Electrotechnology Practice

A practical approach

Steven Hanssen
Jeffery Hampson

6th Edition

Electrotechnology Practice: A Practical Approach
6th Edition
Steven Hanssen
Jeffery Hampson

Portfolio manager: Sophie Kaliniecki
Product manager: Sandy Jayadev
Content developer: Stephanie Davis
Senior project editor: Nathan Katz
Cover designer: Regine Abos (Studio Regina)
Text designer: Rina Gargano (Alba Design)
Permissions/Photo researcher: Debbie Gallagher, Brendan Gallagher
Editor: Sylvia Marson
Proofreader: James Anderson
Indexer: Julie King
Art direction: Linda Davidson
Cover: Getty Images/cravetiger
Typeset by KnowledgeWorks Global Ltd.

Any URLs contained in this publication were checked for currency during the production process. Note, however, that the publisher cannot vouch for the ongoing currency of URLs.

Sixth edition published in 2023

National Library of Australia Cataloguing-in-Publication Data
ISBN: 9780170458863
A catalogue record for this book is available from the National Library of Australia.

Cengage Learning Australia
Level 5, 80 Dorcas Street
Southbank VIC 3006 Australia

Cengage Learning New Zealand
Unit 4B Rosedale Office Park
331 Rosedale Road, Albany, North Shore 0632, NZ

For learning solutions, visit **cengage.com.au**

Printed in China by 1010 Printing International Limited.
3 4 5 6 7 27 26 25

Brief contents

Contents

Guide to the text

As you read this text you will find a number of features in every chapter to enhance your study of electrotechnology and help you understand how the theory is applied in the real world.

CHAPTER-OPENING FEATURES

Refer to the **Introduction** for a contextualised summary of the chapter.

Work health and safety

This chapter provides electrotechnology workers with knowledge and skills about work health and safety fundamentals.

Electrotechnology workers will gain an overview of work health and safety knowledge and skills that will allow them to implement safe procedures for working in the electrotechnology industry. This chapter provides underpinning knowledge for the unit UEECD0007 from the UEE training package.

Identify the key concepts you will engage with through the **Learning Objectives** at the start of each chapter.

LEARNING OBJECTIVES

Effective communication
- Outline communication skills electrotechnology workers use.
- Provide techniques for overcoming barriers to effective communication.
- Describe written procedures and work instructions and provide applications for use.

Electrotechnology work environment
- Describe possible hazards at a worksite.
- Recognise various safety signs.
- Outline the importance of good housekeeping in the workplace.
- Identify a range of fire extinguishers suitable for a specific type of fire.

Legal and ethical issues
- Outline the general aims and objectives of WHS legislation.
- Outline the WHS responsibilities, rights and obligations of those in a workplace.
- Explain the principle of 'duty of care'.

Risk mitigation
- Identify the elements of risk management.
- Describe the components of a risk assessment.
- Explain the requirements for accurate WHS record keeping.

Chemicals in the workplace
- Describe the components of a safety data sheet.
- Name likely chemical hazards, state effects on health and outline control measures.

Confined spaces
- Define a 'confined space' and outline its potential hazards.

Physical and psychological hazards
- Name common non-electrical hazards in the workplace, state effects on health and outline control measures.

Manual handling
- Define 'manual handling'.
- Describe correct lifting procedures.

Working at heights
- Identify hazards, equipment and precautions when working at heights.

Working with electricity
- Discuss the effects of electric shock and describe precautions to minimise those effects.
- Recognise the various protective devices intended primarily for the protection of conductors and equipment.

FEATURES WITHIN CHAPTERS

All-important calculations are worked through step by step in easy-to-follow **Example boxes**. Worked examples show the process undertaken to solve a problem using the calculations.

EXAMPLE 9.3

Convert 70 °C to Kelvin.

$$K = °C + 273.13$$
$$= 70 + 273.15$$
$$= \mathbf{343.15\ K}$$

Practice what you have learned in the text using **Exercise boxes**.

EXERCISE 3.3

- **a** Which units of measurement (kilometres, metres and millimetres) would you use to measure:
  - **i** The direct distance between Brisbane and Sydney?
  - **ii** The length of an AFL ground?
  - **iii** The diameter of a copper wire?
  - **iv** The thickness of wire insulation?
- **b** Estimate the depth, width and height of a miniature circuit breaker.
- **c** The length of a lighting circuit cable is 50 m correct to the nearest metre. What are the upper and lower tolerances for the length of the cable?
- **d** The distance between the workshop and the customer's premises is 18 km, correct to the nearest 0.5 km. What is the percentage tolerance for this distance?

Switch On boxes highlight important tips and hints.

SWITCH ON

Safety with adhesives

Some adhesives have components that are flammable and have toxic hazards. Some adhesives may affect the eyes and skin if direct contact occurs and cause irritation or allergic reactions. The vapours from the adhesive may cause respiratory irritation. Care must be taken when solvents are used, and evaporation occurs to complete the bonding. Ventilation must be provided in order to minimise contamination of the workplace atmosphere. Always keep adhesives away from heat, sparks and open flame because some solvents are flammable. The primary hazard with hot-melt adhesives is burns. With all adhesives always wear eye and face protection. To prevent repeated, or prolonged contact with the skin always wear appropriate gloves. Before using any adhesive, refer to the safety data sheet of that particular adhesive.

The advantages of using adhesives are:

- Dissimilar materials can be joined.
- The bond is continuous.
- On loading, there is a more uniform stress distribution.
- Electrolytic corrosion is prevented.

The limitations of using adhesives are:

- An increase in service temperature decreases the bond strength.
- Bonded surfaces are usually difficult to dismantle.
- There is a need to prepare the surface.

Check your understanding of the content by answering the **Review Questions** as you progress through the chapter.

REVIEW QUESTIONS

1. To what does the term 'work environment' refer?
2. What are two main criminal issues that affect a workplace?
3. What is a standard work procedure?
4. What may constitute a hazard in the workplace?
5. Explain the meaning of the term 'risk'?
6. Name the five broad categories of workplace hazards.
7. What are key activities in the prevention of accidents occurring from worksite hazards?
8. Existing or potential hazards can be prevented or controlled by control measures. List the six control measures in order of priority.
9. Outline the aim of work environment safety signs.
10. What do mandatory signs indicate?
11. Describe the symbolic shape of a restriction sign.
12. What do danger signs indicate?
13. Describe the symbolic shape of a warning sign.
14. Name the four elements required to sustain a fire.
15. Which extinguisher types are suitable for extinguishing class E fires?
16. How is a fire extinguished by a wet chemical extinguisher?
17. PASS is a useful mnemonic for using a fire extinguisher. To what do the letters refer?
18. How frequently should fire extinguishers be serviced?
19. How does a fire blanket extinguish a fire?
20. Who is responsible for workplace housekeeping?

END-OF-CHAPTER FEATURES

At the end of each chapter you will find several tools to help you to review, practise and extend your knowledge of the key learning objectives.

Review your understanding of the key chapter topics with the **Chapter Review**.

CHAPTER REVIEW

1.1 Effective communication

- Being an effective communicator provides a means for establishing a trusting, collaborative relationship with customers.
- Forms of communication for electrotechnology workers include verbal, listening, non-verbal, written, compassion, cultural awareness, trust, and providing instruction.
- Barriers to effective communication include physical, social and psychological.

1.2 Electrotechnology work environment

- A worksite means the place of employment, base of operation or location of workers.

1.3 Legal and ethical issues

- In general, occupational health provides a broad framework to protect the safety, health and welfare of all persons within a workplace.
- The objectives of WHS in Australia are structured around eight core values.
- Duty of care is based on the principle of reasonableness, and is founded in the law of negligence.
- Negligence occurs when there is a failure to exercise a degree of reasonable care under the circumstances, which results in an unintended but foreseeable injury to another party.
- Induction is imparting information to a prospective staff member.

Test your knowledge and consolidate your learning through **Trial Exam**.

TRIAL EXAM

For Chapter 1 knowledge assessment, please complete the following trial exam.

1 The form of listening that requires complete attention and engagement is:
   a listening with purpose
   b contingent listening
   c passive listening
   d aural communication

2 The positive emotion that allows people to show that they care and are willing to help is:
   a indifference
   b coldness

10 What legislation provides a broad framework which codifies the duties of care that are owed under common law?
   a *Wiring Rules*
   b material safety data sheets
   c work and safety procedures
   d Work Health & Safety Acts

11 Which of the following statements imposes a clear obligation on all persons to ensure safe work environments?
   a not to put others at risk
   b safety instructions are only for me

END-OF-BOOK FEATURES

Check your work using the **Answers to the exercises**.

Answers to the exercises

Chapter 1

Exercise 1.1

a single 9 m industrial
b single 5 m domestic

Exercise 1.2

a 2.5 m
b 3.75 m

Exercise 1.3

Arm paralysis

Chapter 3

Exercise 3.1

c 49 m–51 m
d ± 2.78 %

Exercise 3.4

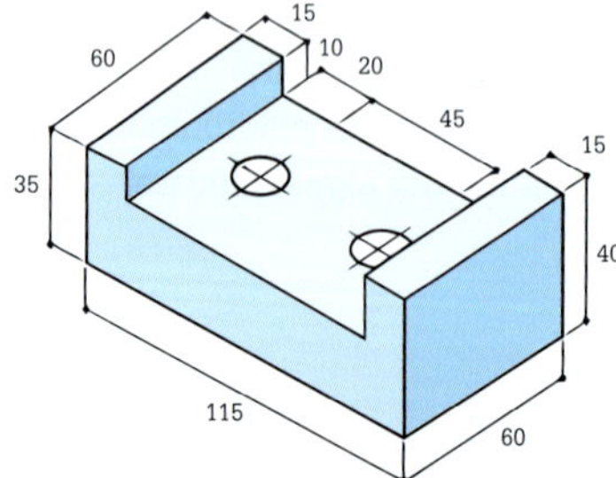

Guide to the online resources

FOR THE INSTRUCTOR

Cengage is pleased to provide you with a selection of resources that will help you prepare your learning and assessments. Contact your Cengage learning consultant for more information.

MINDTAP

Premium online teaching and learning tools are available on the MindTap platform – the personalised eLearning solution.

MindTap is a flexible and easy-to-use platform that helps build student confidence and gives you a clear picture of their progress. We partner with you to ease the transition to digital – we're with you every step of the way.

The *Cengage Mobile App* puts your course directly into students' hands with course materials available on their smartphone or tablet. Students can read on the go, complete practice quizzes or participate in interactive real-time activities.

Series *MindTap* for Hanssen and Hampson's *Electrotechnology: Principles and Practice* is full of innovative resources to support critical thinking, and help your students move from memorisation to mastery! Includes:

- Hanssen and Hampson's *Electrical Trade Principles and Electrotechnology Practice* eBooks
- Chapter reviews
- Concept checks
- Videos
- Trial exams
- Labelling activities
- Online chapters.

MindTap is a premium purchasable eLearning tool. Contact your Cengage learning consultant to find out how *MindTap* can transform your course.

INSTRUCTOR RESOURCES PACK

Premium resources that provide additional instructor support are available for this text. Premium resources include PowerPoints, Test Banks, artwork from the text and Online-only chapters.

These resources save you time and are a convenient way to add more depth to your classes, covering additional content and with an exclusive selection of engaging features aligned with the text.

The Instructor Resource Pack is included for institutional adoptions of this text when certain conditions are met.

The pack is available to purchase for course-level adoptions of the text or as a standalone resource.

Contact your Cengage learning consultant for more information.

SOLUTIONS MANUAL

The **Solutions Manual** provides detailed answers to every question in the text.

MAPPING GRID

The **Mapping Grid** is a simple grid that shows how the content of this book relates to the units of competency needed to complete the Certificate III in Electrotechnology Electrician (UEE30820).

Preface

The sixth edition of *Electrotechnology Practice: A Practical Approach* continues the tradition of previous editions and its companion *Electrical Trade Principles: A Practical Approach* in that it is written to support students in their study of electrotechnology practices. The focus of this edition is the essential knowledge and skills that relate directly to electrical work. The text reflects the changing needs of the electrotechnology industry and presents a broad-based expression of knowledge, which is essential for effective workplace participation. It is also a very practical reference text for anyone interested in practical aspects of the electrical industry.

The text utilises graphical ways of working with ideas and presenting information: to this end the text uses over 1000 illustrations to convey various concepts and real-world aspects of electrotechnology practices. The illustrations provide variety and meaning to the subject material, which aids the learner in retaining knowledge and engages the electrotechnology student.

Individuals recognise and process information in different ways and attain understanding at different rates. This text addresses the uniqueness of individual learning processes. The text is student-centred in that it helps learners in ways that are effective for them, and matches individual needs. The structure of the text leads the learner from what they know to what they need to know, and provides examples and exercises coupled with topic summaries and review questions. For the teacher, the structure of the text makes it well-suited for blended learning.

This sixth edition is available in print and digital format. As an eBook, the text enables a more personal ownership of the learning experience. For example, the eBook format enables students to search for words and phrases, highlight important data and bookmark key pages. In addition, students can add their notes on the topic and explore additional interactive activities. Colleges, trade learning centres and schools find that they have access to a relevant and up-to-date depository of electrical content suitable for the delivery of many varied electrical certificates and diplomas. For instructors of these courses, the eBook allows for additional notes.

This text provides a basis for developing a sound understanding of the skills and associated knowledge underpinning electrotechnology work. In addition, the content makes the development of essential skills and knowledge as flexible and accessible as possible, while ensuring the delivery of appropriate instruction that is linked to industry standards. The content in this sixth edition is updated to reflect essential capabilities required by existing electrical courses contained in the National Electrotechnology Training Package – UEE20.

The text follows the uniform structure and system of delivery as recommended by the nationally accredited vocational educational and training authorities. Relevant elements and performance criteria along with range of conditions are available from Training.gov.au, online at https://training.gov.au.

This self-paced text is ideal for various modes of delivery, including workplace-based learning, classroom learning, blended workplace and classroom learning, and distance learning in the classroom or online through the eText format.

Topics such as 'Select wiring systems and cables ' and 'Control and protection of electrical installations ' align with the latest Australian standards AS/NZS 3000:2018 and AS/NZS 3008.1.1:2017. On-the-job fault-finding skills are an essential facet of all electrical activities, which is reflected in the topic 'Design, install and verify compliance and functionality of general electrical installation'.

We, the authors, believe that this text is designed to bring students of the electrotechnology industry closer together by providing understandable concepts.

Special thanks are due to Sandy Jayadev (Content Manager, Vocational Education and Training), Stephanie Davis (Content Developer) and Nathan Katz (Senior Project Editor) at Cengage Learning, and Sylvia Marson (Copyeditor), for their work, support and advice.

Steven Hanssen
(Assoc.DipEng(Elec) TAFE NSW, BT UTS)

Jeffery Hampson
(Cert IV in Training and Assessment, Dip T (TAFE) BEd)

About the authors

Jeffery Hampson and Steven Hanssen each have more than 35 years of teaching experience in the VET sector. Jeff has taught electrical trades students at SkillsTech Australia (TAFE Queensland) as a leading vocational teacher. Steven is currently a head teacher of electrical trades in TAFE NSW Sydney Region.

Acknowledgements

The authors wish to thank the many dedicated VET educators and industry experts who provided valuable feedback and willingly shared their knowledge in the preparation of this new edition of *Electrotechnology Practice: A Practical Approach*. Their helpful suggestions and robust critique, concerning many aspects of the technical content of this text, have ensured that the content remains up to date and reflects current industry best practice.

Cengage would like to thank the following lecturers and industry experts for their valuable feedback for this new edition of *Electrotechnology Practice: A Practical Approach*:

- James Slattery – TAFE SA
- Andrew Smith – TAFE SA
- Paul Mansfield – TAFE SA.

Every effort has been made to trace and acknowledge copyright. However, if any infringement has occurred, the publishers tender their apologies and invite the copyright holders to contact them.

Work health and safety

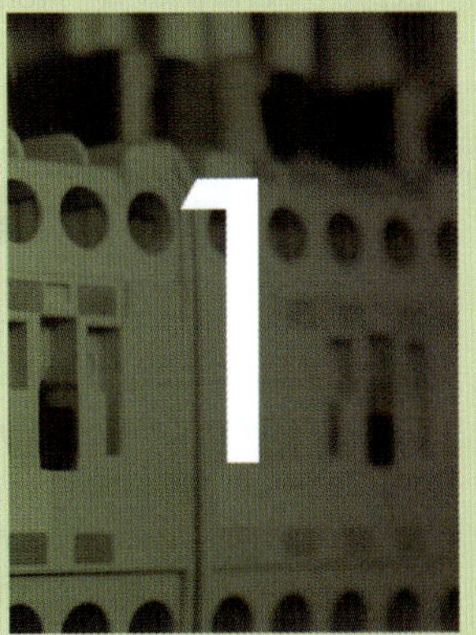

This chapter provides electrotechnology workers with knowledge and skills about work health and safety fundamentals.

Electrotechnology workers will gain an overview of work health and safety knowledge and skills that will allow them to implement safe procedures for working in the electrotechnology industry. This chapter provides underpinning knowledge for the unit UEECD0007 from the UEE training package.

LEARNING OBJECTIVES

Effective communication

- Outline communication skills electrotechnology workers use.
- Provide techniques for overcoming barriers to effective communication.
- Describe written procedures and work instructions and provide applications for use.

Electrotechnology work environment

- Describe possible hazards at a worksite.
- Recognise various safety signs.
- Outline the importance of good housekeeping in the workplace.
- Identify a range of fire extinguishers suitable for a specific type of fire.

Legal and ethical issues

- Outline the general aims and objectives of WHS legislation.
- Outline the WHS responsibilities, rights and obligations of those in a workplace.
- Explain the principle of 'duty of care'.

Risk mitigation

- Identify the elements of risk management.
- Describe the components of a risk assessment.
- Explain the requirements for accurate WHS record keeping.

Airborne contaminants

- Name likely airborne contaminants, state effects on health and outline control measures.

Chemicals in the workplace

- Describe the components of a safety data sheet.
- Name likely chemical hazards, state effects on health and outline control measures.

Confined spaces

- Define a 'confined space' and outline its potential hazards.

Physical and psychological hazards

- Name common non-electrical hazards in the workplace, state effects on health and outline control measures.

Manual handling

- Define 'manual handling'.
- Describe correct lifting procedures.

Working at heights

- Identify hazards, equipment and precautions when working at heights.

Working with electricity

- Discuss the effects of electric shock and describe precautions to minimise those effects.
- Recognise the various protective devices intended primarily for the protection of conductors and equipment.

1.1 Effective communication

Being able to communicate effectively is an important skill for electrotechnology workers. Having effective communication skills is essential for team collaboration with your co-workers as well as with other trades that may be onsite. It is also important in providing excellent customer service.

Workers who take the time to listen to and understand their customer's concerns are better prepared to address issues as they arise, resulting in better customer relations. On the other hand, poor communication, or a lack of communication, can lead to customers misunderstanding instructions, such as failing to follow correct procedures for the operation of equipment and devices. It can also lead to a breakdown in workflow within the team, resulting in delays in receiving goods or completing scheduled work on time. Being an effective communicator provides a means for establishing a trusting, collaborative relationship with customers and co-workers.

Forms of communication

Figure 1.1 Summarises some of these effective communications skills.

FIGURE 1.1 Effective communication skills

Verbal communication

Excellent verbal communication is fundamental to most work environments. For verbal communication to be effective, you should try to speak with clarity, precision and honesty. It is important to know with whom you are communicating and speak appropriately according to the person's age, culture and level of technical literacy. Do not let your own circumstances set the tone and do not allow emotions to affect your communication. You can:

- encourage customers to communicate by asking open questions such as, 'Can you tell me more about that?'
- be respectful when addressing the customer by using inclusive language; ask how they would like to be addressed, such as their preferred name and pronouns, and use inclusive language to respectfully engage with people
- speak in clear, complete sentences and avoid using jargon.

Listening with purpose

Listen with purpose in order to understand the other person's experience. This form of listening requires complete attention and engagement. This skill is important not only for workers, but also for supervisors to build trust and commitment with the staff they are supervising. Listening includes both verbal and non-verbal communication skills. For example:

- acknowledging with a nod of the head, but not interrupting the other person
- engaging by facing the person and maintaining eye contact
- show understanding with the use of a few words, such as 'I understand' and 'go on'.

Non-verbal communication

The use of non-verbal communication, such as body language, eye contact, facial expressions, gestures, posture and tone of voice, is important in establishing rapport. Simply smiling can go a long way in developing meaningful communication. You can:

- maintain eye contact and nod your head to show that you are interested in what the customer is saying
- use open body language such as facing the person speaking and keeping an open (but not slouched) posture.

Written communication

Written communication skills are also essential for effective day-to-day communication. Electrotechnology workers may be required to complete a range of written forms and documents, such as injury notifications, risk assessment and control forms, checklists, work activity records, work tags, and electrical work and test documentation. **Figure 1.2** shows some of the written documentation encountered in electrical work. It is important for written documents to be accurate and up to date. Some handy tips:

- take notes while discussing a customer's requirements so that you do not forget anything
- write legibly and clearly, using simple language
- ensure that all notes and documents have accurate dates and times recorded.

OHS Risk Assessment and Control Form | Risk assessment completed by:

Document number	Initial issue date	Current version	Current version issue date	Next review date

Risk assessment title:

Step 1: Identify the activity

Describe the activity

Describe the location

Step 2: Identify who may be at risk by the activity

A number of persons may be at risk from the activity. This may affect the risk controls suggested.

Steps 3 to 7: Identify the hazards, risk, and rate the risks

1. An activity may be divided into tasks. For each task identify the hazards and associated risks.
2. List existing risk controls and tabulate the risk rating.
3. Additional risk controls may be required. If required, re-rate the risk.

C = consequence
L = likelihood
R = risk rating

Tasks	Hazards (Step 3)	Associated risks (Step 4)	Existing risk controls	Risk rating with existing controls (Step 5)			Additional risk controls required (Step 6) Apply the hierarchy of control	Risk rating with additional controls (Step 7)		
				C	L	R		C	L	R

Step 8: Documentation and supervisor approval

Completed by: Authorised by: Date

Step 9: Implement the additional risk controls identified

Indicate briefly what additional risk controls (see Step 6) were implemented, when and by whom.

Risk control	Date	Implemented by:
Risk control	Date	Implemented by:

Step 10: Monitor and review the risk controls

It is necessary to monitor risk controls and review risk assessments regularly.

Review date	Reviewed by:	Authorised by:
Review date	Reviewed by:	Authorised by:

Documentation

It is a requirement that legal and advisory documentation that supports the risk assessment be listed. Acts, Regulations, Codes of Practice and Standards.

HAZARD CHECKLIST | Version 1 10/09

Site: .... Date: / /
Conducted by: .... Position: ....

No.	HAZARD	Status Yes	No	NS
1.	Slippery or uneven ground surfaces			
2.	Impact from vehicles			
3.	Impact from machines			
4.	Hot or cold surfaces			
5.	Falls from mezzanine floors			
6.	Dangerous surfaces (rough/sharp etc.)			
7.	Hand tools			
8.	Power tools			
9.	Chemicals			
10.	Dusts			
11.	Pressurised systems			
12.	Lighting			
13.	Noise			
14.	Vibration			
15.	Fire			
16.	Radiation			
17.	Lifting			
18.	Biological hazards			
19.	Ergonomics			
20.	Electromagnetic fields			
21.	Explosion			
22.	Electrical energy			
23.	Workplace violence/stress etc.			

Other/Further details:

Signature of person conducting inspection:

Copies provided to:

FIGURE 1.2 Some forms of written communication

Compassion

Conveying compassion is an essential communication skill in many work areas. Compassion is a positive emotion that allows people to show that they care and are willing to help. Showing compassion has the ability to spread harmony in any environment, making it an ideal communication tool in a workplace. It can be a stressful time for customers when electrical issues disrupt their day-to-day conduct or impact their business and productivity. Focusing on compassion allows people to recognise and appreciate others sincerely. Through communicating compassionately, you may be able to more rapidly and effectively identify the root cause of the issue.

You can practise being compassionate in the work environment by:

- listening with purpose and without judgement
- noticing when others are having a tough time and trying to make them feel more comfortable
- accepting criticism
- being careful when expressing opinions to someone to avoid hurting their feelings
- be conscious of working *with* your customers to minimise disruptions to their work as much as practically possible.

Cultural awareness

It is likely that in your everyday work you will be with people who come from a wide range of social, cultural and educational backgrounds. It is important to be aware of, sensitive to, and respectful of these differences. For example, people for whom English is not their first language may experience feelings of frustration and isolation from not being able to fully express themselves in English. Be sensitive to their feelings, speak at a comfortable pace and volume, use pauses, avoid complicated or technical sentences, and clarify your understanding by repeating or summarising what was said. Acknowledgement of cultural diversity, appreciation of the value in this diversity, and use of culturally-inclusive language is important in establishing respectful relationships.

Trust

It is important to inspire trust in customers and co-workers by listening with purpose and taking every complaint and concern seriously. It takes time to build trust and the following steps can foster a trust-based relationship:

- be willing to admit mistakes
- always tell the truth
- share information openly

Providing instruction

At the completion of a task it is often necessary to provide clear instruction on how something works. An effective communication strategy is to ask the customer to repeat the information back to you. This method improves the level of understanding developed by the customer. For example, you can say:

- 'Can you tell me how I need to start this motor?'
- 'I have told you a lot of information. Now I want you to repeat it back to me to make sure you remember everything.'
- 'Let's review what we just discussed. Can you explain it to me in your own words?'

Overcoming barriers to effective communication

On occasions the message is not received the way the sender intended. Ineffective communication can result in poor customer interactions and relationships. To overcome these, it is necessary to understand the types of communication barriers faced in day-to-day communications. **Figure 1.3** shows three major barriers to effective communication.

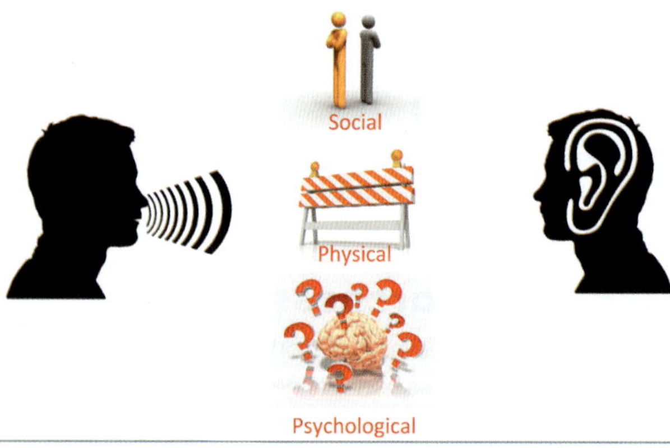

FIGURE 1.3 Barriers to effective communication

Physical barriers

The environment in which you communicate with a customer can make a noticeable difference in effective communication. Hectic, loud and distracting environments can increase stress. To create a safe and comfortable environment, try mitigating worksite noise whenever possible.

Social barriers

Social barriers include differences in language, religion, culture, age and customs. Acknowledging and respecting each customer's cultural background can help avoid prejudice and facilitate clear and effective communication. It is also prudent to tailor your communication strategies depending on the prior knowledge and experience of the person with whom you are conversing. For example, a 16-year-old may be more familiar with technology compared to a 70-year-old.

Psychological barriers

Anxiety and stress are examples of psychological barriers. To assist in reducing these barriers to effective communication, it helps to take extra time to listen, empathise, and be supportive.

Written procedures and work instructions

A procedure is defined as a documented process for undertaking quality activities. The procedures are used as a reference where they provide guidance and consistency when workers perform tasks. **Figure 1.4** specifies what procedures should be carried out.

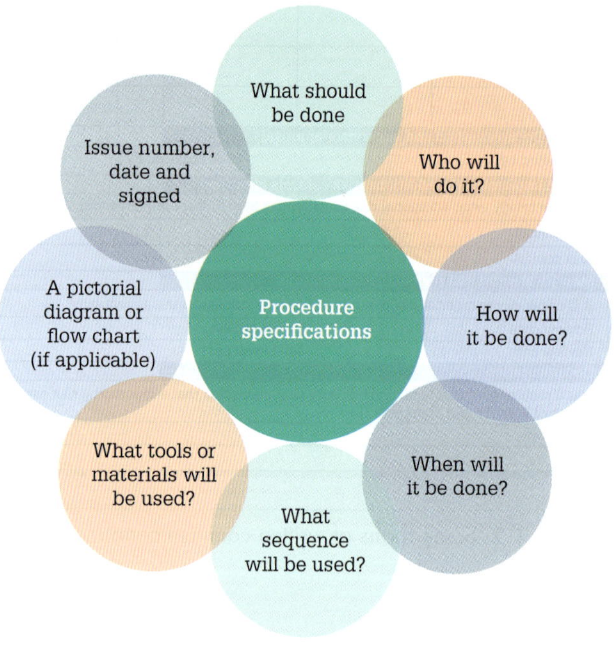

FIGURE 1.4 Procedure specifications

Written procedures are important devices, and like any device have significance when properly used. A procedure for contingent testing is illustrated in **Figure 1.5**.

Work instructions

Work instructions (as shown in **Figure 1.6**) are documents that help in the organisation of tasks usually performed by an individual worker. Producing work instructions is necessary when the procedure is complex and involves many tasks. If the procedure requires few tasks, then the work instructions can be included within a procedure document without side-tracking from the individual worker's understanding of the overall procedure.

There is no perfect format for all work instructions. The types of work instructions that work best are developed over time through trial and error. Individual workers process information differently.

Work instructions should accommodate all types of learners. Some workers are visual learners and process illustrations faster. Other workers may process written text better. For some workers a flow chart is suitable while

ACE Electrical

Sustained Voltage testing

Issue Number 1 **Issue Date 07/2018**

The Principle behind this test is to determine if a sustained voltage remains present.

Note: This procedure is to be performed by a Licensed Electrical Worker only.

If a non-conclusive test result is obtained (higher than 20V a.c. or 50V d.c.) when testing to prove de-energised, further testing is to be conducted to determine whether it is from a sustained voltage source. The procedure for measuring sustained voltages is as follows:

Testing is performed between each phase and earth using a multimeter.

1) This method can only be used when the measured voltage is:
   i. 50 volts a.c. or lower, or ii. 120 volts d.c. or lower.

Note: If the voltage is greater than these, go to Step 7 below. Otherwise go to Step 2) below.

2) Connect a resistance of 20k Ω in parallel with the voltmeter when measuring the voltage.

Note: This is done to place a load in the circuit to check if the source voltage is sustained at this load.

3) If measuring a.c. voltage, the sustained voltage is above 20V, go to Step 5.
4) If measuring d.c. voltage, the sustained voltage is above 50V, go to Step 5.
5) If an acceptable test is not obtained then substitute the resistor connected in Step 2 with 3k Ω resistance and repeat the sustained voltage measurement.

Note: This is again done to place a load in the circuit to check if the source voltage is sustained at this load at this lower resistance.

6) If the sustained voltage remains above the values listed in Steps 3 and 4 then the circuit is to be considered as not proven de-energised – go to Step 7.
7) Report any instances of circuits not being able to be proven de-energised immediately so that further investigation can be conducted into the cause.

Always refer to AS/NZS 3000:2018 Wiring Rules

C. R. Meecham.
Manager

FIGURE 1.5 Procedure for contingent testing

other workers may be totally lost trying to understand a flow chart.

A flow chart (as shown in **Figure 1.7**) is a graphical or symbolic representation of a process. Each step in the process is represented by a different symbol and contains a short description of the process step. The flow chart symbols are linked together with arrows showing the process flow direction.

Day-to-day communications

Every electrical enterprise depends on suppliers to provide goods and services. Motivating those suppliers to provide their best service can be a challenge.

Selecting suppliers

When selecting suppliers an enterprise must decide first on what their main concern is from their supplier; for example, dependability of supply, quality, bulk buys, price, discounts, credit, payment plans, returns policy option or a mixture of these. Once this has been done a list should be established which details ideal suppliers. This list can then be used to contact the suppliers' sales representatives. Qualities to look for with sales representatives are knowledge of products and personal commitment to clients through regular meetings based on flexible schedules. Always have a secondary supplier just in case the primary supplier cannot meet the enterprise's requirements in time. Using local suppliers can also help in reducing on-hand supplies.

Providing quality customer service

Good communication between workers and customers is vital for a business to operate efficiently and effectively. Treat each communication as a way both to exchange information and to build a relationship. By communicating properly with customers it can help motivate them and encourage them to contribute ideas that you can use. They may be able to provide you with some great information as to what they like and dislike about the enterprise. This will allow the enterprise to make any necessary adjustments so as to provide a better service.

Work Instructions for Equipment Checking

Disconnect all equipment and connections from the power source prior to the inspection.

Description (Use a tick for sound, a cross for attention needed)	
Check that the equipment has no external damage.	☐
Visually check cords, plugs and socket-outlets and other accessories for obvious damage:	
Discolouration (may indicate exposure to heat, chemicals or moisture)	☐
Cuts, abrasions, twists in the outer sheath of the cord	☐
Electrical tape or other tapes attached to the cord (look underneath)	☐
Damage to the insulation on the pin of the plugs	☐
Physically check:	
Run the flexible cords through the hand to detect any twisting of the inner cores or damage to the outer sheath	☐
Lightly jiggle plugs connected to sockets to ensure they are not loose.	☐
Flexible cords are effectively anchored to equipment.	☐
Extension leads – connect the plugs/sockets of the lead together to check that the terminals have not spread.	☐
Check where applicable:	
Operating controls are in good working order (secure, aligned and appropriately identified	☐
Covers, guards and the like are secured in the manner intended by the manufacturer	☐
Ventilation inlets and exhausts on electrical equipment are unobstructed	☐

FIGURE 1.6 Work instructions

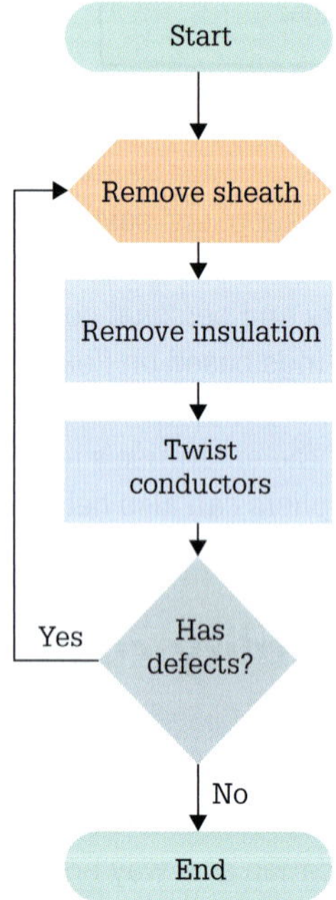

FIGURE 1.7 Flow chart

Good customer service, looking after the customer base and its needs, will probably be the biggest factor determining how well the enterprise performs when compared to its competitors. If customers complain, do not ignore them. Instead talk to the customers and proactively try to sort any issues out. Indeed, encourage customers to provide you with valuable feedback, by giving them a call and asking them what they think of the services.

REVIEW QUESTIONS

1. Identify two outcomes for electrotechnology workers being able to communicate effectively.
2. Which form of communication is fundamental to most work environments?
3. In what forms can non-verbal communication occur?
4. Name three barriers to effective communication.
5. How is it possible to develop trust when communicating?
6. What are work instructions?
7. What is a flow chart?
8. Why is good communication between workers and customers vital for a business?

1.2 Electrotechnology work environment

The work environment, or worksite, is the place of employment, base of operation or location of workers. It includes all of the employer's buildings or facilities located within the same building and their parking facilities.

In today's changing times, safe premises, buildings and security address not only natural disasters but also crime, violence in the workplace and acts of terrorism. Security means the protection of the premises, the employees working there, visitors and assets.

The most common threat in the work environment is workplace violence. Workplace violence includes cases of unhappy, angry workers causing harm to a building, its assets or other workers, and includes domestic issues that carry over into the workplace. This can have fatal consequences if these workers are not identified and counselled before the situation escalates.

Criminal issues include stealing and sabotage to buildings, assets or documents (paper or data). Persons engaged in these activities are usually disgruntled workers, opportunists or hackers. Our era has seen an increase in terrorist-related violence. Workplaces now have the potential to be used as weapons of mass destruction.

When developing a practical plan of action, the workplace environment must be examined from several viewpoints: operations, physical security and valid data-collecting processes. A practical plan of action theoretically leads to safer environments for workers and the surrounding community.

Standard work procedure

Standard work procedure means implementing specific, efficient plans of action for each task or process undertaken in a workplace. It starts with identifying the generally accepted safe and sound way to perform a particular task, then developing methods and procedures. These standard procedures become an effective baseline against which improvement actions are measured.

Standard work procedures enable each worker to learn and follow best practices that help them complete each task or process to a high level of efficiency with safety.

Observation and analysis of each process enables the identification of workplace activities that are not safe or are inefficient and that must therefore be eliminated or improved. This can be anything from a worker engaged in excessive lifting to having to work live. The process is then documented as a carefully planned step-by-step sequence of actions, applied and monitored to ensure it is implemented as planned. A flow chart for the establishment of standard work procedures is illustrated in **Figure 1.8**.

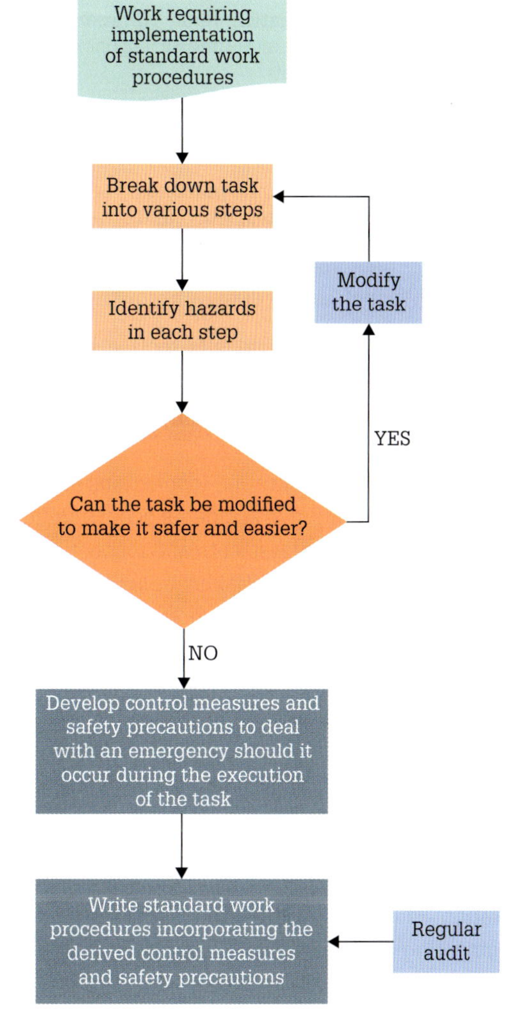

FIGURE 1.8 Standard work procedure flow chart

Hazards at the worksite

A hazard can be work practices, procedures or anything that has the potential to harm the health or safety of a person. Risk is a measure of the probability of a specific harmful effect in particular circumstances. It is important to distinguish between hazard and risk. Worksite hazards occur:

- in the work environment
- as a result of the use of machinery, tools and materials
- as a result of unsuitable work systems and procedures.

Hazards encountered at a worksite can be classified into five broad areas; see **Table 1.1**.

TABLE 1.1 Examples of worksite hazards

Hazard category	Example
Physical	noise, radiation, light, vibration, temperature, humidity, ergonomic (movement)
Chemical	poisons, dusts, lead, solvents, resins, glues, fluxes
Biological	viruses, plants, parasites, vermin, insects, mites, wood and other plant material (allergies), infections (tuberculosis), viruses (from needlestick injuries)
Mechanical/ electrical	slips, trips and falls, tools, electrical equipment
Psychological	fatigue, violence, bullying, stress

Worksite risk assessments and inspections are key activities in the prevention of accidents occurring from worksite hazards. Risk assessments and inspections:

- identify existing and potential hazards
- increase worker awareness, leading to the prevention of worksite accidents and illnesses
- ensure compliance with standards and regulations.

Existing or potential hazards identified by worksite risk assessments and inspections can be prevented or controlled by the following six levels of control measures, in order of priority.

1 Elimination – stop whatever is causing the hazard.
2 Substitution – use a lower hazard alternative.
3 Isolation – separate use from the rest of the workplace.
4 Engineering controls – install equipment that will reduce exposure or risk.
5 Safe work practices – change the way people work.
6 Personal protective equipment (PPE) – gloves, goggles, ear plugs and respirators, for example, can reduce worker contact and exposure to the hazard. PPE is always the last resort, but in some worksite situations, it may be the most practicable.

When assessing workplace hazards and risks always consider and document the probability of an event, length of exposure to the hazard or risk and the consequences.

Work environment safety signs

The aim of work environment safety signs is to regulate and control safety-related behaviour, to warn workers and members of the general public of health and safety hazards and to provide emergency information, including fire protection information. The Australian standard setting out requirements for the design and use of safety signs is AS 1319:1994 *Safety signs for the occupational environment*. The standard specifies several sign classifications and layouts as follows.

Prohibition signs

Prohibition signs, as illustrated in **Figure 1.9**, indicate that an action or activity is not permitted. Their designated symbolic shape is a red circle with a diagonal red slash through it. This is usually superimposed over a black pictograph; for example, a person, to indicate what specific activity is referred to. The background is white and any text is black.

FIGURE 1.9 Prohibition sign

Mandatory signs

Mandatory signs, as illustrated in **Figure 1.10**, indicate that an instruction must be carried out. Their symbolic shape is a blue circle. A white pictograph such as hearing protection is superimposed on the blue circle to indicate the activity that is mandatory. The background is white, and any text is black.

FIGURE 1.10 Mandatory sign – hearing protection must be worn

Restriction signs

Restriction signs, as illustrated in **Figure 1.11**, place a numerical or other defined limit on an activity or use of a facility. Their symbolic shape is a red circle, but without the diagonal slash as in prohibition signs. This would also have a black pictograph or other legend inside the circle, a white background and any text in black.

FIGURE 1.11 Restriction sign

Danger signs

Danger signs, as illustrated in **Figure 1.12**, warn of a particular hazard or hazardous condition that is likely to be life threatening. Their symbolic shape is the word DANGER in white on a red oval, which is surrounded by a black rectangle. This usually forms a heading for a white background on the sign. Alternatively, it may occupy the left side of a horizontal sign. Any text is in black.

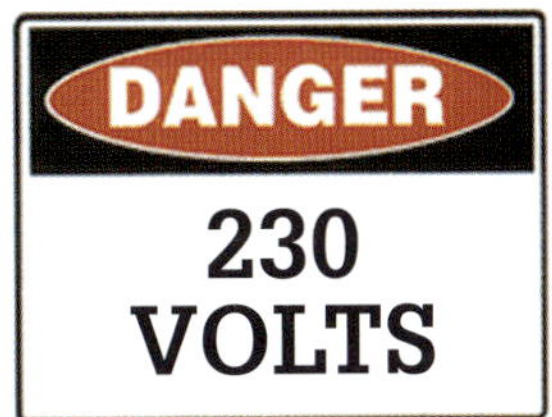

FIGURE 1.12 Danger sign

Warning signs

Warning signs, as illustrated in **Figure 1.13**, warn of a hazard or hazardous condition that is not likely to be life threatening. Their symbolic shape is a black triangular outline. A black pictograph usually appears inside the triangle to indicate the specific detail, for example, an information symbol. The sign background is yellow with any text in black.

FIGURE 1.13 Warning sign

Emergency information signs

Emergency information signs (see **Figure 1.14**) indicate the location of, or directions to, emergency-related facilities such as exits, safety equipment or first aid facilities. The background is green and any text or pictograph is white.

FIGURE 1.14 Emergency information sign

Fire signs

Fire signs, as illustrated in **Figure 1.15**, advise the location of fire alarms and firefighting equipment. The background is red and any text or pictograph is white.

FIGURE 1.15 Fire sign

The type of work environment safety sign used should be suitable for the intended application and workers should be informed of its purpose.

Work environment safety signs should be located where the message is legible, and where they attract the attention of, and are clearly visible to, all workers by being placed at eye height. Signs should be located against a contrasting background so they are more obvious, thereby reducing the risk of them becoming obscured by stacked materials or other visual obstructions.

For maximum effectiveness, work environment safety signs should be maintained in good condition, kept clean and well illuminated.

Fires

Fire protection in the workplace requires appropriate means of extinguishing local fires in all locations where workers are employed. These locations should have adequate safety notices to instruct, warn and guide workers about possible fires. Portable fire extinguishers apply an extinguishing medium that cools burning fuel, displaces or eliminates oxygen or stops the chemical reaction so a fire cannot continue to burn. When the safety pin is removed and the handle of an extinguisher is activated, a canister of high-pressure gas is triggered, forcing the extinguishing medium through a tube and out the nozzle. For fire to exist, four elements must be present at the same time:

1. some sort of fuel or combustible material
2. oxygen to sustain combustion
3. heat to raise the combustible material to its ignition temperature
4. a chemical reaction.

Not all fire extinguishers can be used effectively on all types of fires. Some fires involve combustibles such as paper, some involve liquids and others involve energised electrical equipment. Different types of fire extinguishers are distinguished from each other by their colour scheme and are designed to extinguish different classes of fire. Fire extinguishers empty quickly, anywhere from 8 seconds to 60 seconds. Fire extinguishers (see **Figure 1.16**) are classified by the type of fire they best extinguish.

FIRE CLASS	WATER AIR-PRESSURISED WATER	WET CHEMICAL	FOAM	DRY CHEMICAL POWDER	CARBON DIOXIDE (CO_2)
A Ordinary combustibles (wood, paper, plastics, etc.)	YES	YES	YES	YES	NO
B Flammable combustible liquids	NO	NO	YES	YES	YES
C Flammable gases	NO	NO	NO	YES	NO
E Fire involving energised electrical equipment	NO	NO	NO	YES	YES
F Fire involving cooking oils and fats	NO	YES	YES	YES	NO

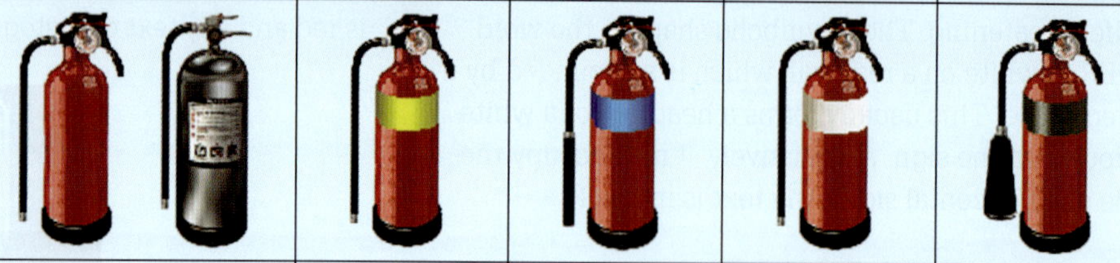

FIGURE 1.16 Fire extinguishers and classifications

Water is one of the most commonly used extinguishing agents for class A fires. Another extinguisher used for class A fires is the air-pressurised water (APW) extinguisher. An APW can be recognised by its large silver container. APWs are two-thirds filled with water then pressurised with air. APWs extinguish fire by cooling the surface of the fuel to remove the heat element of the fire.

Wet chemical extinguishers utilise an aqueous solution discharged in a fine spray to the surface of class F fires (see **Figure 1.16**). Wet chemical extinguishers extinguish fire by smothering the surface of the fuel to separate the oxygen element of the fire from the fuel. The wet chemical also cools the surface of the fuel to remove the heat element of the fire.

Foam extinguishers contain a solution of aqueous film-forming-foam (AFFF) concentrate and water. When the extinguisher is operated, the solution is discharged through the nozzle that is designed to excite air to produce a foam discharge. Foam extinguishers extinguish fire by smothering the surface of the fuel to remove the oxygen element of the fire. The foam also limits the release of flammable vapours to prevent any re-ignition of the fire.

Dry chemical extinguishers coat the fuel with a thin layer of fire-retardant powder. Dry chemical extinguishers extinguish fire by smothering the surface of the fuel to separate the oxygen element of the fire from the fuel. The powder also works to disrupt the chemical reaction.

Carbon dioxide (CO_2) is a non-flammable gas placed under extreme pressure within a CO_2 extinguisher. CO_2 extinguishers extinguish fire by displacing the oxygen element of the fire. Because of its high pressure, pieces of dry ice are also emitted from the extinguisher which has a cooling effect on the fire.

There is another fire class, class D, for combustible metals. Some ships are made with magnesium steel to make the ship lighter and therefore faster in the water. However, if enough heat is generated the metal will burn.

Fire extinguishers should be serviced every six months. They are checked to make sure they are charged and nothing is missing on them. You can check the yellow metal tag on an extinguisher to see when it was last tested. Written on the front of each type of fire extinguisher are instructions for its correct use. To use a fire extinguisher properly all you have to remember is PASS.

SWITCH ON

P – Pull the safety pin
A – Aim at the base of the fire
S – Squeeze the trigger
S – Sweep at the base of the flame from side to side

Fire blankets, as illustrated in **Figure 1.17**, extinguish fire by smothering the surface of the fuel to remove the oxygen element of the fire. They can be used for small class A and small cooking fat fires but are mainly used to wrap around workers if their clothes catch alight. Fire blankets are either a flame-retardant-treated woollen material or a combined Proban® cotton and Aramid (nylon fibre) flame-retardant textile.

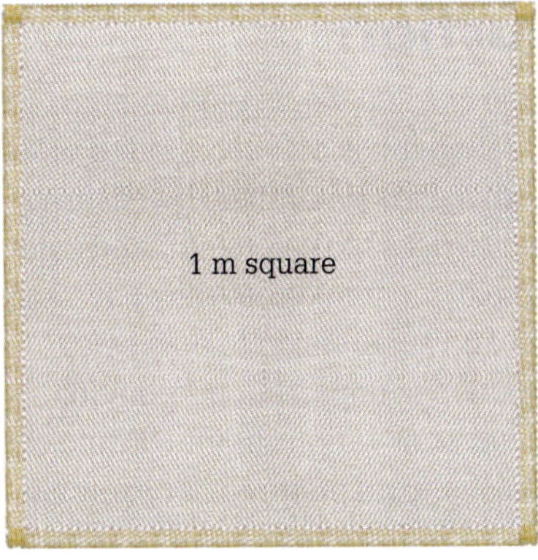

FIGURE 1.17 Fire blanket

Housekeeping

Housekeeping is not just cleanliness; it is a significant factor in creating a safe, healthy workplace for workers. The development of a good housekeeping culture in a workplace is a team effort and should be the desire of every worker. Best-practice procedures make housekeeping a standard part of working. Good housekeeping also raises awareness and highlights the importance of sustainability, environmental responsibility and hygiene concerns.

Where a workplace has been allowed to become cluttered and polluted, poor work practice procedures and frequent accidents often result. Good housekeeping means that:

- work areas are free from rubbish and obstructions
- surfaces are safe and suitable
- surfaces are free from slip/trip hazards
- appropriate waste bins are available
- stock/material is stored safely
- aisles are unobstructed and clearly defined with adequate lighting. There should be good vision at corners and the aisles must be wide enough for the processes carried out.

SWITCH ON

Workplace housekeeping is a task of both employers and workers, for the benefit of both. Good industrial housekeeping creates a working environment in which workers can do their job correctly, professionally and in safety.

REVIEW QUESTIONS

1. To what does the term 'work environment' refer?
2. What are two main criminal issues that affect a workplace?
3. What is a standard work procedure?
4. What may constitute a hazard in the workplace?
5. Explain the meaning of the term 'risk'?
6. Name the five broad categories of workplace hazards.
7. What are key activities in the prevention of accidents occurring from worksite hazards?
8. Existing or potential hazards can be prevented or controlled by control measures. List the six control measures in order of priority.
9. Outline the aim of work environment safety signs.
10. What do mandatory signs indicate?
11. Describe the symbolic shape of a restriction sign.
12. What do danger signs indicate?
13. Describe the symbolic shape of a warning sign.
14. Name the four elements required to sustain a fire.
15. Which extinguisher types are suitable for extinguishing class E fires?
16. How is a fire extinguished by a wet chemical extinguisher?
17. PASS is a useful mnemonic for using a fire extinguisher. To what do the letters refer?
18. How frequently should fire extinguishers be serviced?
19. How does a fire blanket extinguish a fire?
20. Who is responsible for workplace housekeeping?

1.3 Legal and ethical issues

In force across Australia is the national harmonisation law that is a significant reform to WHS legislation. This reform is reflected in new federal Acts called the Work Health and Safety Acts (WHS Acts).

All workers in the electrotechnology industry should read the WHS legislation, which must be available to them in the workplace.

In general, work health (which includes psychological health as well as physical health) and safety provides a broad framework incorporating legislation (which codifies the duties of care that are owed under common law), policies, procedures, obligations and practical means that aim to protect the safety, health and welfare of all persons within a workplace. A workplace is any place where work is being performed. The harmonised principles of the WHS Acts are:

1. All persons in a workplace must be given the highest level of health and safety protection that is sensibly feasible.
2. Those who manage or control work activities that give rise, or may give rise, to risks to health or safety are responsible for removing or minimising health and safety risks, so far as is sensibly feasible.

3 Employers and self-employed people should develop a 'hands-on' approach and take sensible workable measures to ensure health and safety in their business activities.
4 Employers and workers should exchange information about workplace risks to health or safety and actions that can be taken to eliminate or reduce those risks.
5 Employees are entitled, and should be encouraged, to be represented on health and safety issues.

The objectives of WHS are structured around the following eight core values:

1 to secure and promote the health, safety and welfare of people at work
2 to protect all people at a place of work against risks to health or safety arising out of the activities of persons at work
3 to promote a safe and healthy work environment for all persons at a workplace to protect them from injury and illness
4 to provide for consultation and cooperation between employers and employees in achieving the principles of the relevant state or territory WHS Act or regulation
5 to ensure that risks to health and safety at a workplace are identified, assessed and eliminated or controlled
6 to provide a legislative framework that enables a living standard of WHS to take account of future changes in technology and work practices
7 to deal with the impact of particular classes or types of dangerous goods and plant at, and beyond, places of work
8 to develop and promote community awareness of WHS issues.

Every worker in Australia and New Zealand has a right to healthy and safe work and to a work environment that enables them to live a socially and economically productive life.

Responsibilities, rights and obligations

The WHS legislation in Australia and New Zealand places an absolute duty on employers and controllers of workplaces (including directors and managers) to provide a safe and healthy workplace for employees and visitors to the workplace. The various Acts impose clear obligations on all persons to ensure safe work environments.

Employers are to provide work environments that ensure the health and safety of workers. Employers must also maintain the various classes or types of dangerous goods and plant and systems of work under their control without risking the health and safety of any person.

Workers have obligations not to put others at risk and to obey the reasonable instructions of their employer in relation to WHS.

Persons other than employers and workers (visitors or unwelcome persons) must not put others and workplaces at risk and must obey the WHS instructions specific to the workplace location they have entered.

Induction

Induction is a legislative requirement. The WHS Act states that a manager has an obligation to protect the health and safety of workers and others by ensuring that they are not exposed to risks to their health and safety arising from their employment and that they have enough information, training and supervision to stay safe.

Induction is imparting information to a prospective staff member at various points during the familiarisation process for a work program; for example, at an interview and at the local workplace. All new staff should be given a general orientation program that includes providing information which is specific to the local workplace and all relevant safety information.

Induction is normally carried out by a nominated staff member who provides information about processes and activities regarding the work area including health and safety matters. All new staff must be informed about the hazards and the possible risks and know how to avoid or minimise the risks.

Health and safety committees

Health and safety committees are set up in workplaces to help resolve health, safety and welfare issues that arise in the workplace. The committee is a representative group of the employer and workers that meets in a cooperative way to improve systems which assist proposed changes to the workplace, workplace policies, and practices or procedures that could affect the health, safety or welfare of any person in the workplace.

Health and safety committees should:

- develop safe systems of work and safety procedures
- analyse accidents and causes of notifiable occupational diseases, and make recommendations to prevent recurrences
- review risk assessments
- examine safety audit reports
- consider reports submitted by safety representatives
- monitor the effectiveness of health and safety training
- monitor and review the adequacy of health and safety communication within the workplace.

Safety inspectors

A workplace health and safety inspector may enter any workplace to monitor its compliance with the WHS Act and to exercise their powers while they are in that workplace. Usually the inspector will be visiting to undertake a health and safety inspection. However, they could also visit after an accident that may have been caused by work activities. After entering the workplace the inspector has the power to:

- search any part of the workplace
- carry out inquiries, examinations, surveys and investigations, including taking measurements,

photographs and samples with respect to the degree of risk at the workplace or the standards of health and safety existing at a workplace
- inspect and copy documents
- make inquiries into the circumstances and probable causes of workplace incidents
- take any person, equipment or materials into the workplace to assist in the exercising of a power
- require any person in the workplace to give reasonable help
- require a person to produce specific documents
- issue improvement or prohibition notices
- seize evidence of a WHS offence
- seize anything dangerous or otherwise used to commit a WHS offence.

Note that it is an offence to obstruct, threaten or interfere with a WHS inspector who is exercising their powers under the legislation.

Providing first aid

The following information does not represent legal advice and all first aiders should seek legal advice from a law practitioner.

For the most part, it is reasonable to assume that a person experiencing an emergency incident wants to have aid provided to help recover or survive. However, for a conscious person consent needs to be obtained before commencing any intervention. Consent is implied in emergency situations where a casualty is unconscious or injuries prevent the casualty from giving consent. The first aider must also maintain confidentiality with respect to the emergency incident.

Duty of care

Duty of care is based on the principle of reasonableness, which underlies the law of negligence. A person has to conduct themselves in their working life in a way that is considered reasonable for someone in their position to do. Reasonable conduct may be thought of as that which is 'acceptable, average, equitable, fair, fit, honest, proper, right, tolerable or within reason' (*Collins English dictionary and thesaurus*, HarperCollins Publishers).

There is a general duty of care on employers of the workplace to ensure the health, safety and welfare at work of all employees and others who come to the workplace.

There is also a general obligation on workers to take care of others and cooperate with employers in matters of health and safety. A worker must also cooperate with the employer or other person so far as is necessary to enable compliance with the WHS Act.

Negligence occurs when there is a failure to exercise a degree of reasonable care under the circumstances, which results in an unintended but foreseeable injury to another party.

Four important issues must be proven in court:

1. That there existed a duty to act on the part of the defendant towards the plaintiff.
2. That the duty was breached; that is, not fulfilled.
3. That the circumstances that led to the breach were foreseeable.
4. That the breach of duty was the proximate cause of an actual injury to the plaintiff.

Reasonably prudent individual

Upon re-creation of the events that took place during the administering of first aid, the behaviour of the first aider will be compared to what the reasonably prudent person in similar circumstances would have done. This is a matter of trying to establish in court, generally through the testimony of expert witnesses, what the standard is for persons with similar competencies.

Good Samaritan

The Good Samaritan doctrine is a legal principle that prevents a first aider who has voluntarily helped a victim in distress from being successfully sued for 'wrongdoing' with respect to the administration of first aid only. Its purpose is to keep people from being reluctant to help a stranger in need for fear of legal repercussions if they make some mistake in treatment. This doctrine was primarily developed for first aid encounters and every country has its own adaptation of it, the crucial points being the same.

Someone who makes the decision to intervene must perform to the level of knowledge and competence that could be reasonably expected of a similarly trained person in that situation. In other words, you need not do anything, but if you do you must do your best according to your training.

Unlike a passer-by, who may or may not choose to provide assistance when unexpectedly encountering an emergency and a person in need, those workers, such as electricians, with an occupational requirement to be trained in first aid and cardiopulmonary resuscitation (CPR) are expected to provide care in the workplace until skilled medical assistance arrives.

Personal protective equipment

Personal protective equipment (PPE) refers to garments, equipment or barrier substances designed to be worn by a person to protect them from exposure to risks of injury or illness. Different types of PPE may be used depending on the type of hazard. Note that all PPE must be approved by Australian standards. See **Table 1.2** for some common examples of PPE.

TABLE 1.2 Some common types of PPE

Clothing	Flame-resistant clothing (100% cotton clothing) that covers the whole body (neck to wrists and ankles) must be worn by all electrical workers involved with **de-energised electrical work activities**. Flame-retardant clothing that covers the whole body (neck to wrists and ankles) must be worn by electrical workers involved with **live work activities**.	Source: Shutterstock.com/M2020
Insulated gloves	Insulated gloves for working on low-voltage equipment are to be rated to the highest voltage expected when performing the task. The gloves must comply with AS 2225:1994 *Insulating gloves for electrical purposes.*	Source: Shutterstock.com/AleksandrN
Safety footwear	Safety shoes/boots must comply with the requirements of AS/NZS 210.2:2000 *Occupational protective footwear – Requirements and test methods.* The shoes/boots selected should have minimal synthetic material in their construction and must have a full leather upper. When in service the shoes/boots must be in good condition and are not to have any exposed metal, such as steel toe-caps.	Source: Shutterstock.com/BW Folsom
Face shields	Face shields are to cover the full face and have no exposed metal parts and have an electrical rating suitable for the task being performed.	Source: Shutterstock.com/Sinchai24

REVIEW QUESTIONS

1. Outline the purpose of national work health and safety harmonisation law.
2. What is the definition of a workplace?
3. What does the WHS legislation in Australia and New Zealand place on employers and controllers of workplaces (including directors and managers)?
4. Outline the obligations placed on workers under WHS laws.
5. What is meant by the term 'induction'?
6. Name three powers of a workplace health and safety inspector.
7. What is a reasonable assumption for a person experiencing an emergency incident?
8. On what principle is duty of care based?
9. When administering first aid, to whom will the behaviour of the first aider be compared if an explanation is required?
10. What doctrine was primarily developed for first aid encounters?
11. To what does personal equipment refer?
12. What checks are necessary for personal protective equipment?

1.4 Risk mitigation

Risk mitigation is an obligation to health and safety from all persons engaged in an electrical business. Risk mitigation is a consultative process that embeds an effective risk management culture throughout the enterprise. Risk is inherent in all electrical tasks and a competent worker and employer should be able to discern predictable risks and their consequences. Once identified, steps must be undertaken to remove or limit the dangers that affect health and safety.

For electricians, risk management must be implemented in the field; therefore, the process of conducting a risk assessment must be tailored to the particular task. The important thing you need to decide on is whether a hazard is important and whether you have it contained by adequate safeguards that minimise the risk.

The specific elements of risk management are shown in **Figure 1.18**.

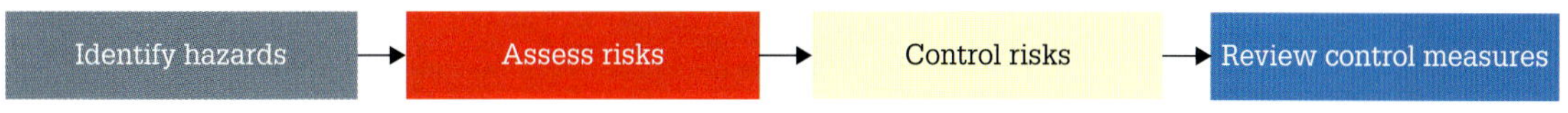

FIGURE 1.18 Risk management elements

Conducting a risk assessment

The first element when conducting a risk assessment is to define the relationship between yourself, the task and its environment. Am I suitably trained, with experience of the equipment or job on which I will be working, in order to establish safe working practices? Identify other stakeholders that are linked to the work such as peers, assistants or other persons (such as machine operators) who are associated with the work and establish a communication and consultation dialogue with these individuals. The information acquired from these sources will assist you in determining if you have the ability, personnel and equipment to manage the risks faced by the task. It is important that perceptions of risk be acknowledged and documented and the reasons for them understood and addressed.

The second element of risk assessment is the discovery of the health and safety risks to be managed (what can happen). This process is vital because a likely risk not recognised at this stage is excluded from further examination. Consultation with other workers is essential in order to bring light to potential hazards. The aim is to generate a comprehensive list of events that might affect the health and safety of personnel linked to the task. Walking around and looking at what could reasonably be expected to affect health and safety can produce a 'risk list'. Risk recognition is an action or series of actions taken to identify hazards and to measure the risk or probability that something will occur because of the hazard. In assessing risk, the harshness of the consequences is noted. Once a list of possible events is identified and documented it is then necessary to review the possible causes and scenarios initiating a particular event (how it can happen and who might be harmed).

Risk analysis

The purpose of risk analysis is to ignore the trivial acceptable risks and to concentrate on significant hazards, their consequences and the probability that the consequences could result in serious harm to persons. Once risks are identified and assigned a classification (high: potential to kill; medium: severe injury; small: minor injury) a decision needs to be made as to whether existing safety measures and controls and procedures are adequate or if more are required. The aim of risk analysis is to make all risks to health and safety small impact by adding to control measures as necessary.

Risk action

Risk action involves identifying the range of options for containing risk, assessing those options, preparing a safe course of action plan and implementing it. The following list illustrates the risk response options:

- Evade the risk by determining not to proceed with the task.
- Reduce the likelihood of the consequences.
- Reduce the effects.

Course of action plan

Responsibility for the control of risk ought to be borne by informed, competent personnel who are equipped to manage the risk. The successful delivery of an action plan requires an efficient management system, which specifies the safety methods chosen, and allocates individual responsibilities and accountability for actions, and checks them against specific details. It is essential to observe the work process to ensure that unforeseen conditions do not change risk ratings. When the work has been completed, it is essential that the effectiveness of the course of action plan is reviewed to see if the measures taken were successful. All features of the risk management process are documented in order to substantiate that a methodical method of risk identification has been implemented. An example of a risk evaluation chart is illustrated in **Figure 1.19**.

A risk assessment and control form is shown in **Figure 1.20**.

LEVEL OF EXPOSURE TO A HAZARD CAUSING INJURY					CONSEQUENCES
1	2	3	4	5	
Almost certain	Likely	Possible	Maybe	Rare	
Expected	Expected	Expected	Expected	Probably	Death, permanent disablement
Expected	Expected	Expected	Probably	Probably	Serious bodily injury
Expected	Probably	Probably	Might occur	Might occur	Medical treatment
Probably	Probably	Might occur	Could occur	Could occur	First aid
Probably	Might occur	Could occur	Could occur	Could occur	No injuries

EXPECTED–ACT NOW. Do something to manage these risks immediately.

PROBABLY–ACT AS SOON AS POSSIBLE. Do something to manage the risks assessed.

MIGHT–PLAN to manage these risks.

COULD–OK for now. Review if any equipment/people/materials/work methods or procedures change.

Source: http://www.hr.unsw.edu.au.

FIGURE 1.19 Risk evaluation

Record keeping

Record keeping is critical to the risk management process. Appropriate recording of the workplace health and safety risk management process demonstrates that an enterprise has been actively working to ensure workplace health and safety. Keeping records will also help the business to maintain track of what has been done and is planned to do. Effective records should enhance the success of the risk management process. Enterprises with five or more employees must record the significant findings of their risk assessment.

The records should show that the process has been conducted appropriately and must include information about the hazards and associated risks at the workplace or worksite. The detail and extent of recording will depend on the size of the workplace or worksite and the potential for major workplace health and safety issues. Information recorded could include:

- How hazards were identified at the workplace or worksite:
  - When and where hazard identification was carried out.
  - Summary of identified hazards.
- How the risks associated with the workplace hazards were assessed:
  - Summary of identified hazards.
  - Whether there is any risk associated with each hazard identified.
- How decisions on control measures to manage exposure to the risks were made:
  - What risk assessment method was used?
  - What new measures have been identified to control any risk?
  - What actions are regarded as not reasonably practicable and the reason for this?
  - What are the reasonably practicable risk control measures for implementation?
- How the effectiveness of the measures was monitored and reviewed; and any checklists and worksheets used in working through the workplace health and safety risk management process:
  - Timeline and person responsible for the implementation of the practicable risk control measures.
  - Who was involved in the hazard identification, risk assessment and risk control processes?
  - Who was consulted in the hazard identification, risk evaluation and risk control processes?

OHS Risk Assessment and Control Form		Risk assessment completed by:		
Document number	Initial issue date	Current version	Current version issue date	Next review date

Risk assessment title:

Step 1: Identify the activity

Describe the activity

Describe the location

Step 2: Identify who may be at risk by the activity

A number of persons may be at risk from the activity. This may affect the risk controls suggested.

Steps 3 to 7: Identify the hazards, risk, and rate the risks

1. An activity may be divided into tasks. For each task identify the hazards and associated risks.
2. List existing risk controls and tabulate the risk rating.
3. Additional risk controls may be required. If required, re-rate the risk.

C = consequence
L = likelihood
R = risk rating

Tasks	Hazards (Step 3)	Associated risks (Step 4)	Existing risk controls	Risk rating with existing controls (Step 5)			Additional risk controls required (Step 6) Apply the hierarchy of control	Risk rating with additional controls (Step 7)		
				C	L	R		C	L	R

Step 8: Documentation and supervisor approval

Completed by:	Authorised by:	Date

Step 9: Implement the additional risk controls identified

Indicate briefly what additional risk controls (see Step 6) were implemented, when and by whom.

Risk control	Date	Implemented by:
Risk control	Date	Implemented by:

Step 10: Monitor and review the risk controls

It is necessary to monitor risk controls and review risk assessments regularly.

Review date	Reviewed by:	Authorised by:
Review date	Reviewed by:	Authorised by:

Documentation

It is a requirement that legal and advisory documentation that supports the risk assessment be listed. Acts, Regulations, Codes of Practice and Standards.

FIGURE 1.20 Risk assessment and control form

Maintenance of appropriate records will assist an enterprise to improve safety continuously and in addition demonstrates compliance with obligations under the WHS legislation. The assessment will have to be reviewed as new machinery, substances, procedures or personnel are introduced, as these could lead to new hazards. Make note of the date the assessment was considered, but the existing record should only be altered if there is a significant change in risk or precautions.

SWITCH ON

Promotion of principles and responsibilities

Workplace safety principles and risk management cannot operate in isolation from workers. A supportive workplace culture is essential to ensuring that every worker with risk management responsibilities feels confident raising, discussing and managing risks. A positive workplace culture will also include evaluation and rewards for individuals demonstrating risk management competencies.

EXAMPLE 1.1

Activity: Hammer drilling into concrete. Consequently, using the Hazard Checklist shown in Figure 1.21 you would say yes for number 8, 10, 13 and 14. In addition, in the further details you should identify specific hazards as:

- Hazards identified: Eye hazards, respiratory hazards, vibration hazards.

HAZARD CHECKLIST | Version 1 10/09

Site: .... Date: / /
Conducted by: .... Position: ....

No.	HAZARD	Status		
		Yes	No	NS
1.	Slippery or uneven ground surfaces			
2.	Impact from vehicles			
3.	Impact from machines			
4.	Hot or cold surfaces			
5.	Falls from mezzanine floors			
6.	Dangerous surfaces (rough/sharp etc.)			
7.	Hand tools			
8.	Power tools			
9.	Chemicals			
10.	Dusts			
11.	Pressurised systems			
12.	Lighting			
13.	Noise			
14.	Vibration			
15.	Fire			
16.	Radiation			
17.	Lifting			
18.	Biological hazards			
19.	Ergonomics			
20.	Electromagnetic fields			
21.	Explosion			
22.	Electrical energy			
23.	Workplace violence/stress etc.			

Other/Further details: ____________________

Signature of person conducting inspection: ____________________

Copies provided to: ____________________

FIGURE 1.21 General hazards checklist

Risk assessment for electrical work

All electrical work undertaken by electrical personnel must follow risk management processes. Risk management includes the following suggested steps:

- Identify all work activities.
- Establish work tasks.
- Identify electrical hazards relating to individual work duties.
- Determine the most appropriate control(s) for the dangers.
- Select the workers involved in implementing the identified controls.
- Analyse the risks associated with the hazards and controls.
- Obtain work permits based on the analysis undertaken to perform the work activity.
- Provide a tool box meeting (documented) and walk-around for the safety observer.
- Implement controls.
- Monitor and review effectiveness of checks.

REVIEW QUESTIONS

1. What type of process is 'risk mitigation'?
2. Where is risk management implemented for electricians?
3. What are the specific elements of risk management?
4. What is the purpose of risk analysis?
5. Outline the elements of risk action.
6. Why should all features of a risk management process be documented?
7. How can an enterprise demonstrate they have been actively working to ensure workplace health and safety?
8. What is essential to ensuring that every worker with risk management responsibilities feels confident raising, discussing and managing risks?

1.5 Airborne contaminants

Airborne contaminants include:

- asbestos particles
- glass and rock wool particles
- fibreglass particles
- silica dust
- wood dust
- Legionella
- bird droppings.

Asbestos

Asbestos is the general name used for a group of naturally occurring mineral silicate fibres. The three types of asbestos are blue, brown and white. The breathing in and retention of asbestos fibres causes several diseases: asbestosis, lung cancer and mesothelioma. There is no safe known level of exposure to asbestos and there is no known cure for its diseases.

Asbestos can be found in premises in forms such as roofing sheets, as shown in **Figure 1.22**, wall sheeting (fibro), sprayed-on finishes (on ceilings) and as lagging. Older-type switchboards using zelemite boards contain asbestos. Some old ceramic fuse holders and switch gear include a woven asbestos which acts as an arc shield.

FIGURE 1.22 Asbestos roof sheet

In those premises where it is assumed that asbestos material exists, an asbestos audit must be undertaken. Smaller fibres, less than 3.5 microns in length, can be inhaled and deposited in the lung where they can accumulate and cause fibrotic changes, while larger fibres cause skin irritation due to mechanical action.

Note: Lebah, zelemite and asbestos electrical panels have been confirmed to contain chrysotile (white asbestos). Electrical panels installed before 1988 would also be considered to contain chrysotile unless proven otherwise.

Safe Work Australia requires persons who manage or control a workplace to protect anyone that works with asbestos. This includes:

- having an asbestos register
- having an asbestos management plan
- controlling asbestos in the workplace
- holding the correct training and licensing
- monitoring workers' health.

The asbestos register must be prepared, maintained and kept at the workplace. There is no requirement to maintain an asbestos register if the building was constructed after 31 December 2003 or there is no asbestos identified in the workplace. The asbestos register:

- lists any asbestos identified or is assumed to be present at the workplace
- documents the date when the asbestos was identified
- details the location, type and condition of the asbestos
- must be maintained to ensure current information
- must be transferred to the employer or business when there is a change of management or workplace controller.

Other information that may be contained in an asbestos register includes:

- details of asbestos assumed to be present
- analysis results confirming existence of asbestos
- details of areas within the workplace that are inaccessible.

Figure 1.23 shows some signs used to warn of asbestos.

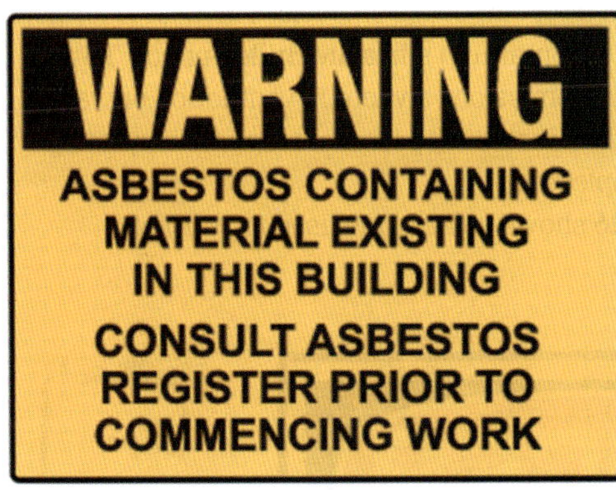

FIGURE 1.23 Asbestos warning signs

Wherever possible, asbestos-containing materials should be labelled. A label stating that the building contains asbestos can be affixed to the main electrical switchboard.

If a workplace has identified the presence of asbestos, it must prepare an asbestos management plan, which:

- identifies the location of asbestos
- documents decisions and the reasoning behind the decisions about the management of asbestos at the workplace
- stipulates the procedures for handling incidents and emergencies involving asbestos
- contains current information
- is reviewed at intervals not exceeding five years
- is available to any worker or any health and safety representatives who represent workers at the workplace
- contains information, consultation and training responsibilities to workers carrying out work involving asbestos.

Glass and rock wool

Glass wool and rock wool products (also called mineral wool) are manmade fibrous insulation materials used in the thermal or sound insulation of commercial, industrial and domestic buildings, plant and equipment.

Glass wool, as illustrated in **Figure 1.24**, is a fibrous product formed by either blowing or spinning a molten mass of glass.

FIGURE 1.24 Glass wool

Rock wool, as shown in **Figure 1.25**, is a fibrous product manufactured by a process of blowing or spinning from a molten mass of rock. In Australia, this is usually a mix of basalt and slag.

FIGURE 1.25 Rock wool

Dust from these products may cause discomfort of the nose, throat and respiratory tract, in particular for those suffering from upper respiratory or chest complaints such as hay fever, asthma or bronchitis.

Glass and rock wool manufactured in Australia is now biosoluble, which means that the material dissolves in bodily fluids and is rapidly cleared from lung tissue. If you must disturb glass or rock wool insulation, wear gloves, a long-sleeved shirt, long pants and safety goggles. It is also prudent to use respiration protection and engage in safe work practices to minimise exposure.

Fibreglass

Fibreglass or glass fibre is a material made from extremely fine fibres of glass. It is used as a reinforcing agent in conjunction with epoxy and polyester resins for many polymer products. The resulting composite material is properly known as fibre-reinforced polymers (FRPs) and it is made of panels. The danger with these materials when worked is that the fibre – whose size is similar to that of asbestos – can be inhaled and lodged in the lungs. Another hazard associated with fibreglass is skin and eye irritation. The reinforcing fibres can rise to the surface with wear and become a risk factor for skin burns. The fibre is ridged and can tear the skin. Some Type X switchboard insulating panels are an FRP product.

Switchpanel, also known as Typanel, is a kraft-paper-based substrate phenolic laminate with a phenolic resin and an anti-tracking melamine surface on one side. Other substrate materials for panels include canvas, linen or glass fibre. These substrates are impregnated with 30 per cent or more of thermosetting phenolic resin and cured.

Silica dust

Typical products such as clay bricks, concrete, tiles and fibre cement products contain silica. Silica is a class 1 carcinogen (cancer-causing substance) and the breathing in of the silica duct can cause silicosis. The dust is formed when products containing silica are drilled or cut. To control the dust emissions a vacuum ventilation or water spraying method including personal respiration systems should be employed.

Figure 1.26 shows some signs used to warn of silica dust.

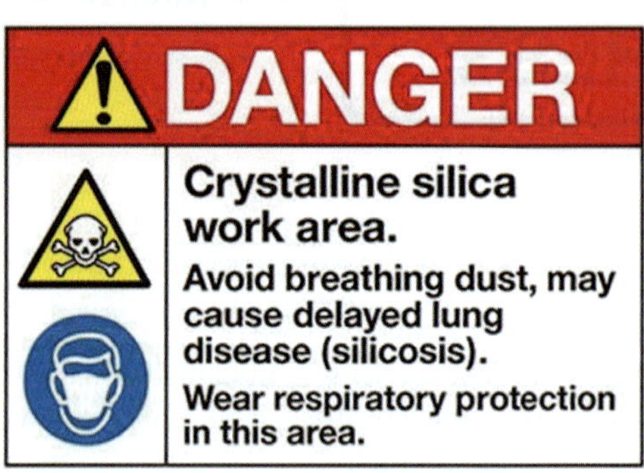

FIGURE 1.26 Silica dust warning signs

Airborne crystalline silica can bioaccumulate in the lungs and cause disease of the respiratory system. Large bioaccumulated loads of crystalline silica in the lung can cause silicosis, which is a build-up of connective tissue in the lung. Silicosis is an irreversible and progressive condition that in stages may not be noticeable. Severe silicosis can result in poor oxygenation of blood, difficulty in breathing and death. Evidence suggests that smokers are more susceptible to the long-term effects of silica dust exposure.

For silicosis to develop one needs to be exposed to substantial airborne quantities of respirable crystalline silica for prolonged periods.

The workplace exposure standard (WES) for respirable crystalline silica was changed to an eight-hour time weighted average (TWA) of 0.05 mg/m^3 in December 2019 and is in effect in all jurisdictions.

Wood dust

Wood dust can cause cancer and other symptoms depending upon the type of natural wood, or manmade material such as medium-density fibreboard (MDF), and any preservatives contained therein. The dust results from work activities such as sawing, drilling and sanding. Personal respiration systems should be employed.

If you are in the field and are unsure of the material you are working with, a drill attached to a vacuum cleaner fitted with a high-efficiency particulate air (HEPA) filter via an appropriate dust extraction unit should be used.

Legionella

Legionella is a bacterium surviving in water (at a particular temperature with nutrients) that can cause severe infections in persons exposed to sources containing the bacterium. When present in the bio-film (the slime on the surfaces that are contacting the water) Legionella is hazardous. An electrician could be exposed to the bacterium when working on air-conditioning systems including their ductwork and cooling towers. Spas, showers and fountains are other possible sources of Legionella. Legionnaires disease causes fever and flu-like symptoms and in severe cases pneumonia. Personal respiration systems should be employed.

Bird droppings

Bird droppings may present a risk of disease to electricians working in premises where a large number of birds are roosting, as **Figure 1.27** shows. The greatest risks arise from organisms that flourish in the droppings, feathers and nesting debris in roof cavities and eaves. Infection is most likely when dust containing the organisms or parasites is inhaled or ingested (via dirty hands when eating).

Source: Mark Saville

FIGURE 1.27 Bird droppings can represent a risk of disease

REVIEW QUESTIONS

1. What diseases are attributable to the breathing in and retention of asbestos fibres?
2. Where can asbestos be found in some premises?
3. Describe the effects that exposure to dust from glass wool and rock wool products may have on humans.
4. What does biosoluble mean in relation to glass wool and rock wool products?
5. What type of material has particle sizes similar to that of asbestos?
6. Name the disease attributable to the breathing in of silica dust.
7. State one method used to control emissions of silica dust.
8. Name one type of work activity that results in wood dust.
9. Where could an electrician be exposed to Legionella bacterium?
10. How can infection from bird droppings occur?

1.6 Chemicals in the workplace

The word 'chemicals' includes substances such as paint, glue, cleaning agents, fuel, pesticides and solvents.

In order to ensure chemical safety in the workplace, information must be available about the identities and hazards of the chemicals. All persons involved with chemicals in the workplace should read AS 2508 *Safe storage and handling information cards for hazardous materials*. Dangerous goods are solid, liquid or gaseous substances that may cause fires, explosions, rapid chemical reactions, immediate health risks (such as poisoning) or long-term effects such as cancer.

Dangerous goods can kill or injure workers. Their effects are usually sudden, evident and can be violent towards the worker. Dangerous goods are classified into nine classes according to the danger characteristics as illustrated in **Figures 1.28**, **1.29** and **1.30**, which comply with the Australian Dangerous Goods (ADG) code. The globally harmonised system of classification and labelling (GHS) applies to hazardous chemicals at workplaces. The ADG code applies when transporting dangerous goods.

It is important to know which dangerous goods produce toxic gas; which are highly flammable; which are dangerous when wet; or which are dangerous when they come into contact with air.

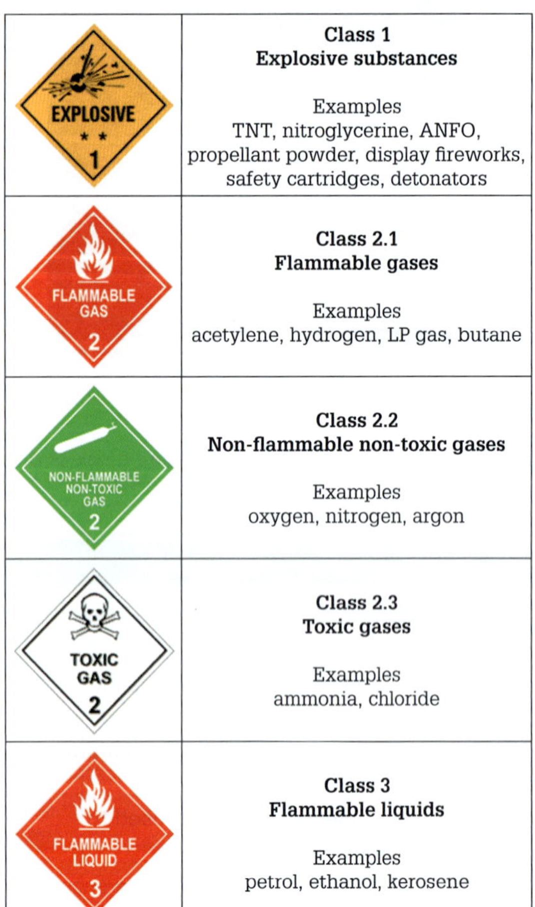

EXPLOSIVE 1	**Class 1** **Explosive substances** Examples TNT, nitroglycerine, ANFO, propellant powder, display fireworks, safety cartridges, detonators
FLAMMABLE GAS 2	**Class 2.1** **Flammable gases** Examples acetylene, hydrogen, LP gas, butane
NON-FLAMMABLE NON-TOXIC GAS 2	**Class 2.2** **Non-flammable non-toxic gases** Examples oxygen, nitrogen, argon
TOXIC GAS 2	**Class 2.3** **Toxic gases** Examples ammonia, chloride
FLAMMABLE LIQUID 3	**Class 3** **Flammable liquids** Examples petrol, ethanol, kerosene

FIGURE 1.28 Classification diamonds 1

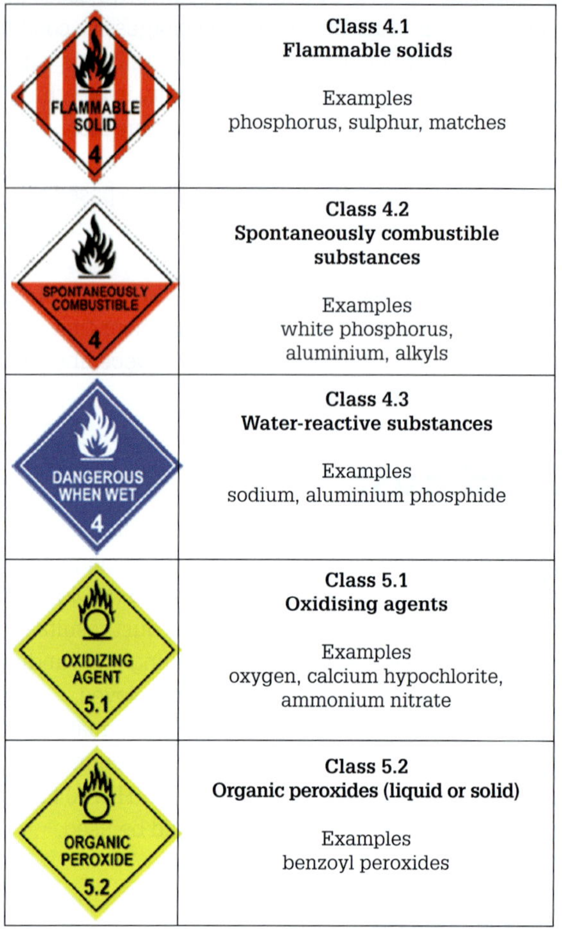

FLAMMABLE SOLID 4	**Class 4.1** **Flammable solids** Examples phosphorus, sulphur, matches
SPONTANEOUSLY COMBUSTIBLE 4	**Class 4.2** **Spontaneously combustible substances** Examples white phosphorus, aluminium, alkyls
DANGEROUS WHEN WET 4	**Class 4.3** **Water-reactive substances** Examples sodium, aluminium phosphide
OXIDIZING AGENT 5.1	**Class 5.1** **Oxidising agents** Examples oxygen, calcium hypochlorite, ammonium nitrate
ORGANIC PEROXIDE 5.2	**Class 5.2** **Organic peroxides (liquid or solid)** Examples benzoyl peroxides

FIGURE 1.29 Classification diamonds 2

Hazardous substances are chemicals and other substances that can cause injury, illness or disease. The health effects may be acute or chronic. Acute illnesses usually occur rapidly as a result of short-term exposure, and are of short duration. Acute illnesses can include irritation, burns, vomiting, anaphylaxis, and drowsiness. Chronic illnesses generally occur as a result of long-term exposure, and are of long duration – symptoms may develop only years after repeated and prolonged exposure. Chronic illnesses can include cancer, asthma, dermatitis, anaemia, chronic bronchitis and other diseases.

Dangerous goods and hazardous substances can enter the body of a worker through inhalation, absorption through the skin and ingestion.

A substance is deemed hazardous if:

- it is listed on the National Occupational Health and Safety Commission's (NOHSC) *List of designated hazardous substances*
- it meets the criteria in the NOHSC *Approved criteria for classifying hazardous substances*.

Some dangerous goods are also hazardous substances, including toxic substances in Class 6 (poisons) and corrosives in Class 8.

Manufacturers of chemicals are required to supply accurate data to enable the identification of those substances that are hazardous or dangerous.

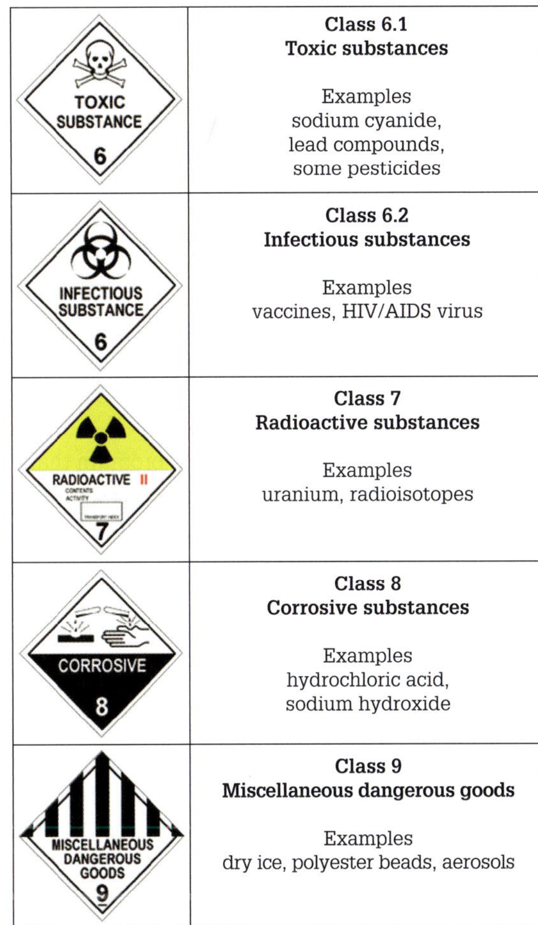

TOXIC SUBSTANCE 6	**Class 6.1** **Toxic substances** Examples sodium cyanide, lead compounds, some pesticides
INFECTIOUS SUBSTANCE 6	**Class 6.2** **Infectious substances** Examples vaccines, HIV/AIDS virus
RADIOACTIVE II 7	**Class 7** **Radioactive substances** Examples uranium, radioisotopes
CORROSIVE 8	**Class 8** **Corrosive substances** Examples hydrochloric acid, sodium hydroxide
MISCELLANEOUS DANGEROUS GOODS 9	**Class 9** **Miscellaneous dangerous goods** Examples dry ice, polyester beads, aerosols

FIGURE 1.30 Classification diamonds 3

Safety data sheets

Safety data sheets and labels indicate whether or not the chemical substance or material is considered a hazardous substance.

Safety data sheets contain information about the substance such as:

- a statement indicating whether it has been classified as hazardous to health in accordance with NOHSC criteria
- the contents
- what it should be used for and how to use it safely
- its health effects
- first-aid instructions
- advice about safe storage and handling

Make sure the safety data sheet is not out of date (i.e. older than five years).

An example of a safety data sheet for mineral turpentine (see **Figure 1.31**) is shown in the appendix to this section.

Source: Recochem Inc., Australia

FIGURE 1.31 Mineral turpentine

Safety data sheets are the primary source of chemical hazard information and must be readily accessible to workers. They should contain all pertinent physical and health hazard information, exposure limits, precautions for safe handling and use and applicable control measures including PPE use requirements.

Chemical substance labelling is also an effective method used to communicate chemical hazard information. Labels are a direct visual reminder of the hazards presented by a chemical substance and they include all appropriate environmental and health hazard warnings.

How to read a safety data sheet

The safety data sheet summarises knowledge of the health and safety hazard information of the material and how to handle and use the product safely in the workplace. Each user should read the safety data sheet and consider the information in the context of how the product will be handled and used in the workplace, including in conjunction with other products. The following information will help you translate a safety data sheet.

Acronyms

- AICS: Australian Inventory of Chemical Substances
- CAS number: Chemical Abstracts Services Registry Number
- Hazchem code: Emergency action code that provides information to emergency services
- UN number: United Nations Registry Number
- ADG code: Australian dangerous goods code.

Format

The Australian Safety and Compensation Council (ASCC) has developed minimum standards for the format and layout of a safety data sheet. For each product, the safety data sheet should provide:

- the name of the product
- the name, address and telephone number of the manufacturer or distributor
- the chemical components
- hazard and precautionary statements, hazard pictogram, and signal word
- health effects (short and long term)
- fire and explosion information
- requirements for safe handling and ways of controlling exposure to the substance
- first aid information
- storage and transport requirements
- spills and disposal information
- emergency information for firefighters
- contact information for further details.

For more information on each section of a safety data sheet, refer to the Appendix at the end of the chapter.

Storage procedures

When storing chemical substances:

1. Make a register of every chemical substance to be stored. The register must contain details of all dangerous goods and hazardous substances currently used, stored or handled on the premises.
2. Obtain from the supplier a safety data sheet for each chemical substance.

Storage procedures involve the following:

- manage your stock – only keep minimum amounts
- label shelves and storage cabinets with a segregation scheme so that chemical substances can be put away in the right place quickly
- refer to safety data sheet for specific chemical substance incompatibilities
- ensure that the storage cabinets are locked
- do not store liquids above solids in case of contamination in the event of a spillage
- limit the size of containers where possible to ≤5 L/kg
- always store chemicals on spill trays – kitty litter trays are ideal
- do not overload shelves
- do not store containers on the floor
- dispose of outdated chemical substances including all portable LPG cylinders that are not in test (i.e. 10 years).

All workers must be given a training program that informs the workers not only of the hazards of the chemical substances in their work area but also how to use the information generated in the hazard communication program.

Labelling

If you transfer chemical substances into a secondary container, ensure that the new container is properly labelled. This means that the full name of the chemical substance, appropriate risk and safety phrases as per the safety data sheet, dangerous goods class and subsidiary class diamond(s) if suitable are recorded. Also, if the chemical substance is classified as hazardous then the word 'Hazardous' must be printed on the label.

REVIEW QUESTIONS

1. Explain what is meant by the term 'Dangerous goods'.
2. Explain what is meant by the term 'Hazardous substance'.
3. How can dangerous goods and hazardous substances enter the body of a worker?
4. Name three examples of Class 3 Flammable liquids.
5. Name two examples of Class 4.3 Water-reactive substances.
6. Name two examples of Class 8 Corrosive substances.
7. What do safety data sheets and labels indicate?
8. What information should a register of chemical substances contain?

1.7 Confined spaces

A confined space is an enclosed, or partly enclosed, area which is not planned, or intended, principally as a place of work as **Figure 1.32** shows. In addition, a confined space may have a restricted means for entry and exit. Types of confined space are outlined in **Table 1.3**.

TABLE 1.3 Confined spaces

Confined space type	Example
Open spaces	drains, trenches and pits
Tunnel-type spaces	sewers, pedestrian tunnels, ductwork, shafts and lift wells
Building space	unventilated ceiling space as illustrated in Figure 1.32, basements and cold rooms, storage rooms, closets, plant rooms and crane cages
Moving space	lifts
Tanks and vessels	storage tanks, pressure vessels, vacuum vessels, boilers and silos

FIGURE 1.32 Confined ceiling space

The potential hazards that could be expected when working in confined spaces are:

- low oxygen
- mechanical motion
- non-ionising radiation or ionising radiation
- asphyxiants (dusts) and poisoning (inhalation/skin)
- electric shock
- noisy environment
- fire/explosion
- manual handling
- head or eye impact
- constrained movement and entrapment including burial beneath solids
- heat stress or cold stress
- falling objects
- drowning
- claustrophobia
- slips, trips or falls

The major hazard when working in confined spaces is an unsafe level of oxygen – depleted or enriched (less than 19.5% or greater than 23.5% in volume). Note that excess carbon dioxide in a confined space depletes oxygen levels.

A space that would not normally be regarded as a confined space may become a confined space by the nature of the work that is done within it, for example welding, which can cause atmospheric contamination.

Where entry into a confined space is necessary, appropriate safety procedures must be followed so that the risk associated with entry into and performance of work within the confined space is minimised as far as is reasonably practicable. Before working in a confined space, suitable emergency response and first aid procedures and provisions must be planned and established.

Figure 1.33 shows sample signage and entry permit for working in a confined space.

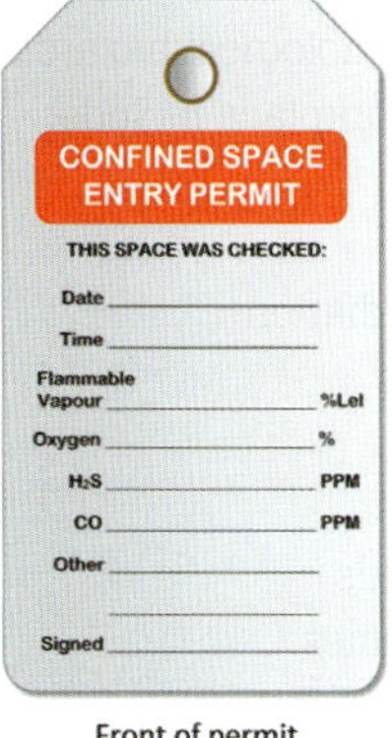

Front of permit

Back of permit

FIGURE 1.33 Confined space signage and entry permit examples

Entry means the action by which an authorised person passes through an opening to a permit-to-enter-required confined space. Entry includes consequent work activities within the confined space and has occurred as soon as any part of a person's body breaks the plane of an opening to the confined space.

There must always be a safety observer available at the entrance to the confined space whose task is to provide information to assist rescue workers. This person should never enter the confined space unless trained.

Portable gas detectors

Without appropriate gas detection, hazardous atmospheres within confined spaces may significantly affect the health of any entry person. Many airborne contaminants cannot be detected by smell or vision and can only be sensed with special detectors.

A range of Standards Association of Australia approved portable gas detectors is available for confined space applications to evaluate contamination levels. Confined space entry gas detectors are capable of continuous measurement (with alarms) of many gases such as oxygen, combustibles' LEL (lower explosive limit – a combustible gas or vapour in air which will ignite if a source of ignition is present), carbon dioxide, carbon monoxide, hydrogen sulphide, nitrogen dioxide, sulphur dioxide, chlorine, chlorine dioxide, phosphine, ammonia and hydrogen cyanide. Gas detectors have built-in sample pumps, hose draw with water trap and filter and data-logging functions.

The most common form of sampling the air within a confined space is the sample draw method. The advantage of this method is that the detecting is performed by a competent person outside the confined space. With a sample draw method, a pump moves the sample from the atmosphere (up to 30 m within the space) and draws it through a hollow tube to the sensors.

For personal protection always test the sides of a confined space hatch before opening. This process prevents the lemming effect whereby one entry person blindly follows another in the absence of any real or understood rationale.

Persons who are required to work in confined spaces as part of their employment must be provided with training. Such training would include:

- legislation
- codes of practice and the Australian/New Zealand standard AS/NZS 2865:2009 *Safe working in a confined space*
- confined space hazard identification
- safe work procedures
- confined space entry/exit procedures
- correctly fitting a full-body harness
- breathing apparatus
- gas detection
- rescue
- emergency procedures
- fire fighting and equipment inspection and maintenance.

REVIEW QUESTIONS

1 What is a confined space?
2 Name three examples of open spaces that can be classified as confined space.
3 List four potential hazards that could be expected when working in confined spaces.
4 What is the major hazard faced when working in confined spaces?
5 What is the task of a safety observer?

1.8 Physical and psychological hazards

The physical and psychological hazards that may be encountered in the workplace include:

- industrial noise
- vibration
- ultraviolet (UV) radiation
- overuse syndrome
- stress
- drugs and alcohol
- optical fibre hazards
- temperature extremes.

Industrial noise

Noise is defined as unwanted sound. Industrial noise for workers can become a major industrial disease which threatens the quality of life. Unpleasant health effects of industrial noise include hearing loss, sleep disturbances, job performance reduction and frustrated communication responses (some sounds are still loud, others are not heard and this makes speech difficult to understand). Protecting the hearing and health of workers imposes a wide range of explicit obligations upon employers, employees and manufacturers of machinery and equipment.

Noise levels are measured in decibels (dB(A)) and exposure to noise is based on the amount a person receives during a typical day. Hearing loss may begin with continuous daily noise over 75 dB(A) over several years. The level of continuous noise which is classified as excessive is 85 dB(A). Some noises, such as blasts, explosions and heavy hammering, are so loud they can damage the sensitive hair-like nerves in the cochlea of the inner ear, causing immediate permanent damage. Sound energy that is too loud or intense can damage and break off the hair-like nerve cells. As they are nerves they do not grow back once they are injured and they cannot be repaired. To avoid damaging these sensitive nerves, workers need protection from workplace noise.

A noise problem in the workplace can be mitigated by applying any combination of the following three control measures:

1 Replacing or modifying the noise source to eliminate or reduce the noise output. This is the best way to prevent occupational noise-induced hearing loss.
2 Using sound absorbing screens to weaken the sound conduction path thereby reducing the noise level reaching the worker. The louder the sound, the less

time it takes to cause a hearing loss. Therefore, the noisier the workplace, the fewer exposures and the shorter the number of years it will take to produce a significant hearing loss.

3 Varying the worker's exposure either through limiting the exposure time or by using personal protective equipment. Using personal hearing protectors will reduce noise by an amount between 15 dB and 25 dB.

Personal protectors

Figure 1.34 shows the basic types of personal protectors for noise. These are the insert type and the ear-muff type, and there are advantages and disadvantages to both. For the insert type the advantages are that they are small and easy to carry. The disadvantages are that they can get dirty when removed and inserted with dirty hands. Muff-type protectors are technically better for reducing noise levels.

FIGURE 1.34 Hearing protectors

Too much noise in the workplace may cause other effects such as continuous ringing in the ears, stress, high blood pressure and tiredness and, at very high levels, physical pain.

Action must be taken to protect hearing where 75 dB is exceeded at any time, even if only for a short period.

Vibration

Vibration motion can be in one or more directions – up and down, side to side, front to rear or rotational. A worker who regularly and frequently is exposed to high levels of vibration can suffer permanent injury.

Vibration hazards in the workplace usually present themselves in two forms:

1 whole-body vibration, where the body is shaken by a machine or vehicle
2 hand–arm vibration, where the vibration effect is localised to a particular part of the worker's body.

The rapid motion of incorrectly suspended cabins and seats during bumpy rides in tractors, trucks, buses or earth-moving equipment can expose the worker to whole-body vibration, resulting in haemorrhoids, backache and heart problems.

Electrical workers operating hand-held machinery such as rotary hammers, drills, grinders and pneumatic tools and fixed machinery such as bench grinders and lathes may suffer from vibration syndrome. Vibration syndrome is characterised by aches in the arms and shoulders, damage to nerves, muscles and joints and vibration white finger in the hands. This condition, for which there is no cure, starts as tingling or numbness in the fingers and prolonged exposure can lead to gangrene in the fingers and hands. The extent of damage vibration causes to the human body depends on:

- the length of time of exposure
- the frequency (motion) at which the vibration occurs
- the amplitude of vibration.

To limit exposure to vibration, portable hand tools should have recoil damping and air cushioning to soften vibration, and padded handles to reduce transmission. Vibration exposure can be limited by using seat and cabin suspension.

Ultraviolet radiation

Ultraviolet (UV) radiation is a known cause of skin cancer, skin ageing and eye damage and may affect the body's immune system. Therefore, it is essential that all workers exposed to UV radiation at work should be given training so that they understand the risks associated with UV exposure.

Electrical workers who work outdoors are likely to suffer health damage from exposure to solar UV radiation. Other sources of UV radiation at a worksite that electricians may be exposed to include arc welding, UVA radiation tanning lamps, high UVB and UVC emitting lamps used to sterilise work areas and medium- to low-power UVA lamps used in insect control.

UV radiation is known to have unfavourable health effects both in the short and long term. UV radiation is absorbed in the skin and the adverse health effects are mostly confined to the skin and eyes. In most cases it is thought that shorter wavelengths (UVB) are more harmful than longer wavelengths (UVA). Short-term exposure to UV radiation causes reddening of the skin, sunburn and swelling, which may be very severe.

The most serious long-term effect of UV radiation is the induction of skin cancer. The non-melanoma skin cancers are basal cell carcinomas and squamous cell carcinomas. They are relatively common in light-skinned people, although they are rarely fatal. They occur most frequently on sun-exposed areas of the body such as the face and hands and show an increasing incidence with increasing age.

Malignant melanoma is the main cause of skin cancer death. A higher incidence is found in people with large numbers of moles, those with a fair skin, red or blond hair and those with a tendency to freckle, to burn and not to tan on sun exposure. There is no known low exposure limit for UV effects such as skin cancer.

Exposure of the eye to UV radiation causes inflammation of the cornea and the conjunctiva, more commonly known as welder's flash. Symptoms range from mild irritation to severe pain and possibly irreversible damage.

In relation to sources of UV radiation, well-designed engineering, administrative and personal protective controls can keep the risks to a minimum. Engineering controls would include the provision of shade cover or canopies for outdoor work. For indoor work, opaque barriers and UV-radiation-blocking filters would be suitable.

Administrative controls for outdoor workers would include rescheduling outdoor work programs where possible to be performed outside the peak UV radiation period from 10:00 am to 2:00 pm. For indoors the administrative controls would include warning signs and limiting the time during which UV radiation sources are switched on.

Personal protective equipment for outdoor workers should consist of protective clothing that is loose fitting, made of close-weave fabric and provides protection to the neck and the lower arms and legs. Hats should shade the face, neck and ears and have a wide brim (8–10 cm). If hard hats have to be worn they should have attached neck flaps.

The type of sunscreen used should be a minimum broad-spectrum SPF 15 and be applied regularly and liberally to exposed skin. Sunglasses should be close fitting, of a wrap-round design and block at least 99 per cent of the UV radiation.

Overuse syndrome

Occupational overuse syndrome (OS) is a form of injury which affects tendons, joints and muscles in the fingers, hands, wrists and elbows. It is caused by repetitive movements or an uncomfortable working stance stressing the body parts beyond their physical limit. OS is also known as repetitive strain injury, or RSI.

Occupational OS is usually associated with repetitive hand movements such as using a screwdriver or placing conductors in stator slots when rewinding small-frame electric motors, but any part of the body can be affected. Rest is usually the best cure.

To limit exposure to OS in the workplace, work benches should be at waist height so that shoulders can relax and arms can bend gently at the elbows. In addition, all hand tools used should be ergonomically designed. Some tasks such as motor rewinding can only be carried out by hand. Therefore, the intensity, duration or frequency of the activity can be reduced through the use of frequent breaks.

Stress

Workers experience work-related stress when they feel that they are unable to cope with the working environment demands placed upon them. Stress can be the result of understaffing, bullying, harassment or intimidation, long work hours, job insecurity or poor management practices. Work-related stress can lead to mental and physical ill health.

When a worker is exposed to constant, extended work-related stress, they may experience physical and emotional symptoms such as:

- frequent headaches
- feeling frustrated and irritable or angry
- loss of energy and motivation
- changes in appetite and weight
- sleep difficulties
- generally feeling worn out or run down.

Attitude has a lot to do with whether events and occurrences at work produce a feeling of stress. Stress management comes down to finding ways to change your thinking and manage your expectations. Some important ways to adjust your attitude include:

- Be realistic. Don't expect too much of yourself or others.
- Try to be assertive rather than passive or aggressive.
- Be flexible.
- Think positively. Look at each stressful event and occurrence as an opportunity to improve your life.
- Don't take work problems home or home problems to work.

To help reduce work-related stress, workers need to develop a network of friends and family members to create a strong social support group that will be there when serious stress is experienced.

Drugs and alcohol

The rights of persons to drink and take drugs socially are acknowledged, but when work performance suffers or fellow workers are endangered then some action must be taken.

SWITCH ON

Workers should not be adversely affected by alcohol or drug use during working hours

Drug and alcohol abusers in the workplace can be difficult to identify but there are some signs that indicate possible drug and alcohol problems. These include:

- regular and often unexplained absences
- involvement in workplace accidents
- unreliable work patterns and reduced productivity
- lack of concern with personal hygiene
- overreaction to real or imagined criticism.

Physical signs can include tiredness, hyperactivity, dilated pupils, slurred speech, unsteady walk, bloodshot or glassy eyes, persistent cough, hangovers and mood swings.

Drug and alcohol prevention workplace programs seem to be the best way of preventing, detecting and dealing with drug and alcohol abusers. These programs can help individual workers to become healthy and productive again.

Optical fibre hazards

Working with fibre optics can have some risks. However, by following your workplace code of procedures the hazards can be reduced. Always assume that every fibre is active. It is easy to forget to deactivate a fibre before viewing it. Remember, you will not see or feel the danger, which will be enhanced when optical microscopes are used. A serious hazard of optical fibre work is the fibres themselves (see **Figure 1.35**).

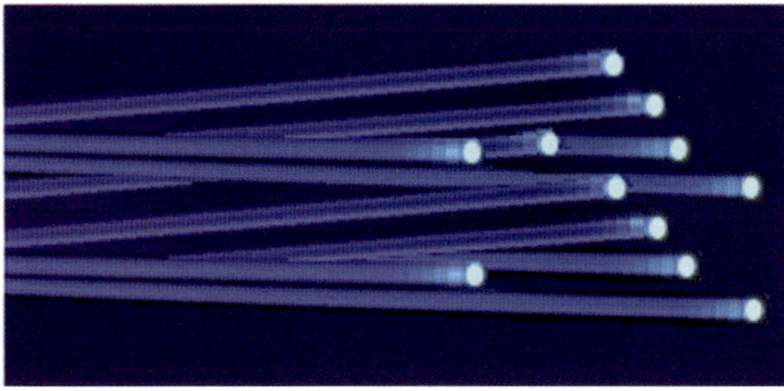

FIGURE 1.35 Optical fibres

Side-shielded safety glasses and disposable latex gloves are recommended personal protective equipment when working with optical fibres. Tiny shards of glass will be present when performing joining or terminating procedures and if any contact is made with the eyes this may cause severe injury. When working with bare fibres, assume at all times that your fingers are covered with glass fragments. It is essential that your hands are washed immediately after leaving the work area. Never eat or drink in your job area. If fibre particles are ingested they can cause internal haemorrhaging.

It is suggested that when working with optical fibres you use a work surface that will allow fibre shards to be visible. If sitting, a smooth-surfaced chair is recommended as it prevents glass shards from standing vertical.

When using UV light to cure adhesives in splices and connectors, UV-rated safety glasses must be worn.

Termination kits using chemicals and hot-melt processes should be used in a well-ventilated area.

All fibre off-cuts must be collected and disposed of in an industry standard sharps container. The container itself must be removed in an approved manner by using a hazardous waste service.

Temperature extremes

Extreme temperatures can adversely affect the human body, which has an ideal core temperature of about 37 °C. Those people most at risk are the elderly, young children and people with certain medical conditions or who are taking certain medications.

Extreme heat

Heat stress occurs when the human body is unable to cool itself sufficiently to maintain a healthy core temperature. The normal process the human body uses for cooling is perspiration. Sometimes this method of cooling is insufficient and the body temperature keeps increasing. Heat-related illness can range from mild conditions such as a rash or cramps to very serious conditions such as heatstroke, which can be fatal.

Contributing factors to heat stress include overexertion in hot weather, exposure to radiant heat from the sun or fires, including bushfires, and exercising or working in hot, poorly ventilated or confined spaces. Some heat-related illnesses include:

- exacerbation of existing medical conditions
- heat rash, which is a skin irritation caused by excessive sweating
- heat cramps, which may occur after strenuous activity in a hot environment, when the body is depleted of electrolytes and water
- dizziness and fainting, usually resulting from reduced blood flow to the brain
- heat exhaustion, which is a serious condition and, if not treated, can develop into heatstroke. Excessive sweating in a hot environment reduces blood volume in the body. Symptoms include paleness and sweating, rapid heart rate, muscle cramps, headache, nausea and vomiting, dizziness or fainting.
- heatstroke, which occurs when the core body temperature rises above 40.5 °C, is a medical emergency that requires immediate attention as the body's internal systems start to shut down. Symptoms include staggering, confusion, fitting or unconsciousness.

Preventing heat-related illness

Prevention is the best way to manage heat-related illness. Preventative measures include:

- drinking plenty of water
- avoiding exposure to heat and staying out of the sun as much as possible
- covering exposed skin, which includes 'slip, slop, slap' (slip on a shirt, slop on sunscreen, slap on a hat) as well as 'seek' shade and 'slide' on sunglasses
- planning ahead so that outdoor activities occur during the cooler parts of the day
- taking frequent breaks and, whenever possible, stay indoors or in the shade
- maintaining air circulation and staying cool, such as by using wet towels, putting feet in cool water and taking cool (not cold) showers
- maintaining energy levels by eating smaller meals more frequently and eating cold meals such as salads.

Monitor for symptoms of heat-related illness and check in on others, especially those more vulnerable to the effects of extreme heat.

Extreme cold

Exposure to extreme cold may result in a condition known as hypothermia, which occurs when the temperature of the human body drops below 35 °C. Hypothermia can occur if the temperature of the environment is less than body temperature, as a body will radiate heat into the surrounding environment. If the body cannot sufficiently compensate for the heat lost to the environment, then the body temperature will begin to drop.

The human body can lose heat through:

- conduction, which is direct transfer from the body to an object that is cooler than the body
- convection, where air or liquid flows across the skin drawing off heat
- radiation, where heat is transferred into the environment
- evaporation, where heat is drawn off when fluid on the skin turns to vapour.

Hypothermia may be classified as mild, moderate or severe.

- The signs and symptoms of mild hypothermia include: pale and cool to touch as blood vessels constrict in the skin; numbness in the extremities (fingers and toes); sluggish responses to questioning, drowsiness or lethargic; shivering; and increased heart and breathing rates.
- The signs and symptoms of moderate hypothermia include: decreasing conscious state; incontinence as a result of an increased workload on the kidneys; shivering ceases; slowed heart and breathing rates and low blood pressure.
- The signs and symptoms of severe hypothermia (below 28 °C) include: unconscious and not responsive; slow and irregular heartbeat; pupils unresponsive to light; muscle rigidity; pulse and breathing may be difficult to detect.

Preventing hypothermia

Exposure to cold weather, even for short periods, can be dangerous. Shivering and feeling cold or numb are warning signs that the body is losing excessive heat.

Prevention is the best way to manage hypothermia. Preventative measures include:

- avoiding prolonged exposure to cold weather
- watching for weather conditions that may increase the risk of hypothermia and acting accordingly
- wearing several layers of clothing to trap body heat
- wearing a waterproof outer layer to stay dry
- wearing gloves, scarves and socks, and have spares to change out when wet
- wearing thermal insulated boots
- wearing warm headgear
- wearing loose-fitting clothes and boots as restricted blood circulation increases risk of hypothermia
- drinking plenty of fluids but avoiding alcohol and caffeine
- eating regularly
- taking regular breaks to reduce the risk of physical fatigue
- monitoring body temperature
- not delaying changing out of wet clothes.

Monitor for symptoms of hypothermia and check in on others, especially those more vulnerable to the effects of extreme cold.

REVIEW QUESTIONS

1. Name four unpleasant health effects of industrial noise.
2. What level of continuous noise is classified as excessive?
3. What are the two forms of vibration hazard?
4. Name two tools likely used by electrical workers that may cause the user to suffer from vibration syndrome.
5. What are three adverse effects of ultraviolet (UV) radiation?
6. What is the main cause of skin cancer death for outdoor workers?
7. What causes occupational overuse syndrome (OS)?
8. How can drug and alcohol abusers in the workplace be identified?
9. Name the recommended personal protective equipment when working with optical fibres.
10. What condition may result from exposure to extreme cold?

1.9 Manual handling

The term 'manual handling' is used to describe any activity requiring the use of the hands or bodily force applied by a person to lift, lower, push, pull, heave, carry, move, support or restrain an object, person or animal. Manual handling also covers activities which require the use of bodily force such as operating power tools or crowbars and repetitive movements such as using a screwdriver or keyboard activity.

Injuries

Injuries most frequently associated with manual handling include:

- back injuries – spine, joints, ligaments, muscles and intervertebral discs
- fractures – to the fingers, hand, feet and toes
- lacerations – to the hands and fingers
- crush injuries – to the fingers
- sprains – to the wrist, thumb and ankle
- strains – to the back, shoulder, arms, hands and fingers
- contusion – bruising to various parts of the body
- hernia – an opening in the wall of a muscle, tissue or membrane that normally holds an organ in place.

Every muscular effort, however slight, involves the spine. When you lift, your back is put under stress, especially the lower spine. The lower spine is very mobile and is able to bend forwards, sideways and backwards but is capable of only very little rotation. Twisting or jerking while lifting and carrying can injure the small facet joints (stabilising joints located between and behind adjacent vertebrae) which guide movement of the back.

Intervertebral discs, which separate the vertebrae (spinal bones), and the ligaments, which hold the vertebrae together, are also at risk. The vertebrae of the spinal column run down the back as shown in **Figure 1.36**, connecting the skull to the pelvis. These bones protect nerves that come out of the brain and travel down the spinal cavity and out to the entire body.

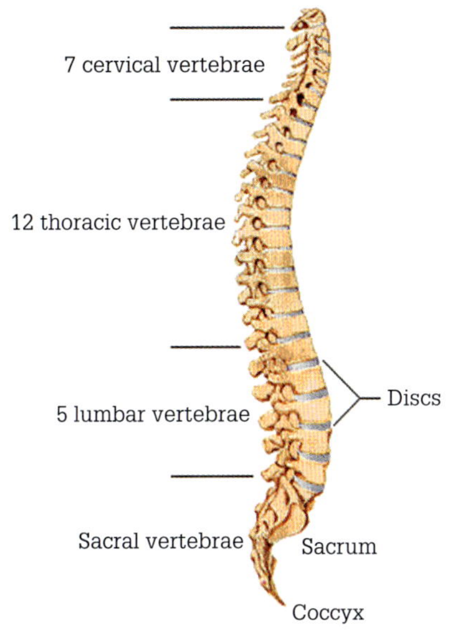

FIGURE 1.36 S-shaped spine

The intervertebral discs are composed of soft gelatinous substances which provide spinal column cushioning. The discs are also surrounded by a strong fibrous ring and with repeated incorrect lifting the discs, fibrous ring or its supporting ligaments may tear or rupture.

In general, back injuries are caused by wear and tear and damage to the joints, ligaments, muscles and intervertebral discs which occur during day-to-day manual handling activities. Manual handling injuries may result from:

- ongoing wear and tear caused by frequent or extended periods of manual handling activity, such as frequent handling of electrical stock throughout the day
- sudden damage caused by extreme or difficult manual handling or awkward lifts such as lifting an electric motor, an air-conditioner or white goods from the ground to a van or utility.

Direct injury can be caused by unexpected happenings such as an electrician walking on irregular ground (construction site) while carrying cartons of electrical cable or equipment and experiencing a trip hazard and falling. Some examples of actions that may cause manual handling injuries are:

- activities involving sudden, jerky or hard-to-control movements such as hammer drilling
- activities involving too much bending, reaching or twisting such as installing cables
- activities where a long time is spent in the same posture or position such as working in confined spaces within roofs
- activities that are fast and repetitious such as coil turns placement in stator slots with respect to motor rewinding
- activities where heavy equipment has to be lifted and carried manually, for example, air-conditioners and stoves
- activities where force is needed to carry out a task such as when using screwdrivers or spanners.

Procedures for lifting

There are several simple procedures to remember when lifting or handling loads.

Plan the lifting activity

Assess the risk factors with respect to the load including:

- type of load – glass, drum, etc.
- weight of load
- size of load
- distance the load has to be carried or moved
- route
- placement requirements.

If in doubt, do not lift alone.

Keep the load close to the body

The further away the load is from your body the more stress is placed on your lower spine. To prevent this type of stress when lifting from a ground position the knees must be bent with the load gripped between the waist and shoulders to ease the strain on the back and the arms. Keeping the load close to your chest with the shoulders back and your bottom out makes you more secure. The spine has a natural S curve. The shoulders are back and the S curve is directly over the pelvis. Make sure you know where the centre of gravity of the load is and keep the heaviest side nearest to your body at waist level.

Maintain balance

Stand square on to the load and as close to it as is comfortable. Have your feet flat on the floor surface with your leading leg forward, preferably facing the direction you're going to move. Ideally you should lift, carry and place in one direction where possible.

Use your legs

Bend your legs slightly, and use your leg muscles to take the weight. Do not turn or twist your body once you have made the lift. Always bend your knees when setting the load down. An illustration of the correct lifting technique is shown in **Figure 1.37**.

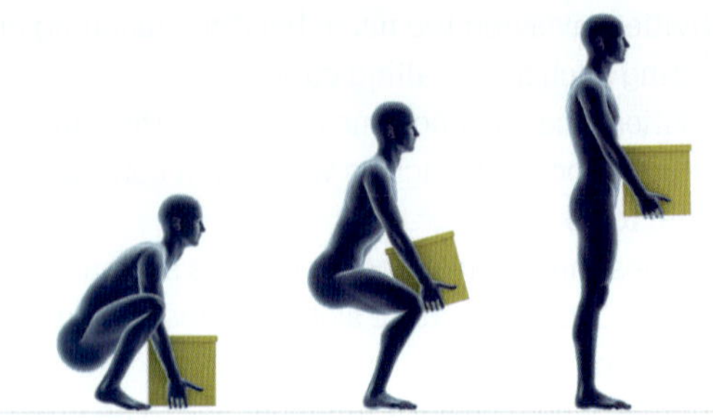

Source: iStockphotos/angelhell

FIGURE 1.37 Correct lifting technique

Some loads can be pushed if the load is on a smooth surface. A recommended method in these situations is to turn your back to the load and push with your legs. Do not pull the load if possible.

Alternative materials-handling aids

Alternative materials-handling aids for carrying or moving loads should be used whenever possible to minimise lifting and bending activity. These alternative materials-handling aids include use of lift truck, scissors lift truck, pallet truck and other mechanical devices – levers, crowbars, gloves.

Implementing proper lifting techniques and other safety measures can significantly reduce your chance of a back injury incident.

All persons involved with manual handling should read the following documents:

- National standard for manual handling, NOHSC:1001 (1990), Commonwealth of Australia
- National code of practice for manual handling, NOHSC:2005 (1990), Commonwealth of Australia
- Information Booklet, Manual handling (1992), Commonwealth of Australia.

REVIEW QUESTIONS

1. What does the term 'manual handling' describe?
2. Name four common injuries associated with manual handling.
3. What in general causes back injuries during day-to-day manual handling activities?
4. List four risk factors with respect to a load that needs to be moved.
5. Name the four simple procedures for safe manual handling.
6. Describe an appropriate method to move a load that is on a smooth surface.
7. Name three alternative materials-handling aids.
8. How can you significantly reduce your chance of a back injury incident?

1.10 Working at heights

Not all electrical installation work can be done on the ground floor or at floor level. To perform installation work above these levels, ladders or scaffolding are required. While ladders are uncomplicated, planning and care are required to use them safely. Scaffolding over 2 m in height requires a permit and must comply with state or territory legislation.

Before work starts, you need to identify all physical locations and tasks that might cause you to fall. Identification is mandatory under the *Code of Practice for the Prevention of Falls in Housing and Construction.*

Ladder hazards

Ladder accidents are usually caused by inappropriate selection, lack of maintenance or unsafe use. Some of the more common hazards involving ladders, such as unsteadiness, falls and electric shock, can be foreseen and prevented. Prevention of ladder accidents requires planning, correct ladder selection, safe work procedures and effective ladder maintenance.

SWITCH ON

Hazard prevention

- Always face the ladder when ascending or descending.
- Ascend or descend one rung at a time.
- Always have three limbs (two arms and one leg or one arm and two legs) on the ladder rungs at one time.
- Never hand-carry equipment or tools on a ladder – use tool belts.
- Never over-reach in order to carry out a task; move the ladder.
- Always place the ladder on firm or level ground.
- Ensure that the support for the top of the ladder is secure.

Ladder types

All ladders should be inspected on delivery, after suspected damage and once every six months. The safety inspection of ladders should look for the following:

- *stiles*: broken, split, cracked, decayed

- *rungs or treads*: missing, broken, split, cracked, decayed, corroded, worn, dirty
- *fittings*: broken, worn, loose, faulty
- *spreaders and ropes*: broken, worn, decayed
- *safety feet*: broken, worn, faulty.

In addition to the above, fibreglass ladders should have a smooth surface that is clean and polished, of uniform colour and without any pits, chips, voids or longitudinal grooves along the stiles. All ladders used by electricians should have the following information permanently marked on the ladder:

- manufacturer's name
- load rating (in kg)
- working length of the ladder.

Step ladders should be used in the fully opened position only and a worker should not stand above the second-top tread. For a single or extension ladder a worker should not stand above the third-top rung.

Portable ladders should comply with the requirements of the following relevant Australian and New Zealand standards:

- AS/NZS 1892.1:1996 *Portable ladders – Metal*
- AS 1892.2:1992 *Portable ladders – Timber*
- AS/NZS 1892.3:1996 *Portable ladders – Reinforced plastic*
- AS/NZS 1892.5:2000 *Portable ladders – Selection, safe use and care.*

Portable ladders are designed for one-man operation and are constructed under two general classes:

- Type I Industrial – heavy duty with a minimum load rating of 120 kg
- Type II Domestic – light duty with a maximum load rating of 100 kg.

Only industrial wooden or fibreglass ladders stamped with the AS/NZS mark must be used by electrical workers when carrying out their activities. Maximum ladder lengths are shown in **Table 1.4**.

TABLE 1.4 Maximum ladder lengths

Construction material	Single	Extension	Step ladders
Metal ladders and reinforced plastic ladders	9 m industrial	15 m industrial	6.1 m industrial
	5 m domestic	7 m domestic	2.4 m domestic
Timber ladders	9.2 m runged	15.3 m	5.5 m industrial
	4.9 m cleated		2.4 m domestic

EXERCISE 1.1

The selection of a ladder must take into account the top three rungs where a person should not stand. Using **Table 1.4**, which metal ladder and reinforced plastic ladder would be suitable for reaching an object that is:

a 10 m above the ground?

b 4 m above the ground?

Step ladders

Step ladders (see **Figure 1.38**) can be used when installing lighting points, fans and other electrical fittings.

FIGURE 1.38 Fibreglass step ladders

Step ladders should be securely fully open. All four legs must be on solid, level ground with the spreaders locked fully open. Never climb on the cross-bracing. Never use a folding step ladder in an unfolded position (leaning it against a wall) unless it is designed for that purpose. The main cause of injuries is from step ladders tipping sideways. Where a potential fall hazard exists (to below where the step ladder is standing) the minimum distance between the step ladder and fall hazard is to be the sum of the person's head height (i.e. from floor/ground) and a 1.5 m 'buffer zone'. For example, if a worker is positioned on a ladder 1 m above the ladder base and the worker is 1.6 m tall, a minimum distance of 4.1 m is required between the base of the step ladder and the fall hazard. If installation or repair work is to be carried out closer to the fall hazard, the worker should be secured in a harness which is not attached to the step ladder. AS/NZS 1891.4:2009 *Industrial fall-arrest systems and devices – Selection, use and maintenance* requires fall protection to be provided where a person could fall a distance greater than 2 m.

SWITCH ON

Note that step ladders, with their broad tread, are safer than rung-type ladders because of their stability and balance.

Single-length and extension ladders

Before working at height, identify all hazards, assess their risks and prepare a safe work procedure. Fall-arrest equipment will be essential for some tasks. An extension and a single-length ladder are illustrated in **Figure 1.39**.

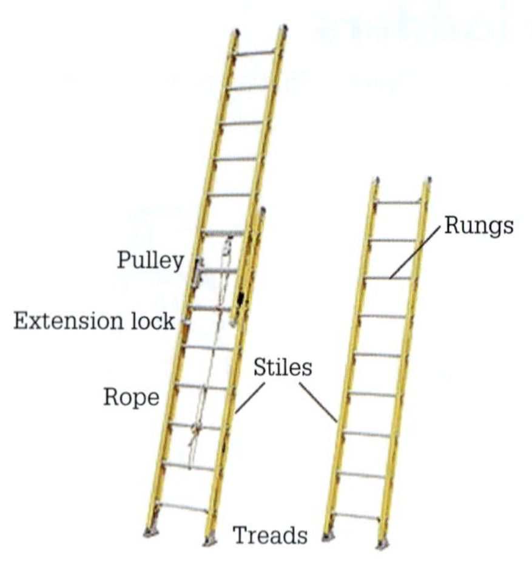

FIGURE 1.39 Extension and single-length ladders

Always use two people to carry and set up a single-length or extension ladder, if possible. Small ladders can be carried parallel to the ground by one person when lifted at the midpoint of the side rail. With two-person ladders the ladder should be carried on the same shoulder. Extension ladders are adjustable in length and they consist of two or more single sections travelling in guides or brackets.

When installing a ladder:

1. Visually inspect the building features to see if anything will impair the safe climbing, descending or moving of the ladder.
2. Place the ladder on its back with the feet butted up against the building. If this is obstructed, a second person can assist in setting up the ladder by placing their foot on the bottom rung and the other foot on the base of one stile.
3. Lift the ladder's top by maintaining pressure against the wall or assistant.
4. Lift the ladder overhead. With your hands extend the ladder to your highest point of reach, work hand over hand pushing the ladder towards the building or assistant. When you reach beyond the middle point of setting up your extension ladder or lift past the middle rung, the increasing force of the weight of the top of the ladder will be working against your efforts.
5. Once vertical, rest the ladder against the building wall, lift the ladder's base with both arms and move the ladder slowly away from the building to obtain the correct ladder pitch (see **Figure 1.40**). Make sure you do not move the ladder into window glass or mark the exterior building surface.
6. Extend the ladder using the ladder's rope and pulley system while stabilising the ladder with your right or left foot on the ladder's bottom rung. The ladder must be close to if not fully vertical in position. Once the ladder is erected to the correct working height, slowly lower the top of the ladder until it rests against the building. Both feet of the ladder and both left and right top stiles must rest equally against the building. To lower the ladder, follow these steps in reverse order.

FIGURE 1.40 Ladder pitch

EXERCISE 1.2

Determine the distance an extension ladder must be from a wall, when measured at the base, if the ladder extends:

a 10 m

b 15 m

Scaffolds

A scaffold is any temporary elevated platform and its structure is used for supporting workers or materials or both. Regulations govern the safe use of scaffolds and these vary from state to state. Scaffolds as shown in **Figure 1.41** must be erected by a construction worker called a scaffolder or rigger whose main job is to erect or dismantle scaffolds (of metal or timber) on construction sites, and who has received appropriate training in this field.

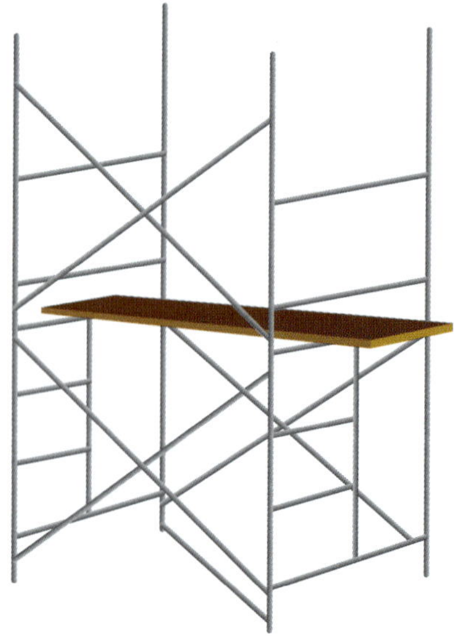

FIGURE 1.41 Scaffold

Unsafe scaffolds endanger workers in many ways. Mechanisms can break, collapse or give way. Planks, boards, decks or handrails can fail and it is possible for entire structures to collapse. Even on sound scaffolds, workers can slip, trip or lose their balance, and without appropriate protection they don't have to fall far to get injured. In many cases, scaffold accidents are the result of untrained or incorrectly trained workers.

Tools, materials or equipment can fall from a scaffold and strike workers below. Therefore, the scaffold must be barricaded or a toe board must be placed along the edge of the scaffold platform. Panelling or screening must protect persons below when tools, materials or equipment are piled higher than the top edge of the toe board. Alternatively, guardrail systems, canopies or catch platforms may be installed to retain materials.

Workers on scaffolds more than 2.4 m above a lower level must use fall protection. A personal fall-arrest system consists of an anchorage, connectors and a body harness. It may also include a lanyard, deceleration device or lifeline. An emergency rescue plan is also an essential requirement with any personal fall-arrest system.

Elevated work platforms

An elevated work platform (EWP) consists of a telescoping device, scissor device, boom lift as illustrated in **Figure 1.42** or articulating device used to raise or lower workers, equipment and material to and from a work location above the EWP's support surface.

The operator/worker may independently control the height and reach of the EWP from within the worker box. Owing to the EWP's particular operation, risks to workers can arise from:

- lack of training in proper use
- no pre-planning for activities has been carried out
- not using additional safety equipment such as a safety belt or harness
- over-reaching by operators from the worker box
- lack of suitable maintenance or repair
- use on unstable ground or in extreme weather conditions such as wind and storms.

FIGURE 1.42 Installing fibre-optic cables as an aerial system, using a boom lift

REVIEW QUESTIONS

1. Which items of equipment is an electrician likely to use when working above ground or floor level?
2. What does the prevention of ladder accidents require?
3. When should ladders be inspected?
4. All ladders used by electricians should have what information permanently marked on the ladder?
5. Which ladder type should be used in the fully opened position?
6. What is the main cause of injuries from step ladders?
7. What must workers do before working at height?
8. Determine the distance an extension ladder must be from a wall, when measured at the base, if the ladder extends 12 m.
9. What is a scaffold?
10. What is required for workers on scaffolds more than 2.4 m above a lower level?

1.11 Working with electricity

While electricity allows us to experience a more comfortable existence, it is not without its hazards. As such, it is important to adopt work practices and attitudes that promote working safely with electricity.

Common electrical hazards

There are three categories of common electrical hazards:

- electric shock
- arcing
- toxic gases associated with the arcing hazard.

Electric shock

Low-voltage contact can cause severe shock and burn injuries and even death. The shock experience can be received by direct contact with a live part, arcing or tracking (equipment leakage currents associated with appliances such as ranges).

Electric shock can also be encountered with earth conductors when the earth conductors are carrying a total load current because of an open-circuited main neutral.

Electric shock can occur with neutrals. For example, if the neutrals are connected at a common terminal, as can exist with some equipment, when the circuit being worked on is isolated another circuit may be alive, allowing the disconnected neutral to become live when disconnected at the common terminal.

Burns

There are five possible burns that a person who has suffered electric shock could experience.

1. Point of contact and exit burns that are local and can be just to the skin or deeper, reaching to the bone.
2. Arc burns that can be very extensive, especially if the person experiences a high-voltage flashover. Arcs are extremely hot. The person will be physically scarred from the experience and may lose limbs.
3. Radiation, similar to sunburn, from the intensity of the arc.
4. Burns from the vaporised metal of a conductor that can saturate the face or hands.
5. Finally, high voltage allows a fault current to flow with impunity into the body of a person, causing contact burns that can punch through the layers of the skin destroying the tissue below. Burns caused on the body of a person are sterile and heal very quickly.

Electric shock is not an isolated event and many persons will experience one at some point in their life. In Australia, the majority of electric shocks occur at the mains potential of 230 V and 400 V. Most fatalities occur at these voltages and high-voltage injuries are mainly from burns. Due to the number of material variables within the person, it is not possible to state the minimum current that will result in death. Muscle tissue does play a part and it appears the current threshold is lower for women than men.

Arcing

A short-circuit fault can cause an intense, determined and rapidly intensifying arc of electrical energy to be established within a very small interval in time. This energy when released suddenly will result in terrible burns on anyone within the arc's range.

Toxic gases

The arcing and burning associated with a low-voltage electrical fault may cause toxic gases to be emitted which can have an adverse effect on a target organ. The effect may be severe if carcinogens are present.

Electrical hazards represent a serious, widespread occupational danger; all electricians are exposed to electrical energy during the performance of their daily duties, and electrocutions do occur. Other low-voltage electrical hazards include:

- wiring not in conformance with the Standards Association of Australia standard
- exposed electrical parts
- contact with overhead supply
- defective/inadequate insulation
- improper earthing of equipment
- overloaded circuits (can produce heat or arcing)
- damaged power tools, equipment and testing devices.

Lack of training and an unsafe work environment (wet work surfaces, inclement weather conditions, noise, toxic chemicals and flammables) are common circumstances that can aggravate working on low-voltage tasks.

The physiological effects of current

When working on or near electric circuits capable of delivering high power, electric shock becomes more of a reality and pain is the least significant outcome of electric shock. Therefore, safe working practices must always be implemented. Electric shock is a general term for the excitation or confusion of the purpose of nerves or muscles caused by the passage of an electric current. It is usually painful but is not necessarily associated with actual damage to a person. There are two methods by which a person can suffer electric shock. The first is by direct contact whereby a person comes into contact with a conductor that is normally live and becomes part of the fault path. The second is by indirect contact whereby a person comes into contact with material that has become live under fault conditions.

The effect that electric shock has on the body depends on the magnitude of the fault current, the body parts through which the current flows and the physical condition of the person. Serious electric shock is almost always associated with alternating current and is rare when low-voltage direct currents are used.

Electrical injuries have three categories: nerve immobilisation, fibrillation and tissue destruction.

The common feature of electric shock is a brutal piercing and deadening pain at the points of entry and exit. Muscular action is complex and under the control of nerves. Electric shock is frequently accompanied by an unintentional contraction of muscles associated with the path of current. It is this unconditioned response that can tear tendons and the muscle tissue apart. The motor neurons are living cells capable of carrying an electric signal. Nerves are made up of neurons and they can be stimulated in various ways; it could be thermally, chemically, by pressure, light or a current entering the body. The current's effect on the nerves and the nerve centre of a person is one of prime control and restraint.

As a direct result of electric shock, a person may clench and be unable to release a conductor or an appliance, or if the person has touched a live conductor without clenching, the muscles of the person's back and legs may contract harshly so that the person is unwillingly thrown backwards.

Another possible result of involuntary muscle contraction is that the muscles of the diaphragm and chest may contract and thereby prevent breathing. This could lead to death by suffocation. Death may also occur as a result of current flowing through the respiratory control centres of the central nervous system of a person. However, death is usually caused by direct interference with the action of the heart.

The ventricles of the heart have three states – rest, normal beat and fibrillation. If the heart is violently disturbed by the passage of current it may pass to the fibrillation state. Once there it will exist in this state and death will occur unless action is taken to 'defibrillate' the heart. It is also possible that the experience of electric shock creates fear in a person and the person dies of a heart attack rather than from the electric shock experience.

LV rescue training

Due to the dangers associated with contacting live electrical conductors or electrical equipment, it is essential for those undertaking electrical work receive appropriate training in LV rescue. This training must include safe work methods for removing an electric shock victim from a live electrical situation in accordance with workplace emergency management procedures. See Chapter 2 of this textbook for more detailed information.

Body resistance

The resistance of the body from hand to hand or from hand to foot is variable and depends upon the area of electric contact and whether that area is dry, moist or wet. Australian Standard AS/NZS 60479.2:2000 *Effects of current on human beings and livestock – Special aspects* tables the human body's opposition for different voltages and lists a person's body resistance range as being between 1750 Ω and 6100 Ω at a touch voltage of 25 V. In dry conditions the body can offer some resistance to current but above 30 V the body practically has no impeding effect at all. The body acts like a voltage-dependent resistor, meaning that as the voltage increases the resistance of the body decreases.

Physiological sensations of current

Practically speaking, the magnitude of fault current can be neither predetermined nor discovered after an electric shock. The physiological values of current are suitable only for research on deciding on permissible leakage currents for electrical devices. At around 15–17 mA rms, the so-called 'grip' current, it becomes extremely difficult to release a conductor held in the hand. Reactions to 50 Hz rms current are shown in **Figure 1.43**.

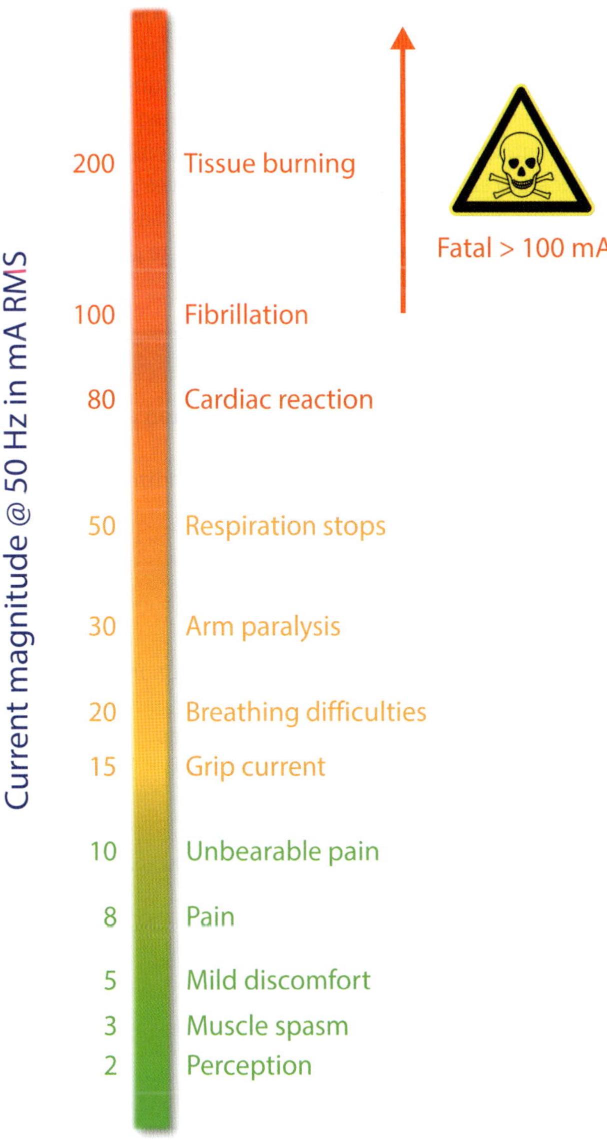

FIGURE 1.43 Possible reactions to 50 Hz rms current (assumed conditions – hand grip of 5 seconds duration)

EXERCISE 1.3

Using **Figure 1.43**, determine the possible effect on a person who has accidentally held a live metallic object causing a current of 30 mA to flow through their body.

Effect of voltage frequency

It is unfortunate that mains voltage is generated at a frequency of 50 Hz, because at this frequency voltage finds resonance with the nerve endings within the human body, producing maximum excitation and therefore control of the nerves. Lower or higher frequencies do not do this. However, if a person's body is exposed to high-frequency electric fields it will be heated up like any other conductor. **Note:** This should not be taken to imply that voltages at frequencies other than 50 Hz are safe.

The fundamental principles

The Australian/New Zealand Standard AS/NZS 3000:2018 *Wiring Rules*, 'Protection for safety' lists 10 fundamental principles for protection against the damaging effects of current. The carrying out of these 10 principles is intended to ensure the safety of persons, livestock and property against dangers and damage that may arise in the reasonable use of electrical installations. The principles are listed in the Switch On box below.

SWITCH ON

10 fundamental principles for protection against the damaging effects of current

1. Protection against both direct and indirect contact by use of extra-low voltage
2. Protection against direct contact
3. Protection against indirect contact
4. Protection by use of residual current devices (RCDs)
5. Protection against thermal effects in normal service
6. Protection against unwanted voltages
7. Protection against over-current
8. Protection against fault currents
9. Protection against over-voltage
10. Protection against injury from mechanical movement

Causes of electrical accidents

Risks for electricians arise in a work situation because it is sometimes difficult to find faults or malfunctions in electrical equipment when the equipment is not operating.

Some common causes of electrical accidents in fault finding or repair include:

- exposed live terminals
- terminals or conductors being live under different conditions of operation of the equipment
- loose or disconnected leads becoming live
- test equipment conducting the potential closer to the electrician
- test equipment being the wrong class for the task (particularly test probes)
- inadequate test points
- incorrect or poorly maintained testing instruments
- inadequate knowledge of equipment or causes of faults
- equipment located in hazardous areas, which often include bolt-on or screw-on covers
- working alone on live equipment or installations.

Protective measures

Fuses and circuit breakers are intended primarily for the protection of conductors and equipment. They prevent overheating of conductors that might otherwise create hazards for installations. They also open the circuit under certain hazardous earth-fault conditions in an MEN (multiple earth neutral) system. These protective devices may also protect persons against electric shock; see **Table 1.5**.

TABLE 1.5 Protective measures

Protective device	Explanation
Fuse	This device will open a circuit when a predetermined excess of current flows. It may be able to be rewired or alternatively may incorporate a wire in a glass case or embedded in insulating powder within a cartridge case.
Circuit breaker	This is a form of switch which opens automatically if the circuit is overloaded; it may operate on either a thermal or a magnetic principle.
Earthing	The external metal casing of electrical equipment, cables and conduit must be earthed if required. This arrangement provides a low impedance path for the return of earth-fault currents. Earthing also ensures that a shorted circuit is automatically disconnected from the supply by drawing sufficient current to blow the fuse or operate the circuit breaker.
Separated extra-low voltage (SELV)	This is a means of protecting persons against electric shock. SELV is defined as not more than 50 V a.c. or 120 V ripple-free d.c. These voltage limits are regarded as user safe. Circuits using voltages to these maximum levels are designed and protected to remain under these safe voltage levels in normal operation and under single-fault conditions.

»

TABLE 1.5 Protective measures (*Continued*)

Protective device	Explanation
Residual current device (RCD)	A residual current device (RCD) or safety switch is a device that immediately trips out the electricity supply in the event of a fault to earth (current leaking to earth). This device can provide protection from harmful electric shocks in situations where a person comes into contact with a live conductor and earth. RCDs are designed to operate within 10–50 μs.
Isolation of electrical supply	When performing electrical work on low-voltage circuits and equipment, all sources of low voltage are to be isolated and locked off. Where isolation switches are unable to be locked, they are to be secured to prevent inadvertent operation.
Securing devices	Securing devices can be used to ensure that the isolation points, when in the open position, require a deliberate action to engage or disengage them. These securing devices should have the facility to allow a padlock to be applied. Warning signs and tags must also be put on the lock or switch to inform other persons of electrical isolation.
Low-voltage systems	Low voltage is defined as voltage exceeding 50 V rms a.c. or 120 V d.c. but not exceeding 1000 V rms or 1500 V d.c.

Permit-to-work system

A permit-to-work system is a formal written system using a systematic disciplined approach to assessing the risks of a task and specifying the precautions to be taken. Examples where a permit-to-work system is required include:

- performing live work
- hot work (work involving welding, thermal or oxygen cutting, heating, grinding and other fire-producing or spark-producing operations)
- roof access
- excavation
- elevated work (a permit is required for all forms of mechanical work platforms and scaffolding)
- high-voltage work
- working in a confined space.

It is also a way of establishing an effective means of communication and understanding between the enterprise personnel requiring the work to be done and the workers or contractors who are going to do the work.

The permit-to-work system:

- specifies the work to be done and the equipment to be used
- specifies the safety precautions to be taken when performing the task
- gives permission for work to begin
- advises occupants of the workplace that work is being performed within their workplace
- provides a check to ensure that all safety considerations have been taken into account, including the validity of permits and certificates and compliance with the workplace safety policies and procedures. On completion of work it provides a mechanism to check that all work has been completed to the satisfaction of the controller of the workplace.

The issue of a permit to work does not in itself make a task safe. It also does not represent permission to do dangerous work and therefore should not be seen as a simple way of eliminating a hazard or reducing risk.

Clearance certificates

A clearance certificate is a permit for entry for workers required to undertake installation or maintenance work in hazardous locations such as laboratories, animal houses, glasshouses, chemical stores and flammable fluid stores. This is to ensure that a defined area is safe prior to work being undertaken.

Isolation permissions

It is the responsibility of all workers and contractors performing work within a worksite to ensure the safety of themselves as well as other occupants of the building or work area. To enable the safe conduct of electrical maintenance and installation work, plant and equipment to be worked on must be isolated to the satisfaction of the controller of the workplace.

If the electricity source is not isolated, the electrician would be working on a live circuit. The electrician would be vulnerable to electrical hazards including electric shock, electrical flashover and short-circuit. The consequences of an electrical accident may be electrocution, serious burns on human bodies, fall from a height, fire and explosions and so on. Therefore, any electrical equipment to be worked on must be isolated from all sources of supply by either:

- opening switches or isolators
- removing fuses or links
- opening circuit breakers
- removing circuit connections.

Prior to commencing isolation work, the controller of the electrical installation must be consulted to confirm that the isolation of supply will not cause a loss of supply to critical medical equipment, information technology equipment or other essential equipment. The following items must be planned in advance prior to commencing isolation work:

1. The time for the isolation to start.
2. The extent of the isolation to ensure the electrical equipment to be worked on is removed from all sources of supply.
3. The operation of control devices which may inadvertently energise the installation or equipment to be worked on.
4. Information provided in a formal written system which will be given to all affected persons regarding the agreed outcomes.

Isolation and tagging must be carried out in accordance with the agreed isolation, safety tagging and locking procedure. By observing these requirements, the risk of injury to workers as well as disruption to building occupants can be minimised.

Once work is completed the electrical supply can be reconnected, provided all parties to the isolation agree that they are satisfied that the work carried out has been done to an acceptable standard and the necessary tests have been carried out. The affected parties must sign off on the permit.

REVIEW QUESTIONS

1 List the three categories of common electrical hazards.
2 What may be the consequences of low-voltage contact?
3 What may the arcing and burning associated with a low-voltage electrical fault produce?
4 State the factors that determine the severity of electric shock.
5 List the three categories of electrical injuries.
6 How can the passage of current through the human body prevent breathing?
7 What does the term 'grip' current refer to?
8 Name five common causes of electrical accidents in fault finding or repair.
9 If the electricity source is not isolated, the electrician would be working on what type of circuit?
10 Who has responsibility to ensure the safety of themselves as well as other occupants of the building or work area?

CHAPTER REVIEW

1.1 Effective communication

- Being an effective communicator provides a means for establishing a trusting, collaborative relationship with customers.
- Forms of communication for electrotechnology workers include verbal, listening, non-verbal, written, compassion, cultural awareness, trust, and providing instruction.
- Barriers to effective communication include physical, social and psychological.

1.2 Electrotechnology work environment

- A worksite means the place of employment, base of operation or location of workers.
- Standard work procedure means implementing specific, efficient plans of action.
- Worksite hazards may be physical, chemical, biological, mechanical/electrical, or psychological.
- The aim of work environment safety signs is to regulate and control safety-related behaviour.
- The type of work environment safety sign used should be suitable for the intended application, and workers should be informed of its purpose.
- Fire protection in the workplace requires appropriate means of extinguishing local fires in all locations where workers are employed.
- Not all fire extinguishers can be used effectively on all types of fires.
- Different types of fire extinguishers are distinguished from each other by their colour scheme.
- Fire blankets extinguish fire by smothering the surface of the fuel to remove the oxygen element of the fire.

1.3 Legal and ethical issues

- In general, occupational health provides a broad framework to protect the safety, health and welfare of all persons within a workplace.
- The objectives of WHS in Australia are structured around eight core values.
- Duty of care is based on the principle of reasonableness, and is founded in the law of negligence.
- Negligence occurs when there is a failure to exercise a degree of reasonable care under the circumstances, which results in an unintended but foreseeable injury to another party.
- Induction is imparting information to a prospective staff member.
- Health and safety committees are set up in workplaces to help resolve health, safety and welfare issues that arise in the workplace.
- A workplace health and safety inspector may enter any workplace and exercise their powers while they are in a workplace.

1.4 Risk mitigation

- The first element when conducting a risk assessment is to define the relationship between yourself, the task and its environment.
- The second element of risk assessment is the discovery of the health and safety risks to be managed (what can happen).
- The purpose of risk analysis is to ignore the trivial acceptable risks and to concentrate on significant hazards, their consequences and the probability that the consequences could result in serious harm to persons.
- Record keeping is critical to the risk management process.

1.5 Airborne contaminants

- Airborne contaminants include asbestos particles, glass and rock wool particles, fibreglass particles, silica dust, wood dust, Legionella, and bird droppings.

1.6 Chemicals in the workplace

- All persons involved with chemicals in the workplace should read AS 2508 *Safe storage and handling information cards for hazardous materials.*
- Hazardous substances are chemicals and other substances that can cause injury, illness or disease.
- Safety data sheets and labels indicate whether or not the chemical substance or material is considered a hazardous substance.
- The safety data sheet summarises knowledge of the health and safety hazard information of the material and how to handle and use the product safely in the workplace.
- When storing chemical substances, you should make a register of every chemical substance to be stored and obtain from the supplier a safety data sheet for each chemical substance.
- All workers must be given a training program that informs them not only of the hazards of the chemical substances in their work area but also how to use the information generated in the hazard communication program.

1.7 Confined spaces

- A confined space is an enclosed, or partly enclosed, area which is not planned, or intended, principally as a place of work.
- Where entry into a confined space is necessary, appropriate safety procedures must be followed.
- Persons who are required to work in confined spaces as part of their employment must be provided with training.

1.8 Physical and psychological hazards

- Noise levels are measured in decibels (dB(A)) and exposure to noise is based on the amount a person receives during a typical day.
- The basic types of personal protectors for noise are the insert type and the ear-muff type, and there are advantages and disadvantages to both.
- Ultraviolet (UV) radiation is a known cause of skin cancer.
- Occupational overuse syndrome (OS) is a form of injury which affects tendons, joints and muscles in the fingers, hands, wrists and elbows.
- Stress can be the result of understaffing, bullying, harassment or intimidation, long work hours, job insecurity or poor management practices. Work-related stress can lead to mental and physical ill health.
- Drug and alcohol prevention workplace programs seem to be the best way of preventing, detecting and dealing with drug and alcohol abusers. These programs can help individual workers to become healthy and productive again.
- Extreme temperatures can adversely affect the human body, which has an ideal core temperature of about 37 °C.

1.9 Manual handling

- The term 'manual handling' is used to describe any activity requiring the use of the hands or bodily force applied by a person to lift, lower, push, pull, heave, carry, move, support or restrain an object, person or animal.
- Alternative materials-handling aids for carrying or moving loads should be used whenever possible to minimise lifting and bending activity.

1.10 Working at heights

- Not all electrical installation work can be done on the ground floor or floor level. To perform installation work above these levels, ladders or scaffolding are required.
- Always use two people to carry and set up a single-length or extension ladder, if possible.
- An elevated work platform (EWP) consists of a telescoping device, scissor device, boom lift or articulating device.

1.11 Working with electricity

- Electric shock is a general term for the excitation or confusion of the purpose of nerves or muscles caused by the passage of an electric current.
- There are two methods by which a person can suffer electric shock: direct and indirect.
- At around 15–17 mA rms, the so-called 'grip' current, it becomes extremely difficult to release a conductor held in the hand.
- Fuses and circuit breakers are intended primarily for the protection of conductors and equipment. A fuse will open a circuit when a predetermined excess of current flows.
- A circuit breaker is a form of switch which opens automatically if the circuit is overloaded.
- A residual current device (RCD) or safety switch is a device that immediately trips out the electricity supply in the event of a fault to earth (current leaking to earth).

TRIAL EXAM

For Chapter 1 knowledge assessment, please complete the following trial exam.

1 The form of listening that requires complete attention and engagement is:
   a listening with purpose
   b contingent listening
   c passive listening
   d aural communication

2 The positive emotion that allows people to show that they care and are willing to help is:
   a indifference
   b coldness
   c awareness
   d compassion

3 Ineffective communication can result in:
   a more efficient work practices
   b poor customer interactions and relationships
   c improved customer interactions and relationships
   d interactive workplace relations

4 What should a practical plan of action for the work environment lead to?
   a good housekeeping
   b a safer environment for workers and the surrounding community
   c provision of emergency information
   d legible environmental safety signs

5 A hazard that can be encountered at a worksite is lead. Lead is classified as :
   a physical hazard
   b mechanical hazard
   c biological hazard
   d chemical hazard

6 For maximum effectiveness, occupational environmental safety signs should be:
   a installed 3 m above a walkway
   b maintained in good condition, kept clean and well illuminated
   c written in several languages
   d replaced at intervals not exceeding 3 months

7 A white sign with a red circle without a diagonal slash through it is a:
   a prohibition sign
   b mandatory sign
   c restriction sign
   d warning sign

8 For fire to exist, what four elements must be present?
   a smoke, heat, fuel, chemical reaction
   b aqueous solution, oxygen, smoke, spark
   c fuel, oxygen, heat, chemical reaction
   d carbon dioxide, fuel, heat, smoke

9 A fire extinguisher should be serviced:
   a monthly
   b 6 monthly
   c annually
   d after use

10 What legislation provides a broad framework which codifies the duties of care that are owed under common law?
   a *Wiring Rules*
   b material safety data sheets
   c work and safety procedures
   d Work Health & Safety Acts

11 Which of the following statements imposes a clear obligation on all persons to ensure safe work environments?
   a not to put others at risk
   b safety instructions are only for me
   c care for yourself only
   d only employers are responsible for safety

12 Who has the responsibility to review risk assessments?
   a the worker
   b the employer
   c the health and safety committee
   d the first aid person

13 The main purpose of a safety inspector is to:
   a observe electrical personnel working in hazardous situations
   b monitor compliance of a workplace with the WHS Act
   c establish a good housekeeping culture in a workplace
   d carry out induction of new electrical staff

14 Who is responsible for PPE approval?
   a the worker
   b the employer
   c Standards Australia
   d the safety observer

15 All electrical workers beginning work in a particular location for the first time must undergo:
   a induction
   b resuscitation skilling
   c housekeeping awareness training
   d observation and analysis procedures

16 The Good Samaritan doctrine is the principle upon which:
   a the permit to work is based
   b the duty of care is based
   c the classification of hazardous materials is based
   d the guideline of fairness is based

17 Risk mitigation is:
   a a process that promotes an ineffective risk management culture throughout the enterprise
   b an obligation to health and safety from all persons engaged in an electrical business
   c a voluntary reporting scheme administered by an overlord
   d a web-based incident reporting system for electrical businesses

18 The specific elements of risk management are:
  a reporting, recording, and rationalising
  b identifying, recording, reporting, and administering
  c alerting, risk minimising, and evaluation of risk classes
  d identifying hazards, assessing risks, controlling risks, and reviewing control measures

19 The breathing in and retention of asbestos fibres may cause:
  a discomfort of the nose, throat and respiratory tract
  b asbestosis, lung cancer and mesothelioma
  c skin and eye irritation
  d silicosis

20 Silicosis is:
  a biosoluble and dissolves in bodily fluids
  b a build-up of connective tissue in the lung
  c a skin disease
  d a bacterium surviving in water

21 A substance that can cause injury, illness or disease is:
  a a dangerous good
  b a harmful substance
  c a hazardous substance
  d an inert substance

22 A safety data sheet provides:
  a a list of designated hazardous materials
  b instruction on the use of the material
  c classification of hazardous materials
  d knowledge of the health and safety hazard information of the material

23 An enclosed, or partly enclosed, area which is not planned, or intended, principally as a place of work is:
  a a confined space
  b an example of a moving space
  c a noisy environment
  d a source of non-ionising radiation

24 The major hazard when working in confined spaces is:
  a muscle cramps
  b physical exertion
  c an unsafe level of oxygen
  d poor illumination

25 An activity which requires repetitive movements, such as using a screwdriver can cause:
  a overuse syndrome
  b stress
  c white knuckles
  d burns

26 Hearing loss may begin with continuous daily noise over:
  a 55 dB(A)
  b 65 dB(A)
  c 75 dB(A)
  d 85 dB(A)

27 Which of the following affords the best protection from exposure to solar UV radiation for electricians?
  a using PPE
  b working between the hours of 11 am and 3 pm
  c applying liberal amounts of coconut oil
  d using spray tanning lotions

28 What can vibration syndrome cause?
  a aches in the arms and shoulders, damage to nerves, muscles and joints
  b loss of hearing
  c loss of sight
  d brain damage

29 When carrying a load, it should be:
  a close to your body
  b at arm's length
  c at head height
  d placed on your shoulder

30 How can you significantly reduce your chance of a back injury incident?
  a ensuring that the load has excess weight
  b carrying heavy loads great distances to build stamina
  c implementing proper lifting techniques and other safety measures
  d taking frequent rest breaks to recover

31 Step ladders are safer than rung-type ladders because:
  a of their stability and balance
  b rung ladders are generally shorter
  c rung ladders have spreaders
  d step ladders are heavier than rung ladders

32 The ladder pitch for a single-length ladder is:
  a equal to a 45° angle
  b equal to a 75° angle
  c equal to a 35° angle
  d equal to an 85° angle

33 Grip current is assumed to be in the range of:
  a 3 to 5 mA
  b 100 mA or higher
  c 15 to 17 mA
  d 20 to 30 mA

34 The formal written system using a systematic disciplined approach to assessing the risks of a task and specifying the precautions to be taken is:
  a touch and go system
  b the fundamental system
  c she'll be all right system
  d permit-to-work system

35 A residual current device (RCD) or safety switch:
  a is a circuit control switch
  b connects to the main earth isolator
  c is a means of arc prevention
  d is a device that immediately trips the electricity supply in the event of a fault to earth

36 There are three categories of common electrical hazards, two of which are electric shock and toxic gases. Name the third category.
  a electrocution
  b thermal heating
  c arcing
  d confined locations

APPENDIX Safety data sheets

This appendix explains the information given in the various sections of a safety data sheet compliant with the Globally Harmonized System of Classification and Labelling of Chemicals (GHS) and then gives an example of a safety data sheet for mineral turpentine.

Section 1: Product identification

Section 1 contains information pertaining to:

- Product name
- Synonyms
- Recommended use and restrictions on use of chemical
- Name, address and phone number of the manufacturer or distributor.

Section 2: Hazards identification

The following indicates how Section 2 of a safety data sheet should be formatted.

Safe Work Australia regulates hazardous chemicals. Some manufacturers will not list the substances as hazardous if they are not on the Australian Inventory of Chemical Substances (AICS).

Some information is withheld here because the material is proprietary. (The word 'proprietary' indicates that a party, or proprietor, exercises private ownership, control or use over an item of property.)

This section of the safety data sheet should identify if the chemical is classified as hazardous or a dangerous good, state the signal word (either warning or danger), and list the appropriate GHS classifications and hazard and precautionary statements. The purpose of the hazard and precautionary statements is to facilitate a standardised approach to risk assessment. Pictograms or hazard symbols may be shown on the safety data sheet as a visual representation of the type of hazard.

Hazard statements

This methodology uses a hierarchical approach to risk assessment and provides a description of the potential risk posed by the chemical. Hazard statements specify the physical, health or environmental hazards of the chemical. A hazard statement may also state the routes of exposure, internal organs susceptible to the chemical, and the severity of the chemical hazard. Each statement is accompanied by a code beginning with the letter H followed by three numerals and is specific to that statement. Individual hazard statements may also be combined on a safety data sheet to give a more precise description of the hazard of a particular chemical.

Precautionary statements

The primary purpose of a precautionary statement system is to predict what accidents, health effects or incidents may occur, how they may happen and how they may be prevented. This methodology also uses a hierarchical approach to safety. Precautionary statements are categorised as general, preventative, response, storage and disposal statements. Each statement is accompanied by a code beginning with the letter P followed by three numerals and is specific to that statement. Individual precautionary statements may also be combined on a safety data sheet to give a more precise description of safety procedures or measures.

Section 3: Composition and information on ingredients

This section is important because the identity of any hazardous chemical is listed. If the material is a mixture of more than one chemical, the safety data sheet will often list the chemical components of the mixture, though this is not essential if the chemical is non-hazardous or is in sufficiently low amounts. The proportion of each chemical should also be stated; however, a range may be given and the total of all components may not equal 100%.

Many chemicals will have a Chemical Abstracts Service (CAS) Registry Number which:

- is a unique numeric identifier
- designates only one substance
- has no chemical significance.

Section 4: First aid measures

This section will give information concerning initial first aid care for a person exposed to the chemical, including whether medical care is required and the immediate and delayed effects of exposure. This section should contain information on:

- what to do if the substance is swallowed
- what to do if there is contact with the eyes
- what to do if there is contact with the skin
- what to do if the substance is inhaled
- first aid facilities
- advice to doctor
- additional relevant information.

Section 5: Fire-fighting measures

This section of the safety data sheet describes measures for fighting a fire involving the chemical or a fire located near the chemical and should contain information on:

- suitable extinguishing equipment
- specific hazards arising from the chemical, such as toxic or flammable gases from combustion
- special protective equipment and precautions for fire fighters
- additional information such as HAZCHEM codes.

HAZCHEM codes provide useful information to emergency services about procedures for handling a fire or spill of a chemical. The HAZCHEM code consists of either two or three characters, the first being a numeral, followed by either one or two letters. Examples are:

- 3WE, 2E, 3YE, 1Z, 1SE, 4X.

Section 6: Accidental release measures

This section states the procedures to follow after the spill or leak of the chemical in order to minimise harm to people or the environment. The procedures to follow may depend on local regulations, which may not be specified on the safety data sheet. This section should provide information on:

- personal precautions, protective equipment and emergency procedures
- environmental precautions
- methods and materials for containment and cleaning up.

Section 7: Handling and storage

This section of the safety data sheet will provide information on how to handle the chemical safely and hygiene practices, such as washing hands after use and removing contaminated clothing. The appropriate storage conditions to minimise the risk of accidental release of the chemical will also be described, including which storage conditions to avoid.

Section 8: Exposure controls and personal protection

This section should detail how the risk of exposure to the chemical can be avoided or minimised through the use of equipment and procedures and may provide information on:

- exposure standards
- biological monitoring
- personal protective equipment, including respiratory protection, hand protection, eye and face protection, and protective clothing
- engineering controls.

Section 9: Physical and chemical properties

This section will provide information on the physical and chemical properties of the chemical and may include the following:

- appearance – colour and form
- odour –can be aromatic, pungent, fragrant, etc.
- melting/freezing point and boiling point
- flammability, including explosion/flammability limits, flash point, auto-ignition temperature and decomposition temperature
- vapour pressure and relative vapour density
- density or specific gravity
- solubility
- pH – corrosive pH scale (shown in **Figure 1.44**).

Remember pH is a logarithmic scale: 7 is neutral, 6 is 10 times stronger than 7, 5 is 100 times stronger than 7, 4 is 1000 times stronger than 7 and so on.

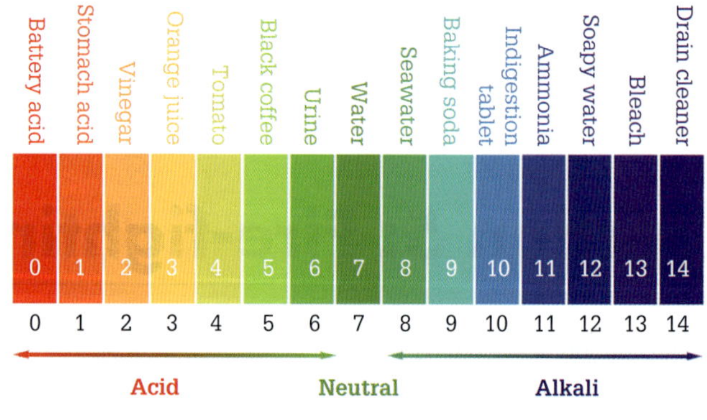

FIGURE 1.44 pH scale

Section 10: Stability and reactivity

This section of the safety data sheet may provide information on:

- chemical reactivity
- chemical stability
- possibility of hazardous reactions
- conditions to avoid
- incompatible materials
- hazardous decomposition products.

Section 11: Toxicological information

This section should provide a description of the known acute and chronic health effects of the chemical and how these effects were determined, such as from animal testing. Where available, this section will give information concerning:

- acute toxicity
- skin corrosion/irritation
- serious eye damage/irritation
- respiratory or skin sensitisation
- germ cell mutagenicity
- carcinogenicity
- reproductive toxicity
- specific target organ toxicity
- aspiration hazard.

Section 12: Ecological information

This section details the potential environmental and ecological harm posed by the chemical following spills or disposal. Where available, information will be provided about the ecotoxicity of the chemical on fish, aquatic invertebrates, algae and microorganisms. The safety data sheet should also describe the movement and fate of the chemical once in the environment, particularly its persistence and degradability, bioaccumulative potential, and mobility in soil.

Section 13: Disposal considerations

This section should provide information on disposal methods though this will depend on local and/or state regulations. Normally a chemical is suitable for disposal by an approved waste disposal agent.

Section 14: Transport information

United Nations number

UN numbers range from UN0001 to about UN3500 and are four-digit numbers representing a particular chemical or group of chemicals (e.g. UN1170 represents ethanol, and UN1300 represents a Class 3 turpentine substitute).

Proper shipping name

The shipping of hazardous materials can pose a serious danger to anyone who might come in contact with the shipment. Therefore, it is critical that the proper shipping name for a particular substance or article is the name used to describe that substance or article during transport. The proper shipping name for ethanol is 'Ethanol', but the proper shipping name for a less common substance that is not individually listed in the ADG will be a generic description; for example, 'Flammable Liquid n.o.s.' (not otherwise specified). In this case, the chemical name of the substance would be included in brackets following the generic description.

Class

The class encodes the general hazard class and subdivision (see **Table 1.6**). Some chemicals may have a subsidiary risk, which occurs where there is a 'sub-risk' attached to the chemical; for example, chlorine gas is classed as toxic first and corrosive second.

TABLE 1.6 Hazard classes

General hazard class	Subdivision
Class 1 Explosive	1.1 Substances with a mass explosion hazard
	1.2 Substances which present a projection hazard but no mass explosion hazard
	1.3 Substances which present both a fire hazard and a minor blast or projection hazard (or both) but not a mass explosion hazard
	1.4 Substances with a minor explosion hazard

»

TABLE 1.6 Hazard classes *(Continued)*

General hazard class	Subdivision
Class 1 Explosive	1.5 Very insensitive substances with a mass explosion hazard
	1.6 Very insensitive articles with no mass explosion hazard
Class 2 Gases	2.1 Flammable gases
	2.2 Non-flammable, non-toxic gases
	2.3 Toxic gases
Class 3 Flammable liquids	
Class 4 Flammable solids	4.1 Flammable solids, self-reactive substances and solid desensitised explosives
	4.2 Materials liable to spontaneous combustion
	4.3 Substances which, in contact with water, release flammable gases
Class 5 Oxidising substances and organic peroxides	5.1 Oxidising agents
	5.2 Organic peroxides
Class 6 Toxic and infectious substances	6.1 Toxic substances
	6.2 Infectious substances
Class 7 Radioactive substances and articles	
Class 8 Corrosive substances	
Class 9 Miscellaneous dangerous substances	

Packing group

The packing group for a chemical indicates the degree of hazard associated with its transportation. The highest group is Group I (greatest danger), Group II is intermediate (medium danger), and Group III chemicals present the lowest hazard (minor danger).

An example of a shipping document is illustrated in **Figure 1.45**.

Section 15: Regulatory information

This section states whether there is other applicable regulatory information for the chemical not already provided in the safety data sheet, such as:

- Montreal Protocol for ozone depleting substances
- The Stockholm Convention for persistent organic pollutants.

Standard for the Uniform Scheduling of Medicines and Poisons (SUSMP), poisons schedule

This section may provide the poisons schedule of the chemical, which provides a means of classifying poisons to identify the degree of control to exercise over their availability to the public.

Australian Inventory of Chemical Substances (AICS)

The AICS is a database of chemicals and identifies the regulatory obligations concerning the importation and manufacture of the chemical, such as requiring a permit.

Dangerous goods initial emergency response guide (SAA/SNZ HB76-2010)

This guide provides emergency response information for dealing with accidents, spills, leaks or fires involving dangerous goods. Information on hazards, protective clothing and emergency procedures is given. Consignors must provide to the prime contractor or driver *emergency information* for each of the dangerous goods in the consignment similar to dangerous goods initial emergency response guide (SAA/SNZ HB76).

ORIGINAL

SHIPPING DOCUMENT Invoice No. : 833485761

THIS LOAD CONTAINS DANGEROUS GOODS

Name and Address of Consignor Jones Transport

Name and Address of Consignee Brown's Store

Place of Despatch Brisbane

Port of Destination Perth

DESCRIPTION						
Proper Shipping Name	**Class Number**	**Subsidiary Risk Number**	**UN Number**	**Packing Group**	**Aggregate Net Quantity**	**Number and Type of Packages**
Propionic Acid	8	-	1848	III	1200 litres	6 × 200 litres plastic drum
Tetrahydrofuran	3	-	2056	II	400 litres	2 × 200 litres drum

Place of Issue Brisbane **Date of Issue** 10/8/19

Authorised Officer T.W. Wormell **Signature** T. W. Wormell

Shipping

A 0977239

FIGURE 1.45 Example of a shipping document

Section 16: Other information

This section will give other information relevant to the interpretation of the safety data sheet, such as the date of preparation, description of changes from previous version, abbreviations and acronyms used, and literature sources of data.

Example information

Other information pertaining to the safe handling of substances may include:

- Outer warning signs – an example of an outer warning sign is shown in **Figure 1.46**.
- Transport guides – an example of a transport guide is shown in **Figure 1.47**.
- Placards at storage locations.
- Format and location of manifests and emergency plans – the dangerous goods manifest should list all dangerous goods and hazardous substances on the site. A basic dangerous goods emergency procedure and generic emergency response guides for different types of goods are shown in SAA/SNZ HB76 *Dangerous goods – Initial emergency response guide* published by Standards Australia.

FIGURE 1.46 Warning sign

EMERGENCY PROCEDURE GUIDE – TRANSPORT

TRADE NAME	UN No	HAZCHEM
Metham Soil Fumigant	3267	2X

Emergency response Guide 36(HB 76-2004)

EMERGENCY CONTACTS

Organisation	Location	Telephone	Ask for
Chemicals Limited	Sydney	1800XXXXXX	Shift Supervisor

HAZARDS

FIRE	May burn but does not ignite readily. Containers may explode when heated. When heated vapours may form explosive mixures with air. Contact with metals may evolve flammable hydrogen gas. Runoff may pollute waterways. May be transported in molten form. Fire may produce irritating, poisonous and/or corrosive gases. Some may decompose explosively(d) or polymerise violently(P) when heated or involved in a fire.
HEALTH	Poisonous, may be fatal if inhaled, swallowed or absorbed through skin. Inhalation, ingestion or contact with substance may cause injury or death. Contact with molten substance may cause severe burns. Runoff from fire control or dilution water may be poisonous and/or corrosive to waterways.

➤— READ OTHER SIDE —➤

EMERGENCY PROCEDURE GUIDE – TRANSPORT

TRADE NAME	UN No	HAZCHEM
Metham Soil Fumigant	3267	2X

Emergency response Guide 36(HB 76-2004)

EMERGENCY PROCEDURES

IF THIS HAPPENS	DO THIS
For All Emergencies	Immediately contact police or fire brigade.
Tanker Accident	Spill or leak area should be isolated immediately for at least 25 m in all directions.
LARGE SPILL	Consider initial downwind evacuation for at least 250 m.
FIRE	When any large container (including rail and road tankers) is involved in a fire, consider initial evacuation for 800 m in all directions.

PROTECTIVE CLOTHING

Wear SCBA and chemical splash suit.
Fully-encapsulating, gas-tight suits should be worn for maximum protection.
Structural firefighter's uniform is NOT effective for these materials.

ADDITIONAL PROCEDURES

Keep unauthorised personnel away.
Keep upwind to higher ground.
Ventilate enclosed spaces before entering.

Police or Fire Brigade Dial: 000

➤— READ OTHER SIDE —➤

FIGURE 1.47 Transport guide

The following is an example of a safety data sheet for low odour turpentine. The full SDS is available for reference at: http://www.recochem.com.au/files/downloads/Low_Odour_Turpentine_v7.pdf.

SAFETY DATA SHEET

SECTION 1 IDENTIFICATION: PRODUCT IDENTIFIER AND CHEMICAL IDENTITY

Product Identifier	**LOW ODOUR TURPENTINE**
Other Names	Turpentine substitute
Manufacturer's Product Code	17062
Recommended Use	Solvent, paint thinner

Details of Supplier/Manufacturer

Company:	Recochem Inc. ABN: 69 010 485 999
Address:	1809 Lytton Road, Lytton, Queensland 4178
Phone:	(07) 3308 5200 Fax: (07) 3308 5201
Website:	www.recochem.com.au

Emergency Telephone Numbers

Business Hours:	(07) 3308 5200
After Hours:	1300 131 001
Poisons Information:	Australia: 13 11 26 New Zealand: 0800 764 766

SECTION 2 HAZARDS IDENTIFICATION

Hazardous chemical	*according to classification by Safe Work Australia*
Dangerous goods	*according to the Australian Code for the Transport of Dangerous Goods by Road and Rail*

Signal Word	**DANGER**

GHS Classification	Pictogram	Hazard statement
Flammable Liquids, Category 3	 FLAME	H226 Flammable liquid and vapour
Aspiration Hazard, Category 1	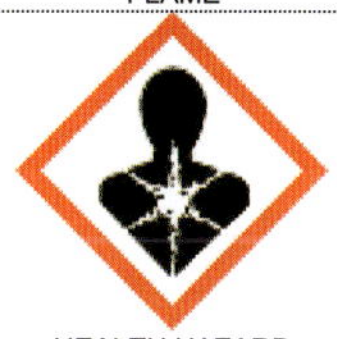 HEALTH HAZARD	H304 May be fatal if swallowed and enters airways

Source: Recochem Inc Australia, http://www.recochem.com.au.

FIGURE 1.48 Example of a safety data sheet

REVIEW QUESTIONS

Look in the Recochem Inc safety data sheet for mineral turpentine and the appendix text to find the answers to the following questions.

1 What are other names for mineral turpentine?
2 List two of the hazard statements for mineral turpentine.
3 What is the first aid treatment for inhalation of mineral turpentine?
4 In the event of a spill, what environmental precautions should be taken?
5 State what eye/face protection should be worn when handling mineral turpentine.
6 What are the engineering controls needed?
7 State the explosion/flammability limits (%).
8 What conditions must be avoided when handling and storing mineral turpentine?
9 State the aspiration hazard posed by mineral turpentine.
10 What is the Australian Dangerous Goods class of mineral turpentine?
11 State the poison schedule.

Cardiopulmonary resuscitation and basic first aid

This chapter provides electrotechnology workers with knowledge and skills required to perform cardiopulmonary resuscitation (CPR) in line with the Australian Resuscitation Council (ARC) guidelines.

Electrotechnology workers will also gain an overview of basic first aid for working in the electrotechnology industry. This chapter provides underpinning knowledge for the unit HLTAID009 as applied to the UEE training package.

LEARNING OBJECTIVES

Regulations and codes of practice

- List the relevant ARC guidelines for providing CPR.
- Outline steps involved in risk mitigation when attending a first aid incident.
- Describe common infection control procedures.
- Outline the purpose of first aid codes of practice.

Legal, workplace and community considerations

- Explain the meaning of duty of care.
- Outline the importance of knowing own skills and limitations when providing CPR.
- Describe how privacy and confidentiality requirements can be satisfied.
- Explain the meaning of consent as applied to first aid.
- Outline stress management techniques and provide a reason for relieving stress.

Considerations for performing CPR

- Describe the process of CPR and describe how the upper airway relates to breathing.
- Outline the purpose and detail the operation of an AED.
- Describe the chain of survival and how it relates to performing CPR.
- Explain how to contact emergency services.

Administering CPR and basic first aid

- Outline the priority action plan and detail its importance in performing CPR.
- Detail the process of performing CPR.
- Describe how to determine a patient's level of consciousness.
- Outline how to treat clinical shock.

2.1 Regulations and codes of practice

First aid is the initial treatment provided to an injured or infirmed person until professional medical treatment arrives or is available. This initial first aid is usually provided by first aiders, people who are trained to provide initial lifesaving treatment to someone who is injured or infirmed.

The aims of first aid are to:

- promote a safe environment
- preserve life
- prevent the worsening of injury or illness
- help promote recovery
- provide comfort to the injured or infirmed.

Australian Resuscitation Council (ARC) guidelines

The Australian Resuscitation Council (ARC) is a voluntary coordinating body representing all major groups involved in the teaching and practice of resuscitation. Under its aims and objectives, the ARC produces guidelines based on consideration of all available scientific and published material, which are accepted by all member organisations before being released. The published guidelines foster uniformity and simplicity in resuscitation techniques and terminology and are an invaluable source of information.

SWITCH ON

The ARC guidelines can be found on their website at https://resus.org.au > Guidelines.

Relevant guidelines are:

- Guideline 5 – Breathing
- Guideline 6 – Compressions
- Guideline 7 – Defibrillation
- Guideline 8 – Cardiopulmonary Resuscitation
- Guideline 10.1 – Basic Life Support.

Minimising risks and potential hazards

A hazard can be a work practice, procedure or anything that has the potential to harm the health or safety of a person. Risk is a measure of the probability of a specific harmful effect occurring in particular circumstances. It is important to distinguish between hazard and risk. Worksite hazards occur:

- in the work environment
- as a result of the use of machinery, tools and materials
- as a result of unsuitable work systems and procedures.

Hazards encountered at a worksite can be classified into five broad areas:

1. physical
2. chemical
3. biological
4. mechanical/electrical
5. psychological.

Worksite risk assessments

Worksite risk assessments and inspections are key activities to prevent accidents occurring from worksite hazards. Conducting regular worksite risk assessments provides evidence that the workplace is safe. After undertaking a worksite risk assessment, employers should make decisions about establishing appropriate prevention and control measures.

When undertaking a risk assessment, it is important to adhere to work health and safety (WHS) legislation and relevant Commonwealth, state or territory regulations and any approved codes of practice pertaining to the control of hazardous substances in the workplace.

Hazards can include objects in the workplace; for example, machinery and dangerous chemicals a person may use when performing their duties. A risk comes about when it is likely that a hazard will cause harm. The level of risk depends on factors such as the frequency of the task, the number of workers involved, and the severity of any likely injury.

Some examples of potential hazards in the workplace include:

- blocked walkways
- faulty electrical equipment or overloaded circuits
- high noise levels
- inaccessible fire extinguishers
- incorrect storage of materials
- poorly maintained equipment or improper use of equipment
- wet or uneven floor surfaces.

A risk assessment matrix, like that shown in **Figure 2.1**, is a tool for evaluating both the probability of an incident occurring and, if the incident were to happen, the severity of the consequences. By aligning the **Likelihood** of a particular risk occurring with the **Most likely consequence** if it were to occur, it is possible to assign a risk ranking between 1 and 25, with a risk ranking of 25 being the worst possible condition. An activity that has a risk considered 'almost certainly to occur' resulting in a 'disastrous' consequence will be classified as high risk (25), while an activity that has a risk considered 'extremely unlikely to occur' resulting in a 'minor' consequence will be classified as very low risk (1).

In order to determine the risk, use the matrix in **Figure 2.1** and:

- determine the most likely consequence of the risk (Consequence Assessment)
- determine the likelihood of the consequence occurring (Likelihood Assessment)
- align the Likelihood and the Consequence Assessments to determine the risk score
- use the risk score to determine the appropriate actions to take in addressing the risk.

Risk assessment matrix

Likelihood / Most likely consequence	Almost certain to occur — Consequence expected to occur on a weekly basis or more frequently	Good chance of occurring — Consequence expected to occur more than once in 3 months but less than once a week	Likely to occur — Consequence expected to occur more than once a year but less than once in 3 months	Unlikely to occur — Consequence expected to occur more than once in 3 years but less than once a year	Extremely unlikely to occur — Consequence has not occurred and is expected to occur less than once in 3 years
Disastrous fatality or extensive damage	High 25	High 24	High 22	Moderate 19	Moderate 15
Critical amputation or major damage	High 23	High 21	Moderate 18	Moderate 14	Moderate 13
Serious more than 1 week off normal duties or serious damage	High 20	Moderate 17	Low 12	Low 9	Very Low 6
Significant less than 1 week off normal duties or negligible damage	Moderate 16	Low 11	Low 8	Very Low 5	Very Low 3
Minor first aid injury or no damage	Low 10	Low 7	Very Low 4	Very Low 2	Very Low 1

FIGURE 2.1 Risk assessment matrix

Risk mitigation

The most important step in managing risks on a worksite involves eliminating the risks as much as is reasonably practicable. The risk should be minimised if elimination is not possible. Employers have a legal obligation to establish appropriate procedures to minimise or eliminate the risk or hazard. If, as a result of undertaking a risk assessment, there is a risk factor identified on the worksite, then employers must establish appropriate procedures to mitigate the hazard or risk. Risk mitigation might include:

- training workers
- providing first aid facilities, such as safety showers and eye wash stations
- providing personal protective equipment (PPE)
- developing and communicating emergency procedures for the worksite.

Hierarchy of control

It is a requirement of many worksites that selection of risk control measures is based on the hierarchy of control model, as shown in **Figure 2.2**. The hierarchy of control ranks risk control measures in decreasing desirability and effectiveness.

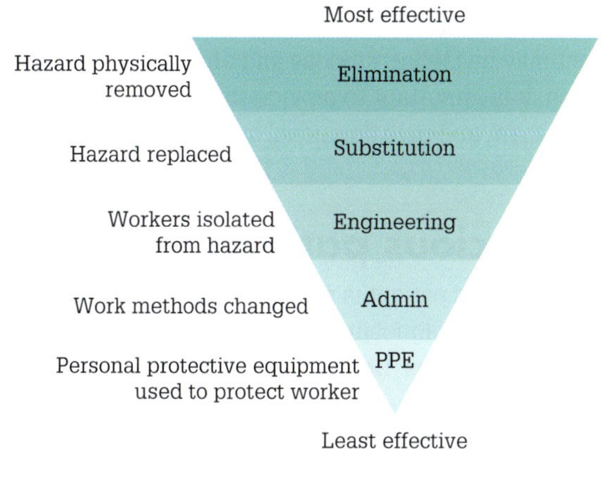

FIGURE 2.2 Hierarchy of control

First aid legislation

Government at both state and federal level have in place first aid legislation, comprising a series of laws and rules that apply to first aiders.

First aid is regulated at the following levels:

- in thc workplace by employers.
- at a state/territory level through state/territory legislation, which may include:
  - Code of Practice
  - Occupational Safety and Health Act
  - Occupational Safety and Health Regulations
- at the national level (Australia) through federal law:
  - First Aid in the Workplace Code of Practice.

It is important for all workers to be aware of the existence of this legal framework and to know where to access appropriate documentation. When performing first aid you should always:

- act in a reasonable manner
- comply with relevant legislation
- follow any applicable codes of practice
- act in accordance with your level of training
- follow any workplace enterprise policies and procedures
- gain consent prior to assisting the casualty
- document the incident and first aid management
- ensure that your first aid certification is maintained and current.

SWITCH ON

Relevant state and territory WHS regulators

- NSW http://www.safework.nsw.gov.au
- VIC http://www.worksafe.vic.gov.au
- QLD https://www.worksafe.qld.gov.au
- TAS http://www.worksafe.tas.gov.au
- SA http://www.safework.sa.gov.au
- WA https://www.commerce.wa.gov.au/worksafe
- ACT https://www.worksafe.act.gov.au
- NT http://www.worksafe.nt.gov.au

Code of conduct

A first aid code of conduct is a set of directions outlining the expected standards, rules and obligations of a person providing first aid. Anyone providing first aid must always treat a patient with respect regardless of their race, age, religion or gender. It is important to bear in mind that the patient may be feeling anxious and afraid of their situation. As such, it is important to provide them with reassurance, to calm them and give emotional support to aid in reducing their anxiety.

Unconscious patient

An unconscious patient is not able to communicate with you to assist in diagnosing their particular injury or illness if not obvious. Depending on the level of patient consciousness, it may be possible to formulate a diagnosis by the patient's reactions to a stimulus or movement.

It is important to treat an unconscious patient as if they are conscious, as they may still be able to hear what you are saying. For example, tell the patient what you are going to do and do not say anything that you would not say if the patient were conscious. This may, for example, be comments about the patient's injuries or physical appearance.

SWITCH ON

When treating a patient:

- be respectful and empathetic (show understanding)
- provide them with reassurance to reduce their anxiety
- develop a rapport and keep them informed of what you are doing and your plan of action
- seek their permission before entering their personal space
- get their assistance and tell them how they can help
- treat them as you would wish to be treated.

Infection control

When administering first aid there is a risk of diseases being transmitted due to the presence of blood and body fluids. Some of these diseases include HIV virus, the various types of hepatitis, and, recently, COVID-19. The transmission of disease when administering first aid is rare and is largely avoidable. The risks can be mitigated during resuscitation by using pocket masks or other barrier devices.

Communicable diseases

Communicable diseases are diseases that spread from one person to another and include:

- colds
- influenza
- COVID-19
- measles
- mumps
- glandular fever
- HIV
- tuberculosis
- some forms of meningitis
- some skin infections
- hepatitis A, B and C.

It is possible that the person providing first aid can acquire the disease through:

- blood
- saliva
- vomit
- urine
- faeces.

The disease may enter the bloodstream of the person providing the first aid through cuts, grazes or the mucous membranes.

Precautions to take before management of patient

Whenever possible:

- cover exposed wounds such as cuts and grazes with a waterproof dressing
- wear disposable nitrile surgical gloves

FIGURE 2.3 Resuscitation barrier device and protective equipment

- wear eye protection such as wrap around safety glasses
- use a pocket mask or other barrier devices if administering CPR
- use antiseptic hand gel or alcohol wipes
- wash hands with warm soapy water for 15 seconds before and after treatment.

Figure 2.3 shows some useful protective equipment.

Precautions to take after management of patient

- If you come in contact with blood or other body fluids, wash your skin thoroughly with soap and running tap water, and sanitising hand gel or alcohol wipes if available.
- If a sharp object, which may be contaminated, punctures your skin, wash the site thoroughly with soap and running tap water, sanitising hand gel or alcohol wipes if available, and seek medical advice as soon as possible.
- If you wore a non-disposable face mask, soak it for 30 minutes in bleach or disinfectant, and then wash with detergent and dry it. Appropriately dispose of any contaminated materials (such as bandages) and restock the first aid kit.

REVIEW QUESTIONS

1. What is first aid?
2. List the five principles of first aid.
3. Name five relevant ARC guidelines for the provision of first aid.
4. What is a hazard?
5. What is risk?
6. Name two key activities in the prevention of accidents occurring from worksite hazards.
7. List four risk mitigation actions.
8. What is a first aid code of conduct?

2.2 Legal, workplace and community considerations

Legal, workplace and community considerations include:

- obtaining consent before administering first aid treatment
- requirements for exercising a duty of care
- behaving respectfully when administering first aid treatment
- being aware of your own skills and limitations
- maintaining privacy and confidentiality regarding any first aid incident in which you are involved
- participating in post-incident debriefing and evaluation
- managing stress.

Consent

Consent is permission or agreement from the patient to be treated by you. When providing first aid it is necessary to obtain consent from a patient, where possible, before applying first aid. Any treatment provided without the patient's consent may constitute assault.

Consent can take one of two forms:

1. Implied (taken as given) consent occurs when the patient is unconscious (or when the patient is unable to verbally communicate with you) and is not able to provide their expressed consent.
2. Expressed consent is when a conscious patient in need of first aid treatment provides their oral or written permission for treatment.

If the patient is under 18 years of age, it is considered to be implied consent, but where possible the person providing first aid should obtain the consent of a parent or legal guardian. If a patient is unable to communicate verbally, then body language and other non-verbal cues can be used. Essentially if the patient does not allow you to assist, that means they are refusing treatment. A person has the right to **refuse** treatment.

SWITCH ON

Before treating a patient:

- obtain their consent where possible before applying first aid
- if under 18 years of age, obtain consent from a parent or legal guardian
- if unconscious, then consent is considered to be implied.

Duty of care

Anyone providing first aid has a duty of care, which is a legal obligation for you to protect yourself and the person to whom you are providing first aid. When the decision to provide first aid to a patient is made and treatment subsequently commences, the person providing first aid is committed to provide a duty of care to the patient.

With regard to the provision of first aid, duty of care means that you will provide reasonable treatment to the casualty to the best of your ability and to the level of training you have. After commencing first aid, you have an obligation to provide duty of care until:

- another or more experienced person takes over the provision of first aid
- medical aid arrives
- you are physically unable to continue with the provision of first aid
- the situation becomes unsafe to continue the provision of first aid.

A duty of care may be breached through either action or inaction (e.g. if you do nothing and the person's condition deteriorates).

SWITCH ON

- When providing first aid it is essential to keep within the scope of your training.
- A duty of care is in place once the provision of first aid has commenced.
- After providing first aid it is necessary to complete required documentation and keep it confidential.
- Maintain your skills and knowledge through regular training.
- Workplace first aid kits and equipment must be maintained.

Respectful behaviour

Another aspect of administering first aid is behaving respectfully, which is the morally correct conduct of a person providing first aid. If you are called upon to administer first aid you will, at all times, behave professionally and respectfully. In this regard, the old adage of treating others as you would like to be treated, holds true.

SWITCH ON

Behaving respectfully means:

- introducing yourself and obtaining the patient's consent where possible before administering first aid
- treating the patient with dignity and respect at all times
- being compassionate and understanding
- speaking in a calm and reassuring voice
- protecting the patient's privacy and confidentiality.

Demonstrating culturally appropriate behaviour is also important when administering first aid. A culturally aware person is able to communicate sensitively and effectively with people who have a different culture, ethnicity, religion, gender, language, and sexuality. Included in cultural awareness and considerations are:

- using appropriate communication
- being mindful of body language
- maintaining eye contact
- appropriate treatment of patients of different gender.

Own skills and limitations

If called on to administer first aid, you are limited to the extent of the training. When providing first aid these limitations can include:

- the scope of the training received, which means your acquired skills and knowledge in terms of procedures, actions and processes you can administer
- your level of confidence, as proficiency generally increases with confidence
- enterprise policies and procedures, which means whenever providing first aid in the workplace, you must be within the boundaries of your enterprise's policies and procedures
- legal framework, which means the law is very strict in what can and cannot be done when administering first aid.

SWITCH ON

First aid providers should:

- maintain their first aid skills and knowledge and keep their first aid certification current by refreshing their CPR skills every 12 months
- keep within the bounds of their skills and knowledge, and the limitations of their training, to manage a patient to a standard of care that is appropriate to their level of training.

Privacy and confidentiality

Privacy and confidentiality are two important considerations when administering first aid. These are legal terms that relate to:

- the patient being free from intrusion
- the first aid provider being constrained from releasing any information about the patient to a third party.

The person receiving first aid treatment has a right to privacy and confidentiality. Anyone providing first aid has a duty and legal responsibility to protect the person's privacy at all times. An unconscious patient may be most vulnerable due to the nature of the injuries as well as their circumstances. It is not uncommon for bystanders to pull out a mobile phone and start taking photos and videos. When administering first aid do not:

- TXT information of the patient to any third party
- communicate details of the incident and the patient to a third party
- take personal pictures of the patient
- leave the person receiving treatment exposed.

Any information and documentation obtained pertaining to a person who is receiving first aid must remain confidential.

SWITCH ON

First aid providers should:

- protect their patient's privacy at all times while administering first aid
- maintain the patient's privacy by not releasing any information to a third party
- in the workplace, abide by the relevant enterprise's policy and procedures
- be familiar with the legislative requirements governing privacy.

Confidentiality of first aid records

As mentioned, confidentiality is a legal requirement for the first aid provider to not divulge information about the patient to a third party or to be negligent with the first aid records. Laws pertaining to who can access first aid records, the extent of this access, and what incidents must be reported vary with each jurisdiction. The following people, however, have the right to access casualty and incident information: ambulance paramedics/officers or a treating doctor, and those investigating workplace illness or injury such as police, coroner, workplace inspection authority, the courts and employer. With the consent of the person who received treatment, access can be granted to the insurance company handling a claim related to the incident, trade union representatives or occupational health and safety committees investigating the incident.

Regardless of who may have genuine access to first aid records, the privacy of the person who received treatment must be respected to the utmost extent possible. The person who has control of these first aid records has a responsibility to ensure that:

- records are only released to people with appropriate authority
- all records are securely stored, such as in a locked filing cabinet or electronic records being protected by password and a range of IT security measures
- the person who received first aid be informed if access was provided to a third party
- records of anyone who accessed particular documents, including when and why, are maintained.

Individual first aid records must be retained at the workplace for the period specified by relevant legislation, which varies by jurisdiction.

SWITCH ON

Requirements for first aid records and documentation include:

- securely storing first aid documentation
- following enterprise policies and procedures
- being familiar with the legislative requirements governing privacy.

Debriefing and evaluation

Two important post-incident processes are debriefing and evaluating. Debriefing is a process of obtaining information from a person following an incident and evaluation is a process of making judgements about the value of the provision of first aid.

Debriefing

A major first aid incident can be a very traumatic experience for the first responders. A debriefing session is an important process that needs to be undertaken post-incident and followed up by an evaluation of the outcomes of that incident. After providing first aid and care of the patient has been transferred to medical aid, there are a number of events that require attention. These include:

- cleaning up the site
- attending to any WHS issues
- completing any reporting requirements
- undertaking a post-incident debrief.

After a first aid incident those who were present will react differently. How people react will depend on the individual and the incident. As such, a post-incident debriefing is a vital part of managing the incident.

The purpose of having a post-incident debriefing is:

- to look after the welfare of individuals involved in the incident by giving them the opportunity to discuss the feelings and emotions they may be experiencing after the incident
- bringing the incident to a close
- providing support to the person providing first aid
- to collect information in the hope of preventing a similar incident from occurring
- identifying any gaps in the emergency action.

A debriefing may involve:

- collecting and documenting all relevant details pertaining to the incident and the effectiveness of incident management processes and first aid administered
- detailing any information communicated by people involved in the incident
- advising of any further assistance that may be available, such as counselling.

A debriefing should include any individual who was involved in the incident. They should be encouraged to discuss the process and its outcomes. It is important, following the debriefing, to make appropriate referrals to counsellors, mediators or industry chaplains as appropriate. Anyone providing first aid needs to consider their own psychological wellbeing.

Evaluating an incident

Evaluation is an important aspect of continuous improvement and development for any organisation or individual. Evaluation may be part of a formal or informal debriefing process. Post-incident evaluation can consider options or strategies that can improve workplace conditions, which in turn prevents future stress and provides methods for eliminating or reducing risks.

Most enterprises have plans in place which are used for evaluating incidents. These may include a First Aid Action Plan, an Emergency Action Plan, a Risk Management Plan and the like. These plans should be assessed for their usefulness on an ongoing basis.

All plans should comply with:

- established first aid principles
- Australian Resuscitation Council (ARC) guidelines
- enterprise policies and procedures
- Australian national peak bodies
- industry standards
- state or territory legislation and regulations.

SWITCH ON

When debriefing and evaluating an incident:

- be understanding and display empathy
- provide advice on what further support or assistance is available
- collect and document all relevant details pertaining to the incident
- detail any information communicated by people involved in the incident.

Managing stress

Stress is a human biological response to certain situations. It is a natural feeling of not being able to cope with specific demands and events. Stress management is about being able to recognise and manage or regulate our level of stress.

Administering first aid in an emergency can be very stressful and emotionally draining. Seasoned first aiders or other first responders can experience disagreeable consequences when attending an emergency situation. As humans we react differently to emergency situations and may display a diverse range of responses to the situation. Some of these responses may not surface until some time after the event. While some people manage stresses by talking, some people withdraw, and others just prefer to physically work off the stress. As stated previously, a post-incident debrief is a crucial part of the incident management process.

SWITCH ON

Watch out for these signs and symptoms of stress:

- feelings of guilt, fear or shame
- perspiration
- anxiety
- increased heart rate
- elevated blood pressure.

Stress management for those involved in administering first aid in an emergency incident may include:

- post-incident debriefing to allow the individual to de-stress
- post-incident evaluation to identify gaps in the Emergency Action Plan
- access to professional services such as counsellors, doctors, a telephone helpline, or religious guidance
- peer support.

Practical application of stress management will largely depend on the particular workplace and the strategies they have available for managing workplace incidents. Not all individuals will require professional post-incident support and, in most cases, an initial debrief will suffice.

Self-care

Self-care is an important part of being able to successfully manage stress. It is widely recognised that good mental health and wellbeing allows us to live our lives in a positive and meaningful way and cope with life's changes and challenges. It is therefore important to look after yourself so you can remain resilient during stressful times. It is important to be able to recognise what works for *you* to ensure that you are indeed looking after yourself. It is important to be aware of your own needs as what self-care techniques work for a colleague may not work or be suitable for you.

Some tips for self-care are:

- Connecting regularly with family and friends and doing things together that you all enjoy.
- Getting regular exercise helps with reducing stress levels as well as improving overall health. Aim for about 20 minutes a day; just a 10-minute walk has health benefits.
- Eating a balanced diet as this helps with energy levels and stress management.
- Asking for assistance and accepting it when it is offered. Talking to someone, whether family member, friend or professional, can help reduce stress.
- Reinforcing your self-confidence by encouraging and rewarding yourself for getting through a difficult period or achieving a goal, however small. Being positive and compassionate towards yourself can be useful in reducing stress.
- Planning something such as a trip with friends or a night at the theatre creates something pleasant to look forward to.
- Getting good quality sleep to allow body and mind to refresh.
- Regularly engaging in an activity that you find relaxing, whether it be gardening, listening to music, going to the beach or reading.

Figure 2.4 shows some self-care activities.

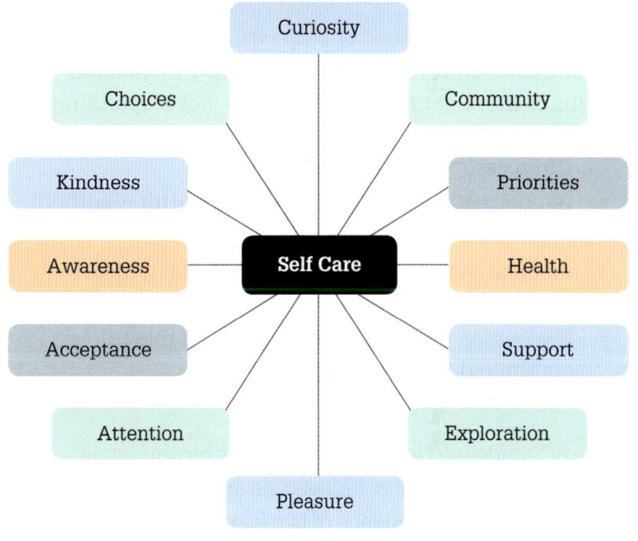

FIGURE 2.4 Self-care

Most workplaces will have in place clear policies and procedures to ensure a safe, risk-free working environment. It is the responsibility of both the employer and employees to ensure all members of staff work collaboratively to create a work environment that endeavours to recognise potential stressors and facilitate or implement immediate action to eliminate or reduce stress and harm.

REVIEW QUESTIONS

1. What is meant by 'consent'?
2. When would consent be implied?
3. Explain what is meant by 'duty of care'.
4. How can a duty of care be breached?
5. What is meant by 'behaving respectfully'?
6. How can a first aider ensure they are compliant with confidentiality of first aid records?
7. Who should be involved in debriefing sessions?
8. What does self-care encompass?

2.3 Considerations for performing CPR

Cardiopulmonary resuscitation, which is commonly abbreviated to CPR, is an emergency procedure that is undertaken on a patient who has no signs of life. The purpose of CPR is to preserve brain functions by manually circulating oxygenated blood through the patient's body until advanced measures are provided to restore spontaneous blood circulation and breathing.

CPR is a process involving manual chest compressions and lung inflation by expelling air into the patient's mouth. The purpose is to compress the heart and pump blood through the body and deliver oxygen to the brain until the heart regains normal sinus rhythm. A patient is in cardiac arrest when the heart stops pumping and the body is starved of oxygen. To be effective, CPR must be administered without delay and with the patient laying on a firm surface.

SWITCH ON

Administer CPR when a patient is **NOT** displaying signs of life: not breathing normally, not responding, and not moving.

CPR is stopped when:

- the patient is revived and starts breathing without assistance
- medical help such as ambulance paramedics arrive and take over
- the person performing the CPR is unable to continue due to physical exhaustion.

ARC guideline No. 6 – Compressions recommends giving thirty (30) chest compressions and two (2) breaths aiming to achieve a cadence of five (5) sets of compressions in two (2) minutes.

Breathing

The body requires oxygen in order to function. The process of moving oxygen into (inhaling) and out (exhaling) of the lungs is known as breathing or ventilation. The breathing process supplies the necessary oxygen to the body and expels waste gases including carbon dioxide from the body (see also **Figure 2.5**). Air is drawn into the lungs when the diaphragm and intercostal muscles expand the chest. Within the lungs oxygen crosses to the blood for transportation around the body. Air is expelled from the lungs when the diaphragm and intercostal muscles relax. A patient who is gasping for breath or breathing abnormally and is unresponsive to stimulus requires resuscitation.

Two classifications are used to describe breathing: either effective or ineffective. Absence of breathing is the worst case. Ineffective breathing may be due to:

- direct depression of or damage to the breathing control centre of the brain
- upper airway obstruction

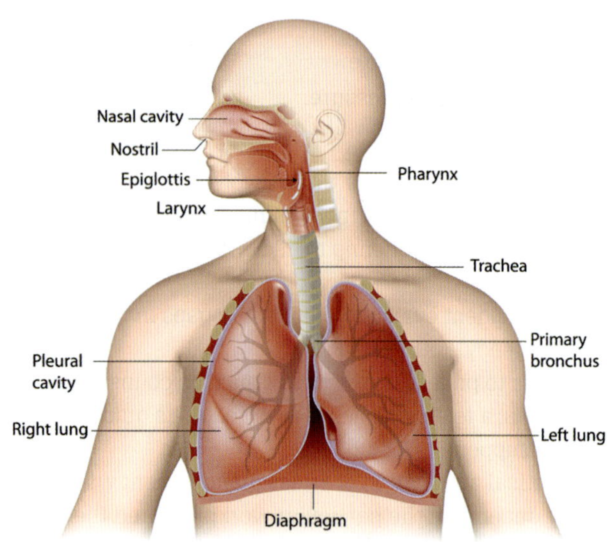

Source: Shutterstock.com/Alila Medical Media

FIGURE 2.5 Respiratory system

- paralysis or impairment of the nerves or muscles that control breathing
- lung impairment
- drowning
- suffocation.

Someone in cardiac arrest may exhibit abnormal gasping (agonal gasps). In this situation:

1. Look for movement of the upper abdomen or lower chest.
2. Listen for the exhaling of air from nose and mouth.
3. Feel the chest and upper abdomen for movement.

Movement of the lower chest and upper abdomen is not indicative of a clear airway. **Figure 2.5** shows the upper respiratory system. A patient may experience impaired or total loss of breathing well before they lose consciousness. Respiratory distress syndrome, where the patient experiences absence of normal breathing, is potentially life-threatening as the lungs cannot provide enough oxygen for the body.

Respiratory distress may be due to asthma, airway obstruction, hyperventilation, croup and epiglottitis. Signs of respiratory distress include difficulty in breathing and the associated psychological distress.

Adults and children have anatomical and physiological differences in airways. A child's airway is narrower and therefore more prone to blockage by blood or secretions. A nasal obstruction can cause respiratory distress in a child as children tend to breathe through their nose. The main response to respiratory distress in a child is to increase the rate and effort of breathing. In infants the trachea is shorter, softer and more pliable and may be distorted by excessive backward head tilt (overextension). It is important therefore when performing CPR on an infant to keep the head in a neutral position with the lower jaw supported at the point of the chin with the mouth maintained open.

Automated External Defibrillator

An Automated External Defibrillator (AED) is an electronic device used in the treatment of a cardiac condition known as fibrillation. An electric shock may cause a person to experience fibrillation, which is a condition where the upper chambers (atria) of the heart beat out of coordination with the lower chambers (ventricles). An AED is able to deliver a therapeutic electric shock to the heart for the purpose of re-synching heartbeat to normal sinus rhythm.

An AED, like that shown in **Figure 2.6**, is a battery-operated device that comprises:

- main unit
- pad-type electrodes
- accessories, including a razor.

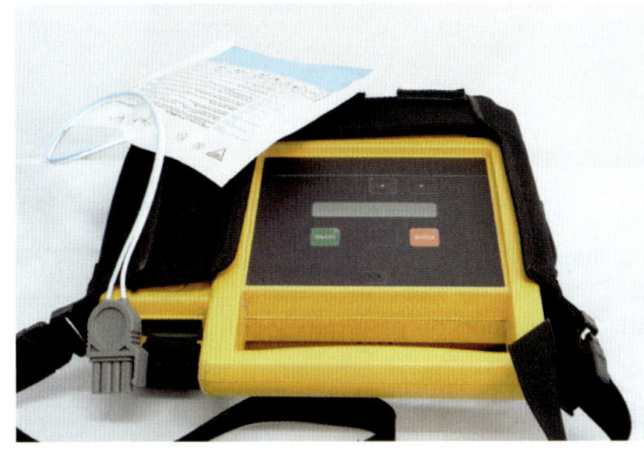

Source: Shutterstock.com/Stoyan Yotov

FIGURE 2.6 Automated External Defibrillator (AED)

When powered, an AED provides verbal instructions through an inbuilt loudspeaker. AEDs are readily available in shopping centres, workplaces and public facilities.

SWITCH ON

Checking the charge status of AED batteries is a function of WHS audits in the workplace.

Basic operation of AED

If two people who are able to render first aid are present, one should go for help and collect an AED (if available), while the other begins CPR on the patient.

While AEDs have printed and verbal instructions, the basic steps in their use are:

1. Open the AED case and power ON
2. Follow the verbal prompts
3. Expose the patient's chest
4. Check for a pacemaker or implant scars (place pads away from the site of the implant)
5. Remove any jewellery and medication patches
6. Place the pads on the patient's bare chest
   a. one pad to right chest wall, below the collarbone
   b. other pad to left chest wall, below the left nipple
   c. check that both pads adhere to the skin (use the razor to remove excess body hair if needed)
7. The AED will check and analyse the patient's heart rhythm
8. The AED will provide a verbal prompt if it is necessary to administer the shock

It is important to ensure that the interval between the start of the cardiac arrest and the delivery of defibrillation is as brief as possible to afford the patient the greatest chance of survival. It is vital to continue CPR except when the AED is administering a shock or the AED detects normal heart rhythm and advises cessation of CPR. It is also important to leave the defibrillator pads in place even if the patient is conscious.

The appropriate use of an AED is on a patient who is unresponsive and unconscious as well as not breathing normally.

ARC guideline No. 7 – Automated External Defibrillation (AED) in Basic Life Support (BLS) is a good source of information.

Chain of survival

The chain of survival, as shown in **Figure 2.7**, documents a series of actions taken to reduce the risk of death from sudden cardiac arrest. A cardiac arrest is when the heart stops beating, which means that the brain and vital organs are denied oxygen. As a result, the casualty becomes unconscious and stops breathing or does not have normal breathing. A cardiac arrest is a medical emergency where every second counts. The step in each link of the chain of survival closely links to the next.

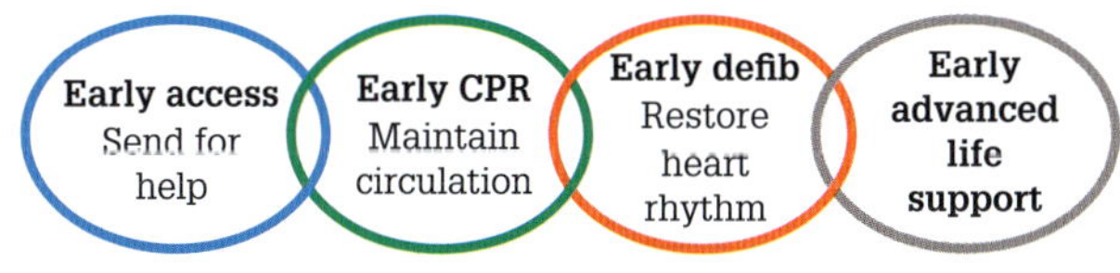

FIGURE 2.7 Chain of survival

The links in the chain of survival are:

1 Early Access

Recognising that a person is experiencing cardiac arrest is a vital first step as some people do not realise the seriousness of the situation until it is too late. It is necessary to identify this in a patient by thoroughly checking their breathing and level of consciousness. After determining that the patient is unconscious and not breathing, it is necessary to call an ambulance without delay to ensure defibrillation and life support arrives as soon as practical. In emergency situations it is safer to err on the side of caution.

2 Early CPR

It is possible to maintain oxygenation of vital organs such as the brain if CPR is commenced within four minutes of the heart arresting. CPR assists in circulating oxygenated blood around the body until the heart resumes normal sinus rhythm.

3 Early Defibrillation
The survival rate from cardiac arrest is enhanced further if CPR is started within four minutes and defibrillation commenced within 8–12 minutes. It is believed that the chances of survival reduce by 9 per cent for each minute without defibrillation.

4 Early Advanced Life Support
The chance of survival is further increased if medical treatment such as giving medication and stabilising the airway occurs. The patient will need to attend a hospital for appropriate care and monitoring to ensure their long-term health and wellbeing.

Making an emergency call

In an emergency, call Triple Zero (000) as shown in **Figure 2.8**.

The Triple Zero (000) service is the most expedient method to get the right resources from emergency services to help you and should be used to contact Police, Fire or Ambulance services in life-threatening or time critical situations.

Calls to Triple Zero (000) are free and can be made from mobile phones, home or work phones or pay phones.

There are a few simple steps in making a Triple Zero (000) call to report an emergency:

1 Ensure your own safety, stay calm and call Triple Zero (000).
2 An operator will ask you if you require Police, Fire or Ambulance. As you are requiring ambulance service, respond with 'Ambulance'. The operator will ask for location information if you are calling from a mobile or satellite phone.
3 An emergency service operator will then take details of the situation.
4 Do not hang up, and be sure to answer all of the operator's questions in a clear and concise manner.
5 Provide details of the specific location, such as street number, street name, nearest cross street, and suburb. In rural areas it is important to provide the full address and distances from landmarks and roads.

Source: Shutterstock.com/Aleksandra Gigowska

FIGURE 2.8 Making an emergency call

6 Stay on the line until the operator has all the information they need.
7 If possible, wait at a prearranged meeting point or in a prominent location. This will assist the Ambulance Service when arriving to quickly locate the emergency.
8 If you make a Triple Zero (000) call while travelling on a motorway or rural road, advise the operator of the direction you are travelling and the last exit or town you passed to assist emergency services to correctly locate the incident.

REVIEW QUESTIONS

1 What is the purpose of CPR?
2 When is CPR administered?
3 When is the administration of CPR ceased?
4 Name the two classifications used to describe breathing.
5 What is an Automated External Defibrillator (AED)?
6 Describe fibrillation.
7 Explain why it is important to ensure that the interval between the start of the cardiac arrest and the delivery of defibrillation is as brief as possible.
8 Name the four links in the 'chain of survival'.

2.4 Administering CPR and basic first aid

First aid is the emergency treatment of an injury or illness at an initial level until skilled medical assistance arrives. The responsibilities of a first aider include swift and efficient treatment of the injury. This means to restart breathing and heartbeat, to stop the bleeding and to treat the body for shock and hypothermia.

A designated first aider should have completed a recognised training course such as those offered by

St John Ambulance or a Red Cross first aid course and have a working knowledge of their first aid manuals. In addition, the first aider should be able to keep concise records of treatment given.

The priority action plan

The priority action plan as suggested by St John Ambulance is a step-by-step strategy for managing the emergency incident. The acronym DRSABCD (Danger, Response, Send for help, open Airway, normal Breathing, start CPR and attach Defibrillator) is used as guidance for the action plan. The acronym allows the first aider as the first responder to identify and deal with initial life-threatening conditions before commencing treatment of non-life-threatening injuries. Refer to **Figure 2.9** for details of each step.

St John

First aid fact sheet

DRSABCD action plan

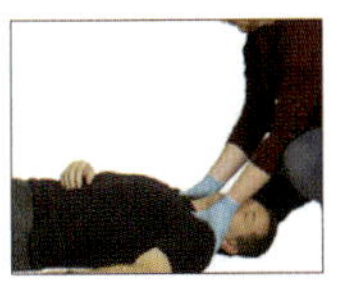

Danger Check for danger and ensure the area is safe for yourself, bystanders and the patient.

Response Check for a response: ask name and squeeze shoulders.

No response? Send for help.

Response? Make comfortable; monitor breathing and response; manage severe bleeding and then other injuries.

Send for help Call triple zero (000) for an ambulance or ask a bystander to make the call. Stay on the line.

If alone with the patient and you have to leave to call for help, first turn the patient into the recovery position before leaving.

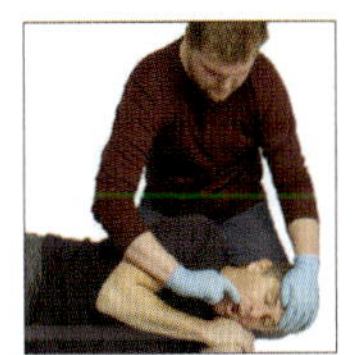

Airway Open the patient's mouth and check for foreign material.

Foreign material? Roll the patient onto their side and clear the airway.

No foreign material? Leave the patient in the position found, and open the airway by tilting the head back with a chin lift.

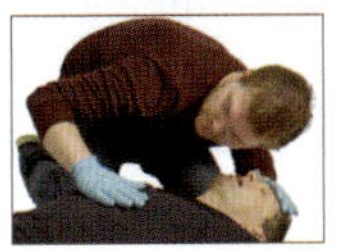

Breathing Check for breathing Look, listen and feel for 10 seconds.

Not normal breathing? Ensure an ambulance has been called and start CPR.

Normal breathing? Place in the recovery position and monitor breathing.

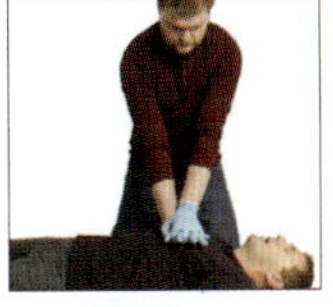

CPR Start CPR – 30 chest compressions followed by 2 breaths.

Continue CPR until help arrives, the patient starts breathing, or you are physically unable to continue.

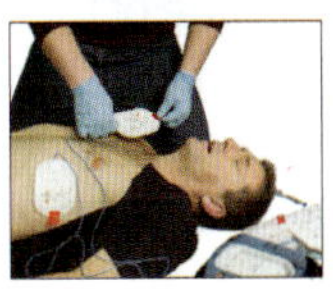

Defibrillate Apply a defibrillator as soon as possible and follow the voice prompts.

You could save a life with **first aid** training • www.stjohn.org.au • 1300 360 455

Source: © St John Ambulance Australia Ltd. All care has been taken in preparing the information but St John takes no responsibility for its use by other parties or individuals. St John encourages first aid training as this information is not a substitute for first aid training.

FIGURE 2.9 DRSABCD Action Plan

Emergency procedure at an accident

When approaching an accident scene:

- Look around for dangers.
- Check casualties and their response.
- In Australia, call 000; in New Zealand, call 111. (or also 112 from mobile phones).
- Care for injuries.

It is necessary to have an order of priority at an accident to ensure the safety of the first aider, then casualties, then bystanders, and to treat the casualties with life-threatening conditions first and then progress to the less serious conditions.

1 Look around for dangers.
   - Is the area safe (e.g. any live electrical wires, leaking fuel and fires)?
   - What does the incident or emergency involve?
   - How many persons are involved in the incident?
   - Is anyone able to assist?
2 Check the casualty's response.
   - You must first look for conditions that are an immediate threat to the casualty's life.
   - Remember COWS (see below).
   - Is the casualty responding to spoken commands or not?
   - If the casualty is not responding, immediately roll them into the recovery position and check their airway.
   - Once the airway is clear, check breathing.
   - If the casualty is breathing, check the circulation for regularity and strength.
   - Check for external bleeding.

One of the most common and intense emotions at the time of a medical emergency for some first aiders is fear (fear of death, fear of failure and fear of negative consequences), which can quickly result in panic.

Casualty examination

Determine the casualty's level of consciousness via gentle touching and loud talking. Do not shake the casualty. Casualty examination for the first aider follows a plan, which is known by the initials of its component parts, 'COWS'. This is used to remind first aid providers of some simple steps that will help to determine a casualty's ability to respond. These are:

- **C**an you hear me?
- **O**pen your eyes.
- **W**hat's your name?
- **S**queeze my hands.

If the casualty is conscious and they express no pain, observe their behaviour for any distress, unusual position or posture and body swelling. Where there is more than one casualty, always give priority to the unconscious casualty. If unconscious, perform DRSABCD.

Checking vital signs

If a person is unconscious, the first step is to check their mouth for any items blocking the airway. These items could include their tongue, food or vomit. If blockages are found, gently roll the person onto their side into the recovery position. Clear any blockages using your fingers, then check for breathing. If no blockage is found, roll the person onto their back and check for breathing. Listen for the sound of the breath, look for movements of the chest or feel for the breath on your cheek.

Rescue breathing

The term 'rescue breathing' has replaced 'expired air resuscitation' (EAR). The guidelines now recommend that full CPR be given to all those requiring resuscitation. If the person is breathing, roll the person into the recovery position. Phone 000 and check the person regularly until medical assistance arrives.

If the person is not breathing, place one hand on their forehead tilting the head back and with your other hand move the chin slightly downwards so that their mouth opens. Seal your mouth over their mouth, pinch their nose gently and blow steadily for two breaths.

If there are no signs of life after two breaths, make sure 000 has been called and commence chest compressions. The rescuer should quickly move to commence compressions but review the casualty's condition if any signs of life have returned (coughing, movement, normal breathing). The important thing is not to delay commencing compressions while you look for signs of life.

Chest compressions

When there are no signs of life present (the casualty is unconscious, unresponsive and not moving), the rescuer should commence CPR. When engaging in chest compressions:

- interruptions to compressions should be minimised
- compressions should be fast and hard
- over-ventilation should be avoided.

The compression ventilation ratio is 30:2 (30 compressions to two ventilations) for infants, children and adults.

Find the lower half of the sternum – you should visualise the 'centre of the chest' and compress at that point. There is no need for measuring. Position the heel of your hands in the centre of the person's chest; interlace your fingers and lift them off the chest. Using the heel of your hand, give 30 compressions. Each compression should depress the chest by about one-third. After 30 compressions take a deep breath, seal your mouth over the person's mouth, pinch their nose and give two firm breaths.

Continue giving 30 compressions followed by two breaths until medical help arrives. If signs of life return, move the person into the recovery position. Continue to monitor breathing and be ready to start CPR again at any time.

CPR should be performed by a single rescuer until other rescuers are available, then it may be performed by two rescuers (one performing chest compressions and one performing rescue breaths).

Figure 2.10, from St John ambulance, details the steps for performing CPR on a patient who is an adult or child over 1 year of age.

First aid fact sheet

CPR adult or child (over 1 year)

CPR is the action of giving 30 compressions followed by 2 breaths. Try to achieve 5 sets of 30:2 in about 2 minutes (or 100–120 compressions/minute).

If unwilling or unable to give breaths, giving compressions only is better than not doing CPR at all.

Give 30 compressions

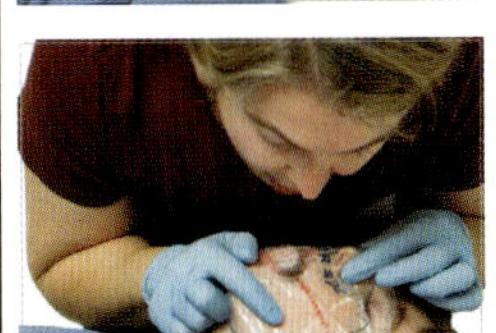

1. Position the patient on their back, on a firm surface, and kneel beside the patient's chest.
2. Locate the lower half of the sternum (breastbone) in the centre of the chest.
3. Place the heel of one hand on the lower half of the sternum and the heel of your other hand on top of the first hand.
4. Interlock the fingers of your hands and raise your fingers.
5. Press down on the sternum.
6. With your arms straight, press down on the patient's chest until it is compressed by about one-third.
7. Release the pressure. Pressing down and releasing is 1 compression.
8. Give 30 compressions.

Giving 2 breaths

1. Open the airway by tilting the head and lifting the chin.
2. With the head tilted backwards, pinch the soft part of the nose closed with your index finger and thumb, or seal the nose with your cheek.
3. Open the patient's mouth by placing your thumb over the chin below the lip and supporting the tip of jaw with the knuckle of middle finger. The chin is held up by your thumb and fingers to open the mouth and keep the airway clear.
4. Take a breath and place your lips over the patient's mouth, ensuring a good seal. Blow steadily for about 1 second, watching for the chest to rise.
5. Turn your mouth away from the patient's mouth. Watch for their chest to fall. Listen and feel for signs of air being expelled. Maintain head tilt and chin lift.
6. Take another breath and repeat the sequence above. This is now 2 breaths.
7. If the chest does not rise, recheck the mouth and remove any obstructions. Make sure the head is tilted and chin lifted, and ensure there is a good seal around the mouth (or mouth and nose).

In a medical emergency call Triple Zero (000)

DRSABCD Danger ▸Response ▸Send for help ▸Airway ▸Breathing ▸CPR ▸Defibrillation

You could save a life with first aid training • **www.stjohn.org.au** • 1300 360 455

FIGURE 2.10 Administering CPR

Level of consciousness

'Level of consciousness' is a term used to describe a person's alertness and understanding of what is occurring in their surrounds. A person's consciousness can vary from being normal, such as being completely alert, talking and making sense, to being deeply unconscious and not responding in any way. Changes in the level of

SWITCH ON

A = alert

V = responds to voice

P = responds to pain

U = unresponsive

consciousness of a person may be due to an injury to the head. An AVPU scale is used to determine levels of consciousness.

Also note the time of response, observe pupils and note any change in AVPU.

To check alertness, ask questions such as:

- What is your name?
- Where are you?
- What day and time is it?
- What happened?

Ask the injured person the above four questions often and note any changes. Note the level of responsiveness. If the person is alert, they are an 'A' on the AVPU scale. If the injured person is unable to answer, ceases to answer or only answers a few of the questions, the first aider needs to assess responsiveness to verbal or painful stimulus.

To assess response to verbal stimulus, speak or yell to the injured person and see whether they open their eyes, move or make some sound. If the injured person responds, they are a 'V' on the AVPU scale. If the injured person is unable to or ceases to respond to verbal stimuli, the first aider needs to assess the person's responsiveness to pain. Gently pinch the person's skin on both sides of the body and watch their face for movement. If the person shows any movement or makes any sounds, they are a 'P' on the AVPU scale.

If the injured person is not responsive to pain, they are considered unconscious and in a coma. This is indicated as 'U' on the AVPU scale. The AVPU information needs to be given to the medical professionals when they arrive.

Clinical shock

Clinical shock is brought on by the body's circulatory system failing to deliver a sufficient supply of blood to all parts of the body following severe injuries. Shock involves the following symptoms:

- anxiety and restlessness or irritability
- rapid weak pulse and fast shallow breathing
- pale, cool, moist skin
- dazed look, nausea or vomiting
- complaint of thirst.

Treatment for shock

Treatment for shock involves the following procedure.

1. Lie the casualty on their back.
2. Elevate the legs by 30 cm to return the blood supply to the brain.
3. Control any bleeding (if necessary).
4. Loosen the casualty's clothing at the neck, chest and waist and cover the casualty to maintain body temperature.
5. Only moisten the casualty's lips with water (no drink is to be given).
6. Comfort and reassure the casualty.

REVIEW QUESTIONS

1. The responsibilities of a first aider include swift and efficient treatment of the injury. What does this involve?
2. Explain the purpose of the priority action plan as suggested by St John Ambulance.
3. Name the acronym used as guidance for the first aid action plan and state the meaning of each letter.
4. List actions to take at an accident in their order of priority.
5. What is the purpose of an AVPU scale?
6. What can bring on clinical shock?
7. COWS is a useful acronym for casualty examination. What do each of the letters in COWS mean?
8. What may cause changes in the level of consciousness of a person?

CHAPTER REVIEW

2.1 Regulations and codes of practice

- First aid is the initial treatment provided to an injured or infirmed person until professional medical treatment arrives or is available.
- The Australian Resuscitation Council (ARC) has Guidelines covering the teaching and practice of resuscitation.
- A hazard can be a work practice, procedure or anything that has the potential to harm the health or safety of a person.
- Risk is a measure of the probability of a specific harmful effect occurring in particular circumstances.
- The most important step in managing risks on a worksite involves eliminating the risks as much as is reasonably practicable.
- A first aid code of conduct is a set of directions outlining the expected standards, rules and obligations of a person providing first aid.

- The transmission of disease when administering first aid is rare and is largely avoidable. The risks can be mitigated during resuscitation by using pocket masks or other barrier devices.

2.2 Legal, workplace and community considerations

- Consent is permission or agreement from the patient to be treated by you. When providing first aid it is necessary to obtain consent from a patient where possible before applying first aid.
- Anyone providing first aid has a duty of care, which is a legal obligation for you to protect yourself and the person to whom you are providing first aid.
- Another aspect of administering first aid is behaving respectfully, which is the morally correct conduct of a person providing first aid.
- If called on to administer first aid you are limited to the extent of your training.
- Privacy and confidentiality are two important considerations when administering first aid.
- Two important post-incident processes are debriefing and evaluating.
- Stress is a human biological response to certain situations. It is a natural feeling of not being able to cope with specific demands and events.

2.3 Considerations for performing CPR

- Cardiopulmonary resuscitation, which is commonly abbreviated to CPR, is an emergency procedure that is undertaken on a patient who has no signs of life.
- The purpose of CPR is to preserve brain functions by manually circulating oxygenated blood through the patient's body until advanced measures are provided to restore spontaneous blood circulation and breathing.
- The process of moving oxygen into (inhaling) and out (exhaling) of the lungs is known as breathing or ventilation.
- An Automated External Defibrillator (AED) is an electronic device used in the treatment of a cardiac condition known as fibrillation.
- The chain of survival documents a series of actions taken to reduce the risk of death from sudden cardiac arrest.
- The Triple Zero (000) service is the most expedient method to get the right resources from emergency services to help you and should be used to contact Police, Fire or Ambulance services in life-threatening or time-critical situations.

2.4 Administering CPR and basic first aid

- A designated first aider should have completed a recognised training course such as those offered by St John Ambulance or the Red Cross and have a working knowledge of their first aid manuals.
- The priority action plan as suggested by St John Ambulance is a step-by-step strategy for managing the emergency incident using the acronym DRSABCD as guidance for the action plan.
- It is necessary to have an order of priority at an accident to ensure the safety of the first aider, then casualties, then bystanders, and to treat the casualties with life-threatening conditions first and then progress to the less serious conditions.
- The compression ventilation ratio is 30:2 (30 compressions to two ventilations) for infants, children and adults.
- Clinical shock is brought on by the body's circulatory system failing to deliver a sufficient supply of blood to all parts of the body following severe injuries.

TRIAL EXAM

For Chapter 2 knowledge assessment, please complete the following trial exam.

1 When approaching an incident or accident scene, the first action is to:
  a Ask bystanders what happened and check if anyone has called Triple Zero (000).
  b Check for potential hazards or dangers and make sure that the area is safe for you, the casualty and others.
  c Collect all available first aid resources and bring them to the scene.
  d Immediately start to check the casualty's airway, breathing and circulation.

2 Once a first aider starts providing first aid, they have a duty to provide care to the best of their ability until:
  a the patient becomes confused and starts telling you that they do not need any help.
  b they determine that there are enough bystanders who should be able to manage the situation.
  c they have been relieved by a more qualified person or their own safety is at risk.
  d they have called Triple Zero (000) and advised the casualty to wait until the ambulance arrives.

3 To ensure privacy and confidentiality of first aid records you would:
  a destroy all first aid records 28 days after the date of the incident.
  b send first aid records to the person receiving treatment within 7 days of the incident.
  c store them in the first aid kit at work for quick access.
  d store them in an appropriate location, which is only accessible by authorised staff.

4 It is important to participate in a debriefing session after a major first aid incident to:
 a ensure that all first aid supplies have been restocked and ready for next use.
 b comply with Australian Resuscitation Council Guidelines.
 c satisfy Safe Work Australia's requirements for dealing with major incidents.
 d review the situation and identify if there is need for further support or education.

5 The most appropriate action to take if you are unable to gain the consent of a casualty because they are not responsive is to:
 a call Triple Zero (000) and let the operator provide consent.
 b commence first aid as the casualty's consent is assumed in this situation.
 c use the casualty's phone to contact one of their family members and obtain consent from them.
 d wait for the casualty to become lucid and then ask for their consent.

6 A colleague tells you they are unable to sleep properly and are having flashbacks after responding to a serious first aid incident. The most appropriate assistance to provide the colleague is to:
 a reassure them that this a normal response to a serious incident and not to worry about it.
 b encourage them not to think about the incident and get them to think about something more positive.
 c ask them to share their concerns with you and use reflective listening techniques.
 d suggest that they seek professional support and consider using stress-management techniques.

7 When you attend a first aid incident involving minor bleeding, what standard infection control precautions should the first aider take?
 a Find out if the patient has an infectious disease to assist in determining the appropriate precautions.
 b Wash the wound with cool running water for 20 minutes before touching the wound.
 c Wear disposable gloves and wash hands before and after treating the casualty.
 d Wait until all bleeding has stopped before managing the casualty.

8 What is the primary focus of the 'chain of survival'?
 a Provide five minutes of CPR before sending for help
 b Assess the chance of survival before commencing CPR so as not to waste energy
 c The earlier someone calls for help and starts CPR the better the chance of survival
 d Survival from a sudden cardiac arrest depends on how quickly emergency services can attend.

9 In Australia, emergency services are contacted by using a phone and dialling:
 a Triple Zero (000)
 b Triple Nine (999)
 c Triple One (111)
 d Nine-One-One (911)

10 The term used to describe a casualty who does not respond to talk or touch is:
 a disorganised
 b conscious
 c delirious
 d unconscious

11 Cardiopulmonary resuscitation (CPR) commences with:
 a 2 rescue breaths
 b 30 rescue breaths
 c 2 chest compressions
 d 30 chest compressions

12 It is recommended that when providing chest compressions to an adult casualty, place:
 a one hand at the top of the chest and depress to about one-third chest depth.
 b two fingers over the centre of the chest and compress at a rate of 120 per minute.
 c two hands over the centre of the chest and depress to about one-third chest depth.
 d two hands at the end of the sternum and depress about half chest depth.

13 While an AED is analysing the patient's heart rhythm and before administering a shock, you should ensure that:
 a a spare battery is available.
 b nobody is touching the patient.
 c the AED is within test.
 d the patient is placed in the recovery position.

14 In which of the following situations should the first aider stop performing CPR?
 a After completing 2 minutes of uninterrupted chest compressions
 b After a defibrillator (AED) has delivered the third shock
 c If a bystander taps you on the shoulder and says 'he's gone, mate'
 d When health care or emergency services personnel are ready to take over.

15 When a casualty is unconscious and either lying face down or on their back, they are at risk of developing:
 a an airway obstruction and/or breathing difficulty.
 b reduced flow of blood to the vital organs.
 c low blood pressure and severe shock.
 d hyperthermia and seizures.

Fabricate, assemble and dismantle utilities industry components

This chapter provides electrotechnology workers with the knowledge and skills to interpret mechanical drawing, sketching, detail and assembly drawings. They will also gain an overview of hand and power tools.

LEARNING OBJECTIVES

Mechanical drawing

- Outline the importance of drawing standards
- Identify various line types
- Apply dimensioning techniques
- Use appropriate drawing scales with the metric system
- Recognise a variety of machining symbols
- Apply standard drawing abbreviations
- Apply sketching techniques to record technical data

Workshop planning

- Outline methods used to work safely in an industrial work environment
- List typical non-electrical workplace hazards
- List and explain the qualities and function of various materials used in the electrotechnology industry
- Outline the workshop planning process

Measuring and marking-out

- Explain the reasons for measuring and marking-out
- List and explain the qualities and function of common measuring and marking-out tools
- Explain how accurate measuring and marking-out can reduce waste

Holding and cutting

- Identify holding and cutting tools
- List and explain the qualities and function of common holding and cutting tools
- Outline appropriate safety procedures for using holding and cutting tools

Drills and drilling

- List and explain the qualities and function of common drills used in the electrotechnology industry
- Outline appropriate safety procedures for drilling

Tapping and threading

- List and explain the qualities and function of common taps and dies used in the electrotechnology industry
- Outline appropriate safety procedures for tapping and threading

General hand tools

- List and explain the qualities and function of common hand tools used in the electrotechnology industry
- Outline appropriate safety procedures for using common hand tools

Joining techniques

- List and explain the purpose of various joining techniques such as soldering, brazing, hard soldering and welding
- Outline appropriate safety procedures for soldering, brazing, hard soldering and welding

Power tools

- Outline the application of common power tools used in the electrotechnology industry
- Explain the operation of common power tools used in the electrotechnology industry

Sheet metal work

- List and explain the purpose of common sheet metals used in the electrotechnology industry
- Explain various techniques used in the electrotechnology industry to fabricate items from sheet metal
- Outline common methods used in the electrotechnology industry to join sheet metal

High accuracy measurement (low tolerance measurement)

- Explain the meaning of tolerance
- Outline the qualities and explain the function of micrometers and Vernier calipers

Dismantling and assembly techniques

- Outline a safe and systematic method for disassembling and reassembling electrotechnology components

3.1 Mechanical drawing

Mechanical drawing is the primary method of communication between all people involved in the design and manufacture of components, buildings and constructions as well as engineering projects.

Drawing standards

Engineering drawings and other technical drawings have to be prepared in a way that workers in a specific industry can recognise. These techniques are standards or conventions. Drawings are made to standard conventions so that:

- they all use the same symbols, lines, dimensioning techniques and so on
- they can be understood throughout Australia or internationally.

All drawings should follow the rules of:

- Australian Standard AS 1100.101:1992/Amendment 1:1994 *Technical drawing – General principles*
- Australian Standard AS 1101 series
- Australian/New Zealand Standards AS/NZS 1102 series 1997.

The most common system of paper sizes in Australia is from the ISO standard. Drawings are usually produced in standard-size sheets, ranging from A0 to A4. Standard sheets not only aid storage but it is also important in the photocopying of drawings. The A0 size, as shown in **Figure 3.1**, is defined as having an area of one square metre. Square metre size allows paper weights to be expressed in grams per square metre. An A0 sheet (1188 mm × 841 mm) can be divided up evenly into the various other sizes simply by halving the sheet along the long side in each case.

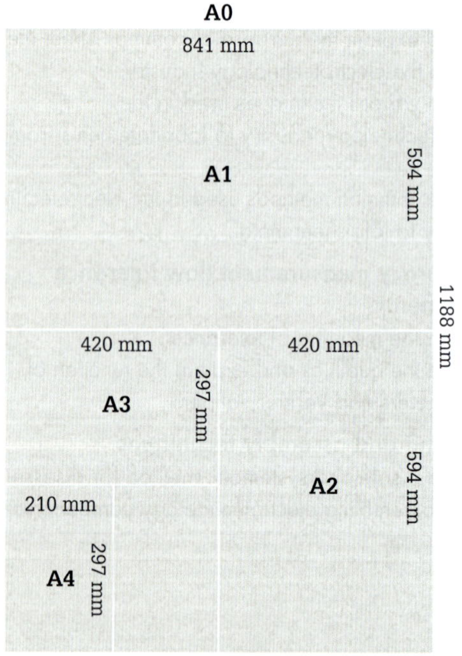

FIGURE 3.1 Drawing sheet sizes

As the sides of every size in the series are in the ratio of 1: √2 (= 1 : 1.414), each size is half the area of the next larger size, as shown in **Table 3.1 (a)**.

TABLE 3.1 (a) ISO standard paper sizes

Drawing sheet size	Size in millimetres
A0	1188 × 841
A1	841 × 594
A2	594 × 420
A3	420 × 297
A4	297 × 210
A5	210 × 148
A6	148 × 105

TABLE 3.1 (b) Grid reference system

Details of grid references					
Detail	A0	A1	A2	A3	A4
Number of vertical zones designated 1, 2, etc.	16	12	8	8	4
Number of horizontal zones designated A, B, etc.	12	8	6	6	4

Drawing sheets A5 and A6 do not have grid references as they are used for notepads and postcards.

AS 1100.101 also requires that necessary information is shown on every drawing. The location of that information is also standardised, as presented in **Figure 3.2**. In the past, the small arrowheads in the centre of the outside of the margins were 'camera marks' used to align cameras when it was common practice to photograph drawings and store them on microfilm.

Drawings may be divided into zones by a grid reference system based on horizontal numbered (L to R) and vertical lettered divisions (A at the top) to assist in readily locating a particular dimension or feature, especially in large drawings (**Table 3.1 (b)**). Note that the letter I is not used, and letters are read from the bottom of the drawing.

Most drawing sheets also have:

- a margin or border
- a title block
- a projection symbol (see **Figure 3.3**)
- differ only in the position of the plan, front and side views
- a list of materials and parts
- a format that can be revised when necessary.

The intention of an engineering drawing is to convey all the necessary information about how to make the part. For many engineering components, the needed information cannot be communicated with a single view of one side of the part. Rather than using a number of drawing sheets with different side views of the part, several aspects can

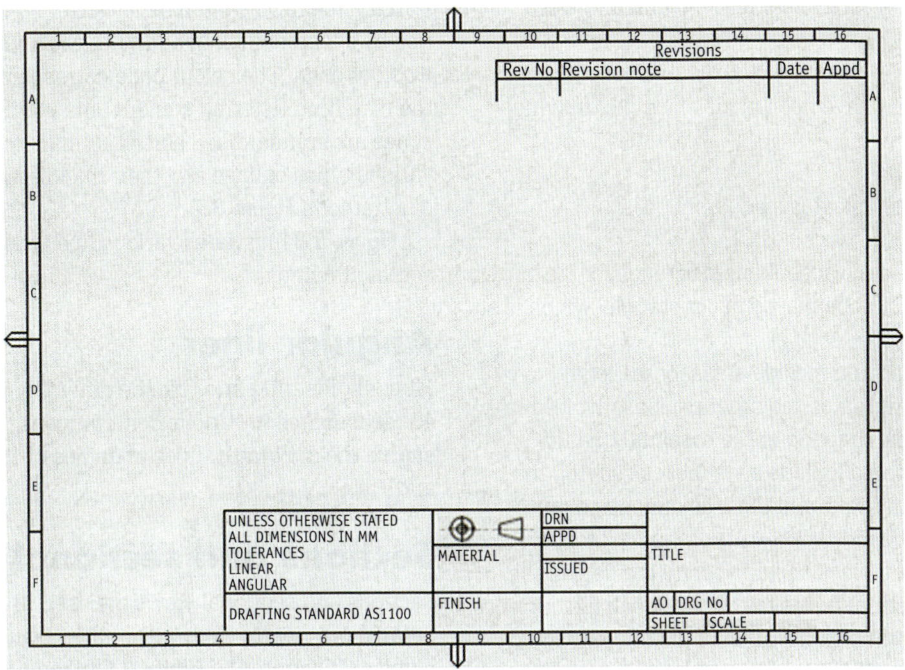

FIGURE 3.2 Basic information on drawing sheet

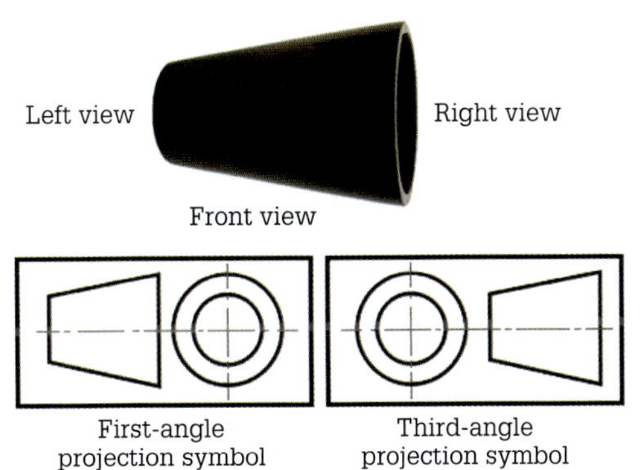

FIGURE 3.3 Projection symbols

be combined as a single drawing using either first-angle or third-angle projection.

Drawings by themselves are not enough to tell the reader everything they need to know. Each drawing requires information about materials, joining methods, tolerances, and instructions for the manufacturer.

Lines in drawings differ in thickness and in the way they are drawn, depending on the size of the paper and the job to be done. However, each line and each thickness must conform to national and international standards.

Written information on a drawing is always in standard lettering. The standard regulates the shape and size of letters and numerals. Symbols are used for items such as dimensions, radius, diameter, tolerance, surface textures, weld details, and methods of projection.

Title block information

A characteristic title block (see **Figure 3.4**) includes the border and the various sections for providing quality, administrative and technical information. The title block contains all the information that enables the drawing to be interpreted, identified, and archived.

The title block should include sufficient information to determine the type of drawing, for example, overall arrangement or detail. It should also clearly describe in a specific way what the drawing represents.

The basic requirements for a title block located at the bottom right-hand corner of a drawing are:

- the name of the company
- the name of what is drawn
- the drawing number for storage and reference purposes
- the sheet number in a set of drawings
- the name of the person who drew the drawing
- the name of the individual who reviewed the drawing
- the issued date

DESIGN ENGINEER	DATE 30.06.19	PROJECT: FLUID PUMP STATION VARIABLE SPEED DRIVES 22.5–240 kW	DRAWING NUMBER SD-806/05 Sheet 04 of 04
DESIGN MANAGER	DATE 30.06.19		
DRAWN	ELECTRICAL DESIGN	PLAN TITLE: SWITCHBOARD CABINET DETAILS GENERATOR TERMINATION ENCLOSURE GENERAL ARRANGEMENT	MARCH 2019
SCALE	NTS		

FIGURE 3.4 Typical title block

- the size of the original sheet that the drawing was drawn on
- the scale
- any changes that have been made since the drawing was initially drawn
- projection symbols
- material that the object is made from
- grade of finish
- tolerance to state the acceptable size range for the parts.

These items should be placed in a rectangle, which is at the most 170 mm wide.

The title block should include areas for the legal signatures of the originator and other persons involved in the production of the drawing to the essential quality.

The drawing should include a symbol classifying the projection, the main scale, and the true dimension units if millimetres are not used.

The drawing title block should specify the date of the first revision. In distinct boxes to the title block the current revision with an outline explanation of the review should be shown (see **Figure 3.5**). On completion of each drawing revision, an additional correction box should be completed to provide a detailed history of the drawing.

1	SECTION ON AA DELETED		01.07.19
ISSUE	AMENDMENT DETAILS	INITIALS	DATE

FIGURE 3.5 Typical revision box

Line types

If all lines on a drawing were equally thick, the drawing would be confusing and difficult to interpret, as the outlines would not stand out from the dimension lines. By varying the thickness and style of lines on a drawing you can provide meaning that is otherwise difficult to express in words. To make sure everyone interprets drawings the same way, the use of each line type and thickness is defined in AS 1100.101:1992 *Technical drawing – General principles*. Technical drawers use different thicknesses of line to differentiate line 'meanings'.

The ISO defines a set of standard metric line widths for drafting, as shown in **Figure 3.6**.

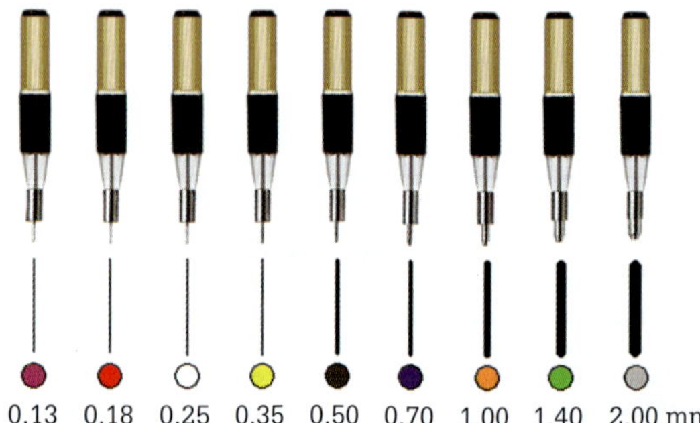

FIGURE 3.6 Standard metric line widths (pens are used to create ink lines on the drawing sheet)

Like the ISO A and B series sheet sizes, the pen sizes rise by a factor of $\sqrt{2}$. In addition, each width is assigned a colour code. The colour code corresponds to that for the matching lettering stencil. Line widths and colour codes are standardised across all manufacturers. The various types of lines and their thickness requirements are illustrated in **Figure 3.7**.

Figure 3.8 illustrates various line types applied to actual drawings.

Angular lines

Most engineering lines can be drawn by using 30°–60° and 45° set squares in various combinations. The combinations enable the drawing of lines at angles of 15°, 30°, 45°, 60° and 75°.

Sections and sectional views

Sections and sectional views are used to show hidden detail more clearly. They are created by using a cutting plane as illustrated in **Figure 3.8** to reduce the object. A section displays the outline of the subject matter at the cutting plane. Hatching is used in order to aid readability, and on sections and sectional views substantial areas should be hatched.

Dimensions

The outline of the object indicates the shape of the object; the dimensions indicate the size of the object. Only those dimensions necessary for the manufacture of the object should exist on the drawing. All dimensions should be shown on the drawing once only and applied to the view where the detail is seen strongest as an outline. Dimensions on engineering drawings are expressed in millimetres. Points to consider relating to dimensions are:

- Place dimensions well clear of drawings and one another.
- Place dimensions where they will be best understood (in relation to a datum).
- Use a 0.50 pen for the numbers and arrowheads.
- Show dimensions once only.
- Keep dimensions from the actual view.
- Dimension all circles and arcs radially.
- Avoid crossing dimension lines.
- Avoid dimensioning hidden details.
- Dimension along the dimension lines placing sizes so that they will read on the bottom and right-hand side of the sheet.

Dimension lines and projection lines

Projection lines are used to indicate the extremities of a dimension. They are drawn up to 1 mm from the outline of the object. Dimension lines are used to label a specific dimension. They have one or more arrowheads, which are typically 3 mm long × 1 mm wide. Thin lines (0.18 mm, 0.25 mm or 0.35 mm) are used for both projection lines and

Type of line (mm)	Line thickness			Application
	A0	A1	A2 A3 A4	
Continuous–thick	0.7	0.5	0.35	Visible outlines Border lines
Continuous–thin	0.35	0.25	0.18	Fictitious outlines Imaginary intersection of surfaces Dimension lines, projection lines, intersection lines and leaders Hatching Outlines of revolved sections Adjacent parts and tooling Fold and tangent bend lines Short centre lines
Continuous—thin, freehand or ruled with zig-zag	0.35	0.25	0.18	Indication of repeated detail Break lines (other than on an axis)
Dashed—medium	0.50	0.35	0.25	Hidden outlines
Chain—thin	0.35	0.25	0.18	Centre lines Pitch lines Alternative position of moving part Path lines for indicating movement Features in front of a cutting plane Developed views Material to be removed
Chain—thick at ends and at change of direction, thin elsewhere	0.7 0.35	0.5 0.25	0.35 0.18	Cutting planes
Chain—thick	0.7	0.5	0.35	Indication of surfaces to meet special requirements

FIGURE 3.7 Types of lines and their thickness requirements

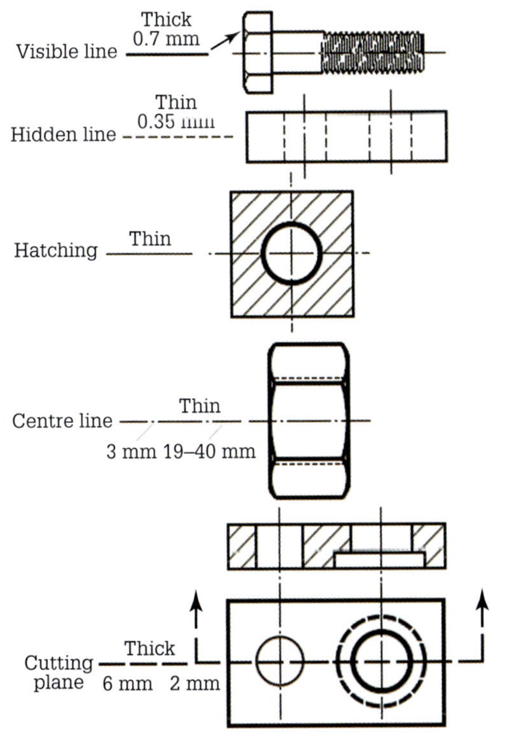

FIGURE 3.8 Drawings showing various line types

dimension lines. Dimension lines and projection lines are illustrated in **Figure 3.9**. (**Note:** In **Figures 3.9** and **3.11**, the datum is the reference from which measurements are made.)

Linear dimensions

Linear dimensions are in millimetres. To avoid having to specify 'mm' after every dimension, a label such as 'all dimensions are in mm' or 'unless noted otherwise (UNO) all dimensions are in mm' is written in the title block.

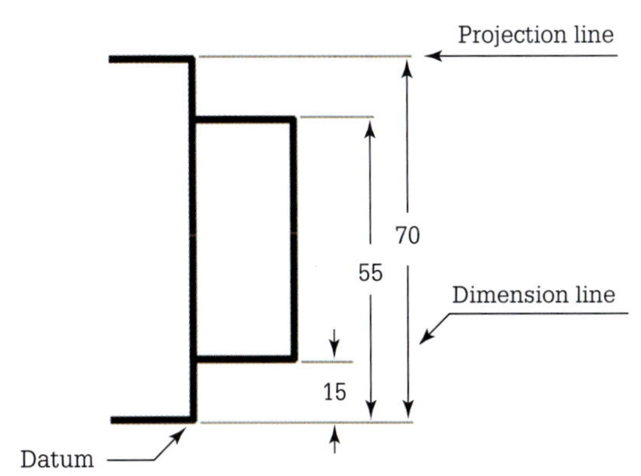

FIGURE 3.9 Dimension lines and projection lines

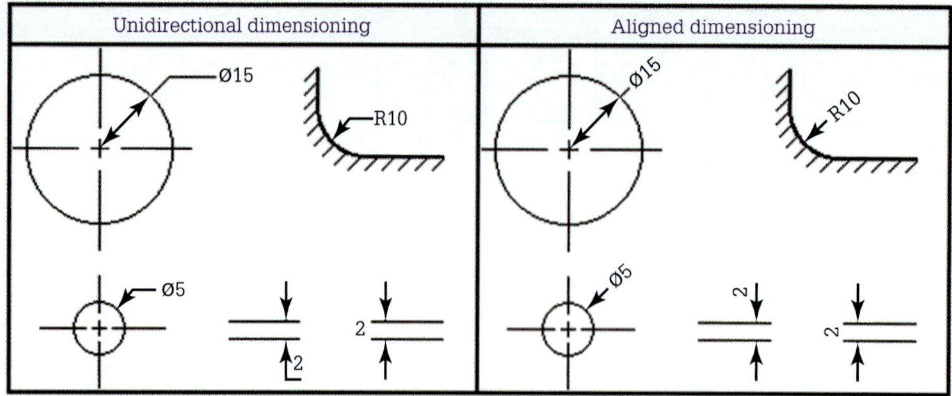

FIGURE 3.10 Dimensioning methods

If the dimension is less than one, a leading zero should be used before the decimal point, for example, 0.5.

Angular dimensions

Angular dimensions are usually specified in decimal degrees, degrees and minutes, or degrees, minutes and seconds.

e.g. 28.5° 28°30' 28°30'16''

If the angle is less than one degree, a leading zero should be used.

e.g. 0.5° 0°30'

Dimensioning methods

Two methods of dimensioning, as shown in **Figure 3.10**, are in common use:

1 unidirectional where the dimensions are printed horizontally; and
2 aligned, where the dimensions are printed parallel to their dimension line. Aligned dimensions should always be readable from the bottom or the right of the drawing.

Overall dimensions

When several dimensions make up an overall length, the overall size can be shown on these piece dimensions (see **Figure 3.11**). When specifying an overall dimension, one or more non-critical component dimensions must be omitted.

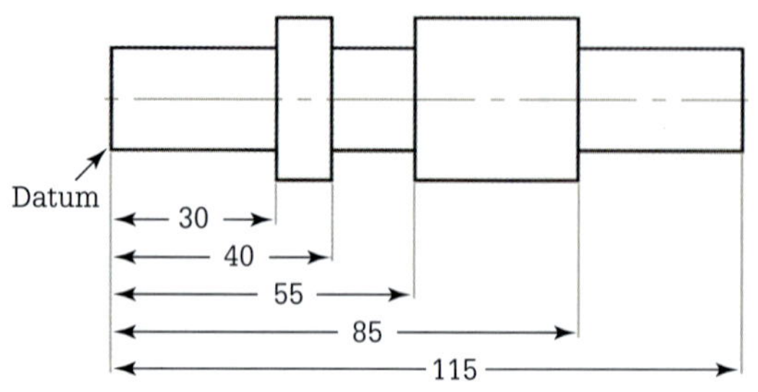

FIGURE 3.11 Overall dimensions

Auxiliary dimensions

When all of the component dimensions must be specified, an overall length may still be defined as an auxiliary dimension. Auxiliary dimensions are never toleranced and are shown in brackets (see **Figure 3.12**).

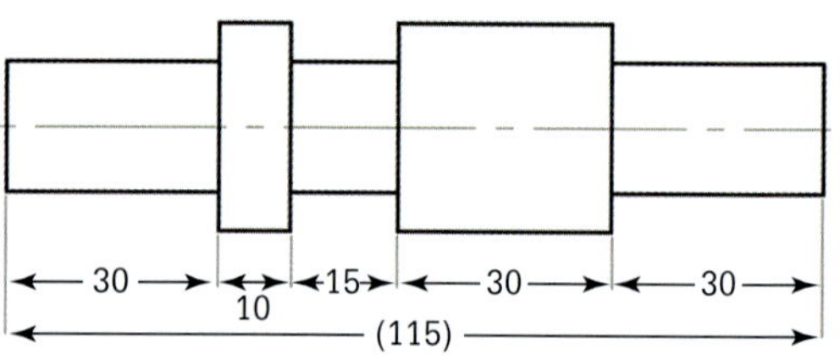

FIGURE 3.12 Auxiliary dimensions

Dimensions not to scale

Dimensions that are not to scale are underlined (see **Figure 3.13**).

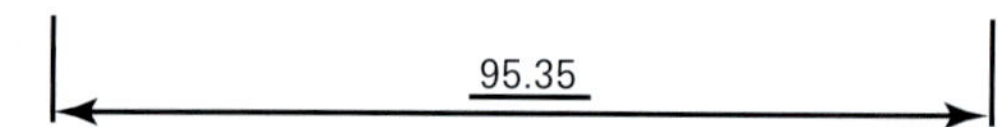

FIGURE 3.13 Dimensions not to scale

Dimensions not complete

When a dimension line cannot be entirely drawn to its normal termination point the free end is terminated by a double arrowhead (see **Figure 3.14**).

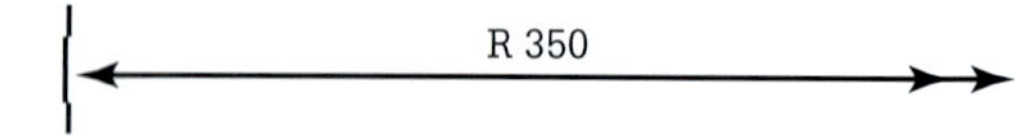

FIGURE 3.14 Dimensions not complete

Scales

The actual size of an object may vary from 1 mm to thousands of mm. Small objects should be drawn at their natural size. Larger ones should be drawn to a recognised scale so that they will fit on the drawing sheet.

Large details, structures and machine parts are drawn smaller than actual size, while tiny things such as instrument parts are drawn larger than their correct size.

The scales recommended for use with the metric system are shown in **Figure 3.16**. Decimal multiples of these base scales are also used, for example, 1:100, 1:250, 20:1.

EXERCISE 3.1

Draw in the required line types on the drawings shown in **Figure 3.15**.

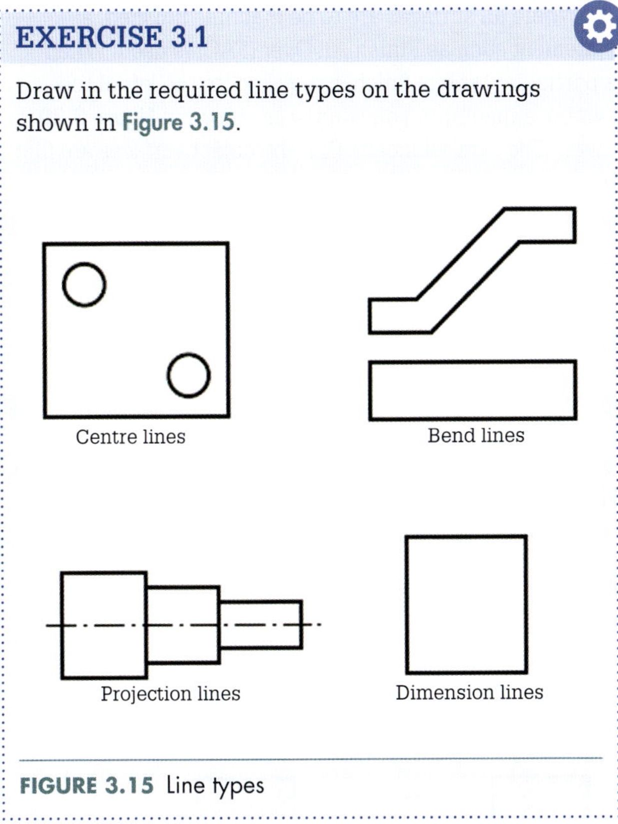

FIGURE 3.15 Line types

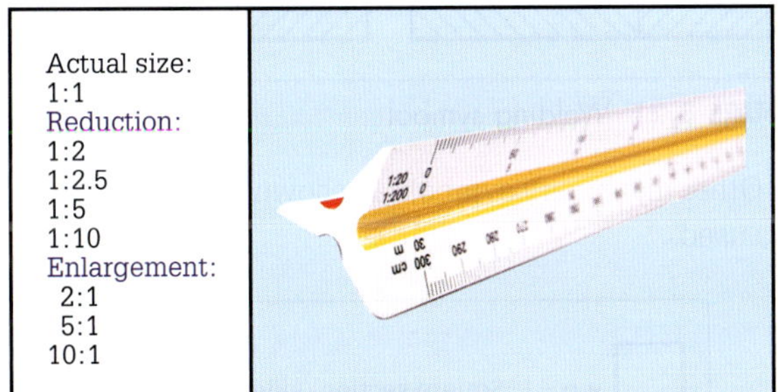

Source: Shutterstock.com/Bokic Bojan

FIGURE 3.16 Metric scales

Indication of scale

The scale should usually be noted in the title block of a drawing. When more than one scale is used, they should be shown close to the views to which they refer, and the title block must read 'scales as shown'. If a drawing uses predominantly one scale it should be noted in the title block together with the wording 'or as shown'.

Use of scales

The choice of scale depends on two elements:

1. the extent of the object to be drawn
2. the quantity of detail that needs to be shown.

If the distance shown in **Figure 3.17** represents 15.35 m, it will be to a scale. If scaled dimensions and written dimensions differ, then the written dimension should always be used.

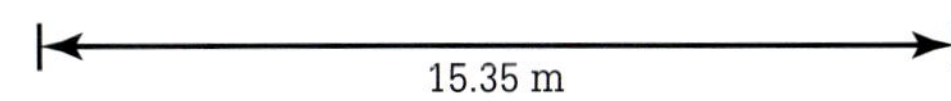

FIGURE 3.17 Written dimensions

The scale shows how much smaller the object or house plan is compared to its original size. Refer to **Table 3.2** for examples of measurements when drawn to scale.

TABLE 3.2 Scales

Scale	Unit of measurement on the drawing	Unit of measurement on the actual object
1:20	1 mm	20 mm
	2 mm	40 mm
	10 mm	200 mm
	100 mm	2000 mm, or 2 m
1:10	100 mm	1000 mm, or 1 m
1:100	100 mm	10000 mm, or 10 m
1:50	100 mm	5000 mm, or 5 m
1:200	150 mm	30000 mm, or 30 m

The **tolerance** is taken to be 0.5 of the unit being measured. For example, the tolerance of 1000 mm is ± 0.5 mm or the upper limit is 1000.5 mm and the lower limit is 999.5 mm.

SWITCH ON

It is simply a matter of multiplying the scale measurement by the scale ratio.

- 1:(10) × 100 mm = 1000 mm, or 1 m
- 1:(50) × 100 mm = 5000 mm, or 5 m
- 1:(200) × 150 mm = 30000 mm, or 30 m

EXERCISE 3.2

Refer to **Figure 3.18** to answer the following questions:

a When measured on the diagram, dimension A is 55 mm. What is the actual length of this side if the scale is 1:10?

b When measured on the diagram, dimension Z is 70 mm. What is the actual length of this side if the scale is 1:5?

c When measured on the diagram, dimension J is 12 mm. What is the actual length of this side if the scale is 1:20?

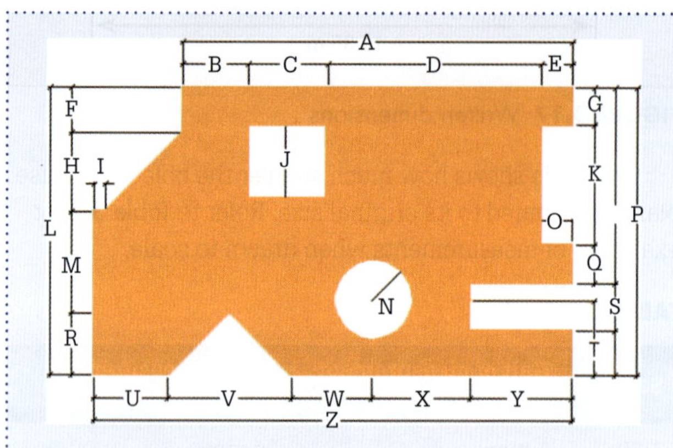

FIGURE 3.18 Scale measurement

EXERCISE 3.3

a Which units of measurement (kilometres, metres and millimetres) would you use to measure:
  - **i** The direct distance between Brisbane and Sydney?
  - **ii** The length of an AFL ground?
  - **iii** The diameter of a copper wire?
  - **iv** The thickness of wire insulation?

b Estimate the depth, width and height of a miniature circuit breaker.

c The length of a lighting circuit cable is 50 m correct to the nearest metre. What are the upper and lower tolerances for the length of the cable?

d The distance between the workshop and the customer's premises is 18 km, correct to the nearest 0.5 km. What is the percentage tolerance for this distance?

Machining symbols

Machining symbols are used to communicate meaning in a simple form. Once learned, symbols help in the understanding of exactly what needs to be done. Engineering drawing symbols are drawn to Australian Standard AS 1101 Part 3.

Some symbols are used to describe the level of finish required for particular surfaces of an object. The symbol for a machine finish is illustrated in **Figure 3.19**.

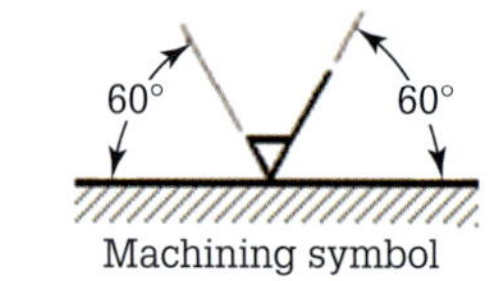

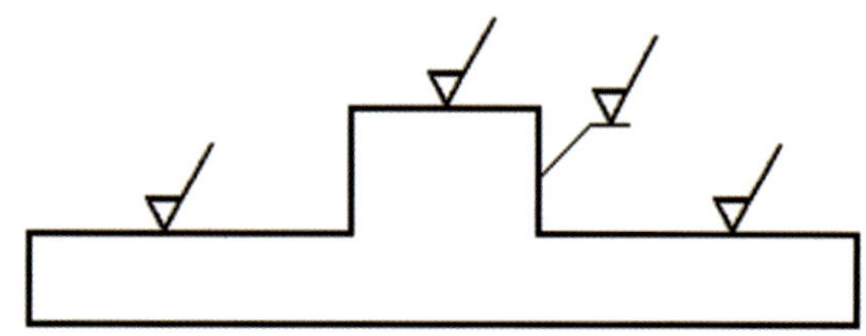

FIGURE 3.19 Machine symbols

Where all surfaces are to be machine finished, a statement such as Finish All Over (FAO) is used. When a particular type of finish is required, to the left of the symbol a machining allowance as shown in **Figure 3.20** is given. The symbol means that the object surfaces are first made bigger by that allowance.

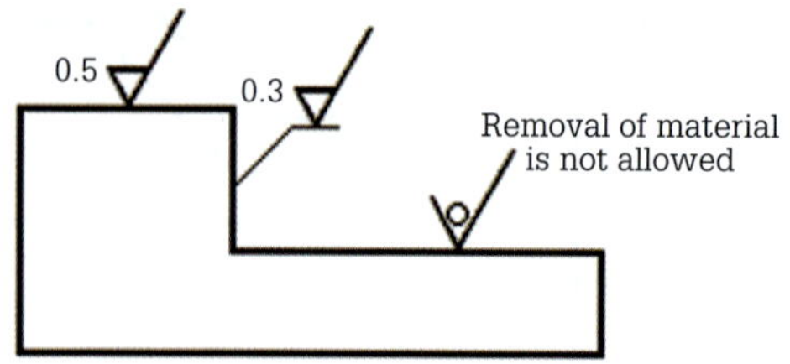

FIGURE 3.20 Machining allowance

When material is not to be machine finished, another symbol (Removal of material is not allowed), shown in **Figure 3.20**, is used.

Welding symbols are also used in many engineering drawings. The symbol for a corner fillet weld is illustrated in **Figure 3.21**.

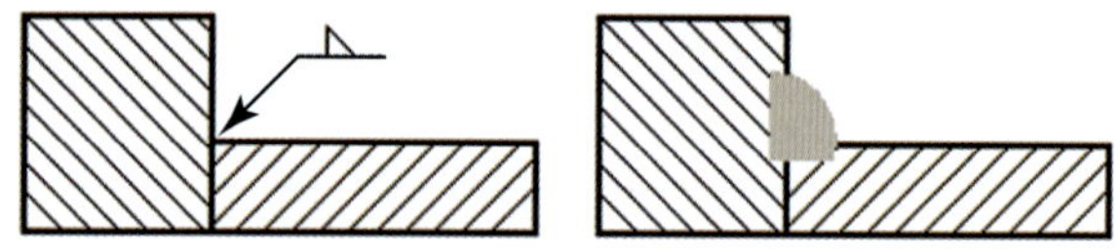

FIGURE 3.21 Welding symbol

Other machinery symbols as shown in **Figure 3.22** are also used.

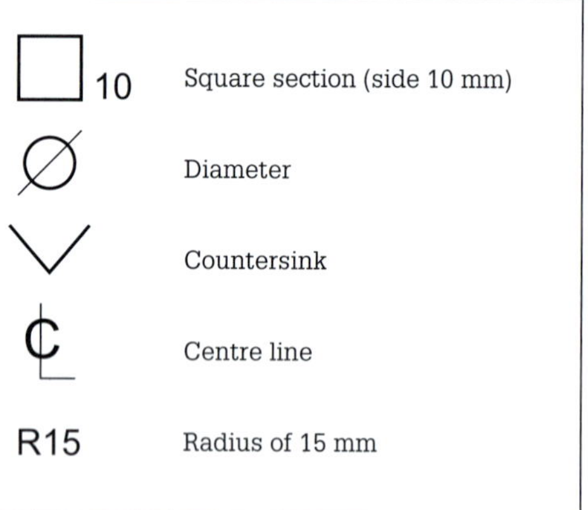

FIGURE 3.22 Additional machinery symbols

Abbreviations

Standard drawing abbreviations, shown in **Table 3.3**, are also used for engineering drawings.

Types of drawing representations

Drawings of an object can be depicted as a detail, assembly, a sequence of side views, called an orthographic projection,

TABLE 3.3 Standard drawing abbreviations

Abbreviation	Term
ASSY	assembly
DWG	drawing
FIN	finish
NS	near side
NTS	not to scale
RAD or R	radius
STL	steel
TYP	typical

or as a figure that looks like the actual appearance, called a pictorial projection.

Detail drawings

A detail drawing is a multiview representation of a single part of a mechanical component that includes a complete and exact description of its form, dimensions and construction. The person who makes the mechanical element must clearly understand the shape, size, material and surface finish of a part, what machinery operations are necessary and what limits of accuracy must be observed from the detail drawing. **Figure 3.23** is an example of a detail drawing.

Assembly drawings

An assembly drawing, illustrated in **Figure 3.24**, is a presentation of the individual components put together, showing each part in its operational position. The separate parts are put together according to the assembly drawings.

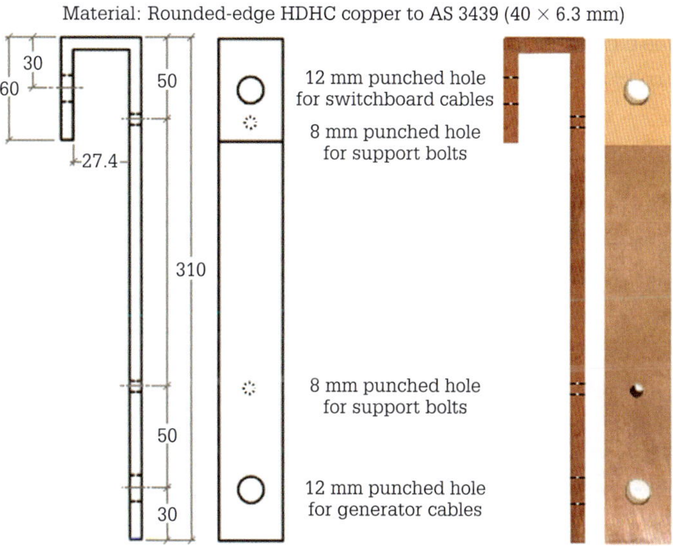

FIGURE 3.23 Detail drawing

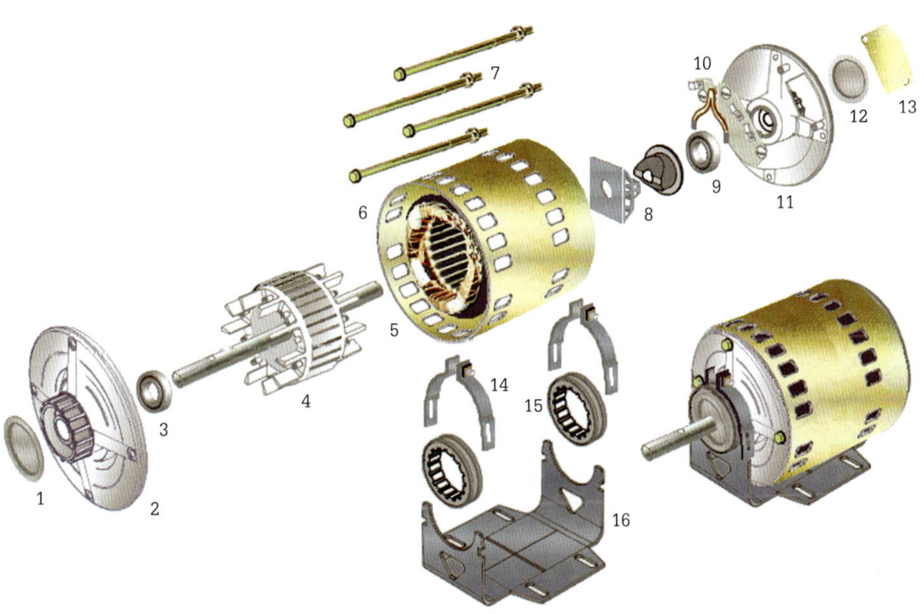

FIGURE 3.24 Assembly drawing of a split phase motor

Assembly drawings should include item letters or numbers representing the various parts. Circles may enclose the part numbers with a leader pointing to the piece. A parts list for the split-phase motor is illustrated in **Figure 3.25**.

ITEM	DESCRIPTION	ITEM	DESCRIPTION
1	Bearing cap DE	9	Bearing NDE
2	Shield DE	10	Cut-out switch
3	Bearing DE	11	Shield NDE
4	Rotor core	12	Bearing cap NDE
5	Stator	13	Terminal cover
6	Frame	14	Mounting clamp
7	Fixing bolt shields	15	Vibration mount
8	Centrifugal switch	16	Frame support

FIGURE 3.25 Split-phase motor parts list

Pictorial views

Pictorial views is a method of drawing objects in a three-dimensional form. There are three different pictorial projection drawing methods for illustrating objects or buildings:

- isometric projection
- axonometric projection
- oblique projection.

We will use a brick, illustrated in **Figure 3.26**, to demonstrate the three pictorial projection drawing methods.

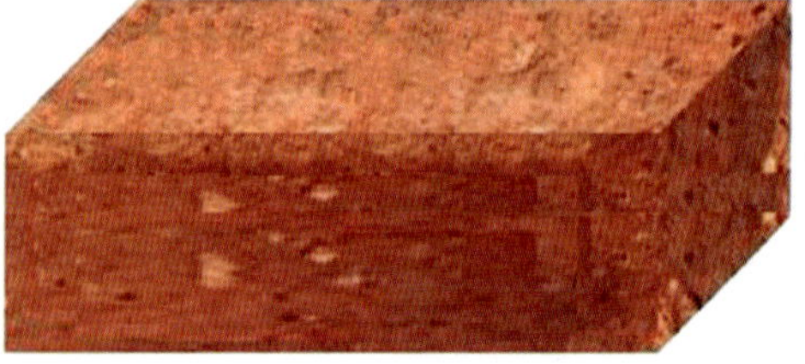

FIGURE 3.26 Brick

Isometric projection

In an isometric drawing, illustrated in **Figure 3.27**, the object's perpendicular lines are drawn vertically, and the horizontal lines in the width and depth planes are shown at 30° to the horizontal. When drawn from these guidelines, the lines parallel to these three axes are at their actual (scale) lengths. Lines that are not parallel to these axes will not be of their exact length.

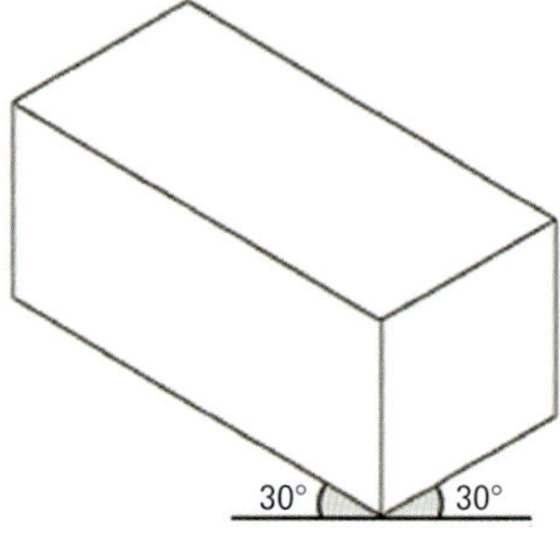

FIGURE 3.27 Isometric projection

EXAMPLE 3.1

Drawing isometric objects using the box method

Draw an isometric box as shown in **Figure 3.27** to the outside dimensions of the object you are drawing as shown in **Figure 3.28**.

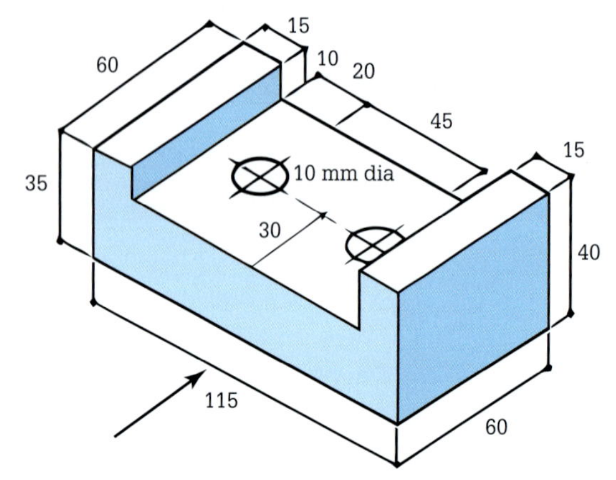

FIGURE 3.28 Isometric object

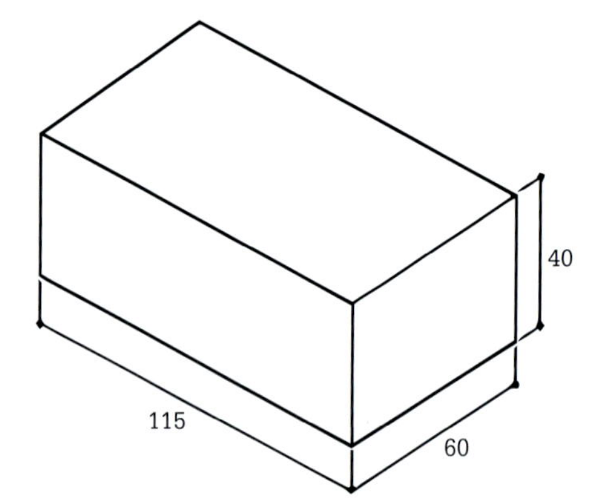

FIGURE 3.29 Isometric object outline

Using the box drawn to the outside dimensions of the object complete the outline as shown in **Figure 3.30**. The finished object should resemble **Figure 3.31**.

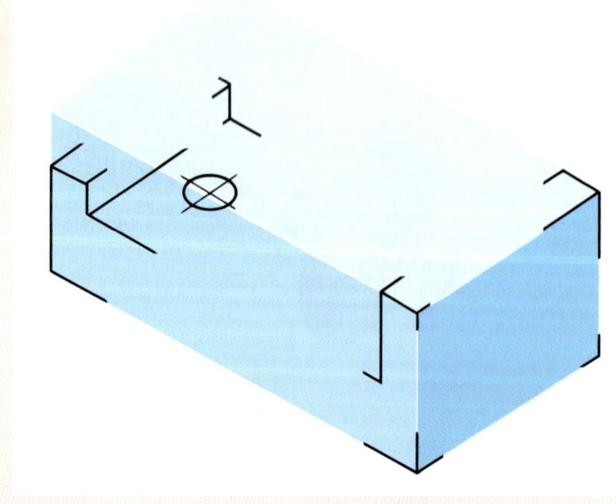

FIGURE 3.30 Completing the outline

»

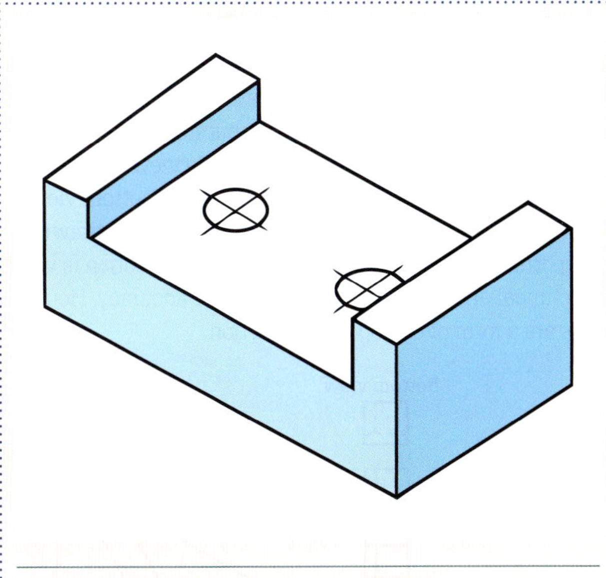

FIGURE 3.31 Finished isometric drawing

Axonometric projection

With axonometric projection, shown in Figure 3.33, all the horizontal lines are drawn at an angle of 45° to the horizontal and all the vertical lines remain vertical.

Oblique projections

Oblique projections, shown in Figure 3.34, can be either:

- cavalier
- cabinet.

In cavalier projections all horizontal lines in the front view of the object are drawn horizontal to give a correct front view. All vertical lines are drawn perpendicular and the side view is drawn full width at 45° to the horizontal.

In cabinet projections all horizontal lines in the front view of the object are drawn horizontal to give an accurate front view. All vertical lines are drawn upright and the side view is drawn at half width at 30° to the horizontal.

EXERCISE 3.4

Draw the isometric objects in Figure 3.32 to the dimensions indicated.

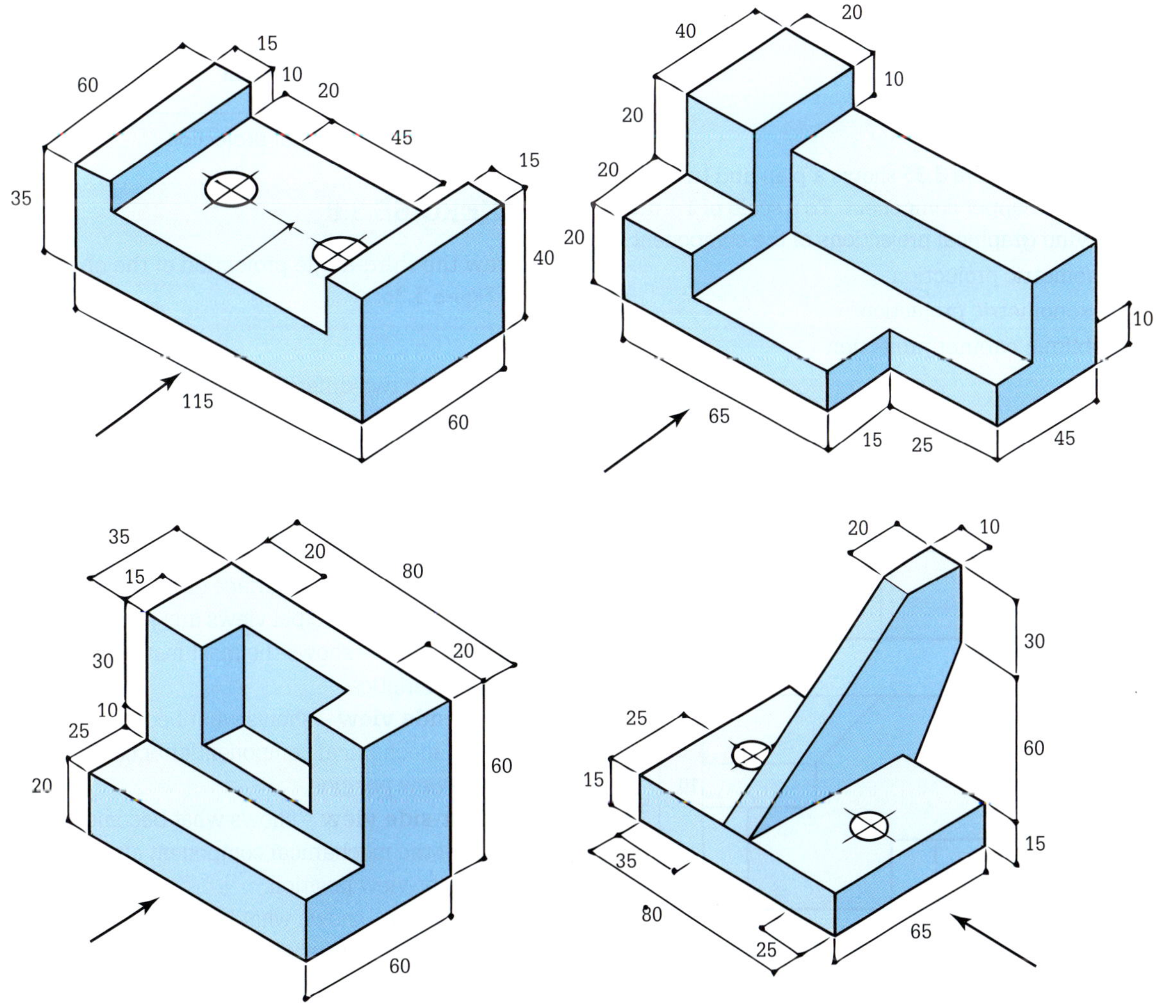

FIGURE 3.32 Isometric exercises

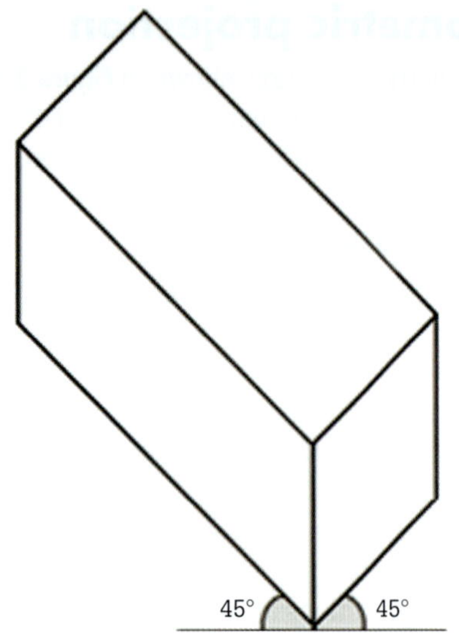

FIGURE 3.33 Axonometric projection

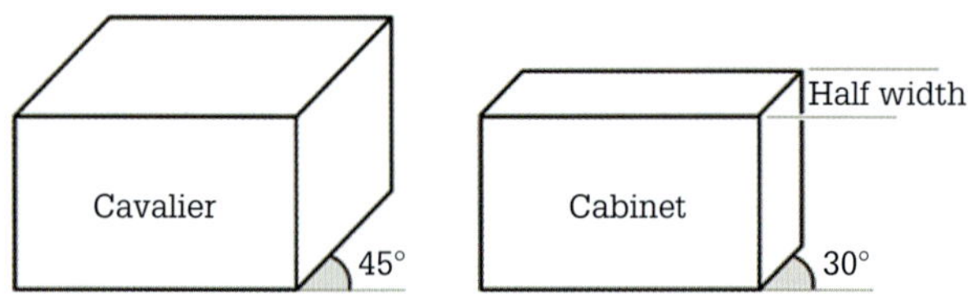

FIGURE 3.34 Oblique projections

EXERCISE 3.5

The drawing in **Figure 3.35** shows a plan and front elevation of a copper component. To a scale of 1:1 draw the following graphical projections of the component.

a An isometric projection

b An axonometric projection

c An oblique cabinet projection

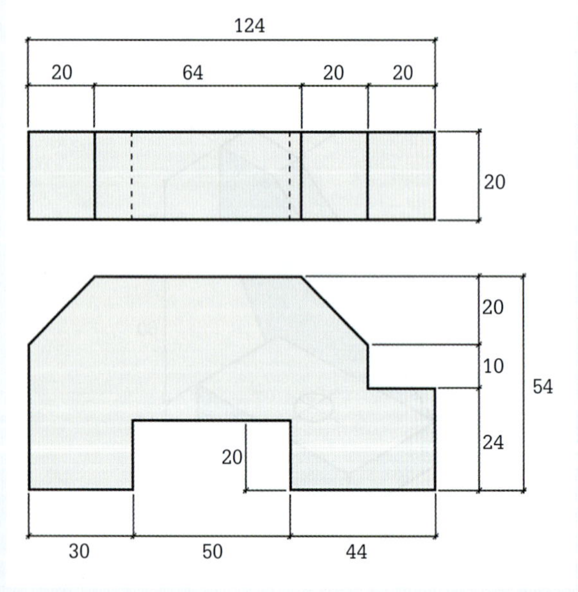

FIGURE 3.35 Copper component

Orthogonal projection

Orthogonal projection, shown in **Figure 3.36**, is a way of drawing an object that shows all the side views of an object drawn as separate distinct drawings but in relationship to one another. The names given to each side view are the plan view, front elevation, side elevations and bottom view. Engineering and building construction objects are drawn using orthographic projection. Orthogonal projection is used in technical and engineering drawings for accuracy. House plans are a form of orthogonal projection.

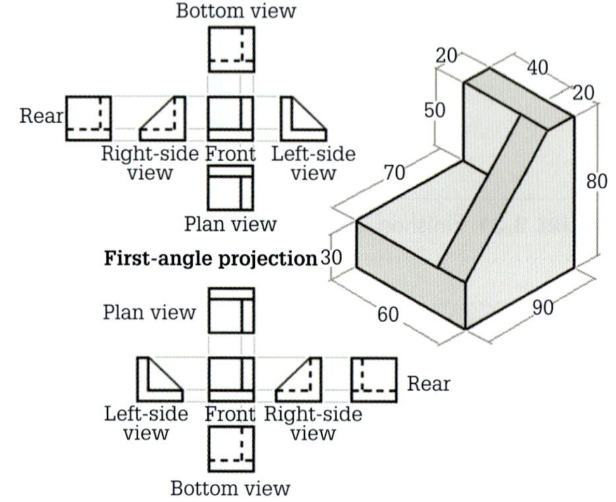

FIGURE 3.36 Orthogonal projection

EXERCISE 3.6

Draw the third-angle projection of the objects shown in **Figure 3.36**.

There are two different ways of presenting orthogonal projection, based on two distinct principles, called first-angle projection and third-angle projection. Third angle is used in Australia, Canada and the United States. First angle is employed in Europe.

Each view represents that which is seen when looking perpendicularly at each surface of the mechanical component. The principal views are:

- **Front view** – shows the most features or characteristics.
- **Left-side view** – shows what becomes of the left side of the mechanical component after establishing the front-view position.
- **Right-side view** – shows what becomes of the right side of the mechanical component after determining the front-view position.
- **Plan view** – shows what becomes of the top of the mechanical component once the position of the front view is established.
- **Bottom view** – shows what becomes of the bottom of the mechanical component once the position of the front view is established.

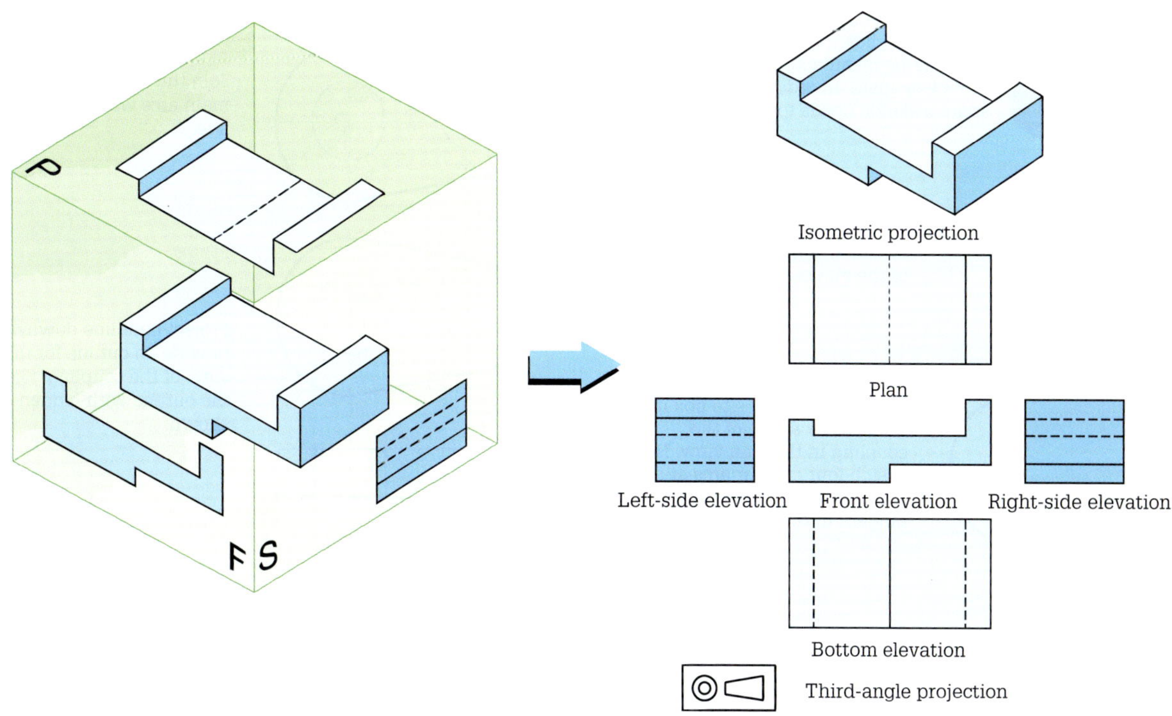

FIGURE 3.37 Third-angle projection

- **Rear view** – shows what becomes of the rear of the mechanical component once the position of the front view is established.

The standard views used in a mechanical drawing orthogonal projection are the top, front and right-side views. A third-angle projection is illustrated in **Figure 3.37**.

EXERCISE 3.7

Draw the first-angle projection of the side block in **Figure 3.38**.

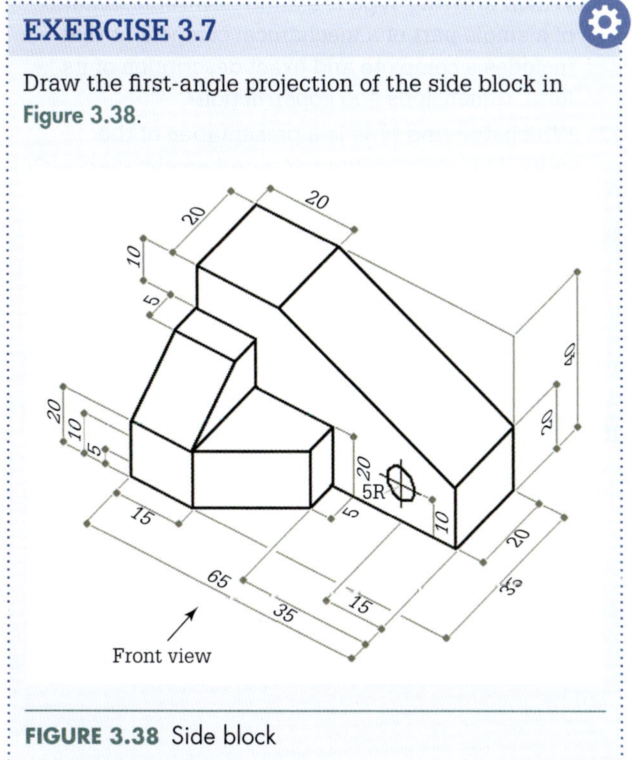

FIGURE 3.38 Side block

Sketching

Freehand sketching is a fast, reliable method used to record data for electrical fitting components and circuit wiring. Sketching is a clear way to record technical data that may be needed later for reassembly, reconnection or reference. Sketching can be used to produce orthogonal projection, detail and assembly drawings and pictorial drawings.

Electricians must read and work from drawings and specifications and make quick, accurate sketches when communicating technical information or ideas. Sketches that you will prepare may be for your use or use by other workers. One of the main advantages of sketching is that few materials are required. Pencil and paper are all you need.

For making dimensional sketches in the field, you will need a measuring tape or rule, depending on the extent of the measurements taken. In freehand pencil sketching, draw each line with a series of short strokes instead of with one stroke. Strive for a free and easy movement of your wrist and fingers. You don't need to be a draftsman or an artist to prepare excellent working sketches.

Printed grids or papers with lightly ruled lines are often used for sketching. The grids are used to keep freehand lines straight. Grids may also be used to create accurately scaled sketches, keeping the drawing in correct proportion. A sketch is not a rough drawing. It should be a clear image sketched in proportion, clearly dimensioned and labelled. A method used to sketch a copper coupling is illustrated in **Figures 3.39** and **3.40**.

This figure shows a copper coupling that needs to be manufactured in order to provide a spare. In order to make the spare a sketch needs to be made.

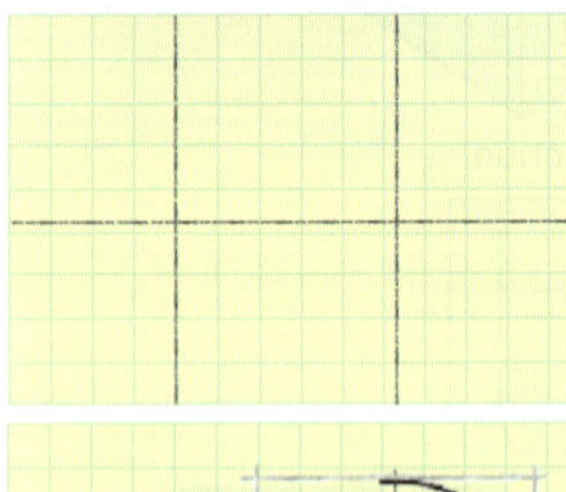

Using sketching techniques we will draw a series of centre lines to define some of the dimensions of the object in the plan view.

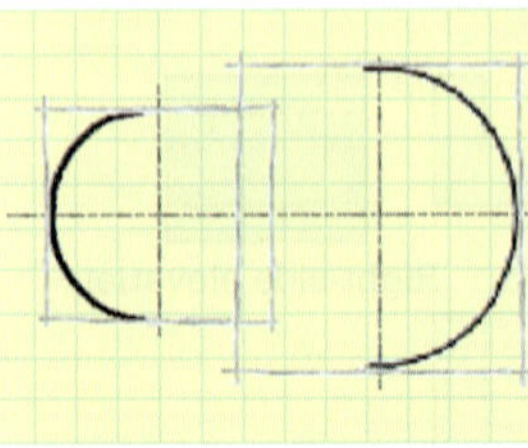

Use very light lines to box in the circular shape of the coupling in the plan view. Next sketch four circular arcs at four points where the arcs are tangent to the lines of the box.

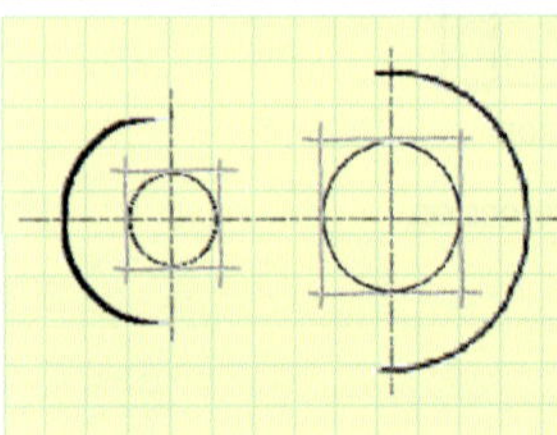

Use very light lines again to box in the circular shape of the holes. Next sketch in the circular shape of the holes.

FIGURE 3.39 Copper coupling: Initial steps

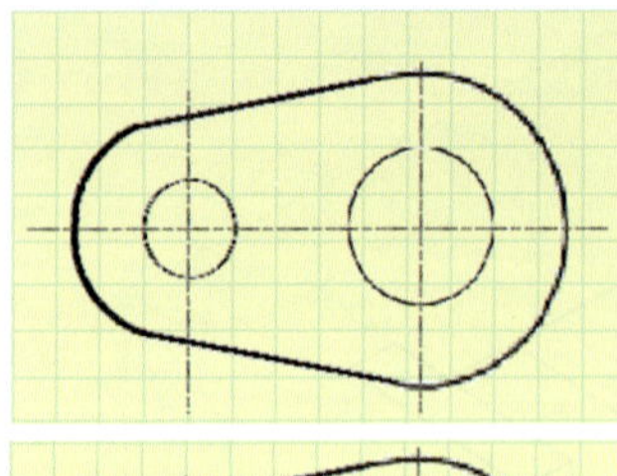

Join the extremities of the main arcs with a straight line.

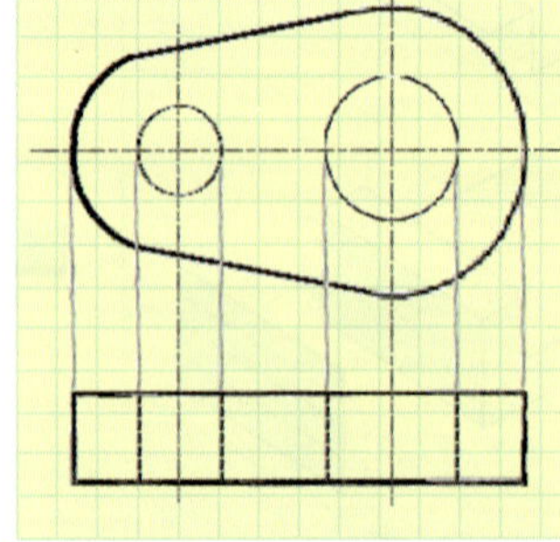

Project light line downwards to provide an outline for the side view of the coupling. Sketch in the outline with hidden detail shown.

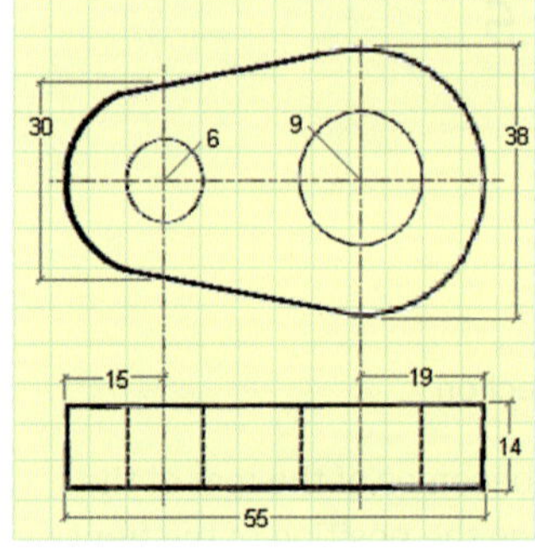

Finally the sketch is dimensioned.

FIGURE 3.40 Copper coupling: Final steps

REVIEW QUESTIONS

1 Why are drawings made to standard conventions?
2 What is the ratio of the two sides of 'A-series' drawing sheets?
3 Draw the projection symbol for first angle projection.
4 Detail the basic requirements for a title block.
5 Which organisation defines a set of standard metric line widths for drafting?
6 State the purpose of a thick chain type of line.
7 List the points to consider in relation to dimensioning.
8 In what units are linear dimensions usually specified?
9 What does the drawing abbreviation 'NTS' mean?
10 What two elements influence the choice of drawing scale?
11 Which drawing type is a multiview representation of a single part of a mechanical component that includes a complete and exact description of its form, dimensions and construction?
12 Which drawing type is a presentation of the individual components put together, showing each part in its operational position?
13 Name the drawing projection where the object's perpendicular lines are drawn vertically, and the horizontal lines in the width and depth planes are shown at 30° to the horizontal.
14 Why is orthogonal projection used in technical and engineering drawings?
15 State the purpose of freehand sketching.

3.2 Workshop planning

An important outcome from good workshop planning is the ability to provide a safe working environment, which in turn can facilitate a better and more efficient workshop.

Working safely

A large part of working safely is identifying hazards and risks in the work environment and implementing suitable control measures.

Non-electrical hazards

Common non-electrical workshop hazards and risks include:

- being struck by moving objects, such as forklifts
- exposure to dangerous goods or hazardous substances
- handling heavy tools and equipment
- injuries from dangerous machinery and equipment (plant)
- injuries from lifting, pushing and pulling heavy loads
- noisy machinery
- slips, trips and falls
- strain from repetitive tasks, such as stripping cable.

Risk management

Involving workers from all levels in health and safety issues can produce a safer workplace for all. Consultation is an important part of risk management. **Figure 3.41** shows a 5-step approach to risk management.

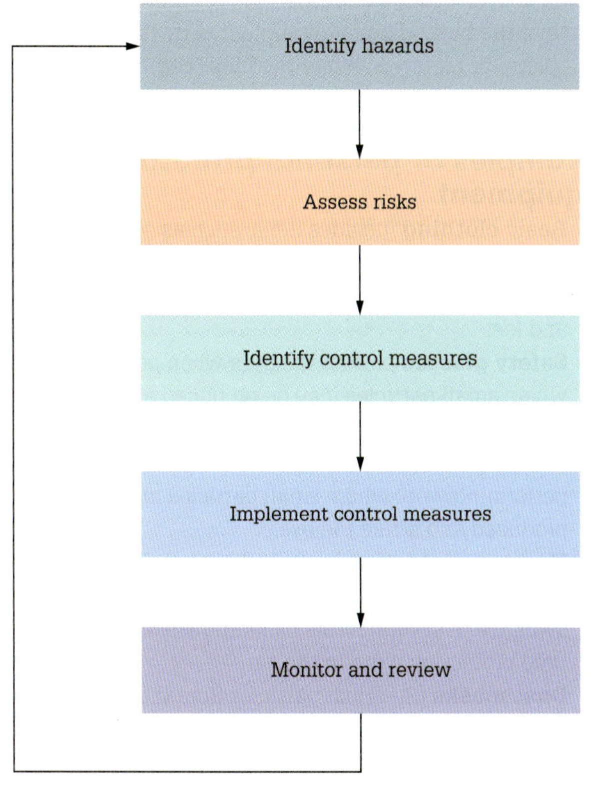

FIGURE 3.41 A 5-step approach to risk management

Legal responsibilities

Under work health and safety legislation, there are specific duties to ensure health and safety in workplaces. Chapter 1 details some of these responsibilities and legal obligations.

Some of the responsibilities expected of employers include:

- Providing and maintain a working environment for employees that is safe and free of risks to health, so far as is reasonably practicable. Employees may include contractors and agency staff.
- Providing employees with the necessary information, instruction, training or supervision to enable them to do their work in a way that is safe and without risks to health.
- Ensuring that the conduct of business does not endanger other people (including visitors, the public and other workers).
- Reporting notifiable incidents to the appropriate authority (WorkSafe).

Some of the responsibilities expected of employees include:

- Taking reasonable care for personal health and safety in the workplace. Employees must also take reasonable care for the health and safety of others who may be affected by what the employee does or does not do.
- Cooperating with the employer about any action they take to comply with the WHS Act or Regulations. This includes using equipment properly, following safe work policies and procedures and attending relevant training.
- Not intentionally or recklessly interfering with or misusing anything at the workplace to support health, safety and welfare.

Implementing safe work methods

Working safely involves the correct use of hand tools and power tools as well as using appropriate personal protective equipment (PPE) correctly.

Safe use of hand tools

Hand tools and power tools often cause accidents in the workplace. These accidents result in injuries that range from the smallest cuts or bruises to amputation or death. The use of some basic safety rules can prevent accidents when using hand tools.

There are four basic rules to follow when using hand tools. They are:

1. choose the correct tool for the job
2. make sure the tool is in good condition
3. use the tool in the correct way
4. store and transport the tool safely.

Select the correct tool for the job

Frequently the incorrect tool is used for a job simply because it is convenient. This happens most often in the case of spanners. An oversized spanner may just grip the nut or bolt head enough to allow you to place force on the spanner. The result can be that the spanner slips and rounds off the nut or you can injure your hand.

Sometimes spanners, wrenches, lumps of steel or wood and the like, are used as hammers. You can add many more examples of incorrect usage to this list. All of these incorrect tool selections often damage the tool or the job and significantly increase the risk of injury.

Make sure the tool is in good condition

You must always check the condition of tools before use. A tool in poor condition can cause severe damage. Things to look for include:

- check the wooden handles of hammers for cracks or loose fit into the head
- check the hammer head for cracks or chips and the correct shape
- chisels and punches and the like should be sharp and have no mushroom heads
- all pliers, shears, and the like must be sharp, and the blades come together as designed
- screwdrivers must be in good condition and the correct size and shape of blade used
- spanner jaws that are damaged due to wear or incorrect use must be thrown out.

Use the tool in the correct way

It is possible to damage the job or injure yourself by using a tool incorrectly. No matter which tool you are using, there will be some basic safety precautions you should follow. Use the following as a basis for your own personal list:

- always use knife, screwdriver and similar tools pointing away from your body
- if necessary, wear protective clothing or fit safety guards to equipment to avoid personal injury
- make sure you are balanced, to avoid stress on your body such as overreaching and stretching.

Store and transport the tool safely

If you use an open type toolbox with shallow trays to store tools, you will be able to see the tools rather than have to rummage through, and possibly cut your hand. The same toolbox will be a handy transport container for the tools.

Safety tips for power tool use

There are a number of safety tips that can reduce the risk of injury to the operator when using portable power tools. These tips apply to electric, pneumatic and fuel driven tools in most cases. Remember that most power tools have a rotating or reciprocating action and the tool often gives a high torque kick when started. Use the following list as a guide:

- remove hanging jewellery such as chains and bracelets
- wear close fitting clothing that will not get tangled in the tool during use
- ensure minimum safety dress — safety glasses are needed in most cases
- check any need for safety boots, ear protection, dust mask, etc.
- tie back long hair
- check that the blade or cutting tool is tight in the power tool
- test for correct position or operation of any safety guards
- make sure you are balanced and on firm footing before you switch on
- make sure the work is secure and will not shift as you apply pressure with the tool
- let the tool reach full speed and listen for correct operating sound
- use variable speed tools at the correct speed for the job
- do not overload tools – if a portable tool slows noticeably, it is being overloaded
- never carry or suspend a portable tool by its lead. Leads are securely adhered to both the plug and the tool body but this anchoring is only strong enough to support the lead, not the tool
- avoid getting the tool wet and take great care that neither the plug, extension socket or the tool lies in water when you are working in a damp situation.

Personal protective clothing and equipment

Personal protective clothing and equipment is essential when using most power tools and some hand tools. There is a vast range of equipment for protecting all parts of the body. The following list is by no means complete, but lets you know the most common forms of protection available.

Examples of personal protective equipment

- **Basic clothing** includes items such as close-fitting clothing (overalls), gloves and steel capped boots. These items give some protection to the body, hands and feet.
- **Safety glasses** protect the eyes when performing work where small particles may be produced and strike the eye, such as when grinding, chiselling or drilling.
- **Face shields** provide full-face protection when performing work where small particles may be produced and strike the eye.
- **Hearing protection** is vital whenever the ambient noise level rises above 80 dBA. Hearing protection is available in several forms including small disposable-type protection worn in the ear.
- **Dust masks** filter dust particles from the air and protect lungs from damage.
- **Breathing respirators** remove dangerous particles, vapours and gases from the air and protect lungs from damage. It is important to realise that in some cases, dangerous gases may displace the air and a respirator will only filter out the poisonous gases.

If there is not enough oxygen to breathe, then death may quickly result. In such cases, sophisticated breathing equipment is needed.

- **Safety helmets** protect your head against falling objects and other similar hazards. Most large building sites require the use of safety helmets to prevent injury.

Source: Shutterstock.com/Ricardo Romero

FIGURE 3.42 Example of PPE suitable for workshop activities

Workshop materials

The major classifications of materials used in the electrotechnology industry are metallic and non-metallic.

Metallic materials

Metallic materials may be pure metals, which are elements, or alloys that are combinations of a metallic element and one or more different elements. Mild steel is one of the cheapest metals but it may not be a suitable replacement for other metals in specific applications. Each metal behaves in a different way when subjected to the effects of heat, chemical action, electricity, magnetism or force.

Metals conduct heat and electricity much better than non-metals. In addition, metals in the solid state are composed of crystals or grains. Sometimes these grains are clearly visible as is the case in broken pieces of metal. The size and distribution of these grains changes due to changes in the shape forced on the metal or changes in temperature. Furthermore, the properties of a metal vary with changes in its grain structure. Forcing a solid metal to change its shape can raise its temperature. Try bending a piece of copper wire back and forth until it breaks. You can feel that the wire is hot at the break.

Strength of metals

The strength or ability of a metal to resist deformation or fracture by a load applied in a particular way is an important property. The greater the compressive strength of a metal then the greater is its ability to resist deformation by a compressive load. The greater the tensile strength of a metal then the greater is its ability to resist fracture by a pulling load. If the applied load is large enough the metal will break or fracture.

Toughness of metals

A suddenly applied load of a given size can have an effect on metal that is different from the effect of the same size load applied steadily. A metal that tends to fracture under impact or shock loads is brittle. A metal that tends to resist fracture under impact or shock is tough. Tough metals have to be strong and ductile to the right degree. The head of an engineer's hammer must be tough and their faces have to be harder than the objects they strike.

Hardness of metals

The harder a metal is, the more it is able to resist deformation and penetration. The softer a metal is, the easier it is to form and machine it to shape. When one material moves against another material, it is likely one will cut, wear or scratch the other.

Ductility of metals

Applying stress to most metals allows them to flow under pressure to some extent without breaking. It is possible to force them to a new permanent shape by plastic deformation. For example, mild steel is easily bent. The more ductile a metal is, the easier it is to deform it plastically without breaking it.

Types of steel

Low carbon steels are alloys of iron and carbon, with possibly traces of other elements, and may be categorised as:

- mild steels, containing up to 0.25% carbon
- structural steels, containing from 0.23% to 0.35% carbon.

Low carbon steels are not as strong and hard as higher carbon steels, but they are generally more ductile, easier to machine and weld, and cost less than other steels. Low carbon steels can generally be carburised, which involves heating the steel to a high temperature in contact with a material that contains carbon. The effect is that extra carbon is absorbed into the surface of the steel resulting in steel with a high carbon exterior and a low carbon core. It is also possible, through further heat treatment, to produce a hard skin on a softer shock resistant core. This process is suitable for the manufacture of gears and dies.

SWITCH ON

Safety note

Some compounds used in carburising contain cyanide, which is a poisonous substance. Anybody performing carburising requires thorough training in the safe use of these compounds.

Medium carbon steels are plain carbon steels containing from 0.35% to 0.5% carbon. Medium carbon steels are stronger, harder and tougher than low carbon steels, but they are less ductile and more difficult to weld.

High carbon steels are plain carbon steels, containing from 0.5% to 0.8% carbon. High carbon steels are less ductile and more difficult to weld and machine than lower carbon steels, but they are stronger, harder and more wear resistant. Heat treatment can improve these properties.

Plain carbon tool steels are high carbon steels of highest quality, containing from 0.8% to 1.5% carbon.

Steel alloys

Adding various elements to the basic steel during manufacture makes alloy steels that exhibit special or improved properties. Steel manufacturers supply charts and books giving the composition of these steels and the correct heat treatment for each of them.

Two examples from a very large range of alloy steels are:

- stainless steel, which contains nickel and chromium, resists corrosion very well
- high-speed steel, which contains tungsten, is suitable for some cutting tools.

Hardening and tempering

Plain carbon steels with a carbon content of more than 0.35% can be brought to their hardest condition by heating to a specified temperature and cooling very rapidly by quenching in a fluid. After quenching, most steel is too hard and brittle to use and must be tempered.

Tempering, which involves relatively low temperatures, generally follows hardening and involves heating the hardened steel to a specified temperature and then cooling. This reduces the hardness and increases the toughness of the steel.

Cast iron

Cast irons easily melt and in this state it is possible to cast into intricate shapes in sand or metal moulds. Cast irons contain between 2% and 5% carbon, with some traces of other elements. A common example is grey cast iron used for machine bases, engine blocks and general engineering castings.

Ordinary cast iron is too brittle to use where there are suddenly applied stresses. A special annealing treatment can produce malleable cast iron, which improves ductility. Some cast iron has alloying elements added to give special properties such as increased tensile strength.

Copper and copper alloys

As copper is a good electrical and thermal conductor and resists corrosion it finds application for electrical wires and busbars, as well as pipes and tanks for liquids. Copper work hardens as is the case when drawing through a die to produce thin wire. Hardening increases its tensile strength but makes it more brittle, so it is usual to anneal copper by heating to red-hot and then quenching it by plunging in cold water.

Aluminium

Aluminium is a good electrical and thermal conductor as well as resisting corrosion and has a density of about one-third that of steel. Alloying aluminium with other elements produces strong, lightweight parts. Magnesium, titanium and beryllium are common alloying elements.

Non-metallic materials

Plastics is a general term used to describe a range of polymers we use as plastics. Some of these are recognisable by their chemical abbreviations such as PET (polyethylene terephthalate) and PVC (polyvinyl chloride). Others are identified by their trade names such as Styrofoam (foamed polystyrene) and Plexiglas or Perspex (polymethyl methacrylate).

Plastics typically used as engineering materials are generally polymers, which are chemical compounds having very large molecules made up of repeating smaller units known as monomers. The size of the molecules together with their physical state and the structures that they adopt, are the principal causes of the unique properties associated with polymers. This includes the ability to be moulded and shaped.

The two major categories of polymers are thermoplastic and thermosetting. Thermoplastics, which include polyethylene, polyvinyl chloride and polystyrene, are capable of being moulded and remoulded repeatedly through the application of heat. On the other hand, thermosetting resins cannot be reprocessed by re-applying heat.

The physical state and morphology of a polymer have a strong influence on its mechanical properties. Properties of plastics include their stiffness, breaking stress, toughness, elongation, poor heat conduction, electrical insulation, dielectric strength and dielectric loss. To obtain particular properties in a plastic material, the polymer may be combined with additives including plasticisers, colourants, reinforcements and stabilisers.

Plasticisers

Plasticisers are used to change the temperature at which the material becomes stiff. They may also change the flammability, odour, biodegradability and cost of the finished product.

Colourants

Colourants are used to obtain the desired colour of the finished product. The ease with which colour is incorporated throughout a moulded article is an advantage of plastics over metals and ceramics which both depend on coatings for colour.

Reinforcements

Reinforcements may be added to the resin to enhance the mechanical properties such as tensile strength and stiffness of a plastic.

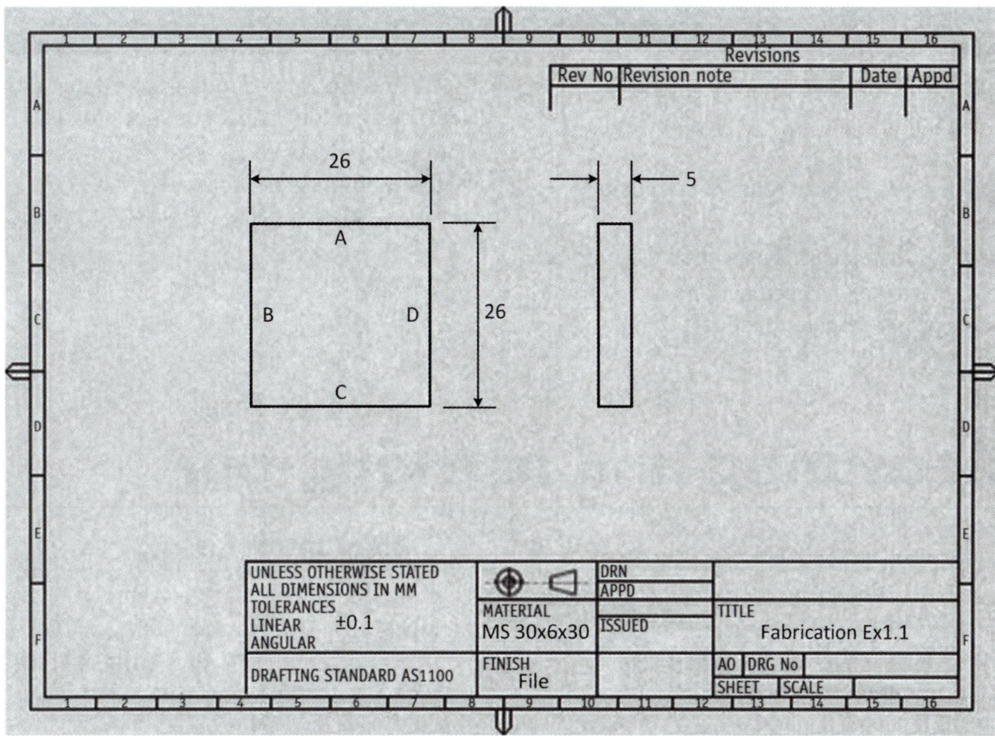

FIGURE 3.43 Sample workshop job

Stabilisers

Stabilisers added to a plastic counter the effects of ageing so that the material's properties change as little as possible throughout its service life. Certain stabilisers may be added to prevent oxidation, whilst others may be added to prevent degradation due to sunlight, ozone and biological agents.

Task planning

Before a task is undertaken, planning of the task should be performed. The planning process should include a risk assessment, collection of appropriate tools and materials, location of a suitable site to perform the task, bins for the collection of waste materials, obtaining plans and drawings for the task, ensuring that all required processes and techniques are understood, and ensuring that the worker has the ability to safely and correctly use the required tools.

Figure 3.44 shows a segment of a job breakdown (planning) sheet used to fabricate the part detailed in **Figure 3.43**.

Job Breakdown Sheet				
	Operation	***Tools or equipment***	***Personal and job safety***	***Calculations***
1.	Remove burrs from sides and scale from face	Bastard cut hand file	File handle fitted eye protection possible laceration	
2.	Scribe line on face 2mm back from milled edge and parallel to it using rule and scriber (side 'C')	150mm steel rule scriber	Eye protection	
3.	Securely clamp stock in bench vice using vice guards for protection	Bench vice vice guards	Eye protection pinch hazard	
4.	Hacksaw just on the waste side of this line keeping cut straight	Hand hacksaw with 24tpi blade	Eye protection possible laceration	

FIGURE 3.44 Sample job breakdown sheet for Figure 3.43

REVIEW QUESTIONS

1 What forms a large part of working safely?
2 List common non-electrical workshop hazards and risks.
3 What are four basic rules to follow when using hand tools?
4 Name the PPE items identified as basic clothing.
5 What are the two major classifications of materials used in the electrotechnology industry?
6 What term is used to describe a metal that tends to fracture under impact or shock loads?
7 Which process reduces the hardness and increases the toughness of the steel?
8 Name the two major categories of polymers.

3.3 Measuring and marking-out

An electrician must have an overall knowledge of engineering drawing and surface developments in order to mark out replacement parts on sheet metal. Consequently, it is essential to have a comprehensive understanding of measurement and marking-out tools, and also of how to test these tools so that a standard of accuracy can be achieved and maintained.

Marking-out

Marking-out is the process of drawing essential information in the form of lines, circles, outlines, centre marks and so on to indicate the position and extent of work to be done and to provide guidance in setting up work.

Accurate marking-off is often necessary on large or irregular shaped workpieces to set them up on a machine prior to machining and is therefore an important skill required of a person working in an engineering environment.

There are four basic reasons for marking-off an engineering workpiece:

1 It allows inspection for faults such as insufficient material for machining, pits or cavities in a casting, distortion of castings or other defects, which would waste the machining time if machining went ahead.
2 It allows for variations in casting shapes and distributes errors evenly.
3 It assists in setting up on the machine.
4 It serves as an approximate guide for machining.

In setting up work, all measurements should be taken from an edge or base line called the datum (reference point). The following marking-out tools enable the transfer of information to the workpiece.

Marking-out tools

Steel rule

A steel rule as shown in **Figure 3.45** is used for measuring items that are flat or straight and as a straight edge for cutting stator insulation sheets. Different rulers have varying degrees of accuracy; some break their measurements down to millimetres while others break down to half millimetres. Make sure you have your eye directly over the measured marking on the rule to prevent parallax error. Parallax error is any superficial dislocation or difference of orientation of an object viewed along two different lines of sight.

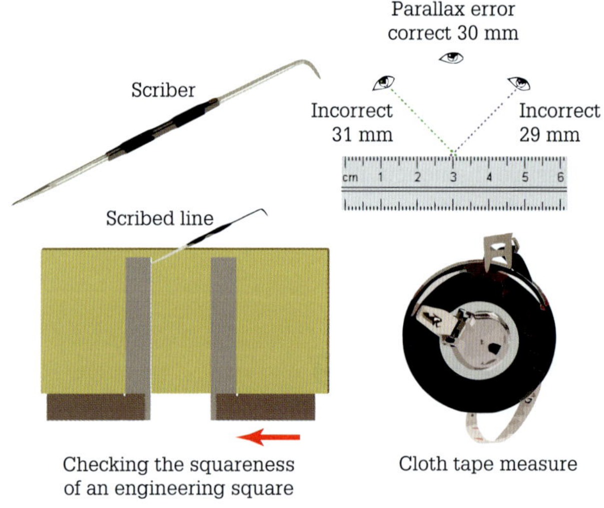

FIGURE 3.45 Marking-out process, tools and accuracy

Scribers

When marking out steel for various tasks, a scriber is used to scratch a fine line on the metal (see **Figure 3.45**). The scriber can be either a short length of sharply pointed, heat-treated steel or a pencil-like marking tool with a carbide tip.

Marking-out media

Scribed marks on galvanised metal are hard to see, so a coating of Prussian blue dye (also known as layout stain or engineer's blue) is used to provide contrast. Lines scribed on the dyed surface stand out sharply and brightly. To ensure that the scribed lines are not erased they should be 'witness marked' lightly with a centre punch at several points along the line. Other mediums used for highlighting markings are printer's ink, chalk and marking pens.

Tape measure

Cloth tape measures, as shown in **Figure 3.45**, can be used to measure the cable route length required for voltage drop calculations. Steel tape measures should not be used near live equipment for safety reasons.

Engineer's squares

An engineer's square (see **Figure 3.46**) is accurate both inside and outside and is the workshop standard for 'squareness'. The engineer's square has a polished precision-ground steel stock and blade for extreme accuracy. Engineer's squares can:

- assist in scribing lines at right angles to a datum edge
- check how flat a surface is
- test accuracy of surfaces at right angles to each other
- test the set-up of a workpiece clamped to an angle plate.

In the workplace, engineer's squares must be checked for accuracy. One method used is to scribe a line from a straight surface then turn the engineer's square over and compare against the scribed line.

Combination square

Combination squares (see **Figure 3.46**) consist of a rule and three stocks namely a:

- square and 45° stock
- centring square
- protractor.

The combination square performs many tasks in marking-off, such as:

- Measuring angles – a combination square can reliably measure 90° and 45° angles. The 45° angle is used commonly in creating mitre joints.
- Determining flatness – when working with wood the first step is to designate a reference surface on a board which is known as the face side. The rest of the workpiece is measured from the face side with the next surface being the face edge.
- Measuring the centre of a circular bar or dowel. The rule is assembled through the centre of the centre square, the two cast iron legs of the centre square are then placed against the outside of the bar (dowel) allowing a centre line to be scribed alongside the ruler. Perform this action at two locations and the intersecting lines will approximate the centre of the bar (dowel).
- Protractor head allows angles to be set and measured between the base and ruler.

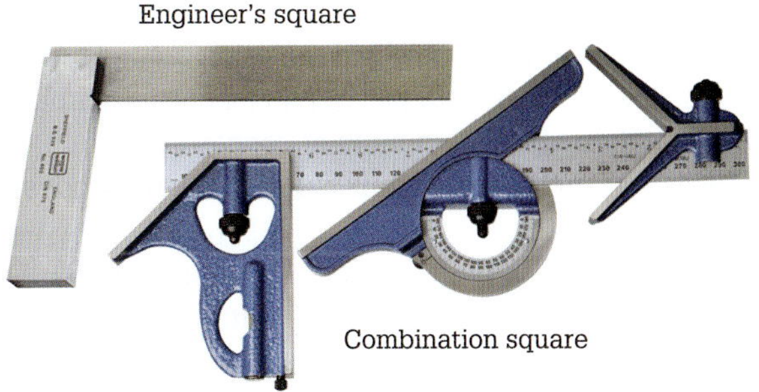

FIGURE 3.46 Engineer's square (left) and combination square (right)

Outside calipers

Outside calipers (see **Figure 3.47**) are used for taking external comparative and transfer measurements. The legs of the caliper are hardened and tempered and fitted with a bow spring or a friction joint. The legs of the bow spring type pivot on a fulcrum spool to assure even tension. The adjusting screw has a self-seating washer and nut to permit quick and positive adjustment.

Inside calipers

Inside calipers (see **Figure 3.47**) are used for measuring inside dimensions such as the size of a cavity or hole.

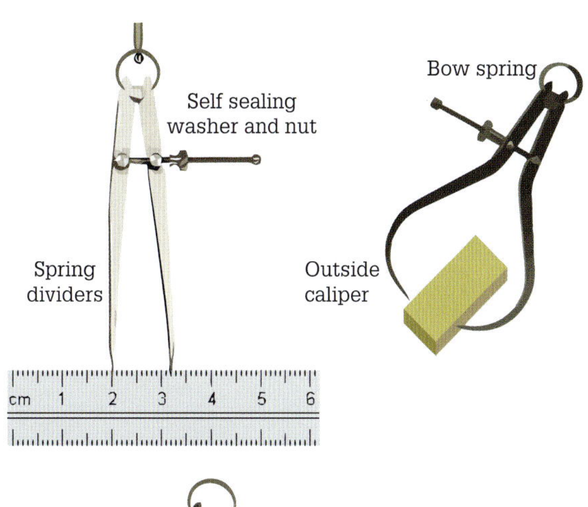

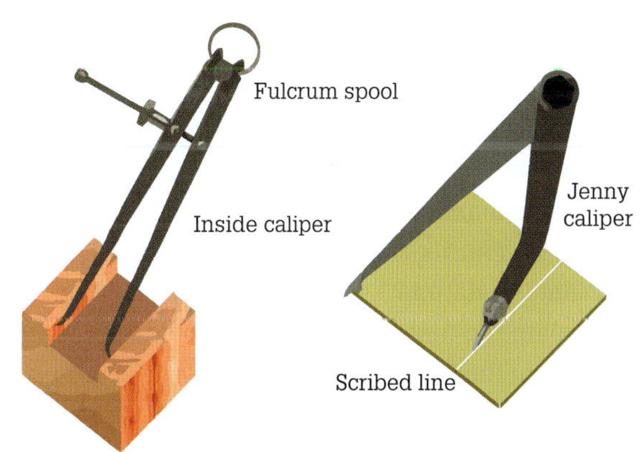

FIGURE 3.47 Calipers and dividers

Jenny calipers

Jenny calipers are used for taking, transferring and marking out a set measurement, the straight leg acting as the datum (reference from which measurements are made) and the bent leg and adjustable scriber point marking the workpiece (see **Figure 3.47**).

Jenny calipers are purposely designed to scribe a line parallel to an edge or to pinpoint the centre of a round or square section of solid steel. They are also known as odd-leg or hermaphrodite calipers.

Spring dividers

Spring dividers (see **Figure 3.47**) are used for marking out circles and for taking transfer measurements.

Vee block

Vee blocks, as shown in **Figure 3.48**, are made from either cast iron or hardened steel and are machined as a pair. They are square on all surfaces and the vee is parallel to and centre with two opposing sides. They are useful to locate round workpieces such as shafts for marking-out.

Vee blocks should never be used to support roughcast or black surfaces without using a protection such as paper or thin sheet steel between the vee block and the rough surface.

Source: Shutterstock.com/Peter Sobolev

FIGURE 3.48 Vee block

Reducing waste

Waste represents a significant loss of natural resources that is reflected in the cost of the final product. The outcomes from waste reduction include a more productive and profitable business that is able to reduce the final cost of their products.

One of the keys to reducing waste when fabricating is to be able to improve the usage of the primary material. This can range from arranging the parts to be cut from the primary material in a manner that best utilises the available material, using off-cuts to produce additional items or batching items together that have similar straight edges or radii.

For fabricators, a certain amount of scrap is unavoidable, but automation and technology provide a tangible method to minimise waste during the fabrication process. Modern computer numerical control (CNC) equipment or laser cutters can maximise the number of items or forms made from a single primary product due to the precision accuracy of the equipment, which can greatly reduce the amount of waste material. Software advances can also improve precision accuracy and assist in maximising the usage of materials.

REVIEW QUESTIONS

1 What is marking-out?
2 Provide four reasons for marking-off.
3 What is a datum?
4 Explain the function of Jenny calipers.
5 What is the function of a scriber?
6 Name four common marking-out media.
7 What is the purpose of an engineer's square?
8 Outline the main difference between an engineer's square and a combination square.

3.4 Holding and cutting

Two common fabrication processes are holding work in a secure device so that it can be cut or shaped. A common method for holding a workpiece involves some type of vice fitted to a workbench. The workbench itself must be strong, rigid and heavy enough for the type of work. The bench top must be at a convenient height for working and be at least 600 mm wide. Common cutting tools are hacksaws, files and snips.

Hacksaw

A hand hacksaw (see **Figure 3.49**) is a stiff-backed, bow-type saw with a replaceable metal cutting blade. There are from 18 to 32 teeth per 25 mm (one inch) in a raker tooth pattern hacksaw blade. The raker design provides a smooth cut and prevents the blade from binding when cutting materials by making a bigger cut than the width of the blade. Always wear safety glasses when using a saw of any kind.

Bimetal hacksaw blades improve cutting performance through improved rake angle and precision ground teeth. They include 14 or 18 teeth per 25 mm for cutting through heavy metals such as aluminium or copper. Cutting stainless steel or thinner metals can occur using blades with 24 or 32 teeth.

The hacksaw spine (U-shaped tube holding the blade) limits how far the blade can cut into a piece of material. The suggested numbers of hacksaw teeth suitable for various material thicknesses are shown in **Table 3.4**.

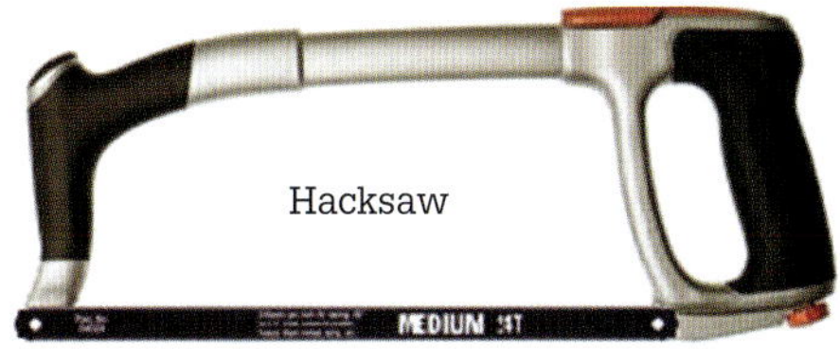

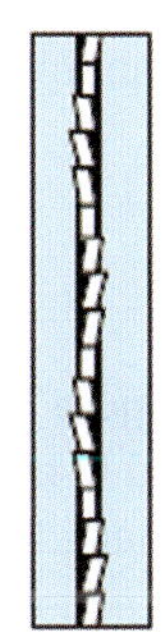

FIGURE 3.49 Types of hacksaws, blades and a raker tooth pattern

TABLE 3.4 Recommended hacksaw blades

Teeth	Material thickness
14	2.4 mm to 25 mm
18	1.6 mm to 8 mm
24	1.6 mm to 3 mm
32	<1.6 mm

The hacksaw blade must be correctly aligned in the frame for the best possible use. The teeth must point forward with at least three teeth in contact as the downward cutting force 'hacks' at the workpiece. The downward pressure should be released on the return stroke.

Regular cutting speed should be between 40 and 50 strokes per minute. After a few strokes a new blade must be re-tensioned. Hard materials require slower speeds. Always 'score' the material to be sawn with several light cuts to provide guidance for the blade. After using a hacksaw, the blade needs to be cleaned of any lubricant and material particles and the tension on the spine released.

Damage to hacksaw blades can be caused by any of the following:

- putting a new blade into an old cut (thinner)
- excessive tension
- work not being held securely
- using an incorrect blade.

Always lubricate the cut if cutting steel. A spray-can of penetrating lubricant could be used to lubricate the cut but it is easier to use a vessel partially filled with auto transmission fluid. Use a small paintbrush to wipe a smear on the stock where you will be cutting.

The round carbide grit hacksaw blade shown in **Figure 3.50** cuts on any side in any direction. The blade is grit coated with tungsten carbide around its 360° cutting surface, which can cut slots, angles and curves in metal, tile, ceramics and wallboard.

FIGURE 3.50 Round carbide grit hacksaw blade

Power hacksaw

The power hacksaw (see **Figure 3.51**) is a blade designed to cut all metal angle, bar, strap, channel, pipe and other shapes of construction steel. This saw is easy to operate and unlike a hand hacksaw it cuts on the return stroke.

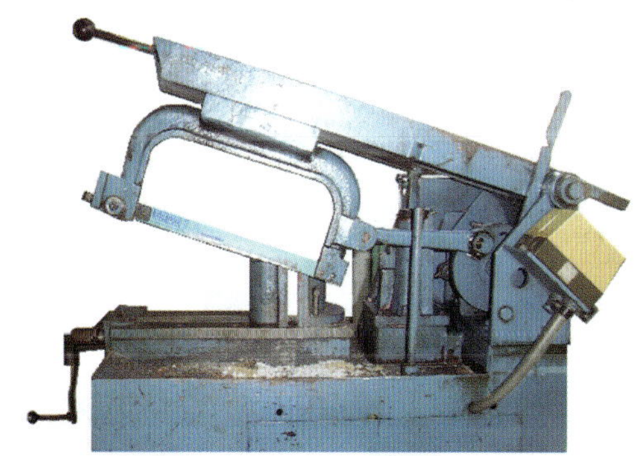

FIGURE 3.51 Power hacksaw

Always use unbreakable bimetal hacksaw blades with power hacksaws. The bimetal construction provides a shatterproof blade that can withstand high feed pressures, giving accurate, fast cutting rates. Most power hacksaws use pump-driven machine oil to lubricate the blade as it cuts the workpiece.

When using a power hacksaw:

- Clamp the workpiece firmly in the vice of the saw, at the required cutting angle. Tension the blade and ensure that as many teeth as possible will come in contact with the workpiece when cutting begins.
- Turn on the power hacksaw and lower the blade until it starts to cut.

Power hacksaws have an automatic shut-off which should stop the saw when the workpiece has been cut through.

Cold chisel

Cold chisels, as shown in **Figure 3.52**, are used for cutting, shaping or removing cold metals softer than the chisel's cutting edge, such as soft iron, aluminium, brass, or copper. Only use a cold chisel if it is in good condition; that is, the cutting edge is sharp and the struck head is not mushroomed or chipped. The cold chisel has a cutting edge at one end, and a struck face at the other to be struck with a ball-pein hammer.

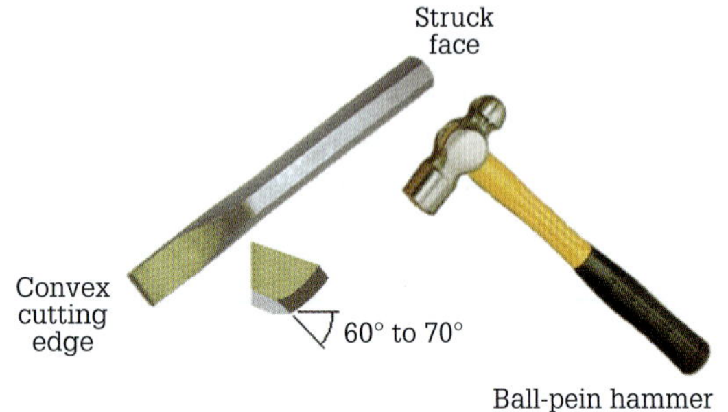

FIGURE 3.52 Cold chisel

For shearing and chipping, hold the chisel at an angle that permits the bevel of the cutting edge to lie flat on the shearing plane. The point angle of the chisel must be 70° for hard metals and 60° for soft metals with a marginally convex cutting edge. Furthermore, always wear safety goggles or a face shield and gloves or place a sponge rubber shield on the shank of the cold chisel before striking the struck face.

Cold chisels come in diverse forms and with different points, as illustrated in **Figure 3.53**.

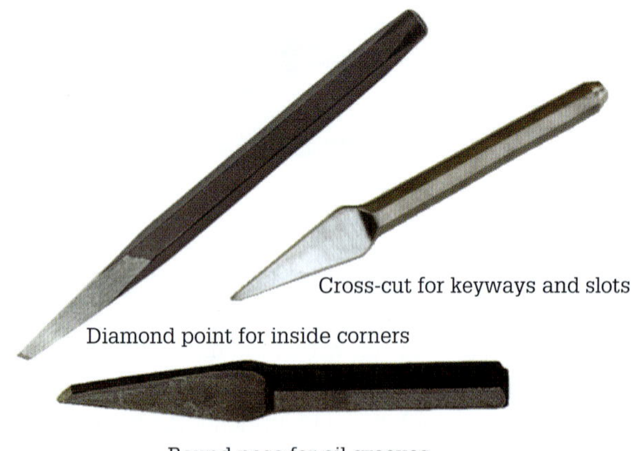

FIGURE 3.53 Various chisel points

Tin snips

Tin snips, illustrated in **Figure 3.54**, are prepared in various shapes and sizes for different tasks and are available for cutting in straight lines, or curved to the left or curved to the right.

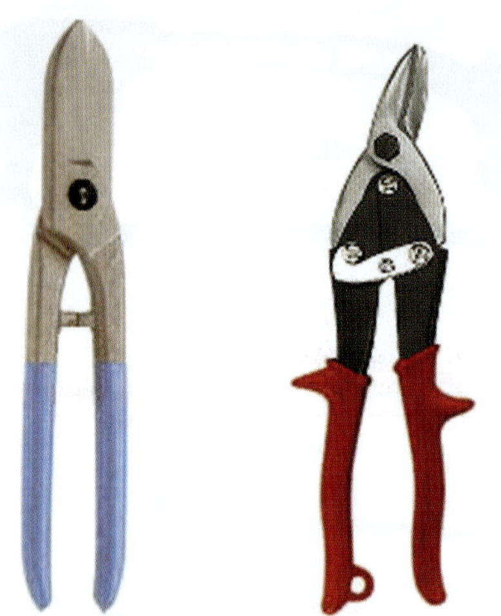

FIGURE 3.54 Tin snips

Cut so that the waste is on the right if you are right handed or on the left if you are left handed. Tin snips are for cutting thin sheet metal only, up to 1 mm.

Always keep the pivot bolt adjusted at all times and oil the blades and pivot bolt after use. Always wear gloves (the edges of cut metal are very sharp) and safety glasses when using tin snips.

Bolt cutter

A bolt cutter, as shown in **Figure 3.55**, is a large-jawed cutting tool that can be used to cut round steel rods and soft bolts or small square-shaped steel bars. The design of the bolt cutter amplifies the squeezing force applied to the cutting edge to cut.

FIGURE 3.55 Bolt cutter

Vice

A common device used for holding and securing work is the vice. Two variations are the bench vice and the machine vice.

Bench vice

Metalworking bench vices as illustrated in **Figure 3.56** are known as engineers' or fitters' vices. The vice is bolted to the top surface of the bench with the face of the fixed jaws just forward of the forward-facing edge of the bench. The bench height should be such that the topmost part of the vice jaws is at or just below the elbow height of the user when standing upright.

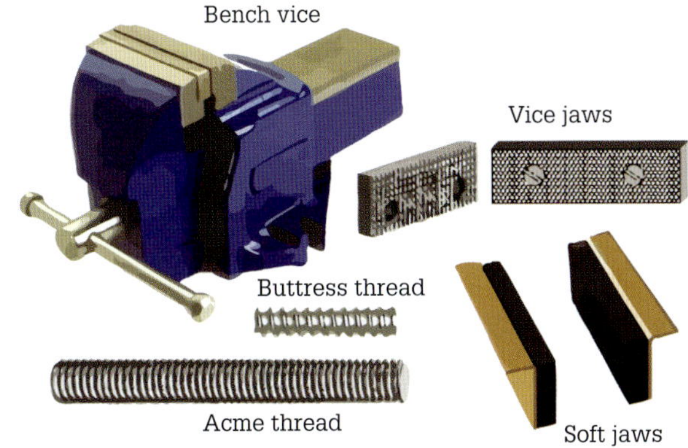

FIGURE 3.56 Bench vice and components

The nut in which the movable jaw screw turns may be split so that, by means of a lever, it can be removed from the screw. Consequently, the screw and movable jaw may be quickly slid into a different position at which point the nut is again closed onto the screw. Many fitters prefer to use the greater precision available from a plain screw vice. The vice may contain other features such as a small anvil on the back of its body, but it is in general better to separate the functions of the various tools.

Vice screws, as shown in **Figure 3.56**, are usually either of an Acme thread form or a buttress thread. Those with a quick-release nut use a buttress thread which is able to withstand a heavy thrust in one direction yet unscrews quickly in the opposite direction. In addition, the Acme thread is a screw thread having a 29° included angle.

Acme threads are a standard screw thread profile, which offer high strength with ease of machining and assembly. They are typically found where large loads require moving or accuracy is required, as in vices or the lead screw of a lathe.

Buttress thread is a thread form designed to withstand axial thrust in one direction only.

The body and jaws of the bench vices found in the workshop are generally made out of cast iron, which is resilient in compression but brittle under shock. As a result, they should not be hammered.

The jaw faces of the vice shown in **Figure 3.56** are made from hardened steel (127 mm × 25 mm). They are also serrated, having a criss-cross pattern which provides a firmer grip on the workpiece. The jaw faces are screwed to the sliding jaw and the fixed jaw so that if they are damaged they can be replaced.

Soft jaws

It is often a problem when holding soft materials that the hardened steel serrated jaws on the bench vice will mark the surface of the work piece. To prevent this occurring, soft jaws as shown in **Figure 3.56** are used. Soft jaws fit over the jaws of the bench vice and are made of a malleable material.

The machine vice

The machine vice (see **Figure 3.57**) is used with a pedestal drilling machine. The slots on the side of the base of the vice enable the operator to screw the machine vice to the drilling machine table.

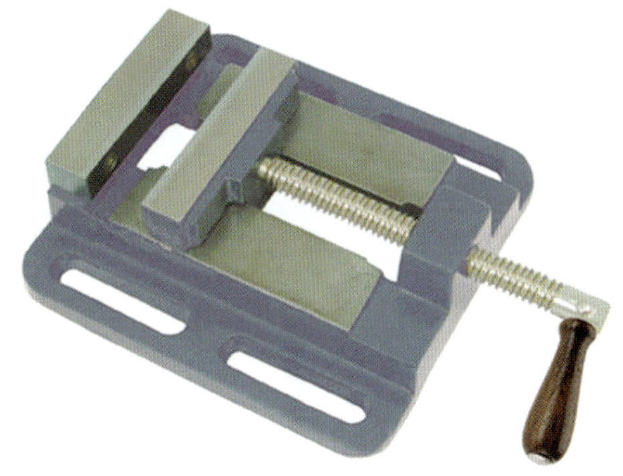

FIGURE 3.57 Machine vice

Files

A hand file is forged from carbon tool steel that has teeth cut, chiselled or punched on its surface. The file is used manually to reduce or smooth metal surfaces called the stock. A hand file and its parts are shown in **Figure 3.58**.

The length of a hand file is measured from heel to point. The heel is where the file begins to taper into the tang. The tang is made of softer metal and can be bent, but the main body of the file is brittle and will snap and break because of its hardness if used as a lever or used as a hammer. The longer the file the further apart are its teeth. Therefore, longer files are coarser than shorter ones. The edge is the narrow cross-section or side of the file and can be smooth without teeth (a 'safe-edge' file) or have single-cut teeth (a flat file). Files can be double-cut or single-cut. The difference between a double-cut and a single-cut file is illustrated in **Figure 3.58**.

The rasp has punched single-cut file teeth on edge for use on timber and some soft metals. There is also a dreadnought file consisting of single curved rows of coarse teeth for filing soft materials such as aluminium.

Most files (see **Figure 3.58**) have three grades of cut: bastard-cut, second-cut and smooth-cut, and for timber there is a rasp. Bastard-cut has the fewest number of teeth per 25 mm length. Second-cut has more teeth per 25 mm and smooth-cut has the greatest number of teeth per 25 mm.

The coarser (double-cut) the grade of the file the rougher will be the finish of the workpiece. The length of the file and the grade of its cut must be well matched with the

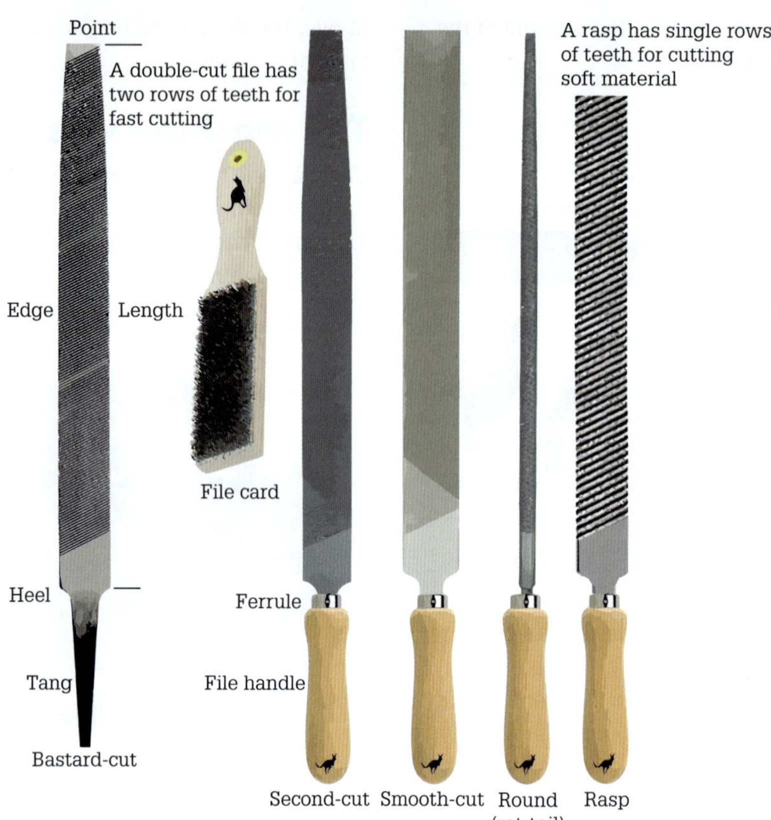

FIGURE 3.58 Hand file types and descriptions

amount of stock to be removed and the quality of finish that is required. A file's shape should be matched to the form of the workpiece from which stock is to be deleted. Hand files come in various shapes:

- Flat files are used for fast removal of stock from plain surfaces.
- Safe-edge files allow the filing of one surface without damaging an adjoining one.
- Half-round files are used for rounded holes, concave surfaces and flat surfaces.
- A rat-tail file is used for enlarging holes and grooves.
- A triangle file is used for corners, flat surfaces and for filing angles on cutting blades.
- A square file is used for square holes.
- A warding file is used for narrow spaces where a flat file cannot fit.

Hand files should be held by both the point and the handle. Keep the point between the tips of the fingers and the thumb of one hand and allow the file handle to rest in the palm of the other hand. Push the file forward with light yet steady downward pressure. Draw back with the same action. Repeat the steps as necessary to remove required stock. Always wear goggles or safety glasses when filing metal.

File handle

Every file must have a handle fitted in order to prevent the tang of the file causing an injury to the hand and to provide a safer grip on the file. Except for screw-on plastic file handles – to fit a file handle (see **Figure 3.58**) insert the tang into the hole in the handle and then tap the handle on a flat surface to seat the file securely. Never use a hammer to drive a file into the handle.

To release the file from the handle gently tap the ferrule (the metal band at the base of the handle) on the edge of a flat surface while you pull in the opposite direction on the handle. Always ensure that wooden file handles are free of splinters and cracks before use.

File card

Files are subjected to clogging – small particles or pins of metal clogging the teeth of the file. A clogged file will cause 'pinning' (scratching) on the workpiece. A file that is chalked before use will help to prevent pinning, but this can also reduce the cutting effect. A file card, shown in **Figure 3.58**, is used to clean the file as needed to remove any material stuck between the file teeth.

Abrasive paper/cloth

For extra smooth surfaces, wrap a piece of emery cloth (abrading fabric) around the file and stroke in the same manner as when removing stock. Glass paper is used to polish motor end-shield flanges which are then smeared with a light coating of grease and attached to the motor frame. The motor shaft is also cleaned and polished with oiled emery cloth. Wet and dry abrasive paper can also be used.

SWITCH ON

Safety procedures

Holding and cutting tools can cause cuts and puncture wounds if they are not handled properly. The safe way to work with these tools is to concentrate on the task at hand. There are several safety rules that can help prevent hazards associated with the use of holding and cutting tools:

- Keep all tools in good condition with regular maintenance.
- Use the right tool for the job.
- Examine each tool for damage before use and do not use damaged tools.
- Operate tools according to the manufacturers' instructions.
- Use the right PPE.

REVIEW QUESTIONS

1. Describe a hand hacksaw.
2. How many teeth would a hand hacksaw blade have when used to cut through thin-walled tubing (<1.5 mm)?
3. What is the recommended cutting speed for a hand hacksaw?
4. What is the purpose of a cold chisel?
5. What is the maximum thickness of sheet metal that can be cut with tin snips?
6. Name the tool used to cut round steel rods and soft bolts.
7. Name two types of vice.
8. At what height should a machine vice be installed?
9. What type of jaws are used to prevent marking occurring when working on a piece using a bench vice?
10. What is a hand file?
11. Between which two points is the length of a hand file measured?
12. Name three grades of file cut.
13. State the purpose of a file handle.
14. What is clogging in reference to a hand file?
15. State the use of glass paper when working on electric motors.

3.5 Drills and drilling

Drilling is an operation to make holes in a workpiece using a rotating cutting tool called a drill. Do not confuse drilling with boring. Boring is the operation of enlarging existing holes. The most common drill in the workshop is the twist drill.

The twist drill is usually made of high-speed steel (HSS). The construction of the twist drill allows the efficient cutting of metals, wood and plastics. Twist drills up to about 12 mm usually have a straight shank. Larger drills usually have a tapered shank known as a Morse taper. This shank fits into a tapered drive and will not slip in a chuck like the straight shank drill.

Twist drills

To make a hole in any material it is necessary to use a twist drill that is correct for the purpose. The drill must be sharpened to the necessary requirements and used correctly. Typical twist drills that an electrician will use are illustrated in **Figure 3.59**.

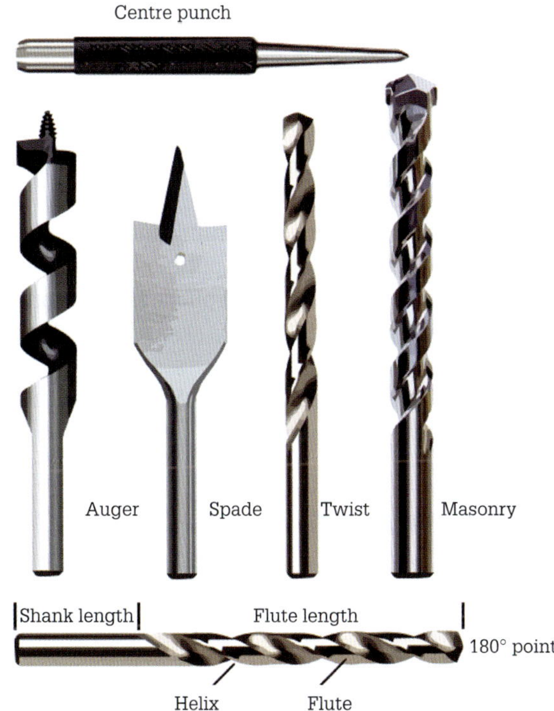

FIGURE 3.59 Twist drills and centre punch

The auger bit is used when drilling larger-diameter holes (25 mm) in timber frame walls. The threaded tip pulls the bit into the wood. This feature makes the auger bit unsuitable for use in power tools and it should only be employed in a hand brace. The spade or flat wood bit is also suitable for drilling larger holes in timber. These drills are for use in power tools, and care should be exercised, as the drills will splinter the timber as they break through. Twist drills are the most commonly used by an electrician with either a hand or a power drill for making holes in metals and plastics. A twist drill and its various parts are illustrated in **Figure 3.59**.

The twist drill does the cutting with the point and the cutting edge. The flutes carry the swarf (the combination of chips and thin spirals of metal that is produced during drilling) from the point and the cutting edges of the top of the hole where it is cast away. The helix provides a positive cutting rake.

Most problems encountered during the drilling operation stem from the drill point having incorrect angles or clearances. Three important considerations when sharpening twist drills are the:

- length and angle of cutting edges
- lip clearance angle
- correct location of point.

There are five types of twist drills: titanium coated, carbide tipped, cobalt bits, carbon-steel and high-speed steel. The two basic types of twist drills commonly used are the high-speed steel (HSS) for metal, and carbon steel for timber. HSS drills are preferable as they can withstand higher operating temperatures than carbon steel twist drills.

Drill sizes are typically measured across the drill cutting edges with a micrometer. Straight shank twist drills are usually available in sizes from 0.75 mm to 13 mm. Some twist drills have a unique self-centring split point design that does not need a centre punch mark to enable drilling. As a rule of thumb, drilling speed for twist drills of less than 4.25 mm diameter is 1200–1500 rpm and for twist drills of 4.25 mm diameter or greater is 900 rpm.

Masonry bits are designed for drilling into brick, block, stone, quarry tiles or concrete. The cutting tip is made from tungsten carbide bonded to a spiralled steel shaft. They are used in a power drill with a slow rotational speed and a hammering action.

A useful item in the measurement of drill sizes is a drill gauge, as illustrated in **Figure 3.60**.

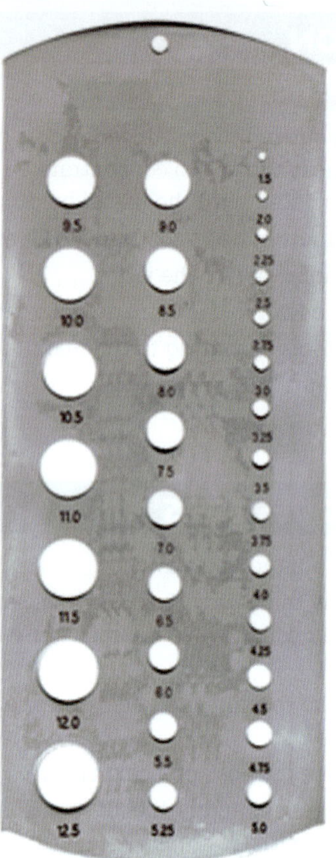

FIGURE 3.60 Drill gauge

Centre punch

The centre punch as shown in **Figure 3.59** is used to locate the centres of circles and arcs when marking out on various materials such as steel and copper. The centre punch is used in conjunction with a ball-pein hammer. When marking out a centre point prior to drilling ensure that the punch mark is deep enough for the drill bit to centre on the mark.

The centre punch point should be finished to 60°, which suits the drill angle. If the centre punch is used for witness marking (defining a line on sheet metal), the point should be sharper, with 40° recommended.

Hole saw

A hole saw or fly cutter as illustrated in **Figure 3.61**, is used for cutting large-diameter holes in sheet metal. The hole saw is employed in a power drill at low rotational speeds because the blade saws its way through the workpiece. The fly cutter uses a tool bit to cut its way through the workpiece and is also used at low rotational speeds. Always exercise caution if using a trammel. Trammels can also be used to cut circular gaskets from various types of sheet materials.

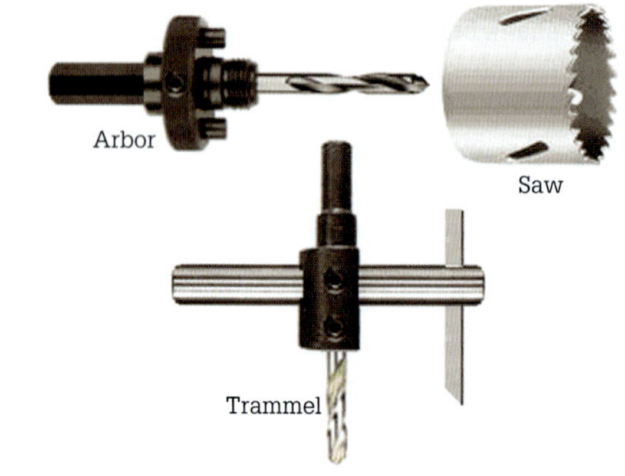

FIGURE 3.61 Hole saw and fly cutter/trammel

Using a hole saw is the fastest and easiest way to make a hole in plasterboard. A hole saw has three parts: an arbor that attaches to a standard drill, a pilot bit, and the hole blade itself.

A common problem when using a hole saw is that it tends to tear out as the saw exits on the opposite side of the material. To prevent this from happening, begin drilling from one side of the material but before the saw goes all the way through go around and finish the hole from the other side. The result is a clean hole on both sides.

REVIEW QUESTIONS

1 Name five types of twist drills.
2 What is the typical range of sizes for straight shank twist drills?
3 What is the point angle for a centre punch?
4 What is the purpose of a hole saw or 'fly cutter'?
5 Name three important considerations when sharpening twist drills.

3.6 Tapping and threading

Two important factors to consider when it comes to tapping and threading are the size and type of thread. The two main types of threads are the imperial and the metric. Metric threads (60°) are designated with a capital M plus an indication of their nominal outer diameter and their pitch:

$$\text{M size} \times \text{pitch}$$

For instance:

$$\text{M10} \times 1.5$$

The pitch is the distance from a point on a screw thread to a corresponding point on the next thread measured parallel to the axis. Metric coarse thread does not have a pitch specified. A coarse thread is just expressed as M10. A tap drill formula for 60° threads is:

Nominal outer diameter of thread − pitch = tap drill size

This realises approximately 68% to 77% of the required thread depth.

EXAMPLE 3.2

Use Table 3.5 to determine:

a the tap drill size for an M8 × 1.0 thread.

$$8 - 1.0 = 7.0 \text{ mm}$$

b the tap drill size for an M10 × 1.5 thread.

$$10 - 1.5 = 8.5 \text{ mm}$$

A metric thread chart is illustrated in Table 3.5.

EXERCISE 3.8

Use Table 3.5 to determine the tap drill size for:

a an M3 × 0.5 thread
b an M10 × 1.25 thread
c an M20 × 2.5 thread

TABLE 3.5 Metric thread chart

Tap size	Primary major dia (mm)	mm per thread	Tapping drill size (mm)
M1.6 × 0.35	1.6 mm	0.35	1.25 mm
M2 × 0.4	2 mm	0.4	1.6 mm
M2.5 × 0.45	2.5 mm	0.45	2.05 mm
M3 × 0.5	3 mm	0.5	2.5 mm
M3.5 × 0.6	3.5 mm	0.6	2.9 mm
M4 × 0.7	4 mm	0.7	3.3 mm
M5 × 0.8	5 mm	0.8	4.2 mm
M6 × 1	6 mm	1	5 mm
M8 × 1.25	8 mm	1.25	6.8 mm
M8 × 1	8 mm	1	7 mm
M10 × 1.5	10 mm	1.5	8.5 mm
M10 × 1.25	10 mm	1.25	8.8 mm
M12 × 1.75	12 mm	1.75	10.2 mm
M12 × 1.25	12 mm	1.25	10.8 mm
M14 × 2	14 mm	2	12 mm
M14 × 1.5	14 mm	1.5	12.5 mm
M16 × 2	16 mm	2	14 mm
M16 × 1.5	16 mm	1.5	14.5 mm
M18 × 2.5	18 mm	2.5	15.5 mm
M18 × 1.5	18 mm	1.5	16.5 mm
M20 × 2.5	20 mm	2.5	17.5 mm
M20 × 1.5	20 mm	1.5	18.5 mm
M22 × 2.5	22 mm	2.5	19.5 mm
M22 × 1.5	22 mm	1.5	20.5 mm
M24 × 3	24 mm	3	21 mm
M24 × 2	24 mm	2	22 mm
M27 × 3	27 mm	3	24 mm
M27 × 2	27 mm	2	25 mm

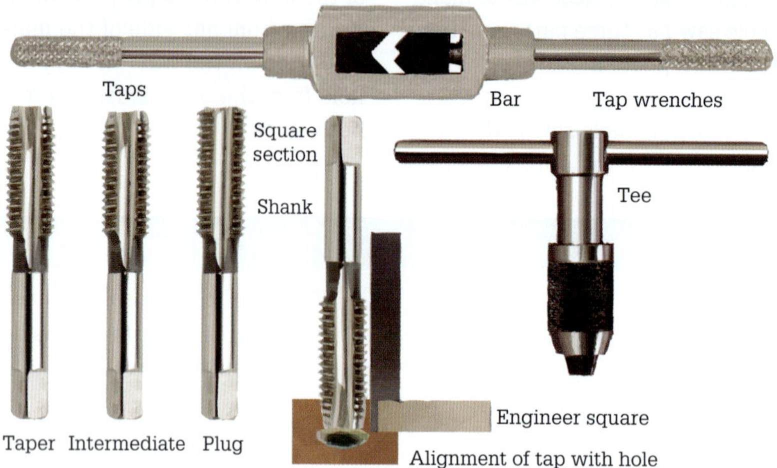

FIGURE 3.62 Taps, tap wrenches and alignment

Taps

Cutting internal screw threads on the inside of holes is a process known as tapping. A tap is a threaded tool which has three flutes that cut the thread to form cutting edges (see **Figure 3.62**). The shank is squared to fit a tap wrench.

The tap chosen to cut the thread must match the pitch and number of threads per mm on the fastener. The number of threads per mm varies depending upon the strength and torque required for the task and the size of the fastener.

Metric taps are specified with a thread pitch. The thread pitch is the distance between threads measured along the length of the tap stated in millimetres. For example, a thread pitch of 2.0 means that the distance between one thread and the next is 2.0 mm.

Taps vary in the shape of the tap as well. There are three types of taps. The taper tap, shown in **Figure 3.62**, has teeth ground to provide a long bevelled starting entry thread leading to the main threading cutters. This form allows easy starting and gradual cutting of the thread. This tap must cut a thread entirely through the workpiece in order to cut all threads to the correct depth. It can also be used to start a thread in a blind hole (a hole having a bottom).

The intermediate tap, illustrated in **Figure 3.62**, has about four cutting teeth ground tapered leading to the main threading cutters. This tap is used after the taper tap and is for finishing 'through hole' tapping in a workpiece and for cutting a thread in a blind hole.

The plug tap, shown in **Figure 3.62**, is not for starting threads. It is used for tapping a full thread to the bottom of blind holes.

A cutting lubricant should always be used on taps to enable smooth cutting and prevent tearing of the thread. When using a tap it should be in line with the hole axis and aligned by using an engineer's square, as shown in **Figure 3.62**.

The tap direction should be reversed each quarter of a turn to break off the cut metal and to prevent tearing the thread. An important point with tapping is to drill a tapping hole of the correct diameter.

SWITCH ON

Hazard

Taps are made from hardened metal and can shatter and break easily if undue force is applied. Always wear eye protection when tapping.

Tap wrench

Tap wrenches can be either bar or tee type, as illustrated in **Figure 3.62**. Bar-type tap wrenches are turned by two hands, while tee-type tap wrenches are used where space limitations make it difficult to use the bar type.

Stock and dies

Dies are used to produce external screw threads on various materials such as steel rods or conduit. There are three main types of dies: adjustable half dies, button dies and die nuts. A button die, a die nut and a stock are illustrated in **Figure 3.63**.

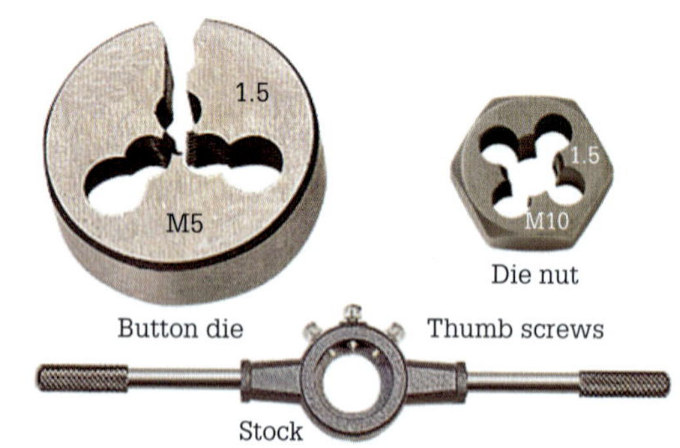

FIGURE 3.63 Stock and dies

Matched-pair half dies have adjustment screws to provide minute control of the depth of material cut from the workpiece. Adjustment screws allow smaller cuts to be taken to reach the correct thread size. The button or split die provides a partial amount of adjustment in the depth it will cut.

Button dies should be unadjusted for the first cut and then gradually reduced to finished thread size. Both of these dies are held in and turned by a stock. The die nut is made from a hexagonal steel stock and is turned by a spanner. Die nuts are of a fixed size and are used to clean rusted and damaged threads or for checking an existing thread. The bar workpiece should be chamfered to assist in starting the die. The workpiece must be securely supported and the die must be held square to the vertical axis of the workpiece. A cutting lubricant should be used and the thread regularly cleaned with a wire brush.

Extractors

Extractors are used to remove broken taps, screws, bolts, studs, pipe and lubrication fittings without damage to the threaded hole. Extractors are either straight flute or helical flute, as illustrated in **Figure 3.64** or are of a particular design for taps.

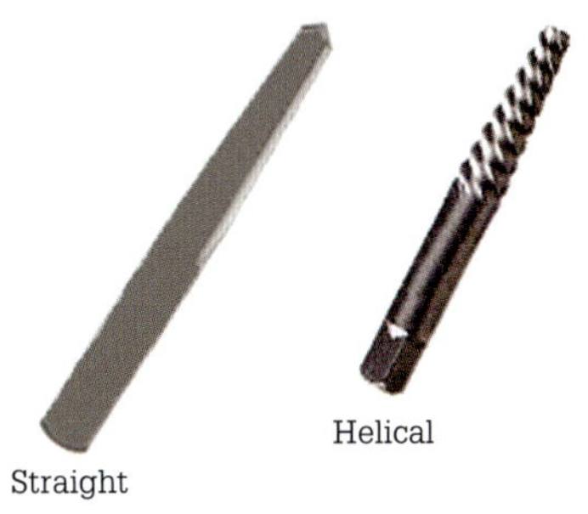

FIGURE 3.64 Extractors

To remove a broken bolt, drill a hole in the broken bolt and insert the extractor in the hole and using a tap wrench turn anticlockwise. The extractor acts like a corkscrew: it grips the sides of the drilled hole and removes the broken bolt on its threads without damaging the threaded hole.

SWITCH ON

Hazard

Extractors are made from hardened metal and can shatter and break easily if undue force is applied. Always wear eye protection when using an extractor.

REVIEW QUESTIONS

1. What are two important factors to consider when it comes to tapping and threading?
2. To what does screw pitch refer?
3. What is a suitably sized tapping drill diameter for an M5 × 0.8 thread?
4. Where would a plug tap be used?
5. Which tool is used to produce external screw threads on various round materials?
6. What tool is suitable for removing a broken bolt?

3.7 General hand tools

Installation work and the disassembly and assembly of electrical equipment require knowledge of basic hand tools and power tools and how to use them correctly. Hand tools are an extension of one's hands and it is the shape of the tool and its handle that determines the way the tool will be used. Most hand tools have a handle that is in line with the working end of the tool. This is why most portable power tools have a pistol grip handle.

Screwdrivers

There are many types of screwdrivers in use today, but the two main types for the electrician are the flat-blade and the Phillips-head screwdriver, as illustrated in **Figure 3.65**.

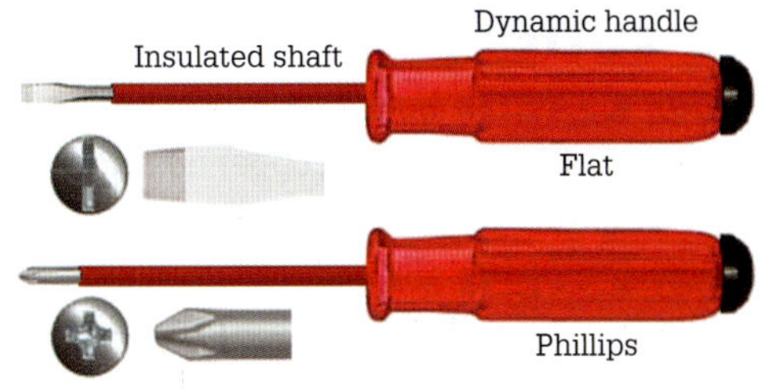

FIGURE 3.65 Flat-blade and Phillips-head screwdrivers

All screwdrivers employed by the electrician should be insulated to withstand 1000 V and be approved by Australian standards (AS 3527.2:1990 *Hand-operated screwdrivers and screwdriver bits – Insulated screwdrivers*).

Never use a screwdriver on an object held in the hand because the screwdriver may slip and puncture or lacerate the hand. Always use a machine vice to hold the object or use some other mechanical means (such as a clamp).

The flat-blade screwdriver has a flat, slightly curved flat-ended tip. Because of this type of design, the flat screwdriver needs to be directly vertical to the screw slot. Otherwise the blade will twist out of the screw slot when turned and strip the screw slot. Flat-blade screwdrivers are measured across the width of the tip in millimetres. A Phillips-head screwdriver has a cross or four-teeth pattern as its blade tip. The purpose of this screwdriver design is to resist stripping the slots in the screw. With the provision for four-teeth grooves in the screw, the Phillips blade holds to the sides of the screw slot firmly, reducing stripping. This characteristic enables more torque to be applied to screw in the screw. Phillips screwdrivers come in various sizes and are referred to by number: # 0, # 1, # 2 and # 3 and so on.

Combination pliers

Combination pliers, as shown in **Figure 3.66**, can be used to cut various sizes of electrical conductors and cable and are ideal for gripping a broad range of objects.

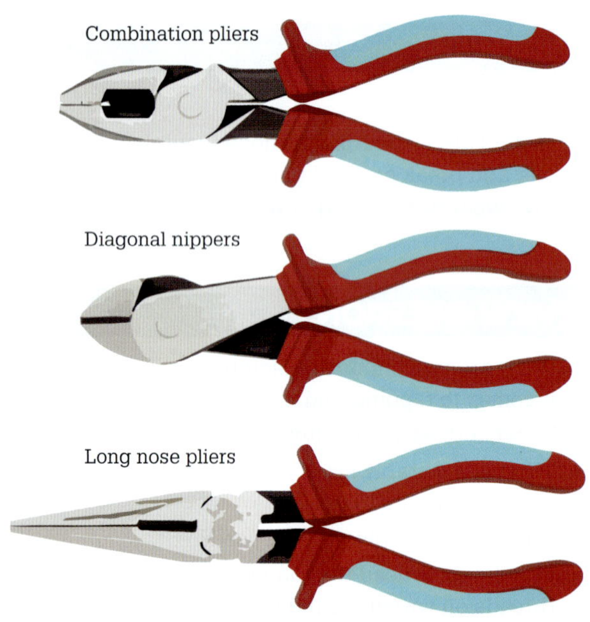

FIGURE 3.66 Combination pliers, diagonal nippers and long nose pliers

Combination pliers have gripping jaws in addition to cutting edges. The wide ends of the jaws are flattened to grip and twist bare electrical conductors together. Combination pliers are not a crimping tool and should not be used to fix various crimp-type terminations. However, there are combination pliers that are designed to crimp lugs onto cables.

Diagonal cutters

Diagonal cutters or nippers, shown in **Figure 3.66**, are also called side cutters. When it comes to cutting, diagonal cutters are more efficient than pliers as the cutting surfaces extend right to the end of the tool and therefore a greater cut pressure can be applied to the conductor.

Diagonal cutters have two cutting blades set diagonally to the joint and handles. To preserve the cutting edge, diagonal cutters should take several cuts to snip off conductors and not one big bite. They should only be used for copper or aluminium cable up to 6 mm^2 in cross-sectional area (CSA). They should not be used to cut steel wire or self-tapping screws.

Always point the conductor ends that are to be cut away from your body because the velocity of small pieces of cut conductors is high and if they strike an eye they will damage it. Consequently, always wear safety glasses when using diagonal cutters.

Long nose pliers

Long nose pliers (see **Figure 3.66**) have machined gripping jaws that are ideal for grasping and twisting. The tapered head design allows access to tight spaces. Long nose pliers usually have a cutting edge.

Wire strippers

When the cable is cut to the correct length, sufficient insulation needs to be stripped off the conductor to accommodate the termination. Two types of wire-stripping tools are a knife and the mechanical stripper, as illustrated in **Figure 3.67**.

FIGURE 3.67 Knife and mechanical wire strippers

One type of wire stripper has spring-opening jaws with an adjusting screw to fit cables up to 5 mm without damaging the conductor. The other type of mechanical wire stripper can remove the sheath from 1.0 mm^2 TPS and 2.5 mm^2 TPS cable. This mechanical stripper also removes the primary insulation from those cables. Nicks and broken strands of stripped conductors are not acceptable. Always cut off the damaged portion, readjust tool and re-strip. Re-twist stranded conductors after stripping in the natural lay of the cable. When using a knife always ensure that the cutting edge faces away from your body.

Cable cutters

A cable cutter, shown in **Figure 3.68**, is designed for cutting stranded aluminium and copper non-armoured cables up to 25 mm^2. The specially designed shape of the cutting blades means that the tool cuts without leaving frayed conductor ends. Cable cutters are not intended for cutting steel wires or SWA cable.

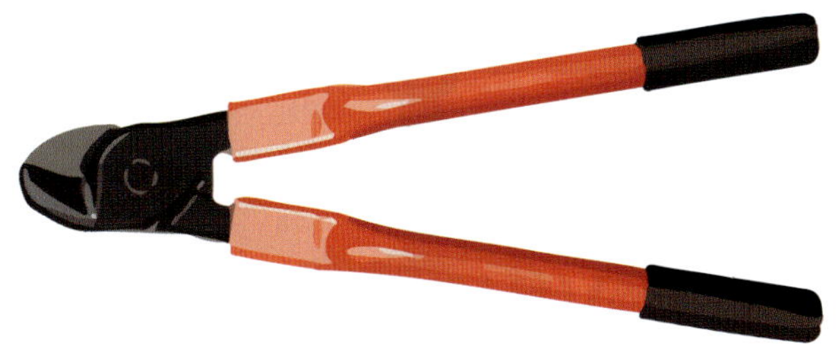

FIGURE 3.68 Cable cutters

Crimp tool

To form a crimp connection, a crimping tool (see **Figure 3.69**) is used to compress a termination tightly onto the cable conductors. A crimp termination provides a reliable metal-to-metal bond for stranded or solid conductors. The finished termination, if correctly formed, is mechanically reliable and electrically sound.

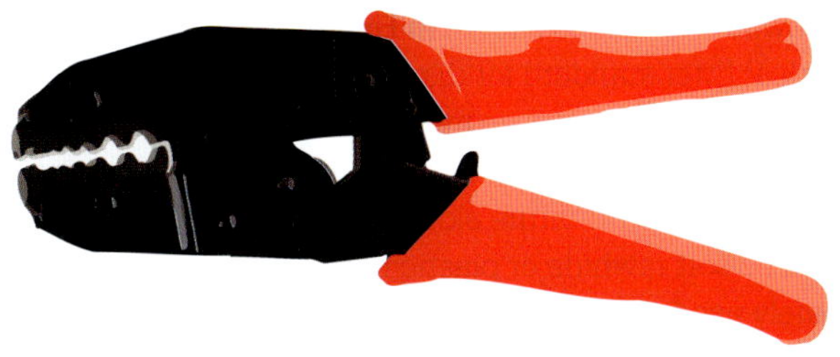

FIGURE 3.69 Hex crimping tool

This type of precision hand crimping tool employs a ratchet action ensuring that correct crimping pressure is applied for reliable indent crimping of cable terminations.

Spanners

Spanners come in all shapes and sizes and are used in the assembly or dismantling of all types of electrical equipment.

Open-ended spanner

The open-ended spanner, illustrated in **Figure 3.70**, is the most common type and may have a single or double end. The head has its jaws offset by about 15° from the axis of the shaft. The offset allows the spanner to be turned over to engage different faces of a nut when working in confined spaces. This type of spanner only grips across two flats of the nut or bolt head. When using this type of spanner, it should be turned so that the lower jaw leads to minimise damage.

FIGURE 3.70 Open-ended spanner

Note that metric spanners (10 mm, and the like) relate to the gap between the jaws in millimetres and therefore equate to the dimension across the flats of the nut.

Ring spanner

The ring spanner, illustrated in **Figure 3.71**, has a completely enclosed head and may have 6 or 12 flats. The size used to describe a spanner is the distance across the flats of the nut or bolt head to be turned. A six-flat spanner is shaped to fit against all sides of a hexagon nut or bolt head to ensure a tight fit. A 12-flat spanner bears upon the corners or faces of a hexagon nut or bolt head and will not slip. The twelve flats permit the turning of the nut where only a short pull of the spanner is possible.

FIGURE 3.71 Ring spanner

Ring spanners are stronger than the open-ended type, allowing greater leverage to be applied to the nut. Ring spanners also have different-sized heads at each end and can be offset in design.

Shifting spanner

A shifting spanner, shown in **Figure 3.72**, has a lower jaw which can be 'shifted' to fit any fastener size within the spanner jaw range. Shifting spanners should only be used if the correct-sized open-ended or ring spanner is not available. Both the fastener and spanner can be damaged if used on tight bolts or nuts. Always turn this spanner towards the adjustable jaw and not the fixed jaw to prevent slippage and to provide maximum turning torque.

FIGURE 3.72 Shifting spanner

Multigrips

Multigrips, as shown in **Figure 3.73**, are adjustable pliers that can be used if you need to undo something round without a hex head. They are used to tighten metallic conduit fittings. (Tip: Protect chrome or other shiny surfaces with a rag before using multigrips.)

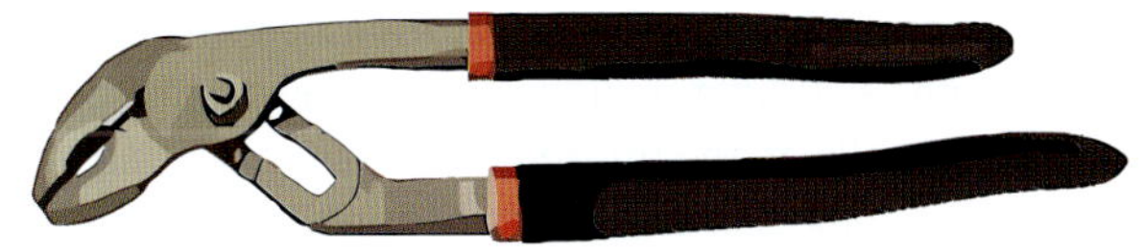

FIGURE 3.73 Multigrips

Vice grips

Among many other styles of pliers are vice grips. Vice grips are used throughout the electrical trade. Vice grips (see **Figure 3.74**) open up with a little tab on the handle. After the vice is opened, an adjustment screw is used to fit the vice to the diameter of the material it has to grip. The curved jaws put tremendous pressure on the object they are gripping, and the hardened teeth are designed to grip and clamp a workpiece from any angle. The vice grip has a quick releasable lever operated by the thumb.

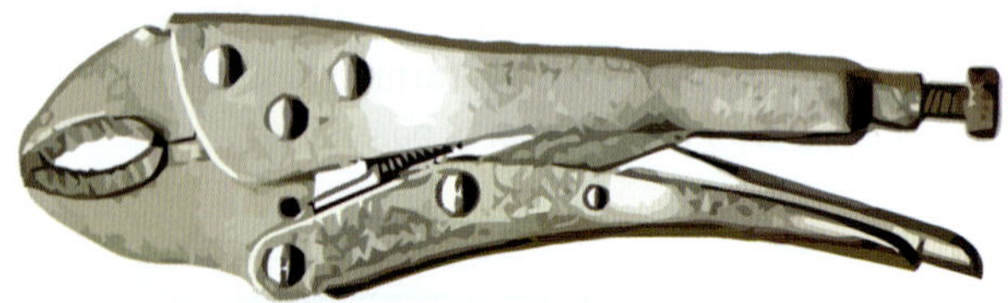

FIGURE 3.74 Vice grips

Allen keys

An Allen, hexagon ball-end driver or Unbrako key, as illustrated in **Figure 3.75**, is a tool used to drive screws and bolts that have a hexagonal socket in the head. Allen keys are measured across-flats (AF), which is the distance between two flat sides of the key.

FIGURE 3.75 Allen keys

Pullers

Two-arm and three-arm pullers (see **Figure 3.76**) provide a mechanical advantage for the extraction, without damage, of seated ball and roller bearings and gears. Bearing pullers usually have self-centring and self-aligning arms to provide leverage and guidance to ensure easy removal of mechanical parts seated on the shaft.

When using a bearing puller, oil the threaded shaft and make sure that the puller is aligned with the bearing shaft. To guarantee a straight pulling force, the jaws of the puller should be parallel to the threaded shaft of the bearing puller. When removing a stubborn bearing, strike the head of the puller screw with two sharp hammer blows. Always wear safety glasses when using a bearing puller.

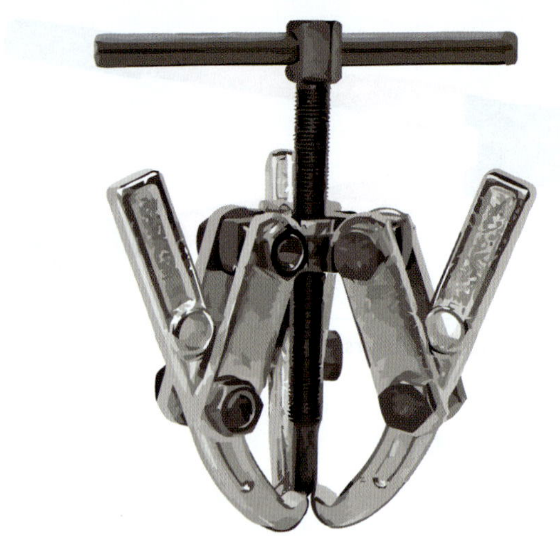

FIGURE 3.76 Three-arm puller

Hammers

The rawhide mallet as shown in **Figure 3.77** is the electrical fitter standard for dismantling and assembling equipment such as electric motors. In the rewind industry, the rawhide mallet is used to assist in the insertion of pre-formed coils into armature slots and for shaping the ends of coils. The soft rawhide or plastic head allows the mallet to be safely used on all types of metals.

FIGURE 3.77 Rawhide mallet, ball-pein and claw hammer

A ball-pein (peining means making metal softer) hammer as illustrated in **Figure 3.77** has a rounded end for working metal into a wanted shape. The other end is used for shaping and striking metal, such as punches and chisels, in metal fabrication (switchboards).

The electrician uses a claw hammer for striking cable pin clips and fixings into various materials.

SWITCH ON

Principles of tool use

When using tools:

- Decide on the right tool for the job.
- Always use ergonomically designed quality-made tools.
- Inspect tools for defects before use.
- Keep cutting tools sharp.
- Maintain tools carefully and store them correctly after each use.
- Carry tools in a tool box.
- Wear safety glasses and well-fitting gloves.
- Keep the work environment clean and tidy.

REVIEW QUESTIONS

1. Name the two main types of screwdrivers.
2. Why is it unadvisable to use a screwdriver on an object held in the hand?
3. What name is also given to diagonal cutters?
4. What is the function of a crimp tool?
5. Name two broad classifications of spanners.
6. What is the purpose of an Allen key?
7. How are Allen keys measured?
8. How should the jaws of a puller be aligned?

3.8 Joining techniques

A joining technique employed by electricians is soldering, typically to join copper wires.

Soldering

Electricians carrying out soldering tasks can face the risk of respiratory hazards due to exposure to soldering fumes and gases. Without correct engineering controls and respiratory protection, the risk of illness is severe. The principal hazards of soldering are heat, fumes and the composition of the solder. Other hazards created by soldering include solder splashes and flying molten solder hazards as well as toxic fumes from cable insulation if subjected to high heat. The following lists PPE required when soldering:

- eye protection (safety glasses with side shields)
- leather gloves for skin protection
- cotton clothing (shirt with long sleeves)
- sufficient mechanical (general and/or local exhaust) ventilation to limit exposure levels.

Soldering should be done in well-ventilated areas so that any fumes produced by the soldering process are dispersed. The soldering process may cause dizziness and headache with possible allergic reactions such as throat and respiratory irritation. When the soldering task has been completed, a thorough hand and face wash with soap should remove any residual particles.

Solder is a low-temperature-melt alloy metal. Soft soldering (sweating) is a joining process using a filler metal with a melting temperature below 427 °C and below the melting point of the parent metal. Soft soldering is the process of linking metals through the application of lead-free-based solder with the aid of a fluxing agent and a heat source. Heat source refers to heat that is either flame or from a soldering iron. Suggested solder to use is one that is 99.3% tin and 0.7% copper. The recommended tip to use if using a soldering iron is one that is iron-clad or nickel-clad. The flux cleans the soldering surface and enables the solder to flow.

Heating of the joint as shown in **Figure 3.78** should begin with the flame perpendicular to the conductors. This preheat will conduct initial heat into the joint for even distribution throughout. The flame should not be moved towards the joint as the insulation of the conductors will burn. Touch the solder to the joint. If the solder does not melt, remove it and continue the heating process. Molten solder will be drawn into the joint by capillary action regardless of the direction by which the solder is being fed.

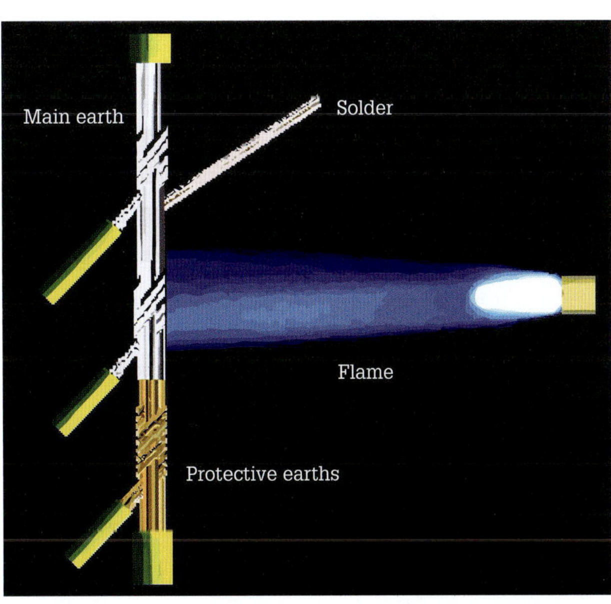

FIGURE 3.78 Soft soldering earthing conductors

After the joint has been completed, natural cooling is preferred. Shock cooling with a water-based rag may cause stresses in the joint that could result in subsequent failure. Once the joint is cool, remove any remaining flux with a wet cloth.

Brazing

The difference between soldering and brazing is temperature. Soft soldering (sweating) is a joining process using a filler metal with a melting temperature below 427 °C and below the melting point of the parent metal. Brazing is also a merging process using a filler metal with a melting temperature of 593 °C and 871 °C but below the melting point of the parent metal. With both soldering and brazing alloys, the parent metal is not melted (fused), but the solder diffuses into the parent metal making a surface alloy with the parent.

Silver solder

Silver soldering, also known as 'hard' soldering or silver brazing, is a joining process in which two or more components are joined by melting and flowing a filler metal (alloy) as shown in **Figure 3.79** into a joint gap by capillary action. Capillary action is the tendency of a liquid alloy to be drawn into small openings such as those between the walls of tubes or fittings without any voids or gaps. The filler metal wets the parent metal. Wetting is the ability of the molten filler metal to bond metallurgically the components together. This bonding enables brazed joints to remain liquid- and gas-tight under heavy pressures, even when the joint is subjected to shock or vibrational types of loading. Hard soldering should be regarded as being potentially hazardous as metal fume is produced by the melting process.

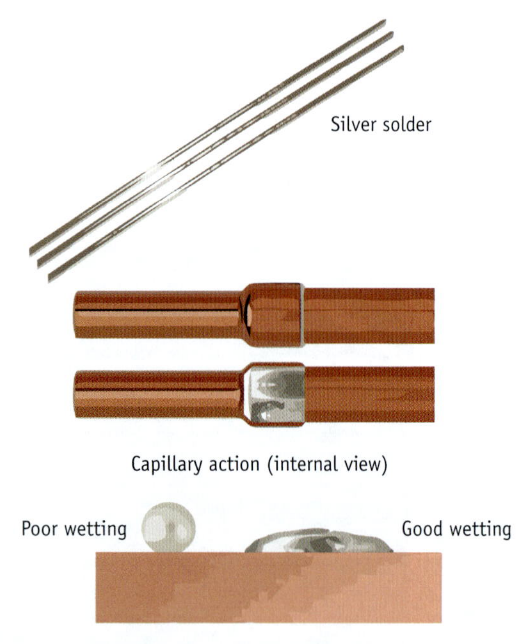

FIGURE 3.79 Silver solder, capillary action and wetting

Materials that can be present in some silver solders are silver, tin, copper, phosphorus, zinc, silicon, manganese and nickel. The various combinations of these metals enable a filler metal to braze copper and copper-based alloys, engineering materials, tungsten carbide and stainless steel. Hard-soldered joints have high strength and ductility with smooth fillets (the alloy deposited and penetrated and fused with the base metal to form a joint) and excellent corrosion resistance.

Heating source

Soft soldering and hard soldering can be carried out using two standard gases – acetylene and oxygen or MAPP gas. Most brazing alloys can be applied with a portable, lightweight torch. Always heat the parent metal (not the alloy itself) then allow the alloy to melt on contact with the pre-heated parent metal. Always make the alloy run to the heat or a cold-soldered joint will be created. A hard-soldered and cold-soldered joint is shown in **Figure 3.80**.

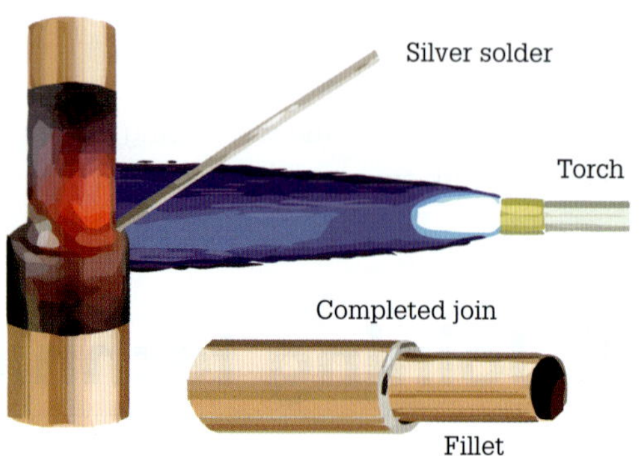

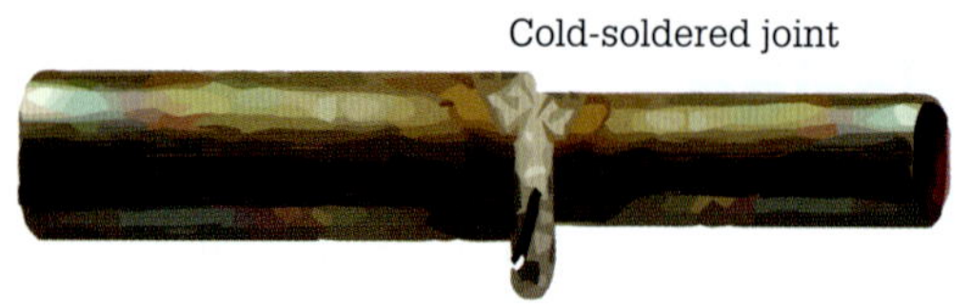

FIGURE 3.80 Hard- and cold-soldering joint

Hard-soldered joints are liquid- and gas-tight, can withstand shock and vibration and are unaffected by normal temperature changes.

Welding

With welding methods, the MIG (Metal Inert Gas) welding process is the simplest of them all to do. It is simple because a MIG welder (see **Figure 3.81**) has a welding wire that continually feeds into the welding puddle and arc.

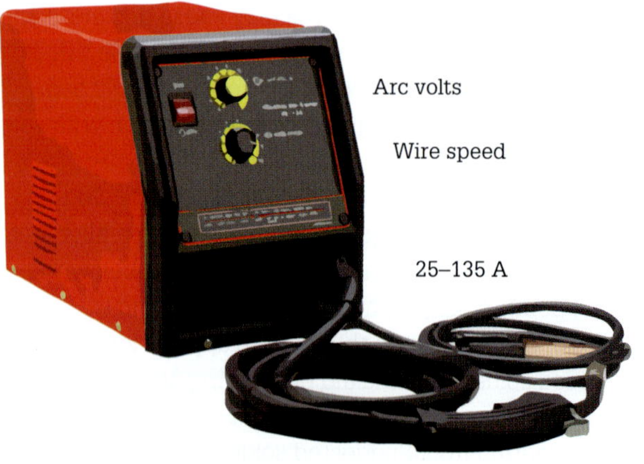

FIGURE 3.81 MIG welder

The setting of the voltage and wire speed is the single most important aspect of tweaking the MIG for the ideal welding arc as illustrated in **Figure 3.82**. Voltage and feed should work together. Otherwise the wire feed can be too slow or too fast, with the wire not emerging quickly enough or regularly striking the parent metal.

Source: Alamy/David Gartland

FIGURE 3.82 MIG arc

Selecting the voltage suitable for the task is essential for good weld penetration and sound welds. The higher the voltage, the hotter the weld puddles. The lower the voltage the colder the weld puddles. Experience with MIG will teach you how many volts to use for different tasks. For sheet metal work, a cold welding puddle should be used. With plate, a hotter puddle would be selected.

Other types of welding include tungsten inert gas (TIG) and arc welding.

Machine screws

One of the simplest methods for joining materials is using machine screws and nuts. Machine screws provide a sound mechanical fastening, which is easy to assemble and remove. When specifying a machine screw for a particular application it is necessary to know:

- head style
- length
- diameter
- thread type
- material
- drive configuration
- finish.

The chosen style of machine screw will depend on appearance, application and strength. A range of machine screws and head types is shown in **Figure 3.83**.

Some screws, such as the flat and oval countersunk heads, require countersinking of the work surface prior to assembly. This usually makes them unsuitable for sheet metal work. The oval head screw is installed in conjunction with a cup washer, eliminating the need for countersinking. Usually the larger the head, the greater the mechanical strength of the screw.

The drive configuration refers to the type of tool necessary to drive the screw. The Allen head screw allows for positive tool positioning and gripping. It finds frequent use in securing items to shafts.

Source: Shutterstock.com/Flipser

FIGURE 3.83 Range of machine screws

Machine nuts

Machine nuts, used with machine screws, provide a sound mechanical fastening, which is easy to assemble and remove. When specifying a machine nut for a particular application it is necessary to know:

- style
- chamfer
- thread type and size
- material
- finish
- width
- thickness.

The hexagonal style is most common and preferable to the square type as it looks better and is less prone to 'rounding'. A double chamfer nut will protect the surface of the work piece. A single chamfer nut will make circular abrasions when the non-chamfered side is in contact with the work piece.

The thread type and size must be the same as the mating screw. A thicker nut is generally stronger than a thin nut.

Machine washers

Machine washers may be of the flat or lock type. Flat washers are in contact with the screw head (or nut) to provide more surface area for the clamping action or to protect the surface of the work piece from abrasion. Lock washers minimise the chance of vibration loosening a screw (or nut).

The spring lock washer uses the temper of the steel to maintain the locking action – compressing the washer causes the spring tension to exert a force that prevents counter-rotation.

Tooth-type lock washers provide maximum gripping action. The teeth tend to flatten in use, so it is necessary to discard the washer and replace it after use.

The inside diameter of a washer must correspond with the clearance hole diameter of the associated screw. The outside diameter of the washer will depend on the specific application.

REVIEW QUESTIONS

1. List the principal hazards of soldering.
2. What PPE is required when soldering?
3. Describe the soft soldering process.
4. What is the main difference between soldering and brazing?
5. Explain the term 'capillary action'.
6. Soft soldering and hard soldering can be carried out using two standard gases. Name these gases.
7. What is the single most important aspect of tweaking the MIG for the ideal welding arc?
8. What is one of the simplest methods for joining materials?

3.9 Power tools

Before you use any power tool, verify that it has an electrical test label to indicate that it has successfully passed an inspection for electrical safety within the previous six months.

Hammer drill

Hammer drills, as shown in **Figure 3.84**, usually have three modes of operation: drilling, hammering only and a combination of normal drilling method with a hammering action. The combination third method enables the hammer drill to bore holes in masonry quickly without burning the hardened bit. Some hammer drills are capable of delivering up to 4200 blows per minute with a 4–22 mm masonry drill bit.

Source: Shutterstock.com/kasarp studio

FIGURE 3.84 Battery-powered hammer drill

Screwdriver

An adjustable clutch screwdriver is shown in **Figure 3.85**. The adjustable clutch screwdriver delivers a range of torque to drive a wide variety of fasteners into soft or hard material.

Source: Shutterstock.com/begun1983

FIGURE 3.85 Adjustable clutch screwdriver

Heat gun

A heat gun (see **Figure 3.86**) is commonly used in the electrotechnology industry for shrinking 'heat shrink' insulation tubing and tapes, drying out wet circuits in MIMS cable and softening PVC conduit in preparation for bending. Some heat guns blow hot air at extremely high temperatures (315 °C to 538 °C).

FIGURE 3.86 Heat gun

When working with heat guns:

- Always wear safety glasses.
- Never obstruct or cover the air inlet grilles.
- Do not touch the metal nozzle.
- Do not point the heat gun at people or animals.
- Do not use a heat gun near flammable materials.
- Always switch the heat gun off before putting it down on any surface.
- Always allow the heat gun to cool off thoroughly when you have finished the task.

Source: Shutterstock.com/ApoGapo

FIGURE 3.87 Angle grinder

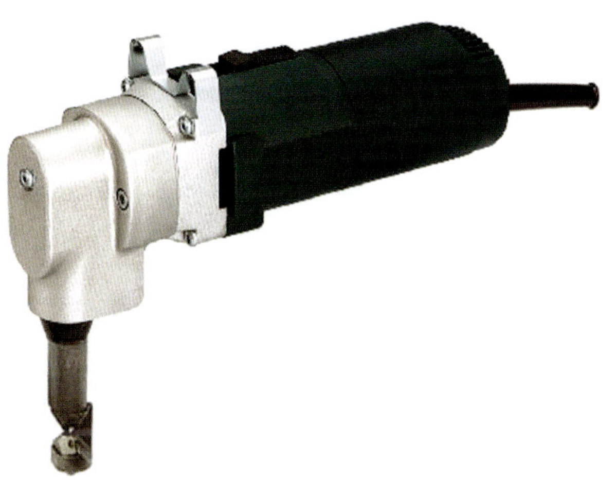

FIGURE 3.88 Nibbler

Angle grinder

The angle grinder is designed to smooth off rough edges of materials. When grinding, the grinding disc must be kept at an angle of 15° to 30° to the workpiece. Even though cutting wheels are available, angle grinders should not be used for cutting materials. Angle grinders in operation can kick back – where the disc is thrust back towards the operator and the discs themselves can shatter. Drop saws, not angle grinders, should always be used for cutting metals. An angle grinder is illustrated in **Figure 3.87**.

Always wear gloves, safety glasses, steel-capped safety boots, knee pads, ear plugs and overalls if using an angle grinder. Never put an angle grinder down until the disc stops turning and then always put it disc upwards. For extra personal protection, the angle grinder should be connected to a residual current device (RCD) protected circuit. Always remove the plug from the socket outlet when changing discs.

Nibbler

A nibbler, as shown in **Figure 3.88**, is the ideal tool for making straight and curved cuts in sheet metal.

Pedestal drill press

The pedestal drill press enables accurate drilling of holes, but reaming, countersinking and boring can also be done.

A pedestal drill press is controlled by a start/stop station and should have an isolation switch as well as an emergency stop station. Many pedestal drill presses have a spindle shaft turned by a V-belt on an overhead stepped cone pulley system driven by a small electric motor. Spindle shaft speeds are usually controlled by changing the V-belt position on the stepped cone pulleys; however, there are pedestal drill presses that have electronic speed control, and others have gear drives. Pedestal drill capacity is typically limited to 13 mm diameter for a Jacobs chuck and higher with Morse taper drill bits.

On V-belt-type pedestal drill presses, there should be a micro-switch that prevents the drill from starting when persons are accessing the head to alter the chuck speed. The feed lever is used to raise and lower the spindle chuck. A workpiece being machined with the pedestal drill press must either be in a vice or clamped to the table or base. A piece of plywood should be used for workpieces that are directly clamped to the table to prevent drilling into the tabletop.

After a part is secured it must be aligned on the spindle by releasing the table lock and swinging the table around the column left or right and up and down by using the elevation crank. A depth gauge next to the feed lever can be used to control the depth of drilling. A pedestal drill press is illustrated in **Figure 3.89**.

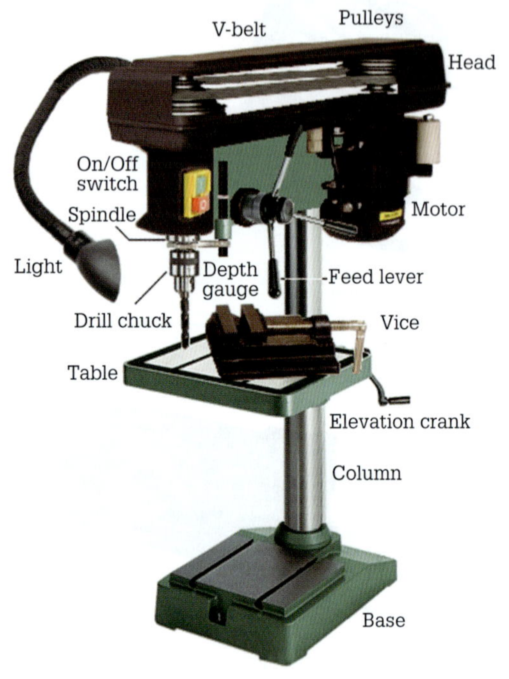

FIGURE 3.89 Pedestal drill press

Always wear safety goggles or safety glasses with side shields and a full-face shield when needed and use ear plugs during extended periods of operation. Do not wear gloves, loose clothing or jewellery that may catch in the chuck or drill bit. Long hair should be in a snood or hair-net to prevent scalping by the pedestal drill press's moving parts. Never hold any workpiece by hand and always make sure that the workpiece is well lit. When drilling is finished always use a brush to remove metal swarf.

Bench grinder

Bench grinders are used for sharpening, shaping and forming metal workpieces such as cold chisels, screwdriver tips, drill bits and various types of cutting edges. The grinding wheel is made of a bonded abrasive. Bench grinders, as shown in **Figure 3.90**, are sized by the abrasive wheel size and electric motor power rating. The maximum speed rating of the abrasive wheel must match the electric motor speed.

Before using a bench grinder the abrasive wheels should be tested by sounding, a process called 'ringing' the wheel. A non-metallic object is used for this test, and the wheel is lightly tapped. If no clear metallic ring is heard then the wheel is unsafe to use. An unsafe wheel will shatter and can cause severe injury or death to the bench grinder operator.

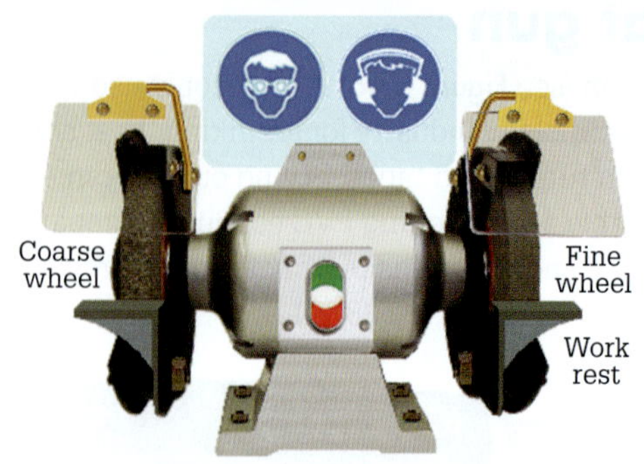

FIGURE 3.90 Bench grinder

Glazing can occur on the grinding wheel when the heat from grinding reacts with a burdened wheel to create a smooth surface on the grinding face. Loading of an abrasive wheel can also occur. Loading takes place when a build-up of swarf in a grinding wheel clogs the spaces between the coarse grains. When working with bench grinders:

- Never stand in front of the abrasive wheel when starting a bench grinder.
- Never grind on the side of the disc or grind soft material such as copper or aluminium.
- Ensure that the work rest is adjusted (slightly above the centre of the wheel with a gap of 1.5 mm between the work rest and the grinding face of the abrasive wheel).
- Dress abrasive wheels that are worn or have grinding faces out of square using a dressing stick or tool which also sharpens the wheel (dressing protrudes the abrasive grains on the wheel surface).
- Always wear appropriate PPE such as hearing protection, safety goggles or face shield, leather apron, respirator if required, overalls and safety boots when operating a bench grinder. Do not wear jewellery and ensure that long hair is in a snood or hair-net.

REVIEW QUESTIONS

1. List the three modes of operation employed by hammer drills.
2. Name common applications for a heat gun.
3. What power tool should be used in preference to an angle grinder for cutting metals?
4. List the PPE required when using an angle grinder.
5. Outline the uses of a nibbler.
6. How should metal swarf be removed from the work piece after drilling?
7. What are the functions of a bench grinder?
8. In relation to grinding, what is loading?

3.10 Sheet metal work

Sheet metals are sheets that are less than 6 mm thick and are produced by reducing the thickness of billets of metal by repeated rolling. They are available as flat sheets or as coiled strips. Sheet metal is widely used for numerous industrial and non-industrial applications in the electrotechnology industry including appliance and switchboard manufacturing, enclosures and hot water tanks (copper). Brass sheet is used for engraving purposes and lamp fixtures while phosphor bronze is used for fuse clips and switch components.

Classification of sheet metals

Sheet metals may be classified as ferrous or non-ferrous.

Ferrous metals

Ferrous metals are those metals containing iron, which include:

- Mild steel sheet is the most common and has a black oxide finish. It is the cheapest sheet metal and is easy to work but readily rusts. Bright steel sheets have the scale removed from black iron by 'pickling' and cold rolling. Mild steel sheet has many applications including car bodies, whitegoods cabinets, pressed steelwork and light fabrication.
- Galvanised iron is steel sheet having a zinc coating. It is available in flat and corrugated profiles. The galvanising produces a bright spangled effect when new and resists corrosion. Sharp bending tends to cause the galvanising to flake. Galvanised iron has many applications due to its corrosion resistance.
- Tinplate is steel sheet with a very thin coating of tin. It has a bright 'silvery' look and is highly resistant to corrosion. Popular applications include food and drink containers.
- Zincanneal is galvanised iron heated in a furnace to 340 °C and rolled. This causes the zinc coating and the surface metal to alloy. This process prevents the surface from flaking when bent.
- Terne plate is mild steel plate having an alloy coating containing 85% lead and 15% tin. It is suitable for panel beating and automotive panels as the coating prevents corrosion of the steel plate.

Sheet metals such as mild steel are usually coated with a variety of finishes to control corrosion, improve durability and provide resistance to abrasion and chemicals. Some of the coatings used for sheet metal are powder coatings, zinc (galvanising), tin, nickel and cadmium plating.

Non-ferrous metals

Non-ferrous metals are metals that do not contain iron and include:

- Copper is soft and easy to work. It resists corrosion and is a good conductor of both heat and electricity. It work-hardens and requires annealing to maintain its workability.
- Aluminium is very soft and resists corrosion. It is a good conductor of both heat and electricity. Popular applications include building panels and roofing. Aluminium foil is a common household item.
- Lead is a heavy metal, which is very soft and easy to work. It is very durable in damp situations and finds popular use as damp course and flashing.
- Brass is an alloy of copper and zinc, making it stronger and harder than copper. It is easy to work and resists corrosion.
- Duralumin is an alloy of copper and aluminium, which hardens with age to become much harder than aluminium but decreases ductility. A common use is for aircraft panels.

Tools for sheet metal work

Various hand and power tools are available for working with sheet metals. Some of these are discussed below.

Aviation snips

The most common hand tools for sheet metal work are the tinsnips. Aviation snips (or shears) are a type of tinsnips that are made of tool steel and used for cutting sheet metal by hand. They are available in a variety of shapes and sizes. Aviation snips are not generally suitable for cutting metal thicker than 1 mm. Thicker metals need cutting by machine or by hand using a hacksaw or cold chisel.

There are three styles of aviation snips, as shown in **Figure 3.91**:

- straight snips for cutting straight lines or outside curves
- right curved snips for cutting inside curves
- left curved snips for cutting inside curves

Universal snips have heavy section blades but narrow cutting edges and can cut without obstruction from the cut edge or waste.

Using aviation snips

When using aviation snips:

- grip the handles between thumb and first finger and allow first and second fingers to curl around the lower handle
- place the third and small fingers between the handles but rest them on top of the lower handle

Source: Shutterstock.com/Yanas

FIGURE 3.91 Aviation snips

- squeeze the hand closed and check that the handles and blades close
- open the hand and check that the handles part and the blades open ready for cutting
- place the sheet metal into the blades, with the junction of the blades aligned with the cutting line
- squeeze the hand closed – do not allow the blades to close completely or the metal immediately ahead of the snips will distort
- open the hand and move more metal into the blades (alternatively, push the snips further along the metal)
- repeat until finished.

Guillotine

The guillotine is a tool for cutting pieces of sheet stock to size. It has a bed to place the sheet metal, a hold bar that usually operates as a safety guard to keep fingers away from the cutting blade. The cutting blade is very sharp and operates in a shearing action. For small work, the foot-operated guillotine is sufficient. For larger, heavier work, power driven guillotines are more suitable.

Many guillotines have back gauges and a stop. These are extremely useful when cutting a quantity of stock to the same size. The back gauge is like a ruler to measure the distance from the back edge of the blade. **Figure 3.92** shows a treadle (manually operated) guillotine.

To make a cut using a treadle guillotine:

- place the sheet metal onto the bed of the guillotine and slide it against the side gauge
- slide the sheet metal under the blade to the required depth (either to the stop or to the cutting line)
- sight along the blade to make sure the cutting line is parallel with the blade

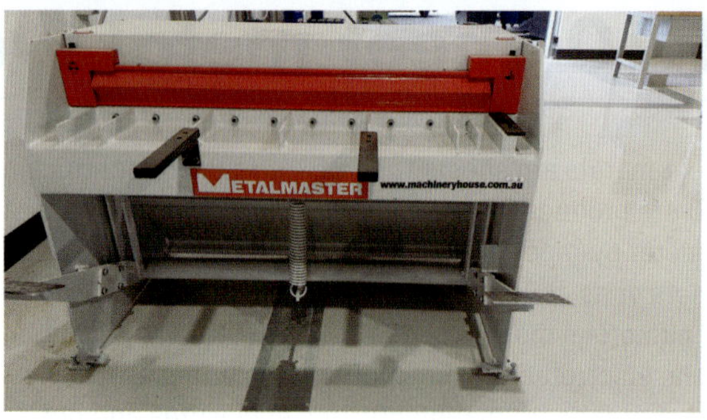

FIGURE 3.92 Treadle guillotine

- hold the metal firmly in position against the side gauge with hands and fingers well clear of the hold bar
- operate the foot treadle.

The exact procedure will vary with each guillotine. Do not attempt to use a guillotine until you have received proper instruction in its operation.

SWITCH ON

Safety considerations

When using a guillotine, be sure to keep hands and fingers well clear of the hold bar and blade. Keep other foot clear of the foot treadle to prevent injury when operating the foot treadle. Be careful of sharp edges on the cut metal.

Sheet metal bender

The sheet metal bender is a tool used to bend sheet metal. Sheet metal benders are known by various names that describe their configuration – pan brake, box brake, finger bender, and the like. **Figure 3.93** shows a small bench-mounted sheet metal bender. Before performing any bending operations, it is necessary to check and adjust the:

- finger seating
- clamping pressure
- finger clearance
- wing stop angle.

When using a sheet metal bender, observe the manufacturer's specifications for bending detail. The position of the bend line between the fingers and the wing break is all-important. The fold line on the material must bisect the finger clearance. The finger clearance must equal the thickness of the material to bend.

Joining sheet metal

Always wear leather gloves when handling sheet metal to avoid cuts from sharp edges. Sheet metal can be joined with a variety of fasteners such as pop rivets,

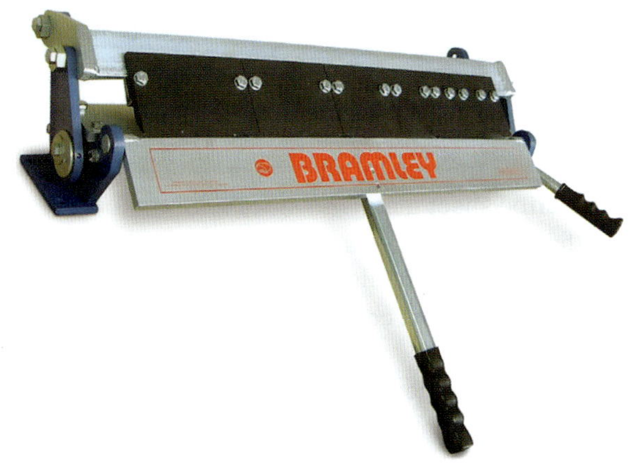

FIGURE 3.93 Bench-mounted sheet metal bender

metal threads, self-tapping screws or drive screws (see Figure 3.94). When using sheet metal always consider the actual construction sequence and assembly in order to avoid creating waste.

- Pop rivets provide a simple yet effective method of joining sheet metal. They are very effective where only one side of the join is accessible.
- Metal thread screws can be screwed into the sheet metal to be joined and tightened with nuts.
- Self-tapping screws cut their thread when driven into sheet metals that have been drilled with a pilot hole to accept the screw.
- Drive screws are used to fix sheet metal in the form of labels to thicker metal such as an electric motor frame.

FIGURE 3.94 Fastener types

Making mitres

Figure 3.95 (A) shows the necessary steps for making mitres in trunking.

Figure 3.95 (B) shows the necessary steps for making a tee joint in trunking.

Mark where the mitre corner will occur (point A) on the back of the trunk; using an engineer's square and scriber, draw a scribed line across the outside of the back of the trunk.

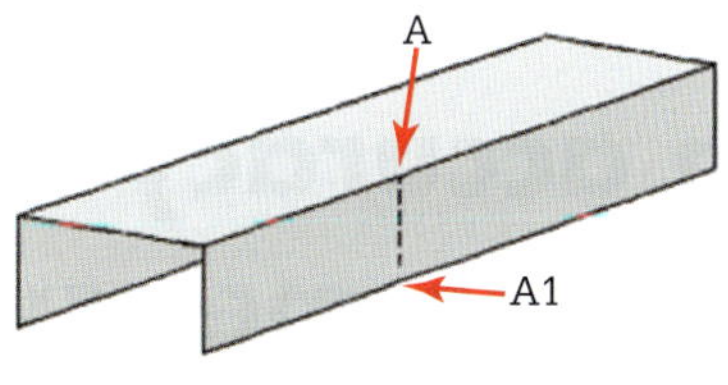

On the back of the trunk place a further two scribe marks, each equal to the width of the trunk (at points B and C) to extend beyond the trunk corner. The central mark B should be extended completely around the outside of the trunk using an engineer's square and scriber.

On the base of the trunk, scribe a line to connect the continuous line from B1 to both A and C. Scribe a line 10 mm away from, but parallel to, A–B1, and extend this across the back of the trunk. This will produce a 'lap' when the trunk mitre is formed. Scribed line B1 should also extend up the back of the trunk to A1.

Using tin snips cut along the line on one side, and on the other side cut on the parallel line which is 10 mm in from the original line connecting A–B1. You will cut out the complete V marked into the trunk. Also remove excess material at A and B1 to allow the outside of the trunk to fold easily, and the tags to overlap.

Fold the trunk around until the back sections touch each other. Fold the tag inside the back trunk. Rivet the join. To produce an external trunk mitre, reverse your V line marks on the trunk so that point B1 is exchanged for point B and the face of the trunk is cut out and not the back. Then follow the above steps.

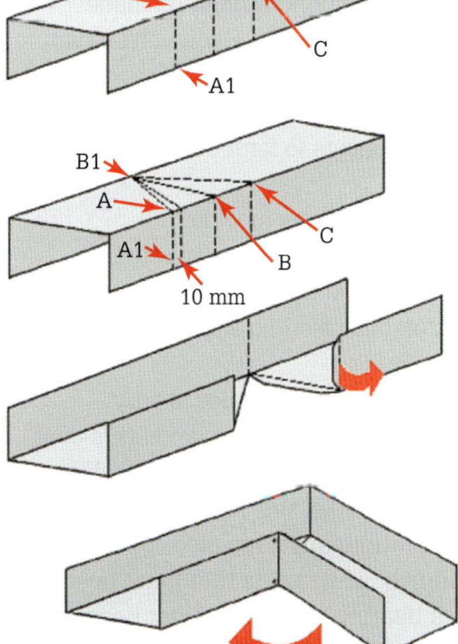

FIGURE 3.95 (A) Making mitres in trunking

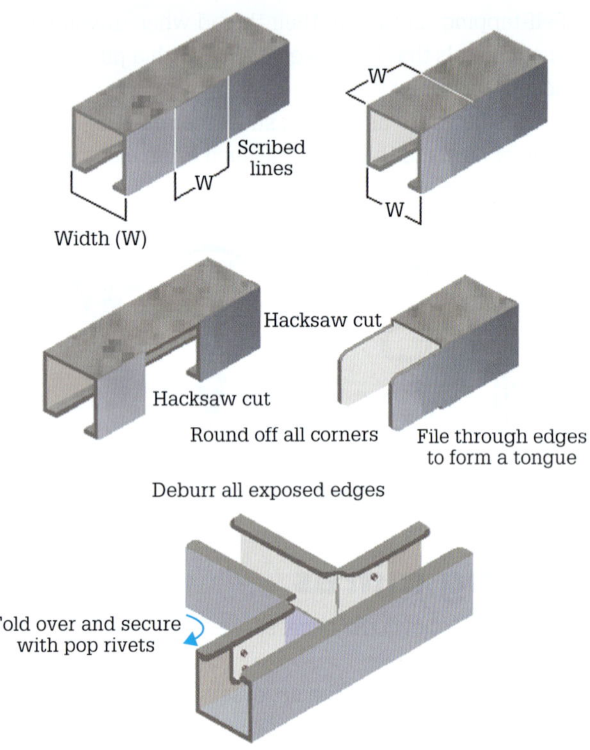

FIGURE 3.95 (B) Forming a tee joint

REVIEW QUESTIONS

1. How is sheet metal made?
2. In what format are sheet metals available?
3. What are the two broad categories used to describe sheet metals?
4. Why are sheet metals such as mild steel coated with a variety of finishes?
5. What is the most common hand tool for sheet metal work?
6. State the function of a metal working guillotine.
7. What is the purpose of the back gauge on a metal working guillotine?
8. Name three types of fasteners that can be used to fix sheet metal.

3.11 High accuracy measurement (low tolerance measurement)

The cost of manufacturing a particular component is made up of several factors. One of these is the degree of precision to which the component must be made. Making engineering components usually involves two separate processes:

1. Forming and machining the component to shape.
2. The assembly of components to form the finished product.

In manufacturing, limit systems provide:

- guides to the size parts must be, if they are to fit together and function correctly
- standard methods of dimensioning engineering drawings.

To make a component to an exact size is very expensive both in money and time. To save on this cost, most parts are allowed a tolerance on the actual size of the part's features. For example, the outside dimension of a part that does not fit into anything else can be made to a wide tolerance, while parts that must be a running fit, such as a bearing and shaft, must be made to a fine tolerance.

Figure 3.96 illustrates how the cost of manufacturing a component increases with the degree of required precision.

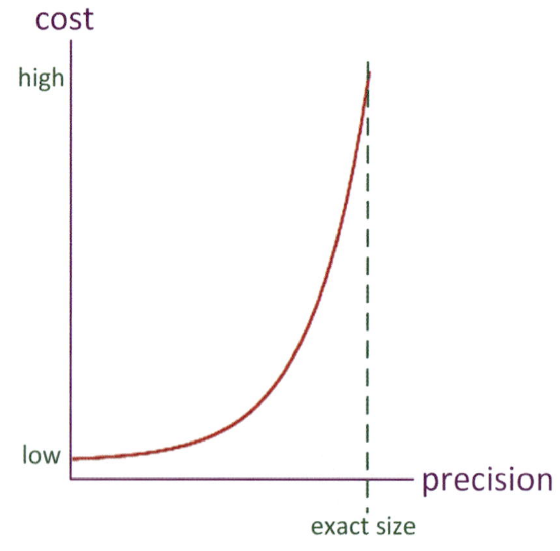

FIGURE 3.96 Component cost increases with degree of precision

Tolerance is the total amount by which a dimension can vary around the ideal or design size, without affecting its function. The maximum and minimum sizes of each dimension are given and these are known as the limits of size and are documented as a plus and minus amount. For example, a dimension might be 20 mm, with a tolerance of plus 0.5 mm and minus 0.5 mm, which is shown on a drawing as 20 ±0.5 mm.

Often it is necessary to measure dimensions to a higher degree of precision than can be attained using an engineer's rule. Two tools used for precision measuring are the micrometer and the vernier caliper.

Micrometer

Micrometers are used where a measurement requiring a level of precision less than 0.5 mm is essential. Micrometers provide precise, quantitative measurements of a workpiece's attributes such as thickness, depth, height, length, inside or outside diameter, roundness or bore. The most common type of micrometer is that used for external measurements, as shown in **Figure 3.97**.

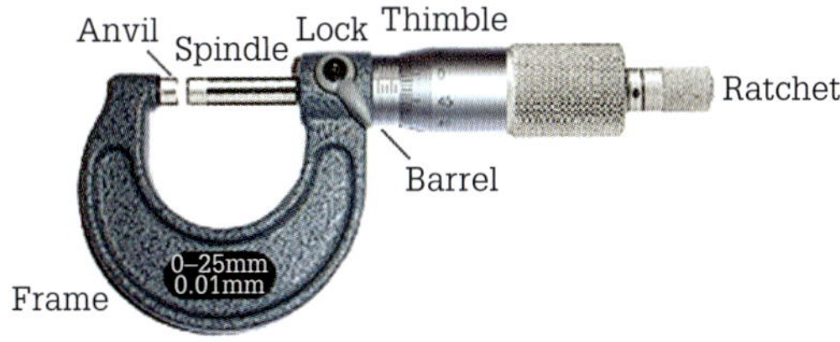

FIGURE 3.97 Outside micrometer 0–25 mm

An outside micrometer consists of a flat U-shaped frame with an anvil fixed at one end and a moving spindle attached onto the thimble. By turning the thimble on its screw thread inside the barrel the distance between the measuring faces of the anvil and the spindle is controlled. The spindle thread has a pitch of 0.5 mm so that one revolution of the thimble moves it and the spindle 0.5 mm. The barrel has a datum line and fixed significant markings on it and is attached to the frame. The datum line is graduated with millimetre divisions above and half millimetres below. The thimble has 50 equally spaced divisions and, when one revolution moves it forward 0.5 mm, one division represents 0.01 mm.

In **Figure 3.98**, the first significant figures are taken to the main scale reading to the left of the thimble: 14 mm. The remaining digits are taken from the scale reading that lines up with the datum line on the barrel: 32 on the thimble scale. Thus, the reading is 14.32 mm (14 mm + 0.32 = 14.32 mm).

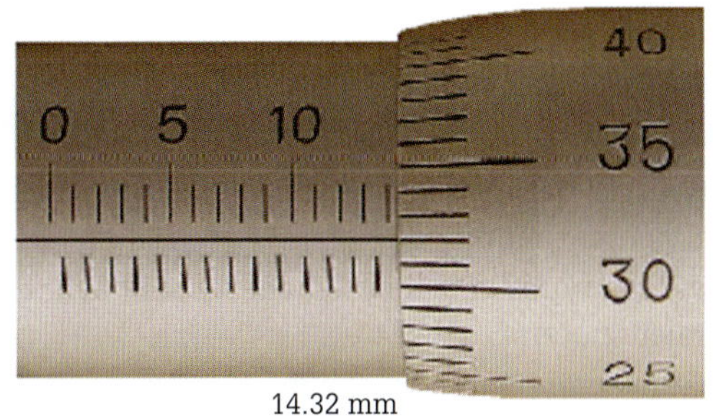

FIGURE 3.98 Micrometer measurement

EXERCISE 3.9

Determine the micrometer measurement in **Figure 3.99**.

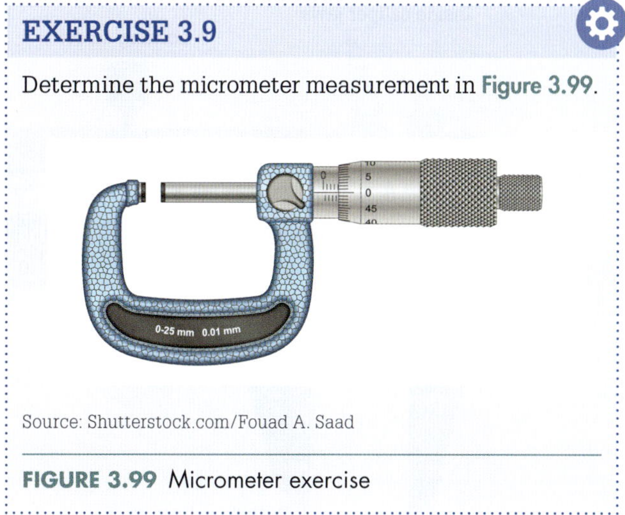

Source: Shutterstock.com/Fouad A. Saad

FIGURE 3.99 Micrometer exercise

The ratchet ensures that the spindle is adjusted on the workpiece with the same degree of pressure each time a measurement is made. A lock provides a method of fixing the spindle after setting the micrometer so that the micrometer has use as a measuring gauge. Micrometers are available in vernier scale (**Figure 3.99**), digital (**Figure 3.100**) and dial variations.

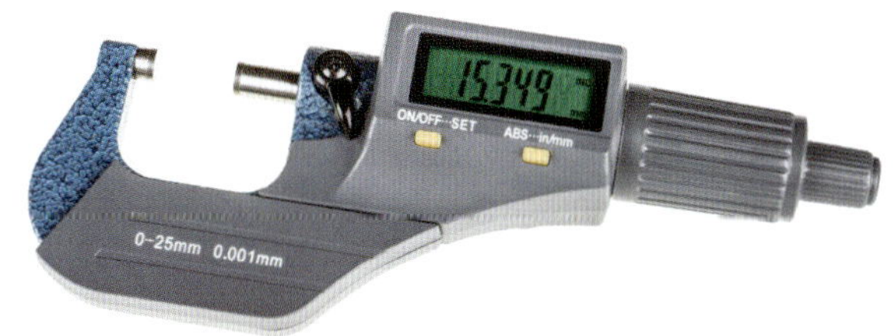

Source: Shutterstock.com/KPixMining

FIGURE 3.100 Digital micrometer

Vernier caliper

A vernier caliper, as shown in **Figure 3.101**, will make measurements within the confines of its graduated beam.

The vernier caliper can measure the outside diameter of a workpiece, the inside diameter of a hole or the deepness of a hole. One jaw of the caliper is fixed, and the other jaw moves and is connected to the vernier. A vernier has two graduated scales, a main scale like a ruler and a vernier scale which permits readings to the nearest fraction of a division on the main scale. A measurement is made by combining the readings from the two scales.

When using a vernier caliper ensure that the vernier caliper correctly reads zero when the jaws are closed. Then close the jaws around the workpiece – the jaws should exert a firm pressure. When the locking screw is tightened the caliper can be detached from the workpiece and the measurement read. In addition, the main scale can be read to the nearest millimetre.

The main scale works just like a ruler: the 0 mark on the vernier is compared to the main scale and the result is written down. In our example this is main scale significant figure '2'. If you look at **Figure 3.101** you will see that the distance between the divisions on the vernier

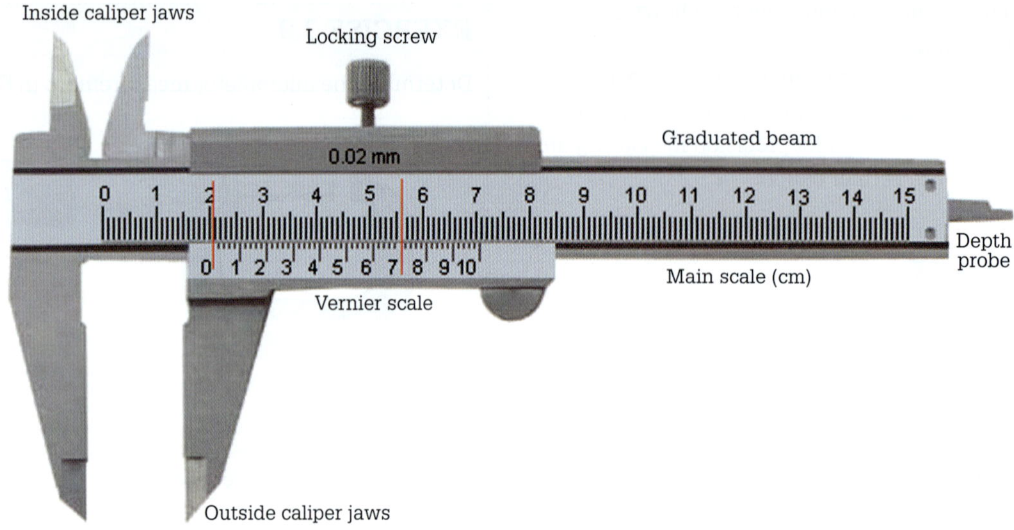

FIGURE 3.101 Vernier caliper

scale are not the same as the divisions on the main scale. The zero (0) line on the vernier occurs between 20.00 and 21.00 on the vernier scale, but the 10 line on the vernier occurs between 70.00 and 71.00 on the main scale. Thus, the distance between the divisions on the vernier are 90% of the space between the divisions on the scale. Such spacing guarantees that one of the vernier markings will line up exactly with a measurement mark on the main scale. Decide which vernier mark comes closest to matching any main-scale mark; in our example the vernier mark is seven. Combine the two readings to give the final length of 20.70 mm.

In **Figure 3.102** the first significant figures are the main scale reading to the left of the vernier zero: 32 mm. The remaining digits are taken from the vernier scale reading that lines up exactly with any main scale reading: 1 (in blue) on the vernier scale. Thus, the reading is 32.10 mm (32 mm + 0.1 = 32.10 mm).

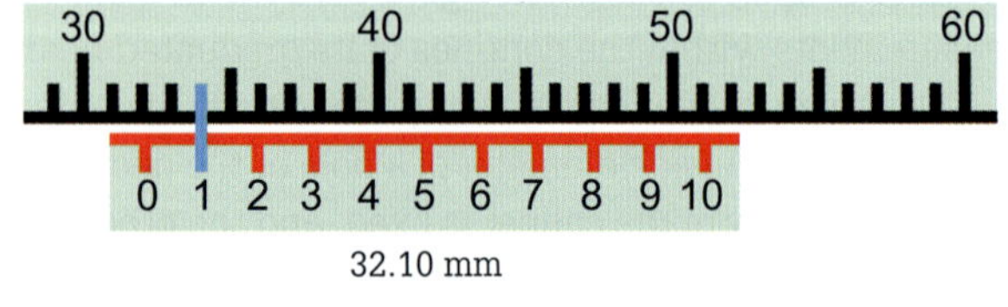

FIGURE 3.102 Vernier measurement

EXERCISE 3.10

Determine the vernier measurement in **Figure 3.103**.

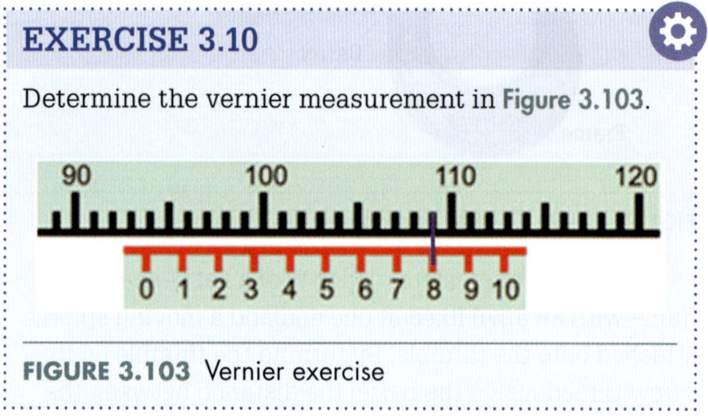

FIGURE 3.103 Vernier exercise

Vernier calipers are available in vernier scale (**Figure 3.101**), digital (**Figure 3.104**) and dial variations.

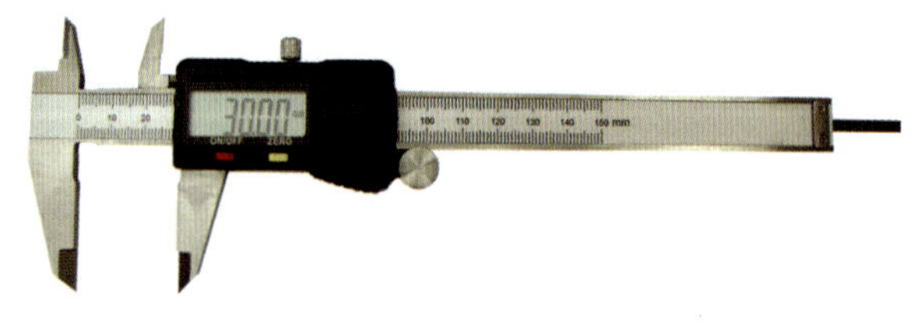

Source: Shutterstock.com/Ilona Koeleman

FIGURE 3.104 Digital vernier caliper

REVIEW QUESTIONS

1 Name the two processes involved in making engineering components.
2 What is meant by the term 'tolerance'?
3 Name two tools used for precision measuring.
4 What quantitative measurements of a workpiece's attributes can be made with a micrometer?
5 How is the datum line on the barrel of an external micrometer graduated?
6 Name the types of measurement readout available on micrometers.
7 What measurements can be made with a vernier caliper?
8 What reading should a vernier caliper indicate when the jaws are closed?

3.12 Dismantling and assembly techniques

When working on a task, it is best to keep the dismantling techniques and tools as simple as possible. The development of a dismantling plan is an important first step in carrying out certain tasks. As part of that plan, the worker needs to know what result is required before starting on a dismantling technique. What needs to be complete to satisfy the WHS, *Wiring Rules* and other regulations? Does the work require a total overhaul of an installation or machine or just components? Once these questions and others are answered then some dismantling techniques will be eliminated.

Dismantling is not just taking things apart using a casual approach. Worker safety is always the primary objective. Before selecting a dismantling method, the following five points should be considered.

1 Prior to starting any dismantling operations, all energy sources (electrical, pneumatic or hydraulic) should be disconnected and tagged or locked out from the work area.
2 The dismantling technique chosen should take into consideration the types of PPE required to perform the tasks and what are the hazards.
3 The location of the work to be accomplished needs to be considered in selecting a dismantling technique.
4 A method to capture and control waste such as abrasive particles, cleaning liquids, decommissioned cables, machine parts and materials is also critical.
5 What mechanical tools must be used to complete the task?

Experienced workers will be more efficient and safer using techniques that are familiar to them.

Assembly requires coordination of various parts, tools, fixtures and workers with a necessary element being that the results of the assembly can be tested. Assembly is the reverse of disassembly. With installations and machines, electrical workers should be able to diagnose and solve problems using their skills and knowledge.

Advances in technology have created some environmental challenges. With electrical and electronic equipment and materials, their decreasing end-of-life rate results in disposal problems.

Provision of impermeable areas should be made at worksites for depollution, dismantling and preparation of elimination. This procedure will minimise the impact on the environment and facilitate the reuse, recycling and recovery of waste materials for maximum value.

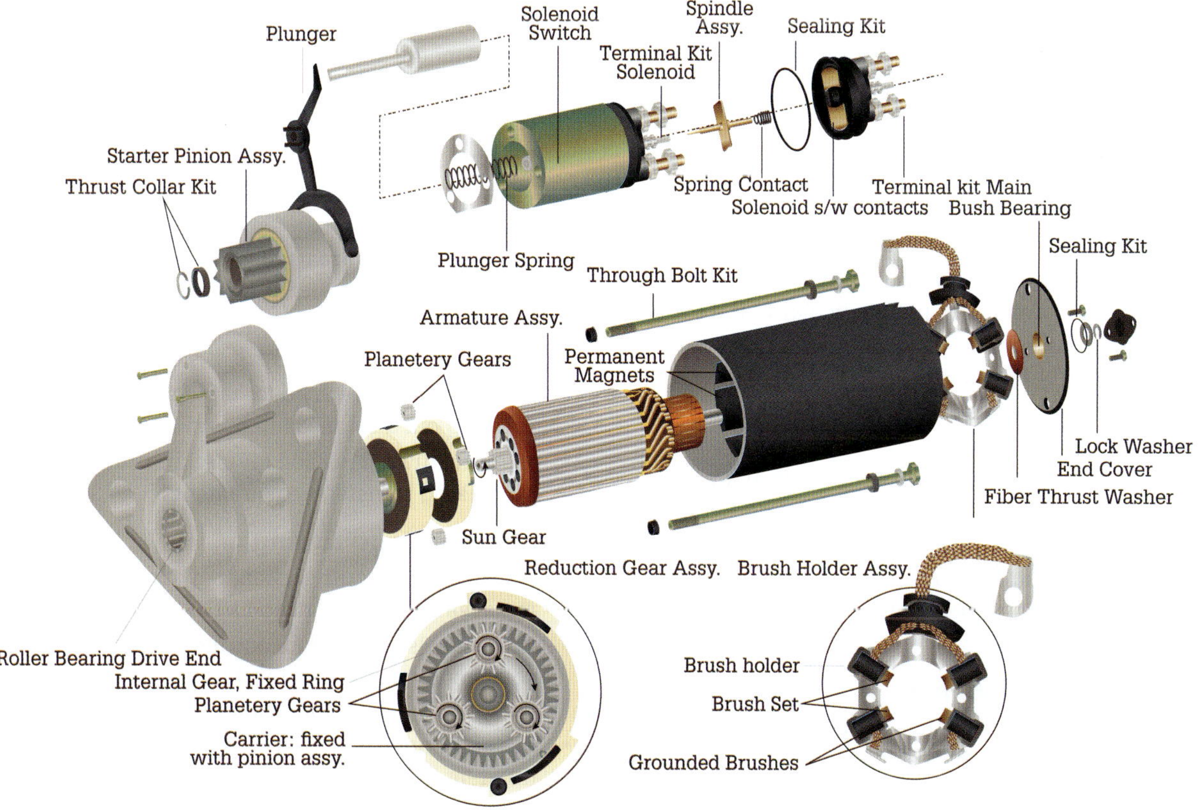

Source: Shutterstock.com/Pankaj_Digari

FIGURE 3.105 Exploded view of starter motor

Dismantling procedure

When it is necessary to dismantle a machine, it is important to follow a set plan. Wherever possible, use the manufacturer's information as a guide to dismantling any machine. Failure to do so could cause damage to machine components and unnecessary expense.

In the absence of manufacturer's details, you may be able to dismantle relatively simple machines by following a general procedure developed after thoroughly inspecting the machine. It is important to mark components such as end shields and frame to ensure correct alignment when reassembled. It is very important to document what you do at every step. This will be of great importance when it comes time to reassemble the machine. Remember it could be several months before you reassemble.

Another important facet is to draw an exploded diagram of the machine as shown in **Figure 3.105**. This does not need to be a work of art, but remember a picture is worth a thousand words.

CHAPTER REVIEW

3.1 Mechanical drawing

- Mechanical drawing is the primary method of communication between all people concerned with the design and manufacture of machine components, buildings and constructions and engineering developments.
- Engineering drawings and other technical drawings have to be done in a way that workers in the specific industry can recognise.
- Lines in drawings have a diverse thickness and are unalike in the way they are drawn, depending on the size of the paper and the job to be done.
- A title block contains the border and the many sections for providing quality, administrative and technical information.
- A detail drawing is a multiview representation of a single part of a mechanical component that includes a complete and exact description of its form, dimensions and construction.
- An assembly drawing is a presentation of the mechanical components put together, showing all parts in their operational positions.
- In an isometric drawing the mechanical component's vertical lines are drawn perpendicularly, and the horizontal lines in the width and depth planes are shown at 30° to the horizontal.
- Orthogonal projection is a way of drawing an object from different directions.
- The standard views used in a mechanical drawing orthogonal projection are the top, front and right-side views.
- Freehand sketching is a fast, reliable method used to record data with respect to mechanical components and circuit wiring.

3.2 Workshop planning

- An important outcome from good workshop planning is the ability to provide a safe working environment, which in turn can facilitate a better and more efficient workshop.
- A large part of working safely is identifying hazards and risks in the work environment and implementing suitable control measures.
- The major classification of materials used in the electrotechnology industry is metallic and non-metallic.
- Plastics is a general term used to describe a range of polymers we use as plastics.
- The planning process should include a risk assessment, collection of appropriate tools and materials, location of a suitable site to perform the task, bins for the collection of waste materials, obtaining plans and drawings for the task, ensuring that all required processes and techniques are understood, and ensuring that the worker has the ability to safely and correctly use the required tools.

3.3 Measuring and marking-out

- Marking-out is the process of laying out essential information in the form of lines, circles, outlines, centre marks and so on.
- Tape measures can be used to measure the cable route length required for voltage drop calculations.
- An engineer's square is accurate both inside and outside and is the workshop standard for 'squareness'.
- Outside calipers are used for taking external comparative and transfer measurements.
- The inside caliper is used for measuring inside dimensions such as the size of a cavity or hole.
- Jenny calipers are used for taking, transferring and marking out a set measurement.
- A spring divider is used for marking out circles and for making transfer measurements.

3.4 Holding and cutting

- A hacksaw is a stiff-backed, bow-type saw with a replaceable metal cutting blade.
- Cold chisels are used for cutting, shaping or removing cold metals that are softer than the chisel's cutting edge, such as soft iron, aluminium, brass or copper.
- Tin snips are made in various shapes and sizes for different tasks and are available for cutting in straight lines or curves to the left or curves to the right.
- A bolt cutter is a large-jawed cutting tool which can be used to cut round steel rods and soft bolts or small square-shaped steel bars.

- Most files have three grades of cut: bastard-cut, second-cut and smooth-cut, and for timber there is a rasp.
- Every file must have a handle in order to prevent the tang of the file causing an injury to the hand and to provide a safer grip on the file.
- A file card is used to clean the file as needed to remove any material wedged between the file teeth.
- For extra smooth surfaces wrap a piece of emery cloth (abrading fabric) around the file and stroke in the same manner as when removing stock.

3.5 Drills and drilling

- The centre punch is used to pinpoint the centres of circles and arcs when marking out on different materials such as steel and copper.
- To drill a hole in any material the correct type of drill bit must be used.
- There are five types of twist drills: titanium coated, carbide tipped, cobalt bits, carbon steel and high-speed steel.
- Drill sizes are typically measured across the drill cutting edges with a micrometer.
- A hole saw or fly cutter is used for cutting large-diameter holes in sheet metal.

3.6 Tapping and threading

- Cutting internal screw threads on the inside of holes is a process known as tapping.
- Dies are used to produce external screw threads on various materials such as steel rods or conduit.
- Extractors are used to remove broken screws, bolts, studs, pipe, and lubrication fittings without damage to the threaded hole.

3.7 General hand tools

- Hand tools are an extension of one's hands, and it is the shape of the tool and its handle that determines the way the tool will be used.
- All screwdrivers employed by the electrician should be insulated to withstand 1000 V and be approved by Australian standards.
- Combination pliers have gripping jaws in addition to cutting edges.
- Diagonal cutters have two cutting blades set diagonally to the joint and handles.
- Two types of wire-stripping tools are a knife and the mechanical stripper.
- A cable cutter is designed for cutting stranded aluminium and copper non-armoured cables up to 25 mm^2.
- A crimping tool is used to compress a termination tightly onto the cable conductors.
- Spanners come in all shapes and sizes and are employed in the assembly or dismantling of all types of electrical equipment.
- Multigrips are adjustable pliers that can be used if you need to undo something round without a hex head.
- Vice grips are ideal for tightening, clamping, twisting and turning.
- An Allen, hexagon ball-end driver or Unbrako key is a tool used to drive screws and bolts that have a hexagonal socket in the head.
- Pullers provide a mechanical advantage for the extraction, without damage, of seated ball and roller bearings and gears.
- The rawhide mallet is the electrical fitter standard for dismantling and assembling equipment such as electric motors.
- A ball-pein (peining means making metal softer) hammer has a rounded end for working metal into a wanted shape.

3.8 Joining techniques

- Soldering should be done in well-ventilated areas so that any fumes produced by the soldering process are dispersed.
- Solder is a low-temperature-melt alloy metal.
- The difference between soldering and brazing is temperature.
- Silver soldering, also known as 'hard' soldering or silver brazing, is a joining process in which two or more components are joined by melting and flowing a filler metal (alloy) into a joint gap by capillary action.
- With welding methods, the MIG (Metal Inert Gas) welding process is the simplest to do.
- The setting of the voltage and wire speed is the single most important aspect of tweaking the MIG for the ideal welding arc.

3.9 Power tools

- Before you use any power tool, verify that it has an electrical test label to indicate that it has successfully passed an inspection for electrical safety within the previous six months.
- Hammer drills usually have three modes of operation: drilling, hammering only and a combination of normal drilling method with a hammering action.
- Always wear gloves, safety glasses, steel-capped safety boots, knee pads, ear plugs and overalls if using an angle grinder.
- Bench grinders are used for sharpening, shaping and forming metal workpieces such as cold chisels, screwdriver tips, drill bits and various types of cutting edges.

3.10 Sheet metal work

- Sheet metals are sheets that are less than 6 mm thick and may be classified as ferrous or non-ferrous.
- The most common hand tools for sheet metal work are the tin snips.
- The guillotine is a tool for cutting pieces of sheet stock to size.
- Always wear leather gloves when handling sheet metal to avoid cuts from sharp edges.
- Sheet metal can be joined with a variety of fasteners such as pop rivets, metal threads, self-tapping screws or drive screws.

3.11 High accuracy measurement (low tolerance measurement)

- Tolerance is the total amount by which a dimension can vary around the ideal or design size, without affecting its function.
- Micrometers are used where a precision measurement requiring an accuracy of less than 0.5 mm is essential.
- The vernier caliper can measure the outside diameter of a workpiece, the inside diameter of a hole or the depth of a hole.

3.12 Dismantling and assembly techniques

- When working on a task, it is best to keep the dismantling techniques and tools as simple as possible.
- The development of a dismantling plan is an important first step in carrying out certain tasks.
- Prior to starting any dismantling operations, all energy sources (electrical, pneumatic or hydraulic) should be disconnected and tagged or locked out from the work area.
- Assembly requires coordination of various parts, tools, fixtures and workers with a necessary element being that the results of the assembly can be tested.

TRIAL EXAM

For Chapter 3 knowledge assessment, please complete the following trial exam.

1 Engineering drawings and other technical drawings have to be prepared in such a way that workers in a specific industry can recognise. This might be:
   a sketches or photos
   b standards or conventions
   c roughs or drawings
   d criteria or documentation

2 The sides of every size in the A series are in the ratio of:
   a 1:√1.414
   b 1:√2
   c 1:1.73
   d 1:2

3 The size of an A0 drawing sheet is:
   a 297 mm × 210 mm
   b 420 mm × 297 mm
   c 841 mm × 594 mm
   d 1188 mm × 841 mm

4 On the set of engineering drawings where would you find the projection symbol?
   a in the title block
   b in the materials list
   c in the revision box
   d in the symbols

5 What data should be shown on the drawing once only?
   a hatching
   b dimensions
   c projection lines
   d tolerance

6 A line when measured direct from a drawing measures 50 mm. If the drawing is scaled 1:100, what is the actual length this dimension represents?
   a 500 m
   b 50 mm
   c 5000 mm
   d 500 m

7 A multiview representation of a single part of a mechanical component that includes a complete and exact description of its form, dimensions and construction is called:
   a an assembly drawing
   b a pictorial drawing
   c an isometric drawing
   d a detail drawing

8 With drawing, what is the meaning for the term 'datum'?
   a all measurements should be taken from the edge or base line
   b all measurements are taken from the margin or border
   c all measurements are taken from the title block
   d all measurements are based on a grid reference system

9 An accuracy of 0.01 mm is possible with measurements using a:
   a steel rule
   b micrometer
   c Jenny caliper
   d scribed mark

10 Name the caliper purposely designed to scribe a line parallel to the edge.
   a outside
   b inside
   c Jenny
   d vernier

11 Name the engineering instrument that provides a measurement that is made by combining the readings from two scales.
   a steel rule
   b distance-measuring sensor
   c dial guage
   d vernier caliper

12 What type of tool makes use of a raker pattern?
   a hacksaw blade
   b pliers
   c screwdriver
   d cold chisel

13 To cut thin stock less than 1.6 mm, what teeth size hacksaw blade should be used?
a 18
b 14
c 32
d 24

14 What type of tool has a cutting edge at one end and a struck face at the other?
a pliers
b diagonal nippers
c cold chisel
d tin snips

15 A tool that amplifies the squeezing force applied to the cutting edge to cut is a:
a vice
b cold chisel
c trammel
d bolt cutter

16 A hand file is measured from:
a heel to the point
b edge to tang
c point to tang
d edge to heel

17 A flat file can be cleaned using:
a soap and water
b a file card
c a bannister brush
d emery cloth

18 Glass paper can be used to:
a mask work surfaces
b polish motor end-shield flanges
c clean the file
d pin the workpiece

19 What type of tool has flutes?
a a hole saw
b a metal thread screw
c a button die
d a twist drill

20 The tool used for witness marking is:
a a hole punch
b a pin punch
c a centre punch
d an auger

21 Into which material would an auger bit be suitable?
a brick
b timber
c sheet metal
d concrete

22 What type of tool uses an 'arbor'?
a tap wrench
b vice grips
c hole saw
d micrometer

23 The tap drill size for an M10 × 1.5 tap is:
a 10.0 mm
b 7.0 mm
c 5.0 mm
d 8.5 mm

24 The tool used for tapping a full thread to the bottom of blind holes is:
a a taper tap
b an intermediate tap
c a plug tap
d a die nut

25 What type of tool is used to remove a broken stud that is flush to a surface?
a vice grips
b extractor
c multigrips
d pliers

26 Before using any power tool in the workplace you should:
a check to see if it is industrial quality
b note if it has double insulation
c verify that it has a current electrical test label
d confirm its power rating

27 The most appropriate power tool for cutting large metal stock is the:
a drop saw
b nibbler
c guillotine
d angle grinder

28 What type of tool is tested by sounding?
a file
b ball-pein hammer
c the wheel of a grinder
d open-ended spanner

29 The tool used to drive screws and bolts that have a hexagonal socket in the head is the:
a Allen key
b Phillips-head screwdriver
c combination pliers
d pozi-drive screwdriver

30 When tightening metallic conduit fittings, what type of tool should be used?
a shifting spanner
b pliers
c multigrips
d vice grips

31 When dismantling an installation or machine, what is the primary objective?
a to obtain the tools
b speed
c keep cost low
d worker safety

32 What is the difference between soldering and brazing?
a temperature
b wetting
c speed of the joining process
d size of the filler metal

33 When using a MIG welder, increasing the voltage:
a produces a cooler weld puddle
b produces a hotter weld puddle
c increases the wire feed speed
d decreases the wire feed speed

4 Fix and secure electrotechnology equipment

This chapter provides electrotechnology workers with knowledge and skills about fixing and support devices and techniques. This chapter provides underpinning knowledge for the unit UEECD0020 as applied to the UEE training package.

LEARNING OBJECTIVES

Fasteners and fixings
- Recognise various types of fasteners and fixings.

Stud wall attachment devices
- Identify different types of attachment devices.

Fasteners
- Identify different types of powder-actuated fasteners.
- Select different types of threaded fasteners.
- Recognise various types of washers.
- Distinguish different types of keys.
- Demonstrate knowledge of various types of non-threaded fastening devices.

Fixings
- Identify various types of fixings.

Fixing adhesives and tapes
- Recognise various types of adhesives and tapes.

4.1 Fasteners and fixings

In general, a fastener connects together permanently two items or parts, while a fixing attaches a movable thing or part to a permanent one. Both terms are used to mean the same thing. For example, a metal screw is a fastener and can be used to attach wallboards to a metal stud frame. The same type of the metal screw can also be used as a fixing to attach a mounting block to a brick wall.

Fasteners and fixings cover a broad range of attachment devices or processes: bolts, circlips, nails, rivets, screws, staples, pins, anchors, chemical processes and welds, both permanent and tack (temporary fix), and all items associated with their use (nuts, washers, plastic plugs, and the like.).

The choice of an attachment device depends on several factors, such as:

- sustainability
- function
- properties – stiffness, strength, service life
- requirements of the device – magnitude and direction of forces
- installation process
- aesthetic considerations
- cost.

Sustainability

Sustainability and sustainable work practices are important issues with many enterprises across every industry segment making moves to reduce their environmental impact. The broader construction industry, according to the Department of Climate Change, Energy, the Environment and Water (DCCEEW) is responsible for about half of all solid waste generated globally.

On a brighter note, the DCCEEW also noted that of an estimated 19 million tonnes of construction and demolition waste created in Australia annually, about 55 per cent is recovered and recycled.

Some simple steps to reduce waste and act as an ethical custodian of the planet include:

- sustainable designs – plan the work with sustainability in mind, which may involve:
  - choosing fixing and fastening materials made from sustainable resources
  - storing material onsite so that it is protected from damage and secured from theft (copper is frequently stolen)
  - reduce the amount of waste, which may include using appropriately sized supports to avoid off-cut waste
- order appropriate quantities – limit the amount of material purchased, which may involve:
  - carefully calculating required quantities in advance
  - designing a cabling plan to assist in identifying the required fixings and fastenings
- purchase sustainable materials – these are healthier for the environment than their non-sustainable counterparts. Wherever possible:
  - use recycled materials or materials that can be recycled – the choice of materials can greatly assist in reducing waste
- return excess material – rather than merely discarding unused materials, which end up in landfill, consider:
  - returning surplus material to the supplier
  - recovering leftover fixings and fastenings that can potentially be used on other jobs.

No matter the size and scope of the job, it is important to consider all possible ways to reduce waste on every project. A modern version of an old adage is very appropriate – 'look after the cents and the dollars will look after themselves'. Meaning it is possible to actively participate in environmentally sustainable work practices by adopting small changes to work practices. By following best sustainability practices, it is possible to minimise waste and do your part to protect the environment.

Nails

Nails are the most common, practical metal devices for joining two pieces of timber together. Nails prevent separation of the joined timber, and they also prevent movement of each piece of wood past each other, as shown in **Figure 4.1**.

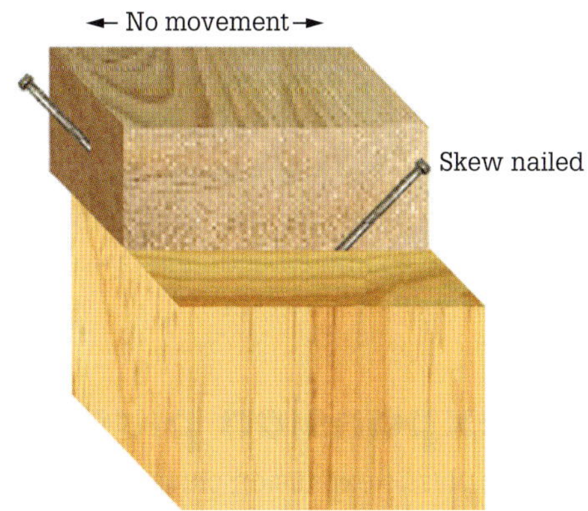

FIGURE 4.1 Using nails to join timber

Nail length should be chosen so that half to two-thirds of the nail length penetrates the timber receiving the nail point. Large-diameter nails can support greater loads than small-diameter nails. Blunt-point nails reduce nail-induced timber splitting that occurs with diamond-point nails.

Always pre-drill (0.8 × diameter of the nail) when using nails in hardwoods to prevent splitting. Various types of nails are illustrated in **Figure 4.2**.

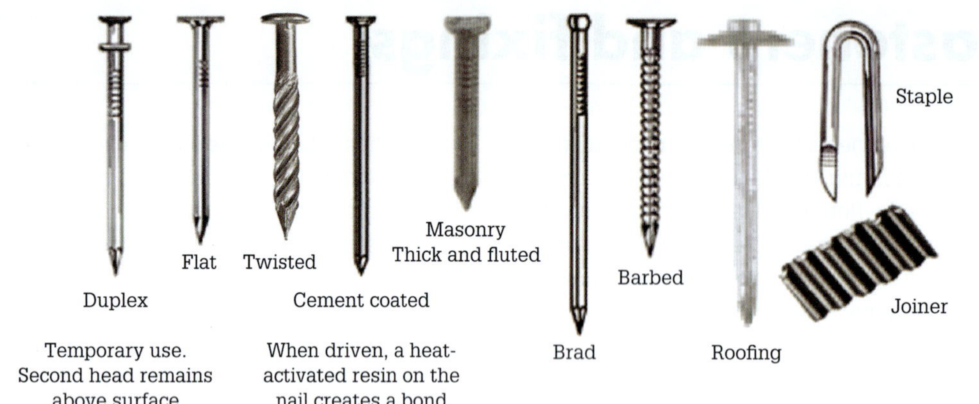

Source: Masonry: Alamy stock photo/trekandshoot

FIGURE 4.2 Various types of nails

Masonry screws

Masonry screws are a reliable and cost-effective method for attaching a broad range of fixtures such as saddles directly to concrete and brick. They are available with Phillips or hex head styles, as shown in Figure 4.3.

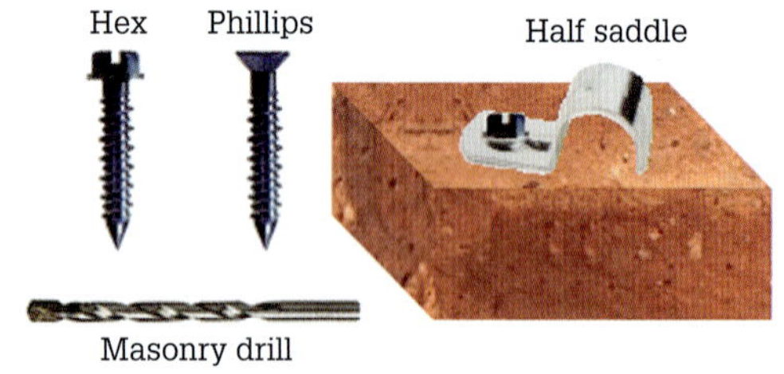

FIGURE 4.3 Masonry screws

To install these screws, drill a pilot hole with a hammer drill and masonry drill to the depth the masonry screw penetrates plus 5 mm to allow for displaced material. After making the hole, drive the masonry screw home. Masonry screws cut a thread into the brick or concrete as they are being screwed. Correct pilot-hole diameter and a 30 to 45 mm embedment is essential to ensure maximum pull-out strength.

Plastic expansion plugs

Various types of plastic expansion plugs are used to fix accessories such as mounting blocks to masonry walls as shown in Figure 4.4. To install a plastic expansion plug, drill a pilot hole at the marked location with a hammer drill and masonry drill to the depth necessary to support the load. Tap the plastic expansion plug into the hole with a hammer.

FIGURE 4.4 Plastic expansion plugs

Masonry expansion anchors

Masonry expansion anchors range from light-duty to very heavy-duty fixings; these are often used in the electrical industry and can be used in concrete, brick or stone. Hammer-set anchors are light-duty expansion anchors, as shown in Figure 4.5. These anchors are quick to install: just drill a hole with a hammer drill and masonry drill to the required depth and hammer the anchor home. Once installed, hammer-set anchors can be difficult to remove.

FIGURE 4.5 Hammer-set expansion anchor

Sleeve anchors

Expanding sleeve anchors, as shown in Figure 4.6, are designed for light- to heavy-duty pull-out and shear capacities in all types of masonry. Hole depth can be determined by taking into account the required embedment and thickness of the material to be fastened plus 5 mm. It is important to drill the correct diameter hole, otherwise the whole anchor spins in the hole and the anchor does not fasten. The sleeves of the anchor push against the sides of the hole and get tighter as the nut or flange nut (combined nut and serrated washer) is tightened.

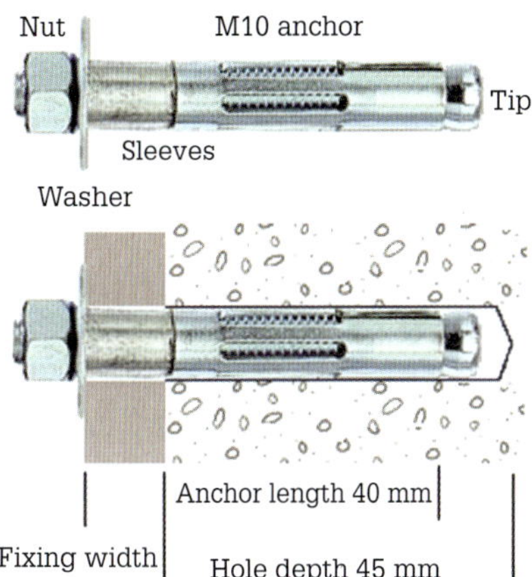

FIGURE 4.6 Sleeve anchor

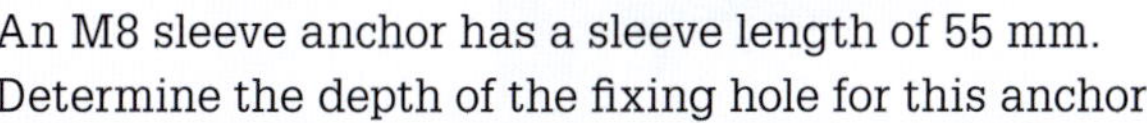
EXERCISE 4.1

An M8 sleeve anchor has a sleeve length of 55 mm. Determine the depth of the fixing hole for this anchor.

The tip of the sleeve anchor must protrude from the bottom of the anchor sleeve to ensure proper expansion. In hard masonry, a short sleeve may be suitable for maximum holding power. In softer masonry, longer sleeves provide superior holding strength. Multiple short sleeves increase the friction force mechanism used by most expansion anchors to resist tension loads. Sleeve anchors are used to fasten electrical equipment such as motors or switchboards to the floor but are also used on walls and concrete ceilings. It is the diameter of the anchor that determines its loading capacity when supporting a load fixed vertically. When installed on the ceiling, it is the depth of penetration that affects the size of the load that a sleeve anchor can support. If you over-tighten the nut, the anchor may break the masonry around the hole.

Wedge anchors

Wedge anchors, shown in **Figure 4.7**, are also designed for light- to heavy-duty pull-out and shear capacities in all types of masonry. The wedge anchor uses a one-piece clip to ensure uniform holding capacity that increases as tension is applied.

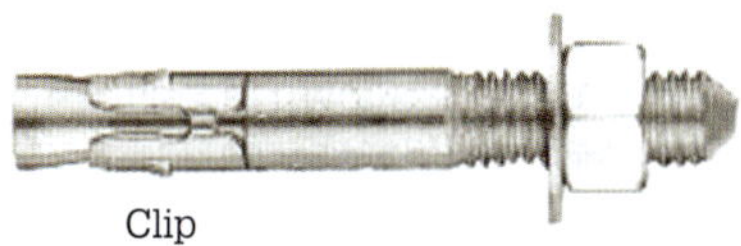

FIGURE 4.7 Wedge anchor

Over-tightening can cause wedge anchors to break free. Use a socket wrench or ring spanner to tighten the nut.

Chemical anchors

A chemical anchor, illustrated in **Figure 4.8**, is a threaded rod that is set in liquid epoxy or acrylic resin and activator. The liquid chemicals become hard when mixed to provide a locking mechanism between the threads on the rod and the rough internal sides of a hole in the masonry. Some chemical anchors use a plastic sleeve cage.

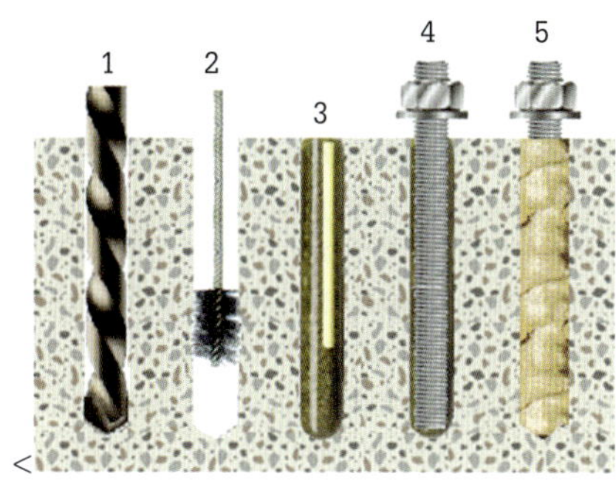

FIGURE 4.8 Chemical anchors

When using chemical anchors:

- Drill the correct-sized hole to the specified depth using a hammer drill and a masonry drill.
- Clean the hole with a brush and remove debris using a hole blower. A chemical anchor's bonding performance is severely reduced if the holes are not cleaned properly.
- Insert the correct size of chemical capsule or inject chemicals without creating air pockets. Avoid breathing the vapours and prevent skin contact by wearing PPE, which includes gloves and safety glasses when injecting the chemicals. Ensure that you read and understand the safety data sheets (SDSs) supplied with chemical anchors.
- Drive or twist the stud to the bottom of the hole following the manufacturer's requirements.
- Do not move the stud before the resin has finished curing.

As chemical anchors do not place any expansion stress on the surrounding masonry, they can be put near the external edges of the masonry materials if required. Chemical anchors have excellent pull-out strength and are used where cyclic uplift loads (motor starting) and vibrating loads are developed by various types of machinery.

REVIEW QUESTIONS

1. Explain the difference between a fastener and a fixing.
2. Name four factors to consider when selecting an attachment device.
3. Which fastener is the most common, practical metal device for joining two pieces of timber together?
4. Name two head styles for masonry screws.
5. What is a common use for plastic expansion plugs?
6. Outline the procedure for installing a light-duty masonry expansion anchor.
7. An M12 sleeve anchor has a sleeve length of 80 mm. Determine the depth of the fixing hole for this anchor.
8. What is a chemical anchor?

4.2 Stud wall attachment devices

A stud wall frame consists of a horizontal bottom plate fixed to the floor or floor bearer and a horizontal top plate to tie the studs together. Studs are the vertical members; noggings are the horizontal pieces between the studs. Bracing (metal angle or timber) is required to provide strength across the length of the wall frame to prevent wall racking and a sideways distortion of the structure in windy conditions. A timber stud wall frame is illustrated in **Figure 4.9**. It is common to line the internal side of the frame with a plasterboard wallboard covering.

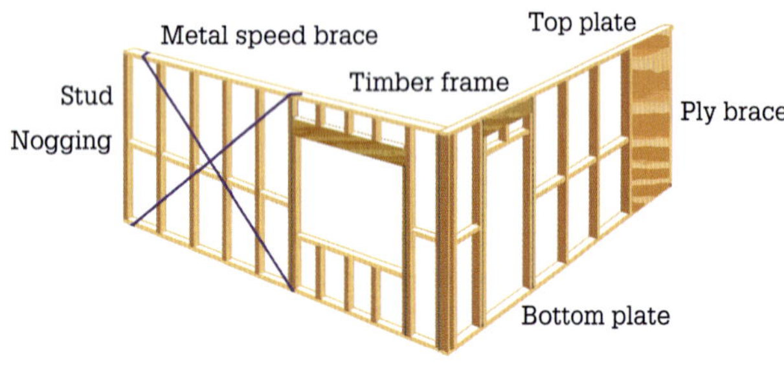

FIGURE 4.9 Timber stud wall

Fixings for plasterboard

The most robust fixings are those that provide a tight hold over a large area of the plasterboard. Fixing strength of plasterboard (12.5 mm thick) anchors is expressed as kilogram-force (kgf). The kilogram-force is an SI unit of force. The kilogram-force is equal to the mass of one kilogram multiplied by acceleration due to gravity. Therefore, one kilogram-force is equal to 1 kg × 9.80665 metre per second^2 = 9.80665 kilogram × metre per second^2 = 9.80665 newton (10 N) (1 kN ≈ 100 kg supported load).

Threaded plasterboard anchor

Threaded plaster board anchors as shown in **Figure 4.10** are available in both nylon and metal. They have a light-duty holding strength (below 10 kgf) and are used as a hanging fixture for clocks and to fix some wall-mounted luminaires.

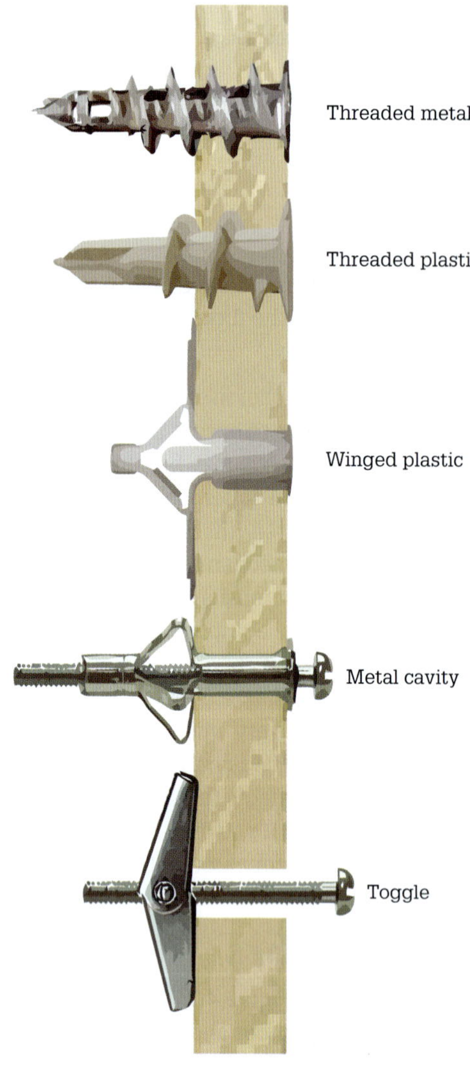

FIGURE 4.10 Various plasterboard fixings

Winged plastic anchors

Winged wallboard anchors, shown in **Figure 4.10**, have a light-duty holding strength (below 10 kgf) and are used for similar applications to threaded wallboard anchors.

Metal cavity anchors

These plasterboard anchors as shown in **Figure 4.10** have anti-rotation teeth to prevent the fixing from spinning while tightening. They also have anchor wings to increase their grip area on the plasterboard. These anchors may be rated to 20 kgf for fixing lighting track, luminaires and heaters.

Toggle anchors

Toggle anchors (see **Figure 4.10**) are sized by either of two methods: by the diameter of the metal thread or by the length of the metal thread. Long metal threads enable wider equipment to be fixed to the wallboard.

Toggle anchors are the strongest of all the wallboard anchors, and they can also be used to attach some fittings to ceilings. Toggle anchors may be rated to 20 kgf for fixing lighting track, luminaires and space heaters. When used in ceilings, recommended loading ability of the toggles has to be reduced by a factor of 0.8. Toggle anchors are available in gravity-type toggle for fixing into walls and spring-type toggle, which are suitable for use when fixing to ceilings.

However, a more efficient and stronger fixing method for ceilings and hollow walls is to fix a nogging, as shown in **Figure 4.11**, between the ceiling joists or wall studs and then fix the fitting to the nogging through the plasterboard via screws or metal threads.

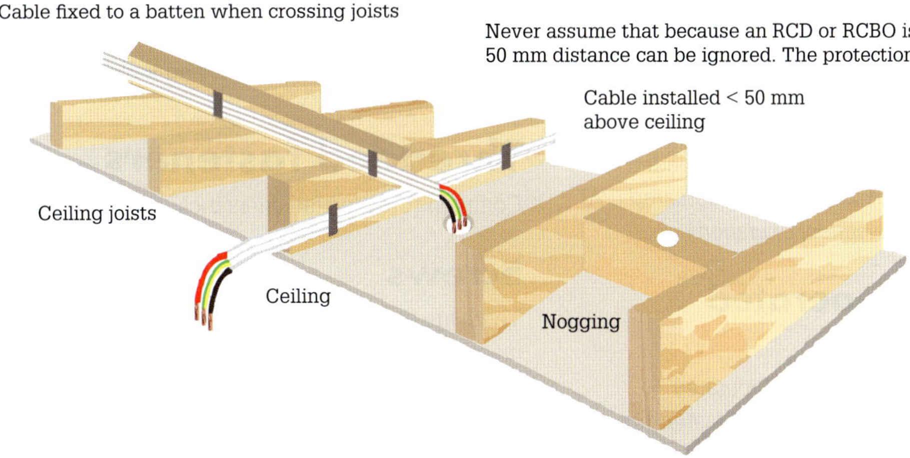

FIGURE 4.11 Nogging between ceiling joists

REVIEW QUESTIONS

1. Describe the construction of a stud frame wall.
2. What internal lining is commonly used for timber stud wall frames?
3. What constitutes a robust plasterboard fixing?
4. Name two applications for a winged wallboard anchor.
5. State the approximate holding strength for a metal wallboard cavity anchor.
6. Name two applications for a metal wallboard cavity anchor.
7. What is the strongest method for fixing to a ceiling?
8. Name the two types of toggle anchor.

4.3 Fasteners

A fastener can be broadly defined as a hardware device used to mechanically join or attach two or more objects together. The purpose of a fastener is to create non-permanent joins. This means that it is possible to remove or dismantle the objects without damaging the joining components.

Powered setting tools

When it is necessary to install a number of fasteners, it may be prudent to use a powered setting tool. Two styles of powered setting tool are the powder-actuated setting tool and the gas-powered setting tool.

Powder-actuated setting tool

Powder-actuated fastening systems use an explosive power tool to force all types of fasteners into steel, concrete, brick and rock to fasten equipment and machinery such as conduit, switchboards and electric motors. Although this fastening system is simple to use, there are safeguards and protections that must be observed. All electrical workers who may operate powder-actuated tools must be trained in their operation. Potential operators must pass an examination and receive a certified operator card to become a certified operator of powder-actuated fasteners. A powder-actuated tool and various fasteners are shown in **Figure 4.12**.

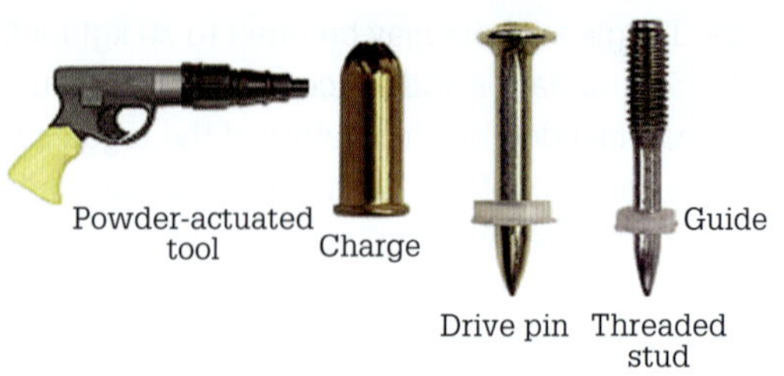

FIGURE 4.12 Powder-actuated tool and fasteners

The fasteners used in powder-actuated tools are factory-made from superior steel and heat treated to create a very hard ductile fastener. These properties are necessary to permit the fastener to penetrate masonry and steel materials without breaking. In addition, the fastener is equipped with a guide that aligns and retains the fastener in the tool as it is being driven. The more commonly used fasteners are drive pins and threaded studs. (Refer to AS/NZS 1873.1:2003 *Powder-actuated (PA) hand held fastening tools – Selection, operation.*)

Expanding-gas-operated setting tool

Gas technology uses an expanding-gas-operated electronic-controlled setting tool for driving fastening elements such as hardened pins into wood, steel, and concrete. No special training is required to use this tool. This tool as illustrated in **Figure 4.13** can fix electrical cable and, with the utilisation of an adaptor, conduit saddles.

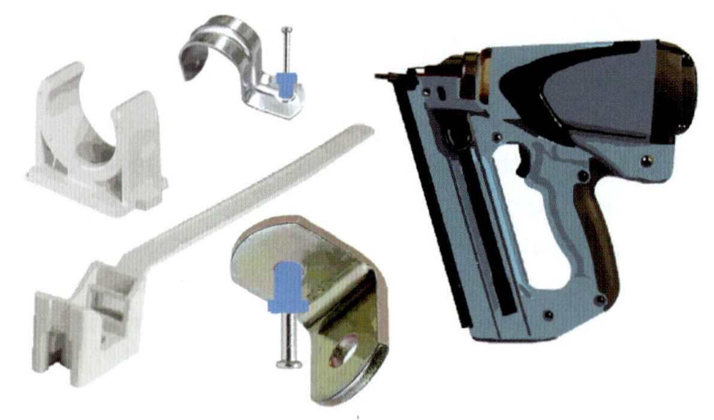

FIGURE 4.13 Gas-powered setting tool

Threaded fasteners

Common threaded fasteners include screws and bolts.

Screws

There are different types of threaded fasteners used in installation work: screws with needle or self-drilling points and metal threads. The shank is the section below the head, and the thread can conclude as a needle or a self-drilling point. Screws are available with different heads and slot configurations, as shown in **Figure 4.14**.

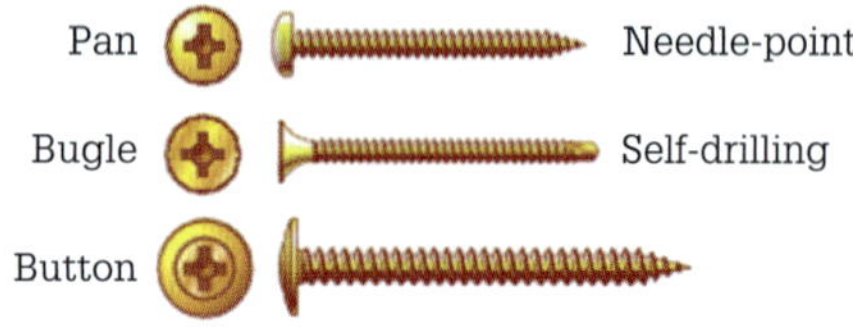

FIGURE 4.14 Yellow zinc-plated Phillips head screws and standard drivers

The needle-point screw is capable of self-drilling and fixing through light-gauge metal up to 0.6 mm without drilling a pilot hole. Self-drilling screws are designed to drill through steel up to 6.4 mm thick. This type of fastener drills, threads and fastens in one action.

The most commonly used screws are described in **Table 4.1**.

TABLE 4.1 Commonly used screws

Screw	Description
Pan head	These provide a flat bearing surface for the head and protrude above the surface. They are often used for attaching metal fittings and electrical accessories.
Bugle head	This is the most standard threaded fastener. Designed mainly for the installation of plasterboard, the bugle head is used for general electrical accessory installation (mounting blocks) and has the widest range of sizes. The tapered head allows it to be driven flush in soft materials without having to be countersunk.
Button head	These are similar to the pan head but have an extensive flat bearing surface under the curved low-profile head. The flat underside is ideal for metal-to-metal surfaces that are also flat.
Metal threads	Metal threads or machine screws (see **Figure 4.15**) have a thread along the length of the shank. They are often used with a nut, or they are screwed into a threaded hole.

FIGURE 4.15 Metal threads

Bolts

A bolt is the word used for a threaded fastener, designed to be used in conjunction with a nut. Bolts provide stronger joints than either screws or metal threads.

The reason is that as the joint is secured by tightening the nut on the bolt, the load in most cases becomes entirely a shear force. A standard bolt has a hex head (as shown in **Figure 4.16**) and a smooth shoulder area beyond the threading.

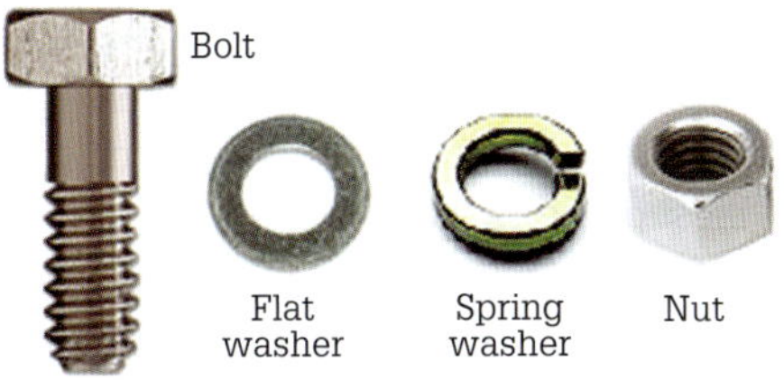

FIGURE 4.16 Bolt and fittings

Bolts are available in many different types and sizes, depending on which fastening problem you are trying to solve.

The thread pitch is inscribed between the diameter and length (both measured in millimetres) and is separated on each side by an ×.

Normally the diameter of the bolt is preceded by a capital M to designate metric.

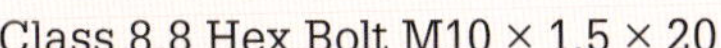

EXAMPLE 4.1

Class 8.8 Hex Bolt M10 × 1.5 × 20

This is a hex head bolt with a diameter of 10 mm, a 1.5 thread pitch, measuring 20 mm long, and made of Class (grade) 8.8 steel.

There are some variations of this abbreviation and sometimes the material and head style, and type appear after the size.

EXAMPLE 4.2

Hex Bolt M10 × 1.5 × 20 CL 8.8

Types of bolts

- Anchor bolts – extended connections with handle
- Carriage bolts – domed top with square under head
- Flange bolts – ridge around the bolt head
- Hex bolts – hex head of various materials
- Hex tap bolts – trimmed hex head with no flange
- Lag bolts – screw-like pointed end
- Machine bolts – fine or coarse threads
- Shoulder bolts – allow slide or pivot
- Square neck bolts – square or domed top with square under head

Anchor bolts

Anchor bolts, as illustrated in **Figure 4.17**, are designed for anchoring structural supports to concrete foundations. Such structural supports include building columns, column supports for highway signs, street lighting and traffic signals, steel bearing plates and similar applications.

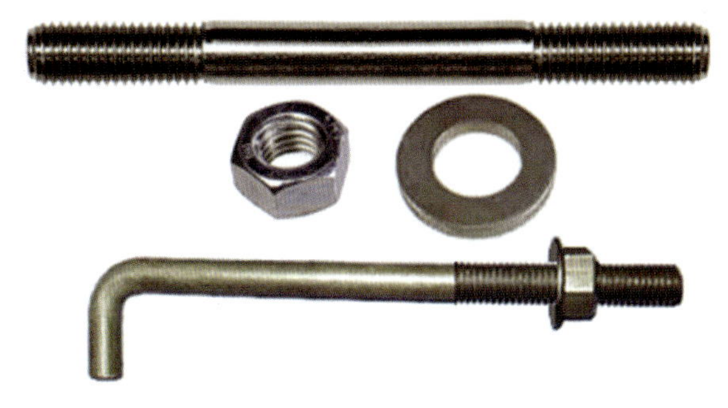

FIGURE 4.17 Anchor bolts

They can be straight or bent, headed and headless, carbon, carbon-boron, alloy or high-strength low-alloy steel anchor bolts (also known as anchor rods).

Carriage bolt

Carriage or coach bolts, shown in **Figure 4.18**, are commonly used in wood. They have a domed top and a square section under the head. The square portion under the head of the carriage bolt pulls into the wood as the nut is tightened for a very secure fit.

FIGURE 4.18 Various types of bolts

These are available in a variety of diameters and are made from a variety of types and grades of steel.

Flange bolts

Flange bolts (see **Figure 4.18**) are named for the edge or skirt around the bolt head. They are designed to provide the holding power as if a washer was installed. The skirt acts as a distributor of the entire clamping load.

Hex bolts

Hex bolts (see **Figure 4.18**) are a very common bolt used in construction or repair and are available in a large variety of sizes and diameters. The best way to choose a hex bolt is to decide which material the bolt is made from to best suit your needs.

Hex bolts are also known as hex-cap screws or machine bolts. They are available with a standard amount of threading or fully threaded depending on the length of the bolt.

Lag bolts

Lag bolts, as shown in **Figure 4.18**, are often referred to as coach screws. Lag bolts are not like other bolts as they look like a screw rather than a bolt; they have a pointed end instead of the more typical blunted end found on other bolts. They are used to attach fittings to timber structures.

Lag bolts need pilot holes to be drilled to ensure that the bolt head does not snap when turning torque is applied. As a rule of thumb, the diameter of the pilot hole should be ½ the diameter of the lag bolt when using softwood and ¾ the diameter for use in hardwood.

EXAMPLE 4.3

A 10 mm lag bolt requires a pilot hole size of 5 mm for softwood and 7.5 mm for hardwood.

EXERCISE 4.2

Determine the pilot hole size for a 12 mm lag bolt for use in softwood and hardwood.

Shoulder bolt

Shoulder bolts, as shown in **Figure 4.18**, are often used as pivot points. These bolts permit that to which they are connected to move, slide and even twist. Shoulder bolts can also be used as a point where bearing surfaces join other surfaces.

The shoulder feature may also act as an alignment feature in some applications.

Machine bolts

Machine bolts, shown in **Figure 4.18**, are made with hexagonal or square heads. Bolts are ordered by diameter, length, and grade.

The shank diameter, the diameter of the unthreaded portion of the bolt below the head, is a little larger than the nominal or standard size of the bolt. Consequently, a hole that is to accept a machine bolt must be drilled marginally bigger than the shank diameter.

Machine bolts are manufactured with fine and coarse threads that extend in length from twice the diameter of the bolt plus 5 mm to twice the diameter plus 10 mm (depending on the length of the bolt). Furthermore, they come in many different types including stainless steel, brass, silicon bronze and zinc plated.

There are several grades of machine bolts, as shown in **Figure 4.18**. The lowest grade is 4.6 and is known as commercial grade (low-carbon steel) with other grades being 4.8, 5.6, 5.8 and 6.8. Next is structural grade 8.8 (medium-carbon steel, quenched and tempered) followed by 10.9 (medium-carbon steel, quenched and tempered) and the highest grade 12.9 (alloy steel, quenched and tempered), which is known as a high-tensile bolt. The first number represents the ultimate tensile strength (UTS) of the bolt in 100s of MPa: 400 MPa, 500 MPa, 600 MPa, 800 MPa, 1000 MPa and 1200 MPa. The second number represents the point at which the bolt permanently stretches. The commercial grade bolt stretches to 60% of its UTS, the structural category at 80% and the high-tensile bolt at 90%.

The strength of the assemblage depends on the diameter of the bolt or the thread engagement of the screw. Thread engagement is the distance a screw extends into the threaded hole. The minimum thread engagement should be a distance equal to the diameter used; preferably one to one-and-a-half times the screw diameter.

Washers

Washers, shown in **Figure 4.19**, are used to allocate the fastening pressure over a larger area, and prevent defacement (damaging the surface). They also provide a broader bearing surface for bolt heads and nuts. Plain washers are used in the assembly of nuts and bolts to provide a smooth surface for the nut or bolt to turn against. Lock washers (see **Figure 4.19**) reduce the chance of a bolt or nut from loosening under vibration.

FIGURE 4.19 Plain and lock washers

Washers prevent stress-related damage to the bolt and minimise loosening of the bolt. Specifications for washers are given by inside and outside diameter, thickness and type.

Keys

The key is a small piece of shaped metal embedded to some extent in the shaft and to a degree in the hub to prevent rotation of a gear or pulley on the shaft. The key is placed in a groove cut on the shaft, called the key seat, and the other part fits into a groove cut in a hub, called the key way. Therefore, after assembly, locked together by a key, the shaft and hub rotate together. There are several different types of keys; see **Table 4.2**.

TABLE 4.2 Types of key

Type	Description
Square key	The width of a square key, shown in **Figure 4.20 (a)**, is usually one-quarter of the shaft diameter. One-half of the key is fitted into the shaft and one-half is mounted into the hub.
Woodruff key	A woodruff key as illustrated in **Figure 4.20 (b)** is semicircular in shape and fits a key seat of the same shape. The rectangular section of the key fits a key way in the mating part.
Pratt and Whitney key	The ends of the Pratt and Whitney key, shown in **Figure 4.20 (c)**, are rounded, and this key fits into a slot in the shaft that has the same shape.
Gib head key	The Gib head key, shown in **Figure 4.20 (d)**, is identical to the square key but has a head to provide easy removal.

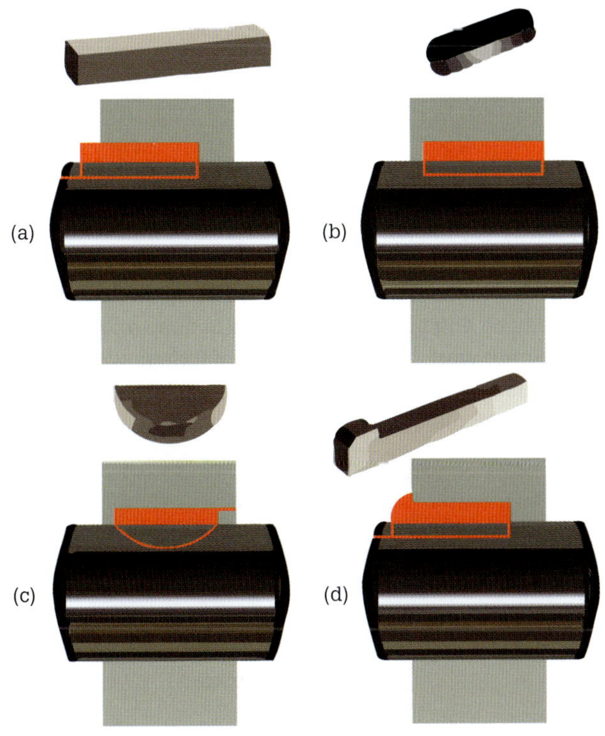

FIGURE 4.20 Various types of keys

Non-threaded fastening devices

Non-threaded fasteners are a large group of fastening devices. Some of these are discussed below.

Dowel pins

Dowel pins (see **Figure 4.21**) are made of treated alloy steel and are used in assemblies where parts must be accurately positioned and held in complete relation to one another. They guarantee faultless alignment and simplify quicker disassembly and assembly of parts.

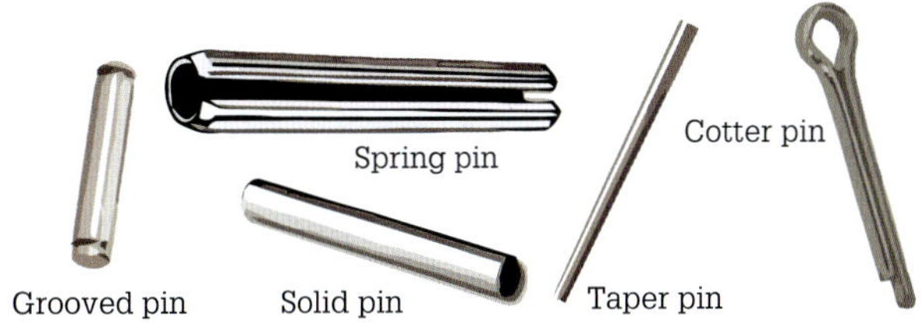

FIGURE 4.21 Dowel and cotter pins

Cotter pins

Cotter pins, as illustrated in **Figure 4.21**, are fitted into drilled holes in shafts to prevent parts from slipping or spinning off. The legs of the pin are bent out to retain the pin when in use.

Retaining rings

Retaining rings or circlips, shown in **Figure 4.22**, are internal or external devices used to keep parts from slipping or sliding apart. While most retaining rings need a groove to seal them in position, some types are self-locking and do not require the use of a groove. Retaining rings are installed or removed by circlip pliers.

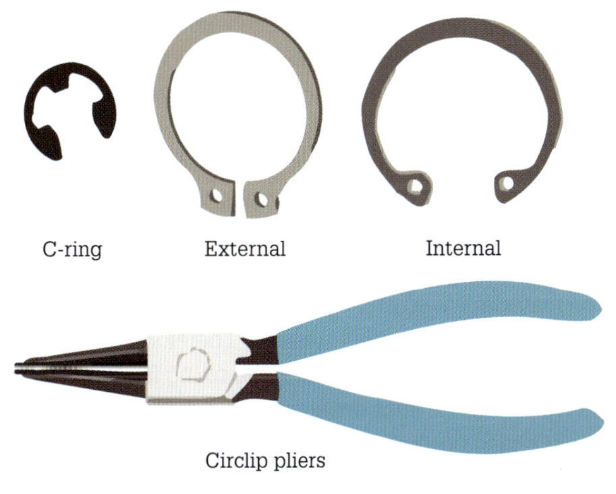

FIGURE 4.22 Retaining rings and circlip pliers

REVIEW QUESTIONS

1 What is a 'fastener'?
2 Name two types of powered fastener setting tool.
3 Describe the physical properties of a metal thread screw.
4 Name three fittings commonly used with bolts.
5 What is the function of a lock washer?
6 What is the function of a key?
7 What is the function of a retaining ring?
8 What tool is needed to install or remove a retaining ring?

4.4 Fixings

An important rating to consider when selecting a fixing is its load rating.

Load rating of fixings

The load ratings of fixings are expressed in kilograms (kg) or newtons (N) as either static load limit or ultimate static load limit.

1 kilogram (mass) = 9.8 newtons (force)

Static loads are any loads that remain in a stationary position for extended periods of time. By contrast, static load limit is a rated stationary load limit for a fixing and integrates a substantial safety margin which is a factor of 4:1 (sometimes 5:1). For example, if a fixing must carry a load of 5 kg the fixing used is designed to carry 20 kg (or 25 kg).

Ultimate static load limit is the highest rated stationary load limit (the number of kilo newtons per square millimetre) allowed which, if exceeded, may cause failure of the fixing.

As a rule of thumb, in order to ascertain how many fixings you'll need to a load, divide the total weight of the load by the static load rating of the fixing.

There are two other expressions of loads – shear and tensile. Shear loads are created when the weight of the fixed equipment exerts force parallel to the surface of the base material, while tensile loads are created when the fixed equipment exerts force perpendicular to the base material, for example, a ceiling surface to which a suspended load is attached. Always determine the weight of suspended loads before attaching to a ceiling fixing.

With the equipment that is suspended in exterior locations the effects of shock loading must be considered. Shock loading is the magnification of the stationary weight of the material. Wind and seismic forces can impose several times the stationary weight. Therefore, both the fixing and the structure to which the fixing is attached must be capable of supporting several times the weight of the suspended equipment. Reference should be made to a structural engineer for interpretation before proceeding with any suspension of the material.

Installing electrical equipment properly begins with understanding how the walls and ceilings of structures are built. There are different methods and fixings that are designed for the various types of walls and ceilings. Choosing what fixing to use can be confusing, but it comes down to two main factors:

1 the weight of the equipment that you want them to support
2 the type of wall, floor and ceiling material.

Table 4.3 summarises the tensile strength for a number of common bolts.

TABLE 4.3 Examples of bolt tensile strength

Bolt class	Tensile strength (MPa [N/mm^2])
4.6	400
5.8	500
A2/A4 – 50	500
A2/A4 – 70	700
A2/A4 – 80	800
8.8 ≤ M16	800
AS 1252 ≤ M16	800
8.8 ≥ M16	810
AS 1252 ≥ M16	815
B7/L7	830
10.9	1000
12.9	1200

EXAMPLE 4.4

Masonry expansion anchor M6 × 8 × 90 mm with flange nut.

If the anchors had to support a 50 kg switchboard measuring 450 mm wide by 900 mm deep and fixed to a concrete wall, then the number of fixings would be:

Tensile stress area (TSA) of the anchor is 20.737 mm^2 (calculated from area of fixing, which is

$$TSA = \frac{\pi d^2}{4} = \frac{\pi \times 8^2}{4}$$

A Class 4.6 bolt has a tensile strength of 400 MPa, so to calculate the ultimate tensile strength (UTS):

$$UTS \ (in\ newtons) = Tesnile\ Strength\ (in\ MPa) \times area\ (in\ mm^2)$$

$$= 400 \times 20.737$$

$$= 8.295\ \text{kN}$$

»

To convert force in kN to kg, multiply the kN by 101.972 (which represents the reciprocal of acceleration due to gravity multiplied by 1000)

$$\text{kilograms} = \text{kN} \times 101.972$$
$$= 846 \text{ kg}$$

Applying a safety margin of 4:1 then an M6 × 1 Class 4.6 bolt could carry a load of 211.5 kilograms shear weight vertical force flat to the wall. The horizontal force taken at 0.6 times the vertical force would be 127 kg.

The actual tension force on the bolt is the width divided by the height, multiplied by the mass of the object. Therefore, the tension created by the mass:

$$0.45 \div 0.9 = 0.5$$
$$0.5 \times 50 \text{ kg} = 25 \text{ kg}$$

Now 127 divided by 25 = 5.08. To support the load securely six M6 masonry anchors would be suitable. Additional anchors may be warranted, not because of weight, but because of the size of the fixture.

However, industry practice uses four; therefore, each bolt would safely hold a load of 40 kgf.

The load values quoted in fixings catalogues are for loads applied only to the orientation (usually vertical) described in the catalogue or instruction sheet. A fixing with an ultimate static load rating of 21 kg safely holds 7 kg at its static load rating.

EXERCISE 4.3

Class 5.8 Masonry expansion anchor M8 × 10 × 90 mm with flange nut having a tensile strength of 500 MPa is used to support a motor starter measuring 500 mm wide by 1000 mm tall and weighing 100 kg to a masonry wall. How many anchors are required to adequately support the starter?

Note that the loads apply only to the fixing; the base material to which the fixing is attached should be evaluated separately. Furthermore, the properties of the base material play a pivotal role when selecting a suitable fastener/anchor and determining the load it can hold.

Factors affecting the selection of fixings

Environment

A fixing's resistance to weathering, ageing and rotting is a major factor when selecting for specific locations. The fixing may have to be installed in aggressive environments such as chemical industries or in marine areas where wind-blown salt spray is of concern. Other sites may have a high amount of dust or may drip water or oil and so on, which may affect the fixing and thereby reduce its holding force. Ambient temperature can affect the fixing by causing expansion, which effects elongation of fixing holes.

Impact damage

Fixings could be placed in a location where they may suffer impact damage, for example, damage caused by ladders propped up against the fixed equipment during maintenance or repair conditions.

Fixings may also have to contain impact loads. Impact loads are loads that change suddenly. Impact loading may be generated onto a fixing either by the collision of a moving body or by the sudden application of force or motion on the equipment or building fabric the fixing is attached to. An example of an impact load is the starting impact load of electric motors. The starting impact load varies according to the type of motor-starting method used.

Dynamic or vibrating loads

Fixings may also have to support dynamic or vibrating loads. Moreover, dynamic or vibrating loads are loads that are always changing. For example, if you mount the bracket on a concrete wall in order to install an air-conditioner, the load on the fixings is a dynamic load. The strength rating of the fixings reduces because of the constant vibration of the load.

SWITCH ON

Due to the inexact quality of concrete, a recommendation is that the fixing chosen requires a rating for about four times the weight it will carry if it is going to tolerate a static load and eight times the weight if it is going to support a dynamic or impact load.

Fixings intended for fire-resistive construction must be evaluated for load resistance during fire exposure.

Base material

When selecting suitable fixings for equipment, the type of the base material must be considered; see **Table 4.4**.

TABLE 4.4 Examples of fixings for different base materials

Base material	Fixing
Brickwork	masonry nails, masonry screw anchors, masonry expansion anchors, plastic expansion plugs, sleeve anchors, wedge anchors and powder-actuated fasteners
Concrete	masonry nails, masonry screw anchors, masonry expansion anchors, plastic expansion plugs, sleeve anchors, wedge anchors, chemical anchors and power-actuated fasteners
Plasterboard	threaded wallboard anchors, winged plastic anchors, toggle anchors
Steel	powder-actuated fasteners, threaded fasteners and bolts
Timber	coach bolts, threaded fasteners, bolts, and nails

Applications for fixings

Masonry screw anchors are used for attaching various types of components such as conduit saddles, ducts and cable trays, and other shear-type loads to concrete and masonry. Shear loads act at right angles to the axis of a fixing and directly on the face of the structural material. Furthermore, shear performance is governed mainly by the shear strength of the fastener material. Note that hardened fasteners experience performance problems in corrosive environments. Accordingly, use masonry screw anchors in dry, non-corrosive environments only.

Plastic expansion plugs are a very popular form of fixing that are available in a range of colour-coded sizes to suit different gauges of screws. This flush surface anchor can be used for shear-type loads such as conduit saddles, mounting blocks, and luminaires.

Plastic expansion plugs and their screws are a lightweight fixing that should not be overloaded, as they are only capable of holding a maximum of 12 kg in ideal conditions. Plastic expansion plugs are inadequate where heat is involved or where the hollow brickwork is the base material.

Masonry expansion anchors, sleeve anchors, and wedge anchors are dependent on the size of the stud for their fixing rating. For example, an M6 expansion anchor in 30 MPa concrete (concrete strength is measured in megapascals (MPa)) has a shear working load of 2 kN, while an M16 expansion anchor in the same concrete will have a shear working load of 10.5 kN. These anchors are ideal for installing heavy equipment and machinery, as well as light fittings in concrete and most types of masonry.

Chemical anchors are particularly suited for use close to the edge of concrete or masonry building fabric, or where there may be a possibility of cracking if an expansion anchor is used. These anchors are ideal for installing heavy equipment and machinery in concrete and most types of masonry. They have superb tensile strength properties, which makes them ideal for suspending equipment from concrete ceilings. For example, an M10 chemical anchor in 30 MPa concrete has a shear working load of 7.1 kN.

Threaded plasterboard anchors, winged plastic anchors, and toggle anchors are only suitable for light shear loads. These anchors are useful for fastening electrical fittings such as mounting blocks, socket outlets and light switches to plasterboard.

Fixings must be of a type suitable to the location in which they are used. Where fixings are to be used externally or exposed to the weather, stainless steel or brass is preferred, and plain steel not accepted. Where fixings are used internally, cadmium plated fixings are acceptable. All fixings, fastenings, and supports must be of adequate strength and size and arranged to ensure the installation against mechanical failure under standard conditions of use and wear and tear.

Suggested fixings for an installation

Fixings must be secure in the parent material and appropriately sized to suit the mass, dimensions and locality of the equipment or accessories being fixed. For outdoor exposures as well as in damp situations including coastal regions (due to salt spray), preferred fixings are hot-dipped galvanised or if available, marine grade 316 stainless steel. The following suggested 'fixings' minimum requirements should be considered.

Enclosures

Switch panel enclosures, load centres, meter panel enclosures, general purpose enclosures, distribution boards and boxes (for example, door chimes) must be secured with a minimum of four fixings of the required length. Possible fixings, to suit the parent material are:

- button-head cadmium-plated steel needle point 9 g × 30 mm
- M5, round head slotted metal thread of the required length for use with a nut and washer or in a tapped hole
- M8 × 41 mm masonry anchor.

Switchboards

Floor-mounted proprietary or custom switchboards should be mounted on hot-dipped galvanised plinths to AS 1650 to a minimum thickness of 85 mm. The switchboards should be fixed to a concrete floor with M12 hot-dipped galvanised 8.8 or 10.9 structural bolts with washer at the front and rear. The bolts should be placed at both ends and at intervals not exceeding 1500 mm along the length of the switchboard.

Wall-mounted switchboards should be installed using the appropriate number of fixings (minimum of 4).

Suggested fixings are:

- To timber: use M12 to M20 hexagon drive head A2 grade stainless steel or spun galvanised coach screws of suitable length including a flat washer.
- To masonry and concrete: use M12 to M20 stainless steel sleeve anchor marine-grade 316 of proper length.

Surface-mounted accessories

Accessories such as surface switches, mounting flanges, mounting blocks, pedestal boxes, backing plates and mounting rings, sockets, batten holders and ceiling roses and extension rings should be fixed to the parent material with a minimum of two fixings for lighter equipment to four fixings for heavier equipment. Suggested fixings are:

- Steel cadmium-plated Bugle head screws, 7 g × 25 mm or 30 mm
- Pan-head screws 7 g × 25 mm or 30 mm
- Wafer-head screws, 9 g × 25 mm or 30 mm
- Washer-head screws, 12 g × 25 mm or 30 mm.

For pedestal boxes, where fixing to concrete, use expansion anchors M5 (minimum) of suitable length.

For outdoor exposures as well as in damp and industrial atmospheres and coastal areas, fixings should be marine-grade 316 stainless steel.

Surface-mount luminaires and fans

Luminaires and fans must be securely fixed to structural members of the building frame, or securely fixed by hangers or brackets.

The minimum size of fixing for luminaires, hangers or brackets for various surfaces should be as follows:

- Fixing to timber: steel cadmium plated
- Wafer-head screws, 9 g × 30 mm
- Washer-head screws, 12 g × 30 mm
- To masonry and concrete: use M6.5 (min) sleeve or expansion anchors of suitable length.

For fixing to hollow, concrete blocks use M5 zinc-plated spring toggles with washers under head of the screw.

The minimum size of fixing (always consider mass in kilograms of the object to be fixed) for fans, hangers or brackets for various surfaces should be as follows:

- To timber: M6 round-head metal thread bolted with a cadmium washer and nut.
- Metal frames: M6 round-head metal thread tapped or bolted with a cadmium washer and nut.
- Fixing to concrete: expansion anchors M5 (min) of suitable length.

Cable trunking, tray, and skirting ducts

Cable trunking, trays and skirting ducts ≤ 100 mm wide should be attached with two fixings suitable for the parent material at intervals of not more than 1000 mm.

Conduits

PVC and steel conduits 16 mm to 32 mm should be fixed as follows:

- To masonry and concrete walls an M6.5 × 25 mm masonry anchor should be used. The drill-hole depth must be deeper than the anchorage depth to provide space if the screw emerges from the anchor. For conduits used in protected locations, nylon nail-in anchors 6.5 × 25 mm or metal nail-in pin anchors can be used.
- To timber: where directly fixed, use steel button- or mushroom-head cadmium-plated needle point 7 g × 30 mm screws. As the timber can split when drilling the pilot hole, the size of the pilot hole must not exceed the minor diameter of the screw thread. An alternative to drilling pilot holes is to use a type 17-point screw.
- To sheet metal: use a cadmium-plated, self-tapping, slotted pan-head screw 4 g × 12.7 mm.
- To steel plate: M5 round-head slotted metal thread of the required length for use with a nut and washer or in a tapped hole.

Conduits 40 mm and greater should be fixed as follows:

- To masonry and concrete walls an M8 × 41 mm masonry anchor should be used.
- To timber: where directly fixed, use steel button- or mushroom-head cadmium-plated needle point 7 g × 35 mm.
- To sheet metal: use a self-tapping, slotted pan-head screw, 12 g × 12.7 mm.
- To steel plate: M6 round-head slotted metal thread of the required length for use with a nut and washer or in a tapped hole.

Cable ties

Cable ties can be tightened by hand or a special tool and should be tightened adequately to hold cable together and to fix cables to supports. In addition, care is necessary to avoid tight twisting of the cable, tearing of the outer jacket or cutting or wearing through due to abrasion of the cable.

It is recommended that only Velcro-type hook and loop cable ties as illustrated in **Figure 4.23**, should be used. However, nylon/zip-style cable ties can also be used. All cable ties should be of sufficient width and suitable length to allow adequate security of the tie. Cable ties used externally must be UV resistant.

Sources: iStockphotos/BirgitKorber; iStockphotos/Winai_Tepsuttinun

FIGURE 4.23 Velcro hook and loop cable tie; nylon zip-type cable tie

REVIEW QUESTIONS

1. How is the load rating of fixings expressed?
2. Describe the term 'shear load'.
3. A certain load has a mass of 5 kg. What rating of fixing is required to support this load, taking into account the applicable safety margin of 4:1?
4. What are the two considerations when selecting a fixing?
5. Why is ambient temperature a consideration in the selection of a fixing?
6. What are impact loads?
7. What type of anchors are particularly suited for use on plasterboard?
8. What type of applications are suited to chemical anchors?
9. How many fixings are recommended to fix a switch panel enclosure?
10. Which cable ties are suitable for external installations?

4.5 Fixing adhesives and tapes

An adhesive is a substance capable of holding material together through surface attachment. Trowel or cartridge can be used to apply some adhesives. A cartridge gun is illustrated in **Figure 4.24**.

FIGURE 4.24 Cartridge gun

When surfaces of the material are bonded together by interfacial (a surface forming a common boundary) forces they are said to demonstrate good adhesion. Proper adhesion requires very close surface contact. Adhesion strength is the force necessary to pull the adhesive completely from the surface. Good adhesion and cohesion are needed to achieve high-performance bonds. Poor adhesion and cohesion are illustrated in **Figure 4.25**.

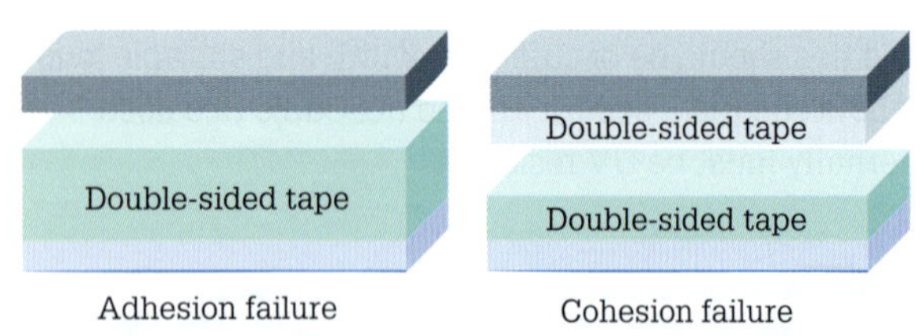

FIGURE 4.25 Adhesion and cohesion failure

Adhesive forces hold two materials together at their surfaces while cohesive forces exist between molecules of the same material.

Double-sided adhesive tape

Double-sided adhesive tape consists of a pressure-sensitive adhesive pre-applied to a special release liner. The tape is applied to the surface, and the liner is peeled off leaving a clean, dry strip of adhesive for joining lightweight materials.

Double-sided pressure-sensitive self-adhesive tapes as shown in **Figure 4.26** are variously employed for a temporary and permanent bonding of a broad range of materials. The advantage of self-adhesive tapes is their ease of handling and the rapid realisation of the adhesive bond. A tape cutter and tape dispenser are shown in **Figure 4.27**.

FIGURE 4.26 Double-sided self-adhesive tape

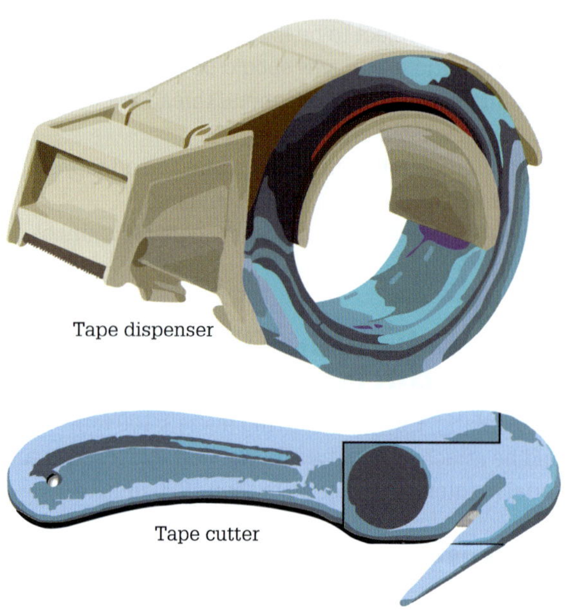

FIGURE 4.27 Tape cutter and dispenser

Double-sided tapes are used indoors. They can be used to bond metal to metal or dissimilar materials. Surfaces to which the tape is to be applied should be as clean and dry as possible with any loose material removed. Some double-sided adhesive tapes can support a continuous load of 5 kg/25 mm^2.

For an electrician, double-sided tapes can be a practical solution for mounting small devices or DIN 35 mm rail (shown in **Figure 4.28**) without the necessity of drilling holes in a switchboard cabinet.

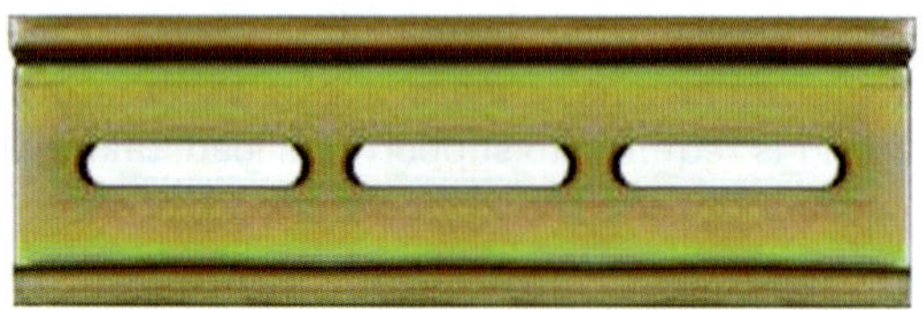

FIGURE 4.28 DIN rail

The tape can also be used for fixing PVC mini-trunking and cable tie bases (see **Figure 4.29**) when fixing cable to ceilings.

FIGURE 4.29 Cable tie base

Cable tie bases can be used for supporting 20 mm and 25 mm conduit and electrical power, data, and fire cables. Some of these adhesive fasteners have a static load limit ranging from 5 kg to 15 kg.

For high-strength bonds, paint, oxide films, oils, dust, mould and all other surface contaminants must be completely removed from the surfaces to be bonded. The amount of surface preparation required depends on the required bond strength and the environmental conditions.

Particular advantages of double-sided tapes are:

- uniform distribution of mechanical loads
- dampening of vibrations and noise
- absorption of impact
- dissimilar materials can be joined.

Potential limitations are:

- stress cracking caused by the adhesive
- disassembly.

Epoxy adhesives

Epoxy adhesives as shown in **Figure 4.30** use an epoxy resin plus a hardener.

One-part or two-part epoxy adhesives readily bond most plastics, metal, ceramics, wood and some rubbers. In general, epoxy adhesives provide excellent adhesion, chemical and heat resistance, excellent mechanical properties and superb electrical insulating properties. Furthermore, epoxy adhesives have exceptionally high strength at service temperatures with greater impact, peel and bond strength properties than are attainable with other adhesives.

A two-part epoxy adhesive must be thoroughly mixed together with a spatula using the recommended mixing ratio. This adhesive usually gels in two hours, with initial bonding being achieved in six to eight hours and maximum bond strength after three days. When using epoxy adhesive avoid touching the hardener and the resin and breathing in fumes.

Epoxy adhesives can be used to encapsulate electric motor windings and transformer windings. Epoxy adhesive is also used to fix anchor bolts into concrete. Epoxy adhesive allows the anchor bolts to withstand higher impact loads.

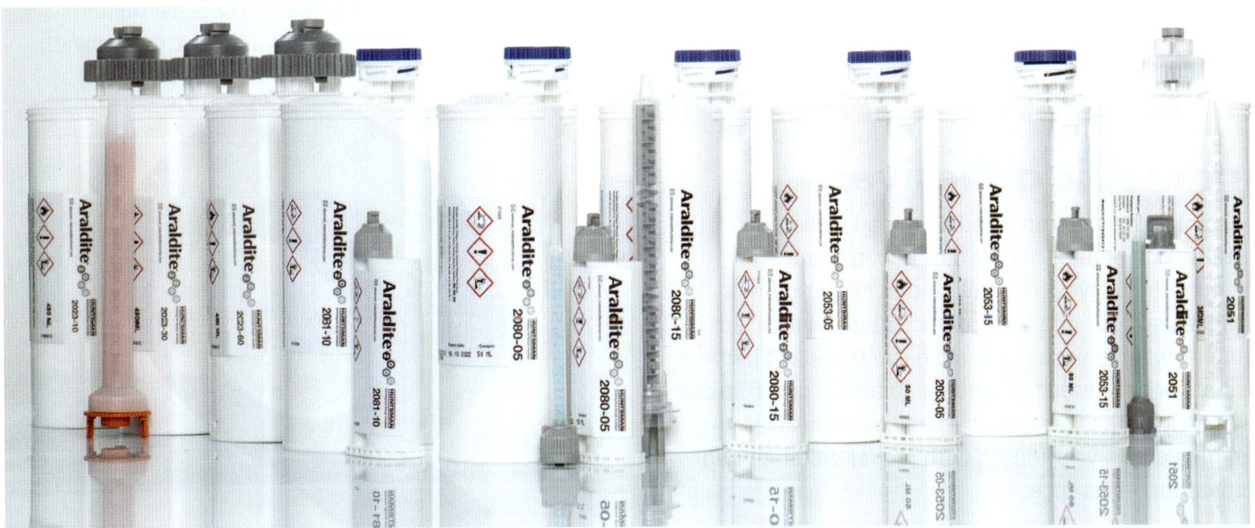

Source: Reproduced courtesy of Huntsman Advanced Materials (Australia) Pty Ltd.

FIGURE 4.30 Epoxy resin and hardener

UV-curable adhesives

Some adhesives are UV- and light-curing, and are designed to adhere to a broad range of materials such as metals, glass and most plastics. These adhesives are formulated for bonding, coating and encapsulating. Applications include bonding and potting fibre-optic cables, connectors and terminations, cementing and coating optical parts and coating or encapsulating electro-optic and laser components. In addition, UV- and light-curing adhesives are ideal for applications that require resistance to thermal cycling under vibration, shock stresses or environmental exposure.

Hot-melt adhesives

Hot-melt adhesives (see **Figure 4.31**) also adhere to a broad range of materials when melted and cooled. These adhesives are applied in molten form to one surface. Once the surfaces being bonded are pushed together the heat in the adhesive quickly dissipates into the bonded parts and the adhesive solidifies, forming an almost instant bond. Furthermore, hot-melts are environmentally safe and non-volatile. Conduit saddles and other electrical fittings can be fixed with hot-melt adhesive.

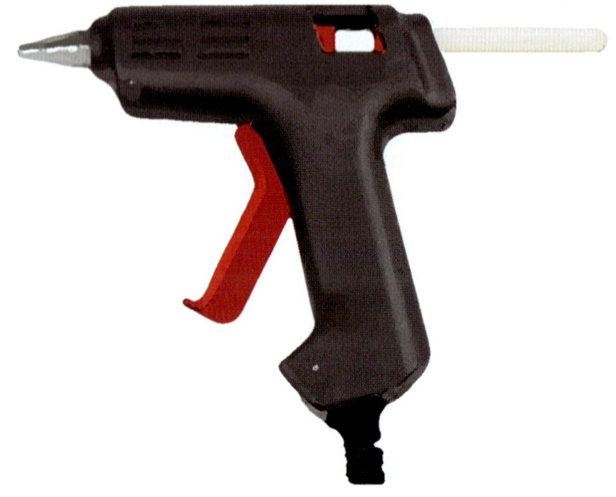

FIGURE 4.31 Hot-melt gun and glue stick

Anaerobic adhesives

Anaerobic adhesives are used as thread lockers, retaining compounds and as gasket compounds. Anaerobic adhesives remain liquid in air but cure in the absence of air and the presence of metal ions. For example, when an anaerobic adhesive is applied between a nut and a bolt thread, it hardens to form a cross-linked plastic with adhesion ability. Anaerobic adhesives offer high adhesive strength with proper vibration and temperature resistance.

SWITCH ON

Safety with adhesives

Some adhesives have components that are flammable and have toxic hazards. Some adhesives may affect the eyes and skin if direct contact occurs and cause irritation or allergic reactions. The vapours from the adhesive may cause respiratory irritation. Care must be taken when solvents are used, and evaporation occurs to complete the bonding. Ventilation must be provided in order to minimise contamination of the workplace atmosphere. Always keep adhesives away from heat, sparks and open flame because some solvents are flammable. The primary hazard with hot-melt adhesives is burns. With all adhesives always wear eye and face protection. To prevent repeated, or prolonged contact with the skin always wear appropriate gloves. Before using any adhesive, refer to the safety data sheet of that particular adhesive.

The advantages of using adhesives are:

- Dissimilar materials can be joined.
- The bond is continuous.
- On loading, there is a more uniform stress distribution.
- Electrolytic corrosion is prevented.

The limitations of using adhesives are:

- An increase in service temperature decreases the bond strength.
- Bonded surfaces are usually difficult to dismantle.
- There is a need to prepare the surface.
- Health and safety is a responsibility.

Self-amalgamating tape

Self-amalgamating or self-fusing tape as shown in **Figure 4.32** is suitable for the permanent waterproofing of electrical cables and cable joints. These tapes are used to join and repair a varied range of power and distribution cables up to 46 kV. The tape is also UV resistant and provides a non-deteriorating dust- and dirt-proof sheath.

FIGURE 4.32 Self-amalgamating tape

REVIEW QUESTIONS

1. With regards to fixing adhesives, what is needed for proper adhesion?
2. What is the difference between adhesive forces and cohesive forces?
3. Provide two advantages of double-sided tapes.
4. List some common uses of cable tie bases.
5. List some common uses of epoxy adhesives.
6. What electrical accessory may be fixed with hot-melt adhesive?
7. Name the adhesive that can be used as thread lockers, retaining compounds and as gasket compounds?
8. List four advantages of using adhesives.
9. State two limitations for adhesives.
10. What is a use of self-amalgamating tape?

CHAPTER REVIEW

4.1 Fasteners and fixings

- In general, a fastener permanently connects together two items or parts, while a fixing attaches a movable thing or part to a permanent one.
- Plastic expansion plugs are used to fix accessories such as mounting blocks to masonry walls.
- Expanding sleeve anchors are designed for light- to heavy-duty pull-out and shear capacities in all types of masonry.

4.2 Stud wall attachment devices

- A stud wall frame consists of a horizontal bottom plate nailed to the floor and a horizontal top plate to tie the studs together.
- Toggle anchors are the strongest of all the wallboard anchors, and they can also be used to fix some fittings to ceilings.

4.3 Fasteners

- Powder-actuated fastening systems use an explosive power tool to force all types of fasteners into steel, concrete, brick and rock to fasten equipment and machinery such as conduit, switchboards and electric motors.
- There are different types of threaded fasteners used in installation work: screws with needle or self-drilling points and metal threads.
- Metal threads have a thread along the length of the shank.
- A bolt is the term used for a threaded fastener, with a head, designed to be used in conjunction with a nut.
- Washers are used to apply the fastening pressure over a larger area, and prevent defacement.
- A lock washer avoids a bolt or nut loosening under vibration.
- The key is a small piece of metal embedded to some extent in the shaft and to a degree in the hub to prevent rotation of a gear or pulley on the shaft.
- Dowel pins are used in assemblies where parts must be accurately positioned and held in complete relation to one another.
- Cotter pins are fitted into drilled holes in shafts to prevent parts from slipping or turning off.

4.4 Fixings

- The load ratings of fixings are expressed in kilograms (kg) or newtons (N) as either static load limit or ultimate static load limit.
- A fixing's resistance to weathering, ageing and rotting is a major factor when selecting for specific locations.
- Accessories such as surface switches, mounting flanges, mounting blocks, pedestal boxes, backing plates and mounting rings, sockets, batten holders and ceiling roses, and extension rings should be fixed to the parent material with a minimum of two fixings.
- Cable bundling should be tightened by hand or a special tool and should be tightened just sufficiently to hold cables together and to fix cables to supports.

4.5 Fixing adhesives and tapes

- An adhesive is a substance capable of holding material together by surface attachment.
- Double-sided pressure-sensitive self-adhesive tapes are used for temporary and permanent bonding of a broad range of materials.
- For high-strength bonds, paint, oxide films, oils, dust, mould and all other surface contaminants must be completely removed from surfaces to be bonded.
- Self-amalgamating or self-fusing tape is suitable for permanent waterproofing of electrical cables and cable joints.

TRIAL EXAM

For Chapter 4 knowledge assessment, please complete the following test exam.

1 The fixing strength of plasterboard anchors is expressed as:
   a kilograms
   b newtons per second
   c kilogram-force
   d millimetres per second

2 Which of the following anchors is the strongest of all the wallboard anchors?
   a winged plastic anchor
   b plastic threaded anchor
   c metal threaded anchor
   d toggle anchor

3 Which of the following are suitable for fixing into timber? (Select multiple responses if appropriate.)
   a wood screws
   b lag bolts
   c self-tapping screws
   d self-drilling screws
   e metal thread screws
   f hollow wall anchors
   g toggle devices

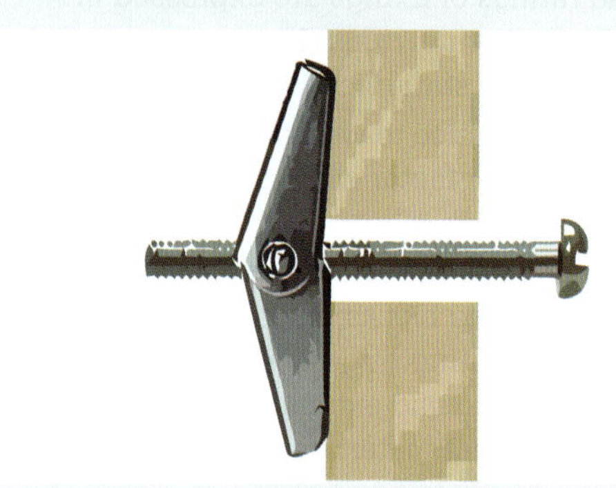

FIGURE 4.33 Figure for Question 4

4 The fixing device shown in **Figure 4.33** is a:
   a lag bolt
   b self-tapping screw
   c self-drilling screw
   d metal thread screw
   e winged plastic wall anchor
   f toggle anchor

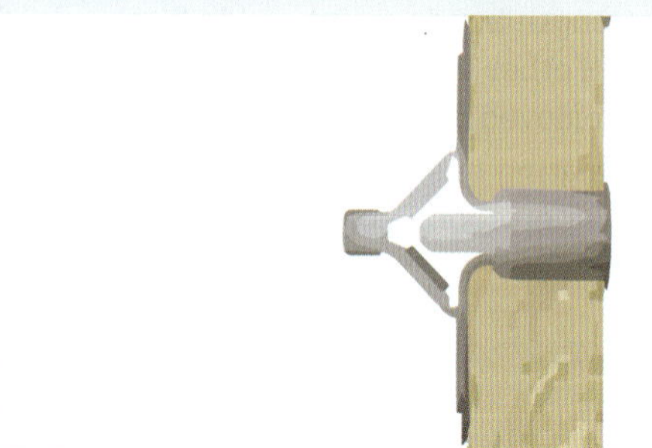

FIGURE 4.34 Figure for Question 5

5 The fixing device shown in **Figure 4.34** is a:
   a lag bolt
   b self-tapping screw
   c self-drilling screw
   d metal thread screw
   e winged plastic wall anchor
   f toggle anchor

FIGURE 4.35 Figure for Question 6

6 The fixing device shown in **Figure 4.35** is a:
   a lag bolt
   b self-tapping screw
   c self-drilling screw
   d metal thread screw
   e winged plastic wall anchor
   f toggle anchor

7 Which of the following are suitable for fixing close to the edge of a brick? (Select multiple responses if appropriate.)
   a wall-plug
   b expanding concrete fixing device
   c gas-powered fixing tool
   d powder-actuated fixing tool
   e chemical anchor
   f masonry nail
   g masonry screw
   h expanding nylon anchor

FIGURE 4.36 Figure for Question 8

8 The fixing device shown in **Figure 4.36** is a:
   a plastic expansion plug
   b expanding concrete fixing device
   c chemical anchor
   d masonry nail
   e winged plastic wall anchor
   f toggle anchor

Masonry: Alamy stock photo/trekandshoot

FIGURE 4.37 Figure for Question 9

9 The fixing device shown in **Figure 4.37** is a:
 a wall-plug
 b expanding concrete fixing device
 c chemical anchor
 d masonry nail
 e winged plastic wall anchor
 f toggle anchor

10 A bolt is the term used for a:
 a metal thread
 b threaded fastener, designed to be used in conjunction with a nut
 c fastener where the thread is cut along the full length of the shank
 d threaded stud

11 To prevent a nut from loosening under vibration:
 a a flat washer is used
 b a woodruff key is used
 c a cotter pin is used
 d a lock washer is used

12 What type of non-threaded fixing assures perfect alignment and facilitates quicker disassembly and assembly of parts in exact relationships?
 a dowel pin
 b nail
 c plastic plug
 d retaining ring

13 How should ties for cable bundling be tightened?
 a by use of pliers
 b by hand
 c by twisting the tie
 d by using a torque wrench

14 When surfaces of the material are joined together by interfacial forces, they are said to demonstrate:
 a good stability
 b good cohesion
 c good adhesion
 d good consistency

15 A fixing that is known for its excellent adhesion, chemical and heat resistance, excellent mechanical properties and superb electrical insulating properties is:
 a self-amalgamating tape
 b hot-melt adhesive
 c double-sided adhesive tape
 d an epoxy adhesive

16 Self-drilling screws are designed to drill through:
 a steel up to 6.4 mm thick
 b masonry
 c plasterboard
 d timber

5 Drawings, diagrams and electrical schedules

This chapter provides electrotechnology workers with essential knowledge and skills pertaining to the use of drawings, diagrams, cable schedules, industry standards, codes of practice and specifications applicable to the electrical industry. This chapter deals with the interpretation of common electrical diagrams, which include schematic, wiring, and architectural diagrams as well as other forms of technical communication. This chapter provides underpinning knowledge for the unit UEECD0051 from the UEE training package.

LEARNING OBJECTIVES

Architectural drawings

- Describe the characteristics and function of various architectural drawings
- Use an architectural floor plan to mark up the location of electrical accessories.
- Use a site plan to mark up the location of electrical services.
- Recognise common electrotechnology symbols used on floor plans.
- Recognise and use appropriate drawing scales.

Building construction drawings

- Explain common types of residential building construction.
- Identify common building constructional elements.
- Outline appropriate cable routes in various parts of a residential building.
- List the various stages of building construction.
- Identify stages of construction where an electrician will be onsite.

Circuit diagrams

- Explain the function of circuit diagrams.
- Detail conventions and common symbols used in circuit diagrams.

Electrical drawings

- Name and explain the purpose of common electrical drawings and diagrams.
- Use a wiring schedule.
- Develop an electrical specification schedule.

Wiring diagrams

- Explain the purpose of wiring diagrams.
- Outline the process for converting a circuit diagram to a wiring diagram.
- Develop a cable schedule.

Job specifications

- Outline the purpose, format and content of a typical job specification.

Electrical licensing

- Name the regulations applicable to the performance of electrical work.
- Outline the scope of work covered by electrical licensing.

Technical standards

- Explain the difference between performance-based and prescriptive requirements.
- Outline the purpose of technical standards.
- Summarise the roles of various bodies in the development of standards.

5.1 Architectural drawings

An electrician must be familiar with all the common terms used with the various building construction methods. This knowledge is necessary for the correct interpretation of building plans and specifications.

Location plan

Every building begins with a declaration of essentials, established by the end-user and designer, from which an initial location is acquired. A location plan aims to indicate the geographical location of the site(s) that is of interest.

A location plan shows a 'bird's eye view' of the development site indicating the location of the block of land. The purpose of the location plan, shown in **Figure 5.1**, is to display the relationship between the building(s) and other structures outside the building site location. The location plan reveals the locality and borders of the site you are planning to work on and is identified with an outline in red. It may show lot numbers, plan numbers and a north point.

FIGURE 5.1 Location plan version 1

The location plan can be drawn using various scales such as 1:1250 or 1:2500. However, satellite images can be used for location plans as shown in **Figure 5.2**. In addition, the location plan may identify where the major public utilities' inlets such as electricity supply, potable

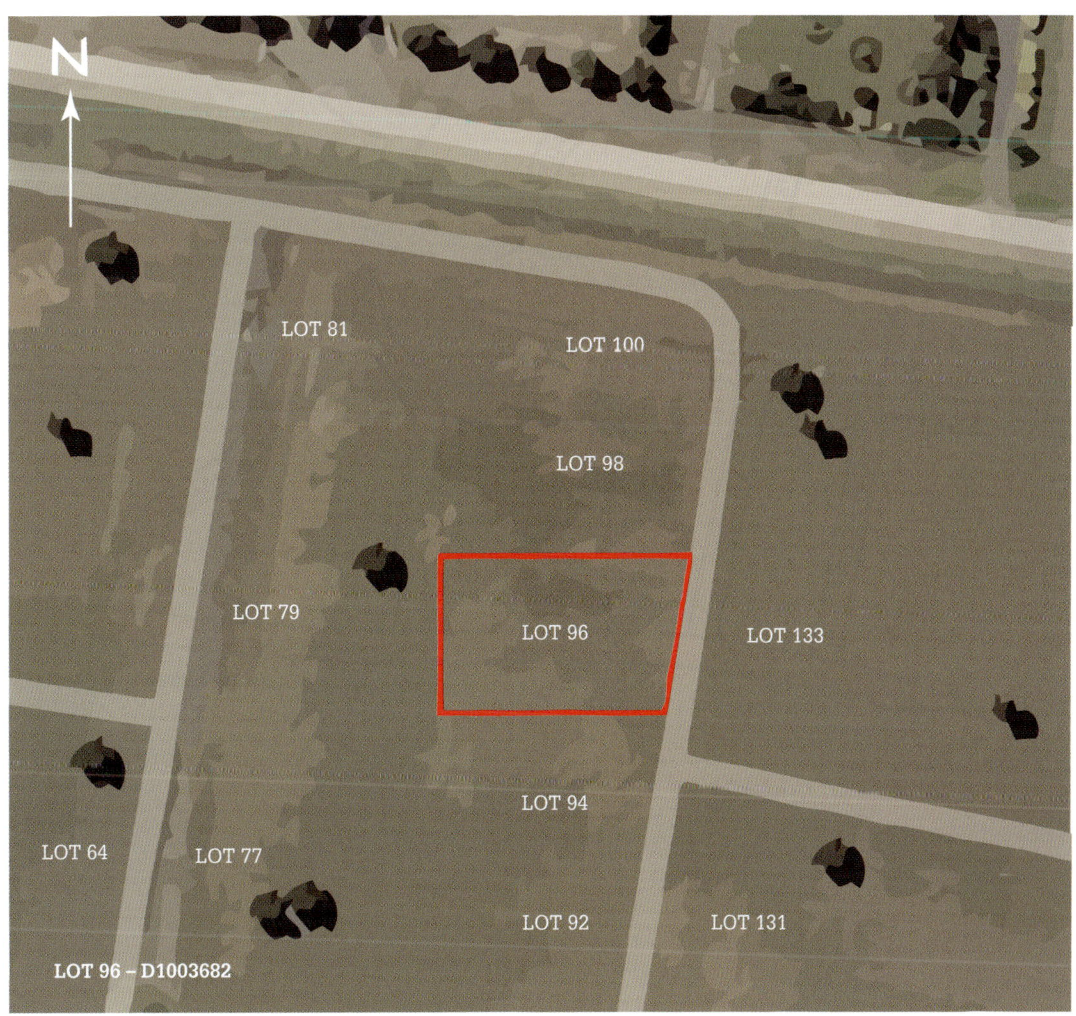

FIGURE 5.2 Location plan version 2

water supply main, gas supply main, telecommunication facilities, and sewerage lines are positioned. The location plan should be detailed enough to identify roads and buildings on land adjacent to the application site to confirm that the precise location of the application site is distinct. The site must be edged or shaded in red, and any other adjacent land owned or controlled by the end-user edged or shaded blue. In addition, the north point must be shown on the plan.

A location plan can be:

- A map produced using standard drawing practice either by hand or computer-aided design (CAD).
- A copy or photocopy of a published map with hand-drawn or marked comments.
- A copy or photocopy of an aerial photograph as shown in **Figure 5.2** with handwritten notes or symbols.

Note: Photocopied maps ought to include symbols that are easily recognisable.

EXERCISE 5.1

In **Figure 5.3** identify with a red circle the location 6 Elm Avenue and with a blue circle the location 15 Field Road.

Lot map

The first step in site plan development is the lot map. This map, which is generally furnished by a surveyor, is a land plan that delineates the property lines with their bearings, dimensions, street names and any existing easements.

FIGURE 5.3 Location plan exercise

The lot map forms the foundation of all future information and site development. An example of a lot map is shown in **Figure 5.4**.

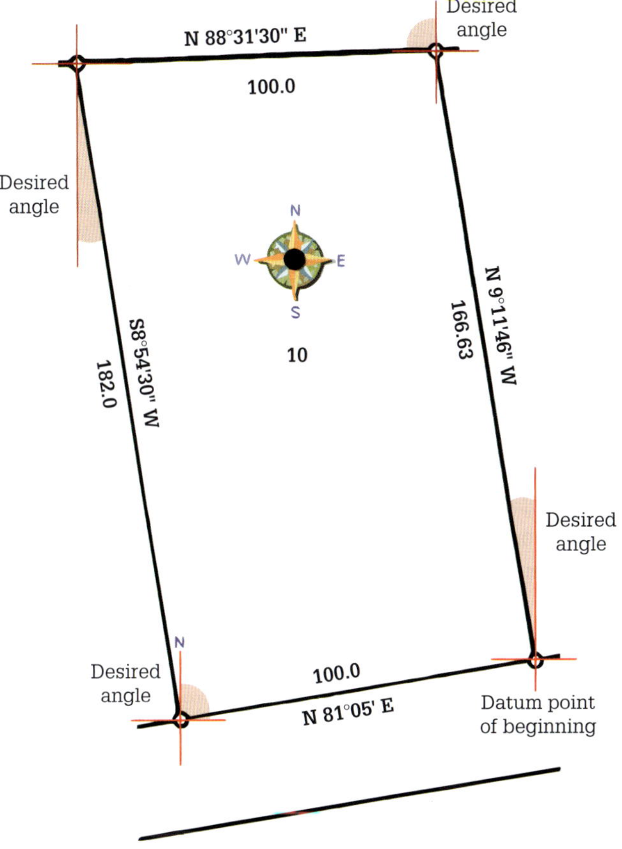

FIGURE 5.4 Lot map with line bearings and property lengths

The property line bearings are described in degrees, minutes, and seconds; the property line dimensions are noted in metres.

Lot lines are arranged by polar coordinates; that is, each line is defined by its length and the angle relative to the true north or south. Polar coordinates use compass direction, degrees, minutes, and seconds. A lot line may read N 80° 30″ 12′ W.

Figure 5.4 shows a lot map with given property lines, bearings and dimensions. To lay out this map graphically, start at the point labelled datum. From the datum, you can outline the lot line in the northwest quadrant with the given dimension. The next bearing falls in the northeast quadrant, which is illustrated by superimposing a compass at the lot line intersection. From there, you can outline the outstanding lot lines in the same way the previous lot lines were defined, finishing at the datum.

EXERCISE 5.2

Using the following dimensions (1 cm = 2 m) and bearings, lay out the lot lines graphically. From the datum go N 5° W for 30 m, then W 3° S for 20 m, then S 1.5° E for 30 m. Join the end of this lot line to the datum and measure its length and bearing.

Soil report

Understanding the type and permeability of surface and subsurface soils at the site is vital to the design of building foundations. Moreover, this ensures that there is no adverse impact on other properties due to water infiltration.

Soils and geology surveys evaluate the bearing qualities of soil conditions such as soil (organic, sandy, reactive clays), moisture content, expansion coefficient and the soil-bearing pressure. These properties of soils can vary from lot to lot and can change considerably within a plot of ground. Site investigations may include test borings at several locations on the site as illustrated in **Figure 5.5**.

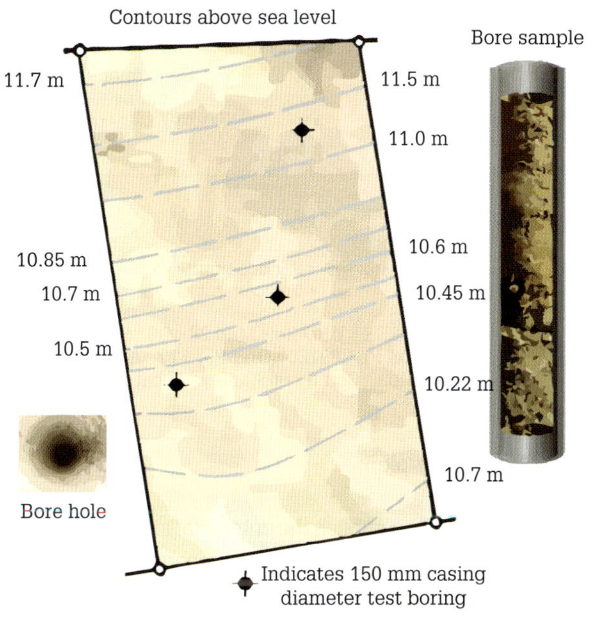

FIGURE 5.5 Contours and boring

These borings are then plotted on the lot map, with an allocated test boring identification and a written or graphic report. When there are geological concerns and soil instability, the particular problem areas are an important primary consideration for foundation design in the proposed project process.

With the completion of the lot map, there is now an accurate plot of earth for locating building setbacks and other factors such as easements and contours that impact on the property. In relation to the architectural working drawings, this part of the drawings is called the site plan.

Site plan

A site plan as illustrated in **Figure 5.6** sets the course for the project. A site plan enables the electrical installer to coordinate with other selected project services for location and trenching for these services. Coordination prevents the crossing of water, sewer and natural gas service lines and distribution mains by the various services.

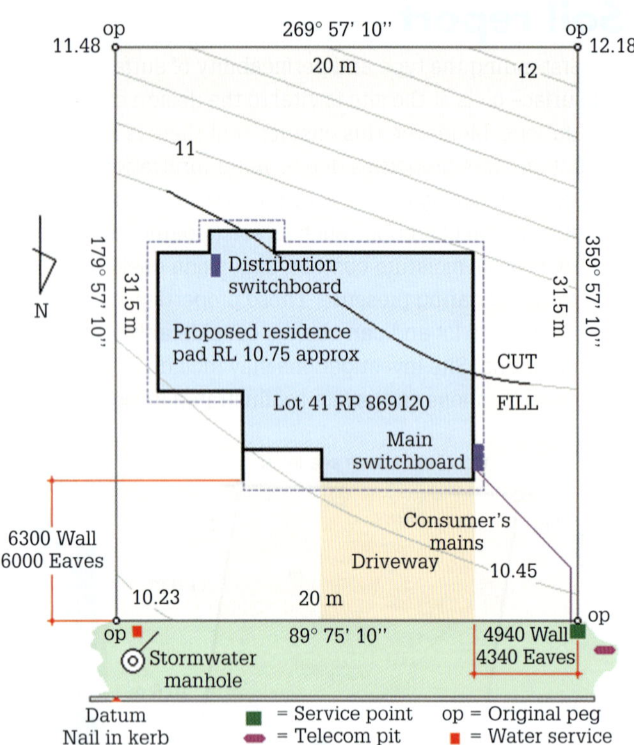

FIGURE 5.6 Site plan

Local councils establish the criteria for site plan submission and the procedures to be followed in preparing an application. The site plan must show the location of all structures on the property and is usually drawn at a scale of 1:500. The following particulars must be indicated on the site plan:

- Location of the proposed structure (external walls should be in thick line and eaves marked with a dashed line).
- Setbacks from boundaries. These distances should always be to the outermost projection of the structure (gutter).
- Position of any existing buildings.
- Property description (lot and registered plan number).
- Location of the road and its name.
- Location and dimensions of any easements.
- Total floor area of the proposed structure.
- Site coverage (total area of all structures on the property).
- Scale of plans (not less than 1:200) with north point shown.
- Contours of the property (every 0.5 m) must be shown accurately on the site plan and be in terms of the same datum. The datum is to be indicated on the plan.
- Building platform level and finished floor level (for dwellings).
- Positions where the heights of retaining walls differ.
- Driveway location and grade (where it meets the road).
- Location of swimming pool fences, gates and filtration equipment (including pipelines).
- Location of sewerage lines, septic trenches and effluent disposal areas.
- Extent of paving and concrete paths.
- Size and location of downpipes and methods of disposal.
- Location of trees and other obstacles located outside the boundary of the property in council's reserve.
- Service point, consumer's mains and main switchboard location and distribution boards.

EXERCISE 5.3

Using a circle, identify the following on the site plan of Figure 5.7: the consumer's mains, the distribution board, contour 5.0 m and the length of the western boundary.

FIGURE 5.7 Site plan exercise

Standard floor plan for domestic buildings

The standard floor plan format for domestic dwellings as illustrated in **Figure 5.8** is chosen by people who are deciding upon the layout. The standard floor plan option is available as a starting point from which they can develop the interior design and tailor it to suit their own personal requirements.

Some standard floor plans as shown in **Figure 5.8** are drawn on either A3 or A2 size sheets. These plans could include dimensioning, labelling of all rooms and bath, toilet and shower locations.

A typical floor plan is shown in **Figure 5.9**. Note that the floor plan displays the outside shape of the building and the arrangement and shape of the rooms. A floor plan can also include the type, size and location of the doors and windows; the location of utility installations; and the location of stairways.

A detailed standard floor plan, as illustrated in **Figure 5.10**, shows all the dimensions.

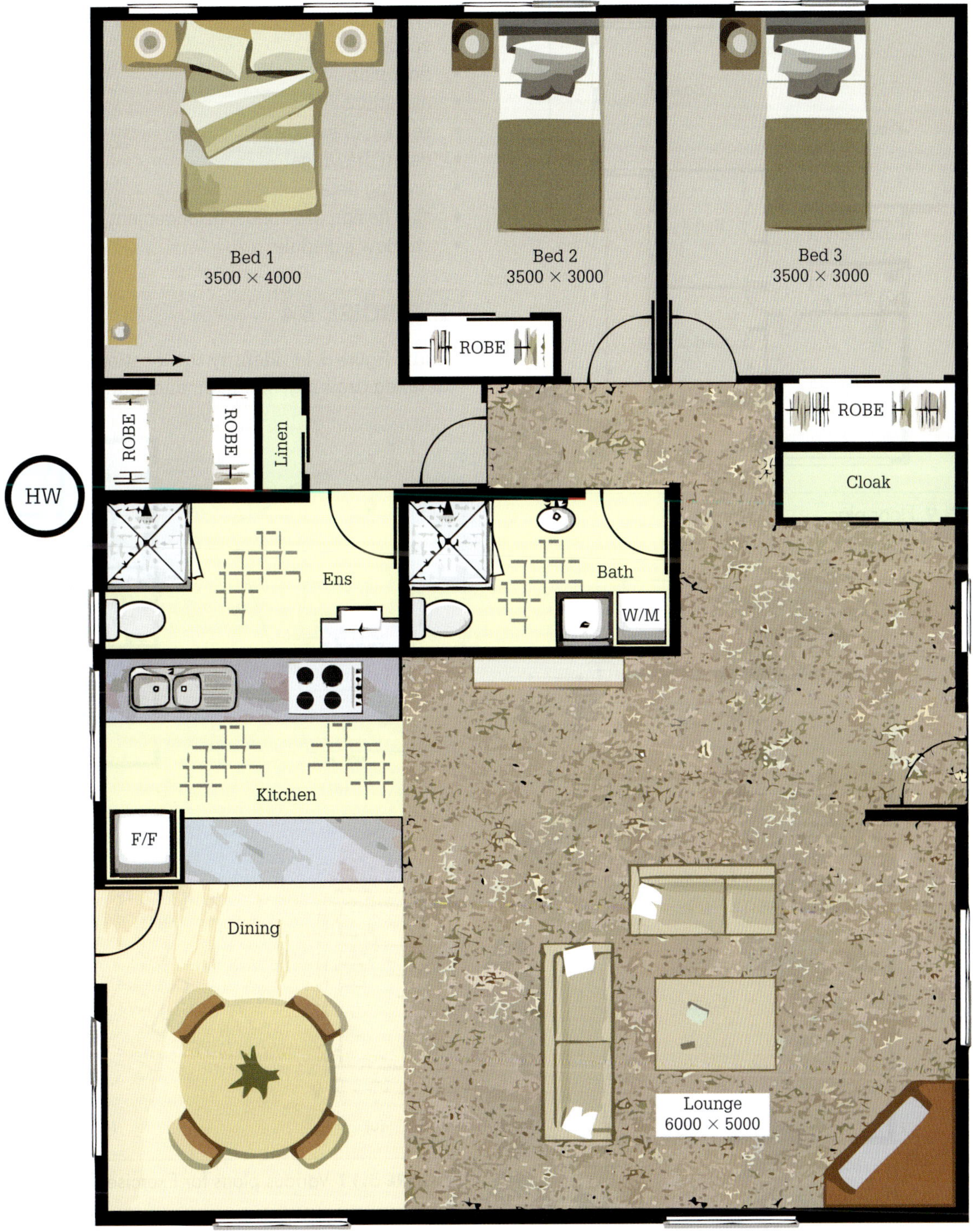

FIGURE 5.8 Standard floor plan

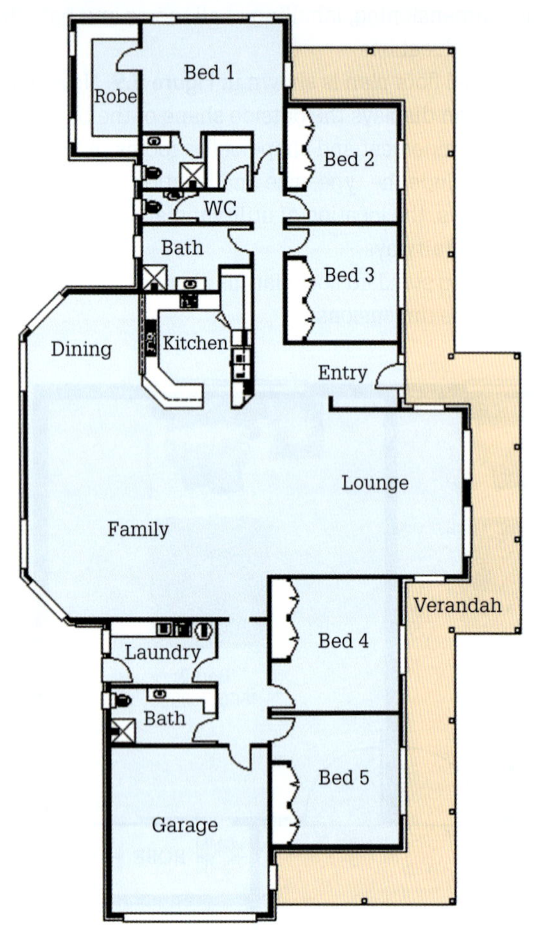

FIGURE 5.9 Floor plan

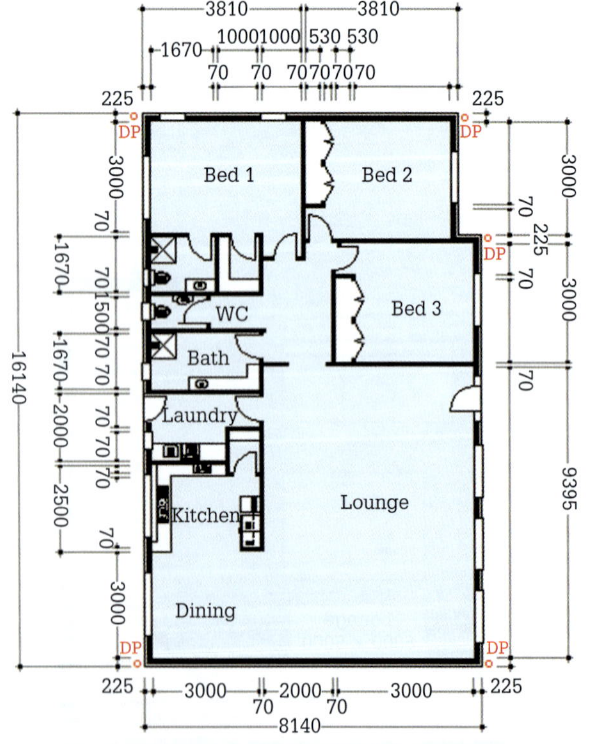

FIGURE 5.10 Detailed floor plan

Working drawings for domestic building work

The working drawings set are the primary format for structural building designs. All working drawings must be arranged to meet the requirements of the Building Code of Australia and the residential timber framing standard, AS 1684, *Residential timber-framed construction* Series (1, 2 and 3), covering cyclonic and non-cyclonic areas, which specifies the safety requirements for all domestic building work in Australia. Included in the set are:

- site plan
- specifications
- floor plan
- elevations
- section drawings
- detail drawings
- bracing detail
- slab design or subfloor arrangement
- window schedule.

EXERCISE 5.4

Using **Figure 5.11**, identify the location plan, floor plan detailed drawing, site plan, and standard drawing.

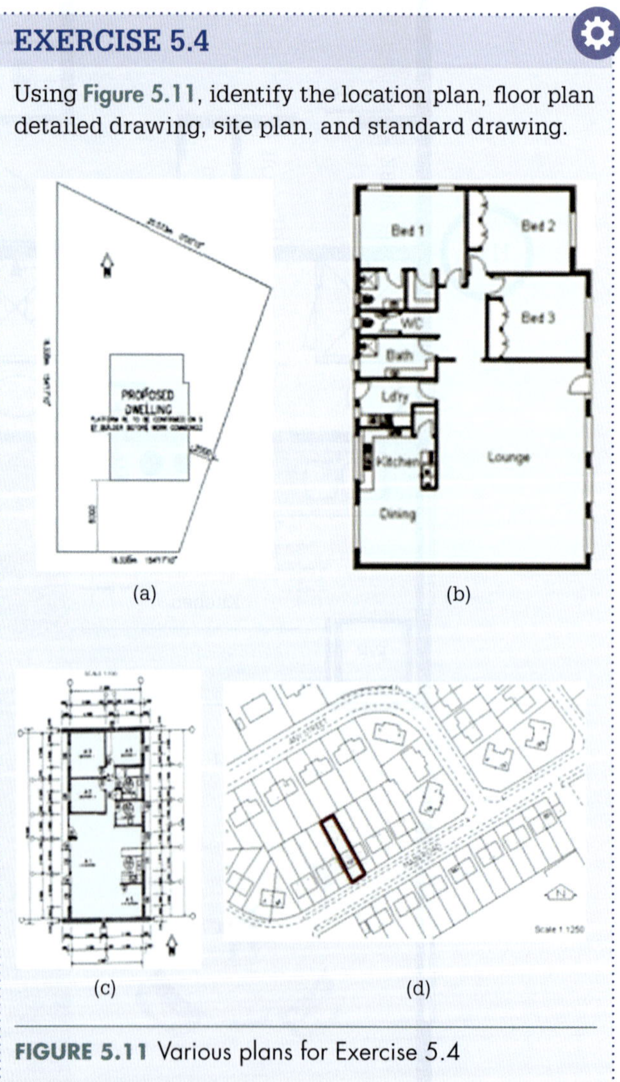

FIGURE 5.11 Various plans for Exercise 5.4

Symbols

Derived electrical symbols and their meanings can be sourced from AS/NZS 1102.102:1997 *Graphical symbols for electrotechnical documentation*.

Symbols are a technical language device that has particular meaning. Usually visual symbols act as shortcuts that convey one or more messages, which are known to both the sender and the recipient. When symbols are standardised, a common language is provided, which orders and standardises electrotechnical documentation throughout Australia and New Zealand.

The following symbols are used on architectural floor plans to show the location of the switches (**Figure 5.12**) and luminaires (**Figure 5.13**) required in an installation as detailed in an electrical schedule as indicated in **Table 5.1**.

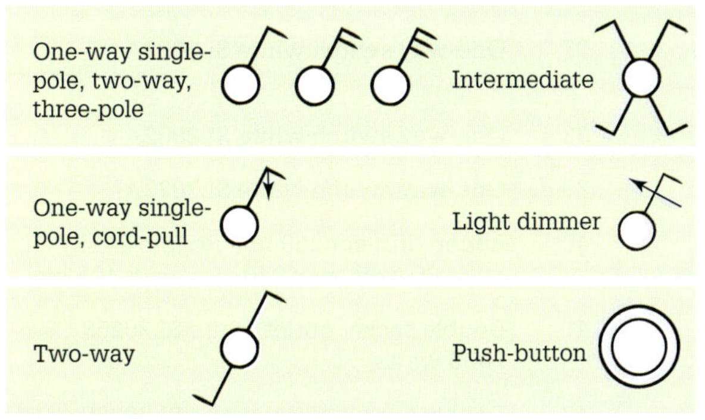

FIGURE 5.12 Electrical symbols – switches

Lighting diagrams as illustrated in **Figure 5.14** show the actual position of all luminaires and control switches using Australian symbols. The lines connecting the luminaires to their designated switch(es) are shown as dotted curved or straight lines.

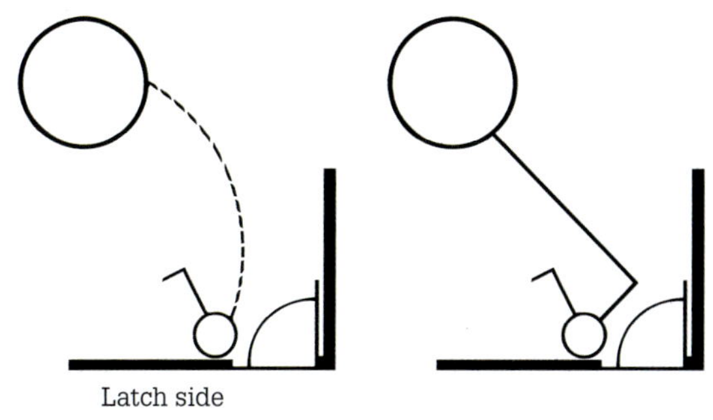

FIGURE 5.14 Lighting diagram

If there is more than one switch at that location, then the switch symbols are drawn beside each other.

Australian symbols – power

The following symbols are used on architectural floor plans to show the location of socket outlets (**Figure 5.15**) required in the installation as detailed in an electrical schedule.

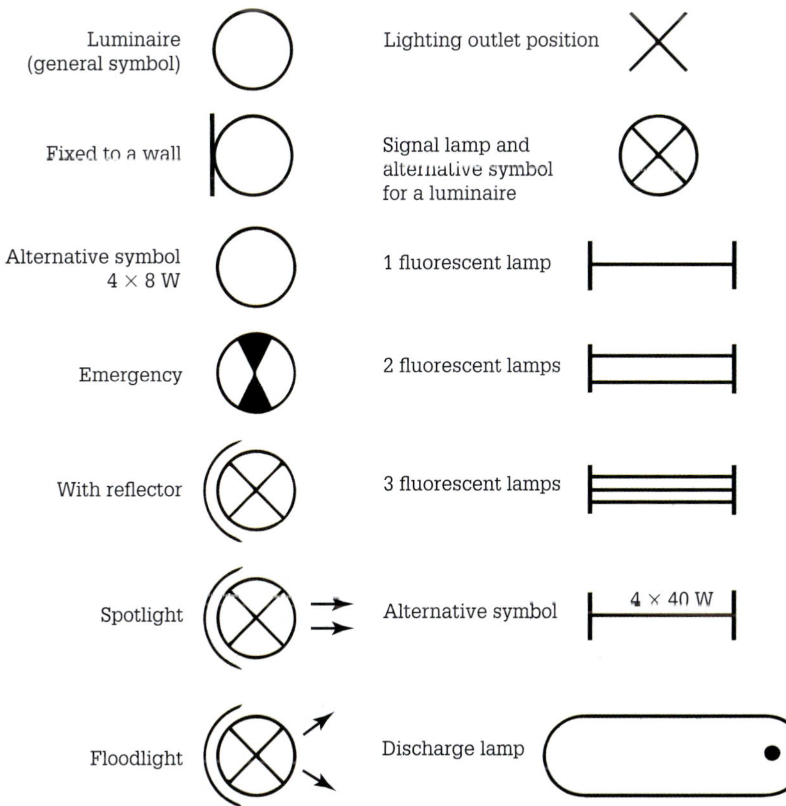

FIGURE 5.13 Electrical symbols – luminaires and lamps

TABLE 5.1 Electrical specification schedule

Location	No.	Appliance	Location	No.	Appliance
Underground	1	Consumer's mains	WC	31	One-way switch white SL2032A
Exterior	2	Switchboard 230 series enclosure and 4RCBE120/30 1 mod, 1 pole MCB/RCD	Hall ceiling	32	Round diffuser 250 mm Reggio 24E27 GLS lamp 60 watt max 240 mm dia × 105 mm
Lounge	3	Double socket outlet white SL2025S 2000 watts	En-suite	33	Single socket outlet white SL2015S 1000 watts
Lounge	4	Double socket outlet white SL2025S 2000 watts	En-suite ceiling	34	Round diffuser 250 mm Reggio 24E27 GLS lamp 60 watt max 240 mm dia × 105 mm
Dining ceiling	5	Round diffuser 250 mm Reggio 24E27 GLS lamp 60 watt max 250 mm dia × 105 mm	Walk-in-robe ceiling	35	Round diffuser 250 mm Reggio 24E27 GLS lamp 60 watt max 240 mm dia × 105 mm
Dining wall	6	Clock 230 V Independence Studios clock with thermometer and hygrometer	Hall	36	Two-way switch white SL2031VA
Kitchen	7	Double socket outlet white SL2025S 2000 watts	Bed 1	37	One-way switch white SL2032A
Kitchen ceiling	8	Fluorescent batten 1/36	Bed 1	38	One-way switch white SL2032A
Dining	9	One-way switch white SL2032A	Bed 1	39	One-way switch white SL2032A
Dining	10	One-way switch white SL2032A	Bed 1 ceiling	40	Round diffuser 250 mm Reggio 24E27 GLS lamp 60 watt max 240 mm dia × 105 mm
Exterior	11	Single socket outlet white WPBS0115 IP53 15 amp	Bed 1	41	Double socket outlet white SL2025S 2000 watts
Exterior wall	12	Exterior luminaire Egea wall bracket EL compact fluorescent lamp 23 W	Bed 1	42	Double socket outlet white SL2025S 2000 watts
Kitchen	13	Double socket outlet white SL2025S 2000 watts	Bed 2	43	Double socket outlet white SL2025S 2000 watts
Kitchen	14	Double socket outlet white SL2025S 2000 watts	Bed 2 ceiling	44	Round diffuser 250 mm Reggio 24E27 GLS lamp 60 watt max 240 mm dia × 105 mm
Lounge	15	Two-way switch white SL2031VA	Bed 2	45	One-way switch white SL2032A
Lounge	16	Double socket outlet white SL2025S 2000 watts	Bed 2	46	Double socket outlet white SL2025S 2000 watts
Kitchen	17	One-way switch white SL2032A	Bed 2	47	Double socket outlet white SL2025S 2000 watts
Pantry ceiling	18	Round diffuser 250 mm Reggio 24E27 GLS lamp 60 watt max 240 mm dia × 105 mm	Bed 3	48	Double socket outlet white SL2025S 2000 watts
Laundry	19	Single electric mains pressure hot water Rheem 400 L 3.6 kW	Bed 3	49	One-way switch white SL2032A
Laundry	20	Single socket outlet white SL2015S 1000 watts	Bed 3 ceiling	50	Round diffuser 250 mm Reggio 24E27 GLS lamp 60 watt max 240 mm dia × 105 mm
Laundry	21	One-way switch white SL2032A	Bed 3	51	Double socket outlet white SL2025S 2000 watts
Laundry ceiling	22	Round diffuser 250 mm Reggio 24E27 GLS lamp 60 watt max 240 mm dia × 105 mm	Bed 3	52	Double socket outlet white SL2025S 2000 watts
Lounge	23	One-way switch white SL2032A	Exterior wall	53	Exterior luminaire Tyr lamp: E27 electronic fluorescent 20 W max IP54 rated
WC	24	Bathroom heater Heatstar 450 watt THP 450S	Lounge	54	One-way switch white SL2032A
WC	25	Single socket outlet white SL2015S 1000 watts	Lounge	55	Two-way switch white SL2031VA

»

TABLE 5.1 Electrical specification schedule (*Continued*)

Location	No.	Appliance	Location	No.	Appliance
Lounge	26	Intermediate switch white SL2031VA	Lounge ceiling	56	Round diffuser 250 mm Reggio 24E27 GLS lamp 60 Watt max 240 mm dia × 105 mm
WC ceiling	27	Round diffuser 250 mm Reggio 24E27 GLS lamp 60 watt max 240 mm dia × 105 mm	Lounge	57	Double socket outlet white SL2025S 2000 watts
WC	28	One-way switch white SL2032A	Lounge ceiling	58	Round diffuser 250 mm Reggio 24E27 GLS lamp 60 watt max 240 mm dia × 105 mm
Hall	29	Two-way switch white SL2031VA	Lounge	59	Double socket outlet white SL2025S 2000 watts
Hall ceiling	30	Round diffuser 250 mm Reggio 24E27 GLS lamp 60 watt max 240 mm dia × 105 mm	Hall and dining ceiling	60	Ionisation smoke alarm

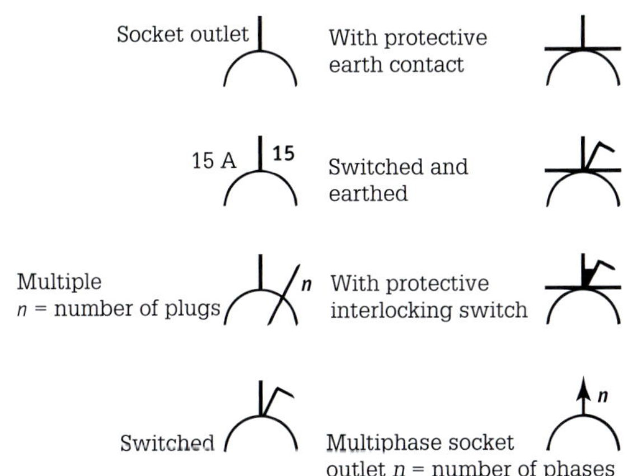

FIGURE 5.15 Electrical symbols – power

Drawing scales

A scale drawing is a reproduction diagram of a building or object but at a standardised fraction of its original size. When designing a building, you use a particular scale to draw diagrams. For example, when you need to draw a plan of a building, reduce the size of the diagrams to 1/50 or 1/100 of its actual size; that is, you use a 1:50 or 1:100 scale. In contrast, when you need to draw a small machine part, you draw it many times larger than its actual size. Refer to **Table 5.2** for standard drawing scales.

The standard measurement units used on architectural drawings are the SI units millimetres (mm) and metres (m). The traditional three-sided scale rule as used by architects is shown in **Figure 5.16**. The scale on the rule is selected so that by placing the rule's edge on a reduced-scale drawing the scale of the drawing (in mm) may be changed directly into the dimensions of the object (in metres).

FIGURE 5.16 300 mm metric architect scale

TABLE 5.2 Standard drawing scales in metric format

Standard scales – metric	
Enlargement	**Scale factor**
2:1 (enlarged 2 times)	0.5
5:1 (enlarged 5 times)	0.2
10:1 (enlarged 10 times)	0.1
50:1 (enlarged 50 times)	0.02
100:1 (enlarged 100 times)	0.01
Normal	
1:1 (full size)	1.0
Reduction	
1:10 (reduced 10 times)	10.0
1:20 (reduced 20 times)	20.0
1:50 (reduced 50 times)	50.0
1:100 (reduced 100 times)	100.0

EXAMPLE 5.1

The scale used on a floor plan is 1:100. If a room is measured from the plan at 50 mm × 55 mm, what is the actual size of the room?

Solution:

As the scale is 1:100, we need to multiply each measured value by 100, which gives us:

$$50 \times 100 = 5000$$

$$\text{and } 55 \times 100 = 5500$$

so, the actual dimension of the room is 5000 × 5500.

EXERCISE 5.5

Use a scaled ruler to help answer the questions for the floor plan shown in **Figure 5.17**. Scale is 1:200. Find the:

a Length of the house in m
b Width of the house in m
c Total floor area of the house in square metres
d Area of Bed 1 in square metres
e Width of Bed 1 window, in mm
f Dimensions of Lounge in m

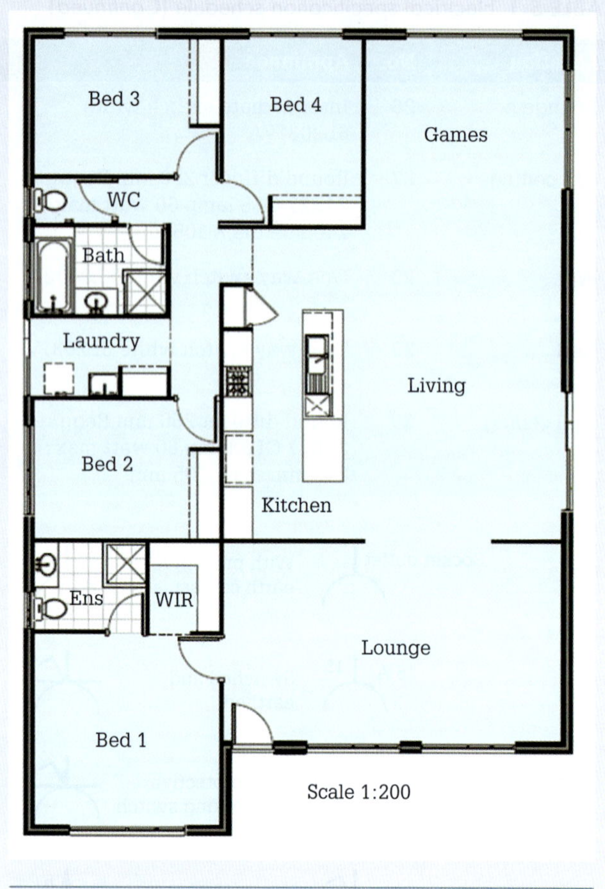

FIGURE 5.17 Floor plan skills practice

The scale should usually be noted in the title block of the drawing. When more than one scale is used, they should be shown close to the views to which they refer, and the title block should read 'scales as shown'.

Installation diagram

The installation diagram, illustrated in **Figure 5.18**, is a topographical layout, which provides the location and type of switches, socket outlets, lighting points, switchboard, consumer's mains and water heater. The topographical layout shows the lighting point(s) and their controlling switch(es) in their relative physical positions with connection lines between them.

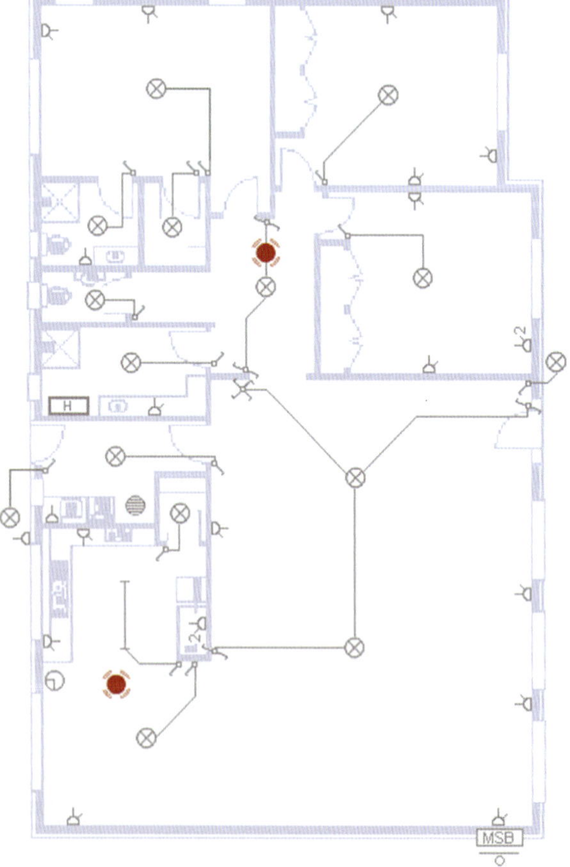

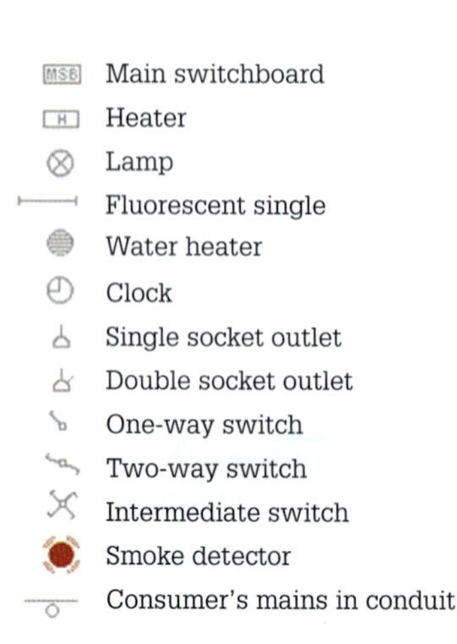

FIGURE 5.18 Installation diagram

REVIEW QUESTIONS

1 Which drawing forms the first step in site plan development?
2 What must the site plan show?
3 To what scale is the site plan typically drawn?
4 Which drawing is available to a customer as a starting point from which they can develop the interior design and tailor it to suit their own personal requirements?
5 Name the drawings typically included in the working drawings set.
6 What is the purpose of visual symbols on a diagram?
7 What is detailed in lighting diagrams?
8 Name the type of drawing that is a reproduction diagram of a building or object but at a standardised fraction of its original size.
9 What is an installation diagram?
10 What type of plan displays the outside shape of a building and the arrangement and shape of the rooms?

5.2 Building construction drawings

Building construction follows an ordinal procedure. The stages of building construction include:

- setting-out
- drainer, foundation trenches, temporary power pole meter box position, underground conduit
- foundation poured
- slab, thickening beam (heating systems or cables may have to be installed) poured or bearers, joists and flooring erected
- wall frames erected
- roof framing, fascia and barge boards fitted
- roof tiled or sheeted
- windows and external doors installed
- outer lining erected
- soffits fitted
- plumbing installed
- electrical fit-out (rough-in, or first fix) installed
- ceilings and walls lined, and cornice fitted
- bath, shower, WCs and kitchen installed
- doors, architraves, and skirting fitted
- plumbing finished off
- painting
- electrical finish off or second fix.

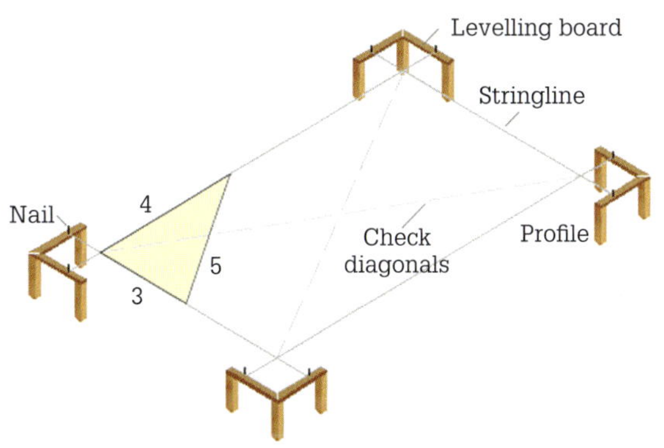

FIGURE 5.19 Setting-out

Setting-out

Setting-out, as shown in **Figure 5.19**, means installing profiles to provide a means of accurately aligning, squaring and levelling the position of the building on the site.

Footings

Footings are designed to suit the building structure and subsoil type. Concrete footings, illustrated in **Figure 5.20**, serve two main purposes. First, they help 'spread' the load from the building structure they support. Second, they create a level stable surface on which to build the structure.

Depending on the type of soils that the building structure sits on, the concrete footings vary in width and

Sources: iStockphoto/ezza116; Alamy/Jason Meyer

FIGURE 5.20 Footings

depth. Every kind of soil has an acceptable load that it can withstand without settling and causing movement in the building structure. Consequently, the width of the footing is sized in order to reduce the pressure on the soil.

Brick veneer

Brick veneer construction is a single brick covering tied across the air space to a timber or steel frame with wall ties embedded in the mortar joints. The timber or steel frame does all the structural load-bearing work by supporting all the wall and ceiling linings and the roof trusses and roofing material. Various types of anchors are suitable for securing the bottom plate to the slab floor. The footings and wall details of a low-set brick veneer slab floor house is shown in **Figure 5.21**.

The slab edges are rebated to prevent water from the environment entering onto the slab floor. Heavy rain can saturate the brick veneer allowing water to enter the 45 mm cavity between the brick and timber or steel frame. The damp-proof course catches this moisture and carries it outside through the weep holes. TPS cables must not rest on the wall ties because the cable acts as a bridge and directs water to electrical fittings, the timber or metal frame, and the wallboard. In some jurisdictions, conduit carrying underground consumer's mains must enter the cavity between the brick veneer and the stud wall. Consequently, the conduit passes through the reinforced foundation and up into the rebate in the slab floor. Brick veneer is the most popular house construction in Australia. A brick veneer building with a wooden floor is illustrated in **Figure 5.22**.

Cavity brick

Cavity brick (see **Figure 5.23**) is two brick walls separated by a 45 mm cavity and tied to each other with wall ties. The inner brick wall carries out the same structural load-bearing work as a timber or metal frame of a brick veneer construction.

With this type of construction, the electrician has to work closely with the bricklayer so that wall boxes used for switches and socket outlets are placed in the correct locations according to the electrical plan.

Timber frame with external linings

Exterior cladding for external linings for a light timber or metal frame house can be fibre cement sheet as shown in **Figure 5.24**, or stucco or weatherboards.

Timber frames can be constructed from hardwoods, Radiata pine or cypress pine. The frame has anchor rods through it to tie the frame together in high-wind conditions. As there is no cavity between the external lining and the stud frame, the electrician must drill holes (max. 25 mm diameter) through the timber members in order to install TPS wiring. All TPS cables may require mechanical protection if installed within 50 mm of the exterior or interior lining surface. (Refer to AS/NZS 3000:2018 *Wiring Rules – External influences* and *Protection against mechanical damage*.)

Timber wall frame

Figure 5.25 shows a timber wall frame with the names of the various components that make up the frame.

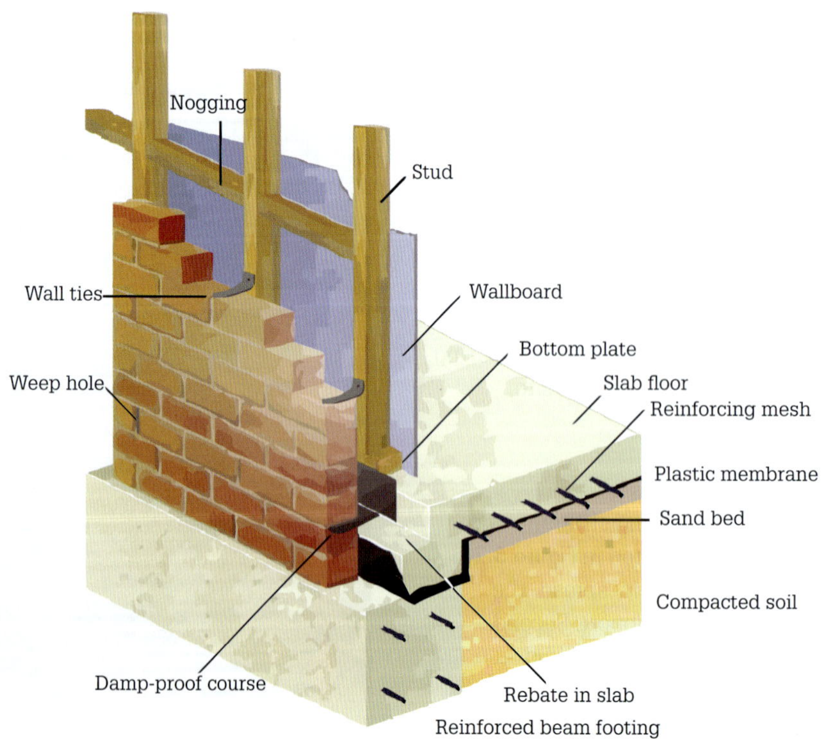

FIGURE 5.21 Low-set brick veneer walls, slab floor, and footings

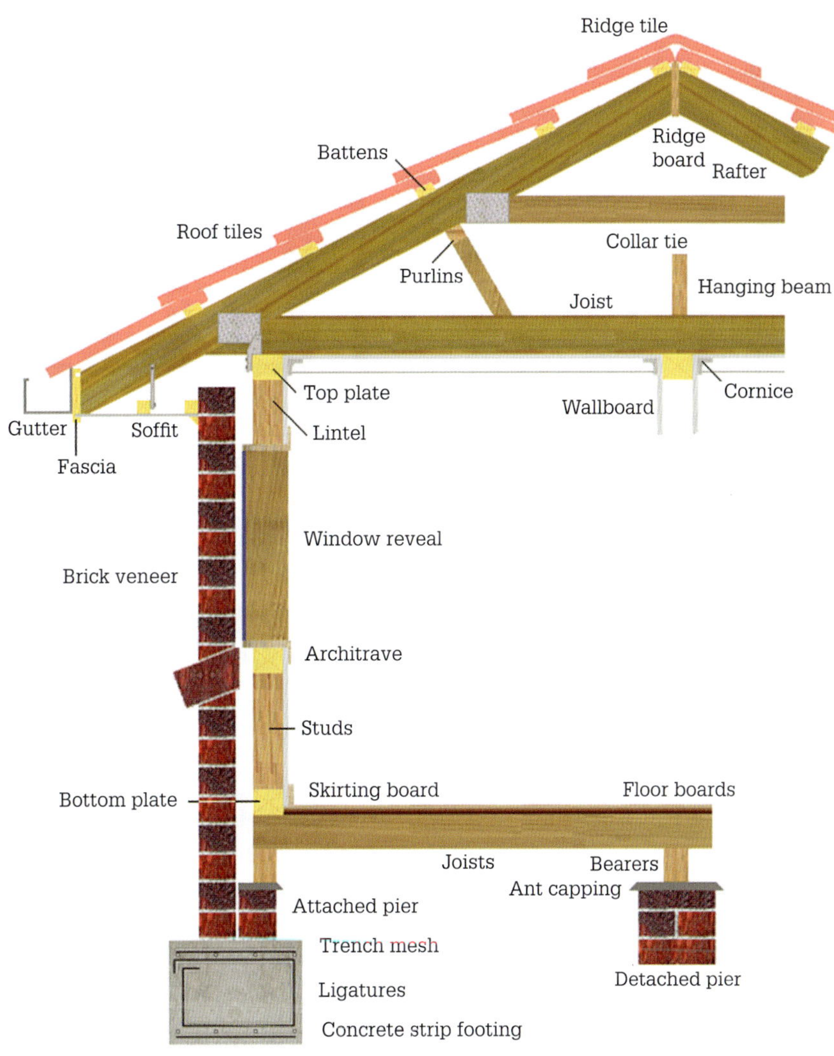

FIGURE 5.22 Brick veneer with a wooden floor

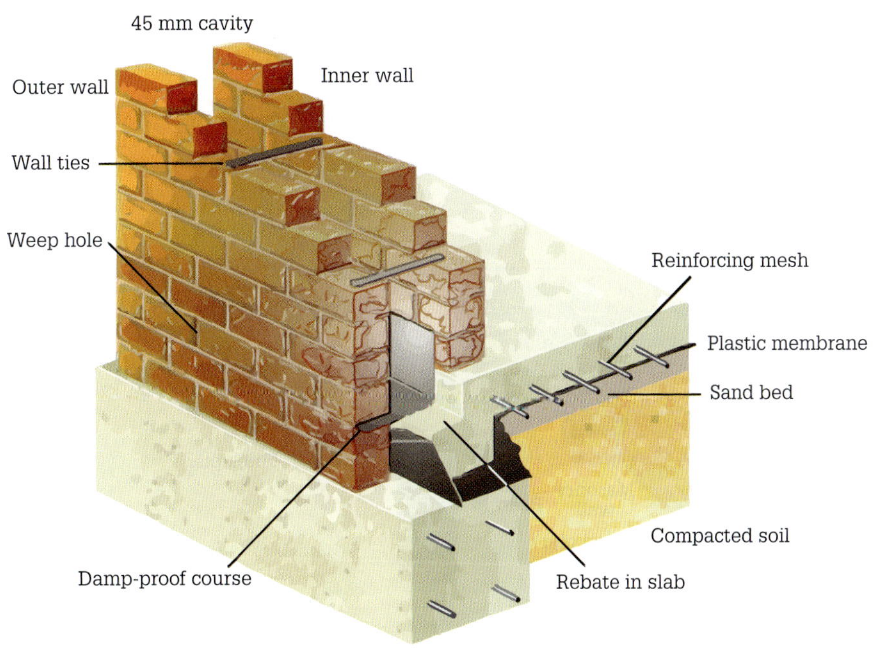

FIGURE 5.23 Cavity brick walls, slab floor, and footings

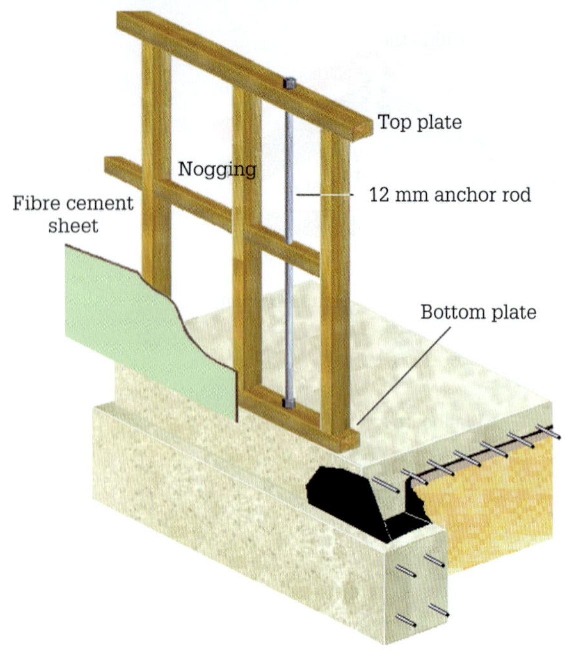

FIGURE 5.24 Light wall frames with external linings

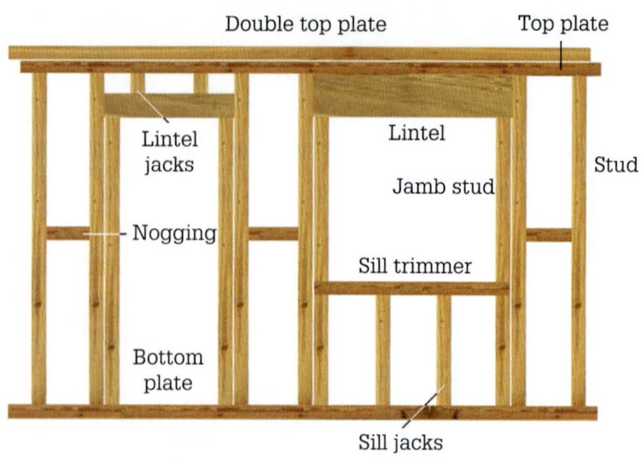

FIGURE 5.25 Timber wall frame

Steel wall frame

Figure 5.26 shows a steel wall frame construction.

Source: Shutterstock.com/Lev Kropotov

FIGURE 5.26 Steel wall frame construction

Timber floors

Timber floors are used where strip footings, piers, steel columns or concrete stumps and engaged piers are used to support the subfloor timbers called bearers, as shown in **Figure 5.27**.

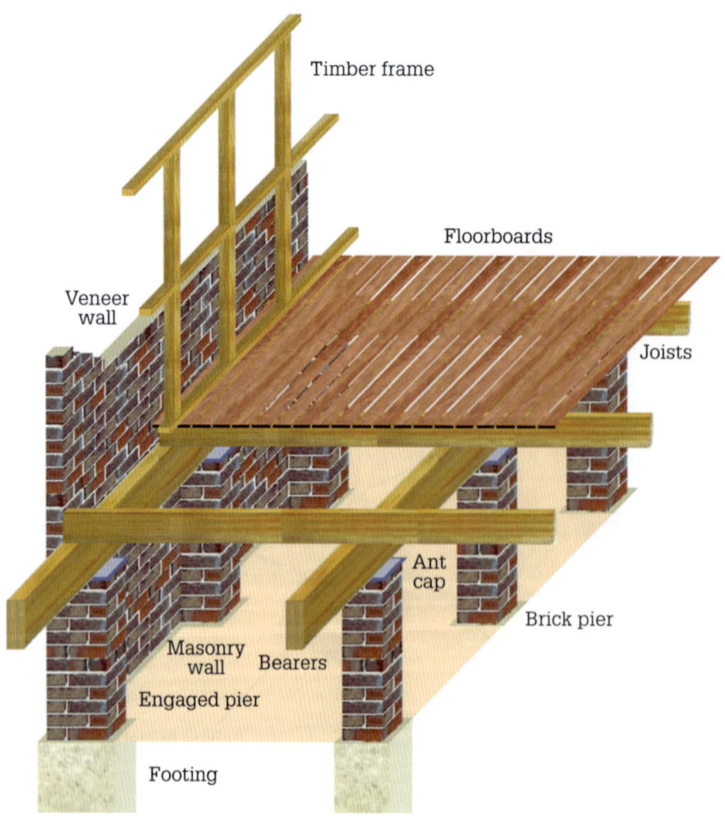

FIGURE 5.27 Timber floors, bearers, joists, and piers

The 110 mm masonry wall is constructed with 350 mm × 350 mm engaged piers up to bearer level. All piers contain a 12 mm reinforced bar taken up from the footings. A 12 mm threaded rod is welded to the reinforcing bar and made through the bearer to fix the bearer to the pier. The hollow pier core is filled with concrete. This construction provides superior holding capabilities for the frame of the house.

The joists are connected to the bearer via a joist strap or a triple grip connector. The stud walls are fixed to the joist with a 12 mm bolt through the bottom plate and joist. All TPS cable clipped to the joists must be further than 50 mm to the underside of the floorboards.

Roof

The illustrations in **Figure 5.28** show common roof types such as skillion, gable, Dutch gable, box gable and hip roof with valleys that are used throughout Australia.

Roof truss

Many roof trusses today are prefabricated for most types of roofing designs such as hip and valley, gable end, Dutch gable, and box gable. A standard prefabricated truss and its various components are illustrated in **Figure 5.29**.

Prefabricated trusses are constructed using truss plates. The erected truss is secured to the top plate with triple grips or metal straps. TPS cable can be fixed to the bottom cord as long as it is not within 50 mm of the ceiling lining surface. Ideally all TPS cable should be fixed to the side of a batten which sits on top of the bottom cords. This method meets requirements for cables in accessible spaces and also lifts the TPS cables above any ceiling insulation that may be installed.

TPS cable in the roof cavity can rest on the ceiling surface in areas that are classified as unlikely to be disturbed. Ideally, even in these locations TPS should be clipped to a timber batten. (Refer to AS/NZS 3000:2018 *Wiring Rules – Selection and installation of wiring systems and for cables in contact with thermal insulation*.)

Eaves

The eaves are the roof overhang of a building. The soffit is the underside of the eaves as illustrated in **Figure 5.30**. TPS cables can be run in the space above the soffit without fixing as they are not likely to be disturbed. Ideally they should be clipped to a timber batten. (Refer to AS/NZS 3000:2018 *Wiring Rules – Selection and installation of wiring systems – Wiring systems likely to be disturbed*.) Note that the cables may have to be de-rated due to the elevated ambient temperatures in the roof space.

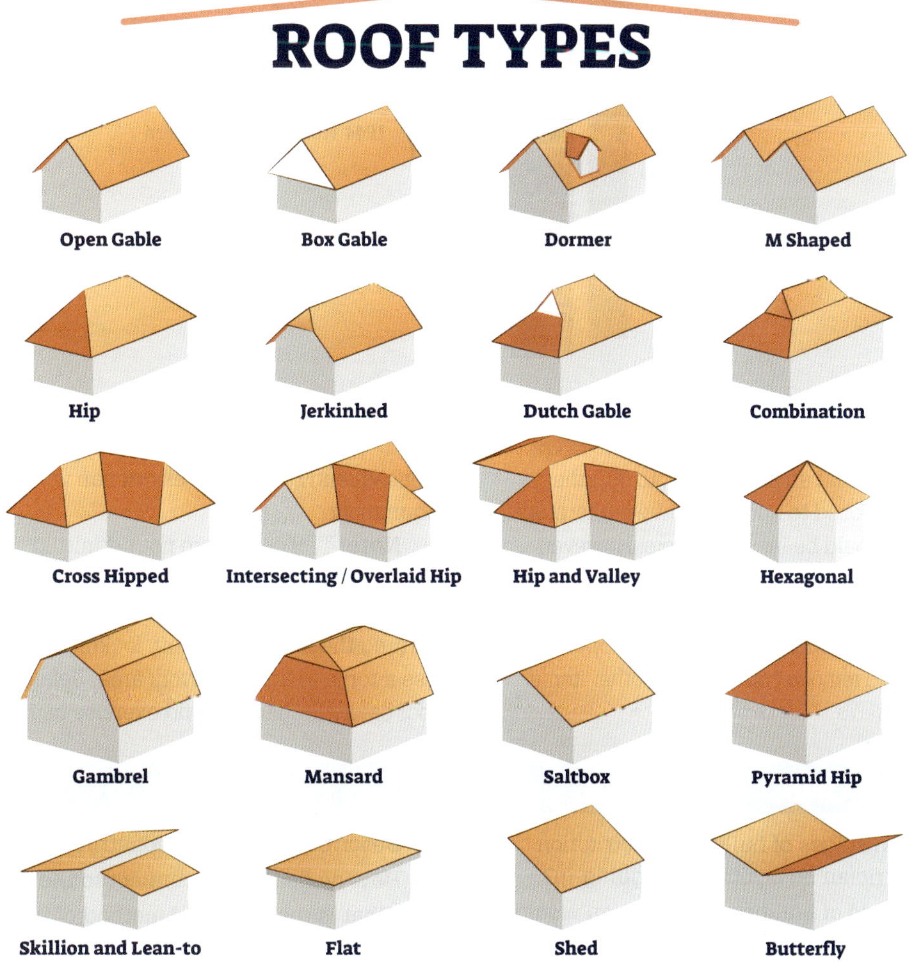

Source: Shutterstock.com/VectorMine

FIGURE 5.28 Common roof types

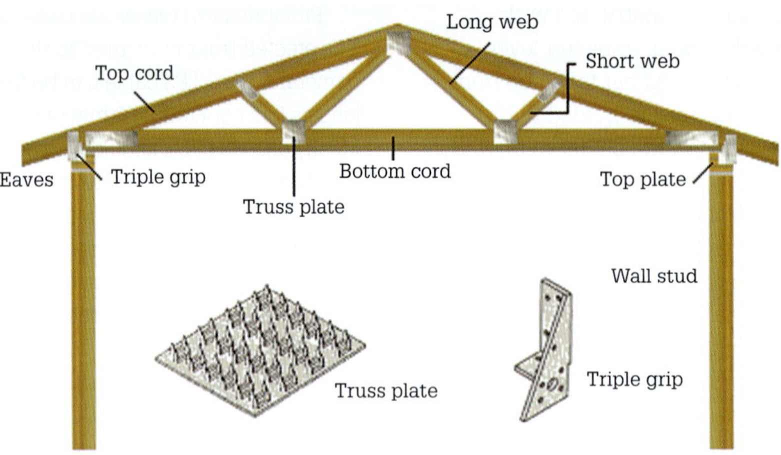

FIGURE 5.29 Truss components

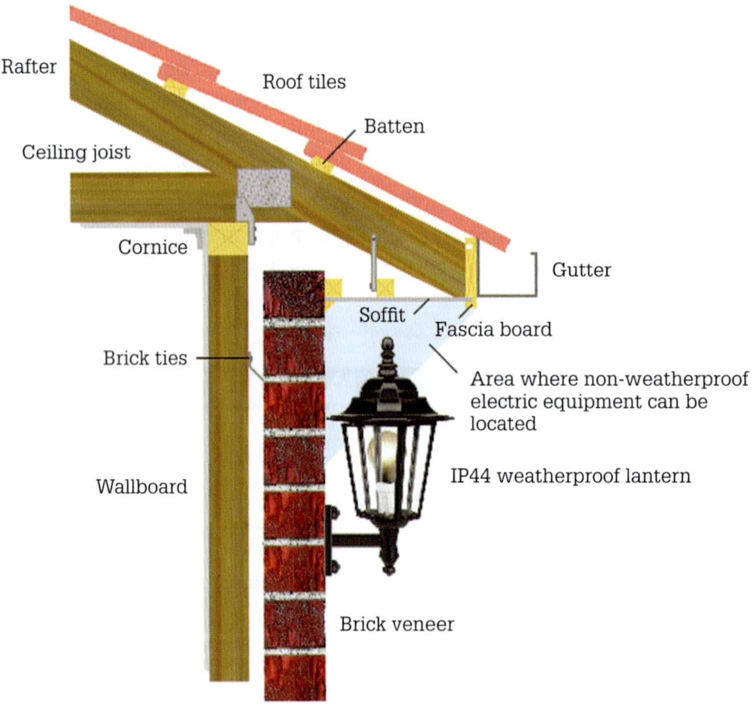

FIGURE 5.30 Eave details

The area underneath (at an angle of 40° to 60°) the soffit is not usually exposed to rain and non-weatherproof electrical equipment could be installed there. However, the direction of rain is dependent on the intensity of wind that means that no area under a soffit is regarded as weatherproof. In addition, there are owners of properties that hose their soffit to remove cobwebs and other items. Consequently, a prudent electrician would use weatherproof equipment.

Note: TPS cables are not permitted between the roof tile or metal roof sheeting and the rafter. If you must enter an established roof cavity always isolate the installation by turning the main switch off, locking out and tagging and use a torch to light the work area. Before entry, inspect the roof cavity for hazards and exercise caution at all times.

REVIEW QUESTIONS

1 Between which stages of building construction would an electrician be onsite for roughing-in?
2 After which stage of building construction would an electrician be onsite for finishing-off?
3 What is meant by 'setting-out'?
4 Describe brick veneer construction.
5 Why are slab edges of brick veneer construction rebated?
6 Describe cavity brick construction.
7 How do floor joists connect to the bearer?
8 What is the preferred method of installing TPS cables in a trussed roof?
9 What is a soffit?
10 Detail the precautions to take before entering an established roof cavity.

5.3 Circuit diagrams

Circuit diagrams are an important part of everyday electrical work activities. All electrical workers need to be conversant with obtaining information from these diagrams.

Circuit symbols

Circuit diagrams (see **Figure 5.31**) show by means of single lines and circuit symbols, like those shown in **Table 5.3**, the flow of electric power or the sequence of operation of the circuit. The circuit diagram should follow a logical progression from input to output.

It is possible to combine these symbols to form what is known as a circuit (or schematic) diagram. A circuit diagram only shows connections and not the physical location of components. The use of standard symbols makes reading and understanding the operation of circuits much simpler.

Circuit representation

Figure 5.31 shows how circuit components can be represented using symbols in a circuit diagram.

Note the orientation of the circuit symbols in the circuit diagram of **Figure 5.31**. By convention:

- the power source such as a battery is drawn to the left
- a load device such as a lamp is drawn to the right
- control and protective devices are drawn on the top line, which connects the positive of the power source to the load
- the return current path does not contain control devices and is drawn on the bottom
- control and protective devices such as switches and circuit breakers are orientated such that the 'hinge' side of the blade is closest to the load device and the switch blade operates in a clockwise direction.

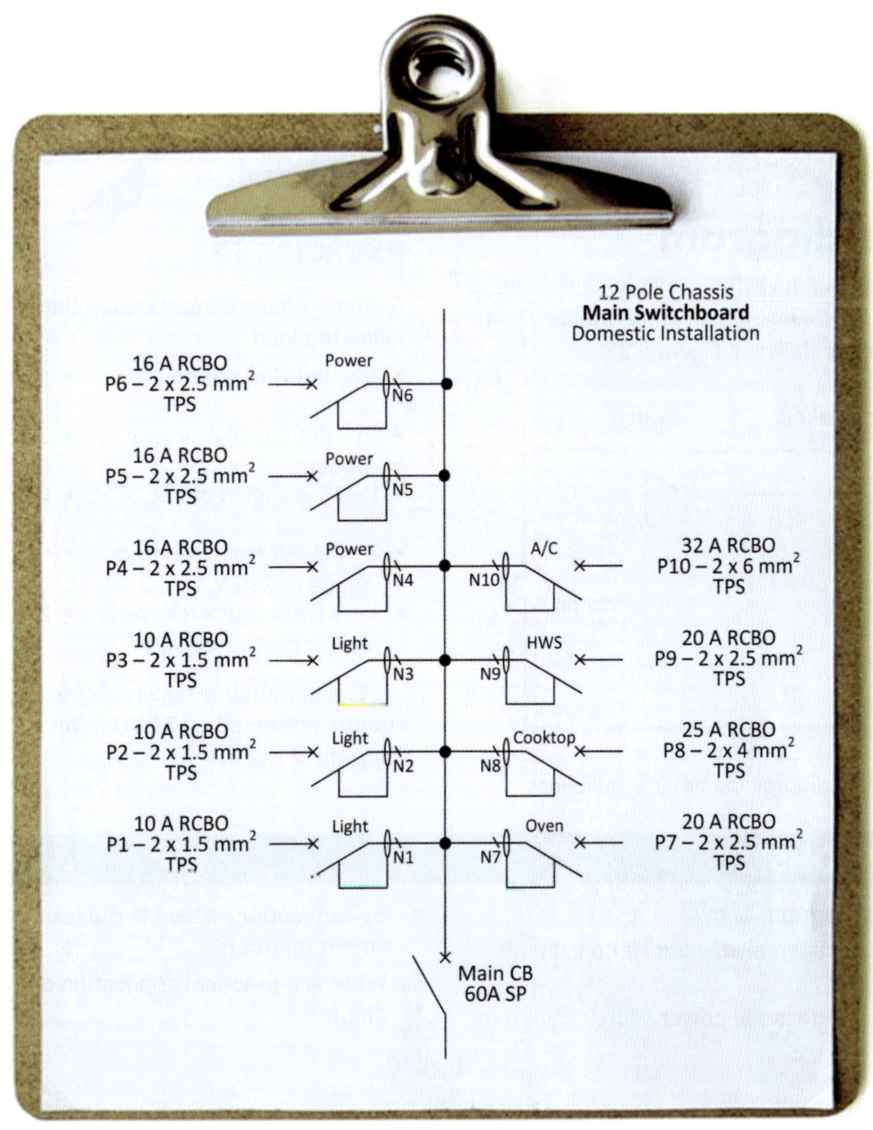

FIGURE 5.31 Using symbols to represent components in a circuit diagram

TABLE 5.3 Circuit symbols and function

Component	Symbol
Cell	+
Battery	+
Conductor	
Joined conductors	
Control switch	
Filament lamp	
Fuse	
Circuit breaker	

Single-line diagram

Related to the circuit diagram is the single-line diagram, which can be used to represent circuit configurations in an electrical installation, as shown in **Figure 5.32**.

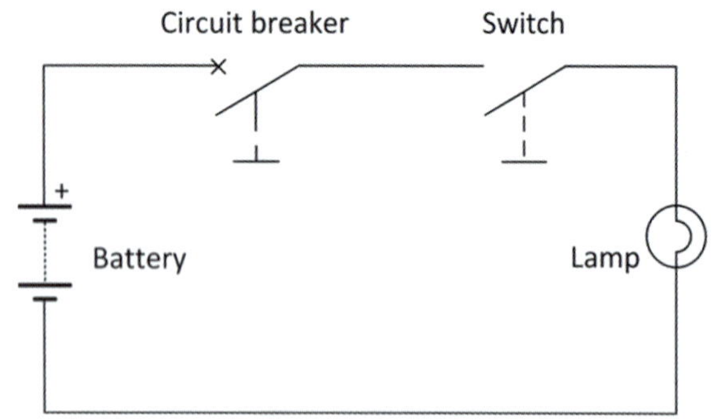

FIGURE 5.32 Single-line diagram domestic installations

The single-phase 230 V domestic installation depicted in **Figure 5.32**, has the following load:

- 38 – lighting points
- 4 – 200 W exterior lights
- 15 – single 10 A socket outlets
- 9 – double 10 A socket outlets
- 1 – 15 A single-phase socket outlet
- 1 – 4.2 kW cooktop @ 230 V
- 1 – 5.2 kW oven @ 230 V
- 1 – 4.8 kW controlled water heater
- 1 – 14 kW single-phase ducted air-conditioner rated at 23.6 A.

EXERCISE 5.6

From the single-line layout diagram in **Figure 5.32**, answer the following questions:

- **a** How many single-pole (SP) circuit breakers are installed on the main switchboard?
- **b** How many triple-pole (TP) circuit breakers are installed on the main switchboard?
- **c** What is the current rating of the main circuit breaker?
- **d** How many power circuits are supplied from the main switchboard?
- **e** How many circuits are protected by RCBOs?
- **f** What type of circuit does 'P10' supply?

EXERCISE 5.7

A single-phase domestic installation has the following load.

▪ 18 – lighting points	▪ 1.5 mm² protected by 10 A RCBO
▪ 15 – double 10 A socket outlets	▪ 2.5 mm² protected by 16 A RCBO
▪ 1 – 15 A socket outlet	▪ 2.5 mm² protected by 16 A RCBO
▪ 1 – 8.4 kW free-standing range @ 230 V	▪ 6 mm² protected by 32 A RCBO
▪ 1 – 4.8 kW quick recovery water heater @ 230 V	▪ 2.5 mm² protected by 25 A RCBO

The installation consists of five circuits, all circuits protected by RCBOs. Develop a single-line diagram of the installation.

REVIEW QUESTIONS

1. What do circuit diagrams show?
2. How is circuit operation made simpler on a circuit diagram?
3. By convention, where is the power source shown in a circuit diagram?
4. By convention, where is the load device shown in a circuit diagram?
5. What is a practical application of a single-line diagram?

5.4 Electrical drawings

There are a number of diagrams used in the electrical industry. Each of these have a specific purpose. It is important for an electrician to be able to read information presented in these diagrams.

Block diagrams

Block diagrams, illustrated in **Figure 5.33**, are a method of explaining complex systems in a simple form. The purpose of the block diagram is to show how the components of the circuit relate to each other and therefore the individual wiring connections are not shown. In block diagrams, various items are shown in a geometrical figure (square or rectangle) and labelled to indicate its purpose.

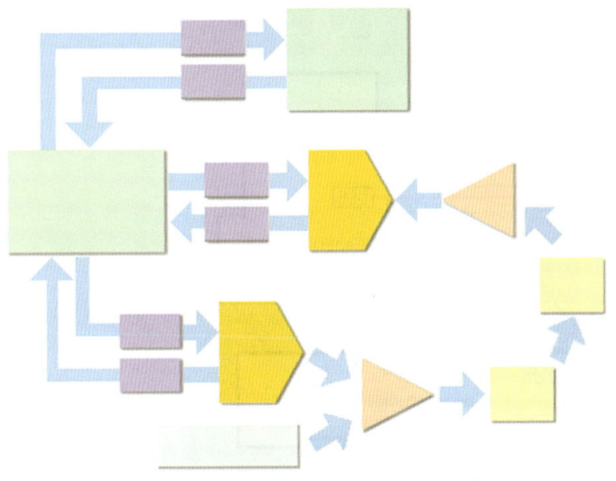

FIGURE 5.33 Block diagram without labels

How to construct a block diagram

- Identify the process. Define the start and end point for the system process to be inspected.
- Identify the key process contributors. Identify all essential elements that participate in the process operations.
- Outline the diagram. For each step, describe the activity, place it in a small box and locate the box in the order the events occur.
- Label the diagram. Above each box diagram write the name of the process being examined.
- Indicate input and output. Just outside of each block list the input that activates the process and outside the bottom of the diagram list the output that ends the process.
- Connect the process steps. Connect the boxes with arrows to show the sequence of activities.
- Verify the accuracy. Consult with others familiar with the system process to verify that the block diagram accurately reflects the system process operation.

With electrical installations, block diagrams are used to design complete circuits by breaking them down into smaller blocks. Each block performs a particular job and is labelled appropriately. A block diagram shows how the blocks are connected together. Only the inputs and outputs of each block are shown. No details of what occurs within each block are given. This way of looking at circuits is called the systems approach. A complete electrical installation (see **Figure 5.34**) is made up from labelled blocks that are joined by arrows. The arrows indicate the direction of flow and inputs to and outputs from the blocks.

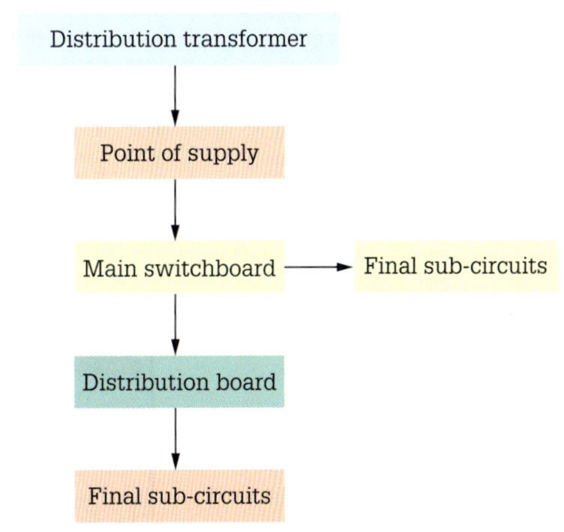

FIGURE 5.34 Installation block diagram

With this incomplete knowledge of the electrical installation, we can apply a little logical fault finding. If there is no output from the distribution board but we can measure the output from the central switchboard then a component on the distribution board is faulty. With more knowledge, we can break the blocks down into smaller blocks and finally the wiring within the blocks will be understood.

Routes of underground feeder cables

All joints in underground cables and terminations are made either by means of compound-filled boxes or by means of approved epoxy-resin pressure-type jointing kits. Where cables are cut and not immediately made-off, the ends must be sealed immediately to prevent the ingress of moisture.

The routes of underground feeder cables are accurately recorded on an electrical site plan as illustrated in **Figure 5.35**. All exterior fixtures should also be located on the site plan. As each section of the feeders is completed an insulation resistance test must be carried out to confirm the integrity of the laid cable.

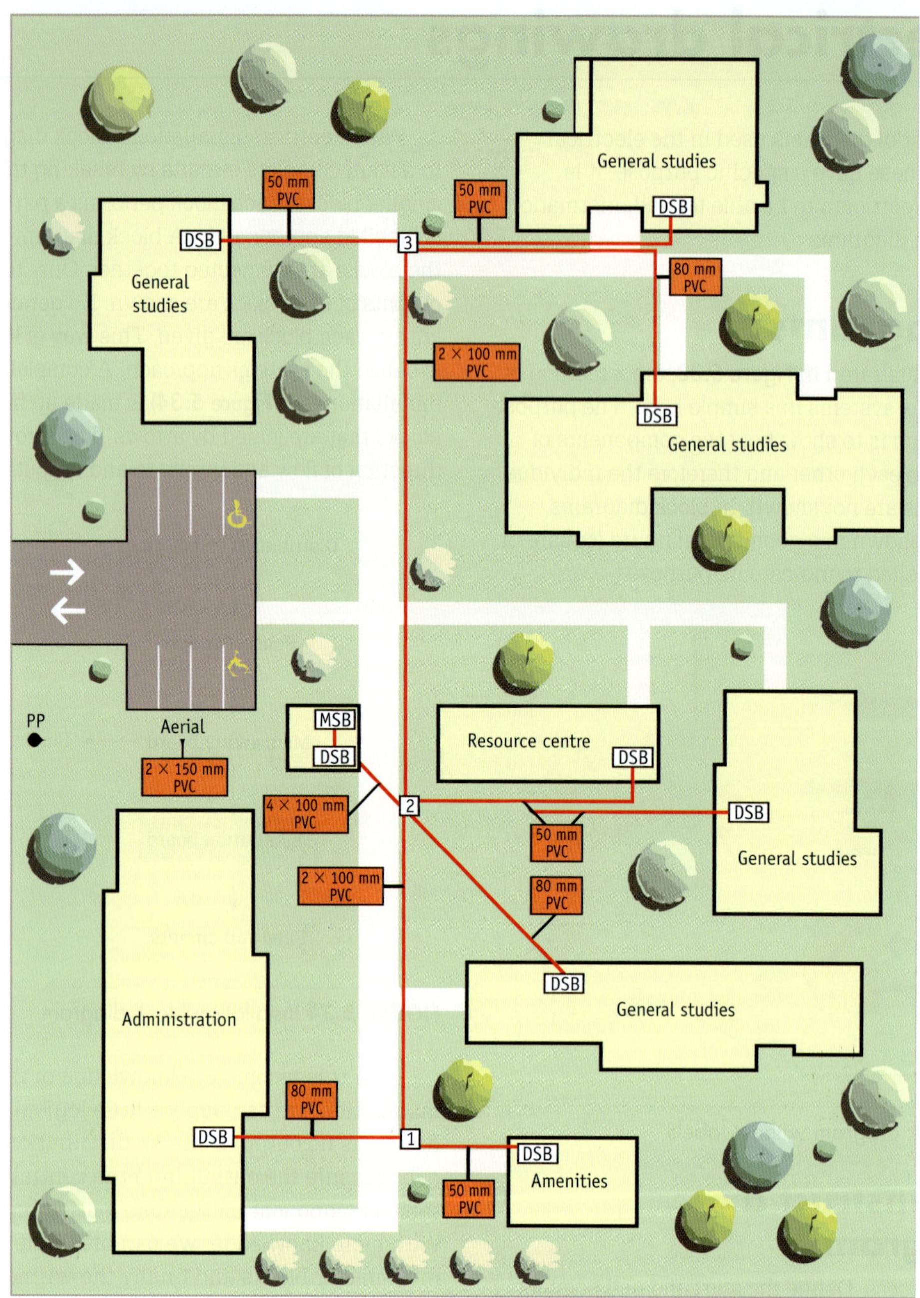

FIGURE 5.35 Underground feeders – electrical mains and sub-mains site plan

Backfilling of the trenches for the underground feeders is made with a grading of the material to ensure settling without voids. The grade material should be tamped down every 150 mm.

After installing the conduits and backfilling, the electrician must precisely mark the location of underground cables with route markers consisting of a marker plate set flush on a concrete base. The marking plates occur at each change of direction, termination point and building entry point. Marker plates indicate the direction of the cable run by means of direction arrows on the marker plate and indicate the distance to the next marker.

Electrical reticulation diagram

The electrical services reticulation diagram uses a single-line diagram to indicate the interconnection of switchboards within an electrical installation. An electrical services reticulation layout for a large complex is illustrated in **Figure 5.36**.

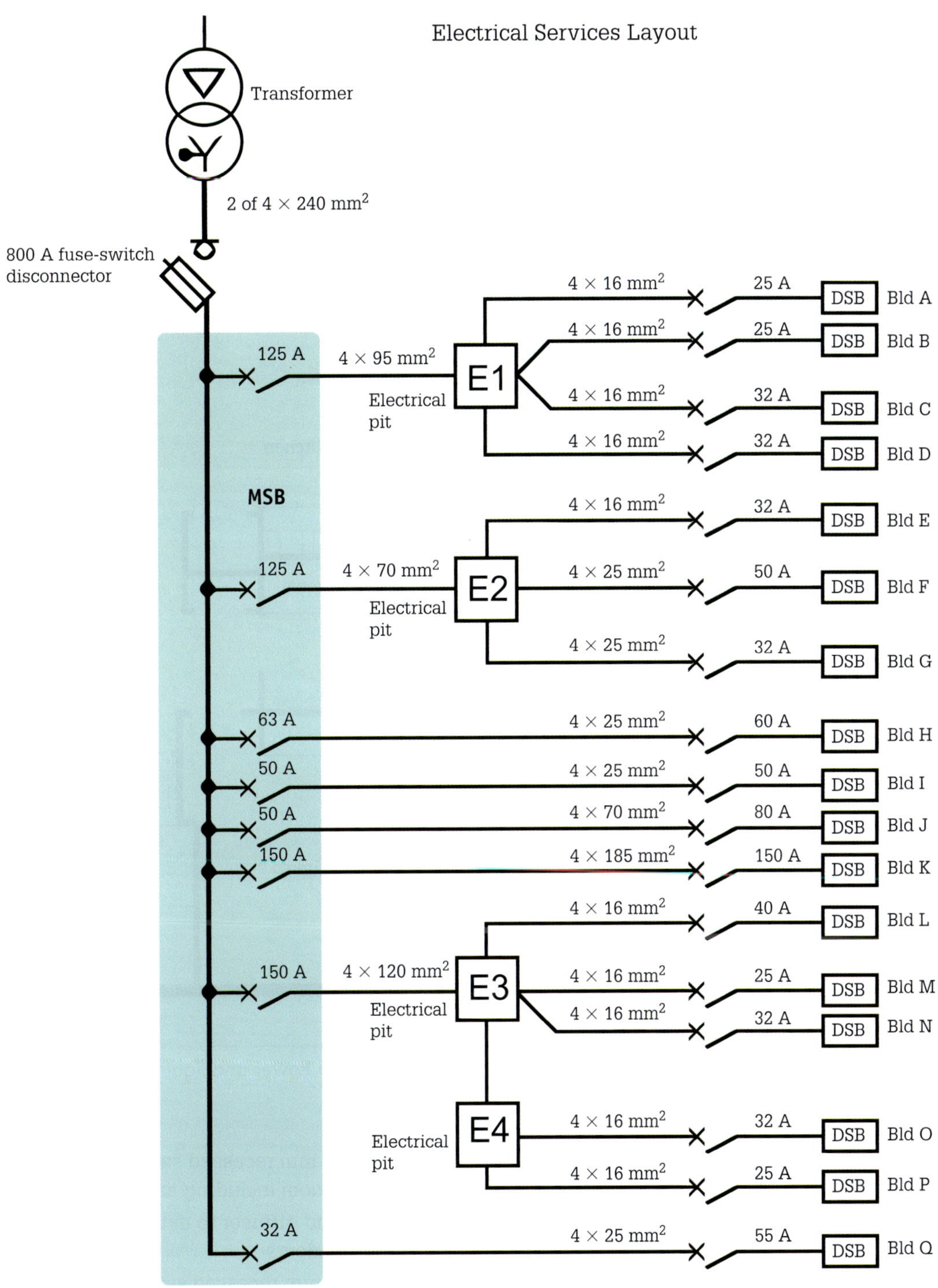

FIGURE 5.36 Electrical reticulation services layout

Wiring schedule

A wiring schedule tabulates information about cables for a particular installation. The schedule can include the number of phases, circuit protection type and rating, the quantity and size of the cables, and circuit identification.

An illustration of a portion of the circuit wiring schedule for a carpentry/joinery workshop is shown in **Figure 5.37**.

Line diagrams

Line diagrams, illustrated in **Figure 5.38**, are a simplified notation for representing a power system. Instead of representing each conductor with a separate line or terminal, only one conductor is represented. Symbols on the diagram do not represent the location of the electrical equipment.

Figure 5.38 shows line diagrams for various lighting circuits where power is looped at the light and is controlled by various switch configurations. The slashes through the lines represent the number of conductors between the components.

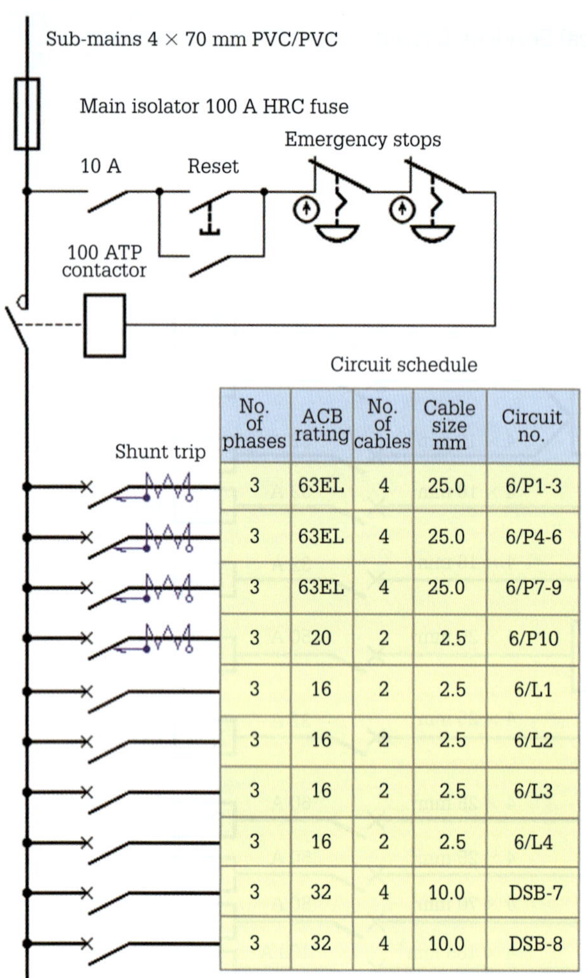

Circuit schedule

No. of phases	ACB rating	No. of cables	Cable size mm	Circuit no.
3	63EL	4	25.0	6/P1-3
3	63EL	4	25.0	6/P4-6
3	63EL	4	25.0	6/P7-9
3	20	2	2.5	6/P10
3	16	2	2.5	6/L1
3	16	2	2.5	6/L2
3	16	2	2.5	6/L3
3	16	2	2.5	6/L4
3	32	4	10.0	DSB-7
3	32	4	10.0	DSB-8

FIGURE 5.37 Portion of the circuit wiring schedule

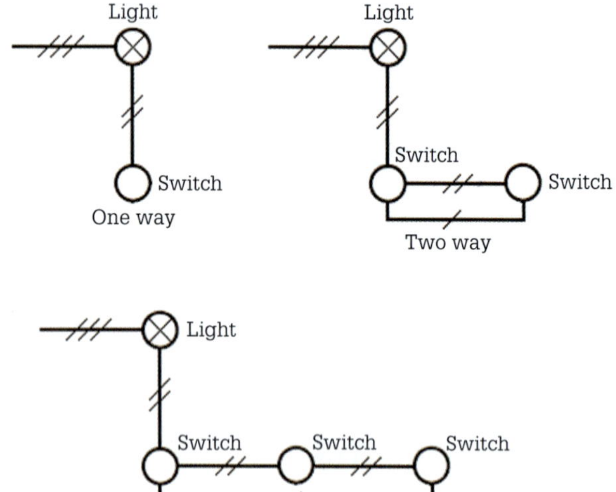

FIGURE 5.38 Lighting line diagrams

EXERCISE 5.8

Using Figure 5.39, show the power and lighting layout using electrotechnical symbols.

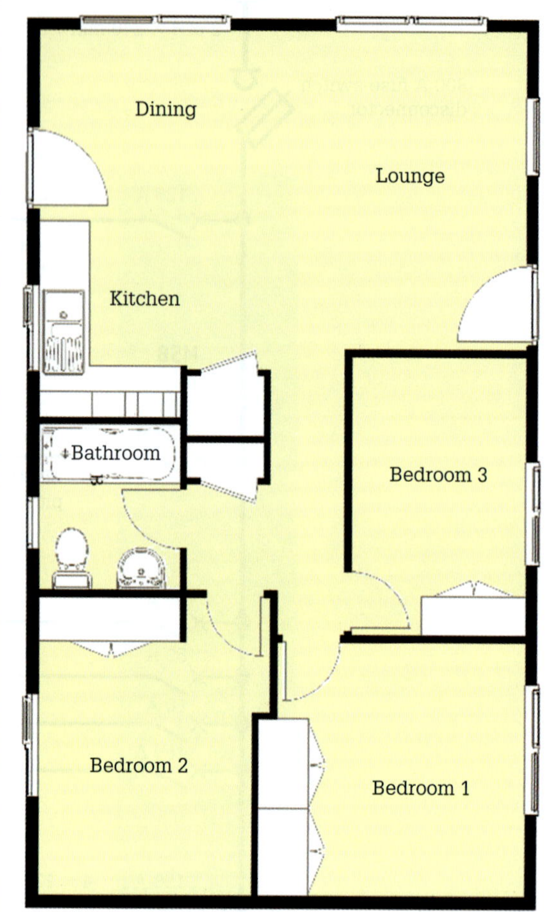

FIGURE 5.39 Power and lighting layout exercise

Lights:

- Two 100 mm recessed satin chrome light fittings to each room including area outside bathroom
- One round diffuser to exterior walls near both external doors
- One 100 mm recessed satin chrome light fitting to pantry

Socket outlets:

- 2 × double socket outlets per room including area outside bathroom
- Single socket outlet for range hood connection (kitchen)
- Single socket outlet for the dishwasher
- Single socket outlet for the microwave

Phone points:

- One point to the kitchen
- One point to bedroom 1

TV:

- One point to the lounge room

Smoke alarms:

- Smoke alarms hard wired (three required).

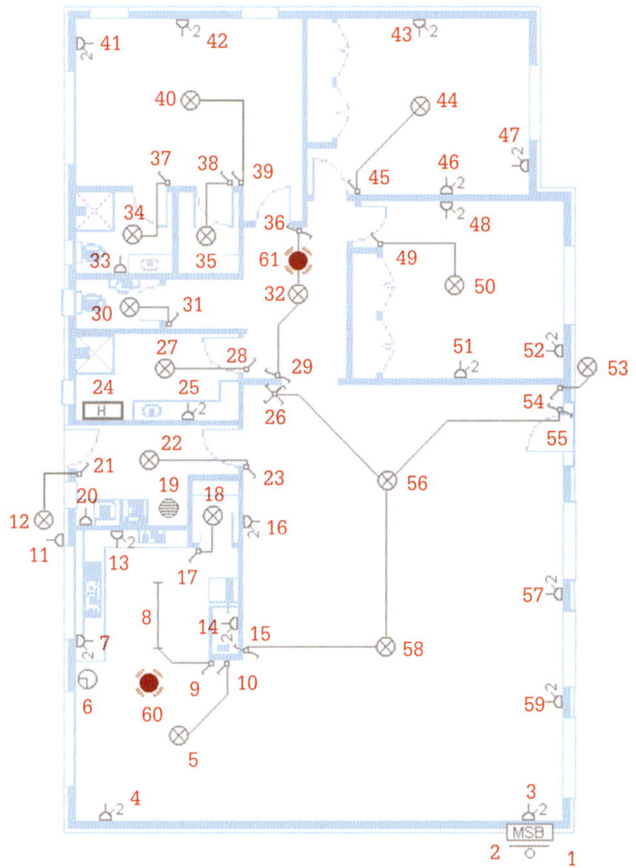

FIGURE 5.40 Installation diagram with item numbers

Electrical specification schedule

The item numbers in **Figure 5.40** refer to a typical electrical specification schedule used to ensure that the installation is supplied with all the equipment required as shown in **Table 5.1**.

EXERCISE 5.9

a Using the floor plan template of **Figure 5.41**, produce a lighting diagram. Insert the symbols for the switches and luminaires. Draw in the connecting lines to their appropriate switches. Include a compact fluorescent luminaire in each room space and one fluorescent tube luminaire in the kitchen. A spotlight is required for the washing machine and a luminaire fixed to the outside wall is to the right-hand side of the sliding doors. Each room space must have its luminaire controlled by a wall switch located on the latch portion of the entrance to the room space. A two-way switch is installed in the laundry area opposite the entrance and near the bedroom door.

b Using the floor plan template of **Figure 5.42**, produce a topographical socket outlet layout diagram. Insert the symbols for the socket outlets. Except for the bathroom (one socket outlet) every room space must have two double 10 A socket outlets.

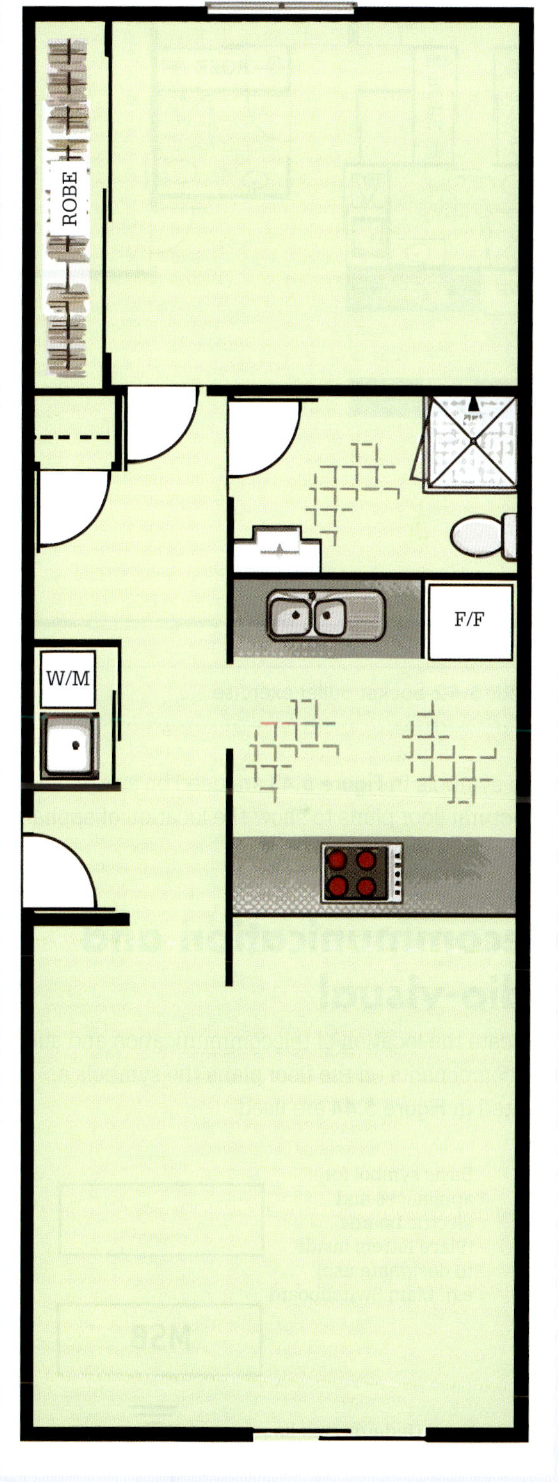

FIGURE 5.41 Lighting exercise

»

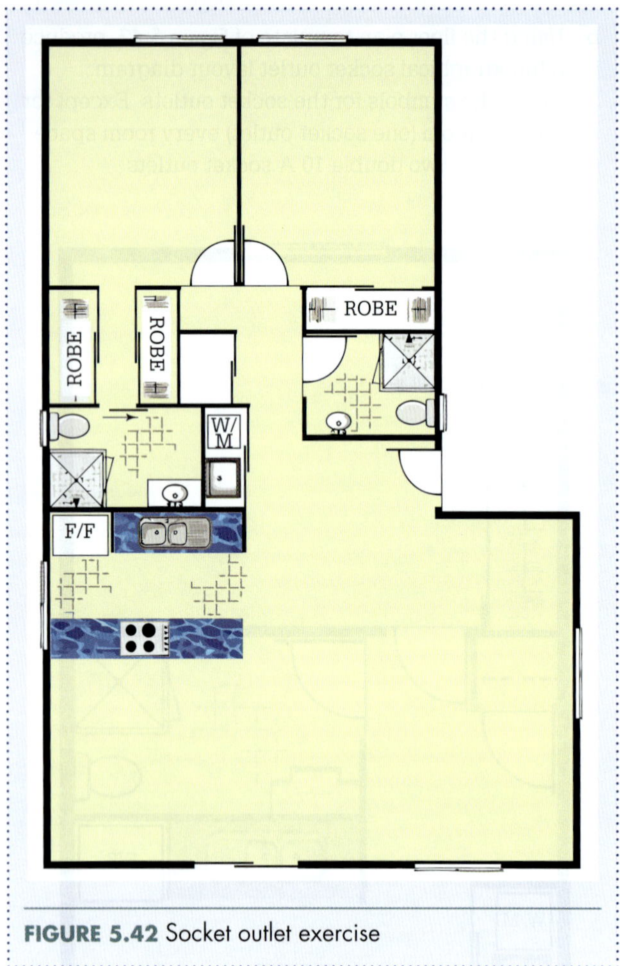
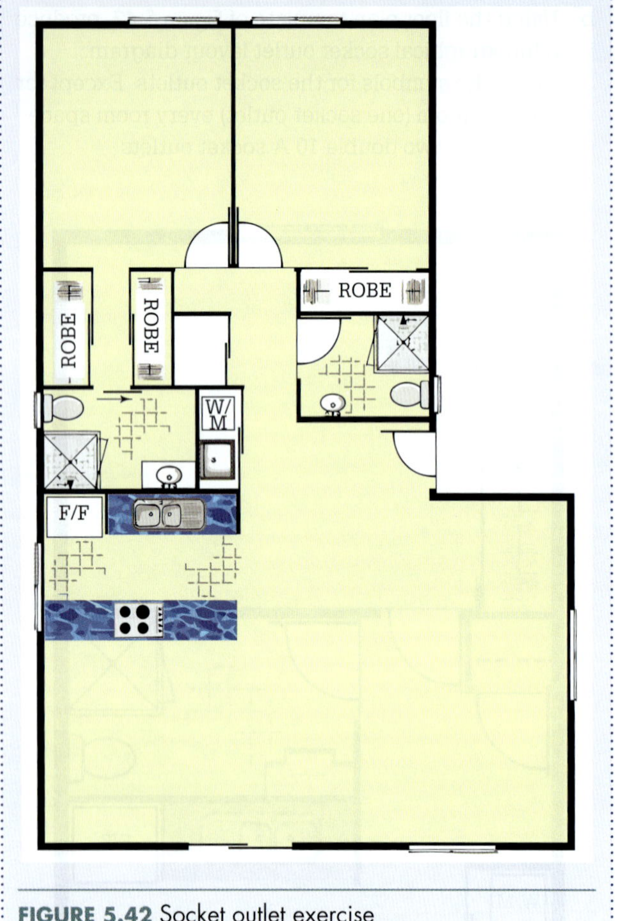

FIGURE 5.42 Socket outlet exercise

The symbols in **Figure 5.43** are used on site plans and architectural floor plans to show the location of appliances and necessary conductors.

Telecommunication and audio-visual

To indicate the location of telecommunication and audio-visual components on the floor plans the symbols as illustrated in **Figure 5.44** are used.

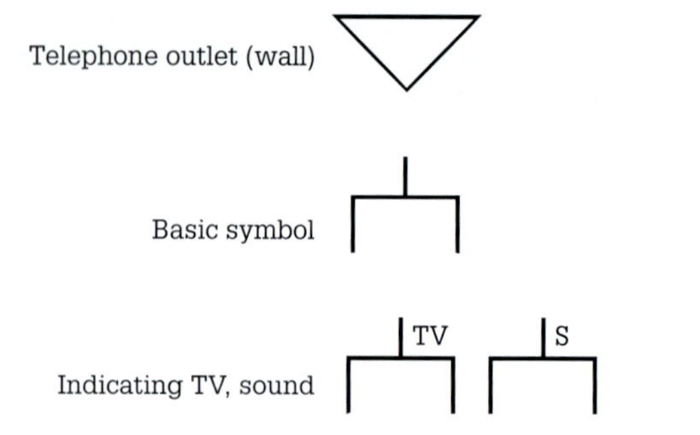

FIGURE 5.44 Telecommunication and audio-visual symbols

Additional types of working electrical drawings

A reticulation diagram for distribution boards within an installation together with their associated feeder cables sizes is illustrated in **Figure 5.47**.

An electrical services layout for lighting and power is depicted in **Figure 5.48**.

A single-line diagram of a single busbar riser feeding sub-mains distribution boards is shown in **Figure 5.49**.

A diagram for a commercial building showing cable rising mains is depicted in **Figure 5.50**.

Particular conductors such as neutral and protective earths have their special identification as shown in **Figure 5.51**.

A surface socket can be regarded as a lighting point under certain conditions as illustrated in **Figure 5.52**.

Figure 5.46 illustrates various electrotechnical diagrams to explain a light controlled by a single switch and protected by a circuit breaker.

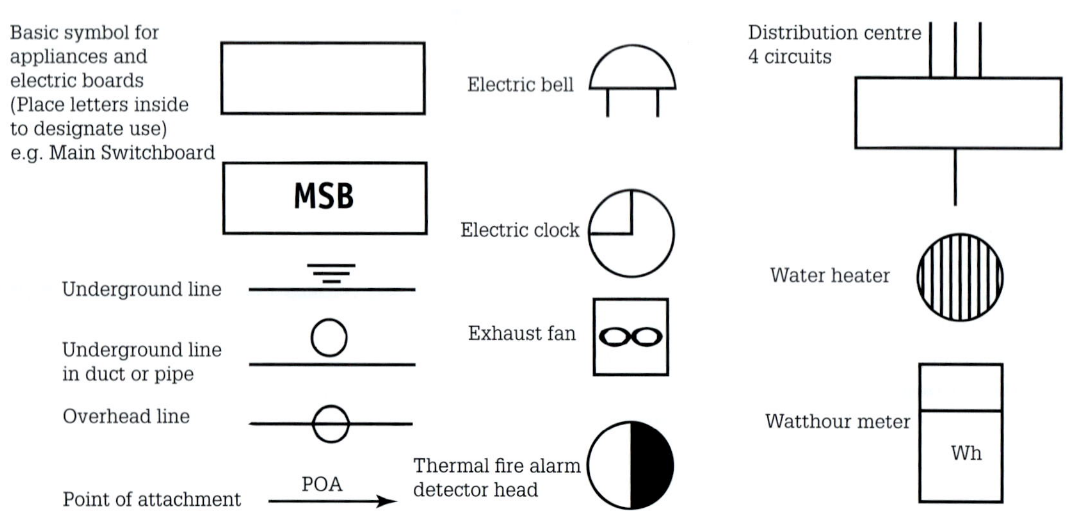

FIGURE 5.43 Appliances, boards and line symbols

EXERCISE 5.10

Write the grid reference for the location of the electrotechnical symbols in **Figure 5.45** in the Table provided in the figure.

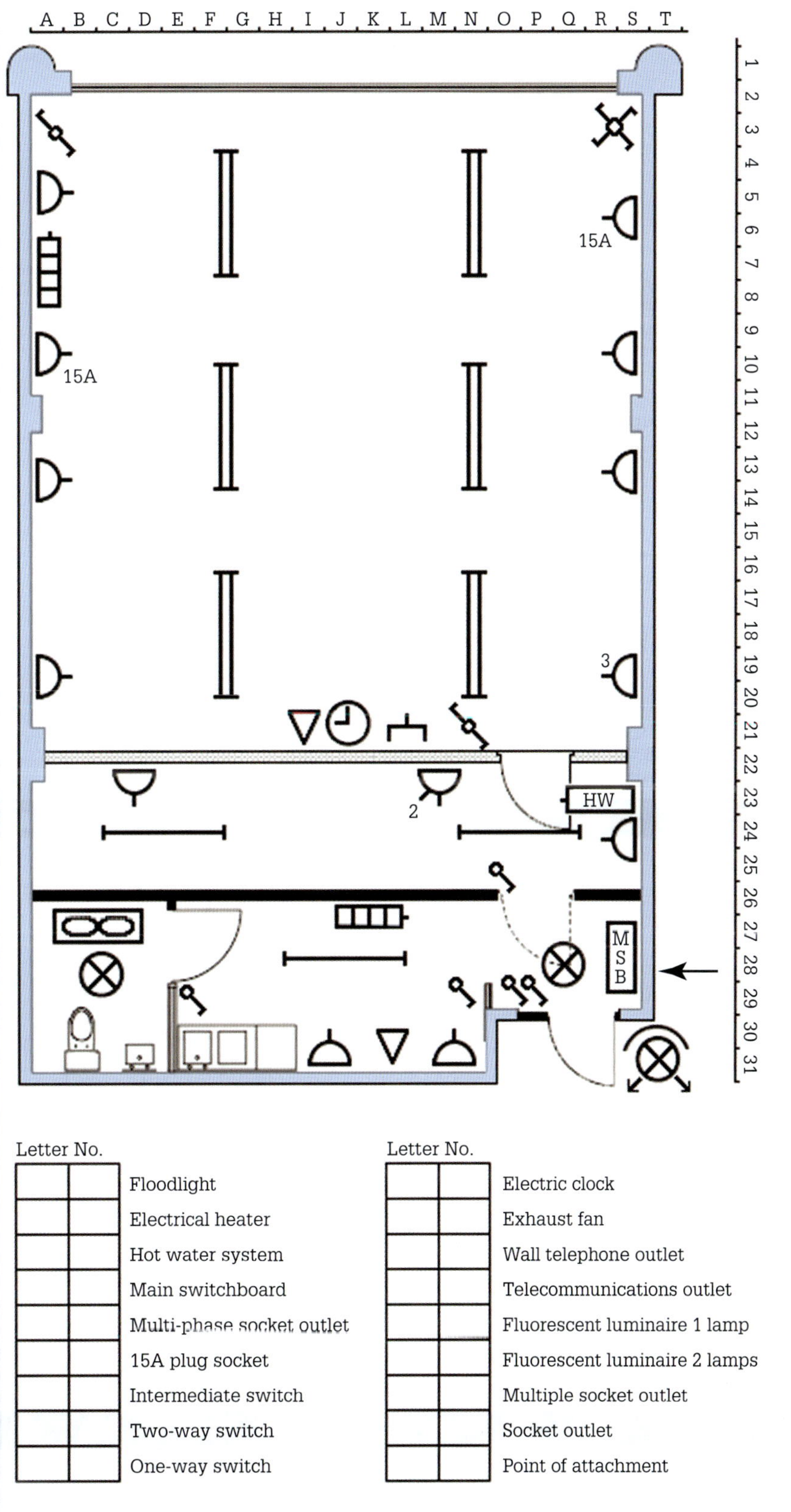

Letter	No.		Letter	No.	
		Floodlight			Electric clock
		Electrical heater			Exhaust fan
		Hot water system			Wall telephone outlet
		Main switchboard			Telecommunications outlet
		Multi-phase socket outlet			Fluorescent luminaire 1 lamp
		15A plug socket			Fluorescent luminaire 2 lamps
		Intermediate switch			Multiple socket outlet
		Two-way switch			Socket outlet
		One-way switch			Point of attachment

FIGURE 5.45 Electrotechnical symbol exercise

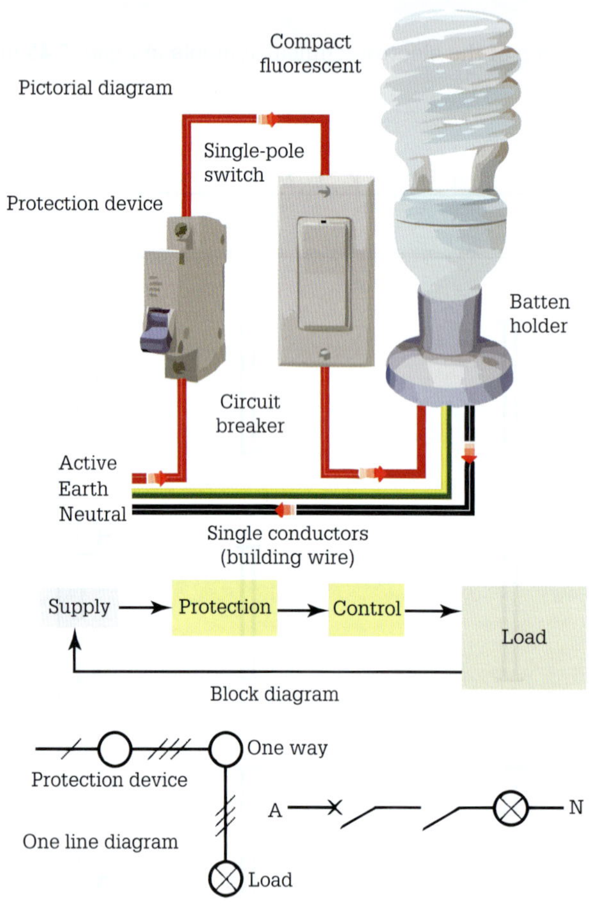

FIGURE 5.46 Various electrotechnical diagrams

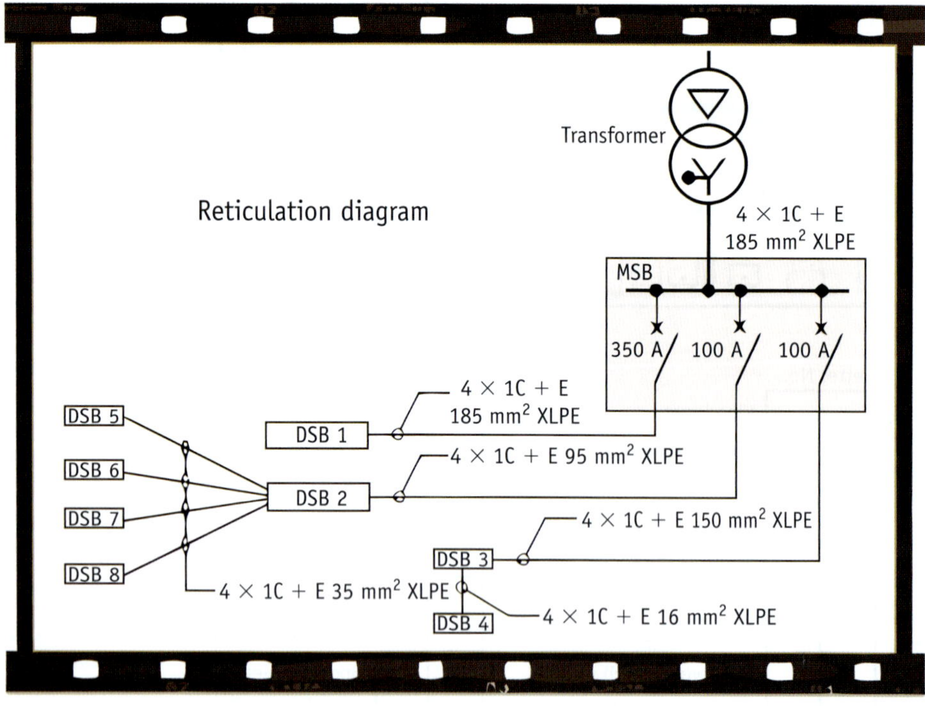

FIGURE 5.47 Reticulation diagram

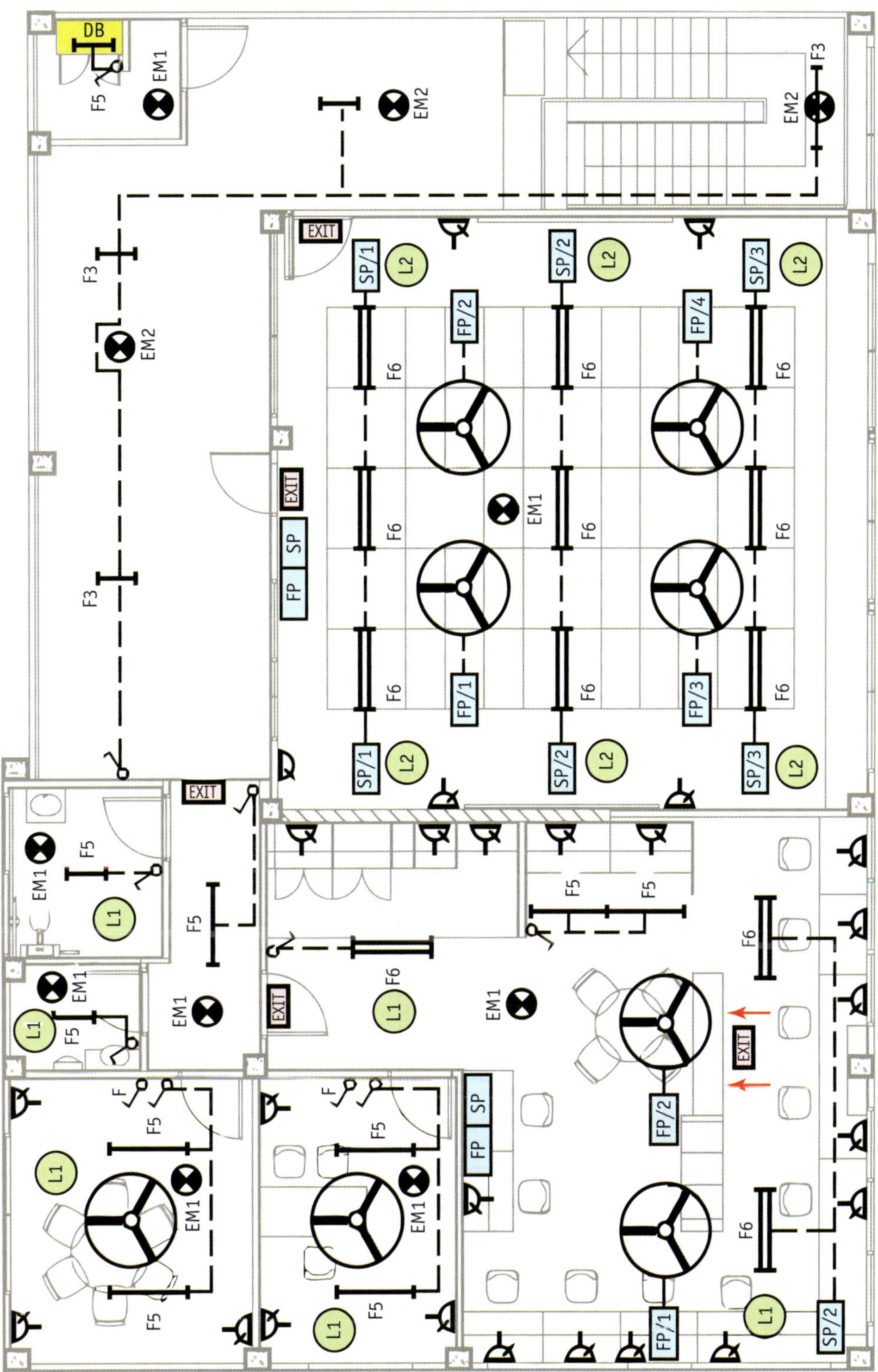

FIGURE 5.48 Electrical services layout

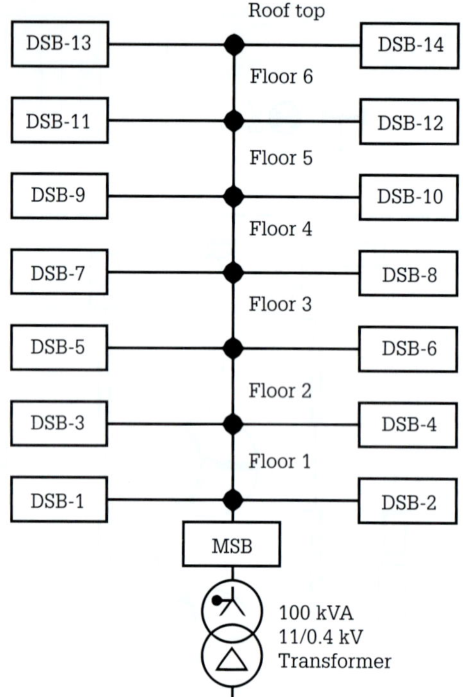

FIGURE 5.49 Line diagram of a single-feed busbar rising main

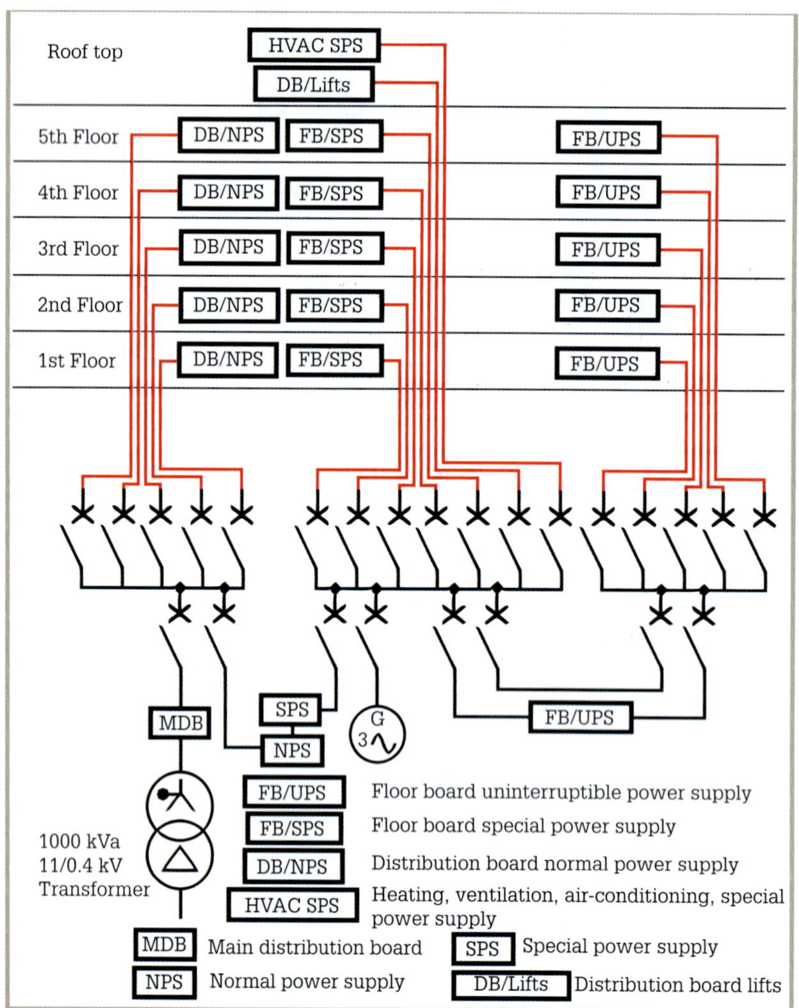

FIGURE 5.50 Electrical services – rising mains commercial building

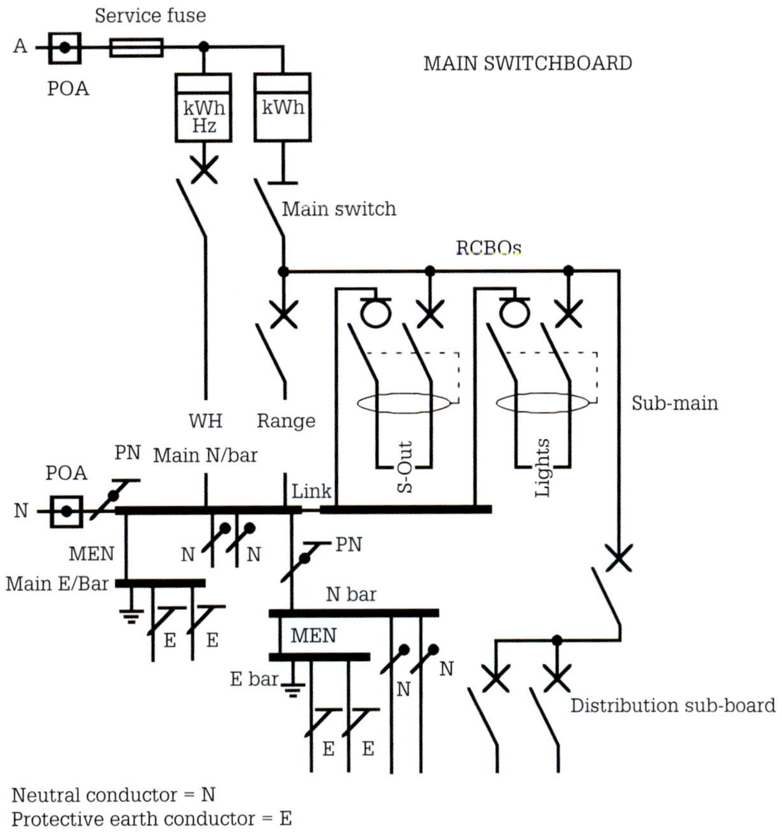

FIGURE 5.51 Specific conductor symbols

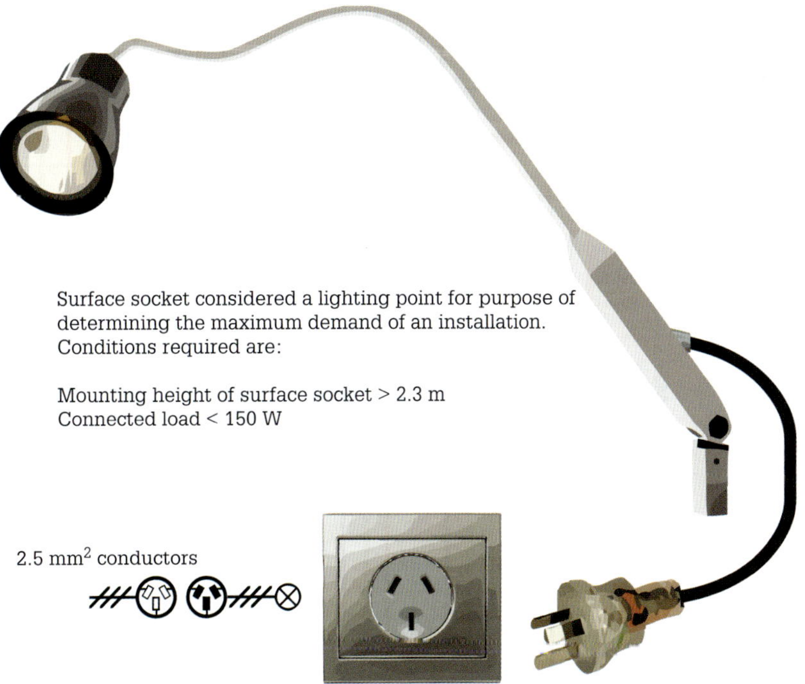

FIGURE 5.52 Surface socket lighting point

REVIEW QUESTIONS

1 Which diagram provides a method of explaining complex systems in a simple form?
2 Name the diagram used to accurately record the routes of underground feeder cables.
3 How is the direction of underground cable runs indicated on a marker plate?
4 Which diagram uses a single-line diagram to indicate the interconnection of switchboards within an electrical installation?
5 What is a wiring schedule?
6 Which diagram is a simplified notation for representing a power system?

5.5 Wiring diagrams

A wiring diagram or drawing is a detailed diagram of a circuit showing all of the conductors and where they are connected, connectors, links, bars and electrical devices of the circuit. The wiring diagram can also identify the conductors by numbers, letters or colour coding. Wiring diagrams are necessary to fault-find and repair electrical installation circuits. A wiring diagram for a domestic installation is shown in **Figure 5.53**.

It shows the electrical components and the interconnecting conductors which are colour coded. Wiring diagrams are always shown with components in their de-energised circuit condition.

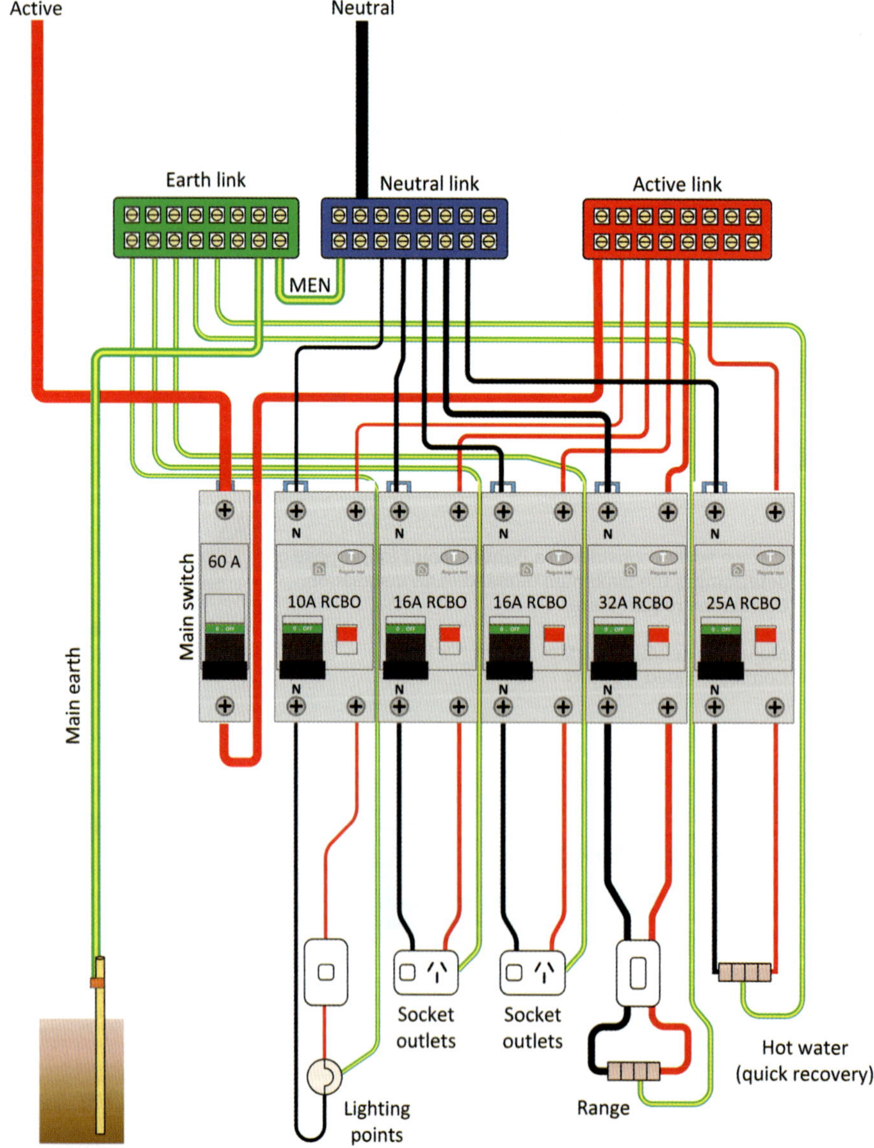

FIGURE 5.53 Domestic installation wiring diagram

Converting circuit diagram into a wiring diagram

A wiring diagram, as shown in **Figure 5.53**, can assist with the physical installation of the wiring. A wiring diagram is usually developed from the associated circuit diagram, which should detail wire numbers and terminal numbers. Numbering all of the wires and device terminals facilitates the development of a wiring diagram using a 'join the dots' process. Each point that shares the same number is electrically common and needs to be electrically continuous. Use straight lines and only connect wires at terminal points on equipment. It is important to ensure that all connections occur at equipment terminals. In practice, it is best to only connect a maximum of two wires to any one point and do not effect (make) a mid-wire splice or join.

Circuit diagrams are easy to read and make it relatively easy to logically troubleshoot a circuit. Wiring diagrams show how equipment physically interconnects.

Cable schedule

A cable schedule is a table containing information about cables in a particular installation. The following is the type of information that a cable schedule could include:

- isolator type and rating
- circuit number
- protection device type and rating
- phase (line) and neutral conductor sizes
- earth conductor size
- number of points
- length of the circuit
- circuit name.

Without a cable schedule (see **Figure 5.54**) you cannot easily estimate the cable cost and total length of cable required.

Cable schedules are used for commercial and industrial installations and rarely for domestic installations. Schedules should be typed on white card and installed in clear acrylic faced holders fixed to the back of switchboard doors or on a wall adjacent.

Circuit no.	Core size mm^2	Current (A)	Route length	Application
MSB C/B 1	1.0 T&E flat	10	30 m	Lighting
MSB C/B 2	1.5 T&E flat	16	20 m	Lighting
MSB C/B 3	2.5 T&E flat	16	25 m	Power circuit
MSB C/B 4	4.0 T&E flat	32	15 m	Air/C
MSB C/B 5	6.0 T&E flat	40	8 m	Range

FIGURE 5.54 Cable schedule

EXERCISE 5.11

Calculate the route length of each lighting circuit in the location diagram shown in **Figure 5.55** and develop a cable schedule. Allow 2 m of cable for each switch drop. Use a scale 1 mm in the diagram = 100 mm actual size.

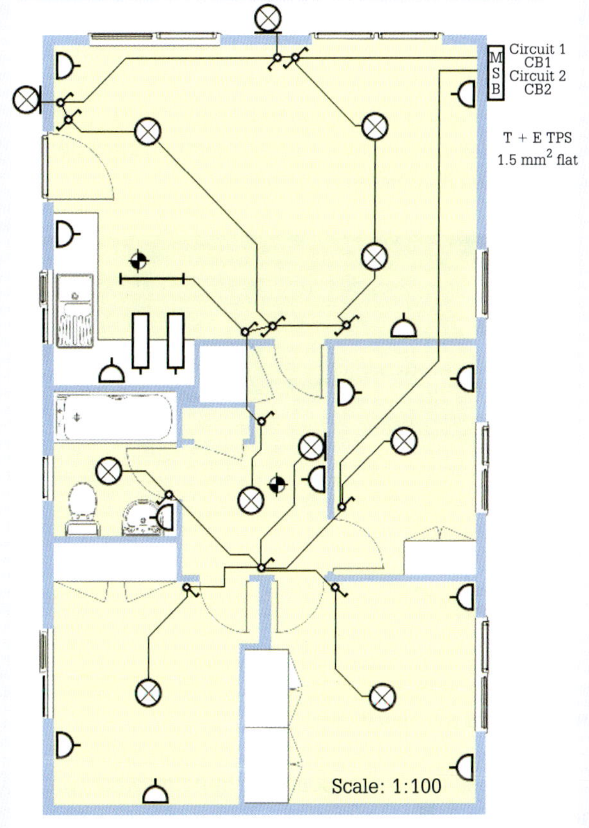

FIGURE 5.55 Location diagram exercise

REVIEW QUESTIONS

1. Which drawing is a detailed diagram of a circuit showing all of the conductors and where they are connected, connectors, links, bars and electrical devices of the circuit?
2. What information is provided on a cable schedule?
3. In what installations are cable schedules used?

5.6 Job specifications

Job specifications, or simply 'specs', detail the work and workmanship needed to complete a building project, which includes the electrical works. Job specifications form part of the formal contract process (**Figure 5.56**). By using a job specification it is possible to reduce the number of variations or extras on a building project. A job specification details the acceptable level of quality for the project. For a specification to be enforceable, it must be included or referenced in the contractual agreement between the customer and the builder.

Source: Shutterstock.com/Lost_in_the_Midwest

FIGURE 5.56 Job specification

A job specification for domestic building work would include the following sections:

- Preliminaries
- Common requirements
- Termite management
- Demolition of existing structure
- Site management
- Earthworks
- Service trenching
- External – fences and barriers
- External – gardening
- Pavement base and sub-base
- Concrete pavement
- Segmental pavers – sand bed
- Concrete
- Brick and block construction
- Light steel framing
- Light timber framing
- Sheet flooring and decking
- Roofing
- Cladding
- Windows and glazed doors
- Doors and access panels
- Overhead doors
- Door hardware
- Glass components
- Insulation and sarking membranes
- Lining
- Joinery
- Miscellaneous appliances and fixtures
- Rendering and plastering
- Waterproofing – wet areas
- Ceramic tiles
- Resilient finishes
- Carpets
- Engineered panel floors
- Timber flooring
- Floor sanding and finishing
- Painting
- Mechanical design and install
- Hydraulic design and install
- Electrical design and install.

NATSPEC

Construction Information Systems Limited, or NATSPEC (its trading name), is a not-for-profit organisation that is owned by the design, build, construct and property industries through professional associations and government property groups. NATSPEC's primary service is a comprehensive national specification system endorsed by government and professional bodies. NATSPEC, the National Building Specification, is for all building structures with specialist packages for architects, interior designers, landscape architects, structural engineers, service engineers and domestic owners.

For example, NATSPEC uses templates in which specifications are prepared covering electrical services systems, performance requirements and standards. The quality of an electrical installation project is dependent on the documentation provided. NATSPEC template specifications are a multipurpose document, whose function is to define the quality required for an installation and the processes necessary for achieving it accurately. NATSPEC's role includes the selection of materials, but it also provides the requirements for the minimum level of workmanship required for the installing work.

REVIEW QUESTIONS

1 What is detailed in a job specification?
2 What service is provided by NATSPEC?
3 What is needed to make a job specification enforceable under law?

5.7 Electrical licensing

Electrical licensing requirements vary between each jurisdiction, but a common element is that the requirements for electrical licensing are established in legislation.

The Electrical Safety Act

The Electrical Safety Act is the legislative framework for electrical safety. The purpose of the Electrical Safety Act is to protect persons and property from electric shock, fire and other hazards resulting from electricity. The Electrical Safety Act sets out a framework that establishes consultative structures, standards, and safety management systems, imposes obligations and provides a system of licensing and penalties.

The purpose of the following legislation is to set out precautions to be taken against the risk of death or personal injury from electricity during work activities.

- Electrical Safety Acts in 2006 (New South Wales), 1998 (Victoria), 2002 (Queensland), 1996 (South Australia), 1945 (Western Australia), 1997 (Tasmania), 2005 (Northern Territory), 1971 (Australian Capital Territory), 1992 (New Zealand) and amendments.
- Electricity regulations in 2001 (New South Wales), 2009 (Victoria), 2002 (Queensland), 1997 (South Australia), 1947 (Western Australia), 2005 (Northern Territory), 2004 (Australian Capital Territory), 1997 (New Zealand) and amendments.
- Work Health and Safety Acts in the various states, territories, and New Zealand.
- Work health and safety regulations in the various states, territories, and New Zealand.

Work Health and Safety Acts, Electrical Safety Acts and electricity regulations provide a number of means to ensure that the hazards associated with the installation, operation and maintenance of electrical installations are actually controlled. Legislation makes mandatory the technical standard AS/NZS 3000:2018 *Wiring Rules*.

Legislation has also established the Electrical Certificate of Compliance scheme as a means of industry self-regulation requiring compliance with the standards.

Electrical safety regulations

Electricity safety regulations identify specific ways for an electrical entity, employer, designer, manufacturer, importer, supplier, installer, repairer, person in control, worker and other persons to meet their safety obligations under the Electrical Safety Act.

Ministerial notices

Ministerial notices are used in urgent circumstances. Such a notice would describe the manner in general terms of how to carry out electrical safety obligations according to the circumstances. Ministerial notices are published in the government gazette.

Codes of practice

Codes of practice provide practical advice on how to meet the electrical safety obligations established under the Electricity Act. The objective of codes of practice (i.e. 'Safe Electrical Work for Low Voltage Electrical Installations' and 'Electrical Practices for Construction Work') is to outline principles and methods of safe work, organisation and performance required to be undertaken when carrying out electrical installation and maintenance work. Codes of practice provide competent persons working on low-voltage electrical installations or systems with:

- the principles of safe working practices
- recommended procedures for safe working practices.

Technical standards

Standards are intended to advise electrical workers on the procedures necessary to ensure compliance with legislation. Current technical standards include AS/NZS 3000:2018 *Electrical installations – buildings, structures and premises* (*Wiring Rules*). This standard is the most applicable technical standard for electrical installations and is mandated by the legislation. It calls up a number of attendant standards.

Technical standards contain the latest available knowledge on how to install and test an electrical installation. The standards are constantly reviewed and improved as a result of working experience and through the development of better equipment and fittings.

Responsible person(s) and electrical installations

The owner(s) and operator of an electrical installation have a general duty of care. These persons must take reasonable steps to ensure that the installation complies with current obligatory standards and that the installation is safe and safely operated. However, the people who are responsible for carrying out the electrical installation work are:

- The licensed electrical contractor (as an employer and controller of the electrical work) who must, therefore, comply with the specific requirements of the legislation to ensure safe electrical installation and safe electrical equipment.
- A prescribed class of persons: the licensed electrical worker who is engaged to carry out the work.
- For electrical installation work: the licensed person who signs the Certificate of Compliance.

It is an electrical contractor, licensed electrical worker and the signatory to the Certificate of Compliance who has statutory obligations to implement safe systems of work to prevent persons from being exposed to electrical hazards. Where installation work is required, it must be undertaken by a licensed electrical worker only.

Licensing requirements

A state- and territory-based licensing system is in place in Australia to ensure that only competent workers who have met certain standards engage in electrical work. By 'competent worker' is meant a person who has acquired, through training, qualifications, experience or a combination of these, the knowledge and skill enabling that person to inspect and test electrical work. 'Electrical work' is defined as the manufacturing, constructing, installing, testing, maintaining, repairing, altering or replacing of electrical equipment or parts.

The following classes of electrical work licences may be issued, depending upon the regulator, in each state or territory.

Australian Capital Territory

- Electrical contractor's licence: permits electrical contracting.
- Electrician's licence unrestricted: holder can perform any class of wiring work as an employee.
- Restricted electrician's licence: allows the disconnection and connection of electrical equipment that is incidental to a worker's main occupation or trade.
- Permit: allows a worker to work under the supervision of an unrestricted licence holder to gain experience in that field.

New South Wales

- Electrical contractor's licence: permits electrical contracting.
- Qualified supervisor certificate: holder can carry out and supervise all types of electrical work or carry out restricted electrical work.
- Restricted electrician's licence: allows the disconnection and connection of electrical equipment that is incidental to a worker's main occupation or trade.

Victoria

- Electrical contractor's licence: permits electrical contracting.
- Unrestricted licence: holder can carry out every class of electrical wiring work in Victoria.
- Restricted licence Class 'B': holder can carry out every class of electrical wiring work in Victoria while under the supervision of an unrestricted licence holder.
- Restricted licence Class 'S': holder can carry out restricted electrical work in Victoria as endorsed on the licence: plumbing/gasfitter equipment installer.
- Restricted licence Class 'N': holder can carry out the installation of cold-cathode discharge lighting systems in Victoria.
- Restricted licence Class 'R': holder can carry out electrical work on premises in Victoria of which the holder is the bona fide occupier.

Queensland

- Electrical contractor's licence: permits electrical contracting.
- Electrical licence: holder can carry out every class of electrical wiring work in Queensland as endorsed on the licence: electrical mechanic, electrical fitter, electrical fitter/mechanic, electrical linesman and electrical jointer.
- Permit: allows the holder to work in the electrical trade while awaiting results of the application or while undergoing training.
- Restricted electrical licence: allows the disconnection and connection of electrical equipment that is incidental to a worker's main occupation or trade: refrigeration/air-conditioning equipment installer.

Tasmania

- Electrical contractor's licence: permits electrical contracting.
- Electrical mechanic licence: holder can perform any class of wiring work as an employee.
- Interim electrical mechanic's permit: holder can carry out every class of electrical wiring work in Tasmania while under the supervision of an electrical mechanic licence holder.
- National restricted electrical licence: allows the disconnection and connection of electrical equipment that is incidental to a worker's main occupation or trade: instrumentation/process control equipment installer.
- Restricted electrical permit: holder can carry out commissioning, fault-finding and replacement of components and work on equipment wiring of refrigeration equipment.
- Electrical wiring works permit: holder can perform electrical wiring work in the holder's or immediate family's home or non-profit organisation without the need for a contractor's licence.

South Australia

- Electrical contractor's licence: permits electrical contracting.
- Electrical worker registration: holder can perform any class of wiring work as an employee.
- Restricted electrical worker: allows the disconnection and connection of electrical equipment that is incidental to a worker's main occupation or trade: communication equipment and laboratory/scientific equipment installer.

Western Australia

- Electrical contractor's licence: permits electrical contracting.
- 'A' Grade electrical mechanic and/or electrical fitter: holder can perform any class of wiring work as an employee.
- 'A' Grade electrical fitter: holder can perform any class of fitting work without supervision.
- Permit: allows a worker to work for a nominated employer under supervision of an 'A' grade licence holder to gain experience in that field.
- 'C' Grade electrical worker's licence: this is issued to an engineering tradesperson, and electrical or electrical mechanic apprentices/trainees.

Northern Territory

- Electrical contractor's licence: permits electrical contracting.
- Electrical worker's licence: holder can perform any class of electrical work as an employee as endorsed on the licence: electrical mechanic, electrical fitter, electrical cable jointer and electrical linesman.
- National restricted licence: holder is limited to electrical work on apparatus and appliances and replacement of fuse links at voltages not exceeding 650 V: commercial and industrial equipment installer.

An electrical contractor's licence allows the holder to establish the business of an electrical contractor and to contract for or to carry out for fee or reward electrical installation work. Electrical installation work means the installation or maintenance of electric cables, accessories or equipment in any electrical installation. The holder of a contractor's licence must maintain civil liability insurance against personal injury and damage to property in connection with the electrical installation work.

An electrical mechanic's licence is an open electrical work licence which allows the holder to perform all electrical installation work and to maintain or repair any electrical equipment when employed by or as an electrical contractor. The electrical mechanic's licence does not allow the holder to 'contract' to the general public or industry to carry out electrical work. The licence does enable the holder to perform electrical work for themselves or their relatives at premises owned or occupied by them. It also allows a contract to the general public to carry out some restricted electrical work.

Other electrical qualifications

An electrical linesperson qualification allows the holder to perform all electric line work. The electrical work to which this qualification applies involves installing or maintaining an overhead line, street lighting and the testing of such lines.

An electrical fitter's qualification allows the holder to perform all electrical equipment work that is physically isolated from the mains supply. The electrical work to which an electrical fitter's qualification applies involves the rewinding and repairs to an electric motor, power tools, transformers, switchgear, various electrical articles and the wiring of a switchboard. In addition, the qualification allows the holder to operate and test the electrical equipment.

SWITCH ON

Both the electrical mechanic and the electrical fitter have the single title 'electrician'. However, a common practice is to use the title 'electrician' for the electrical mechanic.

An electrical joiner's qualification allows the holder to install, join and terminate high- and low-voltage cables that require specialised knowledge or skill.

A restricted electrical work licence enables the holder to perform particular types of electrical work incidental to a particular trade or calling. For example, a refrigeration mechanic's restricted electrical licence allows the holder to install and maintain refrigeration equipment. It does not allow them to carry out any electrical installation wiring from the switchboard. They can disconnect and reconnect fixed wired equipment.

A plumber's restricted electrical licence enables the holder to disconnect and reconnect the wiring to a hot water storage device. A gas fitter's restricted electrical licence may allow the holder to work on the 'electrical' components of gas/electrical appliances. All restricted electrical licences do not enable the holder to perform electrical work outside of the conditions that are stated in the licence.

An electrical work training permit allows an applicant to undertake specific training and gain the experience needed to meet the eligibility requirements for an electrical work licence.

All cabling work that is connected or is intended to be connected, either directly or indirectly, to the telecommunications network requires the cabler to be registered – open or restricted or directly supervised by an open registered cabler. Cablers working on the fire alarm, security alarm, data and lift cables industries are also required to be registered cablers.

A competent person trained in testing and tagging work can inspect, test and tag plug-in-type electrical equipment. Test and tag persons are trained in the use of a portable appliance tester (PAT) and have the knowledge and skills enabling them to perform an inspection and testing task correctly.

Ideally all electrical workers should be trained and competent to carry out rescue breathing (rescue breathing is a component of cardiopulmonary resuscitation (CPR)) and basic first aid.

REVIEW QUESTIONS

1 What is set out under the Electrical Safety Act?
2 What is identified under electricity safety regulations?
3 When are Ministerial notices used?
4 Name the documents that provide practical advice on how to meet the electrical safety obligations established under the Electricity Act.
5 Who has statutory obligations to implement safe systems of work to prevent persons from being exposed to electrical hazards?
6 What work is allowed under an electrical fitter's qualification?
7 What work is allowed under an electrical joiner's qualification?
8 What work is allowed under a plumber's restricted electrical licence?

5.8 Technical standards

A technical standard is a living document that establishes benchmarks of good practices and a common language to provide a level of quality that is respected. These living documents were created by consensus and are continually revised and updated to meet changing requirements and emerging technologies. Standards outline conditions and require actions designed to make sure that products and systems, goods and services are safe, reliable and fit for purpose. Standards provide order, conformity, and convenience, ease of choice, economic efficiency, quality and universality.

Standards are written by organisations such as Standards Australia, which was set up for that purpose. Standards associations serve both the public good and the electrical industry. The members of the associations through various committees contribute their time, experience and technical knowledge to the development of standards. This process empowers the standard with integrity and technical credibility.

Standards bodies

The Australian Standards Association functions as a coordinating body, providing a structure and a forum for developing standards. Technical committees are created using a balanced approach, which means that each committee is structured to take advantage of the combined experience and expertise of its members – with no single industry group dominating.

The technical committee considers the views of all members, including the public, and develops the details of each standard by a consensus process. Significant agreement among committee members based upon established procedures, rather than a simple majority of votes, is required before a draft standard is finalised. When a draft standard has been agreed upon, it is submitted for public review and amended if necessary. The document is then given a final review and published as an Australian standard.

Standards Australia represents Australia on two international standardising bodies: the International Organization for Standardization (ISO – based on the Greek word for equal) and the International Electrotechnical Commission (IEC).

The ISO is a network of the national standards institutes of 100-plus countries with a secretariat in Switzerland. The objective of the ISO is to simplify the international harmonisation and merger of industrial standards. This goal is achieved through consensus agreements between the various standards institutes. International standards are developed by technical committees to define the characteristics and framework that products and services are expected to meet when trading across the world. A simple example of ISO standardisation is the ease of choice between lamp caps for light bulbs: Swan bayonet or Edison screw.

The IEC is the international standards and assessment body that prepares and publishes international standards for all electrical, electronic and related technologies. Through its members, the IEC promotes international cooperation to standardise the design and construction of electrical devices and equipment. This ensures that the product is reliable, safe and dependable. A simple example of IEC standardisation is the range of 'Clipsal' brand products.

Use of standards

Standards are voluntary documents that establish specifications, procedures and guidelines. The purpose of these is to ensure products, services and systems are safe, consistent and reliable. Standards deal with a wide variety of subjects, which can include construction, energy, consumer products and services, the environment, and water utilities.

Essentially, standards fall into three categories:

- International Standards, which are developed by ISO, IEC and ITU for countries to adopt for national use. Standards Australia embraces the development and adoption of international standards.
- Regional Standards, which are prepared by a specific region, such as the European Union's EN standards.

- National Standards, which are developed either by a national standards body, such as Standards Australia, or other accredited bodies. Any standards developed under the Australian Standard® name were created in Australia or are adoptions of international or other standards.

Statutory regulations may call upon Standards such as the *Wiring Rules*, which effectively makes the standard part of the regulation.

Wiring Rules

Wiring Rules is the most widely circulated Australian and New Zealand standard and is used nationally throughout the electrical industry. It is a regularly revised performance-based standard providing outcomes and a comprehensive method for compliance. Adherence to the *Wiring Rules* ensures safety and suitability in all types of installation work.

The Australian/New Zealand Standard (*Wiring Rules*) provides minimum requirements that relate to electrical installations. The requirements of this standard are enforceable by legislation throughout Australia, its territories and in New Zealand.

The standard is developed by a consensus-based process among various stakeholders, including the Association of Consulting Engineers Australia, Australian Building Codes Board, Australian Industry Group, Communications, Electrical and Plumbing Union, Consumers' Federation of Australia, Electrical and Communications Association (Qld), Electrical Regulatory Authorities Council, ElectroComms and Energy Utilities Industries Skills Council, Energy Networks Association, Engineers Australia, Institute of Electrical Inspectors, National Electrical and Communications Association and Telstra Corporation Limited and their equivalent organisations in New Zealand. In addition, the standard has been agreed to by government regulators and is enforced by regulation.

Arrangement of the technical standard

The AS/NZS 3000 Standard is a document containing two parts. Part 1 covers essential elements necessary to meet regulatory requirements. This part includes the scope, application and fundamental principles and provides a method for acceptance of installation design practices that may not be addressed in Part 2. Part 2 covers installation practices that achieve compliance with Part 1. **Table 5.4** provides an overview of what the various sections of the *Wiring Rules* cover.

It is useful to think of the *Wiring Rules* as a tool, which is an essential part of day-to-day work. It is therefore important for the novice to develop familiarity with the layout of these rules so that pertinent information can be located quickly and accurately. To use a standard, one has to first locate the specific information and then apply that information to a specific situation.

It may be useful to add your own indexing tabs to parts of the standard that are used frequently, such as maximum demand tables. Your own margin notes are also quite helpful in demystifying some content in a manner you understand. To locate specific information in the standard you can:

1. Use the contents at the beginning of the standard, which provides broad context for the information you seek, such as 'switchboards'.
2. Use the index at the back of the standard to locate more specific information such as 'Switchboards – prohibited locations'.

Reading the *Wiring Rules*

Each section of the *Wiring Rules* contains information in the form of 'clauses'. The clauses are identified by a point numbering system. For example, the clauses pertaining to Section 2 commence with the number '2'.

A clause details the specific requirements pertaining to a specific aspect of an electrical installation. For example, *Clause 2.3* in *Section 2* details the requirements for the 'Control of Electrical Installation'. As the degree of specificity of the information increases, that part of the clause is designated with a new heading, which is identified with an additional point and number. For example, *Clause 2.3.3* details the requirements for 'Main Switches'. This is broken down further to provide more specific details as is the case with *Clause 2.3.3.3*, which details the 'Number of main switches'. It follows then that if '2.3' represents a clause, then '2.3.3' and '2.3.3.3' are sub-clauses.

Typefaces

The *Wiring Rules* make use of four different typefaces, which have the following specific purpose:

- **Opening statements are in bold print and define the fundamental principles and requirements.**
- The body of the clause, contained in Sections 1 to 8, appear in normal print and indicate the minimum methods deemed to satisfy the requirements of the clause. Normal print in an Appendix represents informative material for guidance only.
- *Exceptions or variations to requirements are in italic print and generally provide specific examples where the requirements do not apply or where they are varied for certain applications. Italic text can provide specific examples of methods deemed to satisfy the requirements of the clause.*
- Text preceded by the word 'NOTE' and printed in reduced normal print provide explanations and advice.

When reading content in the *Wiring Rules* it is vital to read text in conjunction with the preceding and following paragraphs, as these may contain additional or modifying requirements.

TABLE 5.4 Overview of the contents of the *Wiring Rules*

Part	Section	Outline
1	1	**Scope, application and fundamental principles** Provides definitions of electrical terms used in the Standard and establishes the fundamental principles for safety, performance and responsibilities for ensuring that an electrical installation is safe and operates as intended.
2	2	**General arrangement, control and protection** This section stipulates the minimum requirements for the selection and installation of switchgear and control gear necessary to satisfy the requirements of Part 1.
	3	**Selection and installation of wiring systems** This section specifies the minimum requirements for the selection and installation of wiring systems necessary to satisfy the requirements of Part 1.
	4	**Selection and installation of electrical equipment** This section stipulates the minimum requirements for the selection and installation of electrical equipment, which includes appliances and accessories necessary to satisfy the requirements of Part 1.
	5	**Earthing arrangements and earthing conductors** This section specifies the minimum requirements for the selection and installation of earthing arrangements necessary to satisfy the requirements of Part 1.
	6	**Damp situations** This section establishes the minimum requirements for the selection and installation of electrical equipment in damp situations necessary to satisfy the requirements of Part 1.
	7	**Special electrical installations** This section stipulates the minimum requirements for the selection and installation of electrical equipment in special electrical installations necessary to satisfy the requirements of Part 1.
	8	**Verification** This section sets out the minimum requirements for the inspection and testing to fulfil the fundamental safety principles of Part 1 that relate to the verification of an electrical installation.
Appendices	A	Referenced documents
	B	Circuit protection guide
	C	Circuit arrangements
	D	Minimum sizes of posts, poles and struts for aerial line conductors
	E	Electrical installation requirements in national construction codes
	F	Surge protective devices
	G	Degrees of protection of enclosed equipment
	H	WS classification of wiring systems
	I	Protective device ratings and metric equivalent sizes for imperial cables used in alterations and repairs
	J	Symbols used in this standard
	K	Switchboard requirement summary
	L	(deleted)
	M	Reducing the impact of power supply outages–continuity of supply for active assisted living and homecare medical situations
	N	Electrical conduits
	O	Installation of arc fault detection devices (AFDDs)
	P	Guidance for installation and location of electrical vehicle socket-outlets and charging stations
	Q	DC circuit protection application guide

EXERCISE 5.12

Use the current edition of the *Wiring Rules* to look up the title of each of the following clauses.

a 1.5.9
b 2.2.2
c 2.3.7
d 2.7
e 3.9.10

National requirements

Specific requirements of the *Wiring Rules* have different application in Australia and New Zealand. Requirements that are specific to Australia or New Zealand are identified by the following symbols appearing in the outer margin:

- Applicable in Australia only [A].
- Applicable in New Zealand only [NZ].

Performance-based content

The 2018 edition of the *Wiring Rules* is performance based rather than being based on prescriptive requirements, which means the electrical designer has a great deal of responsibility to design an installation that complies with the standard.

Regulations

There are numerous Australian Standards and regulations that pertain to electrical installations. The purpose of these documents is to ensure the safe design and installation of these systems as well as stipulating required system performance. The Australian Clean Energy Council (CEC) publishes additional guidelines covering the design and installation of grid-connected PV and standalone power systems. Local supply authorities also publish their own local requirements, such as the *Service and Installation Rules of NSW*.

All electrical systems must comply with the *Wiring Rules*. Furthermore, the installation of wiring and equipment must comply with the relevant WHS and safe work practice requirements.

Table 5.5 is not an exhaustive list and only shows some of the applicable Standards. Any electrical installation may be required to comply with any or all of these Standards. Those involved with the design, installation and maintenance of these systems have responsibility to keep up to date with current requirements of their work practices.

TABLE 5.5 Australian Standards pertaining to electrical installations

Standard	Application
AS/NZS 3000:2018 Electrical installations (known as the Australian/New Zealand *Wiring Rules*)	This Standard details the requirements for the design, construction and verification of electrical installations, which includes the selection and installation of electrical equipment that forms any part of the electrical installations. It comprises two parts: Part 1 details the minimum regulatory requirements to ensure that the electrical installation is safe Part 2 details compliant work methods and installation practices aligned with the requirements of Part 1.
AS/NZS 3008.1.1:2017 Electrical installations – Selection of cables Cables for alternating voltages up to and including 0.6/1 kV – Typical Australian installation conditions	This Standard details methods of cable selection for electrical cables and installation methods that are in common use at working voltages up to and including 0.6/1 kV at 50 Hz a.c. for Australian conditions.
AS/NZS 3008.1.2:2017 Electrical installations – Selection of cables Cables for alternating voltages up to and including 0.6/1 kV – Typical New Zealand conditions	This Standard details methods of cable selection for electrical cables and installation methods that are in common use at working voltages up to and including 0.6/1 kV at 50 Hz a.c. for New Zealand conditions.
AS/NZS 4777.1:2016 Grid connection of energy systems via inverters Installation requirements	This Standard details the electrical and general safety installation requirements for inverter energy systems not exceeding 200 kVA intended to supply electrical energy to and from a low voltage electrical installation that connects to the grid.
AS/NZS 4777.2:2020 Grid connection of energy systems via inverters Inverter requirements	This Standard details the specifications, functionality, testing and compliance requirements pertaining to electrical safety and performance of inverters that connect between low voltage energy sources and/or energy storage systems. This includes electric vehicles that operate in vehicle to grid mode as well as standalone inverters that connect to a low voltage electrical installation that may connect to the grid.
AS/NZS 5033:2021 Installation and safety requirements for photovoltaic (PV) arrays	This Standard details the general installation and safety requirements for PV arrays, which includes d.c. array wiring, electrical protection devices, switching, and earthing provisions.

»

TABLE 5.5 Australian Standards pertaining to electrical installations *(Continued)*

Standard	Application
AS/NZS 3010:2017 Electrical installations – Generating sets	This Standard details the minimum safety requirements for the use of generating sets that supply electrical energy at potentials exceeding 50 V a.c. or 120 V d.c.
AS/NZS 4509.2:2010 Stand-alone power systems System design	This Standard provides requirements and guidance pertaining to the design of standalone power systems having energy storage at extra-low voltage and used to supply extra-low and/or low voltage electrical energy in a domestic installation.
AS 3011.1:2019 Electrical installations – Secondary batteries installed in buildings Vented cells	This Standard specifies minimum requirements to ensure safety from fire and electric shock from the installation of vented lead-acid batteries and vented alkaline batteries.
AS 3011.2:2019 Electrical installations – Secondary batteries installed in buildings Sealed cells	This Standard specifies minimum requirements to ensure safety from fire and electric shock from the installation of sealed secondary batteries.
AS/NZS 5139:2019 Electrical installations – Safety of battery systems for use with power conversion equipment	This Standard specifies general installation and safety requirements for battery energy storage systems located in a dedicated enclosure or room and supplies electrical energy to other parts of an electrical installation via power conversion equipment.

National Construction Code

The National Construction Code of Australia (NCC) is a nationally uniform approach to technical building requirements for the design and construction of buildings and other structures. Its purpose is to meet community needs for health, safety and amenity (comfort and quality) within buildings. The code is also a living document because it must be maintained to incorporate the results of research and changes in technology and to reflect current best practice. A simple example of the NCC in action is the introduction of enhanced sound impact insulation in buildings to reduce residential amenity problems caused by sound transmission (from TVs) between dwellings.

Electricians need to have a working knowledge of the NCC to ensure that their installation work does not breach any requirements of the building code.

REVIEW QUESTIONS

1. What is a technical standard?
2. Standards are described as living documents. What does this mean?
3. Standards Australia represents Australia on two international standardising bodies. Name these bodies.
4. Essentially, standards fall into three categories. What are these categories?
5. How are exceptions to a requirement documented in the *Wiring Rules*?
6. What is the National Construction Code of Australia?
7. Why do electricians need to have a working knowledge of the National Construction Code of Australia?
8. Which Standard details the minimum safety requirements for the use of generating sets that supply electrical energy at potentials exceeding 50 V a.c. or 120 V d.c.?

CHAPTER REVIEW

5.1 Architectural drawings

- A location plan aims to indicate the geographical location of the site(s) that is of interest.
- Soils and geology surveys evaluate the bearing qualities of soil conditions such as soil (organic, sandy, reactive clays), moisture content, expansion coefficient and the soil-bearing pressure.
- A site plan enables the electrical installer to coordinate with other selected project services for location and trenching for these services.
- A floor plan can include the type, size and location of the doors and windows; the location of utility installations; and the location of stairways.
- All working drawings must be arranged to meet the requirements of the Building Code of Australia and the residential timber framing standard.
- Symbols are a technical language device that has particular meaning.

- A scale drawing is a reproduction diagram of a building or object but at a standardised fraction of its original size.

5.2 Building construction drawings

- Setting out means installing profiles to provide a means of accurately aligning, squaring and levelling the position of the building on the site.
- Footings are designed to suit the building structure and subsoil type.
- Brick veneer construction is a single brick covering tied across the air space to a timber or steel frame with wall ties embedded in the mortar joints.
- Cavity brick is two brick walls separated by a 45 mm cavity and linked to each other with wall ties.
- Exterior sheathing used for external linings for a light timber or metal frame house can be fibre cement cladding, stucco or weatherboards.
- Roof types such as skillion, gable, Dutch gable, standard hip, and hip roof with valleys are used throughout Australia.
- Prefabricated trusses are constructed using truss plates.
- Eaves or soffit are the roof overhang of a building.

5.3 Circuit diagrams

- Circuit diagrams show by means of single lines and circuit symbols the flow of electrical energy.

5.4 Electrical drawings

- Block diagrams are a method of explaining complex systems in a straightforward manner.
- Line diagrams are a simplified notation for representing a power system.

5.5 Wiring diagrams

- A wiring diagram is a detailed diagram of each circuit installation showing all of the conductors and where they are connected.

5.6 Job specifications

- Job specifications or simply 'specs', detail the work and workmanship needed to complete a building project, which includes the electrical works. Job specifications form part of the formal contract process.
- A job specification details the acceptable level of quality for the project.
- Construction Information Systems Limited or NATSPEC (its trading name), is a not-for-profit organisation that is owned by the design, build, construct and property industries through professional associations and government property groups.

5.7 Electrical licensing

- Electrical licensing requirements vary between each jurisdiction but a common element is that the requirements for electrical licensing are established in legislation.
- The purpose of the Electrical Safety Act is to protect persons and property from electric shock, fire and other hazards resulting from electricity.
- Codes of practice provide practical advice on how to meet the electrical safety obligations established under the Electricity Act.

5.8 Technical standards

- A technical standard is a living document that establishes benchmarks of good practices and a common language to provide a level of quality that is respected.
- The Australian Standards Association functions as a coordinating body, providing a structure and a forum for developing standards.
- *Wiring Rules* is the most widely circulated Australian and New Zealand standard and is used nationally throughout the electrical industry.
- The National Construction Code of Australia (NCC) is a nationally uniform approach to technical building requirements for the design and construction of buildings and other structures.

TRIAL EXAM

For Chapter 5 knowledge assessment, please complete the following trial exam.

1 The plan that aims to indicate the geographical place of the site(s) that is of interest is the:
   a location plan
   b site plan
   c lot plan
   d set-out plan

2 The plan that delineates the property lines with their bearings and dimensions is the:
   a aerial plan
   b lot map
   c contour plan
   d location plan

3 The plan that enables the electrical installer to coordinate with other selected project services for location and trenching for these services is the:
   a architectural working drawing
   b service point plan
   c site plan
   d set-out plan

4 'Footings' are:
   a a single brick row with mortar joints
   b a placing of the feet
   c a part of the construction that spreads the load from the building structure above to create a level stable surface on which to build the structure
   d a set of profiles to provide a means of accurately aligning, squaring and levelling the position of the building on the site

5 'Setting-out' refers to:
  a a single brick row with mortar joints
  b erecting the wall frames
  c a part of the construction that spreads the load from the building structure above to create a level stable surface to build the structure
  d a set of profiles to provide a means of accurately aligning, squaring and levelling the position of the building on the site

6 The type of building construction that requires the electrician to work closely with the bricklayer so that wall boxes used for switches and socket outlets are placed in the correct locations is:
  a cavity brick
  b brick veneer
  c timber frame with external linings
  d metal frame with weatherboards

7 The drawings required to be configured to meet the requirements of the Building Code of Australia and the residential timber framing standard are:
  a standard floor plans
  b working drawings
  c detailed floor plans
  d sectional drawings

8 The type of detailed diagram that shows all of the conductors and where they connect is the:
  a single line diagram
  b connection diagram
  c wiring diagram
  d detail diagram

9 The diagram that uses a systems approach to viewing circuits is the:
  a circuit diagram
  b line diagram
  c wiring diagram
  d block diagram

10 The type of schedule that contains information about cables in a particular installation is the:
  a placement schedule
  b cable schedule
  c installation schedule
  d fit-out schedule

11 The purpose of the Electrical Safety Act is to:
  a establish a nation-wide legislative framework for all forms of electrical licensing
  b ensure compliance with the relevant state electrical licensing laws
  c establish a nation-wide repository of electrical codes of practice
  d protect persons and property from electric shock, fire and other hazards resulting from electricity

12 Documents that provide practical advice on how to meet the electrical safety obligations established under the Electricity Act are:
  a codes of practice
  b ministerial notices
  c safety regulations
  d technical standards

13 The Electrical Certificate of Compliance scheme is:
  a an industry code of practice
  b an example of a ministerial notice
  c documented under various safety regulations
  d a means of industry self-regulation requiring compliance with the standards

14 In order to ensure that electrical work does not adversely affect a building structure, electricians should have a good working knowledge of:
  a Ministerial notices
  b National Construction Code
  c NATSPEC
  d national licensing requirements

15 An organisation whose primary service provides a comprehensive national specification system in template documentation is:
  a Standards Australia
  b NATSPEC
  c The National Construction Association
  d FairWork Australia

Terminating cables and cords

This chapter provides electrotechnology workers with essential knowledge and skills pertaining to wiring systems types, enclosures, applications and terminations. Electrotechnology workers will gain an overview of TPI, TPS and other cable installing methods that will allow them to implement safe procedures for working in the electrotechnology industry. This chapter provides underpinning knowledge for the unit UEEEL0023 from the UEE training package.

LEARNING OBJECTIVES

Cable types and terminations

- Outline the types and applications of common electrical cables.
- State the Australian and international colour codes used to identify conductors.
- Describe the construction of common electrical cables.
- Outline common techniques for terminating cables.

Cords, cables and plugs

- Identify the components of plugs and sockets.
- Describe the construction of flexible cords.

Thermoplastic sheathed (TPS) wiring systems

- Describe the construction of thermoplastic sheathed cables.
- Outline common installation methods for thermoplastic sheathed cables.

Circular TPS wiring systems

- Describe the construction of circular thermoplastic sheathed cables.
- Outline common installation methods for circular thermoplastic sheathed cables.

Non-metallic enclosures

- Describe different non-metallic wiring systems.
- Explain the process of setting and installing non-metallic enclosures such as PVC conduit and trunking.

Metallic enclosures

- Describe different metallic wiring systems.
- Explain the process of setting and installing metallic enclosures such as conduit, tray, ladder, and trunking.

Fire protection cabling and systems

- Outline the characteristics of different fire protection cable types including Pyrolex, Radox, and MIMS.

Steel wire armoured cables

- Outline the construction of and installation methods for steel wire armoured cable.

Trailing cables and catenary systems

- Identify equipment used with trailing cable and catenary systems.

Industry standards and testing requirements

- Outline the importance of industry standards.
- Apply cable testing procedures.

6.1 Cable types and terminations

The medium used to convey electric current from a source of supply to an electrical load is, in most cases, referred to as a cable. A cable comprises a conductor with or without surrounding insulation. The conductor provides a conductive path of extremely low resistance to allow the movement of electrons. Insulation is a material that resists the flow of electrons through it. The insulation enables conductors to be arranged close together without a short-circuit occurring.

Cable insulation

Various materials are used for cable insulation. The choice of insulation depends on the required properties. A standard colour code is used to identify the function of particular cables.

Insulation colour code (installation wiring)

Insulated conductors are identified by a colour code that indicates their intended purpose. **Table 6.1** shows the colours recommended in Australia and New Zealand for 230 V single-phase and 400 V three-phase fixed wiring installations.

TABLE 6.1 Colours recommended for insulated conductors

Purpose	Colour single-phase	Colour three-phase
Earth	Green/yellow	Green/yellow
Active	Red	Red (L_1), white (L_2), dark blue (L_3)
Neutral	Black	Black

Note: Actives can be any alternative colour except black, light blue, green, yellow and green/yellow.

Coloured heat shrink and fixed, or elastic sleeving can also be used to identify cable cores (refer to AS/NZS 3000:2018 *Wiring Rules*).

Types of insulation

Some of the more common cable insulants are listed below.

Elastomer

Elastomers are synthetic compounds or rubber materials that provide a high degree of flexibility and elasticity. The term 'vulcanisation' is often used to describe the cross-linking of all elastomers. Cross-linking is used with elastomers to improve their strength when used for cable insulation.

Polyvinyl chloride

Polyvinyl chloride (PVC) is a general-purpose thermoplastic polymer used for wire insulation, cable insulation, sheaths and conduit. PVC is mainly used in low-voltage applications because it loses its mechanical properties at temperatures above 160 °C and becomes brittle at low temperatures. Its maximum temperature ratings range from 60 °C to 105 °C. PVC also has excellent moisture and chemical resistance together with flame-retardant properties. However, when PVC does burn the resulting smoke is toxic.

Polyethylene

Polyethylene is a thermoplastic material having excellent electrical properties such as a low and stable dielectric constant and a very high insulation resistance. Cross-linked polyethylene (XLPE) is a thermoset polymer that has a broad range of operating temperatures, excellent bending ability and impact and abrasion and flame resistance. XLPE can be formulated to have flame-retardant abilities and unique moisture-resistant properties.

Silicone

Silicone is a thermosetting elastomer noted for its high heat-resistance properties. It is very soft thermoset insulation. Silicone has low moisture absorption, good weather resistance, good radiation resistance, low mechanical strength and reduced abrasion resistance.

Mineral powder

Mineral powder is insulating filler. The mineral powder used for this purpose is compressed magnesium oxide (MgO), which is an inorganic mineral in which current-carrying conductors are embedded. Magnesium oxide is hygroscopic, which means that it retains moisture under certain humidity and temperature conditions. It has, however, a high thermal conductivity.

Cellulose

Conductors can be covered or surrounded with successive wrappings or layers of cellulose (paper) insulating material, impregnated with an insulating compound such as mineral oil, long recognised as an insulating material. Cellulose insulation deteriorates over time and high operating temperatures. Deterioration occurs primarily because moisture is produced by deterioration of the cellulose component of the insulation. Cellulose insulation is layered on conductors in spiral strips to the required thickness.

Cross-linked elastomer compounds

The process of cross-linking is the setting up of chemical links between the molecular chains in the polymer. Polymers are substances (produced by the conversion of natural products or by synthesis from primary chemicals coming from oil, natural gas or coal) whose molecules have high molar masses (molar mass is the weight of one mole (or 6.02×10^{23} molecules) of any chemical compound) and incorporate a long-chain structure of many repeating units. Synthetic polymers have a broad range of properties and uses. The materials commonly called plastics are all synthetic polymers. Cross-linking can result in improved insulation performance in a polymer, including:

- higher operating temperatures
- lower permeability
- improved chemical resistance
- increased elasticity
- improvement in strength

The two main types of plastics are thermoplastics and thermosets. These two cross-linked polymers are cured or set using heat or pressure. Thermoplastics (such as polyvinyl chloride) soften on heating and harden on cooling, while thermosets (such as epoxies, XPLE, and neoprene) may soften on heating but do not melt or flow.

Choice of cable insulation

The choice of insulation depends on the environmental conditions under which it has to operate. For most insulation, the hazards that frequently need to be addressed are excess:

- temperature
- moisture
- mechanical abuse, including impact and vibration
- chemical abuse, including hydrocarbons
- overload and short-circuit
- solar and thermal radiation.

Insulation abbreviations

R–EP–90 type of cable insulation is a cross-linked rubber (R) compound based on the ethylene propylene (EP) co-polymer suitable for a maximum continuous operating temperature of 90 °C. Other insulation abbreviations are shown in Table 6.2.

TABLE 6.2 Insulation abbreviations

Abbreviation	Meaning
R–EP–90	Rubber–ethylene propylene–90 °C
R–S–150	Rubber–silicone–150 °C
V–75	75 °C rated PVC
V–90	90 °C rated PVC
HFS–90–TP	Halogen-free sheath–90 °C–thermoplastic
HF–110–R	Halogen free–110 °C–rubber (sheath)
R–CPE–90	Rubber–chlorinated polyethylene–90 °C
R–HF–110	Rubber–halogen free–110 °C (insulation)
V–90HT	90 °C rated PVC–105 °C for restricted periods
X–HF–90	XLPE–halogen free–90 °C
X–90	Cross-linked polyethylene–90 °C
PVC	Polyvinyl chloride
PILC	Paper insulated lead covered

Note: The use of the higher-temperature insulation compounds does not permit a higher current-carrying capacity. AS/NZS 3008.1.1:1998 (AS/NZS 3008.1.2:1998 for New Zealand) recommends a conductor temperature of 75 °C for current-carrying-capacity calculation.

Cable terminology

Numerous terms are used to describe the function of specific parts of a cable. Some of these are discussed below.

Conductors

An electrical conductor is a material that conducts electric charges freely when a voltage is applied across the material. Most conductor materials are metals.

Conductor material

Conductor materials include copper, copper alloys, aluminium, and aluminium alloys (stranded or solid) because of their high conductivity. Copper is usually the preferred material for conductors, and they can be plain or tinned (electroplated with a nickel alloy). For a given load demand, copper conductors are of smaller diameter than those of aluminium.

In Australia and New Zealand, wire area is measured in square millimetres (mm^2). A thick wire carries more current because it has less electrical resistance over a given length. Socket outlet and lighting wiring is typically 2.5 mm^2 and 1.5 mm^2 copper conductors respectively. Until recently metals were used for telecommunication conductors but now a majority of telecommunications for long distances uses optical fibres to convey signals as pulses of light.

Stranding

Stranded conductor constructions enable a cable to be more flexible than a cable constructed of solid conductors. An increase in cable diameter is associated with the use of stranded wires, and resistance and weight are also affected. A 2.5 mm^2 stranded conductor uses 7 × 0.67 mm diameter wires in a round configuration as compared with a 2.5 mm^2 solid core, which has a 1.78 mm diameter conductor.

Insulation type

Insulation materials are dielectrics that are poor conductors of electricity. Most dielectric materials are solid. Examples include paper, plastics and the oxides of various metals. In the context of a cable, the dielectric is the insulant material that encloses the conductor core of the cable, separating it from the surrounding environment. PVC is one such insulant and has been used as insulation for conductors for many years in Australia and accounts for around 50% of the Australian power cable market.

Properties that determine the insulation value of the material include dielectric strength, permittivity and resistance to tracking (leakage along the surface of the insulation). The term 'insulation', when applied to cables, means that part of the cable that is relied upon to insulate the conductor from other conductors or conducting parts or from earth.

Core filling

Cable filler is Aramid yarn, nylon, cotton and other materials used in multiple conductor cables to occupy the hollow spaces formed by the assembly of insulated conductors to form the core of the desired shape. These types of fillers are called 'dry-core' fillers. Some dry fillers swell when wet to provide an impervious barrier to further moisture penetration.

Some cables use filling compounds such as petroleum jelly, waxes or a thixotropic gel (when shaken changes into liquid) to space the conductors and to flood any voids within the cable and the armour layers if required. These compounds have water-blocking properties to provide waterproofing for the interior insulated conductors of the cable to prevent internal shorting. Moisture affects the performance of cable insulation under the influence of high-voltage stress. High-voltage stress can degrade the insulation and cause water trees to form leading to internal shorting. A cross-section of a cable showing various fillers is illustrated in **Figure 6.1**.

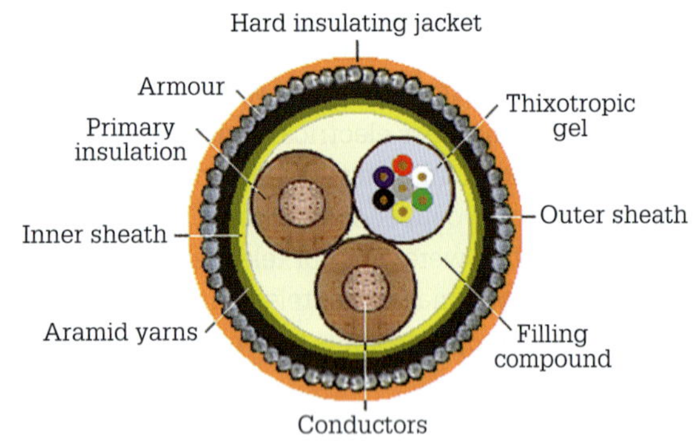

FIGURE 6.1 Cross-section of an armoured cable

The only perfect barrier against the ingress of moisture is a continuous radial and longitudinal metal sheath and sealed tails.

Screens

Screens, when connected to earth, keep electrical interference away from conductors carrying data signals. When the screen in foil form is wound round and round the signal conductors, it is called a wrapped screen. When the screen strands are woven together, it is known as a braided screen. When the screen strands are spiral wound around the signal conductors, it is called a spiral screen. Cable screens must cover the entire length of the cable with 360° coverage. Braided and spiral screens can consist of 50 or 60 strands of copper wire. **Figure 6.2** shows various cable screens.

Round metal copper tube can also be used as a screen.

Hard jacket

A hard jacket (see **Figure 6.3**) over a cable provides additional mechanical protection for the cable. For example, a hard, shiny nylon jacket with suitable chemical additions offers protection from the formic acid that ants and termites use to attack underground cables.

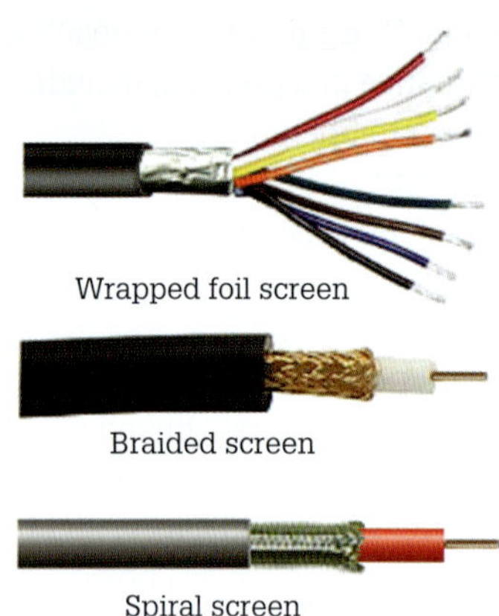

FIGURE 6.2 Cable screens

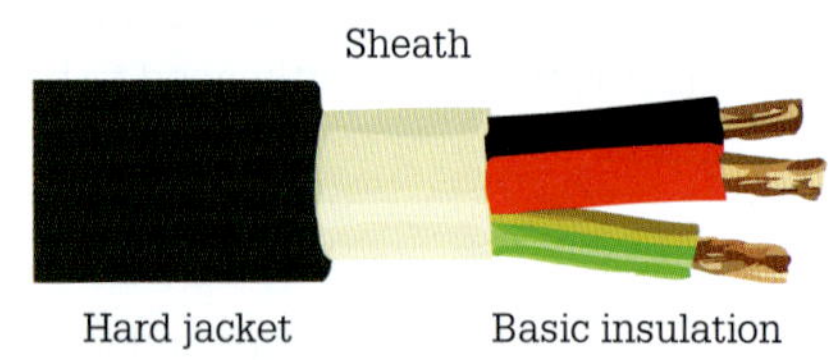

FIGURE 6.3 Hard jacket over a sheath

Voltage rating

The voltage rating of a given cable or insulated conductor is the peak voltage that may be continuously applied without degradation of the insulation causing a safety hazard. It is also called working voltage.

The voltage rating 450/750 V indicates the voltage rating of a cable or conductor insulation as shown in **Figure 6.4**. The number '450' indicates the voltage rating of the insulation to earth. The number '750 V' indicates the voltage rating of the same types of insulated conductors.

PVC basic insulation

750 V

450 V

Earth

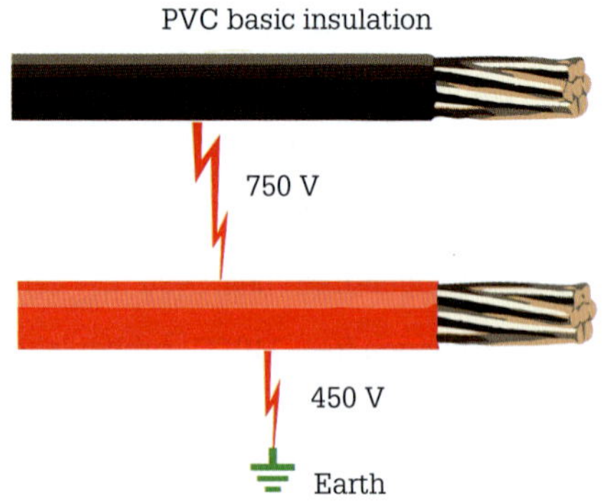

FIGURE 6.4 Insulation voltage rating 450/750 V

The higher the voltage supplying a cable, the thicker is the protective insulation in order to withstand the greater electrical 'pressure' of the higher voltage. A minimum thickness of insulation is required to offer a reasonable level of resistance to mechanical damage. If enough insulation were used to insulate to 230 V only, the insulation covering the conductor would be easily damaged. Therefore, voltage ratings also reflect a safety margin.

Temperature rating

The temperature rating of a cable or insulated conductor is the maximum temperature at which the insulating material may be safely maintained in continuous operation without deterioration of its basic properties. The figure 'V–90' stamped onto TPS cables means that the insulation is designed for an operating temperature of 90 °C in wet or dry locations. If cables with different temperature classifications are contained within the same enclosure, the overall cable rating is determined by the lowest insulation rating. The mixing of various types of cables creates safety issues relating to potential overheating of conductors and insulation and reduces the service life of cables within the enclosure. Cable operating temperature has an effect on the resistance of the cable. Most modern PVC and TPS cables have a recommended maximum operating temperature of 75 °C. The cables have higher resistances at their operating temperature ratings than at ambient temperature.

Sheath

A sheath is a protective outer covering, shown in **Figure 6.5**, which can be applied to cables. A sheath is used in situations where cables need protection from mechanical damage, moisture, oil, fuel, hydrocarbons, solvents, acids, and a variety of different chemicals. Sheaths fit the basic insulation of the cable closely but do not adhere to it.

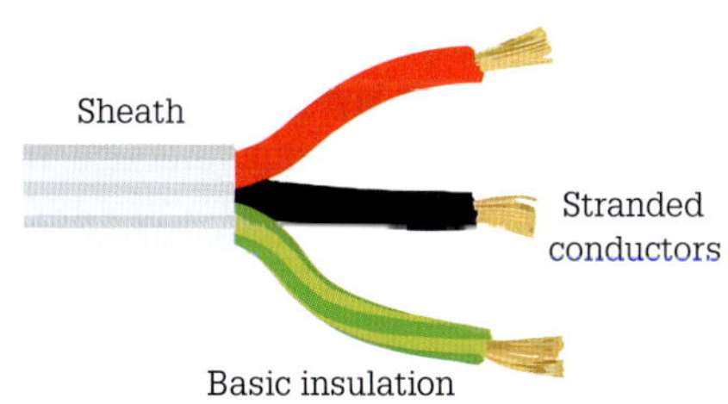

FIGURE 6.5 Thermoplastic sheath

Metals such as copper, aluminium, lead, stainless steel and bronze are also used as cable sheaths. Except for lead, metal sheaths offer the highest degree of mechanical protection when potential crush, shear, and sidewall pressure events occur. Sheaths also provide reinforced insulation that allows the cables to be installed without further enclosure.

Armour

Armour is a mechanical protection for the conductor's insulation. The armour can be a metallic layer of aluminium or steel wire, flat metallic tape or interlocked galvanised steel or aluminium wire.

Serving

Serving or protective coverings can be applied over cable insulation for mechanical, chemical, electrical requirements or all three. It is not unusual for a cable to have several types of protective coverings to meet the demands of a particular application.

Serving can be any protective wrapping such as wire, fibre, tapes or a PVC coating applied over an insulated conductor. Serving materials such as PVC provide protection against chemical corrosion of armour, lead and copper sheaths.

Conductors

Conductors are defined as materials that readily allow the flow of electrons. Metals are good conductors while insulators are not. The most common conductor material used by electrical workers is copper, with aluminium being the only other general-purpose metal used.

Conductor material

Copper and aluminium are commonly used for electrical conductors. Of these two materials, copper is a better electrical conductor. The resistance of a length of copper conductor is relatively small. An aluminium cable would have about 1.6 times the resistance of a copper cable having the same dimensions, owing to the higher resistivity of aluminium. Therefore, the energy losses in the aluminium cable are higher than in the copper cable.

For an aluminium conductor to have the same energy losses as a copper conductor, the aluminium conductor must have a greater CSA. The larger CSA reduces the resistance of the aluminium conductor. AS/NZS 3000:2018 *Wiring Rules* recognises this effect of CSA with aerial conductors. The nominal minimum CSA of a copper aerial conductor is 6 mm^2, while that of an equivalent aluminium aerial is 16 mm^2 (see Clause 3.12.2.2). Strength is also a consideration besides resistance with aerials because aluminium will stretch under its own weight. Therefore, a greater CSA is recommended. In relation to main earth conductors, a 6 mm^2 copper conductor is suitable for a 16 mm^2 copper active conductor. For an aluminium main earth conductor, AS/NZS 3000:2018 *Wiring Rules* requires a minimum 16 mm^2 stranded conductor. Aluminium conductors can be used as other earthing conductors but if they are 10 mm^2 or less they must be a solid conductor.

The smallest nominal size of an active conductor allowed for a copper conductor according to AS/NZS 3000:2018 *Wiring Rules* is 0.5 mm^2 because of its tensile strength. The smallest nominal size of an active conductor allowed for an aluminium conductor is 6 mm^2 (any smaller size is easily broken). The increased diameter of an aluminium conductor is necessary in order to carry the same amount of current as an equivalent copper conductor. Aluminium and copper conductors are illustrated in **Figure 6.6**.

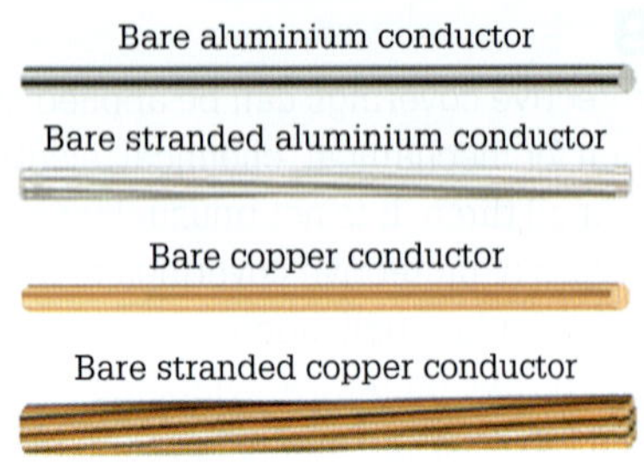

FIGURE 6.6 Aluminium and copper solid and annealed stranded conductors

The density of copper (8.9 kg/cm^3) is 3.3 times higher than that of aluminium (2.7 kg/cm^3). So, for identical linear resistance, an aluminium conductor is twice as light as a copper conductor. Taken on a mass basis, the conductivity of aluminium is about twice that of copper (0.22 kg of aluminium has the same conductive capability as 0.45 kg of copper). This, as well as a greater strength per unit weight, gives aluminium, when reinforced with a galvanised steel wire or aluminium-coated high-tensile steel wire, some advantage over copper as a conductor material when used as a long-span aerial.

Aluminium expands 38% more than copper for a given temperature rise. When terminated with connectors designed for copper and exposed to load currents, some of the aluminium will be extruded from the connector. Subsequently, when de-energised, the aluminium shrinks and oxides form in the voids. The next load cycle experiences increased resistance in the connection which causes more temperature rise, more extrusion of the aluminium and higher resistance in the connection. The sequence repeats itself until the connection fails. Because aluminium expands more than copper, when aluminium is connected to copper terminations it will be placed under stress and the aluminium conductor will experience creep or cold flow.

Creep, or cold flow, is the slow accumulative deformation of a material under stress. Creep is a factor that has caused heating at the connections and terminations of aluminium conductors. A pigtail is used to overcome the problems of creep. This procedure is labour intensive as it requires a short length of copper conductor to be crimped to the aluminium conductor using aluminium/copper-type connectors. Because of creep, aluminium conductors less than 16 mm^2 are not used in Australia.

Copper conductors

Copper used for electrical conductors is electrolytically refined to produce a copper content of 99.9%. Bare copper wire can be hard-drawn solid, hard-drawn stranded, annealed solid or annealed stranded wires. Hard copper wire is drawn from copper wire rod in ambient atmosphere. This process produces a wire that has high conductivity together with high tensile strength. Annealed copper wire is obtained by heating hard-drawn copper wire in a vacuum furnace. This process produces a copper wire that has greater flexibility and conductivity than hard-drawn wire.

Annealed copper conductors are used in stranded form to provide greater flexibility than single or stranded hard-drawn copper conductors. Aside from their appearance and rigidity, hard-drawn and annealed conductors have the same qualities and act identically when they are carrying the same magnitude of current. All conductors for use in Australia must be tested for elongation and tensile strength, conductor resistance and their diameter tested with a micrometer.

Stranded conductor designs for electrical purposes were developed as a means of overcoming the rigidity of hard-drawn solid conductors. For any given conductor size, the greater the number of strands with a corresponding decrease in individual strand size, the more flexible and costly the conductor. For example, seven individual wires of 0.67 mm each are equivalent to a solid 1.78 mm wire even though the stranded conductor is larger in overall diameter. Both forms of conductor, stranded and solid, are regarded as being equivalent to a nominal conductor area of 2.5 mm^2 and have the same current-carrying capacity. Seven individual wires (or strands) are used for 1.0 mm^2, 1.5 mm^2 and 2.5 mm^2 stranded conductors. Australian/New Zealand Standard AS/NZS 1125:2001 *Conductors in insulated electric cables and flexible cords* specifies conductors by their CSA and the number of wire strands.

Due to their flexibility, stranded cables are easier to draw-in and terminate. They also lessen possible conductor breakage when repeatedly flexed.

There are specific numbers of strands called unilay (a multilayer of twisted laid wires with the same direction and same lay length for each layer), as shown in **Figure 6.7**, which lend themselves to round configurations, that is, 7, 19 and 37 (7 = 1 + 6, 19 = 7 + 12 and 37 = 19 + 18).

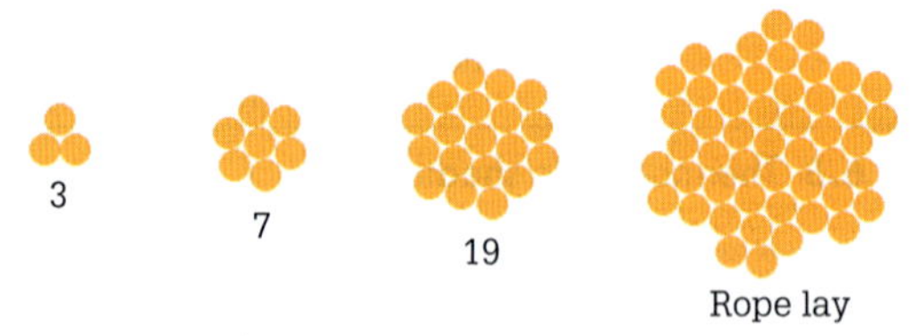

FIGURE 6.7 Three strands, unilay, and rope-lay stranding

Unilay is a round stranding with a minimum of two lays of the same-diameter wires and same lay length arranged to reduce the external diameter of single wires under a simple round strand. Normally greater than 37 wires, rope-type constructions are utilised consisting of 7-wire or 19-wire stranded groups. A 7-group rope-lay stranding is shown in **Figure 6.7**. When compared with a solid conductor of the same length and current, the resistance and weight of the stranded wires are greater. However, this depends on the number of strands and lay length used. Other types of wire grouping include:

- Bunched – wire strands of any number twisted together in the same direction without regard to the geometric arrangement.
- Unidirectional concentric – a central wire surrounded by one or more layers of twisted laid wires with the same direction of lay. It has the advantage of much greater flexibility than unilay.
- Equilay – multi-layers of twisted wire, with the direction of lay reversed for succeeding layers as shown in **Figure 6.8**.

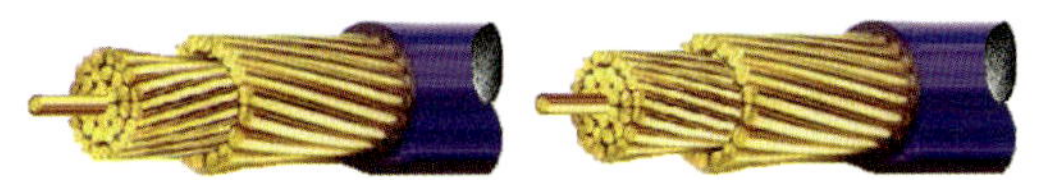

FIGURE 6.8 Equilay stranding and unidirectional wire grouping

The nominal minimum CSA of a bare copper conductor is 6 mm^2; however, a smaller CSA can be used depending upon current-carrying capacity and voltage drop (refer to AS/NZS 3000:2018 *Wiring Rules*, Table 3.3).

Flexible conductors

Some flexible conductors have many fine wires specially made (<0.31 mm diameter) to allow ease of movement. For example, a single-insulated 2.5 mm^2 conductor for a 25 A flexible cord consists of 50 wires each having a diameter of 0.25 mm. **Figure 6.9** shows an insulated flexible conductor used for welding purposes (available in sizes ranging from 10 mm^2 to 240 mm^2).

FIGURE 6.9 Flexible welding conductors

The CSA of single flexible stranded conductors can be as small as 0.5 mm^2 and these are suitable for currents up to 3 amperes.

Current-carrying capacity

A load current which is higher than the current-carrying capabilities of the connected cable can result in the cable overheating and cause a serious safety risk. AS/NZS 3000:2018 *Wiring Rules* Clause 3.4 refers the current-carrying capacity (CCC) of cables to the AS/NZS 3008.1 series. The CCC of a conductor must not be less than the rating of the circuit protective device. **Table 6.3** gives typical values for conductor size in relation to a suitable protective device rating, which is dependent on installation conditions, and load maximum power.

TABLE 6.3 Conductor size in relation to protective device rating and maximum power

Conductor size	Protective device rating	Maximum power of 230 V
1.0 mm^2	6–10 A	2300 W
1.5 mm^2	10–16 A	3680 W
2.5 mm^2	16–20 A	4600 W
4.0 mm^2	25–32 A	7360 W
6.0 mm^2	40 A	9200 W
10.0 mm^2	50 A	11 500 W
16.0 mm^2	63–80 A	18 400 W

Termination of conductors

Conductors must be terminated by one of the following methods:

- Sweated lugs of the appropriate size for the CSA of the conductor used.
- Compression-type lugs. Compression connectors and associated dies for the compression tool must be the correct type and size. The tool must be so designed that the proper compression force must be applied before it can be released.
- Pinch screw-type terminals that do not spread conductors.
- Clamp-type terminals.

For all single connections, when using methods 3 and 4, conductors of up to and including 2.5 mm^2 should be doubled back on themselves to prevent the conductor from being weakened by the screw or clamp pressure and breaking.

Terminating cables

Proper termination of cables is essential to the system's reliable performance because the quality and reliability of electrical connection depend primarily on the terminating technology. Terminating technology concerns various devices that are attached to the ends of wire conductors to facilitate connection to equipment or machinery terminals (refer to AS/NZS 3000:2018 *Wiring Rules*, 'Electrical connections').

Mechanical terminations

One method of mechanical termination consists of corrosion-resistant flat metal surfaces between which the conductor or conductors are clamped. An illustration of this method of termination is shown in **Figure 6.10**, and a practical example is the termination of a flexible cord to a three-pin plug.

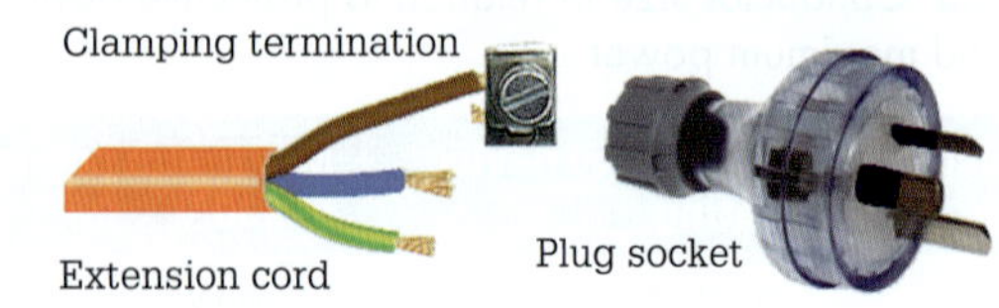

FIGURE 6.10 Clamping-type mechanical terminations

When cutting back insulation to bare the conductor, remove just enough insulation so that the stripped conductor fits the termination exactly. Take particular care not to cut any part of the conductor, since even a slight nick weakens it and eventually causes the conductor to break. In addition, the reduced CSA could result in a high resistance or hot joint. Wire strippers or a diagonal cutting knife are very useful for removing insulation effectively.

Clamping pressure keeps the connection secure. However, weak clamping or continual current overload of the cable causes excessive heat, which over time weakens these types of terminations. The weakening of the termination results in increased contact resistance causing additional heating. The extra heat could cause the insulation to melt. Note that the various cable insulations have different temperature ratings.

Another type of mechanical termination is a cylindrically shaped piece of metal called a barrel. These types of terminations help to eliminate twisting and breakage of stranded wire during connections, and they can improve interconnect performance. Some tunnel-type terminations provide a back wall to prevent over-insertion of the conductors and the possibility of the end of the conductor from shorting. Tunnel-type mechanical terminations are shown in **Figure 6.11**. Examples include the main neutral bar, the earth bar, the connectors on circuit breakers and one- or two- screw tunnel connectors.

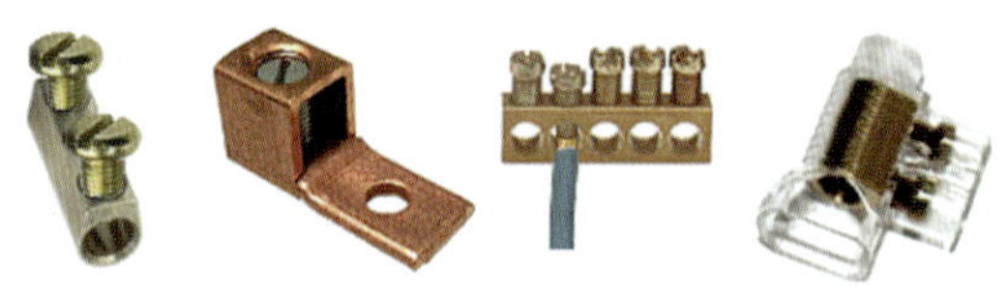

FIGURE 6.11 Tunnel-type mechanical terminations

Other mechanical terminations use a nut on a U-shaped threaded stud to compress the conductors, or the conductor is directly placed under a screw and pinched against a flat plate (see **Figure 6.12**).

Insulation-piercing connectors (IPCs) provide a fast, convenient and reliable method of connecting overhead low-voltage mains to services or other mains. The correct size spanner must be used when installing the IPC (see **Figure 6.13**) and a pull test performed on the conductor to ensure that the connection is sound.

FIGURE 6.12 Line tap and screw mechanical terminations

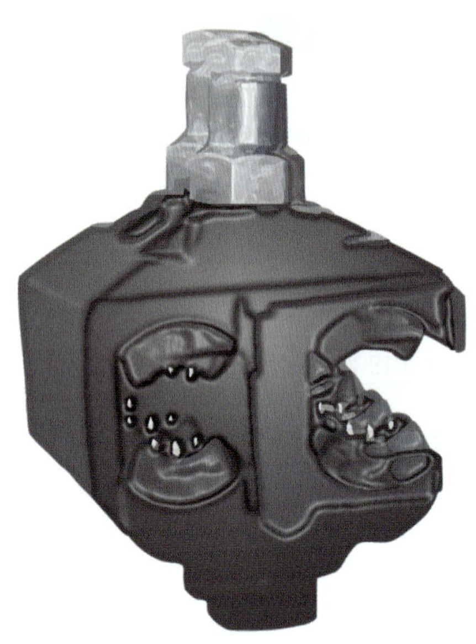

FIGURE 6.13 Insulation-piercing connector

Compression crimp terminators

Compression crimp terminators (see **Figure 6.14**) come in various sizes and configurations. Crimped terminators are called solderless connectors or lugs and are very common in appliances. One reason is that it is easier and faster to make terminations using these lugs.

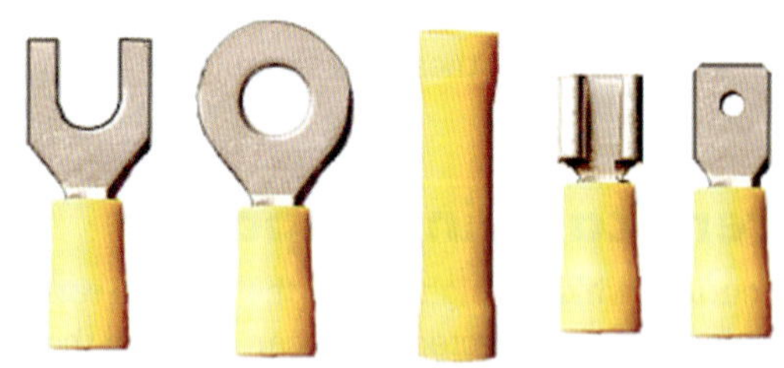

FIGURE 6.14 Compression crimp terminators

In a real compression crimp termination, the CSA of the conductors is only slightly reduced and all voids are practically eliminated. Compression crimping is regarded as a cold weld.

A compression crimp terminator consists of a tin-plated soft-copper sleeve which is crimped over stranded conductors, a PVC insulating sleeve if required and a terminal connection such as a ring, spade, splice or push-on. The crimp terminator must be matched to the conductor size for a reliable crimp connection. Insulated crimp terminators are colour coded for the various sized conductors. The red types are suitable for cable from 0.4 mm^2 to 1.0 mm^2, the blue for cable from 1.1 mm^2 to 2.5 mm^2 and the yellow for cable from 2.5 mm^2 to 6.5 mm^2.

An approved crimping tool (do not make a crimp with pliers or a hammer) suitable for the terminators being used compresses the soft-copper sleeve around the bare stranded conductors to form a good mechanical and electrical connection. A correctly installed crimp terminator should not come apart if an attempt is made to pull the conductors free.

Always cut the insulation from the end of the conductor to the correct length as specified by the compression crimp manufacturer. If you do it correctly, the insulation hits the barrel of the connector when the conductors are inset at the correct length. If the exposed conductor length is correct (1.6 mm past the sleeve) there are no conductor whiskers exposed on the terminal connection. Nicking or cutting conductor strands when removing the primary insulation reduces the tensile strength of the crimp joint. These types of crimp lugs not only crimp the conductor but the conductor insulation as well, hence the requirement to use the correct crimping tool.

Compression crimping may be carried out with hand-operated tools, power tools or automated power tools. Uninsulated crimp terminators and hex and oval crimp tools are illustrated in **Figure 6.15**.

FIGURE 6.15 Uninsulated crimp terminators and crimp tools

Millivolt testing

With compression crimp terminators used on cables greater than 10 mm^2 a millivolt test should be performed at the crimp termination to determine its electrical characteristics. A weak compression termination creates a high-resistance connection resulting in a voltage drop across the crimp termination. This voltage drop can cause poor system performance and connected equipment and machinery to malfunction.

Poor connections do not improve. The heat and possible arcing created causes accelerated oxidation within the crimp termination which in turn causes the resistance to increase further.

Solder terminations

Soldering is a process by which molten solder is used to create a metallic conductive path between metal parts and then allowed to cool. Solder is composed of metal or metal alloys that melt below 425 °C. The essential characteristics of solder include melting temperature, hardness, strength and ease of bonding of the metals being soldered. A wide variety of lead-free solders are available including tin–silver–copper–antimony (Sn–Ag–Cu–Sb), tin–silver (Sn–Ag), tin–antimony (Sn–Sb) and tin–bismuth (Sn–Bi). **Table 6.4** lists various solder alloy melting temperatures.

TABLE 6.4 Solder alloy melting temperatures

Solder alloy	Solid °C	Liquid °C	Eutectic
Sn–Bi	138	138	Yes
Sn–Ag–Cu–Bi	198	215	
Sn–Ag–Cu	186	218	
Sn–Ag	221	221	Yes
Sn–Ag–Cu–Sb	219	219	Yes
Sn–Cu	227	227	Yes
Sn–Sb	232	290	

On the application of heat, solder alloys start to melt at their 'eutectic' (lowest melting point) temperature and, depending upon the solder alloy composition, the alloy will continue to melt and become liquid. A eutectic alloy is composed of one or more metals and has a definite melting point and no intermediate 'plastic' stage. An example of a eutectic solder is tin–silver. Non-eutectic compositions have a semi-liquid temperature range where the solder can be worked due to its plastic nature. An example of a non-eutectic solder is tin–antimony.

On the removal of the heat, the solder alloy starts to solidify and may enter a plastic stage and then become solid at the eutectic temperature. Movement of any metal part being soldered results in an unreliable soldered joint.

Lead-free solders require a melting temperature above 215 °C in order for the solder to change phase from solid to liquid. In addition, the soldering iron tip needs a temperature of 260 °C for successful heat transfer and a long dwell time when in contact with the item to be soldered in order for the solder to flow (called wetting) correctly.

Flux

Flux such as rosin (extract from pine sap) is a chemical cleaner that removes oxidation from metal surfaces so that an effective wetting surface exists to allow a solder-to-metal bond to be made. Oxidation is created in some metals when they are exposed to oxygen. The oxygen and the metal form a non-wetting surface on the metal preventing soldering. Because of the variety of solder alloys available, application of the correct flux becomes critical for the promotion of 'wetting' during the soldering process.

Flux enables all solders to flow at their lowest melting temperature and also allows the metals to be soldered at higher temperatures without burning away. Before using any flux for the first time a material safety data sheet (MSDS) must be read as the user of the flux is responsible for facilitating safety. Generally speaking, completed lead-free joints have a grainy whitish finish as shown in **Figure 6.16**.

FIGURE 6.16 Solder alloy termination finish

An important factor in thermal transfer for soldering with lead-free solders is correct soldering iron tip selection. The right tip should have a similar width to the object being soldered. Flat tips produce a larger contact area with a termination than do round tips and therefore tend to transfer heat more efficiently.

Seamless tubular copper terminations used for solder connectors can be plain or tinned as shown in **Figure 6.17**.

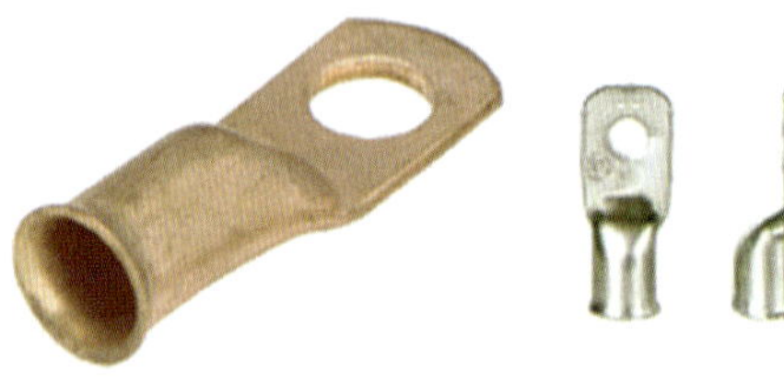

FIGURE 6.17 Solder connectors

Infrared thermal imaging as shown in **Figure 6.18** is an ideal method for inspecting terminations of critical electrical equipment because it reveals hot spots (middle circuit breaker) under actual load conditions.

Joins, taps and splices

A join is the tying together of two conductors so that their union is mechanically and electrically sound. A tap is the connection of the end of the cable to some point along the run of another cable. A splice is the interlacing of stranded conductors so that their union is mechanically and electrically sound.

Joins

To make a tail join, cross the conductors between a pair of pliers as shown in **Figure 6.19** and twist them together for a distance of at least 25 mm and cut the excess wire off.

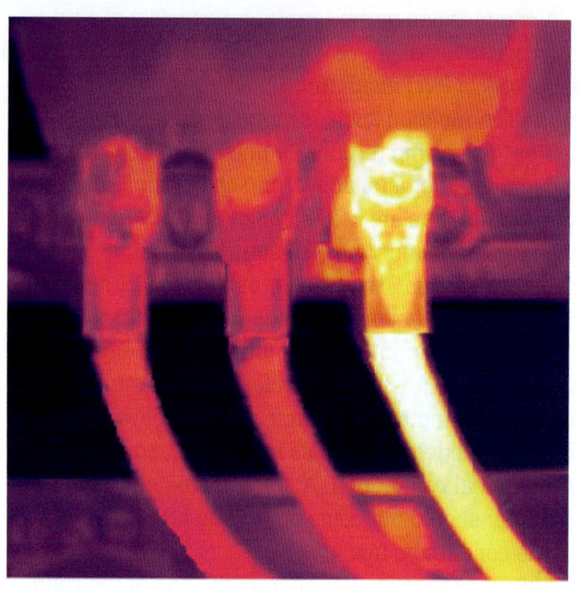

FIGURE 6.18 Infrared thermal imaging

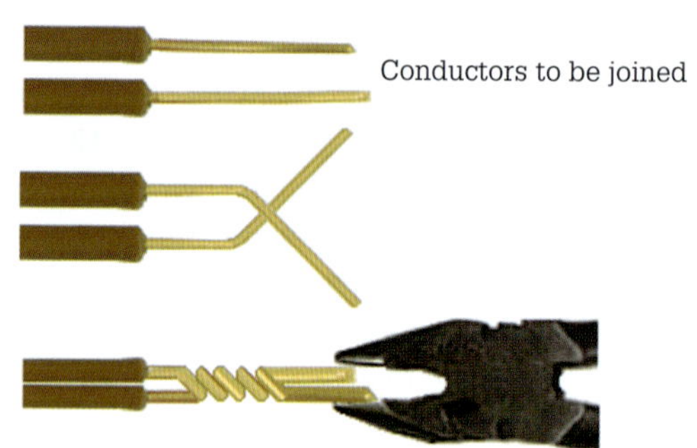

FIGURE 6.19 Tail join

This type of join is suitable for service where there is no mechanical stress. It is used where conductors are to be connected to a switch, socket outlet, and J boxes. Joining two solid conductors together as a single cable as illustrated in **Figure 6.20** provides both mechanical and electrical strength. This type of join is compact and can also be soldered with high-temperature solder if required and must be insulated to the same temperature rating as the conductor insulation.

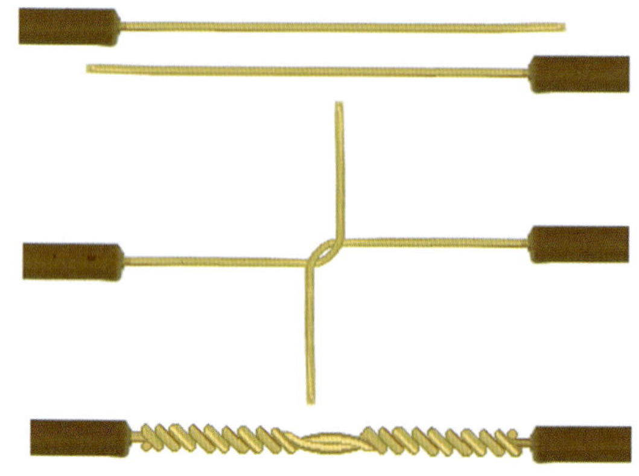

FIGURE 6.20 Stages for joining two solid conductors

Joins in TPS cable must be staggered, as shown in **Figure 6.21**, so that the joins do not come opposite each other.

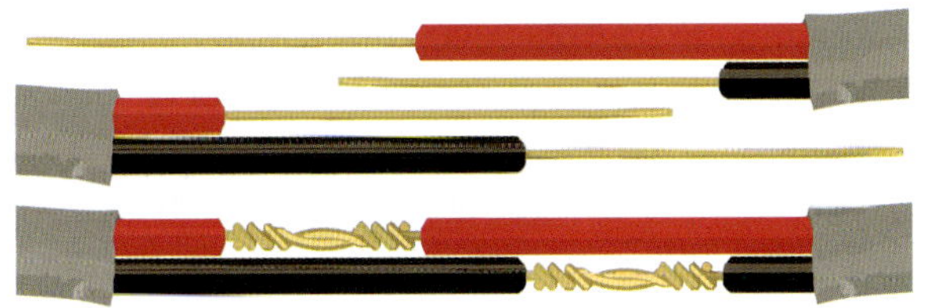

FIGURE 6.21 Joins in TPS cable

This type of join is also compact and should be soldered with high-temperature solder to withstand possible fault currents and insulated to the same temperature rating as the conductor insulation. The twists around the conductors must be in unison and flat, similar to the fingers of your hand when it grips a bar.

Splices

'Y' splices are recommended for connecting protective earths to the main earth conductor, as shown in **Figure 6.22.**

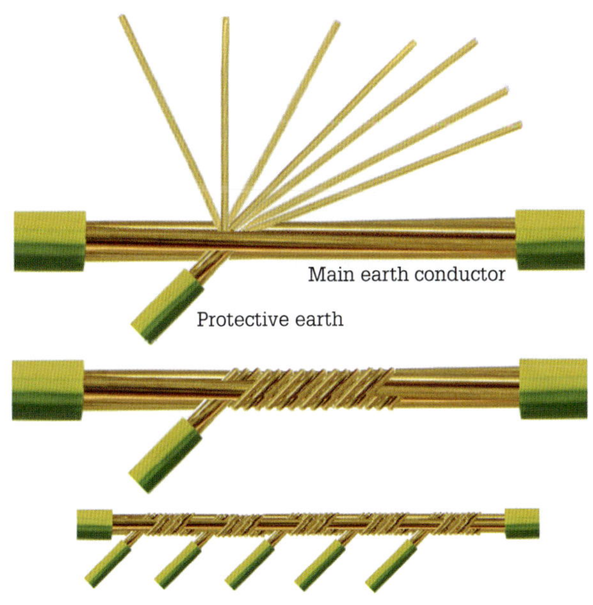

FIGURE 6.22 Protective earths connected to main earth conductor

With a 'Y' splice the protective earth conductors are interwoven through the main earth conductors and the conductor strands are then twisted in unison around the main earth conductor. By joining the protective conductor in this manner, the join is both mechanically and electrically sound independently of any soldering procedure. This type of join is also compact and must be soldered and then insulated to the same temperature rating as the main earth conductor insulation.

'T' splices are constructed by dividing the seven-strand conductor into a group of three and a group of four conductors and wrapping the strands in unison tightly around the main earth conductor, as shown in **Figure 6.23**.

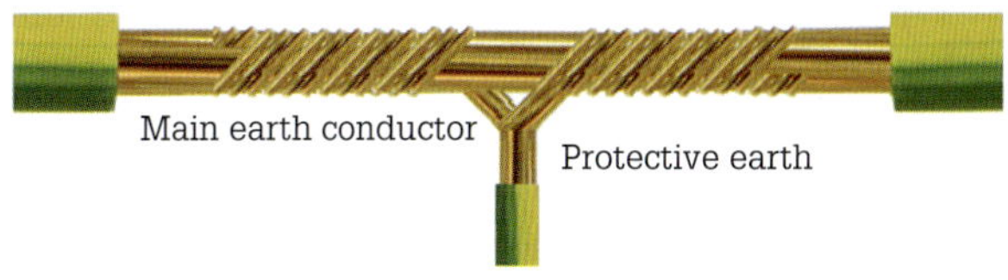

FIGURE 6.23 'T' splice

A 'T' splice is also compact and must be soldered and insulated to the same temperature rating as the main earth conductor insulation. Two-stranded cables up to 4 mm^2 in CSA can also be connected together as a single conductor length using a running splice as shown in **Figure 6.24**.

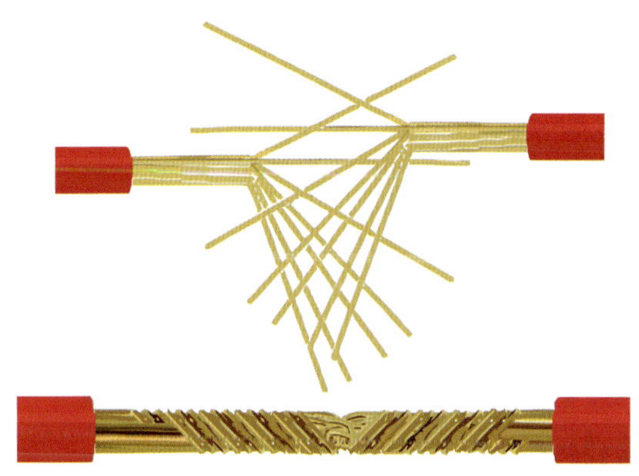

FIGURE 6.24 Running splice for stranded conductors

This splice is created by interweaving the strands of both cables together. The strands of each cable are fanned out at an angle of about 30°. One strand of each cable passes between two strands of the other. The strands are then multiple wrapped in unison around the straight conductors of the opposite cable. By joining stranded conductors in this manner, the join is both mechanically and electrically sound independently of any soldering procedure. This type of join is also compact and can be soldered and then insulated to the same temperature rating as the main earth conductor insulation.

REVIEW QUESTIONS

1. Name the colours recommended in Australia and New Zealand to identify 230 V single-phase fixed wiring installations.
2. Name the general-purpose thermoplastic polymer used for wire insulation, cable insulation, sheaths and conduit.
3. List the properties of silicone that make for good cable insulation.
4. Which of the two materials, copper and aluminium, has the lower resistance for conductors having the same dimensions?
5. What is the purpose of a hard cable jacket?
6. What does the 'voltage rating' or 'working voltage' of a cable indicate?
7. In what situations would a cable be sheathed?
8. What is the smallest nominal size of an active conductor allowed for a copper conductor according to AS/NZS 3000:2018 *Wiring Rules*?
9. Which conductor expands more – copper or aluminium – for a given temperature rise?
10. Why are annealed copper conductors used in stranded form?
11. Name two types of conductor terminations.
12. Name the advantage of using insulation-piercing connectors.
13. Outline the process of soldering.
14. What is an ideal method for inspecting terminations of critical electrical equipment?
15. State the difference between a conductor join and a conductor tap.

6.2 Cords, cables and plugs

A cable is a single-insulated conductor (called a cable core) or an assembly of two or more insulated conductors separated from each other by additional insulation or a protective sheath formed or wound around the group of conductors (refer to AS/NZS 3000:2018 *Wiring Rules*). There are four classes of cables used for installation work:

1. thermoplastic-insulated cables
2. thermoplastic-sheathed cables
3. mineral-insulated metal-sheathed cables
4. armoured cables.

Thermoplastic-insulated cables

Thermoplastic-insulated (TPI) cables, shown in **Figure 6.25**, are described as 450/750 V single-core annealed copper/PVC building wire. The 450 V indicates the voltage rating of the insulation to earth, and the 750 V indicates the voltage rating of the same types of insulated conductors. They are used for switchboard and control panel wiring arranged and contained in capped slotted PVC trunking and for fixed wiring within other types of enclosures. The PVC insulation is a V–90 (temperature rating) category. Colours include red, brown, black, blue, white and green/yellow. Stranded conductor sizes range from 1.0 mm^2 to 120 mm^2 with solid conductors available up to 2.5 mm^2 (1.78 mm). The minimum installed bending radius of TPI cables ranges from 10 mm for 1.0 mm^2 to 70 mm for 120 mm^2 cable.

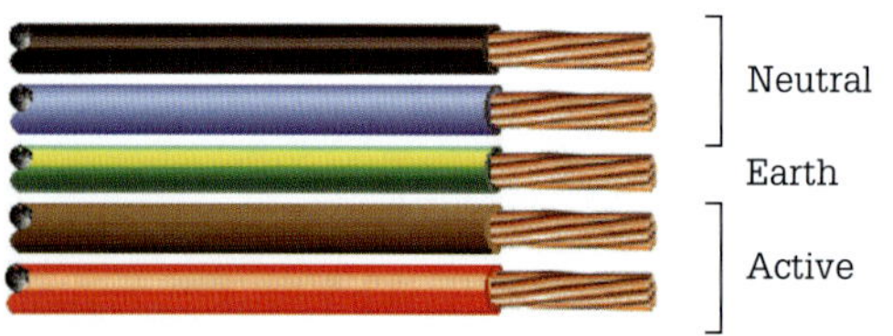

FIGURE 6.25 TPI cables

TPI cables installed in slotted PVC trunking and in steel conduit are illustrated in **Figure 6.26** and **Figure 6.27**.

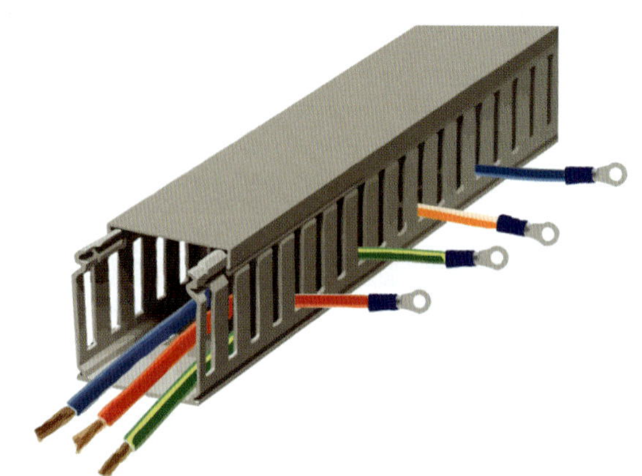

FIGURE 6.26 TPI cables in slotted trunking

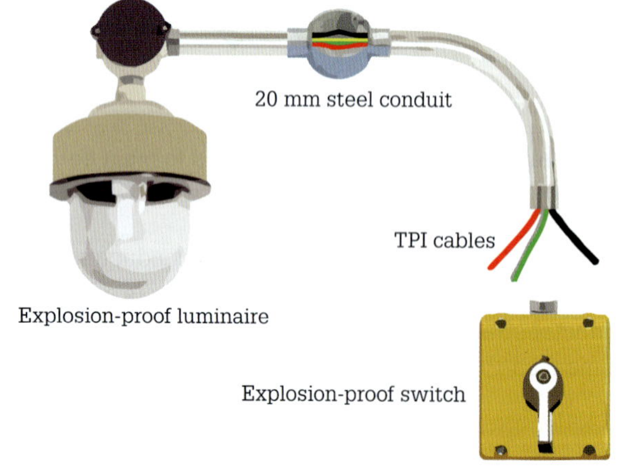

FIGURE 6.27 TPI cables in steel conduit

Flexible cords and cables

A cord consists of a high strand count of fine-gauge high-strength alloy conductors up to and including 4 mm² CSA with no conductor exceeding 0.31 mm in diameter and having up to five cores. Flexible cords, shown in **Figure 6.28**, are rated at 230/400 V with some heavy duty flexible cords rated at V–90.

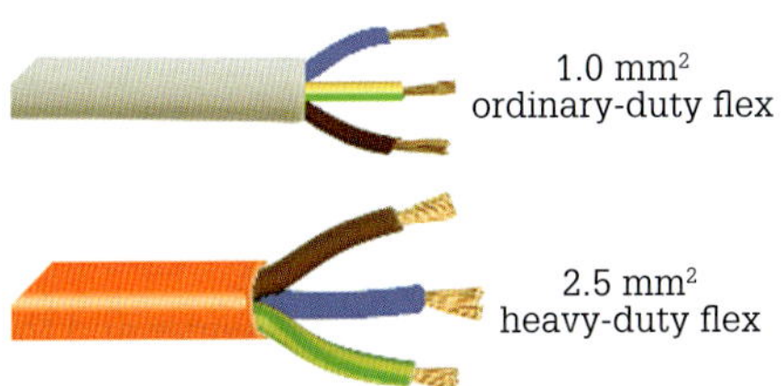

FIGURE 6.28 Flexible cords

Flexible cords are ideal for external wiring of appliances where the temperatures to which the cord is exposed do not exceed 105 °C. Flexible cords are also used as extension cords and on 'pendants' (hanging fixtures such as lamps, luminaires and socket outlets), as trailing cords, for connection of portable lamps and connection of stationary equipment to facilitate frequent interchange. Flexible cords can also be used as exposed fixed wiring if they are of the heavy-duty type or installed in an appropriate wiring enclosure if they are of the ordinary-duty type (refer to AS/NZS 3000:2018 *Wiring Rules*).

When flexible cords are used as extension cords, they have a maximum allowable length. A maximum length is necessary because of the voltage drop that occurs across the cord length. For example, a 1.5 mm² flexible cord loaded to 10 A has a maximum length of 35 m, while a 4.0 mm² flexible cord used for the same current load of 10 A could be up to 100 m in length.

Maximum extension cord length depends on the conductor CSA: the larger the conductor size, the greater the distance can be from a socket outlet to the appliance (see **Table 6.5**).

TABLE 6.5 Conductor CSA and extension cord length

Extension cord length rating in amps	Conductor area in mm²	Maximum in metres
10	1.0	25
	1.5	35
	2.5	60
	4.0	100
16	1.5	25
	2.5	40
	4.0	65
20	2.5	30
	4.0	50

Cables with flexible conductors with a CSA of less than 6 mm² and a core count greater than five cores, as shown in **Figure 6.29**, are classified as flexible cables. Flexible cables are rated at 450/750 V.

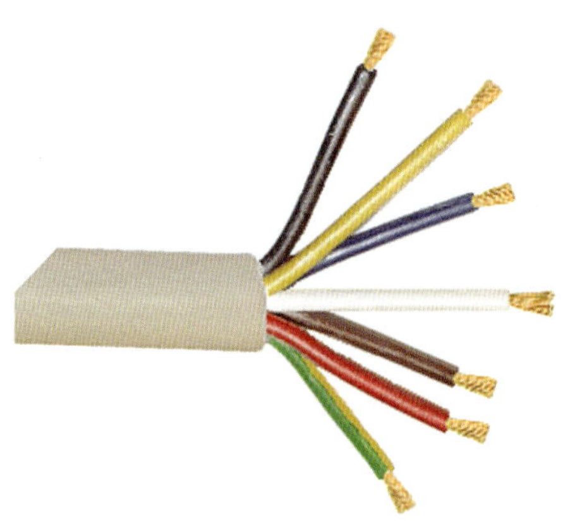

FIGURE 6.29 Flexible cable

Both flexible cords and cables have two sets of insulation; one called the basic insulation, which covers the conductors, and an outer supplementary insulation called the sheath. The supplementary insulation in many flexible cords and cables has properties that allow the cord or cable to be resistant to water, oil, ozone and chemicals. The combination of two sets of insulation is referred to as double insulation.

Flexible cables are for use with control on machinery, cranes, hoists, conveyors, control panels, assembly and production lines, equipment building and processing equipment and as cable for security applications. Fire-resistant (FR) flexible cable constructed to halogen-free, low-smoke, low-toxicity and flame-retardant specifications is used for emergency power supplies and fire alarm systems.

Cable colour code

The standard insulation colours for fixed and flexible cables designed to Australian standards are as follows.

For fixed cables (450/750 V) installation wiring:

- active cores: recommended – red, white and dark blue (three phase) and red or brown for single phase with white as switch wire; alternative – any other colour except black, light blue, green, yellow and green/yellow
- neutral core: black or light blue
- earth: green/yellow.

For flexible cords and cables see **Table 6.6 (A)** and **Table 6.6 (A)**. Functional earthing conductors are typically coloured white or pink.

TABLE 6.6 (A) Single phase

Function	Core colour
Earth	Green/yellow
Active	Brown
Neutral	Light blue

TABLE 6.6 (B) Three phase

Function	Core colour — current Australian	Core colour — European
Earth	Green/yellow	Green/yellow
Phase 1	Red	Brown
Phase 2	White	Black
Phase 3	Blue	Grey
Neutral	Black	Light blue

Single-phase plugs and sockets

A plug top (see **Figure 6.30**) is a device which may be engaged with a cord extension socket and which is designed for the purpose of connecting to a socket outlet any electrical equipment to which the plug top is attached by means of a flexible cord or cable (refer to AS/NZS 3000:2018 *Wiring Rules*).

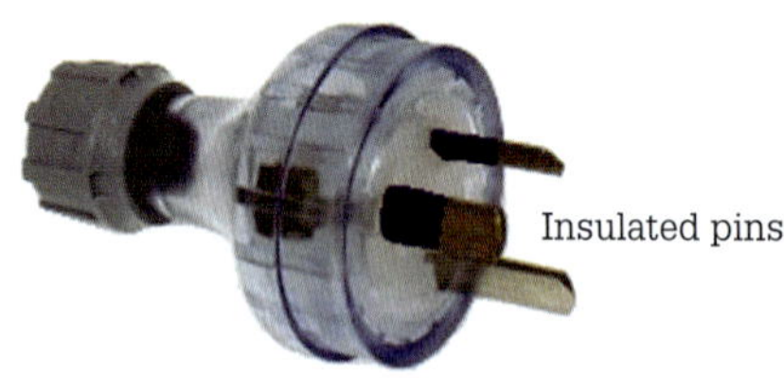

FIGURE 6.30 Transparent three-pin 10 ampere 230 V plug top

The transparent three-pin 10 ampere plug top is also used in New Zealand and Papua New Guinea. This plug has a flat earthing blade and two inclined flat blades for active and neutral connections. The flat blades called pins, measure 6.5 mm by 1.6 mm and are set at 30° to the vertical. The active and neutral pins are insulated to provide electrical safety for persons using a partially inserted plug top. Possible contact with energised pins is, therefore, limited.

The earthing pin is longer than the other two pins to ensure safety. This blade arrangement means that the earth is the first conductor connected to a socket outlet and the last conductor disconnected when the plug top is removed from the socket outlet. Nominal cable sizes of 0.75 mm^2 to 1 mm^2 are used with this size plug top. There is an unearthed version of this plug top as well, with only the two flat, inclined pins as shown in **Figure 6.31**.

FIGURE 6.31 Transparent two-pin 10 ampere 230 V plug top

A plug top with a slightly longer, wider and thicker earth pin is used for equipment drawing up to 15 A (see **Figure 6.32**). Nominal cable sizes of 1.5 mm^2 to 2.5 mm^2 are used with this plug top. Socket outlets supporting this pin also accept 10 A plug tops. Additionally, there exists a 20 A plug top, in which all three pins are oversized.

FIGURE 6.32 Transparent three-pin 15 ampere 230 V plug top

A cord extension socket (see **Figure 6.33**) is a device connected to a flexible cord or cable with which a plug top may be engaged for the purpose of connecting to a socket outlet (refer to AS/NZS 3000:2018 *Wiring Rules*).

The purpose of the shroud with a protective skirt is to provide protection from contact with the plug top pins when the plug top begins to withdraw from the socket. This safety feature ensures that the plug top pins disconnect from the socket terminals before a gap appears between the plug and socket. All plug tops and extension sockets are made from impact-resistant hard plastic and they have a moulded grip to facilitate disconnection.

FIGURE 6.33 Transparent three-pin 10 ampere 230 V cord extension socket

In Australia and New Zealand, all portable electrical appliances intended to be operated at a mains voltage of 230 V must be fitted with an approved plug top. Additionally, items such as extension leads and replacement appliance cords intended for connection directly to a socket outlet must also be equipped with an approved plug top.

REVIEW QUESTIONS

1 List the four classes of cables used for electrical installation work.
2 What are typical applications for thermoplastic-insulated (TPI) cables?
3 Describe the characteristics of an electrical cord.
4 What is the maximum length of an extension cord comprising 1.5 mm^2 and rated at 10 A?
5 Name applications for flexible cables.
6 Name the colours recommended in Australia and New Zealand to identify 230 V single-phase conductors in a flexible cord.
7 Which pin is the longer in a three-pin 10 A 230 V plug top?
8 State the purpose of the protective skirt around a cord extension socket.

6.3 Thermoplastic sheathed (TPS) wiring systems

Thermoplastic sheathed wiring systems may be of single-core or multi-core configuration.

Single-core double-insulated cables

Single-core annealed copper double-insulated TPI cables, known as single double-insulated (SDI) 450/750 V grade cables (see **Figure 6.34**), are also available. These cables are suitable for consumer's mains, sub-mains, and final sub-circuits; however, some authorities may require XLPE. SDI cables can be used unenclosed, enclosed in conduit, buried directly or in underground enclosures. Colours include red, brown, blue and black with a PVC sheath colour of white. Stranded conductor sizes range from 1.5 mm^2 (7/0.50 mm) to 630 mm^2 (127/2.52 mm) with solid conductors available up to 2.5 mm^2 (1/1.78 mm). The minimum installed bending radius of SDI cables ranges from 15 mm for 1.0 mm^2 to 40 mm for 16 mm^2 cables. The insulation jacket of any SDI cable should be neatly trimmed, with minimal edge flash and no mechanical damage to the conductor or insulation. In addition, the conductor stranding lay (twist pattern) should be undisturbed. **Figure 6.35** illustrates 16 mm^2 SDI cable.

FIGURE 6.34 Single double-insulated (SDI) cable

FIGURE 6.35 16 mm^2 SDI cables for consumer's mains

Thermoplastic-sheathed multi-core cables

Thermoplastic-sheathed (TPS) cables, also known as tough plastic sheathed cables, as illustrated in **Figure 6.36**, are rated as 450/750 V flat TPS plain annealed copper cable. These cables have two insulations – basic over the conductors and the sheath as the supplementary – TPS cable is classed as having double insulation (see Clause 1.4.73c). This means that TPS cable meets the minimum standard for safety as required by AS/NZS 3000:2018 (see Clause 1.5.4.3).

The following factors affect insulation quality:

- ozone and partial discharge resistance
- weather resistance
- oil resistance
- water resistance
- resistance to chemicals
- resistance to solvents
- abrasion resistance
- combustion propagation resistance
- insulation resistance
- electric strength.

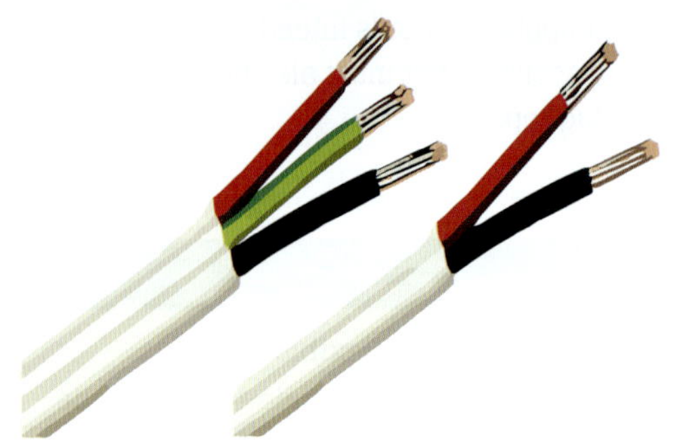

FIGURE 6.36 TPS 1.5 mm² twin and 2.5 mm² twin + earth cables

The performance of TPS (V–90) rated cables satisfactorily addresses these factors.

TPS cables are used for wiring of socket outlets, lighting, and heating loads in domestic, commercial and industrial installations.

There are three main types of TPS cables: two core (twin), two cores + earth and three core + earth. The PVC insulation in new TPS wiring installations is V–90 category with a white PVC sheath rated at the same temperature. The sheath is designed as an easy-peel sheath to allow access to the basic insulation. Depending upon the type of cable, primary insulation colours include red, black and green/yellow. Stranded conductor sizes range from 1.0 mm² to 16 mm² with solid conductors available for 1.0 mm² and 2.5 mm². The minimum installed bending radius of TPS cables ranges from 15 mm for 1.0 mm² twin to 40 mm for 16 mm² cables. The label on a TPS reel and the printing on the cable should have the following details.

For 1 mm² 2C + E – 1/1.13 Red, Blk, G/Y – 450/750 V

For 1.5 mm² 2C + E – 7/0.5 Red, Blk, G/Y – 450/750 V

For 2.5 mm² 2C + E – 7/0.67 Red, Blk, G/Y – 450/750 V

For New Zealand, some TPS cables may have the earth core in the outer position while Australia only has the earth core in the middle position.

In some circumstances, rodents such as rats and mice and marsupials such as possums may cause damage to TPS cable and may expose the conductors. As the insulation for TPS cable has a defined temperature rating (e.g. V–90), excessive heat above the rating may soften the insulation and create a condition called plastic flow.

The use of V–90-rated TPS does not permit a higher current-carrying capacity of the enclosed conductors. AS/NZ 3008.1.1 recommends 75 °C for current-carrying-capacity calculation.

Due to their sheath, TPS cables can be installed unenclosed or installed in various enclosures such as conduit or buried directly in the ground if they are not subjected to possible mechanical damage.

Installing TPS cables

An illustration of a TPS cable run from above the ceilings down to switches and socket outlets is shown in **Figure 6.37**.

FIGURE 6.37 TPS cable run

Where TPS cables are chased (cut) into rendered masonry or concrete walls the cables are not required to be enclosed. However, if TPS cables are to be installed in concrete slabs they must be contained within a suitable enclosure.

An illustration of TPS wiring requirements within a timber stud wall is shown in **Figure 6.38** and in **Figure 6.39** (refer to AS/NZS 3000:2018 *Wiring Rules*).

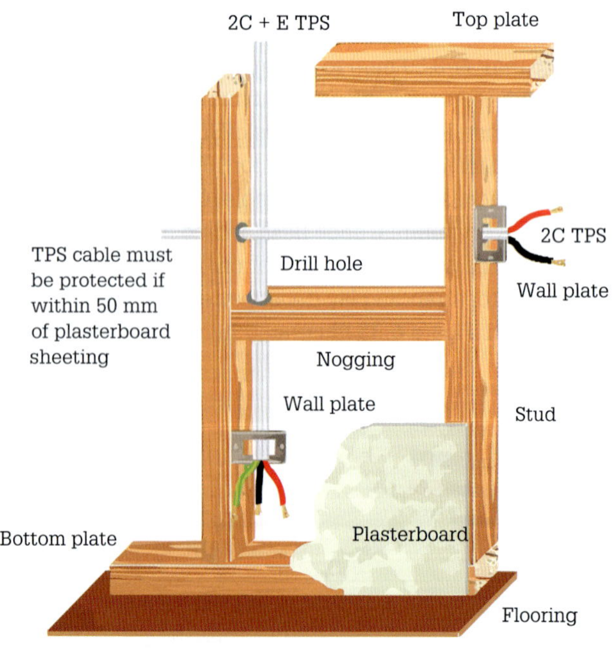

FIGURE 6.38 TPS wiring within a stud wall

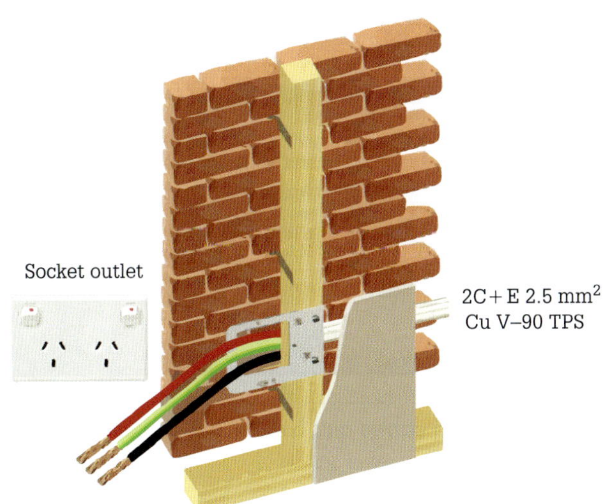

FIGURE 6.39 Flat TPS in a wall cavity

A TPS cable should be run in a manner that eliminates the possibility of strain on the cable itself or on the cable termination. TPS cables should not make contact with items likely to become hot, such as hot water pipes. Cable bending radius should not be less than the maker's recommendation and in any case should not be less than six times the overall cable diameter.

Where TPS cables are in accessible locations they must be supported by cable clips, cable ties or cable saddles depending on the installation method. In accessible locations in a ceiling, shown in **Figure 6.40**, they should be clipped to the side of the joist (but not within 50 mm of the ceiling fabric to prevent penetration by a nail); where they are laid across joists they should be clipped to the side of, for example, a 50 mm × 30 mm batten which is fixed to the joists.

You can remove the sheath to allow single insulated wires in the enclosure but the outer sheathing of the TPS cable must enter some distance into the enclosure. Any exposed TPS cables should be suitably protected or shielded against weathering.

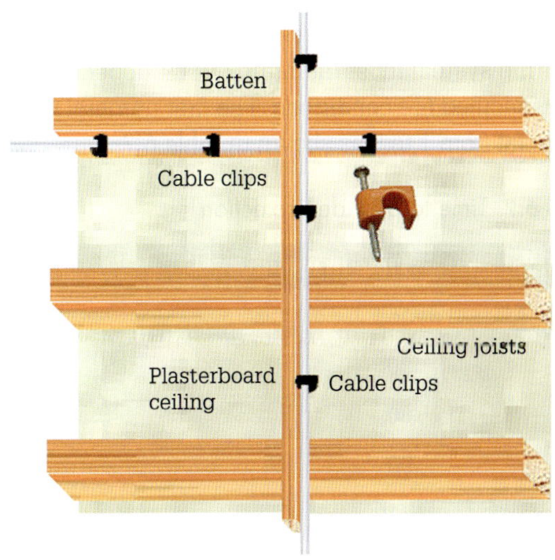

FIGURE 6.40 TPS wiring in an accessible ceiling

TPS cables and thermal insulation

When TPS cables are installed where thermal insulation will be placed, as illustrated in **Figure 6.41**, the cables may be partially or completely surrounded by the thermal insulation. In these circumstances the cables may need to be de-rated because the cables have a reduced ability to dissipate heat. Refer to AS/NZS 3008.1.1:2017 *Electrical installations – Selection of cables*, Section 3, for installation conditions.

For example, referring to Table 10, column 15, of AS/NZS 3008.1.1, the twin plus earth 1.5 mm^2 copper V–90 TPS lighting circuit cable of **Figure 6.41** has a current-carrying capacity of 14 A when partially surrounded and, in column 19, the same cable has a 9 A capacity when completely surrounded.

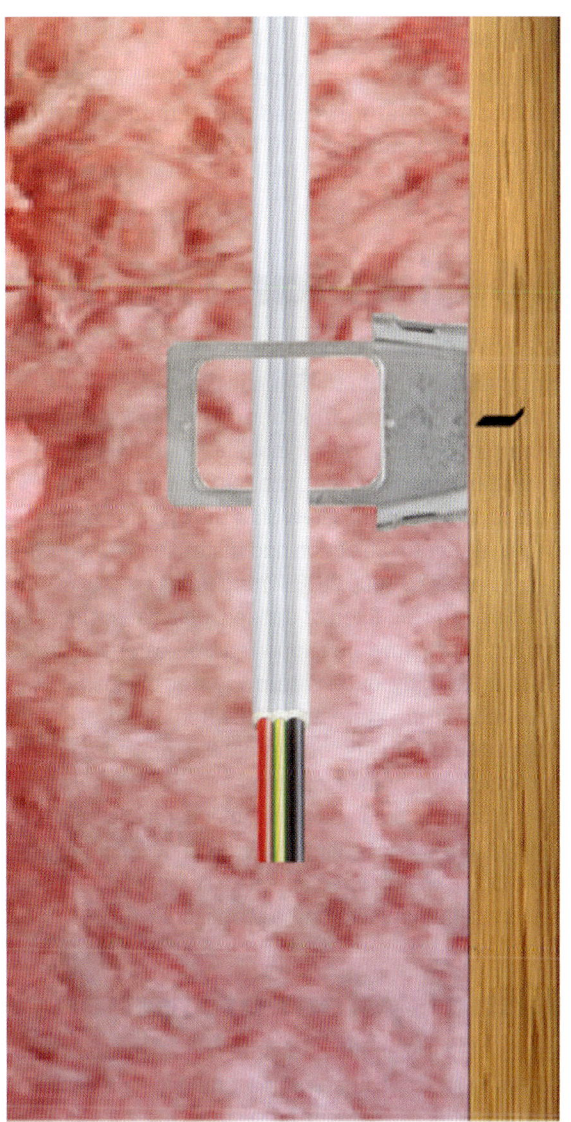

FIGURE 6.41 TPS cables and thermal wall insulation

For a 10 A circuit protection device the value of 14 A for the partially surrounded cable is satisfactory. For the 9 A capacity the cable would need to be protected with a 6 A protective device. If this was unsatisfactory, a 2.5 mm^2 twin plus earth TPS cable must be used which has a rating of 13 A (see column 19).

As building insulation can be installed after installation work is complete, a prudent electrician would assume a worst-case situation and install 2.5 mm^2 at the first fix.

Note: Labour costs to install 2.5 mm^2 at first fix are the same and there is only a slight increase in cable costs.

Figure 6.42 shows bulk thermal ceiling insulation, TPS cables and data cable installed above what is a raked ceiling.

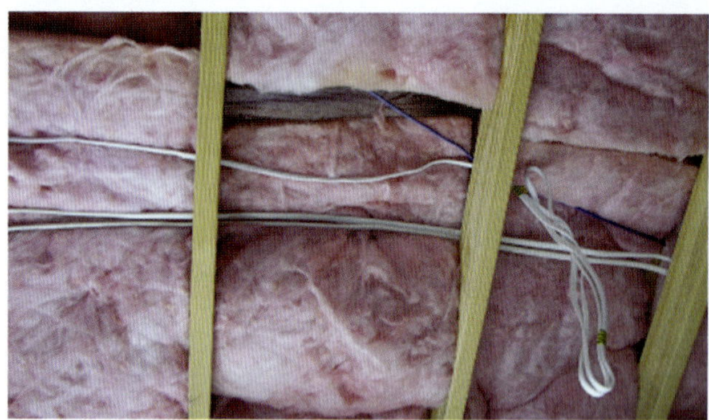

FIGURE 6.42 TPS cables and bulk ceiling insulation

REVIEW QUESTIONS

1 For what is SDI an abbreviation?
2 What is the minimum installed bending radius of 16 mm^2 SDI cable?
3 List four factors that affect insulation quality of TPS cables.
4 Name the three main types of TPS cables.
5 What temperature is used to determine current-carrying capacity of V-90 TPS cables?
6 Is it necessary to enclose TPS cables in a wiring enclosure when installed in rendered masonry or concrete walls?
7 Outline a suitable installation method for TPS cables installed in an accessible space.
8 Why is the current-carrying capacity of a cable reduced when it is installed in thermal insulation?

6.4 Circular TPS wiring systems

Wiring systems refer to the various types of cables, such as insulated, sheathed, armoured, MIMS and fire resistant, and the different methods of installation that can be used. The central consideration in the selection of a wiring system is the degree of safety that the system chosen provides. (Refer to 'Selection and installation of wiring systems', AS/NZS 3000:2018 *Wiring Rules*.)

Cables can rest, fixed or unfixed, on a flat surface, be in a wiring enclosure, be supported by a catenary system and buried directly, or be within a wiring enclosure in the ground. Wiring systems are affected by factors such as temperature, various sources of heat, moisture and humidity, foreign bodies, corrosive or polluting substances, impact vibration, flora and fauna, mechanical stresses and various hazardous locations. AS/NZS 3000:2018 *Wiring Rules* provide the requirements necessary for selection of a wiring system that is fit for the purpose.

Wiring system protection

Wiring system reliability under fire conditions must be maintained for a specified period, and dependability must be obtained when wiring systems are subjected to mechanical impact of a particular degree. To meet these wiring system requirements the standard AS/NZS 3013:2005 *Electrical installations – Classification of the fire and mechanical performance of wiring system elements* was established.

The classification and protection of wiring systems use an alpha-numerical code; for example:

WS 24

The first two letters, WS, stand for wiring system. The first numeral (2 = 30 minutes) represents the minimum time the wiring system must maintain circuit dependability under specified fire conditions. The first numeral is sequential from 1 to 5 as shown in **Table 6.7**.

TABLE 6.7 Time to provide protection

Numeral	Time in minutes
1	15
2	30
3	60
4	90
5	120

The second numeral (4 = very heavy), as shown in **Table 6.8**, represents the degree of impact force and cutting load to which the wiring system can be subjected without losing circuit reliability. The second numeral is sequential from numbers 1 to 5 and represents an impact test and saddle impact test from 2.5 joules to 5000 joules.

In combination with the above, the numbers 1 to 5 also represent a cutting test of 0.3 kN to 5.00 kN. Where the letter 'X' replaces a numeral, it means that this degree of protection does not apply.

TABLE 6.8 Degree of impact force and cutting load

Numeral	Impact result
1	Light
2	Moderate
3	Heavy
4	Very heavy
5	Extremely heavy

EXAMPLE 6.1

1 WS X2

X = no value of fire protection

2 = protected against Moderate impact

2 WS 3X

3 = 60 minutes of fire protection

X = no value of mechanical protection

3 WS 44

4 = 90 minutes of fire protection

4 = protection against Very heavy impact

EXERCISE 6.1

a Determine the WS rating of a wiring enclosure that is required to provide 30 minutes of fire protection and protection against very heavy impact.

b Determine the WS rating of a wiring enclosure that is required to provide 60 minutes of fire protection and protection against extremely heavy impact.

Circular TPS cable

Circular TPS cables are rated as 450/750 V circular TPS plain annealed copper cable. They are used for commercial and industrial wiring to motors, heating and lighting, and in installations where glands are required (see **Figure 6.43**). The PVC insulation is V–90 category with an orange or black PVC sheath rated at V–90. Colours include red, brown, blue, black and green/yellow. Stranded conductor sizes range from 1.5 mm^2 to 6 mm^2. The minimum installed bending radius of circular TPS cables ranges from 40 mm for 1.5 mm^2 twin to 55 mm for 6 mm^2 cable. Due to their sheath, circular TPS cables can be installed unenclosed or installed in various enclosures such as conduit or buried directly in the ground if they are not subjected to possible mechanical damage.

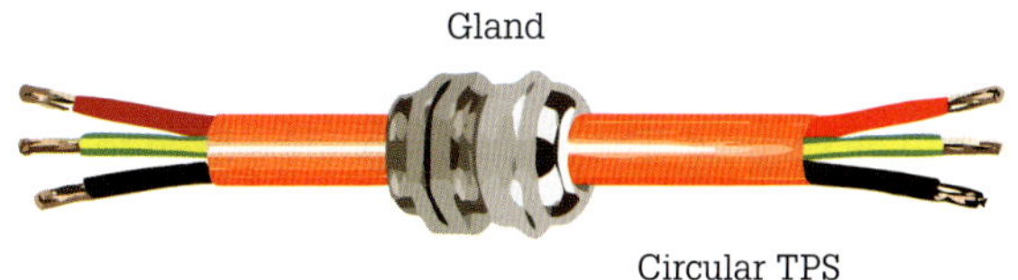

FIGURE 6.43 Circular TPS cable with PVC insulation fitted with a gland

Note: Polystyrene and polyurethane thermal insulation materials if in contact with PVC cables cause degradation of the plasticiser in the PVC cable sheath, leaving the insulation hard and brittle. The cable must be separated from these active materials by a material barrier such as conduit.

Cable ladder systems as illustrated in **Figure 6.44** are used for the support of a combination of suitable segregated power cables, electrical equipment and communication system installations. Cable ladder provides support for any cables installed upon it but, because it is not completely enclosed, in some working environments it may not offer sufficient mechanical protection. For this reason unsheathed, single-insulated (TPI) power cables should not be installed on ladder.

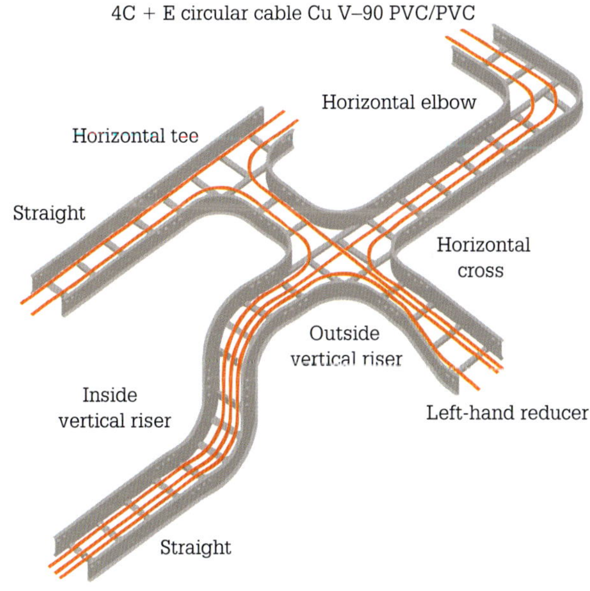

FIGURE 6.44 Circular TPS installed on cable ladder

Circular cable below 100 mm in diameter installed on horizontal ladder should be fixed at 250 mm centres while on vertical ladders fixing should occur at 400 mm centres. Fixing devices for cables include cable ties, strapping, steel, aluminium or moulded cleats and rung slot patterns. Some installation specifications may require that cable exiting the ladder be installed in conduit fixed to the ladder.

Circular TPS cable on tray

Fixing devices as shown in **Figure 6.45** for circular TPS cables on tray include cable ties, saddles and other clamping methods. Saddles for securing multi-core circular cables to the cable tray should be shaped to the form of the cables to be secured. The saddles should be fixed to the cable tray by means of corrosion-resistant round-headed screws, nuts and washers. The shanks of the screws should not extend beyond the nuts by more than three threads.

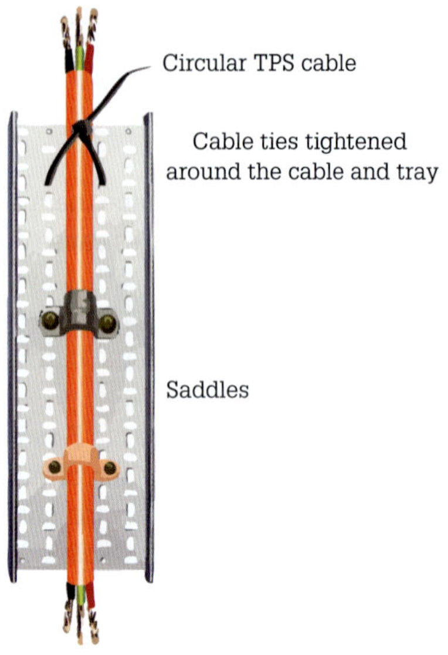

FIGURE 6.45 Circular TPS on tray

Cable saddles or cable clamps with a suitably thick insulated sleeve can be provided along the entire cable route with their spacing in accordance with the manufacturer's recommendation.

REVIEW QUESTIONS

1 To what does the term 'wiring system' refer?
2 List common cable installation methods.
3 Determine the WS rating of a wiring enclosure that is required to provide 60 minutes of fire protection and protection against very heavy impact.
4 List appropriate support methods for circular TPS installed on cable ladder.
5 Outline the fixing requirements for circular TPS cables, not exceeding 100 mm in diameter, installed on cable ladder.

6.5 Non-metallic enclosures

Common wiring enclosures include rigid/flexible steel, light/heavy-duty UPVC, flexible PVC and their associated fittings; trunking such as metal, PVC and their associated fittings; cable tray/rack such as standard/heavy-duty galvanised steel ladder, perforated cable tray and their associated fittings. (Refer to 'Selection and installation of wiring systems', AS/NZS 3000:2018 *Wiring Rules*.)

Wiring enclosures provide mechanical protection for cables and electrical safety to persons and property and provide suitable and accessible enclosures for the conductors. A well-designed electrical enclosure system should have sufficient capacity for future expansion and therefore would be readily adaptable to changing installation needs.

Note: The installation of wiring enclosures is considered electrical work and must be undertaken by a licensed electrician (see Clause 3.10.3.1). Wiring enclosures for consumer's mains must not be installed by the drainer on building sites.

Unplasticised PVC conduit

Unplasticised PVC (UPVC) conduit provides easy installations; it is quickly and easily joined and provides labour savings when compared to rigid metallic conduit. UPVC conduit is available in spigot-socket form with an integral solvent weld joint. UPVC conduit is also referred to in the electrical trade as PVC conduit with the 'U' assumed. Both forms are used in the text.

The standard Australian finished lengths of UPVC conduit is 4 m or 6 m. Various diameters are available such as: 16 mm, 20 mm, 25 mm, 32 mm and 40 mm. Conduit is available in various grades as defined in the Australian standards. These grades are:

- light-duty oval conduit (grey in colour and suited for use in chased walls)
- medium-duty conduit (grey in colour and suited to above-ground service and in concrete slabs): WS X1
- heavy-duty conduit (orange in colour and used for underground applications): WS X2.

Underground conduit (in some Australian states) carrying service cable for single domestic and small installations is a minimum of 50 mm heavy-duty UPVC conduit for two-wire service (600 mm cover) and a minimum of 80 mm heavy-duty UPVC conduit for four-wire service (850 mm cover). Only one proprietary/set bend is allowed at not less than 450 mm rising to the meter box or other terminating position.

UPVC conduit exposed to solar radiation, if not treated for exposure, must be protected with a coating of light-coloured water-based acrylic paint. Medium-duty (grey) UPVC conduit and heavy-duty (orange) UPVC conduit and associated fittings offer both high-impact and high-tensile strength, are non-conductive and have excellent chemical resistance. Examples of UPVC conduit and fittings are illustrated in **Figure 6.46**.

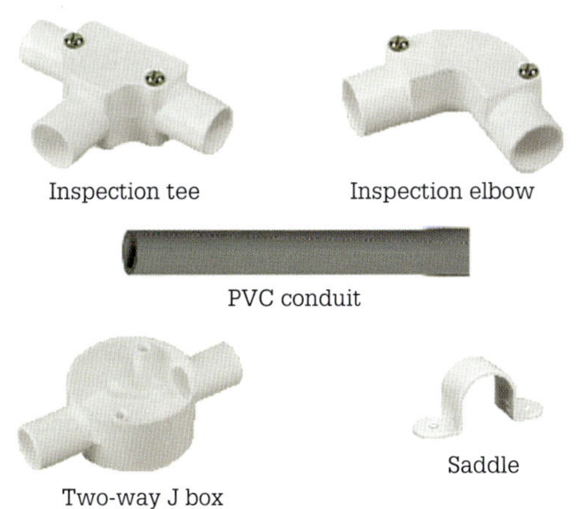

FIGURE 6.46 UPVC conduit and fittings

When cutting UPVC conduits always wear safety glasses with side shields to prevent particles from entering the eye. After cutting UPVC rigid conduit square with a conduit cutter or a hacksaw, sharp edges or burrs from inside the conduit must be removed and a slight chamfer should be applied to the cut end. Kinked or wrinkled bends in UPVC rigid conduit are not acceptable.

UPVC conduit and fittings are meant to be assembled or joined by means of solvent cementing of the integral housings or UPVC couplings. To join conduit and associated fittings apply a liberal amount of solvent cement to the mating surfaces, slide together and give a quarter turn to ensure the solvent is spread evenly on the surfaces. Hold together for a few seconds until the joint is made rigid.

Solvent cements contain chemicals that dissolve the surface of the UPVC, softening it. As the chemicals evaporate, they leave a UPVC resin behind that fuses the mating surfaces. Usually the solvent-cemented joint will be strong enough to install immediately but will take up to 24 hours to cure. As with any chemical product, contact with unprotected bare skin and inhalation of vapour should be avoided.

Conduit runs

Install visible conduit runs so as to blend with the architectural features of the building or structure. Run all types of conduit parallel to walls, floors and ceilings except where it is desirable to follow the architectural layout. Always design the layout of conduit to minimise the number of fittings and sets. Conduit runs should be installed so as to avoid all other pipe systems and services. Install the conduit so that electrical cables are drawn into the conduit only at draw-in boxes, switchboards, socket outlet positions and switch positions. All junction boxes and fittings should be of adequate size to enable the TPI cables to be neatly diverted from one conduit run to another. Where conduits are exposed to view they should be given one coat of primer and two finishing coats of a colour that matches the surroundings.

No wiring should be installed until all the conduit runs complete with draw wires (2.5 mm minimum diameter with 600 mm tails) have been installed.

The starting ends of conduit runs should be temporarily plugged with a soft mastic compound to prevent the ingress of dirt while the rest of the conduit system is installed. Where possible, conduit runs should be in long straight lengths and provided with expansion joints as required. When used, all conduit fittings should be installed in a true horizontal or vertical plane.

Conduit should be fastened to the side of a timber batten where installed across rafters or joists in accessible ceilings to prevent damage should someone step on it.

Select conduit to suit the environment in which it is to be installed. Consider the material performance of the conduit in environmental conditions such as the pH of the environment, temperature and humidity, acids, alkalis, metal salts, organic solvents, aromatic hydrocarbons, ultraviolet (UV) light, water, salt spray and any other factors. Correct selection is important in order to prevent corrosion, deterioration, galvanic action, UV effects and chemical attack. For example:

- Do not use UPVC conduits in areas normally exposed to sunlight or other sources of UV radiation (see Clause 3.3.2.11) unless rated 'T'.
- Do not use steel conduits for above-ground use in coastal areas or other potentially corrosive environments.
- Do not use UPVC conduits where chemicals such as esters and chlorinated hydrocarbons are present.
- Do not use ABS conduit where organic solvents are present.

Environmental concerns also include flora and fauna (see Clauses 3.3.2.9 and 3.3.2.10). For example, flora which includes microorganisms such as mould, bacteria and fungi and the detached parts of plants may cause gradual deterioration of the conduit, causing corrosion and eventual breakdown of cable insulation. **Figure 6.47** shows conduit buried by flora.

FIGURE 6.47 Conduit buried by flora

Native plants, fruit trees and berry-producing shrubs attract black fruit bats (flying foxes) whose manure will quickly erode steel conduit. Conduit can be subjected to attacks from fauna; for example, small animals may cause impact damage and insects and termites may enter the conduit run and attack the insulation of cables. Some soils called acid sulphate soils (ASS) contain sulphuric acid which will attack steel conduit.

Installation of wiring systems can cause environmental damage. For example, in some areas the installation of long runs of underground services may affect flora species such as the black ironbox and the zamia palm, which are threatened species. In addition, fauna species such as the powerful owl, the squatter pigeon and the yellow chat may be disturbed by works associated with underground services.

Conduit should be installed in a manner that prevents formation of moisture traps. Where water or condensation (see Clause 3.3.2.3 and Clause 3.10.2.3) may form in a conduit run, the conduit run must be sloped for drainage, as shown in **Figure 6.48**. The conduit must also be drilled in the underside at each low point to allow the moisture to weep from suitably located drainage points.

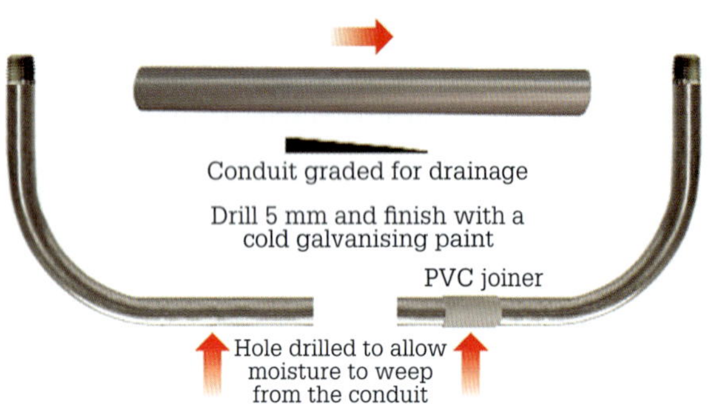

FIGURE 6.48 Conduit graded for drainage

In some installations UPVC conduit adaptors must be used when metallic pipe has to be drilled for drainage points when installed underground (metal conduit cannot be used underground because its wall thickness is too small but metallic pipe can). In addition, a bonded earth (for continuity of a conductive enclosure) may be necessary to connect the two sections of metallic pipe separated by a UPVC conduit adaptor. However, this is not required if the metallic pipe contains insulated and sheathed cables only. Internally wipe out all UPVC conduit and metallic pipe runs to remove moisture before commencing the wiring installation. For UPVC conduits and metallic pipe installed in trenches where the moisture inside cannot escape, a drainage sump as illustrated in **Figure 6.49** is constructed within the trench.

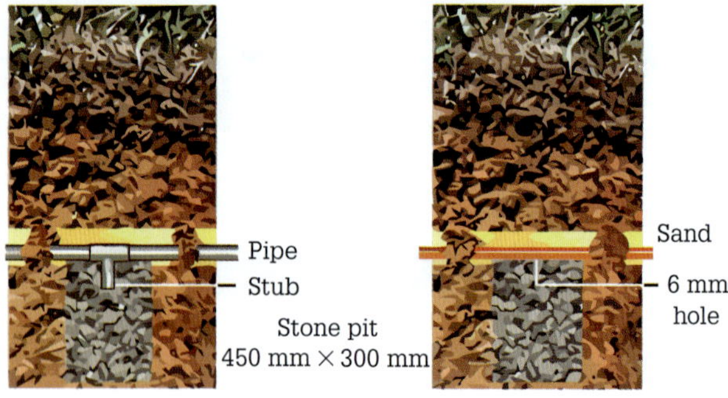

FIGURE 6.49 Drainage sump for UPVC conduit and metallic pipe

Suggested fixing of conduits

Conduits should be neatly run and securely fastened by means of approved saddles. Saddles should be fixed at each change of conduit direction and all boxes carrying light fittings, switches, sockets and other accessories must be rigidly and securely fixed independently of saddles.

It is preferable to fix conduits using double-sided stainless steel saddles. Where space does not permit the use of double-sided saddles an alternative type of saddle (half saddle) can be used. PVC, zinc or galvanised plated saddles or spring clips are also used to fix conduits, as shown in **Figure 6.50** (see Clauses 3.3.2.8, 3.10.3.2 and 3.3.2.12).

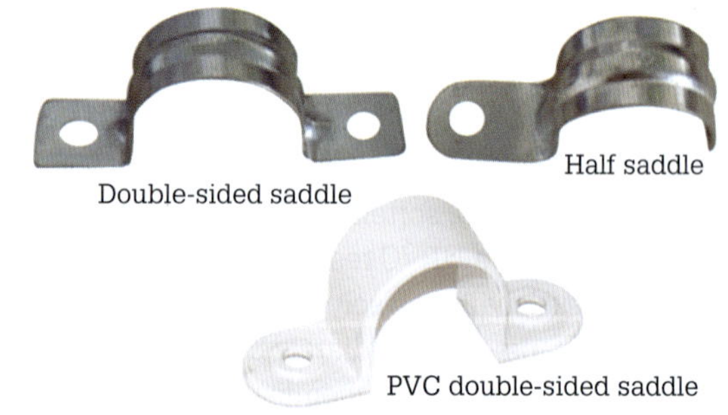

FIGURE 6.50 Conduit saddles

Conduits should be grouped as far as possible and all supports, clips, saddles and brackets spaced to prevent appreciable sag. Also, group saddles in symmetrical positions.

On horizontal runs (walls and ceilings) 20 mm and 25 mm conduits should be supported by not less than one saddle every 900 mm in addition to the support provided by any structure, box or fitting. Conduit should be saddled to the structure of the building within 150 mm of each set, bend or other conduit fittings.

For 32 mm and larger conduits, saddles may be placed not more than 1.2 m apart. Sets, bends and terminations should, in all cases, be supported by two saddles not less than 150 mm and not more than 200 mm from any set, bend or termination.

For vertical runs, conduit should be secured at a maximum distance of 1200 mm.

All purpose-made supports for conduit should be of stainless steel, galvanised steel or aluminium and should be neatly finished, degreased and primed with one coat of zinc chromate and finished with two coats of epoxy-resin-based paint in colours suitable to the surface upon which the conduit is fixed.

Medium-duty conduits

This conduit may be used above ground in positions not exposed to mechanical damage and embedded in concrete slabs. All UPVC conduit is supplied in spigot-socket form with an integral solvent weld joint as shown in **Figure 6.51**. UPVC conduit is available from 16 mm to 150 mm. Standard lengths may be 4 m or 6 m.

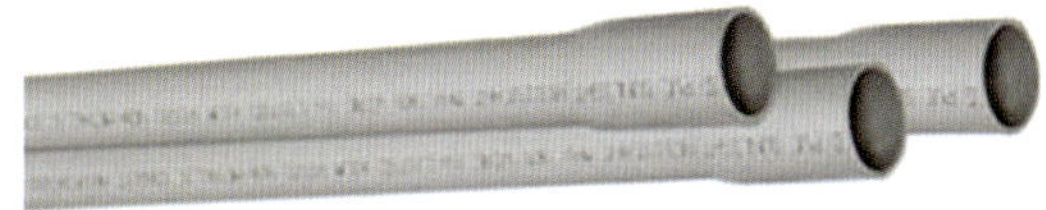

FIGURE 6.51 Medium-duty conduit

There have been installations where underground sheathed cables in medium-duty (grey) conduit were installed as a category 'A' system. Digging up the conduit and replacing it or providing extra protection is an expensive exercise. As per AS/NZS 3000:2018, Table 3.5, column 7, sheathed cables installed in medium-duty, corrugated and flexible conduit are classified as a category B system and must be provided with additional protection as described by Clause 3.11.4.3, AS/NZS 3000:2018.

When medium-duty conduit is fixed to a PVC wall box, a screwed PVC adaptor and circular lock nuts must be used unless the conduit enters the wall box via a moulded conduit entry. The method of supporting medium-duty UPVC conduit must allow for the longitudinal expansion and contraction of the conduits.

For nominal sizes of UPVC medium-duty conduit, 16 mm to 63 mm refer to outside diameter, whereas sizes 80 mm to 150 mm refer to the approximate internal diameter.

Note: According to AS/NZS 2053, rigid non-metallic electrical conduit must meet certain mechanical requirements under specified tests (resistance to compression, resistance to collapse, impact resistance, etc.). AS/NZS 2053 also specifies certain wall thicknesses (2.0 mm, 2.1 mm) pertaining to each duty rating.

Heavy-duty UPVC conduits

These conduits are used underground or above ground where an increased level of mechanical protection is required.

The colours for heavy-duty conduits are as follows:

- orange – enclosing all other cables
- red – enclosing fire alarm cables.

Heavy-duty conduit, shown in **Figure 6.52** is primarily used for underground applications. If installed in accordance with the specific clauses of AS/NZS 3000:2018, it can be used for this application without further protection. This conduit is available in 4-m orange (EO) lengths.

FIGURE 6.52 Heavy-duty conduit (orange)

Trenching for enclosures

Prior to excavation work, location of existing services needs to be determined.

Trenches for wiring enclosures need to be 600 mm deep and 300 mm wide. If chased into rock the excavation needs to be 100 mm square to allow for a 50 mm covering of pea-gravel aggregate concrete. Excavations within 600 mm of existing services should be carried out by hand digging. The other service providers must be consulted for requirements.

Before enclosures or cables are laid, the bottom of the trench must be graded evenly, cleared of loose stones and a 50 mm layer of washed sand added over the full width and length of the trench and tamped. Where the level of the trench bottom has to change, the slope should not be greater than 1 in 12.

After enclosures or cables have been laid a further layer of washed sand should be added over the full width of the trench and tamped, to provide finally not less than 50 mm cover over enclosures or cables. If soil is used the soil should have passed through a sieve with a maximum mesh of 12 mm; for mineral-insulated cables the maximum mesh size must be 6 mm so as not to damage the sheath.

Trenches must be backfilled in layers and each layer should be rammed. The first two layers should be 100 mm deep and rammed by hand; the remaining layers should be not more than 200 mm deep and power ramming may be used. Warning tapes and covers should be included as specified by AS/NZS 3000:2018 *Wiring Rules*.

Where applicable, topsoil and turf should be placed and the final level should be level with or not more than 25 mm above the adjacent ground level. Refer to AS/NZS 3000:2018 *Wiring Rules* for underground wiring systems and categories.

As they are sheathed, SDI and TPS cables can be installed unenclosed or installed in various enclosures such as heavy-duty UPVC conduit, metallic pipe or buried directly in the ground where not subjected to possible mechanical damage.

SDI and TPS cables are suitable for category B underground wiring systems (buried direct in the ground) at a depth of 600 mm measured to the top of the cable and must be laid on a 50 mm bed of washed sand and covered by another 50 mm. In addition, an orange marker tape needs to be provided, with the markings 'danger electric cables below,' 250 mm above the cable.

Wherever trenches exist where other workers or members of the public could accidentally stumble or fall into them, safety barriers, warning signs and suitable night lighting must be provided.

All heavy-duty conduits must be installed at a minimum depth of 600 mm below the finished surface in a trench not more than 300 mm wide. In general, electrical orange heavy-duty UPVC conduit should be used for underground runs. They should be embedded and totally covered in 50 mm of clean washed sand. If more than one conduit is placed in the same trench they should be spaced to a minimum of 50 mm apart.

Orange marker tapes, with the markings 'danger electric cables below', must be provided 300 mm above the conduit and laid continuously along the route of the cable. Trenches should be backfilled with an allowance made for subsidence and compacted after one week.

Cable markers with 'E' or 'Electric cable under' stamped on them should be placed where the cable changes direction and where the cable enters or leaves a building (see **Figure 6.53**).

FIGURE 6.53 Cable marker

Cable pits should be used where underground wiring enclosures change more than two different directions and 50 mm drains should be provided at all cable pits. The soil surrounding the enclosures and cable pits should be treated to protect against termite attack, with a notice stating this displayed in the switchboard.

Drawing-in cables

Installed enclosures should be cleaned internally with a swab or other types of pull-throughs before cables are drawn in.

As cables are installed in conduits, care should be taken to avoid dirt or other particles being drawn into conduits by the cables either by adherence to the surface of the cable or by the motion of the cable at the conduit entry. Suitable bell mouths should be used at the cable entry ends of conduits greater than 50 mm to prevent cable twists from entering the conduit. Cable pulling usually requires two people – one to pull the draw wire and the other to feed the cables into the conduit.

Cables must be pulled from a conduit in as straight an angle as possible to prevent damage to the cable insulation. A cable can also kink as it comes off the reel, causing stress which can significantly impact on the cable's performance. At any expansion joints adequate slack must be left in the cables.

After the cable has been routed through the pathway to the various termination points, more stresses are applied when removing the insulation jacket and terminating the conductors. All of these installation factors can change the physical properties of the cable, which in turn can degrade the cable's electrical performance. Therefore it is critical to maintain the physical integrity of the cable throughout the installation.

When cables are pulled into conduits or other enclosures, the cable drum must be positioned to unreel the cable with the natural curvature (the cable is drawn from the top of the drum) as shown in **Figure 6.54**. By unreeling with the natural curve of the cable into the conduit no undue pressure will be placed on the mouth of the conduit.

FIGURE 6.54 Cable drum mounted on stand

During installation the cable should be carefully examined for any sign of damage as it leaves the drum. This is particularly important with the outer cable layers, where drum transportation may have caused damage.

Draw-in points

Draw-in points provide access to conduit runs and facilitate placing wires in long conduit runs. For surface-run conduit systems, draw-in points need to be provided at suitable intervals not exceeding 12 metres for straight runs and at intervals not exceeding 8 metres or the equivalent of 2 × 90° bends for runs including directional changes.

Where used, draw-in boxes should be of adequate size to enable cables to be drawn in easily and without damage. For 20 mm conduits, round boxes should be used at the draw-in points.

Outlet boxes for fittings can be used as draw boxes. Draw-in boxes should have neoprene gaskets to prevent moisture entering the conduit run. Where pull boxes are located in exposed locations or in floors, they must be fully sealed against the entry of moisture. All wall boxes and draw-in boxes should be fixed independently of the conduit system.

For some underground conduit systems, a sufficient number of accessible cable pits must be provided to enable underground draw-in boxes to be installed. These draw-in boxes must have gasketed covers and be sealed to prevent moisture entry.

Materials used to lubricate cables while drawing in cables to conduits must be non-conductive, non-abrasive and non-hygroscopic. Suggested lubricants are powdered soapstone, talc or silicone grease. The maximum length of any drawing in is dependent on the type of cable, the conduit type and configuration, the lubricant and the lubrication techniques. Never use petroleum-based products as a lubricant.

All conduits in a switchboard enclosure must be installed in a neat and orderly manner, evenly spaced, level and plumb.

High-impact conduits

There are a number of high-impact non metallic wiring enclosures available. Some of these are discussed in the following.

Acrylonitrile–Butadiene–Styrene (ABS) conduit

Conduit made from ABS (see AS 3690 *Installation of ABS pipe systems*) can be used where high-impact resistance combined with high corrosion resistance is required. ABS is very tough and resilient, has high-impact strength, has good chemical resistance, is non-toxic and has a recommended ambient temperature range of –30 °C to +70 °C. ABS conduit is installed according to the same requirements that relate to the installation of PVC conduit. An ABS conduit run must be supported on straight runs with conduit clips or saddles at intervals of not more than 1000 mm. Conduit clips and saddles must be installed so as to allow longitudinal movement of the conduit. ABS conduit must be cleaned and joined with a cleaner and adhesive formulated for ABS, adopting the procedures recommended by the manufacturer.

When ABS conduits are exposed to the weather they will suffer some minor degradation of the exposed surface. The degradation results in a reduction of surface gloss. The effect of long-term exposure to sunlight over prolonged periods has minimal effect on the physical properties.

Solvent cementing ABS to PVC is not recommended (see AS 3691 *Solvent cement and priming (cleaning) fluids*) for use with ABS pipes and fittings. A rubber ring socket or a mechanical connection such as a threaded connection or a flange is recommended where such joints are necessary.

Halogen-free conduit (HFT)

HFT conduit, shown in **Figure 6.55**, has fire-resistant properties (no toxic or corrosive gases are released) that make this conduit suitable for use in fire risk areas and chemical environments. Being 'halogen free' means that these conduits have no waste disposal issues. The material can be reused and recycled safely, as well as used in landfill or incinerated without any danger to the environment. HFT conduit is highly resistant against chemical attack and its suitability for use in these environments requires reference to a manufacturer's chemical resistance chart.

FIGURE 6.55 HFT conduit

UPVC conduit, fittings and joints

All conduits and fittings must comply with AS/NZS 2053 *Conduits and fittings for electrical installations* and the conduits, fittings and accessories used in an installation must be of an approved type and manufacture. UPVC conduits are widely used, primarily because of their low cost and great ease of fabrication.

UPVC conduit should be fixed with PVC (see **Figure 6.56**) or metallic saddles at a maximum of 600 mm spacing. The fastenings should not be over-tightened to permit thermal movement of the conduit. All horizontal runs of conduit should be secured at a maximum distance of 900 mm and vertical runs should be secured at a maximum of 1200 mm. For high ambient temperatures or where rapid changes in temperature are likely to be encountered this distance should be reduced.

Conduits smaller than 20 mm in diameter should not be used except for fire alarm systems where 16 mm in diameter is an accepted size; 25 mm diameter conduit is acceptable for underground installations. Conduits to be installed should be in long lengths (to reduce the number of joins), straight and smooth, with no burrs and sharp edges. In addition, never use offcuts to fabricate long lengths of conduit. UPVC conduit containing scratches or scores deeper than 10% of the conduit wall thickness should not be used.

FIGURE 6.56 UPVC conduit saddle with spacer

Conduit should be cut square, of adequate length to allow conduits to be butted firmly together at joints and hard against the shoulders of accessories such as elbows, tees, J boxes and adaptors. All burrs must be removed from the ends of cut lengths using a burring reamer or rat tail file before installing. Only slow-acting solvent cement especially formulated for watertight conduit fitting joining should be used.

UPVC conduit should not be used in hazardous locations or where subject to physical damage. It should not be used for any surface-mounted installation where contact temperatures or ambient temperatures can exceed 50 °C. If the temperature is less than 50 °C UPVC conduit can be secured in position using spacer-bar-type saddles giving 3 mm minimum clearance from the surface.

Note: Conduit lying directly on a surface will be subjected to higher temperatures than a conduit supported above the same surface.

The number of single-core PVC insulated non-sheathed cables (TPI) run in UPVC conduit should be such as to permit easy drawing of the cables (see Appendix C10, AS/NZS *Wiring Rules*). The actual number of cables drawn into any conduit should not be greater than the number given in AS/NZS *Wiring Rules*, Table C 10. The area of cables including earth wires (see Clause 5.5.5.2) that may be enclosed in any one conduit or pipe must comply with the requirements of AS 3008:1:1. Before a conduit (PVC or metal) is selected to accommodate TPI cables, the number and size of cables to be installed must be known. The number of cables that can be installed in a circular conduit is determined from the ratios of the CSA of the enclosure and the CSA of the cable as follows:

$$\text{Number of cables} = \frac{\text{Internal CSA of enclosure}}{\text{CSA of cable}} \times \text{SF}$$

The space factor (SF) is as follows:

- for one cable in enclosure: 0.5
- for two cables in enclosure: 0.33
- for three or more cables in enclosure: 0.4.

When cables exit conduit or trunking the conductor's insulation must not be damaged (see Clause 3.10.3.5).

Bending UPVC conduit

Circular conduit not exceeding 25 mm in diameter can be bent 'cold' using a bending spring, as shown in **Figure 6.57** (see Clauses 3.9.6 and 3.10.3.4 for cable requirements and bending). Simply insert the correct-sized bending spring into the conduit and bend around your knee to the desired angle. It is often necessary to over-bend to allow for creepage (spring-back).

UPVC conduits must be set as required using correctly sized springs to form radius bends while maintaining the effective diameter and shape of the conduit. Conduit sets must not be distorted by kinks, flattening and wrinkling

EXAMPLE 6.2

1 Use Table C10 of AS/NZS 3000:2018 to determine the maximum number of 1.5 mm^2 single-core V90 sheathed cables that can be safely installed in a 20 mm heavy duty UPVC conduit.

Solution: 4

2 Use Table C10 of AS/NZS 3000:2018 to determine the minimum size of medium duty UPVC conduit that can safely accommodate seven 4 mm^2 single-core V90 sheathed cables.

Solution: 32 mm

EXERCISE 6.2

a Use Table C10 of AS/NZS 3000:2018 to determine the maximum number of 2.5 mm^2 single-core V90 sheathed cables that can be safely installed in a 20 mm medium duty UPVC conduit.

b Use Table C10 of AS/NZS 3000:2018 to determine the minimum size of heavy duty UPVC conduit that can safely accommodate four 2.5 mm^2 single-core V90 sheathed cables.

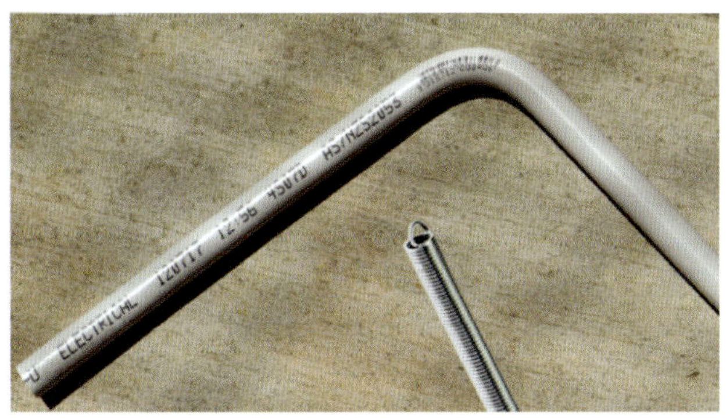

FIGURE 6.57 Conduit bent using a bending spring

because these distortions effectively reduce the internal diameter of the conduit (see Clause 3.10.3.5).

However, some installation specifications do not permit the bending of UPVC conduit by cold means. Where heat is required, the conduit area to be bent (called a set) may be heated by brisk rubbing. On larger conduit, heat can be supplied by careful application of a hot air gun or by placing the conduit in a commercial conduit heater (called a hotbox). The conduit set may be sponged with cool water until it hardens. Use only a flameless heat source such as a heat gun. Do not use an open flame to heat the conduit. Conduit which has heating marks must not be used. When bending UPVC conduit, a minimum radius six times the conduit diameter is recommended, as illustrated in **Table 6.9**.

TABLE 6.9 Bending recommendations

Conduit outside diameter	Minimum radius
16 mm	96 mm
20 mm	120 mm
25 mm	150 mm
32 mm	200 mm
40 mm	240 mm
50 mm	300 mm
63 mm	380 mm

A hotbox, shown in **Figure 6.58**, using the oven principle heats the conduit and softens it so that it can be bent to the desired shape. Before heating the section of conduit, plug both ends to trap air in the conduit. The air, heated in the hotbox, expands to prevent kinks or dislocation of the conduit when it is bent.

All fittings for UPVC conduit should be of the same material as the conduit and all joints should be made with the manufacturer's recommended solvent cement. It is recommended that you use solvent cement that is of a contrasting colour to the conduit, as shown in **Figure 6.59**. This allows visual indication that every fitting and joint has been melded to the conduit.

EXAMPLE 6.3

A 25 mm conduit has to be bent around a 100 mm flange. Using **Table 6.9**, determine the minimum bending radius for the conduit.

Solution: 150 mm

EXERCISE 6.3 

A 40 mm conduit has to be bent around a 220 mm flange. Using **Table 6.9**, determine the minimum bending radius for the conduit.

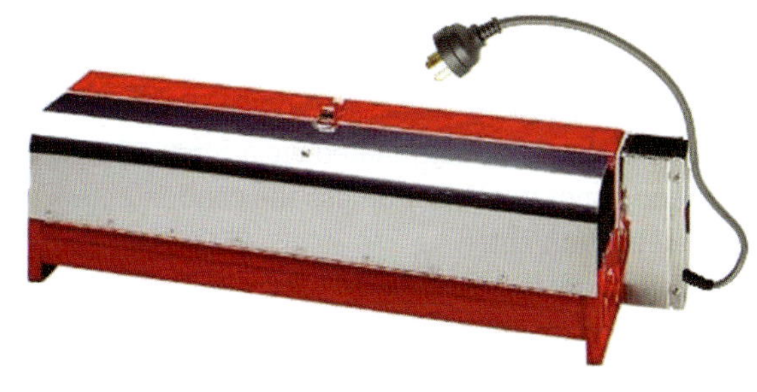

FIGURE 6.58 Conduit hotbox

UPVC conduit installation

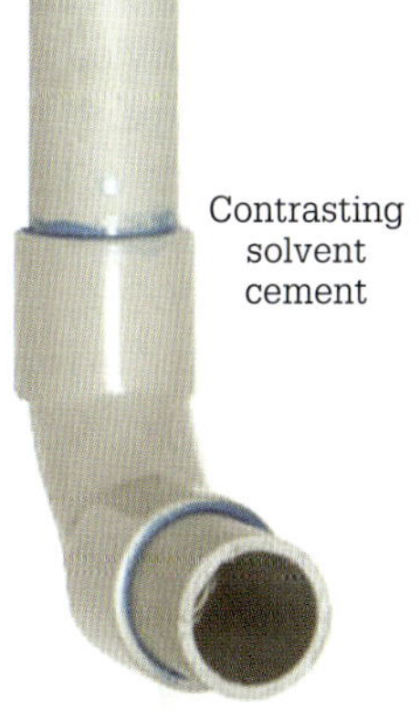

FIGURE 6.59 Conduit solvent cement

UPVC conduit must not be installed in locations where long-term exposure to sunlight or other sources of UV radiation occur unless specifically permitted in any installation contract documents (see Clause 3.3.2.11). Where permitted under an installation contract, sunlight-resistant UPVC conduit and fittings complying with AS/NZS 2053 must be run in locations exposed to direct sunlight. Short runs of heavy-duty UPVC conduits from underground sub-mains may be surface run where they enter an existing building if it is not practicable to conceal them, provided that they are suitably protected from mechanical damage and sunlight. AS/NZS 2053.1 requires that conduits suitable for use in direct sunlight be marked with the letter T.

The conduit run should be completely installed before wiring is drawn in. The wiring should be drawn in at draw-in boxes, inspection fittings, outlets and switchboards. For 20 mm conduits it is suggested that round boxes be used at the draw-in points and for 25 mm or larger conduits rectangular boxes should be used.

All UPVC conduits that are terminated into metal boxes and enclosures (draw-in boxes, joint boxes, switchboards, ducting, etc.) should be terminated via a screwed PVC adaptor and conduit lock nut unless the conduit enters via a moulded conduit entry.

Caution needs to be exercised to prevent accumulation of water, dirt or concrete in the conduit(s) during the construction stage of the building. Therefore, all UPVC conduits must be capped during this stage (see Clause 3.3.2.4).

Insulated, unsheathed cables, if installed within roof cavities, must be enclosed in a wiring enclosure such as UPVC conduit throughout their entire length (see Clause 3.10.1.1).

Except for lighting and power circuits, control, security and fire circuits must be placed in separate conduits.

Expansion joints

Expansion joints connect sections of conduit and provide allowance for movement due to service load, shock or thermal expansion. As thermal expansion (<14 °C) and contraction have significant effects on long straight runs of conduit, expansion joints are used to accommodate changes in length (see Clause 3.10.3.8).

Expansion fittings as illustrated in **Figure 6.60** should be fitted on all straight runs (6 m long) of PVC conduit, except those embedded in concrete or wall chases. Expansion joints should be installed in all UPVC runs, at intervals not exceeding the manufacturer's recommendations.

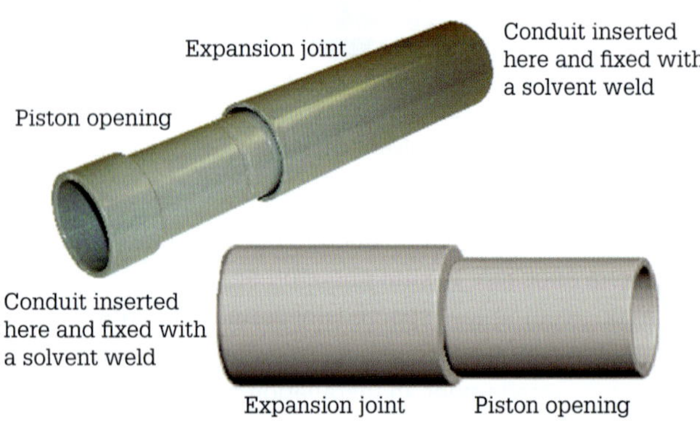

FIGURE 6.60 Conduit expansion fittings

Conduit saddles/clips must be installed close to expansion fittings to provide support. All saddles/clips associated with the straight conduit run must be fitted so as to allow the conduit to move freely while expanding or contracting.

Install expansion couplings wherever control joints occur in a building slab. Each expansion coupling in a building slab may consist of a short length of flexible conduit, to join the two UPVC conduits across the joint, covered by a larger UPVC conduit. The larger conduit should extend 150 mm from the flexible conduit in both directions and must be securely fixed to prevent movement and the annular gaps sealed to prevent concrete entry while the slab is poured.

Do not use bellows-type expansion couplings as shown in **Figure 6.61** in floor slabs or similar installations.

If galvanised steel conduit is used in a building slab, any conduits traversing expansion joints in concrete must be provided with stranded earth wires, which should be bonded to the first conduit boxes in runs on each side of the joint.

FIGURE 6.61 Bellows-type expansion coupling

Conduits in roof spaces

Where conduits are run in roof spaces they can be installed below the ceiling insulation and sarking. Where PVC conduits are installed in roof spaces and are likely to be disturbed (walked on) they must be mechanically protected. In roof spaces, conduits should be run in parallel paths with a minimum of crossovers and be adequately secured by means of saddles.

When designing the layout of conduits in roof spaces, the wiring system according to AS/NZS 3000:2018 *Wiring Rules* must be protected against the effects of heat due to external sources (see Clause 3.3.2.2). Therefore, the conduit-run design must account for expansion due to temperature changes within the roof cavity. An increased temperature condition accelerates the ageing process of PVC conduit and conductor insulation.

Note: When using temperature de-rating tables (AS 3008.1) for cables, the electrician must consider the temperature inside the conduit and not the ambient temperature surrounding the conduit. If choosing between UPVC conduit and metallic conduit in high-temperature environments remember that the higher thermal conductivity of steel removes heat more quickly from the enclosed wiring than will PVC conduit.

Embedded in concrete or wall chases

Light-duty oval conduit comes in a range of low-profile sizes specifically for use in chased walls or floors, as shown in **Figure 6.62**. The low profile allows a minimum depth chase to be made yet still allows adequate space and protection for the cables and a flush wall finish.

Oval conduits should be completely concealed within the building fabric when installed; the outer faces of

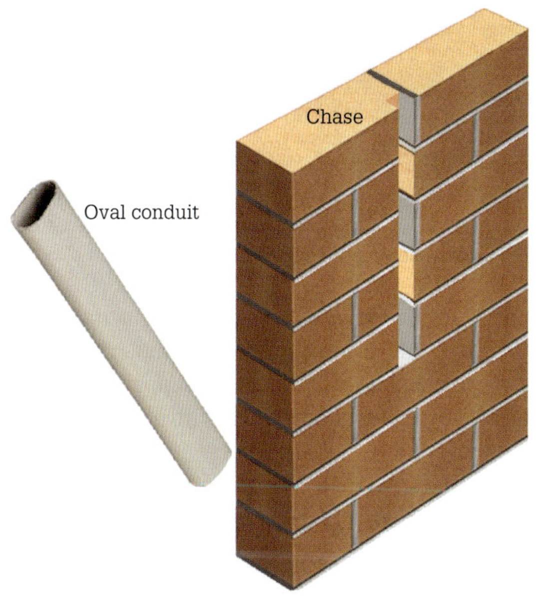

FIGURE 6.62 Light-duty oval conduit

conduits in brick, concrete or cement-rendered walls should not be less than 13.0 mm back from the finished rendered surface. Chasing into single brick walls should not exceed 35 mm. Cables should not be embedded in plaster, concrete, mortar or other finishes unless they are in conduit and capable of being fully withdrawn and replaced.

Conduits installed within concrete render or hard plaster wall finishes must be securely fixed to prevent movement or vibration. The secured fixing is carried out by placing at least 10 mm of render or plaster over the conduit and its fittings.

Penetrations

It is recommended to use mortar of a 3:1 sand cement mix with surfaces 'bondcreted' prior to applying the mortar for repairs to all penetrations in brick, cement blocks and concrete.

However, gunnable sealants and expandable foam are also used. **Figure 6.63** demonstrates a penetration using a non-fire-rated elastomeric seal but with poor workmanship exposed.

Penetrations through damp courses (DPCs) or the touching by conduit of damp courses should not occur.

FIGURE 6.63 Penetration

A DPC as shown in **Figure 6.64** is a course, or layer, of impervious material in a wall or floor to prevent the migration of moisture.

FIGURE 6.64 Damp-proof course

Penetrations through waterproof membranes

Conduits which enter a building at ground level should run under the waterproof membrane and vertically penetrate the membrane and the concrete slab at the appropriate position. Proprietary sealing must be provided between the conduits and the waterproof membrane.

Penetrations through external or existing structures

Where conduits pass through external walls or existing ground floor slabs it is suggested that a penetration 10 times greater than the conduit diameter should be provided, as illustrated in **Figure 6.65**.

FIGURE 6.65 Penetration space

The penetration space around the conduit must be made waterproof and, if in a floor slab, the penetration must be made waterproof and termite-proof by using a proprietary sealing method and termite poison. It is suggested that holes can be cut through concrete and masonry in new and existing structures with a diamond core drill or concrete saw.

In damp, corrosive and exposed environments

In all such environments all conduits, conduit fittings and metal enclosures should be hot-dip galvanised, as shown in **Figure 6.66**; conduit saddles should be of the 'stand-off' type giving not less than 12 mm clearance from building surfaces and be hot-dip galvanised after fabrication. Conduit runs and equipment should be fixed to building structures by means of zinc-coated fasteners.

Inspection-type conduit fittings should be dispensed with as far as possible, but when necessary must be provided with sponge neoprene gaskets to the covers. Inspection covers, saddles, equipment enclosures and so on should be secured by means of brass machine screws with round heads. Where corrosive atmospheres exist (see Clauses 3.3.2.12 and 3.3.2.5) the complete conduit installation including all fittings and supports must be lightly sanded and painted overall with epoxy-resin-based or bitumastic paint.

FIGURE 6.66 Metal switchboard enclosure

In slabs on ground

Conduit must not be installed in a manner that results in penetration of damp-proof courses, waterproof membranes and the like. Conduits must not be run in the concrete topping of any structural concrete structure. This is because they may impair the structural strength of the concrete. Wherever possible, always use preformed bends for all conduits laid in concrete slabs. Conduits run in the sub-base under floor slabs must be heavy-duty UPVC.

Do not run corrugated conduits in concrete slabs because this type of conduit can be crushed by the weight of the concrete. Additionally, the corrugated conduit can be moved and stretched by the weight of the concrete.

Conduits in slabs must be securely fixed using wire ties, as shown in **Figure 6.67**, to the reinforcing mesh, passing above a single layer of mesh or between double layers of mesh.

Where conduits are cast in concrete they should generally be installed in the centre of concrete slabs and above the first layer of reinforcing bars, as shown in **Figure 6.68**.

FIGURE 6.67 Conduit-fixing wire tie

FIGURE 6.68 UPVC conduits placed between reinforcing bars

FIGURE 6.69 Deep pattern J box for ceiling outlet

Ideally, conduits in slabs should be at least 75 mm apart. Terminations should be in deep pattern J boxes for ceiling outlets as shown in **Figure 6.69** embedded in concrete and steel wall boxes for all flush socket outlets and switches.

Conduit should be adequately sealed and protected during a concrete placement. In addition, precautions must be taken to ensure that no movement of conduit or distortion of the conduit walls or fittings occurs (see Clauses 3.3.2.6 and 3.3.2.8). Note that in **Figure 6.69** only one tie has been used to fix the conduit to the deformed steel bar. This is incorrect: it must be a double tie.

Where an expansion or construction joint occurs with a concrete slab, an expansion joint must be made in the conduit run with flexible couplings to 600 mm on either side.

Embedments in concrete such as conduits require that the concrete is vibrated and consolidated around them; otherwise honeycombing or voids can exist under the conduit which could collapse if the area above the conduit is placed under load. Avoid or minimise crossover of conduits within a slab. If a crossover cannot be avoided, the intersection angle should be greater than 30°. Tie together conduits at the point of any crossover.

A minimum horizontal clearance of 75 mm should be maintained between conduits in slabs.

The minimum cover over conduits should not be less than the conduit diameter in interior locations. In exterior locations the conduit should be embedded at least 50 mm below the slab surface. This ensures that the conduit is not subject to weather conditions (moisture reaching the conduit). Conduit requirements in a slab are illustrated in **Figure 6.70**.

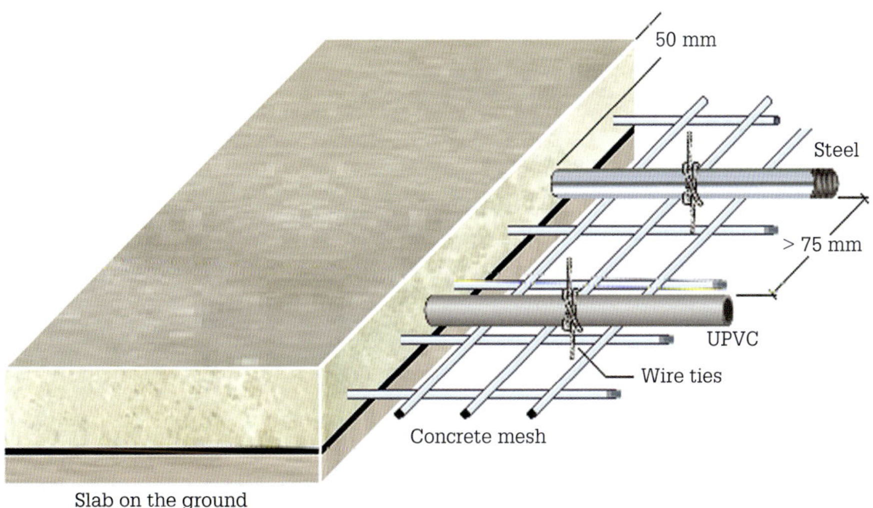

FIGURE 6.70 Conduits in a concrete slab

Where groups of conduits are brought to distribution board positions or the like, they should be fixed with spaces between them equal to one conduit diameter, to permit an adequate fill of concrete. Inspect all conduits prior to pouring concrete, then supervise the concrete pour to ensure that the conduits are not displaced, broken or damaged. After concrete has been placed or render or plaster restored, the conduit stubs projecting through floor slabs, as shown in **Figure 6.71** (for a minimum of 100 mm), and walls must be protected against damage.

FIGURE 6.71 Conduit projecting through a concrete slab

In suspended slabs

A suggested maximum diameter of conduits in suspended slabs is 25 mm. If a number of conduits are to be run in the slab, the conduits must be spaced not less than 75 mm apart. Conduit boxes should be of a suitable depth to finish flush with the underside of the slab, as shown in **Figure 6.72**.

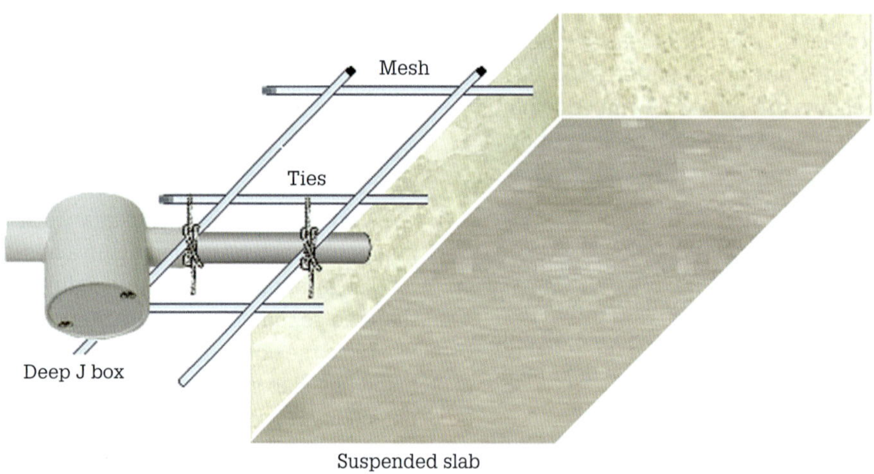

FIGURE 6.72 Deep J box in a suspended slab

In concrete columns

Unless otherwise indicated by the installation designer and depending on the dimensions of the column, up to a maximum of four conduits is acceptable in a column. No more than two conduits should cross any one face of the column and all conduits must be placed centrally, as shown in **Figure 6.73**. Bends in conduits entering the columns should be at 90°. Structural columns must not be chased for conduit installation.

If conduits are not required to enter a column, any conduits within the slab should not be installed through the area one metre around a column.

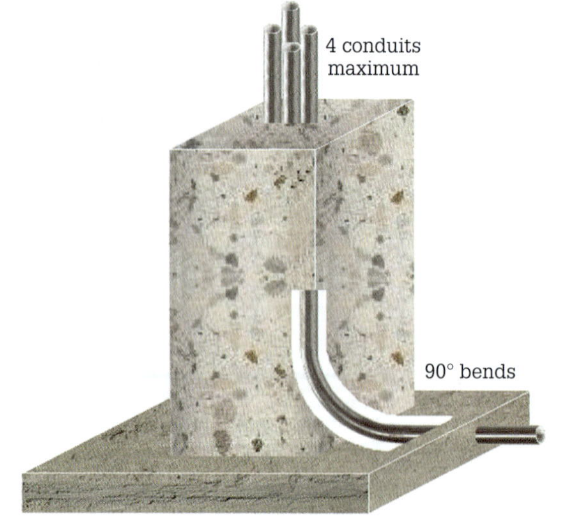

FIGURE 6.73 Conduit in a column

In hollow-block walls

Locate conduits in what will become core-filled sections of precast hollow-block-type walls as illustrated in **Figure 6.74**.

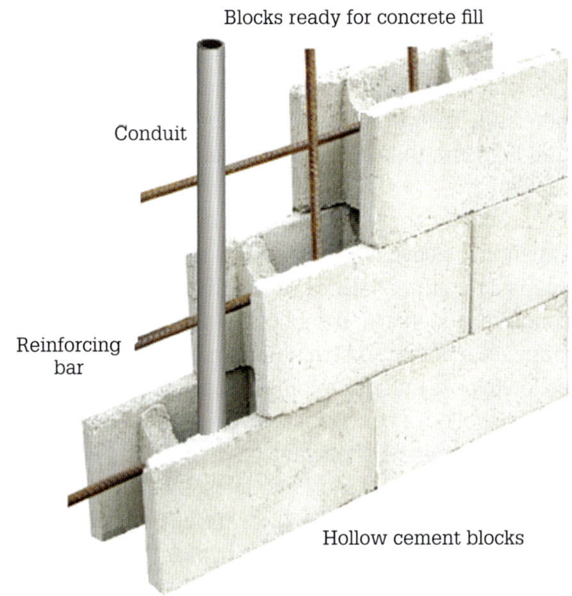

FIGURE 6.74 Conduit in cement blocks

For future use

It is suggested that all conduits installed for future use must be provided with polypropylene draw cords. A length of cord 1000 mm long must be left securely fixed at the ends of each run. These conduits must be capped, sealed against the weather, insects and vermin and labelled.

Non-metallic conduit

Corrugated PVC conduit as shown in **Figure 6.75** may be used as a service conduit (up to 1 m) to allow the cable to enter a cavity or meter box. Where the corrugated conduit is exposed, it must be protected by a steel guard. This conduit is also used where an awkward bend is required and standard bends are not suitable.

Corrugated non-metallic conduit is also used for the following applications:

- in lieu of non-metallic flexible conduit
- in short lengths for the interconnection of sections of light fittings
- for the interconnection of rigid PVC conduits around structural members
- in short lengths for the extension of rigid PVC conduit to light fittings.

Flexible PVC conduit

Flexible PVC conduit, also called electrical hose as shown in **Figure 6.76**, should be used between fixed conduit and equipment likely to be moved or subject to vibration; for example, connecting electric motors, stoves or water heaters. Its length should be as short as possible (<1200 mm).

FIGURE 6.75 Corrugated PVC conduit and wall box

FIGURE 6.76 Flexible conduit

Flexible conduits should be run exposed and be so positioned that they are adequately supported and not susceptible to mechanical damage. The ends of flexible conduit must be securely fixed to the fixed conduit and the equipment at both ends. Use proprietary fixing systems (plain to flexible coupling adaptor sets) for flexible conduit. Note that the adaptor sets can reduce the conduit size at the end of a run, making it difficult to draw through the cables.

Transparent conduit

For high-security situations where cables carry highly sensitive or confidential data, transparent conduit should be used. Transparent conduit allows ready identification of any interference to cables along the run. Normal clear conduit cement is suitable for bonding of joints. Note that transparent conduit as shown in **Figure 6.77** is not UV stabilised and should not be used in direct sunlight.

FIGURE 6.77 Transparent conduit

Fixing of enclosures

The fixing requirements of conduit saddles, enclosures and accessories are shown in **Table 6.10**.

TABLE 6.10 Fixing requirements

Location	Type of fixing
Concrete mesh	Cable ties
Buried in plaster or render	Saddles
Above false ceilings	Spacer bar saddles
Surface	Distance saddles
Fixing of saddles, enclosures and accessories	
Building fabric	**Type of fixing**
Structural steelwork	Purpose-made clamps
Non-structural steelwork	Set screws and nuts
Concrete, brick or building blocks	Plugs and screws
Hollow blocks and pot floors	Butterfly spring toggle bolts or gravity bolts
Timber	Wood screws

Note: No shot firing should be used and no drilling or welding of structural steelwork must be done to affix any enclosures or fittings of any wiring system unless approved by an engineer. Bolts, nuts, washers and screw fixings should be cadmium or zinc electroplated or purpose made for the enclosure or accessory. Examples of types of fixings are shown in **Figure 6.78**.

FIGURE 6.78 Examples of fixings

Fixings and fastenings provided and erected to secure all conduits should be of a proprietary type most suitable in design for the requisite application. Fixings and fastenings must be secure and adequately sized to suit the type, weight, size, shape and environmental location of the conduit being fixed. Fixings and fastenings must be composed of entirely compatible materials throughout. For example, steel fixings and fastenings must not be fitted with brass washers. All screws must be not less than 4 mm diameter and inserted at least 25 mm into woodwork or 50 mm into solid brick or concrete. Fixings to concrete or brick should be via expanding-type metal fixings, the hole for which should be no greater than the recommended diameter. All fixings and fasteners should be galvanised.

Conduits less than 32 mm diameter may be fixed as follows:

1. To masonry, concrete, cement-rendered walls: Ø6 expanding sleeve masonry anchor to accommodate an M4.5 mm thread. However, for 20 mm and 25 mm conduits indoors nylon anchors with mushroom stainless steel expansion nails can be used. Fixing to hollow blocks and mortar should be avoided. If fixing to hollow blocks cannot be avoided, the conduit should be fixed by means of galvanised, round-head screws and spring-loaded butterfly toggles. Block-work fixings should not be made in the joint. Fixings into concrete and concrete block-work can be by means of metal expansion inserts fitted into hammer-drilled holes. Conduit saddles can also be secured to these types of building walls using tapered screws driven into metal, hard plastic, nylon or fibre plugs.
2. To timber: where directly fixed, wood screw No. 8 × 20, round head. Where fixed through fibrous cement, plaster, plasterboard and so on the screws should be of sufficient length to allow for the additional thickness.
3. To sheet metal: No. 8 × 13 mm long galvanised binding head, self-tapping screws.
4. To steel: M5 round-head screws of appropriate length tapped or bolted.
5. To sheet metal: round-head 4 mm diameter set screws and nuts or 7G × 13 mm long galvanised binding head, self-tapping screws (sometimes these are not acceptable – always refer to contract details).

Conduits greater than 40 mm and up to 100 mm diameter may be fixed as follows:

1. To masonry, concrete, cement-rendered walls: Ø8 expanding sleeve masonry anchor to accommodate an M8 thread. The requirements for fixings and fasteners for these sized conduits are the same as for 20 mm, 25 mm and 32 mm conduits.
2. To timber: where directly fixed, steel wood screw No. 12 × 35 mm, round head. Where fixed through fibrous cement, plasterboard and so on to timber the screws need to be of sufficient length to allow for the additional thickness.
3. To sheet metal or steel studs: No. 12 × 13 mm long galvanised binding head, self-tapping screws (sometimes these are not acceptable – always refer to contract details) and for steel studs 7G × 20 mm self-tapping screws.
4. To steel: M6 round-head screws of appropriate length tapped or bolted.

Special care should be taken to ensure accurate placement of all holding down bolts, concrete inserts, sleeves and any special fixing devices. Fixings and fasteners used in external locations and wet areas should be stainless steel.

Cable trunking

The term 'trunking' is used to describe a similar range of enclosures (see Clause 1.4.97 and Clause 3.10.3.9 AS/NZS 3000:2018 *Wiring Rules*). Trunking is available in three types: metal, PVC and multiple compartments.

Trunking is a preformed channel having a removable lid in which cables are installed to protect them from mechanical damage and to provide a route. Trunking enclosure systems can be installed on any flat exposed (walls) or concealed (in ceilings, under floors) surface. Trunking must be accessible and complete along its entire length (see Clause 3.10.3.9).

The advantage of trunking is that new circuits can be easily accommodated as long as de-rating of bunched cables is met. Generally, groups of TPI cables should be installed to avoid de-rating as far as practical. If a number of cables are installed together and each is carrying current, they will all experience an increase in temperature. Because of this, cables installed in groups are de-rated to carry less current than if the same cables were fixed to a solid surface, which can dissipate heat more easily.

All fixing methods must provide a smooth internal surface for the cables, for example, flat or domed headed fixings for joining sections, with the smooth head surface on the inside and nuts on the outside. Screws must not project through the nut more than 3 mm.

Never use self-tapping screws to fix trunking because the screws may penetrate cables. Where metal trunking is cut for installation purposes the cut surfaces should be protected with cold galvanising paint.

Trunking must be installed with the cover fixing screws accessible and the covers should be easily removable (see Clause 3.10.3.9). Trunking needs to be adequately supported in accordance with AS/NZS 3000:2018 *Wiring Rules*, the manufacturers' recommendations and load tables (a rule of thumb ensures that the maximum deflection of the fully loaded cable trunk between support points does not exceed 10 mm). Where they span open areas, it is recommended that the trunking should be capable of supporting an additional load (animals may use the trunking as a crossing point) at mid-span without permanently deforming.

Cables should be installed in trunking neatly and systematically to provide for free circulation of air and dissipation of heat. The maximum number of cables that may be enclosed in one trunking should be such as will permit installation of the cables without damage. Adequate support must be provided for vertically installed cabling ensuring that the weight of cables is sufficiently supported. Adequate support can be provided by retaining clips, which should be installed to retain the cables at intervals not exceeding 1000 mm in all locations except where cable trunking is run horizontally with covers uppermost.

The ends of cable trunking installed outdoors must be sealed to prevent the ingress of soil, water and any other contaminant. The seal should enable the installation of additional cables or the withdrawal of existing cables. Large trunking can be sealed using block-work and weak mortar; smaller openings can be sealed using a weak vermiculite/cement mixture. Where trunking connects to weatherproof or hoseproof (i.e. IP43 to IP56) enclosures, the entry must not compromise the IP rating of the equipment being connected.

Electrical continuity of metal trunking must be maintained at each trunk joint by a copper link (tinned copper for galvanised trunking) fixed on the outside of the trunking, secured by round or cheese-head screws, nuts and shake-proof washers. For continuity, this is only required if single-insulated wires are carried by the trunking. It is important to note that for all trunking, the practical capacity of trunking, be it a duct or a conduit, is considerably less than the theoretical capacity. This is because the helix of the cable is normally retained even when the cables are unwound, thereby causing the cables to wrap round one another and occupy a larger area of the trunking than the theoretical value.

TPS cables and trunking

TPS cable is the simplest wiring system to use for surface wiring – exposed or concealed – because it is quickly and easily installed. **Figure 6.79** shows flat TPS installed in trunking.

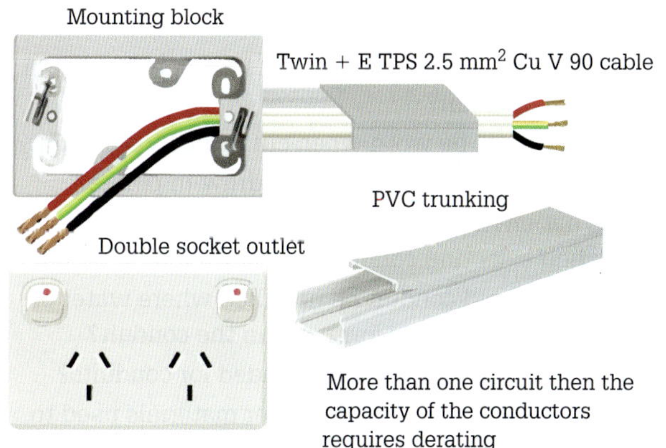

FIGURE 6.79 Flat TPS installed in PVC trunking

TPI cable can also be installed in trunking but the trunking cover must be effectively retained and if readily accessible a tool must be used to remove the cover.

PVC trunking

Trunking is usually grey in colour and of rigid UPVC moulding with removable clip-on covers. The trunking should be supported at a maximum of 600 mm centres and not used where exposed to mechanical damage, high temperatures, weather or moisture. Wiring system classification for mechanical protection is WS X1.

PVC fittings where possible should be high-impact heavy gauge with an all-insulated, clip-on lid.

Robustly constructed, PVC trunking and lids made from heavy-gauge material help to avoid the sagging of the trunking between supports and also prevent warping. Slotted trunking, as shown in **Figure 6.80**, makes the installing of a large number of cables with many terminations a tidy, quick and simple operation. This form of trunking finds application in control panels, switchboards and distribution boards.

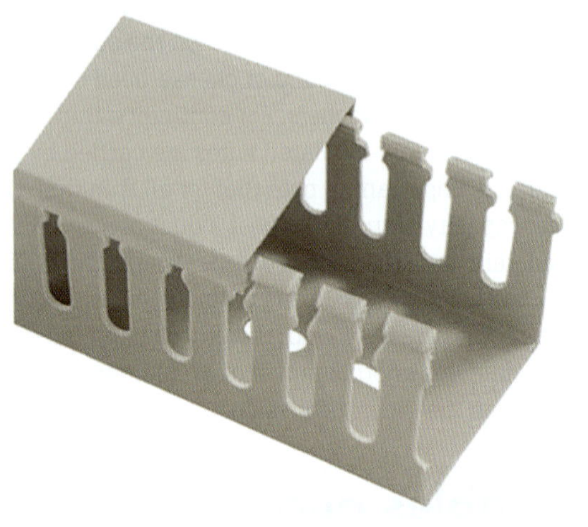

FIGURE 6.80 Slotted PVC trunking

Where changes of direction occur only approved pattern manufactured bends, tees or blank ends should be used. However, conduit branch-offs can be made with suitable clearance in the walls of the trunking and male PVC bushes fixed direct into couplings on the end of the outgoing conduit. Due allowance should be made for the expansion of PVC trunking at a maximum rate of 6.5 mm per 3500 mm run, and in this respect holes for fixing screws must be slotted. Do not use PVC trunking where it can be exposed to sunlight or other sources of UV radiation for extended periods of time.

Mini PVC trunking is ideal where building lease restrictions prevent structural or below-the-surface changes to existing wiring.

Fixings for trunking

Trunking 100 mm wide should be fixed rigidly to building structures at intervals not exceeding 1500 mm. Trunking wider than 100 mm should be fixed at intervals of not more than 1000 mm with two screws per fixing. Suggested fixings are steel wood screws, No. 12, round head; cadmium-plated or cadmium-plated M6 round-head screws; or No. 12 galvanised, binding-head, self-tapping screws.

REVIEW QUESTIONS

1. State the standard Australian finished lengths of UPVC conduit.
2. What minimum bend radius is needed for 20 mm UPVC conduit?
3. What must be worn when cutting UPVC conduit?
4. What is provided by conduit draw-in points?
5. What environmental conditions are of concern when selecting conduit for a particular location?
6. How should conduit be installed where water or condensation may form within the conduit?
7. What fastening is recommended for conduits?
8. What are the requirements for materials used to lubricate cables while drawing into conduits?
9. Where is medium-duty conduit typically used?
10. Trenches for wiring enclosures must be dug to a particular size. State the size.
11. Where can conduit made from ABS be used?
12. Where can HFT conduit be used?
13. When cables exit conduit or trunking, what must not be damaged?
14. Why is it sometimes necessary to over-bend UPVC conduit?
15. Why should a contrasting colour solvent cement to the conduit be used?
16. When are expansion joints used?
17. What must be considered for the design of a conduit run in a roof cavity?
18. Where is low-profile light-duty oval conduit used?
19. What sealant is recommended for repairs to all penetrations in brick, cement blocks and concrete?
20. When conduits pass through external walls or existing ground floor slabs, what size penetration is recommended?
21. With what must all conduits installed for future use be provided?
22. Why is corrugated conduit not suitable in concrete slabs?
23. Why is it necessary for an electrician to supervise a concrete pour for a slab containing conduit?
24. Name the type of conduit that should be used between fixed conduit and equipment likely to be moved or subject to vibration.
25. State the purpose of transparent conduit.

6.6 Metallic enclosures

For AS/NZS 3000:2018 *Wiring Rules* compliance read 'Enclosure of cables' and AS/NZS 2053 *Conduits and fittings for electrical installations.*

Rigid metal conduit is a metal tube of circular cross-section with a running thread on both ends. Each length of rigid metal conduit comes with a straight threaded coupling for joining lengths of conduit, as shown in **Figure 6.81**.

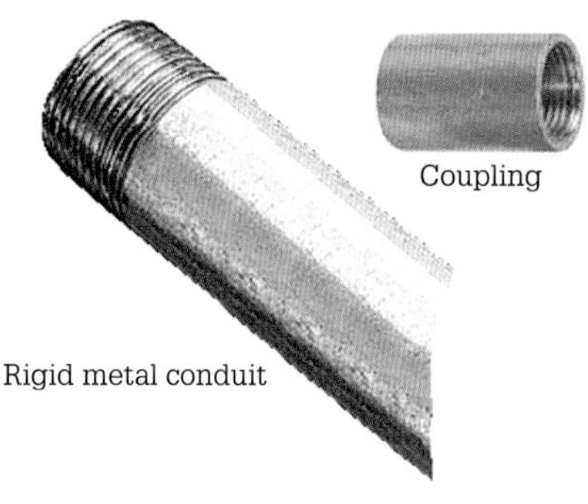

FIGURE 6.81 Metal conduit and coupling

Use hot-dipped galvanised steel conduits to AS 1074 (medium wall thickness) and AS 4792 where high-impact resistance (see Clause 3.3.2.6) is required and the environment in which the conduit is located is not highly corrosive to steel.

Galvanised steel conduit can be depended upon for a long service life. Supplementary coatings can be applied to the exterior of the conduit where additional corrosion protection is needed.

All steel conduits must have screwed joints and screwed terminations. Screw threads of nominal sizes of conduit of 16 mm to 63 mm are metric conduit threads with a 1.5 mm pitch, complying with AS 1721. Sizes 80 mm to 150 mm are British standard pipe (BSP) threads complying with AS 1722. Conduit threads must be cut clean and of sufficient length (10 mm) to permit conduits being butted firmly together at joints and hard against the shoulders of conduit fittings. The ends of cut lengths should be bevelled internally (see Clause 3.10.3.5) and all burrs removed with a burring reamer; all swarf should be removed before installing. Make sure that field-cut threads are coated with an approved electrically conductive, corrosion-resistant aluminium paint.

The steel conduit threads provide a mechanical interface to maintain the connections between conduit sections. The mechanical interface also provides electrical continuity that enables the conduit to be used as an earth conductor. This electrical continuity can also be used to confirm that the threads are appropriately engaged (see Clause 3.10.3.3). If the connection is loose, the electrical resistance will be high and the loose joint can be detected and identified. Thread filler must not be used to reduce looseness.

The standard Australian finished length of rigid metal conduit with coupling is 4 m in various outside diameters from 16 mm, 20 mm, 25 mm, 32 mm, 40 mm, and so on.

A complete rigid metal conduit wiring enclosure system usually requires elbows, bends, tees, junction boxes (one way, two way, three way, four way), and saddles and other fittings. Different conduit fittings and junction boxes are illustrated in **Figure 6.82**.

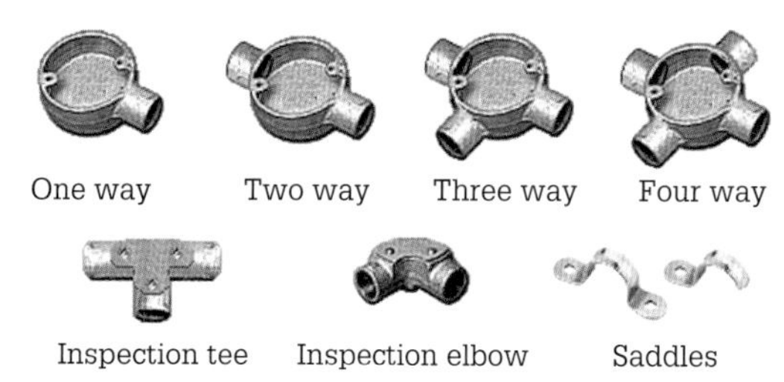

FIGURE 6.82 Rigid metal conduit fittings and junction boxes

Elbows, tees and bends permit an angled change in direction and junction boxes (J boxes) provide an enclosed space for cable termination. Conduit is required to be supported adequately and saddles are used for that purpose. All metallic conduit fittings are made from the same grade of mild steel as the rigid steel conduit. They are galvanised for corrosion protection. Exterior and interior surfaces as well as all threads have a galvanised coating of zinc.

Rigid metal conduits find application with TPI cables in industrial locations such as lift wells and where heavy mechanical damage is possible (see Appendix H4, AS/NZS 3000:2018 *Wiring Rules*, 'Mechanically protected wiring systems'). Metallic conduit has, however, in many installations been superseded by rigid PVC conduit or armoured cables.

Steel conduits may be required to be electrically and mechanically continuous (see Clause 5.5.4.2). This requirement depends on the type of cable enclosed in the conduit. If this requirement is applicable, then all conduits must be securely bonded to their terminating points by lock nuts; complete electrical and mechanical continuity throughout every conduit run must be ensured and each run must be securely bonded to earth. 'Proprietary' means identifiable by naming manufacturer, supplier, installer, trade name, brand name and catalogue or reference number.

Rigid metal conduit systems should be installed on the 'draw-in loop system'. Electrical continuity tests must be carried out prior to the installation of draw cords. Conduit runs should be completely erected and fixed before drawing in any conductors. The ends of conduit entering an enclosure should be finished so as to prevent damage to the cables (such as bushed). All conductive wiring enclosures such as metallic conduit requiring a protective earthing conductor must have a conductor of not less than 2.5 mm^2. Entry to weatherproof equipment and the like by steel conduit should be sealed so as to prevent the ingress of moisture. Horizontal and vertical steel conduit runs that are exposed to weather, or where moisture would be retained between the conduit and the wall, must be installed using saddle spacers or other means of fixing clear of the surface.

Where conduits are installed vertically, they must be installed in such a manner as to avoid damage to themselves and cables contained therein (see Clause 3.9.5). Joints and any damaged coatings must be painted with cold galvanising paint a minimum of 50 mm either side of the joint or damaged area.

During building operations and/or installation of electrical work, the open end of all conduits must be capped to prevent the ingress of moisture and foreign matter.

Conduit sets and bends

When you are installing a conduit run you will need to make bends to go over or around obstacles. These bends must be made without reducing the inside diameter of the conduit at the bend (refer to AS/NZS 3000:2018 *Wiring Rules*). You will make most bends on the job as part of the installation procedure. Changes in direction of a conduit run are achieved by setting the conduit. Sets are defined as a change in direction in a conduit of other than 90°. **Figure 6.83** illustrates some common sets that may be used in a conduit installation.

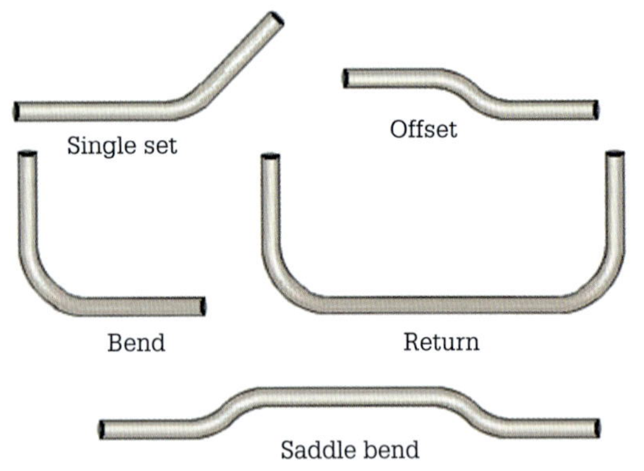

FIGURE 6.83 Conduit sets and bends

To determine offsets, Pythagoras' theorem is applied, as shown in **Figure 6.84**.

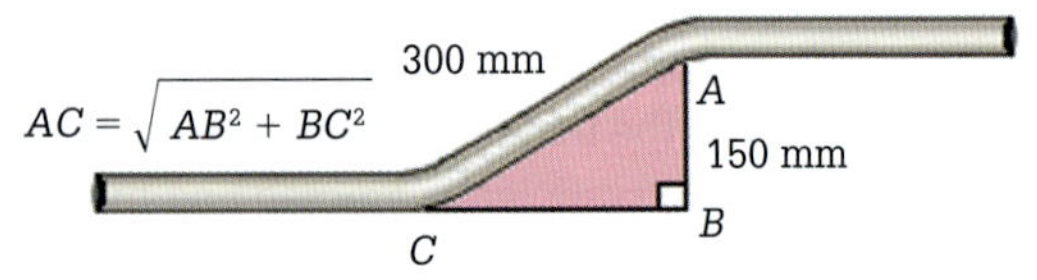

FIGURE 6.84 Applying Pythagoras' theorem

With trigonometry, the ratio of one side to another for a given angle allows us to determine an unknown side or angle of a triangle, if we know at least one side and two angles or two sides and one angle. One of these ratios states that for a 30/60 right-angled triangle the side opposite the 30° angle (the shortest side) divided by the hypotenuse (the longest side) equals 0.5. This is called a sine ratio. This ratio is why a conduit bending tool should have a 30° bending mark.

If you want to make a quick 150 mm offset bend in a length of conduit you make two marks 300 mm apart and, aligning the datum mark of the bender on each of these marks, make two opposite bends to the 30° line. Half of the long 300 mm hypotenuse is 150 mm. The conduit stands up in a 150 mm offset.

Sets in rigid metal conduit are usually made cold using a conduit bender. The set must not cause cables, when inserted, any damage or reduce the internal diameter of the conduit.

Steel conduit is ductile, which means that it bends easily. It is also soft (mild steel) so that it can be cut and threaded easily. In addition, it is either plated with zinc or galvanised.

Cutting steel conduit

Steel conduit should be secured in a pipe vice as illustrated in **Figure 6.85** to allow for ease of cutting.

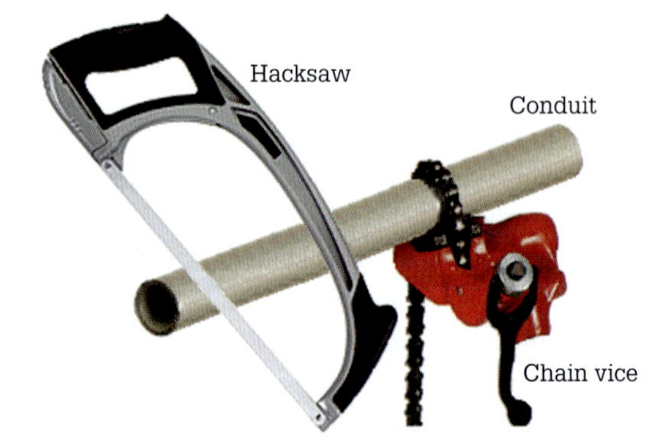

FIGURE 6.85 Cutting steel conduit

Steel conduit is normally cut using a hacksaw. Before cutting conduit with a hacksaw you should inspect the blade and replace it if required. A blade with 32 cutting teeth per 25 mm is recommended for conduit. If the blade needs replacing make sure that the teeth of the replacement blade point towards the front of the saw.

To cut, place the middle of the hacksaw blade on the conduit mark where the cut is to be made. Position the saw so that the end of the blade is pointing slightly downwards and the handle is pointing slightly up. Push forwards and backwards gently until the cut is started. Make long even strokes until the cut is completed. If the cut is correctly done, the end of the conduit should be at right angles to the sides of the conduit and level. Clean the conduit and hacksaw to remove swarf and ream the conduit (as illustrated in **Figure 6.87**) to remove internal sharp edges.

A conduit cutter as shown in **Figure 6.86** can also be used to cut steel conduit.

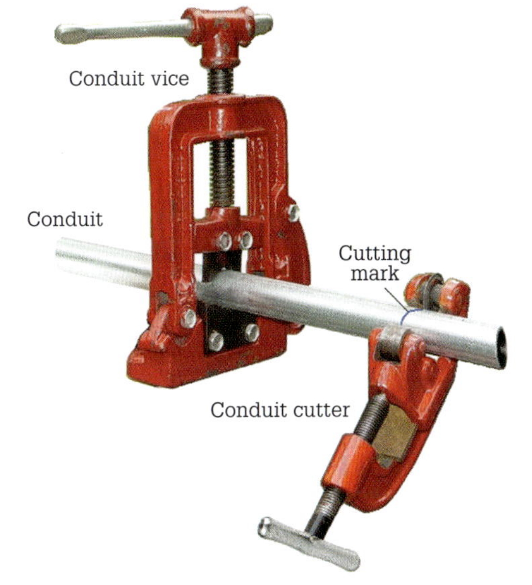

FIGURE 6.86 Conduit cutter

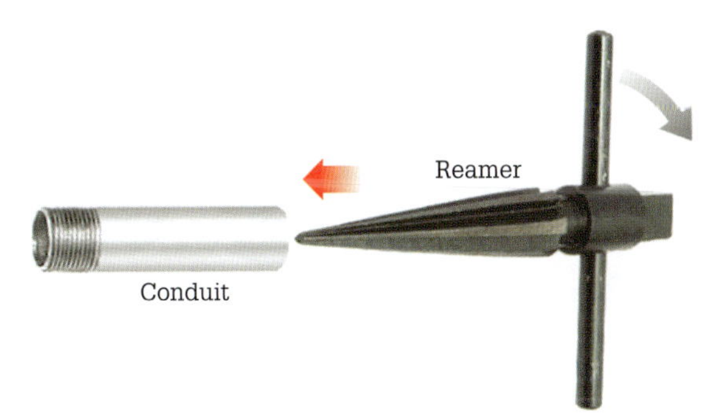

FIGURE 6.87 Conduit reamer

To use a conduit cutter first secure the conduit in a pipe vice. Open the cutter and place it over the conduit with the cutter wheel on the conduit mark. Place some cutting oil on the conduit and begin cutting. Tighten the cutter by turning the screw handle. Tighten the cutter handle a one-quarter turn for each full turn around the conduit. When the cut is almost finished, stop cutting and snap the conduit to finish the cut. This reduces the ridge that can be formed on the inside of the conduit. Clean the conduit and cutter to remove oil and ream the conduit to remove the ridge.

When the conduit is cut, the inside edge will be sharp. This edge will damage the insulation of the conductors when they are installed. To avoid this, the inside edge must be filed smooth and rounded or reamed using a reamer (illustrated in **Figure 6.87**).

Steel conduit will also require a thread to be cut with a stock and die, shown in **Figure 6.88**, so that it can be joined to other lengths of conduit or finished within equipment.

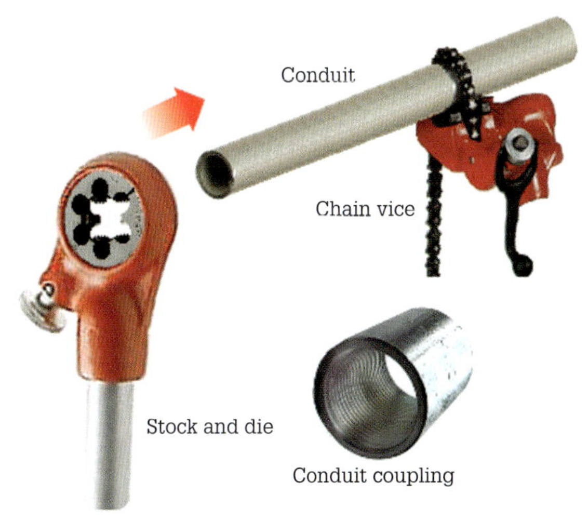

FIGURE 6.88 Cutting a thread

Using a stock and die

1. Chamfer the end of the conduit with a file; this should be as deep as the thread to be cut.
2. Mount the conduit in a chain vice, making sure that the chamfered side of the die is down; this is the side with the markings.
3. Hold the die stock with one hand cupped over the centre; apply inward pressure as you turn the die with the stock until the cut has started.
4. Apply cutting fluid and turn the die stock with one hand; after each revolution reverse the die one-half turn to break the swarf if you are not using a ratchet-type stock as this stock maintains pressure on the die preventing it from fouling.
5. After a few threads have been cut check the fit with a conduit coupling and adjust the die if necessary.

Various tools required for threading rigid metal conduit are shown in **Figure 6.89**:

1. Pipe vice
2. Conduit stock and metric 1.5 mm die; use thread cutting lubrication when cutting thread

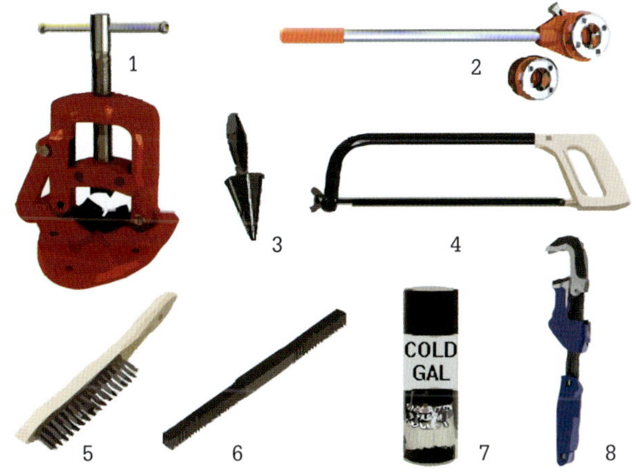

FIGURE 6.89 Tools required for conduit threading

3 De-burring tool to ream conduit
4 Hacksaw to cut the length of conduit (do not use a pipe cutter as it damages the galvanising)
5 Wire brush to clean thread
6 Thread file
7 Cold galvanising to protect thread and any damage to conduit galvanising (see AS/NZS 3000:2018 *Wiring Rules*, 'Presence of corrosive or polluting substances')
8 Stillson wrench to join conduit lengths.

Hand benders

Conduit hand benders, shown in **Figure 6.90**, are practical to use at an installation because they are convenient, portable and require no electrical-mechanical assistance. Hand benders are designed with a specific shape for the controlled bending of metallic conduit without deformation or crushing of the walls of the conduit. Conduit hand benders are fitted with long, rigid handles to enable sufficient force to be applied through the hands and one foot to the conduit in order to bend it to the required shape. Foot pressure is necessary to keep the conduit in the groove of the bender and to prevent kinks.

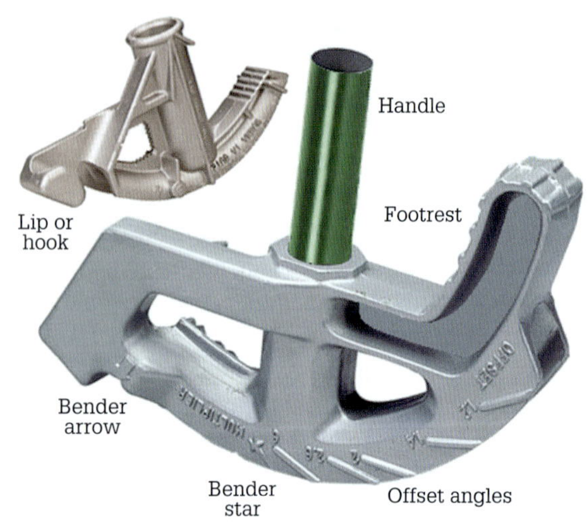

FIGURE 6.90 Conduit hand benders

Bending can also be done in the air, illustrated in **Figure 6.91**, by applying force as close to your body as possible and positioning the conduit length to be bent with both hands or placing it under the armpit for greater downwards bending force.

Hand conduit benders can be used to make various bends in smaller-size conduit (16 mm). The manufacturer of the conduit hand bender may provide documentation indicating starting points, offset angles, distance between offsets, gains and other important values associated with the bender.

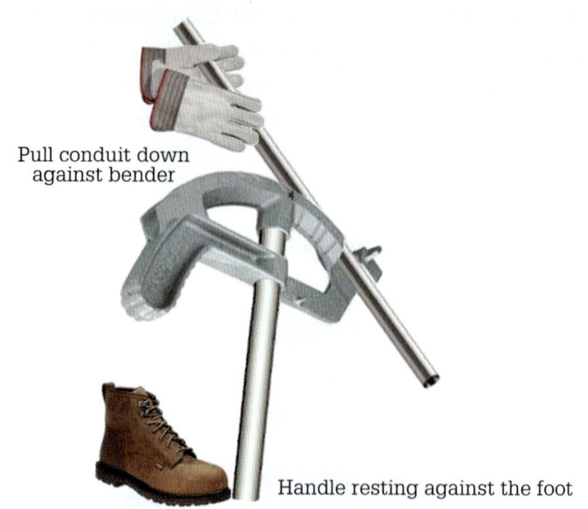

FIGURE 6.91 Bending conduit in the air

Mechanical conduit benders

Mechanical conduit benders, illustrated in **Figure 6.92**, make bending conduit easy. The machine does the work, so electricians don't have to bend conduit manually. The mechanical benders are often mounted on a wheeled frame so they can be rolled to different locations.

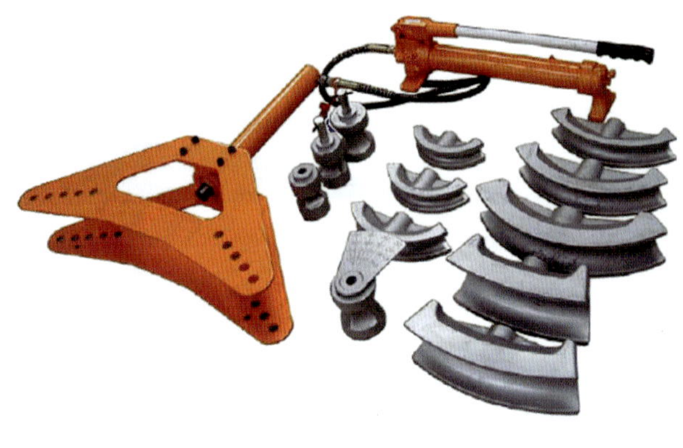

FIGURE 6.92 Mechanical conduit bender

Bending steel conduit

An effective way to practise bending conduit is to use a length of solid copper wire and bend it to resemble the bends required for the conduit fitting work, as illustrated in **Figure 6.93**.

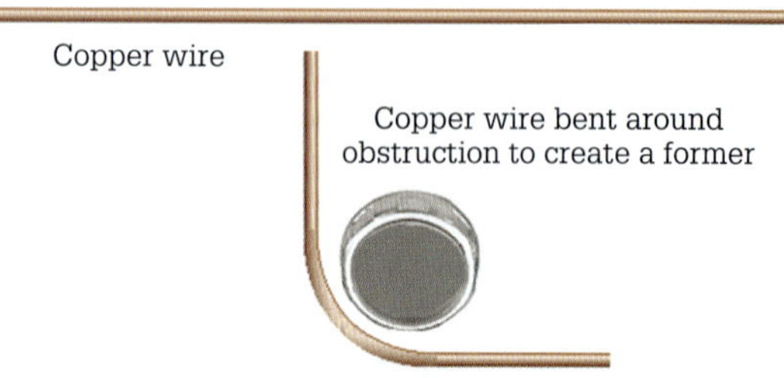

FIGURE 6.93 Copper wire guide

The formed copper wire gives perspective on how to bend the finished conduit. To make the bend accurately, draw the contour of the bend with chalk on a flat surface such as a concrete floor, using the formed copper wire as a guide. Then, as you form the bend match the bend in the conduit with the chalk outline.

Wiring Rules requirements when bending conduit

Refer to Clauses 3.10.3.4, 3.10.3.5 and 3.10.3.6, AS/NZS 3000:2018 *Wiring Rules*.

When bending conduit, the radius of the bend, as shown in **Figure 6.94**, must not cause damage to the cable insulation or to the conductors. In addition, the bend should not substantially reduce the internal diameter of the conduit.

Where cables, flexible cables or flexible cords pass through conduits, the conduit ends must be provided with a bush, shown in **Figure 6.95**, secured to the conduit. If there is no bush, the conduit ends must be filed (a rat tail file can be used) to remove burrs and the exit wall finished flush with the lock nut. The purpose of this rule is to ensure the structural integrity of the cables that exit the conduit.

The radius of the bend must not cause damage to the cable insulation or the conductors.

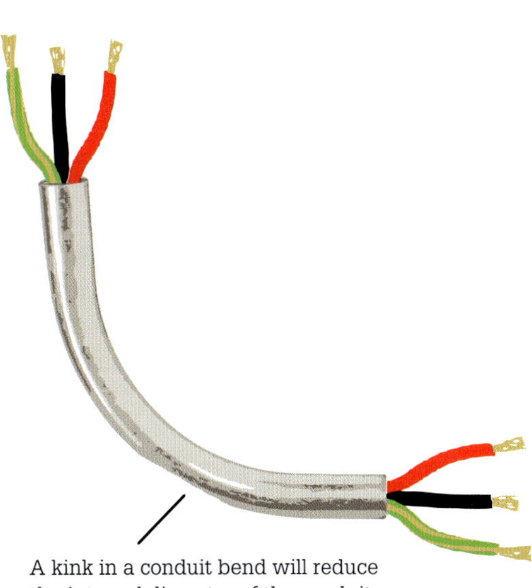

FIGURE 6.94 Radius of conduit bend

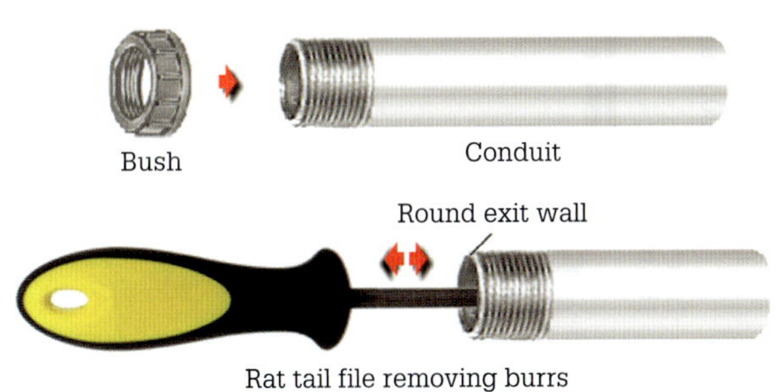

FIGURE 6.95 Removing burrs

FIGURE 6.96 Conduit termination

Terminations of conduit in electrical equipment, shown in **Figure 6.96**, must fully protect the enclosed cables from insulation or conductor core damage.

Hickey

A hickey (**Figure 6.97**) should not be confused with a hand bender because it functions quite differently.

A hickey is an incremental bending device. First, a small bend of about 10° is made. Then the conduit is moved to a new position and another small bend is made.

This process is continued until the bend is completed. A 90° bend can be produced by making a total of nine bends at 10° each.

Manufacturing a 90° bend using a hand bender

With conduit work the 90° bend is probably the most common bend that the electrician will make. Before beginning to make the 90° bend there are two measurements that must be known:

1 Rise or stub distance
2 Take-up distance of the bender.

The required rise is the height of the stub. The take-up is the amount of conduit the bender will use to form the bend. (**Note:** Different brands of hand benders may have a

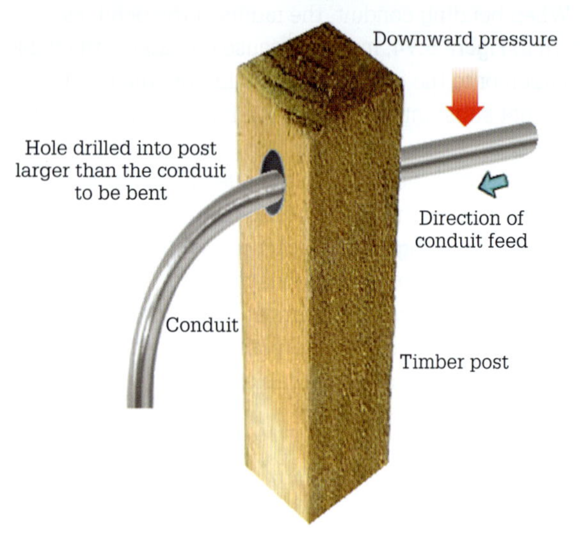

FIGURE 6.97 Timber hickey

different take-up distance.) The take-up is the radius of the groove in the head of the bender measured to its centre.

Once the take-up distance has been determined subtract it from the stub height distance. Mark that new measurement with a permanent marker on the conduit all the way around. This mark will indicate the point at which you will begin to bend the conduit. Most benders have a mark, like an arrow, to indicate the starting point for a bend. Line up the starting point on the conduit with the starting point on the bender. **Figure 6.98** illustrates the take-up (150 mm) required to achieve a 200 mm stub distance with 20 mm steel conduit.

Note: The illustrations in **Figure 6.98** show a hand bender and 16 mm screwed conduit.

Once you have lined up the bender and conduit start marks, use one foot to hold the conduit steady. Keep your heel on the floor for balance. Apply pressure on the bender footrest with your other foot. Make sure you hold the bender handle level with both hands, as far up as possible, to get maximum leverage. Then bend the conduit in one smooth motion, pulling as steadily as possible. You should now have a 90° bend with a 300 mm height which contains a 200 mm stub. After finishing the bend, check to make sure you have the correct angle (use an engineer's square for 90° bends) and bend length, as shown in **Figure 6.99**.

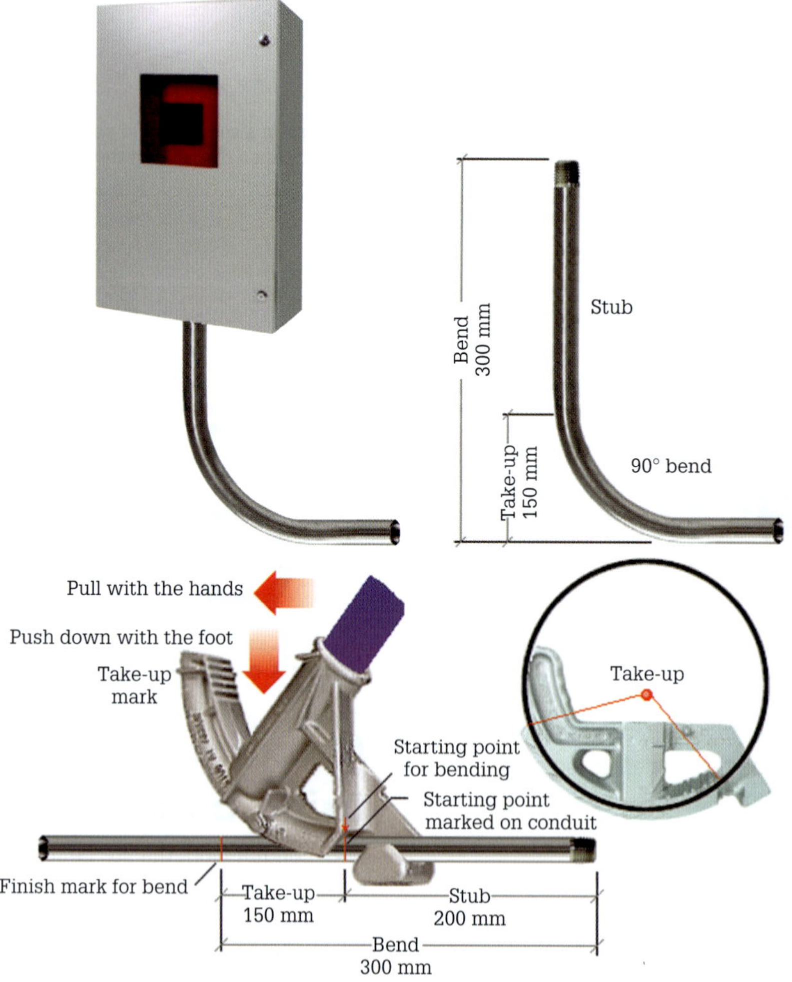

FIGURE 6.98 Take-up on conduit

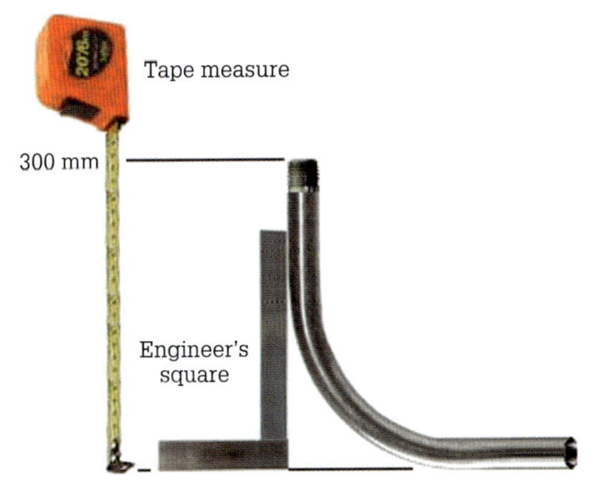

FIGURE 6.99 Checking bend angle

Gain

The gain is the distance saved by the arc of a 90° bend. Knowledge of the gain allows the electrician to pre-cut, de-burr and pre-thread both ends of the conduit before bending. Note that it is easier to work with conduit while it is straight.

When working with conduit, you cannot create a 90° corner at a reference point where the conduit length changes direction. If that occurred, the internal CSA would be severely decreased and the conduit would be kinked. However, with a bend the conduit is not going all the way to the reference point. Therefore you will gain some conduit length.

Figure 6.100 demonstrates that the overall bend length for conduit with a 90° bend is less than the algebraic sum of the vertical distance (A) and the horizontal distance (B) when measured square to the corner. The bend length for a 90° bend can be determined from the following equation:

$$\text{bend length} = (A + B) - \text{gain}$$

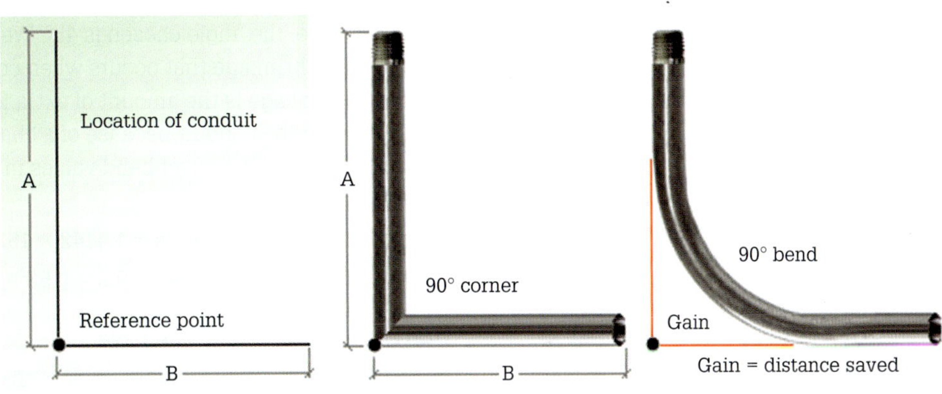

FIGURE 6.100 Conduit gain

Back-to-back 90° bends

A back-to-back bend consists of two 90° bends made on the same length of conduit. When making a back-to-back bend first determine the distance (shown in **Figure 6.101**) between the outside of the holes where the conduit will be fixed. In our example this is 600 mm.

After the first bend is made, as shown in **Figure 6.102**, measure to the point where the back of the second bend will be.

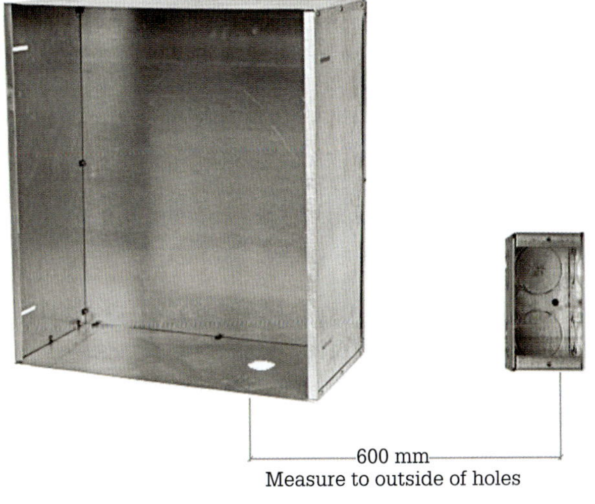

FIGURE 6.101 Dimensions for a back-to-back bend

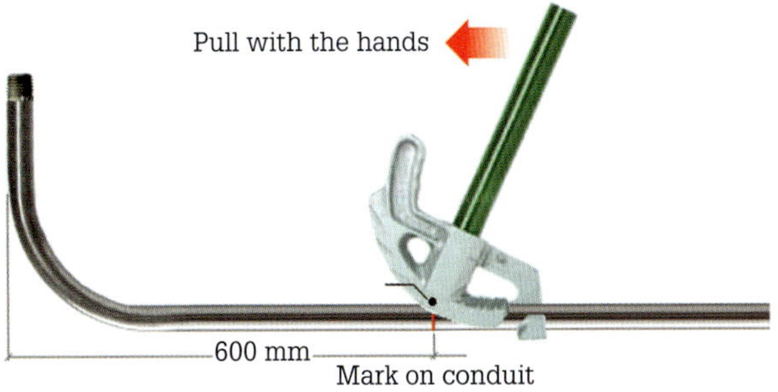

FIGURE 6.102 Bending the second bend

Mark the conduit at the measured distance and then place the conduit mark in line with the centre of the bender (star point) and bend the conduit up to 90°. This will result in a back-to-back bend 600 mm wide, as shown in **Figure 6.103**.

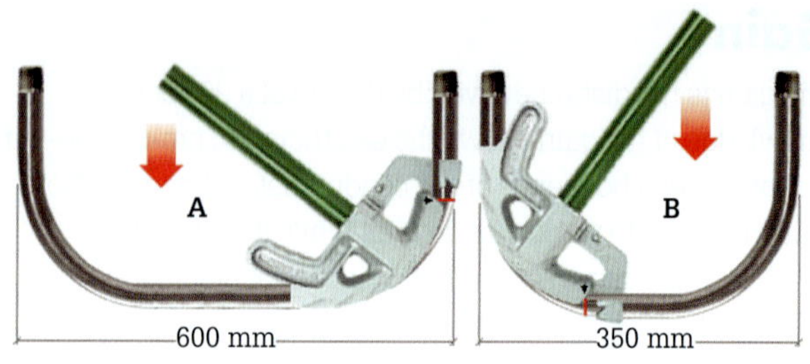

FIGURE 6.103 Back-to-back bend

If the back-to-back bend is a narrow U shape, as shown in Figure 6.103B, you can subtract the take-up from the measured distance between the fixing holes for the conduit. The conduit must be bent on the arrow mark on the bender in the opposite direction to that in Figure 6.103A.

The finished back-to-back bend is illustrated in Figure 6.104.

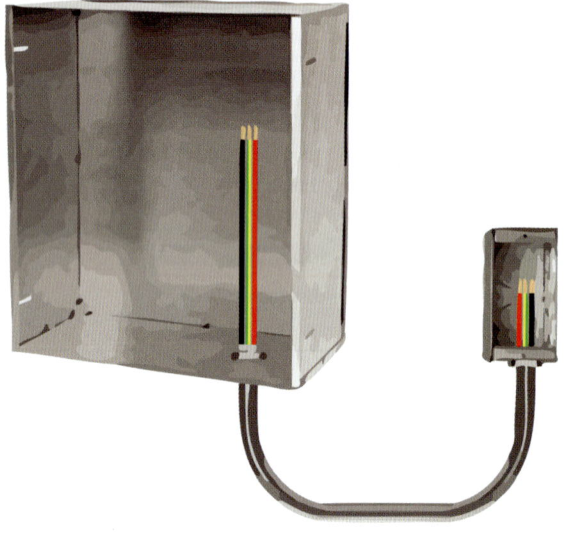

FIGURE 6.104 Installation of back-to-back bend

Making offset bends

Many installations require the conduit to be bent so that it can pass over obstructions such as rolled steel joists (RSJs), changes in floor, ceiling or wall levels, other service pipes or enter switchboards and J boxes. Conduit bends used for these purposes are called offsets. To construct an offset as shown in Figure 6.105, two equal-angled bends of less than 90° are manufactured a specific distance apart.

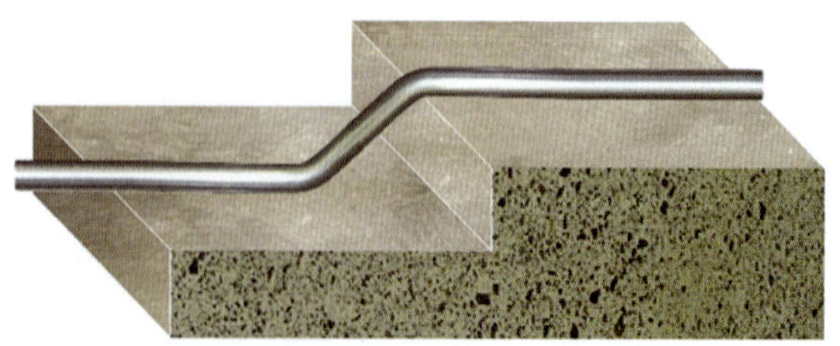

FIGURE 6.105 Offset bend

Determine the height of the offset and the distance the offset occurs from the end of the conduit, as illustrated in Figure 6.106. In the figure the offset is 915 mm from the end of the conduit and is 153 mm high.

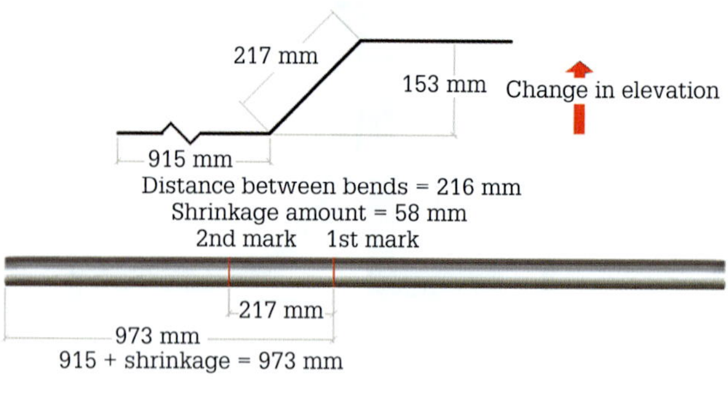

FIGURE 6.106 Dimensions for an offset bend

Choose the angle that will be used for the offset bends. In the example, the angle chosen is 45°. We must now consider the shrinkage that occurs when conduit offsets are bent. Shrinkage is the amount of extra length needed to be added to the conduit because of a change in elevation (rise or fall). Using the offset bend values in Table 6.11 we see:

Distance between bends = 1.4142 × 153 = 216.3 mm

Shrinkage amount for 45 = 9.5 × (153 > 25) = 58 mm

TABLE 6.11 Offset shrinkage table

Angle	Shrinkage of conduit per 25 mm rise	Distance between bends: multiply the offset depth by:
10°	1.6 mm	5.7588
22.5°	4.8 mm	2.613
30°	6.35 mm	2.0
45°	9.5 mm	1.4142
60°	12.7 mm	1.1547

When marking out the conduit, place the first mark on the conduit (all way around) a distance of 915 mm plus the shrinkage amount of 58 mm. This mark will be 973 mm away from the conduit end. Place the second mark (all way around) 217 mm back from the first mark. Bend the conduit on the floor at the first mark in the direction shown in Figure 6.107 to create a 45° bend.

With the bender still on the conduit, turn the whole job over to make an air bend. Slide the bender to the second mark as shown in Figure 6.107. Using your hands and with the conduit under your armpit bend another 45° angle on the conduit. Figure 6.108 illustrates the offset second bend. This bend will give a 153 mm offset 915 mm away from the end of the conduit.

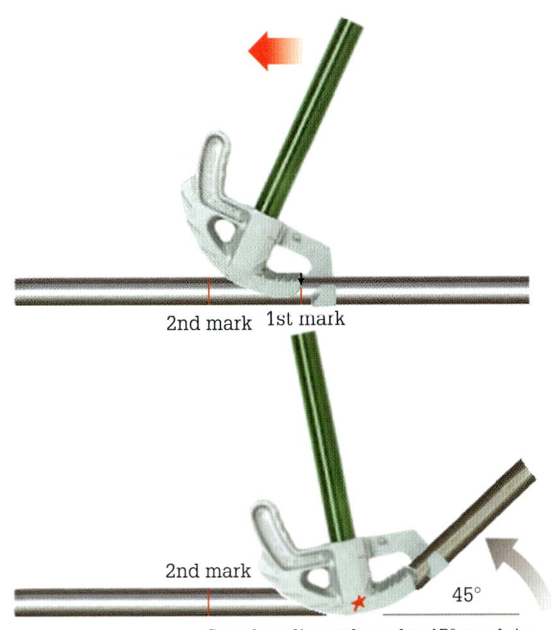

FIGURE 6.107 Bending an offset

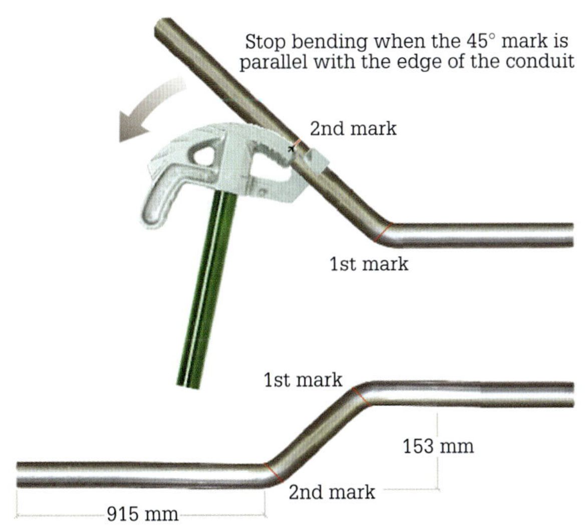

FIGURE 6.108 Completed offset bend

Three-set saddle bend

A saddle bend is used to go around obstacles such as service conduits, as shown in **Figure 6.109**.

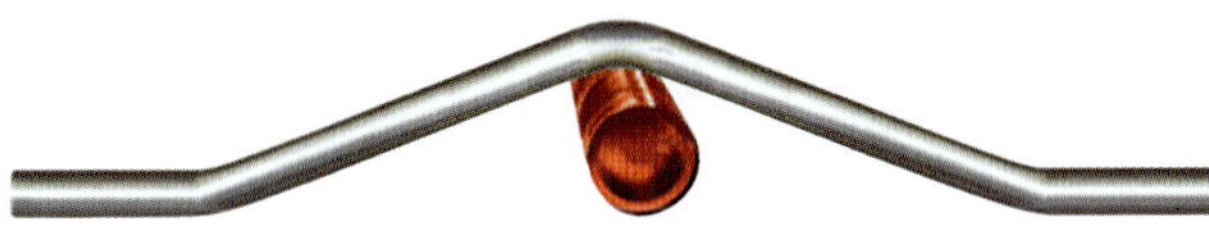

FIGURE 6.109 Three-set saddle bend

When engineering a saddle bend allowance must be made for shrinkage. The centre of the saddle will be reduced for each 25 mm of saddle depth. For example, if the pipe the three-set saddle bend has to go over is 102 mm in outer diameter, there is shrinkage of the conduit on each side of the bend as shown in **Figure 6.110**.

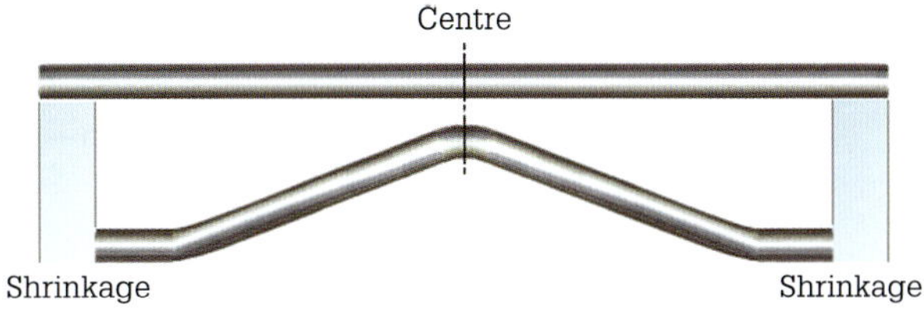

FIGURE 6.110 Shrinkage when bending conduit

Shrinkage is dependent upon the angle of the bend. When making three-set saddle bends, the following steps should apply:

1 Determine the height of the saddle bend and the distance the bend occurs from the end of the conduit to the centre of the saddle bend, as illustrated in **Figure 6.111**. In **Figure 6.111** the centre of the saddle is 1080 mm from the end of the conduit and must clear a pipe 102 mm in outer diameter. The first step is to determine the angle that will be used for the offset bends. Most three-set saddles are bent using 22.5° for the two outside offset bends and 45° for the centre bend. Using **Table 6.11**, calculate the distance between the bends and the shrinkage amount.

$$\text{Distance between bends} = 2.613 \times 102 = 267\text{ mm}$$

$$\text{Shrinkage amount for } 22.5° = 4.8 \times (102/25) = 20\text{ mm}$$

2 Place a mark (all way around) (called the first mark) on the conduit a distance of 1100 mm from the end of the conduit. Place a second mark (all way around) 267 mm to the right of the first mark and place a third mark 267 mm (all way around) to the left of the first mark.

Bend the conduit at the first mark using the star mark in the direction shown in **Figure 6.112** to a 22.5° bend. Stop bending when the conduit is parallel to the 22.5° mark on the bender.

3 Leave the bender on the conduit and slide the bender to the second mark on the conduit, shown in **Figure 6.113**. Turn the conduit through 180° and align both the conduit mark and the bender's arrow mark. Turn both the conduit and the bender over 180° and make an air bend. Bend the conduit until it is parallel with the 45° bender mark.

4 Remove the bender and turn the conduit in the opposite direction. Place the bender on the conduit and align the bender arrow with the third mark on the conduit, as shown in **Figure 6.114**.

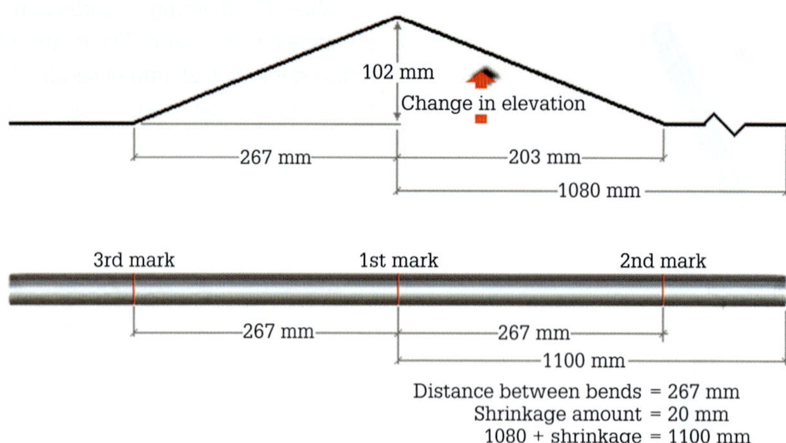

FIGURE 6.111 Dimensions for a three-set saddle bend

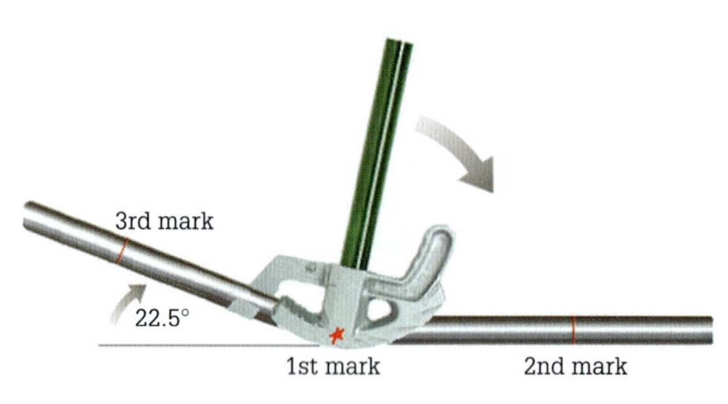

FIGURE 6.112 First bend in a three-set saddle bend

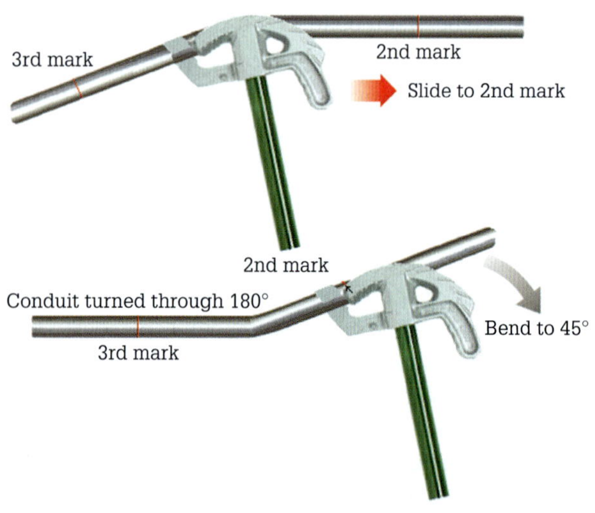

FIGURE 6.113 Second bend in a three-set saddle bend

Bend the conduit until it is parallel with the 22.5° mark on the bender. The final bend completes the set, as shown in **Figure 6.115**.

This will give a three-point saddle bend with the first bend 915 mm away from the end of conduit and 102 mm high.

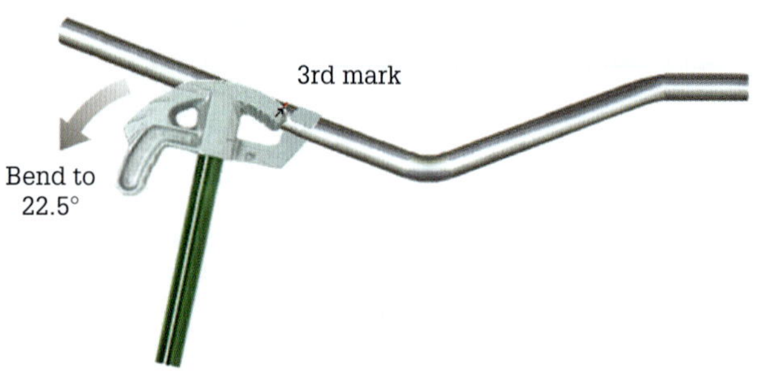

FIGURE 6.114 Final bend in a three-set saddle bend

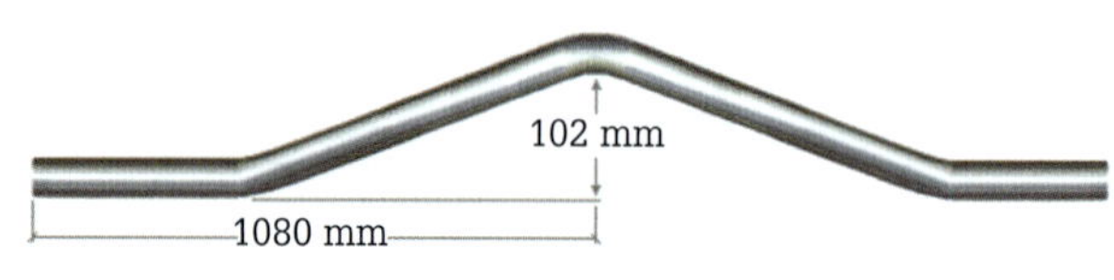

FIGURE 6.115 Completed three-set saddle bend

Four-set saddle bend

A four-set saddle bend can be hard to manufacture. This is because all four bends must be on the same plane as shown in **Figure 6.116**. The four-set saddle is two offsets manufactured back-to-back with the width of the object it has to cross between them.

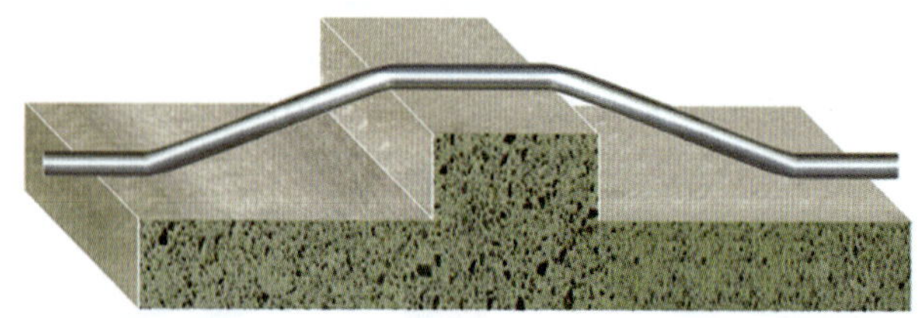

FIGURE 6.116 Four-set saddle bend

When making four-set saddle bends, the following steps should apply:

1 Determine the height of the four-set saddle bend, its width and the distance the rise occurs from the end of the conduit. These dimensions are shown in **Figure 6.117**.
2 Choose the angle to use to make the bends. Calculate the distance between the bends and the shrinkage. For our example we will use all 22.5° angles. Therefore, using **Table 6.11** calculate the distance between the bends and the shrinkage amount.

   Distance between bends = 2.613 × 50 = 131 mm

   Shrinkage amount for 22.5° = 4.8 × (50/25) = 10 mm
3 Place a mark (all way around), called the first mark, on the conduit a distance of 610 mm from the end of the conduit. Place a second mark (all way around) 131 mm to the left of the first mark and place a third mark 300 mm (all way around) to the right of the first mark. Then, place a fourth mark (all way around) 131 mm to the right of the third mark. To assist with alignment of the bends draw a line (use a permanent marker) along the length of the conduit (see **Figure 6.117**) and on the bender as shown in **Figure 6.118**.

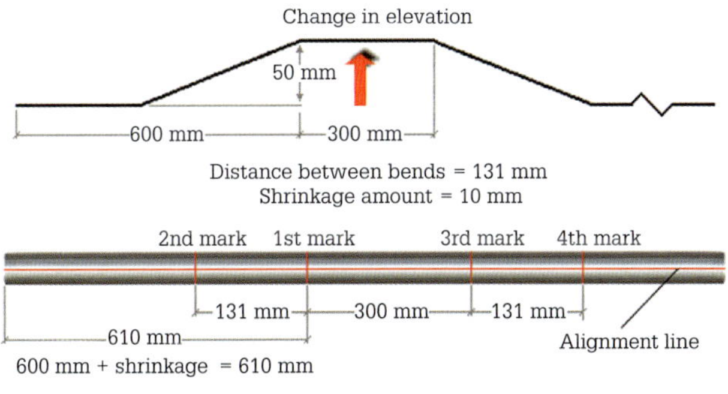

FIGURE 6.117 Dimensions for a four-set saddle bend

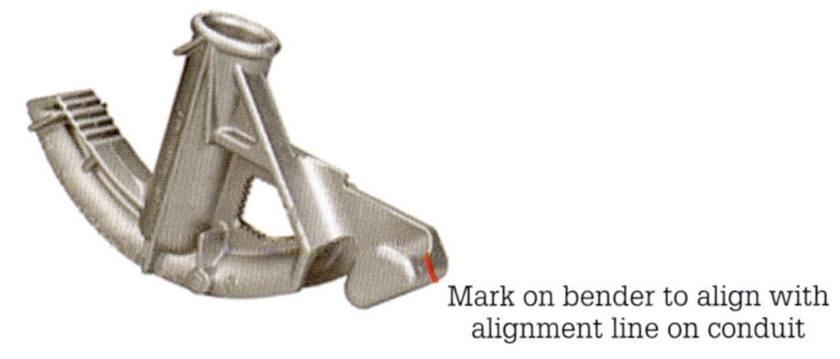

FIGURE 6.118 Alignment mark on bender head

Bend the conduit on the floor at the first mark in tho direction shown in **Figure 6.119** to create a 25.5° bend.

4 Leave the bender on the conduit and slide the bender to the second mark on the conduit. Turn the conduit through 180° and align both the conduit mark and the bender's arrow mark, shown in **Figure 6.120**. Turn both the conduit and the bender over 180° and make an air bend. Bend the conduit until it is parallel with the 22.5° bender mark.

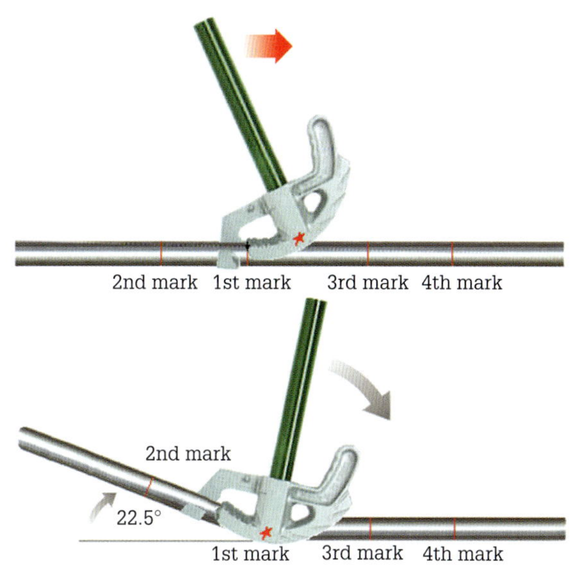

FIGURE 6.119 First bend in a four-set saddle bend

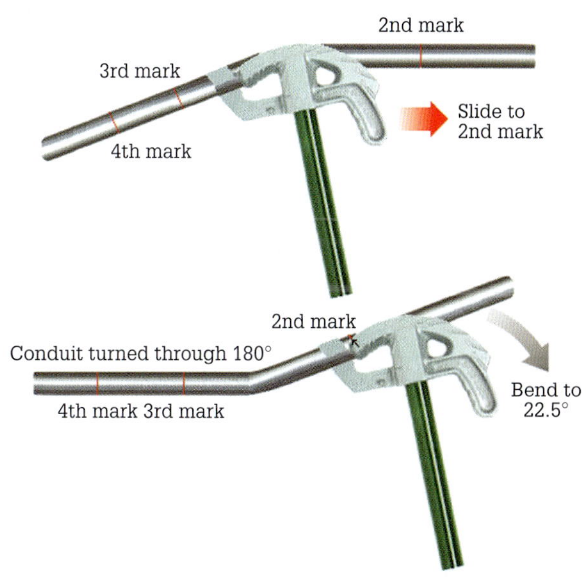

FIGURE 6.120 Second bend in a four-set saddle bend

5 Then take the bender off the conduit and turn the conduit through 180° so that the ends of the conduit are in opposite directions. With the bender on the conduit align the bender's arrow mark with the third mark on the conduit and bend a 22.5° angle as shown in **Figure 6.121**.
6 After the third bend is made slide the bender to the fourth mark as shown in **Figure 6.122** and turn both the conduit and the bender vertically through 180°. Make an air bend when the fourth mark is aligned with the arrow on the bender. The conduit must be bent to 22.5°.

   The final bend completes the set, as shown in **Figure 6.123**.

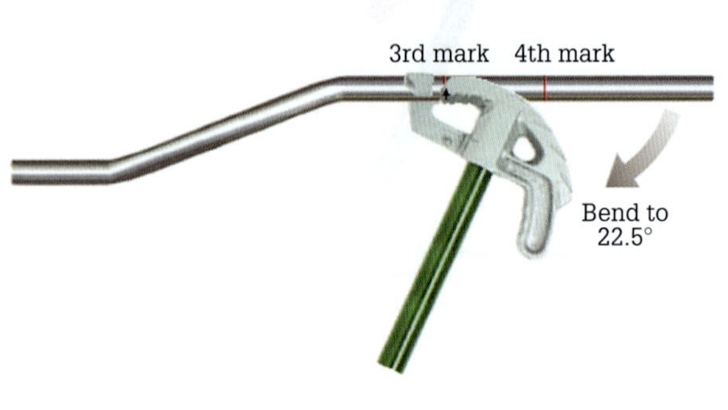

FIGURE 6.121 Third bend in a four-set saddle bend

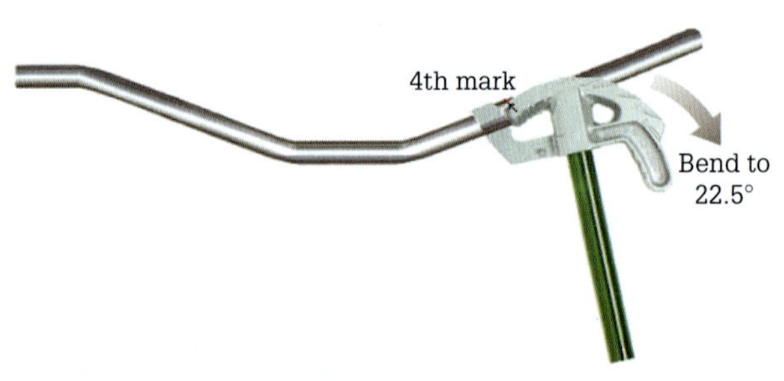

FIGURE 6.122 Final bend in a four-set saddle bend

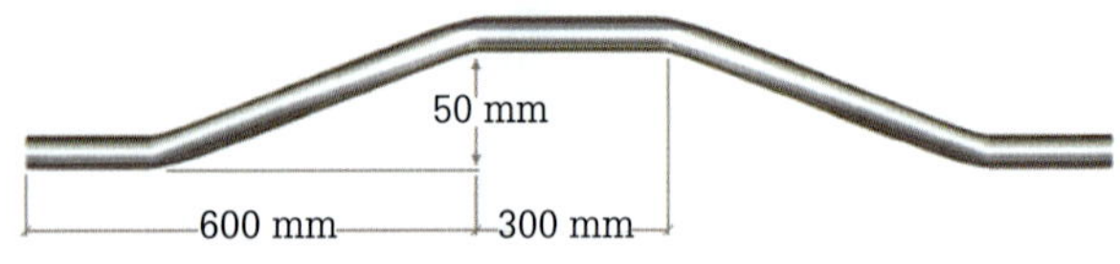

FIGURE 6.123 Completed four-set saddle bend

This will give a four-set saddle bend 50 mm high and 300 mm wide, 600 mm away from the end of the conduit.

Parallel offsets

Sometimes several lengths of conduit must be bent around a common obstruction. In this situation, parallel offsets are constructed. **Figure 6.124** illustrates a conduit run of three parallel offsets. Each offset has been bent using 22.5° bends. Conduit 2 is spaced 72 mm (centre-to-centre) from conduit 1 and conduit 3 is spaced 72 mm (centre-to-centre) from conduit 2. All three conduits are shown with their ends aligned.

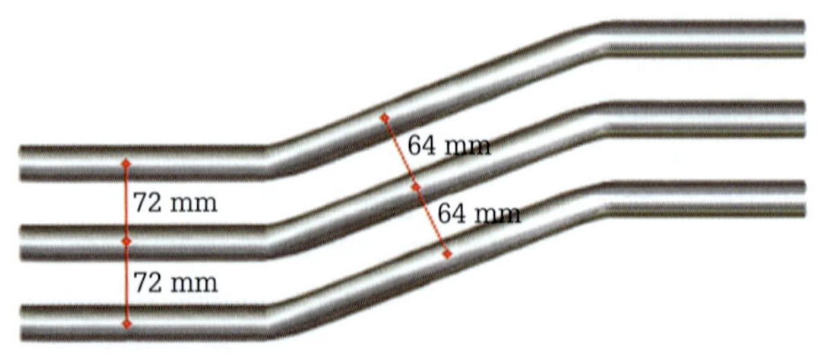

FIGURE 6.124 Incorrectly aligned parallel offsets

Although this may look acceptable in the example in **Figure 6.124**, notice that the three conduits have only 64 mm (centre-to-centre) of separation at the centre section of the offsets, as compared with the 72 mm (centre-to-centre) separation on the ends. This difference between the spacing increases as the degree of the bend used to make the offset is increased. To prevent this difference occurring, an adjustment must be made when bending each conduit length.

We start with the centre of the first bend of conduit A, as shown in **Figure 6.125**.

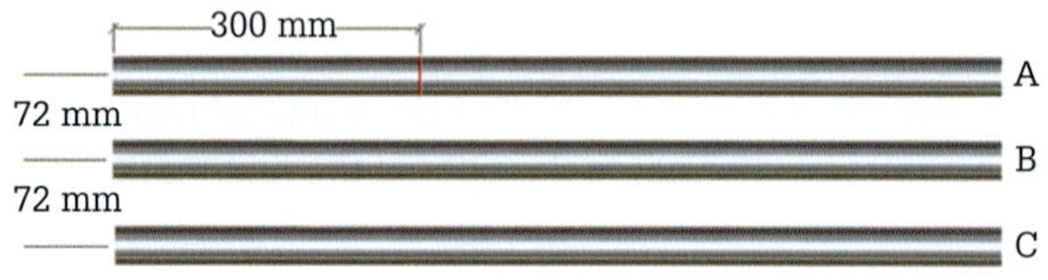

FIGURE 6.125 Dimensioning conduit A

The other two conduits, B and C, must have their bend centreline moved further away from the end of the conduit, as shown in **Figure 6.126**. The amount to add is calculated using the following formula:

Amount added = distance of bend centre for conduit A from the end of the conduit + centre-to-centre spacing × tangent of half the offset angle

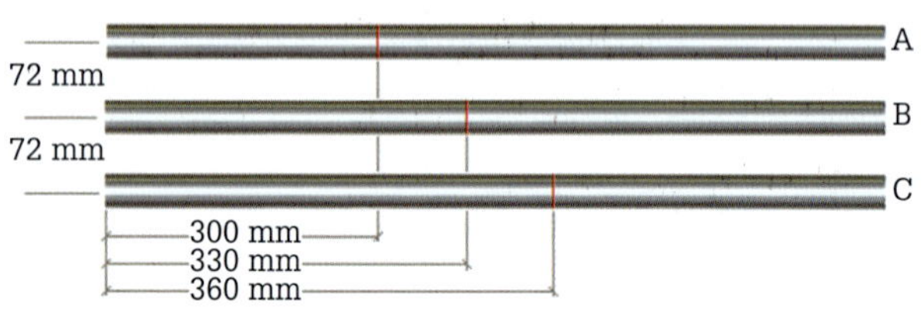

FIGURE 6.126 Dimensioning conduits B and C

For example, **Figure 6.125** shows three conduits laid out ready for marking. The angle of the offset for all three conduits is 22.5°. The centre-to-centre spacing is 72 mm and the start of conduit A's first bend is 300 mm from the conduit end.

The starting point of conduit B will be:

Amount added = distance + centre-to-centre spacing × tangent of half the offset angle

$$300 \text{ mm} + [72 \times \tan(0.5 \times 22.5°)]$$

$$300 \text{ mm} + (72 \times \tan 11.25°) = 300 \text{ mm} + (72 \times 0.414)$$

$$300 \text{ mm} + 30 \text{ mm} = 330 \text{ mm}$$

The starting point of conduit C will be

$$330 \text{ mm} + 30 \text{ mm} = 360 \text{ mm}$$

By using these starting points and bending an offset bend for each conduit length, the final conduit run will look like **Figure 6.127**.

It is recommended that conduits be run directly from an exit point to the termination point with a minimum number of sets as shown in **Figure 6.128** (the number of sets should not be greater than the equivalent of 360° or 4 × 90° bends). It is very difficult to pull conductors through bends that total greater than 360° and in addition the conductors will experience high levels of strain. However, exceptions do exist.

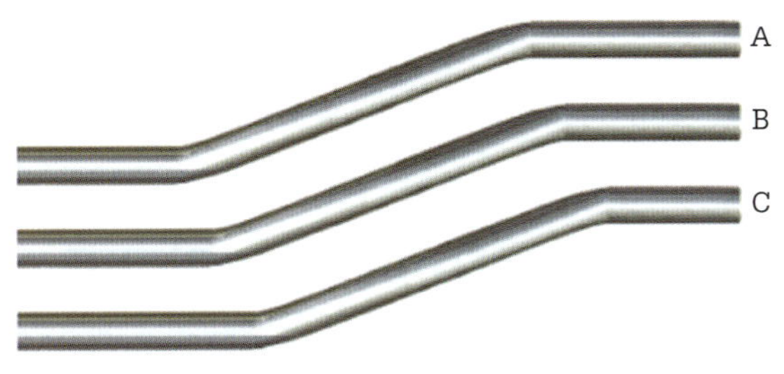

FIGURE 6.127 Completed parallel offsets

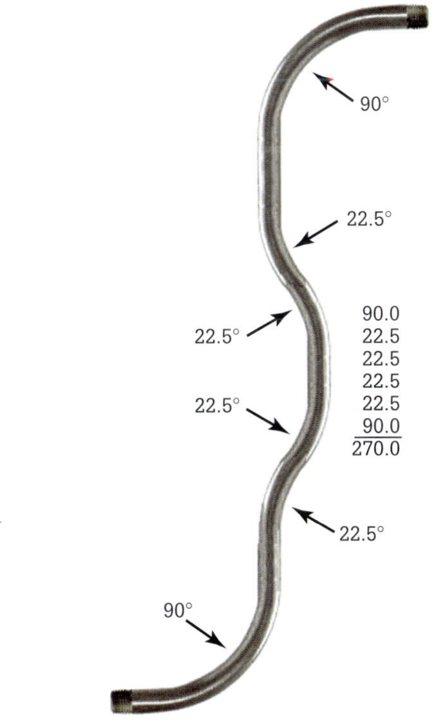

FIGURE 6.128 Conduit sets

Flexible metallic conduit

Flexible metallic conduit as shown in **Figure 6.129** is helically wound from strip electro-galvanised steel. It is designed to be used at the end of a conduit run where movement, vibration or removal of equipment or machinery is a factor when considering a wiring system.

FIGURE 6.129 Flexible metallic conduit and coupling enclosing a cable

Flexible metallic conduit is also available with a PVC coat which is a flexible polymeric material to provide additional protection. Where conductors enter or leave a rigid metal or flexible metal conduit, a bush or a gland with a smooth well-rounded surface must be provided to protect the conductors.

Metal trunking

Metal trunking as shown in **Figure 6.130** should have a demountable metal cover which is usually fixed to the ducting by galvanised metal thread screws to enable access to cables. It is suggested that if choosing ducting for a wiring system, 50% spare space should be allowed for future cable additions.

With any trunking, manufacturer's bends, tees, adaptors and coupling pieces must be used. Radial or throat-pattern fittings are normally used.

All trunking work should run neatly and squarely on walls and ceilings. Trunking should be installed so as to be inconspicuous by running in corners and parallel to wall and ceiling edges. All mitred corners must have provision for earth bonding connections. Trunking must be adequately supported (usually at 1200 mm centres) to prevent deformation and installed so that the covers are accessible. All holes or cut-outs provided for cable entry/exit must be provided with a grommet securely fixed in position. Cables installed must be adequately supported in vertical trunking by cleats and retaining clips.

Metal trunking including joins must be electrically continuous and must be earthed at one end. The wiring

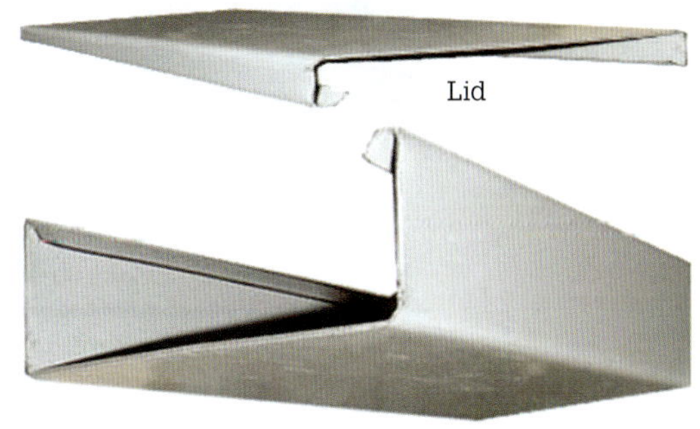

FIGURE 6.130 Metal trunking and cover

system classification for mechanical protection of metal trunking is WS X1 if the trunking is less than 1.6 mm and WS X2 if greater than 1.6 mm (see Appendix H in AS/NZS *Wiring Rules*, Table H5.3).

In-floor cable trunking

An in-floor cable trunking distribution system, when properly designed (total access or junction box access), is a good method for distributing power cables. A system designed with total access throughout its entire length to enable easy installation and maintenance of cables is known as the in-floor cable ducting system. In many new buildings, particularly those with open plans, in-floor cable management systems are particularly effective for large, open areas such as offices with modular partitions, schools and retail stores. This system is installed prior to a concrete pour or it can be chased into an existing concrete floor. Access points for cables and service fittings are provided along the length of the trunking.

An in-floor cable trunking is a cable enclosure system that is embedded in a floor with the cover of the tray flush with the floor level, as shown in **Figure 6.131**.

As a recommendation the size of the under-floor trunking should be such that the CSA of all the cables accommodated within an in-floor trunking does not exceed 30% of the CSA of the trunking. The in-floor cable trunking can be either high-impact UPVC (unplasticised PVC) or galvanised metal or steel of welded construction and of adequate thickness.

The advantages of an in-floor trunking system are that the cables are well protected and concealed in the trunking, so that interruption of service caused by physical damage to cables is minimised. In-floor trunking is also easy to work on. Disadvantages can include impaired access to the trunking and inspection boxes if machines and other items cover them. In addition, it is possible for fluids to seep through the lid of the trunking and cover the cables.

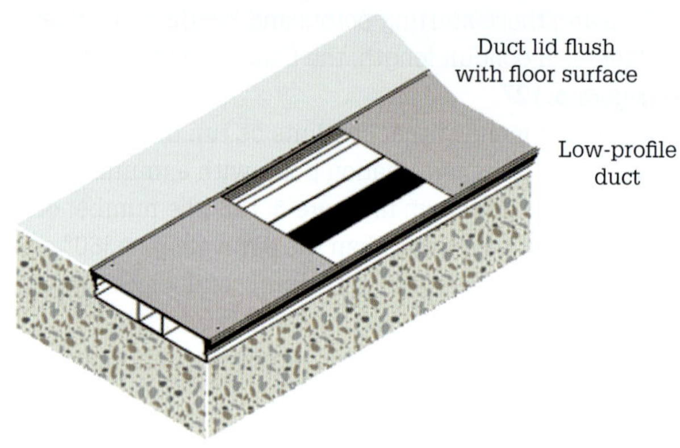

FIGURE 6.131 In-floor cable trunking

Multi-compartment trunking

Multi-compartment trunking can be aluminium or galvanised steel, of proprietary manufacture and should have two or three compartments. Where used to extend an existing installation the trunking should match that installation. It can also be mounted at any height along a wall.

Multi-compartment trunking allows access to the cables wherever it is needed. A photo of this type of trunking attached to a wall is shown in **Figure 6.132**.

The advantage of multi-compartment trunking is that it can be used for power, telephone and internal communications and for PC networking. A disadvantage is that the installation of this system can be made difficult by the columns, doors and windows in a building. Metal multi-compartment trunking also needs to be earthed. All socket outlets require an earth pigtail from the earth terminal to the metal surround as shown by one method of connection in **Figure 6.133**.

FIGURE 6.132 Multi-compartment trunking

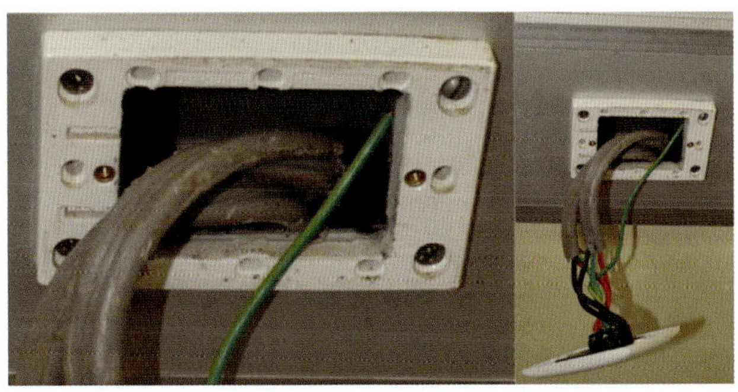

FIGURE 6.133 Socket outlets with earth pigtail

Cable tray and ladder

Cable trays and ladders offer an efficient way to manage a wiring system while keeping it accessible, ordered and secure. These systems have high structural strength and can accommodate large numbers of cables. Cable tray is a rigid structural support system used to fasten securely or support cables and conduits. Cable tray systems are available in a variety of configurations with a diversity of sizes and cover and support options.

Solid cable tray, perforated cable tray, wire tray, cable ladder and all accessories should be proprietary items from a single manufacturer whose range includes splice connections, expansion splices, covers, risers, crossovers, reducers, bends, support brackets and all other accessories that could be used.

It is sometimes necessary to select a cable tray or ladder wiring system without knowing the maximum cable weight. An effective approach is to choose a possible size of tray or ladder and then to calculate approximately the maximum cable capacity (kg/m) which is capable of being contained within it. The cable capacity can be calculated by applying the following equation:

$$\text{Maximum cabling capacity (kg/m)} = \text{cable laying area (m}^2) \times 2800 \text{ (kg/m)}$$

Solid, perforated and wire tray

Solid, perforated and wire (basket) tray as illustrated in **Figure 6.134** is another method used to protect, support and route cables, usually overhead.

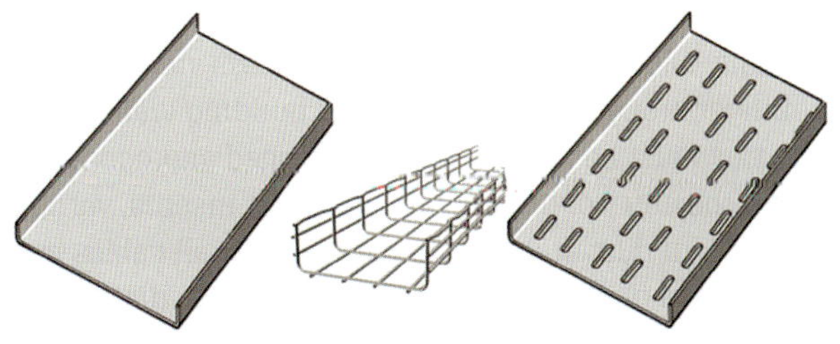

FIGURE 6.134 Solid, perforated and wire cable tray

Perforated cable tray is common for heavier-weight cables and offers excellent protection, while wire tray is typical for light-weight cable runs. Perforated cable trays can support a higher fill capacity than the solid tray due to better airflow around the cables. The perforations also enable the securing of the cables using cable ties rather than saddles as with solid trays. Cable trays offer ease of cable installation because the cables are laid in place rather than pulled as with a conduit enclosure system.

Unlike cable ladder, the type of cable tray together with the electrical load requirements and the type and voltage rating of cables that will use the tray determine the allowable fill for each cable tray. Cable trays should not be filled in excess of 40–50% of the inside area of the tray. If cable trays are overfilled, excessive heat will build up in and around the current-carrying conductors. Excessive heat can cause the cable insulation to break down, leading to potential short-circuit, shock hazards or fire.

To determine the required width of a perforated or wire cable tray (shown in **Figure 6.135**):

- Use the manufacturer's data to determine the total diameter of the cables.
- Multiply the diameter of the cables by the number of cables.
- Allow a space between the cables equal to one cable diameter.
- Sum these dimensions to obtain minimum tray width.

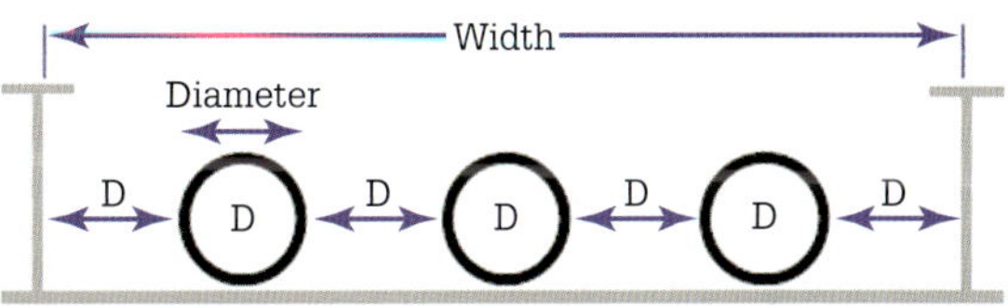

FIGURE 6.135 Width of tray

Power cables under short-circuit conditions (especially three-phase, single-conductor cable arrangements) are subject to significant forces as a result of magnetic fields. For three-phase, single-conductor cables, these forces cause violent whipping of the individual conductors, frequently resulting in inadequately supported cables jumping out of their cable tray systems. Such unrestrained cable movement can cause cable damage, damage to surrounding equipment and possible injury. To prevent the cable movement, cables must be configured correctly on the tray.

The three-phase single conductor cable installation configuration shown in **Figure 6.136a** is technically superior to that shown in **Figure 6.136b**.

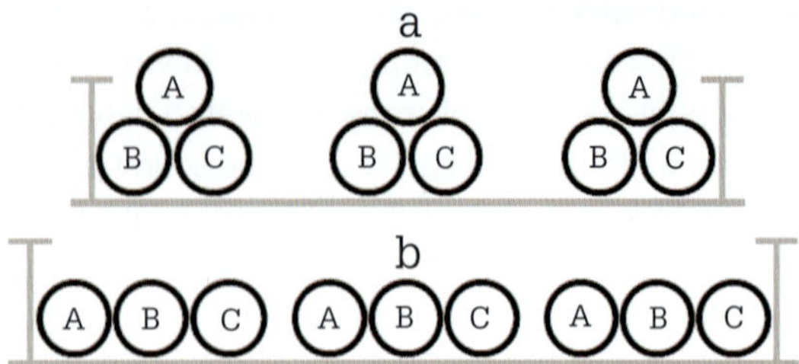

FIGURE 6.136 Cable configuration

The installation of the single cables in the cable tray as per **Figure 6.136a** is preferred as the cables are in an arrangement (trefoil) where they are equilaterally spaced (refer to AS/NZS 3008 1.1 for the recommended method of installation). If each of the phase conductors has the same resistance and reactance, the load currents and the phase voltages for connected equipment will be balanced, assuming that all the loads are balanced three-phase loads. Motors that are supplied with unbalanced three-phase voltages experience additional heating due to the voltage unbalance. A voltage unbalance of even a small percentage can significantly affect the service life of a motor.

The current-carrying capacity of a cable can be greater when installed on a cable tray, particularly when compared with conduit containing many (more than three) conductors. In certain installations, material cost will be reduced for a cable tray system due to the space factor and greater current-carrying capacity of installed cables.

Trays should be fixed to steel brackets and hangers with a deflection not exceeding 10 mm when fully loaded, with consideration to future installation of cables. Tray ends should be bushed or have a grommet to protect cables. Trays should be mounted at 2100 mm or higher above finished floor level and above other services at right angles or parallel to the building structure or according to an installation specification. Access must be provided to cable trays with a minimum of 300 mm above and 600 mm on one side if supporting TPS cables. A perforated cable tray carrying three-phase sub-mains and earthing conductor and other circuit cables is illustrated in **Figure 6.137**.

FIGURE 6.137 Cables covered in dust on perforated cable tray

As a guide, the minimum thickness for cable tray should be:

- steel 1 mm up to 150 mm wide; 1.6 mm for aluminium
- steel 1.2 mm from 150 mm up to 300 mm wide; 2.0 mm for aluminium
- steel 1.6 mm above 300 mm wide; 2.6 mm for aluminium.

Cable tray can be fixed to building structures by stand-off supports at intervals of approximately 1250 mm, supports being provided in any case at each end of each length.

Cable tray wider than 100 mm should be fixed at intervals of not more than 1000 mm with two screws per fixing. The clearance behind the trays must be such as to permit easy access to fixing nuts. Covers in accessible locations must be screw-fixed or of the clip-on type removable only with the use of tools. Cable tray installed externally should be aluminium or stainless steel. Openings through roofs and external building walls for cable tray must be plugged, sealed, flashed and hooded to prevent the entry of driving rain, seepage, insects, vermin and dust. For dust and pressurisation seals, all cables must be cleaned of dust and greased for a minimum distance of 100 mm either side of the cable tray penetration. In dusty environments, the top surfaces of all supports, conduits and cable sheaths surfaces where dust can accumulate will require regular clean-up housekeeping. Excessive amounts of dust act as a thermal barrier which may not allow the installed cables to dissipate internal heat safely. This condition may result in the accelerated ageing of the conductor insulation.

Ladder cable tray

Cable ladder consists of two folded steel or extruded structural-grade aluminium side rails with cable support rungs between the two rails. The rungs are either welded or fitted to the side rails usually 300 mm apart. Cable ladder is typically used in feeder applications for longer runs of multiple cables or of higher current-carrying capacity and weight. Ladder cable tray as shown in **Figure 6.138** is available as a straight length up to 4 m together with 90°, 45° bends, inside and outside risers, tees, reducers and crosses.

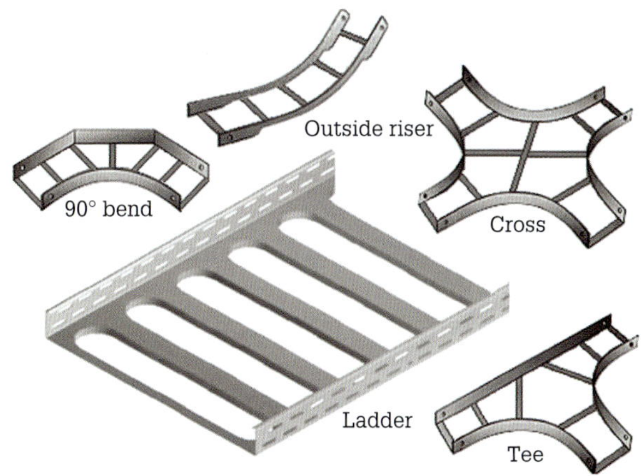

FIGURE 6.138 Ladder cable tray and fittings

Due to its load (mass) bearing capacity ladder cable tray is generally used in installation work for heavy-duty mains and sub-mains power distribution support. The rungs provide a convenient anchor for tying down cables in vertical runs or where the positions of the cables must be maintained in horizontal runs. Cables may exit or enter through the top or the bottom of the ladder. A ladder cable tray provides for the maximum free flow of air, dissipating heat produced in current-carrying conductors.

Ladder tray is easily cut, shaped and fitted together to route cable in either the horizontal or the vertical planes. Ladder tray must be adequately sized to support the cables without bunching. It is suggested that the cable tray or ladder be predrilled to accept bonding conductors. All earth bonding should be complete before the installation of cable commences. No cable crossovers should occur, except where cables enter or leave the trays. In order to prevent short-circuit forces separating cables, cables or groups of cables, depending on their size, must be securely attached to the tray or ladder using proprietary nylon cable ties as shown in **Figure 6.139**, straps or retaining clips. A suggestion for vertical runs is that all cables be secured on every second rung of the cable ladder. On horizontal runs all cables should be secured on every third rung on the cable ladder mounted horizontally and saddled on every second rung on cable ladders mounted vertically edgewise.

All holes for fixings and any other purpose must be machine drilled. Cutting or blowing holes with gas cutting

FIGURE 6.139 Nylon cable tie

equipment must not be carried out. Where it is necessary to cut or weld galvanised ladder, the exposed metal must be deburred, wire brushed and cold galvanised. Ladder sections that are bolted together must not have the threaded portion of the bolt protruding into the ladder section.

Horizontal cable ladders running on the flat and edgewise should be supported at 1500 mm maximum spans. Bends, tees and similar fittings must be supported strictly in accordance with the manufacturer's recommendations. It is suggested that, as a minimum size, cable ladder supports should be constructed of 75 mm × 75 mm × 6 mm angle.

Cable tray and ladder supports can be fixed to the building structure or fabric by means of ≥ 8 mm threaded rod hangers attached to hot-dipped galvanised U-brackets, or by means of proprietary brackets. Where supported by hangers or angle brackets, it is suggested that a minimum distance of 150 mm should be allowed above and 600 mm on one side of the tray or ladder. Surface fixing to ceilings or walls or by angle brackets out from walls are the recommended methods. Cable ladder joins should be within 150 mm of a support except for spans greater than 3.0 m, which should have joins at the centreline of the support.

Cable ladder fittings and supports must be installed so that they cannot be bent out of position or distorted during the installation period. Where misalignment or distortion occurs, all damage should be rectified and additional supports provided.

The minimum clearance to cable ladders, including the underside of supports, should be 2100 mm above floors, platforms, walkways and stairs. Vertical spacing between horizontal cable ladders should not be less than 400 mm bottom to bottom. The side walls of the cable ladder should not be cut for the installation unless absolutely necessary and then have additional supports installed to support this weakened section.

Cable ladders must be provided with spacing between:

1. High-voltage cables
2. Low-voltage power cables
3. Control cables
4. Instrumentation (analogue) and communication cables also provided with barriers to separate them from the other cables
5. Fire system cabling
6. Intrinsically safe (IS) cabling – used in non-incendive circuits as a protection technique to minimise the risk of ignition in hazardous locations. (These cables are identified by their light-blue outer sheath colour which allows identification of the cable along its length. If an

intrinsically safe cable is on the same ladder as non-intrinsic cables it must be separated by an intermediate layer of insulation material.)

Cables should leave the cable tray or ladder in such a manner that no cable is in contact with the side rails. Always provide rounded support surfaces under cables where they leave trays or ladders. However, it is suggested that cable should leave the tray or ladder in conduit, which must be securely anchored to the tray or ladder with a minimum of two anchors. It is suggested that cables be installed in one layer on all ladders and trays with a minimum of 20% spare capacity when the installation work is complete.

Where cable tray or ladder spans open areas it should be capable of supporting an additional load (60–75 kg) at mid-span without permanent deformation.

Mineral-insulated metal served (MIMS) cable is arguably the safest type of cable for use in hazardous locations, so there are few limitations on its use. However, MIMS cable cannot be used where exposed to physical damage or corrosive conditions unless protected by a suitable non-metallic outer jacket. Where MIMS cables are used, the cable tray or ladder should be PVC dipped or PVC served.

REVIEW QUESTIONS

1 What do steel conduit threads provide?
2 State the standard Australian finished length of rigid metal conduit.
3 What type of tool is used to cut a thread on conduit?
4 How should joints and any damaged coatings in steel conduit be treated?
5 What is meant by the term 'set' in relation to steel conduit?
6 What is the typical method used to make sets in rigid metal conduit?
7 Which hacksaw blade is recommended for cutting steel conduit?
8 State a typical application for flexible metallic conduit.
9 Explain the term 'gain' in relation to setting steel conduit.
10 Why should the number of sets in a conduit run be not greater than the equivalent of 360°?

6.7 Fire protection cabling and systems

Heat- and fire-resistant cables are designed to withstand long-term heat, or in some cases heat and fire or fire only. The term 'fire resistant' means that the cable is capable of performing its intended functions under the heat and other conditions likely to be experienced due to a fire at its particular location for a defined length of time. Commonly used types of heat/fire-resistant cables are:

- V–105 PVC cable, temperature rating 105 °C (heat resistant only) used for internal wiring of hot appliances (e.g. fluorescent lights).
- PVC-based cables using halide-based chemicals, such as chlorine, as their fire suppressants. When PVC burns it gives off a halide gas, which rapidly absorbs oxygen thereby starving the fire, causing it to self-extinguish. PVC is a very efficient fire retardant, but its by-products can be hazardous. In high concentrations, chlorine gas is quite toxic and when combined with oxygen and water vapour creates a by-product called hydrochloric acid, which is also potentially hazardous.
- Glass-fibre cables, temperature rating 150 °C (heat and fire resistant) used for internal wiring of hot appliances (e.g. stoves and boilers).
- Mineral-insulated metal-shielded (MIMS) cables, temperature rating 250 °C (heat and fire resistant) used for building risers, lift wells, boiler rooms and slab heating.
- Radox® cable (fire resistant only) insulation which offers excellent resistance to thermal, chemical, electrical and mechanical loads. The electron beam cross-linked insulation is mechanically robust, does not melt and is resistant to most media, impregnation resins and insulating varnishes. Short-term exposures of the cable to temperatures up to 280 °C have no adverse effect on the insulation. The high flame retardance prevents fire propagation in case of fire. Radox® cable is used extensively in fire alarms and security systems.
- Pyrolex™ Ceramifiable® cable, shown in **Figure 6.140**, has an insulation layer that hardens into a protective ceramic shield when exposed to fire. This feature allows the cable to maintain circuit integrity.

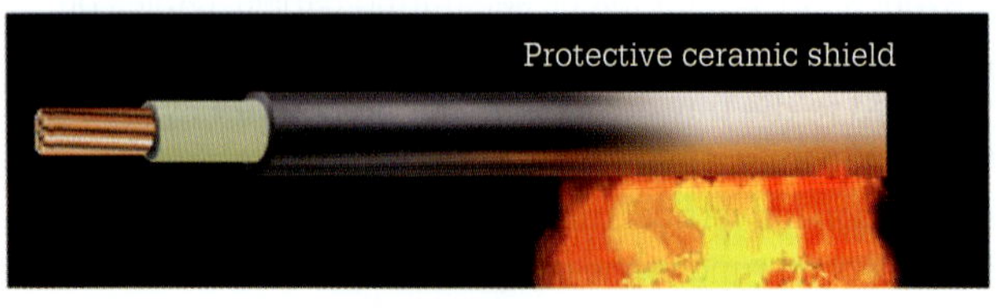

FIGURE 6.140 Pyrolex™ Ceramifiable® cable

Cables for fire detection and fire alarms must always be installed above normal cable and cable support systems. This ensures that other cabling systems affected by heat or fire will not damage the fire detection and fire alarm wiring and support system. Heat- and fire-resistant multi-core circular, copper conductor, mica glass taped with X-90 XLPE insulation and HFS–90–TP sheathed, 0.6/1 kV cables fixed to cable tray with a gland fitted are illustrated in **Figure 6.141**.

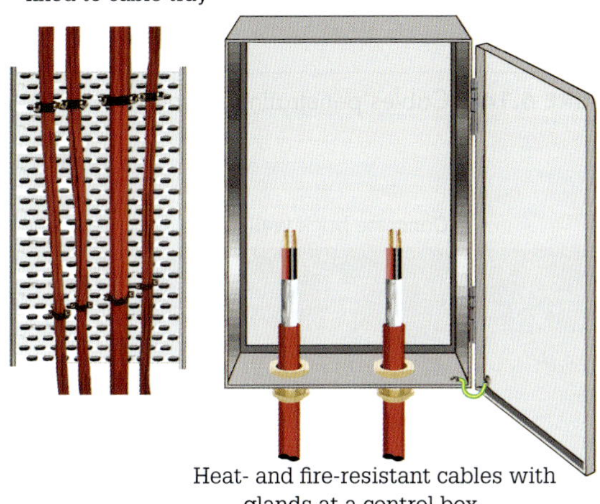

FIGURE 6.141 Heat- and fire-resistant cables with gland attached and fixed to cable tray

Although standard cable cleats, cable joints and cable glands may be used, it is important to ensure that because the cable is required to maintain circuit integrity in a fire, any wiring support and fixing used must also be capable of withstanding that fire.

Fires involving electrical equipment

The risk of a fire involving electrical equipment in a building depends upon the risk of damage that can occur to the cables and the points of termination. Terminations will in time have dust and/or moisture build-up which can cause arcing between terminals. This is why there must be no exposed conductor core beyond the ferrule of the termination. Also, insufficient tightening of terminals can cause local heating because of the high-resistance termination resulting in a potential fire. All cables installed in wall cavities must have a drip loop to prevent possible moisture from collecting on the switch or socket outlet terminations.

Special attention should always be given to all ceiling luminaires and their terminations. Because of their location on the ceiling, a fire can rapidly spread unnoticed through the roof cavity (refer to AS/NZS 3000:2018 *Wiring Rules*, 'Lighting equipment and accessories').

Fire integrity

Refer to AS/NZS 3000:2018 *Wiring Rules*, 'Fire integrity'.

With the introduction of the Building Code of Australia (BCA), building designers now have a system for defining the ability of a wall to resist the devastating effects of fire. Section C of the BCA defines the type and class of buildings and uses three building attributes to express fire-resistance levels (FRLs) as a triple rating in terms of minutes.

A one-hour fire wall is defined as having a fire-resistance level of 60/60/60 where each number represents the number of minutes the building attribute can resist the fire and maintain its function. The building attributes are:

- *Structural adequacy* – This is the ability of a wall, floor or ceiling to continue to perform its structural function for the fire-resistance period.
- *Integrity* – This is the ability of a wall, floor or ceiling to maintain its continuity and prevent the passage of flames and hot gases through cracks in the wall during the fire-resistance period.
- *Insulation* – This is the ability of the wall, floor or ceiling to provide sufficient insulation such that the side of the wall away from the fire does not exceed a predefined temperature during the rated period. At this temperature (a rise of 140 °C over the ambient temperature or a maximum of 180 °C) surface finishes and furnishings in contact with or near the wall may combust.

To maintain the fire integrity of wall, floor or ceiling, electricians must ensure that penetrations are kept to a minimum and well-sealed with fire-resisting material.

Penetrations through fire-rated walls and floors

Without penetration seals, a fire may spread and destroy other parts of an installation (see Clause 3.9.9.3, 'Penetration of fire barriers', AS/NZS 3000:2018 *Wiring Rules*). A fire that is tracking along a cable may pass through a penetration that has been sealed with a non-approved sealant and will possibly burst out the sealant in the process. A fire can quickly consume a building simply by finding its way through any penetration no matter how small.

If a penetration requires a fire rating, the use of a fire-rated mortar, mastic, fire collar (a fire collar is a prefabricated penetration seal system for PVC conduits) or cable intumescent (to swell up as in heat) coating seal is required. All approved penetration seals are required to achieve a three-hour fire rating. The penetration seals must also achieve a three-hour temperature rating; however, where this is not possible due to heat transfer through the cable core, cable coating is provided to prevent the cable from igniting on the other side. Conduits, trunking and cable risers must be sealed at both ends.

A gypsum plaster and vermiculite mortar is designed to stop the spread of smoke and fire through openings in fire-rated walls and floors. A fire-rated acrylic (water-based) gun-grade mastic with limited joint movement capability is also used for the sealing of gaps around conduit, cables, trunking and troughs which penetrate fire-rated walls and floors.

Voids can be filled using Dow Corning® 3-6548 sealant in compliance with the manufacturer's specifications and recommendations. Conduit must be installed concentrically or eccentrically within the fire-rated mortar or mastic system.

In fires, PVC conduits provide a fuel source unless treated with the appropriate seal; they allow a fire to spread easily by the conduit giving way when heated or through the inside of the pipe itself or along the cables inside. To prevent the spread of fire in PVC conduits, an intumescent seal (fire pillows, intumescent paint) is required. Intumescent seals provide a highly efficient barrier against fire, hot smoke and fumes. On reaction to fire they expand many times from their original volume and maintain positive pressure between the walls of the conduit to prevent loss of the seal. Where sealing is not possible due to heat transfer through the cable core, intumescent paint is used to prevent the cable from igniting on the other side.

Note: Polyethylene-insulated unsheathed cables must be installed in fire-rated enclosures (see Clause 3.10.1.2).

Fire protection provided to cables penetrating walls is illustrated in **Figures 6.142** and **Figure 6.143**.

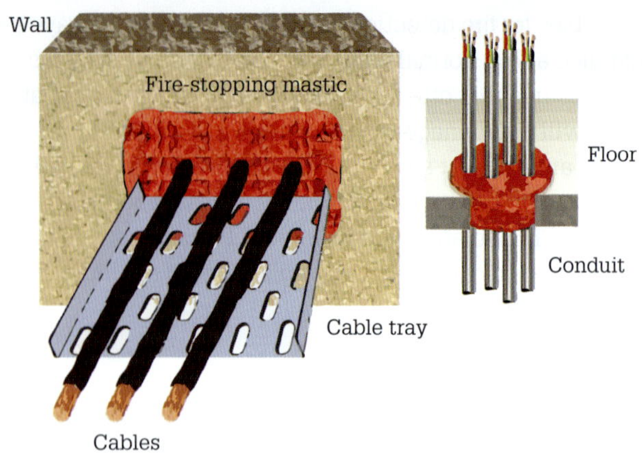

FIGURE 6.142 Cables penetrating a wall

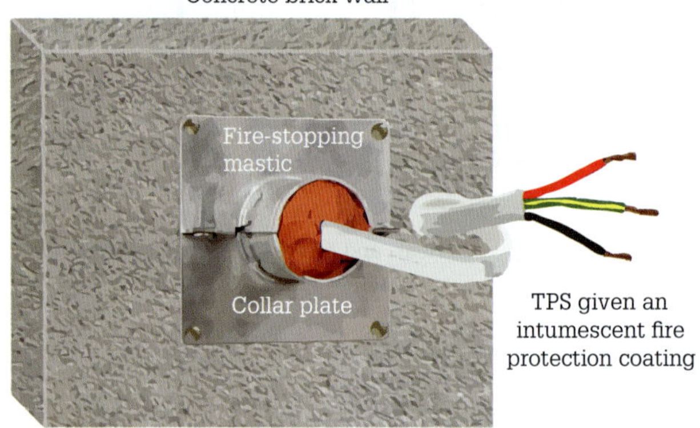

FIGURE 6.143 TPS cables provided with a coating and mastic penetrate a wall

Mineral-insulated metal-sheathed cable

Mineral-insulated metal-sheathed (MIMS) cable consists of a sheath, a conductor and a powdered insulant, as shown in **Figure 6.144**. The seamless outer sheath is made from malleable metal in a wide range of diameters, containing single or multiple conductors. Easily formed or bent, MIMS cable can accommodate virtually any shape. The outer sheath (made of copper, aluminium, stainless steel and other materials) protects the inner conductor from oxidation and aggressive environments. In locations subject to chemical environmental hazards a serving (available in many colours) over the sheath provides protection.

Compressed magnesium oxide is an organic mineral insulation, which gives the cables excellent high-temperature dielectric strength. MIMS cable does not burn nor does it support combustion. The sheath has a melting point of 1083 °C and the insulant melts at 2800 °C. However, the cable (V–90 HT or HFS–90–TP) is limited to a maximum operating temperature of 100 °C (refer to AS/NZS 3000:2018 *Wiring Rules*). The metal sheath is impervious to water, oil and gas and can withstand flattening, twisting and bending hazards.

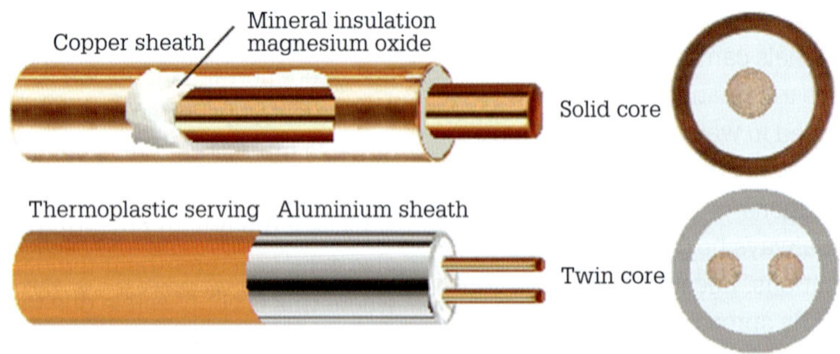

FIGURE 6.144 Mineral-insulated metal-sheathed (MIMS) cables

MIMS cable does not require a separate circuit protective earth conductor as the metal sheath serves this purpose, providing a low-resistance earth continuity path. The internal conductors of multi-core MIMS cables must not be used as protective or bonding earth conductors.

A MIMS wiring system that uses the sheath as a protective earth and a neutral is called an earthed sheath return (ESR) wiring system. An ESR is a restricted system in Australia because circuits cannot be protected by RCDs (refer to AS/NZS 3000:2018 *Wiring Rules*). MIMS cable is used in installations that are subjected to:

- high-temperature conditions (oil refineries, furnaces)
- possible explosions (fire alarm, petrol pumps)
- corrosive conditions (chemical works) – a serving over the sheath is required
- hygiene demands (food-processing plants, breweries and dairies).

MIMS, depending upon voltage class, is used for mains, sub-mains and sub-circuits and in voice communication installations and floor heating systems. The cable is rated in two voltage classes:

1. 0.6 kV/0.6 kV for two, three and four core only, with cable sizes and current rating depending upon core from
   1.0 mm^2 to 4.0 mm^2 and 20 A to 40 A (sub-circuits only).
2. 1.0 kV/1.0 kV for single, two, three, four and seven core, with cable sizes and current rating depending upon core from 6.0 mm^2 (70 A) to 400 mm^2 (1000 A). At this voltage rating MIMS cables, if fitted with suitable take-off boxes and surge diverters on the cable ends if they emanate from low-voltage overhead mains, can be used as consumer's mains.

Installing MIMS

When cutting MIMS cable, a fine-toothed hacksaw should be used. After cutting, the sheath should be ringed and then removed with an appropriate stripping tool such as a rod or a sheathmaster tool. After stripping and exposing the conductor tails and cleaning them, a pot must be fitted by screwing it onto the MIMS sheath. Then, an insulation test (500 V) must be performed between conductors and between each conductor and the cable sheath. If the test reading is less than 200 MΩ, moisture is present in the insulation and must be driven out from each cable end (over a 300 mm length) by applying heat to the cable. The insulation resistance will be restored once the cable has cooled.

The insulating filler, magnesium oxide, is hygroscopic, which means that it will absorb and retain moisture under certain humidity and temperature conditions. When MIMS cable is not properly sealed and is subjected to these conditions, moisture penetrates the insulation thereby reducing the insulation resistance. Test the cable again and if a correct reading is obtained seal the pot with compound. Install pot seal and crimp using an approved crimper (flat plate or an MI crimp).

Coloured sleeving of a sufficient thickness of insulation to suit the particular application can now be placed over the conductor tails. A compression cable gland (standard or flameproof) can be fitted to the MIMS if required. The sequence of completing the seals and conductor tails is illustrated in **Figure 6.145**.

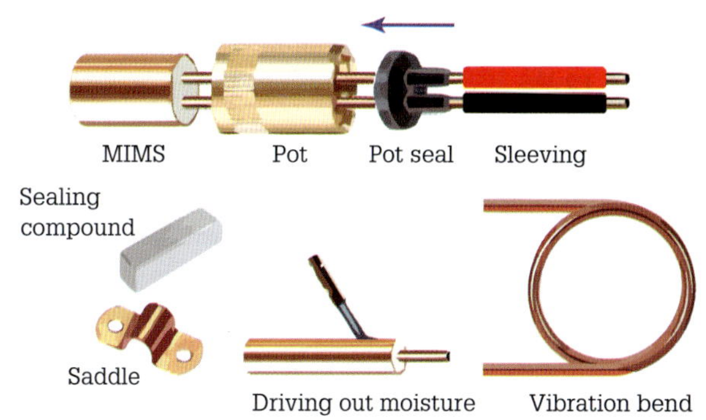

FIGURE 6.145 MIMS sealing sequence

For single-core MIMS, a self-adhering tape or heat shrink is used to seal the conductor tails. If the MIMS is sleeved, any damage must be repaired (some installation specifications require replacement of the length of MIMS regardless of the size of the damage). If the MIMS is connected to an electric motor or across an expansion joint, a vibration bend (helix, S or double S) in the cable must be made (see **Figure 6.145**). Bends (> 6× radius of the cable diameter) can be made by hand or by using a bending lever. The cable can also be straightened by using a roller straightener made for that purpose.

All saddles used with MIMS cables must be of the same material as the sheath to prevent galvanic action. Where MIMS cable is installed on the surface in damp or humid conditions and/or buried in the ground, concrete render, hard plaster wall finishes or concrete floor, it must be provided with PVC serving. MIMS cable installed in exposed locations should be protected to a height of 2 m above floor level.

The passing of single-core MIMS cables through ferrous materials must be avoided or, where this is not possible, the manufacturer's instructions for the limitation of eddy currents should be followed. Single-core cables can run in various cable groupings to minimise the effects of eddy currents. A three-phase run in trefoil grouping (see **Figure 6.146**) must maintain this formation throughout its length and only diverge to facilitate the termination of the cables (refer to AS/NZS 3000:2018 *Wiring Rules*).

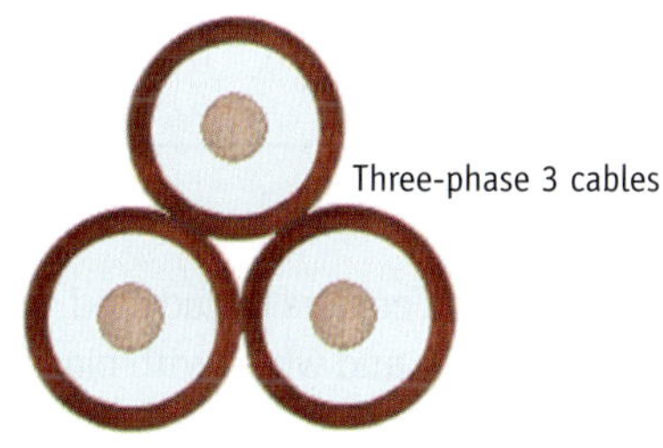

FIGURE 6.146 Single-core MIMS in trefoil grouping

MIMS cable has a wiring system (WS) rating – WS 52W, WS 53W and WS 54W – depending upon the cable grouping.

MIMS cable fixings

The location and type of fixings for MIMS cable are shown in **Table 6.12**.

TABLE 6.12 Location and type of fixings for MIMS cable

Location	Type of fixing
Structural steelwork	Purpose-made clamps or fasteners
Non-structural steelwork	Set screws and nuts
Concrete, brick or building	Plugs and screw blocks
Hollow blocks	Toggle-type screw fixings
Timber	Screws

Note: No shot firing must be used and no drilling or welding of structural steelwork must be done to affix any enclosures or fittings of any wiring system unless approved by an engineer. Bolts, nuts, washers and screws should be cadmium or zinc electroplated. Saddles (must be copper only) or MIMS cable must be served to remove the possibility of corrosion with sheath.

REVIEW QUESTIONS

1. What does the term 'fire resistant' mean?
2. Name three fire-resistant cables and list typical applications for each.
3. Where are cables supplying fire detection and fire alarm systems placed in relation to normal cable and cable support systems?
4. Which cable has an insulating layer that hardens into a protective ceramic shield when exposed to fire?
5. If a cable penetration requires a fire rating, what can be used to seal the penetration?
6. Describe the construction of MIMS cable.
7. Name two voltage classes for MIMS cable.
8. Why is it necessary to seal MIMS cable?
9. In what installation circumstances must MIMS be provided with PVC serving?
10. What must occur with a run of MIMS if it is to be connected to an electric motor?

6.8 Steel-wire armoured cables

Steel-wire armoured (SWA) cable (see **Figure 6.147**) is rated as 0.6/1.0 kV circular SWA plain annealed copper cable. Aluminium armour (for single core) is also used. SWA cables are designed with mechanical protection and are therefore suitable for internal and external use where mechanical damage is a real possibility and for direct burial. SWA cable is used for mains, sub-mains and auxiliary control cables and in installations where glands are required.

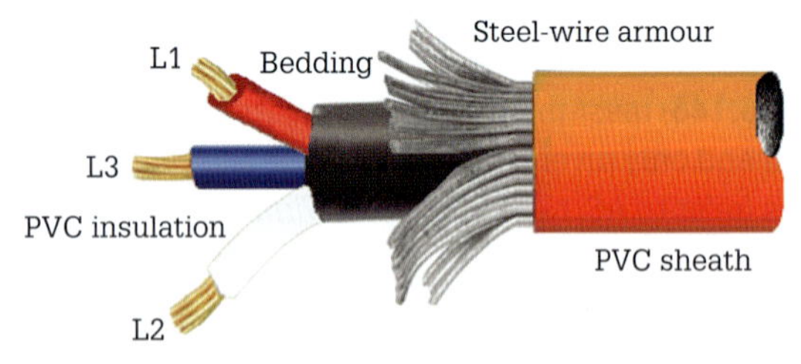

FIGURE 6.147 SWA cable

Conductor insulation colours include red, brown, blue, black, green/yellow and white with black numbers. Stranded conductor sizes range from 1.5 mm^2 to 25 mm^2 and above. The minimum installed bending radius of SWA cables ranges from 6× cable diameter for cables less than 16 mm^2 to 8× cable diameter for cables 25 mm^2 and larger.

Due to their armouring, SWA cables are classified as a category A wiring system and can be installed unenclosed or installed in ducts or buried directly in the ground, concrete, plaster or cement render. If chased in rock with no enclosure SWA cable is classified as category C (refer to AS/NZS 3000:2018 *Wiring Rules*).

SWA cable is composed of inner cores (solid or stranded plain annealed copper) that are individually sheathed (PVC or XLPE), then covered by an initial overall PVC or XLPE bedding, followed by the protective galvanised steel-wire armour with a final PVC or XLPE outer sheath (orange or black) to provide protection to the armour.

XLPE insulation is used in installations where fire, smoke emission and toxic fumes will create a potential threat. The inner cores may number two, three, four or five depending upon the application. All SWA cable must use glands for termination of the cable, as illustrated in **Figure 6.148**.

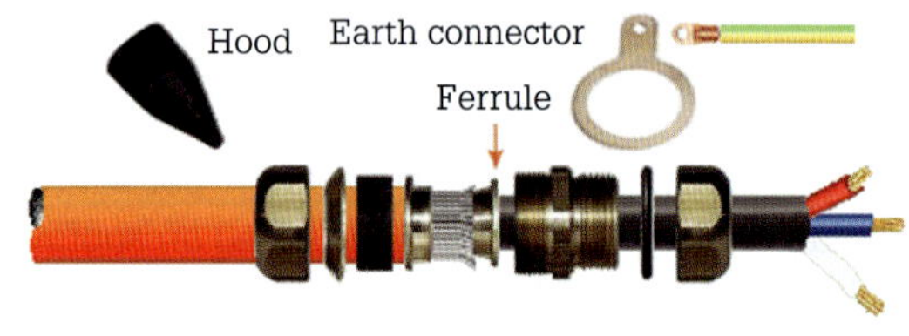

FIGURE 6.148 Gland termination of SWA cable

Note: A ferrule is placed under the armouring to prevent damage to the cable by the fixing procedure. An earth connector must be provided at one end of an SWA run and a hood over the gland may be required.

For armoured cables the gland at one end must have an earth bond attachment, and for cables with conductors larger than 35 mm² the earth termination must be integral with the body of the gland. Glands used with aluminium-armoured cables must be of aluminium to prevent galvanic corrosion.

SWA fixings

Fixings for SWA cable are shown in **Table 6.13**.

TABLE 6.13 Fixings for SWA cable

Building fabric	Type of fixing
Structural steelwork	Purpose-made clamps
Non-structural steelwork	Bolts, washers and nuts
Concrete, brick or building blocks	Expanding anchors
Timber	Coach bolts

Note: No shot firing should be used and no drilling or welding of structural steelwork must be done to affix any enclosures or fittings of any wiring system unless approved by an engineer. Bolts, nuts, washers and screw fixings should be cadmium or zinc electroplated.

Neutral screened cable

Neutral screened cables can consist of one or more insulated inner cores containing plain annealed stranded copper conductors (see **Figure 6.149**). The insulation of the inner PVC core V–90 is surrounded by concentric plain annealed copper conductors. A 3.2 mm thick V–90 PVC sheath for unenclosed underground use covers these conductors.

Neutral screened cable is used for above-ground applications and as underground consumer's mains or sub-mains in domestic or light industrial installations. The cable can also be used in unenclosed situations where the neutral/earth screen can provide protection against the hazards of electric shock. Contact with the active conductor is limited because contact with the neutral screen must be made first. The neutral is always at earth potential. If both conductors are breached the circuit protection device will operate.

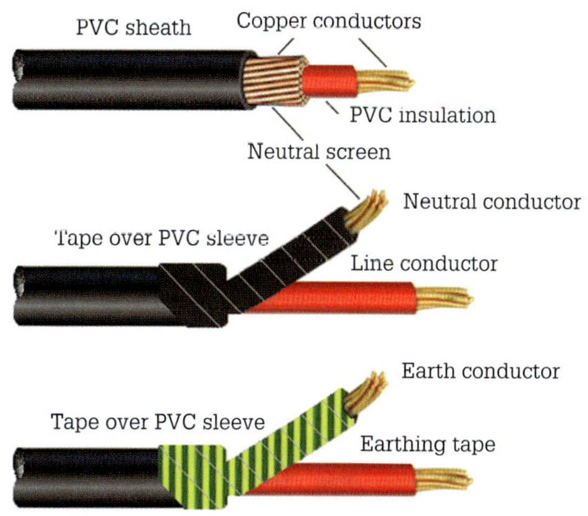

FIGURE 6.149 Preparing neutral screened cables for termination

The inner core is the line conductor. The outer conductor or screen can be either the neutral conductor or a protective earth conductor. The screen is twisted to form either the neutral or the protective earth conductor. If the screen is used as a protective earth conductor then a green/yellow sleeve and earthing tape must be used. All tape used must be of the same temperature rating as the cable.

If buried directly into the ground a neutral screened cable is classified as a Class A underground wiring system. If the cable is installed in conduit it is also classified as Class A (refer to AS/NZS 3000:2018 *Wiring Rules* and the thickness of the sheath insulation). Single-core neutral screened cables are available as 4 mm² (37 A), 6 mm² (46 A), 10 mm² (64 A) and 16 mm² (85 A).

REVIEW QUESTIONS

1 State typical applications for SWA cable.
2 What underground wiring system is applicable to SWA cables?
3 Name the cable insulation necessary where fire, smoke emission and toxic fumes would create a potential threat.
4 In regard to neutral screened cable, which conductor is the line conductor?
5 In neutral screened cables, how is contact with the active conductor limited?

6.9 Trailing cables and catenary systems

Trailing cables are used in applications where standard cables would experience high mechanical stress. These cables experience frequent movement with various types of stress and therefore have a tough, durable, hardwearing sheath. A trailing cable used for crane control is illustrated in **Figure 6.150**.

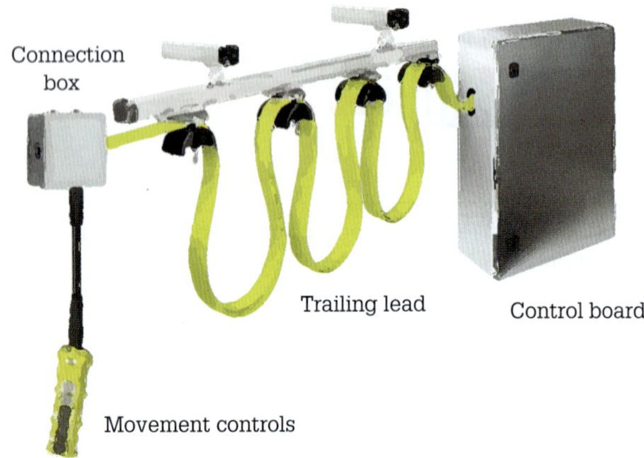

FIGURE 6.150 Trailing cable

These cables are used as control and power cables for conveyors, lifting systems and cranes. In addition, special trailing cables are used to supply a range of surface mining equipment including pontoon pumps, electric drills, draglines, shovels, and the like. Typical voltages for trailing cables range from 400 V to 22 kV.

Some trailing cables contain earth screens and a small conductor called the pilot at the very centre of the cable. The pilot conductor signals protection relays to detect cable breakage during operation or prior to operation.

Catenary supported cables

If a cable cannot be supported for its length when suspended between two fixed points, it will hang freely under the influence of gravity. The shape the cable will take what is known as a catenary curve. A catenary system is an overhead wiring system (CWS) containing one or more cables suspended between two fixed points and hung on galvanised, stainless steel or compacted aluminium alloy support wires. A simple catenary above a suspended ceiling is a fixed-anchored, span-wire construction as shown in **Figure 6.151**.

An outdoor simple catenary wiring system is illustrated in **Figure 6.152** with a cable strainer which is used to tension the support wire.

Catenaries can be used indoors for a trailing cable system as well as outdoors. Catenary wiring systems (CWSs) securely fixed to anchored eye bolts via thimbles and grips may be used where a cable tray is unsuitable indoors or the number of cables does not warrant the use of a cable tray.

Catenaries are normally used for short spans with some CWSs under tension. When cables are released or tensioned (via a strainer) it must be done in a way that controls the effect on the structural stability of the CWS. In particular, a CWS should not be cut and allowed to fall. This practice may cause a hazardous whip action. In addition, any sag must not present a thoroughfare hazard. Catenary cable tails can be terminated directly on structures at each end of the catenary span or continued on to be terminated at equipment.

The support wires for a CWS, depending upon the mass of electrical cable, are usually seven strands of 2.0 mm galvanised steel wire or seven strands of 0.5 mm compacted aluminium alloy wires. Cables supported by a catenary should consist of stranded or flexible cable, using double insulation. In some installations it may be necessary to install a bonding earth to the support wire of the catenary. Cables exposed outdoors must be suitable for exposure to direct sunlight. Some cables have a catenary wire attached to their insulation for easy installation. Relevant *Wiring Rules* clauses for catenaries are illustrated in **Figure 6.153**. (Always refer to the primary source of data AS/NZS 3000:2018 *Wiring Rules*.)

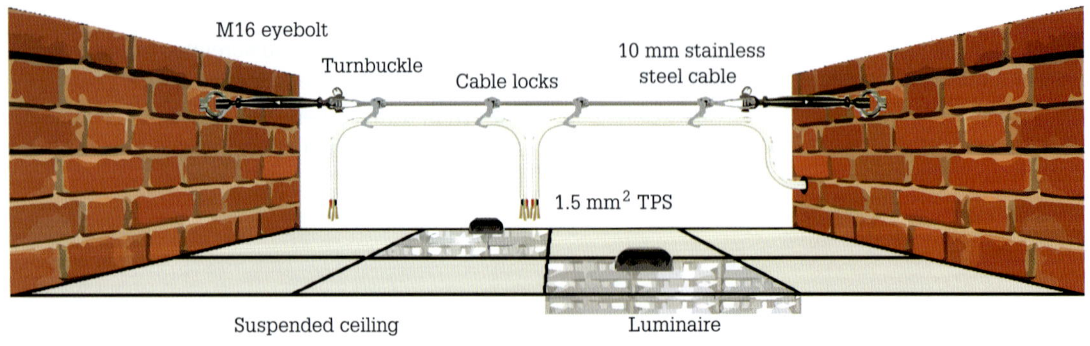

FIGURE 6.151 Simple catenary wiring systems

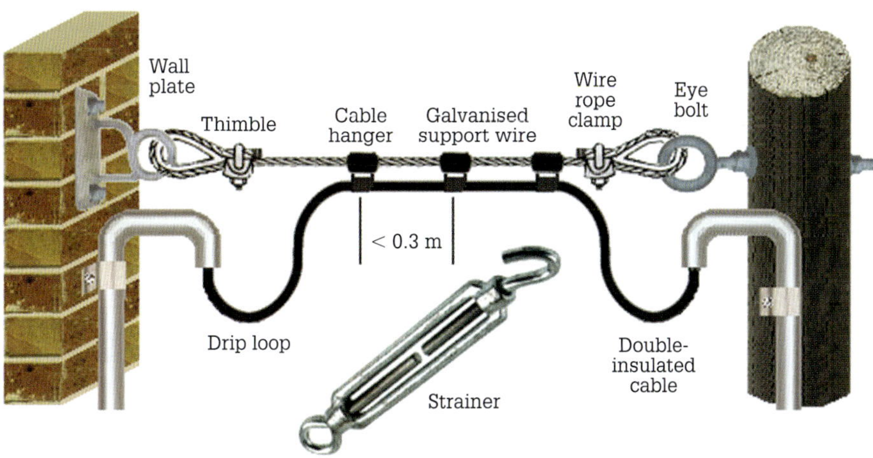

FIGURE 6.152 Simple catenary outdoor wiring systems

Clause
1.4.99
2.3
3.2
3.13
5.3

FIGURE 6.153 Relevant clauses for catenaries

SWITCH ON

Hazard

Before installing any catenary wiring system, ensure that any metal where the catenary is to be located such as conduit, ducts, trunking, structural beams, roofing iron, downpipes, and the like are not live. The Occupational Health and Safety (Fall Prevention) Regulations 2003 has specific provisions for controlling the risk of falling when working at heights which are based on a prescribed hierarchy of controls. Where there is a risk of a person falling 2.0 m or more, emergency procedures must be in place to enable the rescue of a person in the event of a fall.

Pendants

Pendants bring electric power to a work station or work area without the use of extension leads and are usually attached to a support chain. Chain-supported pendant socket outlets are illustrated in **Figure 6.154**.

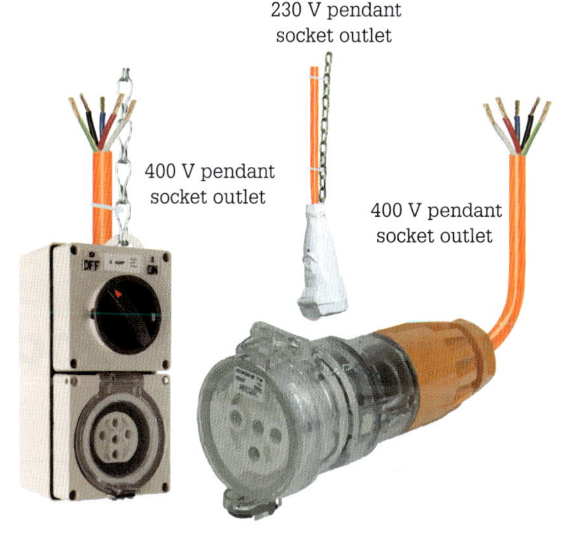

FIGURE 6.154 Pendant socket outlets

Lighting looms

Lighting looms can prevent the installing problems for cables above suspended ceilings as shown in **Figure 6.155**.

FIGURE 6.155 Cables installed above a suspended ceiling

Lighting looms are flat TPS cables installed in the space between a roof structure or concrete flooring and a suspended ceiling. For more information on installing cables in a suspended ceiling see Clause 3.9.3.2 of AS/NZS 3000:2018. A suspended ceiling is shown in **Figure 6.156**.

FIGURE 6.156 Suspended ceiling

The loom circuit provides a series of surface sockets for the luminaires as shown in **Figure 6.157**. Loom circuits are usually prefabricated on site to suit the layout required. This method of wiring is suited to projects where the lighting layout is similar for many floors.

Several looms can be pre-constructed for normal lighting, emergency lighting and security lighting.

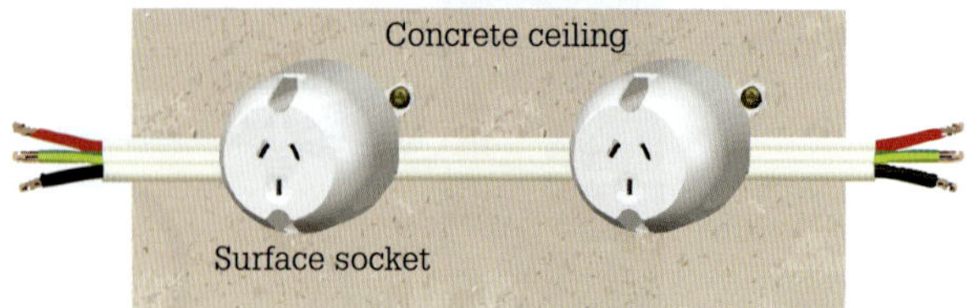

FIGURE 6.157 Lighting loom for surface sockets

Before installing lighting looms always check drawings to confirm their locations and requirements. Always label each loom with its circuit number and distribution board.

REVIEW QUESTIONS

1. State typical applications for trailing cables.
2. What is the purpose of a pilot conductor in a trailing cable?
3. Describe a catenary wiring system.
4. What type of cables are supported on catenary wires?
5. Describe a pendant wiring system.

6.10 Industry standards and testing requirements

Industry standards, such as the National Construction Code, and manufacturers' data play an important part in the selection and installation of wiring systems.

Manufacturer's specifications

The *Wiring Rules* notes that suppliers of wiring systems should provide installers with detailed fixing and support requirements to ensure requirements of AS/NZS 3013. These requirements should include details as to the correct orientation of the wiring system. Installers also have an obligation to comply with manufacturers' instructions for fixing and support requirements of wiring systems.

Specific clauses in the *Wiring Rules* that call on manufacturer data and specifications include:

- 3.7.2.1.2 Aluminium conductors
- 3.7.2.3.2 Crimp joints (compression joints)
- 3.9.5 Wiring systems installed vertically
- 3.9.6 Change of direction.

Sample cable data

Table 6.14 and **Table 6.15** show sample physical and electrical characteristics for a fictitious multi-core electrical cable.

TABLE 6.14 PVC Cable 0.6/1 kV 1.5 mm² 2-40 C+E Physical Characteristics

Product code	Cable						Bending radius – minimum installed (mm)
	Nominal Conductor CSA (mm²)	Core count	Nominal insulation thickness (mm)	Overall diameter		Mass (kg/100 m)	
				Minimum (mm)	Maximum (mm)		
MYCBL01	1.5	2	0.75	9.5	10	14	40
MYCBL02	1.5	3	0.75	9.9	10.5	16.5	50
MYCBL03	1.5	4	0.75	10.9	11.4	21.6	50
MYCBL04	1.5	5	0.75	12.8	13.5	23	60

TABLE 6.15 PVC Cable 0.6/1 kV 1.5 mm^2 2-40 C+E Electrical Characteristics

Product code	Core count	Current rating			Electrical rating	
		Spaced unenclosed (A)	Buried direct (A)	Underground duct (A)	Max d.c. resistance @ 20°C (Ω/km)	Reactance per core (Ω/km)
MYCBL01	2	20	28	23	13.8	0.115
MYCBL02	3	17	25	20	13.8	0.115
MYCBL03	4	17	25	19	13.8	0.115
MYCBL04	5	15	23	17	13.8	0.115

EXAMPLE 6.4

Use the information in Table 6.14 and Table 6.15 to answer the following.

1 Determine the weight a catenary system would need to support if it was required to support 20 m of cable type MYCBL03.

Solution: $\frac{21.6 \times 20}{100} = \mathbf{4.32\ kg}$

2 Determine the current-carrying capacity of cable type MYCBL03 when installed in an underground enclosure.

Solution: **19 A**

EXERCISE 6.4

Use the information in Table 6.14 and Table 6.15 to answer the following.

a Determine the weight a catenary system would need to support if it was required to support 30 m of cable type MYCBL04.

b Determine the current-carrying capacity of cable type MYCBL04 when installed unenclosed and spaced from other cables.

Testing

The *Wiring Rules* requires that wiring systems are tested to ensure safety and proper functioning. Refer to Chapter 16 of this textbook for information relating to testing of wiring systems that form part of an electrical installation.

Flexible cords, plugs, and sockets

Testing of flexible cords and associated plugs and sockets involves visual and electrical testing.

Visual testing

Visibly check the plug top and extension socket for:

- bent or broken pins, as shown in Figure 6.158
- exposed wiring strands or exposed primary insulation at plug top or extension socket grip
- plugs and sockets which should be appropriate for Australian and New Zealand usage.

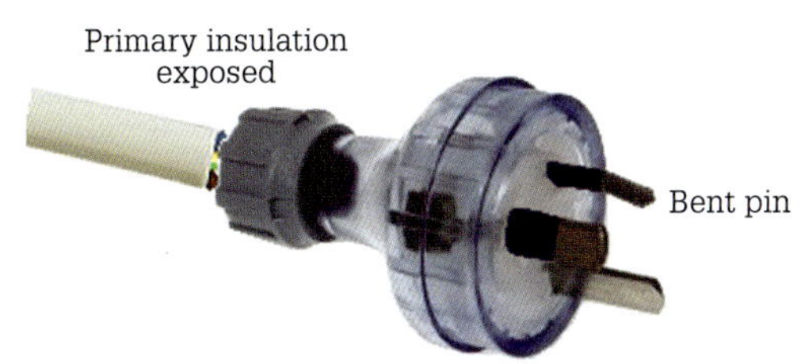

FIGURE 6.158 Visual inspection of plug top

Flexible cord is selected to match the appliance (selection must consider length, voltage drop, current rating, number of cores, outside diameter, shape, insulation quality to suit voltage, operating temperature, environmental conditions).

- Check the cord or cable for nicks, cuts, joins/insulation tape and frayed insulation.
- Verify that strands are not cut from conductors, that no stray strands are protruding, that strands are twisted and doubled over where necessary and terminal screws are tightened.
- Check to see that shroud/cord protector is still correctly fitted and fit for service.
- Is the plug and socket housing serviceable and are all fixing screws and covers secure?

Some imported cord extension sockets have been found to pose a serious threat that could result in electrical shock and fire. It had been discovered that the socket contacts did not adequately grip their corresponding plug contact. This type of high-resistance contact can cause an electrical fault that may result in overheating, melting of the housing and shrouds, fire and possible electric shock. Every cord extension socket, therefore, should be tested for its mechanical gripping ability.

Electrical testing

Electrical tests (see Figure 6.159) check polarity, continuity and insulation resistance.

Many types of electrical equipment provide an internationally accepted mains connector designated IEC 60309 type plugs and socket as illustrated in Figure 6.160.

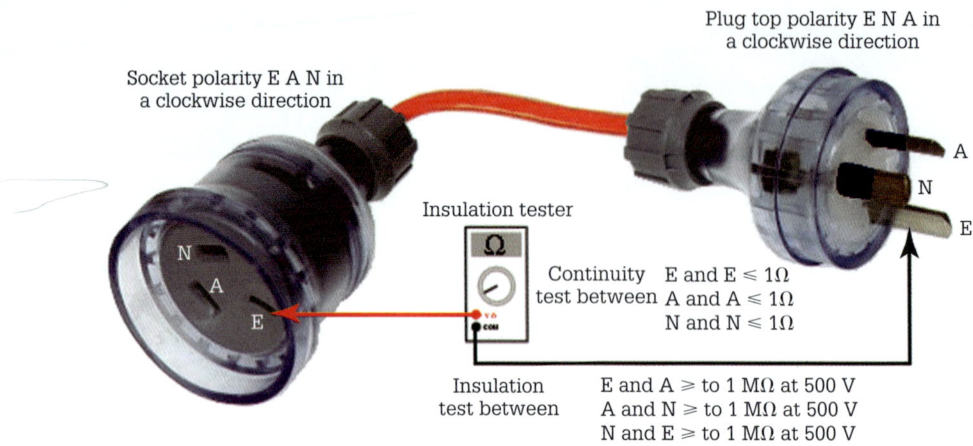

FIGURE 6.159 Polarity, continuity and insulation resistance tests

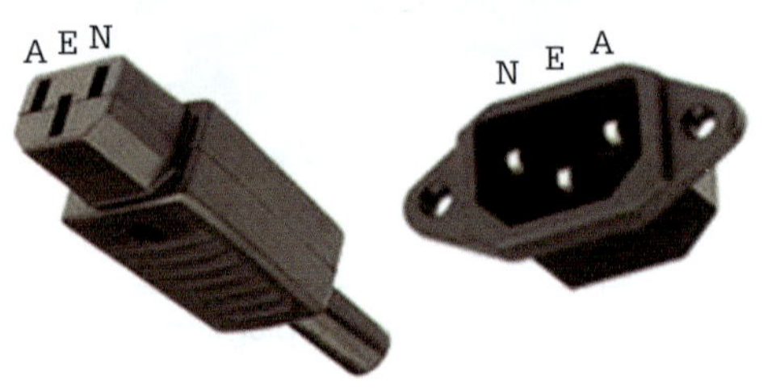

FIGURE 6.160 IEC plug and socket

Using an appliance tester

A portable appliance tester, like the 'Safe T Check' appliance tester shown in **Figure 6.161**, will put the appliance cord or extension lead through an automated series of insulation, earthing and continuity tests to assess whether the appliance cord or extension lead is fit for use.

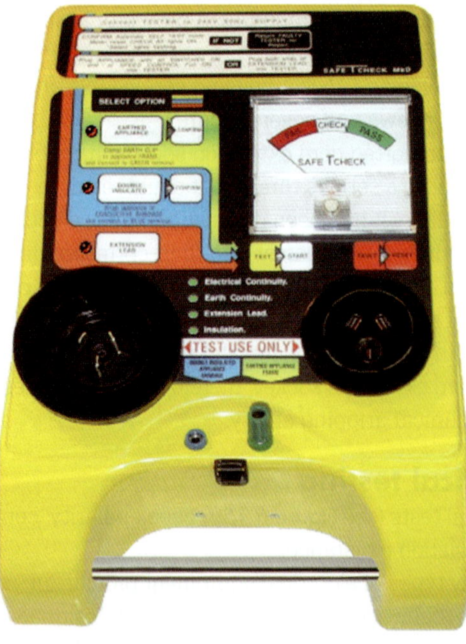

FIGURE 6.161 Safe T Check appliance tester

The procedure is as follows:

1. Plug the tester into a socket outlet and turn it on.
2. Ensure the tester is in the 'self-test' mode (the meter reads 'check').
3. Plug the ends of the extension lead into the plug and the socket provided on the front panel of the tester.
4. Select 'extension lead test' on the membrane pad. The tester now automatically sets up test parameters to test the extension lead.
5. Start tests by pressing 'start' on the membrane pad. Test results are indicated via meter readings and LED fault displays.

Testing installation wiring

After TPS has been installed and the second fixing completed, electrical testing must be carried out. Each circuit must be tested according to the AS/NZS *Wiring Rules*, Clause 8.3. Relevant clauses are listed in **Figure 6.162**. The relevant clauses explain testing requirements for earth resistance, insulation resistance, polarity, earth and neutral, short-circuits, interconnections, earth loop impedance, operation of RCDs and marking.

Clause
2.9.5
8.3.5
8.3.6
8.3.7
8.3.8
8.3.9
8.3.10

FIGURE 6.162 Testing TPS circuits

REVIEW QUESTIONS

1 What must suppliers of wiring systems provide installers?
2 Name the broad types of tests required to be undertaken on flexible cords and associated plugs and sockets.
3 What do you visibly check the plug top and extension socket for?
4 Name the electrical tests required to be undertaken on flexible cords and associated plugs and sockets.
5 What tests does a portable appliance tester perform on the appliance cord or extension lead?

CHAPTER REVIEW

6.1 Cable types and terminations

- Insulated conductors are identified by a colour code which indicates their intended purpose.
- Elastomers are synthetic compounds or rubber materials that provide a high degree of flexibility and elasticity.
- Polyvinyl chloride (PVC) is a general-purpose thermoplastic polymer used for wire insulation, cable insulation, sheaths and conduit.
- Polyethylene is a thermoplastic material having excellent electrical properties such as low and stable dielectric constant and a very high insulation resistance.
- Silicone is a thermosetting elastomer noted for its high heat resistance properties.
- Conductor materials include copper, copper alloys, aluminium, and aluminium alloys (stranded or solid) because of their high conductivity.
- Stranded conductor constructions enable a cable to be more flexible than a cable constructed of solid conductors.
- Screens when connected to earth keep electrical interference away from the signal conductors.
- The voltage rating of a given cable or insulated conductor is the peak voltage that may be continuously applied without degradation of the insulation causing a safety hazard. It is also called working voltage.
- The temperature rating of a cable or insulated conductor is the maximum temperature at which the insulating material may be safely maintained in continuous operation without deterioration of its basic properties.
- A sheath is a protective outer covering that can be applied to cables.
- Armour is a mechanical protection for the conductor's insulation.
- Serving or protective coverings can be applied over cable insulation for mechanical, chemical and electrical requirements or all three.
- Annealed copper conductors are used in stranded form to provide better flexibility than single or stranded hard-drawn copper conductors.
- Some flexible conductors have many fine wires specially made (< 0.31 mm diameter) to allow ease of movement.
- One method of mechanical termination consists of corrosion-resistant flat metal surfaces between which the conductor or conductors are clamped.
- Crimped terminators are called solderless connectors or lugs and are very common in appliances.
- Soldering is a process by which molten solder is used to create a metallic conductive path between metal parts and then allowed to cool.
- Infrared thermal imaging is an ideal method for inspecting terminations of critical electrical equipment because it reveals hot spots under actual load conditions.
- A join is the tying together of two conductors so that their union is mechanically and electrically sound.
- A tap is the connection of the end of the cable to some point along the run of the other cable.
- A splice is the interlacing of stranded conductors so that their union is mechanically and electrically sound.

6.2 Cords, cables and plugs

- Thermoplastic-insulated (TPI) cables are described as 450/750 V single-core annealed copper/PVC building wire.
- A cord consists of a high strand count of fine-gauge high-strength alloy conductors up to and including 4 mm^2 CSA with no conductor exceeding 0.31 mm in diameter and up to five cores.
- Cables with flexible conductors of less than 6 mm^2 CSA and with a core count greater than five cores are classified as flexible cables.
- A plug top is a device that may be engaged with a cord extension socket and is designed for the purpose of connecting to a socket outlet any electrical equipment to which the plug top is attached by means of a flexible cord or cable.
- A cord extension socket is a device connected to a flexible cord or cable with which a plug top may be engaged for the purpose of connecting to a socket outlet.

6.3 Thermoplastic sheathed (TPS) wiring systems

- Single-core annealed copper double-insulated TPI cables are also known as single double-insulated (SDI) 450/750 V grade cables.
- Thermoplastic-sheathed (TPS) cables, also known as tough plastic sheathed cables, are rated as 450/750 V flat TPS plain annealed copper cable.
- Where TPS cables are in accessible locations they must be supported by cable clips, cable ties or cable saddles depending on the installation method.
- When TPS cables are installed where thermal insulation will be placed they may need to be de-rated because the cables have a reduced ability to dissipate heat.

6.4 Circular TPS wiring systems

- Wiring systems refer to the various types of cables, such as insulated, sheathed, armoured, MIMS and fire resistant, and the different methods of installation that can be used.
- Wiring systems can be classified according to their capacity to preserve circuit reliability under fire conditions for a specified period and maintain circuit dependability under mechanical impact of a particular degree.
- Circular TPS cables are rated as 450/750 V circular TPS plain annealed copper cable.
- Due to their sheath, circular TPS cables can be installed unenclosed or installed in various enclosures such as conduit or buried directly in the ground if they are not subject to possible mechanical damage.
- Fixing devices for circular TPS cables on tray include cable ties, saddles and other clamping methods.

6.5 Non-metallic enclosures

- Wiring enclosures provide mechanical protection for cables and electrical safety to persons and property and provide a suitable and accessible enclosure for the conductors.
- Conduits should be neatly run and securely fastened by means of approved saddles.
- Heavy-duty conduit is primarily used for underground applications.
- Prior to excavation work, location of existing services needs to be determined.
- 'Proprietary' means identifiable by naming manufacturer, supplier, installer, trade name, brand name and catalogue or reference number.
- Materials used to lubricate cables while drawing in to conduits must be non-conductive, non-abrasive and non-hygroscopic.
- Circular conduit not exceeding 25 mm in diameter can be bent 'cold' by using a bending spring.
- Expansion joints connect sections of conduit and provide allowance for movement due to service load, shock or thermal expansion.
- Where conduits are run in roof spaces, they can be installed below the ceiling insulation and sarking.
- Light-duty oval conduit comes in a range of low-profile conduit specifically for use in chased walls or floors.
- Penetrations through damp-proof courses (DPCs) or the touching by conduit of damp courses should not occur.
- Where conduits are cast in concrete they should generally be installed in the centre of concrete slabs and above the first layer of reinforcing bars.
- A suggested maximum diameter of conduits in suspended slabs is 25 mm.
- Corrugated PVC conduit may be used as a service conduit (up to 1 m) to allow the cable to enter a cavity or meter box.
- Flexible PVC conduit should be used between fixed conduit and equipment likely to be moved or subject to vibration.
- A lubricant such as dry soap or talc placed on the insulation or blown into the conduit will assist the drawing of the cables.

6.6 Metallic enclosures

- Mechanical conduit benders make bending conduit easy.
- A hickey is an incremental bending device.
- With conduit work the 90° bend is probably the most common bend that the electrician will make.
- The gain is the distance saved by the arc of a 90° bend.
- A back-to-back bend consists of two 90° bends made on the same length of conduit.
- It is recommended that conduits be run directly from an exit point to the termination point with a minimum number of sets (the number of sets should not be greater than the equivalent of 360° or 4 × 90° bends).
- Where metal trunking is cut for installation purposes the cut surfaces should be protected with cold galvanising paint.
- Metal trunking (including joins) must be electrically continuous and must be earthed at one end.
- An in-floor cable trunking distribution system, when properly designed (total access or junction box access), is a good method for distributing power cables.
- Cable trays offer an efficient way to manage a wiring system while keeping it accessible, ordered and secure.
- Trays should be fixed to steel brackets and hangers with a deflection not exceeding 10 mm when fully loaded, with consideration to future installation of cables.
- Cable ladder consists of two folded steel or extruded structural-grade aluminium side rails with cable support rungs between the two rails.

- A ladder cable tray provides for the maximum free flow of air, dissipating heat produced in current-carrying conductors.

6.7 Fire protection cabling and systems

- The term 'fire resistant' means that the cable is capable of performing its intended functions under the heat and other conditions likely to be experienced due to a fire at its particular location for a defined length of time.
- Pyrolex™ Ceramifiable® cable has an insulation layer that hardens into a protective ceramic shield when exposed to fire.
- If a penetration requires a fire rating, the use of a fire-rated mortar, mastic, fire collar (a fire collar is a prefabricated penetration seal system for PVC conduits) or cable intumescent (to swell up as in heat) coating seal is required.
- MIMS cable consists of a sheath, a conductor and a powdered insulant.
- When MIMS cable is not properly sealed, moisture penetrates the insulation, thereby reducing the insulation resistance.
- MIMS cables going to an electric motor require a vibration loop.
- The passing of single-core MIMS cables through ferrous materials must be avoided or, where this is not possible, the manufacturer's instructions for the limitation of eddy currents should be followed.

6.8 Steel wire armoured cables

- Armour is a mechanical protection for the conductor's insulation.
- Serving or protective coverings can be applied over cable insulation for mechanical, chemical and electrical requirements or all three.
- Neutral screened cable is used for above ground applications and as underground consumer's mains or sub-mains in domestic or light industrial installations.

6.9 Trailing cables and catenary systems

- Trailing cables are used in applications where standard cables would experience high mechanical stress.
- Pendants bring electric power to a work station or work area without the use of extension leads.
- A catenary is an overhead wiring system (CWS) containing one or more cables suspended between two fixed points and hung on galvanised, stainless steel or compacted aluminium alloy support wires.

6.10 Industry standards and testing requirements

- Industry standards, such as the National Construction Code, and manufacturers' data play an important part in the selection and installation of wiring systems.
- The *Wiring Rules* notes that suppliers of wiring systems should provide installers with detailed fixing and support requirements to ensure requirements of AS/NZS 3013.
- The *Wiring Rules* requires that wiring systems are tested to ensure safety and proper functioning.
- Electrical tests for extension leads are polarity, continuity and insulation resistance tests.

TRIAL EXAM

For Chapter 6 knowledge assessment, please complete the following trial exam.

1 Aluminium earthing conductors less than 10 mm^2 must be:
  a soldered to their connection point
  b a solid conductor
  c a stranded conductor
  d bonded to a copper tail

2 Pigtails are used with aluminium conductors to overcome the problem of:
  a creep
  b resistivity
  c tensile strength
  d stepping

3 What type of cable is easier to draw-in and terminate?
  a MIMS
  b hard-drawn copper
  c solid
  d stranded

4 A round stranding with a minimum of two lays of the same-diameter wires and same lay length arranged to reduce the external diameter of single wires under a simple round strand is:
  a cable-lay
  b three strands
  c rope-lay stranding
  d unilay stranding

5 Name the termination that is regarded as a cold weld.
  a clamped terminations
  b compression crimping
  c barrel terminations
  d soldered terminations

6 An important factor in thermal transfer for soldering with lead-free solders is correct:
  a melting temperature of the solder
  b hardness of the soldering tip
  c correct soldering iron tip selection
  d constant movement of the metal being soldered

7 The recommended splice for connecting protective earths to the main earth conductor is:
a 'T' splice
b running splice
c tail splice
d 'Y' splice

8 Which of the following conductor colours is suitable as an alternative for the active conductor in fixed wiring?
a white
b light blue
c green
d black

9 A thermoplastic-insulated (TPI) cable is described as 450/750 V single-core annealed copper/PVC building wire. The voltage rating of the insulation to earth for this cable is:
a 415 V
b 450 V
c 230 V
d 750 V

10 A wiring system is rated as 'WS 32'. This means it can withstand:
a fire for 5 minutes and has a degree of impact of light
b fire for 60 minutes and has a degree of impact of moderate
c fire for 60 minutes and has a degree of impact of very heavy
d fire for 120 minutes and has a degree of impact of very heavy

11 What is the main type of wiring used for switchboards and control panels?
a armoured cable
b MIMS
c SDI cable
d TPI cable

12 A TPS cable reel is labelled '2C + E – 7/0.50 Red, Blk, G/Y – 450/750 V'. State the conductor size.
a 1.5 mm^2
b 2.5 mm^2
c 4.0 mm^2
d 6.0 mm^2

13 At what temperature are cables rated for current-carrying capacity calculation?
a 90 °C
b 75 °C
c 110 °C
d 55 °C

14 Name the simplest wiring system to use for surface wiring.
a TPI cable
b cables in conduit
c ducted wiring systems
d flat TPS cable

15 If glands are required then a suitable cable would be:
a circular TPS
b TPI
c flat TPS
d busbar

16 The wiring system that provides for the maximum free flow of air, dissipating heat produced in current-carrying conductors, is:
a PVC trunking
b cable ladder
c UPVC conduit
d steel conduit

17 How should conduit runs be installed?
a only through the floor slab
b with proprietary fittings only
c so as to avoid all other pipe systems and services
d within confined spaces

18 State the colour of heavy-duty UPVC conduit used for enclosing cables other than fire alarm cables.
a orange
b white
c grey
d red

19 If more than one conduit is placed in the same trench they should be spaced to a minimum of:
a 50 mm apart
b 20 mm apart
c 100 mm apart
d 10 mm apart

20 A suitable material used to lubricate cables while drawing in to conduits is:
a oil
b dry soap
c grease
d lard

21 State the type of conduit used for enclosing cables in high-security situations.
a flexible conduit
b metallic conduit
c transparent conduit
d oval conduit

22 Which of the following conduits is specifically for use in chased walls or floors?
a heavy-duty conduit
b ABS conduit
c flexible conduit
d light-duty oval conduit

23 Where conduits pass through external walls or existing ground floor slabs it is suggested that a penetration:
a 10 times greater than the conduit diameter is provided
b 5 mm larger than the conduit diameter is used
c twice the conduit diameter is used
d five times greater than the conduit diameter is provided

24 'Setting' conduit means:
a fixing the conduit to a surface
b bending the conduit
c cutting the conduit at an angle
d curing the conduit-fixing cement

25 When working with conduit what is meant by the term 'gain'?
 a the take-up distance
 b the offset when bending
 c the distance saved by the arc of a 90° bend
 d the distance between bends

26 The number of sets in a conduit run from an exit point to the termination point should not be greater than the equivalent of:
 a greater than 360° or 6 × 90° bends
 b 90° or 1 × 90° bends
 c 180° or 2 × 90° bends
 d 360° or 4 × 90° bends

27 If choosing ducts for a wiring system, how much spare space should be allowed for future cable additions?
 a 50%
 b 10%
 c 25%
 d 30%

28 What type of cable does not burn or support combustion?
 a TPI
 b TPS
 c round armoured
 d MIMS

29 What type of cable can provide protection from electric shock?
 a neutral screened cable
 b armoured cable
 c TPS cable
 d TPI cable

30 Use Table C10 of AS/NZS 3000:2018 to determine the maximum number of 1.5 mm^2 single-core V90 sheathed cables that can be safely installed in a 20 mm medium duty UPVC conduit.
 a 3
 b 4
 c 5
 d 6

7 Document and apply measures to control WHS risks

This chapter provides electrotechnology workers with essential knowledge and skills in applying control measures for dealing with non-electrical hazards and low voltage, extra-low voltage and high-current hazards. In addition, electrotechnology workers will gain knowledge of how to perform efficient and safe isolation and testing and tagging procedures. This chapter provides underpinning knowledge for the unit UEECD0016 from the UEE training package.

LEARNING OBJECTIVES

Risk management and assessment

- Outline the principle and purpose of risk management.
- Explain the process of conducting a risk assessment.
- Outline methods used for hazard identification.
- Show appropriate methods of recording hazards and assessing risk.

Recognising and assigning a level of risk

- Explain how a risk matrix is used to identify risk level.
- Identify the likelihood of an incident occurring.

Identifying risk control measures

- Explain the hierarchy of control.
- Outline what constitutes a reasonable control measure.
- Describe monitoring and review processes.

Documenting control measures

- Explain the purpose of job safety analysis, safe work method statements, and risk registers.
- List relevant industry standards applicable to work health and safety.

Construction site hazards

- Detail common construction site hazards and appropriate control measures.

Hazards associated with high voltage

- Outline hazards of working near high voltage plant.
- Explain the meaning of the terms touch voltage, step potential, induced voltage, and stored energy.
- Outline the requirements for safety services.

Hazards associated with extra-low voltage and low voltage

- Outline hazards of working on or near extra-low voltage and low voltage equipment and circuits.
- List protection measures applicable to low voltage installations.

Disconnecting and reconnecting electrical equipment

- Explain the isolation and tagging procedure for electrical work.
- Explain the hazards associated with alternate supplies.

7.1 Risk management and assessment

All electrical work has the capacity to cause harm. It is important, therefore, to have an understanding of the risks involved in this work and of suitable control measures. The greatest dangers associated with electrical work are electric shock, arc flashes and arc blasts, which can attain temperatures of 19 000 °C.

Principle and purpose

Risk management is an obligation to health and safety from all persons engaged in an electrical business. Risk management is a consultative process that embeds an effective risk management culture throughout the enterprise. Risk is inherent in all electrical tasks and a competent worker and employer should be able to discern predictable risks and their consequences. Once identified, steps must be taken to remove or limit the dangers that affect health and safety.

For electricians, risk management must be implemented in the field; therefore, the process of conducting a risk assessment must be tailored to the particular task. The crucial thing you need to decide on is whether a hazard is important and whether you have it contained by adequate safeguards that minimise the risk.

Figure 7.1 shows the specific elements of the risk management process.

Source: Shutterstock.com/dashadima

FIGURE 7.1 Risk management process

Responsibilities

The people who have a role in managing work health and safety (WHS) risks include:

- persons conducting a business or undertaking, who are referred to as PCBUs
- those who design, manufacture, import, supply and install plant, substances or structures
- officers.

Workers and other persons at the workplace also have duties under the WHS Act. These responsibilities include taking reasonable care for their own health and safety at the workplace.

A PCBU has a requirement to eliminate risks in the workplace. If it is not possible to eliminate these risks then the risks need to be minimised so far as is reasonably practicable. PCBUs have a duty to consult workers about work health and safety.

Company directors and those who hold similar positions are referred to as 'officers'. Officers have a duty to exercise due diligence to ensure the PCBU complies with the WHS Act and WHS Regulations. To this end, they must take reasonable steps to gain an understanding of the hazards and risks associated with the operations of the business or undertaking. They must also ensure that the business or undertaking possesses and uses appropriate resources and processes to eliminate or minimise risks to health and safety.

Workers have a duty in taking reasonable care for their own health and safety and to not adversely affect the health and safety of other people in the workplace. Workers need to conform with any reasonable instructions, as far as they are reasonably able, and to cooperate with reasonable health and safety policies or procedures advised to them. If the business or undertaking provides workers with personal protective equipment (PPE), the workers must so far as they are reasonably able, use or wear this PPE in accordance with the information and training provided.

Conducting risk assessments

It is important to have a clear understanding of what comprises risks and hazards.

SWITCH ON

- **Risk** is defined as the possibility that harm (death, injury or illness) may occur when exposed to a hazard.
- **Hazard** is defined as a situation or thing that has the potential to harm a person.

Figure 7.2 provides an example of the relationship between a hazard and the associated risk.

Hazard	Risk	Contributing factors
Electricity	Electric shock from contacting live terminals	Poor maintenance Insulation failure Not wearing PPE

FIGURE 7.2 Hazard and associated risk example

The first step in the risk management process is identifying potential hazards involved with the work. This involves finding things and situations that could potentially cause harm to people. Hazards generally arise from the various aspects of the work and their interaction, which may include the:

- physical work environment
- equipment, materials and substances used
- performance of work tasks
- design and management of the work.

It is possible to identify hazards by examining the workplace and how work is performed. Workers, manufacturers, suppliers and health and safety specialists are excellent sources of information as well as relevant records and incident reports. **Table 7.1** lists some common types of workplace hazards.

TABLE 7.1 Examples of common hazards

Hazard	Example	Potential harm
Biological	Microorganisms	Hepatitis, Legionnaires' disease, Q fever, HIV/AIDS or allergies
Electricity	Contact with live electrical conductors	Electric shock, burns, damage to organs and nerves leading to permanent injuries or death
Extreme temperatures	Heat and cold	Heat can cause burns and heat stroke or injuries due to fatigue, cold can cause hypothermia or frost bite
Gravity	Falling objects People slipping, tripping or falling	Fractures, bruises, lacerations, dislocations, concussion, permanent injuries or death
Hazardous chemicals	Acids, hydrocarbons, heavy metals, asbestos and silica	Respiratory illnesses, cancers or dermatitis
Machinery and equipment	Being caught in moving parts of machinery or struck by moving vehicles	Fractures, bruises, lacerations, dislocations, permanent injuries or death
Manual tasks	Sustained or awkward postures, high or sudden force, repetitive movements or vibration	Musculoskeletal disorders such as damage to joints, ligaments and muscles
Noise	Exposure to loud noise	Permanent hearing damage
Psychosocial	Bullying, violence and work-related fatigue	Psychological or physical injury or illness
Radiation	Sunlight, welding arc flashes, micro waves and lasers	Burns, cancer or blindness

After identifying hazards, it is necessary to consider how the potential harm can be lessened. **Figure 7.3** summarises some possible actions.

Source: Shutterstock.com/NicoElNino

FIGURE 7.3 Risk assessment considerations

Electrical hazards

Hazards associated with electrical equipment or installations may arise from:

- the age of electrical equipment and electrical installations
- changing or modifying the design of equipment or installation
- designing, constructing, installing, maintaining and testing of electrical equipment or electrical installations
- electrical equipment operated in hazardous areas
- inadequate or malfunctioning electrical protection
- location of electrical equipment
- portable electrical equipment operated in harsh environments
- work undertaken on or near electrical equipment or electrical installations, such as electric overhead lines or underground electric services.

Electromagnetic radiation may also present a potential hazard for workers with some medical conditions; for example, pacemakers.

Hazard identification

Hazard identification means looking for those things that have the potential to cause harm. Harm may be for the short term or the long term, and can have a negative impact on the health and safety of personnel, property and the environment.

Some methods to assist in identifying workplace hazards include:

- **Inspecting the worksite** – Conduct a walk-through of the worksite, being observant of the work environment and work processes. During the walk-through, document any general housekeeping concerns. Be mindful of unseen hazards and those that could affect workers' health over an extended period of time, such as workplace bullying or shift work resulting in stress or fatigue.

- **Having a conversation with those doing the work** – Allow workers to voice their health and safety concerns, near misses or unreported incidents. Confidential surveys are useful tools to identify problems that are less obvious, such as harassment and workplace bullying.
- **Reviewing available information** – Consider information from a range of sources to identify other hazards. Sources can include:
  - enterprise-specific information from local records, which may include any documented incidents, sick leave or worker complaints
  - information from industry regulators, industry associations, trade unions and technical specialists
  - instructions and datasheets for plant, process and chemicals provided by manufacturers and suppliers, such as safety data sheets (SDSs)
  - workers' compensation data for the industry.

As **Figure 7.4** shows, some hazards are not so obvious. This is why it is prudent to involve others in identifying hazards.

Source: Shutterstock.com/Antonio Guillem

FIGURE 7.4 Some hazards are not so obvious

SWITCH ON

The Electrical Regulatory Authorities Council (ERAC) publishes annual data on electrical fatalities and accidents. Visit their website at https://www.erac.gov.au for the latest reports. Fortunately, the 20-year trend is continuing downward, but in the 2019–2020 financial year there were eight fatalities in Australia. This highlights the importance of undertaking a thorough risk assessment prior to commencing any electrical work.

Recording hazards

The benefits of maintaining good records include:

- assisting in targeting training for workers, managers and supervisors for the significant workplace hazards
- demonstrating to stakeholders that WHS obligations are being met as well as complying with legal requirements
- providing a basis for decisions relating to WHS and risk assessments at a later date
- providing evidence as to how and when decisions related to safety issues were made
- serving as a basis for review of risks subsequent to any changes in legislation or business activities.

The size of the enterprise and the potential for major safety issues will largely dictate the detail required in and the extent of the records. Nevertheless, it is essential to maintain information pertaining to:

- identified hazards, assessed risks and control measures implemented, which will include hazard checklists and forms, worksheets and assessment tools
- hazards, incidents, near misses and injuries advised by workers
- details of how and when nominated control measures were implemented, monitored and reviewed
- persons consulted
- records of staff training, currency of the training and refresher requirements
- proposed changes in the workplace or activities.

Legal requirements

The Work Health and Safety Regulations require specific record-keeping requirements (**Figure 7.5**). At a minimum this includes:

- risk assessments and safe work method statements (SWMS)
- incident reports
- health monitoring results
- inspections and modifications to registered plant
- training and licensing records.

It is essential that PCBUs are aware of and comply with all legal requirements and ensure that all WHS records are accessible and available when required.

It is important that all key safety documents are easily accessible to everyone at the workplace. For auditing purposes, it is ideal to have a central collection or single manual containing pertinent documentation. An online documentary repository is ideal.

The following documentation should be available:

- WHS policies displayed on noticeboards
- a register of the chemicals used on the worksite, which is located in their main storage area together with the SDSs
- safe work procedures located near each piece of equipment.

Source: Shutterstock.com/Nattawit Khomsanit

FIGURE 7.5 Recording hazards

Currency of documents

It is important to keep documentation up to date and remove out-of-date safety policies, plans and procedures from circulation. Safety documentation should be reviewed annually or earlier if:

- the current documents do not effectively manage the risk
- new hazards or risks are identified
- changes to the work environment, business or key personnel occur
- new equipment or chemicals are introduced into the workplace
- consultation indicates a need to review the documentation
- workers request a review of the documentation
- an incident occurs.

Do not dispose of or overwrite existing documents, as they may be required for legal or 'knowledge preservation' purposes. It is important, therefore, to ensure that there is no confusion, and no one can use out-of-date documents by:

- clearly marking them as obsolete with wording such as 'replaced' with a date
- relocating printed and online documents to a physical or online archive area as appropriate.

All staff need to be advised that there are new versions of documents that should now be used. Remove and replace any WHS documentation on noticeboards and the like.

REVIEW QUESTIONS

1 What type of process is risk management?
2 List the four steps in the risk management process.
3 What work health and safety obligations are placed on workers?
4 Define the term 'risk'.
5 What is the first step in the risk management process?
6 Explain what is meant by 'hazard identification'.
7 How frequently should safety documentation be reviewed?
8 Why should existing safety documents not be disposed of or overwritten?

7.2 Recognising and assigning a level of risk

A risk assessment generally focuses on identifying a level of risk, which is based on considering the consequences of the risk, given the existing control measures, and the likelihood of those consequences occurring.

A risk assessment that is based on consequence and likelihood events enumerates levels of risk. The risks are then ranked according to the level of risk, from high to low. This provides a means for identifying which risks require attention in an order of priority. **Figure 7.6** shows a commonly used risk assessment matrix for prioritising risks based on likelihood and consequences.

Risk level matrix

Risk assessment matrices, as shown in **Figure 7.6**, are common tools used by a variety of industries and professions for evaluating risk. The purpose of a risk assessment matrix is to assist in determining the likelihood and consequences of various health and safety hazards and risks occurring. The matrix is usually represented in a table or graph format. This risk assessment matrix offers four ratings:

- **very low likelihood + very low consequence** – risks identified in this area are either effectively controlled or are not prevalent enough to warrant concern
- **low likelihood + low consequence** – risks identified in this area are either satisfactorily controlled or are not widespread enough to warrant concern
- **moderate likelihood + moderate consequence** – risks identified in this area are usually controllable within reason and represent the absolute maximum level of acceptable risk in the workplace but should have appropriate control measures implemented
- **high likelihood + high consequence** – risks identified in this area require robust control measures to reduce the risk to reasonably practicable health and safety standards.

The matrix of **Figure 7.6** lists various levels of consequence and likelihood to attain a ranking level of the total risk, which appears in the matrix as a number between 1 (very low) to 25 (very high).

Likelihood of incident

The likelihood of an incident occurring is based on probability and so it is not possible to say with certainty when an incident may occur, only the probability of an incident occurring. If it is determined that the probability of an incident occurring is high, then some action must be undertaken to reduce the risk. On the other hand, if it is determined that the probability of an incident occurring is

Risk assessment matrix					
Likelihood / **Most likely consequence**	Almost certain to occur	Good chance of occurring	Likely to occur	Unlikely to occur	Extremely unlikely to occur
	Consequence expected to occur on a weekly basis or more frequently	Consequence expected to occur more than once in 3 months but less than once a week	Consequence expected to occur more than once a year but less than once in 3 months	Consequence expected to occur more than once in 3 years but less than once a year	Consequence has not occurred and is expected to occur less than once in 3 years
Disastrous fatality or extensive damage	**High 25**	**High 24**	**High 22**	**Moderate 19**	**Moderate 15**
Critical amputation or major damage	**High 23**	**High 21**	**Moderate 18**	**Moderate 14**	**Moderate 13**
Serious more than 1 week off normal duties or serious damage	**High 20**	**Moderate 17**	**Low 12**	**Low 9**	**Very Low 6**
Significant less than 1 week off normal duties or negligible damage	**Moderate 16**	**Low 11**	**Low 8**	**Very Low 5**	**Very Low 3**
Minor first aid injury or no damage	**Low 10**	**Low 7**	**Very Low 4**	**Very Low 2**	**Very Low 1**

FIGURE 7.6 Commonly used risk matrix

low, it may be decided to ensure that the risk remains low through administrative controls. Estimating the likelihood of an incident occurring is a primary source of uncertainty in the risk analysis.

The following examples present common work activities.

EXAMPLE 7.1

Consider gaining access to a worksite.

There is the possibility that uneven surfaces may cause a worker to slip and fall, or materials stored on the ground may cause a worker to trip. These are representative of general hazards rather than high-risk job procedures.

The likelihood of one of these events occurring is variable and depends largely on the condition of the site and general awareness of the worker. The consequences of these events again are variable and could be worse if the worker were to be carrying tools or materials.

Total risk is a function of both likelihood and consequence. As such, while an event may be rare it can still carry moderate consequences such as injury requiring medical attention.

- This activity would have a low-risk rating.

EXAMPLE 7.2

Consider working at a significant height, such as from a mobile platform.

There is a possibility that a worker could fall from the mobile platform.

This event is seen as highly likely to occur if no mitigations are implemented. As such, the task itself is likely to lead to a significant injury or death, unless appropriate safety procedures are implemented. It is therefore essential to carefully mitigate these hazards and review the control measures in order to keep workers safe.

The consequences of this event are severe to both the worker involved and the enterprise. The severity of a death at work should not be understated.

- This activity would have a high-risk rating.

REVIEW QUESTIONS

1. What is the purpose of a risk assessment matrix?
2. Why are risks ranked?
3. What needs to happen if it is determined that the probability of an incident occurring is high?
4. What is a primary source of uncertainty in the risk analysis?

7.3 Identifying risk control measures

A common method used for identifying risk control measures is through the hierarchy of control.

Hierarchy of control

Figure 7.7 shows a simplified version of the hierarchy of control to that shown in Chapter 2. The hierarchy of control is a system for controlling workplace risks that utilises a step-by-step approach to eliminating or reducing risks. It ranks risk controls from the highest level of protection and reliability through to the lowest and least reliable protection.

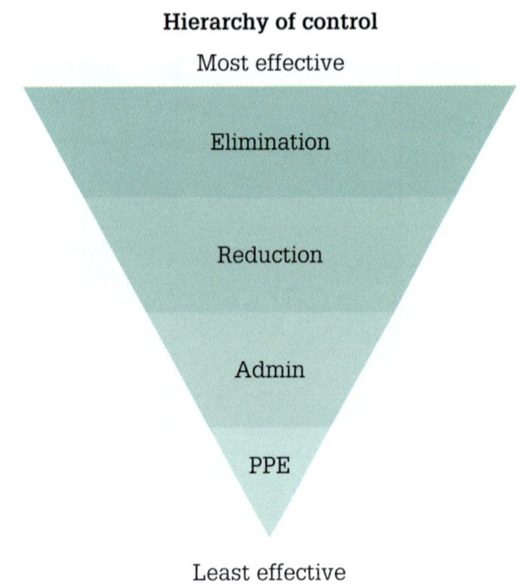

FIGURE 7.7 Simplified hierarchy of control

Control measures

The highest level of control in the hierarchy is totally eliminating the hazard and risk, which is followed by reducing the risk through substitution, isolation and/or engineering controls, then reducing the risk through administrative controls. The lowest level of control involves reducing the risk through the use of PPE.

Eliminating the risk

Eliminating the hazard and its associated risk is the most effective control measure. The most appropriate way to eliminate a hazard is by not introducing the hazard in the first place. For example, undertaking work at ground level eliminates the risk of falling from a height. Hazard elimination is possibly more practical at the design or planning stage where there is more scope to design a product, process or work environment that eliminates the hazard. Other examples of eliminating hazards are disposing of unwanted chemicals rather than storing them onsite and not storing products on the floor where they present a trip hazard.

It may not always be possible to eliminate a hazard. In this case it is imperative to eliminate, as far as possible, risks associated with the hazard.

Risk reduction

Risk reduction is achieved through substitution, isolation or engineering controls of the hazard. If it is not possible to eliminate risks associated with the hazard, then the risks should be minimised by:

1. Substituting the hazard with something safer; for example:
   - using battery tools rather than mains-powered tools
   - using mild detergent and hot water instead of chemical cleaners
   - using water-based paints in place of solvent-based paints
2. Isolating the hazard; for example:
   - installing guard rails around access holes
   - using barriers to isolate pedestrians from moving vehicles
   - using remote controls to operate machinery
3. Controlling the hazard through engineering means, such as a mechanical device or process; examples of engineering controls include:
   - placing guards around moving parts of machinery
   - using mechanical devices such as trolleys or hoists to move heavy loads.

Administrative controls

Administrative controls are work methods or procedures intended to reduce exposure to a hazard. Examples of administrative controls include:

- having documented procedures for safe operation of a particular machine
- implementing appropriate safety signs to warn workers of a hazard
- limiting the time a worker is exposed to a hazardous task.

Personal protective equipment (PPE)

The use of PPE represents the lowest level of control and includes anything workers use or wear to minimise risks posed to their health and safety. PPE may include:

- body protection such as aprons and body suits
- eye protection such as face shields and goggles
- hand protection such as gloves
- head protection such as hard hats
- hearing protection such as ear muffs and earplugs
- high-visibility clothing
- respiratory protection such as face masks and respirators
- safety footwear
- safety harnesses
- sunscreen.

Monitoring and reviewing

Safety measures should be monitored and reviewed regularly. This is particularly important with PPE, which limits exposure to the harmful effects of a hazard but only if workers wear and use it correctly.

Hazards are not controlled at the source through the use of administrative controls and PPE risk reduction. Administrative controls and PPE tend to be least effective measures in minimising risks as they rely on human behaviour and supervision.

It is recommended to only use administrative controls and PPE:

- as a last resort when no other practical control measure is available
- as a temporary measure while instigating a more effective risk control
- to increase the efficacy of higher-level control measures.

It is important to consider a range of risk control measures and select those that most effectively eliminate the hazard. If it is not reasonably practicable to eliminate the hazard, then it is necessary to minimise the risk. Risk reduction may be achieved through a single control measure or through a combination of various control measures that combine to provide the highest level of reasonably practicable risk reduction.

The choice of appropriate control measures should involve consultation with workers and their health and safety representatives (HSRs).

The use of checklists, like that shown in **Figure 7.8**, is an invaluable tool for the ongoing monitoring and review of implemented safety control measures.

Electrical safety checklist

Use this checklist to help identify electrical hazards in your workplace.

Electrical hazards	Yes	No	N/A
Are electrical leads, plugs, sockets and switches in good condition? (Check for fraying and nicks)	☐	☐	☐
Are walkways and other floor areas free from electrical leads lying across floors?	☐	☐	☐
Have you ensured there are no double adaptors used and power boards have over-current protection?	☐	☐	☐
Are electrical leads and power boards tagged where necessary and within date?	☐	☐	☐
Is the location of powerlines and cables (overhead / underground / behind walls) checked before digging, drilling, using cranes or ladders, or erecting scaffolding? For information on underground infrastructure, contact DialBeforeYouDig nationally on 1100 or https://www.byda.com.au/.	☐	☐	☐
Are residual current devices protecting portable electrical equipment?	☐	☐	☐

FIGURE 7.8 Safety checklist

REVIEW QUESTIONS

1. What system is the hierarchy of control?
2. What is the most effective risk control measure?
3. It may not always be possible to eliminate a hazard. What should occur in this case?
4. How is risk reduction achieved?
5. What are administrative controls?
6. Which control measure represents the lowest level of control?
7. Why do administrative controls and PPE tend to be measures least effective in minimising risks?
8. Which item forms an invaluable tool for the ongoing monitoring and review of implemented safety control measures?

7.4 Documenting control measures

Some useful tools to assist in documenting control measures are:

- a job safety analysis
- safe work method statements
- risk registers.

Job safety analysis

A job safety analysis (JSA) is a formalised procedure, which is usually documented on a JSA form, and details the steps necessary to complete a specific job or task. It identifies potential incidents and hazards that could arise while undertaking the job or task.

The first step in writing a JSA is to document the steps needed to complete the work. This is a procedural process that lists all the individual steps from beginning to end. A JSA is a step-by-step process that anyone should be able to follow to complete the task. As such, it is important to use clear and concise language.

It takes time and effort to break a task down into its component steps, but it is a critical process to enable an analysis of the overall risk associated with the task. Another vital step is to consult those who perform the particular task to review the documented steps to ensure that all steps are captured and listed in the correct order. **Figure 7.9** summarises the steps involved in undertaking a job safety analysis.

FIGURE 7.9 Steps involved in job safety analysis (JSA)

Determining hazards and risks

The next step after documenting the steps involved in the task is to identify any hazards that may be present at each step of the task. When identifying risks it is wise to consider the entire work environment in order to determine any possible hazards that may exist while completing the task.

After identifying the hazards, it is necessary to assess the risk of injury presented by each identified hazard. Using a risk matrix can aid in assessing the probability and severity of the hazard to determine an overall risk rating. It is important to use the enterprise's Risk Ranking Matrix as it may differ from that presented here. It is essential to determine the risk level presented by the hazard before applying any control measures. This means that each step in the JSA will reflect the initial risk rating.

Identify control measures

After determining the risk rating, the next step is to identify the practical control measures that can be implemented to avoid potential workplace injuries posed by the hazard. At this point it is prudent to refer to the hierarchy of control (see **Figure 7.7**) and implement the highest level of control that is practicable to enable the performance of the job task.

It is not uncommon to implement a range of control measures. While the use of PPE is considered a last resort when all other measures are impractical, it is often implemented in conjunction with other higher-level control measures.

Determine the residual risk

After selecting appropriate control measures, it is possible to determine the level of residual risk, which is the risk ranking with the control measures implemented. Once again refer to the hierarchy of control (see **Figure 7.7**), which should indicate a lower probability and severity rating. If this is not the case it is essential to determine how the task can be performed more safely through the introduction of additional control measures.

Figure 7.10 illustrates the structure of a JSA.

Communicate job safety analysis

Before finalising a JSA it is important to have the document checked by workers and supervisors to ensure that all critical aspects of the job are covered and appropriate controls are in place to make the job safe. Honest feedback about the selected control measures is critical to ensure that they are practicable and do not introduce additional hazards.

JSAs need to be living documents that capture information about risks, document controls and inform workers of hazards in their job tasks and the safest system of work. A review of the relevant JSA should happen following a workplace injury to gauge whether the JSA had not adequately addressed the hazards and appropriate control measures. Regular reviews of JSAs is essential to effective risk management and to offer insight into the hazards workers face in their jobs and tasks they perform.

Maintaining a good JSA program is an ongoing and evolving process. As noted, if a workplace injury occurs, a review of the relevant JSA is required to see if it had a shortcoming that may have contributed to the incident.

Having up-to-date JSAs that are reviewed regularly by all those involved goes a long way to creating a more inclusive safety culture.

<table>
<tr><th colspan="6">JSA Details</th></tr>
<tr><td>Work activity description:</td><td colspan="2"></td><td>Location and Date</td><td colspan="2"></td></tr>
<tr><td>Name of business undertaking the activity:</td><td colspan="2"></td><td colspan="3" rowspan="4">JSA authorised by:
Name: ____________
Position ____________
Signature: ____________
Date: ____________
Project: ____________</td></tr>
<tr><td>Required plant and equipment:</td><td colspan="2"></td></tr>
<tr><td>Maintenance checks required (for example testing and tagging of electrical equipment):</td><td colspan="2"></td></tr>
<tr><td>Required hand tools:</td><td colspan="2"></td></tr>
<tr><td>Required materials:</td><td colspan="5"></td></tr>
<tr><td>Required PPE:</td><td colspan="5"></td></tr>
<tr><td>Certificates, permits and/approvals required:</td><td colspan="5"></td></tr>
<tr><td>Relevant legislation, codes, standards, SDSs and the like applicable to this activity</td><td colspan="5"></td></tr>
<tr><td colspan="6">JSA – Action steps</td></tr>
</table>

Step	Job step details (steps involved in the job)	Potential hazards	How to control risks (Hierarchy of Control) 1. Elimate, e.g.: eliminate task, remove hazard 2. Substitute e.g.: replace with less hazardous process 3. Isolate e.g.: enclosures, restricted access 4. Engineering e.g.: guarding, separation, redesign 5. Administrative e.g.: SWP, training, schedule 6. Personal Protective Equipment (PPE) e.g.: gloves	Risk Level (from Risk Matrix)	Name of Person(s) responsible for implementing controls

This job safety analysis has been developed through consultation with our workers and has been read, understood and signed by all workers completing the works:

Print Names:	Position:	Signature:	Dates:

Review No	01	02	03	04	05	06	07	08
Initial:								
Date:								

FIGURE 7.10 Example of job safety analysis (JSA) structure

Safe work method statements

A safe work method statement (SWMS) is a document used within the construction industry in Australia. Australian health and safety regulations require an SWMS to be created for high-risk construction work (HRCW). HRCW comprises a number of different types of work, such as working-at-height, trench work, and working near live electrical equipment.

The term SWMS is used exclusively in construction and so may be referred to by other industries as safe operating procedure (SOPs) and safe work instructions (SWIs). Regardless of the name, they form a documented set of directions for a task or activity, which provide workers with the agreed safest and most efficient way of performing a job or task. The effectiveness of the SWMS relies on the worker following the requirements as set out in the document.

Typically, an SWMS will have the following sections:

- Introduction, which summarises the purpose of the particular SWMS
- Scope, which details to whom the SWMS is applicable
- References, which provides source documents used in the preparation of the SWMS

SWMS Details			
PCBU Details: [Name, ABN, Office address, Phone]		Principal contrator (PC): [Name, ABN, Office address]	
Work activity: [Description of job]		Work location:	
		Manager:	
High risk construction work: [list work from WHS Regulations]		Contact phone:	
Workers consulted about this SWMS?			
Person responsible for ensuring compliance with SWMS:		Date SWMS provided to PC:	
Person(s) responsible for reviewing the SWMS:		Date SWMS revised:	
Date SWMS received:		Signature:	
Worker name:		Date received:	
Workers signature:			

Task Details		
List the Tasks	List hazards and risks	List the control measures
	Identify the hazards and risks that may cause harm to workers or the public.	Describe what will be done to control the risk. What will you do to make the activity as safe as possible?

FIGURE 7.11 Example of safe work method statement (SWMS) structure

- Definitions, which provides a list of words commonly used throughout the SWMS and the context in which they are written
- Details of the procedure, which lists what is required for the safe and efficient completion of the task
- Responsible persons, which lists the people at the worksite who are responsible for implementing and maintaining the SWMS
- Reviewers, which lists the people at the worksite who are responsible for reviewing and authorising the SWMS.

SWMSs are typically developed on completion of a risk assessment and consider the work environment in which the task will be undertaken. Like JSAs, an SWMS needs to be reviewed at regular intervals and updated as necessary. **Figure 7.11** illustrates the structure of an SWMS.

Source: Shutterstock.com/docstockmedia

FIGURE 7.12 The risk register

Risk registers

A risk register is a record of all the identified risks, the likelihood and consequences of a risk occurring, the actions implemented to reduce those risks, and who is responsible for managing these risks. It may be a single document, a paper-based file as shown in **Figure 7.12**, or an electronic document. Placing the risk register on an internal intranet allows workers access to the information and enables effective reviews of the documents following incidents or changes in legislation.

The purpose of a risk register is to compile information pertaining to workplace risks in a single location which is easily accessible. Another advantage of a risk register is that information is presented in a clear and consistent manner that makes it easy for people to understand and therefore provide feedback.

A risk register review can assist in implementing improvements in design and work practices in order to achieve the desired health and safety outcomes of the enterprise.

Industry standards

Where required under WHS laws, a duty holder must comply with the relevant Australian Standard to ensure that they satisfy their legal obligations. It is important to understand that Standards themselves are not laws and there is no general requirement to conform to a Standard. There is, however, a requirement to comply with WHS laws, which may reference particular Standards. Where WHS laws require compliance with specific Standards, then a failure to do so may constitute a breach of the WHS laws.

Standards provide guidance on specific matters and as such a Standard may be considered in a court of law when determining compliance with the WHS laws. In this case a court may rule that conforming to a Standard is relevant and that it was reasonably practicable for a person to conform to the Standard in the given circumstances.

Figure 7.13 illustrates the role Standards play in relation to WHS regulations.

Table 7.2 lists some Standards applicable to work health and safety.

FIGURE 7.13 The role of Standards

TABLE 7.2 Australian Standards pertaining to work health and safety

Standard	Application
AS/NZS 1269.1:2005 Occupational noise management – Measurement and assessment of noise emission and exposure	Covers the requirements for, and provides guidance on, the types of occupational noise and procedures for performing noise assessments.
AS/NZS 1891.1:2020 Industrial fall-arrest systems – Harnesses and ancillary equipment	Covers the requirements for the materials, design, manufacture, testing and marking of full-body, combination, and lower-body harnesses that are intended for working at height.
AS 2397:2015 Safe use of lasers in the building and construction industry	Covers the safety requirements for the use of lasers in the building and construction industry.
AS 2832.1:2015 Cathodic protection of metals – Pipes and cables	Covers the requirements for the cathodic protection of buried or submerged metallic pipes and cables.
AS/NZS 3012:2019 Electrical installations – Construction and demolition sites	Establishes the minimum requirements for the design, construction and testing of electrical installations located on construction and demolition sites. It also covers the in-service testing of portable, transportable and fixed electrical equipment used on construction and demolition sites.
AS 4801:2001 Occupational health and safety management systems	Establishes the requirements for an occupational health and safety management system, which may be used for auditing and certification purposes.
AS ISO 31000:2018 Risk management – Guidelines	Provides generic guidelines for organisations on risk management.
SA/SNZ HB 89:2013 Risk management – Guidelines on risk assessment techniques	Provides general guidance to assist in the selection and application of procedures that may be applicable to various risk assessment activities.
HB 158:2010 Delivering assurance based on ISO 31000:2009 – Risk management – Principles and guidelines	Provides guidance for internal auditors and other assurance providers on using the risk management process to develop a risk-based assurance strategy and program, plan an assurance engagement, report the assurance program, and design suitable controls.
HB 327:2010 Communicating and consulting about risk	Provides guidance on establishing risk management communication and consultation.
AS 1885.1–1990 Measurement of occupational health and safety performance	Covers suitable strategies for the recording of workplace injury and disease.

International Standards

Australian Standards reflect Australian workplaces and generally contain information that is more relevant for Australian operating conditions. An international Standard may be considered in aggregation with the equivalent Australian Standard. International Standards may offer additional useful information, which may afford a better overall level of safety compared to the Australian Standard equivalent.

REVIEW QUESTIONS

1 Describe the job safety analysis (JSA).
2 What is the first step in writing a JSA?
3 Define 'residual risk'.
4 Why is it important to have the JSA document checked by workers and supervisors before finalising it?
5 What is a safe work method statement (SWMS)?
6 What is a risk register?
7 Outline the purpose of a risk register.
8 What is the likely consequence of failing to comply with specific Standards where WHS laws require compliance?

7.5 Construction site hazards

A hazard can be work practices, procedures or anything that has the potential to harm the health or safety of a person. Construction site hazards occur:

- in the work environment
- as a result of the use of machinery, tools and materials
- as a result of unsuitable work systems and procedures.

Common hazards encountered on a construction site include:

- manual and mechanical handling (see Section 1.9 in this textbook)
- noise (see Section 1.8 in this textbook)
- dusts (see Section 1.5 in this textbook)
- gases (see Section 1.7 in this textbook)
- chemicals (see Section 1.6 in this textbook)
- working at heights (see Section 1.10 in this textbook)
- working in confined spaces (see Section 1.7 in this textbook)
- harmful airborne contaminants (see Section 1.5 in this textbook)
- physical and psychological hazards (see Section 1.8 in this textbook)
- harmful materials (see Section 1.8 in this textbook).

Hazards when working with machines

Hazards to be aware of when carrying out maintenance or breakdown electrical work concern 'mechanical movement'.

Injuries from the mechanical movement of electrically actuated equipment range from hand and finger lacerations to the more severe (amputations, crushes and fractures), including death.

It is necessary to de-energise electrically actuated equipment or machines, and to lock or tag them out, prior to servicing, maintenance or an emergency event to protect yourself adequately from the hazards of unexpected energisation, start-up or release of stored energy. It should be noted that locks and tags by themselves do not control hazardous energy. It is the isolation of the equipment from the power source and the following of the established procedures for de-energisation and re-energisation of the equipment that controls the power. By using established procedures you are implementing a method that cannot readily be removed, bypassed, overridden or otherwise defeated.

Note: Pushbuttons, and foot, selector, limit, flow, temperature, pressure and photoelectric switches and other control-circuit-type devices, are not energy-isolating devices.

All parts of equipment or machines that move in the course of the performance of work may contribute to accidents causing injury and damages. All electrically actuated rotating, linear and reciprocating machine movements can be dangerous. Examples include:

- belts and pulleys
- gear wheels
- shafts and spindles
- flywheels
- grinding wheels
- slides and cams
- chain and sprocket gears
- power presses
- guillotines
- milling cutters
- saws
- drills.

In addition, the mechanical potential energy of all portions of the equipment or machine is set so that the opening of pipes, tubes, hoses or actuation of any valve, lever or button will not produce a movement which could cause injury.

Typical hazards to look for when identifying dangerous mechanical movement from electrically actuated equipment and machine parts include:

- entanglement
- friction and abrasion
- cutting
- shear
- stabbing and puncture
- impact
- crushing
- flying particles
- protrusions
- drawing in.

It takes far more than merely locking out electrical energy to achieve maximum protection for electricians. The locking-out procedure should include all energy sources:

- fluids under pressure
- compressed gases
- energy stored in springs
- potential energy from suspended parts
- any other source that may cause unexpected mechanical movement.

The source of the energy under control must let the machine 'complete its stroke' or 'come to rest'. After all energy sources have been made safe, the equipment is in zero mechanical state. Zero mechanical state provides maximum protection against unexpected mechanical movement.

Coherent optical hazards

The name 'laser' is an acronym for 'light amplification by stimulated emission of radiation'. A laser is a device that creates a uniform and coherent light very differently from an ordinary light bulb or the sun. Coherent light, as illustrated in **Figure 7.14**, is produced by a standard stimulus which triggers a beam of photons that have the same frequency and are in phase with each other. Two light waves are coherent if the crests and troughs of one wave are aligned with the peaks and troughs of the other.

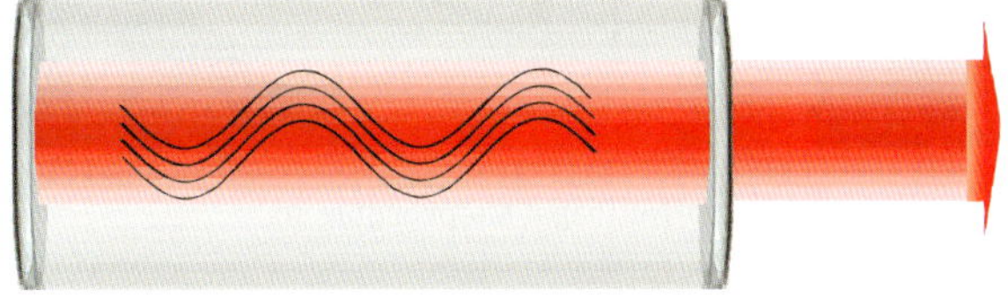

FIGURE 7.14 One colour light (monochromatic and waves in phase (coherent))

By collimating or focusing a particular wavelength of light, energy can be transmitted great distances and remain a focused spot of energy. The movement of light in a laser is shown in **Figure 7.15**.

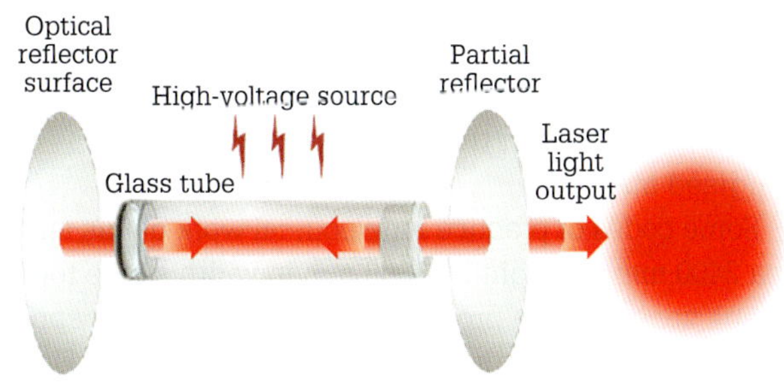

FIGURE 7.15 Movement of light in a laser

Within the cavity of the laser device a beam of light is reflected back and forth along the glass tube, until the waves of light become coherent. A beam of laser light will not spread and diffuse.

Lasers emit beams of non-ionising radiation, at wavelengths from the ultraviolet to the infrared. Thermal burns from visible or infrared laser beams are possible as well as photochemical burns from ultraviolet laser beams. In addition, the skin and eyes are at risk of injury. For lasers in the visible or near infrared part of the spectrum there is a particular likelihood of damage to the retina of the eye, resulting in permanent visual impairment.

Lasers are divided into a number of classes depending on the power or energy of the beam and the wavelength of the emitted radiation. Laser classification is based on power output, wavelength and exposure duration (the laser's potential for causing harm).

Safety Class 2M: visible low-power lasers

2M is a new class of laser with little risk to eyes and no risk to skin.

Our eyes are sensitive to light radiation that lies in a small area of the electromagnetic spectrum categorised as visible light. This visible light corresponds to a wavelength range of 400–700 nanometres (nm) and a colour range from violet to red. The human eye is not capable of seeing radiation with wavelengths outside the visible spectrum. Laser products classified as 2M emit visible output below 1 mW in the 400–700 nm range and can be seen by the eye. Class 2M lasers pose relatively little risk to eyes and no risk to skin under standard conditions.

The eye's natural aversion response (blinking) to bright light prevents damage to the eye. Blinking is a natural involuntary response that causes the individual to blink and avert their head thereby terminating the eye exposure. However, users should not incorporate optics that could collect more light and concentrate the output into the eyes. Lasers are classified on the basis of their accessible emission. For each laser class there is an 'accessible emission limit' (AEL), expressed in watts or joules. This type of laser has little risk to the eyes and no risk to skin. However, it presents a severe glare hazard. A Class 2M product label is illustrated in **Figure 7.16**.

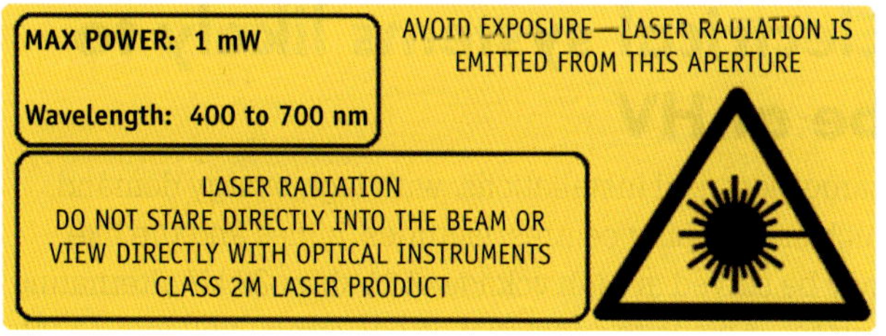

FIGURE 7.16 Laser label

Laser pointers and barcode scanners that are Class 2 laser products are level and orientation instruments used in civil engineering.

Laser beam instruments used in a worksite must conform to Australian standards and should only be used with the following precautions:

- Operators must have completed a recognised course in laser safety.
- Up to 2M Class lasers only are to be used on any construction site.
- Lasers are to be positioned so as not to be at the eye level of persons.
- Warning signs are to be erected indicating the use of a laser in the area.

REVIEW QUESTIONS

1. Name common hazards encountered on a construction site.
2. Name the hazards to be aware of when carrying out maintenance or breakdown electrical work.
3. Describe coherent light.
4. What do lasers emit?

7.6 Hazards associated with high voltage

The term 'high voltage' applies to electrical equipment that operates at more than 1000 V a.c. rms or 1500 V d.c. The state or territory Electrical Safety Act, regulations, AS/NZS 3000:2018 *Wiring Rules* and codes of practice and HV isolation and access procedures manuals must always be consulted when any high-voltage work is intended to be undertaken. These provide advice on ways to discharge safety obligations.

Control measures for HV

Isolation of electrical circuits is a basic safety procedure that protects workers, electrical production and consumption devices or service lines. This means that the power is turned off and isolated at the conduction source so that no energy can enter or leave those devices or service lines.

SWITCH ON

Isolation procedures must include a warning to all workers that the equipment in a particular area is being worked on and must not be operated or the area entered.

Electrical systems likely to be at HV

Large electrical installations, with high energy demand, such as shopping centres, factories, mines and the like may be served at high voltage of 11 kV or 33 kV alternating current. An onsite substation (transformer) will reduce this potential to the standard 230/400 V a.c.

Some machinery, such as pumps, mills, rollers and the like may have 11 kV 3-phase motors as the prime mover.

Onsite HV equipment may include:

- transformer
- switching devices
- protecting devices such as, fuses, isolators, circuit breakers, and protective relays
- control panels (see **Figure 7.17**)
- lightning arrestors
- potential and current transformers, and metering
- auto re-closers
- various ancillary equipment.

Terminology

Terminology associated with HV installations includes:

- touch voltage or potential

Source: Shutterstock.com/Nutthapat Matphongtavorn

FIGURE 7.17 HV control panel

- step potential
- induced voltage
- stored energy.

Touch voltage or potential

Touch voltage (also referred to as step potential) is caused by a fault current flow in a conductive medium establishing a potential difference between the feet of a person with the earth contact point and the hand, head or other parts of the body in contact with the energised material.

Some codes of practice for electrical work define an area 2400 mm in height by 1000 mm in width as the touch voltage zone where a prospective fault voltage may appear at the point of contact and any conductive material. **Figure 7.18** illustrates the fault current path for touch voltage.

FIGURE 7.18 Touch voltage zone

Step potential

Electrical current called a ground fault flows into the ground when anything such as a tree branch, person, metal ladder or the metal arm of a cherry picker accidentally makes contact with an exposed live conductor. Current can also flow into the ground or through objects to the ground if a storm causes a power line to fall.

The current will take many paths back towards the source as it flows through the ground away from the entry point. The current radiates out in concentric circles into the surrounding soil like ripples on a water surface, as shown in **Figure 7.19**.

FIGURE 7.19 Concentric nature of current in the ground

There are also potential differences created between each ring of the 'ripple' and these potentials decrease in magnitude the further they are from the centre. Where the current enters the ground, there is a rise in potential created relative to any circuit-earthing medium further away from the current entry point. In good moist soil there would be less resistance and the potential difference would drop very quickly before the current travels very far from the point of entry.

On a high-resistance surface such as sand or rock, the potential gradients would fall off more slowly and move further from the point of contact. Therefore, there would be less potential difference across each ring. A voltage gradient across a displaced conductor and earth is illustrated in **Figure 7.20**.

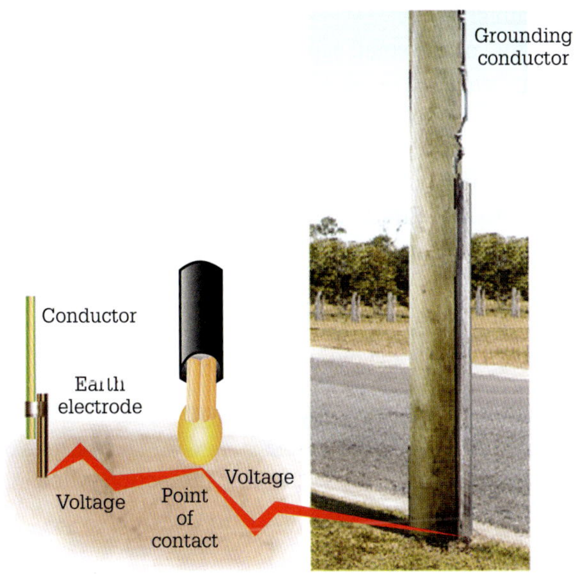

FIGURE 7.20 Voltage gradient

The magnitude of the current flowing to ground will depend upon the soil resistivity at the point of contact. Due to the difference in potential between the rings it is possible for a person, as they move towards or away from the point of contact, to 'step' between high- and low-voltage potentials.

As the human body is usually a better conductor of current than the ground, the fault current will flow from the feet through the body. This process is referred to as 'step potential'. A step potential refers to having one foot on a high-voltage ring near the point of entry and the other foot on a lower-voltage ring further away from the point of entry. Step potential is illustrated in **Figure 7.21**.

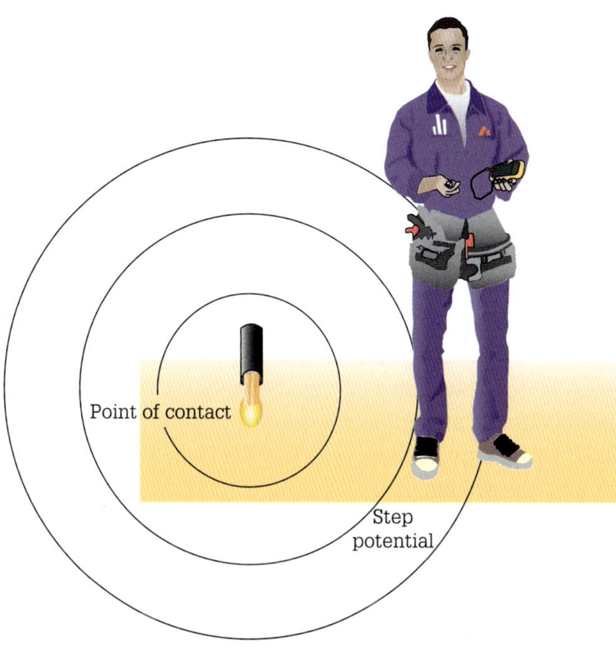

FIGURE 7.21 Step potential

A step potential is defined as the voltage difference between two points on the ground separated by the distance of one pace or 1 metre. Depending upon the fault current, zero voltage is assumed to be approximately 10 metres from the point of contact.

Induced voltage

A potential can be impressed on nearby conductors and conductive material through transformer action and capacitive coupling. Any conductive equipment and structures located within the designated HV hazard zone need to be solidly bonded to earth as shown in **Figure 7.22**.

Source: Shutterstock.com/RachenStocker

FIGURE 7.22 Earthing for HV installations

The permit-to-work system is used extensively for HV works. Before being issued an HV 'access authority', it is necessary to ensure appropriate earthing and bonding of the relevant HV plant and equipment. This arrangement is commonly referred to as access authority earths identified in the Power System Safety Rules. Portable earthing devices may be used where fixed earthing switches are either not available or not suitable for the planned work. When used, portable earths (see **Figure 7.23**) can connect to:

- transmission lines
- substation apparatus
- switchboards
- cables.

The purpose of HV earthing equipment is to provide a conductive path for unforeseen electrical current to flow safely to earth. Current flowing to earth will cause HV protection systems to operate, which will isolate the source of supply from the segment undergoing work. This arrangement also affords some protection from induced voltages and currents.

It is important to bear in mind that cables have capacitance and therefore possess the ability to store an electric charge. In cases where an HV system could potentially discharge through the earthing system, it is essential to allow sufficient time between de-energising the conductors and applying the earthing system. This will allow any energy stored on the cables to safely dissipate through the system. It is important to disconnect cable segment from the source of supply first and allow any stored energy to dissipate through downstream equipment.

Source: Shutterstock.com/angus reid

FIGURE 7.23 Portable earthing for HV installations (orange cables)

Stored energy

Capacitive energy storage is the most common form of electrical energy storage. Capacitors are found throughout the power utility industry and in almost every electronic instrument or appliance and with some electric motors. Disconnected circuit cables if left *in situ* in an installation can store energy as a result of electromagnetic induction. The stored energy in long cable runs presents as a capacitor and can be an unexpected hazard. Procedures must be in place to ensure proper discharge of this energy.

In the utility industry, capacitors are used to correct the power factor, compensating for the inductive nature of loads. In electronic instruments, capacitors are employed in the d.c. power supplies to reduce the a.c. ripple after rectification. In the utility industry, many capacitors are often assembled in series and/or parallel capacitor banks to create large energy storage reservoirs. Upon switching, these capacitor banks can generate enormous voltages and currents in nanosecond to millisecond pulses. With motors, capacitors are used to provide additional starting or running capacity.

There are several severe hazards associated with the use of high-energy capacitors, including mechanical explosion, toxins, fire and especially electric shock.

A high-energy density capacitor can store over 100 kJ. A single such device can deliver well over 1 kA into a short-circuit. Any energy storage capacitor with more than a few joules (16 J could kill) of stored energy can generate significant discharge current if contacted by a person. Small capacitors, such as power filter capacitors in televisions or mains-type instruments, can stop the heart due to muscle response. In comparison, larger energy storage capacitors or capacitor banks would almost invariably produce a fatal shock if contacted by a person.

One unique feature of capacitors is their ability to store energy for relatively extended periods of time after the power source is removed. Even though all power sources have been disconnected a capacitor still presents a severe hazard. Even if a capacitor is fully discharged, it may recharge due to a phenomenon known as dielectric memory. A 0.01% recharge of a 50 kJ energy storage capacitor would result in 50 J of stored energy.

Bleed resistors

Bleed resistors may be on any capacitors larger than 0.1 μF if charged to higher than 230 V. Do not rely on the fact that the bleeders did discharge the capacitors; they can fail too. Capacitors should be checked by discharging with a high-value resistor before touching circuits containing capacitors.

High-voltage small capacitors can be very dangerous because energy in a capacitor is proportional to CV^2. Since the energy is proportional to V^2, a high-voltage capacitor can store substantially more energy per unit volume than a lower-voltage, but higher-capacitance component.

Due to the extreme shock hazard of energy storage capacitors, it is important to isolate capacitors and capacitor banks with enclosures or barriers to prevent accidental contact. Care must be taken to avoid the recharge of the capacitors after the initial discharge.

SWITCH ON

Caution: Discharging is the only protection when working with capacitors.

Issues related to HV installations

Work on, near or associated with high-voltage production and consumption devices and service lines should be carried out only by the authority of certified access permits and switching schedule sheets. Written lockout, isolation and switching events are an essential part of high-voltage procedures before any work is planned in de-energised production and consumption devices and service lines. These systems are developed by a review of appropriate one-line up-to-date drawings for the electrical system. From these pictures primary electrical sources can be identified in addition to all possible sources of alternative feeds.

After a draft lockout, isolation and switching schedule has been prepared it is recommended that it be checked out in the field by at least one other suitably trained and experienced person. Safety earthing locations and methods should also be part of the lockout and isolation procedures and their place in the field noted. In addition, areas that require safety barriers to be erected should have the type of restriction recorded and their location noted.

A switching schedule sheet defines the sequence of each operation for the isolation, testing and earthing of the production and consumption devices and service lines to be worked on. In order to obtain an access permit, the compilation of a written lockout, isolation and switching schedule demonstrates to entities in control of the high-voltage system that a safe system of work has been established.

In addition, a list of personnel who will work in the high-voltage location needs to be compiled. Lists enable the person in control to restrict their tasks within the bounds of their experience, responsibilities and authorisation. Before a permit is issued, the authority will:

- research and document the request
- document the actual switching process
- identify any people or business that will be interrupted.

All individuals required to enter the work area should sign the permit-to-work access. At the completion of work all signatories must be informed before the earths are removed and reverse switching sequence engaged. They ought to sign another form giving written agreement for the reconnection to source of supply. This form should state that the installation is cleared of all earthing leads, labour and tools and that it is safe for reconnection to the supply at a particular time.

After completion of work and prior to re-energising the system, a designated person should:

- ensure that all personnel are clear of conducting elements and have been notified that the system will be energised
- check that the temporary earthing devices and all other safety equipment are removed
- prove the system's insulation by testing the system at a specified voltage to earth
- ensure that the covers and doors of all equipment are closed and the entire installation is safe and secure
- remove the safety tags and locks from the switching equipment
- sign the agreement-to-reconnect form and then energise the system at the agreed time.

All signatories to the form can then revoke the permit-to-work authority.

Safety clearances

Details of high-voltage clearances (exclusion zones) with respect to untrained, instructed, authorised and safety observer personnel are described in Electrical Work codes of practice and the Electrical Safety Regulations. Examples of clearances to be observed with respect to exposed parts for high voltage up to 33 kV phase-to-phase are:

- 3000 mm exclusion zone for exposed parts of an untrained person
- 700 mm exclusion zone with consultation with the individual in control of exposed part for authorised and instructed persons.

Safety services

Safety services are methods, such as shown in **Figure 7.24**, or an element that functions to ascertain a crisis, or is supposed to activate because of the crisis. Refer to Clause 1.4.104 AS/NZS 3000:2018. These include

- fire suppression systems
- building evacuation
- fire detection systems
- smoke detection systems
- warning systems
- extinguishing systems
- evacuation systems
- safety of persons using lifts.

It is necessary to implement the requirements of Clause 7.2 in AS/NZS 3000:2018 when installing electrical elements as part of a safety service. All safety services must be installed to prevent their malfunction in the event of a crisis, especially where the services are dependent upon regular supply.

Examples of safety services such as fire and smoke control equipment are illustrated in **Figure 7.25**.

AS/NZS 3000:2018 does not recognise smoke alarms in domestic installations as safety services (refer to Clause 4.6). However, these devices must be installed as per requirements of the Building Code of Australia (BCA).

Evacuation equipment

Equipment used for emergency evacuation purposes such as sound systems and intercom systems, as well as emergency evacuation lighting systems, are not covered under AS/NZS 3000:2018.

However, the BCA requires high-rise buildings to have a sound and intercom system for emergency evacuation purposes. The installation is required by the BCA when a building height exceeds 25 m in actual height. The installation of emergency equipment must comply with AS 1670.4-2004.

A sound and intercom system installation includes a master control panel, emergency warden intercom points at each emergency exit, manual call points and warning sirens or speakers dispersed throughout the building.

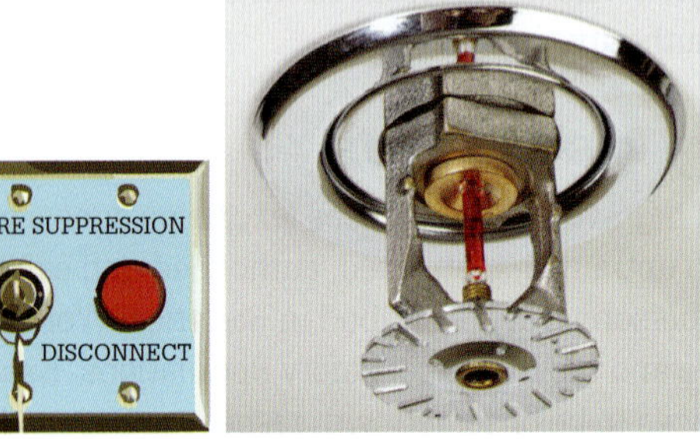

Sources: Shutterstock.com/Justin Kral; Shutterstock.com/H N Y

FIGURE 7.24 Safety services

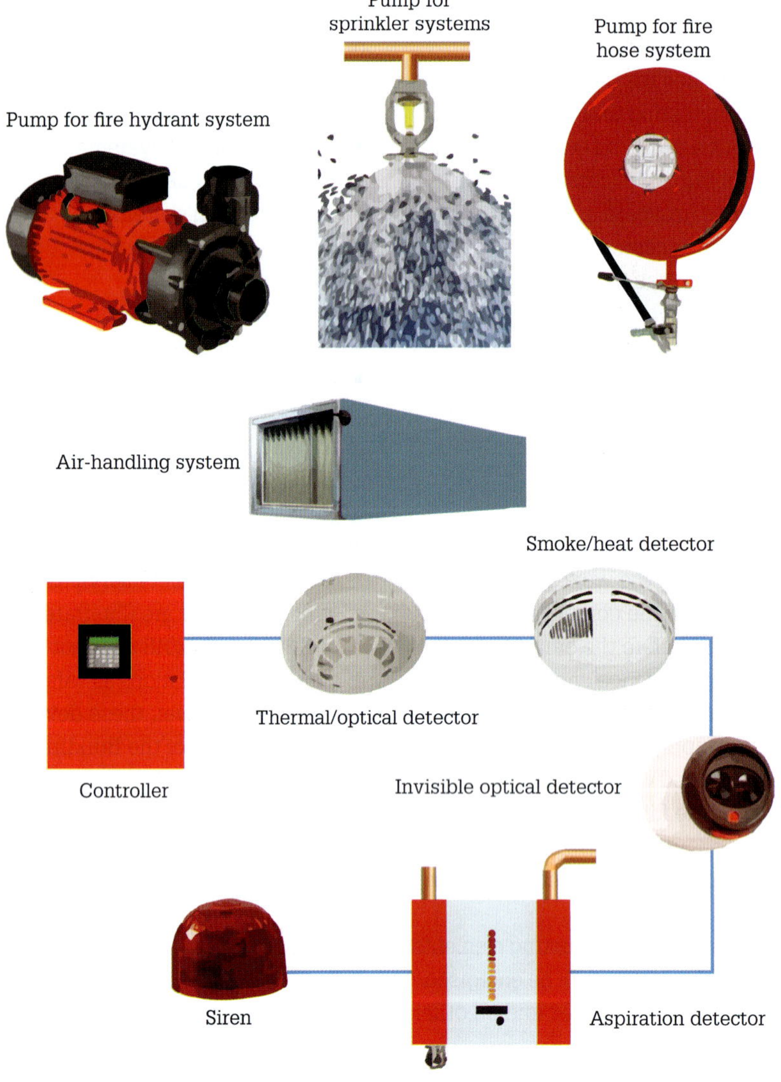

FIGURE 7.25 Fire and smoke control equipment

Lifts

AS/NZS 3000:2018 requires that lifts as illustrated in Figure 7.26 meet certain requirements if supply fails. It makes no recommendations as to their use during an emergency.

It is suggested that lifts for emergency evacuation should only complement an evacuation strategy. An effective strategy embodies fire-isolated stairwells as the main means of egress, sprinkler-protected buildings together with trained fire wardens and tested emergency evacuation systems.

FIGURE 7.26 Safety service – lifts

Special requirements

Requirements for main switches, supplies and alternative power supplies are illustrated in Figure 7.27.

Emergency and standby power systems are intended to provide an alternative source of power if the regular source of power should fail. Any stoppage of the normal electrical supply during emergency conditions could create hazards or hinder rescue and firefighting operations. Generators are by far the most prevalent source of power for emergency and standby power systems. Battery systems are also used for some safety services. Battery systems that are part of the emergency system must be periodically maintained.

This alternative power source usually supplies emergency system loads where the battery source has adequate capacity.

Batteries or emergency generators also provide power for emergency illumination. This form of illumination must include all required means of egress lighting, illuminated exit signs and any other lighting form as necessary to provide required illumination.

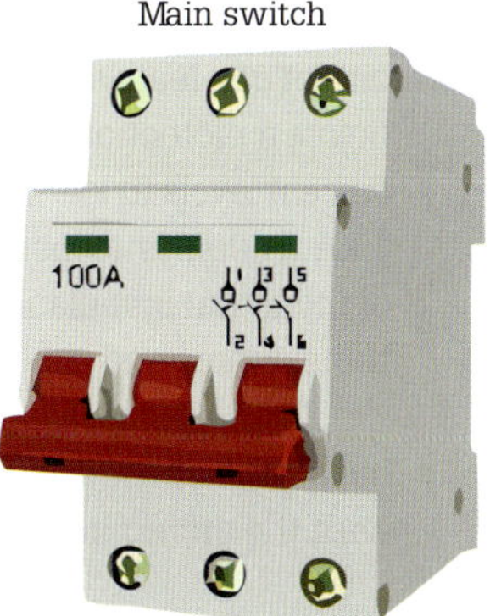

FIGURE 7.27 Special requirements

REVIEW QUESTIONS

1 Define the term 'high voltage'.
2 What is isolation of electrical circuits?
3 What may onsite HV equipment include?
4 Explain how touch voltage (also referred to as step potential) is caused.
5 Explain how a ground fault is produced.
6 What determines the magnitude of the current flowing to ground in the event of a fault?
7 Define 'step potential'.
8 Why is it important to isolate capacitors and capacitor banks with enclosures or barriers to prevent accidental contact?
9 What is the purpose of emergency and standby power systems?
10 What form of illumination must include all required means of egress lighting, illuminated exit signs and any other lighting form as necessary to provide required illumination?

7.7 Hazards associated with extra-low voltage and low voltage

The *Wiring Rules* require that every electrical installation comprise a number of circuits as necessary to:

- avoid danger and minimise inconvenience in the event of a fault
- facilitate safe operation, inspection, testing and maintenance.

Arrangement of circuits

The creation of different circuits in an electrical installation makes it possible to:

- limit the effects of a fault to the circuit concerned
- simplify fault-finding
- carry out maintenance work, inspections, testing or circuit extensions without interrupting the whole electrical installation. In general, the following circuit groups are used:
  - lighting – indoor, outdoor, communal
  - socket outlets – 10 A, 15 A, 20 A
  - cooking – ranges, ovens, cooktops
  - fixed-space heating – air-conditioning, ventilation
  - water heating – instantaneous, storage, spa, swimming pool and laundry equipment
  - safety – emergency lighting, fire protection, security and for uninterruptible power supplies
  - power circuits for fixed plant – lifts, motors, kilns and welding machines.
- control circuits for fixed plant.

The number and type of circuits required are determined by the load on the circuit, the location of the loads, any seasonal variations affecting the loads and any special conditions.

Refer to AS/NZS 3000:2018 *Wiring Rules*, 'electrical installation circuit arrangement'.

Electrical systems likely to be at LV and ELV

Extra-low voltage (ELV) is defined as a voltage that does not exceed 50 V a.c. or 120 V ripple-free d.c.

Low voltage (LV) is defined as voltage exceeding 50 V rms a.c. or 120 V d.c. but not exceeding 1000 V rms or 1500 V d.c. It is possible that some electricians believe that low voltage contact is less hazardous than high-voltage contact. They may believe that an error made in working on a live low voltage system means only an arc flash combined with the operation of the circuit protection device. This perception is incorrect.

Several factors make 'live' low voltage equipment hazardous. One of these factors is the small creepage and clearances between low voltage terminations when using tools or test instruments (see **Figure 7.28**).

(**Note:** A training person must be 1000 mm away from an exposed live part if with a trained person.)

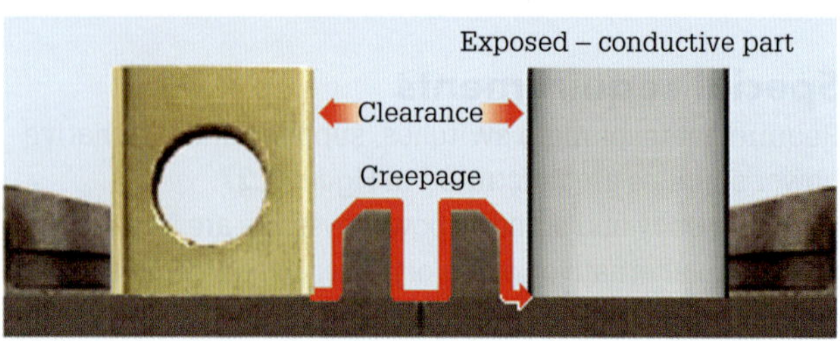

FIGURE 7.28 Clearance and creepage

Creepage distance and clearance distance have relevance for printed circuit boards, transformer insulators and between primary and secondary windings of an isolation transformer (where reinforced insulation is required). In addition, inadequate separation between high-voltage and low voltage cables or terminations can result in hazardous induced voltages and arcing at the terminations.

Another factor is the circuit on which a task is being performed. Under short-circuit conditions a low voltage circuit can feed large amounts of energy in the fault condition before a protection device can operate. The short-circuit fault current levels in low voltage installations can range from 1 kA for domestic installations to over 50 kA for commercial or industrial installations because of the distribution transformers and cable parameters feeding the installation.

There are three categories of common electrical hazards: electric shock, arcing and toxic gases associated with the arcing hazard.

Low voltage installations or systems hazards

Typical sources of low voltage hazards that, by themselves or grouped with other hazard sources, could lead to electric shock and injury include:

- voltages between phases
- voltages between phases and earth – during single phase to earth faults, significant-sized voltages can exist between an unearthed metallic pipe and the general mass of earth
- live exposed conductors
- voltages between live exposed conductors and conductive materials
- voltages across undischarged capacitors
- voltages on disconnected conductors – mainly neutrals and unused conductors that have not been physically removed from service
- induced voltages – due to electromagnetic coupling
- exposed live conductors

- multiple-supply sources
- high impedance neutral (MEN system)
- electrical testing – typical and in damp situations and hazardous areas
- circuits inadvertently becoming live – omission of safety tags and lockouts
- circuits with insulation damage – lack of mechanical protection.

A low voltage circuit is not de-energised until it has been placed into an electrically safe work condition. This means that the circuit must be:

- tested for voltage presence
- isolated from energised parts
- locked
- tagged
- tested again to ensure absence of voltage
- operation of tester is confirmed.

For d.c.-powered equipment, although the voltage levels are usually not hazardous, energy hazards (>5 joules) may still exist.

Hazardous contact

The AS/NZS 3000:2018 *Wiring Rules*, regulations and codes of practice distinguish two kinds of hazardous contact: indirect contact and direct contact.

Indirect contact

Indirect contact refers to a person or livestock coming into contact with an exposed conductive part, which is not usually live, but which has become live accidentally (due to insulation failure or other condition). All electrical installations must comply with the principles of protection for safety of which indirect contact is a significant element (refer to AS/NZS 3000:2018 *Wiring Rules*).

A fault current could raise the exposed conductive part to a voltage likely to be hazardous. The rise in voltage, called the touch voltage, could be at the source of a touch current through a person coming into contact with this live exposed conductive part (see **Figure 7.29**). Note that touch voltage is the voltage that could be 'touched' should the exposed conductive part become live.

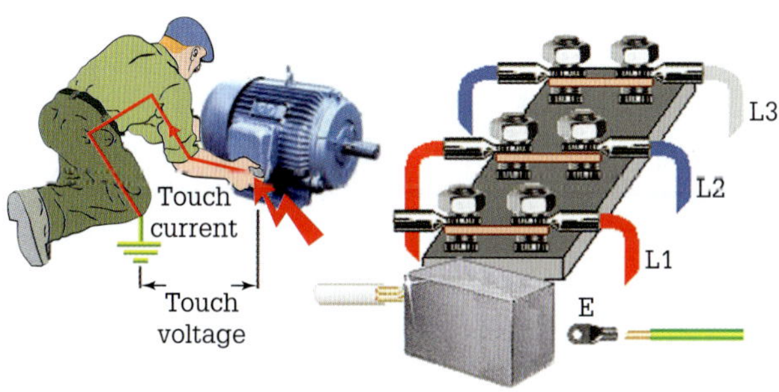

FIGURE 7.29 Touch voltage and touch current

In the example illustrated, due to poor workmanship, both the protective earth and a phase conductor have come away from their lug terminals. This has caused the motor frame to become an exposed conductive part and enabled the touch voltage to rise to 230 V.

Note: Two separate events had to happen for the motor frame to become live. The higher the value of touch voltage the greater is the risk of a hazardous touch current occurring if a person comes into contact with the exposed conductive part and earth.

Possible exposed and unearthed conductive parts used in the manufacturing process of electrical equipment must be separated from the live parts by 'basic insulation'. Failure of the basic insulation will result in the exposed conductive part being live.

Direct contact

Direct contact refers to a person or livestock coming into contact with an exposed conductive part which is usually live and becomes part of the fault path, as shown in **Figure 7.30**.

Measures of protection against direct contact include:

- protection by the insulation of the live part, for example, the insulation surrounding conductors
- protection afforded by barriers or enclosures
- protection provided by use of an RCD.

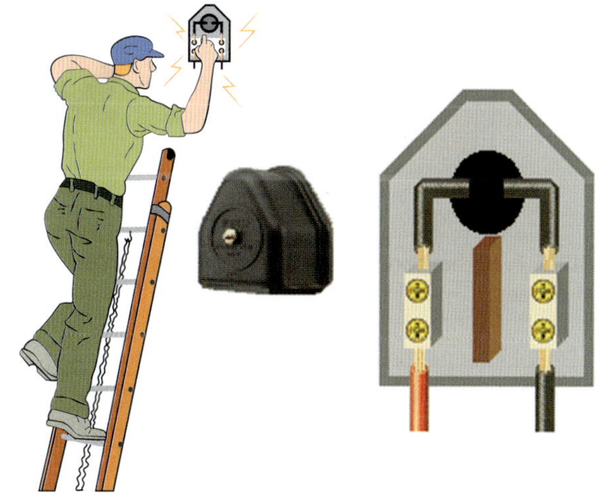

FIGURE 7.30 Direct method: touching live mains in a mains connection box

Protection measures

Various measures are used to protect against the hazard of indirect contact, and include automatic disconnection of the supply of the connected equipment via circuit breaker, fuses, or residual current devices within touch voltage/time safety requirements.

1 The use of Class II insulation materials. *Double insulation* comprises both basic and supplementary insulation, as shown in **Figure 7.31**. Double-insulated hand-held electric equipment carries the square-in-square mark (Class II symbol). Approved Class II equipment does not require a protective earth. Although this design method reduces the risk of a touch voltage existing, a shock hazard can still be present.

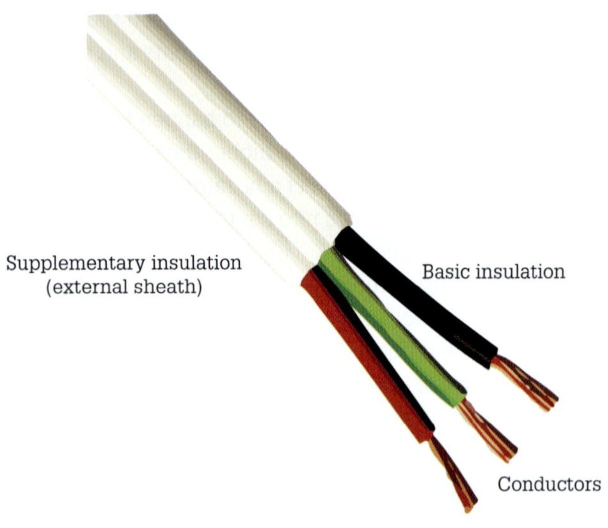

FIGURE 7.31 TPS cable showing basic and supplementary insulation

Reinforced insulation is a single-insulation system, which provides a degree of protection against electric shock equivalent to that of double insulation (see **Figure 7.32**).

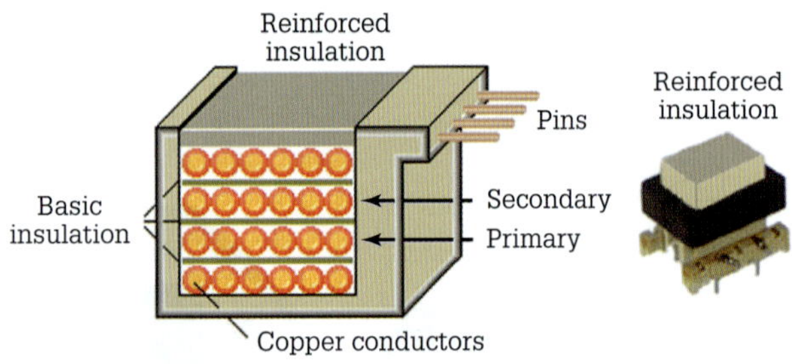

FIGURE 7.32 Section view of a reinforced insulation transformer

Double insulation and reinforced insulation mean that, in general, failure of two independent sections of insulation has to occur before any exposed metalwork can become connected with live conductors. It is important to confirm on a regular basis that the Class II standard is maintained.

2 Equipotential bonding of all extraneous conductive parts within the installation. Equipotential bonding arrangements are intended to lessen the risk associated with the possibility of voltage differences occurring between the exposed metallic parts of fittings and fixed wired appliances and other exposed metal. Equipotential bonding will be used for metallic parts not directly associated with the electrical installation but which may become live as the result of a fault. Exposed metal fittings as shown in **Figure 7.33** that may require equipotential bonding include:
   - water pipes
   - gas pipes
   - sink bench tops
   - metal shower trays
   - urinals
   - exposed structural steel work
   - metal sheathed roofs
   - metallic service brackets (risers).

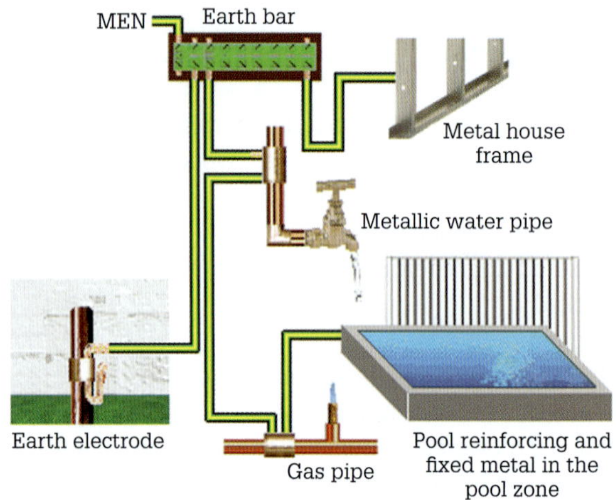

FIGURE 7.33 Equipotential bonding connections

Note that PVC plumbing for water reticulation may mean that some exposed metal fittings might not need equipotential bonding.

All equipotential bonding connections must be mechanically and electrically sound.

3 Non-conducting location, out of arm's reach or the placement of barriers. The possibility of touching a live exposed conductive part while at the same time touching an extraneous conductive path at earth potential is very unlikely, as shown in **Figure 7.34**. The walls and floor must be non-conductive.

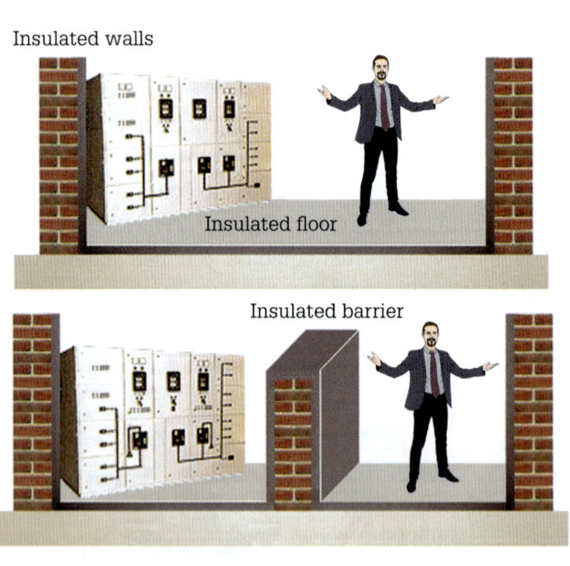

FIGURE 7.34 Protection by out of arm's reach or the placement of barriers

4 Electrical separation of the potential hazard from the rest of the installation by means of an isolating transformer, as shown in **Figure 7.35**.

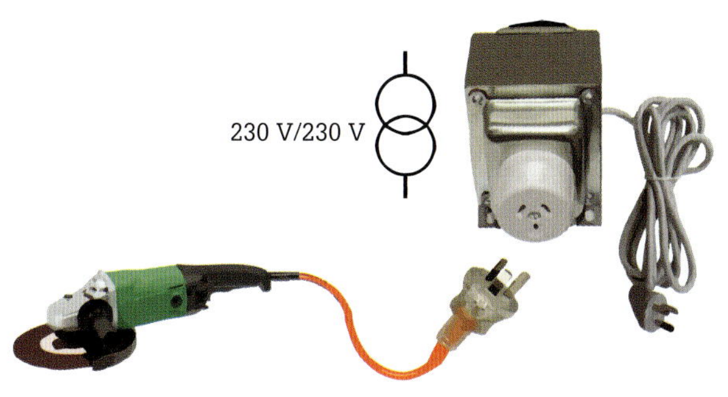

FIGURE 7.35 Class II isolation transformer

5 Use of a residual current device (RCD) as illustrated in Figure 7.36 in the circuit feeding the connected equipment.

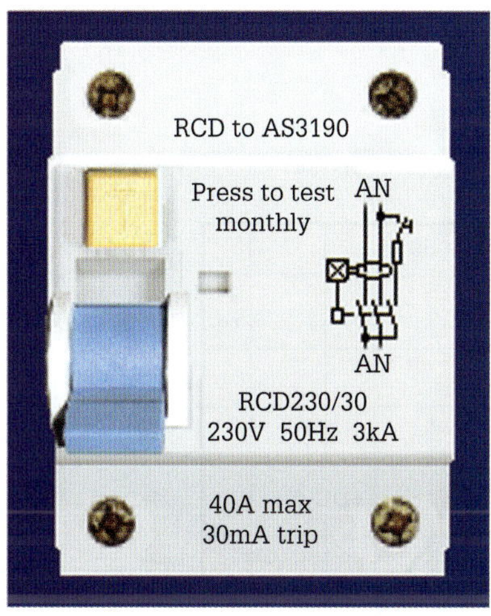

FIGURE 7.36 Residual current device

RCDs provide protection against indirect contact by automatically disconnecting the supply. As the current required to trip an RCD is small, it is not dependent on the very small fault-loop impedance necessary to operate a protective device such as a circuit breaker.

Effects of current on a person

The severity of the electric shock is determined by the type of current. In general, direct current that has no frequency but may be intermittent or pulsating is less dangerous than alternating current. The effects of alternating current on a person depend primarily on the frequency. Currents of 50 Hz find resonance with nerves in the body that make this frequency more dangerous than high-frequency currents, and three to five times more dangerous than direct current of the same voltage and amperage. Alternating current produces severe involuntary muscular contractions, often locking the hand to the current's source; prolonged exposure may result in serious burns if the voltage of origin is high. Direct current tends to cause an involuntary contraction, often throwing the victim away from the current's source. Both a.c. and d.c. currents can cause seizures, ventricular fibrillation and muscle paralysis. Other effects of electric shock can include thermal (burning of skin tissue and organs) and electrochemical damage (from d.c.) (see Chapter 1).

Critical path for electricity

The critical path of electricity through a person's body is through the chest cavity. As shown in Figure 7.37, current flowing from one hand to the other, from a hand to the foot or from the head to either foot will pass through the chest cavity paralysing the respiratory or heart muscles, initiating ventricular fibrillation and/or burning vital organs. The most common entry point for current into a person is the hand, followed by the head. The most common exit point is the foot. The effect of a.c. and d.c. current on the human body is illustrated in Figure 7.38.

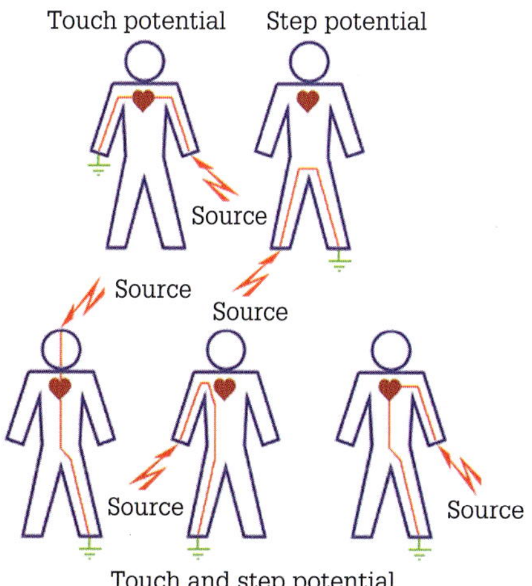

FIGURE 7.37 Possible current paths through a person

Electric shock

Electric shock is not an isolated event and many persons may experience one at some point in their life. Most electric shock injuries occur at voltages above 50 V a.c. or 100 V d.c. However, sometimes the voltage can be much lower, especially in damp situations. In Australia, the majority of electric shocks occur at the mains voltage of 230 V and 400 V respectively. Fatalities occur most often at these voltages and high-voltage injuries are mainly from burns. Due to the number of material variables within the person it is not possible to state the minimum current that will kill. Age and health of the individual and muscle tissue play a part and it appears that the current threshold is lower for women than men. The degree of risk depends not only on current but also on time – the higher the current or the longer the time of shock, the greater the danger.

The relative physiological effects of exposure to increasing a.c. or d.c. currents flowing through the body on a 68 kg male and a 52 kg female are shown in Table 7.3.

It is assumed that ventricular fibrillation that alters the heart's normal rhythmic pumping action can be initiated by a current flow of 500 mA or greater a.c. or d.c. for 10 seconds or more through the chest cavity.

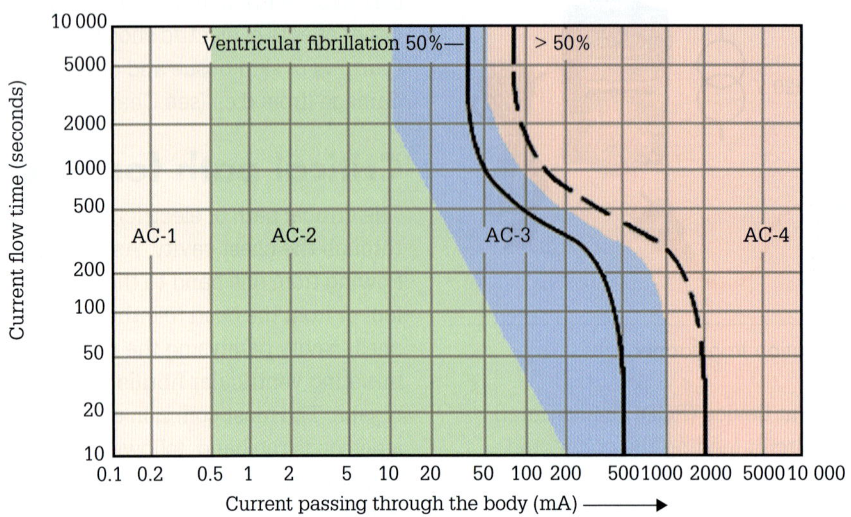

Zones 1 to 4 correspond to different levels of effect

ac/dc-1: no perception

ac/dc-2: perception

ac/dc-3: healable effects

ac/dc-4: possible irreversible effects

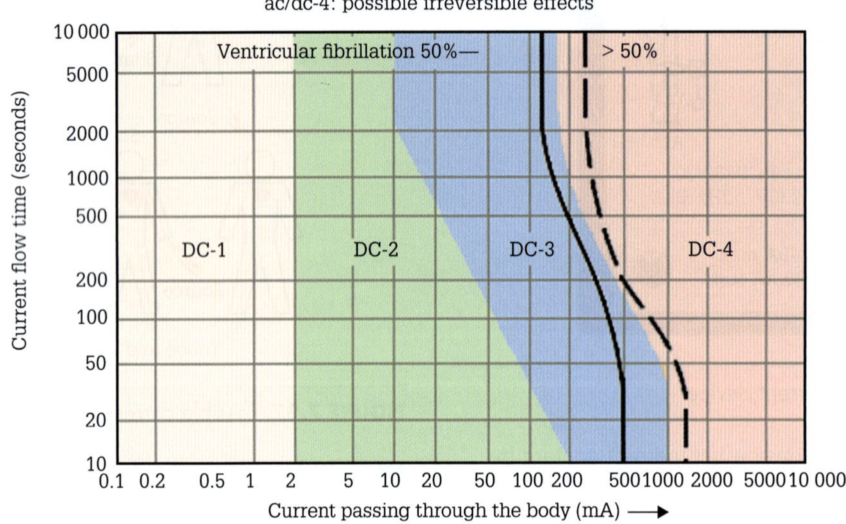

FIGURE 7.38 Effect of a.c. and d.c. current on the human body

TABLE 7.3 Physiological effects of exposure to increasing a.c. or d.c. currents

	d.c. (mA)		a.c. (mA) 50 Hz		a.c. (mA) 10 000 Hz		
Effect	**M**	**F**	**M**	**F**	**M**	**F**	**Severity**
Slight sensation	1	0.6	0.4	0.3	7	5	
Perception threshold	5.2	3.5	1.1	0.7	12	8	
Shock not painful	9	6	1.8	1.2	17	11	
Shock painful	64	41	9	6	55	37	Spasm
Muscle cramps	75	51	15	10.5	75	50	Possibly fatal
Respiratory arrest	200	109	30	19	180	95	Frequently fatal
Cardiac arrest			4000	4000			Possibly fatal
Organs burn			500	5000			Fatal if a vital organ

The maximum current that can cause the muscles of the arm to contract but that allows a person to release their hand from the current's source is termed the let-go current. For d.c., the let-go current is about 75 mA (this is not a definable value) for a 68 kg man; for a.c., it is about 10 mA to 15 mA depending on muscle mass. Electrocution from d.c. current is far less likely than with a.c. current because with d.c. current it is easier to remove the grip on live parts and because d.c. current has a less significant effect on the cardiac system. Direct current, however, should always be treated with care. Burns are the most common form of injury caused by high d.c. current.

RCDs

Protection from a.c. leakage currents likely to cause electric shock is provided by RCDs. To be effective, the RCD must operate very quickly at a low earth leakage current. Those designed to protect human life are engineered to trip out with an earth leakage current of 10 mA within 40 ms (type I), and at a higher earth current of 30 mA they will trip in less than 300 ms (type II) and 40 ms at 150 mA. These limits are well inside the designed safety zone within which ventricular fibrillation would not be expected to occur. These types of RCDs will provide protection against both direct and indirect contact with live parts. The operating characteristics of RCDs are specified in AS/NZS 3190:2011 *Approval and test specification – Residual current devices.*

Also refer to AS/NZS 60479.1:2010 *Effects of current on human beings and livestock – General aspects.*

Hazards of electric arc flashes

Arc flash as shown in **Figure 7.39** is an explosive release of large amounts of heat and ultraviolet (UV) light energy at the point of a fault. It occurs when insulation or isolation between energised conductors is breached or can no longer withstand the applied voltage. The arc flash is a product of short-circuit current and arcing time. Before the arc is quenched, the electrical energy input to the high-current arc plasma is transferred to the surroundings by conduction, convection and radiation, and is also consumed in melting and vaporising the electrode metal at the arc point.

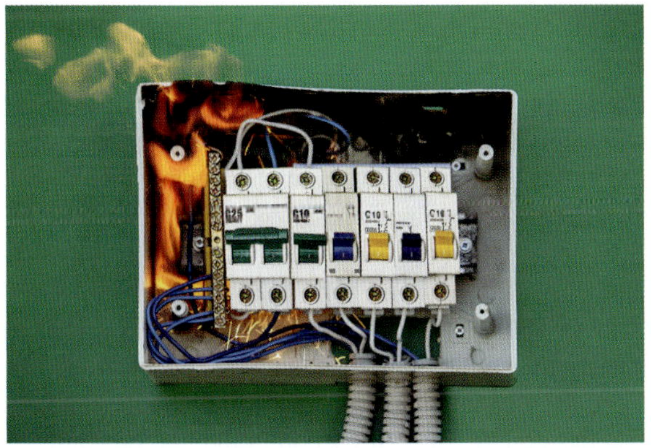

Source: Shutterstock.com/Grigvovan

FIGURE 7.39 Arc flash from a switchboard

Heating effect of arc flashes

A massive amount of concentrated radiant energy that can reach more than 2760 °C can explode outwards from the point of the fault. Arc flashes cause rapid heating of the surrounding air creating a pressure wave of hot gases and molten metal together with sound pressure waves and an intense burst of light. For enclosed equipment a substantial part of the energy is also converted to pressure rise. Exposure to such an event can cause death or severe radiation burns by direct heat and by igniting clothing, together with possible hearing and eyesight damage.

Many different events including the following can cause the arc fault:

- faulty insulation
- improper earthing
- unguarded live parts
- accidental contact with electrical systems
- failure to de-energise electrical equipment when it is being repaired or inspected
- dropped tools
- unsafe work procedures.

A hazardous arc flash can occur with any electrical device, regardless of voltage, in which the energy is high enough to maintain an arc. Likely places where this can happen include:

- main switchboards and sub-boards
- metal-clad switch gear
- transformers
- motors
- motor starters and control cabinets
- fused switches.

The above places should be marked with a clearly visible warning label similar to the one shown in **Figure 7.40** to notify service personnel of potential electric arc flash hazards.

FIGURE 7.40 Arc flash hazard warning label

The label should be located in a place that is readily visible and readable from some distance. The flash protection boundary and its units, the incident energy at the estimated working distance and its corresponding risk category number must be clearly printed on the label. Additional information that is useful that could be printed on the label is the potential size of the incident energy level and the risk category number.

Temperature of the arc

System voltage, arc impedance and available short-circuit current determine instantaneous arc energy. Total arc energy is the immediate arc energy times the arc duration. The amount of energy draw from the supply will in turn determine the temperature of the arc. Conductive vapours help to maintain the arc and the term of the arc is primarily determined by the time it takes for over-current protective devices to open the circuit.

Hard-drawn copper conductors subject to arc temperatures will expand as a vaporised metal to approximately 67 000 times their normal state. For example, 16.39 cm^3 of copper vaporises into 1 098 000 cm^3 of vapour. The superheated surrounding air combined with the expanded copper causes a pressure wave called an arc blast. Direct effects of an arc blast are devastating. The total force on a person standing in front of an open switchboard may exceed a 454 kilogram-force (47.3 N). Such forces may crush an individual's chest or even propel them into equipment, walls and so on, causing additional trauma.

Safety

All electrical work conducted on low voltage electrical installations must be in accordance with the employer's electrical safety hazard management procedure and the guidelines for AS/NZS 4836:2011 *Safe working on low voltage electrical installations*. A flash hazard analysis (exposure level) needs to be done before a person can work near energised equipment to determine the type of protective clothing required.

Mitigation of potential arc flash hazards can be achieved if low-impedance transformers, insulated bus and current-limiting fuse and circuit breaker protection are used. However, the only way to prevent an electric arc flash hazard is always to de-energise and lockout/tagout all electrical equipment prior to performing fault-finding and maintenance activities.

Protective clothing

The main concern with an arc flash hazard is the arc temperature and the flash flame causing the ignition of clothing. At approximately 960 °C for 0.1 s (just over one cycle with 50 Hz supply), exposed human skin experiences a third-degree burn (rendered incurable). The energy threshold for the commencement of a second-degree burn is 5.0 J/cm^2 (1.2 calories/cm^2).

Electricians should wear and be properly trained in the safe use and function of the PPE suitable for the possible electrical hazard which they may face. Training must be in relation to the safe use of hard hats, face shields, flame-resistant neck, face and chin protection, hearing protectors, flame-retardant suits, insulated rubber gloves with leather protectors and insulated leather footwear. PPE for arc flash hazards is required for every part of the body and must be sufficient for protection against the potential arc flash level at the point of work.

Synthetic materials, such as polyester, nylon and synthetic cotton blends, are flammable and must not be used as arc flash PPE. Also, clothing made from natural materials, such as cotton, wool or silk, can ignite under certain conditions. Only clothing that states the ATPV (Arc Thermal Performance Value) or EBT (Energy at Break Through) arc flash rating on its label is suitable for arc flash PPE. The arc flash rating in J/cm^2 or calories/cm^2 is a measure of how much heat energy the material can withstand before the wearer experiences the onset of a second-degree burn. The higher the rating the greater will be the protection. If the clothing does not have an arc rating it is not arc flash protection. Some overalls have an inherent flame-resistant property with an arc rating of 4.8 ATPV and this property cannot be degraded by laundering.

Extra-low voltage (ELV) circuits

Refer to 1.4.128 of AS/NZS 3000:2018 *Wiring Rules*, 'Extra-low voltage electrical installations'.

Voltage is regarded as a difference of potential that exists between all active conductors and all active conductors and earth. Extra-low voltage is a voltage not exceeding 50 V a.c. or 120 V ripple-free d.c. Extra-low voltage is a measure designed to protect people and livestock from direct and indirect contact with live parts without automatic disconnection of the supply.

Two systems are used for this purpose – separated extra-low voltage and protected extra-low voltage. The source of these types of supply must comply with the requirements of AS/NZS 3000:2018 *Wiring Rules*.

Separated extra-low voltage (SELV)

SELV systems are used where an inherently safe system of supply is required. These systems ensure that a failure of the primary insulation does not result in an increased risk of electric shock. This is achieved by electrical separation of the SELV output from the source of energy, typically using a safety isolating transformer.

Protection by SELV is used in high-risk situations where the operation of electrical equipment presents a serious hazard; for example, swimming pools. This method of protection requires supplying electrical energy from a safety isolating transformer, which utilises double or reinforced insulation. For damp situation zone 0 applications, the nominal voltage must not exceed 12 V a.c. or 30 V ripple-free d.c.

Protection

The following conditions of use must be applied when using SELV to ensure effective protection against indirect contact:

- No live conductor at SELV potential must be connected to the earth.
- Exposed conductive parts of SELV circuits or equipment must not be connected to the earth,

other circuits, and their exposed conductive parts or extraneous conductive parts.

- In SELV systems where the voltage does not exceed 25 V a.c. or 50 V ripple-free d.c. there is no need to provide protection against direct contact hazards.
- Each SELV circuit must be segregated from all other wiring systems including other SELV systems. Segregation is not required where the SELV conductor insulation is equivalent to double insulation.

  Segregation is also not required if the SELV insulation is of the same voltage rating as the highest voltage cable rating with which the SELV cable is enclosed.
- Socket outlets and plugs for an SELV wiring system must not have an earth-pin contact and must be of a design that prevents connection to a different voltage level.

Direct and indirect protection

Direct and indirect protection of an SELV wiring system is provided by:

- the extra-low voltage
- minimal risk of accidental contact with a higher voltage
- no return path through a protective earth that a current could take in case of contact.

Safety

The safety of an SELV system is demonstrated by:

- an extra-low voltage system
- a small risk relating to accidental contact with extra-low voltage
- the absence of a return path for any fault current through earth if a person or livestock comes into contact with extraneous metal components of the extra-low voltage system.

Protected extra-low voltage (PELV)

This type of wiring system is used where low voltage is required, but the situation is not as high risk as is the case with SELV wiring conditions. A PELV power supply includes two protective measures (double insulation) against direct and indirect contact with exposed conductive parts by means of a safe separation of the primary and secondary transformer windings (such as IP2X or an insulation rating of 500 V a.c. for one minute). An earthed conductive shield placed between the primary and secondary windings is also effective. The shield collects any leakage current from the primary winding and discharges it to the earth. This prevents any leakage current entering the secondary winding. For this reason, no separate protective earth conductor is required in a PELV system. Connected equipment and exposed metallic parts in a PELV wiring system can be earthed, but this is not a requirement.

If the circuits of a PELV system require earthing, then the secondary circuit of the transformer cannot exceed 25 V a.c. or 60 V ripple-free d.c. and must have an integrated earth connection to enable connection of the protective earth conductors. PELV wiring systems are used in current supply in automation, machine and process control applications.

Isolated supply

Isolated supply has the same significance as 'separated circuit'. Isolation is the electrical separation between two circuits (source and output) in proximity to each other. If an isolated supply is required, three methods are available (refer to Clause 7.4 AS/NZS 3000:2018):

- motor/generator output
- isolated inverter
- isolating transformer.

Transformers provide electrical isolation while allowing the magnetic coupling of electrical power from primary winding to secondary winding. An isolation transformer as illustrated in **Figure 7.41** eliminates any electrical continuity between mains power and the load connected to the transformer output. The mains power is fed to the primary side of the transformer, and the load is connected to the secondary side. Isolation transformers are available in both single-phase and three-phase versions.

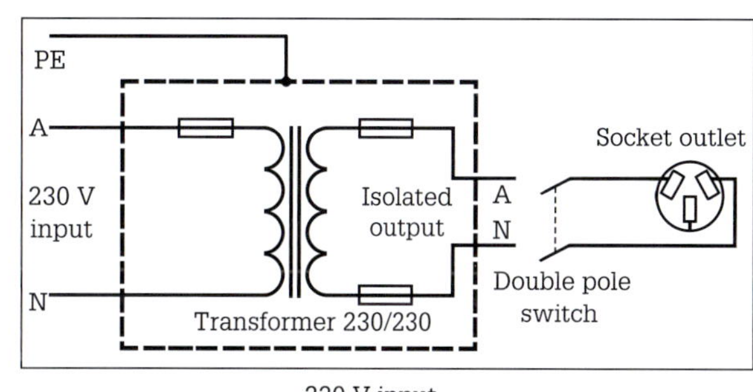

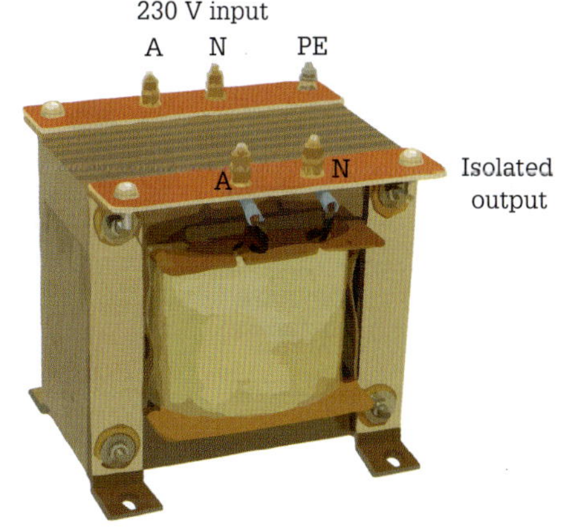

FIGURE 7.41 Isolation transformer

The isolation transformer completely isolates the load from the mains earth provided that there is not an earth connection on the load side of the transformer. As isolating transformers eliminate electrical continuity between circuits they are used for the following applications:

- to isolate shore power from a boat so that galvanic action (electrolysis) cannot take place as the shore earth is isolated from the boat earth
- in hospitals for power supply in body- and cardiac-protected areas
- to avoid tripping of switchboard protection devices when testing portable RCD units.

REVIEW QUESTIONS

1 What factors dictate the number and type of circuits required?
2 Define 'low voltage' (LV).
3 Name the three categories of common electrical hazards.
4 To what does indirect contact refer?
5 To what does direct contact refer?
6 What measures are used to protect against the hazard of indirect contact?
7 What is an arc flash?
8 Where can a hazardous arc flash occur?

7.8 Disconnecting and reconnecting electrical equipment

Considerations for the disconnection and reconnection of electrical equipment are:

- implementing appropriate isolation and tagging-off procedures
- alternative sources of supply.

Isolation and tagging-off procedures

Lockout, isolation and tagging procedures are established to protect persons and property in a workplace location from any electrical hazard, either energy or mechanical in nature, while inspections, repairs or maintenance work are being carried out. Samples of lockout and tagging devices are illustrated in Figure 7.42.

FIGURE 7.42 Lockout and tagging devices

The items in Figure 7.42 include danger tags, barrier tape, hasp with lock and a circuit breaker locking device.

Tags

Two tagging classifications are used in electrical installations to indicate that electrical and non-electrical equipment are isolated. The classifications are a personal danger tag and an out-of-service tag. The personal danger tag is red and black on a white background and is used for labelling electrical and non-electrical equipment. The out-of-service tag is black with a yellow background. This tag is used to identify equipment or machinery that is faulty or not suitable for use and has been taken out of service. Examples of both tagging classifications are shown in Figure 7.43.

FIGURE 7.43 Tag classifications

If working on-site with plug-in type appliances, attach a danger tag to the plug or lock the plug in a suitable enclosure to prevent it being inserted into a live socket outlet.

When required to carry out installation or maintenance work on electrical production and consumption devices or service lines, these devices or lines must be isolated or de-energised so that they are safe to access. Tools generally used for isolating electrical production and consumption equipment from an electrical service line include isolators, circuit breakers, fuses, switches and links.

Only authorised persons with current rescue breathing and CPR certification should apply lockout and tag out devices. All persons must be authorised through training and qualification on the employer's lockout procedures for the equipment and machinery they are assigned to work on. This training must be completed before they can perform any service or maintenance.

Training must include understanding of any required testing instruments used, safe operation of the equipment being tested, and the use of the lockout devices and warning tags provided by the employer. The following steps provide a brief outline of approved application procedures.

1 Notify all persons affected that the equipment requires service or maintenance and is scheduled for shutdown, and lockout or tagout procedures will be implemented.
2 Use established procedures to identify the type, magnitude and hazards of the equipment's energy source. Make sure you know that the proper methods for controlling the energy source are engaged:
   a If the equipment is currently operating, shut it down using normal shutdown procedures.
   b Isolate the equipment from its energy source by activating and energy-isolating devices and prove that all load and control circuits are inactive.
   c Dissipate any stored residual energy using methods such as earthing or bleeding.
   d Ensure that all persons are away from the equipment. Then confirm again that successful isolation has been implemented.

SWITCH ON

Caution: After verifying isolation, check the testing device for correct operation and then make sure that all controls are in the 'off position' and lock and tag.

Although switching off, tagging and locking an isolating switch, fuse or service circuit breaker is necessary, the isolation of the electrical production and consumption devices or service line needs to be proven effective and that in fact it is de-energised.

The 'attempt to operate' isolation procedure is not an effective method of verifying that a circuit is de-energised because mechanical failure with switch contacts cannot be determined by the 'attempt-to-operate' method.

Another procedure that is used is the 'testing for dead' isolation procedure. This process involves turning any isolating switch off and locking the switch, followed by testing all active conductors for the presence of any electrical potential.

This procedure, like the 'attempt-to-operate' procedure, also has flaws. There may be control circuits that can energise the service lines under predetermined conditions.

Confirm, with absolute certainty, that equipment is disconnected from electrical energy by proving that every contact on the load side that feeds the load is indeed not activated. Note that due to the complexity of operations no single solution is applicable to all situations – a range of options must be considered when selecting solutions to mitigate electrical hazards.

Alternate supplies

There are a number of technologies and methods used to produce usable electrical power. We are seeing a shift away from a reliance on burning fossil fuels as the source of this energy towards renewable energy sources using technologies such as wind generators and photovoltaic panels. Common technologies used for alternate electrical power supplies are:

- engine-driven alternators and generators
- photovoltaic panels
- wind-driven alternators and generators
- personal or micro hydroelectric systems.

See Chapter 12 of this textbook for a thorough treatment of working with alternate supplies.

REVIEW QUESTIONS

1 Why are lockout, isolation and tagging procedures established?
2 What does an out-of-service tag indicate?
3 What colours identify the personal danger tag?
4 Who should apply lockout and tag out devices?
5 What should be done once verification of isolation has occurred?
6 List common technologies used for alternate electrical power supplies.

CHAPTER REVIEW

7.1 Risk management and assessment

- The greatest dangers associated with electrical work are electric shock, arc flashes, and arc blasts.
- Risk management is a consultative process that embeds an effective risk management culture throughout the enterprise.
- Once identified, steps must be undertaken to remove or limit the dangers that affect health and safety.
- Workers have a duty in taking reasonable care for their own health and safety and to not adversely affect the health and safety of other people in the workplace.
- Hazard identification means looking for those things that have the potential to cause harm.
- It is important that all key safety documents are easily accessible to everyone at the workplace.

7.2 Recognising and assigning a level of risk

- A risk assessment generally focuses on identifying a level of risk, which is based on considering the consequences of the risk, given the existing control measures, and the likelihood of those consequences occurring.
- The likelihood of an incident occurring is based on probability and as such it is not possible to say with certainty when an incident may occur, only the probability of an incidence occurring.

7.3 Identifying risk control measures

- A common method used for identifying risk control measures is through the hierarchy of control.
- The highest level of control in the hierarchy is totally eliminating the hazard and risk, which is followed by reducing the risk through substitution, isolation and/or engineering controls, then reducing the risk through administrative controls.
- Risk reduction is achieved through substitution, isolation or engineering controls of the hazard.
- The use of PPE represents the lowest level of control and includes anything workers use or wear to minimise risks posed to their health and safety.
- Safety measures should be monitored and reviewed regularly.

7.4 Documenting control measures

- A job safety analysis (JSA) is a formalised procedure, which is usually documented on a JSA form, and details the steps necessary to complete a specific job or task.
- A safe work method statement (SWMS) is a document used within the construction industry in Australia.
- The effectiveness of the SWMS relies on the worker following the requirements as set out in the document.
- A risk register is a record of all the identified risks, the likelihood and consequences of a risk occurring, the actions implemented to reduce those risks, and who is responsible for managing the risks.
- Standards provide guidance on specific matters and as such a Standard may be considered in a court of law when determining compliance with the WHS laws.

7.5 Construction site hazards

- Hazards to be aware of when carrying out maintenance or breakdown electrical work concern 'mechanical movement'.
- Pushbuttons, and foot, selector, limit, flow, temperature, pressure and photoelectric switches and other control-circuit-type devices, are not energy-isolating devices.
- Lasers emit beams of non-ionising radiation, at wavelengths from the ultraviolet to the infrared.
- Thermal burns from visible or infrared laser beams are possible as well as photochemical burns from ultraviolet laser beams.

7.6 Hazards associated with high voltage

- The term 'high voltage' applies to electrical equipment that operates at more than 1000 V a.c. rms or 1500 V d.c.
- Isolation of electrical circuits is a basic safety procedure that protects workers, electrical production and consumption devices or service lines.
- Large electrical installations, with high energy demand, such as shopping centres, factories, mines, and the like may be served at high voltage of 11 kV or 33 kV alternating current.
- Touch voltage (also referred to as step potential) is caused by a fault current flow in a conductive medium establishing a potential difference between the feet of a person with the earth contact point and the hand, head or other parts of the body in contact with the energised material.
- Electrical current called a ground fault flows into the ground when anything such as a tree branch, person, metal ladder or the metal arm of a cherry picker accidentally makes contact with an exposed live conductor.

7.7 Hazards associated with extra-low voltage and low voltage

- The number and type of circuits required in an electrical installation are determined by the load on the circuit, the location of the loads, any seasonal variations affecting the loads and any special conditions.
- Low voltage (LV) is defined as voltage exceeding 50 V rms a.c. or 120 V d.c. but not exceeding 1000 V rms or 1500 V d.c.
- Inadequate separation between high-voltage and low voltage cables or terminations can result in hazardous induced voltages and arcing at the terminations.
- Indirect contact refers to a person or livestock coming into contact with an exposed conductive part, which is not usually live, but which has become live accidentally (due to insulation failure or other condition).
- Direct contact refers to a person or livestock coming into contact with an exposed conductive part which is usually live and becomes part of the fault path.

7.8 Disconnecting and reconnecting electrical equipment

- Lockout, isolation and tagging procedures are established to protect persons and property in a workplace location from any electrical hazard, either energy or mechanical in nature, while inspections, repairs or maintenance work are being carried out.

- When required to carry out installation or maintenance work on electrical production and consumption devices or service lines, these devices or lines must be isolated or de-energised so that they are safe to access.
- The 'attempt to operate' isolation procedure is not an effective method of verifying that a circuit is de-energised because mechanical failure with switch contacts cannot be determined by the 'attempt-to-operate' method.
- Common technologies used for alternate electrical power supplies include engine-driven alternators and generators, photovoltaic panels, wind-driven alternators and generators, and personal or micro hydroelectric systems.

TRIAL EXAM

For Chapter 7 knowledge assessment, please complete the following trial exam.

1 The first element in a risk management process is:
  a identifying potential hazards
  b assessing the risk
  c controlling the risk
  d reviewing control measures

2 It is possible to identify hazards by:
  a undertaking a thorough skills audit
  b reviewing personnel records
  c reviewing control measures
  d examining the workplace and how work is performed

3 A hazard is defined as:
  a an occurrence
  b an injury
  c an illness
  d something that has the potential to cause harm

4 A risk is:
  a the possibility that harm may occur when exposed to a hazard
  b the consequence of an injury
  c the likely outcome of a hazard
  d something that has the potential to cause harm

5 A risk assessment generally focuses on:
  a documenting risk outcomes
  b assessing workers' ability to perform tasks
  c identifying a level of risk
  d identifying hazards

6 The purpose of a risk assessment matrix is to assist in:
  a determining where to lay blame in the event of an incident occurring
  b determining the likelihood and consequence of various health and safety hazards and risks occurring
  c allocating hazard identification colour coding to specific tasks
  d identifying those members of a work team most likely to suffer an injury

7 Using numbers from 1 to 4, list the preferred order for the hierarchy of hazard control.
  [ ] Administrative
  [ ] Reduction
  [ ] Elimination
  [ ] PPE

8 Hazards are **not** controlled at the source through the use of:
  a reduction strategies
  b administrative controls combined with PPE
  c hazard elimination
  d reduction and elimination strategies

9 A formalised procedure, which details the steps necessary to complete a specific job or task is:
  a a safe work method statement
  b a risk register
  c considered unnecessary for electrical work
  d a job safety analysis

10 The risk ranking with the control measures implemented is:
  a the tolerable risk
  b residual risk
  c expected hazard
  d controlled through the use of PPE

11 Low a.c. voltage is defined as:
  a voltage less than 12 V
  b voltage below 50 V but above 30 V
  c voltage between 12 V and 30 V
  d voltage exceeding 50 V but not above 1000 V

12 What factor makes live low voltage equipment hazardous?
  a the lack of reinforced insulation
  b the design of the equipment
  c the small creepage and clearances between low-voltage terminations
  d no circuit protection for the low voltage

13 Three categories of electrical hazards are:
  a electric shock, arcing and capacitive coupling
  b electric shock, arcing and toxic gases
  c arcing, toxic gases and live exposed conductors
  d toxic gases, electric shock and voltages on unused conductors

14 The systems used where an inherently safe system of supply is required is:
a HV
b LV
c SELV
d PELV

15 Which laser class has little risk to eyes and no risk to skin?
a Class 1M
b Class 2M
c Class 3R
d Class 4

16 A serious hazard with lasers is:
a the long distance they project light
b the high audible frequencies they produce
c damage to the retina of the eye
d their visible light output

17 What colours are used for a personal danger tag classification?
a red and black on a white background
b black with a yellow background
c red with a yellow background
d black with a white background

18 The colours used to represent a personal danger tag are:
a black with a yellow background
b red and black on a white background
c red with a yellow background
d white and black on a yellow background

19 What is meant by the term 'touch voltage zone'?
a the work area
b a zone near live terminals
c a working area where it is safe to touch energised conductors
d an area 2400 mm in height by 1000 mm in width where a prospective fault voltage may appear

20 The most common entry point for current entering a worker is the:
a head
b leg
c face
d hand

Lighting circuits, equipment and controls

This chapter provides electrotechnology workers with essential knowledge and skills pertaining to basic illumination concepts, how the various lamps operate, the function of additional devices and equipment and the purpose of special lighting. This chapter provides underpinning knowledge for the unit UEEEL0009 from the UEE training package.

LEARNING OBJECTIVES

Wiring methods
- Explain the difference between the loop at the light and the loop at the switch wiring methods.
- Use diagrams to show implementation of the loop at the light and the loop at the switch wiring.
- Explain one-way, two-way, and intermediate switching.
- Outline the purpose of a switching chart.

Lighting control
- Explain the function of lighting control.
- Detail common methods employed in lighting control.
- Outline the operation of occupancy sensors.

Emergency and evacuation lighting
- Explain the notion of emergency lighting.
- Detail the requirements for effective emergency and evacuation lighting.
- Describe the features of common emergency and evacuation lighting systems.

Principles of lighting
- Outline the process of seeing.
- Explain the meaning of common lighting terms.
- Perform basic lighting calculations.
- Explain the inverse square law and law of reflection for lighting.

Types of luminaires
- Identify basic luminaires.
- Explain the function of a luminaire.
- Outline the basic requirements for electromagnetic compatibility.

Lamp types
- Identify common lamp types.
- State the principles of operation of various lamps and luminaires.

Energy-saving lighting
- Explain why LED lamps are an important and significant advancement in illumination.
- Outline the operation of optical fibre lighting systems.
- Explain the operation of cold-cathode fluorescent lamps.

Fire protection
- Detail the requirements for fire protection systems.
- Summarise the operation of common thermal and smoke detectors.

Requirements for luminaires
- Outline the general requirements for the installation and operation of luminaires.
- Apply the principle of lighting design.

Testing and fault finding
- Outline the requirements for testing lighting circuits.
- List common faults that occur in lights.

8.1 Wiring methods

The most commonly used wiring system for lighting circuits is unenclosed flat TPS cable. **Figure 8.1** illustrates TPS wiring for lighting circuits at the first-fix stage of installation work with long tails for ease of switch connection. On a cable wall, a drip loop prevents moisture from flowing down the cable and onto the connector. Note the mark on the timber stud indicating the position where the wall bracket is required to be fixed.

FIGURE 8.1 TPS wiring at the first-fix stage

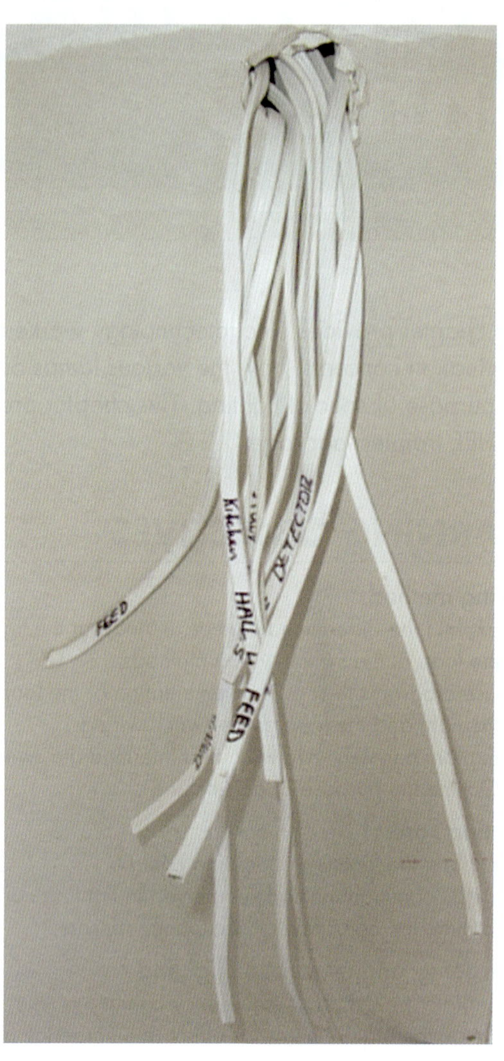

FIGURE 8.2 Identification of TPS lighting circuits

Figure 8.2 illustrates TPS wiring for lighting circuits after the plasterer has finished and has kindly put a hole in the wallboard for the wiring to exit. Plasterers do not have to do this, so it is important to be on good terms with the other trades. Otherwise the electrician finds blank walls. Always mark the floor with an arrow or another symbol directly below where the cable is to exit from the wallboard. Note how the installation electrician has identified each cable by writing on them. In addition, always indicate dimensions on the location plan in case the floor has been sprayed, obliterating all marks.

One-way switched lighting circuits

A one-way switched lighting circuit consists of a single-pole, double-throw (SPDT) switch connected in series to one or more luminaires connected in parallel. A looping at the switch method of wiring a light circuit is illustrated in **Figure 8.3**. Lighting circuits can be run in 1.5 mm^2 cable (nominal minimum cross-sectional area (CSA) of conductors for lighting points) protected by a 10 A circuit breaker or 1.5 mm^2 cable protected by a 10 A or 16 A circuit breaker if installation conditions allow.

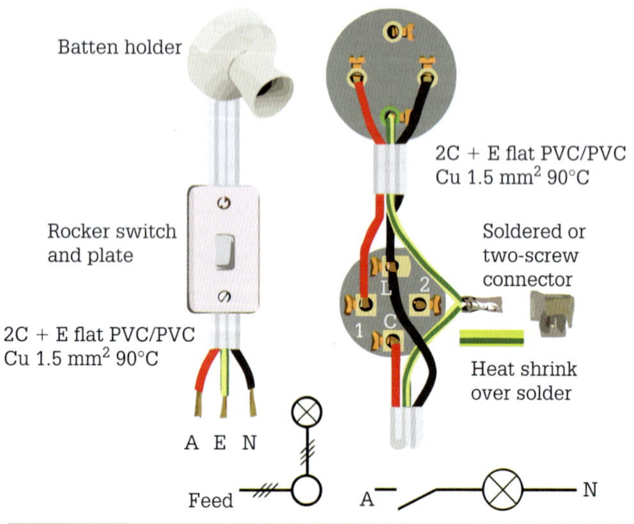

FIGURE 8.3 Looping at the switch method one-way lighting circuits

A batten holder is regarded as an electrical accessory and is used to provide a detachable connection for a lamp.

In **Figure 8.3**:

- the active conductor terminates in the common terminal (C) of the switch
- the neutrals terminate in the loop terminal (L) of the switch
- the switch wire (the conductor that provides a controlled active to the light) terminates in terminal 1 of the switch (terminal 2 is not used).

The earthing conductors are twisted together, soldered and insulated with either earth heat-shrink tubing or earth tape. Alternatively they can be terminated (unsoldered) in a two-screw BP connector. (Refer to AS/NZS 3000:2018 *Wiring Rules* for soldered connections and earth connections.) **Note:** If soldering the earth conductors of XLPE insulated PVC 90° sheathed 450/750 V cable with a gas torch, the insulation encourages a flame as minimal heat only is required to support combustion.

The switch wire, neutral and earth conductors of the cable from the switch then terminate in the relevant terminals at the batten holder, ceiling rose or light fitting. All lighting points in the installation must be provided with an earthing conductor.

Looping at the switch means that the circuit brings the power to the switch first and then to the load. The next light circuit is connected from the common and the loop terminal at the switch and also to the earth two-screw BP connector. In addition, an earth pigtail from the earth termination at each switch must be connected to the building frame if a metal frame construction is used. This method of wiring can be used where there is limited space for the lighting point such as cathedral ceilings and external circuits where the wiring runout and back is over a long distance.

Figure 8.4 shows how the tails of each final sub-circuit TPS cable are prepared for termination. **Note** that the electrician who prepared the terminations used a soldering method to terminate earths and neutrals and then used PVC tape to insulate which demonstrates poor practice. Another example of poor practice is the second tail on the bottom of the figure which is unacceptable because the twisting of the conductors could produce a hot spot which might result in a fire.

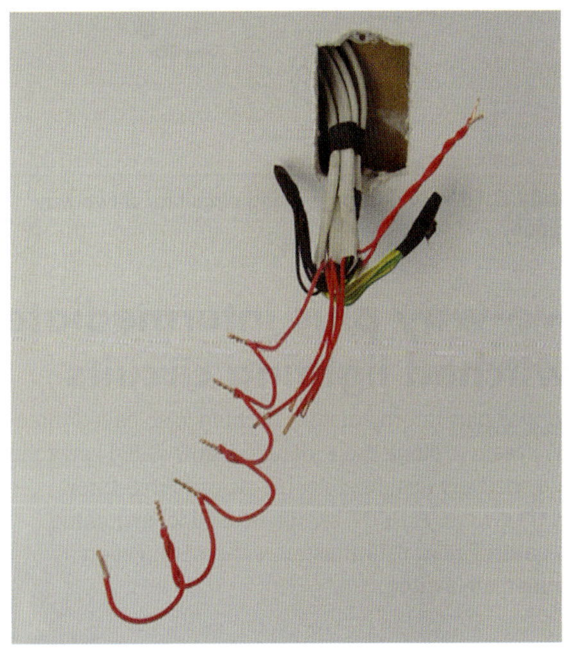

FIGURE 8.4 TPS cable tails prepared for termination

Looping at the light is the most commonly used wiring method for lighting circuits using TPS cables and is illustrated in **Figure 8.5**. This circuit brings power to the batten holder; ceiling rose or luminaire terminals first and then to the switch. Power for the next light circuit is taken from the loop (line active), line neutral and earth terminals at the batten holder, ceiling rose or luminaire terminals. No earth at the switch means no pigtail is available for metal frame buildings at the switch position. Alternative arrangements for earthing the metal frame must be used.

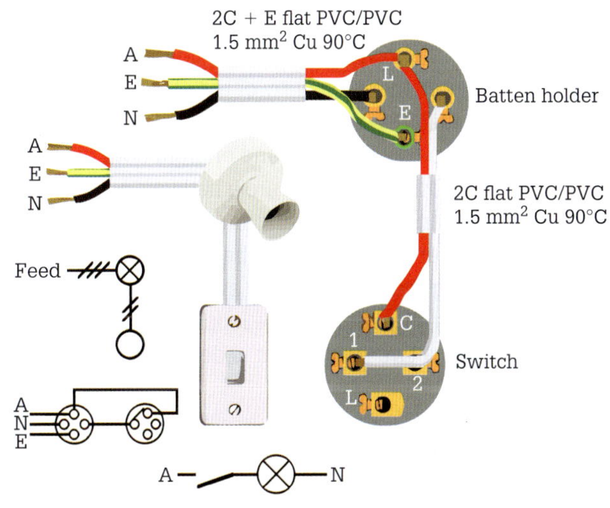

FIGURE 8.5 Looping at the light – one-way switching

A two-core, flat 450/750 PVC insulated, V-90 PVC sheathed TPS cable having 1.5 mm^2 copper conductors is the preferred method of installation. However, two SDI cables between the lighting point and the switch could be used. Two lights controlled by one switch is illustrated in **Figure 8.6** while **Figure 8.7** illustrates the looping of power from one light to another with separate control.

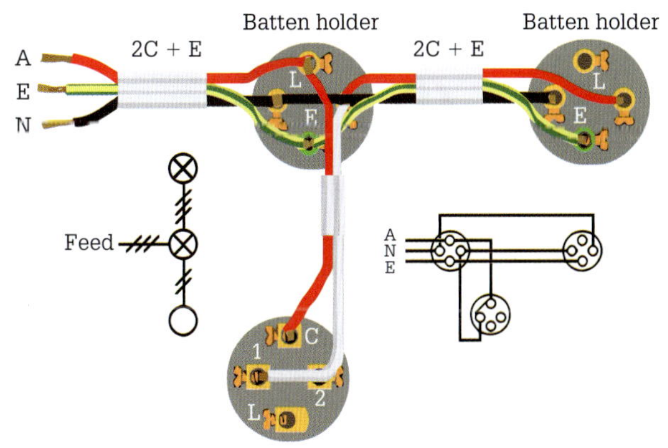

FIGURE 8.6 Two lights controlled by one switch

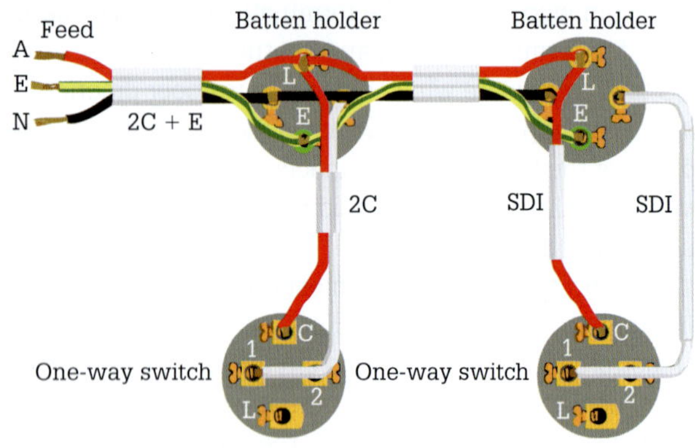

FIGURE 8.7 Looping from one lighting point to the next

Two-way switched lighting circuits

A two-way switched lighting circuit consists of two single-pole, double-throw (SPDT) switches and one or more luminaires. This circuit allows the light fitting to be switched on and off independently of the other switch. An example of such a system is in a stairwell where the light fittings can be switched on and off from both upstairs and downstairs. Other examples are a hallway, a room that has two entrances and a bedroom with the main light switch and another switch near the bedhead. A two-way circuit illustrating the looping at the switch method of wiring is shown in **Figure 8.8** while looping at the light method of wiring a two-way circuit is depicted in **Figure 8.9**.

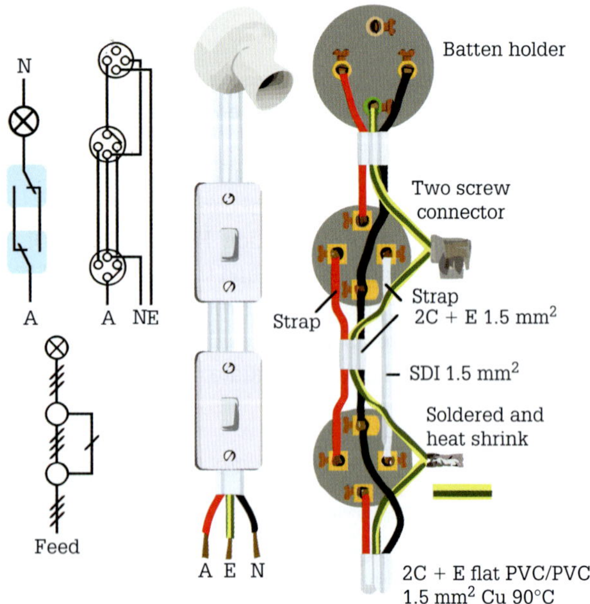

FIGURE 8.8 Looping at the switch method – two-way circuits

Existing light switches in an installation can be rewired to provide for a two-way addition as shown in **Figure 8.10**.

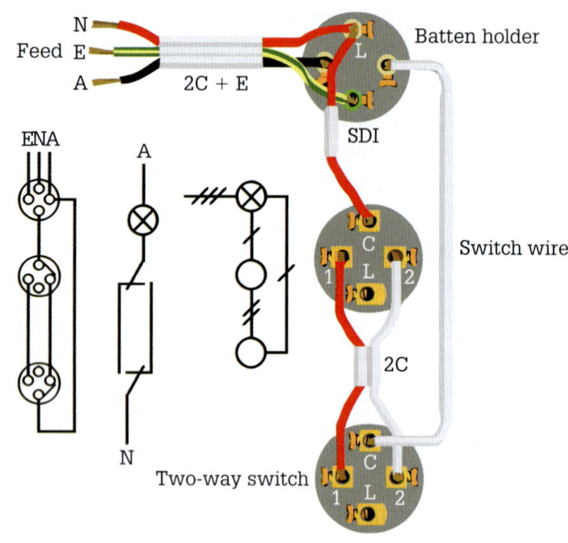

FIGURE 8.9 Looping at the light method – two-way circuits

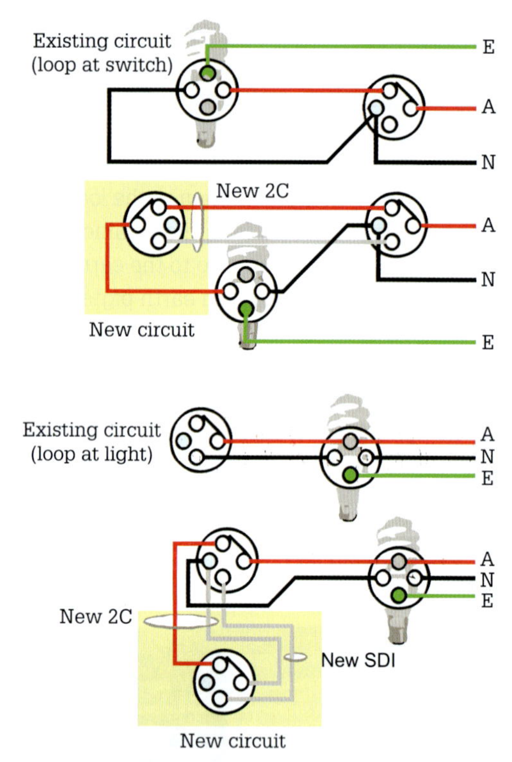

FIGURE 8.10 Rewiring an existing circuit for a two-way addition

Two-way plus intermediate switched lighting circuits

Two-way plus intermediate switching (also called three-way) uses two single-pole, double-throw (SPDT) switches and a specially designed double-pole, double-throw (DPDT) switch. A DPDT switch is used for switching two paths. **Figure 8.11** illustrates the wiring of an intermediate switch.

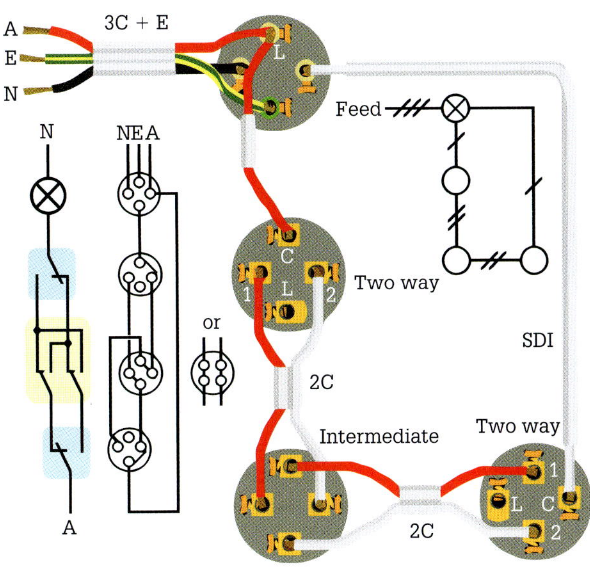

FIGURE 8.11 Two-way plus intermediate switching

Terminals	Toggle up	Toggle down
1–2		
1–3		✓
1–4	✓	
2–3	✓	
2–4		
3–4		✓

✓ = bridge

FIGURE 8.12 Switching chart

The addition of the intermediate switch in **Figure 8.11** provides luminaire on/off control from more than two positions. Multiple numbers of intermediate switches can be added to the two SPDT two-way switches. An example of where intermediate switching can be used is with stairwells whereby the luminaires can be controlled from switches on each floor level.

Switching chart

Note that in the circuit diagram in **Figure 8.11** there are two possible bridging arrangements. This is because manufacturers such as HPM and Clipsal have different internal bridging arrangements for their intermediate switches. Testing the switch with a continuity tester using the terminal numbers on the switch and a switch chart as shown in **Figure 8.12** determines what links are made between the various terminals. A switching chart is a useful tool when an unknown type of switch is encountered.

Converting from a circuit diagram to a wiring diagram

It is important to be able to convert a circuit diagram to a wiring diagram in order to build the circuit correctly. A wiring diagram shows the physical relationship of all the components, as well as the information needed to install a circuit.

EXERCISE 8.1

Convert the circuit diagram in **Figure 8.13** to a wiring diagram using loop at the light method.

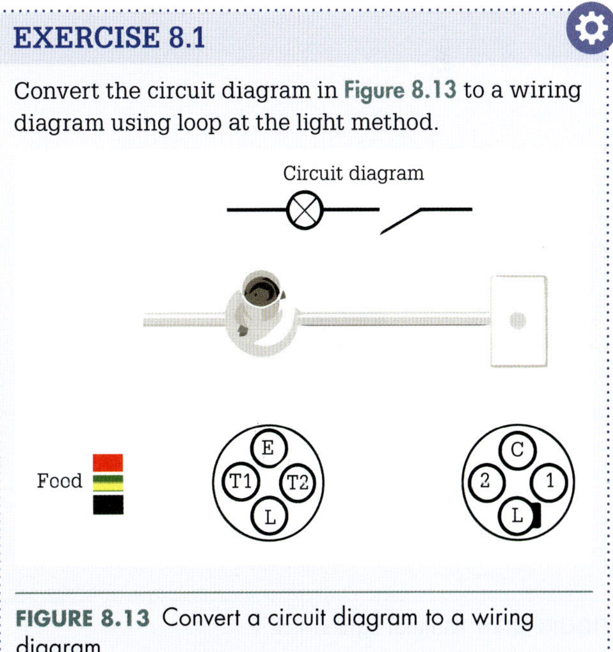

FIGURE 8.13 Convert a circuit diagram to a wiring diagram

REVIEW QUESTIONS

1 Draw a circuit diagram showing the looping at the switch method for a one-way lighting circuit.
2 Draw a circuit diagram for a looping at the light method of wiring for a two-way circuit.
3 Draw a circuit diagram for a two-way plus intermediate switching.

8.2 Lighting control

The overall purpose of a lighting control system is to reduce energy consumption while providing a productive visual environment. This requires:

- making available the right amount of light
- distributing the light where it is needed
- delivering a light when it is needed.

Lighting controls can:

- dim the luminaires in reply to daylight (daylight harvesting/constant light)
- dim the luminaires in reply to lumen depreciation of the luminaire and room surfaces (maintained luminance)

- deliver the right light level for different tasks and movement areas
- offer presence detection to turn lights off if the area is unoccupied
- prevent the artificial lighting from turning on if daylight is abundant
- be linked to time-based controls to turn lights on or off depending on the time of day.

Manual light switches

The simplest form of control of the luminaire is by manual switching and comprises an on/off switch adjacent to the entry to the room/building. It is suggested that the switch mechanisms of manual switches (shown in **Figure 8.14**) be capable of switching a minimum of 10 A.

FIGURE 8.14 Manual light switch

Manual switches are available as one gang (means number of switch mechanisms) one way, two ways or intermediate with a flush (flat on the wall) finish. Manual switches can be mounted on walls or columns to enable switching as required. Each enclosed room should have separate manual switches.

When installed in industrial and commercial installations, the manual switches should be provided with permanently fixed, easily understood labels identifying the area of lights covered by each switch.

For each functional area, manual switches should be positioned in group gang panels as shown in **Figure 8.15**.

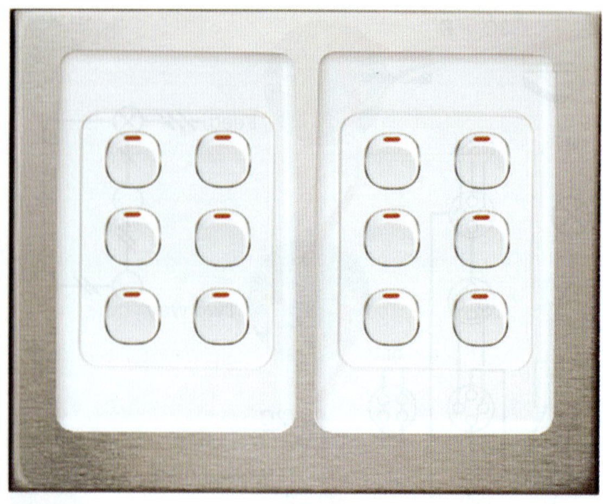

FIGURE 8.15 Group gang panel manual switches

Key switches

Key switches (shown in **Figure 8.16**) require a physical or an electronic key to operate. They can limit the use of lights to authorised personnel.

FIGURE 8.16 Key switch and fingerprint switch

Card switch

The card key switch (shown in **Figure 8.17**) turns electrical circuits on or off when a card key is inserted or removed from its slot. The switch unit consists of a normally open and normally closed isolated relay, allowing it to interface with third-party energy and lighting management control systems.

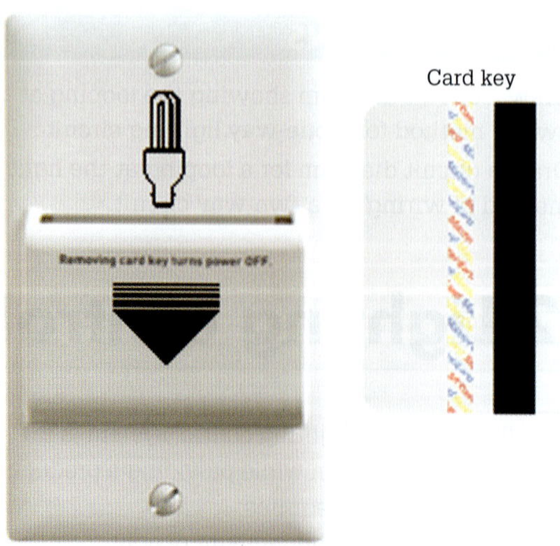

FIGURE 8.17 Card switch

Master on switching

A master on switch (see **Figure 8.18**) is used with two-way plus intermediate switching when control over the 'on-time' for the luminaires is required. This may be at a motel where to attract clients the exterior luminaire would be mastered 'on' thereby preventing individual room occupiers from turning out the exterior light. Control over the exterior luminaire is given back to the room occupier whenever the master on switch is turned 'off'.

Note that the master 'on' switch is connected across terminals 1 and 2 of the two-way switch. When the master on switch is activated these two terminals are shorted together.

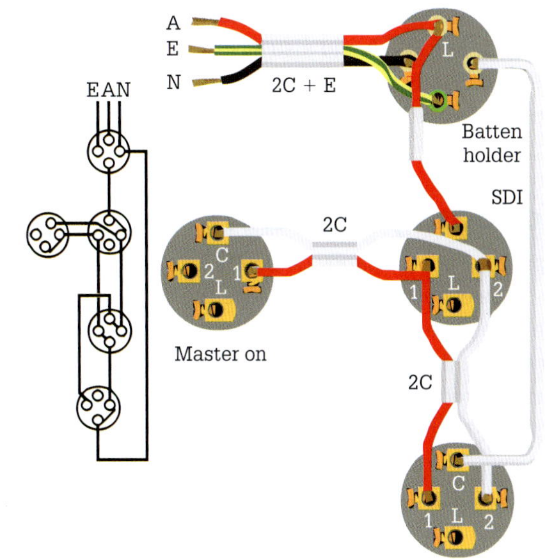

FIGURE 8.18 Master on switching

Master off

Master 'off' switching (see **Figure 8.19**) using looping at the light method is simply an SPDT switch connected to the active conductor upstream of the first two-way switch.

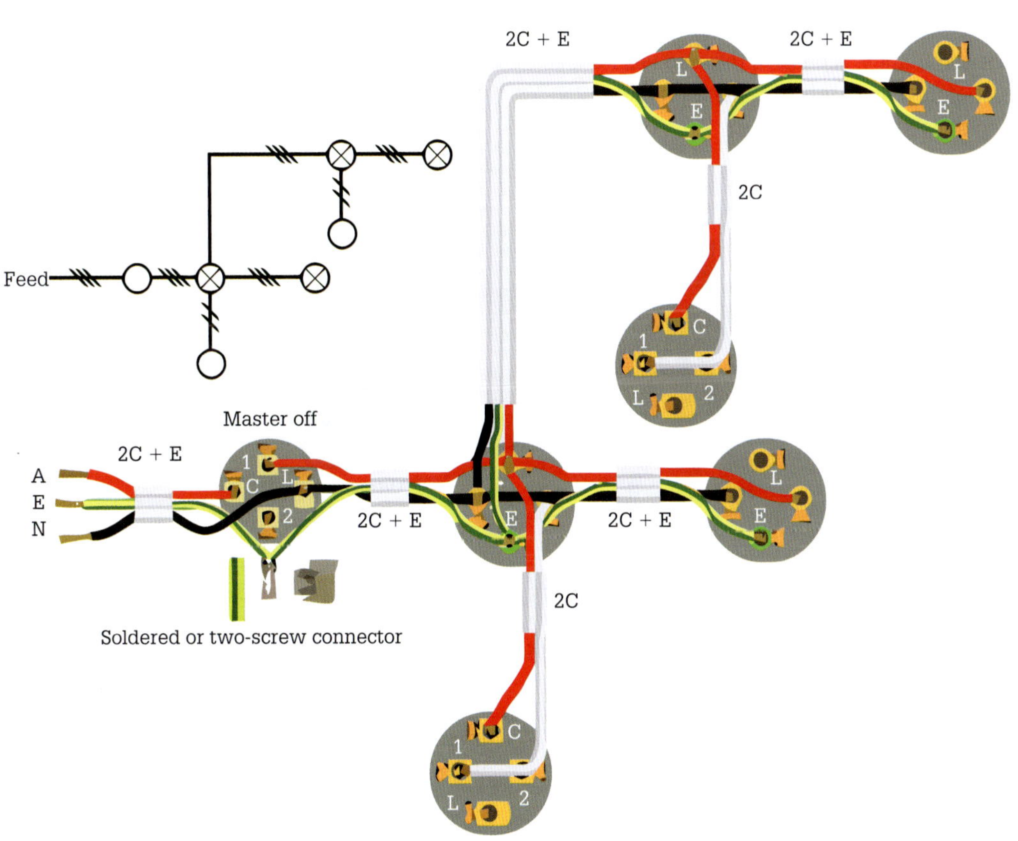

FIGURE 8.19 Master off switching

For commercial premises

A master off switch rated to carry the load current (a contactor can also be used) is usually located at the entrance point to premises where persons enter and leave. In many installation applications, the master off switch is a key-locking switch which prevents unauthorised use. An example of a master off switch is in some restaurants where it is used to turn off all lighting except security lighting within the restaurant at closing.

Programmable control (PC)

Many office lighting systems require 365-day programmable checks on each floor distribution board to control power to office lighting circuits. A programmable control provides prior warning to occupants that lamps are to be automatically switched off. The PC is also interlinked with the security systems so that except for safety lamps all lamps are switched off after the security system is set.

With a sensor-operated PC, shift workers are detected and given reassuring illuminance of egress paths and common areas. Gradual illumination closure occurs after all movement has ceased, and a final sweep of all fields is completed. The PC should also have an easily accessible, visible and clearly labelled after-hours request switch that provides one hour of additional lighting in case of emergency.

Occupancy detectors

The purpose of occupancy detectors is to turn luminaires on or off as occupants enter or leave the monitored room or space. These devices can be used alone or together with other control devices in a coordinated manner and are very effective in reducing energy consumption.

Broadly there are two types: infrared and ultrasonic/microwave. The devices incorporate motion detection; PE (ambient light) detection and PIR (passive infrared) receive capability. The most popular types of sensors are discussed below.

Passive infrared (PIR)

These sensors are used for security purposes in small areas with infrequent occupancy. A passive infrared sensor (see **Figure 8.20**) responds only to infrared energy radiated by the object being sensed.

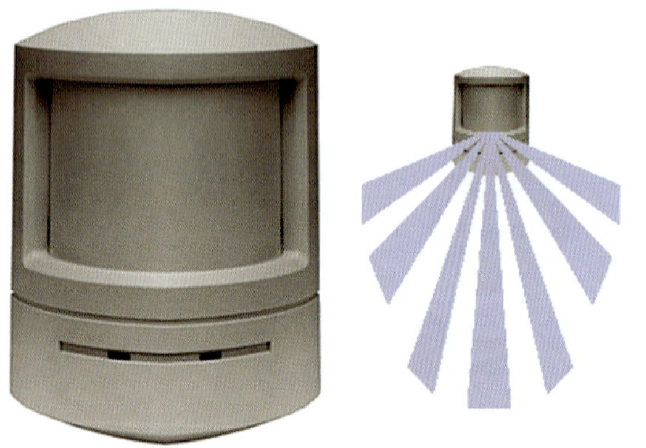

FIGURE 8.20 Passive infrared sensor

All material objects emit electromagnetic radiation, the intensity of which depends primarily on the object's temperature. This phenomenon is known as blackbody radiation. Human beings emit infrared energy through blackbody radiation. Because the amount of radiation depends on the temperature, it is sometimes called thermal radiation or heat radiation. The skin temperature of a person is approximately 34 °C, which is usually higher than background temperatures. As a person walks past the sensor, their higher skin temperature is detected by the PIR sensor which becomes aware of the infrared energy with its pyroelectric material. When subjected to this infrared radiation the pyroelectric material generates a small electric signal. An amplifier circuit boosts the low signal and feeds it to a differential comparator. The comparator looks for a difference in the signal from prior readings of the space to trigger an output. To reduce false activations, an infrared filter which only responds to certain wavelengths is used. After installing PIR sensors a 'walk through' test as shown in **Figure 8.21**, must be carried out to determine the sensor's alignment and check that the head is pointed towards the movement to be detected.

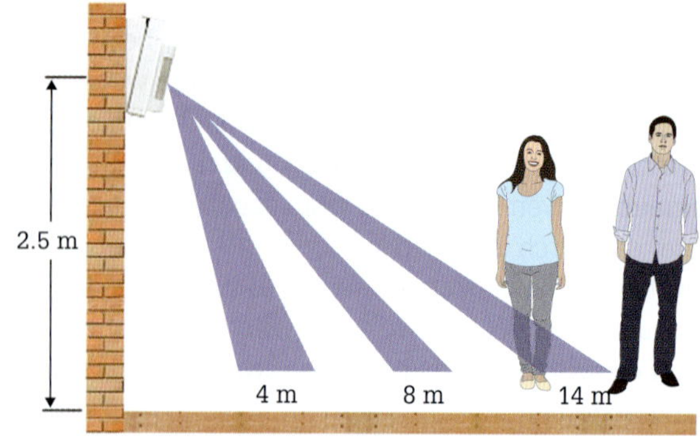

FIGURE 8.21 'Walk through' test

Ultrasonic sensing

Ultrasonic sensors are based on the speed at which sound travels through air. These sensors measure distance by emitting ultrasonic frequency sound waves (a ping), then calculating the time required for the echo to return. The frequency of the echo waves when an object is moving is different from the frequency when the object is stationary (the Doppler effect). If the sensor does not detect movement, it does not activate. An ultrasonic sensor is illustrated in **Figure 8.22**.

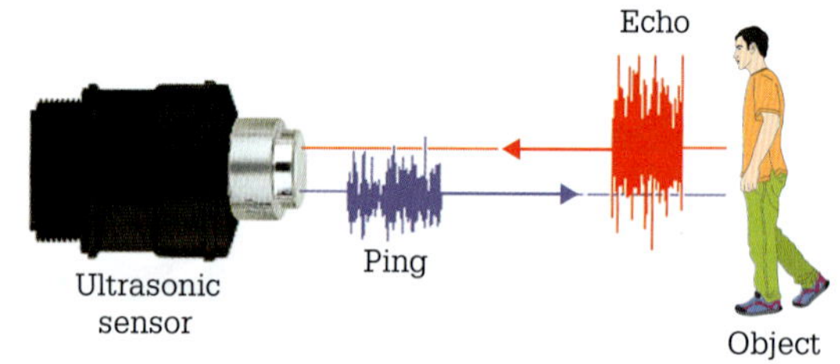

FIGURE 8.22 Ultrasonic sensor

Microwave sensing

Microwave sensing, shown in **Figure 8.23**, is used for large internal spaces (aisles, corridors, halls) or external areas with infrequent occupancy.

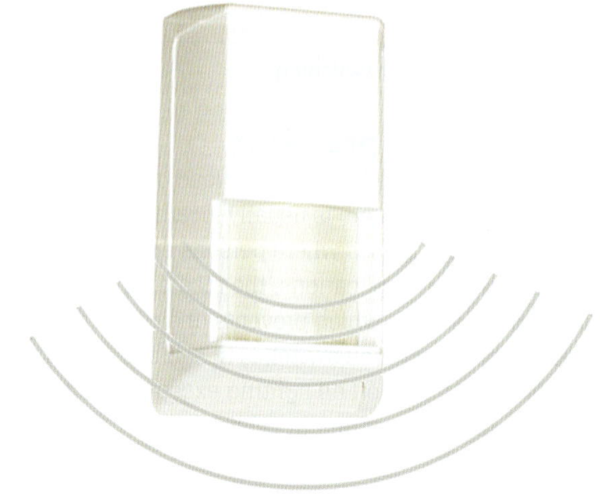

FIGURE 8.23 Microwave sensor

The microwave sensor uses a radar signal (5.8 GHz) which causes microwaves to bounce off an object. A moving object within the view of the microwave sensor returns a frequency higher or lower than the sent frequency. The frequency of the returning microwaves is then measured and a signal to turn on a luminaire occurs when it detects a difference. Microwave sensors can detect motion from a long distance away. They can even detect a slight movement through doors; sheets of glass or thin walls and so can be located anywhere. The disadvantage of a microwave sensor is that it is very sensitive and can be disturbed by electrical devices such as a light and a radio.

REVIEW QUESTIONS

1 What is the purpose of a lighting control system?
2 Name the simplest form of lighting control.
3 What should be provided for manual switches when installed in industrial and commercial installations?
4 State the purpose of an occupancy detector.
5 When would a microwave sensing detector be installed to control lighting?
6 When would a PIR sensing detector be installed to control lighting?

8.3 Emergency and evacuation lighting

Exit and emergency lighting must comply with the Building Code of Australia, AS 2293 Parts 1 and 2 *Emergency evacuation lighting in buildings*. Exit lighting should be illuminated at all times, while emergency lighting must operate automatically and provide sufficient illumination for safe evacuation of all areas when required. The emergency lighting system should be designed with a minimum illumination time of not less than 90 minutes (via a separate battery pack). A separate control system at each relevant distribution board must be provided for testing of the exit and emergency lighting powered from that distribution board. In addition, all emergency and exit lighting systems should be inspected every 12 months with maintenance recorded in a log book.

The term 'emergency lighting' includes four categories of lighting, each incorporating a separate energy supply, as follows:

- emergency escape lighting
- illuminated emergency exit signs
- high-risk-task area lighting
- standby lighting.

Only two of these four categories are addressed in the AS/NZS 2293 standards – namely emergency escape lighting and illuminated emergency exit signs.

Emergency lighting and exit signage and directional exit signage required throughout a building are governed by the National Construction Code (NCA).

The objective of emergency and exit lighting is:

- In an emergency, to safeguard occupants from injury by having adequate lighting, adequate identification of exits and paths to travel to exits and being made aware of the emergency; and that
- A building is to be provided with adequate lighting on failure of normal artificial lighting during an emergency and adequate means of warning occupants to escape and to manage the evacuation process and to identify exits and paths to travel to an exit.

To meet these conditions, exit signs and emergency lighting must be installed in buildings.

Exit signs

Exit signs are designed to minimise the risk of death or injury to persons during an emergency because of a failure to locate an exit.

The signs must be plainly visible at all times and located on, above or adjacent to the specified designated exits and doors. They must also be situated in positions where exits are not readily apparent, that is, corridors, hallways and lobbies.

When a fire alarm sounds, exit signs must direct persons to a safe place or open area.

On failure of the electrical supply to the normal lighting, emergency luminaires and exit signs must be automatically energised from their emergency supply. In the event of a power failure, they must operate on battery power for a minimum period of 90 minutes.

Exit signs are to be mounted between 2 metres and 2.7 metres from floor level.

Where viewing distances exceed 24 metres the exit legend should be increased by the factor:

$$\text{Exit legend dimensions} = \frac{d}{24}$$

where d – viewing distance in metres

The emergency 'Exit' luminaire has been changed to the international standard of pictograms. This new international standard displays an arrow pointing the way and a man hurrying to the nearest way out. This is a universal sign that is to be adopted worldwide. In accordance with AS/NZS 2293.1:2005 Part 1 *Emergency escape lighting and exit signs for buildings – System design, installation and operation*, exit signs and direction arrows as shown in **Figure 8.24** should be white and the background green.

FIGURE 8.24 Exit signs

Where exit signs are installed there must be an unobstructed path of egress to the exit. All illuminated exit signs should incorporate a maintained emergency lighting lamp, a self-contained power pack and electronic control gear. The signs must also be capable of being fixed on ceilings or walls or suspended.

The National Construction Code 2019 (NCC 2019) does not explain how the fittings must be installed; instead it refers electrical installers to AS/NZS 2293.1 for the installation compliance.

It is a requirement of the AS/NZS 2293.2 *Inspection and maintenance* standard that all emergency lighting and exit signs in public buildings be adequately tested and maintained. These systems are required to be tested at least every six months by a qualified electrician and the details of testing, date and person responsible for the work entered into a log book. The log book should also be located on the premises so it can be available for inspection upon request.

Testing procedures

Maintenance and testing procedures vary with the type of system.

Maintenance procedures for single-point systems

Every six months:

- Ensure there have been no power interruptions for 16 hours prior to test.
- Check fluorescent lights for blackened ends. If blackened ends are found, both tube and starter are to be replaced.
- Perform a manual discharge test (simulated power failure). To comply with the requirements of AS 2293, the emergency luminaire must provide the correct illumination for a period of at least 120 minutes on initial installation and 90 minutes thereafter.
- Restore exit or emergency light to normal condition and check that the battery charger operation indicator functions correctly.
- Perform an automatic discharge test. Visually check that the unit operates correctly by means of the indicator at each unit or indicator panels.
- Replace or repair all units that FAIL.
- For indirect lighting systems, check that all reflective surfaces around the exit or emergency light are in good condition and that beam lights do not cause disability glare.
- Record all test results and action taken in a log book.

Every 12 months:

- Carry out the above six-monthly testing and inspection procedures.
- Clean all light-emitting and light-reflecting surfaces.
- Record all test results and action taken in a log book.
- This log book must be kept on site and must be available for inspection and audit when required. The log book is to include the following:
- Schedule of all emergency luminaires and exit signs showing location, circuit and switchboard origin. It is also useful to include manufacturer and lamp type.
- Provision for recording details of tests carried out as detailed in AS/NZS 2293.2 Section 3 at each nominated interval.
- Provision for recording details of any corrective action taken.
- Provision for recording the names of the persons responsible for carrying out the maintenance work and the date of completion of the work.

Maintenance procedures for central systems

Checks should be carried out on the system every six months. The checks should cover:

- batteries
- battery chargers
- inverters
- distribution and control equipment
- lamps
- the wiring system
- six-monthly procedures.

Emergency lighting

Emergency lighting is intended to provide a safe path of 'way out' by illuminating the escape route from a building when a power failure or a disturbance in the supply network occurs. The emergency luminaire(s) must be installed as a lighting circuit and obtain supply from a permanent feed as well as a backup battery. An emergency lighting luminaire is shown in **Figure 8.25**.

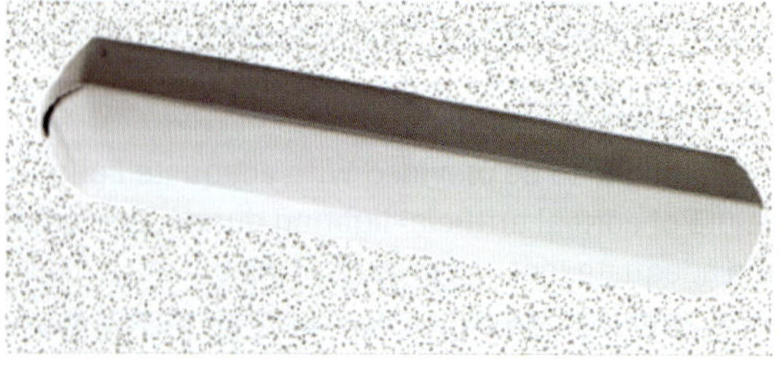

FIGURE 8.25 Emergency lighting luminaire

Every emergency lighting system must operate automatically and give sufficient illumination without unnecessary delay for the safe evacuation of all areas of a building. Emergency lighting luminaires operate only in the event of an emergency and are intended to supply a codified degree of ambient illumination under emergency conditions. To comply with the requirements of AS/NZS 2293, the emergency luminaire must provide the correct illumination for a period of at least 90 minutes. The percentage of voltage drop acceptable for extra-low-voltage systems is usually 5%. For safety reasons and codified requirements, it is essential that the emergency lighting is routinely inspected, regularly serviced and maintained.

Emergency illuminance

The current standard of AS/NZS 2293.1 is based on a ceiling or upper-level lighting system that provides a minimum lighting level directly to the floor surface. The calculations of these are based on a non-smoke condition and Australia requires 0.2 lux while New Zealand uses 1.0 lux. This means that the spacing of the emergency luminaire must be calculated to ensure that horizontal illuminance at floor level is not less than the recommended standard. While the standard provides a reasonable means of finding exits and illuminating pathways, it cannot be maintained under all conditions of a fire where smoke has entered the evacuation path.

Electromagnetic compatibility

Under AS/NZS 4051:1998, emergency lighting luminaires must comply with disturbance levels between 9 kHz and 30 MHz. This is because some emergency lighting luminaires may cause interference to radio communication systems.

Centrally supplied emergency luminaires

A centrally supplied emergency lighting system is a dedicated lighting system in which a number of emergency slave luminaires or exit signs or both without their own batteries are supplied from a common power source. Centralised emergency lighting systems use a central unit composed of a collective bank of batteries (110 V) connected to a d.c./a.c. inverter or a standby generator.

Advantages of centrally supplied emergency luminaires are:

- Maintenance and routine testing is simpler with only one location to consider.
- Battery service life is excellent, between 5 and 25 years depending upon battery type.
- Emergency luminaires are able to operate at relatively high or low ambient temperatures.
- Large batteries provide lower cost per unit of power.
- Emergency luminaires are usually less expensive.

Disadvantages of centrally supplied emergency luminaires are:

- High initial capital equipment expense.
- Higher installation and wiring costs with fire-resistant cable to each slave luminaire.
- Failure of a battery or wiring circuit, which could disable a large part of the system.
- Requirement for a 'battery room' to house cells and charger circuits and so on may also require ventilation of acid gases.
- Voltage drop on slave luminaire connected furthest from the battery bank could cause problems.

Single-point emergency luminaires

A single-point emergency lighting system uses only self-contained emergency luminaires and exit signs. A single-point luminaire is an emergency luminaire or exit sign with all control gear and battery pack housed within the luminaire. All self-contained, non-maintained, single-point fittings must be provided complete with nickel cadmium batteries with dual rate battery charge, charger and electronic control gear. The number of series-connected cells used in single-point emergency luminaires is kept to a minimum to maximise storage efficiency, reliability and cost. The emergency luminaires should be either 10 W quartz halogen or a T5 8 W fluorescent. Luminaires subject to water, dust or insect penetration must be rated at IP65.

Some single-point self-contained emergency luminaires and exit signs use a computerised testing and monitoring system. Each luminaire is complete with its own battery, charger, mains failure relay, d.c./a.c. inverter (for fluorescent tubes) and microcontroller-based communications/monitoring circuit. The whole emergency lighting system is monitored and controlled by a master controller via a communication network and software.

Single point type emergency luminaires should have nickel-cadmium batteries with dual rate battery charger, a test button and a battery charging LED.

Advantages of single-point emergency luminaires are:

- The cost of installation and equipment is low.
- Standard cable insulation may be used, as failure of mains supply due to cable burning through will automatically satisfy the requirement for a luminaire to be energised.
- Maintenance costs are low – periodic test and general cleaning only required.
- Emergency system integrity is improved, with each luminaire independent of the other.
- Emergency system can easily be extended with additional luminaires.

Disadvantages of single-point emergency luminaires are:

- They have limited environmental operating range (batteries may be harmfully affected by a fairly high ambient temperature).
- Battery life is limited depending upon type.
- Testing requires observation of each luminaire.

Maintained emergency luminaire

A maintained emergency luminaire contains one or more lamps, all of which are illuminated during normal mains supply. In the event of a mains power failure the lamps remain illuminated by automatically being switched to the internal emergency supply (battery, inverter). A maintained emergency luminaire lighting system is illustrated in **Figure 8.26**.

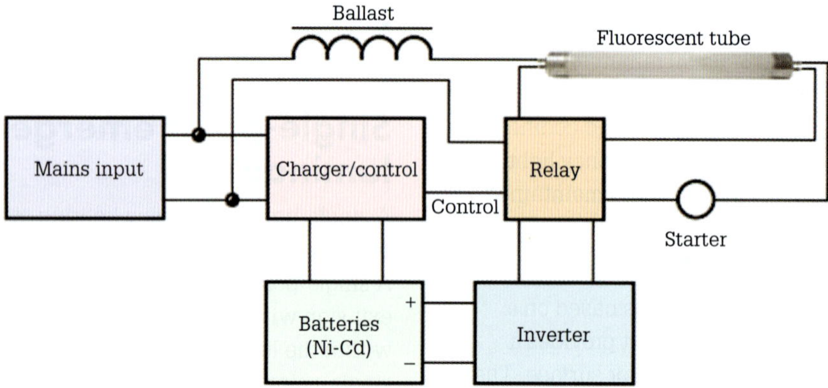

FIGURE 8.26 Maintained emergency lighting system

Non-maintained emergency luminaire

A non-maintained luminaire, shown in **Figure 8.27**, contains one or more lamps which are not illuminated during normal mains power supply, but are automatically powered by an internal stored separate battery upon failure of the normal mains supply. High-temperature nickel-cadmium batteries are used for the power supply because of their long service life (7 to 10 years) and maintenance-free capabilities.

FIGURE 8.27 Non-maintained emergency luminaire

Sustained emergency luminaire

A sustained luminaire is an exit or emergency luminaire with two or more lamps where at least one lamp operates in non-maintained mode and only illuminates when normal supply fails. The lamp in non-maintained mode is illuminated by automatically getting switched to an internally stored battery. The other lamp operates when the regular 230 V power supply is available. The reason for using sustained emergency luminaires is to make sure that the emergency lamp operates when an emergency situation occurs.

Emergency and exit lighting circuits that use the maintained, non-maintained or sustained systems should be wired from distribution boards and labelled with engraved plastic labels. In addition, a key-operated auto-off manual switch should be mounted adjacent to the switchboard for testing of the circuits (to simulate a power outage).

Other types of signage that can be used in conjunction with the emergency illuminance standard requirements are:

- LED-illuminated signs operated by the fire system
- linear light optical fibre systems
- high-density LED stand-alone floor signs and lights
- photo-luminescent signage
- auditory way-finding systems.

Recommended locations

Recommended locations for emergency exit luminaires are:

- at exit doors
- at changes of direction
- at changes of floor level
- on stairs
- at escalators
- in ablution blocks (water closets (WC), toilets)
- in switch rooms
- in covered car parks.

Exit doors

An emergency exit luminaire must be located at each exit door as shown in **Figure 8.28**. Exit signs must be mounted at a height between 2000 mm and 2700 mm above floor level, or immediately above the door if higher than 2700 mm.

FIGURE 8.28 Exit luminaire, emergency lighting and 0.2 lux

Always ensure that the exit luminaire does not exceed the maximum viewing distances recommended. The emergency escape route must always be clear of obstacles. There must be a minimum illuminance level of 0.2 lux (Australia) at floor level on the centre line of permanently unobstructed escape routes. An emergency luminaire must be placed above fire stations as shown in **Figure 8.29**.

FIGURE 8.29 Emergency luminaire above fire station

Intersecting corridors

An emergency luminaire must be installed within 2000 mm of the intersecting corridors as shown in **Figure 8.30**.

FIGURE 8.30 Emergency luminaire near intersecting corridors

Change in direction

An emergency luminaire must be installed within 2000 mm of the intersection at a change of direction as shown in **Figure 8.31**.

FIGURE 8.31 Exit luminaire near change of direction

Change in floor level

An emergency luminaire must be installed at each change of floor level, on the small side as shown in **Figure 8.32**.

FIGURE 8.32 Emergency luminaire near change of floor level

Stairways

Emergency luminaires must be installed to ensure that each flight of stairs at floor level including the intermediate landing receives direct light from the luminaire as shown in **Figure 8.33**.

FIGURE 8.33 Emergency luminaire near stairs

Escalators

Emergency lighting must be provided above escalators to enable users to get off them safely.

Ablution blocks (water closets (WC), toilets)

Ablution blocks incorporating two or more cubicles should have emergency lighting. Emergency lighting should also be provided in handicapped ablution facilities.

Switch rooms

Emergency lighting must be installed in control, plant or switch rooms that have the provision to provide normal and emergency lighting to the premises.

Covered car parks

Covered vehicle areas are part of the premises and must have exit and emergency lighting.

Spacing of emergency luminaires

In Australia the spacing of emergency luminaires is based on illuminance calculations and manufacturer's photometric data where the light reaches the floor directly from the emergency lamp, as shown in **Figure 8.34**. Minimum illuminance level for emergency lighting in Australia is 0.2 lux at floor level on the centre line of permanently unobstructed escape routes.

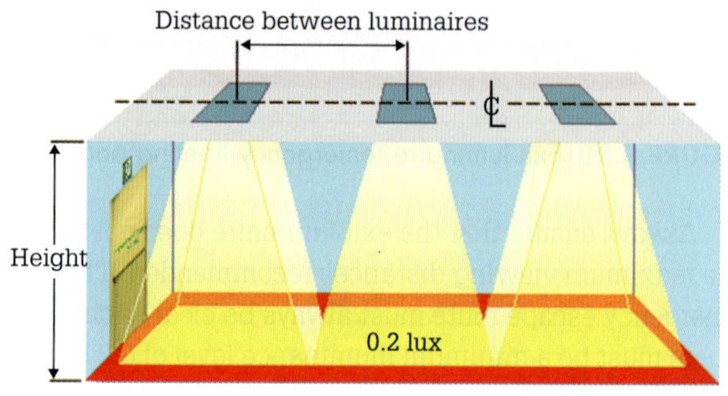

FIGURE 8.34 Siting emergency luminaires

For spacing requirements of emergency luminaires, Table 5.1, 'Maximum spacings for class A, B, C, D and E luminaires' in AS 2293.1:2005 *Emergency escape lighting and exit signs for buildings* must be consulted.

The luminaires can be mounted transversely or axially as shown in **Figure 8.35**.

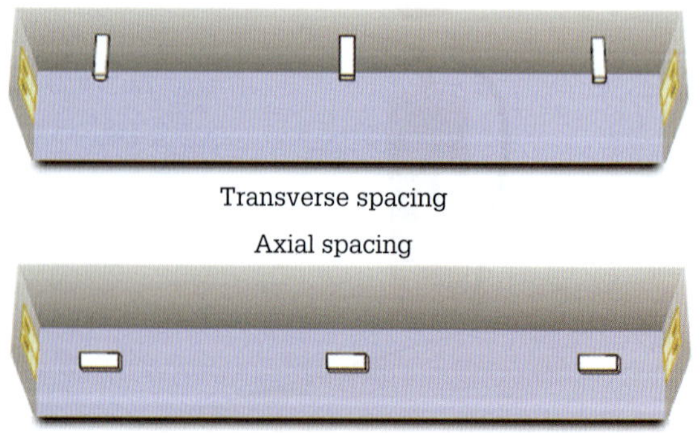

FIGURE 8.35 Transverse and axial spacing

Figure 8.36 illustrates various mandatory locations for exit signs and emergency luminaires.

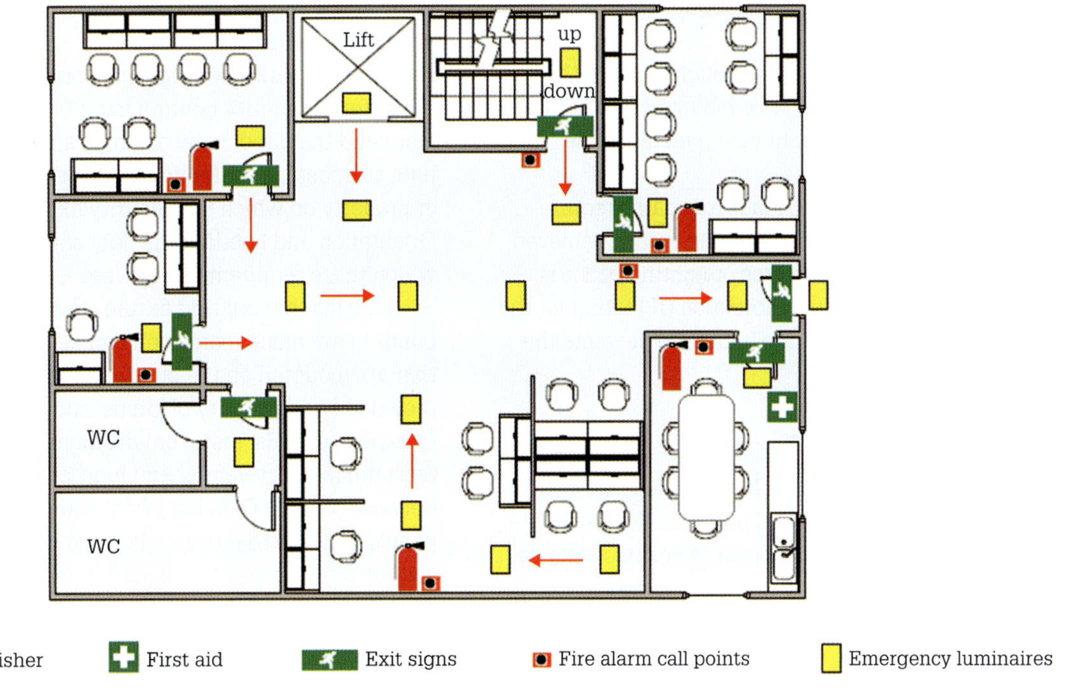

FIGURE 8.36 Location of exit signs and emergency luminaires

Testing of exit signs and emergency luminaires

Testing of exit signs and emergency luminaires must include out-of-hours tests, to demonstrate the emergency and evacuation system's performance. Tests include the following:

- Testing components for correct function and operation.
- Demonstration of illumination performance on site, to at least the level stated in the manufacturer's performance specification for that device.
- Testing the operation of battery discharge test and control test switch functions, including discharge and restoration.
- Demonstration of system functions under mains fail condition.
- Demonstration of the operation of the battery and charger including a full discharge/recharge over the designated time. The batteries must be capable of operating each lamp at its rated output continuously for at least 2 hours during completion tests and 1.5 hours of subsequent tests. As a suggestion, always indelibly mark each battery with its date of manufacture or installation.

Fault-finding

Faults to luminaires are created by either of two conditions: component failures or electricity supply interruptions. Some lighting problems are evident and are located by seeing, hearing and smelling. However, some electrical faults are hidden and can be located by tugging on secondary connecting conductors or by twisting the lamps gently back and forth in their sockets. If this does not locate the problem, the system must be tested for short-circuits and open and earthed circuits.

External lighting

External lighting, illustrated in **Figure 8.37** (bollard, flood and post top), should be arranged to provide compliance with the requirements of AS 1158. Adequate external lighting should be provided to deter vandalism and criminal activities.

FIGURE 8.37 External lighting, bollard, flood and post top

External lighting other than security lighting should be controlled from dusk to dawn using a photocell and operated by movement detector switching.

External lighting of buildings

Vehicle access driveways, roads and paths associated with a particular installation should be illuminated. External lighting should be provided to operate during regular hours of darkness.

Where paths run adjacent to, through or beneath buildings, the luminaire should be mounted on the external wall of the building. White light (using metal halide fittings) should be employed.

Vehicle access driveways, roads and paths require separate illumination. This illumination can be achieved by using pole-mounted luminaires or lighting bollards. As a minimum, the International Protection (IP) rating for the luminaire must indicate that the luminaire prevents the access of insects.

Security lighting

Security lighting is exterior lighting installed solely to enhance the safety of people and property. Security luminaires should remain on at night to ensure safety at the following locations:

- all entrances and exits
- staircases
- in some offices (suggested one luminaire per 200 m^2)
- above the security panel
- in the carpark area (suggested one luminaire per 200 m^2).

Exterior security lighting must be designed to control glare and the direct view of lamps and must prevent light trespass by confining illumination to the premises or property on which the security fixtures are located. Orientation and hooding the light source are the means by which these requirements are met.

It seems that lighting fixtures that are directed at a building are much more useful for security than fixtures that are mounted on the building. Security lighting can be provided from a variety of lamps, such as T8 fluorescents, halogens, par 38s, metal halide, mercury vapour, high-pressure mercury vapour and high-pressure sodium. Because of the golden light from sodium lamps some people may feel that golden lighting is 'less safe'. However, sodium lamps cause less light pollution, emit high levels of light output and have low emissions in the UV end of the spectrum. This also means that the light is not actively attracting insects into the premises.

REVIEW QUESTIONS

1. With what code must exit and emergency lighting comply?
2. State the minimum time an emergency lighting system must provide adequate illumination.
3. What is the purpose of exit signs?
4. At what height must exit signs be mounted?
5. Describe the pictogram for an exit sign.
6. Name the type of lighting that is intended to provide a safe path by illuminating the escape route from a building should a power failure occur.
7. Identify the type of circuit and how emergency luminaires must connect.
8. State the Australian minimum lighting level required for an emergency luminaire directly to the floor surface based on a non-smoke condition.
9. Describe a centrally supplied emergency lighting system.
10. What is a single-point emergency luminaire?
11. State the required rating for single-point emergency luminaires subject to water, dust or insect penetration.
12. Describe the operation of a maintained emergency luminaire.
13. Identify three locations suitable for security lighting.
14. Exterior security lighting must prevent light trespass by confining illumination to the premises or property on which the security fixtures are located. How is this achieved?

8.4 Principles of lighting

Photons (light energy), like electrons associated with an atom, exhibit both wave and particle behaviours. A photon is created in its entirety when emitted from a particular location and it disappears totally when it is absorbed by an object's surface. Light has a consequence on the health, wellbeing, comfort and self-esteem of human beings. The visual system is the most accessible of the human senses in that eyes sense visual comfort. Visible photons are capable of stimulating the eye's retina, resulting in what we call sight. The image of the human eye is shown in **Figure 8.38**.

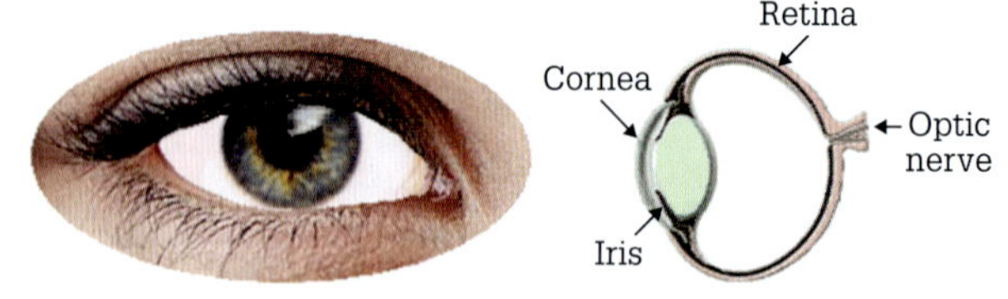

FIGURE 8.38 The human eye

It is the luminance and the colour of the surfaces of objects that the retina perceives.

Types of human vision

Human vision is about how the 'eye' sees or, more accurately, responds to the various wavelengths of visible light and how the cerebral cortex interprets these into images. White results from a mixture of blue, green and red, as shown in **Figure 8.39**.

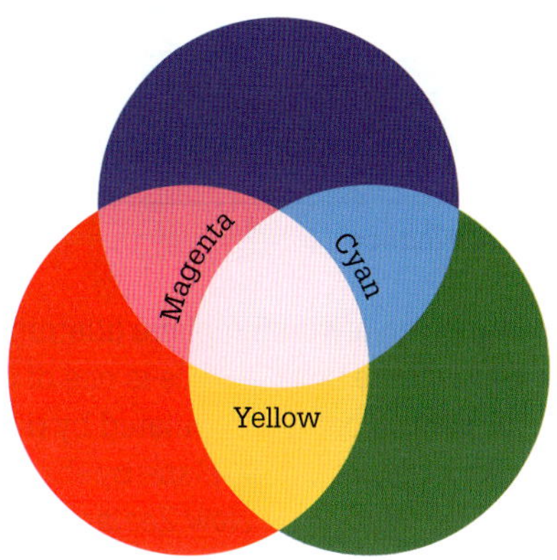

FIGURE 8.39 Colour mixture

To see clearly, we need daylight. Photopic (daylight) vision is the scientific term for the human colour vision under normal lighting conditions during the day or night.

Photopic vision is characterised by high sharpness and colour perception. In photopic vision, the various wavelengths of visible light appear in their corresponding colours. The two physical variables, luminance and wavelength, are transformed into three sensory variables: brightness, hue (the attribute of a colour that allows it to be classified as red, yellow, blue, etc.) and saturation.

Scotopic vision is the monochromatic vision of the eye in low luminance. In this range, the human eye uses rods to sense light. The retina contains two kinds of photoreceptors, rods and cones. Since the rods have a sole absorption maximum of about 1700 lumens/W at a wavelength of 507 nm, scotopic vision is colour blind. Mesopic vision is a combination of photopic vision and scotopic vision in low but not quite dark lighting situations. Mesopic vision results in a very healthy appearance of bluish colours around twilight and dawn.

The electromagnetic spectrum

Radiant energy travels through space in the form of waves made up of vibrating electric and magnetic fields. These waves have both a frequency (how many complete waves pass a given point in one second) and a wavelength (measured in metres). It is these values that distinguish the individual forms of radiant energy from each other. The electromagnetic spectrum is shown in **Figure 8.40**.

A narrow band of the electromagnetic spectrum is visible light, which is composed of different wavelengths (380–780 nanometres) called colours. The colour

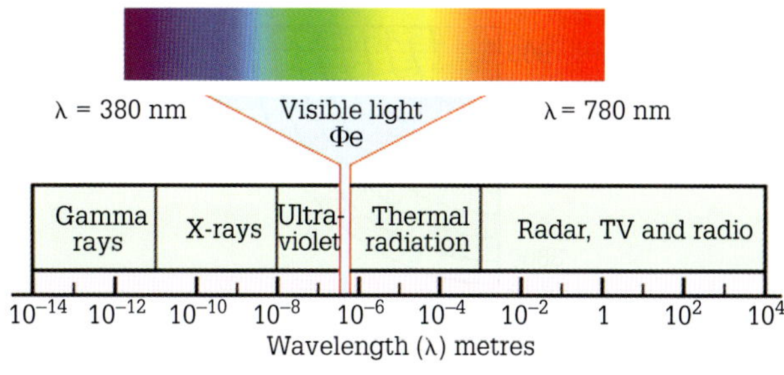

FIGURE 8.40 Electromagnetic spectrum

components of the visible spectrum fall approximately within the following wavelength regions:

- Red: 610–700 nm
- Orange: 590–610 nm
- Yellow: 570–590 nm
- Green: 500–570 nm
- Blue: 450–500 nm
- Indigo: 430–450 nm
- Violet: 400–430 nm.

An even balance of these colour light waves creates white light. The light that we perceive is being reflected from particles floating in the atmosphere. All objects that exist absorb specific wavelengths of light and reflect others.

The wavelengths that are reflected stimulate the eye's retina with a photosensitive pigment that the cerebral cortex accepts to be the colour of the object. The eye's sensitivity is a function of wavelength and is greatest at a wavelength of about 5.6×10^{-7} m (yellow-green) or 560 nanometres.

Visible light is produced between 4.1×10^{14} Hz and 7.5×10^{14} Hz frequencies. A black object absorbs all the colours of the visible spectrum and reflects none. Therefore, black is the absence of light. A white item reflects all visible wavelengths and absorbs none.

Terminology

There are a number of special terms used in lighting. Some of these are discussed in the following text.

Luminous intensity

Luminous intensity is the intensity of the light within a small solid angle and in a specified direction. The symbol I is used to represent luminous intensity although I_V is used in some countries. Luminous intensity is defined as the total number of lumens (see below) emitted from an object's surface in a given direction.

The unit of measurement of luminous intensity is the candela (cd), from which all other units of light are derived. Power measurements are taken at various angles around a light source in order to plot a luminous intensity distribution curve. These curves show the luminous intensity of the light source and light fittings in any direction. Candelas are plotted against angles from the vertical downward direction. A luminous intensity distribution curve is shown in **Figure 8.41**.

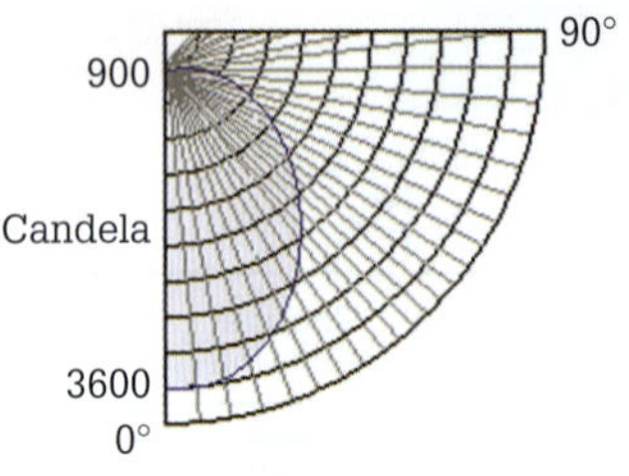

FIGURE 8.41 Luminous intensity distribution curve

Lamps can have the same light output but different distribution curves depending on their style of fitting.

Luminous flux

Luminous flux describes the perceived light power emitted by a light source. It is the quantity of light regardless of the direction that is received or emitted by the surface of an object or a lamp and is measured in lumens (see **Figure 8.42**). Luminous flux may be represented in equations with the symbols F or ϕ. The symbol ϕ is used to express a measure of the rate of flow of radiant energy within the visible region of the electromagnetic spectrum. Luminous flux is measured in lumens, which is abbreviated to lm.

Source: Shutterstock.com/Koretskyi

FIGURE 8.42 Luminous flux

Illuminance

The distribution or density of light on a horizontal surface is called illuminance. Illuminance is also defined as the density of luminous flux (ϕ) incident per unit area (m^2) of the surface. Illuminance is represented in equations with the letter E. Illuminance (how bright the surface appears) is measured in lux (lx). The amount of illuminance varies according to the difficulty of a visual task. Ideal illuminance is the minimum amount of lux necessary to allow a task to be performed comfortably and competently without eye strain. Typical illuminance values are 1 lux, full moon; 10 lux, street lighting; 100 to 1000 lux, work space lighting; and 100000 lux, plain sunshine at noon. One lux of illuminance is one lumen of light distributed over an area of one square metre (1 lux = 1 lumen/m^2), as shown in **Figure 8.43**.

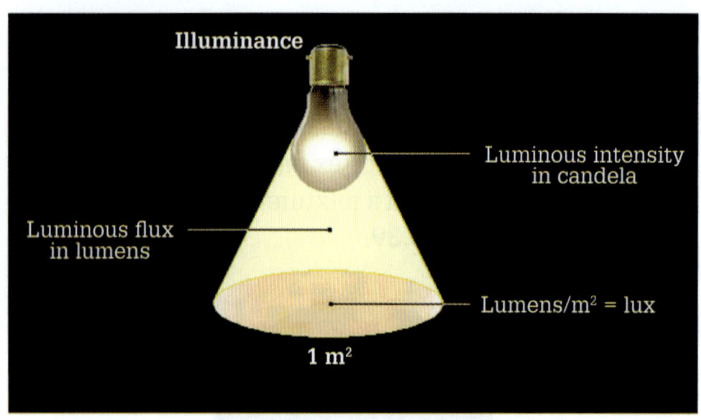

FIGURE 8.43 Illuminance

The illuminance on a working plane (illuminating a horizontal surface where work is done is one of the most common lighting tasks) can be calculated by dividing the luminous flux incident on the plane by the area of the working plane. The Australian/New Zealand Standard AS/NZS 1680 series of standards gives lux levels for many different industries and should be consulted in this respect.

Illuminance is calculated using the following equation:

$$E = \frac{\phi}{A}$$

where E = Illuminance in lux (lx)
ϕ = Luminous flux in lumen (lm)
A = Area of illuminated surface in m^2

EXAMPLE 8.1

Calculate the illuminance on a work plane if 2 × 15 W LED luminaires having a combined output of 1500 lm that falls uniformly on an area of 8 m^2.

$$E = \frac{\phi}{A}$$

$$= \frac{1500}{8}$$

$$= 187.5 \text{ lx}$$

EXERCISE 8.2

Calculate the illuminance on a work plane if 2 × 18 W LED luminaires having a combined output of 2000 lm that falls uniformly on an area of 10 m^2.

Luminance

Luminance is a measure of the luminous intensity (brightness) of the surface of an object in a given direction per square metre of projected area of the surface. Luminance is measured in candelas per square metre (cd/m) and has the symbol L.

Luminance is a directional concept and therefore the luminance at a point on the surface of an object may vary with direction, as illustrated in **Figure 8.44**.

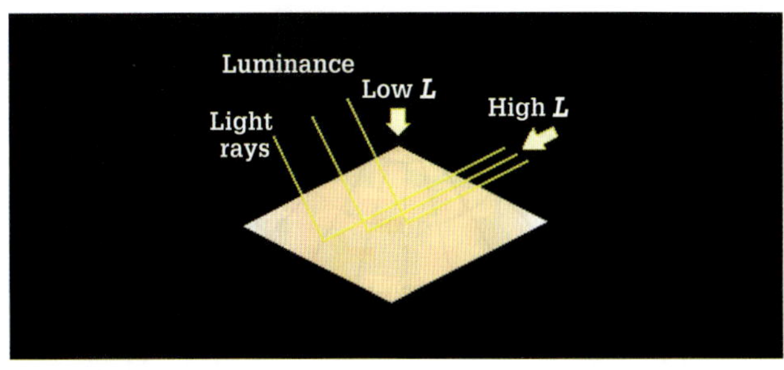

FIGURE 8.44 Luminance varies with the direction

Glare

If the luminance is not controlled, it can produce levels of glare that can spoil or prevent a task being performed. Glare is the contrast between a bright object and a darker background and results in visual discomfort that impairs vision.

Australia uses a glare index system which takes into account the actual installation constraints, as well as a luminaire. This system is a qualified measurement of the visual discomfort of a placed observer who is looking straight ahead. Room reflectance has a chief measure in this calculation. If the floor or walls are too dark, the background luminance is reduced due to there being less reflected light. As the glare calculation involves background luminance, a brighter background will give lower glare figures.

There are two types of glare: discomfort that does not fundamentally weaken the vision of objects and disability glare that prejudices vision without inevitably causing discomfort.

Glare can also be classified as direct or indirect (reflected). Direct glare results from the light source itself, while reflected glare is displayed on the object. Direct glare can be so powerful that, for a considerable period after it has been removed, no article can be seen. Glare in this situation is referred to as a blinding glare.

Indirect glare is a reflection of luminance that partially or wholly conceals the details to be seen on the surface of the object. Direct glare and reflected glare are illustrated in **Figure 8.45**.

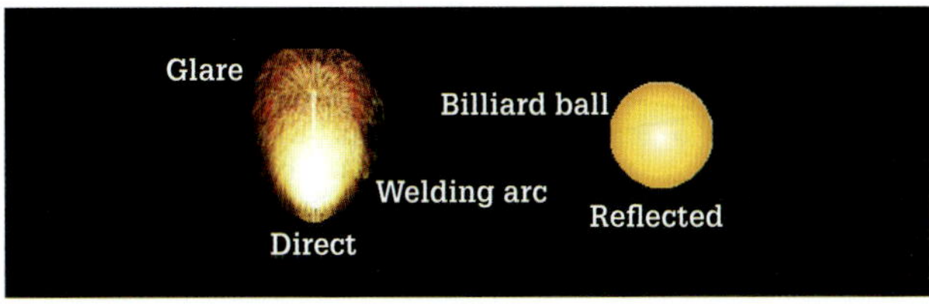

FIGURE 8.45 Glare

Luminous efficacy

This concept indicates how well the light source transforms electrical power into useful light output. It is calculated by dividing the light output in lumens by the electrical input in watts. The value thus obtained is measured in lumens per watt. Luminous efficacy is calculated using the following equation:

$$\eta = \frac{F}{P}$$

where η = luminous efficacy in lumens/watt (lm/W)
F = luminous flux in lumens (lm)
P = true power dissipated by the lamp in watts (W)

All new buildings and renovations must have 80% of their total floor area fitted with energy-efficient lamping. The minimum efficacy allowed for a suitable lamp is 27 lm/W. High-energy-consuming lamps can be used but only in 20% of the floor area.

EXAMPLE 8.2

What is the luminous efficacy of a 50 W, 600 lm lamp?

$$\eta = \frac{F}{P}$$

$$= \frac{600}{50}$$

$$= 12 \text{ lm / W}$$

EXERCISE 8.3

What is the luminous efficacy of a 75 W, 1000 lm lamp?

Measuring illuminance

Lux meters, shown in **Figure 8.46**, are used for lighting surveys. They measure the amount of luminous flux incident on an illuminated surface or a specified plane such as the working plane.

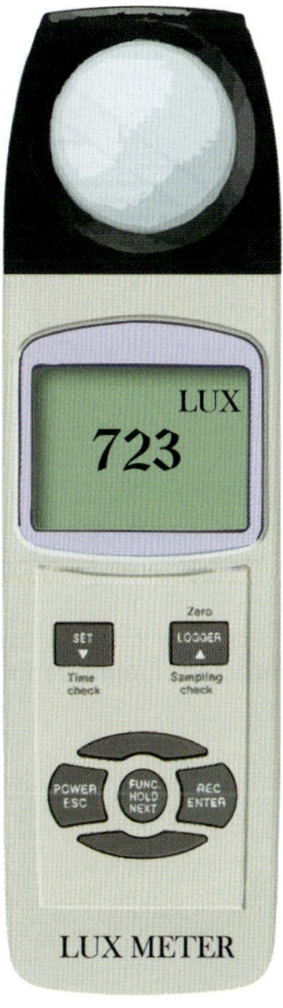

FIGURE 8.46 Lux meter

When measuring a large surface or room, it is recommended that the space be divided into sections or a grid. Take illuminance readings at each grid section and average all of the illuminance readings to determine the average illuminance level for the space.

Luminous intensity distribution curve

A luminous intensity distribution curve, shown in **Figure 8.47**, is essentially the *x*, *y* points of a graph mapped onto a circular grid where each point is defined by the radial coordinate *r* and the angular coordinate *t*. This allows the relationship between two points to be most easily expressed in terms of angles and distance.

A circular grid format allows the visualisation of both the positioning and the light distribution of a light fixture. The luminous intensity dispersal of a light fixture rests upon reflector design, shielding type and lamp-ballast selection. It is presumed that the light fixture position is at the junction of two axes (horizontal and vertical) and that 0° (nadir) is under the light fixture. Other angles, which denote the various locations of a lux meter as it is placed in a circular pattern around the light fixture, are marked on the circular grid as well.

A vertical distribution curve is acquired by taking measurements at several angles of elevation in a vertical plane through the lamp centre. The vertical curve is presumed to denote an average such as would be obtained by rotating the lamp or light fixture about its vertical axis. Each lamp and lamp–luminaire fixture combination has a unique luminous intensity distribution curve. A luminous intensity distribution curve perpendicular to the lamp axis of the luminaire is shown in **Figure 8.48**.

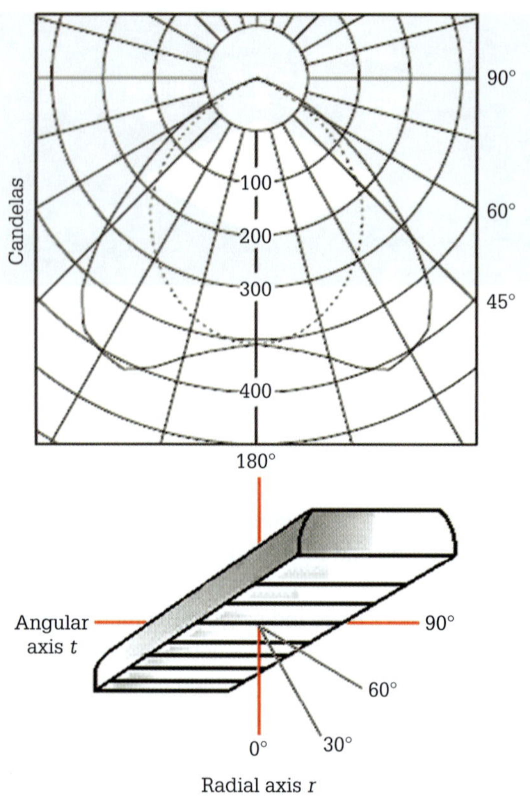

FIGURE 8.47 Radial and angular coordinates

A luminous intensity distribution curve parallel to the lamp axis of the luminaire is shown in **Figure 8.49**.

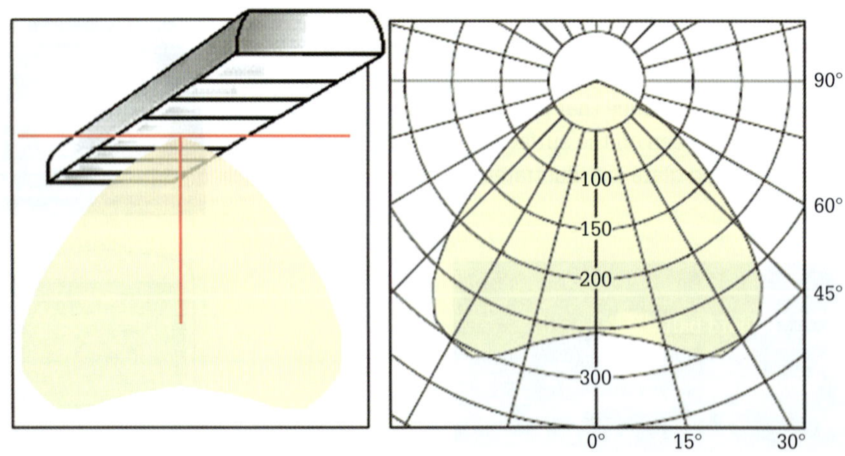

FIGURE 8.48 Perpendicular luminous intensity distribution curve

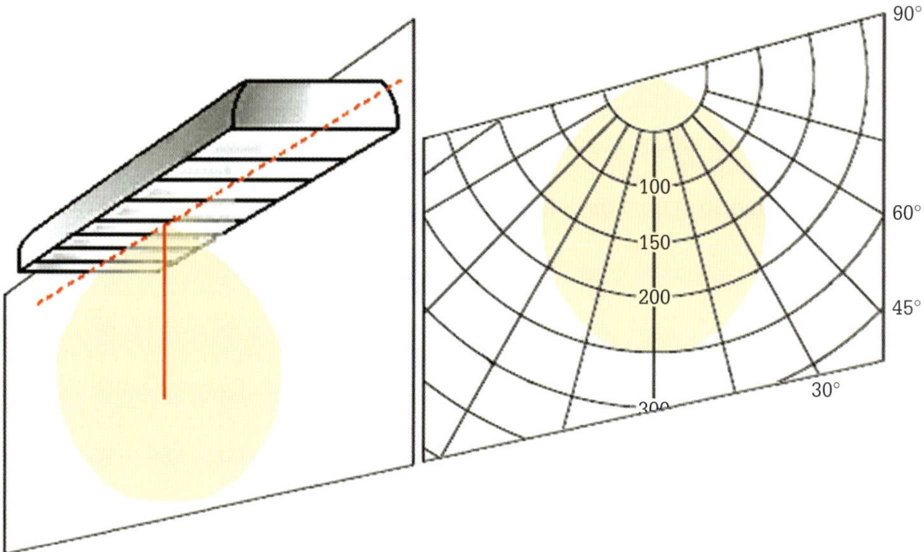

FIGURE 8.49 Parallel luminous intensity distribution curve

If the distribution of the light is symmetrical in all directions around the radial axis then only one luminous intensity distribution curve is necessary. In fact, only one-half of that curve is usually shown (see **Figure 8.50**).

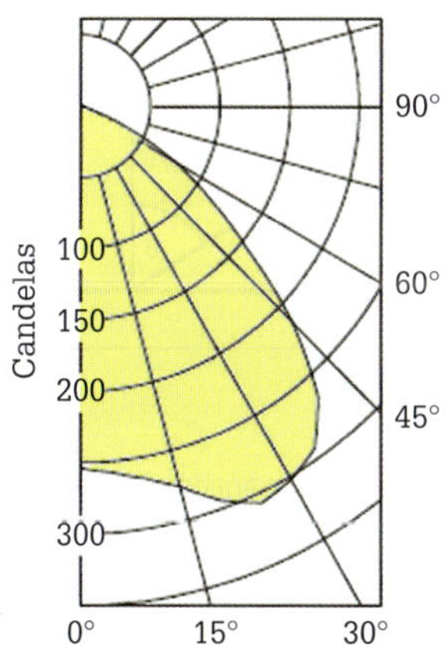

FIGURE 8.50 One-half of a luminous intensity distribution curve

From the distribution curve, a rough estimate of the luminous intensity of a particular angle can be determined. For example:

- On the 0° plane, the measured value is 210 cd.
- On the 45° plane, the measured value at 45° falls between 200 cd and 250 cd and is about 220 cd.
- On thc 30° plano, the measured value is 300 cd.

The inverse square law

The inverse square law tells us that if light is distributed from a point source then the illuminance on the surface of an object at 90° to the point source is inversely proportional to the square of the distance between the point source and the surface of the object. If you have a lamp fixture (point source) and you measure the illuminance with a lux meter at 5 m as 240 lux at the beam centre then at 10 m the illuminance is 60 lux (one-quarter of 240; doubling the distance results in a quartering of the illuminance) at the beam centre. This law is illustrated in **Figure 8.51**.

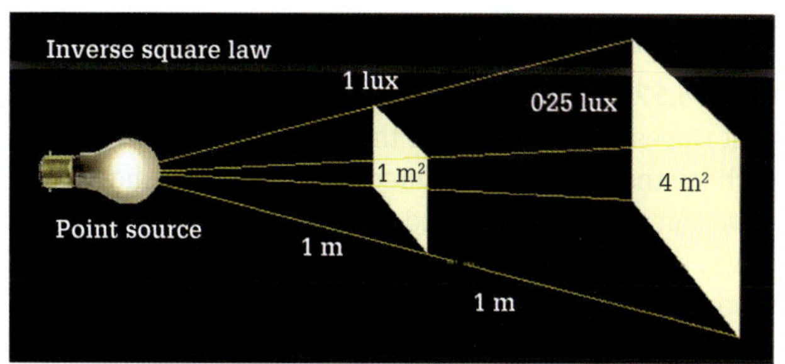

FIGURE 8.51 Inverse square law

The amount of light falling on a surface is the reciprocal, or inverse, of the square of the distance.

The following equation can be used to determine the illuminance if the distance and the luminous intensity of the point source are known.

$$E = \frac{I}{d^2}$$

where E = illuminance in lux (lx)
I = luminous intensity in candela (cd)
d = distance from the point source in metres (m)

EXAMPLE 8.3

This example considers the illuminance at a single point on a horizontal surface from a single luminaire directed straight down. A point light source produces a luminous intensity of 65 lux. Find the illuminance at a point at a distance of 1 m, 5 m and 10 m from the lamp.

$$E_{1m} = \frac{I}{d^2} = \frac{65}{1^2} = 65 \text{ lx}$$

$$E_{5m} = \frac{I}{d^2} = \frac{65}{5^2} = 2.6 \text{ lx}$$

$$E_{10m} = \frac{I}{d^2} = \frac{65}{10^2} = 0.65 \text{ lx}$$

EXERCISE 8.4

A point light source produces a luminous intensity of 180 lux. Find the illuminance at a point at a distance of 5 m from the lamp.

Law of reflection

Figure 8.52 shows light waves approaching a surface. The light waves are referred to as the incident wavefront, and each ray that points in the direction in which the light waves are travelling is called the incident ray. The reflected waves are the waves that bounce off the boundary and head back, and the reflected ray is the ray that points in the direction in which the reflected wavefront is travelling. At the point of incidence (the point where the incident ray strikes the surface), a standard line is drawn.

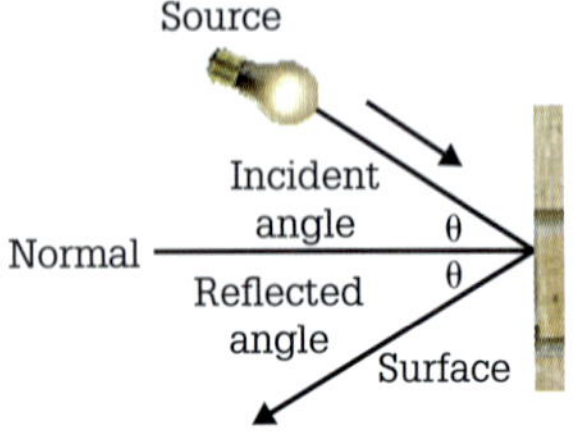

FIGURE 8.52 Law of reflection

The standard line is always drawn perpendicular to the surface at the point of incidence. The standard line creates a variety of angles with the light rays; these angles are necessary and are given individual names. The angle between the incident ray and the normal is the angle of incidence. The angle between the reflected ray and the normal is the angle of reflection. The fundamental law which governs the reflection of light is called the law of reflection and states that:

SWITCH ON

When a light ray reflects off the surface, the angle of incidence is equal to the angle of reflection.

Cosine law

Effective illuminance is proportional to the angle of incidence of the luminous flux on the surface of an object, whereas the inverse square law applies to illuminance at right angles to the surface. The cosine law applies when there is an angle between the direction of the luminous flux and the perpendicular to the object's surface.

If the surface is turned so that rays strike at an angle, the illuminated area increases in size and the illuminance decreases. The ratio of the perpendicular illuminated area in lux to the new illuminated area is equal to the cosine of the angle through which the surface has been moved. Therefore, the illuminance falls by a factor of the cosine of the angle. This is the cosine law of illuminance. If a surface illuminated to 250 lux is turned through an angle of 60°, then the illuminance falls to half or 125 lux because the cosine of 60° is 0.5. An illustration of the cosine law is shown in **Figure 8.53**.

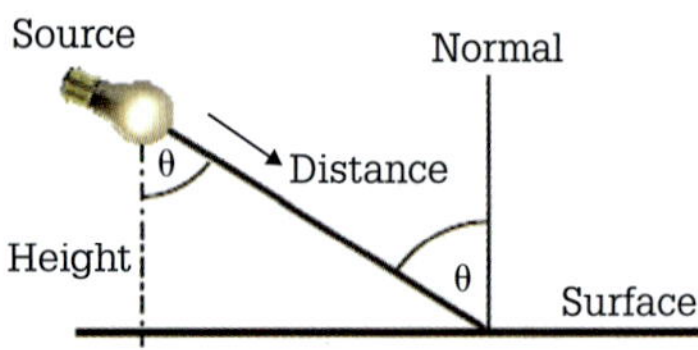

FIGURE 8.53 Cosine law

If the surface is not perpendicular to the direction in which the light travels, then the inverse square law requires changing with a cosine correction. If the light rays fall obliquely, then the same luminous flux from the source spreads over a larger area which is inversely proportional to the cosine of the angle between the surface normal and the rays. Consequently, the inverse square law is modified to the cosine law as shown in **Figure 8.54**.

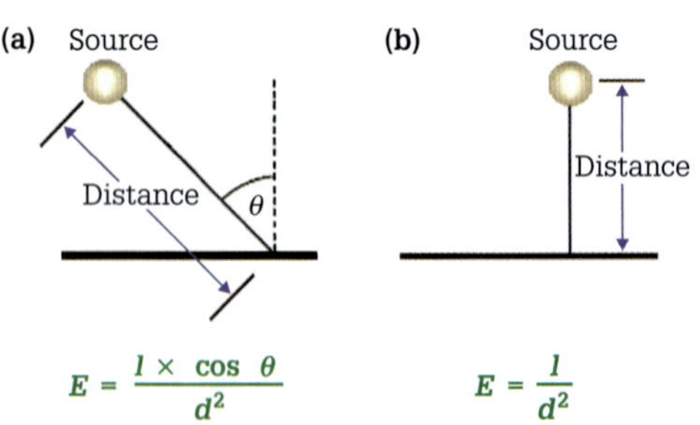

FIGURE 8.54 (a) Cosine law and (b) inverse square law

In **Figure 8.54**:

E = illuminance in lux (lx)

I = source intensity in candela (as taken from a candela distribution curve) (cd)

d = distance in metres (m)

θ = angle of incidence (degrees)

For example, a 300 W lamp gives 259 candelas at the vertical angle at a light level of 108 lux. The light level at the farthest point of 10 m when the lamp is mounted at 55° is then 1.485 lux.

Colour

It is important to understand that every colour has a unique vibrational frequency and wavelength. The wavelength or range of wavelengths that reaches the retina produces the sensation of colour. Colours at shorter wavelengths are called cool (violet and blue) colours, while those at longer wavelengths are called warm colours (orange and red). Colours are just one of the many ways that we take in meaning from the environment.

The colour of any given surface is not an intrinsic quality; it is a radiant energy affected by both the characteristics of the object's surface and the quality of the luminous flux that illuminates it. Consequently, the colour of a given surface may vary depending on the quality of luminance and its spectral distribution (different colours are present to differing degrees). The colours associated with the surfaces of objects in our consciousness are the colours that surfaces are perceived to have by natural luminous flux.

Paper that is white reflects the entire visible luminous flux spectrum and the reflected luminous flux is white radiant energy. If the white paper is illuminated with blue light only, the retina receives only that luminous flux frequency and the paper is now perceived as being blue. This effect is illustrated in **Figure 8.55**.

FIGURE 8.55 Colour effects

Throughout the day, the spectral distribution of natural luminous flux is always changing according to the prevailing weather patterns. As a result, the surface colours of objects change to differing degrees when illuminated by luminous flux from a given source compared with the colours when illuminated by some reference source such as natural daylight. The expression granted to this effect of colour change is colour rendering. Colour rendering is also used to refer to the manner in which different types of illumination can cause colours to look different. The better the colour rendering of an artificial luminous source, the less the difference between the colour of the surface of an object when illuminated by it and the colour when illuminated by natural luminous flux.

The colour-rendering index (CRI)

The change in colour by differing degrees from the primary reference can be tabulated as a number called a colour-rendering index (R_a) (**Table 8.1**). This figure indicates the extent to which the appearance of colours is distorted under the luminous source.

TABLE 8.1 R_a values for various groups

Group	Colour-rendering index	Application
1a	$R_a > 90$	Wherever accurate colour matching is required – picture galleries
1b	90–80	Good colour matching is required – homes
2a 2b	80–70 70–60	Average colour matching is required – schools
3	60–40	Some distortion of colour is acceptable – heavy industry
4	40–20	Distortion of colour acceptable – outdoors

The value given to a primary reference colour is 100. Numbers less than 100 indicate the level of colour distortion. This R_a value can be negative for extreme colour variance from the primary reference colour.

The retina only perceives colours, and functions best in luminous environments. If the environment is dark, the retina perceives the environment in black and white. If the intensity of the luminous flux reaching the retina is the same at every wavelength then it perceives the colour in its lightest mode. This is because the lighter the surface of an object the greater its luminance.

Correlated colour temperature (CCT)

The correlated colour temperature (CCT) is a depiction of the colour emitted from a lamp, based on the colours in Kelvin (K) given off a black body (pure carbon) if heated to various temperatures (as the black body is heated its perceived colour changes, first to red, then yellow, then white and finally blue). The CCT enables us to define the colour appearance of lamps in terms such as warm, cool or cold.

The colour of lamps with a CCT of less than 3500 K is similar to the colour of pure carbon heated to that temperature and is usually described as having a warm appearance. Lamps with a CCT of 3500 K to less than 5000 K are considered to have a cool appearance while lamps having a CCT of greater than 5000 K are said to have a cold appearance.

At sunrise and sunset, when the sun is low, the colour temperature is low. Neutral daylight has a colour temperature of about 6500 K. Colour temperature is higher in the middle of the day and can be as high as 15 000 K. The higher the colour temperature the whiter and bluer the light source appears.

Warm light has more frequencies of the colour red than frequencies of other colours in the visible spectrum, while cooler colours have more frequencies of the colour blue. **Figure 8.56** shows the effects of colour temperature of a gypsum wall.

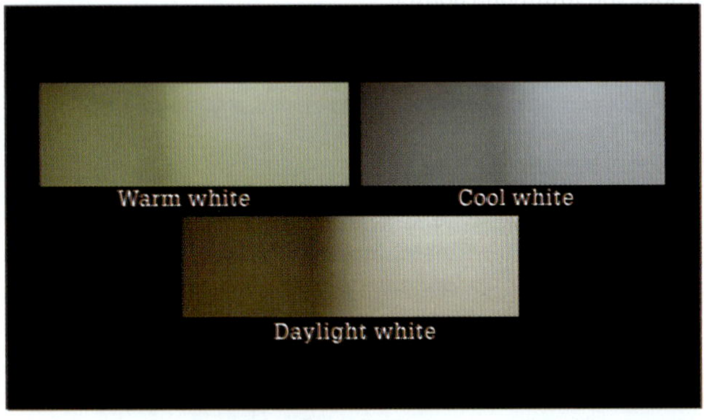

FIGURE 8.56 The colour temperature of a gypsum wall with white luminous flux

Visual comfort

Visual comfort is a product of the environment and illuminance. The goal of lighting is to create an appropriate visual environment so that the visual task originating from the activity performed embodies visual accuracy, contrast sensitivity and speed. In order to meet the requirements of visual comfort, the following items need to be incorporated:

- dimensions of the work area
- height of the work plane
- luminaire mounting height
- reflectance from the ceiling, walls and floor
- photometric data on the fitting to be used
- lumen output times the number of lamps.

Obtaining exact information on the parameters of the environment enables a person to distinguish in the area concerned the dimensions of space, illuminance, colours and the spatial positions of the details.

It is recommended that visual comfort design be in accordance with AS/NZS 1680:2006 *Interior lighting set*. Other documents that need to be consulted are building regulations, building codes and Worksafe requirements.

REVIEW QUESTIONS

1. Why is the visual system the most accessible of the human senses?
2. Describe the basic process involved in human sight.
3. What does the term 'photopic vision' mean?
4. Into which part of the electromagnetic spectrum does visible light fit?
5. What is 'luminous intensity'?
6. What is the distribution or density of light on a horizontal surface called?
7. Name the instrument used to measure the amount of luminous flux incident on an illuminated surface or a specified plane such as the working plane.
8. What is the unit of measure for luminous flux?
9. Describe glare.
10. What type of glare prejudices vision without inevitably causing discomfort?
11. What is the luminous efficacy of a 60 W, 150 lumen lamp?
12. A point light source produces a luminous intensity of 65 lux. Find the illuminance at a point at a distance of 2 m from the lamp.
13. Describe colour rendering.
14. What is the recommended colour-rendering index for homes?
15. Define the expression 'correlated colour temperature'.
16. State the effect that is a product of the environment and illuminance?

8.5 Types of luminaires

A luminaire is a complete lighting device consisting of a lamp or lamps, lamp holders, and visual elements as well as terminals and any necessary control gear and a means for connecting to a power source. Luminaires are also called 'fixtures'. The luminous intensity of a light source is identified volumetrically by its luminous flux and intrinsically by colour rendering and colour temperature.

Fixtures have several illumination and practical purposes. Their illumination purposes are as follows:

- to distribute and direct the luminous flux
- to absorb some of the luminous fluxes.

Their practical purposes are:

- to attach the light source
- to protect the light source from the environment.

Functional luminaires are classified according to the manner of light distribution employed. Light distribution is defined as the ratio of luminous flux radiated vertically with respect to a horizontal plane through the luminaire. This is illustrated in **Figure 8.57**.

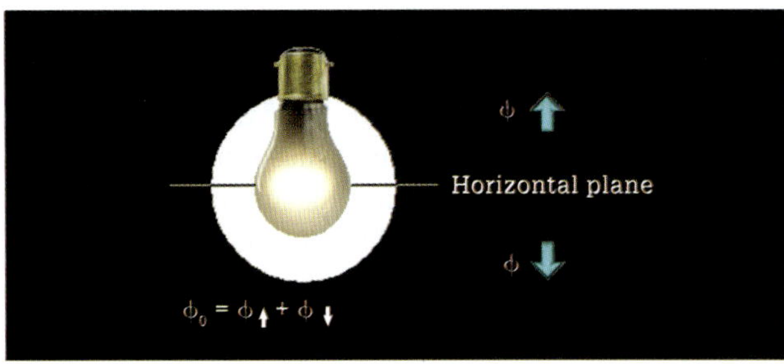

FIGURE 8.57 Light distribution

The classifications used for luminaires are:

- direct lighting
- semi-direct lighting
- general diffused lighting
- semi-diffused lighting
- indirect lighting.

The classifications, together with their symbols, percentage light distributions and distribution curves, are shown in **Figure 8.58**.

Classification	Symbol	Upward light (%)	Downward light (%)	Luminous intensity distribution curve
Direct		0–10	90–10	
Semi-direct		10–40	60–90	
General diffusing		10–60	40–60	
Semi-diffused		60–90	10–40	
Indirect		90–100	0–10	

FIGURE 8.58 Classification of luminaires according to their light output

As seen in **Figure 8.58**, there is a significant difference between upward and downward luminous flux. Luminous flux radiated downwards by the luminaire strikes the working plane (desk top). Luminous flux radiated upwards can only reach the working plane after one or more reflections. As such, only a small percentage of this luminous flux can reach the working plane. The distribution of the luminous flux by a luminaire is characterised by its luminous intensity distribution curve.

In addition to the luminaire's classification, other factors that affect the choice of a luminaire for a particular purpose include the reflection factor of a surface, the absorption factor and, with a bright surface, its transmission factor. Aesthetics also play a role in the selection process.

Efficiency of a luminaire

The luminaire radiates only part of the luminous flux of a light source into the environment and absorbs the rest. The effectiveness of a luminaire in radiating luminous flux is called the efficiency of the luminaire, which has the symbol η_L.

The efficiency of a luminaire is the ratio of the luminous flux emitted by a luminaire to the total luminous flux produced by the light source.

Glare rating

In order to have visual comfort, glare-free conditions are necessary. This is achieved by controlling the luminance emitted by luminaires at angles between 40° and 90° from the line of sight. Luminance within this angular area comes into the outer peripheral field of view, creating a personal glare zone as shown in **Figure 8.59**.

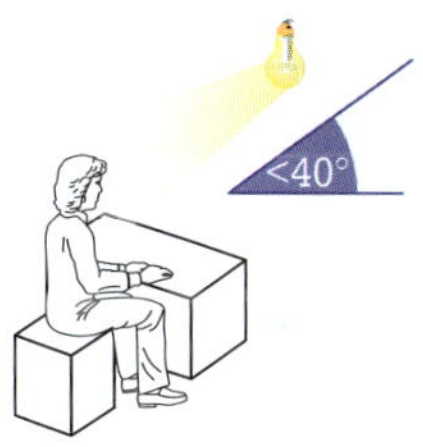

FIGURE 8.59 Glare angle less than 40° from line of sight

A potential source of glare on visual display terminals is high-angle glare of a luminaire candela distribution curve (see **Figure 8.60**). This glare needs to be controlled by shielding the light source from view, cutting off the light angle at about 30°. This leaves a 35° cushion between the high-intensity zone and the glare-producing zone of 70° to 90°.

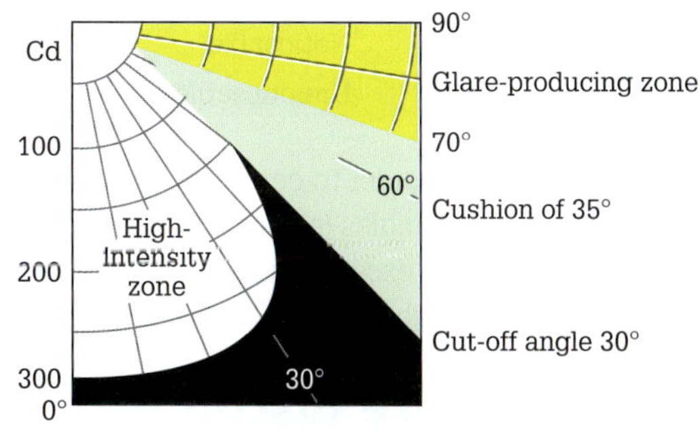

FIGURE 8.60 Location of high-angle glare

AS/NZS 1680 provides two alternative methods for limiting the degree of discomfort glare experienced by persons in interior environments. These methods are the 'luminaire selection' system and the 'glare index (GI)' system. The GI method seems ideal as it can be used to identify glare locations and optical directions within an internal lighting environment.

The De Boer glare rating process measures on a scale from 90 to 10 the visual discomfort of an observer whose optical axis (line of sight) is 90° from the nadir (point below the observer). The illuminated environment is divided into areas and an observer is placed in each of these positions. This information is then fed into a computer program that rotates the observer's *x*- and *y*-coordinates and height on the ground through 360° and calculates the glare at each angle for each of the positions and prints out the glare area.

This method also takes into account the dimensions of the illuminated environment and reflectance as well as the type of luminaire being used. The assumed worst-case position at the middle of the back wall of the general office environment illuminated by 320 lux at a work plane height of 700 mm with a glare rating of 50 is illustrated in **Figure 8.61**.

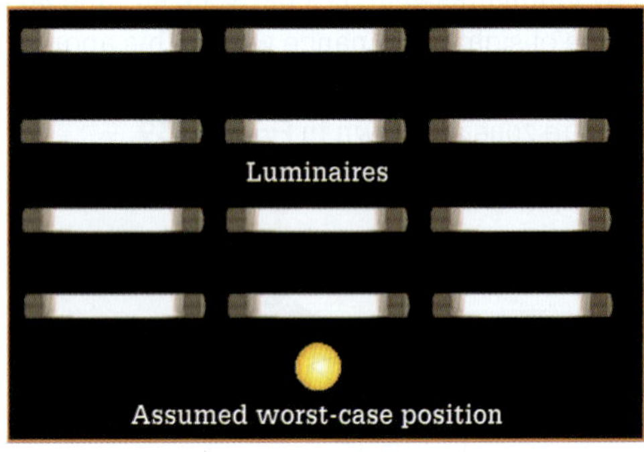

FIGURE 8.61 Worst-case position

The De Boer glare rating scale is a means of comparing the glare rating of various lighting installations and is shown in **Table 8.2**.

TABLE 8.2 De Boer glare rating

90–80	Unbearable
70–60	Disturbing
50–40	Just admissible
30–20	Noticeable
10	Unnoticeable

As a rule of thumb, in order to control glare, light should not be emitted at angles (above 40°) that project illumination at a distance exceeding four times the luminaire's mounting height.

Testing for glare discomfort

One way to check for glare discomfort is to go to the back of the room and stand at the centre and note any bright lights in the peripheral vision. By using your hand to shield your eyes indicate whether you sense an immediate improvement in your comfort. This should be done two to three times at the same position in the room. If there is an immediate sense of improved comfort by eliminating the lights from the peripheral vision then glare discomfort exists.

Controlling light angles

There are three devices that can be used to control the angle at which the light can leave the luminaire. These are the prismatic diffusers, parabolic diffusers and metal or plastic louvres. These devices absorb luminous flux, which is converted to heat and dissipated.

Prismatic diffusers

Prismatic diffusers are made from acrylic or polycarbonate and have myriads of bright prisms on the outside of the panel that enable the luminous flux to be diffused over a wide area. These diffusers lower the light source intensity in the direct glare zone of 40° to 90°. A prismatic panel is shown in **Figure 8.62**.

FIGURE 8.62 Prismatic panel

Prismatic diffuser panels are used with Troffer bar (T-bar) one-piece frame and gyprock recessed fixtures such as the K12, 19 and 15. A recessed T-bar diffuser is illustrated in **Figure 8.63**.

FIGURE 8.63 Prismatic diffuser

Silver-tint versions of the diffuser reduce luminous flux output by up to 25% but deliver better glare control. The luminous intensity distribution curve for 2×36 W, 2850 lm fluorescent tubes covered by an acrylic diffuser is illustrated in **Figure 8.64**.

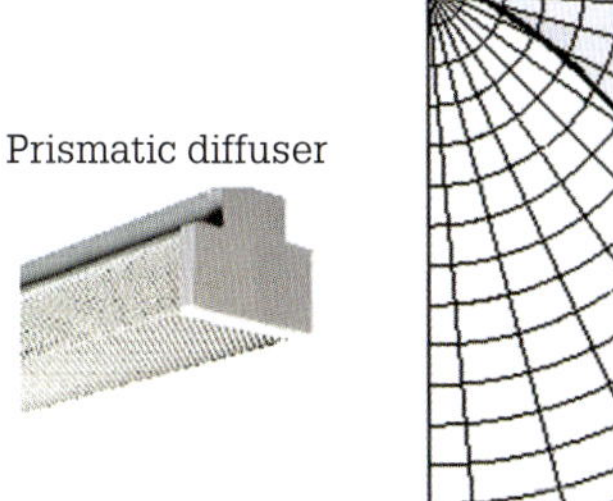

FIGURE 8.64 Prismatic diffuser and luminous intensity distribution curve

Note how the luminous flux is distributed downwards and that the cut-off angle is 30°, meaning that there is a much-reduced output in the direct-glare zone of 40° to 90°.

Parabolic diffusers

These diffusers use a particular kind of open grid pattern called an 'egg-crate' type. Egg-crate diffusers, because of their honeycomb frame of rectangular apertures, can reduce the luminous flux output from the fitting much more than the prismatic diffuser. A parabolic diffuser is shown in Figure 8.65.

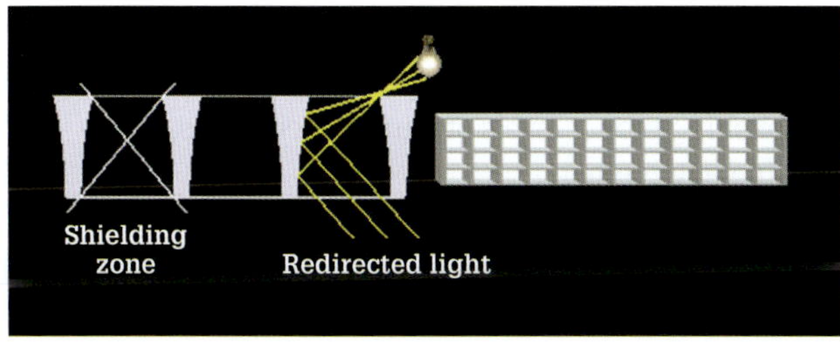

FIGURE 8.65 Parabolic diffuser

Egg-crate diffusers are mainly used in interior environments where very little glare requirements are necessary. These diffusers can reduce glare by concentrating the luminous flux downwards. This feature can, however, create hotspots for areas directly below the luminaire. The shielding zone is the horizontal angle at which the lamp is obscured.

Louvres

Louvres can control the direction of the luminous flux without absorbing a significant amount of the luminous intensity delivered by the fitting. Many louvres limit the angle at which the flux can leave a luminaire by using a grid-like mechanical pattern. A louvre is shown in Figure 8.66

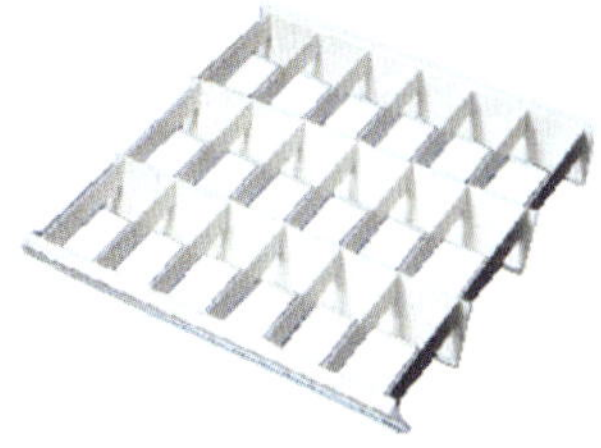

FIGURE 8.66 Louvre

Fluorescent tube sleeves

Fluorescent light filter sleeves are ideal for use with under-shelf and overhead fluorescent lighting. These sleeves reduce direct and reflective glare by up to 80% on the working plane. A fluorescent sleeve is shown in Figure 8.67.

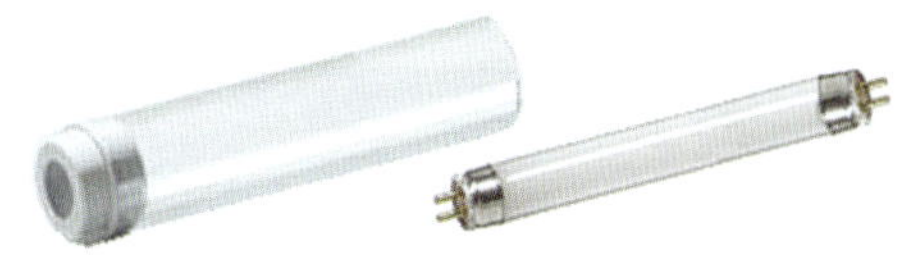

FIGURE 8.67 Fluorescent sleeve

Reflectors

Whenever luminous flux strikes an opaque surface, not all the flux is reflected. Some of the flux is absorbed. The reflected flux can be either directly reflected, which is called 'specular reflectivity' or diffused in all directions. Some reflectors use a reflective film or a painted metallic surface to reflect the flux out of the luminaire.

Other types of reflectors use a silver-coated aluminium or polished anodised aluminium surface to reflect the flux. The brilliant white-powder-coated painted reflector employs heat-resistant paint with a high level of titanium dioxide pigment that produces an extremely reflective surface. The polished anodised aluminium diffuses the reflected light and makes the reflection uniform.

Reflectors are engineered specifically to focus an extremely directional light out of the fixture cavity. The use of reflectors means that the desired illumination in a particular environment can be achieved with fewer luminaires. Floodlights use a parabolic reflector where the luminous source is placed at the focal point of the reflector. This allows the rays of flux to be reflected parallel to one another, concentrating the flux into a tight beam. If the source is moved towards the reflector, the luminance spreads. Moving the source away from the reflector will cause the rays of flux to converge. Several reflectors are shown in Figure 8.68.

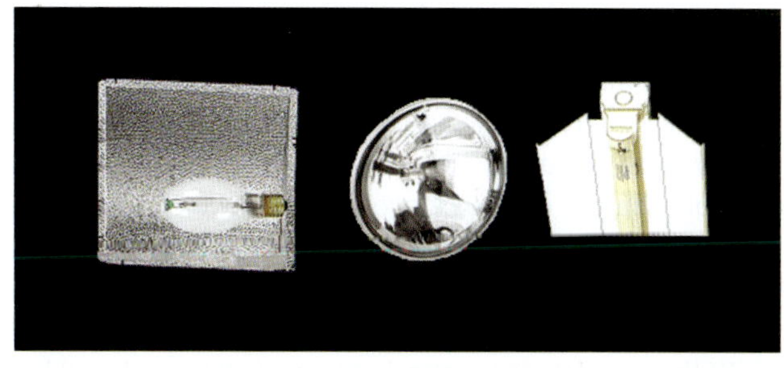

FIGURE 8.68 Reflectors

Electromagnetic compatibility (EMC)

Australian standards require that luminaires must comply with regulations concerning the electromagnetic interference produced by the fixture. Electromagnetic interference is illustrated in **Figure 8.69**.

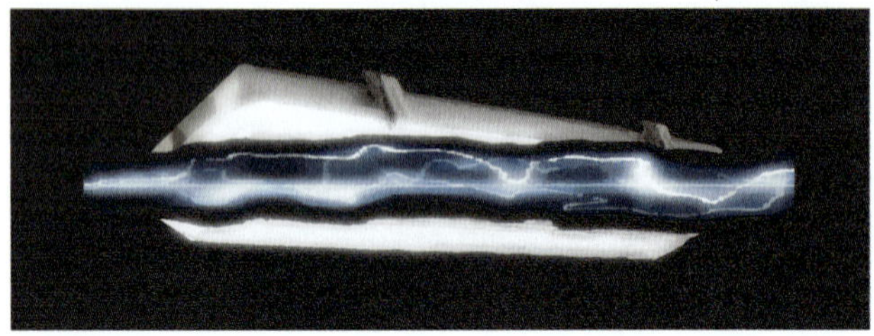

FIGURE 8.69 Electromagnetic interference

A compliance trademark registered to the Australian Communications and Media Authority (ACMA) and Electrical Regulatory Authorities Council (ERAC) is being introduced for luminaires and indicates that the luminaire is compliant with the regulations.

Additional factors

Other factors that affect luminaires and their use are the type of lamp selected, mounting height of the luminaire, the number of luminaires required, room reflectance, height of the working plane and the maintenance factor associated with the lamp and the luminaire.

Decorative luminaires

Decorative luminaires are available in a vast variety of styles, sizes, finishes, diffusers and lamping configurations. They can also be designed to meet specific architectural needs. They can be manufactured in brass, aluminium, steel, stainless steel, iron and copper, using the finest crystals, faux-alabasters, hand-blown glass and acrylics. These uniquely designed luminaires are used in kitchens, hallways, and bathrooms and ballrooms, lobbies and restaurants.

Maintenance

The accumulation of dust, dirt and other materials on lamps, reflectors and diffusers can decrease the light leaving a luminaire by 30% or more. This grime also reduces the amount of light reflected from walls, ceilings and other surfaces in a working environment. Reflectors and diffusers may discolour over time, further reducing the light output. Dirt traps heat within the luminaire, decreasing lamp efficacy and reducing lamp and ballast life. Therefore, the need to clean luminaires periodically is an important part of lamp maintenance. Yearly testing with a light meter is a valuable aid to establishing proper levels of illuminance for a safe working environment.

REVIEW QUESTIONS

1. What comprises a luminaire?
2. How are functional luminaires classified?
3. Provide a definition for 'light distribution'.
4. State the five classifications of luminaires.
5. Draw the symbol of the light output from a general diffusing luminaire.
6. What is necessary for visual comfort?
7. Name three devices that are used to control the angle at which the light can leave the luminaire.
8. State the purpose of a fluorescent tube sleeve.
9. What is the purpose of a reflector?
10. Why is maintenance of luminaires necessary?

8.6 Lamp types

Artificial light sources called lamps are devices producing luminous flux. These devices convert a portion of electric energy into visible radiation.

Light sources can be categorised into two types of lamps or luminaires:

1. incandescent
2. luminescent gaseous discharge.

With incandescent lamps, luminous flux is produced by the heating of a filament (thin wire) to white heat (2800 K) causing it to glow or incandesce. The light generated by white heat contains every wavelength of visible white light.

There are two types of incandescent luminaires available:

1. tungsten incandescent
2. tungsten halogen.

In luminescent gaseous discharge lamps, when an electric current is passed through the pressurised gas the electrons emitted from the filament collide with the gas atoms and the atoms absorb more energy. When these atoms decay to their stable state they release photons of ultraviolet (UV) energy which, when striking a phosphor coating on the inside of the lamp bulb or tube, reradiate the UV energy as visible light. The light generated by luminescent means depends on the composition of the materials used in making the lamp.

There are five types of luminescent gaseous discharge lamps found in industry:

1 fluorescent lamps
2 mercury lamps
3 mercury tungsten blended lamps
4 metal halide lamps
5 sodium lamps.

In terms of the practical application of light sources, lamps can be divided into three main categories.

1 construction
2 operation
3 technical data.

Tungsten incandescent lamps

Incandescent lamps are manufactured with a tungsten filament in a bulb (globe) filled with a noble gas (a gas such as argon that inhibits the evaporation of the tungsten filament). A general light service (GLS) incandescent bulb is shown in **Figure 8.70**.

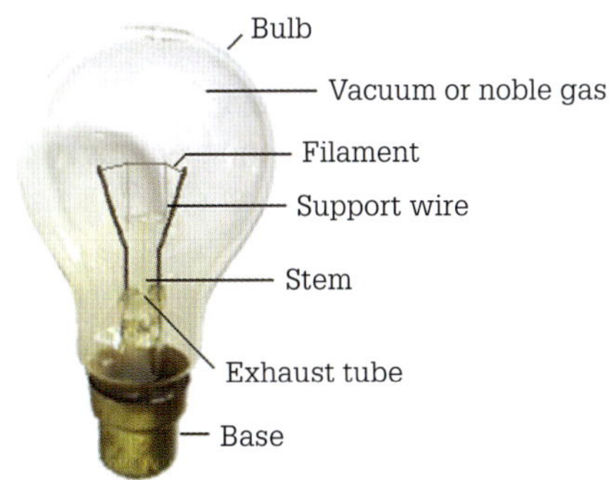

FIGURE 8.70 GLS incandescent bulb

Filaments

Tungsten (symbol W) is the metal used as a filament for incandescent lamps (**Figure 8.71**). It has a high melting point (3422 °C), very high tensile strength at high temperatures, low evaporation rates, and excellent radiation ability together with low vapour pressure at high temperatures. Due to the varying electrical characteristics of incandescent lamps, tungsten filaments are manufactured in different dimensions and various shapes. They can be a straight wire, coiled, double-coiled (most efficient) and flat. Molybdenum support wires and connecting lead-in wires are added to lamps with long filaments to prevent filament sagging when the lamp is energised. Tungsten, molybdenum or dumet conductors are used as the connecting lead-in wires. These wires are used because their expansion rate mirrors that of the glass bulb thereby ensuring a reliable metal-to-glass hermetic seal.

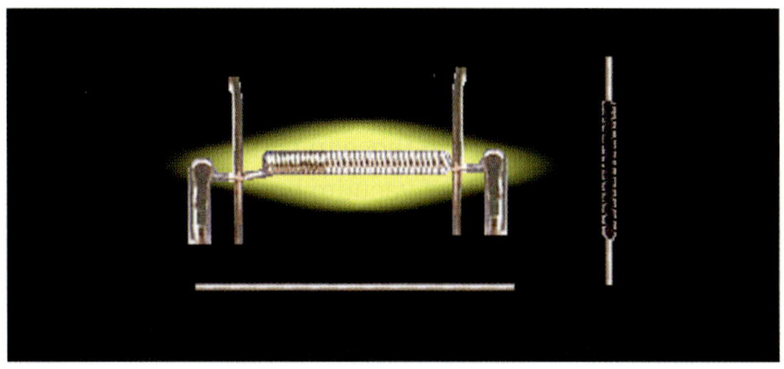

FIGURE 8.71 Filaments

The filament of an incandescent lamp is simply a temperature-dependent resistor. When electrical energy is applied, it is converted to heat in the filament. The filament's temperature rises until it reaches a point where the amount of heat leaving the filament is the same as the quantity of heat being generated in the filament. The hotter the filament becomes, the more efficient it is as a lamp. The better filaments are coiled or double-coiled to secure a bright, uniform light. The upper limit of filament temperature is the melting point of tungsten.

Higher-wattage incandescent lamps tend to be more efficient than lower-wattage lamps. One reason for this is the fact that filaments with larger cross-sectional areas can be operated at a higher temperature, which is better for radiating visible light.

When tungsten leaves the filament body at incandescent temperatures by evaporation, it produces pitting and makes the filament thinner and weaker in certain places. These thinner areas have greater electrical resistance and the current drawn from the filament makes them hotter. When the filament temperature at the narrowest spot reaches the melting point of tungsten (3422 °C) the filament open-circuits at that spot. An electrical arc forms across the gap and the lamp flares up brightly for an instant until the gap widens enough to prevent current flow, and burnout occurs.

The glass envelope

The lead-free glass envelope or hard glass sleeve is called a bulb and encapsulates the filament. The enclosure itself is either processed so that a vacuum (air evacuated) exists inside the glass envelope, or it is filled with low-pressure noble gases such as argon and nitrogen or krypton and xenon. These gases are used because they do not react chemically with the tungsten filament as their electron shells are filled and have nothing to gain in terms of stability. However, the inclusion of gas increases the energy heat loss of the lamp due to conduction and convection.

A vacuum-type bulb provides a stable environment to prevent the loss of heat from the filament to the bulb by convection; a bulb filled with noble gas provides a steady atmosphere to assist in the retardation of tungsten evaporation. The glass itself absorbs about 5–12% of the illuminance from the filament. The maximum filament temperature of a vacuum bulb is 2100 °C.

With a coiled filament in a bulb filled with noble gas the temperature limit is increased to 2500 °C resulting in more luminosity. Tungsten evaporation rates in vacuum bulbs are much higher than those in noble gas bulbs. As a consequence, the glass of vacuum bulbs is gradually darkened as a result of deposits of tungsten that have evaporated from the filament, in contrast to the glass of bulbs with noble gases where the reduced rate of tungsten evaporation results in little blackening of the glass during the life of the lamp. Additives are also included with the noble gas bulbs in order to control gas-borne impurities chemically. Because the evaporation rate has been controlled, the filament can be operated at a higher temperature thereby providing a higher luminous intensity for the same life cycle as the vacuum type.

The glass envelopes may be clear, frosted (pearl), coloured glass, silica sprayed, fused with ceramics or covered with thin coatings of metal to create dichroic globes (dichroic describes the type of coating on the reflectors or front glass). They are made up of dozens of layers of thin materials that have the uncommon assets of selectively reflecting or transmitting certain wavelengths of visible light, infrared radiation and UV. Examples of various incandescent lamps are illustrated in **Figure 8.72**.

FIGURE 8.72 Incandescent lamps – coloured, pearl and insect

Some glass envelopes have neodymium as part of their structure. Neodymium light bulbs enhance the blues and reds and cause colours to appear more vibrant. They do this by absorbing the yellow spectrum from the visible light thereby producing a light that is white and crisp. A neodymium bulb is illustrated in **Figure 8.73**.

FIGURE 8.73 Neodymium bulb

The tin or brass base provides the means for the electrical connection to the filament. There are many different types of bases, but the two main types are the Edison screw (ES) and the Swan bayonet cap (BC) base. These two bases are shown in **Figure 8.74**.

FIGURE 8.74 Base types – Edison screw and a Swan bayonet

Most of the output from an incandescent lamp is in the lower yellow to red frequencies of the visible spectrum, as illustrated in the spectral curve chart (which indicates how much of each wavelength is present) in **Figure 8.75**.

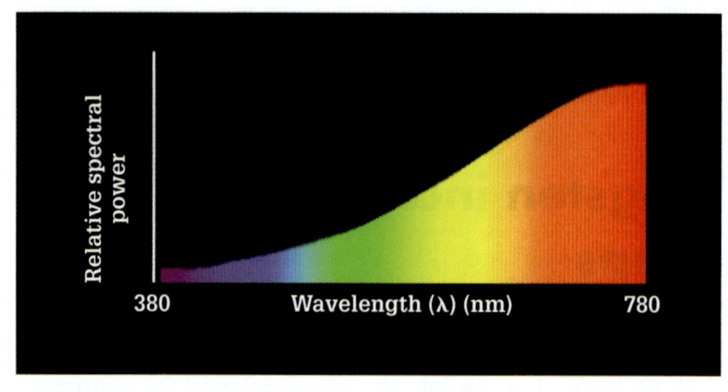

FIGURE 8.75 Spectral curve – incandescent lamp group

Krypton and xenon

Use of the higher molecular weight noble gases krypton (symbol Kr) and xenon (symbol Xe) in a glass bulb increases the strength of the luminous intensity of the filament and improves the lamp's efficiency. This occurs because these gases allow less heat to be conducted to the bulb's surface, which means that more energy leaves the filament by radiation. These higher-fill-pressure gases achieve this because of a specially treated heavy-duty tungsten filament in the bulb. Energy savings of up to 20% are possible when compared to the vacuum and argon-type incandescent lamps. The noble gases krypton and xenon improve the luminous intensity by enabling higher filament temperatures, with xenon performing best (this is used in short-arc projection lamps). The use of these gases also increases the life of the lamp by as much as 20 times. Unlike tungsten halogen lamps, these bulbs can be handled without any special precautions.

Advantages and disadvantages (air-evacuated and argon incandescent lamps)

Incandescent lamps have a low initial cost, instant starting, warm colour temperature and high brightness together with good colour rendition ($R_a > 90$). They require no special control equipment and are easy to dim. Their overall luminous efficacy is the lowest of all the lamps and ranges from 6 lumens to 20 lumens per watt. This means that their running cost per produced lumens per watt is relatively high, due to their low luminous efficacy.

Air-evacuated and argon/nitrogen-filled incandescent lamps have a relatively short life of 1000 hours and are a very inefficient source of light. If the lamp is installed in a horizontal position, the filament sags, resulting in a lower expected life. The 'extra long-life' versions (soft light) of these lamps do not produce as much light as a standard globe. These lamps produce reduced lumens because their lamp filament temperature must be lower to give a longer life.

Less than 10% of the energy input to the lamp produces light, 20% is dissipated as heat (temperatures can exceed 300 °C at the top of the lamp) and the remainder is lost as infrared radiation (vibration frequencies too low to be visible as light).

Voltage variation

The life of an incandescent filament is voltage sensitive. A 10% increase in applied voltage will reduce the filament life to only 30% of its design value, while a 10% decrease in applied voltage will increase its life by 300%, but the light output will decrease by 30%.

Variation in supply voltage to an incandescent lamp causes variation in current and hence filament temperature. This in turn affects the light output and life of the filament. The effect on luminous flux can be approximated using the following formula:

$$\text{Relative flux} = \left(\frac{\text{Applied voltage}}{\text{Rated voltage}}\right)^{3.5}$$

For example, for a 230 V globe subject to 253 V (a 10% increase):

$$\begin{aligned}\text{Relative flux} &= \left(\frac{\text{Applied voltage}}{\text{Rated voltage}}\right)^{3.5}\\ &= \left(\frac{253}{230}\right)^{3.5}\\ &= (1.1)^{3.5}\\ &= \mathbf{1.3959\ (say\ 1.4)}\end{aligned}$$

that is, 140% (an increase of 40%) of the luminous flux it produced at 230 V. The effect on the life of the lamp may be approximated using:

$$\begin{aligned}\text{Life factor} &= \left(\frac{\text{Rated voltage}}{\text{Applied voltage}}\right)^{12}\\ &= \left(\frac{230}{253}\right)^{12}\\ &= (0.909)^{12}\\ &= \mathbf{0.318}\end{aligned}$$

that is, the life expectancy is only about 30% of what it would be at 230 V.

Tungsten halogen

A halogen bulb is like an incandescent globe but with several variations. The noble fill gas argon (krypton is also used) contains traces of a gaseous halogen compound such as iodine or bromine. A halogen is a chemical that forms haloids or salts by pure union with a metal such as tungsten to form, for example, tungsten iodide (one of a number of tungsten halides).

The 'halogen cycle' acts to return evaporated tungsten to the filament. As the tungsten evaporates from the filament, it condenses and accumulates on the bulb surface. The halogen chemically reacts with the tungsten deposit to produce tungsten halides, which evaporate freely. When the halide reaches the filament, the high temperature of the filament causes the halide to decompose, redepositing tungsten on the filament.

The tungsten is redeposited on the coolest part of the filament (commonly the ends of the filament) and as a result the middle portion of the filament eventually thins, creating a weak spot, and finally the filament breaks.

In order for the halogen cycle to work, the bulb surface must be very hot (260 °C), otherwise the chemical reaction between the halogen and the tungsten may not occur because the deposited tungsten is too cool. This means that the glass envelope must be small and made of either hard glass (high strength, heat resistant) or quartz. Since the glass is reinforced, higher fill pressures are employed which means that the filament can now be operated at a much higher temperature (350 °C). The higher temperature results in a higher luminous efficacy.

There are four basic types of tungsten halogen lamps: single ended, double ended, PAR and extra low voltage and these are illustrated in **Figure 8.76**.

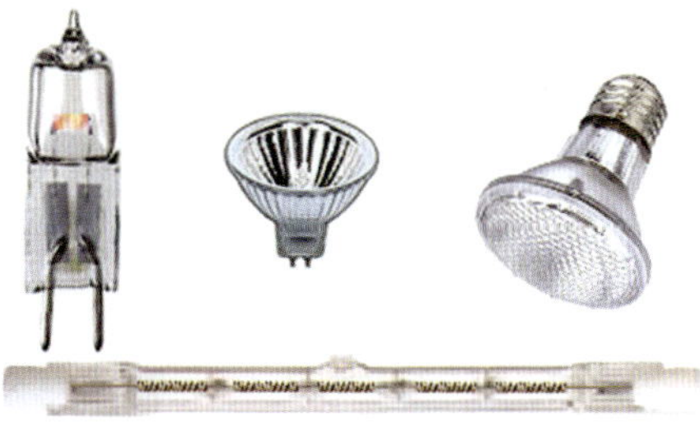

FIGURE 8.76 Halogen bulbs – single ended, extra-low voltage, PAR and double ended

The majority of the light output from a halogen bulb is the same as that of a general lighting service (GLS) incandescent in the lower yellow to red frequencies of the visible spectrum, as illustrated in **Figure 8.77**.

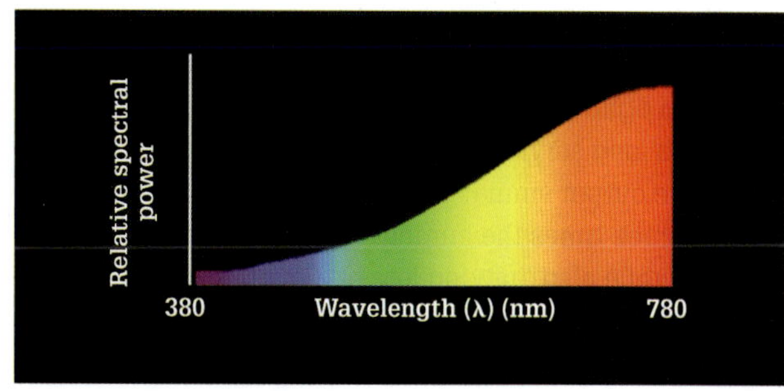

FIGURE 8.77 Spectral curve – halogen lamp group

A halogen lamp's colour temperature is low, 2800–3000 K, so the colour appearance of its light is warm with a slightly more whitish sparkling light. Its colour rendering is excellent with an $R_a > 90$ or 1a. The luminous efficacy of halogen lamps is 13–22 lm/W, meaning that they are 25–30% brighter than an equivalent GLS incandescent of the same wattage. They can give off 95% of their initial light output up to the end of their 2000–3000 hour life.

The double-ended or linear-type lamp must be operated in a horizontal burning position of 4° or less in order to ensure efficient operation of the tungsten cycle. The surface of a quartz halogen lamp must not be touched with bare fingers as the quartz surface is exposed to the salts and oils in the fingers. Such exposure causes a cooler area of the lamp and leads to premature lamp failure.

HID xenon lamps

Unlike halogen lamps, high-intensity discharge (HID) xenon bulbs, shown in **Figure 8.78**, do not have a filament but create light by igniting an electrical discharge between two electrodes.

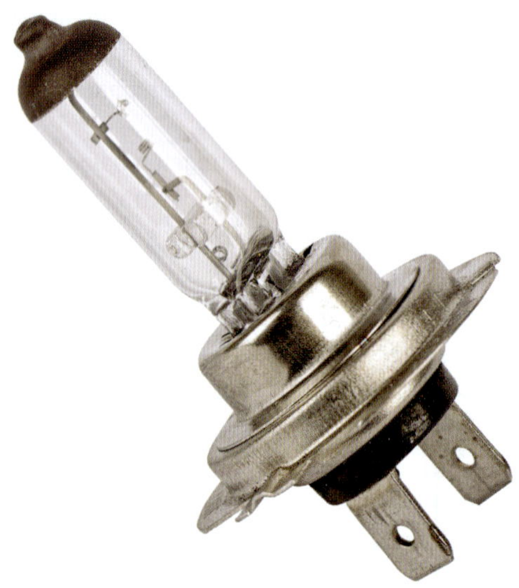

Source: Shutterstock.Com/Neryxcom

FIGURE 8.78 HID xenon bulb

There are distinctive metals contained in the quartz arc tube: high-pressure xenon (99.99%) gas, mercury and metal halide. When voltage is applied to the electrodes from the igniter, an arc is formed between them. The arc activates the xenon gas, which in turn excites the metal halide. The result is the emission of a visible light spectrum.

Xenon lighting is a cooler alternative to halogen and provides a long service life with a clear white light with a bluish hue output similar to daylight. A xenon light bulb delivers three times the luminescence (brightness) of a halogen while also drawing less electrical power and emits no UV radiation. These lamps find extensive use in the automobile industry.

Luminescent gaseous discharge lamps

Luminescent gaseous discharge lamps include the following types of lamps:

- fluorescent
- mercury vapour
- mercury blended tungsten
- metal halide
- sodium vapour.

Fluorescent lamps

Linear fluorescent lamps have a pair of filament electrodes that are hermetically sealed with a very small amount of mercury vapour (5 mg) and a noble gas (argon or krypton) at low pressure inside a glass tube. The inner wall of the tube is coated with a phosphor that produces visible light when excited by UV radiation. A fluorescent lamp is shown in **Figure 8.79**.

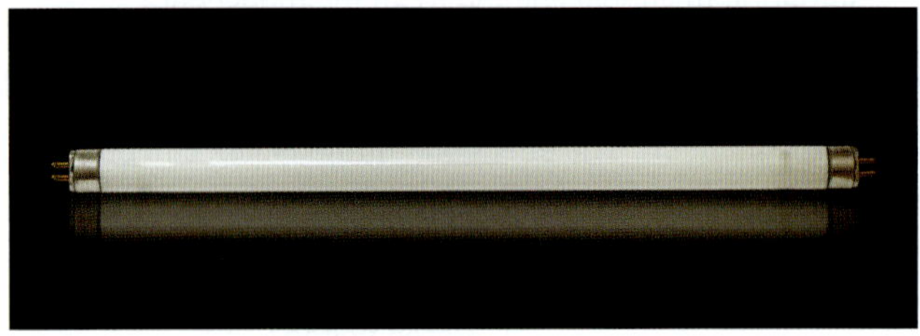

Source: Shutterstock.com/Leonid Andronov

FIGURE 8.79 Fluorescent lamp

Operation

A starter device is incorporated into the circuitry of the fluorescent lamp that at the outset provides current to the electrodes, heating them to the required starting temperature.

Once the electrodes are at the required temperature (hot cathode), the starting circuit opens and a high-voltage pulse initiates an arc across the electrodes, ionising the fill gas. A high temperature is required to ensure that the mercury gas is at the correct pressure so that the UV proton radiation is maximised. When the voltage becomes high enough, the gas nuclei, stressed by the influence of the electrostatic field, decompose into electrons and positive ions. The initially small number of free electrons turns into an avalanche where the electrons in rapidly expanding numbers flow through the gas as a current.

As the high-speed charged electrons move through the tube, they collide with the mercury atoms of the gas inside the lamp, causing their electrons to be excited to a higher energy level.

Mercury is needed because it is the best element to produce UV with high efficiency in a low-pressure discharge lamp. As more electrons collide and more of the mercury gas electrons are being excited to a higher energy level, large quantities of radiation in the form of photons are produced throughout the gas. The gas has now been 'ionised'. Ionisation refers to the amount of energy that is needed to dislodge from a mercury gas atom its most loosely held electron.

The gain in energy that elevated the electron to an excited state is emitted when the electron returns to its normal state. During its reversion or de-ionisation, the electron emits the higher energy in the form of a UV photon with a wavelength of 254 nm. This UV photon radiation energy is absorbed by the phosphor layer on the inside of the glass tube, which then changes the character of the radiation. The energy is transformed from the shorter wavelength into longer wave radiation, including UVB, UVA, visible light and infrared rays, depending on the phosphors used. Finally, the converted radiation fluoresces through the glass of the lamp, which can also be designed to act as a filter and provide additional alteration of the wavelengths. Only 22% of the energy used by the lamp is converted to light. An illustration of this process is shown in **Figure 8.80**.

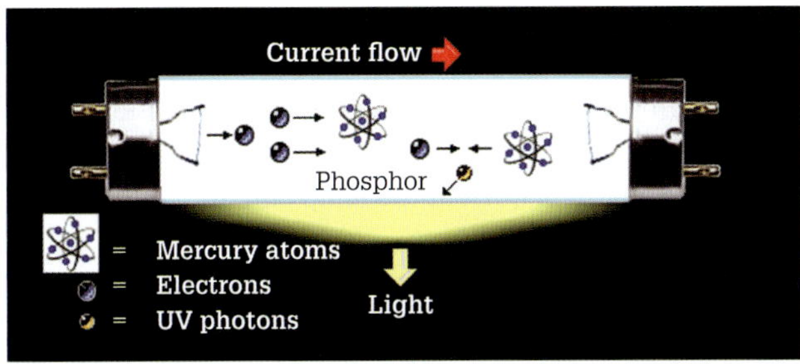

FIGURE 8.80 Radiation process

Phosphors

A phosphor is a synthetic fluorescent substance that has the property of absorbing a 254 nm UV radiation wavelength and emitting it at other wavelengths including visible light. The phosphor produces visible light (full spectrum) by fluorescence.

In a low-pressure standard fluorescent lamp, conversion of the UV proton emitted by the excited mercury into visible light is done using only one type of phosphor powder coating such as zinc silicate, magnesium tungstate or calcium halophosphate that is activated by antimony and manganese. Because only one type of coating is used the visible light produced does not have superb colour-rendering properties (R_a > 40 and < 70, 3 and 2b). The spectral curve for a halophosphor fluorescent lamp is shown in **Figure 8.81**.

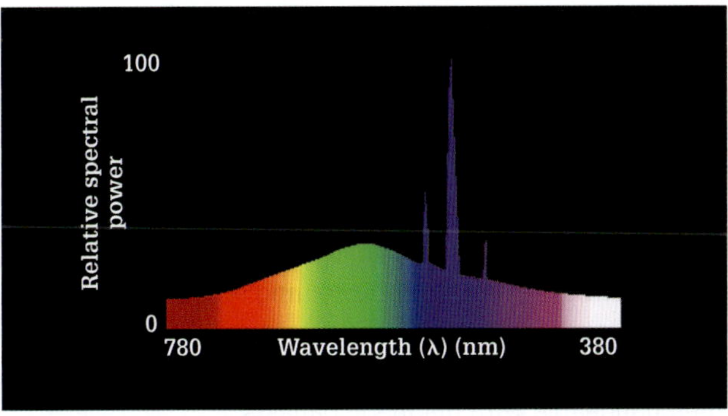

FIGURE 8.81 Spectral curve – halophosphor fluorescent lamp

Triphosphor fluorescent lamps are the standard for fluorescent lighting systems. Triphosphor lamps use rare earth trichromatic phosphors such as aluminate and phosphate phosphors that are activated by divalent europium (blue emissions), trivalent europium (red emissions) and trivalent terbium and cerium (yellow emissions). By using a mixture of these single prime colour phosphors (see **Figure 8.82**), a lamp of different colour temperatures with excellent colour rendition (R_a > 85), good lumen maintenance and high efficacy can be obtained.

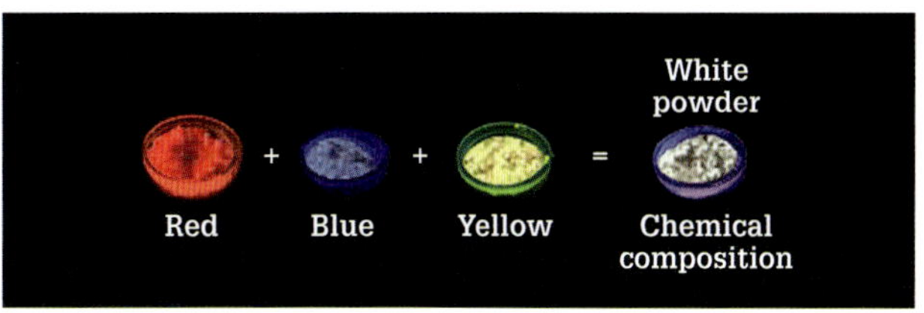

FIGURE 8.82 Triphosphors

The trichromatic system can radiate a visible spectrum very close to that of an incandescent lamp. In addition, triphosphor lamps have five to eight times greater light output than halophosphor lamps and a lifespan of more than 12 000 hours.

Different 'white' and coloured fluorescent lamps are available, each having its own colour characteristics. The triphosphor fluorescent lamps are available in two types: the conventional linear (double-ended) lamp (T8, T5) and a compact (single-ended) lamp. Fluorescent lights are cooler, provide a greater light output for the same energy consumption, are cheaper to operate than incandescent lamps and last longer. Triphosphor lamps meet the 'initial' and 'maintain' efficacy levels as required by the minimum energy performance standards – lamps (MEPS) in AS/NZS 4782.2:2004 *Double-capped fluorescent lamps – Performance specifications – Minimum energy performance standard* (MEPS). The spectral curve for a triphosphor fluorescent lamp is illustrated in **Figure 8.83**.

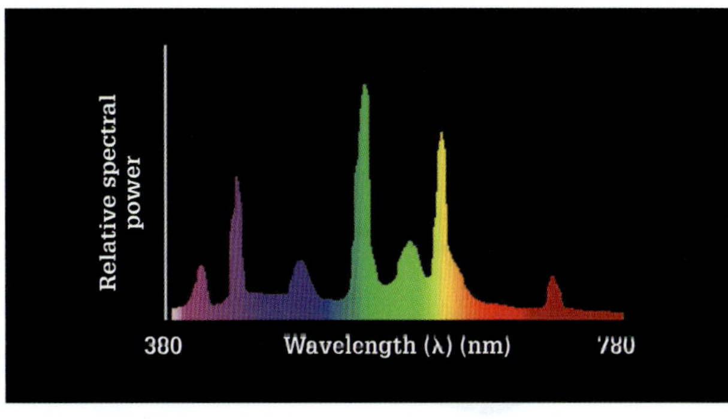

FIGURE 8.83 Spectral curve – triphosphor fluorescent lamp

This curve shows how a triphosphor coating produces pronounced energy bands in the primary colours, enriching its colour-rendering properties compared with the halophosphor fluorescent lamp. These lamps provide a greater light output for the same energy consumption as an equivalent halophosphor fluorescent.

Filaments

The coiled tungsten filaments, referred to as 'hot cathodes', are coated with electron-emissive alkaline earth oxides so that a stream of electrons is boiled off the filaments when the correct voltage is applied. Fluorescent lamps also require additional devices in order to start the ionisation process. These devices include the standard or electronic starter and ballast to ensure the starting and continuity of fluorescence. A capacitor is added for power factor correction only.

Starters and ballasts

The conventional starter switch, also called a 'glow starter', consists of a canister, a small discharge bulb containing a bimetallic switch, neon or helium/hydrogen mix gas and a radio frequency interference (RFI) suppression capacitor. A starter is illustrated in **Figure 8.84**.

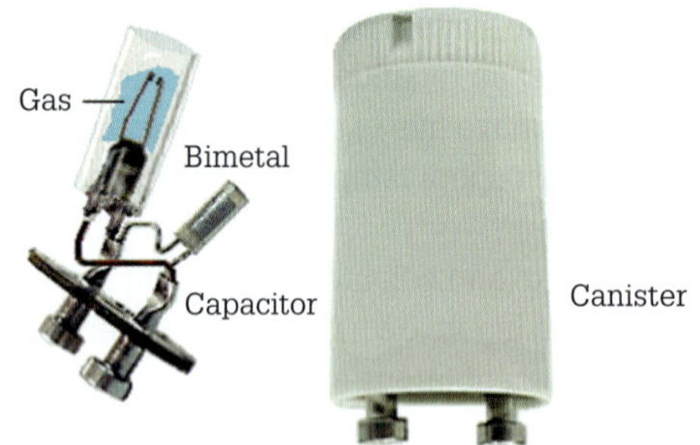

FIGURE 8.84 Starter (4–22 W 220–250 V)

The fluorescent starter is a time-delay switch that is initially in an open state. When power is applied across the fluorescent lamp the gas around the bimetallic switch is ionised (begins to glow) and heats the switch, which begins to close. Once it has closed, the ionised gas around the bimetallic switch is extinguished so that there is no longer any heating and after a period the contacts open.

The closing of the contacts allows maximum current to flow through the fluorescent filaments, heating them, and through the inductive 'ballast', shown in **Figure 8.85**, which produces large amounts of magnetic flux.

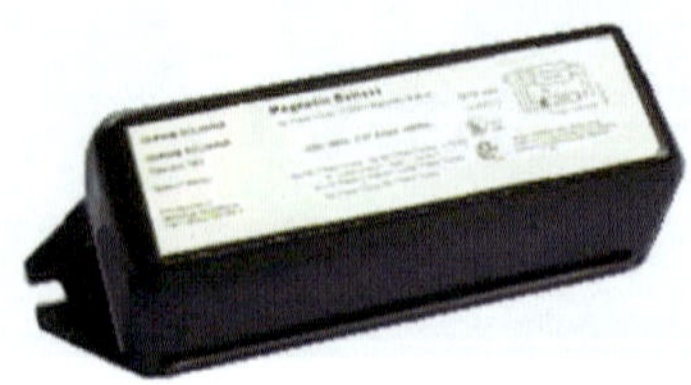

FIGURE 8.85 Inductive ballast

When the starter contacts open, current flow ceases, and the collapsing flux in the ballast produces a largely induced emf of 1000 V or more. This inductive voltage pressure across the tube provided by the ballast enables a stream of electrons to flow through the tube and ionise the mercury vapour. Without the starter, the flow of electrons is not established between the two filaments and the lamp does not ionise. Conventional starters have a random starting pulse production that can cause high lamp filament wear thereby reducing lamp life at every start. While starting they can, if the suppression capacitor is faulty, cause high electromagnetic interference. Also, the starter may keep cycling indefinitely if either it or the fluorescent lamp is defective. Sometimes starters at the end of their life cause flickering of the lamp.

Electronic starters use a programmed ignition pulse suitable for the demand of a fluorescent lamp using conventional ballasts and power factor correction capacitors. They are designed to fit a standard conventional canister. They use thyristor and Zener clamp diode technology to provide a controlled flicker-free start-up.

When a current flows through the fluorescent lamp the noble gas inside becomes warmer. As the gas heats, its electrical resistance decreases, allowing more current to flow, which makes the gas even hotter and allows even more current. This property of the gas is known as negative temperature coefficient (NTC) of resistance. In order to prevent thermal runaway, the impedance of the ballast limits the amount of current the lamp can draw no matter how hot the gas becomes.

For effective operation of the fluorescent lamp, ballasts must generally be fairly closely matched to the lamp in terms of lamp wattage, length and tube diameter. A longer lamp needs a higher inductive voltage for ionisation than a shorter lamp. A connection circuit for inductive ballast is illustrated in **Figure 8.86**.

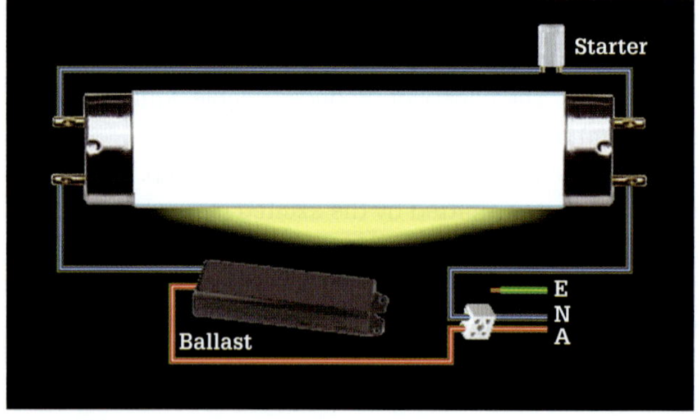

FIGURE 8.86 Inductive ballast connection circuit

Sometimes inductive ballasts will hum. This occurs because loose laminations in the iron core will resonate with the frequency of the voltage applied across the coil ends. Ballasts also consume energy and dissipate the energy consumed in the form of heat. To reduce the amount of energy wasted as heat, low-loss ballasts can be installed.

Electronic ballasts use semiconductor switching power technology to convert incoming 50 Hz power to drive the lamps with high-frequency current of 20 kHz to 40 kHz. Electronic ballasts (see **Figure 8.87**) are directly interchangeable with magnetic ballasts.

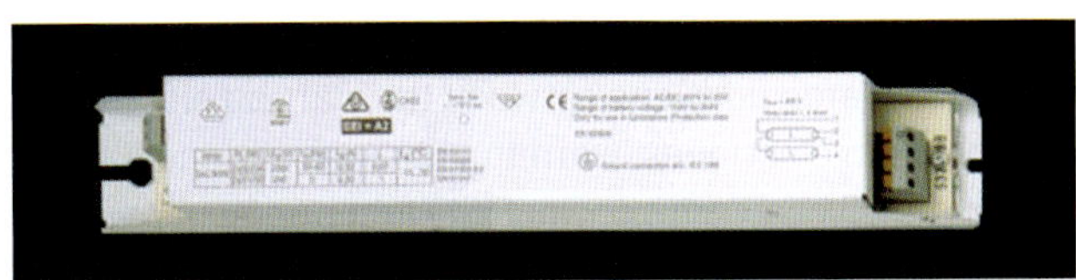

FIGURE 8.87 Electronic ballast

Electronic ballasts are more efficient than inductive ballasts in a number of ways. Most particularly, they waste less power internally than inductive ballasts, saving 3 watts to 8 watts per ballast. Their higher-frequency operation results in an overall lamp-ballast system efficacy increase of 15% to 20% and the absence of lamp flicker.

Selecting the appropriate ballast is the most important part of creating a lighting system with excellent performance. Ballasts for linear fluorescent lamps manufactured in or imported into Australia must comply with minimum energy performance standard (MEPS) requirements set out in AS/NZS 4783.2:2002 *Performance of electrical lighting equipment – Ballasts for fluorescent lamps – Energy labelling and minimum energy performance standard requirements*.

The terms T5 and T8 mentioned above relate to the tube diameters of linear fluorescent tubes. The T stands for the shape of the lamp – 'tubular'. A T5 tube is 16 mm in diameter while a T8 tube is 26 mm in diameter. The numbers 5 and 8 refer to how many eighths of an inch the tube is in diameter: 16 mm = five-eighths and 26 mm = eight-eighths. T8 lamps have been the standard lamp in fluorescent luminaires for many years. The T5 lamps are a recent design with a 20% increase in lumen output over the equivalent T8 lamp. The T5 lamp requires electronic control gear to run, which enhances the energy savings (to approximately 30%) over a magnetically ballasted T8 luminaire. Note that T8 has similar energy performance to T5 when using electronic ballasts. Compared to the T8 the T5s have a higher light output, increased lamp life and improved lumen depreciation. Other advantages of T5 lamps are sustainability focused:

1 38% less glass in construction
2 Less phosphor due to reduced physical size. In addition, T5s use a rare earth triphosphor (RE80) coating that prevents mercury from being absorbed into the phosphor and the bulb glass. This improved coating allows for reduced mercury content in the lamp
3 80% less mercury content
4 Less metal used in casing.

A significant factor that affects lamp light output is ambient temperature. Each lamp system has an optimum ambient temperature at which the lamp produces maximum light output. This is 35° C for T5s, which is 10° higher than that of T8s. This means that in enclosed luminaires the higher-temperature T5 lamps function better than T8 lamps. The higher temperature of T5s makes possible the development of smaller light fittings.

T5 lamps are available in the three standard fluorescent colour temperatures – cool at 4100 K, warm at 3000 K, and neutral at 3500 K. They also have a colour rendering index of 85 and are rated for 20 000 hours at three hours per start.

Fluorescent circuits do fail in service; however, 80% of all faulty fluorescent circuits are due to three faults: the starter, the tube and the ballast.

Stroboscopic effect

The fluorescent lamp extinguishes every time the sine wave passes through zero degrees (100 times per second at 50 Hz), which causes a stroboscopic effect (flicker causing dark intervals). This is not noticeable to the unaided eye but will create undesirable effects on TV or computer monitor screens and in the apparent speed of rotating machinery. Electronic ballast circuits can mitigate this effect somewhat, but standard ballasts cannot. To overcome this effect a lead-lag circuit supplying a twin lamp may be used. One lamp is supplied via the ballast and the other lamp circuit has a capacitor in series with its ballast. This combination creates a 90° phase difference in the currents between the two lamps so that one lamp fills in the dark intervals left by the other lamp. A lead-lag circuit is shown in Figure 8.88.

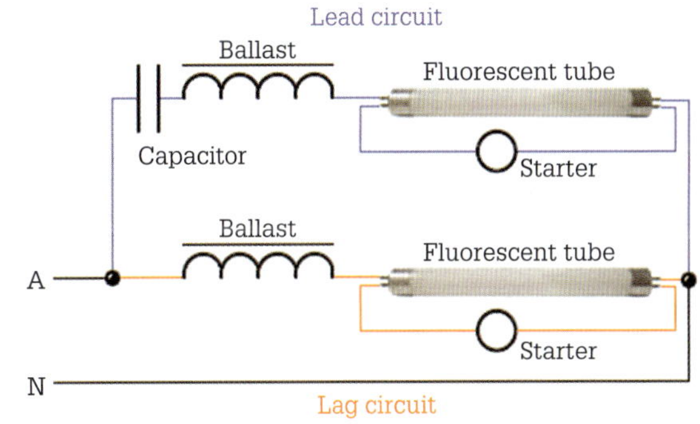

FIGURE 8.88 Lead-lag fluorescent circuit

In a three-phase four-wire installation, rows of fluorescent luminaires as shown in Figure 8.89 can be connected to different phases to create the lead-lag effect.

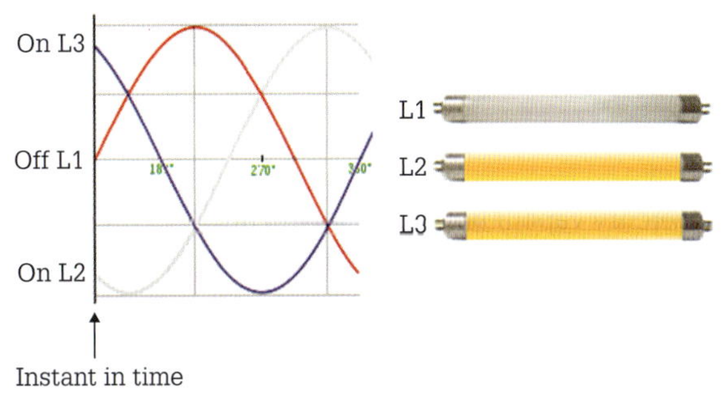

FIGURE 8.89 Lead-lag three-phase fluorescent circuit

Compact fluorescent lamps

Compact fluorescent lamps (CFLs) operate in the same manner as other fluorescents. CFLs can combine the efficacy of fluorescent lamps with the simplicity and expediency of incandescent lamps. These lamps can replace incandescent lamps that are up to four times their wattage, thereby conserving a significant amount of energy. CFLs can have built-in or separate electronic ballast and are only slightly bigger than the incandescent lamps they replace.

CFLs have the advantages of good colour rendition, good efficacy and long life. These lamps can take the shape of an incandescent or can be of circle form or bent or twisted into compact shapes. Unlike normal fluorescents they can be operated in any position. Two CFLs are shown in **Figure 8.90**.

FIGURE 8.90 Compact fluorescent lamps

Mercury lamps

The mercury vapour lamp is the oldest and best-known member of the HID lamps. The mercury vapour lamp is also the least efficient HID lamp, and consequently expensive to operate. Mercury vapour lamps are illustrated in **Figure 8.91**.

FIGURE 8.91 Mercury vapour lamps

Mercury vapour lamps contain an arc discharge quartz tube or a vitreous silica tube (allows UV radiation to be transmitted) containing a starting electrode and two operating electrodes (tungsten and alkaline earth metals) enclosed in an outer glass isothermal envelope.

The discharge tube contains a small amount of mercury and an argon fill gas, while the outer envelope contains a series-starting resistor, support leads, connecting wires and a nitrogen, argon or carbon dioxide fill gas both to prevent oxidation of the arc tube seals and to slow the rate of deterioration of the phosphors if included.

When a voltage is applied across the lamp terminals, an arc discharge develops between the starting electrode and the main electrodes, warming the mercury and fill gas within the discharge tube. Once the argon/mercury vapour has reached sufficient pressure and temperature and the internal resistance of the discharge tube is less than the external resistance, a discharge is struck between the two main electrodes.

The warming period can take up to five minutes; if a restart is required while the lamp is hot, up to a seven-minute delay (for cooling and restart) is possible. The series-starting resistor limits the starting current drawn by the lamp.

Clear mercury vapour lamps produce a bluish-green white light concentrated in five spectral peaks at certain wavelengths directly from the discharge tube. This light gives high distortion of colours (it makes red and orange appear brown) due to the fact that red is lacking in the electromagnetic spectrum (see **Figure 8.92**).

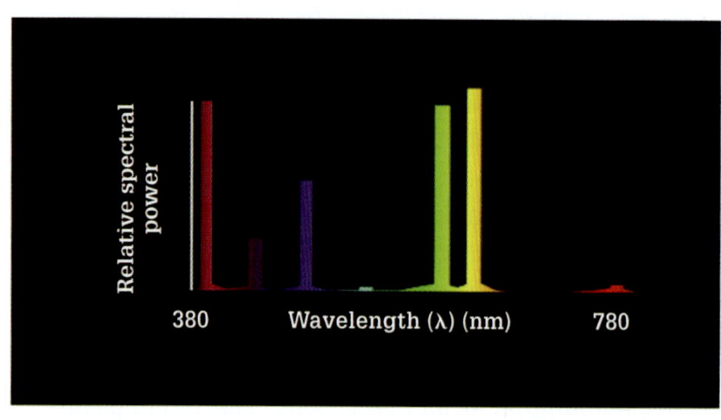

FIGURE 8.92 Spectral curve – mercury vapour lamp

Phosphor-coated lamps improve colour rendition and allow the UV radiation absorbed by the glass of the clear lamp to be converted into visible light.

These lamps have an efficacy in the 25–55 lumens per watt range with a lamp life expectation of up to 24 000 hours and a colour-rendering (R_a) index of 3. These lamps are also voltage sensitive and must be operated within ±5% of their rated voltage. With the exception of self-ballast lamps, mercury lamps need auxiliaries for their operation. They require a ballast to start and regulate the voltage across the lamp, in a similar way to the operation of a fluorescent lamp.

The ballast creates a lagging power factor that must be corrected by connecting a capacitor across the supply. The mercury vapour lamp ballast starts the lamp with a low initial voltage across the electrodes and as the lamp warms the voltage rises until the lamp stabilises.

High-pressure mercury vapour lamps are used as general industrial lighting in high-ceiling environments utilising Highbay fittings and in a wide range of outdoor purposes.

They can be operated in any burning position and have a long life and high reliability in all conditions of operations and a low sensitivity to fluctuation of mains supply voltage.

A circuit diagram of how a high-pressure mercury vapour lamp is connected to the supply is shown in **Figure 8.93**.

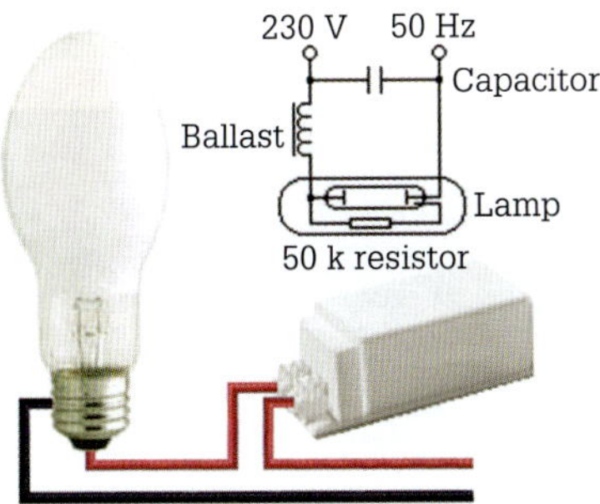

FIGURE 8.93 Circuit diagram – high-pressure mercury vapour lamp

Mercury blended tungsten lamps

Mercury blended tungsten lamps can be identified by the absence of connected ballast and by the heavy-duty tungsten filament encircling the arc discharge tube inside the phosphor-coated isothermal envelope. They are also called 'self-ballasted' or 'colour corrected' or 'blended light' mercury vapour lamps. The tungsten filament located between the two envelopes produces incandescent light and controls the current through the lamp. These lamps have an effective life of 12 000 hours and an efficacy of up to 28 lumens per watt. The spectral curve for a mercury blended tungsten lamp is illustrated in **Figure 8.94**.

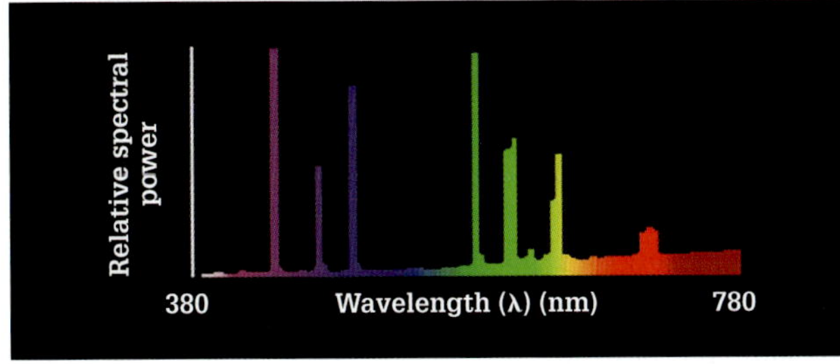

FIGURE 8.94 Spectral curve – blended mercury vapour

Self-ballasted mercury vapour lamps produce a whiter light than an incandescent lamp and can directly replace incandescent lamps in existing installations.

Metal halide lamps

High pressure metal halide lamps are similar in construction to high-pressure mercury vapour lamps but they operate at higher pressures and temperatures. They have two envelopes. The inner quartz arc discharge tube contains metal halides such as dysprosium, thallium, indium and sodium iodides in addition to mercury and the fill gas argon. The outer envelope may or may not have a fluorescent powder coating. Metal halide lamps are illustrated in **Figure 8.95**.

FIGURE 8.95 Metal halide lamps

The principle of operation is that each metallic halide, when vaporised in the discharge tube, produces its distinct electromagnetic spectrum, thereby together covering the entire visible range. Metal halide lamps provide white light in a variety of correlated colour temperatures ranging from 3200 to 5200 Kelvin with a lifespan of 6000 hours to 10 000 hours. They have a colour-rendering index < 65 (R_a = 1a, 1b or 2a). The spectral curve for a metal halide lamp is illustrated in **Figure 8.96**.

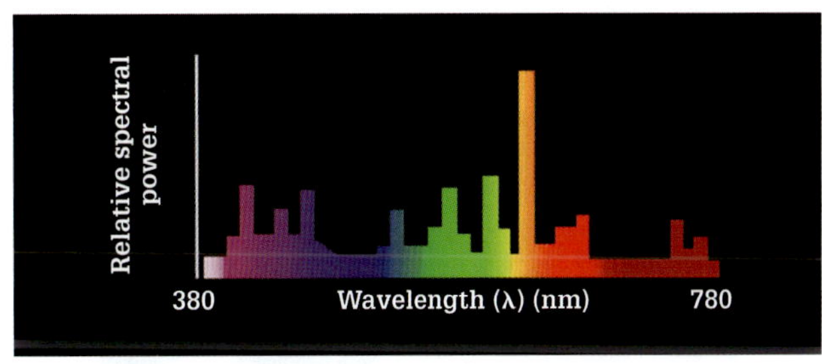

FIGURE 8.96 Spectral curve – metal halide lamp

The spectral curve produced depends on the combination of halides used. Metal halide lamps do not achieve their full light output until after 12 minutes of warm-up time has elapsed. A restart when hot requires only two minutes if an ignitor is connected or seven minutes or longer for an auxiliary-electrode-type lamp. Because of their large quantity of high intense light output, metal halide lamps are used in buildings with high ceilings and other facilities in which lighting is constantly in use for many hours at a time. Light is produced in the arc discharge tube and is started with either the help of an auxiliary electrode (probe start, ignitor) or a short-duration high-voltage starting impulse. A circuit diagram of how a metal halide lamp is connected to an ignitor is shown in **Figure 8.97**.

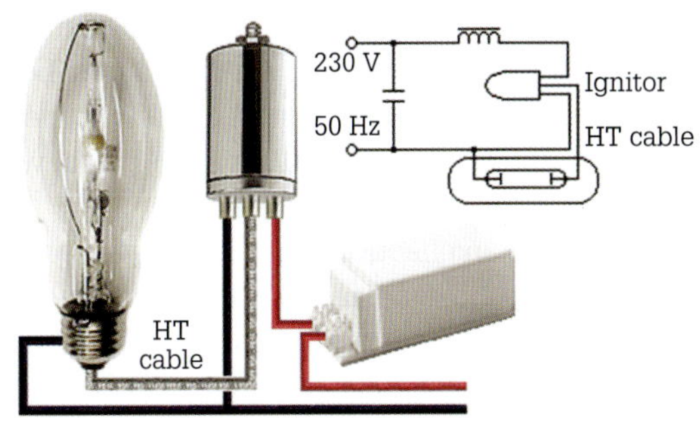

FIGURE 8.97 Connection diagram for a metal halide lamp

Metal halide lamps are used for street lighting, stadium and sports field lighting and for many large-area lighting situations.

Sodium vapour lamps

Sodium vapour lamps come in two varieties, high-pressure sodium and low-pressure sodium. The light emitted from these lamps is monochromatic (comprised of one colour only) totally made up from deep yellow wavelengths at 589.0 nm and 589.6 nm. They are 10 times more efficient than incandescent bulbs; in fact, these are the most efficient lamps made and have a 24000 hour lifespan.

Low-pressure sodium vapour lamps

Low-pressure sodium vapour lamps have the highest efficacy of all the lamps, providing around 180 lumens per watt which gives them an advantage over other types of lamps. The principal reason for the high efficacy is because the colour of the light is close to the maximum sensitivity of the human eye in normal viewing conditions. It also maintains its initial lumen output for the entire life of the lamp. Low-pressure sodium vapour lamps are also known as the SOX (sodium oxide) type. An illustration of a SOX lamp is shown in **Figure 8.98**.

Source: Alamy stock photo/ dpa picture alliance

FIGURE 8.98 Low-pressure sodium vapour lamp

A low-pressure sodium vapour lamp contains either a linear or a U-shaped ceramic discharge arc tube constructed of sodium-resistant lime borate glass inside a clear high-vacuum outer envelope. The outer glass envelope has an internal coating of heat-retaining indium oxide that contributes to the high efficacy of the lamp.

Inside the arc tube there is a mixture of neon and argon starting gases that are used to start the ignition of the lamp together with a small amount of sodium metal. As the sodium metal heats up and vaporises (10 to 12 minutes) in the discharge tube, visible energy is discharged. Low-pressure sodium vapour lamps must be started with either high-reactance autotransformer ballast or an ignitor with ballast to provide the starting voltage pulse.

In order to have an effective lumen life, the arc tube is constructed with small depressions that store the sodium metal. The arc discharge takes place between beehive-shaped triple coil tungsten electrodes mounted at each end of the U-shaped arc tube. These lamps have the poorest colour-rendering characteristics due to their monochromatic light. The spectral curve for a low-pressure sodium vapour lamp is illustrated in **Figure 8.99**.

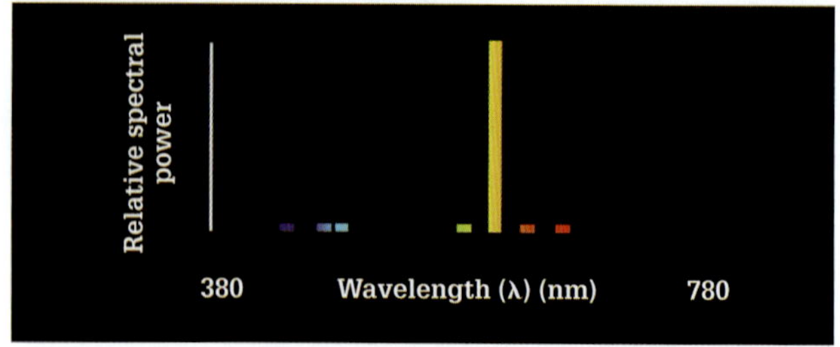

FIGURE 8.99 Spectral curve – low-pressure sodium vapour lamps

Because of their harsh yellow light, low-pressure sodium vapour lamps are suitable for outdoor and some indoor lighting applications where colour rendering ($R_a = 0$) and aesthetics are of low concern. These lamps are most often used as street and security luminaires.

High-pressure sodium vapour lamps

A high-pressure sodium vapour lamp emits a higher-quality broadband light compared to the low-pressure sodium vapour lamp, with an effective life of up to 25000 hours. An orange-white light is produced by the electric current through high-pressure sodium vapour in the arc discharge tube, which is made of sintered aluminium oxide. Sintered aluminium oxide prevents chemical erosion by the sodium and also enables a higher temperature to be established with the discharge tube. High-pressure sodium vapour lamps require a high-voltage starting pulse delivered by an ignitor, while the ballast limits the operating current. The ignitor provides a pulse of at least 2500 volts peak in order to initiate the lamp arc. When the system is energised, the ignitor provides the required pulse until the lamp is lit and automatically stops pulsing once the lamp has started.

White high-pressure sodium lamps are available that produce an excellent quality of light similar to that produced by an incandescent lamp, but these do not have as high a luminous efficacy as conventional golden-white high-pressure sodium vapour lamps. High-pressure sodium vapour lamps are illustrated in **Figure 8.100**.

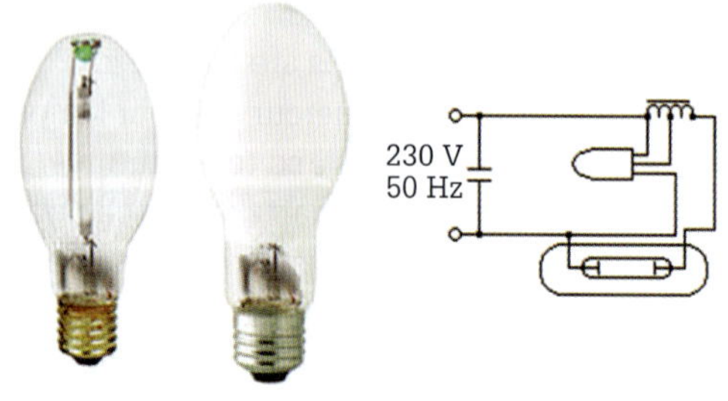

FIGURE 8.100 High-pressure sodium vapour lamps

The spectral curve for a non-white high-pressure sodium vapour lamp is illustrated in **Figure 8.101**.

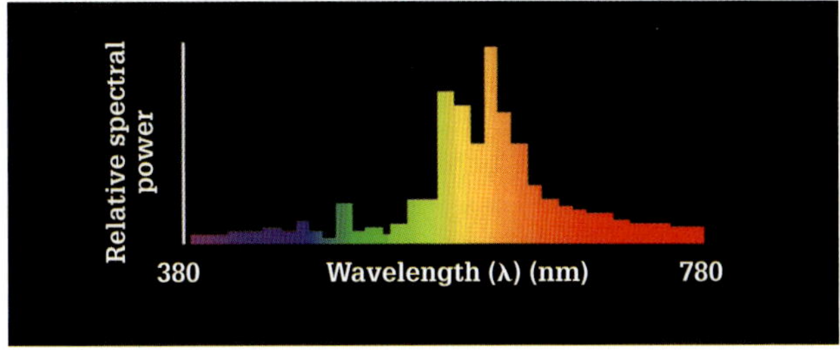

FIGURE 8.101 Spectral curve – high-pressure sodium vapour lamps

Common applications are outdoor lighting around factories and commercial buildings and roadway lighting, while white sodium vapour lamps can be used for shop lighting and building beautification.

Other lamp types

Electrodeless lamps embody a new variety of lamps that produce light by using microwaves or an electromagnetic field to excite the fill gas within a lamp.

A sulphur lamp was established in about 1997 in the United States and according to various sources it was a bright energy-efficient lamp with a light similar to sunlight. The sulphur lamp consisted of an enclosed quartz sphere about 40 mm in diameter containing an argon fill gas, a small amount of sulphur and no electrodes or filaments. Microwaves from a magnetron were used to energise the lamp. The microwaves heated the fill gas exciting the sulphur to release photons of light. The quartz sphere had to be constantly rotated because it would develop a hot spot on the tube as a result of heat convection and melt.

The induction lamp uses a high-frequency generator connected via a coaxial cable to a ferrite core induction coil to establish via a coupling antenna an electromagnetic field within a low-pressure-mix fill gas inside a glass envelope. The interaction of the electromagnetic radiation and the atoms of the fill gas releases UV radiation that is then converted into visible photon energy by a phosphor powder coating on the inner surface of the glass envelope. An induction lamp is illustrated in **Figure 8.102**.

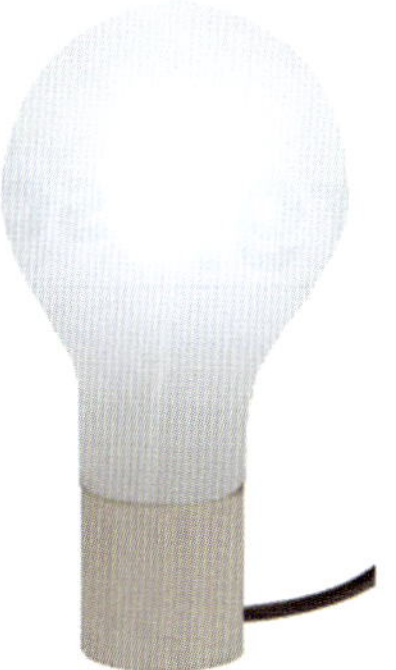

FIGURE 8.102 Induction lamp

The colour rendition (R_a = 82) is similar to fluorescent lamps with a colour temperature of 3000 K. They have an exceptional lifespan of up to 100 000 hours with a lamp efficacy of 70 lm/watt. The spectral curve for an induction lamp is illustrated in **Figure 8.103**.

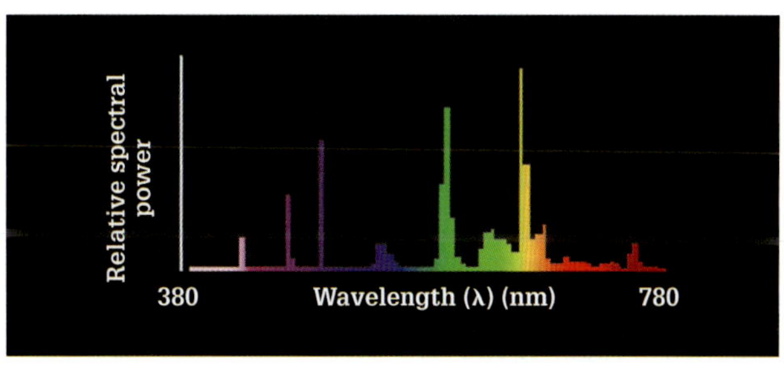

FIGURE 8.103 Spectral curve – induction lamp

These lamps have the potential to replace existing luminaires such as incandescent lamps and high-intensity discharge lamps. They are used both indoors and outdoors in areas requiring long operating hours.

REVIEW QUESTIONS

1. Name the two categories of lamp light sources.
2. Describe the operation of luminescent gaseous discharge lamps.
3. Name five types of luminescent gaseous discharge lamps found in industry.
4. For what is the lighting term 'GLS' an abbreviation?
5. From which metal is the filament of incandescent lamps usually made?
6. In an incandescent lamp, how is the evaporation of the metal filament retarded?
7. Name the two types of bases used for incandescent type lamps.
8. How do the noble gases krypton and xenon improve the luminous intensity of lamps?
9. What effect will a 10% voltage increase have on the life of a 230 V incandescent lamp filament?
10. In relation to tungsten halogen lamps, what is a halogen?
11. Name the four basic types of tungsten halogen lamps.
12. Describe the construction of a linear fluorescent lamp.
13. Why is mercury added inside the tube of a linear fluorescent lamp?
14. In relation to tungsten fluorescent lamps, what is a phosphor?
15. What components are used in the conventional starter switch for linear fluorescent lamps?

»

16 Describe the basic operation of an electronic fluorescent lamp ballast.

17 Describe a common method used with fluorescent lighting to reduce the stroboscopic effect.

18 Describe the operation of a high-intensity mercury discharge lamp.

19 What is an application of high-pressure mercury vapour lamps?

20 Which discharge lamp produces an orange-white light?

8.7 Energy-saving lighting

Possibly the most prevalent energy-saving lamp is the light emitting diode (LED).

Light emitting diodes (LEDs)

LEDs are a significant advancement in illumination. This is because LEDs have a long service life (up to 50 000 hours), no infrared and UV emission, high energy efficiency, low-voltage operation, are environmentally friendly, robust to thermal and vibration shocks and compact in size.

An LED is a semiconductor diode with a junction between an *n*-type (electrons) and a *p*-type (holes) material. When the diode is forward-biased, electrons from the *n*-type region and holes from the *p*-type region meet at the junction and recombine, emitting photons. LEDs are also polarised – anode with respect to cathode. The flat side on the edge of an LED is the cathode side.

LEDs do not produce light by heating a filament or by producing an arc in a discharge tube. They are small solid-state devices that radiate a small brilliant photon when a hole and an electron recombine. An illustration of an LED is shown in **Figure 8.104**.

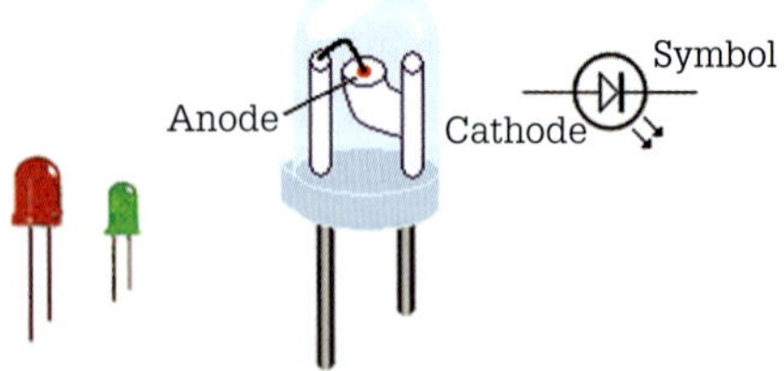

FIGURE 8.104 Light-emitting diodes

The type of material used for the diode construction and the width of the material's band gap determine the colour of the light. LEDs produce deep reds, greens, blues and other colours in monochromatic form directly from the solid-state material. A white LED has a blue LED semiconductor source that illuminates a phosphor to produce white light. LEDs are suitable for a large range of lighting applications that currently use incandescent and fluorescent luminaires. However, LED lights run at high temperatures so large heat sinks need to be fitted to cool them.

Due to their small size LEDs are built into an array to increase lumen output. Individual LEDs have an efficacy of 2–20 lumens per watt with a 3.6 V 20 mA supply input.

Illustrations of LEDs and their applications are shown in **Figure 8.105**.

FIGURE 8.105 LED applications – traffic light and spotlight

A 12 W lamp array utilising LED technology can replace an incandescent lamp used in traffic signals, resulting in significant power savings. LED bulbs, spotlights and strip lights are also available. They can also be used as the lamp for signs, signals and warning lights. Reducing current flow through the solid-state material easily dims LEDs.

LED fluorescent tube replacements

LED fluorescent tube replacements as illustrated in **Figure 8.106** emit the same amount of light as conventional T-series fluorescent lamps, but last up to 10 times longer with low loss of illumination.

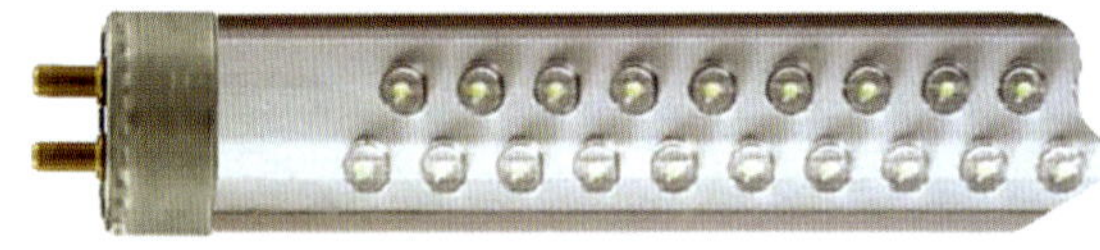

FIGURE 8.106 LED fluorescent tube

LED fluorescent tubes use 5 mm low-decay super bright dual in-line package (DIP) LEDs with 90 lm/W. They have a colour-rendering index of 84 and an average lifespan of more than 50 000 hours while consuming 15 watts. There is no mercury, and no ballast or starter are required. In addition, their power consumption is 15 W while delivering good light quality and good heat dissipation. The aluminium housing acts as the heat sink for the LEDs.

Fibre optics

Fibre optics refers to the branch of science while optical fibre refers to the waveguide.

In an optical fibre system, a halogen lamp injects light into the optical fibres. The light is then transmitted through the fibre by the principle of total internal reflection. Total internal reflection (TIR) is the phenomenon that involves the reflection of the entire incident light off the boundary between two media. TIR only takes place when both of the following two conditions are met:

- a light ray is in the more dense medium and approaching the less dense medium
- the angle of incidence for the light ray is greater than the critical angle (the light ray must hit the walls of the fibre at a minimum angle).

Under this principle, only the visible spectrum (400–780 nm) is transmitted, eliminating heat, and IR (infrared) and UV (ultraviolet) radiation. The result is pure light.

Optical fibre cables consist of glass silica or an optically clear polymer rod called the core through which light is guided by internal reflection. Surrounding this core is a fluoropolymer light-refractive material called the cladding. The purpose of the cladding is to confine all of the light to the core through its refractive nature, as illustrated in **Figure 8.107**.

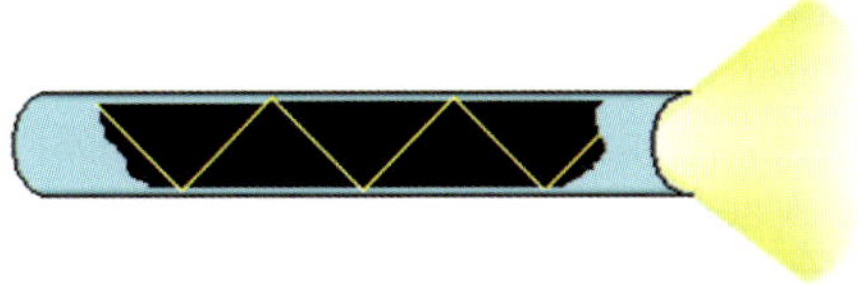

FIGURE 8.107 Optical fibre cable

A lighting system employing optical fibre cable (the ends are called tails) uses a luminaire (150 W metal halide, 50 W tungsten halogen or even a high-brightness LED) coupled to the solid or stranded fibre. The luminous flux emitted from the luminaire is focused to shine on the ends of the fibres at a predetermined angle. The flux is then reflected along each fibre tail where it is emitted directly through its diamond polished end or through a lens to focus the light into a beam. Ellipsoidal aluminium reflectors or dichroic reflectors can be used for high-output applications. These reflectors allow wavelengths longer than 700 nm to pass through to reduce heat, selecting only the visible light.

The cross-sectional area (CSA) of the tails, the intensity of the coupled luminaire and how the light is attenuated (to attenuate is to weaken or to make (or become) thin) as it passes through the fibre determine the light output from the ends of the fibre. Various colour output can be initiated through the use of a rotating colour wheel located in front of the coupled luminaire.

The two types of optical fibre used for lighting purposes are end-emitting fibre and side-emitting fibre. Sidewall optic fibre is illuminated by light being scattered along its entire length and is used for decorative lighting. An example of its use to illuminate the features of a swimming pool is shown in **Figure 8.108**.

FIGURE 8.108 Fibre-optic sidewall lighting on a swimming pool

Side-emitting fibre is mounted with clips or straight tracks while end-emitting fibre can be run in conduit or directly fixed. End-emitting fibre can also be used for decorative purposes such as providing starlight (30 000 or more individual tails) in ceilings and for spotlights, as shown in **Figure 8.109**.

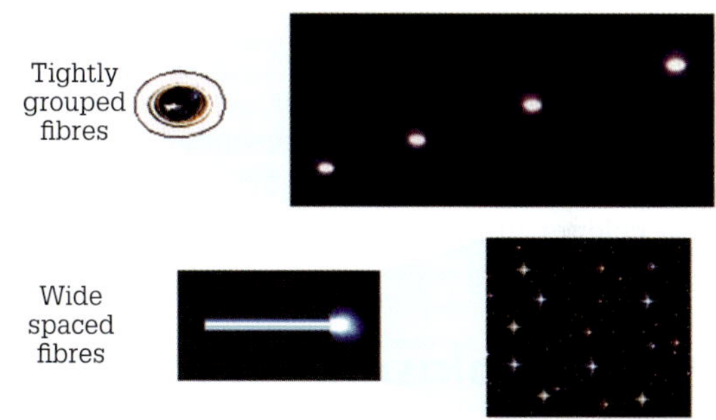

FIGURE 8.109 End-emitting fibres

Cold-cathode lamps

Cold cathode refers to the type of electrode – iron nickel – used by the luminaire. This electrode is not heated in order to heat up the fill gas to establish ionisation. Cold-cathode fluorescents and neon discharge lamps use the cold-cathode principle of operation.

Cold-cathode fluorescent (CCFL) lamps

A cold-cathode fluorescent lamp is basically a fluorescent lamp without the thermal filament. These lamps can be bent and shaped to any configuration and are available in very small lamp diameters. Cold-cathode fluorescent lamps are illustrated in **Figure 8.110**.

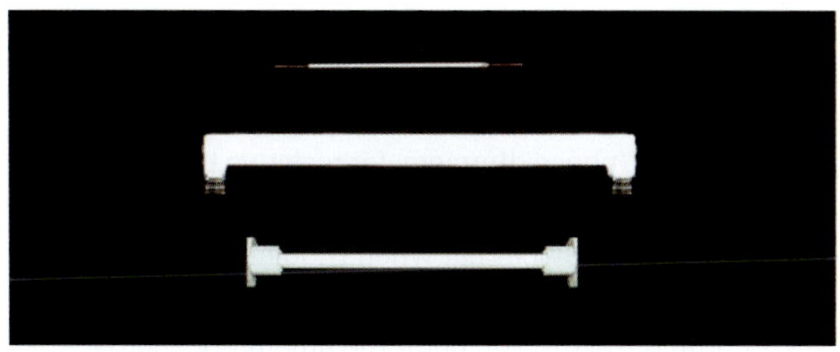

FIGURE 8.110 Cold-cathode fluorescent lamps

Cold-cathode tubes contain a fill gas mix of argon and mercury. Because the mercury is already in vapour form, the fill gas does not require preheating. Lamp ignition is instantaneous and flicker free with operation from both alternating and direct current. These lamps are low-amperage (3 mA to 6 mA), high-voltage (200 V to 700 V rms) input lamps. In some applications they utilise an inverter to generate the high-frequency voltage required from battery power supplies. The high voltage is required to enable electrons to be emitted in an avalanche from the unheated cathode, allowing the current to excite the fill gas inside to produce a glow.

These lamps draw very little power for the luminous flux (up to 80 lumens per watt) they deliver over a lifespan of up to 50 000 hours. Most of the energy drawn by the lamp is converted to visible light (65%) and not to heat (35%).

The cold-cold fluorescent construction allows the lamp to be miniaturised. This makes these lamps ideal for applications such as the document illumination source in photocopiers, facsimile machines and scanners, and in exit signs. These lamps come in different colours with excellent lamp luminance and high colour rendering.

Controlled plasma (CP) lighting

Controlled plasma (CP) lighting technology is an emerging innovative lighting technology based on cold-cathode fluorescent lighting principles. Controlled plasma technology uses an initial current strike voltage to release electrons to create the plasma state within the lamp. It uses only 5 watts, is dimmable and has a lifespan of 20 000 hours. CP lamps have a light output efficiency of 140 lumens/watt; a colour-rendering index of 91 and a cool operating temperature (reduces fire risk and air-conditioning requirements of premises). Due to their triphosphor coating CP lamps are available as a warm white (2700 K) or a cool white (4200 K). This lamp is seen as a replacement for halogen downlights; however, CP lamps are not point-source lights. This is because they have a different directivity of light.

Dimming

Light dimming is achieved by adjusting the voltage across the lamp. Light dimming can be achieved by using variable power resistors and tapped transformers. However, solid-state dimmers are most prevalent. These dimmers operate by controlling the point of switch on and switch off of the voltage waveform applied across the luminaire. Solid-state dimmers (phase dimmers) for resistive incandescent and halogen lamps use the triac thyristor. The triac is a semiconductor device that when triggered can be turned on and off 100 times per second (50 Hz). Therefore, by controlling the turn-on point the amount of energy delivered to a lamp can be controlled. Triac load control is illustrated in **Figure 8.111**.

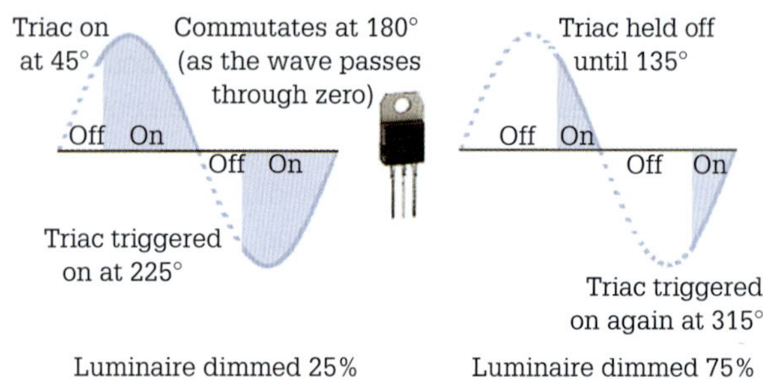

FIGURE 8.111 Triac control of a lamp

Unlike incandescent lamps, fluorescent lamps cannot be properly dimmed with a simple solid-state device such as that used for incandescent lamps. For a fluorescent lamp to be dimmed over a full range without a reduction in lamp life, its electrode heater voltages must be maintained while the lamp arc current is reduced.

Fluorescent lamps can be dimmed with high-frequency dimmable ballast. Dimming ballasts are available as magnetic or electronic devices using insulated-gate bi-polar transistor, integrated circuits (ICs) and power metal oxide semiconductor field-effect transistor (MOSFET) technology. Magnetic dimming ballasts require control gear containing high-power switching devices that condition the input power delivered to the ballasts, while energy management electronic ballasts have a normal line supply and an intelligent control interface card that provides an analogue voltage control signal (1–10 V d.c. potentiometer) for telling the ballast what level to dim to.

Other fluorescent lamp dimmers use a microprocessor-controlled wiring system (digital control) such as the C-Bus system that delivers complete control of a lamp circuit. Dimmers for other types of gas-discharge lighting are also available.

Photosensors and infrared and ultrasonic sensors are sometimes used with dimmers to coordinate electric lighting in various lighting environments.

Daylight

When applying sustainable energy principles, it is important that daylight is incorporated into lighting control systems to gain the benefit of the available daylight. For example, where two or more luminaires are connected to the same lighting circuit and are located close to windows, they should be separately switched from the rest of the room to conserve energy when daylight is available. If there are large window areas, automatic stepless dimming control should be considered.

Daylight sensors

A daylight sensor as shown in **Figure 8.112** is an electronic control device that adjusts the light output of a lighting system based on detected illuminance. These sensors can be used with dimming electronic ballasts to adjust the light output of fluorescent lighting systems in commercial offices.

FIGURE 8.112 Daylight sensor

Light fixtures at the perimeter of floors near windows should have daylight sensors (dawn to dusk) to dim artificial lighting as daylight increases to save energy. Some luminaire manufacturers are installing daylight photocells in suspended and recessed luminaires. The daylight sensor is usually shielded from direct light so that only light reflected from surfaces such as the work plane and walls reach the sensor. Problems with these sensors involve commissioning, locating the sensor which requires accurate placement and calibration for effective operation of the daylight sensor system.

REVIEW QUESTIONS

1 What are the benefits of LED lamps?
2 How is the lumen of an LED lamp improved?
3 Name the lamp that injects light into optical fibres.
4 What is a cold-cathode lamp?
5 What applications suit cold-cathode lamps?
6 How is lamp dimming achieved?
7 What are the requirements for dimming fluorescent lamps?
8 Which natural element needs to be incorporated in lighting control?

8.8 Fire protection

Installation of reliable fire detection systems is essential; therefore, it is crucial that the electrician has an understanding of the types of fire detection technologies available and their limitations.

The prime function of a fire detection system is to identify one or more physical changes in the protected environment which indicate the development of a fire condition. These physical changes can be:

- invisible products released after a fire has started
- temperature
- smoke
- flame.

When a developing fire produces flames, the resulting effect is one of rapid flame and heat growth. Indoors, ceiling temperatures, depending upon the fuel material, can become extremely high within minutes. The heat associated with the flames can also cause other combustible materials to 'flash' (where all combustibles fire at the same time) within an area and immediately place people's lives in danger. Smoke production under these conditions is massive, reducing visibility to zero.

The choice of a fire alarm system depends on the risk of the fire to be detected, type of building structure, the purpose and use of the building, legislation and the local Fire and Rescue Service recommendations. Most fires in a domestic building originate in the living room, followed by the bedrooms, kitchen, and other areas in that order.

Fires that result in death usually occur when people are sleeping and between 7.30 pm and 7.30 am. An efficient fire detection system does save lives and should be installed in all types of domestic premises.

Most of the devices associated with fire detection systems are usually located near or on ceiling surfaces. When a fire develops, hot gases and particles rise directly above the fuel material and invade the ceiling area (see **Figure 8.113**). The ceiling surface causes the hot gas and particle flow to move parallel to the ceiling to other parts of the building.

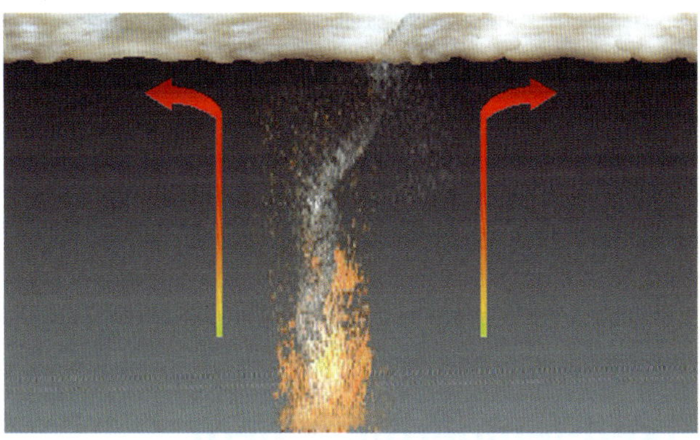

FIGURE 8.113 Hot gases and particles rising to the ceiling surface

In some fire detection systems the detector is used as an alarm device, to sense the changed physical conditions and alert personnel. In other systems the detector is used as a release device which responds to the changed physical conditions by activating fire suppression systems such as CO_2, water or dry chemicals.

Alarms used to alert persons with respect to a fire condition must have a noise level of not less than 5 decibels above average sound level for the protected area (95 decibels is the suggested level for most environments). This requirement should provide adequate audibility. In bedrooms, a minimum degree in the order of 65 decibels to 75 decibels is necessary to wake sleeping persons. It is suggested that an alarm should be placed near the bedhead to wake sleeping children and adults. Flashing lights are also used to warn people of a fire condition.

There are three types of fire detection systems: heat, smoke, and flame detectors. Smoke detectors are the primary domestic detection system with heat detectors being used as an enhancement. Heat detectors should not be used to replace smoke detectors. Flame detectors are an effective outside fire detection system. Devices for each system can be either stand alone or interconnected installations.

Heat detectors

Heat detectors can be classified as:

- fixed temperature
- rate of rise
- rate compensation
- combination
- electronic thermal.

All these heat detectors respond to the thermal energy released by a developing fire. A typical spot-type heat detector is shown in **Figure 8.114**.

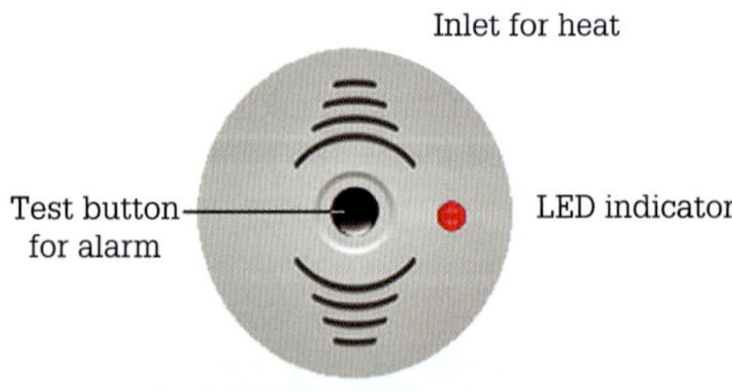

FIGURE 8.114 Spot-type heat detector

Heat detectors are located on the ceiling or high up on walls and respond to the convective nature of the thermal energy of fire. In some models, the detector head is attached to the main body by a bayonet fixing for easy removal for maintenance or replacement.

Fixed-temperature heat detectors

A fixed-temperature heat detector activates an alarm when the temperature of the sensor element reaches a pre-set level, usually between 50 °C and 60 °C. These detectors provide protection from slow-developing fires.

The active elements of fixed-temperature heat detectors can be fusible, pneumatic or bimetallic. A fusible element is a eutectic metal alloy such as bismuth, lead, tin and cadmium that melts at a specified temperature, allowing the alarm contacts to close. A pneumatic element uses air in a chamber that when subjected to heat expands and applies pressure to a flexible diaphragm which in turn activates electrical contacts. With a bimetal element, the width of the gap between the contacts determines the operating temperature: the wider the gap between the contacts the higher the operating temperature setting off the detector. **Note:** All fire detection systems use a normally open electrical contact which closes when sensed physical conditions change.

Fixed-temperature heat detectors have a thermal lag characteristic. This means that the temperature of the environment is higher than the temperature setting of the sensor element when the alarm sounds. Fixed-temperature detectors can provide protection in areas subject to high ambient temperatures such as elevator shafts, switchboard or switchgear enclosures, or in zoned areas such as corridors and stairwells, so that only detectors in the immediate fire area operate.

Some sprinkler systems use liquid-filled bulbs as the fixed-temperature heat detector. When the temperature of the protected area rises, the liquid in the bulb expands until it shatters the glass and allows water to sprinkle over the protected area.

Rate-of-rise detectors

Rate-of-rise detectors are designed to detect sudden increases in ambient temperature. This feature allows these detectors to respond to quickly developing fire. A mechanical-pneumatic element senses the change in temperature and engages with electrical contacts.

Usually, a rate-of-rise detector sounds an alarm when the rise in temperature is 8° C per minute, which is what happens with a fast-developing fire. These detectors can be used in an area where smoke or flames may not reach or to protect areas where the ambient temperature is constant.

Rate-compensation detectors

Rate-compensation detectors use a design method which allows the detector to ignore a rush of warm air in the protected environment to eliminate false alarms that may occur with respect to rate-of-rise detectors momentarily.

Combination detectors

A fixed-temperature detector and a rate-of-rise heat detector are on a par when it comes to protection. However, a combination heat detector provides a higher level of protection because this detector offers the responsiveness of both a fixed-temperature and a rate-of-rise heat detector. Combination detectors are used for monitoring cables associated with cable tray enclosures.

Electronic thermal detectors

An electronic thermal detector uses a thermistor as the sensing element and is either a fixed-temperature or a rate-of-rise electronic detector. The principle of operation is based on the thermoelectric effect. Positive temperature thermistors develop a change in resistance when exposed to increasing changes in temperature. This resistance change is monitored electronically, and the detector responds when the resistance reaches a specific value (fixed temperature) or changes at a sudden rate (rate of rise).

Thermal detectors are usually not acceptable in areas involving people or where high-value thermal contents are housed.

Smoke detectors

As the name implies, these devices are designed to identify a developing fire while in its smouldering or early flame stages. There are two basic types of smoke detectors in use today, and they operate on either an ionisation or a photoelectric principle, with each type having advantages in different applications.

Smoke alarms are the main detection method in areas involving people or buildings with high-value contents and it is a mandatory requirement in the Building Code of Australia (BCA) for all new dwellings and substantial renovations to have smoke alarms.

Most smoke detectors have a small visible indicator which should be seen flashing at roughly one-minute intervals in normal operation. These detectors also begin to make a pulsing 'beep' when the battery is nearing the end of its life. It is important to check that all smoke detectors are working from time to time. This is carried out by pushing the detector's small test button to sound the alarm.

Besides being powered by a 9 V lithium ion battery, other power sources include the mains supply of 230 V a.c. directly (hard wired with battery backup) and an external a.c. to d.c. power supply (typically 12 V or 24 V d.c.). Hard-wired smoke alarms are required in all new and significantly renovated homes. Smoke detectors must be fitted to all homes. With existing homes, battery-operated smoke detectors can be fitted instead of fixed-wired smoke detectors.

Smoke alarms can be installed either as spot detectors or interconnected. Interconnected smoke alarms provide warning to all persons occupying a building at the same time should any one alarm activate. Because children have deeper sleep patterns than adults, smoke alarms in children's bedrooms should be interconnected to parents' rooms.

Recommended smoke alarms carry the Standards Australia Mark or a Scientific Services Laboratory (SSL) label.

Ionisation smoke detectors

These smoke detectors use an open-air ionisation chamber and a few micrograms of americium-241, which is a radioactive material. As the radioactive material decays, it produces a stream of alpha particles. These alpha particles ionise nitrogen and oxygen atoms in the air in the ionisation chamber to produce free electrons and positive ions. On each side of the ionisation chamber is a conducting plate to which is applied a small electric charge.

The charge on the two plates attracts the free electrons and positive ions, causing a small electric current to flow through the ionised air in the chamber. The magnitude of the current flow is measured by an electronic circuit in the smoke alarm.

When polarised smoke or particles from a developing fire enter the ionisation chamber, the free electrons and positive ions combine with these particles and are neutralised. As a result, the chamber current flow between the conducting plates falls and at a pre-set level the measuring circuitry sounds an alarm. The detection chamber ionisation process is illustrated in **Figure 8.115**.

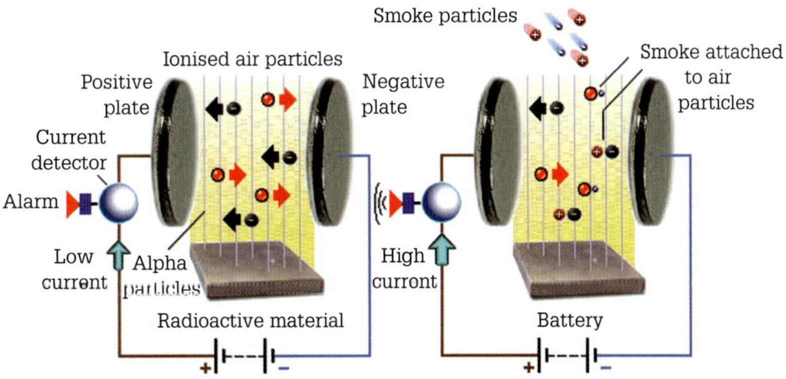

FIGURE 8.115 Detection chamber ionisation process

The ionisation smoke detector is suited to fast- or high-energy-burning fires, which produce little visible smoke but high levels of invisible particles. Ionisation detectors require very low levels of power to operate and so are ideal for use as battery-operated (typically 6 V to 9 V) smoke detectors.

Smoke detectors are placed on ceilings as shown in **Figure 8.116** or high on walls in a manner similar to thermal detectors. Ionisation smoke detectors pose very little risk to persons if used and disposed of according to manufacturers' instructions.

FIGURE 8.116 Ionisation smoke detector on a ceiling

Photoelectric smoke detectors

Photoelectric smoke detectors use a pulsating LED to produce a light beam and a photoelectric cell. Photoelectric smoke detectors are based on the principle that any level of smoke will scatter light beams. The LED is located in a tube and emits light through the air that circulates from the protected area into the smoke alarm. The photoelectric cell is positioned perpendicular to the LED. If the circulating air is free of smoke, the light emitted from the LED is unobstructed and passes through in a straight line, and the photoelectric cell detects the absence of light.

As soon as small amounts of smoke are present in the circulating air, some light beams from the LED start to scatter, enabling the photoelectric cell to detect those light rays (**Figure 8.117**). Once the amount of light detected increases to a set level the alarm is activated.

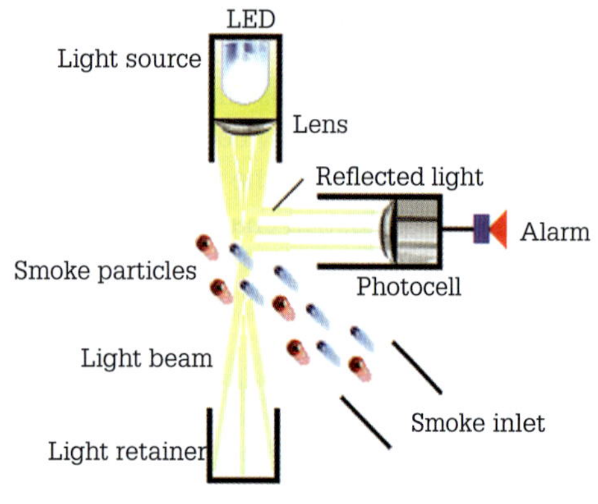

FIGURE 8.117 Detection chamber light beam deflection

Smoke detectors operating on the photoelectric principle respond faster than ionisation detectors to the smoke generated by low-energy or smouldering-type developing fires as these fires produce visible large smoke particles. Photoelectric smoke alarms are suitable for areas likely to be affected by cooking or heating appliances.

Aspirating detectors

Aspirating detectors are smoke detectors that use a small pump to draw samples of the protected environment air through a tube into a detection chamber. The tubes can be located at different height levels within the protected area. This feature enables the detector to pick up on incipient smoke (early smouldering). Incipient smoke does not contain enough heat energy to lift it to the ceiling.

The detection chamber of an aspirating detector is more sensitive than that of conventional smoke detectors. The detector can determine whether a fire is present or whether the smoke is in the air by sampling the density of the smoke in the air. In the detection chamber, the air sample is exposed to a light source. Any light that is scattered by smoke present in the air is detected by sensitive receivers. The device communicates this information to the fire management system and the alarm is sounded.

Installing and positioning smoke alarms

Smoke alarm laws require owners of all homes and units (Class 1a and sole occupancy units in Class 2 buildings) to have installed and to maintain smoke alarms. Class 1a means a single dwelling that is:

- a detached house; or
- one of a group of two or more attached dwellings, each being a building, separated by a fire-resisting wall, including a row house, terrace house, town house or villa unit and cabins in caravan parks.

Class 2 means a building containing two or more sole occupancy units, each being a separate dwelling. A sole occupancy unit means a room or other part of a building for occupation by one owner or joint owners, lessee, tenant or another occupier to the exclusion of all other persons and includes a dwelling. The dwelling must have a kitchen sink and facilities for the preparation and cooking of food, a bath or shower and a toilet and washbasin.

As a legal minimum requirement, a smoke alarm must be installed on the ceiling or, if that is not possible, high on the wall according to manufacturers' requirements. Smoke alarms must be located within any areas (hallways) containing bedrooms and the rest of the house or unit as shown in **Figure 8.118**. If a multiple-storey dwelling does not provide bedrooms on a particular level, smoke alarms must be installed on the predetermined evacuation route from that level.

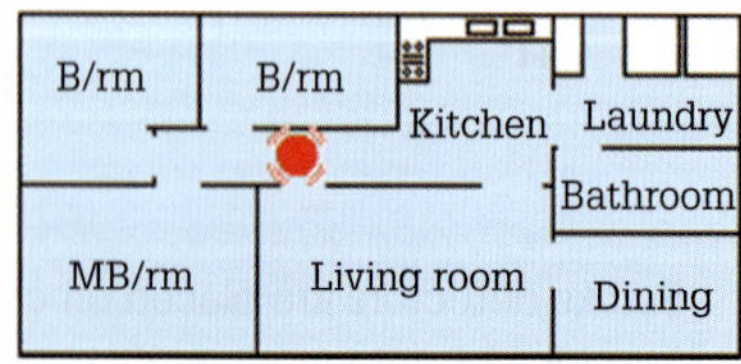

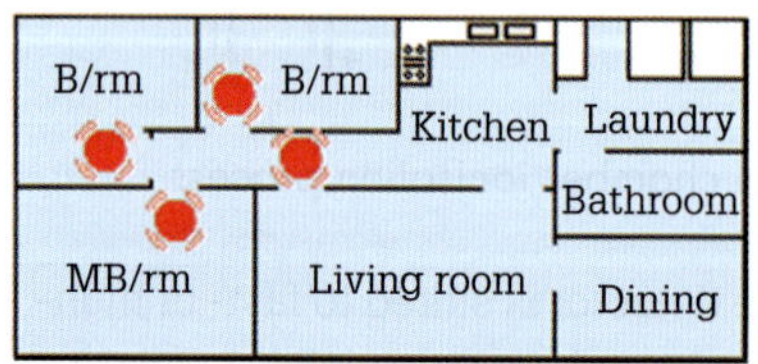

FIGURE 8.118 Minimum and recommended smoke alarm locations

If at least one smoke alarm has not been installed as required by the BCA in an existing dwelling, the owner of the building is guilty of an offence for which a penalty can be imposed.

Service electricians should always seek advice on the selection, placement and maintenance of smoke alarms from the local Fire and Rescue Service fire station.

Flame detectors

Flame detectors detect radiant energy in the form of ultraviolet or infrared radiation over a defined bandwidth. This means that these devices are line-of-sight devices with a 90° to 180° cone of vision, with the highest sensitivity occurring when they are looking directly at the flame, as shown in **Figure 8.119**.

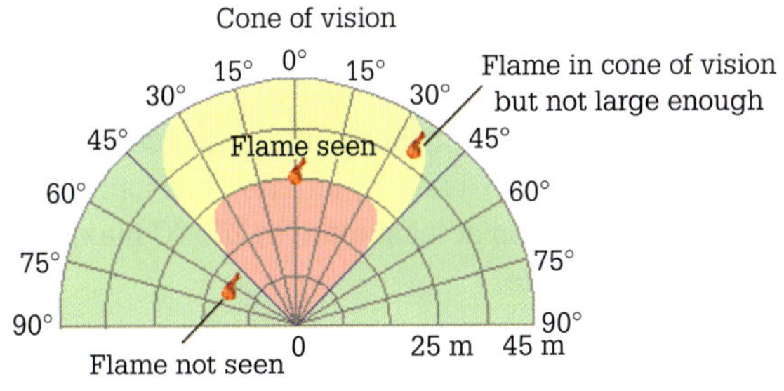

FIGURE 8.119 Flame detector cone of vision

Flame size affects the response of the detectors, and the flame needs to be proportionally greater at maximum distance for reliable detection. For example, a 0.1 m^2 flame at 25 m is seen if it is within the cone of vision. A flame in the yellow area would need to be 0.4 m^2 in order to be seen. When radiant energy indicative of a flaming condition is seen by the detector, it sends a signal to a fire alarm panel. If the detector cannot see the whole area to be protected, additional detectors are required.

Flame detectors are often used for outside fire detection and are the best method of detection for munitions factories, petrochemical industries and liquid fuel and gas fires. Each liquid fuel or gas, when burning, produces a flame with specific radiation characteristics. A flame detection system must be designed for the type of flame that is produced. For example, some fuels such as hydrogen, alcohol and methane burn with no visible flame. A flame detector tuned to the water-band emission frequency of invisible flames sees the fire.

REVIEW QUESTIONS

1 Name four physical changes in the protected environment which indicate the development of a fire condition.

2 What factors determine the choice of a fire alarm system?

3 Where are most of the devices associated with fire detection systems located?

4 Name three types of fire detection systems.

5 Name five classifications of heat detectors.

6 Which heat detector allows the detector to ignore a rush of warm air in the protected environment to eliminate false alarms that may occur with respect to rate-of-rise detectors momentarily?

7 Name the active elements of fixed-temperature heat detectors.

8 What are the two types of smoke detector?

9 Which type of fire detector uses a small pump to draw samples of the protected environment air through a tube into a detection chamber?

10 As a legal minimum requirement, where must a smoke alarm be located in residential dwellings?

8.9 Requirements for luminaires

When installing lighting circuits and their associated components it is important to ensure that the luminaires and their lamps are in working order. Additionally:

- all luminaires and their control device must be level and plumb
- all luminaires' lenses and louvres must be correctly directed and aligned.

Light switches should be located on the latch side of doorways, approximately 1350 mm above the finished floor level. Two-way switching should always be used in hallways, stairwells and in rooms with more than one entrance. Multi-gang switches should only be used as one input feed connected to them. Where adjoining switches on different phases are installed, they should be located at least 150 mm apart. Switching of lighting circuits containing discharge lighting or another lighting with high inductance components may require a light switch to be rated at more than twice the load current due to high switching voltages that may occur. There are specially designed and rated switches available for inductive loads.

All trenches used for installing an exterior lighting system must be backfilled, compacted and restored to the original cover (grass, pavers, and the like).

Fluorescent luminaires installed on concrete ceilings should be fixed to the outlet boxes and with two 6 mm concrete anchors spaced at 75% of the length of the lighting fixture apart. Where cables are run within luminaires, they must be of a heat-resistant quality or protected by heat-resistant sleeving.

The layout of the luminaires as designated on drawings must be followed as far as practicable.

In Australia, the *Wiring Rules* require additional protection by RCDs for all final sub-circuits in residential electrical installations (see Clause 2.6). Where required, RCD protection is installed on the switchboard from where the circuit originates. Alternatively, a residual-current circuit breaker with overload protection (RCBO) could be used. RCBOs are for use against overload, short-circuit and residual current protection in residential, commercial, industrial and mining applications. Lighting circuits in a residential installation are generally rated at 6 or 10 A supply.

Recessed luminaires

Low-voltage recessed ceiling luminaires should be fitted with flexible cords and three-pin plugs. The flexible cord must be three-core, PVC insulated with minimum 0.75 mm^2 conductors and comply with AS/NZS 3191. It should be of suitable length but not greater than 1500 mm. A plug socket, shown in **Figure 8.120**, should be located within 500 mm from the edge of the access aperture to allow the luminaire to be plugged in prior to fixing.

Clause 5.5.4.4, 'Connecting devices' of AS/NZS 3000:2018 *Wiring Rules* needs to be consulted when wiring by a connection in the form of a plug and plug socket. Clause 3.7.2.6, 'Mechanical stress on electrical connections', also has relevance.

Recessed luminaires (shown in **Figure 8.120**) installed in timber-framed ceilings should be fixed to the ceiling members and noggings using the proprietary fixings supplied by the luminaire manufacturer. Low-voltage recessed ceiling luminaires installed in a limited space (e.g. covered way, counter canopy, integrated ceiling/roof) should be wired to a nearby junction box using cable having insulation at least rated at 200 °C maximum operating temperature.

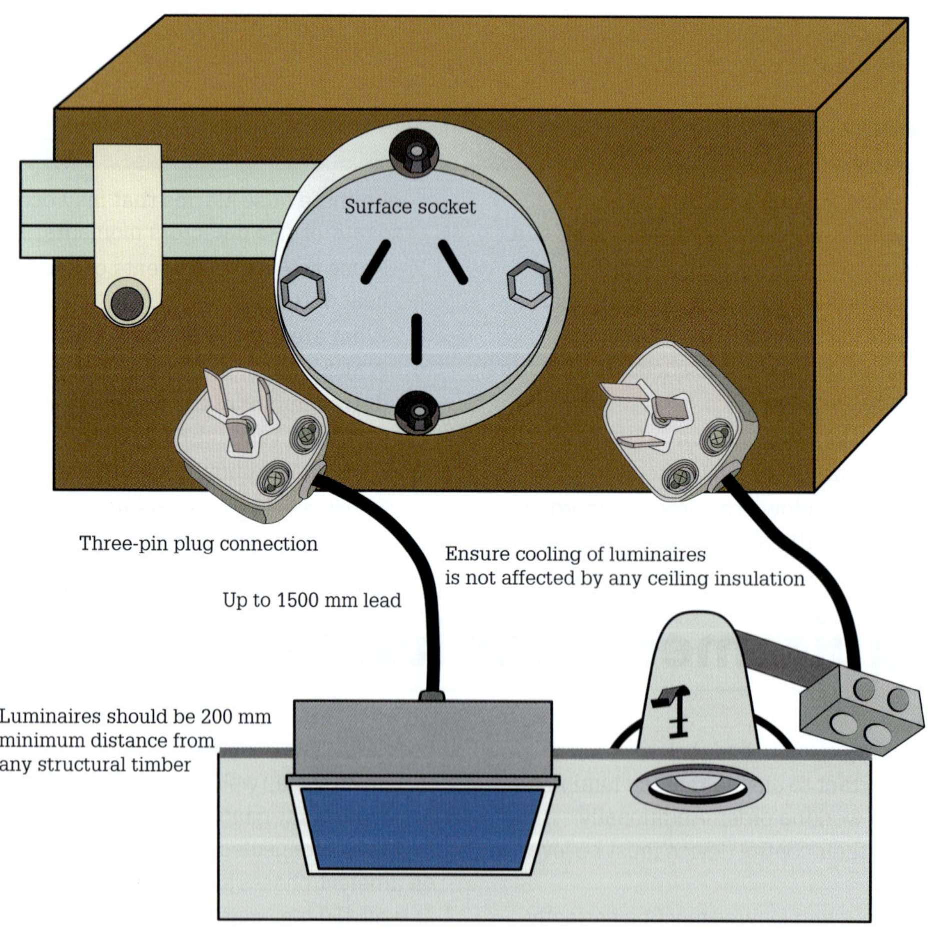

FIGURE 8.120 Recessed luminaires

Recessed luminaires for suspended or modular ceilings should be attached to suspended ceilings with proprietary mounting brackets supplied by the luminaire manufacturer.

Note that the surface socket is fixed so that the three-pin plug is withdrawn in the horizontal plane. Recessed luminaires within false ceilings can also be securely fixed to catenaries.

Refer to AS/NZS 3000:2018, Clause 4.5.2.3 as the primary source for requirements with recessed luminaires.

Lighting track

Lighting track as illustrated in **Figure 8.121** is specialised lighting. Tracks are adaptable, allowing flexibility in sizing, location and fixture styles. Connected lights can be focused on paintings, bench-tops, walls or other room features as required. Refer to AS/NZS 3000:2018, Clause 3.9.7.5 as the primary source for requirements with track systems.

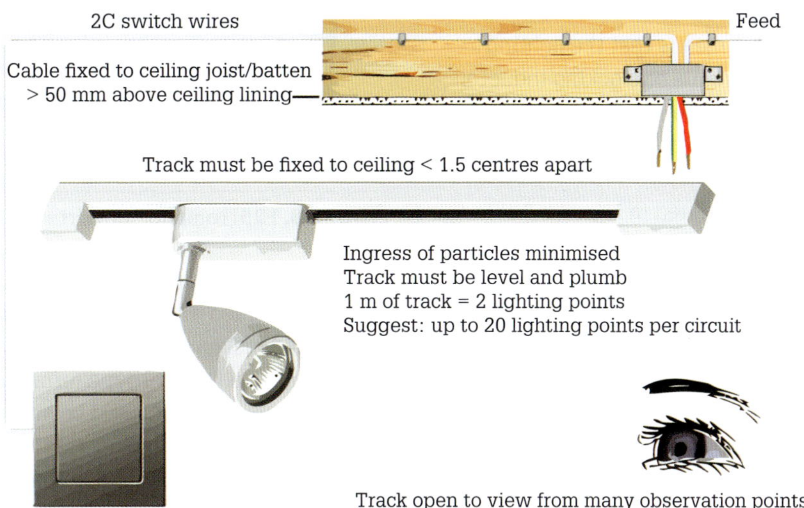

FIGURE 8.121 Lighting track

Festoon lighting

Festoon lighting as illustrated in **Figure 8.122** is used for lighting up pathways, trees, marquees. Lamps are usually spaced one metre apart and 15/25 W coloured harlequin lamps are used.

Refer to AS/NZS 3000:2018, Clause 4.5.1.3 as the primary source for requirements with festoon lighting.

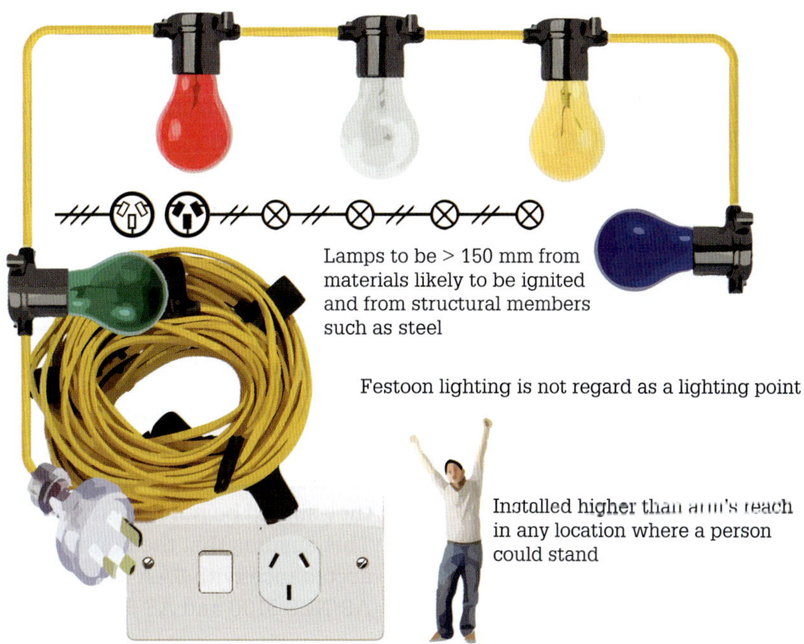

FIGURE 8.122 Festoon lighting

Lighting design

All internal and external lighting for an installation should be designed to provide appropriate lighting quality (intensity, glare, uniformity) and levels which improve the safe movement of persons, personal safety and security.

Luminaires supplied for an installation should be complete with lamps such as LEDs, CFLs, fluorescent lamps, discharge lamps and accessories necessary for their proper functioning. In addition, it is suggested that all discharge luminaires be power factor corrected to a minimum of 0.85 lagging. This means that the lighting circuits incorporate lead-lag circuits or blocking inductors. Finally, all lamps, where appropriate, should be energy-efficiency types. For example, energy-efficient fluorescent lamps should be 230 V grades, general service type with regular bayonet caps (B22) (illustrated in **Figure 8.123**) or with an Edison screw cap.

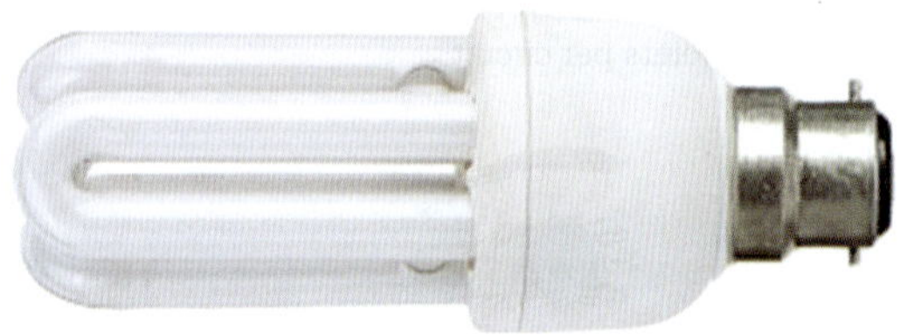

FIGURE 8.123 Compact fluorescent lamp

Edison screw and Swan-type lamp holders and batten holders should comply with AS 3117 and preferably be a ceramic type (better for heat dissipation), shown in **Figure 8.124**, although moulded types with brass-lined shells can be used with lamps rated up to and including 75 watts.

FIGURE 8.124 Ceramic lamp holder

A long skirt lamp holder should be ceramic or moulded glass-filled polyester as illustrated in **Figure 8.125**, complying with AS 3140.

FIGURE 8.125 Lampholder skirt

Suggested features for the screw-type lamp holders are:

- spring-loaded and nickel-plated contacts
- isolated nickel-plated fixed-screw shell
- nickel-plated tunnel-type terminals.

Fluorescent tubes should be straight and of standard lengths, with a bi-pin connection, hot cathodes and a phosphor colour temperature in the range of 4100 to 4300 K.

All fluorescent luminaires should have low-loss lead-lag ballasts.

Mercury vapour lamps should be class MBF/U colour-corrected mercury fluorescent lamps or class MBTF/U where self-ballasted lamps are indicated and should have E27 or E40 screw caps.

Sodium discharge lamps should be high-pressure sodium lamps with tubular or elliptical, frosted or clear glass envelope and have E27 or E40 screw caps.

Where extra-low-voltage lighting is proposed the installation should provide:

- separate transformers for all lamps, unless in a track configuration
- isolation of transformers clear of ceiling insulation
- lamps with low levels of UV emissions
- covered lamps
- screw-fixed lamp holders.

SWITCH ON

Supply of luminaires

To ensure that luminaires required for an installation are of the required quality, style and performance, they should be purchased from a reputable supplier. This ensures that the luminaires are manufactured in accordance with SAA approval and test specification AS/NZS 3100 and AS 60598.1.

Installation requirements

Luminaires must be installed in accordance with Clause 1.7.1, 'Selection and installation of electrical equipment', and Clause 4.5.2, 'Lamps and luminaires', AS/NZS 3000:2018 *Wiring Rules*.

Luminaires need to be installed and arranged in accordance with AS1680.1:1990 *Interior lighting – General principles and recommendations for indoor installations* and AS/NZS 1158.3.1:2005 *Lighting for roads and public spaces for external installations*. In addition, the luminaire must have external identifiers. Refer to AS 3771:1998, Section 1.6.2. The required identifiers include:

- lamp type (essential for re-lamping)
- wattage
- year of manufacture.

Installation of luminaires

Before starting the installing of light fittings, the electrician should have all fixings necessary for their proper installation. In addition, a range of packing pieces should be available to level the luminaires and to prevent distortion.

Lighting should be installed on dedicated circuits and not mixed with general power or permanently connected equipment. However, mixed circuits can be employed under certain conditions. These terms refer to cable CSA and the order of connections. For example, it is allowable to connect a lighting circuit in a socket outlet circuit (a 1.5 mm^2 conductor branching of a 2.5 mm^2 conductor) but you cannot connect a socket outlet circuit in a lighting circuit unless the CSA of the lighting circuit conductors is 2.5 mm^2 or larger.

All light fittings must be readily accessible and must not be mounted on equipment, or a feature (e.g. drops over stoves, plant features, spars, voids) that impedes access or presents a high safety risk. Ideally, all light fittings should be at a height accessible from a 2.0 metre step ladder or lower.

Noggings used to provide a fixing point for luminaires between ceiling joists should be of similar size to timber joists, or of minimum size 75 mm × 50 mm.

Rooms or spaces with only one luminaire should have the luminaire centrally placed for even illumination.

All luminaires must be effectively earthed or in the case of double-insulated fittings must have a suitably terminated earth conductor at each lighting point.

All fixings in outdoor locations should be corrosion resistant.

Where chain suspensions are indicated in a lighting specification, standard chain suspension sets as provided by the supplier of the fluorescent luminaires should be used or, alternatively, electroplated welded steel link chains with a minimum of 3 mm diameter steel should be used. Chains (see **Figure 8.126**) should be hook-mounted from ceiling and lengths adjusted so that the luminaires can hang correctly.

FIGURE 8.126 Suspension chain

Where rod suspensions are indicated in a lighting specification, standard rod suspension sets as provided by the supplier of the fluorescent luminaires should be used or, alternatively, rods should be steel water pipe with gimbal mounting from the ceiling. Rod length should be adjusted so that the luminaires hang correctly.

End-to-end luminaires should be correctly aligned using packing strips where necessary.

Luminaires should not be supported or suspended from J boxes or fittings.

Plastic boxes attached to luminaires for cable protection should not be used as part of the fixing method.

Where a PVC conduit enters a luminaire, a screwed PVC adaptor with circular lock nuts should be used to secure the conduit to the luminaire. Lock nuts are not required for luminaires with screwed conduit entries.

Fluorescent luminaires should be supported by two fixings at each end. One fixing at each end of a narrow fitting is acceptable where a 1.6 mm thick back plate reinforces each fixing screw.

Clean all luminaires installed immediately prior to commissioning. Also, polluted diffusers need to be cleaned to improve light output.

Any damaged painted surfaces on luminaires must be repainted to the same standard as the original paintwork.

Surface-mounted luminaires

Surface-mounted luminaires as shown in **Figure 8.127** should be securely fixed to structural members of the ceilings or walls, or fixed by hangers or brackets which are securely fixed to structural members.

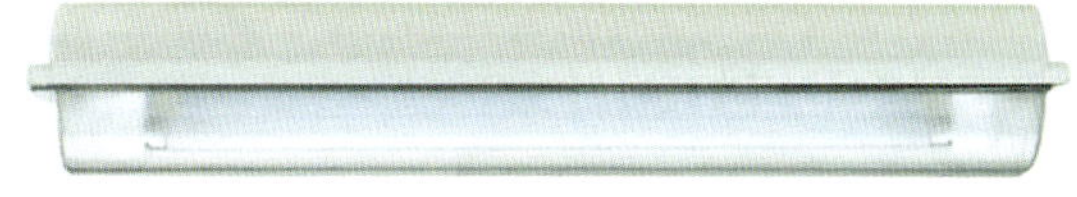

FIGURE 8.127 Surface-mounted luminaire

Recommended minimum size of fixing for luminaires, hangers or brackets for various surfaces are as follows:

1. Fixing to timber: steel wood screw No.10 × 25, round head, cadmium plated (see **Figure 8.128**).

FIGURE 8.128 Steel wood screw

2 Fixing to concrete: screw expanding bolts M5 × 40 (see **Figure 8.129**).

FIGURE 8.129 Screw expanding bolt

3 Fixing to hollow blocks: M5 electro galvanised, round-head screws with spring-loaded butterfly toggles, as shown in **Figure 8.130**.

FIGURE 8.130 Spring-loaded butterfly toggle

All fixings should be fitted with large-diameter (minimum 3 × screw diameter) cadmium-plated washers under the heads of each screw.

Post top luminaires

Post top luminaires are installed outdoors as shown in **Figure 8.131**.

FIGURE 8.131 Post top luminaires

Unless the lighting specification for the installation informs you otherwise, it is suggested that post top luminaires should:

- be mounted on tapered columns
- consist of proprietary brand steel poles, hot dipped galvanised and suitable for base plate mounting on galvanised rag bolt assembly set in a concrete pad
- have poles that are adequately drained and fitted with an approved weatherproof lockable enclosure to house the control gear and fuse, in the lower section of the pole, within 1000 mm of ground level.

Depending upon circuit length, 1.5 mm^2 twin and earth conductor is usually suitable for the final sub-circuit within the post.

A protective earthing conductor, connected to a terminal or suitably insulated and enclosed, must be provided at every lighting point, including transformers supplying extra-low-voltage lighting systems.

REVIEW QUESTIONS

1 At what height above finished floor level should light switches be mounted?
2 In what situation may it be necessary to install a light switch rated at more than twice the load current?
3 What is the recommended cable supplying low-voltage recessed ceiling luminaires?
4 State the focus of lighting design.
5 What type of ballast should be installed in fluorescent lights?
6 State the requirement for placing a single luminaire in a room.

8.10 Testing and fault finding

All electrical circuits must be tested to confirm compliance with the *Wiring Rules*. The basic tests for lighting circuits include:

- earth continuity
- insulation resistance
- polarity
- correct circuit connections.

Refer to Chapter 16 of this textbook for a thorough treatment of installation testing.

Testing lighting circuits

The following is a brief summary of the procedure for testing lighting circuits.

Procedure for testing protective earth conductors

The procedure for testing the resistance of protective earthing conductors for a lighting circuit is:

1. Disconnect the aerial service support equipotential bonding conductor (earth on riser bracket for aerials).
2. Ensure that there are no parallel conductive earth paths. For example, if the water heater is connected there is a possibility that a parallel path exists through the conductive water pipes.
3. At the switchboard disconnect the MEN link.
4. Connect the long lead to the main earth or earthing bar and the other end of this lead to the ohmmeter. Next connect the other ohmmeter lead to the protective earthing conductor at each earthing point for the entire installation in turn and record the results. Remember to deduct the resistance of the trailing lead. Include the following: the PE at each lighting point (see **Figure 8.132**), with all control and isolating switches ON.
5. Disconnect the test leads and reconnect the MEN link and other disconnected earthing and equipotential conductors unless carrying out further tests.

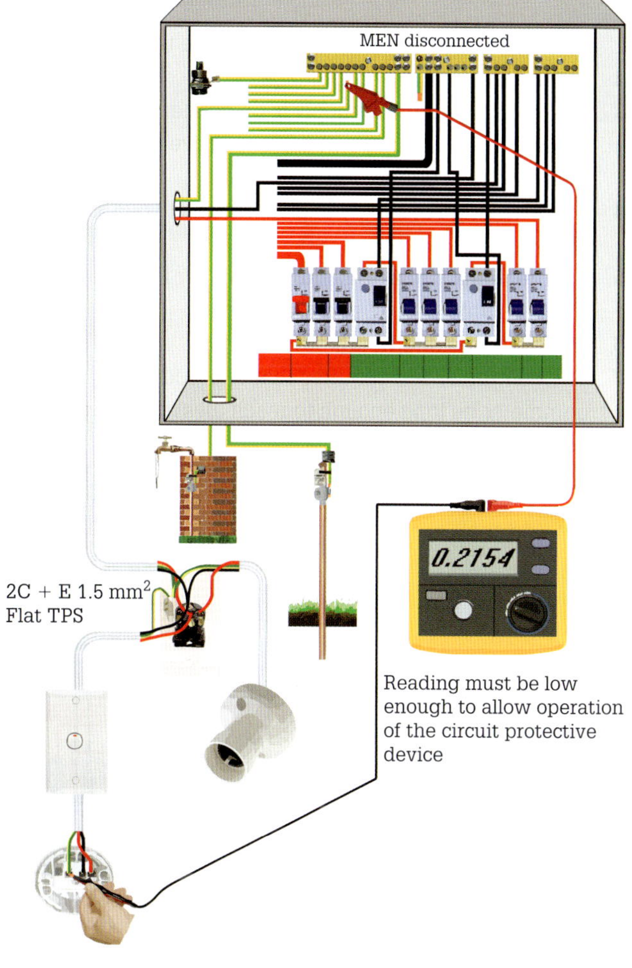

FIGURE 8.132 Resistance of the protective earth at lighting point

Requirements for testing insulation resistance

It is a requirement of Clause 8.3.6.2, AS/NZS 3000:2018 *Wiring Rules* that an insulation resistance test must be carried out on all live conductors (this includes the neutral) and the earth. For an installation this means:

1. The main switch must be off if it is a connected system.
2. All protective devices must be active, with fuse elements in place and circuit breakers on.
3. All circuit switches must be on with electronic controls bypassed.
4. All fixed appliances must remain connected and turned on.

The insulation test must prove that the insulation resistance between live conductors and the earth or any part thereof is not less than 1 MΩ.

Requirements for testing polarity

It is necessary to test an electrical installation for correct polarity to ensure that there is no shock hazard resulting from the incorrect connection of active, neutral and earthing conductors. When undertaking polarity tests ensure:

1. That single-pole switches and protective devices only operate in the active conductor of the connected circuit.
2. That switches and protective devices of multi-phase circuits, with some exceptions, operate in all active conductors of the connected circuit.
3. RCDs switch the active and neutral conductors of the circuit – all live conductors.
4. That the phase sequence of multi-phase socket-outlets is the same for all socket-outlets of the same type within the electrical installation.
5. That the polarity of all flat-pin socket outlets is earth, active and neutral in a clockwise direction commencing from the bottom-most pin.
6. All neutral conductors connect to the neutral bar of the switchboard.
7. The neutral conductor of the consumer's mains connects to the neutral bar of the main switchboard.

Procedure for testing polarity of final sub-circuits

The procedure for testing polarity of final sub-circuits is as follows:

1 Ensure the testing instrument is within test date.
2 Set the tester to the ohms range.
3 Ensure the electricity supply is disconnected and lock and tag the isolation point (never remove a fuse carrier instead of disconnecting).
4 Prove isolation of supply and operation of voltage presence indicator.
5 Ensure items 1 to 4 of 'Requirements for testing insulation resistance' above are proven.
6 Turn 'Off' the circuit protection device.
7 The continuity test requires that one lead of the ohmmeter be long enough to reach all switches, all Edison-type screw lamp holders and socket outlets within the installation. The test is carried out with a long lead connected to the main active conductor at the distribution board and to one terminal of the ohmmeter on its low-resistance scale. The other lead of the ohmmeter is connected in turn to each switch, and Edison goliath lamp holder contacts.

For switches (as shown in **Figure 8.133**):

- With the switch 'On' ≈ 0Ω for common (active) and 0Ω for terminal 1
- With the switch 'Off' ≈ 0Ω for common and ∞ Ω for terminal 1

This assumes that active is connected to common and switch wire to terminal 1. Some practitioners prefer to install active to terminal 1 and switch wire to common to avoid having terminal 2 live when switched off.

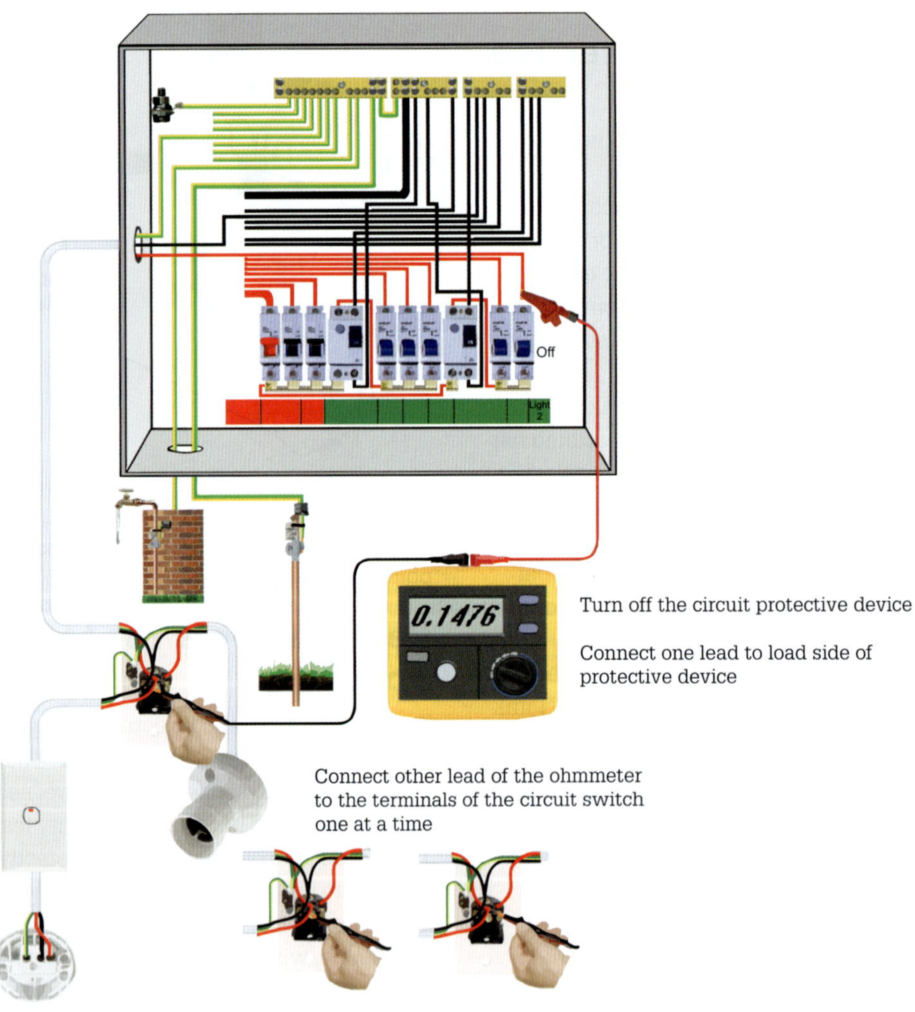

FIGURE 8.133 Polarity and continuity test of typical single-pole rocker switch terminals

For Edison-type goliath lamp holders:

- Outer contact ≈ ∞ Ω
- Inner contact ≈ 0 Ω (switch closed)

The required polarity for an Edison-type goliath lamp holder is centre terminal active and the outer contact or screw base neutral.

Requirements for testing for correct connections

Testing for correct circuit connections requires the use of an ohmmeter. The minimum steps include proving that the active, neutral and protective earthing conductors of each circuit are correctly connected so that there is no:

- short-circuit between the conductors

- transposition of conductors that could result in the earthing system and any exposed conductive parts of the electrical installation becoming energised
- interconnection of conductors between different circuits.

Figure 8.134 illustrates the testing procedure using resistors for each lighting point circuit and at each lighting point under test.

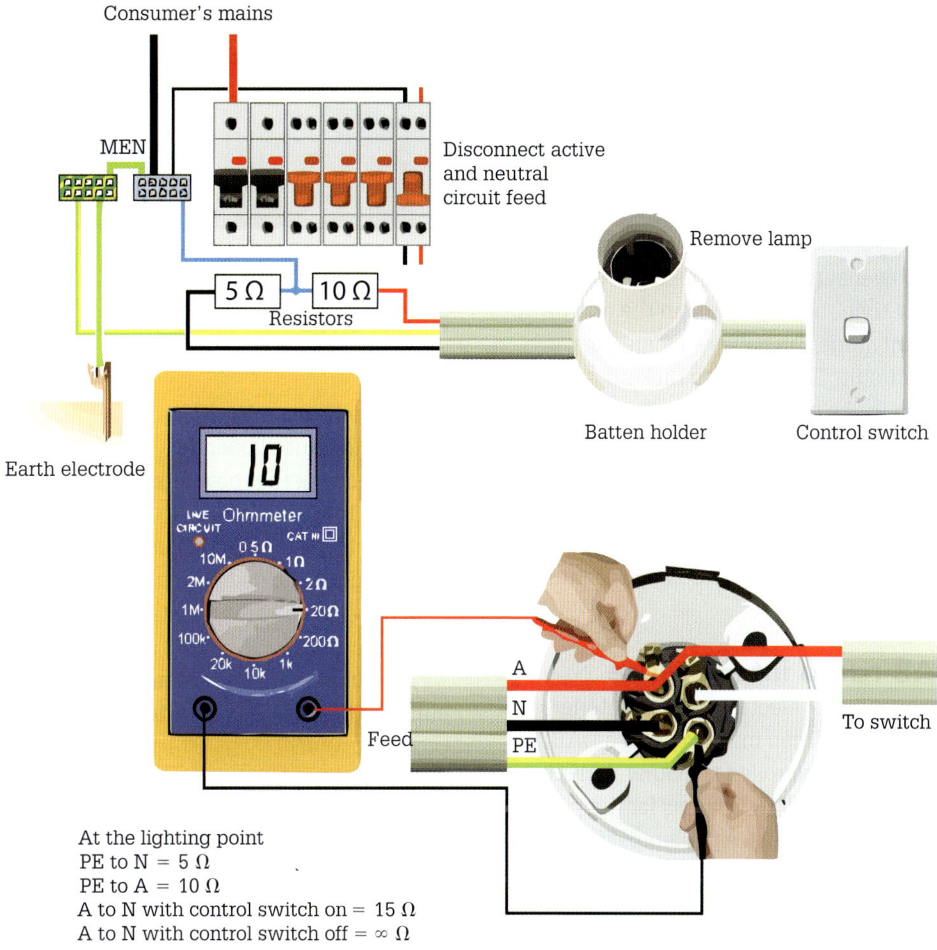

FIGURE 8.134 Connection test – lighting points

Fault finding

Fault finding of electrical installations and circuits is covered in Chapter 16 of this textbook. Following is a brief overview of fault finding LED lights.

LED lighting

Modern LED lamps have a mean time to failure of at least a decade. LED lamps operate quite differently to old-style lamps, which can result in some issues when retrofitting into older circuits. Some issues experienced when switching to LED lamps include flickering, buzzing or not dimming properly.

Preventing premature failure

Many LED lamps have a lifetime of about 25000 hours, which equates to over 20 years if operated for three hours a day. A common reason for premature failure of an LED lamp is due to installing a new LED lamp into a light fitting that still contains some conventional incandescent lamps. While the LED lamp is able to dissipate the relatively small amount of heat they generate, they can struggle to dissipate the additional heat radiated from the higher power incandescent lamps. This situation is easily prevented by replacing all lamps in a fitting at the same time. Not only will the output from the light fitting be more consistent but the risk of the LED lamps overheating is significantly reduced.

Flickering or buzzing

LED lamps can flicker or buzz if the current provided from the LED driver is not constant. LED lamps require a constant current source, so any changes in current can have ill effects on the lamp output. There are a number of possible causes for non-constant current.

1. Incorrect dimmer installed

An inappropriate dimmer can cause flickering and buzzing from an LED lamp. Incandescent lamps used what is known as a 'leading edge' dimmer, which are intended to smoothly dim incandescent-type lamps in a circuit with a range typically between 200 W to 1000 W. LED replacement lamps represent a significantly lower power requirement. For example, replacing four 50 W halogen downlights with four 9 W LED downlights results in a power reduction from 200 W to just 36 W. 'Trailing edge' LED dimmers (around $40) are better suited to the lower power demand of LED lamps. When selecting a dimmer, it is important to check the power rating of the trailing edge dimmer to ensure it suits the rating of the lamps it is to control. It is also prudent to check the LED lamp manufacturer's website for specific recommendations.

2. A high-powered appliance on the same circuit

When an appliance with a switching high power demand, such as a microwave oven, is on the same circuit as low-power LED lamps, it can result in lamp flicker. LED lamps require a much lower voltage than traditional lamps, so they have drivers (power supplies) to reduce the voltage to the LED lamp. Every time the magnetron in the microwave energises, the current surge can cause a momentary voltage drop in the circuit. The observed flickering is the LED driver adjusting to the new input voltage.

3. Loose connections

Loose connections are one of the most common causes of flickering. After eliminating other possibilities check all electrical connections for good contact. Be sure to check the contacts in the lamp holder as elevated temperature can cause springs to fail.

LEDs and DAB radio interference

All electrical appliances emit electromagnetic interference (EMI). In rare cases this EMI can interfere with a DAB radio signal. EMI is tightly controlled through regulation, so if EMI is a problem, check all LED lamps for the regulatory compliance label. Replace the lamps if they do not have this label.

Dull glow when off

Existing light switches may have incurred some damage from switching inductive loads, which resulted in a high resistance circuit between the switch contacts in the off position. This may be due to a build-up of carbon deposits within the switch. As such they allow a small amount of current to flow when switched off. This was not apparent with old-style lamps as the current was never sufficient to appear as light. LED lamps, however, are such low power devices that a small 'leak' of electric current can be sufficient to make them glow.

Fluorescent lighting

Some faults, like that shown in **Figure 8.135**, are easy to see but may prove difficult to diagnose the source.

The following summarises some common problems with fluorescent lighting.

The fluorescent tube refuses to illuminate

This situation can be due to:

- no electrical power due to a tripped breaker or blown fuse
- an open circuit ballast

Source: Shutterstock.com/KYNA STUDIO

FIGURE 8.135 Faulty fluorescent lamps

- a failed starter
- open circuit filament(s).

First check that power is available at the terminals of the light fitting.

Next test the tube by removing it from the fitting and examining the ends of the tube for excessive darkening. If all appears okay then install the tube in a similar fitting and test.

The glow starter is the next item to consider. It is simplest to replace the starter and see if this fixes the problem.

If the tube still does not illuminate at this point it is time to consider the ballast. To check the ballast:

1. Turn off the light at the switch and isolate the circuit using the correct safety procedure.
2. Remove the tube(s) from the fixture and remove the gear tray cover.
3. Check that all of the wires securely connect to the ballast as loose wires could cause the ballast to malfunction. If all is in order at this point, then the problem is likely to be with the ballast itself.
4. As fluorescent lamp ballasts come in a range of shapes and sizes, it is necessary to check the physical measurements and electrical ratings carefully to ensure a suitable replacement is obtained.
5. The ballast is likely to have a wiring diagram on it (or at least in the gear tray) which must be followed when replacing a ballast.

REVIEW QUESTIONS

1. Name the basic tests for lighting circuits.
2. State the minimum acceptable insulation resistance for a lighting circuit.
3. Why is it necessary to test the polarity of a lighting circuit?
4. What are the minimum steps involved in testing for correct connections?
5. What may be a consequence of using an appropriate dimmer to control an LED lamp?

CHAPTER REVIEW

8.1 Wiring methods

- The most commonly used wiring system for lighting circuits is unenclosed flat TPS cable.
- A one-way switched lighting circuit consists of a single-pole, double-throw (SPDT) switch connected in series to one or more luminaires connected in parallel.
- Looping at the light is the most commonly used wiring method for lighting circuits using TPS cables.
- A two-way switched lighting circuit consists of two single-pole, double-throw (SPDT) switches and one or more luminaires.
- Two-way plus intermediate switching (also called three-way) uses two single-pole, double-throw (SPDT) switches and a specially designed double-pole, double-throw (DPDT) switch.
- A switching chart is a useful tool when an unknown type of switch is encountered.

8.2 Lighting control

- The overall purpose of a lighting control system is to reduce energy consumption while providing a productive visual environment.
- The purpose of occupancy detectors is to turn luminaires on or off as occupants enter or leave the monitored room or space.
- A passive infrared sensor responds only to infrared energy radiated by the object being sensed.
- Ultrasonic sensors are based on the speed at which sound travels through air.
- Microwave sensing is used for large internal spaces (aisles, corridors, halls) or external areas with infrequent occupancy.

8.3 Emergency and evacuation lighting

- Exit signs are designed to minimise the risk of death or injury to persons during an emergency because of failure to locate an exit.
- When a fire alarm sounds, exit signs must direct persons to a safe place or open area.
- The spacing of the emergency luminaires must be calculated to ensure horizontal illuminance at floor level is not less than the recommended standard.
- A centrally supplied emergency lighting system is a dedicated lighting system in which a number of emergency luminaires or exit signs or both without their batteries are supplied from a common power source.
- An emergency exit luminaire must be located at each exit door.
- In Australia, the spacing of emergency luminaires is based on illuminance calculations and manufacturers' photometric data where the light reaches the floor directly from the emergency lamp.
- External lighting should be provided to operate during regular hours of darkness.
- Security lighting is exterior lighting installed solely to enhance the safety of people and property.

8.4 Principles of lighting

- Radiant energy travels through space in the form of waves made up of vibrating electric and magnetic fields.
- Luminous intensity in candelas is the intensity of the light within a tiny solid angle, in a specified direction.
- Luminous flux is the quantity of light received or emitted by the surface of an object regardless of direction and is measured in lumens.
- The distribution or density of light on a horizontal surface is called its illuminance and is measured in lux.
- Luminance is the luminous intensity (brightness) of the surface of an object in a given direction per square metre of projected area of the surface.
- Glare is the contrast between a bright object and a darker background and results in visual discomfort that impairs vision.
- Luminous efficacy indicates how well the light source transforms electrical power into useful light output.
- The inverse square law tells us that if light is distributed from a point source then the illuminance on the surface of an object at 90° to the point source is inversely proportional to the square of the distance between the point source and the surface of the object.
- Light waves are referred to as the incident wave, and the ray that points in the direction in which the light waves are travelling is called the incident ray.
- Colour rendering refers to the manner in which different types of illumination can cause colours to look different.
- Visual comfort is a product of the environment and illuminance.

8.5 Types of luminaires

- A luminaire is a complete lighting device consisting of a lamp or lamps, lamp holders, and visual elements as well as terminals and any necessary control gear and a means for connecting to a power source.
- Light distribution is defined as the ratio of luminous flux (ϕ_o) radiated vertically upwards and downwards in relation to a horizontal plane through the luminaire.
- Luminaires are classified as direct lighting, semi-direct lighting, general diffused lighting, semi-diffused lighting and indirect lighting.
- Prismatic and parabolic diffusers and metal or plastic louvres control the angle at which light can leave the luminaire.
- Yearly testing with a light meter is a valuable aid to establishing proper levels of illuminance for a safe working environment.

8.6 Lamp types

- There are five types of luminescent gaseous discharge lamps found in industry: fluorescent lamps, mercury lamps, mercury tungsten blended lamps, metal halide lamps and sodium lamps.
- The lead-free glass envelope or hard glass sleeve is called a bulb and encapsulates the filament.
- The two main base types are the Edison screw (ES) and the Swan bayonet (BC) cap.
- The life of an incandescent filament is voltage sensitive.
- A halogen bulb operates like an incandescent globe.
- Phosphor produces visible light by fluorescence.
- Coiled tungsten filaments are referred to as 'hot cathodes'.
- As the fluorescent lamp extinguishes every time the sine wave passes through zero (100 times per second at 50 Hz) a stroboscopic effect is set up.

8.7 Energy-saving lighting

- LEDs do not produce light by heating a filament or by producing an arc in a discharge tube.
- A main advantage of LEDs is their long life of up to 50 000 hours.
- Optical fibre cables consist of glass silica or an optically clear polymer rod called the core through which light is guided by internal reflection.
- The name 'neon lighting' is used for advertising signs that use a low-pressure gas-discharge tube.
- To improve the power factor, metallised polypropylene capacitors are connected in parallel across the supply.
- Light dimming is achieved by adjusting the voltage across the lamp.

8.8 Fire protection

- The prime function of a fire detection system is to identify one or more physical changes in the protected environment which indicate the development of a fire condition.
- When a developing fire produces flames, the resulting effect is one of rapid flame and heat growth.
- Heat detectors can be classified as fixed temperature, rate of rise, rate compensation, combination, and electronic thermal.
- Smoke detectors are designed to identify a developing fire while in its smouldering or early flame stages.
- Smoke alarm laws require owners of all homes and units (Class 1a and sole occupancy units in Class 2 buildings) to have installed and to maintain smoke alarms.
- Flame detectors detect radiant energy in the form of ultraviolet or infrared radiation over a defined bandwidth.

8.9 Requirements for luminaires

- In simple terms, every lighting system consists of three elements: a lamp, a fixture or luminaire, and a control system.
- All internal and external lighting for an installation should be designed to provide appropriate lighting quality (intensity, glare, uniformity) and levels which improves the safe movement of persons, personal safety and security.
- Before starting the installing of light fittings, the electrician should have all fixings necessary for their proper installation.

8.10 Testing and fault finding

- The basic tests for lighting circuits include earth continuity, insulation resistance, polarity, and correct circuit connections.
- Modern LED lamps have a mean time to failure of at least a decade.

TRIAL EXAM

For Chapter 8 knowledge assessment, please complete the following trial exam.

1 The monochromatic vision of the human eye in low luminance is:
  a scotopic vision
  b photopic vision
  c daylight vision
  d night vision

2 Which of the following colours has a wavelength between 500 and 570 nm?
  a red
  b yellow
  c green
  d violet

3 A term given to the effect of light that bounces off a surface is:
  a reflection
  b radiation
  c point of incidence
  d refraction

4 Light that appears as red would have a wavelength of:
  a 610–700 nm
  b 590–610 nm
  c 570–590 nm
  d 400–430 nm

5 A term given to the perceived light power emitted by a light source is:
a lumen
b illuminance
c luminous flux
d luminance

6 The unit of luminous intensity is the:
a lumen
b candela
c lumen/watt
d luminous intensity

7 The instrument to measure light levels when undertaking a lighting survey is the:
a flux meter
b lux meter
c candela meter
d illuminance meter

8 The term 'optical fibre' refers to:
a LED technology
b colour rendition
c the waveguide
d UV radiation

9 Production of light in a discharge lamp occurs when a gas or vapour is:
a polarised
b ionised
c charged
d stimulated

10 The colour produced by a low-pressure sodium vapour lamp is:
a greenish red
b golden-blue
c yellow
d bluish white

11 The colour produced by a clear mercury vapour lamp is:
a greenish red
b golden-blue
c bluish white
d bluish green

12 A type of electrodeless lamp is the:
a mercury vapour lamp
b HID lamp
c sodium vapour lamp
d sulphur lamp

13 An emergency lighting luminaire which has one lamp connected to the regular supply and one lamp on emergency supply is referred to as:
a maintained
b non-maintained
c sustained
d self-contained

14 What type of occupancy sensor uses a radar signal to detect a moving object within the view of the sensor?
a ultrasonic sensor
b passive infrared sensor
c microwave sensor
d light sensor

15 What type of occupancy sensor uses sound waves to detect a moving object within the view of the sensor?
a ultrasonic sensor
b passive infrared sensor
c microwave sensor
d light sensor

9 Heating equipment and controls

This chapter provides electrotechnology workers with essential knowledge and skills to enable them to evaluate and modify low voltage heating equipment and controls. This chapter provides underpinning knowledge for the unit UEEEL0008 from the UEE training package.

LEARNING OBJECTIVES

Electrical heating control devices

- State methods used to control heat.

Fixed electrical heating appliances

- Discuss the procedures used for space and industrial process heating.

Electrical water heater operation

- Discuss the procedures used for water heating.

Faults in heating equipment and controls

- Recognise various faults that develop in heating devices.

Heating and heat energy

- Discuss the concept of heat.
- Define the terms 'heat', 'temperature', 'heat capacity' and 'heat transfer'.
- State the Law of Conservation of Energy.

Safe working practices

- Understand when to conduct a risk assessment.
- Understand the hazards and safety precautions when working on heating equipment and controls.
- Discuss how to obtain information about heating equipment and controls.

9.1 Electrical heating control devices

In order to manage the thermal communication of a heating appliance, some form of control device is required. This can be achieved by one of three types of control – manual, automatic or programmable.

Manual control

There are several types of manual control devices and their choice depends upon the switching statement the control device executes. The simplest form of a control device is a single-pole single-throw (SPST) switch, as shown in Figure 9.1.

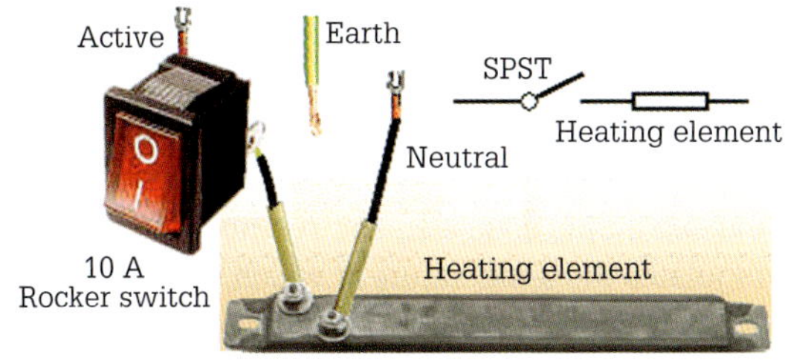

FIGURE 9.1 Manual switch control

Single-pole single-throw switches are available with 10 A to 35 A rated mechanisms and provide a simple on or off switching statement. Several switches could be used to control a bank of heating elements by switching each element on or off individually as required. Another manual switch that was developed from a simple manual switch specifically to control two heating elements is the three-heat switch. These switches have a four-position switching statement – off, low, medium and high as shown in Figure 9.2.

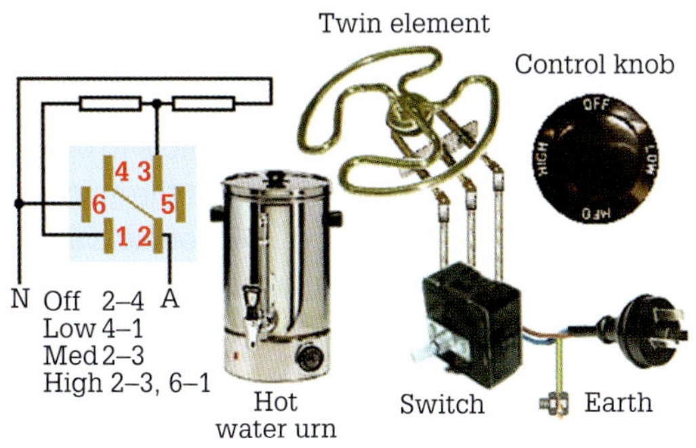

FIGURE 9.2 Three-heat switch

The three-heat switch in Figure 9.2 enables the two elements to be connected in series for low heat at 25% of the possible total power consumed by the heating system. This is achieved by internal connections that allow current to flow from terminal 2 to terminal 4 and then to terminal 1. The next switching statement connects only one element across the supply for medium heat ability that allows 50% of the total power to be used. The last switching statement connects both heating elements in parallel with high heat (100%) power application. Double-pole three-heat switches (see Figure 9.3) are also used for twin heating element control.

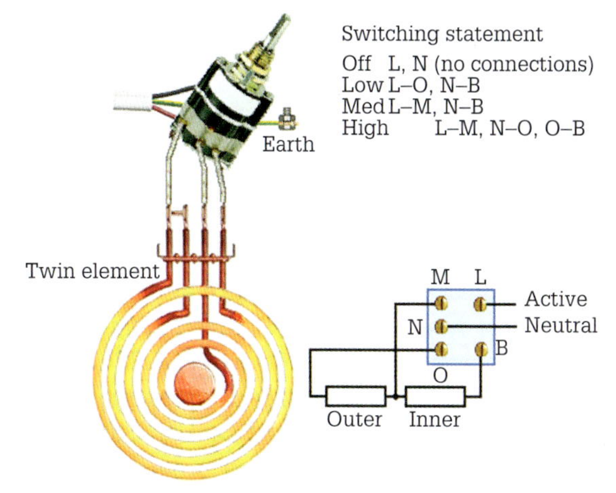

FIGURE 9.3 Double-pole three-heat switch

Automatic control

There are two types of automatic temperature control – thermostatic and an energy regulator device called a simmerstat.

Thermostats

These devices are used whenever a heating system must be maintained within moderately accurate limits. They are designed to hold a thermal load continually at a set-point temperature with very little variation around the set point. To achieve automatic control, these devices vary the time for which current flows to the elements so that the average power is enough to maintain the system at the desired temperature.

There are three main configurations of thermostats – bimetal, strut-and-tube, and bulb-and-capillary.

Bimetal thermostats

The bimetal temperature control is a small-current-operated temperature-sensitive control. Some bimetal thermostats have a conductive bimetal element that carries the circuit current. A bimetal element is shown in Figure 9.4.

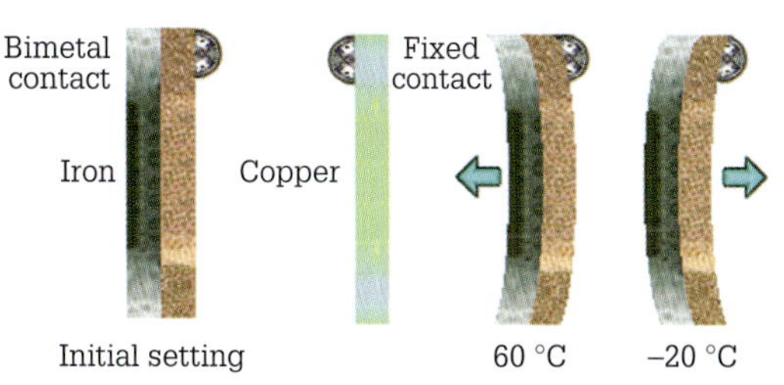

FIGURE 9.4 Bimetal thermostat

The bimetal construction consists of two narrow strips of metal (usually copper and iron) permanently bonded together to make a thin one-piece element. Since these two metals have different coefficients of volume expansion, the bimetal element deforms as its temperature changes. The copper strip has the larger coefficient of expansion, so that it expands more when the bimetal element is heated and contracts more when the element is cooled.

There is only one temperature at which the strip is straight and above or below that temperature the bimetal element expands or contracts, bending itself to one side or the other. Some bimetal elements are circular (disc type) in construction and their expansion or contraction produces a snap-action mechanical force.

The snap-action mechanical power of a current-stressed bimetal element opens or closes the contacts. The snap action of a bimetal assembly provides excellent repeatability of over 100 000 cycles. Once set, this device maintains temperature with a differential of ±5 °C.

The differential is the number of degrees required to make the contacts change position (the relationship of the 'on' and 'off' times or duty cycle). The amount of force needed to make the bending action happen controls the differential. An adjusting screw that sets the distance through which the movable bimetal contact must move determines this force. A narrower differential will give smaller temperature swings but increases cycling rates and causes increased wear on mechanical relays and the heater element. Some bimetal thermostats have no adjustment, and function at a fixed temperature. A variable bimetal thermostat as used with steam irons and a snap-action thermostat are illustrated in **Figure 9.5**.

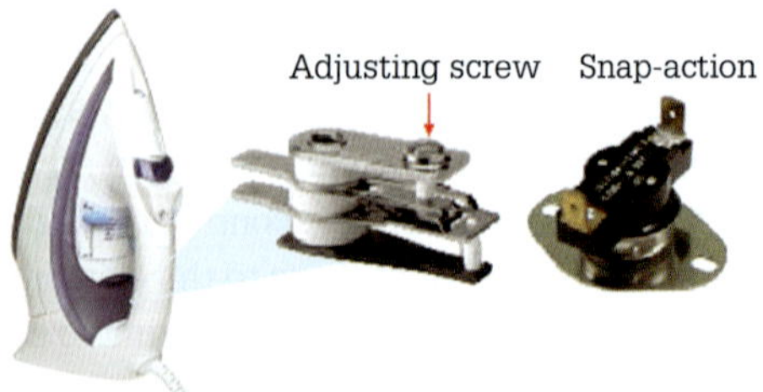

FIGURE 9.5 Bimetal thermostats

Bimetal thermostats are primarily used in household appliances such as steam irons, grills and small electric ovens. The fixed-temperature bimetal thermostats are used as an air or surface temperature protection device in electric motors.

Strut-and-tube thermostats

The strut-and-tube thermostat applies the principle of differential expansion of its two basic parts: the outer shell, the active sensing member made of high-expanding metal, and the strut assembly made of low-expanding metal. Each end of the strut assembly is mechanically connected to the ends of the shell. The principle of differential expansion is applied when a change in force is produced on the low-expansion strut assembly as the high-expanding shell expands or contracts with changing temperature. The response of the outer shell to temperature change is almost instantaneous. A strut-and-tube thermostat is illustrated in **Figure 9.6**.

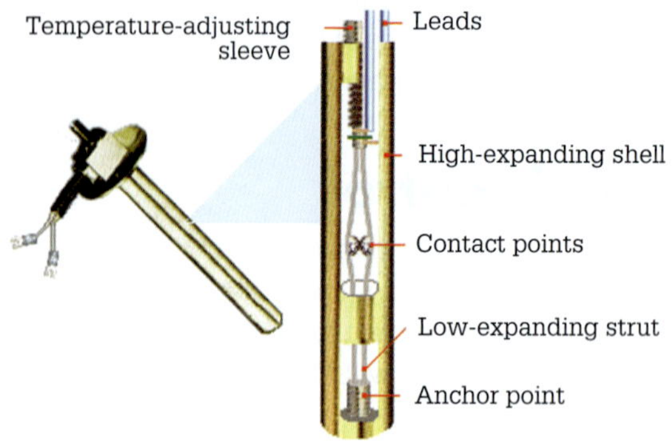

FIGURE 9.6 Strut-and-tube thermostat

A pair of electrical contacts is attached to the strut assembly, and the combined unit is placed under tension inside the expanding shell element. A temperature-adjusting sleeve regulates the temperature at which the contacts 'make' or 'break'. The temperature setting of some strut-and-tube thermostats can be resolved to within 0.05 °C of the required temperature. These thermostats have very high accuracy, with switching differentials of less than 0.1 °C obtainable. This means that every temperature variation, no matter how tiny, causes a corresponding change in the space between the electrical contacts. Therefore, contact action can be brought about by a very minute temperature change. Some strut-and-tube thermostats can control temperatures from –73.3 °C to 593.3 °C.

Bulb-and-capillary thermostats

Bulb-and-capillary thermostats have a sensing bulb at the end of a small capillary tube. A liquid such as mercury is sealed inside the bulb and expands and contracts with temperature change in the sensing bulb. The volume change in the mercury transmits a force up the capillary tube and against an actuator (bellows or diaphragm). The actuator engages the electrical contacts to open or close as required. A set point determines the amount of force required to make the electrical contacts move from the open or closed position to the opposite position. A bulb-and-capillary thermostat is illustrated in **Figure 9.7**.

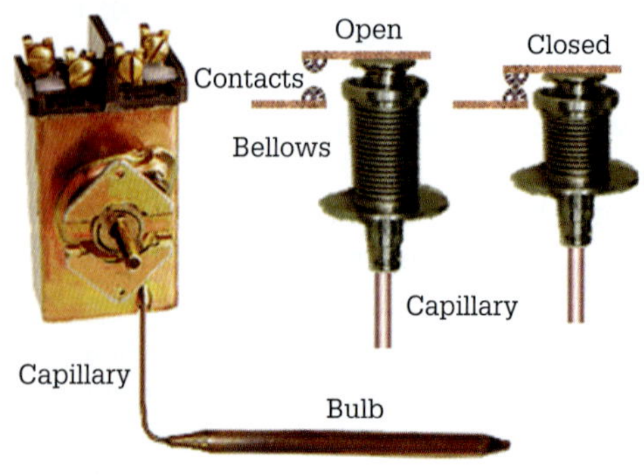

FIGURE 9.7 Bulb-and-capillary thermostat

These thermostats are used for air or liquid heating applications where the load temperature changes very little over time, such as keeping a metal plating bath hot.

The bimetal, the strut-and-tube, and the bulb-and-capillary thermostats are known as closed-loop control sensors. A closed-loop control senses the temperature of the load (called feedback) and opens or closes accordingly.

Simmerstat control (infinite switch)

The simmerstat is a bimetal switch in which the bimetal sensing element is activated by the current flow through its resistance wire element. This element is wrapped around the sensing bimetal element and is connected in parallel with the load that the simmerstat is controlling.

The load is controlled by the cyclic action (on and off) of the bimetal switch that operates according to the heat level dissipated by the switch's internal heating element.

The simmerstat is an open-loop control sensor. These types of sensors do not react to the changes in load temperature. They maintain a set current or energy level that is delivered to the heater or cooler regardless of the temperature of the load. A simmerstat is illustrated in **Figure 9.8**.

The cam enables an infinite number of switching statements (mark space ratios) to exist. By rotating the cam the set-current trip point is created and the cam mechanically places stress similar to the tension in a spring on the bimetal element (high tension = high cycling time, less tension = short cycling time).

Simmerstats are suitable for electric cooking ranges, hot plates, electric stoves and any other electrical heating appliances.

Programmable thermostat control

Programmable heat controllers are used when exceptional accuracy in temperature control is required. These devices are based on microprocessors and thermistor sensors. Most of these programmable heat controllers can store and initiate multiple daily temperature settings and they can adjust the turn on and off times of heating and cooling systems as the outside ambient temperature varies. Programmable heat controllers have viewing capabilities such as LED or liquid crystal displays as well as a data entry membrane switch keyboard. A programmable thermostat controller is illustrated in **Figure 9.9**.

FIGURE 9.9 Programmable thermostat controller

There are five essential operating elements with respect to the programmable thermostat – sensor, input, comparator, output and load. The temperature sensor device, which could be a thermocouple, an RTD or a programmable controller (PLC), sends its information to the input microchip that interprets the information. Once the signal has been decoded the input microchip transfers the information as a temperature value to the comparator (a programmable device). The comparator compares the actual load temperature to this new set-point temperature and then instructs the output microchip in what action to perform. The output microchip implements the instruction by signalling output-temperature-modifying devices such as a relay or timer to switch the load on or off as required. A block diagram of the operating process is shown in **Figure 9.10**.

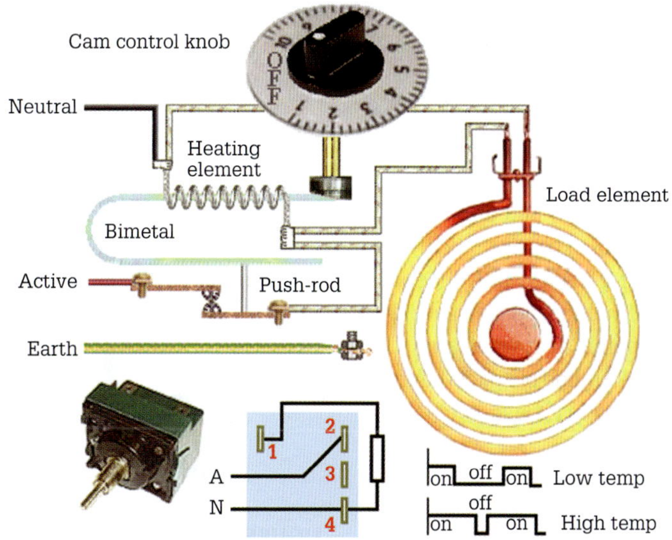

FIGURE 9.8 Simmerstat

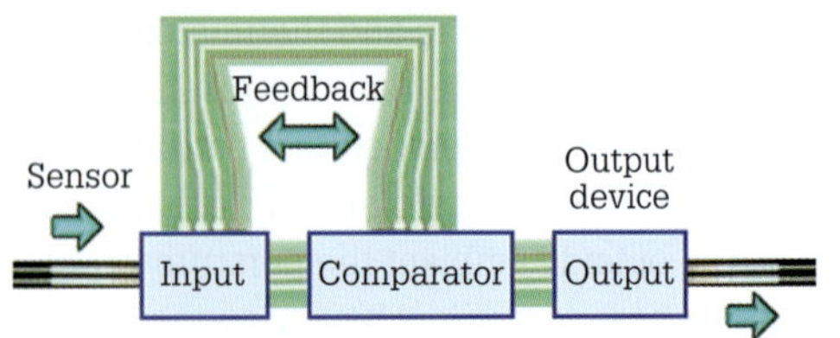

FIGURE 9.10 Block diagram of a programmable thermostat

Testing a thermostat

In order to maintain consistent quality of heat or cold production, it is necessary to perform calibrations on process sensors such as thermostats. A calibration is a matter of qualifying the sensor under test. By knowing the limitations of the sensor, it is possible to maximise the effectiveness of the heating or cooling process loop.

Because most errors in any temperature system are due to the sensor itself an accurate reference temperature must be achieved. This means placing a thermometer – mercury-in-glass or digital – near where the probe of the thermostat is located. This means that both temperature devices are physically exposed to the desired temperature. If the thermostat proves to be more than 15 °C out of step with the thermometer, then any adjustment may not maintain validity and the thermostat should be replaced. Adjustment of most range thermostats is accomplished by removing the range control knob and adjusting a screw in the centre of the hollow shaft. This screw controls the set point. Other types of thermostats have their adjustment at the base of the control shaft. An oven thermostat and a digital thermometer are shown in **Figure 9.11**.

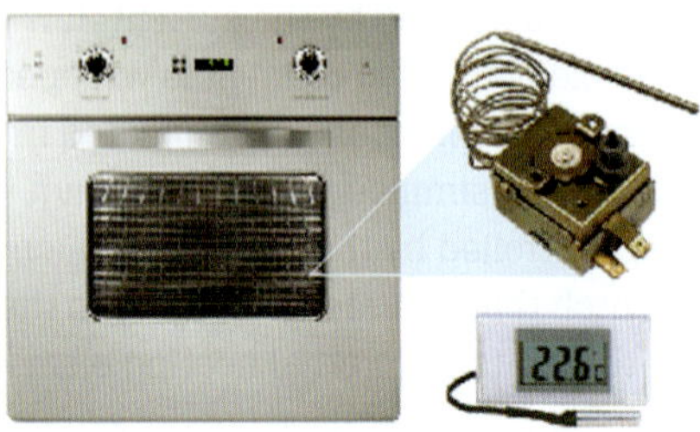

FIGURE 9.11 Oven thermostat and digital thermometer

REVIEW QUESTIONS

1. Name the three types of control used to manage the thermal communication of a heating appliance.
2. Which heating control device has a four-position switching statement?
3. Name the two types of devices that provide automatic temperature control.
4. What applications suit bimetal thermostats?
5. Which temperature control device is used as an air or surface temperature protection device in electric motors?
6. Describe the operation of a bulb-and-capillary thermostat.
7. Name an open-loop control sensor.
8. What type of heat controller is used when exceptional accuracy in temperature control is required?
9. Name the five essential operating elements of a programmable thermostat.
10. What is meant by the term 'calibration'?

9.2 Fixed electrical heating appliances

Fixed electrical appliances include space heating for residential applications and process heating in industrial applications.

Space heating

Commercial and industrial premises and even some residential premises can be cold, draughty and difficult to heat. Poor insulation, high ventilation, excessive outside air filtration and poor draught control lead to chilly indoor environments and reduce work production. Space heating offers direct, controllable and economic heat when and where it is required. There is a broad range of space-heating devices ranging from simple radiators to reverse-cycle air-conditioning and floor heating.

Electrical heaters

Panel, column and architrave oil-filled convection-type space heaters provide a broad range of temperature variation over broad surface areas. An oil-filled column space heater that uses the convective movement of warm air for quick heating is shown in **Figure 9.12**.

FIGURE 9.12 Oil-filled convection space heater

These space heaters are available in various heat settings up to 3600 watts. They have a thermostat that when set automatically maintains the desired temperature.

Floor heating

Heating cables have been developed to meet the specific need for large-area space heating. These cables have a high-temperature capability and low electrical resistance values required for long circuit lengths. The cables are located under concrete, ceramic tiling or soil. There are three types of heating cables: mineral insulated (MI), PTFE (polytetrafluoroethylene) and self-regulating. Mineral-insulated cables have resistance wires embedded in the mineral insulant. The diameter and material of the resistance wires determine the heat output per metre for a given applied voltage. An MI heating cable with a possible operating temperature of 400 °C is illustrated in **Figure 9.13**.

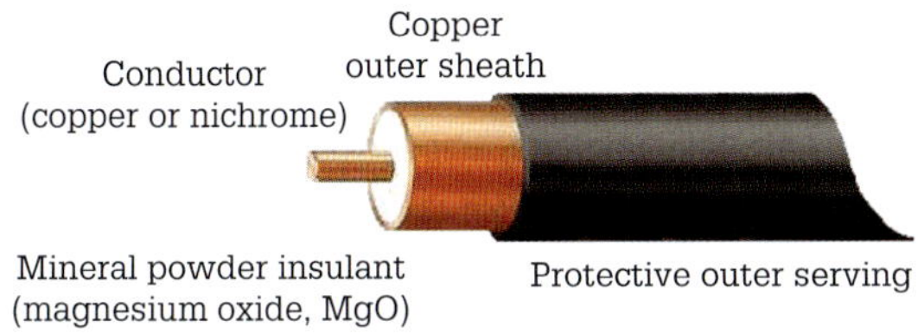

FIGURE 9.13 Mineral-insulated heating cable

The sheath provides excellent heat transfer and mechanical as well as corrosion protection of the element for a long-life installation. Typical applications include wall heating and floor warming.

PTFE-insulated and shielded cables are also supplied with the resistance conductor embedded in the insulating material. The sheath provides excellent heat transfer and mechanical protection of the element but is limited to a surface temperature of up to 260 °C. These heating cables are suitable for wooden floor heating. A PTFE cable is illustrated in **Figure 9.14**.

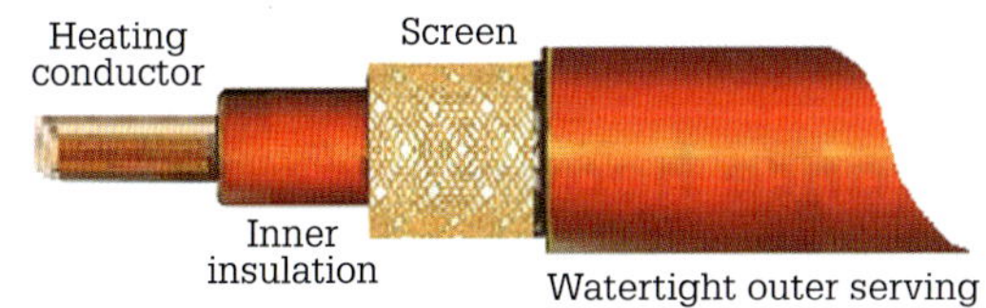

FIGURE 9.14 PTFE heating cable

Self-regulating heating cable consists of a resistive compound suspended in the insulation between the cable conductors. The compound self-limits (increases resistance) with temperature and so reduces power drawn from the supply. A self-regulating heating cable is shown in **Figure 9.15**.

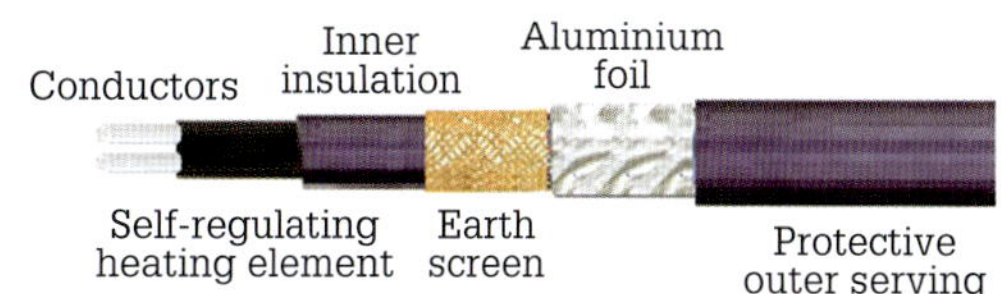

FIGURE 9.15 Self-regulating heating cable

Self-regulating heating cables consist of a semi-conductive polymeric heating element extruded between two parallel copper conductors that serve as the heating element. When the load current flows through the cable the temperature of the conductive core increases.

Because the polymeric heating element has a positive temperature characteristic of resistance, the resistance of the cable increases as the surrounding temperature increases. Since the power drawn by the heating system is a function of temperature, the conductive core responds quickly to any temperature change. These cables cannot overheat and they actually maintain temperatures of up to 65 °C. Both PTFE cables and self-regulating cables are unaffected by oils, lubricants and hydraulic fluids.

The principle of floor warming is to utilise the high thermal capacity of the floor slab as a heat reservoir to store and emit heat on a continuous basis. For the most efficient use, the heat input to the slab is usually via off-peak, low-cost supply. **Figure 9.16** illustrates slab heating using mineral-insulated cable.

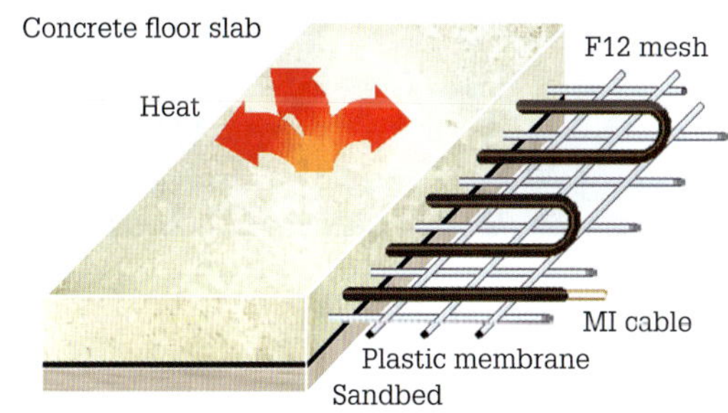

FIGURE 9.16 Slab heating with MI cable

Timber floor heating using PTFE heating cable is shown in **Figure 9.17**.

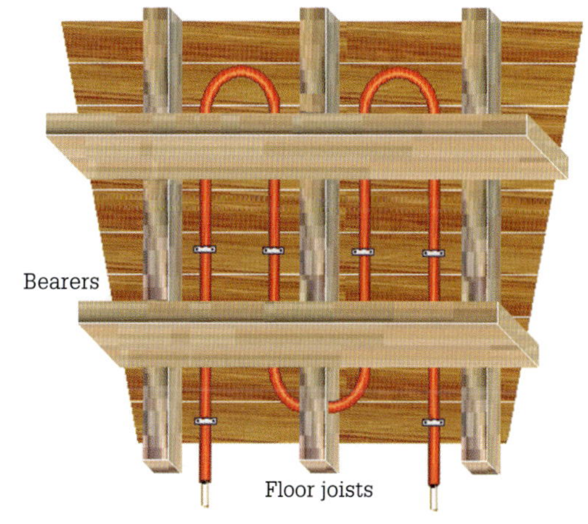

FIGURE 9.17 Timber floor heating using PTFE cable

With tile floor heating, only 50% of the floor area is laid with the heating cable. An under-tile heating system using self-regulating heating cable is illustrated in **Figure 9.18**.

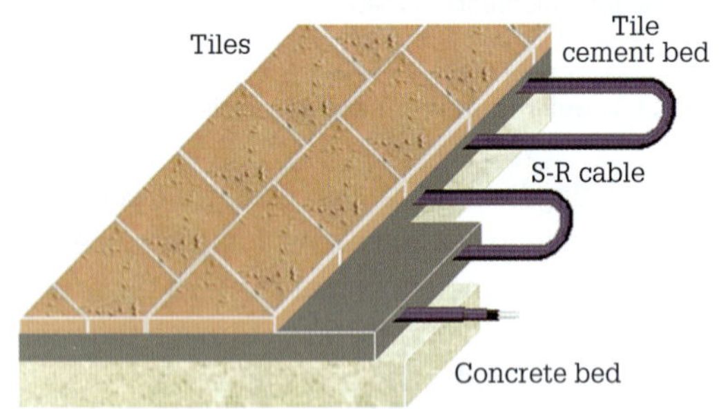

FIGURE 9.18 Under-tile heating using self-regulating (S-R) heating cable

Under-carpet heating is another method used for space heating. There are several methods used – under-carpet heating mats or a very thin (1.0 to 1.5 mm^2 CSA) heating cable. An under-carpet heating system using the cable method is shown in **Figure 9.19**.

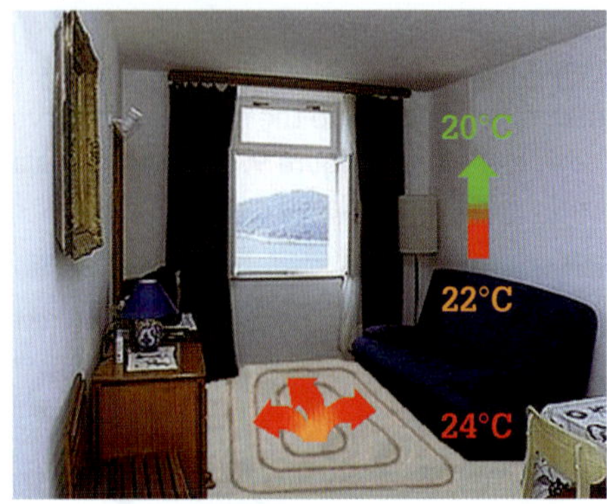

FIGURE 9.19 Under-carpet heating

The mat consists of hard aluminium foil on one side and polyester fabric material on the other side. The heating wires are sandwiched between these materials and because aluminium is an excellent conductor, heat transfer is done very efficiently. The thin cables, once placed on the floor, are covered with protecting scree about 3 mm thick. This is particularly useful and important if carpets are removed and the floor is covered with tiles. This form of space heating provides an extremely quick warming time of between 10 and 20 minutes. Individual automatic thermostatic controls are used to adjust temperature changes while an in-floor temperature sensor provides the correct comfort level required.

Reverse-cycle air-conditioning

Reverse-cycle air-conditioning (see **Figure 9.20**) operates in the same manner as a heat pump.

When the air-conditioner is switched to heating mode, a valve reverses the flow of the liquid refrigerant. Warm air is captured by the refrigerant in the evaporator

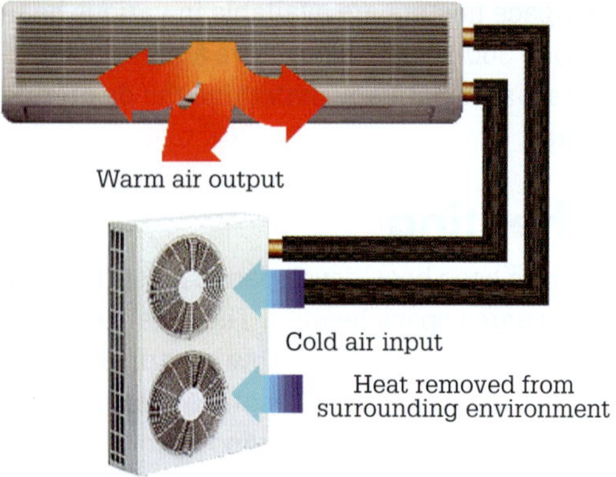

FIGURE 9.20 Reverse-cycle air-conditioner

(outside unit) from the surrounding environment and warms the liquid refrigerant which then becomes a low-pressure vapour. The heat contained as a low-pressure vapour is then transferred via the compressor to the condenser for the inside coil where it becomes a high-pressure vapour. The internal fan draws cool air in through the filtration system over the evaporator coil and transfers the heat from the high-pressure refrigerant vapour into the air that is circulated throughout the room. The refrigerant cools down and changes back into its liquid state where it then passes back to the outside unit to be heated again.

Process heating

Heat plays an indispensable role in a wide variety of manufacturing processes: cooking, softening, melting, drying, curing and fusing. Electric process heating can be delivered in exact amounts at precise temperatures and has various methods of delivery. There are seven categories of process heating methods used for different industrial processes:

- resistance
- induction
- infrared
- dielectric
- arc
- ultraviolet (UV)
- electron-beam heating.

Resistance heating

Resistance heating has been the most economical means of supplying process heat for many applications. Resistance heating is an indirect method of heating that converts nearly 100% of the energy in the electricity to heat. Indirect resistance heating involves passing an electric current through a resistance-heating element and the thermal energy is then transferred by three possible methods: conduction (through physical contact of the element and the material being heated), convection (through the air) and radiation (directly from the elements in the material being heated). At lower temperatures, convection heating is predominant but as the temperature increases radiation becomes more prevalent. The composition of the heating element determines the highest attainable temperature and influences the construction of the heating plant. Elements are made from materials such as iron–chrome–aluminium alloys (Fe–Cr_{21}–Al_4), nickel chromium (Ni–Cr 80:20), graphite, tungsten and silicon carbide. Various resistance-heating elements are shown in **Figure 9.21**.

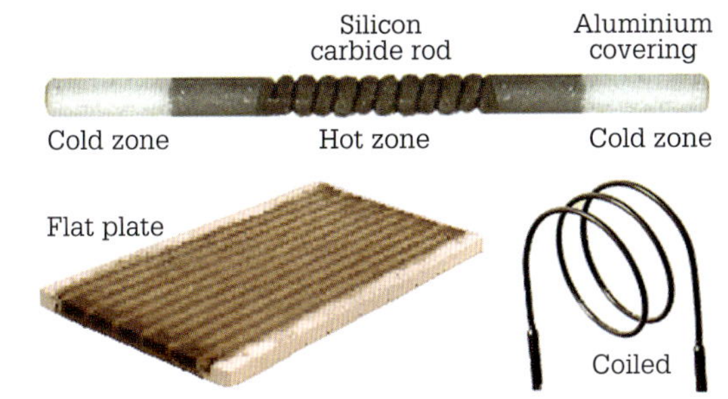

FIGURE 9.21 Various resistance-heating elements

Indirect resistance heating is accomplished by placing the material to be heated in a well-insulated metal container with resistance-heating elements in the walls or floor of the metal container. The inside surface of the metal container is lined with heat-resisting brick, ceramic or fibre batts. The atmosphere may be air, inert gas such as argon or a vacuum, depending on the requirements of the application. Resistance heating is the most simple and therefore the most widely used method of heat generation. Its most significant advantages are wide temperature range, natural method of control, temperature uniformity, flexibility in positioning and design of electric heating elements and low cost.

Induction heating

Induction heating occurs when a substance is heated by induced electric current by being placed within an electromagnetic field. Industries use induction heating for heat treatment, forging, annealing, soldering, pipe welding, thread rolling, removing electrical varnishes and surface coatings from electrical conductors and metal joining. Induction heating is based on the principle of inducing very high values of current on the face of the electrically conductive substances being heated. These substances are exposed to an a.c. electromagnetic field generated by a heating coil. Current is induced into the substance in the same manner that current is induced in the secondary coil of a transformer. The heating coil is energised from a power frequency converter. With ferromagnetic materials, the heating effect is intensive because of the 'skin effect' whereby the current induced in the surface layers is very high. Frequencies of 5 kHz to 30 kHz are useful for thick substances requiring deep heat penetration while higher frequencies of 100 kHz to 400 kHz are valid for smaller substances. Two induction-heating applications are shown in **Figure 9.22**.

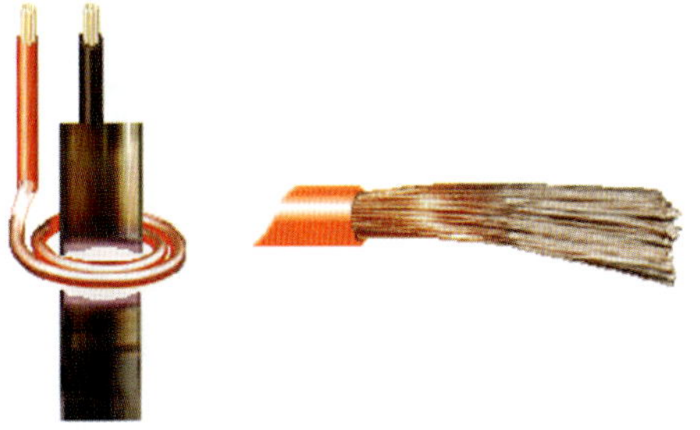

FIGURE 9.22 Induction heating – pipe welding and varnish removal

Infrared heating

Infrared heating is produced by electromagnetic radiation that is generated by a heat source (425–2200 °C) by vibration and alternation of molecules. The infrared region of the electromagnetic spectrum lies between the visible and the microwave radiation at wavelengths from 0.7 µm to 1000 µm. Electric infrared process heating employs electrical resistance to heat an emitting material specifically for the purpose of generating thermal (infrared) radiation.

The infrared energy wave travels at the speed of light and there is little absorption by the air. Optically designed reflectors direct these energy waves onto a substance that then absorbs the heat energy. In industry infrared heating is used to warm food and bake bread and chickens, cure plastics and soften adhesives, dry water- and solvent-based paints and for space heating. The elements of infrared heaters may be glass bulbs, quartz or metallic tubes or ceramic panels. Some infrared heating elements are shown in **Figure 9.23**.

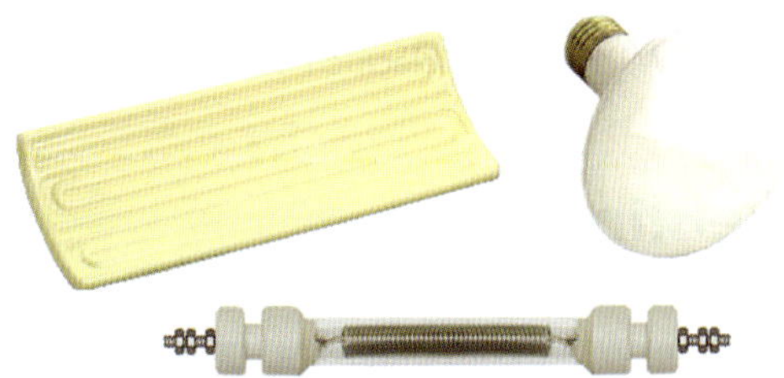

FIGURE 9.23 Infrared heating elements

Dielectric heating

Dielectric heating uses high-frequency electromagnetic waves such as microwaves and radio waves to stimulate molecules in non-conductive substances. The stimulation causes rapid movement of a substance's molecules that in turn causes heat to be produced. This stimulation occurs within the substance being heated when it is placed between two electrodes. Radio waves are used to dry paper and cure glue in the manufacturing of particleboard and plywood. Microwaves are used in the food-processing industry for pre-cooking chicken and thawing deep-frozen foods. In metal foundries dielectric heating is used for the drying and curing of moulds. Dielectric heating of particleboard is shown in **Figure 9.24**.

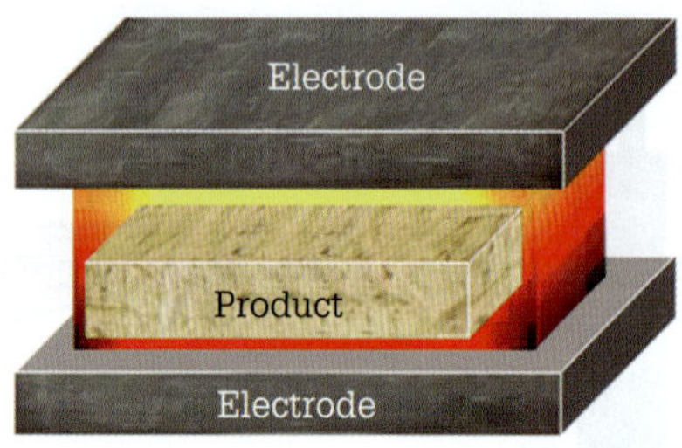

FIGURE 9.24 Particle board manufacturing

With radio-wave dielectric heating, the dielectric substance is placed between two electrodes and the disturbed molecules of the dielectric substance cause friction and therefore create heat. Microwave dielectric heating uses the microwave energy produced by a magnetron tube. This energy wave is transmitted into metal channels called wave guides that direct the 2450 MHz microwaves into the cooking space. Browning of food is achieved by the addition of a resistance element that heats the outside of the food by radiation. An illustration of microwave heating is shown in **Figure 9.25**.

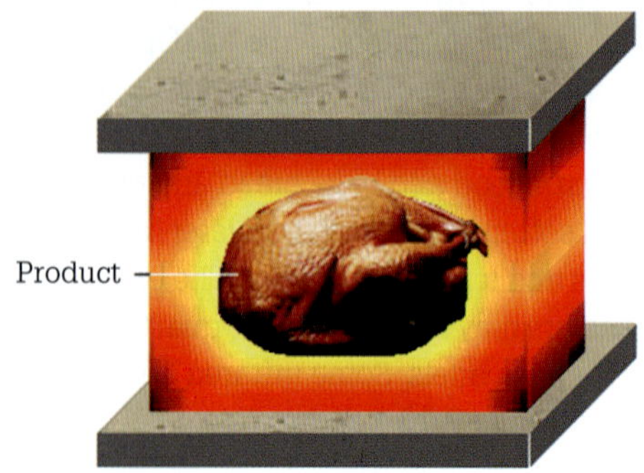

FIGURE 9.25 Microwave dielectric heating

Arc heating

Electric arc heating furnaces are used to refine ores and produce metals by concentrating ores and refining metals through smelting. **Figure 9.26** shows an arc furnace.

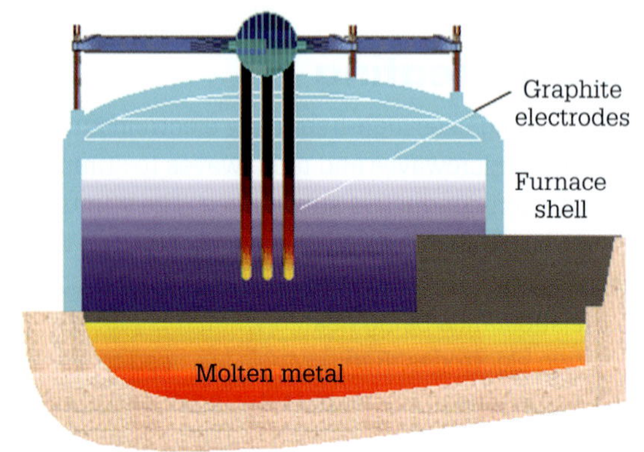

FIGURE 9.26 Arc heating

There are two types of arc furnaces – alternating current and direct current. An a.c. furnace uses three-phase current flowing from a set of three graphite electrodes to another electrode set between the conductive refining metal. Most of the d.c. furnaces have a single carbon electrode called the anode and a carbon cathode at the bottom of the furnace containing the melt. The electrodes are 15–300 mm in diameter and are up to 3 m in length.

The d.c. furnaces are supplied by a high-voltage a.c. supply (up to 22 kV) that is rectified through 6- or 12-pulse thyristor semiconductors and fed to a d.c. reactor then to the carbon electrode. Direct current flows down the graphite rod through the refining metal to the cathode. Electrode currents of up to 120 kA can be realised using d.c. technology, producing up to 150 tonnes of refined metal every heat cycle.

Ultraviolet heating

Ultraviolet (UV) heating is a curing process that uses electromagnetic energy to transform a liquid material to a solid state after it has been applied to another surface. In this process particular chemicals are added to coatings such as inks, polymer coatings and adhesives. When the UV electromagnetic wave strikes the individual coating, the added chemicals react to the wave frequency and molecular bonding occurs. The bonding creates an excellent adhesion and a very durable finish.

UV heating is used to harden polymer coatings on cork and timber floors and in the film coating of various materials. A UV heating lamp is shown in **Figure 9.27**.

FIGURE 9.27 Ultraviolet (UV) heating lamp and UV safety glasses

Electron-beam heating

With electron-beam heating high-velocity electrons are fired at metals and the transformation of each electron's kinetic energy to heat energy at concentrated temperatures heats the metal. Heating a filament in an evacuated cylinder produces an avalanche of electrons. The avalanche of electrons produced is accelerated and focused by various optical devices as the electrons travel to the metal's surfaces. An example of an electron-beam heating process is shown in **Figure 9.28**.

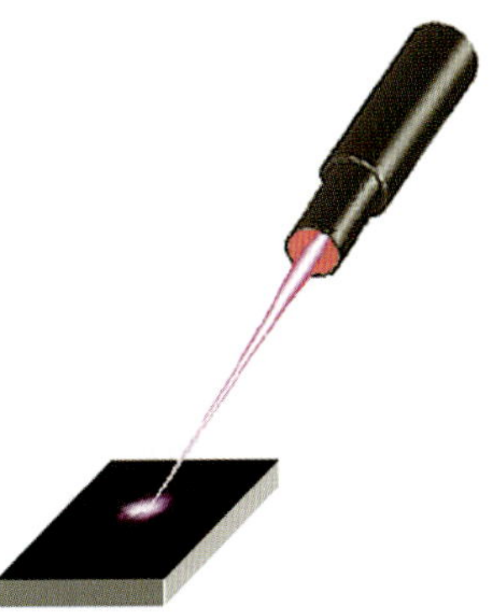

FIGURE 9.28 Electron-beam heating with an electron gun

Cooking appliances

Many of today's appliance manufacturers are striving to make the cooking process faster and more efficient by making available a wide variety of cooking appliances to the consumer. Many appliances are particular in their application, while others can be used for a wide variety of cooking functions. Electric cooking appliances fall into one of the three categories: resistance, induction and electric microwave cooking. The vast majority use resistance-type heating.

REVIEW QUESTIONS

1. Outline the benefits of space heating.
2. What types of cables can be used for floor heating?
3. State the factors that determine the heat output per metre of floor heating cables.
4. Name the manufacturing processes in which heat plays an indispensable role.
5. Name the seven categories of process heating methods.
6. What type of process heating is the most economical means of supplying heat?
7. State the most significant advantages of resistance heating.
8. Describe the induction heating process.
9. Name the two different types of arc furnace.
10. What type of heating method uses microwave and radio waves to stimulate molecules in non-conductive substances?
11. Name the three categories of electric cooking appliances.

9.3 Electrical water heater operation

The water heater is a closed vessel used to supply hot water. The heat generated by resistance elements heats the water. The water heater includes all the controls and sensors necessary to prevent water temperatures from exceeding 60–70 °C. Water heating processes fall into two categories – instantaneous and storage.

Instantaneous water heaters

Instantaneous or tankless water heaters are small cabinets that heat water on demand or instantly as it passes through the heater. They contain no significant water storage, possessing only up to a 6-litre operating holding. These water heaters only use energy when the hot water outlet is turned on and shut down immediately when the outlet is closed.

Originally designed for a one-point use only, units today can supply continuous hot water at flow rates of 7.5 L/min to many points throughout an installation. The apparatus of these heaters includes a cold water inlet, a hot water outlet, a flow switch, a large kW resistance heating element and solid-state controls. These heating units are available in single-phase and three-phase models. The three-phase units can range from 6 kW to 29.1 kW and draw a large operating current (up to 45 A/phase) from the supply. Electricity utilities generally resist the installation of three-phase instantaneous water heaters in domestic, commercial and industrial areas. This is because of the load demands that can be called from a supply system. An illustration of a single-phase instantaneous water heater is shown in **Figure 9.29**.

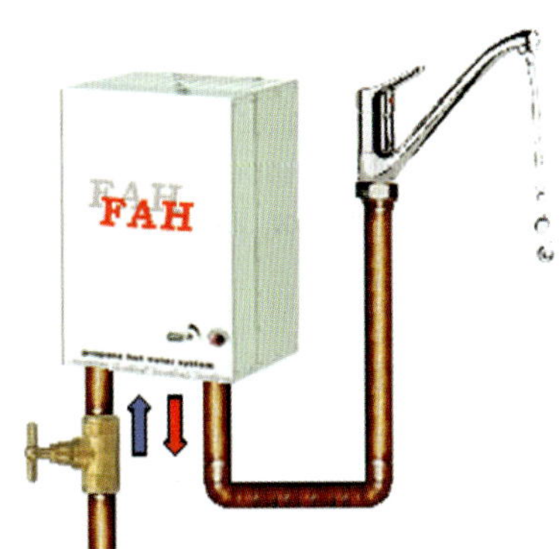

FIGURE 9.29 Instantaneous water heater

The turning on of the kitchen tap in **Figure 9.29** allows mains pressure water to be introduced into the unit, engaging the operation of a water flow switch. The flow switch controls the energy to the electric heating element. Once activated, the element heats the water. The kW rating of the element and the output water flow rate determine the temperature. If the flow of water from the unit is high, the element rating may not be high enough to raise the water temperature to the desired level.

To overcome this problem, some heater units have multi-stepped heating elements that are cycled on and off as the water demand varies. Some heaters are also provided with high-speed switching devices such as thyristors or triacs to vary the flow of current continuously to maintain the desired water outlet temperature in direct proportion to the hot water flow rate demand.

These devices adjust the adequate current flow by delaying the instant in each half-cycle of the supply at which current begins to flow. Instantaneous heaters require no safe tray or drainage outlet, although local regulations should be checked. Illustrations of possible instantaneous water heater components are shown in **Figure 9.30**.

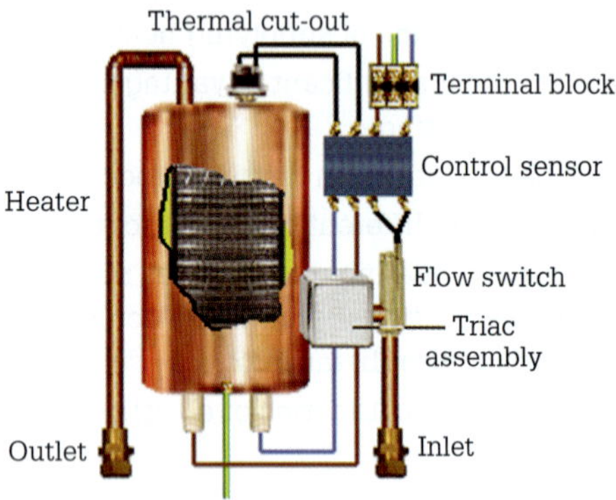

FIGURE 9.30 Instantaneous heater components

Instantaneous water heaters are used in residential, industrial and commercial settings and institutions such as schools, hospitals and nursing homes.

Storage water heaters

Storage tank water heaters or heat exchange units are very common. They store and heat a large volume of water in an insulated (injected polyurethane) copper tank at a thermostatically controlled temperature. The fully enclosed units are mounted at ground or floor level, usually outside the building.

Mains-pressure water heaters

In the case of a mains-pressure unit, the cold water mains pipe is connected directly to the copper storage tank via the cold water inlet. On entering the unit, the cold water is heated by an electric resistance element and then rises to the top of the tank. Turning on a hot water tap allows the heated water to be delivered at the same pressure as the cold water.

The apparatus of these pressurised heaters includes an unvented (water output is at mains pressure) copper or a copper-lined steel, vitreous enamelled (glass-lined) mild steel, stainless steel tank, a combined temperature and pressure valve (PRT – tank connected with discharge pipe), a pressure-limiting valve (PLV – cold water), a non-returning isolating valve and stop cock (NRV – cold water connected), a pressure relief valve (PRV – cold water expansion), an immersion heating element (1.8–4.8 kW – single or double elements), a surface-mounted thermostat incorporating an over-temperature cut-out switch, a sacrificial anode, a cold water inlet and a hot water outlet. An illustration of an unvented mains hot water storage heater is shown in **Figure 9.31**.

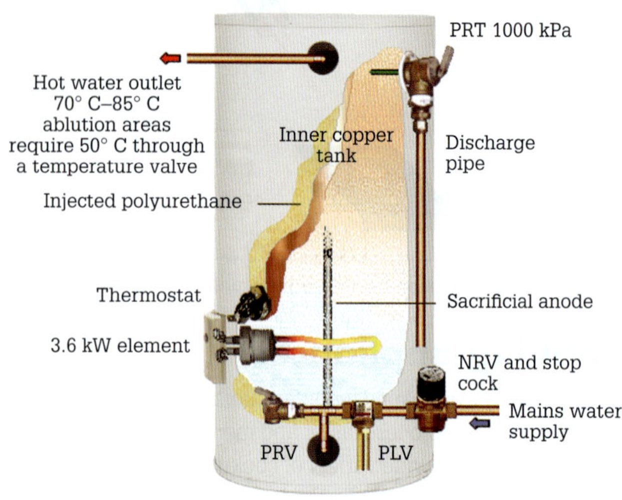

FIGURE 9.31 Unvented mains-pressure hot water storage heater

With mains-pressure storage heaters, the thermostat should be set to deliver the lowest acceptable temperature. Around 65 °C is usually adequate for most consumers. With many storage tanks a sacrificial anode of a metal such as magnesium or aluminium is installed to protect the steel, brass or copper of the tank, element or valves from the corrosive effects of galvanic action.

Mains-pressure storage heaters are generally attached to an off-peak electricity meter that usually operates between 10.00 pm and 7.00 am. This is the most cost-effective way to heat water electrically. When a hot water tap is opened, cold water enters the tank through the NRV and the PLV and the drop in temperature triggers the thermostat and element at the bottom of the tank. When the tap is turned off, the heating element continues to carry current until the thermostat's temperature setting is realised.

Some mains-pressure storage units have an extra element at the top of the tank to provide an enhanced heating capability during the day if all the stored hot water is depleted. This second element is not connected through the off-peak meter, and it operates automatically when the temperature of the stored water falls below a set value. Depending on the size of the tank only the top 40–70 litres is heated, and the unit will 'top up' the hot water as required.

Low-pressure water heaters

Low-pressure or gravity-fed vented (permanently open to the atmosphere) water heaters store water at a pressure lower than mains pressure. The height determines the pressure (pressure = 10 kPa/m height) at which the unit is mounted with respect to the hot water outlets.

The units can be floor or roof space mounted. In addition to their main tank, these units have a cold water header tank controlled by a ball valve either located in the roof space or mounted on the side of the main unit.

This type of storage water heater uses the displacement principle of operation where cold water is delivered to the bottom of the storage tank, displacing the hot water drawn from the top of the tank. The pressure of the hot water at the point of delivery depends on the relative height at which the unit or header is mounted. The low-pressure storage heater with a remote tank is illustrated in **Figure 9.32**.

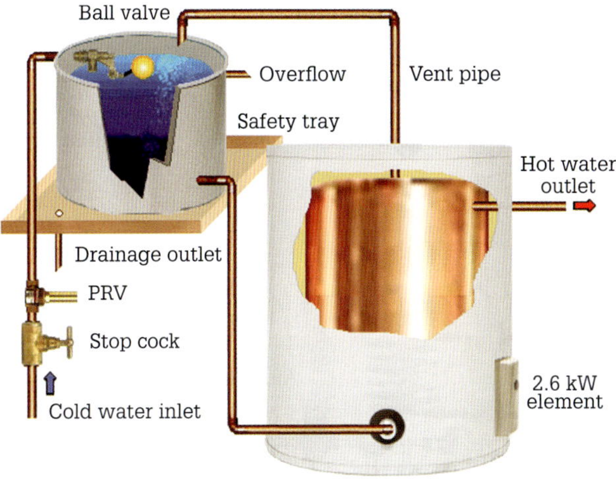

FIGURE 9.32 Low-pressure storage heater

The header tank stores cold water from the high-pressure mains that in turn is fed into the hot water storage tank where the water is heated by the heating element. Hot water is drawn from the top of the tank when required by a delivery point.

The problem with some of these units is that there is the possibility of a partial vacuum being created within the storage tank. A partial vacuum can occur if the vent pipe is blocked or restricted. As a result, the storage tank will partially collapse when the hot water outlet is engaged.

Some storage-type water heaters incorporate a magnetic valve that controls the cold water feed to the storage tank. These types of units are called falling-level storage heaters. An illustration of their principle of operation is shown in **Figure 9.33**. The magnetic valve is installed in the cold water inlet pipe. This valve prevents cold water entering the storage tank until the water in the tank has reached a predetermined temperature.

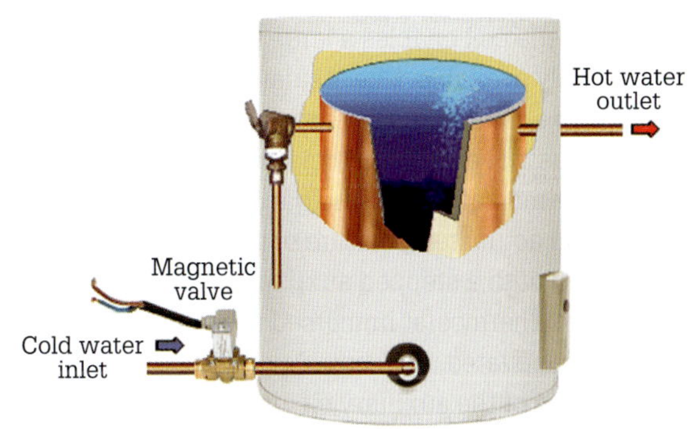

FIGURE 9.33 Falling-level storage heater

Solar water heaters

In solar systems, cold water travels through the roof-mounted solar collector where the water absorbs heat from the sun. Water heating using solar energy occurs during the day, and the solar involvement varies significantly throughout the year depending on the climatic conditions. The apparatus of solar heaters includes the solar collector, insulated storage tank and if required, pump and control valves.

Two types of solar collectors are available – flat-plate collectors and evacuated-tube collectors.

Flat-plate collectors

Flat-plate collectors are the most common collector for domestic water heating. A typical flat-plate collector is an insulated rectangular-type metal box with a transparent cover (similar to a greenhouse) and a black absorber plate. A transparent cover allows the light to strike the absorber plate that then heats up, changing solar electromagnetic radiation into heat energy. The heat energy is transferred to the water passing through the collector piping, heating it to about 70 °C. For greatest efficiency the flat-plate collectors must face directly north. An illustration of a flat-plate solar heater is shown in **Figure 9.34**.

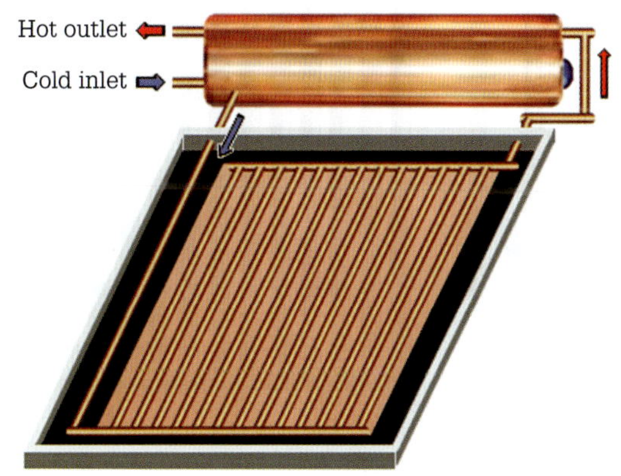

FIGURE 9.34 Flat-plate solar collector

Evacuated-tube collectors

The evacuated-tube collectors consist of rows of parallel transparent double glass tubes, each containing an electromagnetic energy absorber and covered with a solar-sensitive coating. Sunlight enters the tube, strikes the absorber and heats the water flowing through the collector. An evacuated-tube collector is shown in **Figure 9.35**.

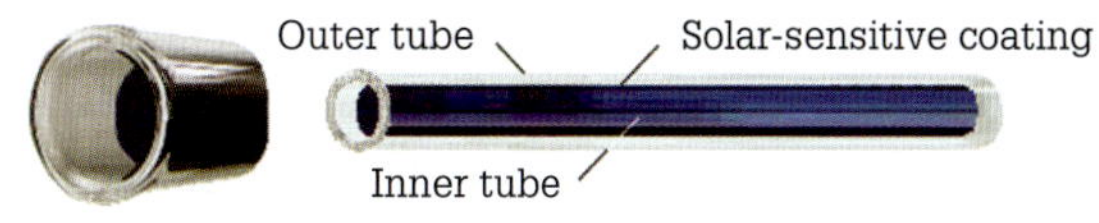

FIGURE 9.35 Evacuated-tube collector

The collectors are manufactured with a vacuum (acting as an insulator to prevent heat loss) between the outer and inner tubes. This allows the water to achieve very high temperatures. The glass tubes can achieve this because of the circular shape of the evacuated tube that allows both direct and diffused sunlight to strike the circumference of the tube.

The heated water can then be stored in a roof-mounted tank or a ground-based unit. If a ground-mounted tank is used, a small pump is required to circulate the water through the elements of the system. Because climatic conditions may vary, many solar water storage systems have an auxiliary heater element (2.4 kW) connected in order to heat the water if required. The auxiliary element can be wired for continuous usage or connected to an off-peak metering system. An evacuated-tube solar heater is shown in **Figure 9.36**.

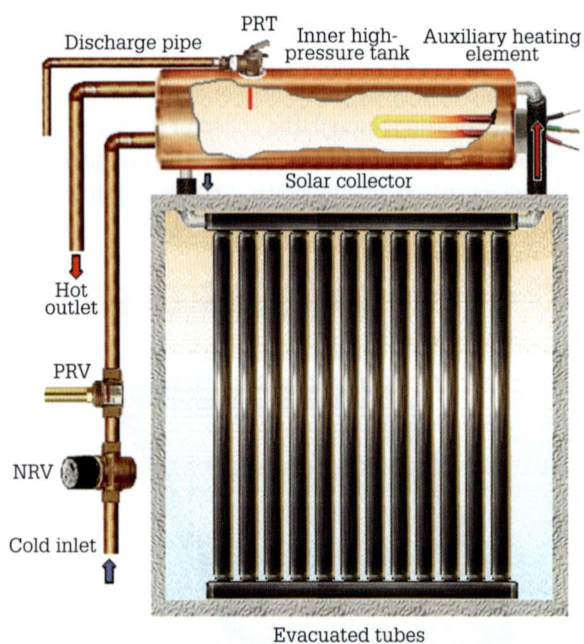

FIGURE 9.36 Evacuated-tube solar heater

Where the storage tank can be mounted above the evacuated tubes the water circulates naturally according to the Thermosyphon or gravity-circulation principle: hot water rises and cold water displacement occurs.

Commercial and industrial applications are designed to supply over 2000 litres of hot water per day.

A typical installation has a group of collectors mounted either on the roof of a building or at ground level together with a closed-circuit heat-exchange unit and several water storage units. The heated water from the solar panels is continuously circulated through the banks of collectors and the heat-exchange unit. The heat-exchange unit also contains auxiliary heating elements in order to provide additional heating.

The solar collector increases the temperature of the stored water temperature by circulation of the heated water through heat-exchange coils. The heat-exchange coils then transfer the acquired heat energy to the circulating water from the storage tanks. An illustration of this process is shown in **Figure 9.37**.

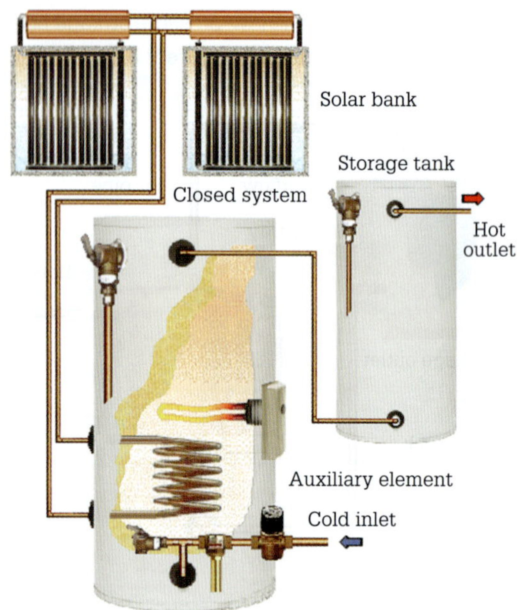

FIGURE 9.37 Industrial and commercial solar water heater

Some areas where solar hot water systems are used include individual residential homes and units, restaurants and canteens and various industrial requirements.

Calorifiers

Calorifiers are cylinders with an internal coil which allows the use of any boiler for hot water production. The calorifier can be either mains-pressure or low-pressure hot water storage systems. A significant amount of heat energy can be transferred to the calorifier, allowing a large production of hot water from a relatively small cylinder. Calorifiers are most suited to applications with moderate to high temperature profiles and can be designed to withstand high pressures.

A central heating boiler is used to heat water passing through a coil contained within the calorifier storage tank. The energy from the coil heats the water stored in the tank. The design of the coil enables a large supply of hot water to be continuously available. The insulation materials in the tank reduce to a minimum the energy lost owing to thermal radiation. Industrial and commercial units have an easily adjustable thermostat, a pressure gauge and a thermometer that shows the actual temperature of the stored water. Some units are fitted with a stepped auxiliary heating element with its control thermostat immersed in the water to be heated. Once activated, the control panel energises the available stages of the auxiliary element until the set-point temperature is reached. Stages are switched on in timed sequence to ensure smooth immersion heater loading.

Safety is paramount for calorifiers and most are fitted with independent high-temperature cut-out devices and an anti-vacuum valve. A low-water cut-out system is also fitted to protect the immersion heater from being energised while the calorifier is empty. A calorifier is shown in **Figure 9.38**.

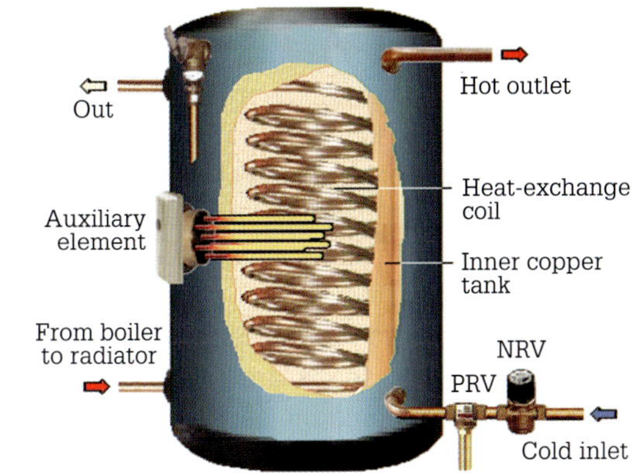

FIGURE 9.38 Calorifier

Small calorifiers are used on ships, in dairies, hospitals and nursing homes, the electroplating and galvanising industry, bottling plants and distilleries for bottle cleaning and pharmaceutical, chemical and fertiliser industries and for heating process feed water for the paint industry.

Heat pump

A heat pump water heater absorbs heat from the surrounding environment and pumps the acquired heat energy into a hot water storage tank. The apparatus for a heat pump includes a compressor, a condenser, evaporator panels, a refrigerant liquid and an insulated pressurised storage tank. The heat pump serves as a heater by absorbing heat from the surrounding environment and pumping it into a closed-system heat-exchanger water storage tank.

All air, even winter air, contains a certain amount of heat. As the fan-assisted outdoor air passes over the evaporator, heat from that air is absorbed by the refrigerant contained inside the copper tubes of the evaporator. The absorption of heat changes the refrigerant from a low-temperature liquid to a low-temperature, low-pressure vapour. The vapour then passes through a compressor where it is compressed into a high-pressure, high-temperature vapour. The hot vapour is pumped into the coil of the heat exchanger in the hot storage tank. As the mains-pressure cold water passes through the heat-exchanger coil, it absorbs heat from the coil. The warmed water is then redistributed through the tank system. The high-pressure, high-temperature vapour condenses back to a low-temperature liquid and is pumped back to the evaporator. The heat pump and its cycle of operation are illustrated in **Figure 9.39**.

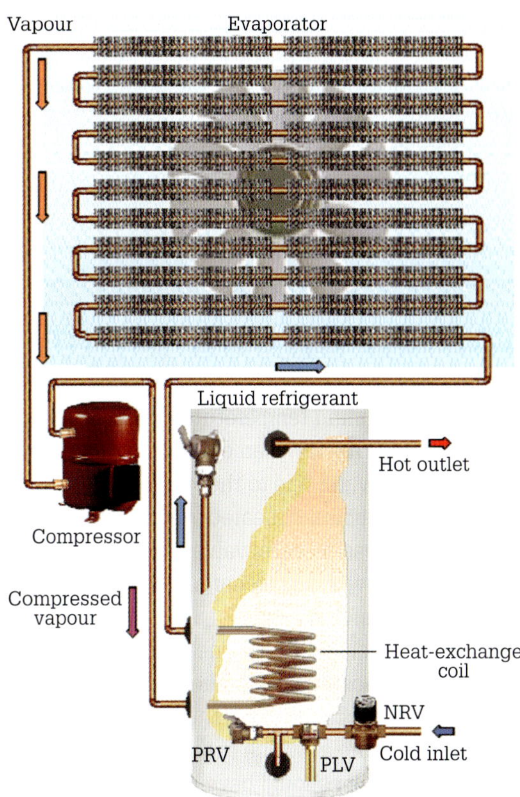

FIGURE 9.39 Heat pump and cycle

Heat pump water heaters operate very efficiently during daylight when the ambient temperature, air movement and solar radiation are at their optimum levels. Heat pumps can be used to heat large volumes of water such as swimming pools and are modified to provide space heating for work and living environments (known as air-conditioners).

Safety precautions

If the water heater's thermostat, which controls the resistive heating element, fails, the pressurised water in the tank might continue to heat and superheat (beyond 100 °C). This causes two problems.

First, since water expands when heated the water pressure in the tank increases as the water is superheated. Water is at its greatest density at 4 °C (where a mass of water occupies the least volume of space). When heat is applied to water at 4 °C, its volume gradually increases as its density decreases until the water boils at 100 °C, assuming open atmospheric pressure. When closed vessels are subjected to expanding water, significant pressures occur. If the pressure exceeds the vessel's maximum pressure threshold (approximately 21.097 kg/cm for some hot water tanks), the tank could rupture or even explode.

Second, the release of superheated water (water heated above 100 °C up to its critical temperature of 374 °C without boiling) causes the water to burst into steam (1 litre of water can produce about 3 litres of steam), causing a sudden increase in volume and release of energy. Lowering the pressure of water lowers the boiling point. There is less pressure above the water to overcome. The superheated vapour plume expands until its pressure equals that of the surrounding atmosphere. The force of superheated steam created within the stored water would exceed the explosion of half a kilogram of trinitrotoluene (TNT). For this reason, pressurised water heaters are required to have incorporated into their system an over-temperature and pressure relief valve to prevent the possibility of an explosion occurring. By controlling the pressure of water, pressure valves control its boiling temperature. Usually, a temperature and pressure relief valve controls the boiling temperature by discharging a quantity of water. However, if an expansion control valve is fitted in the cold water line to the water heater, it may discharge a small amount of heated water instead of the temperature and pressure relief valve. The benefit is that the energy is conserved, as discharged water from the expansion control valve is cooler.

When replacing elements care needs to be taken with respect to the thermostat setting due to the possibility of scalding. It is the responsibility of the installer to set and record the temperature at which the water heater delivers hot water. The installer should have the owner of the system countersign the record.

Tariffs

With respect to hot water tariffs, the purchaser of the energy can select any tariff in the residential, commercial and industrial category range which is appropriate to their requirements and user class. There are in general three types of tariffs: on-demand, economy controlled and off-peak controlled.

On-demand tariffs relate to supply that is available for 24 hours a day and is, therefore, the most expensive. This tariff applies to instantaneous-type hot water systems.

Economy tariffs relate to a supply that is accessed up to 18 hours a day. The hot water system that economy controlled is suitable for includes storage water heaters and solar electric hot water heaters.

The off-peak controlled tariff is suitable for hot water systems (over 125 L) such as the larger type (415 L) storage hot water heaters and the booster element of solar hot water systems. The energy delivered for off-peak heating is over an eight-hour period usually between 10.00 pm and 7.00 am. The heating elements of hot water systems connected for this tariff must be able to heat the water to its specified temperature during this period. All tariffs are billed in units. One unit equals one kilowatt hour (kWh).

REVIEW QUESTIONS

1. Name the two categories of water heating processes.
2. Describe the operation of an instantaneous water heater.
3. Which type of water heater stores and heats a large volume of water in an insulated (injected polyurethane) copper tank at a thermostatically controlled temperature?
4. What type of storage water heater uses the displacement principle of operation?
5. Name the valve that prevents cold water entering the storage tank until the water in the tank has reached a predetermined temperature.
6. Name the two types of solar collectors that are available.
7. State the Thermosyphon principle with respect to solar water heaters.
8. What is a calorifier?
9. Describe the operation of a heat pump.
10. What safety precautions need to be taken when heating elements in a hot water system have been replaced?

9.4 Faults in heating equipment and controls

Some faults that develop in heating devices result from loose termination screws which can generate heat and high temperatures in a small space (less than 2 mm). The free termination becomes a little heating unit itself. Insulation degradation with shorting between conductors or to earth is a possible result. Ineffective and loose connections can be diagnosed by measuring the voltage at the contact point. If a fault exists a small voltage drop is noted.

Mechanically scraped and nicked conductors can reduce the current-carrying effectiveness of a conductor, resulting in localised heating and eventual open-circuiting of the conductor. Arc tracking can occur on insulating surfaces due to salt residue left after condensation has evaporated. This type of fault usually results in an earth condition, especially if the salt on the insulation is between an active conductor and an earth medium. In this situation, heating of the salt occurs which gradually breaks down the conductor's insulation. Heating elements can operate as designed but without adequate control. This means that the heating unit does not turn off once the desired temperature has been reached. This fault condition is usually the result of a malfunctioning thermostat, generally as a result of corrosion from build-up of salts. To set thermostats, just use a screwdriver to move the pointers until they point directly to the desired temperature. Dual-element water heaters have two thermostats, so make sure that both thermostats are set to the same temperature. Other electrical faults that can occur with heating elements include internal element shorts, open circuits, and earth faults.

Wiring requirements of heating equipment

AS/NZS 3000: 2018 *Wiring Rules* must be consulted when installing or when making repairs to heating wiring systems and equipment. Areas of importance include electric duct heaters, heating cable systems, appliances producing hot water or steam, cooking appliances, under-carpet wiring systems and mains switches. Tables C1 and C2 of the *Wiring Rules* should also be consulted with respect to maximum demand requirements.

Fault-finding in water heaters

Water heater elements are rated by their wattage and are sized to specific tanks. **Table 9.1** shows 240 V nominal water heater elements rated by their wattage and sized to specific tanks.

TABLE 9.1 Water heater elements ratings

Heating element	1.8 kW	2.4 kW	3.6 kW	4.8 kW
Service per element	7.5 A	10 A	15 A	20 A

The typical water capacity sizes for storage hot water systems are 25 L, 50 L, 80 L, 125 L, 160 L, 250 L, 315 L and 400 L. The standard wattage ratings of heating units specified in AS/NZS 4692.1:2005 *Electric water heaters – Energy consumption, performance and general requirements* are 1200 W, 1500 W, 1800 W, 2000 W, 2400 W, 3000 W, 3600 W, 4800 W and 6000 W.

To change the water heater element you first need to drain the water out of the tank. After switching off the power, close the water inlet valve, which is normally located above the water heater on the cold water side. If the replacement element is not tightly fitted, it develops a slow leak around it. A terminal block, thermostat and heating element of a single-phase 230V, 3600 W, 400 L mains-pressure water heater is illustrated in **Figure 9.40**.

An isolating switch for the water heater must be installed at the switchboard. A 25-mm flexible or corrugated conduit is required to protect the final sub-circuit conductors to the water heater. The conduit should be connected to the unit with a 25-mm terminator.

FIGURE 9.40 Single-phase mains-pressure water heater wiring connections

The sub-circuit feed is connected directly to the terminal block and earth tab connection. There should be no excess cable tails inside the connection space. To determine if an element is defective, its resistance value needs to be measured. The resistance can be determined from power and voltage ratings by applying the equation:

$$R = \frac{V^2}{P}$$

where R = resistance in ohms

V = supply voltage in volts

P = power in watts

EXAMPLE 9.1 

1 Determine the resistance of a heating element rated at 1.8 kW at a nominal supply voltage of 240 V.

$$R = \frac{V^2}{P}$$

$$= \frac{240^2}{(1.8 \times 10^3)}$$

$$= \mathbf{32\ \Omega}$$

2 Determine the current this heating element would draw from a 230 V supply.

$$I = \frac{V}{R}$$

$$= \frac{230}{32}$$

$$= \mathbf{7.19\ A}$$

EXERCISE 9.1

a Determine the resistance of a heating element rated at 2.4 kW at a nominal supply voltage of 240 V.

b Determine the current this heating element would draw from a 230 V supply.

If the resistance measurement on the suspected element is proven to be sound then a check of the voltage is required. A drop of 10% in voltage reduces the wattage by 20%.

Three-phase water heater elements use a three-phase 400 V a.c. supply, one phase per element for three element water heaters or one phase per two elements for six-element water heaters. Three-phase elements are connected in 'star' configuration as shown in **Figure 9.41**.

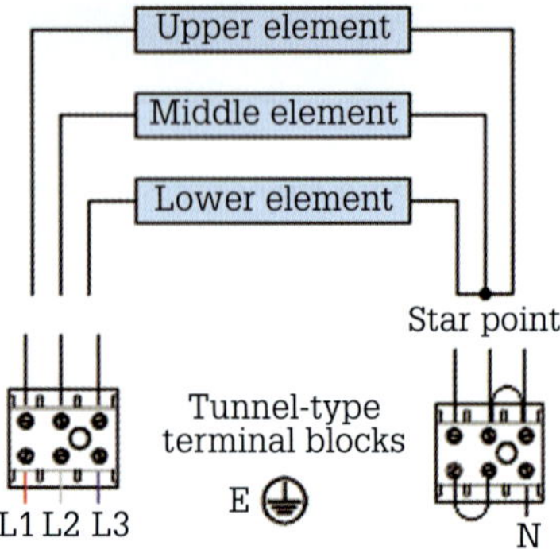

FIGURE 9.41 Three-phase heating elements connected in 'star' configuration

EXAMPLE 9.2

A three-phase 400 V instantaneous water heater is rated at 19.4 kW. Determine the element resistance per phase.

Solution:

As each element has the same resistance, the power dissipated by each element is a third of the total and as they are star-connected, the voltage across each element is 230 V. This yields the equation:

$$R = \frac{3 \times V^2}{P}$$

$$= \frac{3 \times 230^2}{(19.4 \times 10^3)}$$

$$= \mathbf{8.18\ \Omega}$$

EXERCISE 9.2 

A three-phase 400 V instantaneous water heater is rated at 15 kW. Determine the element resistance per phase.

REVIEW QUESTIONS

1 How is it possible to diagnose ineffective or loose conductors at heating element terminations?

2 What is the result of mechanically scraped and nicked conductors supplying a heating element?

3 How are water heater thermostats set?

4 What is a necessary first step in changing a water heater element after isolating the unit?

5 Determine the resistance of a heating element rated at 4.8 kW at a nominal supply voltage of 240 V.

9.5 Heating and heat energy

Heat is the energy transferred by virtue of a difference in temperature between a hot body (physicists use the term 'systems') and a cooler body. Heat can only be identified at the moment it crosses from one body to another. It is a momentary occurrence.

When a hot body is in thermal communication with a cooler body, heat energy is transferred until both bodies become stable or are in equilibrium at the same temperature.

Heat as energy transfer

A refrigerator is an excellent example of heat transference. A refrigerator uses a liquid refrigerant that evaporates readily into gas. Rapid evaporation of the refrigerant in the evaporator (freezer) causes sufficient cooling thereby enabling the gas to remove large amounts of heat. The heat energy from the food is transferred to the cold circulating gas. A compressor is used to suck the gas from the evaporator and recycle the refrigerant through the system. The gas is compressed and discharged into the condenser (black tubes on the back of the refrigerator).

The condenser dissipates the heat energy gained by the gas in the evaporator to the surrounding air and allows the gas to return to liquid form. The liquid is then returned via a capillary tube back to the evaporator thereby completing the cycle. The process of heat as energy transfer is illustrated in **Figure 9.42**.

In SI units the unit of heat is the same unit that is used for other mechanical and electrical forms of energy, the joule (J).

The scientific field of study of the properties of bodies (systems) that have a temperature and of the flow of heat energy from one body to another is called thermodynamics. From this study, three physical laws have been established.

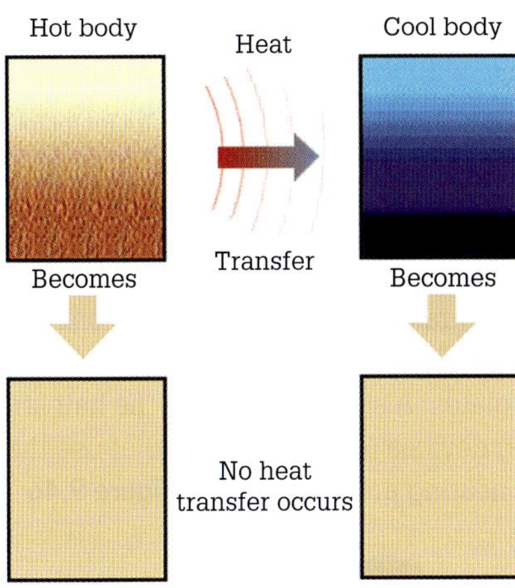

FIGURE 9.42 Heat as energy transfer

The first law of thermodynamics is often called the Law of Conservation of Energy. This law states that the energy can be changed from one form to another, but it cannot be created or destroyed. The total amount of energy and matter in the universe always remains constant.

The second law of thermodynamics states that heat can never pass on impulse from a colder to a hotter body. As a result of this fact, natural processes that involve heat energy transfer must have one irreversible direction. This law also predicts that the heat energy spontaneously disperses if not hindered. A term given to the measurement of the dispersal (before and after effect) is called entropy.

The third law of thermodynamics states that if the thermal motion (kinetic energy) of molecules could be stilled, a state called absolute zero would occur. Absolute zero is a thermodynamic temperature of 0 Kelvin or −273.15 °C.

A theory that aids the understanding of heat as energy transfer is molecular dynamics or kinetic theory. According to this theory, the higher the temperature to which the molecule is subject, the greater is its average kinetic energy. The resulting heat energy is carried by the molecules in the form of vibration or motion of the molecules as they move with different velocities and collide with other molecules.

Therefore, according to this theory an abundance of molecular vibration of a body will feel hot to the touch while a body with no molecular vibration feels extremely cold. An illustration of the heat energy of a vibrating molecule moving from a warm environment to a cooler environment is shown in **Figure 9.43**.

FIGURE 9.43 Heat energy transfer by molecular vibration

Modes of heat transfer

There are three modes of heat transfer:

1 conduction
2 convection
3 radiation.

Conduction

Conduction is the transfer of thermal energy between two solid materials in direct contact. The rate of the transference depends upon the temperature difference between the two solid materials, the thermal conductivity of each material and the distance through which the heat is dissipated.

The thermal conductivity of metals is relatively high and those metals that are the best electrical conductors are also the best thermal conductors. Because metals possess free electrons, their movement throughout the atomic lattice of the metal also conducts heat through the metal. Their rapid movement causes them to travel around more quickly and strike other free electrons which gain more kinetic energy. This effect continues until the two bodies attain the same temperature.

Kinetic energy is, therefore, transferred to the electrons and throughout the metal from the point closest to the heat source towards a point some distance away. Free electrons travel very short distances but because they are vibrating, rapid conduction of heat takes place very quickly. Conduction of thermal energy caused by excessive current flow in a copper busbar is illustrated in **Figure 9.44**.

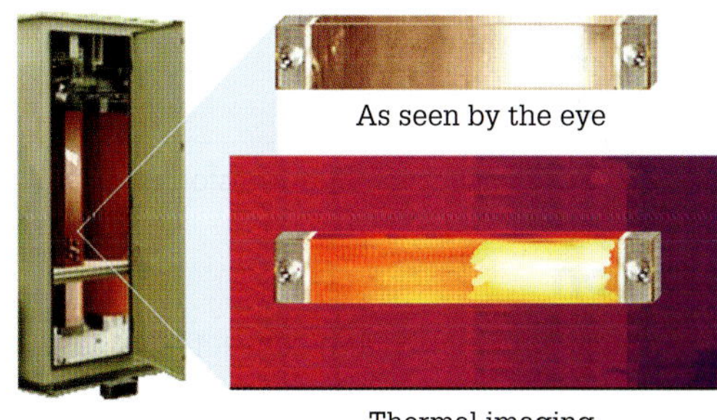

FIGURE 9.44 Heat conduction in a busbar

A copper soldering iron tip is an example of superb heat conduction. A wire-wound heating element heats the copper tip. The copper tip quickly passes on the heat to the solder and the metals to be joined.

Liquids and gases do not have free electrons and the atoms are not as densely packed as in metals. Consequently, heat transference by means of conduction is very limited.

Convection

Convection is based on the principle that a hot fluid (liquids and gases) is less dense than a cold fluid. When a cold fluid is heated, the cool molecules gain kinetic energy and move away from each other. The region surrounding the heated molecules becomes less dense than the area surrounding the other molecules and therefore it rises, displacing the cooler molecules by direct collision. The cool molecules, being denser, fall and move towards the heat source to replace the warmer molecules. This action causes convection currents throughout the fluid.

High-pressure sodium and mercury vapour lamps use convection to dissipate the heat from the ionisation process to the wall of the glass envelope where the heat energy can be transferred to the surrounding atmosphere.

Some electric heaters use the principle of convection to dissipate the heat energy of the elements to the surrounding atmosphere and thereby increase the temperature of an enclosed area.

Oil-filled distribution transformers use the principle of convection to transfer heat via the medium of oil from the windings to the transformer metal case. In a transformer, the heat must pass through the conductors and core iron to the insulation, oil, the steel casing and the paint coating. The heat energy gleaned from the conductors is then transferred to the surrounding atmosphere and dissipated via convection currents. This process is illustrated in **Figure 9.45**.

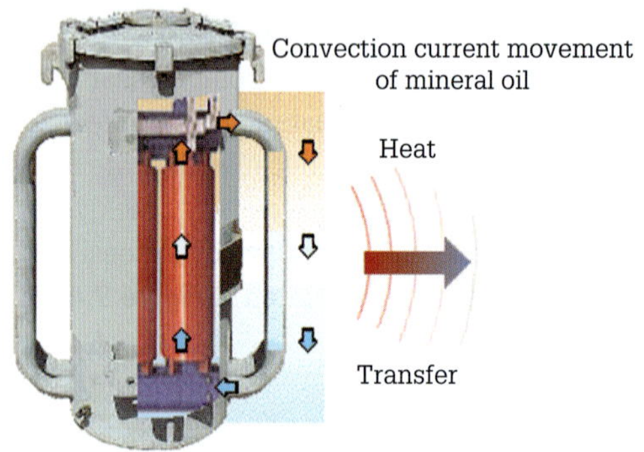

FIGURE 9.45 Heat convection in a transformer

Radiation

Radiation, unlike conduction and convection, does not need a transporting medium such as a molecule to exist. It is an electromagnetic wave like light but with a longer wavelength. It is transmitted at the speed of light through any transparent medium including a vacuum. It can be emitted by the source surface, reflected off a smooth surface or absorbed by the surface. The generated radiant energy is converted into heat energy when bodies in its path absorb it. The heated body can then transfer the heat energy acquired to the surrounding atmosphere (e.g. convection).

Any hot body emits electromagnetic radiation of various wavelengths usually in the infrared region of the electromagnetic spectrum. White heat is seen as illuminance in the visible area of the electromagnetic spectrum.

Radiation heating is used in some oven-type soldering processes. When solder is used in integrated circuits and their connection to packages or circuit boards, the soldered joint needs to be smooth and stable to be reliable. Some convection-type ovens that are used have an unstable atmosphere that heats the flux and solder at differing rates, causing unreliability with respect to the soldering process (heating and cooling occurs from the top down).

Radiation heating using infrared lamp technology is an equilibrium process that provides uniform heating for the soldering process. The infrared rays pass through their emitter tube and only heat the objects at which they are directed and not the surrounding air.

Some hair straighteners use the radiation heating process to provide even heat distribution across ceramic or aluminium plates. A slow slide across the hair straightens it. Two heating devices (infrared lamps) that use the radiation principle of heating are illustrated in **Figure 9.46**.

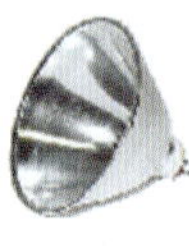

FIGURE 9.46 Infrared heating devices

Radiation is also used for infrared brazing that uses a high-intensity quartz lamp as a heat source. The process is suited to the brazing of very thin materials. The metals to be brazed are supported in a position that enables radiant energy to be focused on the surfaces to be joined.

Temperature

Temperature is a property of a substance and is not a holistic measurement of a substance's internal energy. It defines the direction of heat flow, which is always away from the body that is at a higher temperature and to the body that is at the lower temperature. Temperature is the hotness or coldness of a substance and is the measure of the average kinetic energy of the molecules of a substance. One instrument calibrated to measure temperature is a thermometer.

When the thermometer is put in thermal contact with a substance and reaches thermal equilibrium, then the thermometer displays a quantitative measure of the temperature of the substance. A calibrated and a digital thermometer are shown in **Figure 9.47**.

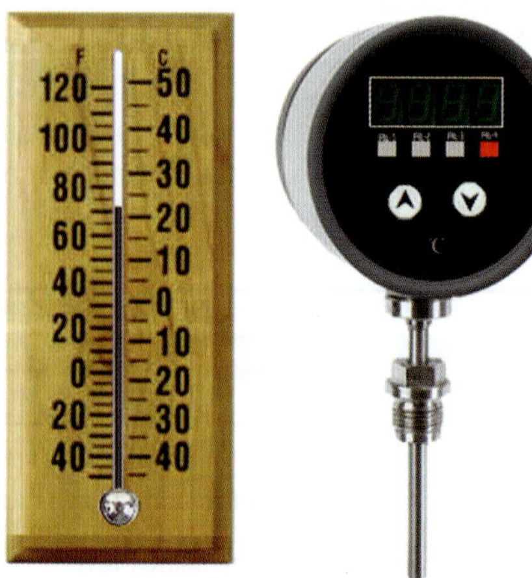

FIGURE 9.47 Thermometers

Calibrated thermometers use an expanding substance such as mercury that enlarges as it gets warmer. Its expansion proportion is linear and can be precisely calibrated. The mercury is allowed to expand into a stem and its expansion is calibrated on the timber plate.

Digital thermometers are temperature-sensing instruments that have a permanent probe and a digital display. They are typically battery powered. The temperature-sensing elements available for digital thermometers include thermocouples, resistance temperature detectors (RTDs) or thermistors. They are easier to read than glass thermometers.

There are also infrared digital thermometers that sense temperature from a distance. The radiation of infrared energy from the surface is associated to the temperature of the body. The infrared thermometer reads the intensity of the infrared energy and converts this reading electronically into an equivalent surface temperature. This inferred temperature is then communicated by a liquid crystal display (LCD).

Temperature scale

The fundamental unit of temperature measure in the Système International d'Unités (SI) has been redefined in terms of the Boltzmann constant, which relates the amount of thermodynamic energy in a substance to its temperature. The Kelvin, which has the symbol K, is the SI unit of thermodynamic temperature. In the Kelvin scale the freezing point of water is 273.15 K, and the boiling point is 373.15 K.

The Kelvin was previously defined as the fraction 1/273.16 of the thermodynamic temperature of the triple point of water. The triple point of water is the distinctive temperature point at which the three phases of water (ice, water and vapour) coexist in equilibrium. It is fractionally higher than the freezing point, being 273.15 K. An illustration of the three states of water is shown in **Figure 9.48**.

Source: Shutterstock.com/CAN BALCIOGLU; Shutterstock.com/Valentyn Volkov; Shutterstock.com/Dziegler

FIGURE 9.48 Three phases of water– vapour, solid, liquid

The triple point of water is the most important defining thermometric fixed point used for the calibration of thermometers that accurately obey known laws. The use of this specified point means that temperature standards around the world can be accurately equivalent. The temperature scale is termed the absolute, the thermodynamic or the Kelvin scale. The Kelvin scale is extensively used by physicists and engineers to define and apply fundamental laws of thermodynamics.

Most temperature scales today are communicated in degrees Celsius (°C), although Fahrenheit (°F) is still in use in some countries. Celsius is an SI-derived unit of temperature and is the scale commonly used. In fact, the boiling point of water at standard atmospheric pressure (99.975 °C) and the temperature at which water freezes and melts (0 °C) classify the Celsius scale. Absolute zero on the Celsius scale is –273.15 °C. Every temperature on the Celsius scale corresponds to a particular temperature on the Kelvin scale. The relation between these scales is:

$$K = °C + 273.15 \text{ and } °C = K - 273.15$$

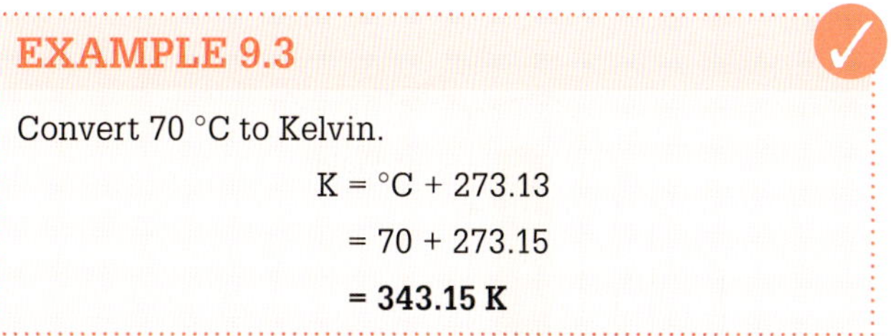

EXAMPLE 9.3

Convert 70 °C to Kelvin.

$$\begin{aligned} K &= °C + 273.13 \\ &= 70 + 273.15 \\ &= \mathbf{343.15\ K} \end{aligned}$$

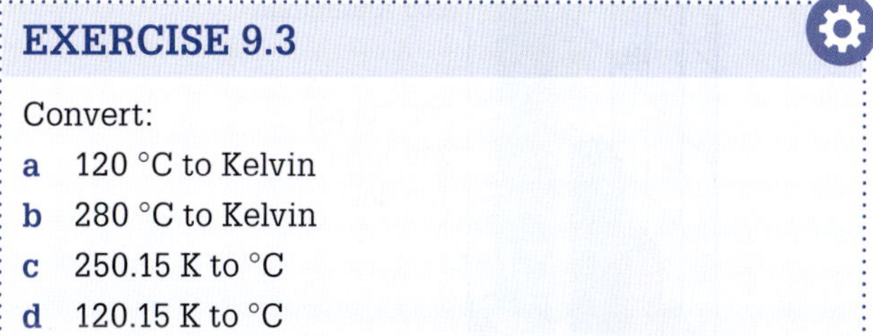

EXERCISE 9.3

Convert:

a 120 °C to Kelvin

b 280 °C to Kelvin

c 250.15 K to °C

d 120.15 K to °C

Temperature can also be distinguished by the colour of the radiation emitted by a hot body. The colour scale of temperature is shown in **Figure 9.49**.

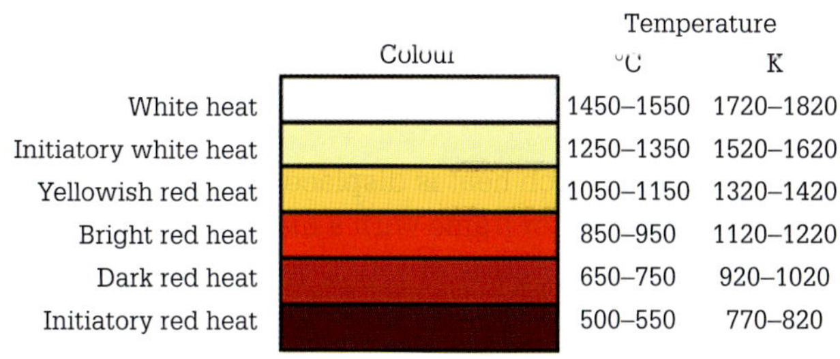

	Colour	Temperature °C	Temperature K
White heat		1450–1550	1720–1820
Initiatory white heat		1250–1350	1520–1620
Yellowish red heat		1050–1150	1320–1420
Bright red heat		850–950	1120–1220
Dark red heat		650–750	920–1020
Initiatory red heat		500–550	770–820

FIGURE 9.49 Colour scale of temperature

Specific heat capacity

When the heat energy absorbed by the body increases, the temperature of the body increases. The specific heat capacity of the body is the amount of heat required per kilogram capacity of the body to raise the temperature by one degree.

In SI units specific heat capacity (symbol *c*) is measured in J/kg•K. The heat gained or lost by a given mass as its temperature changes depends on the relationship between the mass, the change in temperature and the specific heat capacity of the body. The relationship is expressed in the equation:

$$Q = mc\Delta T$$

where Q = quantity of heat in joules

m = mass in kilograms

c = specific heat capacity in J/kg•K

ΔT = change in temperature in Kelvin (note: Since 1 °C = 1 K, temperature changes can be measured in either K or °C)

The specific heat capacity of common substances at constant pressure has been determined and some values are given in **Figure 9.50**.

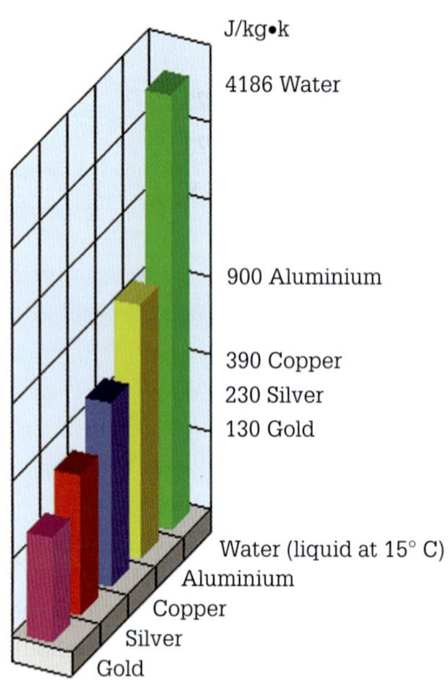

FIGURE 9.50 Specific heat graph

EXAMPLE 9.4

Calculate how much heat is displaced when an aluminium motor frame with a mass of 12 kg cools from a temperature of 110 °C to a temperature of 25 °C.

$$Q = mc\Delta T$$
$$= 12 \times 900 \times (110 - 25)$$
$$= \mathbf{918\ kJ}$$

EXERCISE 9.4

a. A copper busbar with a mass of 5 kg cools from a temperature of 75 °C to a temperature of 22 °C. Calculate the quantity of heat that is lost.

b. If a 0.02 kg silver contact contains 225 J of heat energy after cooling to a temperature of 62 °C, what was the original temperature of the silver contact?

Thermal conductivity

Thermal conductivity is the flow of internal energy from the region of higher temperature to one of lower temperature by the interaction of the adjacent particles, atoms, molecules, ions or electrons in the intervening space. There are four factors that affect the rate at which thermal conductivity occurs.

1. temperature
2. length
3. cross-sectional area (CSA)
4. material.

The rate of heat transfer per unit of time is measured in watts (joules/second) and is denoted by:

$$\frac{\Delta Q}{\Delta t}$$

The rate of heat flow in watts can be determined by applying the following formula:

$$\frac{\Delta Q}{\Delta t} = kA\frac{t_1 - t_2}{l}$$

where k = proportionality constant called the thermal conductivity (J/s.m.°C) which is a characteristic of the substance

l = the distance between the two ends of the object that are at temperatures t_1 and t_2. Where t_1 and t_2 are the initial and final temperatures respectively.

A = the cross-sectional area of the substance

Electrical conductivities (Siemens/m) and thermal conductivities (k) vary for different substances, being greatest for metallic solids, lower for non-metallic solids, very low for liquids and extremely low for gases. Substances with a high conductivity are said to be good conductors. In metallic solids, thermal conductivity almost mirrors electrical conductivity. In metallic solids freely moving electrons transfer not only an electric charge but also heat energy. However, there is a significant difference between them in the amount of energy transfer.

Calculations concerning thermal conductivity can play a role in determining the effectiveness of reverse-cycle air-conditioning especially in room spaces that have a number of glass windows. For example, in **Figure 9.51**, using the above formula, the heat loss from a chamber space that has a glass wall 4.0 m × 2.0 m in area and that is 25 mm thick and has an inner room temperature of 24 °C and an outer temperature of 12 °C is shown.

k of glass = 0.84 J/s.m.°C

$$\frac{\Delta Q}{\Delta t} = kA\frac{T_1 - T_2}{l}$$
$$= 0.84 \times 8 \times \frac{12}{0.025}$$
$$= 3225.6\ \text{J/s}$$
$$= 3225.6\ \text{watts}$$

FIGURE 9.51 Thermal conductivity

This result clearly shows that there is a significant rate of heat loss from the room to the outside environment. Two sheets of glass would substantially diminish the heat loss, thereby reducing the energy drawn over time by the reverse-cycle air-conditioner. Materials that are used to insulate the house, such as building blankets and batts, foils and polystyrene panels, are specified by their *R*-value (thermal resistance). The *R*-value is defined for a given thickness of material as:

$$R = \frac{l}{k}$$

The *R*-value combines the thickness (l) and the thermal conductivity (k) in one number. The higher the *R*-value number, the better the thermal resistance properties of the insulating material. An illustration of *R*8 wall insulation is shown in **Figure 9.52**.

FIGURE 9.52 *R*8 thermal building material

REVIEW QUESTIONS

1 Define the term 'heat'.
2 State the SI unit of measurement of heat.
3 What theory aids the understanding of heat as energy transfer?
4 Name the three modes of heat transfer.
5 Give one practical example of a piece of equipment that uses the principle of conduction in relation to heat developed.
6 Which of the modes of heat transfer is an electromagnetic wave?
7 Why is mercury used as the heat sensitive material in calibrated thermometers?
8 Name the fundamental unit of temperature measurement in the SI system of measurement.
9 What name is given to the distinctive temperature point at which the three phases of water (ice, water and vapour) coexist in equilibrium?
10 Convert 600 K to °C.
11 What is the unit of measure for 'specific heat capacity'?
12 The aluminium fins of an electric motor frame with a mass of 6 kg cool from their operating temperature of 90 °C to the surrounding ambient temperature of 27.5 °C. Calculate the quantity of heat that has dissipated.

9.6 Safe working practices

Fault-finding is a common form of problem solving. Electricians diagnose faulty systems and take direct, corrective action to eliminate any faults in order to return the systems to their normal states.

SWITCH ON

Before undertaking any testing or fault-finding activities on heating equipment and controls it is essential to undertake and document a risk assessment.

Risk assessment

Prior to engaging in fault-finding on heating equipment and controls, it is necessary to undertake a risk assessment. This assessment should:

1 Check condition of PPE such as fire-rated safety clothing, safety footwear and safety glasses.
2 Check/secure area adjacent to work.
3 Check whether an assistant is needed.
4 Consider risks and hazards in work area (see **Table 9.2**). Are relevant permits in place?
5 Be familiar with the replaced component before carrying out any work.
6 Check for availability of diagrams for the heating equipment and controls.

On completion of a job:

1 Tidy up, remove and dispose of redundant materials.
2 Sign off any relevant permits.
3 Sign off job on completion.

Safety precautions when working on heating equipment and controls

When working on or testing heating equipment and controls, it is important to be careful. Whatever type of heating equipment and controls you are testing, whether it be simple or complex, it is paramount to exercise safety.

TABLE 9.2 Risk assessment

Possible hazards	Possible harm to people	Control measures to prevent harm or reduce hazards
▪ Eye injury ▪ Electrocution ▪ Electric shock ▪ Explosion	▪ Burns or fatalities arising from contact with a live conductor. ▪ Eye injury to fatality arising from explosion.	▪ All electrical fault-finding is to be carried out by qualified tradespeople. ▪ Check faults in a planned and concise manner. ▪ Isolate the supply as soon as is practically possible. ▪ Start testing from the source of supply and work by a process of elimination. ▪ Use insulated tools and 'in-service' test equipment.

Working with electrical equipment involves risks that should never be taken lightly. There are a number of safety procedures to follow in order to avoid personal injury, the possible damage to equipment or the danger of fire.

General safety

Before working on any heating equipment and controls, consider following these basic safety precautions to help reduce any hazards.

- Disconnect and isolate the heating equipment and controls which you are testing or repairing from the power source. Never assume that this is the case. Test and test again with a suitable voltage presence indicator to confirm isolation.
- Remove fuses and replace them only after isolating power to the heating equipment and controls.
- Do not reconnect power to heating equipment and controls until work is complete and checked.
- Ensure that all electronic equipment is properly earthed.
- Replace damaged components rather than attempting inappropriate repairs. For example, replace damaged cables rather than repairing with insulating tape.
- Use the correct repair and maintenance tools.
- Reinstate covers after removing them to reduce the risk of electric shock.
- Have safety equipment, such as a fire extinguisher, basic first aid kit and a mobile phone nearby.

Personal safety

It is important to ensure that you are safe when working on heating equipment and controls. Some personal safety precautions to keep in mind are:

- Ensure that the work area is clean, dry, tidy and well ventilated.
- Do not wear loose clothing when working.
- Remove any metallic jewellery such as watches, rings and bracelets from your body.
- Do not handle hot components with bare hands.
- Wear non-conductive footwear when working on power supplies.
- Safely discharge capacitors by using an appropriate load – do not short the terminals.
- Use test equipment rated to Cat III as a minimum, this includes test leads.
- Wear eye protection.

Electric shock

One of the major hazards when working with heating equipment and controls is electric shock. To minimise the risk, you should follow a few safety precautions, including:

- Read any safety procedures that came with the equipment you are about to test or repair.
- Check all wires for solid connections.
- Ensure that all parts of the power supply are securely fixed to prevent accidents.
- Keep water and other liquids away from the heating equipment and controls under repair or test.
- Check for signs of wear, defects and fraying on cables, cords and connectors.
- Wear insulating safety rubber gloves and footwear when testing and repairing heating equipment and controls.

Search for 'electrical safety' at https://www.safeworkaustralia.gov.au for relevant information.

Obtaining information about heating equipment and controls

Obtaining heating equipment and controls is usually a straightforward task. You can nearly always pick up what you want in a single visit to an electrical wholesaler. However, if the customer wants a particular product that has a low turnover rate, the wholesaler may not have it sitting on the shelf and will need to order it directly from the manufacturer for you.

Electrical appliances can, however, present a selection problem. It does happen that you find a particular product is no longer available and a direct replacement does not exist. In this situation you need to draw on your acquired knowledge and experience to select a suitable replacement.

This information is available from appliance nameplates or by searching the manufacturer's website using the product code or model number.

Let us consider the following two space heaters.

EXAMPLE 9.5

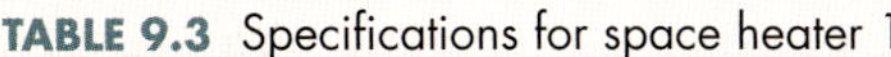

TABLE 9.3 Specifications for space heater 1

Model Name	Space Heater 2400
Model Number	TSH2400
Material	Anodised Alloy
Colour	Black
Cord Length (cm)	80
Heating Area (m^2)	4.5
Corded (Yes/No)	Yes
Indoor Use (Yes/No)	Yes
Outdoor Use (Yes/No)	Yes
Voltage (volts)	240
Wattage (watts)	2400
Amperage (A)	10
Dimensions (mm)	W: 170 H: 50 L: 1350

TABLE 9.4 Specifications for space heater 2

Model Name	Space Heater 3200
Model Number	TSH3200
Material	Anodised Alloy
Colour	Black
Cord Length (cm)	80
Heating Area (m^2)	6
Corded (Yes/No)	Yes
Indoor Use (Yes/No)	Yes
Outdoor Use (Yes/No)	Yes
Voltage (volts)	240
Wattage (watts)	3200
Amperage (A)	13.3
Dimensions (mm)	W: 170 H: 50 L: 1650

In comparing the specifications of the two, you should notice that Space Heater 2 is not an equivalent replacement for Space Heater 1. There are two factors making it unsuitable:

1. Space Heater 2 is 300 mm longer than Heater 1 so it may not fit in the allocated space.
2. Space Heater 2 is rated at 800 watts (or 3.3 Amperes) more than Heater 1 so it may overload the circuit.

Exercise 9.5 will give you practice in identifying suitable replacement appliances.

EXERCISE 9.5

Appliance 1

Electric Storage

250L Electric Unit - Single Element

Number Of Elements:	Single Element
Tank Material:	Vitreous Enamel
Element Size:	3.6 kw
Hot Water Delivery:	250 ltr
System Connections:	Dual Handed
Water Connections:	Rp3/4" (20mm)
Height:	1395 mm
Diameter:	640 mm

Appliance 2

Electric Storage

315L Electric Unit - Single Element

Number Of Elements:	Single Element
Tank Material:	Vitreous Enamel
Element Size:	4.8 kw
Hot Water Delivery:	315 ltr
System Connections:	Dual Handed
Water Connections:	Rp3/4" (20mm)
Height:	1640 mm
Diameter:	640 mm

Is Appliance 2 a suitable replacement for Appliance 1? Give a reason for your answer.

REVIEW QUESTIONS

1. What should be undertaken prior to engaging in fault-finding on heating equipment and controls?
2. Name four possible hazards associated with working on heating equipment and controls.
3. What should be removed from the body prior to repairing heating equipment and controls?
4. What should be reinstated on heating equipment and controls being serviced to reduce the risk of electric shock?
5. When repairing heating equipment and controls, in what condition should the work area be?

CHAPTER REVIEW

9.1 Electrical heating control devices

- Control of heating can be achieved via manual, automatic or programmable means.
- There are two types of automatic temperature control: thermostatic and simmerstat.
- The bimetal temperature control is a small-current-operated temperature-sensitive control.
- The simmerstat is a bimetal switch in which the bimetal sensing element is activated by the current flow through its resistance wire element.
- There are five essential operating elements in relation to the programmable thermostat: sensor, input, comparator, output and load.

9.2 Fixed electrical heating appliances

- There are three types of heating cables: mineral insulated, PTFE and self-regulating.
- Heat plays an indispensable role in a broad range of manufacturing processes: cooking, softening, melting, drying, curing and fusing.
- Process heating methods used for industrial processes are resistance, induction, infrared, dielectric, arc, ultraviolet and electron-beam heating.
- Electric cooking appliances fall into one of three categories – resistance, induction and electric microwave cooking.

9.3 Electrical water heater operation

- Instantaneous or tankless water heaters are small cabinets that heat water on demand or instantly as it passes through the heater.
- Mains-pressure storage heaters are generally attached to an off-peak electricity meter.
- Low-pressure water heaters use the displacement principle of operation.
- Water heating using solar energy occurs during the day and the solar involvement varies significantly throughout the year depending on the climatic conditions.
- Calorifiers are cylinders with an internal coil, which allows the use of any boiler for hot water production.
- A heat pump water heater absorbs heat from the surrounding environment and pumps the acquired heat energy into a hot water storage tank.
- The three types of tariffs are on-demand, economy controlled and off-peak.

9.4 Faults in heating equipment and controls

- Some faults that develop in heating devices result from loose termination screws which can generate heat and high temperatures in a small space.
- Mechanically scraped and nicked conductors can reduce the current-carrying effectiveness of a conductor, resulting in localised heating and eventual open-circuiting of the conductor.
- AS/NZS 3000: 2018 *Wiring Rules* must be consulted when installing or when making repairs to heating wiring systems and equipment.
- Water heater elements are rated by their wattage and are sized to specific tanks.

9.5 Heating and heat energy

- Heat is the energy that is transferred by virtue of the difference in temperature of a hot body with respect to a cooler body.
- Energy can be changed from one form to another, but it cannot be created or destroyed.
- There are three modes of heat transfer: conduction, convection and radiation.
- An instrument calibrated to measure temperature is a thermometer.
- The triple point of water is the most important defining thermometric fixed point used for the calibration of thermometers that accurately obey known laws.
- Four factors that affect the rate at which thermal conductivity occurs are temperature, length, CSA and material.
- Building materials that are used to insulate a house are specified by their *R*-value.

9.6 Safe working practices

- Prior to engaging in fault-finding on heating equipment and controls, it is necessary to undertake a risk assessment.
- Disconnect and isolate the heating equipment and controls which you are testing or repairing from the power source before commencing work.
- One of the major hazards when working with heating equipment and controls is electric shock.
- Obtaining information about heating equipment and controls can usually be done through contacting an electrical wholesaler.

TRIAL EXAM

For Chapter 9 knowledge assessment, please complete the following trial exam.

1 The three-heat switch provides:
   a manual control
   b automatic control
   c programmed control
   d stepped control

2 The control device used to maintain the temperature of a heating system within moderately accurate limits is the:
   a simmerstat
   b enerstat
   c thermostat
   d comparator

3 The heat controller that provides exceptional accuracy in temperature control is the:
   a thermostat
   b enerstat
   c strut-and-tube
   d programmable thermostat controller

4 The diameter and material type used in floor heating cables determines:
   a the heat output per metre
   b the maximum operating temperature
   c the minimum operating temperature
   d suitable applications

5 The most economical means of supplying process heating is:
   a induction heating
   b dielectric heating
   c resistance heating
   d electron-beam heating

6 What type of phenomenon can only be identified at the moment it crosses from one body to another?
   a heat
   b energy
   c gas
   d cold

7 The first law of thermodynamics is often called the:
   a joule law
   b Law of Conservation of Energy
   c physical law
   d kinetic energy law

8 The transfer of thermal energy between two solid materials in direct contact is:
   a radiation
   b convection
   c conduction
   d transference

9 What type of technology uses an equilibrium process to provide uniform heating?
   a radiation heating technology
   b convection-type technology
   c conduction-type technology
   d ultraviolet lamp technology

10 The fundamental unit of temperature measure in the Système International d'Unités (SI) has been defined as the:
   a Kelvin
   b Celsius
   c Absolute
   d Fahrenheit

11 The quantity of heat is calculated from:
   a mc^2
   b $mc\Delta T$
   c klA
   d Rlk

12 What does the *R*-value of thermal insulation indicate?
   a the conductivity for metallic solids
   b the specific heat capacity of a substance
   c the thermal resistance properties of an insulating material
   d the amount of heat energy between a hot body and a cooler body

13 An energy regulator device is called a:
   a three-heat switch
   b infrared switch
   c thermostat
   d simmerstat

14 Which of the following is the purpose of a sacrificial anode?
   a to prevent the corrosive effects of galvanic action
   b to activate an isolating switch for the water heater
   c to reduce the current-carrying effectiveness of a conductor
   d to gradually break down a conductor's insulation

15 Name the device that uses an internal coil which allows the use of any boiler for hot water production.
   a an evacuated-tube solar heater
   b calorifier
   c falling-level storage heater
   d low-pressure storage heater

10 Circuits for socket outlets

This chapter provides electrotechnology workers with essential knowledge and skills pertaining to undertaking electrical work and related activities on circuits supplying low voltage socket outlets. This chapter provides underpinning knowledge for the unit UEEEL0010 from the UEE training package.

LEARNING OBJECTIVES

Circuits for socket outlets

- Identify different types of socket outlets and their purpose.
- State the polarity requirements for single-phase and three-phase socket outlets.
- Determine appropriate cable size and number of socket outlets on a circuit.
- Outline appropriate installation methods for circuits supplying socket outlets.

Final sub-circuits and segregation

- Explain the purpose of a mixed circuit.
- Determine circuit loading for a mixed circuit.

Identifying faults in socket outlet circuits

- Outline testing and fault-finding procedures for circuits supplying socket outlets.

10.1 Circuits for socket outlets

A socket outlet is known colloquially as a power point and may be referred to by electricians as a general-purpose outlet or GPO. The *Wiring Rules* defines a socket outlet (Clause 1.4.110) as a device that has a set of contacts for making a separable connection with the contacts of a plug.

Types

Socket outlets may be single (**Figure 10.1**) or multiple combination outlets such as the double combination outlet shown in **Figure 10.2**.

FIGURE 10.1 Single 10 A single-phase socket outlet

FIGURE 10.2 Double 10 A single-phase socket outlet

Socket outlets are rated in terms of the maximum current they can supply without damage. Typical ratings are 10 A (**Figure 10.1**), 15 A (**Figure 10.3**), 20 A (**Figure 10.4**), 25 A (**Figure 10.5**), and 32 A (**Figure 10.6**). The 15 A rated socket outlet has a wider earth pin than that of the 10 A rated socket outlet. The 20 A rated socket outlet has wider power pins, to provide a greater surface area and hence higher current density, as well as the larger earth pin. A plug can be inserted in a socket of the same or higher rating but not into a socket having a lower rating. As an example, a 10 A rated flat three-pin plug fits in each of the socket outlets, but a 32 A rated flat three-pin plug can only be used with a 32 A socket outlet.

Dedicated circuits supply socket outlets rated above 10 A. **Table 10.1** summarises typical single-phase socket outlet function by rating.

TABLE 10.1 Single-phase socket outlet function by rating

Rating	Function
10 A	earthed domestic appliances and equipment
15 A	caravans, air conditioning, light industrial appliances, and equipment
20 A	plug-in ovens, cooktops, stoves, air conditioning and light industrial appliances and equipment
25 A	plug-in ovens, cooktops, stoves, air conditioning and light industrial appliances and equipment
32 A	plug-in stoves

FIGURE 10.3 Single 15 A single-phase socket outlet

FIGURE 10.4 Single 20 A single-phase socket outlet

FIGURE 10.5 Single 25 A single-phase socket outlet

FIGURE 10.6 Single 32 A single-phase socket outlets

Socket outlets may be configured for horizontal installation as shown in **Figures 10.1** to **10.6**, or vertical installation as shown in **Figure 10.7**.

FIGURE 10.7 Double 10 A single-phase socket outlet – vertical arrangement

Socket outlets are also available in weather-protected styles. **Figure 10.8** illustrates a weather-protected single 10 A single-phase switched socket outlet.

FIGURE 10.8 Single 10 A single-phase weather-protected socket outlet

Polarity

The *Wiring Rules* requires a specific polarisation for socket outlets that accommodate three flat-pin plugs. **Figure 10.9** illustrates the requirement of Clause 4.4.5.

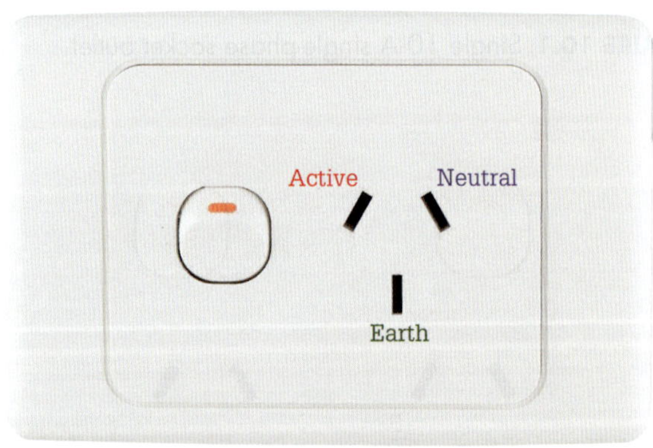

FIGURE 10.9 Polarity of single-phase socket outlet

Pendant-type socket outlet

A switch incorporated in a pendant-type 230 V socket outlet attached to a flexible cord should interrupt all live conductors except where the socket outlet is not dependent on the flexible cable for support (refer to AS/NZS 3000:2018 *Wiring Rules*).

Figure 10.10 shows a pendant-type 230 V socket outlet backed by a chain fixed to the plasterboard ceiling by a hook. The flexible cable is held in place by cable ties attached to the support chain.

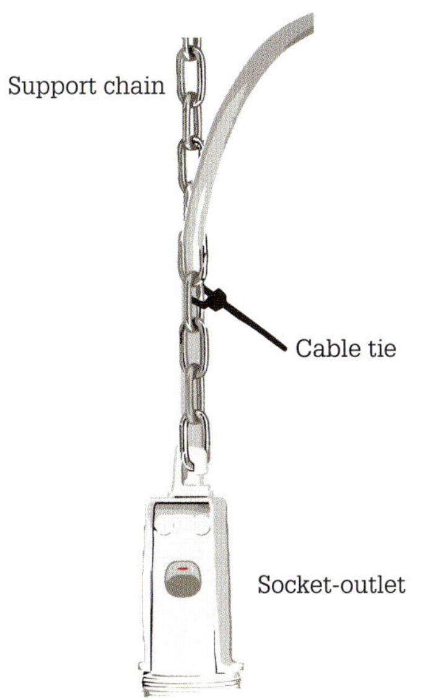

FIGURE 10.10 Pendant-type 230 V socket outlet

Three-phase outlets

Three-phase socket outlets and plugs, shown in **Figure 10.11**, are 10 A, 20 A or 32 A rated, four pin (three-phase and earth) or five pin (three-phase, neutral and earth) with the socket outlets fitted with a triple-pole control switch. **Note:** Australian plugs and flexible cable connected sockets use the colour orange.

FIGURE 10.11 Three-phase permanently wired socket and plug

Only in exceptional circumstances will equipment having a current demand more than 30 A be connected by a socket outlet. Such equipment normally requires a fixed connection. Where three-phase socket outlets are required, each case must be considered on its merits. Five-pin outlets are preferred.

Sockets rated at 10 A and 20 A have the same physical size all over but the 10 A has a keyed slot at the bottom and the 20 A has a flat section. This means that a plug rated at 10 A can fit into a 20 A socket but a 20 A plug cannot fit into a 10 A socket.

Phase sequence (polarity)

The order of the cyclic direction is established by using a three-phase sequence indicator. A three-phase sequence indicator shows the order of the phase sequence. Knowledge of the phase sequence prior to starting electrical motors and other equipment is important. The consequence is an incorrect connection that could cause damage to the equipment. **Figure 10.12** shows an illustration of a three-phase sequence indicator.

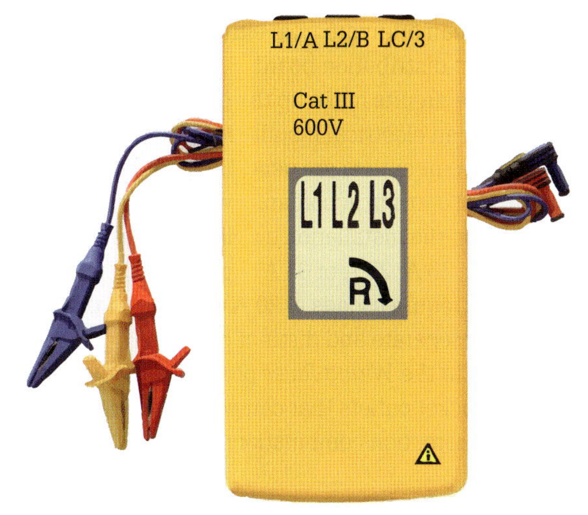

FIGURE 10.12 Phase indicator

Australia follows an anticlockwise phase sequence and the convention for this sequence is ABC or L1, L2, L3 connected left to right on the terminal block. **Figure 10.13** shows the polarity of a 5-pin three-phase socket.

A phase sequence indicator shows the sequence of phase rotation on its digital display. The presence of all three phases is the illumination of three letters and numbers with each phase clearly identified as L1, L2 and L3 in accordance with current practice. An open phase illuminates two phases only. Proper phase sequence displays L1, L2, and L3 while a reverse sequence illuminates L1, L3, and L2.

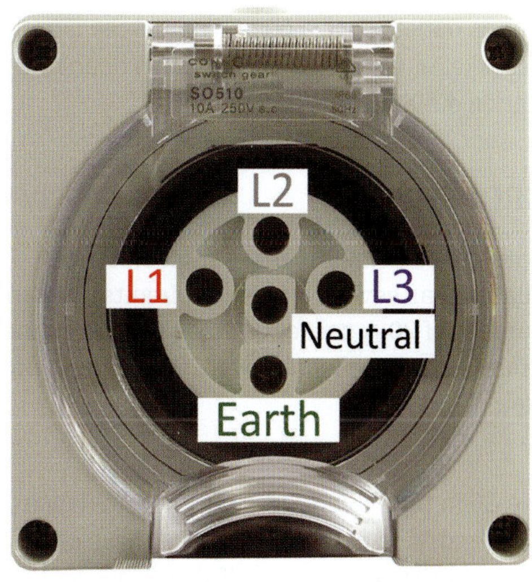

FIGURE 10.13 Polarity of three-phase socket outlet

Circuits for socket outlets

Every installation must be divided into circuits to avoid hazards and to contain the disruption in the event of a fault. In addition, the division of an installation into various circuits makes easier the safe operation of the installation, improves the quality of circuit inspection and aids reliability with testing and makes maintenance simpler. The electrical distribution within an installation ends with final sub-circuits (see AS/NZS 3000:2018). A final sub-circuit is that section of a wiring system that extends beyond the final circuit protection device. Final sub-circuits can originate from either a main switchboard or a sub-main distribution board.

Final sub-circuits make up the greatest portion of the wiring of an electrical installation. Final sub-circuits supplying circuit outlets are generally wired using 2.5 mm^2 TPS twin and earth in accordance with the *Wiring Rules*. Depending on the installation conditions, these may be rated at 16 A, 20 A, 25 A or 32 A. Appendix C5 in AS/NZS *Wiring Rules* requires loads of 20 A or more per phase to connect to a separate and distinct circuit.

Table C9 in the *Wiring Rules* provides guidance on the number of socket outlets that can connect to various cable and circuit protection combinations.

EXAMPLE 10.1

A single-phase domestic residential electrical installation has several final sub-circuits. Determine the number of 10 A socket outlets that can connect to a 2.5 mm^2 circuit protected by a 16 A Type C circuit breaker.

Solution:

From Table C9 a socket outlet is rated at 1 A for all domestic installations. As the circuit is rated at 16 A, it is possible to install 16 socket outlets on this circuit. This is calculated from:

$$Number\ of\ points = \frac{Circuit\ rating}{Rating\ per\ point}$$

Footnote 3 to Table C9 requires that multiple combination outlets have a point count of the number of integral socket outlets. This means that a double 10 A combination outlet counts as 2 points.

EXERCISE 10.1

A single-phase domestic residential electrical installation has several final sub-circuits. Determine the number of 20 A socket outlets that can connect-to a 2.5 mm^2 circuit protected by a 20 A Type C circuit breaker.

Table C6 in the *Wiring Rules* provides guidance on the current-carrying capacity of single-phase cables for common installation conditions. Table C7 in the *Wiring Rules* provides guidance on the current-carrying capacity of three-phase cables for common installation conditions.

EXAMPLE 10.2

Determine the current-carrying capacity of a single-phase circuit comprising two-core and earth, 2.5 mm^2 copper conductors with PVC insulated and sheathed cables, marked V-90 and installed such that they are partially surrounded with thermal insulation.

Solution:

From Table C6, 2.5 mm^2 copper conductors with V-90 PVC insulated and sheathed cables are rated at 20 A when installed partially surrounded by thermal insulation.

EXERCISE 10.2

Determine the current-carrying capacity of a three-phase circuit comprising four-core and earth, 2.5 mm^2 copper conductors with PVC insulated and sheathed cables, marked V-90 and installed such they are partially surrounded with thermal insulation.

Installing TPS cables

A TPS cable should be run in a manner that eliminates the possibility of strain on the cable itself or on the cable termination. TPS cables should not contact items likely to become hot, such as hot water pipes. Cable bending radius should not be less than the maker's recommendation and, in any case, should not be less than six times the overall cable diameter.

Where TPS cables are in accessible locations they must be supported by cable clips, cable ties or cable saddles, depending on the installation method. In accessible locations in a ceiling, shown in **Figure 10.14**, they should be clipped to the side of the joist (but not within 50 mm of the ceiling fabric to prevent penetration by a nail); where they are laid across joists they should be clipped to the side of, for example, a 50 mm × 30 mm batten which is fixed to the joists.

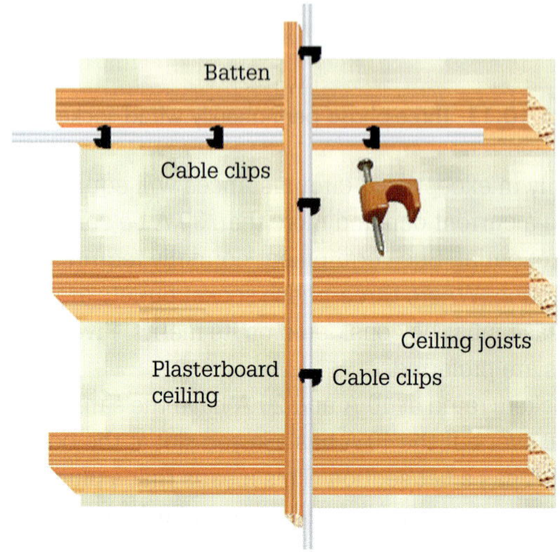

FIGURE 10.14 TPS wiring in an accessible ceiling

It is permissible to remove the sheath to allow single insulated wires in the enclosure but the outer sheathing of the TPS cable must enter some distance into the enclosure. Any exposed TPS cables should be suitably protected or shielded against weathering.

Figure 10.15 shows how TPS cable can supply a socket outlet mounted on a cavity wall.

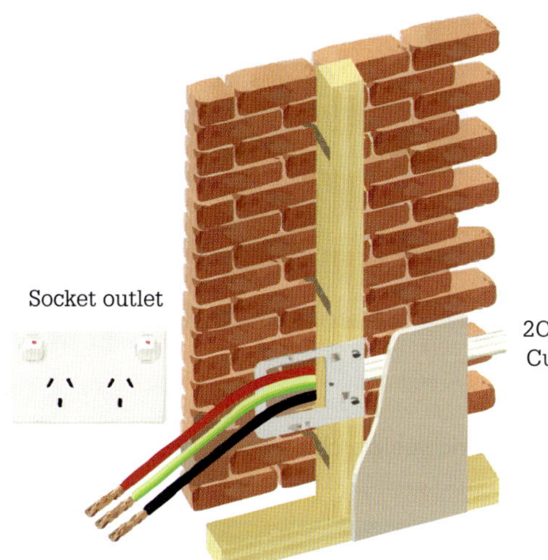

FIGURE 10.15 Flat TPS in a wall cavity

TPS cables and thermal insulation

When TPS cables are installed where thermal insulation will be placed, as illustrated in **Figure 10.16**, the cables may be partially or completely surrounded by the thermal insulation. In these circumstances the cables may need to be de-rated because the cables have a reduced ability to dissipate heat. Refer to AS/NZS 3008.1.1:1998 *Electrical installations – Selection of cables*, Section 3, for installation conditions.

As building insulation can be installed after installation work is complete, a prudent electrician would assume a worst-case situation and install 2.5 mm² at the first fix.

Note: Labour costs to install 2.5 mm² at first fix are the same and there is only a slight increase in cable costs.

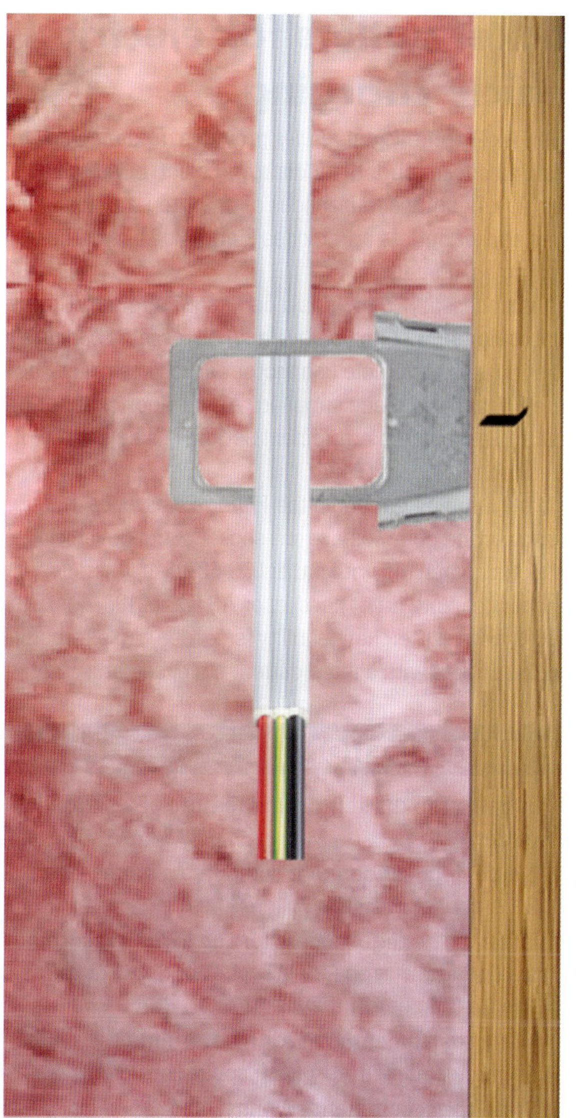

FIGURE 10.16 TPS cables and thermal wall insulation

TPS in plaster and cement render

Where TPS cables are chased (cut) into rendered masonry or concrete walls the cables are not required to be enclosed. However, if TPS cables are to be installed in concrete slabs they must be contained within a suitable enclosure.

TPS cables can be installed in plaster or cement render as illustrated in **Figure 10.17** without the use of an enclosure, provided the installation meets the requirements of Clause 3.3.2.6 and Clause 3.9.4. Note that if a metal wall box is used it should be earthed.

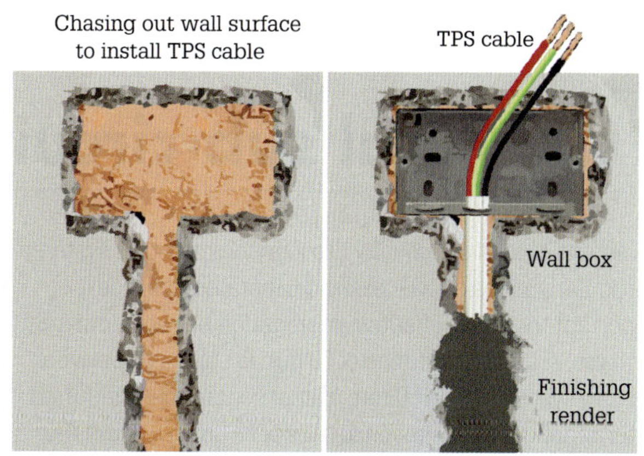

FIGURE 10.17 TPS cables embedded in a plaster wall surface

Socket outlets for electric vehicle charging

Appendix P of the *Wiring Rules* provides guidance on installing socket outlets for charging electric vehicles (EV). The rules recognise four modes of charging: modes 1 and 2 utilise socket outlets and modes 3 and 4 utilise dedicated charging equipment.

It is necessary to provide a dedicated circuit for the connecting point of an EV. It is also necessary to locate socket outlets or vehicle connector as close as practicable to the EV parking place supplied. An outdoor connection point and associated equipment should offer a degree of protection of at least IP4X.

It is a requirement under the *Wiring Rules* for each EV connecting point to have one socket outlet or vehicle connector complying with either IEC 62196-1, where interchangeability is not needed, or AS IEC 62196-2 or IEC 62196-3, where interchangeability is needed. An alternative is socket outlets with a rated current not exceeding 20 A complying with AS/NZS 3112, AS/NZS 3123 or IEC 60309. **Figure 10.18** illustrates a dedicated EV charging point.

Source: Shutterstock.Com/Chesky

FIGURE 10.18 Electric vehicle charging

REVIEW QUESTIONS

1. What is the distinguishing feature of a socket outlet?
2. Name applications that could use a single-phase 20 A socket outlet.
3. What is the requirement for the switch incorporated in a pendant-type 230 V socket outlet that relies on the flexible cord for support?
4. What instrument is used to determine the phase sequence at a three-phase socket outlet?
5. State the recommended phase sequence for Australia.
6. Why are electrical installations divided into circuits?
7. A single-phase domestic residential electrical installation has several final sub-circuits. Determine the number of 15 A socket outlets that can connect to a 4 mm^2 circuit protected by a 25 A Type C circuit breaker.
8. Determine the current-carrying capacity of a three-phase circuit comprising four-core and earth, 4 mm^2 copper conductors with PVC insulated and sheathed cables, marked V-90 and installed so as they are partially surrounded with thermal insulation.

10.2 Final sub-circuits and segregation

As socket outlets are usually associated with other systems, particularly in residential electrical installations, it is important to ensure effective segregation between different systems. A spatial distance of 50 mm is sufficient between cabling systems and 150 mm spatial separation is required at terminations. The reason for this is to minimise the risk of simultaneous insulation failure, such as a nail or screw piercing the insulation, resulting in low voltage appearing on the extra low voltage system and causing a shock hazard. Such extra low voltage systems include television distribution, data, telephone, building control, and security.

Mixed circuit

A mixed circuit is a final sub-circuit that supplies more than one type of load. A common mixed circuit is one that comprises both lighting points and 10 A socket outlets in a residential electrical installation.

Mixed circuits are useful in electrical installations such as granny flats that do not have many lighting points and socket outlets installed. A mixed circuit allows these to be installed on a single circuit. Note that socket outlets are a restricted connection and require a minimum conductor of 2.5 mm^2.

Table C9 in the *Wiring Rules* provides guidance on the number of lighting points and socket outlets that can connect to various cable and circuit protection combinations.

EXAMPLE 10.3

A single-phase domestic residential electrical installation has a number of final sub-circuits. Determine the number of 10 A socket outlets that can be added to a circuit containing 8 lighting points if the circuit is wired using 2.5 mm² copper conductor, V-90 insulated, twin and earth, TPS cables protected by a 16 A Type C circuit breaker.

Solution:

From Table C9 a lighting point is rated at 0.5 A and a socket outlet is rated at 1 A for all domestic installations.

The contribution of the lighting points is:

$$lighting\ demand = number\ of\ points\ 0.5\ \text{A}$$

$$= 8 \times 0.5$$

$$\mathbf{= 4A}$$

As the circuit is rated at 16 A, the available demand for socket outlets is:

$$available\ demand = circuit\ demand - lighting\ demand$$

$$= 16 - 4$$

$$\mathbf{= 12A}$$

it is possible to install 12 socket outlets on this circuit. This is calculated from:

$$Number\ of\ points = \frac{available\ demand}{rating\ per\ point}$$

Footnote 3 to Table C9 requires that multiple combination outlets have a point count of the number of integral socket outlets. This means that a double 10 A combination outlet counts as 2 points.

EXERCISE 10.3

A single-phase domestic residential electrical installation has several final sub-circuits. Determine the number of 10 A socket outlets that can be added to a circuit containing 12 lighting points if the circuit is wired using 2.5 mm² copper conductor, V-90 insulated, twin and earth, TPS cables protected by a 20 A Type C circuit breaker.

REVIEW QUESTIONS

1 Explain why it is important to maintain effective segregation between low-voltage cables and cables supplying other systems such as data.
2 What is a mixed circuit?
3 Provide an application for a mixed circuit.
4 A single-phase domestic residential electrical installation has several final sub-circuits. Determine the number of 10 A socket outlets that can be added to a circuit containing 12 lighting points if the circuit is wired using 2.5 mm² copper conductor, V-90 insulated, twin and earth, TPS cables protected by a 16 A Type C circuit breaker.

10.3 Identifying faults in socket outlet circuits

All electrical circuits must be tested to confirm compliance with the *Wiring Rules*. The basic tests for circuits supplying socket outlets include:

- earth continuity
- insulation resistance
- polarity
- correct circuit connections.

Refer to Chapter 16 of this textbook for a thorough treatment of installation testing.

Testing circuits for socket outlets

The following is a summary of the procedure for testing circuits supplying socket outlets.

Procedure for testing protective earth conductors

The procedure for testing the resistance of protective earthing conductors for a circuit supplying socket outlets is:

1. Disconnect the aerial service support equipotential bonding conductor (earth on riser bracket for aerials).
2. Ensure that there are no parallel conductive earth paths. For example, if the water heater is connected there is a possibility that a parallel path exists through the conductive water pipes.
3. At the switchboard disconnect the MEN link.
4. Connect the long lead to the main earth or earthing bar and the other end of this lead to the ohmmeter. Next connect the other ohmmeter lead to the protective earthing conductor at each earthing point for the entire

installation in turn and record the results. Remember to deduct the resistance of the trailing lead. Include the following: the PE at each socket outlet point (see **Figure 10.19**), with all control and isolating switches ON.

5 Disconnect the test leads and reconnect the MEN link and other disconnected earthing and equipotential conductors unless carrying out further tests.

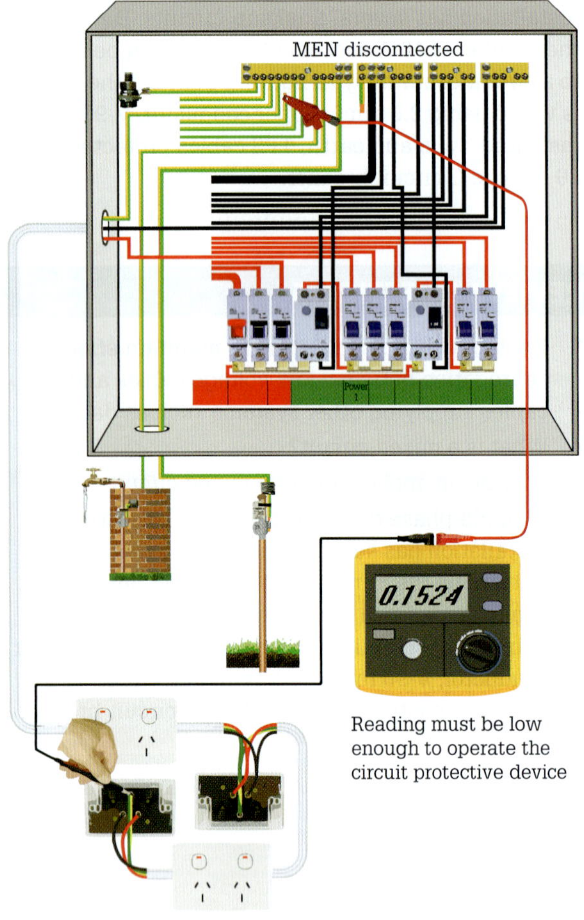

FIGURE 10.19 Resistance test of the protective earth at socket outlet

Requirements for testing insulation resistance

It is a requirement of Clause 8.3.6.2, AS/NZS 3000:2018 *Wiring Rules* that an insulation resistance test must be carried out on all live conductors (this includes the neutral) and the earth. For an installation this means:

1 The main switch must be off if it is a connected system.
2 All protective devices must be active, with fuse elements in place and circuit breakers on.
3 All circuit switches must be on with electronic controls bypassed.
4 All fixed appliances must remain connected and turned on.

The insulation test must prove that the insulation resistance between live conductors and the earth or any part thereof is not less than 1 MΩ.

Requirements for testing polarity

It is necessary to test an electrical installation for correct polarity to ensure that there is no shock hazard resulting from the incorrect connection of active, neutral and earthing conductors. When undertaking polarity tests ensure:

1 That single-pole switches and protective devices only operate in the active conductor of the connected circuit.
2 That switches and protective devices of multi-phase circuits, with some exceptions, operate in all active conductors of the connected circuit.
3 RCDs switch the active and neutral conductors of the circuit – all live conductors.
4 That the phase sequence of multi-phase socket outlets is the same for all socket outlets of the same type within the electrical installation.
5 That the polarity of all flat-pin socket outlets is earth, active and neutral in a clockwise direction commencing from the bottom-most pin.
6 All neutral conductors connect to the neutral bar of the switchboard.
7 The neutral conductor of the consumer's mains connects to the neutral bar of the main switchboard.

Procedure for testing polarity of final sub-circuits

The procedure for testing polarity of final sub-circuits is as follows:

1 Ensure the testing instrument is within test date.
2 Set the tester to the ohms range.
3 Ensure the electricity supply is disconnected and lock and tag the isolation point (never remove a fuse carrier instead of disconnecting).
4 Prove isolation of supply and operation of voltage presence indicator.
5 Ensure items 1 to 4 of 'Requirements for testing insulation resistance' above are proven.
6 Turn 'Off' the circuit protection device.
7 The continuity test requires that one lead of the ohmmeter be long enough to reach all socket outlets within the installation. The test is carried out with a long lead connected to the main active conductor at the distribution board and to one terminal of the ohmmeter on its low-resistance scale. The other lead of the ohmmeter is connected in turn to each active socket.

   For socket outlets, a 10 Ω resistor can be connected to the circuit active and the main neutral bar to provide a load for the ohmmeter. The required polarity for socket outlets, as shown in **Figure 10.20** viewed from the front, is neutral (right-hand side), earth and active in a clockwise direction (earth was proved previously).

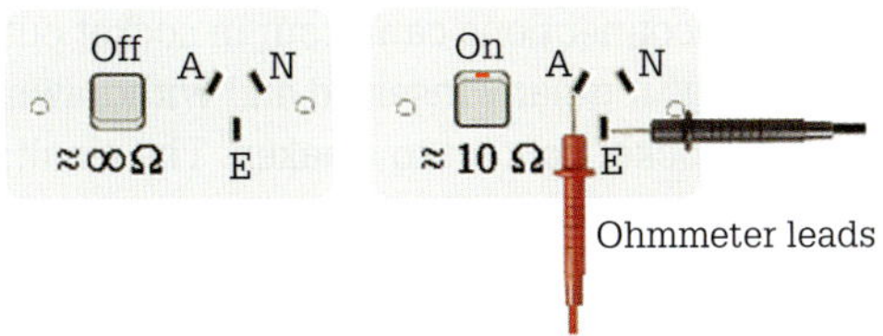

FIGURE 10.20 Typical socket outlets

- Switch Off ≈ ∞ Ω between earth and active terminal
- Switch On ≈ 10 Ω between earth and active terminal

8 Remove the 10 Ω resistor and turn on the main switch if all polarities are correct.

9 Again check the ohmmeter against a 0.5 Ω resistance to confirm the validity of the instrument and record result.

Requirements for testing for correct connections

Testing for correct circuit connections requires the use of an ohmmeter. The minimum steps include proving that the active, neutral and protective earthing conductors of each circuit are correctly connected so that there is no:

- short-circuit between the conductors
- transposition of conductors that could result in the earthing system and any exposed conductive parts of the electrical installation becoming energised
- interconnection of conductors between different circuits.

Figure 10.21 illustrates the testing procedure using resistors for each socket outlet circuit and at each socket outlet under test.

Cable schedule

A cable schedule is a table containing information about cables in a particular installation. The following is the type of information that a cable schedule could include:

- isolator size
- circuit number
- protection device size
- phase (line) and neutral size
- earth size
- number of points
- length of the circuit
- circuit name.

Without a cable schedule (see **Figure 10.22**) you cannot estimate the cable cost and total length of cable required.

Cable schedules are used for commercial and industrial installations and rarely for domestic installations. Schedules should be typed on white card and installed in clear acrylic faced holders fixed to the back of switchboard doors or on a wall adjacent.

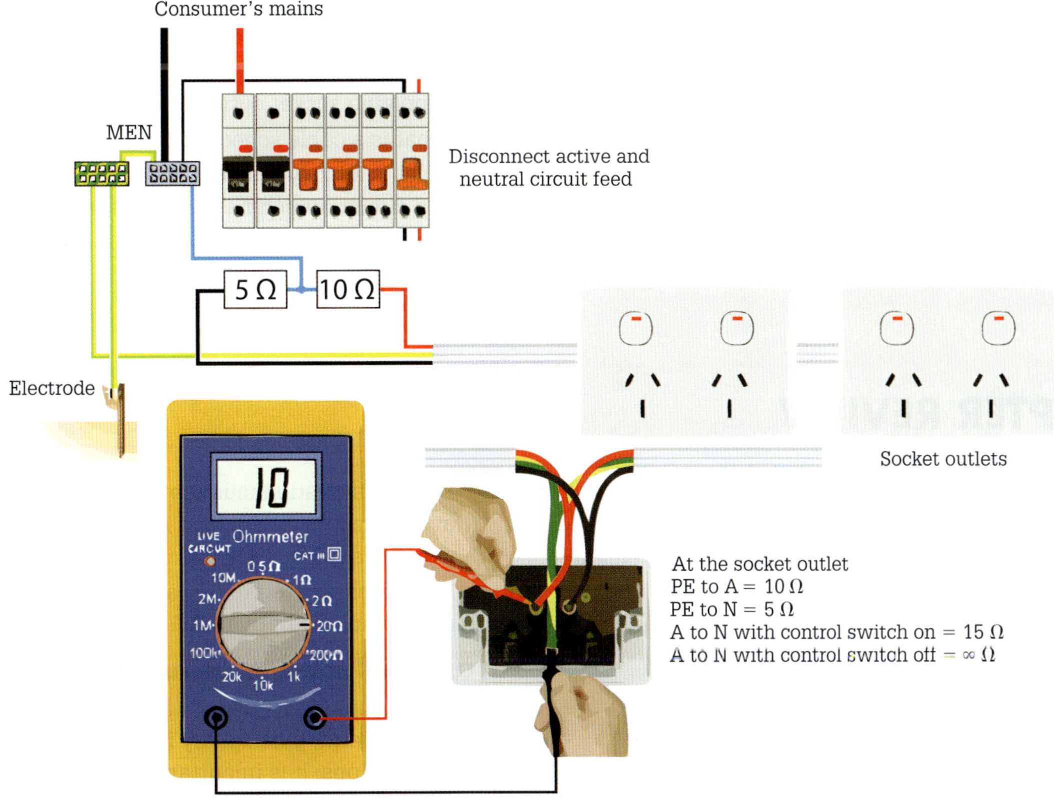

FIGURE 10.21 Connection test – socket outlets

Circuit no.	Core size mm^2	Current (A)	Route length	Application
MSB C/B 1	1.0 T&E flat	10	30 m	Lighting
MSB C/B 2	1.5 T&E flat	16	20 m	Lighting
MSB C/B 3	2.5 T&E flat	16	25 m	Power circuit
MSB C/B 4	4.0 T&E flat	32	15 m	Air/C
MSB C/B 5	6.0 T&E flat	40	8 m	Range

FIGURE 10.22 Cable schedule

Faults in socket outlets

Socket outlets generally provide faultless operation throughout their service life. However, some common faults include:

1. Cracking of the plastic due to being struck by a hard object. The remedy for this is to replace the socket outlet on a like-for-like basis.
2. Loose connections at the screw terminals resulting in wire and terminal burning due to the high resistance contact. The remedy for this is to replace the socket outlet on a like-for-like basis.
3. Control switch fails to operate. This can occur on socket outlets that are in frequent operation. Control switches do have a finite number of operations. The remedy for this is to replace the socket outlet on a like-for-like basis.
4. Loose socket connections can occur in socket outlets that have plugs regularly inserted and withdrawn causing the socket contact to weaken. The resultant high resistance contact can generate sufficient thermal energy to cause fire.

Leakage current

Excessive leakage current in circuits supplying socket outlets can cause the residual current device to operate, which disconnects the circuit from the supply. This problem is possibly attributable more to a plugged-in appliance or two rather than a faulty circuit.

To rule out a faulty circuit it is necessary to remove all plugged-in appliances and perform an insulation resistance test on the circuit. It is important to disconnect all appliances and not merely turn off the control switch as the fault can be on the neutral side.

After clearing the circuit, it is a matter of reinstating each appliance one-by-one and monitoring the leakage current using a tong ammeter. This can be done by clamping the tong around the circuit active and neutral conductors. The resultant reading should be zero ampere. A reading in the order of tens of milliampere can indicate an appliance with an earth fault, which should be checked and repaired or replaced.

REVIEW QUESTIONS

1. List the tests to undertake on a circuit supplying socket outlets to confirm compliance with the *Wiring Rules*.
2. What is the minimum acceptable insulation resistance for a circuit supplying socket outlets?
3. Why is a polarity test undertaken on a circuit supplying socket outlets?
4. What is a cable schedule?
5. What is a likely result of excessive leakage current in a circuit, which has RCD protection, supplying socket outlets?

CHAPTER REVIEW

10.1 Circuits for socket outlets

- A socket outlet is a device that has a set of contacts for making a separable connection with the contacts of a plug.
- Socket outlets may be single or multiple combination outlets.
- Socket outlets are rated in terms of the maximum current they can supply without damage. Typical ratings are 10 A, 15 A, 20 A, 25 A and 32 A.
- Socket outlets are also available in weather-protected and pendant styles.
- Australia follows an anticlockwise phase sequence and the convention for this sequence is ABC or L1, L2, L3.
- Every installation must be divided into circuits to avoid hazards and to contain the disruption in the event of a fault.
- It is necessary to provide a dedicated circuit for the connecting point of an EV.

10.2 Final sub-circuits and segregation

- Socket outlets are usually associated with other systems and, particularly in residential electrical installations, it is important to ensure effective segregation between different systems.
- A mixed circuit is a final sub-circuit that supplies more than type of load.
- Table C9 in the *Wiring Rules* provides guidance on the number of lighting points and socket outlets that can connect to various cable and circuit protection combinations.

10.3 Identifying faults in socket outlet circuits

- All electrical circuits must be tested to confirm compliance with the *Wiring Rules*. The basic tests for circuits supplying socket outlets include earth continuity, insulation resistance, polarity and correct circuit connections.
- A cable schedule is a table containing information about cables in a particular installation.
- Excessive leakage current in circuits supplying socket outlets can cause the residual current device to operate, which disconnects the circuit from the supply.

TRIAL EXAM

For Chapter 10 knowledge assessment, please complete the following trial exam.

1 A device that has a set of contacts for making a separable connection with the contacts of a plug is a:
   a light point
   b socket outlet
   c point
   d load

2 Circuits supplying socket outlets rated above 10 A require:
   a a dedicated circuit
   b a ring feed
   c surge protection
   d higher current density

3 The outlet shown in Figure 10.23 is rated at:
   10 A
   15 A
   20 A
   32 A

FIGURE 10.23 Question 3

4 The flexible cable supplying a pendant socket outlet is often held in place by:
   a cable ties attached to the support chain
   b cable clips to a timber support
   c saddles fixed to a stud or noggin
   d self-amalgamating tape

5 Australian three-phase plugs and flexible cable connected sockets use the colour:
   a black
   b grey
   c blue
   d orange

6 For a three-phase socket outlet, the order of the cyclic direction of the supply is established by using a:
   a voltmeter
   b three-phase sequence indicator
   c tong ammeter
   d oscilloscope

7 Final sub-circuits supplying circuit outlets are generally wired using
   a 1.5 mm^2 TPS twin and earth cable
   b 2.5 mm^2 TPS twin and earth cable
   c 4 mm^2 TPS four-core and earth cable
   d 4 mm^2 TPI cable enclosed in conduit

8 Guidance on the number of socket outlets that can connect to various cable and circuit protection combinations is found in:
   a Table C5 of the *Wiring Rules*
   b Table C7 of the *Wiring Rules*
   c Appendix P of the *Wiring Rules*
   d Table C9 of the *Wiring Rules*

9 A single-phase domestic residential electrical installation has several final sub-circuits. The number of 10 A socket outlets that can connect to a 4 mm^2 circuit protected by a 20 A Type C circuit breaker is:
   a 1
   b 10
   c 20
   d 40

10 Polarity testing of a circuit supplying socket outlets is necessary to ensure that:
   a all switches operate in the neutral conductor
   b all switches operate in the active conductor
   c all switches operate in the earth conductor
   d current flow in the circuit is unidirectional

11 The minimum cable size for an unenclosed cable installed in free-air and supplying a circuit containing 230 V, 10 A single-phase socket outlets protected by a 20 A Type C circuit breaker is:
   a 10 A
   b 16 A
   c 20 A
   d 25 A

12 The recommended number of lighting points that can be added to a 2.5 mm^2, 16 A domestic power circuit that has 5 double socket outlets and 5 single socket outlets is:
   a 2
   b 4
   c 8
   d 12

13 An outdoor connection point and associated equipment for EV charging should offer a degree of protection of at least:
   a IP56
   b IPX3
   c IP4X
   d Category II

14 Where interchangeability is not required, each EV connecting point must have one socket outlet or vehicle connector complying with:
   a IEC 62196-1
   b IEC 62196-2
   c IEC 62196-3
   d AS/NZS3017

15 A single circuit that supplies both lighting points and socket outlets is known as:
   a a light-power circuit
   b a dedicated circuit
   c a mixed circuit
   d an alternate circuit

Control and protection of electrical installations

This chapter provides electrotechnology workers with essential knowledge and skills pertaining to the selection, arrangement and termination of circuits, selecting and installing control and protection devices and systems in electrical installations operating at low voltage. This also includes methods to ensure protection of persons and property, correct functioning, ensuring compatibility with the supply, arranging installation into circuits, selecting and arranging switchgear, and protective devices to meet compliance requirements. This chapter provides underpinning knowledge for the unit UEEEL0003 from the UEE training package.

LEARNING OBJECTIVES

Safety principles

- Outline the safety principles detailed in the *Wiring Rules*.
- Summarise the fundamental requirements for electrical installations to ensure compliance with the *Wiring Rules*.
- State the requirements for safe electrical installation design.

Circuit and control arrangements

- Explain the reasons for dividing an electrical installation into circuits.
- Describe SELV and PELV circuits.
- Explain what is meant by the term 'isolated supply'.

Hazards and risks in an electrical installation

- Outline the harmful effects of an electric current.
- Summarise the physiological effect of an electric current.
- State the risks associated with electric current.

Protection against indirect contact

- Describe the methods of protection used to limit the effects of indirect contact.
- Delineate the earth fault current path.

Earthing

- Describe the fundamentals of the MEN system of earthing.
- Explain why protective earthing conductors (PECs) are necessary.
- Explain how installation faults of equipotential bonding conductors may occur.

Equipotential bonding

- Outline the requirements for equipotential bonding.
- Explain the operation of the single-wire earth return system.

Protection against overload and short circuit current

- Describe the earth fault loop.
- Calculate prospective fault current.
- Determine earth fault-loop impedance.
- Outline the requirements for automatic disconnection of supply.

Devices for automatic disconnection of supply

- Explain the operation of devices used to provide automatic disconnection of supply.
- Select appropriate circuit protection devices.
- Explain the operation of and installation requirements for a core balance earth leakage device.
- Describe the purpose of RCDs.

Protection against over-voltage and under-voltage

- List devices used for over-voltage protection.
- Describe typical lightning and surge protection specifications for an installation.
- Outline the requirements for under-voltage protection.

Control of an electrical installation and circuits

- Describe how control of the installation is achieved.

Switchboards and distribution boards

- Outline the *Wiring Rules* requirements for electrical switchboards.
- Explain the arrangement of switchboard equipment.

11.1 Safety principles

Safety principles require that the design and performance of electrical installations satisfy the requirements of all relevant Acts, regulations, standards and codes of practice publications relevant to the state or territory where the installation is located.

All electrical installations must comply with the prescriptive requirements of AS/NZS 3000:2018 and the local network service provider needs such as their Service and Installation Rules. Additionally, equipment and accessories for electrical installations must comply with all relevant Australian standards for the type of installation into which they are to be installed.

The design of electrical installations should meet design objectives as illustrated in **Figure 11.1**.

Compliant protection arrangements

Compliant protection arrangements are applied to every installation so that any periodic inspection, testing, maintenance and repairs likely to be necessary during the intended life of the installation can be easily and safely carried out. In addition, the reliability of the protective measures for safety remains effective for the life of the installation.

Installation design

The design and installation of electrical installations must provide reasonable safety from danger. To meet this requirement, an installation when properly used, should offer protection from electric shock, physical hazards and fire for all persons who may properly use, maintain, repair, make additions to, test or be exposed to any part of an installation. Furthermore, livestock must be given the same duty of care as persons, and property must be protected from damage as it relates to the function of the installation. The safety principles are illustrated in **Figure 11.2** (refer to Section 1, AS/NZS 3000:2018).

Protection against electric shock

Protection against electric shock must be provided in regular service and in the event of a fault. There are two ways by which persons and livestock can receive an electric shock. These are by 'direct contact' and 'indirect contact'. Protective measures are required against both of these conditions (refer to Clause 1.5.3).

Protection against dangers and damage

All accessories and wiring in an electrical installation must be suitably located so as to prevent electric shock and to provide safe access for operation, maintenance and repair and must be protected against damage. Cables subjected to overload conditions can experience effects such as tempering of the conductor and the reduction of the quality of the insulation. Both of these effects are considered mechanical damage. Basic or primary cable insulation provides protection against electric shock from direct contact of the conductor and the cable must therefore be protected against mechanical damage.

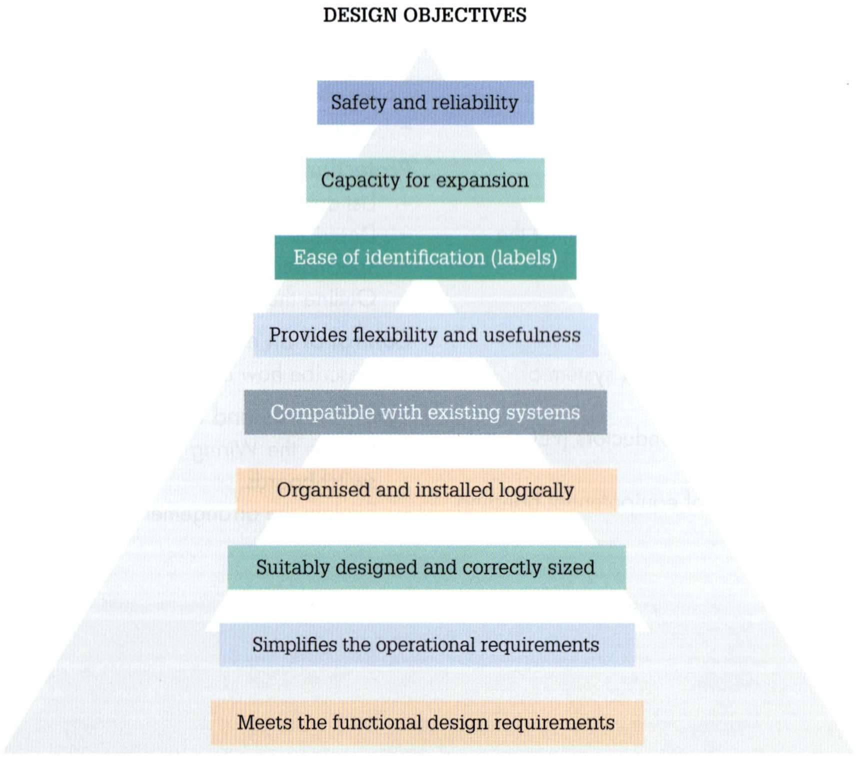

FIGURE 11.1 Design objectives

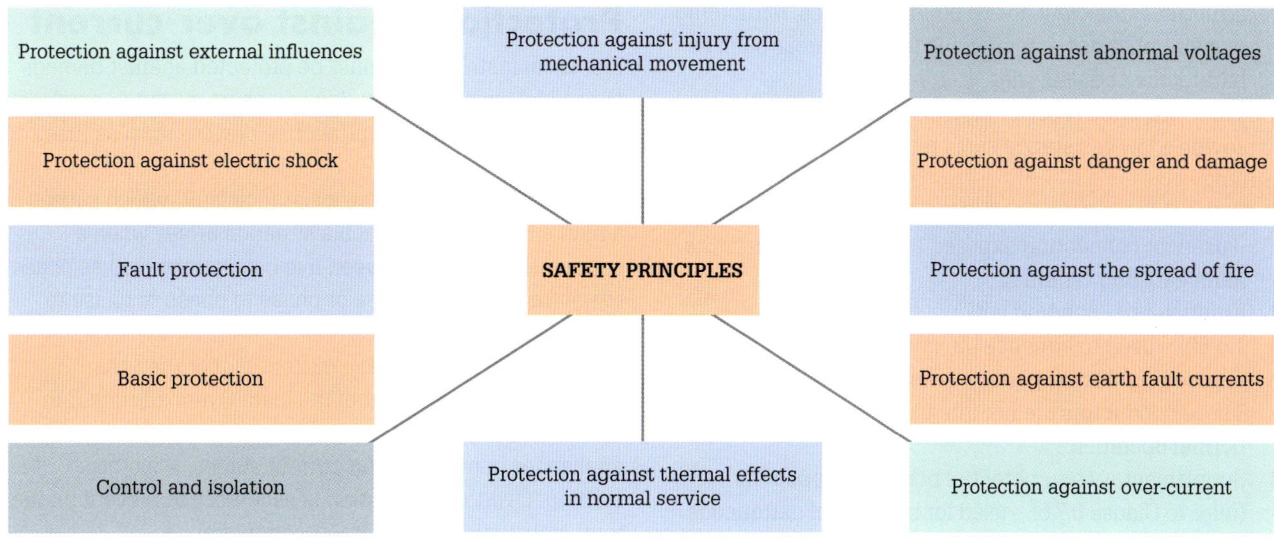

FIGURE 11.2 Safety principles

The protection can be achieved by a protective sheath, MIMS or through the use of enclosures. Corrosion is a form of mechanical damage and metal enclosures and fittings must be protected. Protection is especially important in damp situations where contact between dissimilar metals can result in electrolytic action (refer to Clauses 3.3.2.5 and 3.3.2.6).

Control and isolation

All electrical installations must be provided with a means of isolating or controlling the electricity supply at suitable locations such as at switchboards and at points where appliances or machinery are used. Efficient means of disconnection must be provided and must be easily accessible, with the isolator having a rated breaking current so that the isolator can be switched while a fault is present.

Isolation means the disconnection of electrical energy from every source of supply. Even though the isolation function is carried out by a switch, isolation is intended for operation by electricians who need the connected circuit isolated so as to perform work on the circuit, its accessories or load which would otherwise be live. In a domestic installation, the main switch provides isolation of the connected circuit from the consumer's mains.

Control or functional switching is for disconnecting or connecting loads to a supply, for example, a one-way switch controlling a lighting point. Even though a control switch can turn on or off a load, it is usually not regarded as an isolator because it does not clearly and reliably indicate the isolating position (refer to Clauses 2.3.2.2 and 2.3.7).

Primary protection (protection against direct contact)

All parts of an electrical installation shall be insulated in a way that is appropriate to the function they serve, in consideration of the expected operating environment, so as to prevent danger. Protection by primary insulation is the most usual means of providing basic protection against direct contact and is employed in all installations. Only destruction can remove this type of protection. Protection against direct contact can sometimes be achieved by the provision of obstacles that prevent unintentional bodily approach or unintentional contact with live parts during the operation of equipment in ordinary service. These may be guards, railings or enclosures providing protection to at least IP2X except where larger openings occur during the replacement of parts, such as lamp holders, or fuses, or where larger openings are necessary to allow the proper functioning of equipment.

Another method of protection is to place live parts out of arm's reach, for example, the high placement of bare aerials such as service mains (refer to Clause 1.5.4). Protection by placing out of reach is intended only to prevent unintentional contact with live parts. Two parts are deemed to be simultaneously accessible if they are closer than 2.50 m.

Fault protection (protection against indirect contact)

Contact with accessible exposed and extraneous conductive parts made live by a fault is called indirect contact. Protection must allow a protective device to disconnect the supply before the fault causes a rise in voltage above the conventional touch voltage limit of 50 V. One method of achieving some measure of protection against indirect contact is through a combination of equipotential bonding and automatic disconnection of supply. This involves the bonding together and connection to the general mass of earth:

1. All metalwork associated with the electrical installation termed 'exposed conductive parts'. Examples include conductive sheaths, armours and screens of cables, steel conduit, metal enclosures such as trunking, duct and pipe and the metal cases of appliances and machines.
2. All extraneous conductive parts likely to develop an earth potential. Examples are metal water or gas pipes,

structural steelwork, metal building frames, reinforcing in swimming pools, reinforcing in damp areas such as bathrooms and laundries.

There are three types of protective conductors used to provide a low-resistance earth return path for a fault current:

1. Protective earthing conductors (refer to Clause 5.5.2.1) – used for connecting exposed conductive parts to the main earthing terminal.
2. Equipotential bonding conductors (refer to Clause 1.4.60) – used to equalise the potential between exposed conductive parts or extraneous conductive parts. These conductors are not intended to carry current in normal operation.
3. Supplementary equipotential bonding conductors (refer to Clause 5.7.5) – used for bonding of extraneous conductive parts and their connection to the earthing system to reduce the earth-fault loop impedance.

For the protection to be effective, it is necessary to ensure that automatic disconnection occurs quickly. Automatic disconnection is achieved through an overload protection device. Therefore, it is essential that the earth path for a fault current must be of low impedance to ensure the required disconnection time.

The basic requirement for protection against indirect contact means that the characteristics of the protective device, the earthing arrangements and the impedances of the circuits have to be coordinated so that the magnitude and duration of the voltage appearing on simultaneously accessible exposed and extraneous conductive parts during an earth fault shall not cause danger.

The use of an RCD or RCBO can also reduce the risk of electric shock associated with indirect contact. The RCD or RCBO must have a rated leakage current not exceeding 30 mA and be used in conjunction with the three types of protective conductors above.

Indirect contact protection can be provided by the use of Class 2 equipment. This is equipment having double or reinforced insulation. Class 2 equipment means that the live parts are so isolated by insulation that faults to exposed conductive parts cannot occur.

All electrical installations must be protected against damage caused by excess current due to a fault or overload by suitable protective devices.

Protection against thermal effects in normal service

Cables and fixed equipment must be installed so that surrounding materials, persons and livestock are not at risk of accidental contact with heat. This can be achieved by ensuring that fixed equipment is not accessible and installing suitable guards (refer to Clause 1.5.8 and Clause 4.2). An important consideration is the provision of adequate ventilation for electrical equipment to ensure higher than normal temperatures are not attained in operation.

Protection against over-current

Electrical installations must be protected against damage caused by excess current due to a short-circuit or overload by suitable over-current protective devices such as fuses or circuit breakers.

A short-circuit is a low-resistance fault, which causes a high current to flow. Such a situation arises when a connection 'shorts' between live conductors such as phase to neutral for single phase or phase to phase or phase to neutral for three-phase.

An overload produces a current higher than the normal operating current of the load. Overloads occur in electrically sound circuits and may be caused by faulty appliances, motor starting current surges or motors stalling. Overload protection is necessary to prevent undue temperature rise in the circuit conductors.

Usually, the same protective device is used to protect circuits against both short-circuit and overload currents. A good working knowledge of the prospective short-circuit current is needed in order to select the correct protective device with the suitable breaking current rating (refer to Clause 1.5.9) to adequately protect the circuit.

Protection against earth fault currents

The features of protective devices and the impedance of the earthing system shall be such that, if a low impedance fault occurs anywhere in an electrical installation between an active conductor and a protective earthing conductor or exposed conductive part, automatic disconnection of the supply will occur within a specified time (refer to Clause 1.5.5.3).

Protection against abnormal voltages

In industrial and commercial installations, it is necessary to consider the likelihood of danger occurring from a drop in voltage or loss and subsequent reinstatement of supply. In these situations, it is necessary to provide suitable protection to prevent automatic restarting of machinery. Motor control circuits provide 'no-volts' protection via their 'hold-in' circuit (refer to Clause 1.5.11).

Protection against the spread of fire

Electrical installations must be designed and installed with particular attention given to the risk of fire due to electrical faults and the propagation of fire by using the elements of an electrical system. Fires originating from electrical equipment include:

- The overheating of conductors causing the heating of the insulation or adjacent flammable materials.
- The ignition of flammable materials as a result of electric arcing.

A short-circuit current increases the heat dissipation of a conductor ($P = I^2R$). If the short-circuit current is not cleared quickly, the temperature of the conductors rises significantly resulting in fire.

Protection against injury from mechanical movement

All electrically operated equipment must be suitably located so as to provide safe access for operation, emergency stopping, maintenance and repair. This means that all moving parts of machines must be guarded so as to prevent any unforeseen danger (also refer to Clause 1.5.13).

Protection against external influences

All parts of an electrical installation must be designed in a way that is appropriate to the function they serve, in consideration of the expected operating environment (heat, cold, explosive atmospheres) and other influences such as vibration, dusts or water so as to prevent danger (refer to Clause 1.5.14).

If the installation is engineered to be electrically safe without hazards when used for the purpose intended, then correct functioning has been achieved. Proper functioning also means that any conductors, protective devices, accessories or other electrical equipment have been designed for the conveyance, control and use of electrical energy required for the installation.

Compatibility with supply

To guarantee satisfactory performance of electrical installations, both the safety of supply and the quality of supply need to be considered with equal care. Such factors include:

- feed – single-phase, two-phase, three-phase or d.c. supply
- the voltages, tolerance and frequency of the supply required for the installation
- the current required by the installation from the electricity distributor
- prospective short-circuit current
- the prescribed earthing arrangements
- limits on high-demand loads
- harmonic distortion on the feeders to the installation.

REVIEW QUESTIONS

1. Why are compliant protection arrangements applied to every installation?
2. How can an electrical installation satisfy the requirement of providing reasonable safety from danger?
3. What are two ways by which persons and livestock can receive an electric shock?
4. What is the requirement for all accessories and wiring in an electrical installation?
5. Name the primary method for providing basic protection against direct contact in all electrical installations.
6. When are two parts deemed to be simultaneously accessible?
7. Why is adequate ventilation for electrical equipment an important consideration?
8. What can fires originating from electrical equipment include?
9. How is the correct functioning of an electrical installation determined?
10. Which supply parameters must be considered to guarantee satisfactory performance of electrical installations?

11.2 Circuit and control arrangements

Every installation must be divided into circuits to avoid hazards and to contain the disruption in the event of a fault. In addition, the division of an installation into various circuits makes easier the safe operation of the installation, improves the quality of circuit inspection and aids reliability with testing and makes maintenance simpler. The electrical distribution within an installation ends with final sub-circuits (see AS/NZS 3000:2018). A final sub-circuit is that section of a wiring system that extends beyond the final circuit protection device. Final sub-circuits can originate from either a main switchboard or a sub-main distribution board.

Final sub-circuits make up the greatest portion of the wiring of an electrical installation and are divided into three general groups:

1. fixed-wiring lighting circuits
2. fixed-wiring socket outlet circuits
3. fixed-wiring appliance circuits.

Final sub-circuits are usually rated at 6 A, 10 A, 16 A, 20 A, 25 or 32 A for some fixed-wiring loads.

Appendix C5 in AS/NZS *Wiring Rules* requires loads of 20 A or more per phase to connect to a separate and distinct circuit. Final sub-circuits are illustrated in **Figure 11.3**.

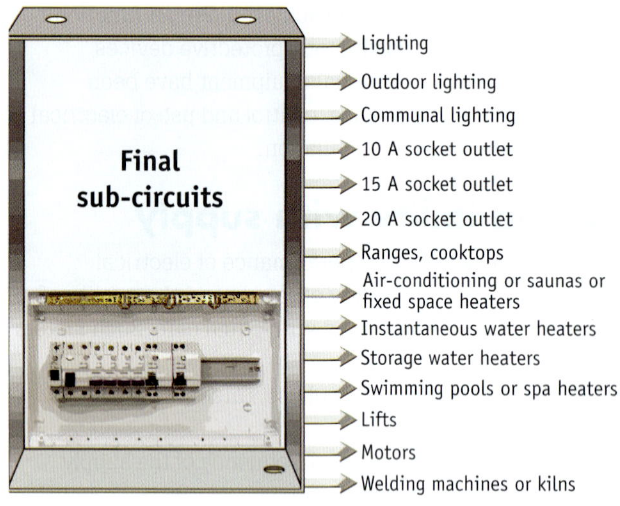

FIGURE 11.3 Final sub-circuits

Other final sub-circuits include safety emergency lighting, fire protection, security and uninterruptible power supplies as well as control circuits for fixed plant.

Circuit arrangement in electrical installations

The creation of separate circuits in an electrical installation makes it possible to:

- limit the effects of a fault to the circuit concerned
- simplify fault-finding
- carry out maintenance work, inspections and testing or circuit extensions without interrupting the whole electrical installation.

The number and type of circuits required are determined by the load on the circuit, the location of the loads, any seasonal variations that affect the loads and any special conditions such as tariff requirements.

Refer to AS/NZS 3000:2018 *Wiring Rules*, 'electrical installation circuit arrangement'.

Table C9 of AS/NZS 3000:2018 provides guidance on the number of points per final sub-circuit in a domestic installation. Suggested number of points for a domestic installation with cables unenclosed in air is illustrated in **Figure 11.4**.

The circuits provided for the different types of loads of the installation also need to be separately controlled and protected so that they are not affected by the failure of other circuits.

Each final sub-circuit must comply with AS/NZS *Wiring Rules* for over-current protection, for isolation and control switching and the current-carrying capacity of the conductors. Further, each final sub-circuit must be separate from other circuits to prevent energisation of an adjacent circuit.

Refer to Chapter 15 of this textbook for a more detailed treatment of sub-circuits.

Extra-low-voltage circuits

Refer to 1.4.128 of AS/NZS 3000:2018 *Wiring Rules*, 'Extra-low voltage electrical installations'.

Voltage is regarded as a difference of potential that exists between all active conductors and all active conductors and earth. Extra-low voltage is a voltage not exceeding 50 V a.c. or 120 V ripple-free d.c. Extra-low voltage is a measure designed to protect people and livestock from direct and indirect contact with live parts without automatic disconnection of the supply.

Two systems are used for this purpose – separated extra-low voltage and protected extra-low voltage. The source of these types of supply must comply with the requirements of AS/NZS 3000:2018 *Wiring Rules*.

Separated extra-low voltage (SELV)

SELV systems are used where an inherently safe system of supply is required. These systems ensure that a failure of

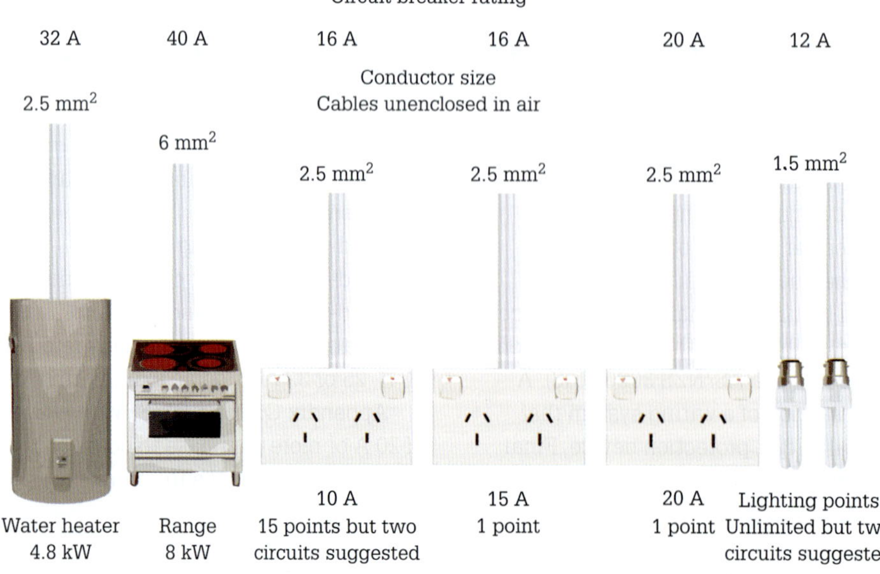

FIGURE 11.4 Final sub-circuit in a domestic installation

the primary insulation does not result in an increased risk of electric shock. This is achieved by electrical separation of the SELV output from the source of energy, typically using a safety isolating transformer.

Protection by SELV is used in high-risk situations where the operation of electrical equipment presents a serious hazard; for example, swimming pools. This method of protection requires supplying electrical energy from a safety isolating transformer, which utilises double or reinforced insulation. For damp situation zone 0 applications, the nominal voltage must not exceed 12 V a.c. or 30 V ripple-free d.c.

Protection

The following conditions of use must be applied when using SELV to ensure effective protection against indirect contact:

- No live conductor at SELV potential must be connected to the earth.
- Exposed conductive parts of SELV circuits or equipment must not be connected to the earth, other circuits, and their exposed conductive parts or extraneous conductive parts.
- In SELV systems where the voltage does not exceed 25 V a.c. or 50 V ripple-free d.c. there is no need to provide protection against direct contact hazards.
- Each SELV circuit must be segregated from all other wiring systems including other SELV systems. Segregation is not required where the SELV conductor insulation is equivalent to double insulation.

  Segregation is also not required if the SELV insulation is of the same voltage rating as the highest voltage cable rating with which the SELV cable is enclosed.
- Socket outlets and plugs for an SELV wiring system must not have an earth-pin contact and must be of a design that prevents connection to a different voltage level.

Direct and indirect protection

Direct and indirect protection of an SELV wiring system is provided by:

- the extra-low voltage
- minimal risk of accidental contact with a higher voltage
- no return path through a protective earth that a current could take in case of contact.

Safety

The safety of an SELV system is demonstrated by:

- an extra-low-voltage system
- a small risk relating to accidental contact with extra-low voltage
- the absence of a return path for any fault current through earth if a person or livestock comes into contact with extraneous metal components of the extra-low-voltage system.

Protected extra-low voltage (PELV)

This type of wiring system is used where low voltage is required, but the situation is not as high risk as is the case with SELV wiring conditions. A PELV power supply includes two protective measures (double insulation) against direct and indirect contact with exposed conductive parts by means of a safe separation of the primary and secondary transformer windings (such as IP2X or an insulation rating of 500 V a.c. for one minute). An earthed conductive shield placed between the primary and secondary windings is also effective. The shield collects any leakage current from the primary winding and discharges it to the earth. This prevents any leakage current entering the secondary winding. For this reason, no separate protective earth conductor is required in a PELV system. Connected equipment and exposed metallic parts in a PELV wiring system can be earthed, but this is not a requirement.

If the circuits of a PELV system require earthing, then the secondary circuit of the transformer cannot exceed 25 V a.c. or 60 V ripple-free d.c. and must have an integrated earth connection to enable connection of the protective earth conductors. PELV wiring systems are used in current supply in automation, machine and process control applications.

Isolated supply

Isolated supply has the same significance as 'separated circuit'. Isolation is the electrical separation between two circuits (source and output) in proximity to each other. If an isolated supply is required, three methods are available (refer to Clause 7.4 AS/NZS 3000:2018):

1 motor/generator output
2 isolated inverter
3 isolating transformer.

Transformers provide electrical isolation while allowing the magnetic coupling of electrical power from primary winding to secondary winding. An isolation transformer as illustrated in **Figure 11.5** eliminates any electrical

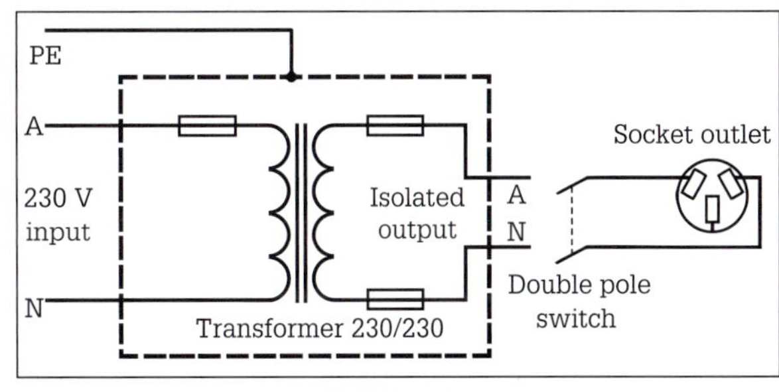

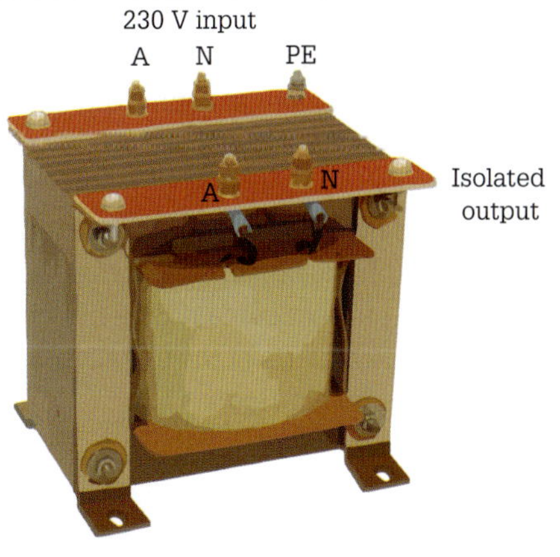

FIGURE 11.5 Isolation transformer

continuity between mains power and the load connected to the transformer output. The mains power is fed to the primary side of the transformer, and the load is connected to the secondary side. Isolation transformers are available in both single-phase and three-phase versions.

The isolation transformer completely isolates the load from the mains earth provided that there is not an earth connection on the load side of the transformer. As isolating transformers eliminate electrical continuity between circuits they are used for the following applications:

- to isolate shore power from a boat so that galvanic action (electrolysis) cannot take place as the shore earth is isolated from the boat earth
- in hospitals for power supply in body- and cardiac-protected areas
- to avoid tripping of switchboard protection devices when testing portable RCD units.

REVIEW QUESTIONS

1. Why are electrical installations divided into circuits?
2. What characterises a final sub-circuit?
3. Name the factors that contribute to the number and type of circuits in an electrical installation.
4. What is the purpose of extra-low voltage supplies?
5. Where would an SELV system find application?
6. How is the safety of an SELV wiring system demonstrated?
7. What is the function of an earthed conductive shield placed between the primary and secondary windings of a PELV transformer?
8. What do transformers provide?

11.3 Hazards and risks in an electrical installation

Every electrical installation is a combination of cables, accessories and energy-converting load equipment installed for a particular purpose within specific environmental conditions. The design of every installation, therefore, must be such that it eliminates or minimises the effects of electric current that may occur within the wiring system. Primarily, the design of the installation must afford protection of the circuit conductors by open-circuiting faulty circuits. If the installation design is not efficient, then the consequences for the circuits are thermal overloading of the conductor insulation and fire. After determining the protection for property, it is necessary to ensure a level of safety for persons and livestock within an electrical installation. Possible hazards that could cause severe injury or damage within an installation are:

- excessive temperatures
- arcing or burning
- explosion
- shock current
- over- and under-voltages
- mechanical movement of connected equipment
- electromagnetic interferences
- variations in the supply feed to the installation.

Persons and livestock must be protected from electric shock hazards (physiological effects of current) arising from direct and indirect contact with live conductors or exposed conductive parts of the installation to prevent the situation represented in **Figure 11.6**.

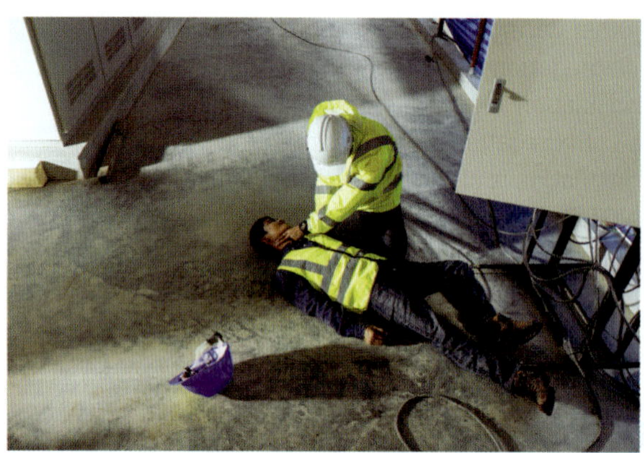

Source: Shutterstock.com/ME Image

FIGURE 11.6 Contact with live parts

Physiological effects of current

Physical contact with live parts of an electrical installation can affect the human body in various ways such as pain, burns or even death.

Electric shock is a general term for the excitation or confusion of the purpose of nerves or muscles caused by the passage of an electric current. It is usually painful but is not necessarily associated with actual damage to a person. There are two methods by which a person can experience electric shock.

The first is the direct method whereby a person comes into contact with a conductor that is normally live and becomes part of the fault path. The second is indirect contact whereby a person comes into contact with material which has become live under fault conditions.

The effect that electric shock has on the body depends on the voltage, magnitude of the fault current, the part of the body through which the current flows, the physical condition of the person, and the duration of the shock. Serious electric shock is almost always associated with alternating current and is rare when low-voltage direct currents are used.

SWITCH ON

Electrical injuries have three categories:
- nerve immobilisation
- fibrillation
- tissue destruction.

See also Chapter 1 of this textbook for information on the physical effects of electric current.

Installation design requirements

Electrical installations in premises must be designed, performance-based and engineered to the requirements of all relevant Acts, regulations and standards and codes of practice publications appropriate to the state or territory where the installation is located.

All electrical installations must comply with the prescriptive requirements of AS/NZS 3000:2018 and the local network service provider requirements such as their Service and Installation Rules. Equipment and accessories for electrical installations must comply with all relevant Australian standards for the type of installation into which they are to be installed.

Safety principles of installation design

Electrical installations must be designed and installed so as to provide reasonable safety from danger. To meet this requirement an installation when properly operated should provide protection from electric shock, physical hazards and fire for all persons who may correctly use, maintain, repair, make additions to, test or be exposed to any part of an installation. In addition, livestock must be given the same duty of care as persons and property must be protected from damage as it relates to the function of the installation.

Protection against thermal effects of electric current

One of the effects of an electric current that is always present is the thermal or heating effect. The heat produced by an electric current is proportional to the square of the magnitude of electric current and the time for which the current is flowing:

$$Heat \propto Current^2 \times time$$

In addition to the two main areas requiring protection for safety – direct and indirect contact – another fundamental principle specified in AS/NZS 3000:2018 *Wiring Rules* where protection is needed to minimise the risk of electric shock and fire:
- protection against thermal effects that may cause persons and livestock to be burnt or flammable materials to ignite. Methods used can be screening, enclosure or guarding.

Protection against injury from mechanical movement

All parts of equipment or machines that move in the course of the performance of work may contribute to accidents causing injury and damage. All electrically actuated rotating, linear and reciprocating machine movements can be dangerous. Examples include:
- belts and pulleys
- gear wheels
- shafts and spindles
- flywheels
- grinding wheels
- slides and cams
- chain and sprocket gears
- power presses
- guillotines
- milling cutters
- saws
- drills.

In addition, the mechanical potential energy of all sections of the equipment or machine is set so that the opening of pipes, tubes, hoses or operation of any valve, lever or button will not yield a movement, which could cause injury.

Typical hazards to look for when identifying dangerous mechanical movement from electrically actuated equipment and machine parts include:
- entanglement
- friction and abrasion
- cutting
- shear
- stabbing and puncture
- impact
- crushing
- flying particles
- protrusions
- drawing-in.

Methods of prevention

To prevent injury from mechanical movement, the installing electrician must ensure that the machine demonstrates the following six defences.

1 **Prevent contact:** A good protection system removes the likelihood of the operator or another worker placing parts of their bodies near hazardous moving parts.
2 **Secure:** Workers should not be able to get rid of or interfere with protective guards easily.
3 **Check falling objects:** Protective guards should ensure that no objects can drop into moving parts.
4 **Efficient design:** The protective guards must be designed so as not to create a new hazard.
5 **Hinder workers:** Guards must not reduce a worker's performance when operating the machine.
6 **Ease of maintenance:** Guards should allow a maintenance worker to carry out work without entering hazardous areas of the machine.

See section 11.1 of this chapter for further information

REVIEW QUESTIONS

1 What constitutes an electrical installation?
2 What is a likely consequence of having an installation design that is not efficient?
3 In what ways can physical contact with live parts of an electrical installation affect the human body?
4 List the factors that determine the severity of electric shock on the human body.
5 With what requirements must all electrical installations comply?

11.4 Protection against indirect contact

A basic safety requirement of AS/NZS 3000:2018 Clause 1.5 is the protection of persons and livestock from 'indirect contact' (refer to Clause 1.4.39) from touching exposed metallic components that have become 'live' due to an earth fault as illustrated in **Figure 11.7**. Therefore, each circuit in an electrical installation must be designed so that automatic disconnection of the circuit occurs to prevent a rise in 'touch voltage' (refer to Clause 1.4.125) to dangerous levels (exceeding 50 V a.c.) within a specified time (refer to Clause 1.5.5.3 and Clause 2.4.2 in AS/NZS 3000:2018).

FIGURE 11.7 Indirect contact

The method of protection used to limit the effects of indirect contact is achieved by a system of equipotential bonding and automatic disconnection of the supply (see Clause 1.5.5.3c). For automatic disconnection to occur, the prospective fault current must be high enough to operate the protective device, particularly where the touch potential exceeds 50 V a.c. (see Clause 1.5.5.3b). The earth fault current flowing through the impedance of the protective earth system causes a voltage to appear on the earthed metal. This voltage is called the prospective touch voltage (see Clause 1.4.125).

Protection from electric shock

There are three methods used to provide protection from electric shock after indirect contact with an exposed conductive part that should not be live and an extraneous conductive part.

1 Ensuring that when an earth fault occurs, and the extraneous conductive part becomes live, it results in the feed being removed from the circuit within a safe time. This means that the circuit needs to be engineered to limit the earth fault-loop impedance.
2 Using earthed equipotential bonding will help to ensure that the resistance between metal parts that can be touched simultaneously is of such a low value that a potential difference exceeding 50 V cannot be maintained. This method of protection cannot be used alone as disconnection of the feed to the circuit is still necessary to protect against other faults such as overheating of the cables.

   Note that if alternative supplies are in use such as an uninterruptible power supply (UPS) or an emergency generator with automatic starting it is possible that a hazardous voltage may be established and maintained.
3 Removing the feed to the circuit before electrocution can occur by use of an RCD or an RCBO protection device.

When a person or livestock touches at the same time, an accessible exposed part and an extraneous conductive part of the installation under earth fault conditions some current will flow. The magnitude of touch current that flows depends on:

- the potential difference across the accessible exposed part and the extraneous conductive part
- the fault-loop impedance of the circuit, which includes the impedance of the person or livestock forming part of the fault circuit.

The use of extra low voltage and electrical separation, as described in the previous section, are also suitable methods for protecting against indirect contact.

Current path

Figure 11.8 illustrates the path fault current will take should a low impedance fault to earth occur.

Class II Protection

To provide protection against indirect contact, Class II equipment or appliances, which do not rely only on basic insulation, can be used. With Class II protection, additional safety precautions, such as double insulation or reinforced insulation, are provided. Therefore, there is no reliance on either protective earthing or installation conditions in the event of an earth fault.

If Class II protection is not used, then protection can be provided by use of residual current devices or a residual current circuit breaker with over-current protection.

Other protection methods include the use of extra-low voltage and electrical separation.

Damp situations

The *Wiring Rules* require that the selection and installation of electrical equipment in damp situations must provide enhanced protection against electric shock, due to the reduction in body resistance and the likelihood of bodily contact with earth potential (refer to Clause 6.1.2). Section 6 of AS/NZS 3000:2018 details the requirements for damp situations within electrical installations.

REVIEW QUESTIONS

1. How does 'indirect contact' occur?
2. How are the effects of 'indirect contact' minimised?
3. What is the method of protection used to limit the effects of 'indirect contact'?
4. List the factors that determine the magnitude of 'touch current'?
5. What categorises Class II equipment?
6. Why do damp situations represent an increased risk for electric shock?

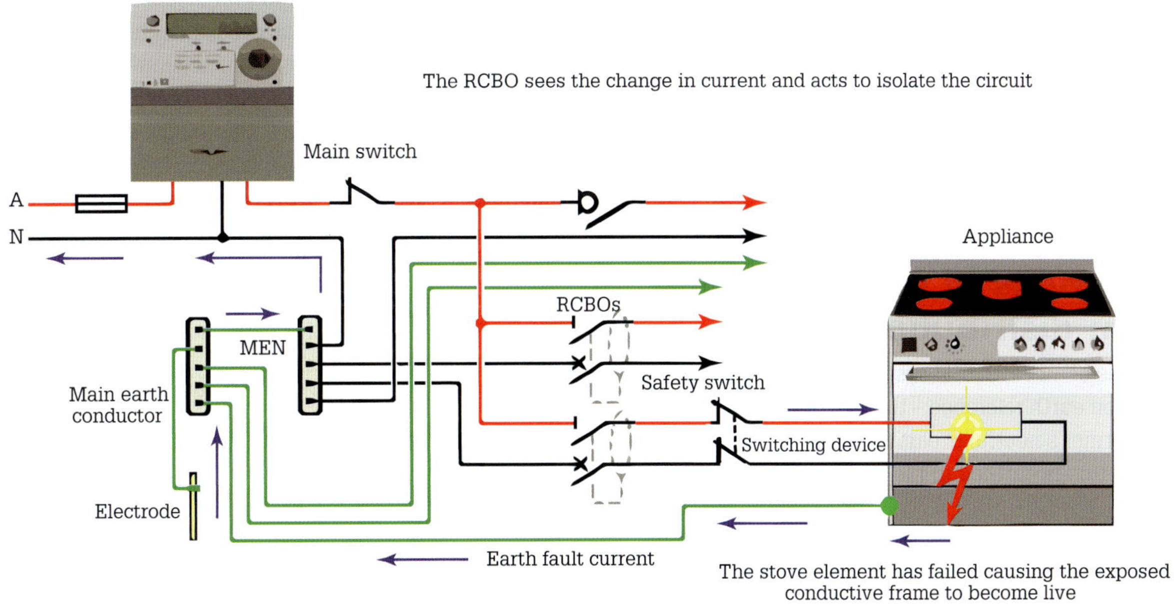

FIGURE 11.8 Earth fault current path

11.5 Earthing

Without exception, the earth is the most important conductor of any electrical installation. An effective earth ensures personnel safety and equipment protection particularly during fault conditions. Section 5 of AS/NZS 3000:2018 *Wiring Rules* deals comprehensively with earthing arrangements and earthing conductors.

Earth

The term 'earth' refers to the general mass of ground or soil. The term 'earthed' (see *Wiring Rules*, Clause 1.4.47) refers to a conductor or metal within the electrical installation that is purposely connected to earth.

The voltages referred to when talking about electrical systems are usually voltages referenced to the general mass of earth. Earth, therefore, represents the reference point, or zero potential point, to which all other voltages refer.

The general mass of ground/soil (earth) is chosen as the zero reference point since it is accessible everywhere. Earth is simply an electrical reference point for the system. When you are standing on the ground, your body is approximately at the voltage potential of the earth. If a conductor or all accessible metal connects to the general mass of earth then, within the electrical installation, the conductor and all accessible metal parts are approximately at ground potential.

There are many methods to distribute electrical energy. The preferred means is by having a four-wire secondary distribution transformer. In order to create a zero reference point, the neutral star point of a four-wire distribution transformer secondary connects solidly to the general mass of earth. The neutral is a return path to complete the electrical circuit and originates from the transformer star point.

The primary purpose of earthing the neutral of a distribution transformer secondary is to provide system voltage stability when imbalance in loading causes neutral current flow. The neutral point of the low-voltage system of the distribution transformer is solidly connected to the general mass of earth via an earth electrode to achieve a very low earth resistance (recommended ≤ 1 Ω). This provides a parallel current path through the general mass of earth ensuring a very low-impedance return path for the current resulting in a very low voltage drop. A secondary purpose is the detection of earth fault currents or earth leakage currents and allowing the protective devices to operate within the desired time.

Advantages of an earth star system distribution

Advantages of an earth star system distribution are:

- Common earth reference – the distribution transformer's neutral is earthed and the consumer's installation is earthed to the same reference voltage point.
- Voltage stability – star-point-connected transformers are earthed primarily to provide voltage stability to a system where an imbalance in loading results in neutral current flow. The neutral star point allows the vector sum of three-phase currents displaced at 120°, nullifying it to zero.
- Low-voltage supply – equipment can be connected to one phase and neutral instead of phase to phase.
- Lower insulation requirements – low voltage means that the insulation requirements for a cable and equipment can be smaller.
- Phase-to-earth faults – if a phase conductor makes contact with the earth the resulting short-circuit earth fault would result in a significantly increased current in the active, causing the protection devices to operate.

Purpose of earthing system

The earthing system in an installation has the following purposes (see *Wiring Rules*, Clause 5.1.2):

1. To minimise the voltage rise (see 'Protection from electric shock' earlier in the chapter) of any earthed metal that may become energised due to the passage of fault currents.
2. To provide a low-impedance path to allow circuit protection to operate when required to clear faults resulting from an insulation failure to earth.

The earthing systems adopted by the majority of power authorities in Australia are the multiple earthed neutral (MEN) system and the common multiple earth system (CMEN). In the MEN system, the high-voltage and low-voltage earth systems are kept separate. The CMEN system is a variation in which high-voltage earths are also bonded to the low-voltage system.

Multiple earthed neutral (MEN) system

The MEN system of earthing (see *Wiring Rules*, Clause 1.4.83) is one in which the distribution neutral conductor is used as a low-resistance return path for earth fault currents and where its potential rise is kept low by having it connected to earth at a number of locations along its length. The neutral conductor is connected to earth at the distribution transformer, at each consumer's installation and specified poles or underground pillars. The resistance of the neutral conductor of the distribution system and the earth must be as low as possible at any location. **Figure 11.9** illustrates a residential MEN earthing system.

The general mass of earth provides a very large, in cross-sectional area (CSA), conductor in parallel with the service neutral(s); as such it serves to create a very-low impedance conductive path back to the source of supply – a distribution transformer. When the installation is earthed according to the AS/NZS 3000:2018 *Wiring Rules*, protection is being provided for persons and animals against the danger of electric shock. A correctly earthed electrical system provides for the electrical isolation of any defective equipment through the operation of protective devices.

With this type of system, the low-voltage distribution transformer neutral point is connected to the ground (soil) both at the transformer and the base of approximately every fourth transmission pole. These grounding conductors attach to earth electrodes embedded in the soil. A timber batten covers the grounding conductor on the pole to provide mechanical protection and also protects people and animals from 'touch potentials'. **Figure 11.10** illustrates the grounding conductor.

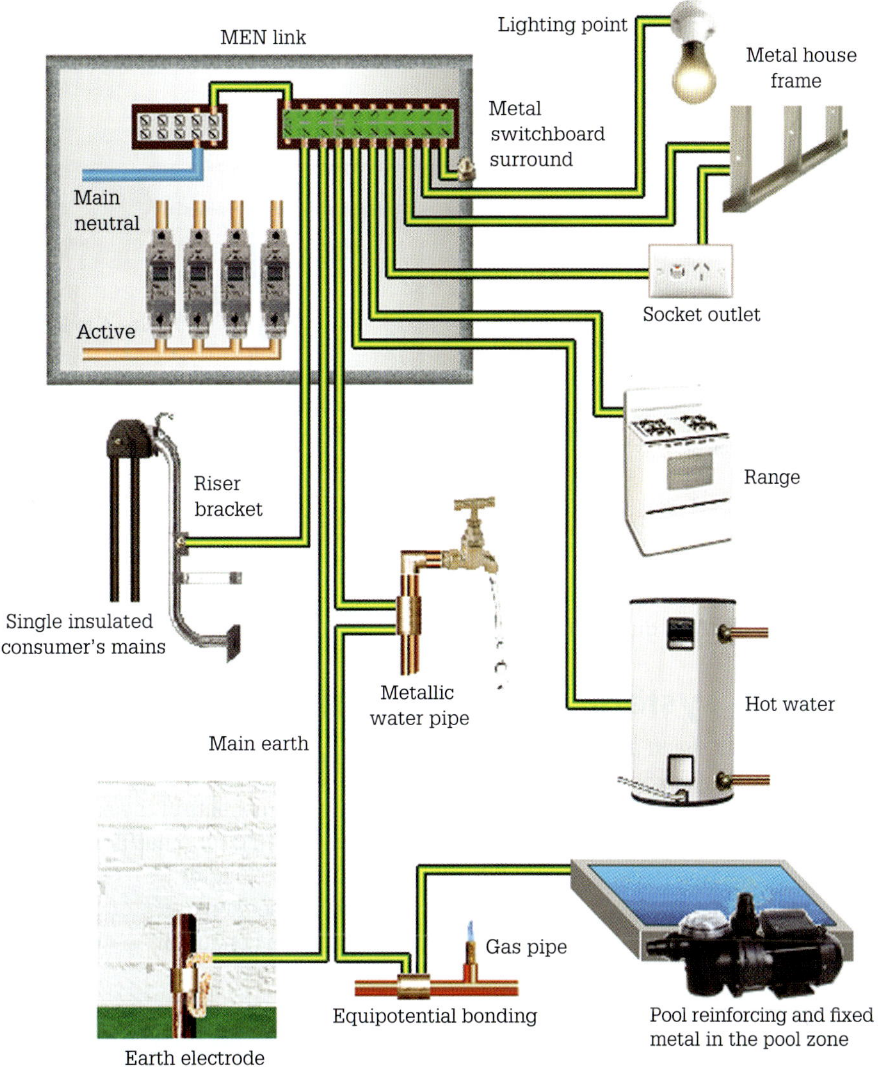

FIGURE 11.9 Residential MEN earthing system

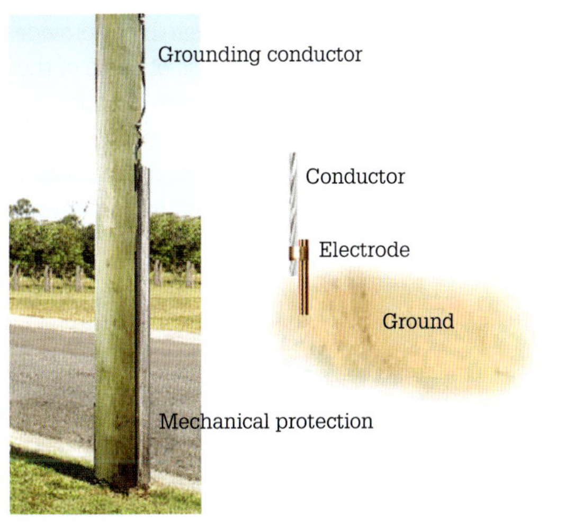

FIGURE 11.10 Grounding conductor

Touch voltage

Touch voltage (see Clause 1.4.125) is a term used to describe the potential difference that occurs across a person or animal by their touching an exposed grounding conductor and the soil at any time when current is flowing. Touch voltage is defined as the potential difference between a person's outstretched hand touching an earthed structure and their foot. In this situation it is assumed a person's maximum reach is one metre.

The MEN system of earthing uses the distribution neutral conductor to carry any fault currents that may flow as a result of the active conductor passing current to earth. The number of parallel paths that the fault current can access in the case of an installation's earth fault occurring assists this flow of fault current. In practice, the number of parallel paths reduces the impedance of the neutral conductor and the general mass of earth to a very small value. The small value of impedance keeps the neutral at earth potential or zero volts. **Figure 11.11** illustrates the many parallel paths created by the distribution neutral and the ground.

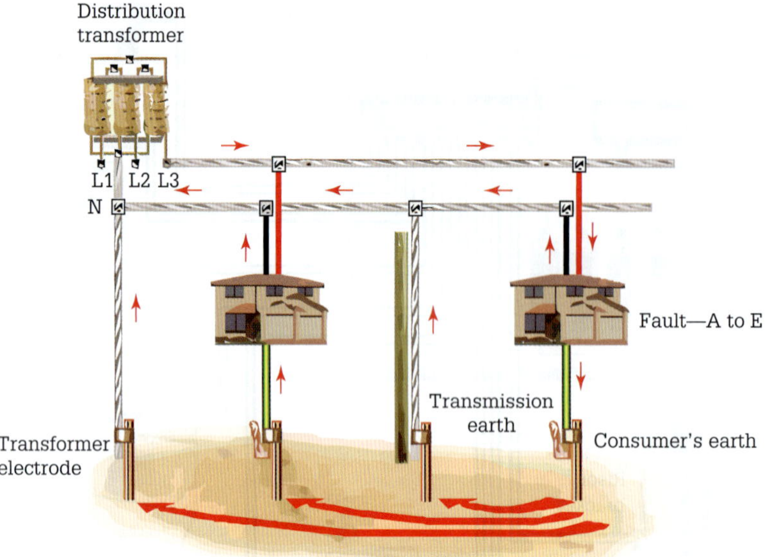

FIGURE 11.11 Parallel paths in an MEN distribution system

Operation of MEN system

Design of the MEN system ensures that if a fault occurs where the active conductor contacted exposed metal, the fault current could flow through the earth conductor as a return path to the switchboard, allowing the protective device to open-circuit, thereby making the circuit safe.

However, reliance on the MEN system alone is insufficient for personal safety. In the event that a neutral was open-circuited or the MEN link was missing, the only path for the current to return to the distribution point is through the ground at the consumer's earth electrode and then to the supply pole transmission earth or other consumer's mains earths.

This path, under very dry conditions, could be a high-impedance path resulting in a rise in potential between the earthing system and all exposed metallic components that are in contact with it. Such a situation could create hazardous touch voltages on exposed metallic conductive parts of the earth circuit.

Installing RCD or RCBO switches in a consumer's installation greatly minimises the effects of touch voltages and their potentially lethal effects.

The effectiveness of the MEN system is reliant on the continuity of the main neutral conductor, the MEN link and the connection of the main earthing conductor to the earth electrode. The best way to connect the earths would be to run all the earthing conductors separately back to a single earth bar. The practical way is to arrange circuit earths as a strict 'tree' structure, with the circuit and bonded earths only connected to the main earth conductor.

Earthing system components

The main components of the earthing system are:

- main earthing terminal/bar
- earthing conductor
- MEN connection
- main earth conductor
- earth electrodes.

Main earthing terminal/bar

Each electrical installation must have a connection point to the earthing system. A terminal (soldered connection) or earth bar (see Clause 5.3.4) as shown in **Figure 11.12** provides this point. The terminal or earth bar provides for the connection to the general mass of earth of the main earth, MEN connection, protective conductors, equipotential bonding conductors and conductors for functional earthing. All connections to the main earth bar must be readily accessible without disturbing any other wiring.

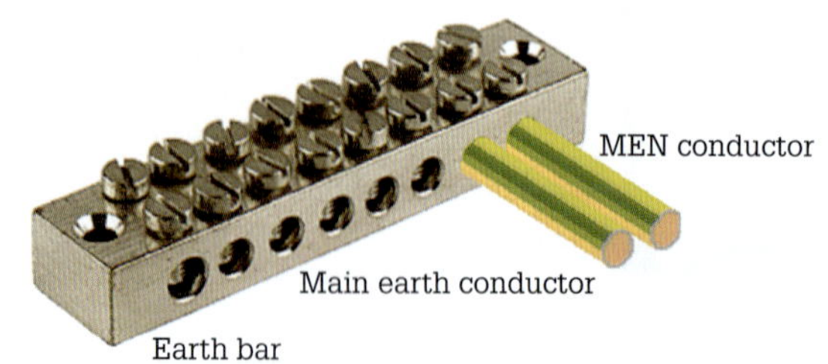

FIGURE 11.12 Earth bar

Earthing conductor

Earthing conductors are usually high-conductivity copper, as shown in **Figure 11.13** (see Clause 5.3.2).

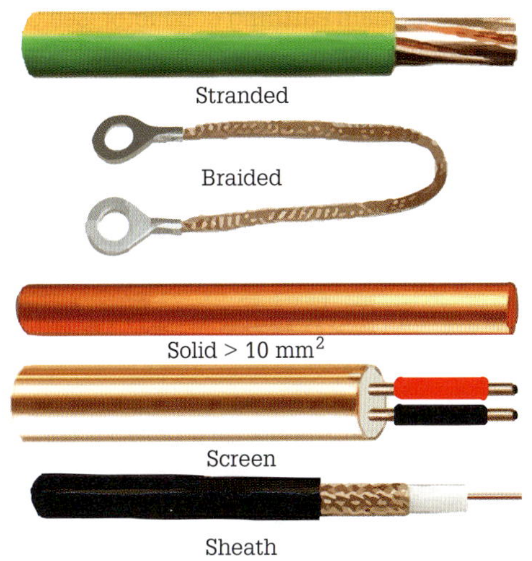

FIGURE 11.13 Copper earthing conductors

Aluminium conductors can also be used as earthing conductors as shown in **Figure 11.14** (see Clause 5.3.2.1.2).

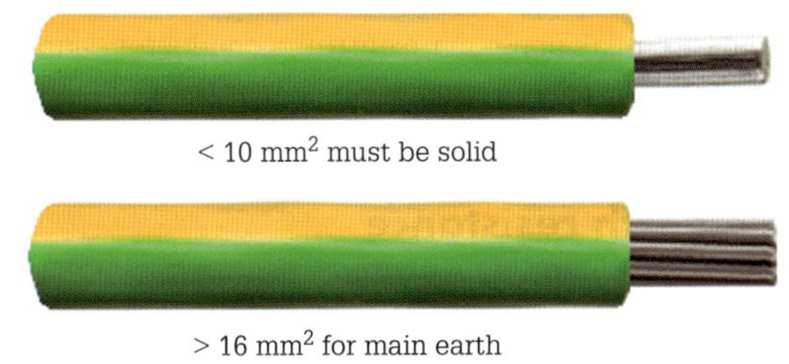

FIGURE 11.14 Aluminium earthing conductors

When aluminium conductors are used as earthing conductors, galvanic corrosion must be prevented, and terminations must be suitable for use with aluminium.

SWITCH ON

Clause 5.3.2.1.3 allows other materials such as stranded steel catenary wire, metal conduit and conductive sheaths to be used as earthing conductors.

When earthing conductors are used in cables, flexible cables or flexible cords (see Clause 5.3.3.4) they must meet minimum CSA requirements. For example, the minimum size of a copper earthing conductor in the arrangement of a single-core insulated cable, flexible cable or flexible cord is 2.5 mm^2.

Joining

There are two main means by which earthing conductors can be connected: soldered connections and tunnel-type connections (see Clause 3.7.2.11). Where soldering is used for the joining or connection of earthing conductors, the earthing conductors must be mechanically joined independently of the solder. **Figure 11.15** shows a method that can be used to meet the requirements.

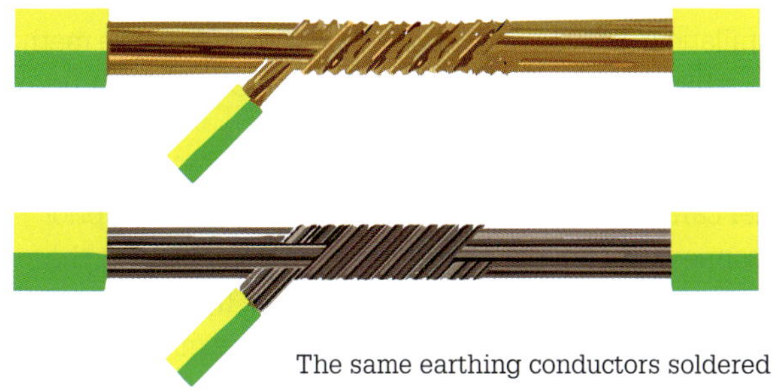

FIGURE 11.15 Connection of earthing conductors

Note how the Y conductors loop under several strands of the cable they are required to be joined to. This method ensures mechanical independence from the solder.

Tunnel-type connections, which include earth bars and links, are usually of the two-screw type but can also be a one-screw design (see Clause 2.10.4.2). Clause 3.7.2.11b also needs to be referred to regarding the type of screws used and also states an exception to the diameter of the one-screw requirement. Tunnel-type connectors are illustrated in **Figure 11.16**.

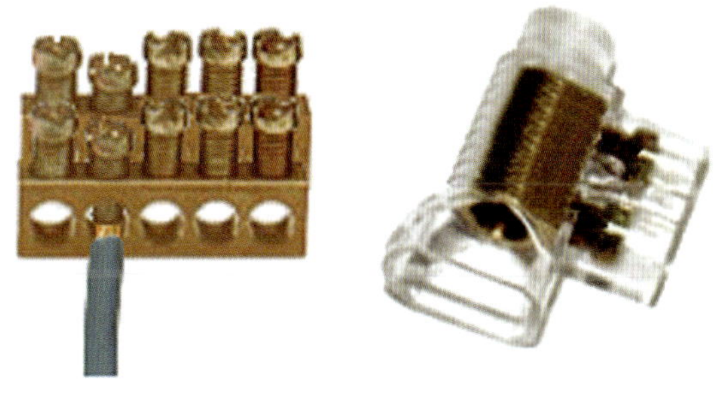

FIGURE 11.16 Tunnel-type connectors

A note under Clause 2.10.4.2 refers to other means of joining or connecting earthing conductors.

MEN connection

The MEN connection is a link between the main earth terminal/bar and the main neutral conductor bar and is made at the main switchboard (see Clause 5.3.5) as illustrated in **Figure 11.17**.

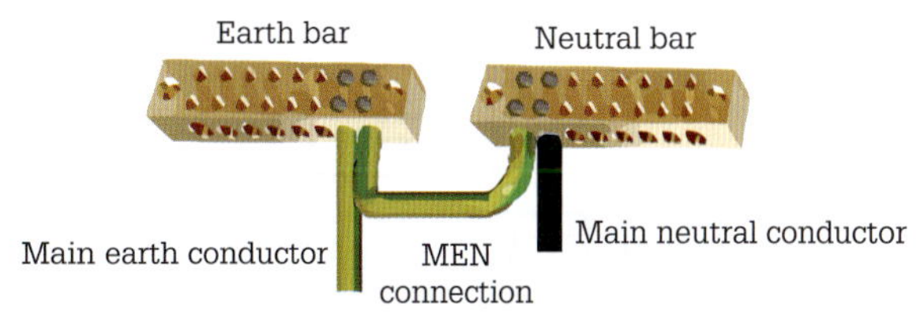

FIGURE 11.17 MEN connection at the switchboard

The current-carrying capacity of the MEN link must not be less than that of the main neutral conductor (see Clause 5.3.5.2 and note the exceptions). If the MEN link is insulated, it must have the colour green or green/yellow (see Clause 5.3.5.3).

Where a switchboard is installed in a separate installation (such as outbuildings and detached portions of an installation), particular attention must be given to the method of earthing and the MEN connection (see Clause 5.5.3.1).

Testing

After carrying out tests on the electrical installation, prior to connection to supply, it is imperative to ensure that the MEN connection is solidly installed. There have been serious electrical accidents where the MEN link was missing and subsequent faults occurred in the installation. For example, when an active-to-earth fault occurs, the earth fault current flows through the MEN link and the low-impedance consumer's mains neutral and the distribution neutral conductor to the source of supply (distribution transformer). The fault current then flows through the installation's protective device, which opens or trips, isolating the circuit. If the MEN link was not in place, the return fault current path would be via the main earth and the general mass of earth to the source of supply. The impedance of the general mass of earth return path could be much higher than the MEN/neutral return path. The higher impedance limits the magnitude of the fault current to a value that may not cause the protective device to operate. If the protective device does not work, a voltage capable of producing an electric shock may remain on the exposed metal of the appliance or equipment. Therefore, ensuring the connection of the MEN link is central to the electrical safety of the installation.

Even in a soundly earthed MEN system, rise in neutral potential can occur due to:

- ineffective earth bonding connections
- voltage drop in the neutral conductor due to single-phase loads
- voltage drop in the neutral conductor resulting from harmonic currents: back emf from single-phase motors after being switched to the supply mains.

Note that in the event of an open-circuited supply neutral or if the polarity of the consumer's mains becomes reversed then the main earth rises to the supply voltage.

Main earth conductor

The main earthing conductor (see Clause 1.4.81) is a conductor connecting the main earthing terminal/connection or bar to the earth electrode, as illustrated in **Figure 11.18**.

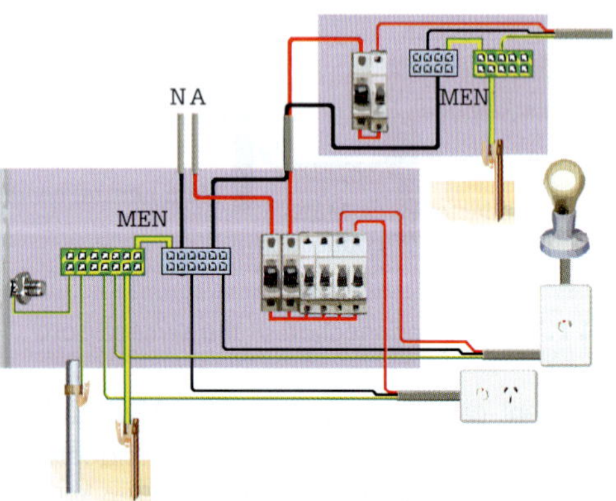

FIGURE 11.18 Circuit diagram of an MEN earth system

The size of the main earthing conductor (see Clause 5.3.3.2) is determined from AS/NZS 3000:2018 *Wiring Rules*, Table 5.1. This table relates the size of the main earthing conductor to the CSA of the largest active conductor of the consumer's mains (see **Figure 11.19**). There are, however, exceptions to this: where double insulation is maintained between the point of supply and the load terminals of the protective devices and where the CSA of the consumer's mains is larger than that required to carry the maximum demand of the installation.

Consumer's mains insulation if Table 5 is used

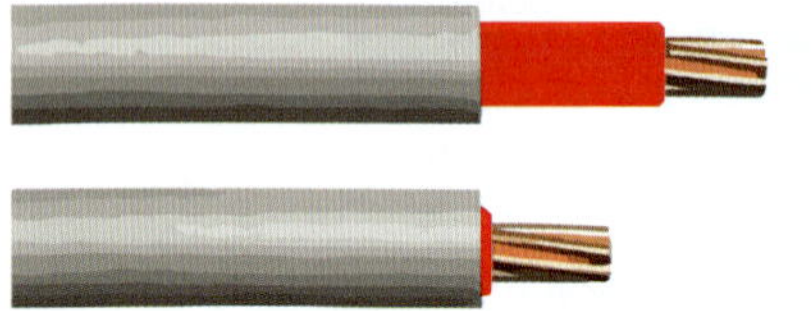

Consumer's mains insulation for the exception

FIGURE 11.19 Consumer's mains

The CSA of a copper main earthing conductor cannot be less than 4 mm^2 and need not be larger than 120 mm^2 (Clause 5.3.3.2).

Main earth resistance

The resistance of the main earthing conductor (see Clause 5.5.1.4) is measured at the main earthing terminal/connection and bar and the earth electrode, including the connection to the earth electrode. The measured resistance between these two points must not be greater than 0.5 Ω.

Labelling of the main earthing connection

The main earth connection should have a permanent label (see Clause 5.5.1.3) attached to the connection to the earth electrode with legible warning against disconnection of the main earth conductor as shown in **Figure 11.20**.

FIGURE 11.20 Main earth conductor warning label

> **SWITCH ON**
>
> The warning label material must not be affected by mechanical impact and environmental conditions and is required to last the life of the installation. In addition, the location of the earth electrode must be identified at the main switchboard (see Clause 5.3.6.4).

Earth electrodes

The purpose of the earth electrode is to connect the neutral of the electrical installation to the general mass of earth (see Clause 5.5.1.2). The principle of earthing is to regard the general mass of earth as a reference (zero) potential. Therefore, all wiring systems connected directly to it will be at zero potential or above it by the level of the voltage drop in the connected system (e.g. the voltage drop across a protective earth carrying fault current). The type of earth electrode should consist of one of the types described in Clause 5.3.6.2 and Table 5.2 of AS/NZS 3000:2018. A copper-clad rod-type electrode is illustrated in **Figure 11.21**.

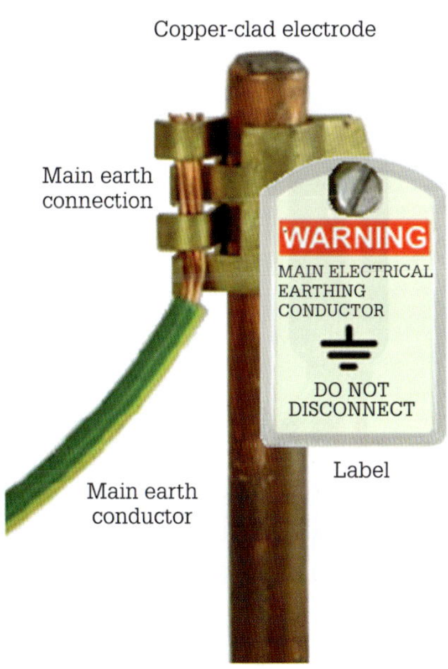

FIGURE 11.21 Copper-clad rod-type electrode

Copper-clad earth rods are manufactured using steel to AS 3679 and hard-drawn copper tube manufactured to AS 1572, with a nominal diameter of 12 mm. All manufactured earth rods are extendable, by using couplers, and are pre-pointed. Rod-type electrodes must be driven to a depth (see Clause 5.3.6.3) of 1.2 m (Australia) and 1.8 m (New Zealand). Strip electrodes have other requirements.

Some additional examples of earth electrodes are:

- an approved earth stake driven to a depth of at least 1.2 m into the ground
- a copper strap of at least 25 mm × 1.6 mm × 3 m buried at a depth of at least 0.5 m in a horizontal trench
- bare copper cable of 25 mm^2 × 3 m buried at a depth of at least 0.5 m in a horizontal trench.

Location of earth electrode

The earthing conductor and its connection to the earth electrode must be located (see Clauses 5.3.6.4, 5.5.1.2, 5.5.5.1, 5.5.5.2 and 5.5.5.3) in moist soil with protection from mechanical damage, and galvanic corrosion must be prevented. Earth electrodes as illustrated in **Figure 11.22** are for an existing residential installation, prior to fit-out and a construction site electrode.

FIGURE 11.22 Earth electrodes

Note the lack of mechanical protection in the first and third electrodes (from left to right) and the required use of a cold galvanising coating on the main earth termination.

Continuity

When undertaking electrical installation work, the requirement for testing of the installation in AS/NZS 3000:2018 *Wiring Rules*, Clause 5.5.4 on continuity, allows for any earth electrodes that are faulty to be identified, especially at the earth clamp connection. A visual check at the earth electrode also allows for the identification of severe corrosion. Repairs need to be undertaken, and the relevant certificate of electrical safety issued.

The PVC pipe surrounding the earth electrode in **Figure 11.23** allows for water to enter the electrode area when surrounded by a concrete slab. Note the lack of mechanical protection.

Resistance of an earthing electrode

The resistance of the earth electrode to the general mass of earth is made up of three essential elements:

- resistance of the electrode
- contact resistance between the electrode and the soil
- resistivity of the soil.

Soil resistivity is the determining element for the resistance of an earth electrode and therefore to what depth it must be driven to obtain low ground resistance. The resistivity of the general mass of earth surrounding an electrode can be enhanced by using an earthing

FIGURE 11.23 Earth electrode surrounded by a concrete slab

compound, illustrated in **Figure 11.24**. The compound is mixed with clean water until a slurry that can be poured is achieved. The slurry is then poured into the area where the electrode is to be installed. Note that the earthing compound reduces the resistivity of the soil only when dissolved in the soil. Using the compound dry does not serve any purpose at all.

FIGURE 11.24 Earthing compound

Protective earthing conductor (PEC)

A PEC (see Clause 5.5.2) is any earth conductor, excluding the main earthing conductor, which is intended to carry earth fault currents and connects any part of the earthing system to the electrical installation or electrical equipment required to be earthed, or to any other part of the earthing system. All PECs must connect directly to the main earth bar or via various methods to the main earth conductor. The symbol in **Figure 11.25** signifies the protective earthing conductor terminal. The tag containing the symbol is placed at the equipment earthing point.

FIGURE 11.25 Protective earthing conductor terminal symbol

Types of PEC

The PEC can take many forms (see Clause 5.3.2.2), such as:

1 A separate conductor which must be green/yellow insulated (see Clause 5.3.2.4 for exceptions).
2 An earth conductor included in a sheathed cable with other conductors.
3 Metallic sheaths, armours and screens of cables, as shown in **Figure 11.26**.

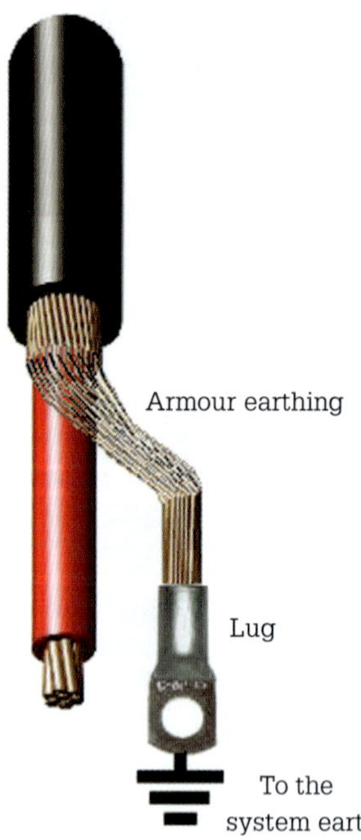

FIGURE 11.26 Armour PEC

4 Conducting cable enclosures such as conduit, tube, pipe, trunking, ducting and similar wiring enclosures (see Clause 5.6.2.4). **Note:** Flexible metallic conduit cannot be used as a protective conductor. This type of conduit cannot be relied upon to retain a low-resistance conducting path.
5 Conductive framework, as illustrated in **Figure 11.27** (see Clause 5.3.2.3c).

FIGURE 11.27 Conductive framework

6 Catenary wires (see Clause 5.3.2.3d) as shown in **Figure 11.28** which illustrates requirements for a catenary PEC.

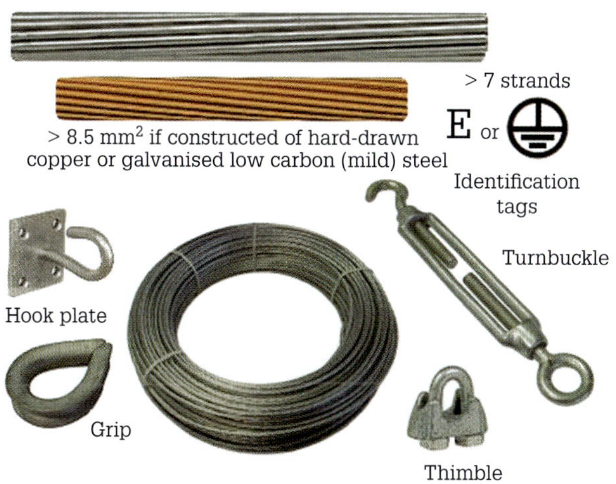

FIGURE 11.28 Catenary wires and fittings

Note that catenaries when used as an aerial earth (see Clause 3.8.3.4) must be identified as such.

The electrical continuity of a protective earthing conductor must be proven. In addition, the electrical continuity must be protected against mechanical, chemical or electrolytic wear and tear.

Size of PEC

The purpose of the protective earthing conductors is to provide a safe path for earth fault current so that the protective device operates to remove dangerous potential differences, which are inevitable under fault conditions. The minimum size of protective earthing conductor required is determined in relation to the CSA of the active-phase conductor supplying the portion of the installation to be protected. Examples of minimum protective earthing conductor sizes from AS/NZS 3000:2018 *Wiring Rules*, Table 5.1, are shown in the following table.

Nominal size of active conductor	Copper protective earthing conductor
10 mm^2	4.0 mm^2
16 mm^2	6.0 mm^2

SWITCH ON

The earthing conductor fixed to the metal frame of the main switchboard is a protective earthing conductor and not an equipotential bonding conductor. A metal switchboard enclosure where *unprotected* consumer's mains enter must be earthed with a protective earthing conductor with a CSA of not less than that of the consumer's mains neutral conductor.

Arrangement of earthing conductors

Examples of earthing arrangements using an earthing bar and a tree arrangement are shown in **Figure 11.29**.

Figure 11.30 illustrates a main switchboard supplying a distribution switchboard at outbuilding 1 which in turn supplies another switchboard at outbuilding 2. Both outbuildings have a separate MEN connection.

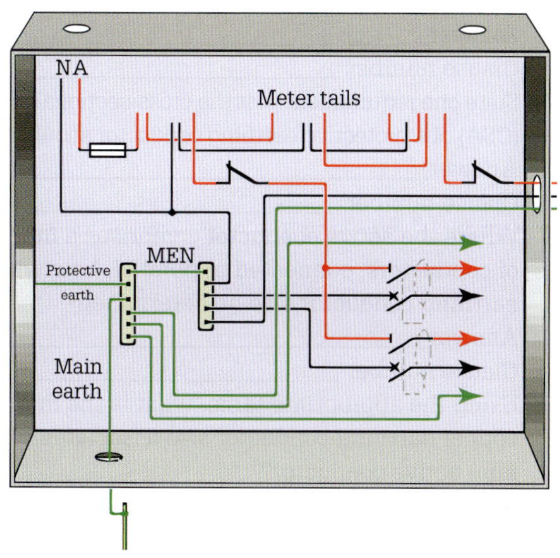

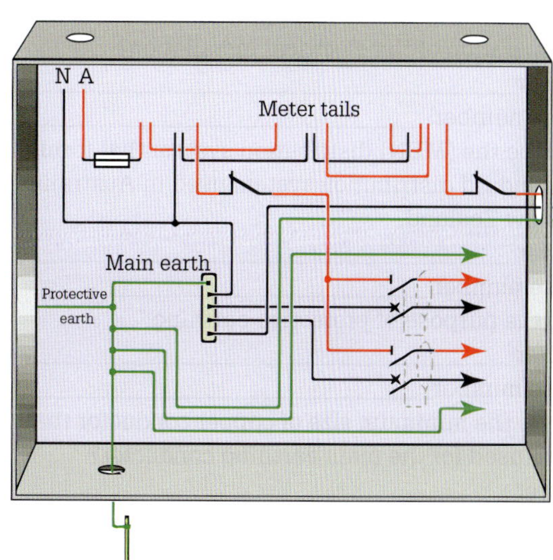

FIGURE 11.29 Examples of earthing arrangements

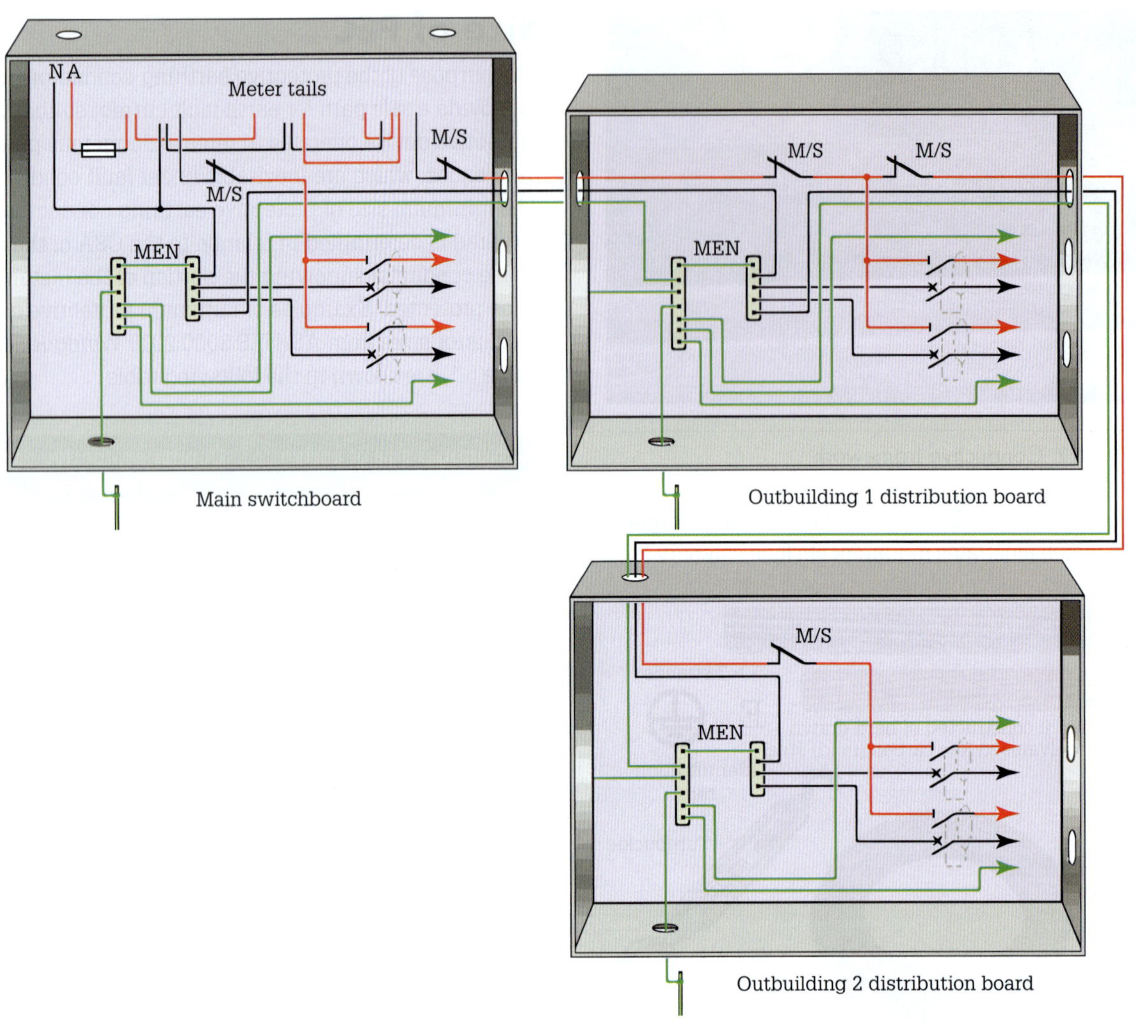

FIGURE 11.30 Separate MEN connections

EXERCISE 11.1

Wiring Rules questions on earthing

Refer to *Wiring Rules*, Section 5, 'Earthing arrangements and earthing conductors' to locate the appropriate rule; summarise the requirement and document the applicable clause number.

1 What are the five earthing functions that selected earthing arrangements must perform to ensure a safe operation of the electrical installation?
Answer ________________
Clause number ________________

2 Describe the (MEN) distribution system that forms the standard distribution system used in Australia and New Zealand.
Answer ________________
Clause number ________________

3 State the purpose of protective earthing.
Answer ________________
Clause number ________________

4 What is the minimum size of copper conductor that can be used for the main earthing conductor?
Answer ________________
Clause number ________________

5 What is the requirement for conductive sheaths, armours and screens of cables if used as a protective earthing conductor?
Answer ________________
Clause number ________________

6 When can a catenary wire be regarded as a protective earthing conductor?
Answer ________________
Clause number ________________

7 State one requirement that the cross-sectional area (CSA) of a protective earthing conductor should ensure.
Answer ________________
Clause number ________________

8 Where the active conductor comprises a number of conductors, connected in parallel, how will the earthing conductor size be determined?
Answer ________________
Clause number ________________

9 Using the appropriate table, what is the minimum copper earthing conductor size associated with a 35 mm^2 active conductor?
Answer ________________
Clause number ________________

»

10 State the **maximum** CSA of a copper main earthing conductor.
Answer____________________
Clause number ____________________

11 State the function of the MEN connection or link within an electrical installation.
Answer____________________
Clause number ____________________

12 How is the CSA of the MEN connection determined?
Answer____________________
Clause number ____________________

13 Where should the location of the earth electrode be identified?
Answer____________________
Clause number ____________________

14 Should the earthing contact of every 230 V socket outlet connect to earth?
Answer____________________
Clause number ____________________

15 Should a protective earthing conductor be provided at every lighting point?
Answer____________________
Clause number ____________________

16 How should an earthing conductor, deemed to be the main earthing conductor, be arranged?
Answer____________________
Clause number ____________________

17 What warning must the permanent label attached to the connection to the earth electrode and the main earthing conductor display?
Answer____________________
Clause number ____________________

18 State the maximum resistance allowed of the main earthing conductor, measured between the main earthing terminal/connection or bar and the earth electrode, plus the connection to the earth electrode.
Answer____________________
Clause number ____________________

19 How should exposed conductive parts of wiring enclosures be earthed?
Answer____________________
Clause number ____________________

REVIEW QUESTIONS

1 Which conductor is classified as the most important in an electrical system?
2 To what does the term 'earth' refer?
3 What is the primary purpose of earthing the neutral of a distribution transformer secondary?
4 Describe the multiple earthed neutral system.
5 Provide a definition for touch voltage.
6 Name the conductors that would terminate in an earth bar.
7 What are two materials commonly used for earthing conductors?
8 There are two common connection methods for joining earthing conductors. What are these methods?
9 What is the MEN connection?
10 What current rating is specified for the MEN link conductor?
11 How is the minimum size of the main earthing conductor determined?
12 What is the maximum size for a copper main earthing conductor?
13 Which earthing system component ensures conductivity with the general mass of earth?
14 What characterises a protective earthing conductor?
15 Outline the purpose of the protective earthing conductor.

11.6 Equipotential bonding

Equipotential bonding (see Clause 5.6) is a method used to reduce potential difference between exposed conductive parts of electrical equipment and extraneous conductive parts of the installation by connecting these parts with cables known as 'bonding conductors'. For example, extraneous metal parts in the pool zone, as illustrated in **Figure 11.31**, must connect together electrically to ensure they are at the same potential. A difference in potential may arise due to currents flowing in the general mass of earth which may not necessarily be fault currents.

Bonding conductors within the electrical installation

Equipotential bonding conductors, with green/yellow coloured insulation, connect the electrical installation earthing system and the metal pipes of other services such as gas and water, as shown in **Figure 11.32**. The bonding of service pipes must occur as near as possible to their point of entry to the building. Clauses 6.6.2.2 to 5.6.2.7 detail these bonding requirements.

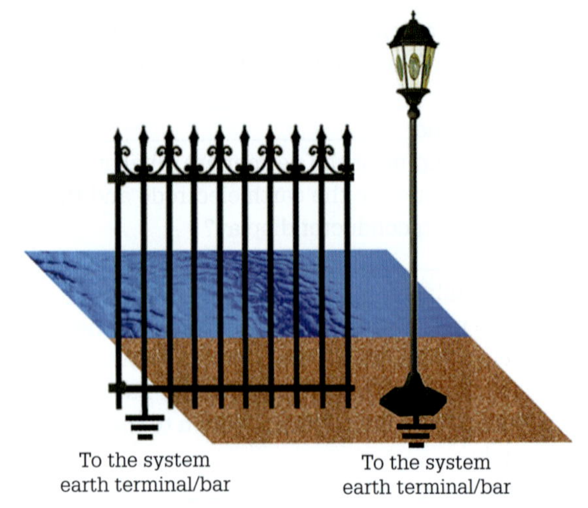

FIGURE 11.31 Extraneous metal parts in a pool zone

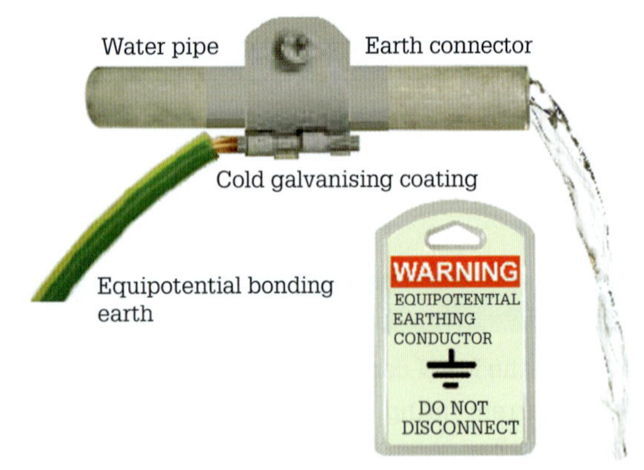

FIGURE 11.32 Equipotential bonding earth

Every bonding connection should have an identifying label. The minimum size of bonding conductors is selected in accordance with Clause 5.6.3.2 and relates to the particular bonding application. For example, a 4 mm^2 copper equipotential bonding conductor is used with bond conductive piping, cable sheaths, and wiring enclosures.

Bonding conductors external to the electrical installation

Bonding conductors, having a CSA not less than 4 mm^2 for copper conductors, connect together extraneous conductive parts. These extraneous parts are exposed conductive materials, which are not associated with the electrical installation, but which may provide a conducting path if not bonded, giving rise to electric shock. Examples within a swimming pool zone include swimming pool reinforcing and metal pool fences.

The resistance of any equipotential bonding conductor should not be more than 0.5 Ω (same as the main earthing conductor). The low value resistance of the bonding conductor ensures that its voltage drop does not prevent the operation of the protective device when carrying fault current.

Installation faults in equipotential bonding conductors

The following installation faults can occur in equipotential bonding conductors:

- The metal cable sheath, metallic wiring enclosure or bare earthing conductor is not separated from or bonded to metal pipes containing flammable liquids as required. (Note the term 'unavoidably in contact' in AS/NZS 3000:2018 *Wiring Rules*.)
- Equipotential bonding conductor of electrical equipment and conductive parts associated with the swimming or spa pool installation not arranged as required.
- Equipotential bonding conductors of the telecommunications equipment not arranged as needed.
- Size of the equipotential bonding conductor associated with metallic piping, cable sheath, wiring enclosure or with a pool is less than the minimum required.
- Resistance of equipotential bonding conductor exceeds 0.5 Ω.
- Equipotential bonding conductor not effectively insulated and identified as an earthing conductor.
- Equipotential bonding conductors are not actually protected against mechanical and chemical influences and electrodynamic forces.
- The equipotential bonding conductor is not attached to the water piping as close as possible to the entry of the water piping.
- Metallic piping systems unavoidably in contact with exposed metal of wiring enclosures are not connected by means of an equipotential bonding conductor.

Functional earthing

Functional earthing (see Clause 5.3.7) refers to the need for certain points in electrical equipment or systems to be connected to the earthing system in order to maintain their proper function. In power systems, an example of a functional earth is earthing of the star point of a three-phase transformer. RCDs are provided with a functional earthing conductor typically coloured white or pink. The purpose of the functional earthing conductor is to provide an alternative signal path for the operation of the RCD if the neutral is disconnected.

Functional earthing conductors do not perform a safety function and are not considered to be part of the earthing system of an electrical installation. A functional earth terminal marking symbol and marking is shown in **Figure 11.33**.

FIGURE 11.33 Functional earth terminal's symbol and marking

Many electronic devices require a protective earth and a functional earth. These different earth requirements result in electromagnetic field problems called noise. Noise can cause interference to electronic equipment that carries data. Telecommunications earth or communications earth conductors are required to be connected to clean low-noise functional earths in order to function correctly and dependably.

Single-wire earth return (SWER) system

The SWER system uses one metal conductor (in most cases a bare conductor) to provide power and the general mass of earth (ground) as a return current conductor, as shown in **Figure 11.34**. As the general mass of earth is used as a conductor, the resistivity of the soil must be within a defined range for the system to conduct effectively. The SWER system is a low-cost and low-maintenance method of supplying power to rural (isolated) areas where the base loads (100 kW to 200 kW) and the population are small. An isolating transformer separates the main feed high-voltage system (33 kV or 11 kV) from the SWER line (19.1 kV or 12.7 kV). The single-conductor SWER line, which can be up to 300 km in length, feeds a single-phase distribution transformer which steps down the line voltage to two single-phase outputs in a 230–0–230 V a.c. arrangement or one circuit of 460 V.

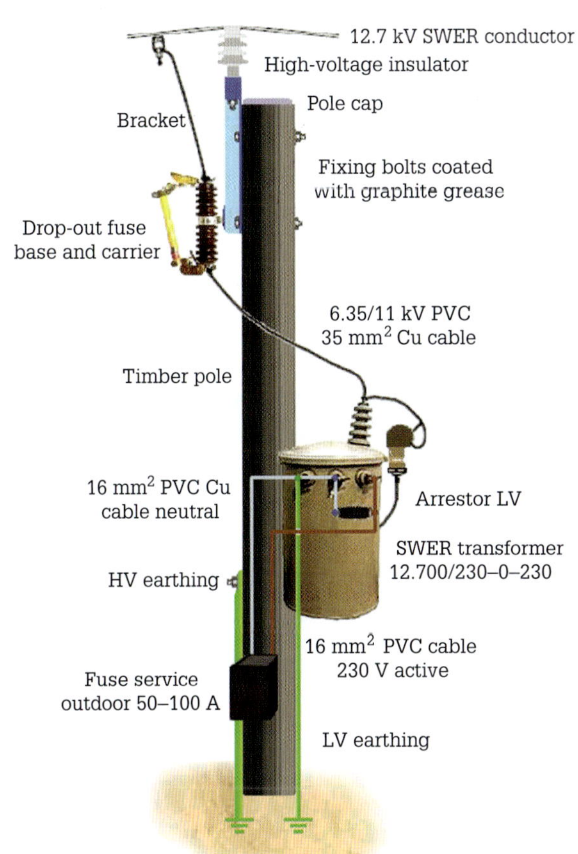

FIGURE 11.34 SWER system

A SWER system consists of a separate high-voltage earthing system and a low-voltage earthing system. Therefore, for each installation there is a low-voltage earth electrode system at the SWER transformer and an earth electrode at the installation's distribution board earthing the low-voltage neutral. The design and impedance of the low-voltage earthing system must be arranged so that the MEN system complies with AS/NZS 3000:2018 *Wiring Rules*. A SWER circuit diagram is illustrated in **Figure 11.35**.

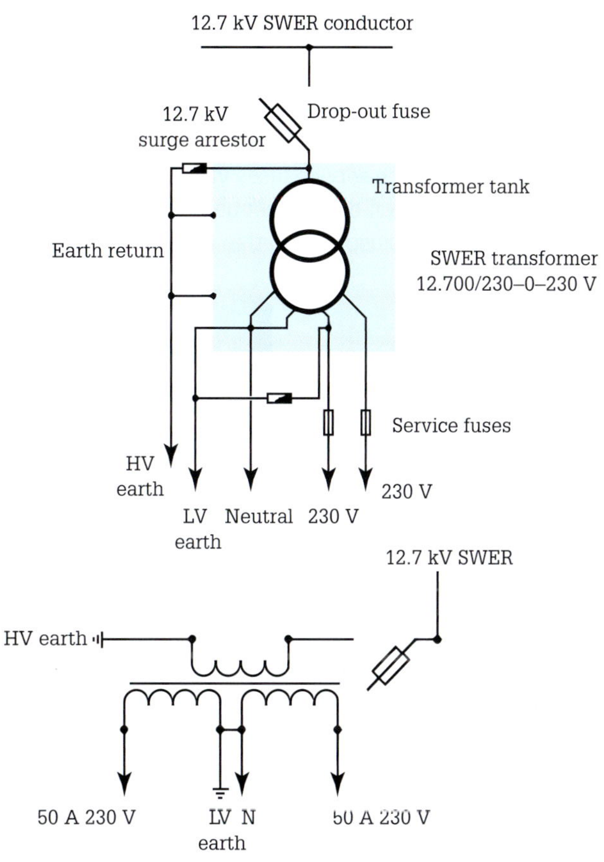

FIGURE 11.35 SWER circuit diagram

A disadvantage of an SWER system is that electrical equipment used in cold rooms, air-conditioners and irrigation pumps may not operate as efficiently as they could on three-phase power and their output may be restricted.

Electric motors connected to SWER supplies must be rated for either 230 V or 460 V 50 Hz single phase. Consequently, rural populations are unable to take advantage of low cost three-phase 400 V electric motors to run their operations. Although 230 V motors are readily available, the sizes are limited to small kW ratings. Single-phase motors of 460 V are non-standard and expensive. In addition, there are voltage regulation problems as SWER consumers increase their load demands. However, this problem may be eliminated as power electronic solutions become more readily available.

An SWER line must be installed at a safe height and if near stockyards, irrigation areas (use of spray rig booms) or other areas where breaching the exclusion zone is possible the SWER line must be insulated and warnings of its presence and potential danger must be provided. In South Australia, Stobie pole (composite concrete and steel pole) or SURELINE® poles are used to carry SWER lines.

SWITCH ON

Where a SWER system for distribution of electricity is used, electricians must be aware of the safety issues and the requirements of the network distributor.

Any conductive metal poles, posts, struts, brackets, stay-wire rope and any other conductive support (as shown in **Figure 11.36**) used for the distribution of the low-voltage supply (230 V) from the service fuses via aerials to the consumer's distribution board must be earthed (see Clause 5.4.5). Note the four exceptions to this clause requirement.

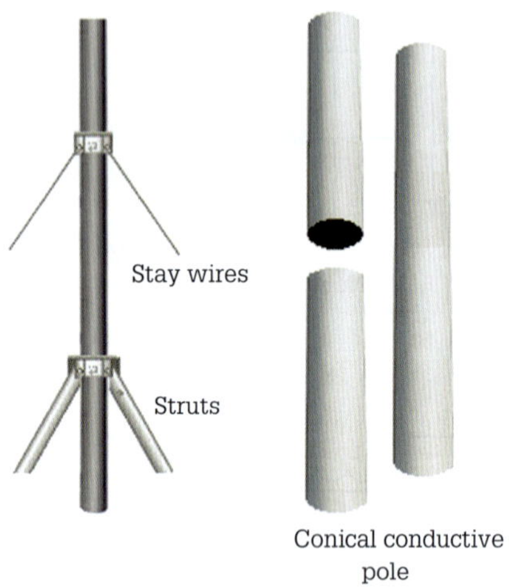

FIGURE 11.36 Conductive poles, struts and stay wires

Stay-wire ropes are either galvanised or stainless steel consisting of a 7 mm to 10 mm strand using seven wires (1 × 7 wire rope).

EXERCISE 11.2

Wiring Rules **questions on equipotential bonding**

Refer to *Wiring Rules*, Section 5, 'Earthing arrangements and earthing conductors' to locate the appropriate rule; summarise the requirement and document the applicable clause number.

1 Is equipotential bonding of extraneous conductive parts included within a protective earthing arrangement for an electrical installation?
Answer____________________
Clause number ____________________

2 State the maximum resistance of equipotential bonding conductors.
Answer____________________
Clause number ____________________

3 State the purpose of equipotential bonding.
Answer____________________
Clause number ____________________

4 When should conductive water piping be bonded to the earthing system?
Answer____________________
Clause number ____________________

5 Is it a requirement that the conductive reinforcing within a concrete floor or wall establishing part of the shower or bathroom be bonded to the earthing system of the electrical installation?
Answer____________________
Clause number ____________________

6 What items must be equipotential bonded in swimming pools and spas?
Answer____________________
Clause number ____________________

7 What is the minimum CSA of equipotential bonding conductors required to connect conductive parts of the shower, bathroom, swimming pool or spa?
Answer____________________
Clause number ____________________

REVIEW QUESTIONS

1. What is the function of equipotential bonding?
2. What colour insulation identifies equipotential bonding conductors?
3. Name non-electrical items that would be equipotential bonded within a swimming pool zone.
4. What is the maximum permitted resistance of equipotential bonding conductors?
5. Describe the purpose of functional earthing.
6. Describe the single-wire earth return system.

11.7 Protection against overload and short-circuit current

Each circuit in the electrical installation must be designed to protect people, livestock and property from harmful effects (see Clause 1.6.1). The method of protection is outlined in section 11.4 'Protection against indirect contact'. Automatic disconnection of the phase conductor must occur within specified times (see Clause 1.5.5.3d) when a short-circuit (phase to earth) of negligible impedance occurs between a phase (active) conductor and exposed conductive parts anywhere in an electrical installation. It is the trip characteristic of the protective device that determines whether or not disconnection occurs within specified times.

Earth fault-loop

The earth fault-loop impedance is the impedance of the phase-to-earth fault current loop starting and ending at the point of the earth fault, as shown in **Figure 11.37**.

The impedance needs to be low enough to allow the fault current to operate the protective device, within a given period. The earth fault-loop impedance path in an MEN system includes the following elements (see **Figure 11.37**):

1 The phase conductor as far as the position of the fault, including service mains, service line, consumer's mains, sub-mains and the final sub-circuit.
2 The protective earthing conductor, including the main earthing terminal/connection or bar and MEN connection.
3 The neutral return path, containing the neutral conductor between the main neutral terminal or bar and the neutral point of the transformer, as well as supply mains, service line and consumer's mains.
4 The path to the neutral point of the transformer and the transformer winding.

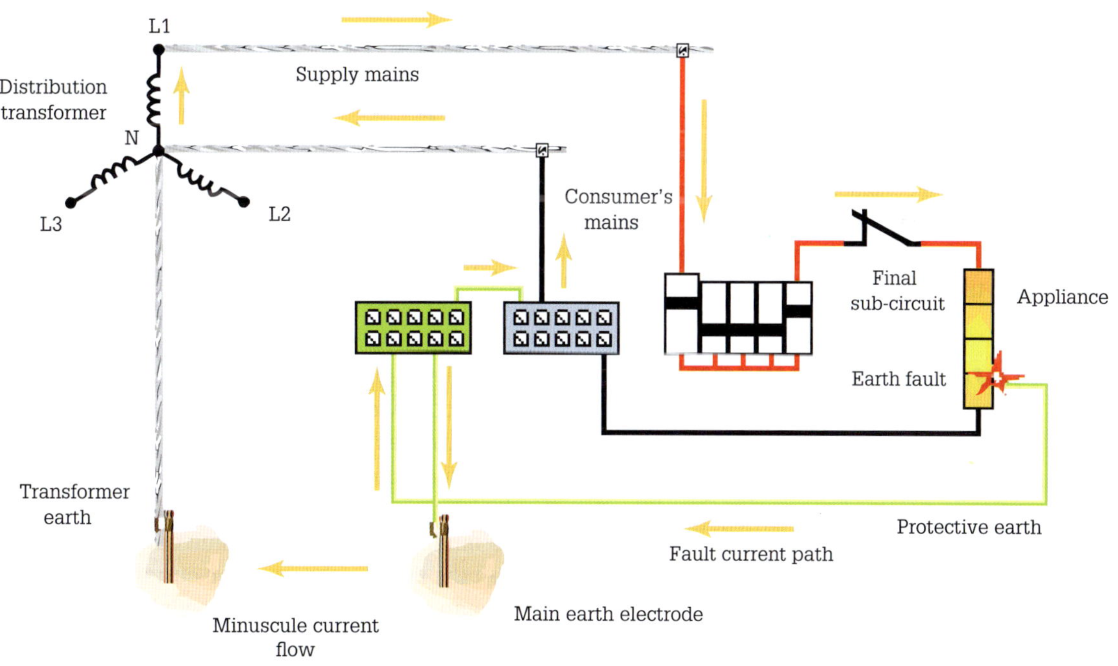

FIGURE 11.37 Earth fault-loop impedance path in an MEN system

Fault-loop impedance test

An earth fault-loop impedance or load test is applied to the installation to ensure continuity and effectiveness of the earthing arrangement. The circuit diagram of the earth fault-loop in an MEN system is shown in **Figure 11.38**.

The main concern with earth fault impedance is the need to disconnect the circuit within a required time when the touch voltage exceeds 50 V a.c. Clause 1.5.5.3d provides the recommended disconnection times for protective devices: 0.4 s for socket outlets and 5 s for other circuits, which include sub-mains and final sub-circuits that supply fixed or stationary equipment. To ensure the protective device operates within the prescribed time, its time–current (tripping) characteristics must be examined.

Supplementary equipotential bonding

In order to ensure that automatic disconnection occurs within the required time should a low impedance fault occur between active and the conductive frame of

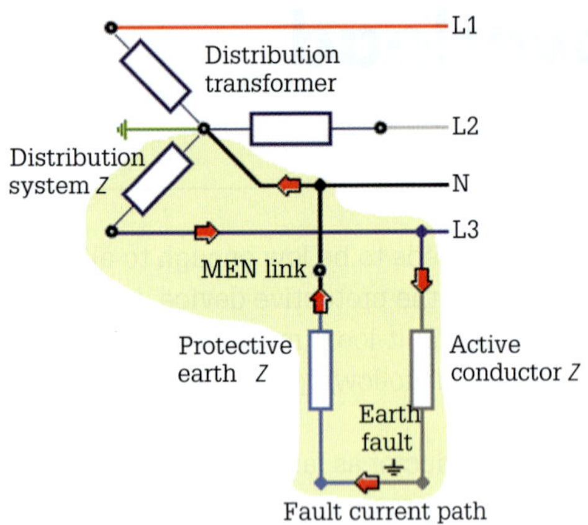

FIGURE 11.38 Equivalent fault-loop circuit for an MEN system

electrical equipment, it is imperative that the associated earthing conductor has a very low impedance. Should it be necessary to reduce the impedance of the earth fault-loop so that the requirements of Clause 5.7.2 are satisfied it may be practical to bond extraneous conductive parts to the earthing system. This effectively increases the CSA of the earthing conductor, thereby decreasing the impedance of the return path offered by the earthing conductor. This in turn will increase the magnitude of the fault current to a value that will allow the over-current protection device to operate in the required time.

If it is not possible to reduce the impedance of the earth fault-loop using supplementary equipotential bonding to an appropriate value, then consideration should be given to using an alternative protective device having a lower automatic operating current that affords disconnection within the required time such as an RCD or RCBO.

Prospective fault current

The electrical installation must be capable of enduring, without destruction, the prospective fault current that can be drawn under a fault condition. The prospective fault current refers to the magnitude of current that would flow at a particular point in the installation if a fault of negligible impedance occurred at that point. The prospective fault current depends on the voltage at the source and the impedance of the upstream circuit.

The prospective fault current is the maximum current that can flow under short-circuit conditions within the electrical installation. The value of the short-circuit current is reliant upon the size and impedance of the transformer and the impedance of the conductors supplying the fault. For example, in domestic installations connected to a low-voltage supply with a 70 A service from a 315 kVA supply transformer, the prospective short-circuit current according to some local electricity distributors at the consumer's terminals is deemed to be 6.5 kA phase to earth and 10 kA between phases for 0.1 s. Commercial and industrial installations are rated as having 25 kA fault levels.

These values for prospective fault current are the maximum fault current likely to be encountered at the point of supply (POS). Higher prospective short-circuit values apply to the consumer's POS closer to supply transformers.

Fault current data

Data concerning prospective short-circuit and fault currents at the POS are obtained from the local electricity distributor. Knowledge of prospective fault current is necessary for an installation designer to provide adequate protection for the installation when choosing conductors, switch gear, switchboards and protective devices.

The selection of switchboard-mounted switchgear (isolating switches, RCDs, MCBs or HRC fuses) requires knowledge of the prospective fault currents that are likely to be present on an installation and the kA rating of the protection devices that is appropriate for these fault currents. A protection device must have a high enough kA rating to open safely without destroying itself under fault current conditions (refer to Clause 2 of AS/NZS 3000:2018).

In domestic installations, the circuit breakers downstream from the service fuses are usually 6 kA with some 10 kA.

Prospective fault level at the secondary of a distribution transformer

Switchboards use a cascading protection method where protection devices with different kA ratings connect in series either on the same switchboard or separate switchboards. These devices provide backup protection for each other by limiting the fault current flowing through the series combination of protection devices.

The let-through energy (I^2t) that passes before a protection device detects the fault is constrained by the breaking capacity of an upstream protection device if a downstream device fails to break the circuit in time. A calculation is used to provide a quick and simple means of determining the prospective fault current.

At the POS, two prospective current fault levels are acceptable. The first is the real fault level determined primarily by the impedance of the distribution transformer supplying the installation (see **Figure 11.39**). The second is that stated by a supply authority (e.g. 10 kA).

The fault level at the transformer terminals may not reflect the actual prospective fault current level at the main switchboard because the additional impedance of the consumer's mains may make a significant reduction.

The equation used for calculating the prospective fault level at the secondary terminals of a distribution transformer as shown in **Figure 11.39** is derived from the equation for 3-phase power:

$$P = \sqrt{3} \times V_L \times I_L \times \cos\phi$$

FIGURE 11.39 Pole-mounted 100 kVA distribution transformer

As the power factor of the load is unknown, it is usual to express the transformer rating as apparent power in kVA or MVA.

Three-phase apparent power

$$VA = \sqrt{3} \times V_L \times I_L$$

For a distribution transformer

$$kVA = \frac{\sqrt{3} \times V_L \times I_L}{1000}$$

Transposing the above equation specifically for the secondary line current at full-load we obtain:

$$I_{LS} = \frac{kVA \times 1000}{\sqrt{3} \times V_L}$$

Transformer fault current

The impedance of the transformer has a major effect on system fault levels. It determines the maximum value of current that flows under fault conditions. Transformer impedance is usually expressed as a percentage of supply voltage to cause full-load current to flow if the secondary connects to a short circuit. A typical value for the impedance of a supply transformer is 5%.

Prospective fault current (PFC)

$$I_{PFC} = \frac{I_{LS} \times 100}{Z\%}$$

Therefore:

$$I_{PFC} = \frac{kVA \times 1000 \times 100}{\sqrt{3} \times V_{LS} \times Z\%}$$

where: V_{LS} = secondary load voltage

$Z\%$ = impedance of the transformer

kVA = transformer capacity

EXAMPLE 11.1

A 150 kVA, 11 kV:230/400 V, three-phase distribution transformer has 5% impedance. Calculate the maximum prospective fault current at the secondary terminals of the transformer.

Prospective fault current

$$I_{PFC} = \frac{kVA \times 1000 \times 100}{\sqrt{3} \times V_{LS} \times Z\%}$$

$$= \frac{150 \times 1000 \times 100}{\sqrt{3} \times 230 \times 5}$$

$$= 7530 \text{ A}$$

EXERCISE 11.3

a A 500 kVA, 11 kV:230/400 V, three-phase distribution transformer has 4.5% impedance. Determine the maximum prospective fault current at the secondary terminals of the transformer.

b A 750 kVA, 11 kV:230/400 V, three-phase distribution transformer has 4.8% impedance. Determine the maximum prospective fault current at the secondary terminals of the transformer.

Impedance of source

In order to determine the downstream fault level, it is necessary to represent the upstream fault level as an impedance. This represents the value of impedance that will cause a current equal to the prospective fault current for the nominal supply voltage. The equation for this is as follows:

$$Z_S = \frac{V_S}{I_{Fault}}$$

where: Z_S = the impedance of the source

V_S = the nominal phase voltage, 230 V

I_{Fault} = the magnitude of the fault current

EXAMPLE 11.2

The supply authority service rules state that the prospective fault current at the POS in a domestic installation is 10 kA. Calculate the impedance at the POS.

$$Z_S = \frac{V_S}{I_{Fault}}$$

$$= \frac{230}{10\,000}$$

$$= 0.023\ \Omega$$

EXERCISE 11.4

The supply authority service rules state that the prospective fault current at the POS in a non-domestic installation is 25 kA. Calculate the impedance at the POS.

Installation fault level

The prospective fault current diminishes as the fault moves further away from the source of supply. The impedance of the circuit conductors combines with the impedance of the source to further reduce the fault current.

Example 11.3 examines the effect the impedance of the consumer's mains has on reducing the fault level at the main switchboard. For this example, consider the equivalent circuit of **Figure 11.40** where the fault loop comprises the impedance of the source (Z_S), the impedance of the active conductor of the consumer's mains (Z_{CMa}), and the impedance of the neutral conductor of the consumer's mains (Z_{CMn}).

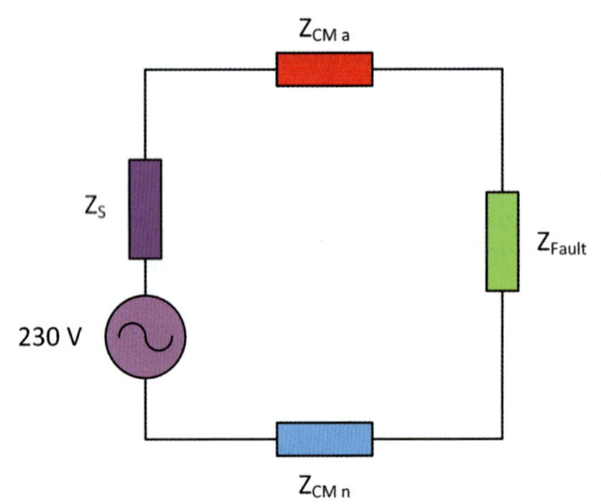

FIGURE 11.40 Equivalent fault-loop circuit for fault at MSB

EXAMPLE 11.3

The supply authority service rules state that the prospective fault current at the POS in a domestic installation is 10 kA, which equates to an impedance at the POS of 0.023 Ω (Example 11.2). Calculate the fault current level at the main switchboard if the impedance of the active conductor in the consumer's mains is 0.005 Ω and the impedance of the neutral conductor in the consumer's mains is 0.005 Ω.

$$I_{\text{Fault MSB}} = \frac{V_S}{Z_S + Z_{CMa} + Z_{CMn}}$$

$$= \frac{230}{0.023 + 0.005 + 0.005}$$

$$= 6.97 \text{ kA}$$

EXERCISE 11.5

The supply authority service rules state that the prospective fault current at the POS in a non-domestic installation is 20 kA, which equates to an impedance at the POS of 0.0115 Ω. Calculate the fault current level at the main switchboard if the impedance of the active conductor in the consumer's mains is 0.006 Ω and the impedance of the neutral conductor in the consumer's mains is 0.006 Ω.

Circuit fault level

For a line-to-earth fault in a final subcircuit, the fault current flows from the source of supply, the installation active conductors, the low impedance fault, and the protective earth conductor, the MEN link, and the supply neutral conductor as shown in **Figure 11.41**. In this situation, some current will also flow through the main earthing conductor, the general mass of earth and the MEN system of nearby electrical installations, creating multiple parallel paths. As environmental conditions are variable, the impedance of this return path could be extremely low (as is the case after a prolonged period of rain). In this case it is better to err on the side of safety and assume that the impedance of the return path is negligible, which will result in a higher value of fault current than might actually occur.

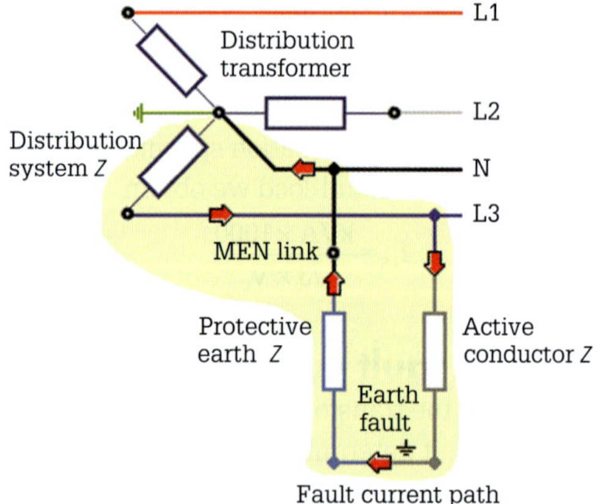

FIGURE 11.41 Equivalent fault-loop circuit for installation with MEN earthing system

Example 11.4 examines the effect the impedance of the final subcircuit conductors have on reducing the fault level at load should a fault of negligible impedance occur between active and earth. For this example, consider the equivalent circuit of **Figure 11.42** where the fault loop comprises the impedance of the source (Z_S), the impedance of the active conductor of the consumer's mains (Z_{CMa}), the impedance of the active conductor of the final sub-circuit (Z_{FSCa}), the impedance of the protective earthing conductor of the final sub-circuit (Z_{FSCpe}), and the impedance of the return path (0 Ω).

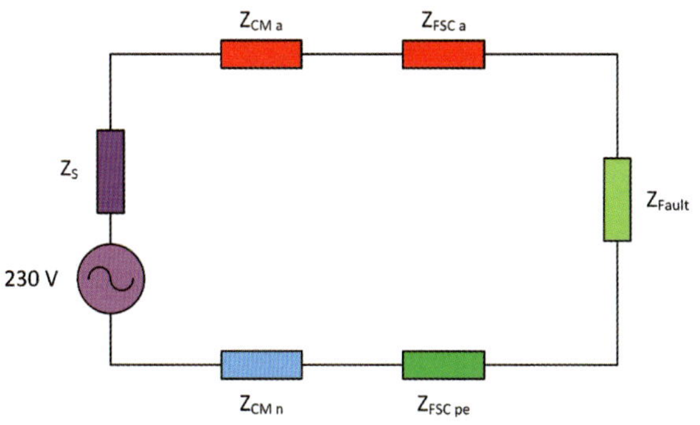

FIGURE 11.42 Equivalent fault-loop circuit for FSC with MEN earthing system

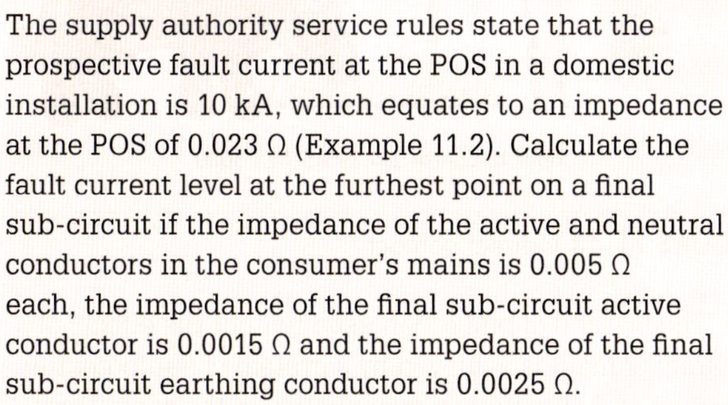

EXAMPLE 11.4

The supply authority service rules state that the prospective fault current at the POS in a domestic installation is 10 kA, which equates to an impedance at the POS of 0.023 Ω (Example 11.2). Calculate the fault current level at the furthest point on a final sub-circuit if the impedance of the active and neutral conductors in the consumer's mains is 0.005 Ω each, the impedance of the final sub-circuit active conductor is 0.0015 Ω and the impedance of the final sub-circuit earthing conductor is 0.0025 Ω.

$$I_{FaultMSB} = \frac{V_S}{Z_S + Z_{CMa} + Z_{FSCa} + Z_{FSCN}}$$
$$= \frac{230}{0.023 + 0.005 + 0.0015 + 0.0025}$$
$$= 7.19 \text{ kA}$$

In this example the expected fault current is slightly higher than that calculated at the main switchboard of Example 11.3. The reason for this is we are assuming the impedance of the return path, from the MEN link, is 0 Ω (owing to the multiple parallel return paths through the general mass of earth and neutral conductors).

EXERCISE 11.6

The supply authority service rules state that the prospective fault current at the POS in a non-domestic installation is 20 kA, which equates to an impedance at the POS of 0.0115 Ω.Calculate the fault current level at the furthest point on a final sub-circuit if the impedance of the active and neutral conductors in the consumer's mains is 0.006 Ω each, the impedance of the final sub-circuit active conductor is 0.0085 Ω and the impedance of the final sub-circuit earthing conductor is 0.0095 Ω.

Determining fault current level

It is important to be able to calculate fault current levels to enable the correct selection of protection devices. A circuit breaker, which is not able to safely isolate the fault current may suffer catastrophic failure. For example, if the fault level at a switchboard was determined to be 6 kA, it would be inappropriate to install 4.5 kA circuit breakers without installing additional upstream fault current limiters.

EXAMPLE 11.5

A factory installation is supplied from a 500 kVA, 11 kV:230/400 V, three-phase distribution transformer having 5% impedance. The main switchboard is supplied by a 60-m length of four-core and earth 185 mm² X-HF-90 cable having copper conductors. This main switchboard supplies a sub-board through a 45-m length of 95 mm² four-core and earth V-90 insulated cable having copper conductors. An MEN connection exists on the low-voltage side of the distribution transformer and the main switchboard. Determine the maximum prospective fault current at the sub-board.

Solution:

1 Calculate the prospective fault current at the distribution transformer:

$$I_{PFC} = \frac{kVA \times 1000 \times 100}{\sqrt{3} \times V_{LS} \times Z\%}$$
$$= \frac{500 \times 1000 \times 100}{\sqrt{3} \times 230 \times 5}$$
$$= 25\,100 \text{ A}$$

2 Calculate the impedance of the distribution transformer:

$$Z_S = \frac{V_S}{I_{PFC}}$$
$$= \frac{230}{25\,100}$$
$$= 0.00916 \ \Omega$$

3 Calculate the impedance of the active conductor of the consumer mains:

$$Z_{CMa} = L_{CM} \times \frac{\sqrt{R_{CM}{}^2 + X_{CM}{}^2}}{1000}$$
$$= 60 \times \frac{\sqrt{0.129^2 + 0.0725^2}}{1000}$$
$$= 0.00888 \ \Omega$$

X-HF-90 is an XLPE cable with a maximum conductor temperature of 90 °C

4 Calculate the impedance of the active conductor of the sub-mains:

$$Z_{SMa} = L_{SM} \times \frac{\sqrt{R_{SMa}{}^2 + X_{SMa}{}^2}}{1000}$$
$$= 45 \times \frac{\sqrt{0.236^2 + 0.0766^2}}{1000}$$
$$= 0.0112 \ \Omega$$

»

5 Calculate the impedance of the protective earth conductor of the sub-mains:

$$Z_{SMPe} = L_{SM} \times \frac{\sqrt{R_{SMPe}^{\ 2} + X_{SMPe}^{\ 2}}}{1000}$$
$$= 45 \times \frac{\sqrt{0.884^2 + 0.0853^2}}{1000}$$
$$= 0.040\ \Omega$$

From AS/NZS 3000:2018, Table 5.1, the protective earth associated with 185 mm² active copper conductors is 25 mm² copper.

6 Calculate the fault level at the sub-board:

$$I_{Fault\,SB} = \frac{V_s}{Z_s + Z_{CMa} + Z_{SMa} + Z_{SMPe}}$$
$$= \frac{230}{0.00916 + 0.00888 + 0.0112 + 0.040}$$
$$= 3.32\ \text{kA}$$

Consequently, all circuit protective devices that are on the main switchboard need to have a fault current rating not less than 3.32 kA (a 4.5 kA rated MCB is readily available).

EXERCISE 11.7

An installation in a factory is supplied from a 400 kVA transformer with 5% impedance. The three-phase consumer's mains consist of 25 m of 185 mm² copper V–90 SDI cables touching each other and the sub-mains consist of 35 m of 70 mm² copper V–90 four-core and earth. An MEN connection exists on the low voltage side of the distribution transformer and the main switchboard. Determine the maximum prospective fault current at the sub-board.

Earth fault-loop impedance

Earth fault-loop impedance is the impedance of the earth fault-current loop, which comprises the active-to-earth loop that commences and concludes at the point-of-earth fault. It is essential that this impedance is sufficiently low to ensure that the supply is automatically disconnected within the specified time if a fault of negligible impedance occurs anywhere in the electrical installation between an active conductor and a protective earthing conductor or exposed conductive part.

Fault-loop impedance considerations

Important considerations for ensuring a safe electrical installation, which has low earth fault-loop impedance are:

- indirect contact
- disconnection time
- maximum circuit length.

Indirect contact

An essential safety requirement demanded by AS/NZS 3000:2018, Clause 1.5, is the protection of persons and livestock from 'indirect contact' (refer to Clause 1.4.39) by touching exposed metallic components that have become 'live' due to an earth fault as illustrated in **Figure 11.43**. Therefore, each circuit in an electrical installation must be designed so that automatic disconnection of the circuit occurs to prevent a rise in 'touch voltage' (refer to Clause 1.4.125) to dangerous levels (exceeding 50 V) within a specified time (refer to Clause 1 in AS/NZS 3000:2018).

FIGURE 11.43 Indirect contact

The method of protection used to limit the effects of indirect contact is achieved by a system of equipotential bonding and automatic disconnection of the supply (see Clause 1.5.5.3c). For automatic disconnection to occur, the prospective fault current must be high enough to operate the protective device, particularly where the touch potential exceeds 50 V a.c. (see Clause 1.5.5.3b). The earth fault current flowing through the impedance of the protective earth system causes a voltage to appear on the earthed metal. This voltage is called the prospective touch voltage (see Clause 1.4.125).

A load test or earth fault loop impedance test is applied to the installation to ensure continuity and effectiveness of the earthing arrangement. The earth fault-loop is the circuit path followed by a fault current due to the effect of low impedance occurring between the active conductor and the protective earth system. The fault-loop impedance for an MEN system includes the impedance of the distribution system and the impedance of the active and protective earth within the installation. If the total impedance of the fault-loop circuit is low, sufficient fault current flows, causing a protective device to operate within its specified fault clearing time.

Time for disconnection of supply

To give the required protection, the disconnection of the supply must occur within the specified time limits given in Clause 1.5.5.3d. These time limits are 0.4 seconds for Class I equipment, portable equipment and socket outlets and 5 seconds for lighting circuits and other circuits including sub-mains and final sub-circuits supplying fixed or stationary equipment. Refer to AS/NZS 3000, Appendix B4.3.

The actual disconnection time is dependent upon the installation conditions and whether a person is likely to be in contact with exposed conductive parts at the instant of an earth fault. Therefore, the disconnection time for socket-outlets is less than that for fixed equipment. It is assumed that socket outlets are used to supply portable and hand-held appliances so there is a higher risk of a person being in contact with the appliance at the time of a zero-impedance earth fault. The 0.4 s time has been chosen based on the likely damp/wet situation of the portable or hand-held appliance. In a damp situation, a person in poor health touching a portable or hand-held appliance would have a greatly reduced chance of survival with a long disconnection time under a zero-impedance earth fault condition. In order to make sure that the circuit protection device will operate within specified time, it is crucial that the fault-loop impedance of the circuit is below the maximum permitted value. A low value of fault-loop impedance results in a higher fault-current with a corresponding faster circuit protection device disconnection time. The speed at which the protective device clears a fault is dependent upon the size of the fault current, which in turn is dependent upon the value of the phase earth-loop impedance (Z_S).

Appendix B4.5 provides information on the determination of the fault-loop impedance.

$$Z_S = \frac{U_0}{I_a}$$

where Z_S = maximum fault-loop impedance (Ω)

U_0 = nominal phase voltage (the phase to earth voltage 230 V)

I_a = prospective short-circuit current causing automatic tripping of the protective device (A)

I_a for fuses is approximate mean values of AS 60269.1.

I_a for circuit breakers is the mean tripping current as follows:

- Type B = 4 × rated current
- Type C = 7.5 × rated current
- Type D = 12.5 × rated current

The various circuit breakers types B, C and D have different time–current curves for tripping. Type B trips faster than type C for a given over-current, and type C faster than type D.

Another way of looking at the various circuit breaker types is that type D requires more current than type C, which requires more current than type B to trip in a given amount of time. Therefore, a circuit protected by a type D needs a lower earth loop resistance than type C, and a type C requires a lower earth loop resistance than one protected by a type B.

EXAMPLE 11.6

A 230 V final sub-circuit supplying fixed equipment protected by a 16 A type C MCB is illustrated in Figure 11.44. Calculate the maximum earth fault-loop impedance allowed in the event of a zero-impedance earth fault and the time it takes to trip the circuit breaker.

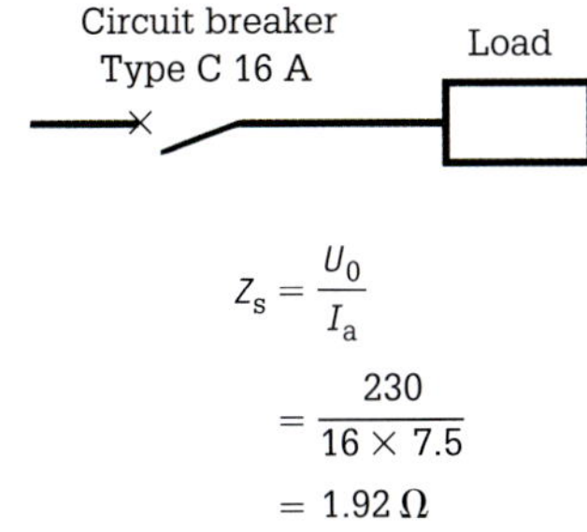

FIGURE 11.44 Example 11.6

A type C MCB trips within 5 to 10 times of the rated load current (I_N). Using some simple calculations and a typical type C characteristic curve as shown in Figure 11.45, the time to trip can be determined for the specified circuit breaker.

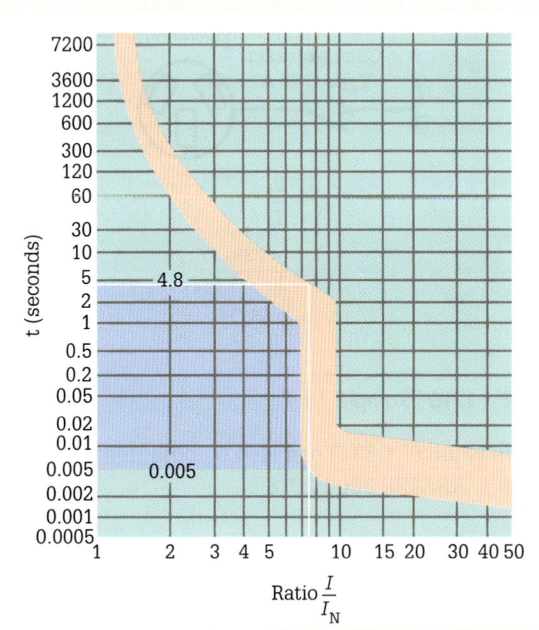

FIGURE 11.45 Tripping characteristics for a type C MCB

On the *x*-axis of the graph there is a ratio $I:I_N$; I is the fault current and I_N is the current which the circuit breaker will carry continuously under specified conditions and on which the time current

»

characteristics are based (e.g. I_N= 32 A trip type C marked 'C32'). To find the tripping time, the ratio needs to be determined then looking at the graph the tripping time can be determined. Therefore, the ratio is:

Using the value of I_S, we can calculate the ratio of $I:I_N$ from:

$$\frac{I}{I_N} = \frac{120}{16} = 7.5$$

The value for I is calculated from: $\frac{230}{1.92} = 120$

The graph ratio $I:I_N$ at the value of 7.5 shows that the circuit breaker takes between 0.005 s and 4.8 s to trip at maximum earth fault-loop impedance. This operating time is well within the time allowed for 230 V circuits protected by type C MCBs to give compliance with 5.0 s disconnection time for a final sub-circuit supplying fixed equipment. An impedance value larger than this will not give the required tripping time.

If a person in indirect contact was the cause of the earth fault then the severity of the electric shock received depends entirely on the impedance of the earth fault-loop.

Instead of the calculation used for the example, Table 8.1 in AS/NZS 3000:2018 could be used to arrive at the same result. For a 16 A circuit breaker, Column 3 for type C circuit breakers gives the same value of 1.9 Ω maximum earth fault-loop impedance. In practice, different circuit breaker manufacturers have different operating characteristics for their circuit breakers. Therefore, if time ranges for a 5 s operating characteristic for a particular circuit breaker are required, reference should be made to the manufacturer's data.

EXAMPLE 11.7

A 230 V final sub-circuit supplying socket outlets protected by a 20 A type C MCB is illustrated in **Figure 11.46**. Calculate the maximum earth fault-loop impedance allowed in the event of a zero-impedance earth fault and the time it takes to trip the circuit breaker.

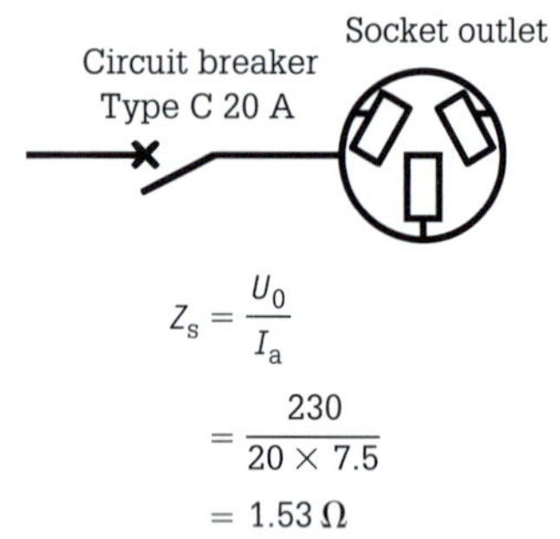

FIGURE 11.46 Example 11.7

This value of 1.53 Ω can also be obtained from Table 8.1, AS/NZS 3000:2018.

As a circuit of socket outlets, the maximum tripping time for the circuit protection device is 0.4 seconds. Therefore:

$$\frac{I}{I_N} = \frac{150}{20} = 7.5$$

The 150 comes from $\frac{230}{1.53} = 150$

From the graph of **Figure 11.45** and the ratio $I:I_N$ it can be seen that the circuit breaker takes 0.005 seconds to trip at a maximum earth fault-loop impedance of 1.53 Ω. However, the maximum time is 4.8 seconds. This operating time is well outside the time allowed for 230 V circuits protected by type C MCBs to give compliance with 0.4 s disconnection time for a final sub-circuit supplying socket outlets. Therefore, the type C circuit breaker is unsuitable, and another circuit breaker must be chosen.

EXAMPLE 11.8

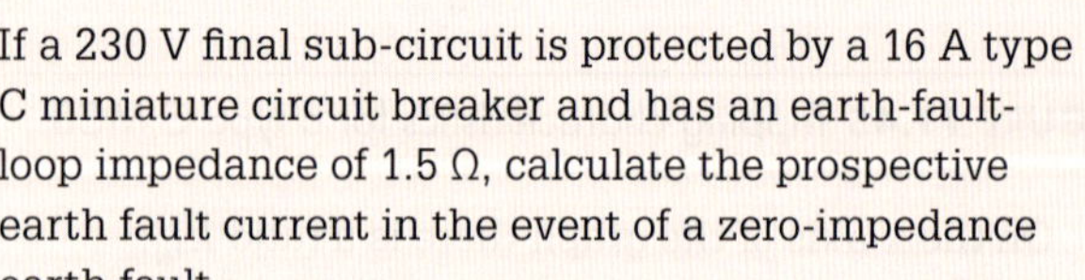

If a 230 V final sub-circuit is protected by a 16 A type C miniature circuit breaker and has an earth-fault-loop impedance of 1.5 Ω, calculate the prospective earth fault current in the event of a zero-impedance earth fault.

$$I_{PFC} = \frac{U_0}{Z_S} = \frac{230}{1.5} = 153.33\text{ A}$$

A type C MCB trips within 5 to 10 times of the rated load current (I_N). Using some simple calculations and a typical type C characteristic curve as shown in **Figure 11.45**, the time to trip can be determined for the specified circuit breaker. On the x-axis of the graph is the ratio $I:I_N$, where I is the fault current and I_N is the current rating of the circuit breaker. To find the tripping time determine the ratio and then read the tripping time from the graph shown in **Figure 11.45**.

Therefore, using the values in the above example, the ratio is:

$$\frac{I}{I_N} = \frac{153.33}{16} = 9.6$$

With I:I_N at the value of 9.6, the graph shows that the circuit breaker takes a maximum of 0.02 s to trip. This operating time is well within the period allowed for 230 V circuits protected by type C MCBs to give compliance with 0.4 s disconnection.

If a person in direct contact was the cause of the earth fault, the severity of the electric shock received depends entirely on the impedance of the earth fault-loop. It can be seen from our example that electrocution is the highest hazard that is associated with low levels of fault current (longer time for the protective device to operate) and not the highest fault levels.

EXERCISE 11.8

a A 230 V final sub-circuit supplying socket outlets is protected by a 20 A type C MCB. Calculate the maximum earth fault-loop impedance allowed in the event of a zero-impedance earth fault and the time it will take to trip the circuit breaker.

b A 230 V final sub-circuit supplying fixed equipment is protected by a 32 A type C MCB. Calculate the maximum earth fault-loop impedance allowed in the event of a zero-impedance earth fault and the time it takes to trip the circuit breaker.

Maximum circuit length – tabulated values

The 5 second and 0.4 second maximum tripping time for the circuit protection device govern the design of single-phase electrical installations. A significant factor arising from the tripping time requirements is the maximum allowable distance from a socket outlet back to the distribution board. This distance has to be determined not only by consideration of the allowable voltage drop but also of the maximum earth fault-loop impedance, either of which may be a critical factor.

If the route length of sub-mains or final sub-circuits is excessive then the fault-loop impedance is greater than the maximum allowed. Appendix B.5, Table B1, shows the maximum circuit lengths for various sizes of conductors and circuit protection devices. If circuit lengths are kept at these distances, then the requirements of earth fault-loop impedance are met.

The information provided in Table B1 is related to the 0.4 second tripping time only. AS/NZS 3000:2018 has not provided a table for the 5 second tripping time. However, it is possible to use Column 4 of Table B1 for type C circuit breakers at the 5 second disconnection time to determine the maximum allowable circuit lengths.

EXAMPLE 11.9

1 A socket outlet circuit with a route length of 55 m is to be installed in a commercial building as shown in **Figure 11.47**. The cable chosen to be suitable is a 2.5 mm² TPS twin plus earth V-90 insulated copper cable clipped to the timber joists in the insulated ceiling space. This cable is protected by a 16 A type C MCB. Determine if the route length meets the earth fault-loop impedance limits.

Using Column 4 of Table B1 of AS/NZS 3000:2018, the maximum length of this circuit, if using 2.5 mm² cables and the 16 A type C MCB, is 85 m. Therefore, the route length of 55 m is well within the limit set by Table B1.

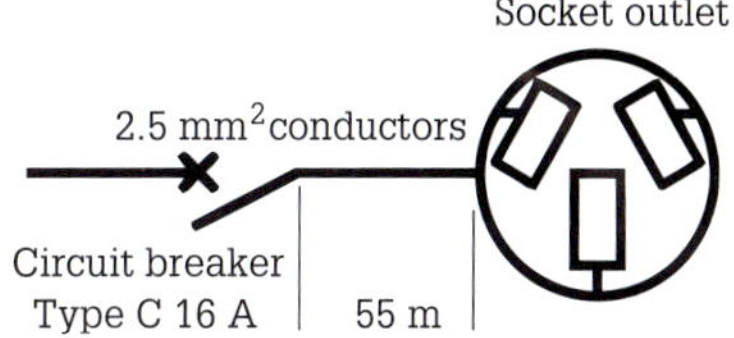

FIGURE 11.47 Socket outlet circuit

2 Determine the maximum length for a 230 V 2.5 mm² twin plus earth TPS cable to be installed as a final sub-circuit supplying fixed equipment when protected by a 20 A type C MCB.

Using Column 4 of Table B1 the maximum length of this circuit, if using 2.5 mm² cables and the 20 A type C MCB, is 68 m.

EXERCISE 11.9

Determine the maximum length for a 230 V 4 mm² twin plus earth TPS cable to be installed as a final sub-circuit supplying fixed equipment when protected by a 25 A type C MCB.

Maximum circuit length – calculated values

The maximum route length of the circuit (circuits deemed to comply with the term Z_{int}) originating from the protective device can be determined using the following equation from Appendix B5.2.2.

$$L_{max} = \frac{0.8 \times U_0 \times S_{ph} \times S_{pe}}{I_a \times \rho(S_{ph} + S_{pe})}$$

where: L_{max} = maximum length in metres

U_0 = nominal phase voltage (230 V)

ρ = resistivity in ohm-mm²/metre

22.5×10^{-3} for copper

36×10^{-3} for aluminium

I_a = trip current setting for the instantaneous operation of a circuit breaker; or the current that assures operation of the protective fuse concerned, in the specified time

S_{ph} = CSA of the active conductor in mm^2

S_{pe} = CSA of the protective earthing conductor in mm^2

Refer to the 'Notes' to this equation in AS/NZS 3000:2018 if using the calculation method.

EXAMPLE 11.10

Calculate the maximum route length of a 230 V 1.5 mm^2 twin plus earth TPS cable to be installed as a final sub-circuit supplying fixed equipment when protected by a 10 A type C MCB as illustrated in Figure 11.48.

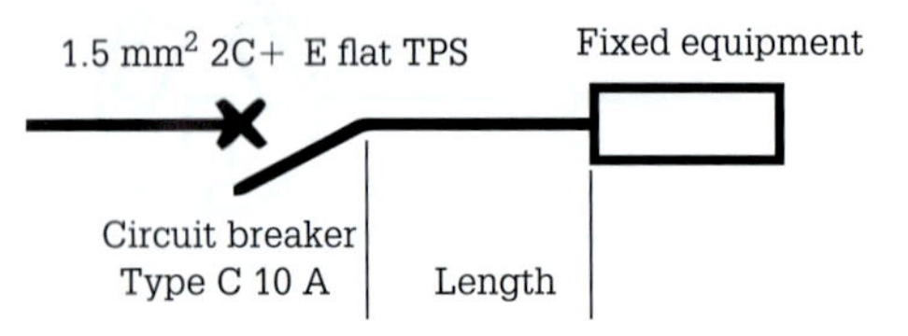

FIGURE 11.48 Maximum route length example

$$L_{max} = \frac{0.8 \times U_0 \times S_{ph} \times S_{pe}}{I_a \times \rho(S_{ph} + S_{pe})}$$

$$= \frac{0.8 \times 230 \times 1.5 \times 1.5}{(7.5 \times 10) \times (22.5 \times 10^{-3}) \times (1.5 + 1.5)}$$

$$= 81.8\,\text{m}$$

Table B1, Columns 1, 2, 3 and 4, confirm the calculation: active is 1.5 mm^2, earth is 1.5 mm^2 with a protective device rating of 10 A using a type C circuit breaker protecting a conductor length of 82 m.

EXERCISE 11.10

Calculate the maximum route length of a 230 V 4 mm^2 twin plus earth TPS cable to be installed as a final sub-circuit supplying fixed equipment when protected by a 16 A type C MCB.

Automatic disconnection

Automatic disconnection must occur when a short-circuit earth fault occurs between the active conductor and the protective earth conductor or other exposed conductive parts of the installation.

When an earth fault happens, the impedance of the fault current path must be low enough to allow sufficient current to flow to guarantee that the protective device operates within a specified time. Refer also to AS/NZS 3000:2018, Appendix B4 'Protection by automatic disconnection of supply'.

The earth fault-loop impedance is the impedance of the phase-to-earth fault current loop starting and ending at the point of the earth fault, as shown in Figure 11.49.

The earth fault-loop impedance path in an MEN system includes the following elements:

1 The final sub-circuit protective earthing conductor, 'E' link, MEN
2 The neutral return path consisting of the terminal on the 'N' link, the main neutral to the POS and then to the transformer neutral point
3 The path through the transformer winding
4 The L1 service line to the POS then through the active consumer's mains to the main switch and RCBO, final sub-circuit to the point of the fault.

The main concern with earth fault impedance is the need to disconnect the circuit within a required time

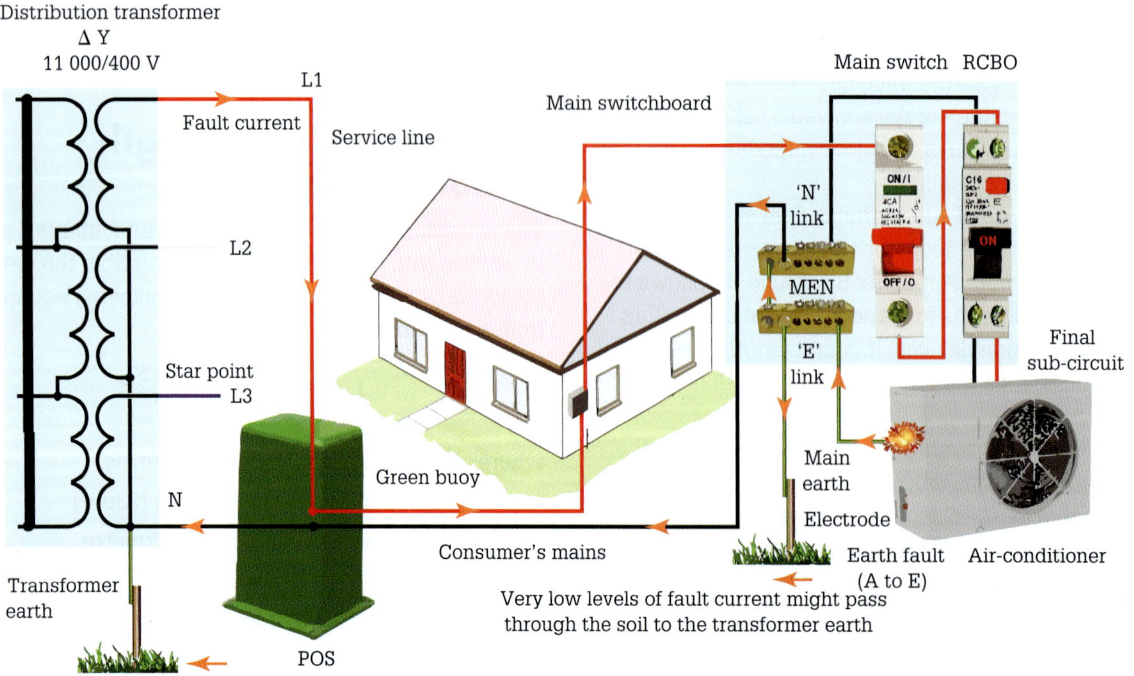

FIGURE 11.49 Earth fault loop

when the touch voltage exceeds 50 V a.c. Recommended disconnection times for protective devices are 0.4 s for socket outlets and 5 s for lighting circuits. To ensure the protective device operates within the allowed time, its time–current (tripping) characteristics must be examined.

REVIEW QUESTIONS

1 When must automatic disconnection of the phase conductor occur?
2 What makes up the earth fault-loop impedance?
3 Why does the earth fault-loop impedance need to be sufficiently low?
4 When may it be practical to bond extraneous conductive parts to the earthing system?
5 A 225 kVA, 11 kV:230/400 V, three-phase distribution transformer has 5% impedance. Calculate the maximum prospective fault current at the secondary terminals of the transformer.
6 The supply authority service rules state that the prospective fault current at the POS in a non-domestic installation is 20 kA, which equates to an impedance at the POS of 0.0115 Ω. Calculate the fault current level at the main switchboard if the impedance of the active conductor in the consumer's mains is 0.010 Ω and the impedance of the neutral conductor in the consumer's mains is 0.010 Ω.
7 To give the required protection, the disconnection of the supply must occur within the specified time limits. What is the specified time for circuits supplying socket outlets?

11.8 Devices for automatic disconnection of supply

A protective device must be placed at the point where a reduction of the cross-sectional area (CSA) of the circuit conductors or another change causes alteration in the characteristics of the installation. Protective devices on installations using a selective coordination protection arrangement called discrimination may be all circuit breakers (CBs) incorporating short-circuit and overload release, circuit breaker–fuse combinations or all fuses (see **Figure 11.50**).

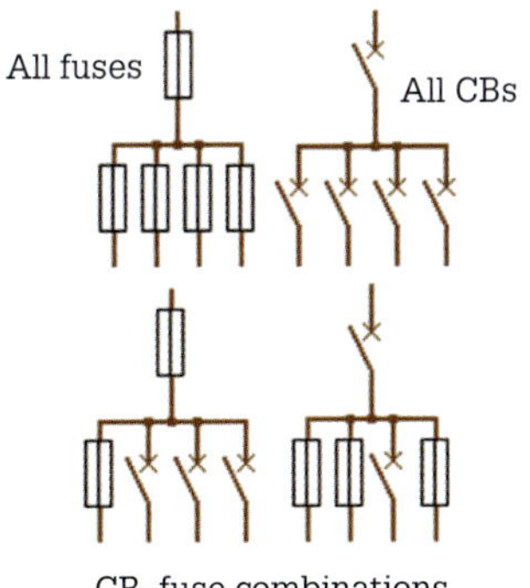

FIGURE 11.50 Possible arrangements of series protective devices on an installation

Series-connected protective devices

An installation comprising some series-connected protective devices is considered a selective coordination arrangement when, in the event of short-circuit or overload, the installation is interrupted only by the device that is immediately upstream of the fault point. The term 'upstream' refers to the proximity to the origin of the installation and 'downstream' refers to the proximity to the load. Selectivity is ensured when the time–current characteristic of the upstream MCB is above the time–current characteristic of the downstream MCB. Selectivity may be total or partial.

All of these types of protective devices must be able to interrupt any over-current including the prospective short-circuit current at the location where the devices are installed. 'Interrupting rating' is defined as the prospective short-circuit current at rated operating voltage that a device is intended to interrupt under load conditions. Where fuses or circuit breakers that don't have sufficient interrupting ratings are used, they can rupture aggressively with consequent hazards to equipment and personnel. Refer to the clause on devices for protection against both overload and short-circuit in AS/NZS 3000:2018 *Wiring Rules*.

Series rated

Series rated is a combination of two short-circuit protective devices that have been tested under the manufacturer's test conditions to work together to clear the fault. Only tested combinations can be used.

Series rated is a combination of circuit breakers or fuses and circuit breakers that can be used at short-circuit levels above the interrupting rating of the load-side (protected) circuit breaker, but not above the interrupting rating of the main or supply-side device. A series-rated combination can consist of fuses protecting circuit breakers or circuit breakers protecting circuit breakers.

For short-circuit protection, the series combination interrupting rating must be higher than the available short-circuit current at the load-side (protected) circuit breaker. Also, the interrupting rating of the supply-side fuse or circuit breaker (protecting) must be greater than the available short-circuit current at its line terminals.

Disadvantage

A disadvantage of a series-rated combination (see **Figure 11.51**) is that for fault conditions the supply-side circuit breaker (protecting) must interrupt at the same time as, and in harmony with, the load-side (protected) circuit breaker. This means that the entire circuit branch loses power because the protecting circuit breaker feeding the branch must open under medium- to high-level short-circuit conditions.

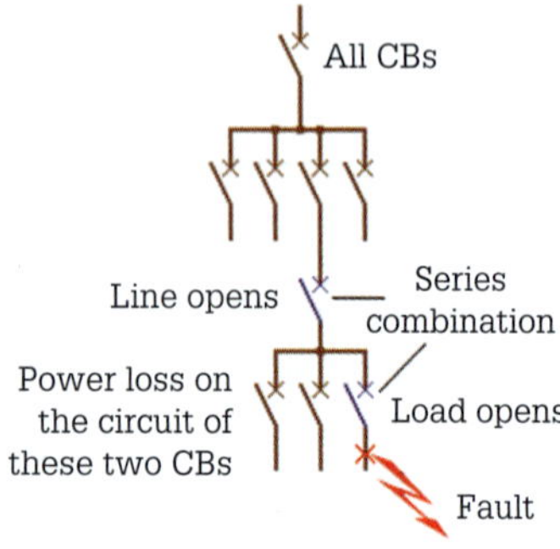

FIGURE 11.51 Series-rated circuit breaker combination

Figure 11.52 shows an installation system with a 100 A HRC fuse and a 32 A sub-main circuit breaker feeding a distribution board containing 16 A circuit breakers. If a fault occurs at point A, the 100 A HRC fuse operates, and the total installation disconnects. If the fault is at point C, the 16 A circuit breaker should operate and not the 32 A circuit breaker or 100 A HRC fuse. A fault at point B should operate the 32 A circuit breaker and not the 100 A HRC fuse. If this takes place, the system has discriminated adequately.

Lack of discrimination would occur if a fault at point C caused operation of the 32 A circuit breaker or 100 A HRC fuse, but not the 16 A circuit breaker. Remember that selective discrimination is ensured when the time–current characteristic of the upstream HRC fuse or MCB is above the time–current characteristic of the downstream MCB. To guarantee discrimination is very difficult, particularly where an installation includes a mixture of types of fuse, or of fuses and circuit breakers. Manufacturers' operating characteristics must be studied to ensure discrimination.

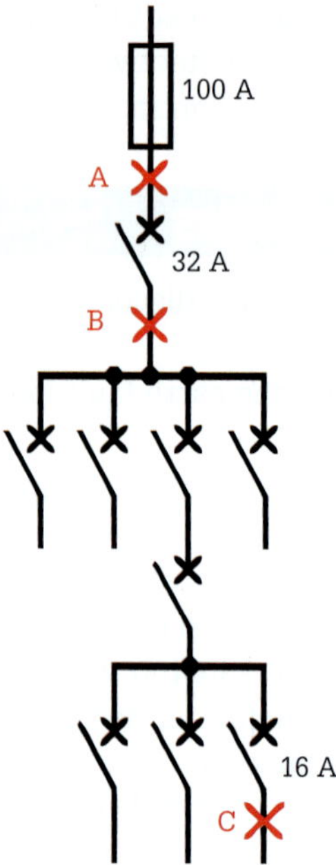

FIGURE 11.52 Installation system layouts for discrimination

Circuit protection devices

Devices used for the automatic disconnection of supply include:

- fuses
- fault current limiters
- circuit breakers.

The continuous load ratings of all electrical equipment are based on thermal considerations, the limiting factor being the maximum temperature at which the insulation can be stressed without any deterioration occurring. Circuit protection devices are designed to protect the insulating materials by stopping current flow or opening the circuit. To do this, the circuit protection device is always connected in series with the circuit it is protecting. Essentially these devices are over-current protection devices. These devices are designed to protect only the conductors from the over-temperature damage by short-circuits and current surges. There are two over-current protection devices: fuses and circuit breakers. There is another protection device called the core balance earth leakage device. Its role is to protect small appliances used in earthed situations and indirectly this device also protects people.

Fuses and circuit breakers also protect against short-circuit currents.

Fuses

The word *fuse* comes from the Latin word meaning 'to melt'. A simple fuse is a short length of wire called the element that is designed to melt and separate when a fault current flows. If this happens we say that the fuse has 'blown'. Fuses are classified as being of an open or closed type. However, the element itself is contained within a ceramic or glass housing. If the wire element itself can be touched the fuse is of the open type, and if the wire element cannot be touched the fuse is of the closed type. It is possible under fault conditions for a circuit to remain energised for the presence of an arc (a stream of incandescent vapour containing volatile metal particles from the element) that can be struck between the ends of a blown fuse.

HRC fuse

Closed-type fuses are called cartridge fuses. A simple cartridge fuse consists of the fuse element surrounded by a transparent glass case so that the fusible element can be visually inspected. A special type of cartridge fuse is a high-rupturing-capacity fuse called an HRC fuse. The HRC fuse is housed in a specially designed carrier and base. The element can consist of parallel and series loops of pure silver wire or a perforated thin copper foil enclosed in a thick ceramic cylinder as shown in **Figure 11.53**.

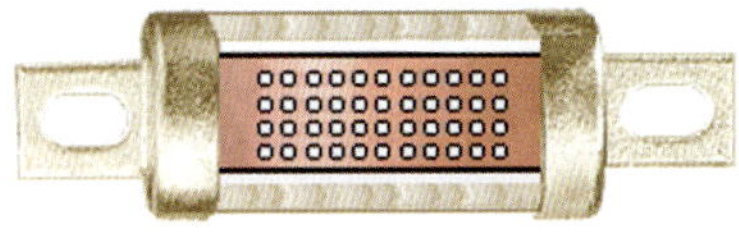

FIGURE 11.53 An HRC fuse showing the paralleling elements

The connection configuration of the element ensures that the element always ruptures between the loops. The void around the element of the cylinder is filled with silica powder that assists in quenching the arc under fault conditions. The quenching occurs when the silica powder turns into glass. An HRC fuse is designed to contain the energy realised when the element blows under fault conditions and can clear a fault in about 0.003 seconds. These fuses are used in industrial and commercial installations where high prospective fault currents could occur. There is a particular type of HRC fuse (red band) designed to protect electric motors (see **Figure 11.54**). As the electric motor can draw up to 10 times its rated load current upon starting, these fuses are designed to handle this momentary start-up current.

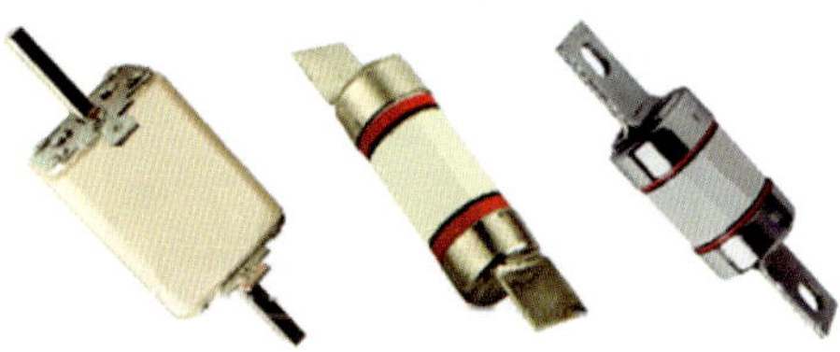

FIGURE 11.54 HRC fuses of 100 A, 20 A and 60 A values

Fuse time–current characteristic curves

Time–current characteristics determine how fast a fuse responds to over-currents. All fuses have an inverse time characteristic by which fusing time decreases as the amount of over-current increases. When properly rated, fuses provide both overload and short-circuit protection for the installation's conductors, cables and connected components. Some fuses must be able to carry 150% (fusing factor of 1.5) of their rated current for at least one hour. For other types of fuses, the fusing factor is 1.25.

The time–current characteristic is used to calculate the melting time of the fuse element at a particular overload. The time–current characteristic curve is drawn in a double logarithmic grid pattern. The vertical axis of the graph represents time in seconds, and the horizontal axis displays a range of fault currents in amperes, as shown in **Figure 11.55**.

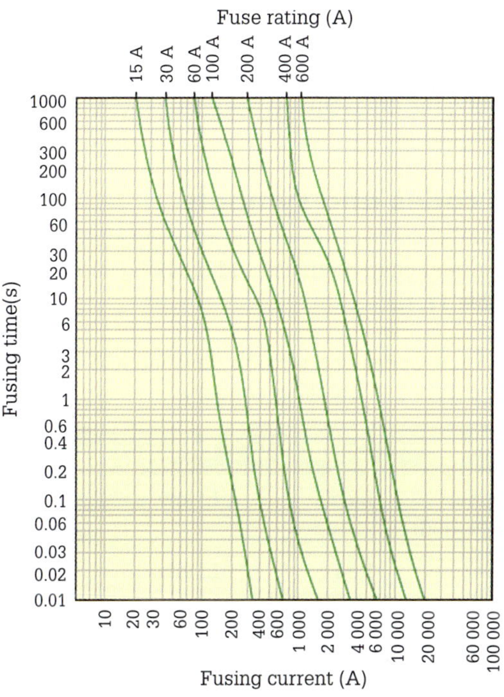

FIGURE 11.55 Fuse time–current characteristic curves

If the inrush current or the expected fault current of the circuit is known, then you locate that current value on the *x*-axis of the graph and draw a line vertically to where the line intersects the curve of the rated fuse. The time to interrupt can then be read on the *y*-axis.

EXAMPLE 11.11

1 Determine the fusing time for a 15 A fuse through which a 200 A fault current flows.

Referring to **Figure 11.55**, on the fusing current axis read off 200 A. Now read vertically until the line intersects the 15 A fuse link curve. Read across to the fusing time in seconds. The answer is 0.1 seconds.

2 Determine the fusing time for a 60 A fuse with a fault current of 1.67 times its rating.

The answer is: Referring to **Figure 11.55** ($60 \times 1.67 = 100$ A) = 300 seconds.

EXERCISE 11.11

Determine from **Figure 11.55** the fusing time for a 30 A fuse through which a 300 A fault current flows.

If the fault current falls to the right of the curve, the fuse opens. Each curve is exclusive for one fuse rating and provides the typical opening time for the amperage rating selected. Time–current characteristic curves are extremely useful in defining a fuse, since fuses with the same current rating can be represented by considerably different time–current curves. The time–current characteristic curve is used to check whether a particular fuse copes with the operating demands of the protected circuit. Some circuits such as motor circuits can draw high inrush currents from the supply. By knowing the value of that inrush current and consulting the time–current characteristic, a suitable fuse can be selected.

Fuse codes and symbols

In Australia, codes and symbols are used for fuse time–current characteristics and breaking capacity (see **Table 11.1**).

TABLE 11.1 Codes and symbols denoting fuse time–current characteristics and breaking capacity

Letter code	Colour code	Breaking capacity symbol
FF (super quick acting)	Black	L: low
F (quick acting)	Red	L: low
M (medium time lag)	Yellow	H: high
T (time lag)	Blue	H: high
TT (super time lag)	Grey	H: high

For example:

- Breaking capacity symbol L: 35 A at 230 V or 10 times the rated current.
- Breaking capacity symbol H: 1500 A at 230 V.

Time-lag fuses have additional thermal inertia designed to tolerate normal initial or start-up overload pulses.

Fault current limiters

A fault current limiter (FCL) is a circuit protective device (isolator) which limits the prospective fault current when a fault occurs. Current limiters fit three basic categories: circuit breakers, fuses and positive-temperature-coefficient (PTC) devices.

One of the most severe problems in the electrical installation is the short-circuit current. During a fault, the short-circuit current might increase up to 100 times (instantaneous fault current) its nominal value and cause irreversible damage to cables and connected equipment. Very fast response times for FCLs are necessary to limit the first cycle of the fault current.

FCLs must be selected to limit instantaneous fault currents to a value of the capacity of the cables and equipment being protected. When selecting suitable FCLs for an installation, the following three parameters are considered:

1. The short-circuit current available from the supply
2. The ratings and characteristics of the connected equipment
3. The rating and characteristics of the protective devices.

Where protective equipment is selected with a fault interrupting capacity that is less than the fault level at the installation point, it must be backed up by a device that will limit fault currents to a level the equipment can withstand safely. Typically an HRC fuse as shown in **Figure 11.56** or a suitably designed circuit breaker is used as the FCL in these situations.

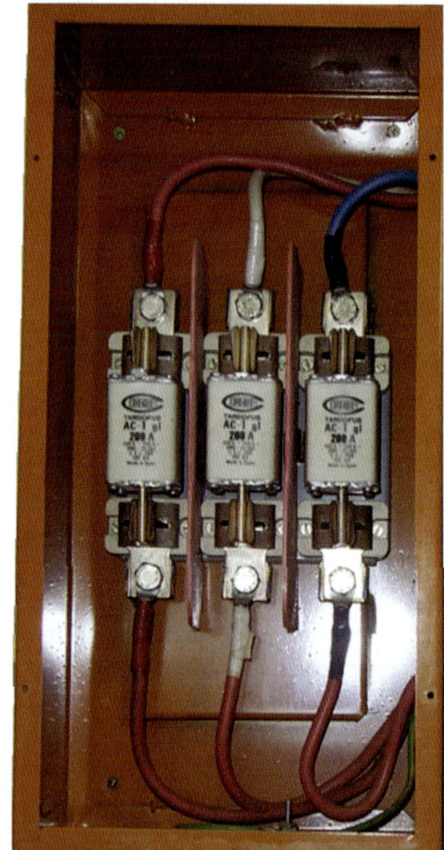

FIGURE 11.56 HRC 200 A fault-current-limiting fuses

The FCLs in **Figure 11.56** are used to protect a three-phase sub-main and connected circuits.

Fuses are the oldest and most common current limiters. The detailed design aspects of current-limiting fuses for fault-current-limiting applications are somewhat different from those used for typical load applications. In particular, when used as an FCL the rating of the fuse element is significantly smaller than the standard load current expected to be carried by the FCL, thus ensuring rapid operation.

In general, where an HRC fuse rated up to 100 A is installed no additional FCLs need to be connected to the circuit(s) protected by that fuse. An HRC service fuse rated up to 100 A provides suitable fault current limiting to an installation.

Circuit breakers

Circuit breakers are electromechanical devices extensively used in domestic, commercial and industrial electrical installations to protect circuits against overcurrent faults. A circuit breaker is a reliable over-current protective device that opens and closes like a switch.

The protecting element of a circuit breaker is referred to as the 'tripping element'. Unlike a fuse, a circuit breaker can be manually reset. A breaker trips a circuit while a fuse opens a circuit. Circuit breakers are principally designed to protect cables. A breaker should be designed to trip in different times at different overloads.

An overload is a circuit condition whereby the load device draws more current from the supply than the conductor's rated operating current. For example, a circuit can be 50% or even 100% or more overloaded. Therefore, the larger the overload, the shorter is the time required to operate the breaker.

This characteristic is known as the inverse time characteristic because as the current increases the time decreases. However, on heavy overloads and short-circuits the tripping characteristic should change from inverse to instantaneous to trip the circuit as quickly as possible. It is also necessary for a breaker to be designed so as to allow its tripping time to vary at different operating temperatures. At high temperatures, it should trip faster than at low temperatures. The effect is known as a temperature-sensitive characteristic as the tripping time will depend on the ambient (air) temperature. Circuit breakers are classified by the type of tripping element they employ.

Tripping releases

Circuit breakers are equipped with bimetallic-based, inverse-time-delayed overload releases and instantaneous over-current releases (electromagnetic short-circuit releases).

Thermal type

The thermal-type circuit breaker relies on the heating effect of the current for its operation and utilises a bimetallic element to transform heat energy into mechanical movement. The thermal element is connected to a latch device that 'latches' the switching unit. As the bimetal heats and bends, it moves the latch that releases the switching mechanism.

The bimetal element can be adjusted so that the breaker trips at any current up to 10 times the rated load value. Even when no current is flowing through the breaker, an increase in ambient temperature affects the element, causing it to move away from the latch. The tripping times of this breaker, therefore, decrease as the ambient temperature increases. More specifically the tripping time increases for temperatures below 30 °C and decreases for temperatures above 30 °C. These breakers are very capable, offering excellent overload protection (see **Figure 11.57**).

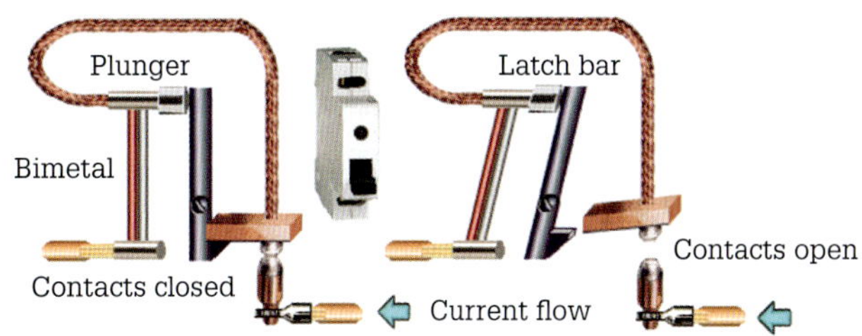

FIGURE 11.57 Thermal circuit breaker showing the bimetallic strip

Magnetic type

This type of circuit breaker release works on an entirely different principle. It consists of an electromagnetic coil and a steel plunger. The switching unit comprises a spring-loaded element on which the latch is mounted. Light overloads do not produce a magnetic field strong enough to overcome the spring force of the switching mechanism.

With heavier overloads, the steel plunger moves into the coil, increasing the magnetic force which is then strong enough to pull the latch piece against the spring and so trip the breaker. Unlike a thermal breaker, the operating times of a magnetic breaker are not affected by temperature. These breakers have excellent short-circuit protection capabilities and are shown in **Figure 11.58**.

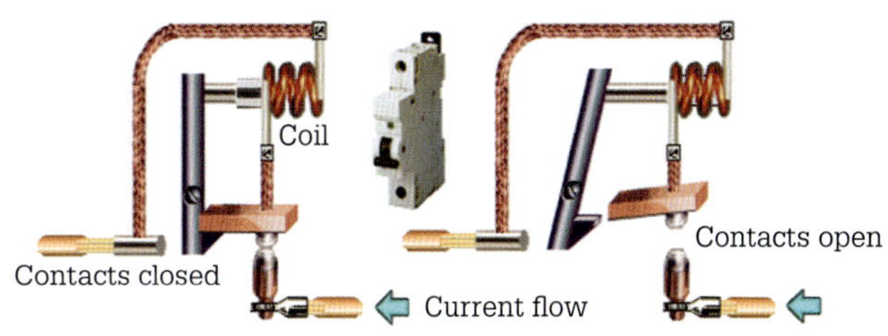

FIGURE 11.58 Magnetic circuit breaker

Combined thermal magnetic type

This breaker is a combination of the other two breakers. On small overloads, only the thermal action operates, offering both variations in tripping times with temperature and time delay. At higher overloads, the magnetic unit kicks in and overrides the thermal action and trips the breaker.

Neither a thermal nor a magnetic breaker alone can guarantee complete protection. Both allow cables to be overloaded under certain conditions. However, the combined breaker has all the desirable features of both the thermal and magnetic types and so protects the cable all the time. In Australia, this type of breaker is required in all domestic installations and is illustrated in **Figure 11.59**.

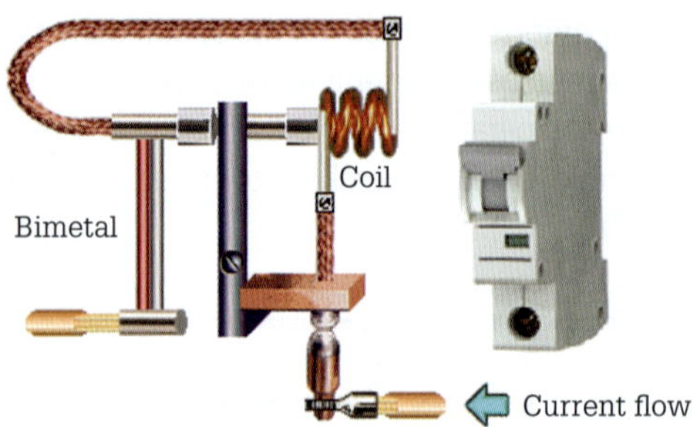

FIGURE 11.59 Combination circuit breaker

Circuit breakers also have trip-free and arc-extinguishing features.

Trip free

On most circuit breakers the manual operating lever indicates whether the breaker is closed, open or tripped. It can be clearly seen in a group of breakers that one has tripped as opposed to which one is off. All operating levers are trip free, which means that if the lever is held or locked in the 'on' position, the breaker still trips in the event of a circuit fault. It also means that the breaker cannot be closed onto a fault. This feature is illustrated in **Figure 11.60**.

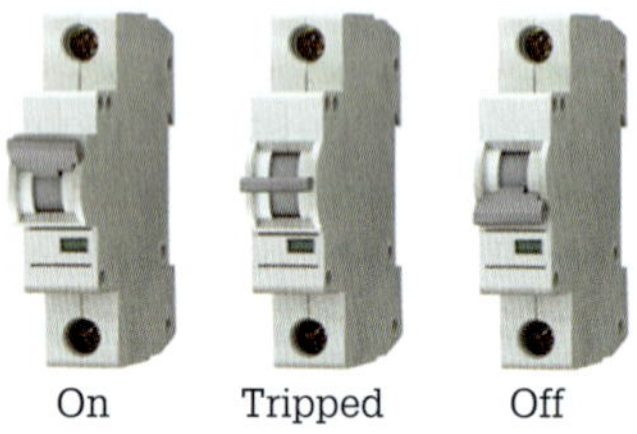

FIGURE 11.60 Trip-free feature

The switching mechanism is independent and manually operated. The contact mechanism is composed of fixed and moving contacts made of silver graphite, which is reliable and has a long life and anti-weld properties.

Arc-extinguishing device

When a breaker operates under a large fault condition, an arc is created between the two contacts as they open. An arc is a stream of incandescent ionised gas containing volatile metal particles from the contacts. This arc can cause pronounced damage if allowed to persist for even a short time. Therefore, some means of extinguishing the arc is required. The device used for arc extinguishing is called an arc chute or a de-ion grid.

This device consists of a group of shaped steel plates that are held with spaces between them. When an arc is struck, heat is established which causes the arc to rise and be drawn into the de-ion grid via an arc guide where the arc is rapidly split into smaller arcs and quenched. A de-ion grid is capable of quenching an arc in 0.005 seconds and is shown in **Figure 11.61**.

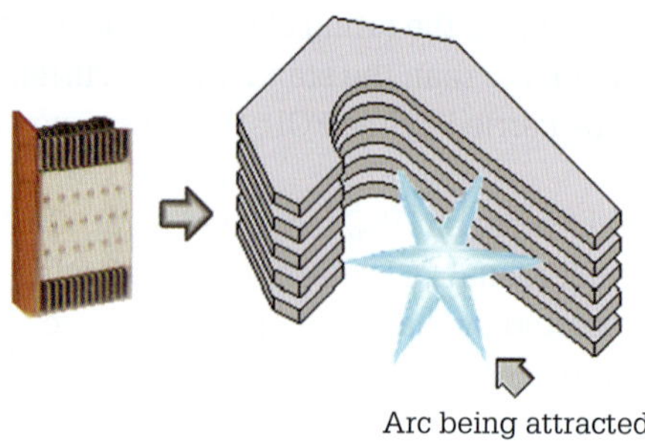

FIGURE 11.61 De-ion arc grid

A sectional view of a combined magnetic and thermal circuit breaker illustrating the main components is shown in **Figure 11.62**.

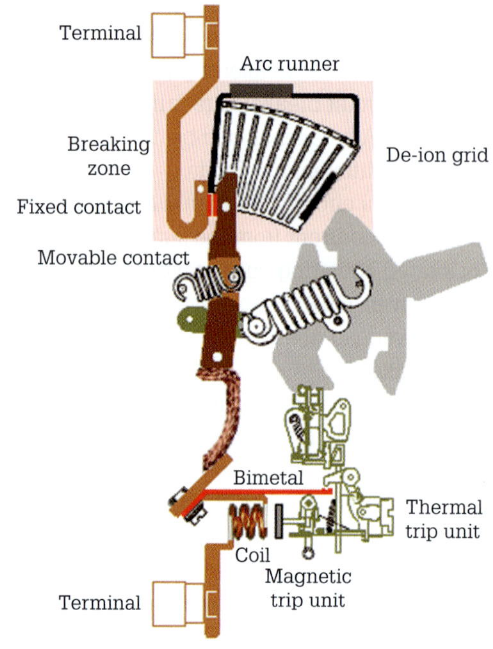

FIGURE 11.62 Sectional view of a combined magnetic and thermal circuit breaker

Miniature circuit breakers

A miniature circuit breaker (MCB) is a protective device that is installed primarily to protect final circuits against overload and short-circuits. As a current-limiting device, MCBs protect conductors, cables and circuit components from damage by reducing the peak let-through current, which causes damaging magnetic forces, and peak let-through energy, which generates heat. An MCB is a small compact device, limited to cables no bigger than 50 mm^2 CSA for a 125 A load.

The MCB is commonly installed on a distribution board and may be a single- or three-phase device. Various types of MCBs are available (types B, C, D). These types relate to the tripping characteristic of the device allowing for maximum system protection. Short-circuit ratings for the three types of MCBs are a minimum of 3 kA and as high as 25 kA. The MCB largely replaces the cartridge fuse and semi-enclosed rewireable fuse (SERF) as a safer alternative.

MCB time–current characteristic curves

The time–current (tripping) characteristic defines the MCB's speed of response (trip time) to various levels of over-current. The time–current characteristic enables the device to be optimally matched to the application. For example, water heater and stove circuits that can tolerate moderate over-currents only are best protected by MCBs with type B trip characteristics.

Key selection criteria for MCBs are the trip-time zones determined at a particular design temperature. Upper and lower limits of the curves show minimum and maximum adjustment tolerances. Unless otherwise stated by the various manufacturers, all thermal and thermal–magnetic circuit breakers carry 100% rated current continuously and trip within one hour at 140% rating. Type B MCBs have rated currents of 6 A, 10 A, 13 A, 16 A, 20 A, 25 A, 32 A, 40 A, 50 A and 63 A.

Type B MCBs

Type B MCBs provide a uniform switching capacity of 10 kA across their 6 A to 63 A rating range. Instantaneous tripping (short-circuit release) occurs between approximately three and five times the rated current (I_N) in 50 Hz systems. The fast trip time of these type B devices, as shown by the curve (rapid magnetic trip time), maximises protection of conductors, control circuits and programmable logic controllers (PLCs) under low short-circuit fault levels that could cause damage.

The type B time–current characteristic, shown in **Figure 11.63**, is designed primarily for use in cable protection applications, making type B MCBs ideally suited for domestic and commercial installations having little or no switching surges (lighting and distribution circuits for electrical heating, water heaters and stoves).

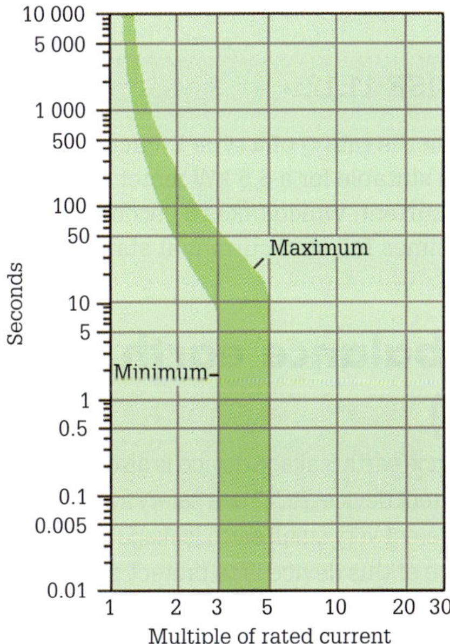

FIGURE 11.63 Type B time–current characteristic

The time–current characteristics of all types of MCBs have a vertical section where there is a broad range of operating times for a definite current. For the type B characteristic in **Figure 11.63**, the range of operating times for a short-circuit is 0.01 s to 13 s. At 0.01 s the release action would be instantaneous magnetic trip. Between 0.01 s and 13 s the release action would be a combination of thermal and magnetic action. Above 13 s the release action would be thermal only.

The operating current during the period concerned is a fixed multiple of the rated current. For example, a type B MCB has a multiple of 5, so a 40 A device will operate over the time range of 0.01 s to 13 s at a current of 5 × 40 A = 200 A.

Type C MCBs

Type C MCBs have rated currents of 6 A, 10 A, 13 A, 16 A, 20 A, 25 A, 32 A, 40 A, 50 A, 63 A, 80 A, 100 A and 125 A. The 80 A, 100 A and 125 A type C MCBs are suitable for fault current levels of up to 25 kA. Instantaneous tripping occurs between 5 and 10 times the rated current in 50 Hz systems. A higher instantaneous trip level (in comparison with type B MCBs) prevents nuisance tripping and components being protected can typically withstand higher fault currents without being damaged.

The type C time–current characteristic, shown in **Figure 11.64**, is designed for medium magnetic start-up currents. Therefore, type C MCBs are best suited for general use in commercial and industrial installations (lighting, socket outlets) and where the use of highly inductive circuits, for example, banks of fluorescent lighting and small motors, can produce switching current surges that would cause a type B circuit breaker to trip.

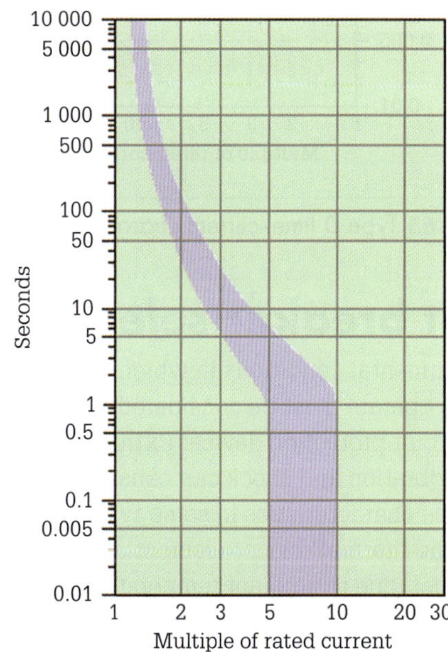

FIGURE 11.64 Type C time–current characteristic

Type D MCBs

Type D MCBs have rated currents of 6 A, 10 A, 13 A, 16 A, 20 A, 25 A, 32 A, 40 A, 50 A, 63 A, 80 A, 100 A and 125 A. Instantaneous tripping occurs between 10 and 20 times the rated current in 50 Hz systems. The higher instantaneous trip level (in comparison with type B and type C MCBs) protects against nuisance tripping due to high transient currents and starting currents of electrical machinery.

The type D time–current characteristic, shown in **Figure 11.65**, is designed for heavy magnetic start-up currents where the components being protected can withstand the higher currents without being damaged. Therefore, type D MCBs are best suited for the protection of equipment such as transformers, large motor starting, some fluorescent lighting, X-ray machines, industrial welding equipment and similar applications where abnormally high inrush currents are experienced.

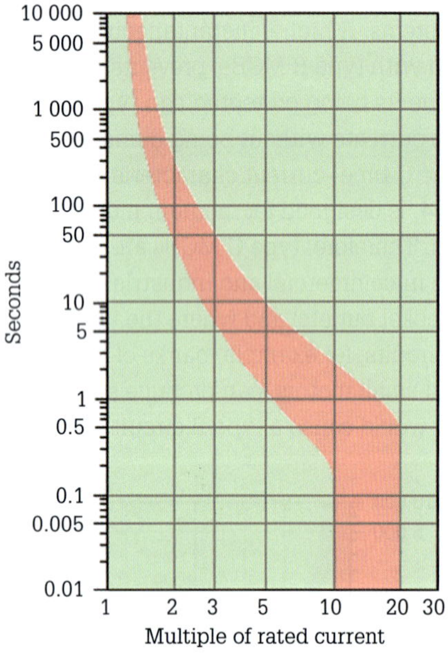

FIGURE 11.65 Type D time–current characteristic

Circuit breaker selection

The environmental conditions in which an MCB is required to operate must be considered when selecting a suitable circuit protection device. Extreme temperature, humidity, vibration and shock can cause undesirable performance characteristics in some types of MCBs. For instance, the thermal-release element can be less reliable when it is hot (due to ambient temperature) than when it is cold.

EXAMPLE 11.12

1 Determine the tripping time of a 16 A circuit breaker passing a current of 32 A for a type B, C and D circuit breaker.

$$\frac{I}{I_N} = \frac{32}{16} = 2$$

where I = actual current

I_N = circuit breaker rating

The answer '2' is the multiple of rated current.

Find the multiple of rated current 2 on each of the circuit breaker time–current characteristic curves shown and trace upwards to read off the minimum times and the maximum times.

Type B = 40 s minimum and 210 s maximum

Type C = 28 s minimum and 155 s maximum

Type D = 40 s minimum and 300 s maximum

2 Determine the rating of a type C circuit breaker that would be suitable for a 5.5 kW motor rated at 10.4 A full load current.

As a rule of thumb, a motor draws six times full-load current at start.

Assuming that a 5.5 kW motor requires up to 8 seconds to start then the starting current for this motor is:

$$10.4\text{ A} \times 6 = 62.4\text{ A}$$

From the type C time–current characteristic above trace across from the 8 second point of the seconds (vertical) axis to the curve to find the tripping time. From this point read down to the multiple of rated current value.

$$\frac{\text{Actual current}}{\text{CB rating}} = \frac{I}{I_N} = 4.8$$

$$\text{CB rating} = \frac{\text{Actual current}}{4.8} = \frac{62.4}{4.8} = 13\text{ A}$$

A 16 A type C circuit breaker would be suitable.

EXERCISE 11.12

Determine the rating of a type D circuit breaker that would be suitable for a 6.5 kW motor rated at 12.3 A full load current, which takes 8 seconds to start and draws 8 times full-load current at start.

Core balance earth leakage (CBEL)

A core balance earth leakage device is also known as a residual current device (RCD) or a safety switch and was created to detect very small earth leakage currents (10 mA). The purpose of this device is to protect people and livestock from electric current passing through the body to the earth. It is this ability that makes the RCD the only protection device that affords protection to people and livestock. RCDs only afford protection from active to earth faults and do not protect against active to neutral faults. Active to neutral faults need to be cleared by over-current protection devices such as circuit breakers. The RCD operates on the principle that the magnetic flux in a differential transformer is balanced under normal conditions. This means that the current flowing in the active is of the same magnitude as the current in the neutral conductor.

The differential transformer uses the effects of magnetism to detect any out-of-balance current in the system. If the current in the return conductor, the neutral, does not equal the current entering the active conductor due to an earth fault, then the system is out of balance. The resultant magnetic field induces a voltage in the secondary (I_{trip} winding) of the toroidal transformer. A semiconductor circuit amplifies this voltage to provide sufficient energy to trip the circuit breaker. Clause 2.6.3.2.2 requires all final sub-circuits in domestic and residential electrical installations in Australia to be protected by RCDs having a maximum rated residual current of 30 mA. RCDs having a maximum rated residual current of 10 mA are used in extremely hazardous areas, including hospitals.

The RCD is either double pole or four poles and switches off actives and the neutral. A core balance earth leakage protection device is illustrated in **Figure 11.66**.

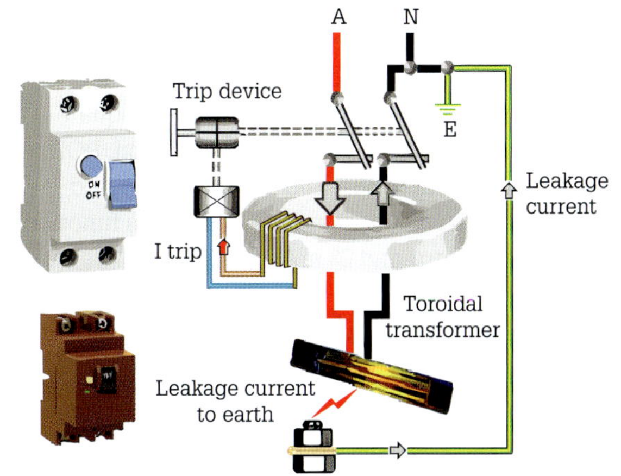

FIGURE 11.66 RCD showing leakage current to earth

RCD characteristics

RCDs are set to operate the circuit protection instantaneously when a current of a pre-set magnitude or more leaks from an active-to-earth either through a living being or equipment. The time–current curves illustrated in **Figure 11.67** indicate the operating current values of the RCD devices, depending on their standardised characteristics.

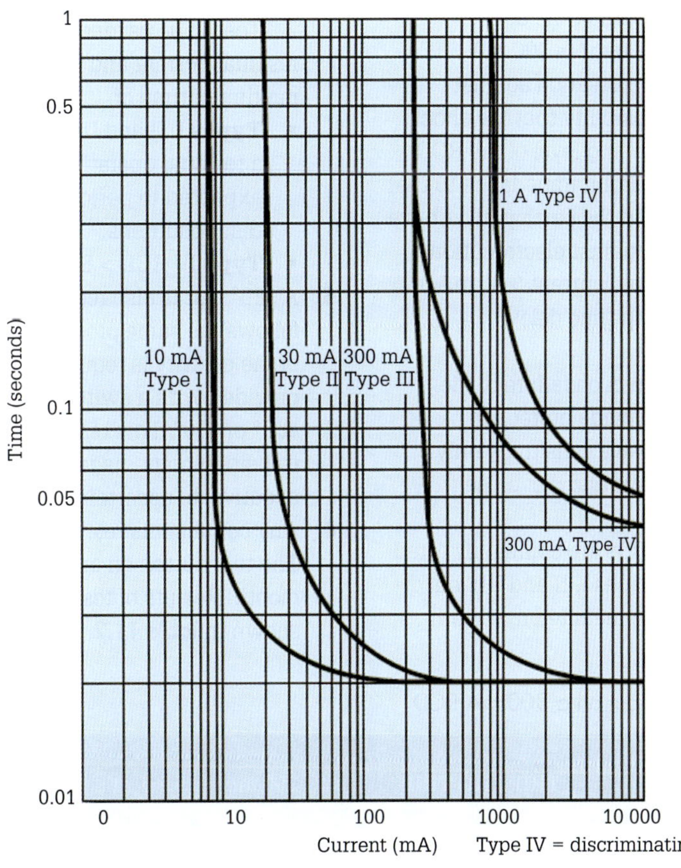

FIGURE 11.67 RCD time–current characteristic curves – general and delayed

It is essential that the RCD detects current that is below the minimum anticipated current through the body. The recommended tripping current for shock protection is a maximum of 30 mA, which will provide adequate protection for the vast majority of the population. RCDs should operate between 50% and 100% of their rated tripping current. In practice, most 30 mA RCDs function at levels between 18 mA and 23 mA.

Classification of RCDs

There are four main types of RCDs: Type I to Type IV. The main differences are the trip times (measured in milliseconds) and the rated residual trip currents ($I_{\Delta}n$) measured in milliampere. Types I and II are used to protect people and animals from electrocution and have maximum leakage current ratings ranging between 10 mA and 30 mA and have different trip times. Types III and IV are mainly used in situations where nuisance tripping may be a problem, and in the protection of equipment and cables; they have maximum leakage current ratings ranging between 30 mA and 300 mA (up to 2000 mA for special Type IV), with different trip times.

- Type I RCDs with rated residual current ($I_{\Delta}n$) ≤ 10 mA.
- Type II RCDs with rated residual current > 10 mA ≤ 30 mA.
- Type III RCDs with rated residual current > 30 mA ≤ 300 mA without selective tripping time delay characteristics. **Note:** RCDs > 30 mA ≤ 100 mA typically give a high degree of protection against electrocution, but there is a possibility that the shock current could fall below the tripping level of the RCD. This could occur if additional circuit impedance to that of the human body is included in the earth path. Above 100 mA there is no protection against electrocution.
- Type IV RCDs with rated residual current > 30 mA ≤ 300 mA with selective tripping time delay characteristics.
- Type FS RCDs automatically open on failure of the supply voltage or continue to provide protection.
- Type AC RCDs have a.c. sine wave sensitivity only.
- Type A RCDs have a.c. sine wave and d.c. pulse sensitivity.

Delayed-response (Type IV) RCDs are commonly fitted upstream of general-type RCDs (Types I, II and III), but general-type RCDs should never be mounted upstream of Type IV. The term 'upstream' refers to the proximity to the origin of the installation, while the term 'downstream' refers to the proximity to the load.

Type testing of RCDs

All RCDs must be tested by an electrician prior to their initial introduction to service and before the return to service after a repair or servicing which could have affected the electrical operation of the device. The tests performed include:

1 $I_{\Delta}n$

Test to be carried out by rated residual current. Rated residual current: value of nominal residual (leakage) current at which positive operation of the RCD or relay is ensured.

2 $\frac{I_{\Delta}n}{2}$ trip / no trip

Test to be carried out at half the rated residual current. Result of the test: the device does not trip.

3 Residual current time ($I_{\Delta}nt$) ≤ 300 ms

Result of the test: the device trips within the maximum time of:

- Type I RCD ≤ 40 ms
- Types II and III RCD ≤ 300 ms
- Type IV RCD > 130 ms ≤ 500 ms

4 5 × $I_{\Delta}nt$

Test to be carried out by five times the rated residual current and the device trips within the maximum time of:

- **Types** I, II and III RCD 40 ms: a Type I RCD with a residual operating current of 10 mA would be expected to perform ≤ 40 ms at 50 mA under type-test conditions.
- **Type** IV RCD > 50 ms ≤ 150 ms

5 A 180° test is also required to be carried out. This test follows the same procedure as above, but a changing of the polarity is required. RCD testers are frequently provided with a switch to start the flow of test current at 0° or 180°. This test is carried out at a different transition point, negative to positive or positive to negative, to ascertain the maximum time taken to trip.

6 Any comment is relevant to the test or the device. The test button on an RCD should be activated every month. Results of tests performed on a 300 mA RCD are shown in **Table 11.2**.

TABLE 11.2 Results of tests carried out on a 300 mA RCD

Location w/shop	Device $I_{\Delta}n$	$\frac{I_{\Delta}n}{2}$ TRIP / NO TRIP	$I_{\Delta}nt$ ≤ 300 ms	5 × $I_{\Delta}nt$ ≤ 40 ms	180°	Comments
Circuit 1	300 mA	No trip	65 ms	35 ms	39 ms	New RCD
Circuit 3	300 mA	No trip	20 ms	24 ms	30 ms	OK
Circuit 4	300 mA	No trip	400 ms	190 ms	–	Fail (tag off)

Installation of RCDs

AS/NZS 3000:2018 requires that where RCDs are installed lighting circuits must be distributed between the RCDs if there is more than one RCD and at least two lighting circuits. For residential installations where there is more than one final sub-circuit, a minimum of two RCDs must be installed and a single RCD cannot protect more than three final sub-circuits.

For Australian installations, Clause 2.6.3.2.2 requires additional protection by 30 mA RCDs for all final sub-circuits in domestic and residential electrical installations. Further, when RCD protection is required, the RCDs must be installed at the switchboard where the final sub-circuit originates. It is important for installation electricians to have a good working knowledge of the requirements of Clause 2.6.3.

Figures 11.68, **11.69**, **11.70** and **11.71** illustrate various installation methods to ensure compliance with AS/NZS 3000:2018.

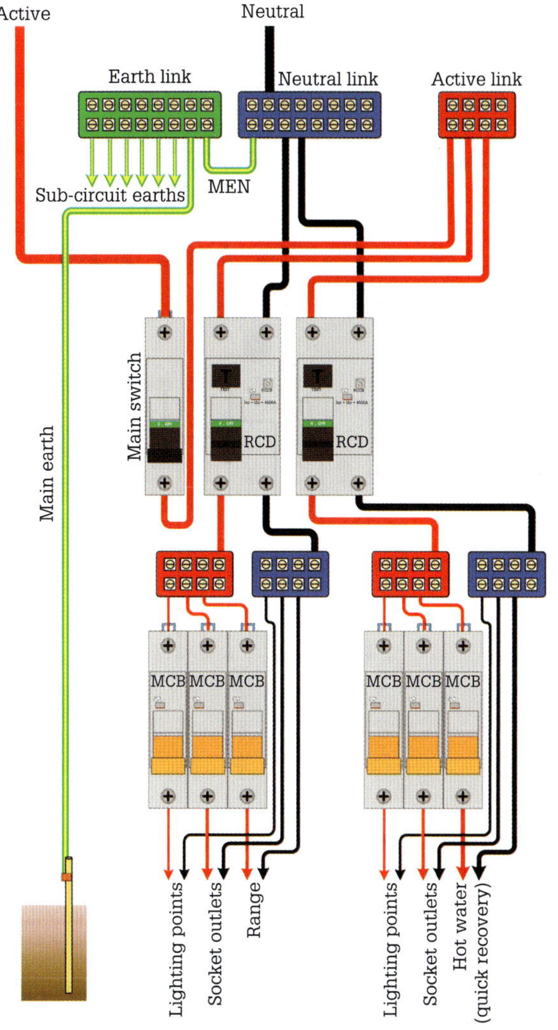

FIGURE 11.69 Installation protection by circuit breakers and two, two-pole RCDs

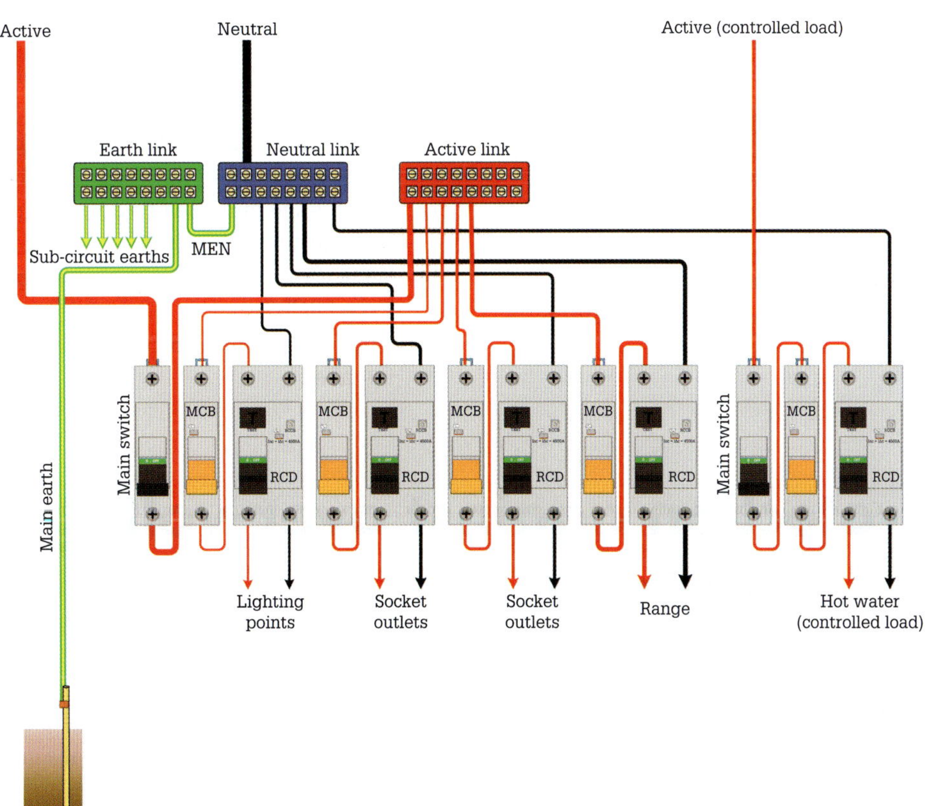

FIGURE 11.68 Installation protection by circuit breakers and individual two-pole RCDs

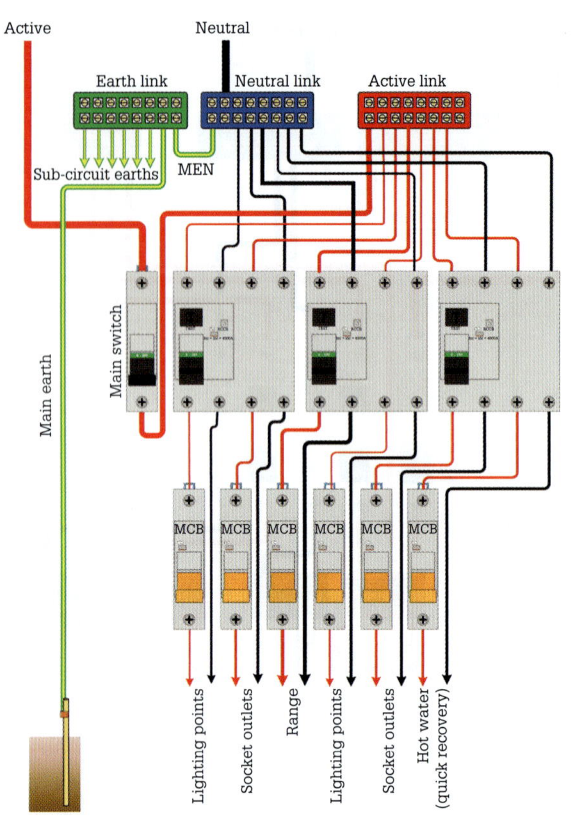

FIGURE 11.70 Installation protection by circuit breakers and three, four-pole RCDs

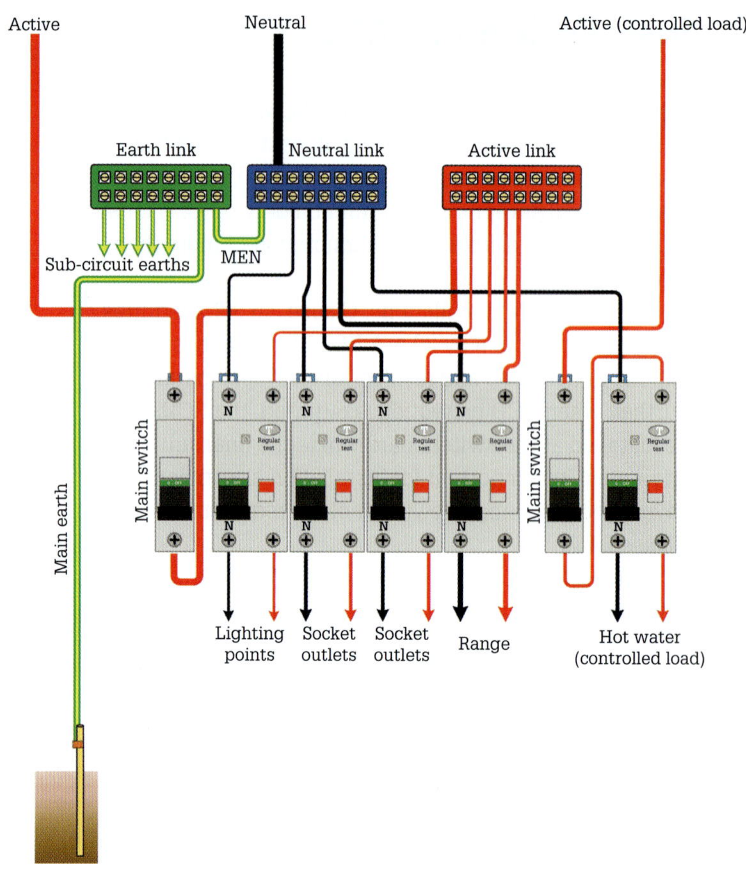

FIGURE 11.71 Installation protection by individual MCB/RCDs (RCBOs)

REVIEW QUESTIONS

1. Where in a circuit are protective devices placed?
2. When is an installation considered to have a selective coordination arrangement?
3. What is a disadvantage of a series-rated combination?
4. Name three devices used for the automatic disconnection of supply.
5. Name a common type of closed fuse.
6. What is the purpose of a fuse time–current characteristic curve?
7. Name three categories of fault current limiter.
8. Name the advantage a circuit breaker offers over a fuse.
9. What does the term 'trip-free' mean in relation to circuit breakers?
10. Name two types of tripping mechanism for circuit breakers.
11. Where would a type B miniature circuit breaker find application?
12. What environmental factors must be considered when selecting a miniature circuit breaker?
13. What is the purpose of a core balance earth leakage circuit breaker?
14. What is the residual trip current range for a Type I RCD?
15. Where are delayed-response (Type IV) RCDs installed?

11.9 Protection against over-voltage and under-voltage

Increasing the system voltage above its upper design limit or decreasing it below its lower design limit may result in hazardous situations. AS/NZS 3000:2018 specifies a nominal supply voltage for Australia of 230/400 V + 10% to −6% (in accordance with AS 60038); and for New Zealand, 230/400 V + 6% to −6% (in accordance with IEC 60038). This means that an acceptable single-phase supply voltage ranges from 216 V to 253 V in Australia and 216 V to 244 V in New Zealand.

A voltage in excess of the upper acceptable voltage is known as over-voltage. Depending on the duration of the over-voltage event, it can be classified as transient – a voltage spike – or permanent, leading to a power surge. Electrical consuming devices are designed to operate at a particular upper supply voltage limit and may suffer considerable damage if the supply voltage is higher than that for which the devices are rated.

A voltage less than the lower acceptable voltage is known as an under-voltage. Depending on the duration of the under-voltage event, it can be classified as a brownout – intentional act for load reduction in an emergency, which lasts for several minutes or hours – or unintentional, causing a sag (or dip) of short-term duration. Different types of electrical load will be affected by sags in different ways. Some loads will be severely affected, while others may not be affected at all.

Over-voltage

Over-voltage, shown in **Figure 11.72**, is defined as a power surge that takes place when the voltage is 10% above rated rms voltage for one or more cycles. Each over-voltage condition has an associated transient (short-lived) voltage wave shape. The wave form arriving at a load and its magnitude and frequency (which depend on the impedance of the circuit) differ considerably. Heat in conductors, cables and equipment above normal conditions may be an outward indication of an over-voltage.

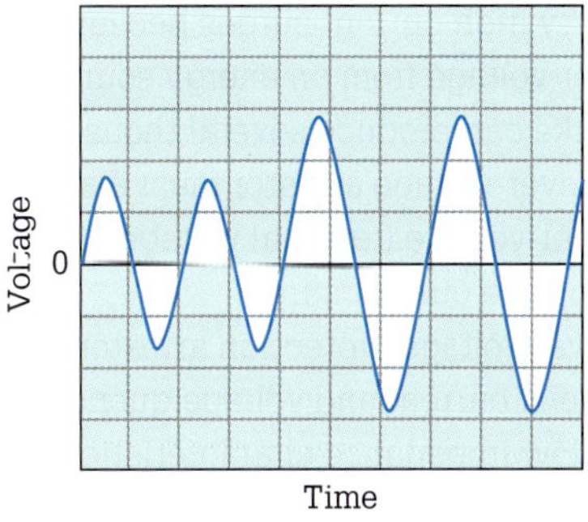

FIGURE 11.72 Over-voltage waveform

Over-voltages are also called swells or spikes. When swells or spikes are short-duration increases above nominal voltage (above 253 V for a 230 V system), they are referred to as surges. Swells and spikes are defined by their voltage gradient and time characteristic. Swells are modest voltage increases lasting from nanoseconds to one second. Spikes are high voltages lasting only microseconds.

Causes of over-voltage

Over-voltage can be caused by:

- rapid reduction in power loads by heavy equipment due to faulty switching operations
- transient switching operations that control inductive circuits such as motors and transformers (incorrect tap settings)
- lightning discharges on or near an electrical system
- energising a capacitor bank.

Faulty switching operations

Faulty switching operations can be caused by a failed power supply unit controller or incorrect wiring in the installation. The relatively high frequency voltage waves that can occur, as a result, also represent dangerous over-voltages. Protection against these is vital. They last for less than half a cycle of the mains voltage. A typical switching transient waveform is illustrated in **Figure 11.73**.

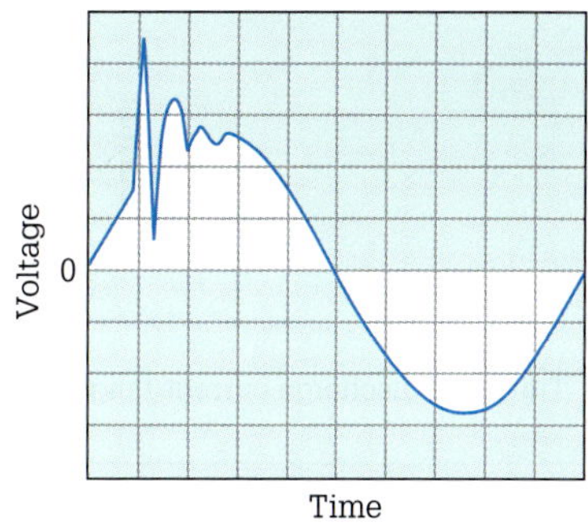

FIGURE 11.73 Switching transient waveform

Transient switching operations

Transient switching operations (e.g. motors and transformers) create over-voltages because the switching action does not operate in sync with the zero point of the a.c. waveform. This means that in many switching operations there is a sudden change of the current value from a high value to zero, such as when a circuit breaker operates to clear a fault. Due to circuit impedances, the sudden change of the current value leads to transient over-voltages with high-frequency oscillations and high-voltage peaks or swells. These transient over-voltages can damage electrical conductors, cables and equipment by conductive (transferred directly) coupling, inductive coupling or capacitive coupling.

Lightning discharges

Bolts of lightning produce very high currents. Therefore, if a direct strike hits transmission lines it can cause a sudden large voltage drop and at the same time generate very high secondary voltages across the transmission lines even in well-earthed systems. This secondary voltage (called a spike) can be coupled in the same manner as transient switching operations.

Indirect lightning strikes can also induce secondary voltages in any electrically energised material. A lightning transient waveform occurring on a single-phase 50 Hz supply line is illustrated in **Figure 11.74**.

The destructive potential of lightning transients is defined by their peak voltage, the residual current and the time of duration of the current flow.

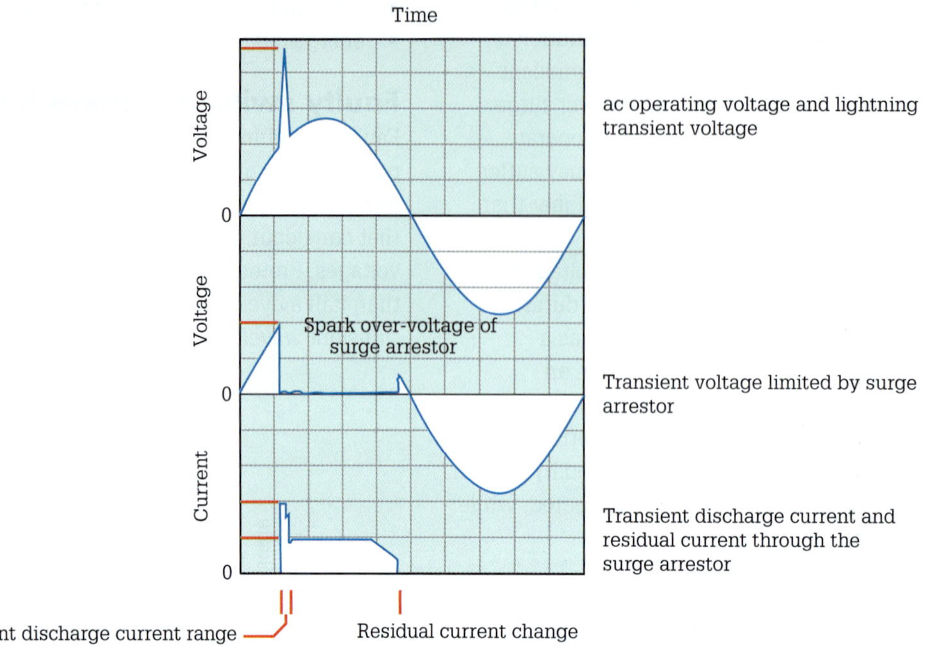

FIGURE 11.74 Lightning transient waveform

Other over-voltage situations

Other over-voltage conditions include:

- Voltage unbalances (single-phase fault on a three-phase system) where the voltages on the phase conductors are different.
- A neutral–earth voltage rise that is caused by poor workmanship when installing an installation's earthing requirements.
- A breakdown of insulation between electrical circuits or systems of different voltages.
- Electrostatic over-voltages in unearthed electrical systems as a result of atmospheric discharges (a discharging cloud) between an overhead line and the ground.

Insulation breakdown

The dielectric strength of the insulation protecting the conductors, cables and equipment determine the voltage ratings of these items. For example, insulation rated at 3 kV may be 2 mm to 2.8 mm thick. The thickness of insulation varies depending on the type of conductor, cable or electrical equipment. Insulation is also affected by temperature and has operating and maximum temperature ratings. As over-voltages are voltages in excess of electrical system design values, they place stress on insulation due to the heating effect causing rapid ageing of the life of the insulation. This in turn results in the premature breakdown of its electrical properties.

Methods of over-voltage protection

Over-voltages are prevented from reaching conductors, cables and equipment in an installation by clamping (extinguishing them before a dangerous value has been reached). To prevent over-voltages, an installation must have installed quick-reaction over-voltage protection arrestors that can respond in nanoseconds. Over-voltage arrestors respond during the high-frequency rising period of the over-voltage and contain the over-voltage by earthing it. Over-voltage arrestors ensure that the function of the circuit or equipment is guaranteed.

Over-voltage protection arrestors must be able to withstand very high current surges because a short-circuited over-voltage from an energy source such as a lightning strike can produce several thousand amperes. An effective over-voltage arrestor must also remove any remaining over-voltage (residual voltage) that is still present during the current surge.

For the over-voltage protection arrestors to discharge their dual role of current and voltage quenching, arrestors must have a low impedance characteristic. Over-voltage protection by the different types of arrestor devices should have the following properties:

- rapid response times
- a short earth conductor path to the ground (soil)
- high current-carrying capacity
- containment of residual voltage
- good resetting time
- long service life.

Types of over-voltage arrestors

When a lightning event occurs near or strikes a transmission or distribution line, a high-voltage impulse waveform appears on the line. If it reaches a transformer or an installation protected by lightning surge diverters, the impulse, if large enough, will travel through the diverter to the ground, impressing on the transformer or installation only that part of the voltage waveform that occurred before it conducted. While transformer insulation must be able to withstand a voltage at least equal to the actuation voltage of the diverter, an installation must also have secondary protection. The secondary protection allows the remaining current impulse from the lightning to be discharged and limited to a voltage level compatible with the installation system.

There are four types of over-voltage arrestors:

1 valve-type arrestor (spark-gap arrestor)
2 varistors (voltage-dependent resistors)
3 gas-filled arrestors
4 suppression diodes (solid state).

Valve-type arrestor

This type of arrestor consists of two types of elements connected in series. The first element is a series of 'spark gaps' followed by the valve element composed of voltage-dependent resistors (VDRs). When a high-voltage impulse waveform appears on the line with the transformer, the impulse flashes over the series-connected air gaps and then discharges to the ground through the voltage-dependent resistors.

After the impulse has spilled over to the ground, the standard supply potential tends to maintain the flashover across the elements of the arrestor. The follow-on current that flows due to normal supply voltage cannot be maintained due to the current-limiting nature of the diverter's elements and the flashover is cleared.

This type of arrestor is of robust construction and its method of assembly is such that adequate contact pressure is maintained at all times on the faces of the series voltage-dependent resistors. The design of the series gaps and voltage-grading resistors is such that the gap setting cannot be affected by vibration, mechanical shock or change in temperature. A valve-type surge arrestor is illustrated in **Figure 11.75**.

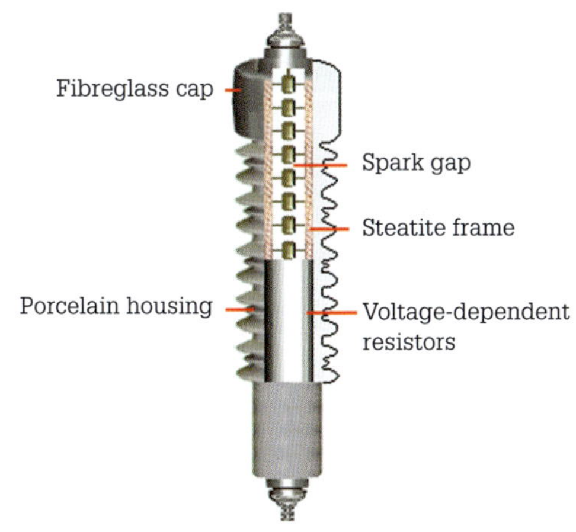

FIGURE 11.75 Valve-type surge arrestor

A 255 V, 75 kA spark-gap-type lightning current arrestor with over-voltage protection capacity of ≤ 3.5 kV for protection of a single-phase consumer wiring system is shown in **Figure 11.76**, along with an arrestor for protection of a three-phase consumer wiring system.

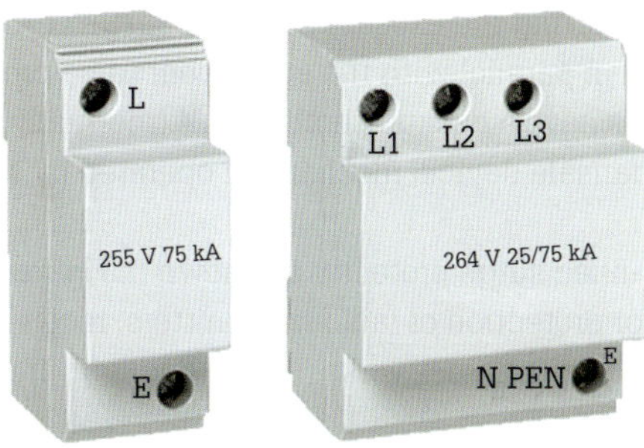

FIGURE 11.76 Single-phase and three-phase lightning current arrestor

Metal oxide varistor (MOV)

This surge protection device is a non-linear resistor device using metal oxide discs arranged in a column. The resistance of a varistor varies automatically in response to changes in voltages appearing across it. During normal conditions, the nominal line-to-ground voltage is applied continuously across the diverter terminals. When over-voltages occur, the arrestor limits the over-voltage through reduced impedance between line and earth. After the passage of the over-voltage condition the diverter returns to its high-impedance state. An illustration of a metal oxide varistor is shown in **Figure 11.77**.

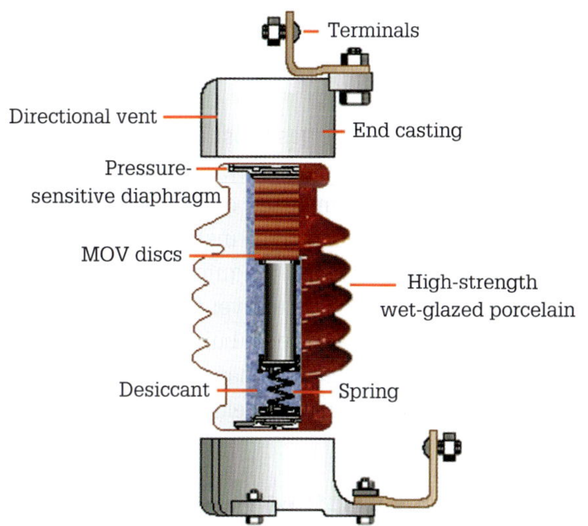

FIGURE 11.77 Metal oxide varistor

Gas-filled suppressor (GDS)

GDSs use an inert gas (argon/hydrogen) pressurised to 1 bar in a replaceable capsule and two or three electrodes. The gas becomes conductive (surge current capability of up to 2500 A) only when high-voltage transients occur. The breakdown voltage of the GDS is related to the gas pressure and electrode. When the striking voltage of the GDS is exceeded, an ionised discharge is developed across the electrodes. As the fault current increases, an arc discharge is produced. The gas presents very high impedance (open circuit) in the normal state. When a voltage spike appears, the gas ionises, presenting a low-impedance path and actually diverting the excess voltage to earth. GDSs are connected in parallel with the equipment to be protected. Some GDSs have a backup short-circuit (thermally operated failsafe device) mechanism operated by solder pellets.

A gas-filled surge protector as shown in **Figure 11.78** is suitable for protection of radio transmitters, receivers, high-frequency local area networks (LANs) and high-frequency coaxial cable.

FIGURE 11.78 Coaxial gas-filled surge protector and gas capsule

During normal operation, a signal can pass through the surge protector to the equipment while the gas inside the capsule remains inert. When lightning strikes the antenna, current (max. 20 kA (8/20 μs)) flows through the coaxial cable to the surge arrestor. When the voltage appearing across the capsule increases and reaches the d.c. spark over-voltage, the gas ionises and becomes conductive. Current is then diverted through the gas capsule earth-bonding conductor to the ground (soil). Once the surge has been discharged the gas capsule returns to its resting state.

The operating life of the gas capsule depends on the number of surges and their intensity. The gas-tube holder unscrews, providing for simple gas capsule replacement. Coaxial surge protectors should be installed in coaxial feeders as near as possible to the equipment to be protected. At transmitter and receiver sites it is desirable to install them at the coaxial cable building point of entry. GDSs for coaxial cable protection have an earth stud to ensure connection to the earth for either cable or direct mounting.

Solid-state equipment protectors

Some solid-state arrestors consist of Zener diodes, which exhibit voltage-limiting characteristics. When a transient over-voltage occurs along the line with a voltage exceeding the reverse-biased voltage rating of the Zener diode, the Zener will conduct and the transient will be clamped at the Zener voltage. In contrast to MOVs, Zeners do not degrade with repeated surges and so will last longer.

However, the MOV has a significant advantage over the Zener: its ability to handle transients of much larger energy content. Zeners need a heat sink in order to cool the rapid build-up of heat which occurs in the P-N junction after it has encountered a transient. Otherwise a significant hot spot is created at the junction. Another advantage of the MOV is its ability to survive much higher instantaneous power.

In comparison with the gas-filled arrestors, the solid-state arrestors are superior in speed and voltage control and have a long life cycle. The main difference between the two devices is clamping time, or the time it takes to remove the surge off the line.

Another type of solid-state arrestor is the silicon-controlled rectifier (SCR). SCRs are used to protect electronic circuits, telecommunication equipment and devices and equipment such as programmable logic controllers from over-voltage and over-current. The SCR channels the extra voltage into the earth wire of the socket outlet and prevents it from flowing through electronic devices; however, the proper equipment voltage is allowed to flow through its regular path.

Arrestor locations

Table 11.3 and **Figure 11.79** illustrate the application of four categories to the wiring of the power system. These four categories represent the majority of places, from the electrical service entrance to the most remote socket outlet, equipment and coaxial cable.

TABLE 11.3 Arrestor locations, types A–D

A	B	C	D
Socket outlets and long sub-circuits (more than 10 m from category B)	Mains distribution boards, sub-boards	Outside	Inside for equipment, coaxial, telecommunication data

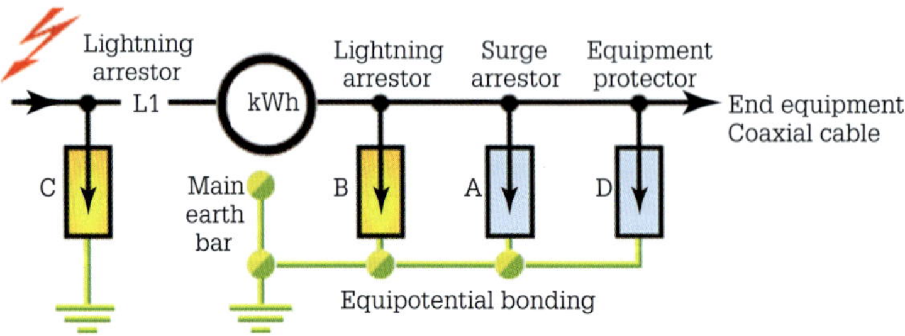

FIGURE 11.79 Arrestor locations

The over-voltage protection device's equipotential bonding conductors must be connected by the shortest possible route to the main earth terminal of the installation. Longer routes reduce the efficiency of the over-voltage protector.

Internal over-voltages

Internal over-voltages are those voltages generated inside electrical equipment. Typical causes of over-voltages are:

- switching on and off motors and current transformers
- switching on and off photocopiers
- switching on switch-mode power supplies
- switching on and off phase-controlled modulators
- tripping a fuse, a relay or an interrupter
- electrical power outages of any kind.

These temporary over-voltages of several kV with a very fast rise time can lead to the destruction of insulation (by creating pin holes in the insulation material), tracks on PC boards and electronic devices such as triacs and thyristors and computer data.

Low-energy secondary-protection MOVs are frequently used in mains protection circuits. Three types of MOV insulation are available: porcelain, polymer and silicon rubber. When configured correctly MOVs provide essential clamping against over-voltage transients. MOVs are usually specified to initiate clamping at an effective voltage of 275 V with a discharge capacity of 20 kA (8/20 μs). The 20 kA is the impulse current. The 8/20 μs means that the first value (8) is the rise time (from 10% to 90% of peak), while the second value (20 μs) is how long it takes the over-voltage to decrease to half its peak value.

Mains filters are also used in conjunction with MOVs by providing reducing filtering to slow down the rate of rise of the residual voltage prior to the MOV clamping becoming active.

A third protective measure is required if any further over-voltage possibility exists between the distribution board and any connected equipment. Surge-protected MOV socket outlets as shown in **Figure 11.80** must be installed in such cases.

FIGURE 11.80 Surge-protected MOV socket outlet

A socket outlet surge protector works by directing any extra voltage into the outlet's earth conductor, thereby stopping it from reaching any equipment. At the same time, the protector allows the standard voltage to remain. **Note:** An MOV provides reliable surge protection but degrades each time. It may even last only once.

Typical lightning and surge protection specifications for an installation

All mains-operated electrical equipment should be protected from lightning and power surges. The protection system must consist of surge protection devices and earthing systems. The protection system should not affect the operation of equipment under normal operating conditions. All surge protection systems must be installed in accordance with the manufacturer's recommendations.

Appendix F in AS/NZS 3000:2018 provides guidance on the selection and installation of surge protection devices (SPDs). In essence, the installation of primary SPDs is near the origin of the electrical installation or at the main switchboard, and installation of secondary SPDs is at switchboards remote from the main switchboard. The selection of secondary SPDs should be such that they are coordinated with the primary SPDs in accordance with the manufacturer's instructions.

When installed, SPDs should be installed after the main switch and before any RCD protection devices. An appropriate fuse or circuit-breaker, which is separate from the SPD, should afford appropriate protection and isolation of the SPD. The SPD should connect between phase and neutral conductors at the main switchboard. When installed, SPDs are required to be legibly and permanently labelled as to their function. **Figure 11.81** shows how a primary surge protection device can be installed in the main switchboard.

Remote equipment

Remote equipment in a separate building or switchboard must be supplied from the unfiltered mains and must have a surge suppression device at its remote termination. These remote devices should have a 500 V let-through voltage.

Surge protection devices should be provided at both ends of the 4–20 mA signal cables and digital data lines that clamp the voltage to no more than 45 V. Each device must be securely bonded to the earthing system. The case of each transmitter and each receiver must be connected to the earthing system. Remote transmitters must use a local earth system.

Surge protection devices are not required if the signal loop:

- does not extend outside the switchboard; or
- does not extend outside the confines of a building.

Surge protection devices must be of the series-connected type, comprising three stages of protection, failsafe operation (short-circuit ability) and common and differential mode protection.

Coaxial antenna cables must be protected by coaxial surge protectors suitable for the frequency of operation of the antenna system. The surge protector must be securely bonded to the ground via an earth-bonding conductor to the earth bar.

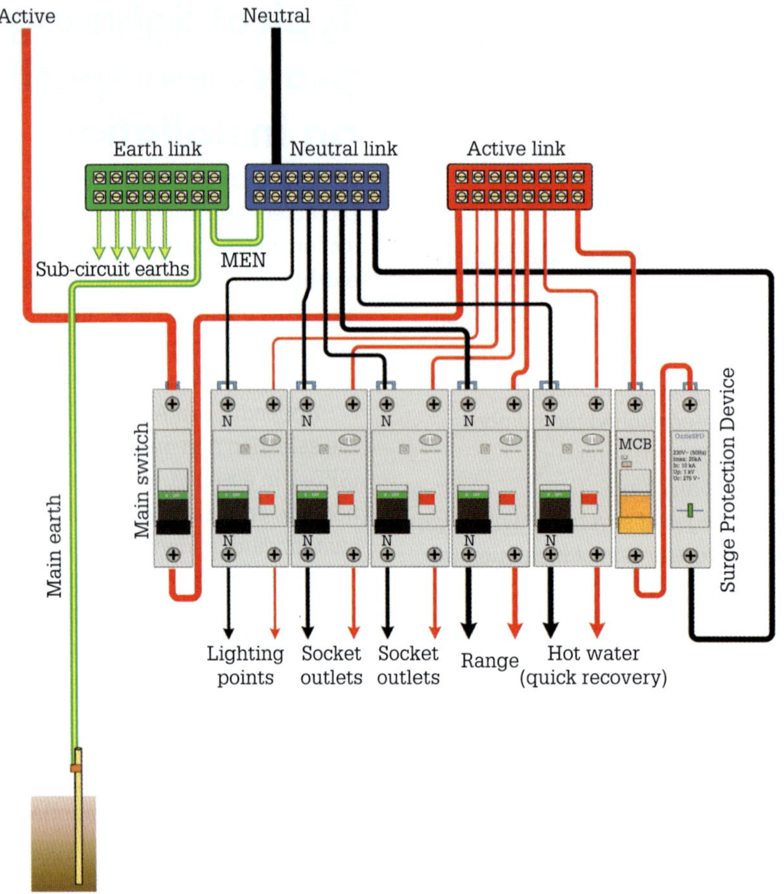

FIGURE 11.81 Installation of primary surge protection

Under-voltage

An under-voltage (see **Figure 11.82**) (also called a brownout) is defined as a decrease in the voltage to between 10% and 90% of the standard voltage for a period of time of greater than 60 seconds. Under-voltage can be the result of switching on or off a load or a capacitor bank. A brownout can also occur if a distribution transformer blows one of its high-voltage fuses. If this happens the single-phase voltages can fall to half their normal value.

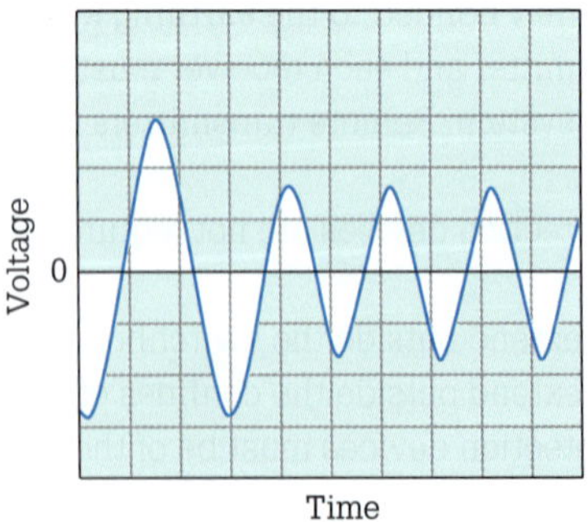

FIGURE 11.82 Under-voltage waveform

Solutions to under-voltage include:

- adding voltage regulators to improve the voltage profile
- adding shunt capacitors or static VAr (volts-amp reactive) compensators to reduce the line current
- increasing the size of the transformer, adding series capacitors or increasing the size of line conductors to reduce the system impedance.

Sags

Power sags, illustrated in **Figure 11.83**, involve voltages below nominal voltage for a short period and are one of the more common forms of power problems experienced by consumers.

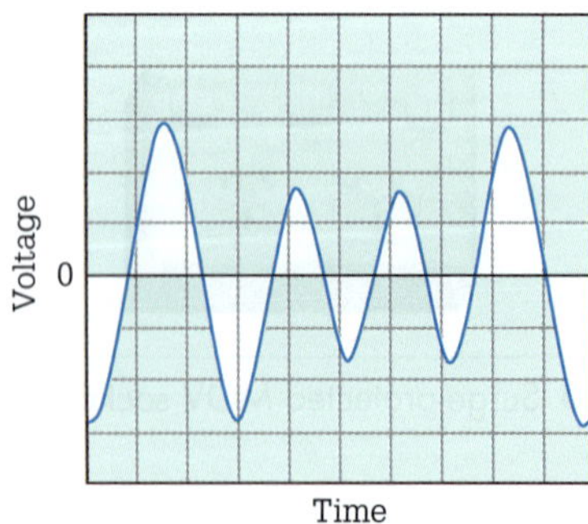

FIGURE 11.83 Sag voltage waveform

Sags can be generated both internally and externally from a consumer's facilities. External causes of sags primarily come from the utility transmission and distribution network. Sags coming from the utility have a variety of causes including lightning, animal and human activity and normal and abnormal utility equipment operation.

Sometimes externally caused sags can be generated by other consumers via the starting of large motors or the switching off of shunt capacitor banks.

Internal sags are generated by similar loads to external sags. The large inrush of current required to start these types of loads lowers the nominal voltage level available to other equipment that shares the same electrical system.

Under-voltage protection

Protective measures must be taken where a drop in voltage or loss of voltage and the subsequent automatic restart of equipment when the voltage is restored could cause a dangerous situation. This may be damage to the installation, equipment or work in progress or injury to the operator of the equipment.

Automatic restart of equipment after a power failure and subsequent restoration of power can be avoided with an under-voltage release main switch. The switch has an under-voltage release facility which means that in the event of a power failure the under-voltage release coil returns the switch to the off position, preventing automatic start-up of machinery on restoration of power. Under-voltage release add-on devices as shown in **Figure 11.84** are available for circuit breakers and RCDs. When the voltage drops below a certain level or in the event of a voltage failure (less than 10% to 35%), the under-voltage release device trips the circuit breakers or RCDs. The circuit breakers or RCDs remain in the off position until the operator manually switches them on again.

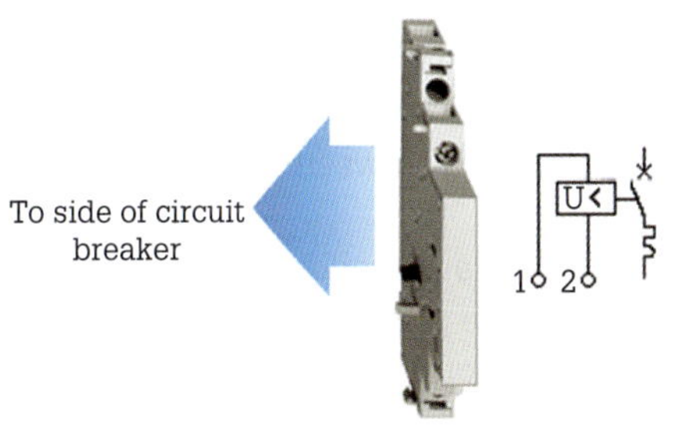

FIGURE 11.84 Sideways-mounted under-voltage release device and symbol

Circuit breakers and RCDs that can have under-voltage release add-on devices are characterised by an active trip-free release, meaning that the contacts cannot be closed when the supply voltage is too low and it is not possible to prevent a release by blocking the rocker (trip-free ability).

Over/under-voltage relays

Over/under-voltage relays (see **Figure 11.85**) provide protection for equipment where either an over- or an under-voltage condition is potentially damaging. They are designed to operate when the operating voltage reaches a pre-set value and drop out when the operating voltage drops to a level below the pre-set value.

FIGURE 11.85 Plug-in over/under-voltage relay

The pick-up voltage setting is user-adjustable from 85% to 115% of the nominal voltage rating. The relay energises when the monitored voltage is above the pick-up setting. The relay de-energises when the monitored voltage (over and under) is below the drop-out setting for a period longer than the drop-out time delay (T), which is fixed during manufacture. A function diagram of an over/under-voltage relay is illustrated in **Figure 11.86**.

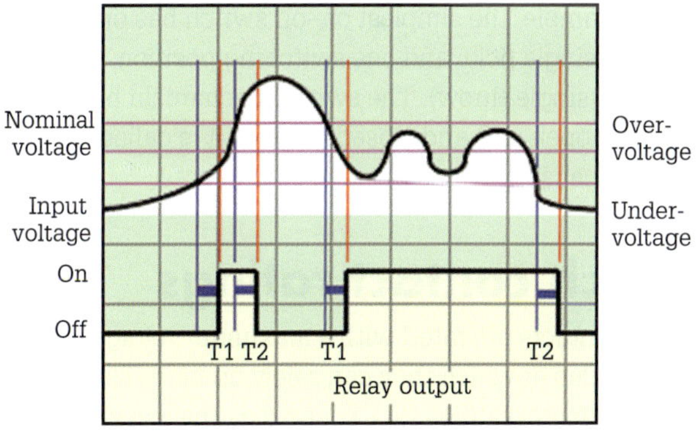

FIGURE 11.86 Function diagram of an over/under-voltage relay

REVIEW QUESTIONS

1. What categorises a voltage spike?
2. Name four causes of over-voltage in a system.
3. What is a common method for protecting an installation against over-voltage?
4. Identify the qualities of an effective over-voltage protection device.
5. Name four types of over-voltage protection devices.
6. What are typical causes of internal over-voltage?
7. How does a socket outlet surge protector work?
8. What is another name commonly used to describe an under-voltage condition that lasts more than 60 seconds?
9. What are three solutions to under-voltage?
10. What are common external causes of voltage sags?

11.10 Control of an electrical installation and circuits

Control of an electrical installation is achieved by means of a mechanical switching device. This device is capable of making, carrying and breaking an electrical current under normal installation conditions.

A switch is a component that allows us to control whether the current flows or not in a circuit. A closed switch has negligible resistance and allows current to flow in the circuit, whereas an open switch has nearly infinite resistance. An open switch prevents current from conducting or passing through the electric circuit.

There are three important features to consider when selecting a switching device:

- **Contacts** (e.g. single pole, double throw)
- **Ratings** (maximum voltage and current)
- **Method of operation** (toggle, slide, key.)

There are also several terms used to describe switch contacts:

- **Pole** – number of switch contact sets.
- **Throw** – number of conducting positions, single or double.
- **Way** – number of conducting positions, three or more.
- **Momentary** – switch returns to its normal position when released.
- **Open** – off position, contacts not conducting.
- **Closed** – on position, contacts conducting; there may be several on positions.

For example, the simplest on–off switch has one set of contacts (single pole) and one switching position, which conducts (single throw). The switch mechanism has two positions: open (off) and closed (on), but it is called 'single throw' because only one position conducts.

Switch contact ratings

Switch contacts are rated with a maximum voltage and current value at which they can safely operate. There may also be different ratings for a.c. and d.c. The a.c. values are higher because the current falls to zero many times each second and an arc is less likely to be maintained between the switch contacts.

For extra-low-voltage electronic tasks, the voltage rating is less significant but the current rating is an important consideration. The maximum current is less for inductive loads (coils and motors) because they cause more sparking at the contacts when switched off.

The current rating corresponds to the maximum allowable current the switch can carry when it is in the closed position. The current rating is based on the physical size of the switch contacts as well as the type of metal used for the contacts of the switches. Some switches have gold- or silver-plated contacts to ensure a very low resistance when closed.

The voltage rating of a switch corresponds to the maximum voltage that can safely be applied across the open contacts without internal arcing. The voltage rating does not apply when the switch is closed since the voltage drop across the closed switch contacts is practically zero.

Devices used for isolation and switching are broadly classified as:

- isolators
- switches.

Isolators

Isolators or switch disconnectors as illustrated in **Figure 11.87** are used for the electrical isolation of electrical systems from all sources of power and to conduct continuous current up to the level of their defined rated current level. They provide an effective method to isolate power in an emergency and should have a facility to lock the device in the OFF position.

Isolators find application as maintenance and repair switches installed locally to hard-wired loads such as motors, air-conditioners, water heaters and as a means to isolate the electrical supply from the loads and thereby protect persons from hazardous situations. These switches also adopt the function of the main switch on a switchboard as illustrated in **Figure 11.88**.

In some jurisdictions there may be a requirement to leave a half pole space between the main switch and the first circuit breaker.

Isolators do not have overload or short-circuit breaking capacity and must be protected from electrical damage. Therefore, the conductors feeding the isolator and then to the load must be protected against overloads and short-circuits by the upstream feeder protective devices located on the switchboard for this purpose. Isolators must be visible and readily available in relation to the load that they are connected to in order to verify visually that their downstream circuit is de-energised.

SWITCH ON

Isolating switch

Some isolating switches and circuit breakers look very similar except for the labelling. So check the circuit breaker before you install it to make sure it is a circuit breaker. Isolating switches do not provide over-current protection.

FIGURE 11.87 Various types of isolators

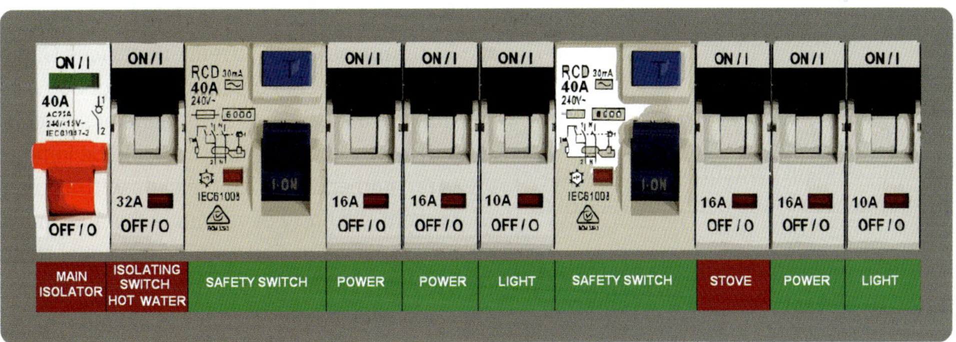

FIGURE 11.88 Main isolator and hot water isolator

Switches

The requirement to install a switching device as shown in **Figure 11.89** is the result of a need to isolate a heating source (such as a range, conventional cooktops and induction cooktops) from inflamed pots and pans. (Refer to AS/ NZS 3000, Clause 4.7.1.)

REVIEW QUESTIONS

1. How is control of an electrical installation achieved?
2. What are three important features to consider when selecting a switch?
3. State the purpose of an isolator.
4. Where would an isolator find application?
5. What is the difference between an isolator and a circuit breaker?

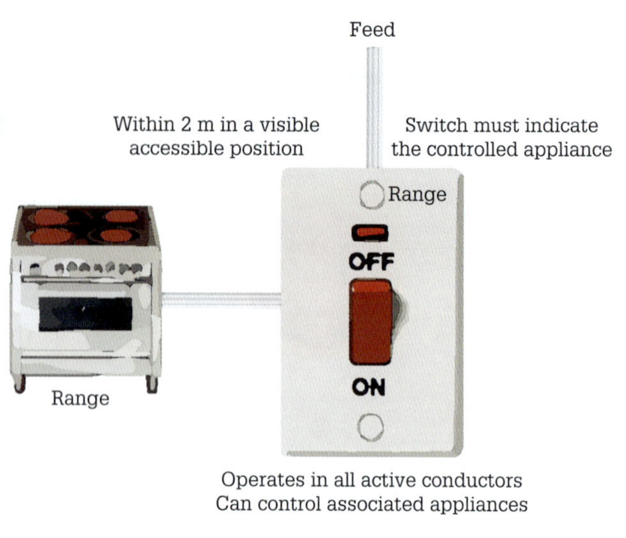

FIGURE 11.89 Switching device

11.11 Switchboards and distribution boards

AS/NZS 3000:2018, Section 2.10, 'Switchboards' specifies requirements for location, construction and mounting of switchboards. Regardless of the number or type of switchboards within an installation, they must all:

- Comply with the unique requirements of the relevant supply authority. Complying ensures that the installation coordinates with their protective and control equipment.
- Be readily accessible in well-ventilated areas, free from excess humidity, dust, and dirt and away from hazardous materials. A main switchboard should be located within easy access to the main entrance of the building.
- Be weatherproof; if installed outdoors, the switchboard must be protected to prevent moisture from entering the enclosure.
- Be constructed of materials (all steel construction or all insulated polycarbonate) able to withstand the environmental constraints and severe operating conditions that may be expected at the installation location.
- Be safely and securely mounted to a permanent structure such as a wall or a floor.
- Have a means to prevent strain/damage to cables (see Clause 2.10.6).
- Have no exposed live parts if the switchboard is domestic. Note Clause 2 for non-domestic boards.
- Include protective and metering devices positioned so as to provide safe and easy access for operation and maintenance.

All circuit breakers are mounted so that the ON/ OFF and current rating indications are clearly visible with covers or escutcheons in position. In addition, align operating toggles of each circuit breaker in the same plane.

Effect of eddy currents

To minimise eddy current effects (see Clause 3.9.10.2), where single-core cables rated over 300 A pass separately through a steel cable entry plate or switchboard surrounds, the entry holes should be joined with slots. It is recommended to fill the slots with epoxy or silicone to prevent the entry of vermin. All mains and sub-mains must be supported or tied within 200 mm of their terminations. This support or tie must be substantial due to stresses that occur under short-circuit conditions.

Spread of fire

In order to limit the possible spread of fire (see Clause 2.10.7) where a busway or cable is brought into a switchboard, the insulated busway/cable entry plate must have a close tolerance cut-out to fit the busbars/cable.

Switchboards must have a lid/door which:

- can be opened without damaging any cables or equipment
- is capable of being locked
- can remain open while an electrician works on the switchboard.

Marking

Switchboards must be clearly marked with numbers or letters to identify each switchboard from other switchboards within the installation.

All switchboard components must be legibly marked (see Clause 2.10.5.1) to indicate their relationship with various sections of the installation. For general light and power distribution boards, provide schedule cards of a minimum size of 200 × 150 mm, with typewritten text showing the following as installed information:

- sub-main designation, rating, and short-circuit protective device
- light and power circuit numbers and current ratings, cable sizes and type and areas supplied
- single-line diagram for the location of underground mains/sub-mains.

Mount schedule cards in a holder fixed to the inside of the switchboard or door, next to the distribution circuit switches. Protect with hard plastic transparent covers.

In domestic installations locate the switchboard and metering panel so as to reduce electromagnetic fields in active use areas.

Switchboards in a domestic installation must not be installed behind locked gates or a door unless they are fitted with the electricity distributor's locking systems.

Types of switchboards

Main switchboards for industrial and commercial installations are of the modular type as illustrated in **Figure 11.90** while the smaller sub-boards are usually of steel construction with insulated busbar systems.

FIGURE 11.90 A modular type switchboard

Domestic switchboards similar to that shown in **Figure 11.91** are common in many installations.

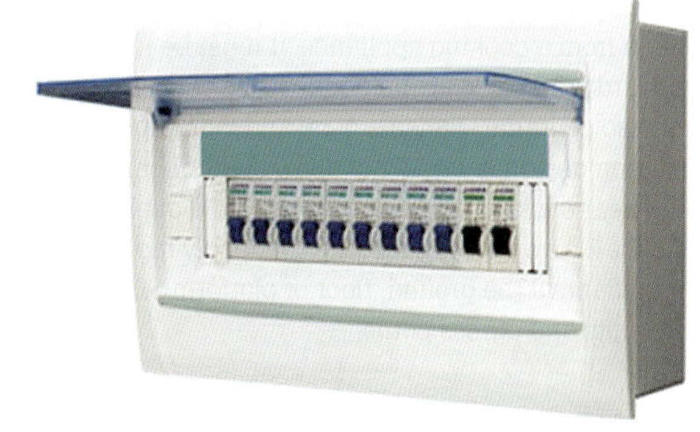

FIGURE 11.91 Domestic switchboard

All the switches and breakers on these switchboards are DIN rail mounted.

Restricted locations for switchboards

Clause 2.10.2.5 details the restricted locations of switchboards. This clause indicates a number of locations where the installation of a switchboard is restricted. These are/or near:

- water containers and cooking appliances
- showers/baths
- swimming pools, paddling pools and spas, fountains and water features
- refrigeration rooms
- fire hose reels
- automatic fire sprinklers
- in cupboards.

If the switchboard is required to be installed in a cupboard as shown in **Figure 11.92**, the cupboard's design should make it impossible for objects to be placed in front of the board.

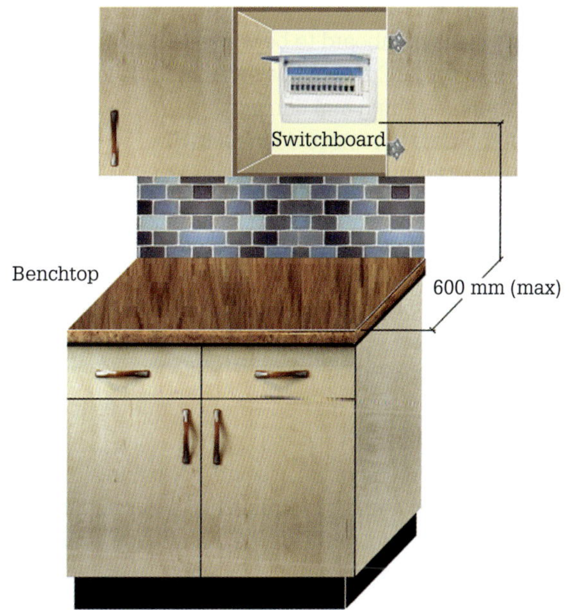

FIGURE 11.92 Switchboard in a cupboard

If the switchboard is located where it could be affected by steam from a cooking appliance, it must be installed in a cupboard with close fitting doors. However, as cooking steam also contains elements of oil and fat a prudent installation designer would not allow switchboards to be located near cooking appliances even if in a cupboard.

Clause 2 gives the restrictions that apply regarding:

- height above the ground, floor or platform
- fire exits and egress paths.

Fixings

Floor-mounted switchboards should be fixed to the floor or other structure with M12 hot-dipped galvanised bolts through the mounting holes provided in the base.

Wall-mounted switchboards should be installed using a sufficient number of fixings. However, not less than four fixings per switchboard should be used. It is suggested that fixings should be:

- To timber:
  - Proprietary assemblies – hot-dipped galvanised steel No. 12 wood screws of suitable length.
  - Custom-built assemblies – 10 diameter coach screws of appropriate length with a full washer under the head.
- To masonry and concrete:
  - Proprietary assemblies – stainless steel 6 mm diameter expanding sleeve masonry anchors of the projecting stud type each fitted with a full washer and nut.
  - Custom-built assemblies – stainless steel 10 mm diameter expanding sleeve masonry anchors of the projecting stud type each equipped with a full washer and nut.

Checking switchboards in the field

1. Isolate the switchboard to be worked on.
2. Inspect installed low-voltage switchboards for fixing, alignment, earthing and damage.
3. Clean interiors to remove debris and dirt.
4. Check the tightness of all accessible mechanical and electrical connections. Re-torque busbar connections.
5. Adjust access doors and operating handles for free mechanical operation.
6. Adjust circuit breaker trip and time-delay settings to values specified.
7. Test operation of RCDs.
8. Repaint scratched or marred exterior surfaces to match original finish.

A main switchboard is an assembly of circuit protection devices and other equipment necessary for the monitoring and control of the whole installation. A block diagram showing the distribution relationship is illustrated in **Figure 11.93** (refer to AS/NZS 3000:2018 *Wiring Rules* for the definition – main switchboard and distribution board).

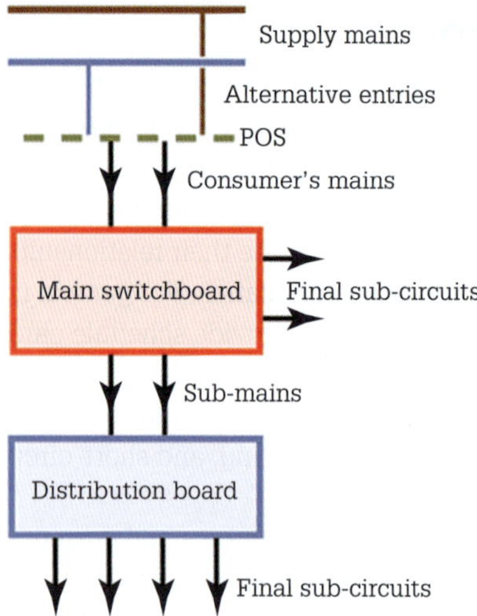

FIGURE 11.93 Installation block diagram

The main switchboard may have a consumer's distribution board mounted on it, or the consumer's distribution board may be in a remote location. A main switchboard containing the energy meter, main switches, RCD and two socket outlets for a domestic three-phase installation is illustrated in **Figure 11.94**. Note the in-service tags on the battery charger and extension lead.

FIGURE 11.94 Main switchboard for three-phase supply (Queensland arrangement)

The consumer's board for the main switchboard in **Figure 11.94** is illustrated in **Figure 11.95**. The consumer's board was located directly on the main switchboard in the garage. The consumer's board has 12 circuit breakers and 6 RCDs/MCBs.

FIGURE 11.95 Consumer's distribution board (not finished)

If the back of a metal switchboard surround is within 50 mm of the back surface of an internal wall lining it may be necessary to install an earthed metal plate between the inner face of the wall and the back of the switchboard. (**Note:** All parts of an electrical installation must be adequately protected against damage – this includes the switchboard.)

Figure 11.96 shows a simple checklist that can be used when inspecting switchboards.

☑ Switchboard check list

- ☐ Is the switchboard installed correctly (location allows heat to be dissipated), including restricted locations and installation at the recommended height?
- ☐ Is the switchboard enclosure fitted with an approved lock which is compatible with the utility's master key system or does it have a working simple self-latching device?
- ☐ Are panels fixed hinged or removable according to their size?
- ☐ Are all exposed conductive parts – switchboard case and associated equipment of the switchboard – earthed?
- ☐ Are the correct protective overload devices fitted?
- ☐ Are proper RCDs fitted?
- ☐ Is the switchboard free from any exposed live parts?
- ☐ Are the phase-to-neutral, phase-to-phase and phase-to-earth clearances and creepage distances correct?
- ☐ Is the fault current rating for circuit breakers and RCDs suitable for the load?
- ☐ Are main switches fitted and clearly labelled and their operation clear and unmistakable? If there is more than one main switch are they grouped together and identified as main switches?
- ☐ Are main switches for emergency systems such as fire and smoke control systems, evacuation equipment and lifts connected on the supply side of all other main switches?
- ☐ Are conductors and control equipment associated with emergency systems segregated from all other wiring systems on the switchboard?
- ☐ Are neutral bars, earth bars, and active links fitted and correctly terminated?
- ☐ There is usually one conductor per terminal, and separate neutral and earth bars have an MEN connection.
- ☐ Is the switchboard wiring clear of any sharp edges or bare live parts and correctly terminated without being subjected to undue stress?
- ☐ Are any cables bent? If so, do the bends comply with requirements?
- ☐ Are all soldered connections insulated to the same grade as the cable insulation?
- ☐ Are the laminated circuit schedule card and route of consumer's mains provided?
- ☐ Are all cables and conduit access holes cut into the switchboard that are not of a close fit sealed with a fire-retardant sealant?
- ☐ Where single-core cables pass through the switchboard surround is there provision to limit eddy currents?
- ☐ If self-tapping screws have been used on the switchboard could they penetrate a cable?
- ☐ Do the metering services meet the local utility's requirements?
- ☐ Can the switchboard door travel through approximately 100° and be held there?

FIGURE 11.96 Switchboard checklist

Control of outgoing circuits

All outgoing circuits must have a control device installed at the originating switchboard.

Sub-mains

A sub-main circuit can be defined as a circuit connected directly from the main low-voltage switchboard to any other sub-main distribution board or from one sub-main switchboard to another. The CSA of any sub-main cable must be able to carry the maximum demand of the portion of the installation it is supplying.

When an electrical supply is required for outbuildings, such as a detached workshop, garage or garden shed, sub-mains are used to supply the power from the main switchboard. A sub-main distribution board is required to be installed in an outbuilding. Sub-circuits are then run from the sub-main distribution board to lights, socket outlets and other fittings.

Each sub-main circuit is required to have its protection device at the main switchboard. In addition, each sub-main circuit may have an earthing conductor connected to the main switchboard earthing bar and taken to the earthing bar on the sub-mains distribution board. Alternatively, an MEN system may be established at the outbuilding with a separate earth stake. RCD protection at outbuildings or on the sub-main circuit is recommended.

Final sub-circuits

The electrical distribution within an installation ends with final sub-circuits. A final sub-circuit is that section of a wiring system that extends beyond the final circuit protection device. Final sub-circuits can originate from either a main switchboard or a sub-main distribution board.

Final sub-circuits make up the greatest part of the wiring of an electrical installation and are divided into three general groups:

1. Fixed-wiring lighting circuits
2. Fixed-wiring socket outlet circuits
3. Fixed-wiring appliance circuits

Final sub-circuits are usually rated at 6 A, 10 A,16 A, 20 A, 25 A or 32 A for some fixed-wiring loads.

Switchboard design and arrangement of equipment

Switchboards can be custom built by the electrician. Alternatively, a switchboard can be a commercially made ready-built board of standard design. Ready-built switchboards are used in domestic installations.

Switchboard design and arrangement of equipment on the panel board depend upon several factors:

- the load demand of the installation
- the number and type of circuits required
- the prospective fault current (with a maximum transformer capacity of 315 kVA the maximum prospective fault current at the POS for a 70 ampere two-wire service is 6.0 kA if the POS is greater than 50 m from the supply transformer)
- fault current rating of the switchboard (residential switchboards should have a minimum fault rating of 6 kA)
- accessibility, location and environmental conditions.

Refer to AS/NZS 3000:2018, Part 2, *Wiring Rules*, 'General arrangement, control, and protection: switch boards and emergency systems'. **Figure 11.97** shows a typical metal domestic switchboard for Queensland including the Bramite® type Z insulating panel board.

FIGURE 11.97 Domestic switchboard and panel board (Queensland)

Switchboards can be made of other material such as plastic or fibreglass but are subject to approval by the utility provider. The meter panel is usually 600 mm high × 600 mm wide.

Switchboard equipment that may be installed on a switchboard includes main switch or switches, links and service fuses, circuit protection devices, RCDs and meters with their associated equipment, as shown in **Figure 11.98**. Where the maximum demand of any installation required to be metered exceeds 100 A per phase, meters that operate in conjunction with current transformers are used. The current transformers must be mounted so that the secondary terminals are always accessible while the switchboard is energised. The red spot on the current transformer is always to face the incoming supply. Note the provision of an insulated entry for a flexible cord through the bottom of the switchboard enclosure.

FIGURE 11.98 Switchboard equipment placement (Queensland)

Adequate (minimum 600 mm) clear working space must be provided in front and to the sides of a metering enclosure. The top of the meter panel must not be mounted more than 2 m and the bottom not less than 0.7 m from finished ground level.

Another consumer's distribution board for Queensland is illustrated in **Figure 11.99**.

Figure 11.100 shows a typical metal domestic three-phase switchboard for New South Wales with **Figure 11.101** showing the individual equipment used.

In Australia, we use an off-peak tariff system for water heating. The water heater is connected to its kilowatt hour meter with a tone-controlled switch (audio PLC). The service provider controls the switching of the water heater and offers a greatly reduced tariff rate. One possible MEN distribution arrangement for a domestic main switchboard is shown in **Figure 11.102**.

Mounting of equipment

The mounting of equipment within a switchboard must comply with AS/NZS 3000:2018 *Wiring Rules* and local energy distributor requirements where applicable. Equipment should be fixed on the mounting chassis plate by means of bolts, washers and nuts or bolts screwed into tapped holes in the chassis plate. In the latter case the

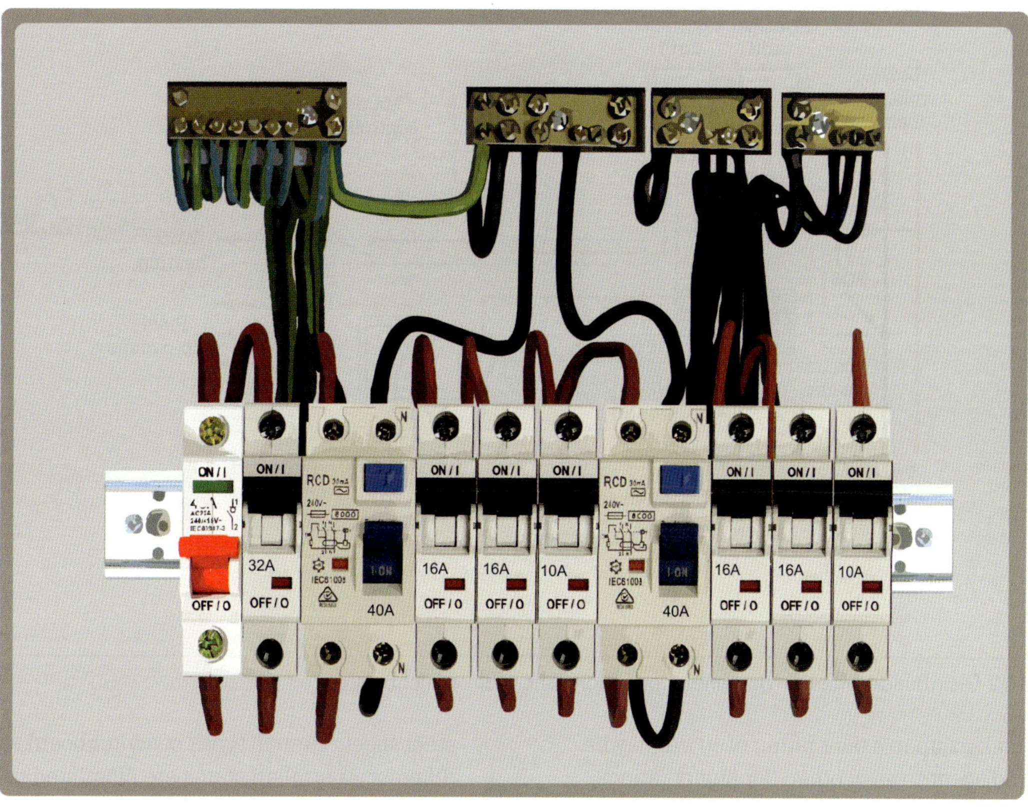

FIGURE 11.99 Distribution board (Queensland)

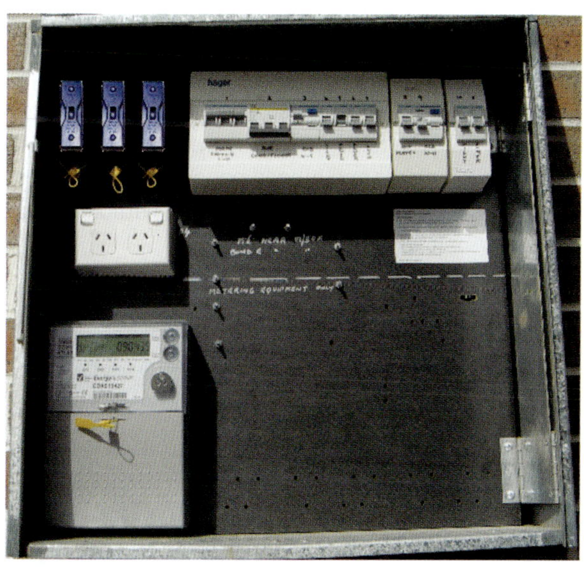

FIGURE 11.100 Three-phase domestic switchboard (New South Wales)

minimum thickness of the chassis plate should not be less than 2.5 mm. Self-tapping screws are not an acceptable method as the screws may penetrate cables. In designing a custom-built switchboard, the following requirements, depending upon local energy distributor specifications, should be adhered to:

- A minimum distance of 50 mm is necessary between any piece of equipment and the frame or internal partitioning. This minimum space is required on all sides of the equipment. A minimum of 75 mm is necessary between horizontal rows of equipment.

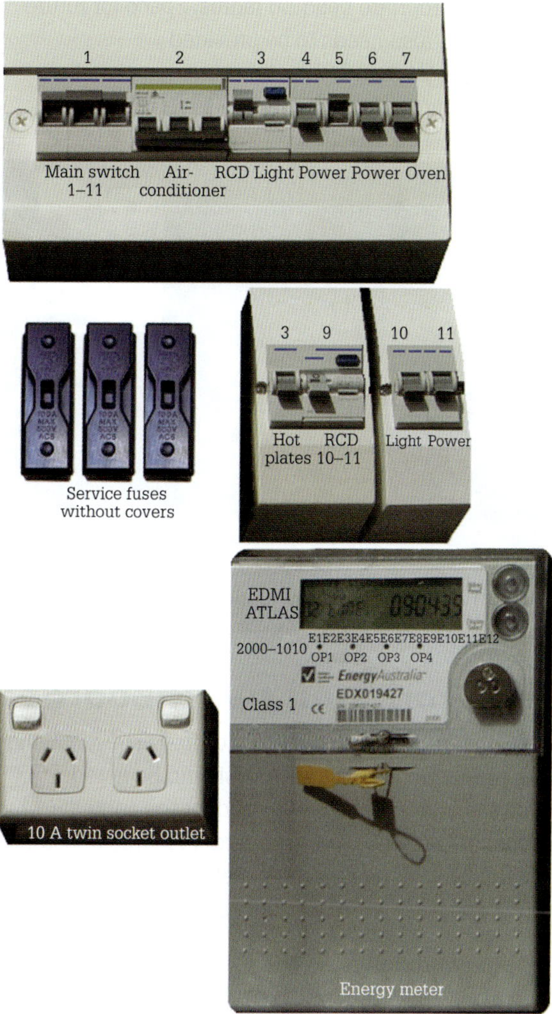

FIGURE 11.101 Switchboard equipment

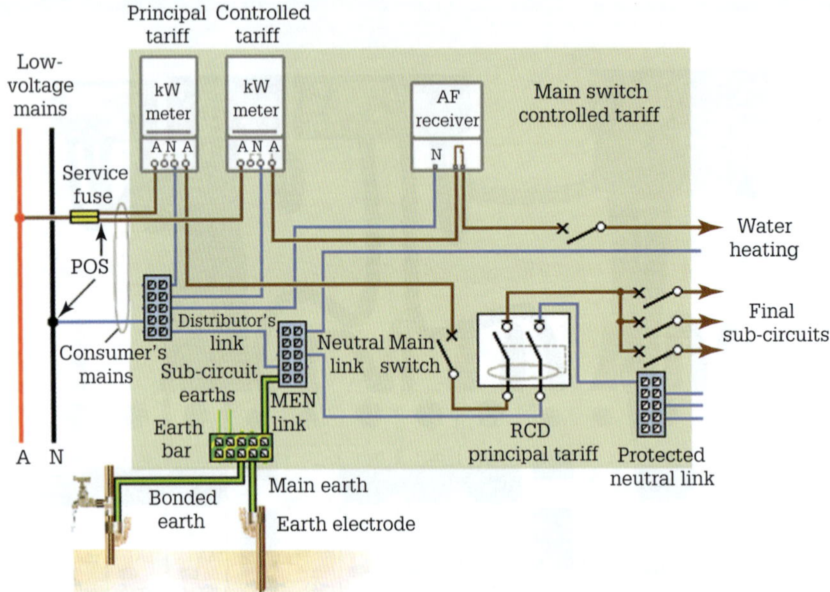

FIGURE 11.102 Distribution MEN arrangement for a domestic main switchboard

- Circuit breakers up to a fault rating of 10 kA may be installed adjacent to each other. For higher ratings, a minimum separation of 40 mm should be allowed between circuit breakers or isolators.
- Sufficient space shall be provided for wiring allowing for the appropriate bending radius. Space for future equipment also needs to be provided.

An older-type distribution board with its labelled simple layout is illustrated in **Figure 11.103**.

FIGURE 11.103 Old-type distribution board used for flats

Metalwork and switchboards

A high standard of metalwork is required. The following points concerning the metalwork should be noted:

1 The thickness of the metal used for the cabinet depends on the size of the board. For large freestanding switchboards the minimum metal size for major components is 2 mm thick furniture-quality bright steel sheet. Various types of switchboard surround for different Australian states are illustrated in Figures **11.104**, **11.105** and **11.106**.
2 All hinged panels must be suitably stiffened and fitted with lift-off hinges.
3 Large lift-off panels are not favoured, but if unavoidable they must be equipped with a means of handling them, such as fitted D handles. Such panels must have a means of support, such as studs or a supporting ledge, for use while the fixing screws are being installed.
4 Escutcheon plates and hinged panels should be fixed in place with a fixing that can be undone without the use of tools.

FIGURE 11.104 South Australian switchboard surround

FIGURE 11.105 Victorian switchboard surround

FIGURE 11.106 New South Wales switchboard surround

Busbars

Busbars are constructed in the form of straps, which are first cut to length and can be provided with sets of punched holes through which bolts are received for mounting the busbars on suitable supports within a switchboard. They can then be connected to each other and for mounting electrical cable connectors thereon.

All busbars should be of solid drawn, high-conductivity copper. However, other types of material such as aluminium, brass and copper-clad aluminium are also used. If joined together busbars should overlap for a distance equal to twice the width of the bar to prevent localised heating. Copper and brass contact surfaces should be tinned (acid-base flux may not be used) or silver-plated and bolted together with cadmium-plated bolts and nuts.

A busbar as shown in **Figure 11.107** is a solid conductor, usually rectangular in cross-section, which forms a link between two or more electrical circuits.

FIGURE 11.107 Busbars

Busbars are usually provided for the following applications:

- distribution of supply voltage (phase or active busbars)
- connection of equipment with ratings exceeding the current rating of 70 mm^2 conductors
- connection of outgoing circuits with current ratings in excess of that allowed for 70 mm^2 conductors
- collector bars for parallel cables
- connection bars for neutral conductors (brass busbars)
- connection bars for earth conductors (brass busbars)
- busbar connections to miniature circuit breakers in domestic switchboards (comb-type copper busbar).

See also Chapter 15 of this textbook for information about electrical switchboards.

REVIEW QUESTIONS

1. What is the requirement for orienting circuit breakers on a switchboard?
2. How are the effects of eddy currents minimised where single-core cables rated over 300 A pass separately through a steel cable entry plate or switchboard surround?
3. How should switchboards be marked?
4. What type of switchboard is commonly used in industrial and commercial installations?
5. What is required if a domestic switchboard is required to be installed in a cupboard?
6. Provide a definition for a sub-main.
7. Provide a definition for a final sub-circuit.
8. What metering is required where the demand exceeds 100 A?
9. Name the equipment that may be installed on a switchboard.
10. What is a busbar?

CHAPTER REVIEW

11.1 Safety principles

- Safety principles require that the design and performance of electrical installations satisfy the requirements of all relevant Acts, regulations, standards and codes of practice publications relevant to the state or territory where the installation is located.
- Compliant protection arrangements are applied to every installation so that any periodic inspection, testing, maintenance and repairs likely to be necessary during the intended life of the installation can be easily and safely carried out.
- The design and installation of electrical installations must provide reasonable safety from danger.
- To guarantee satisfactory performance of electrical installations, both the safety of supply and the quality of supply need to be considered with equal care.

11.2 Circuit and control arrangements

- Every installation must be divided into circuits to avoid hazards and to contain the disruption in the event of a fault.
- Final sub-circuits make up the greatest portion of the wiring of an electrical installation and are divided into three general groups.
- The number and type of circuits required are determined by the load on the circuit, the location of the loads, any seasonal variations that affect the loads and any special conditions such as tariff requirements.
- Voltage is regarded as a difference of potential that exists between all active conductors and all active conductors and earth.
- Extra-low voltage is a voltage not exceeding 50 V a.c. or 120 V ripple-free d.c.
- SELV systems are used where an inherently safe system of supply is required.
- A PELV power supply includes two protective measures against direct and indirect contact with exposed conductive parts by means of a safe separation of the primary and secondary transformer windings.
- Transformers provide electrical isolation while allowing the magnetic coupling of electrical power from primary winding to secondary winding.

11.3 Hazards and risks in an electrical installation

- Every electrical installation is a combination of cables, accessories and energy-converting load equipment installed for a particular purpose within specific environmental conditions.
- Persons and livestock must be protected from electric shock hazards (physiological effects of current) arising from direct and indirect contact with live conductors or exposed conductive parts of the installation.
- Physical contact with live parts of an electrical installation can affect the human body in various ways such as pain, burns or even death.
- Electrical installations in premises must be designed, performance-based and engineered to the requirements of all relevant Acts, regulations and standards and codes of practice publications appropriate to the state or territory where the installation is located.
- Heat produced by an electric current is proportional to the square of the magnitude of electric current and the time for which the current is flowing.
- All parts of equipment or machines that move in the course of the performance of work may contribute to accidents causing injury and damage.

11.4 Protection against indirect contact

- Each circuit in an electrical installation must be designed so that automatic disconnection of the circuit occurs to prevent a rise in 'touch voltage' to dangerous levels within a specified time.
- The method of protection used to limit the effects of indirect contact is achieved by a system of equipotential bonding and automatic disconnection of the supply.
- To provide protection against indirect contact, Class II equipment or appliances, which do not rely only on basic insulation, can be used.
- The *Wiring Rules* require that the selection and installation of electrical equipment in damp situations must provide enhanced protection against electric shock, due to the reduction in body resistance and the likelihood of bodily contact with earth potential.

11.5 Earthing

- The MEN system was designed to ensure that if a fault occurs where the active conductor contacted exposed metal, the fault current could flow through the earth conductor as a return path to the switchboard, allowing the protective device to open-circuit, thereby making the circuit safe.
- The effectiveness of the MEN system is reliant on the continuity of the main neutral conductor, the MEN link and the connection of the main earthing conductor to the earth electrode.
- The main earthing conductor of an MEN earth system connects the main earthing terminal/connection or bar to the earth electrode.
- Earthing conductors are usually high-conductivity copper.
- The MEN connection is a link between the main earth terminal/bar and the main neutral conductor bar.
- The size of the main earthing conductor is determined from AS/NZS 3000:2018 *Wiring Rules*, Table 5.1.

- The minimum size main earthing conductor is 4 mm^2 for copper conductors and 16 mm^2 for aluminium.
- The main earth connection should have a permanent label attached to the connection to the earth electrode.
- The resistivity of the general mass of earth surrounding an electrode can be enhanced by using an earthing compound.
- A protective earthing conductor (PEC) is any earth conductor that is not a main earthing conductor, bonding conductor connecting the earthing system to the electrical installation or the earth conductor of any electrical equipment requiring earthing.
- The purpose of the protective earthing conductors is to provide a safe path for earth fault current so that the protective device operates to remove dangerous potential differences, which are inevitable under fault conditions.

11.6 Equipotential bonding

- Equipotential bonding is a method by which different earthing conductors cause various exposed metallic conductive parts and extraneous conductive parts to be at the same or approximately the same potential.
- Functional earthing refers to the need for certain points in electrical equipment or systems to be connected to the earthing system in order to maintain their proper function.
- The SWER system uses one metal conductor (in most cases a bare conductor) to provide power and the general mass of earth (ground) as a return current conductor.
- An SWER system consists of a separate high-voltage earthing system and a low-voltage earthing system.

11.7 Protection against overload and short-circuit current

- Each circuit in the electrical installation must be designed to protect people, livestock and property from harmful effects.
- The earth fault-loop impedance is the impedance of the phase-to-earth fault current loop starting and ending at the point of the earth fault.
- An earth fault-loop impedance or load test is applied to the installation to ensure continuity and effectiveness of the earthing arrangement.
- In order to ensure that automatic disconnection occurs within the required time should a low impedance fault occur between active and the conductive frame of electrical equipment, it is imperative that the associated earthing conductor has a very low impedance.
- The electrical installation must be capable of enduring, without destruction, the prospective fault current that can be drawn under a fault condition.
- It is important to be able to calculate fault current levels to enable the correct selection of protection devices.
- Automatic disconnection must occur when a short-circuit earth fault occurs between the active conductor and the protective earth conductor or other exposed conductive parts of the installation.

11.8 Devices for automatic disconnection of supply

- A protective device must be placed at the point where a reduction of the cross-sectional area (CSA) of the circuit conductors or another change causes alteration in the characteristics of the installation.
- An installation comprising some series-connected protective devices is considered a selective coordination arrangement when, in the event of short-circuit or overload, the installation is interrupted only by the device that is immediately upstream of the fault point.
- Devices used for the automatic disconnection of supply include fuses, fault current limiters, and circuit breakers.
- The environmental conditions in which an MCB is required to operate must be considered when selecting a suitable circuit protection device.
- A core balance earth leakage device is also known as a residual current device (RCD) or a safety switch and was created to detect very small earth leakage currents (10 mA). The purpose of this device is to protect people and livestock from electric current passing through the body to the earth.

11.9 Protection against over-voltage and under-voltage

- Over-voltage is defined as a power surge that takes place when the voltage is 10% above rated rms voltage for one or more cycles.
- Over-voltage places stress on insulation due to the heating effect, causing rapid ageing of the insulation, resulting in the premature breakdown of its electrical properties.
- Over-voltages are prevented from reaching conductors, cables, and equipment in an installation by short-circuiting (extinguishing them before a dangerous value has been reached).
- The over-voltage protection device's equipotential bonding conductors must be connected by the shortest possible route to the main neutral terminal of the installation.
- An under-voltage (also called a brownout) is defined as a decrease in the voltage to between 10% and 90% for a duration of greater than 60 seconds.
- Power sags involve voltages below nominal voltage for a short period and are one of the more common forms of power problems experienced by consumers.
- Protective measures must be taken where a drop in voltage or the loss of and automatic restart of voltage could cause a dangerous situation.

11.10 Control of an electrical installation and circuits

- Control of an electrical installation is achieved by means of a mechanical switching device. This device is capable of making, carrying and breaking an electrical current under normal installation conditions.
- Switch contacts are rated with a maximum voltage and current value at which they can safely operate. There may also be different ratings for a.c. and d.c.

- Isolators or switch disconnectors are used for the electrical isolation of electrical systems from all sources of power and to conduct continuous current up to the level of their defined rated current level.
- Isolators find application as maintenance and repair switches installed locally to hard-wired loads such as motors, air-conditioners, water heaters and as a means to isolate the electrical supply from the loads and thereby protect persons from hazardous situations.

11.11 Switchboards and distribution boards

- AS/NZS 3000:2018, Section 2.10, 'Switchboards' specifies requirements for location, construction and mounting of switchboards.
- All circuit breakers are mounted so that the ON/OFF and current rating indications are clearly visible with covers or escutcheons in position.
- To minimise eddy current effects, where single-core cables rated over 300 A pass separately through a steel cable entry plate or switchboard surrounds, the entry holes should be joined with slots.
- In order to limit the possible spread of fire where a busway or cable is brought into a switchboard, the insulated busway/cable entry plate must have a close tolerance cut-out to fit the busbars/cable.
- All switchboard components must be legibly marked to indicate their relationship with various sections of the installation.
- Floor-mounted switchboards should be fixed to the floor or other structure with M12 hot-dipped galvanised bolts through the mounting holes provided in the base. Wall-mounted switchboards should be installed using a sufficient number of fixings.
- All outgoing circuits must have a control device installed at the originating switchboard.

TRIAL EXAM

For Chapter 11 knowledge assessment, please complete the following trial exam.

1 The design and installation of electrical installations must provide:
 a reasonable safety from danger
 b ease of maintenance and testing
 c application of prescriptive measures
 d a framework for compliance

2 Two ways by which persons and livestock can receive an electric shock are:
 a direct and inconsequential contact
 b intentional and non-intentional contact
 c direct and indirect contact
 d simultaneous failure of RCD and over-current protection

3 In electrical installations, two major types of risk exist. Shock current is one; name the other:
 a fault currents
 b unwanted voltages
 c mechanical movement
 d excessive temperatures

4 An SELV system finds application where:
 a it is inappropriate to bond to earth
 b there is no MEN connection
 c direct earthing systems are used
 d an inherently safe system of supply is required

5 Transformers provide:
 a electrical isolation
 b electrical continuity
 c high load power factor
 d low load power factor

6 Class II equipment is categorised by:
 a basic insulation only
 b double insulation or reinforced insulation
 c basic insulation and an earth pin
 d either an SELV or PELV supply

7 The conductor classified as the most important in an electrical system is:
 a active
 b bonding
 c earthing
 d neutral

8 The MEN connection is:
 a the link between the main active and main neutral
 b the link between the main earth and main neutral
 c the link between the main active and main earth
 d required for correct equipotential bonding

9 The minimum size for an aluminium main earthing conductor is:
 a 2.5 mm^2
 b 4 mm^2
 c 6 mm^2
 d 16 mm^2

10 A voltage spike is categorised by:
 a an over-voltage of short duration
 b an over-voltage of long term
 c an under-voltage of short duration
 d an under-voltage of long term

11 The prospective touch voltage should not exceed:
 a 25 V
 b 50 V
 c 100 V
 d 230 V

12 Disconnection of the supply for a circuit supplying 10 A socket outlets undergoing an earth fault must occur within:

a 0.4 second
b 1 second
c 5 seconds
d 12 seconds

13 For a type D 50 A circuit breaker the mean tripping current causing automatic disconnection of supply is:

a 4 × rated current
b 6 × rated current
c 7.5 × rated current
d 12 × rated current

14 In what location must protective devices be placed with the circuits of an installation?

a where voltage drops occur
b where a potential short-circuit exists
c where the supply authority regulations suggest
d where a reduction of the CSA of the circuit conductors occurs

15 A circuit protection device must be:

a connected at the end of all circuits
b connected to the earthing circuit
c connected in series with the circuit it is protecting
d connected across the active and neutral circuit conductors

16 Over-current protection is provided by a:

a circuit breaker
b rocker switch
c isolating switch
d residual current device

17 A 200 kVA, 11 kV/400/230 V, three-phase distribution transformer has 5% impedance. The maximum prospective fault current at the secondary terminals of the transformer is:

a 1.66 kA
b 83.3 kA
c 16.66 kA
d 10.0 kA

18 Which of the following demonstrates a defence for preventing injury from mechanical movement?

a protrusions
b safety guards
c entanglement
d drawing-in

19 A partial reduction in rms voltage is called a:

a voltage buck
b voltage drop
c reduced boost
d voltage sag

20 Equipotential bonding and automatic disconnection of the supply is a method of protection to limit the effects of:

a direct contact
b indirect contact
c circulating currents
d excessive insulation temperatures

12 Alternate supplies

This chapter provides electrotechnology workers with knowledge and skills required to identify, shut down and restart systems with alternate power supplies. It also includes information to assist in identifying system configuration, hazards and controlling the associated risks, isolating and testing for isolation, and requirements for system restart. This chapter provides underpinning knowledge for the unit UEEEL0047 from the UEE training package.

LEARNING OBJECTIVES

Working safely with alternate supplies

- Outline procedure for the safe working on installations with alternate supplies.
- Explain the importance of hazard identification and risk control measures.

Types and configurations of alternate supplies

- Describe the main types, arrangements, and configurations of alternative supplies.
- Outline the fundamental requirements for operating alternate supplies.

Safe isolation

- Explain the importance of safely shutting down and isolating an alternate supply.
- Follow a procedure for safely shutting down and isolating an alternate supply.

Battery storage systems

- Describe common battery storage systems used to augment alternate supplies.
- List Standards and regulations that apply to installations with alternate supplies.

12.1 Working safely with alternate supplies

It is important to understand some of the terminology associated with alternate supplies, which are defined in AS/NZS 3000 and associated standards. **Table 12.1** summarises some pertinent definitions.

TABLE 12.1 Definitions pertaining to alternate supplies

Term	Meaning
Alternate supply	A system of electrical supply that is intended to maintain the functioning of all or part of an electrical installation in case of interruption of the normal electrical supply.
Normal supply	The electrical supply to which the electrical installation connects under normal conditions of operation. This is usually a distribution network but may be a private electrical generation system.
Supplementary supply	A system of electrical supply that is intended to operate in conjunction with the normal electrical supply.

This chapter will look at working safely with electricity generation systems and inverters, which are supplemental to the normal supply obtained from an electrical distribution network.

Risk determination

Before undertaking *any* electrical work, it is essential that a thorough a risk assessment be undertaken. The purpose of this is to determine potential hazards associated with the work. There is a range of risks to consider, which include (but are not limited to):

- immediate risk of electric shock to person undertaking the work and others
- fire or explosion from arcs and sparks
- release of poisonous fumes or gas from heat or fire
- oxygen depletion from insulating gases (SF_6) or other gases and fumes displacing air in confined spaces
- electric shock from stored energy sources such as battery banks
- conductors or other electrical equipment becoming live from unexpected sources such as:
  - automatic, inadvertent or remote switching
  - portable generators
  - photovoltaic arrays
  - uninterruptible power supply systems (UPS).

After identifying possible hazards, it is necessary to put in place preventative control measures to mitigate the incidence of potential hazards. The aim of risk mitigation is to eliminate the risk. **Figure 12.1** details the principles of risk control applicable to electrical isolation. If the risk assessment indicates that it is not possible to sufficiently reduce the risk to a level that allows the work to proceed, then the work is not commenced. Chapters 1 and 2 identified the three principles of risk management as:

1. identifying potential hazards
2. assessing and ranking the associated risks
3. applying suitable control measures to mitigate or reduce the risk.

SWITCH ON

Always operate on the premise that **all** electrical conductors and equipment, which includes neutral and earthing conductors, are **live** until proven otherwise.

Always test before touching

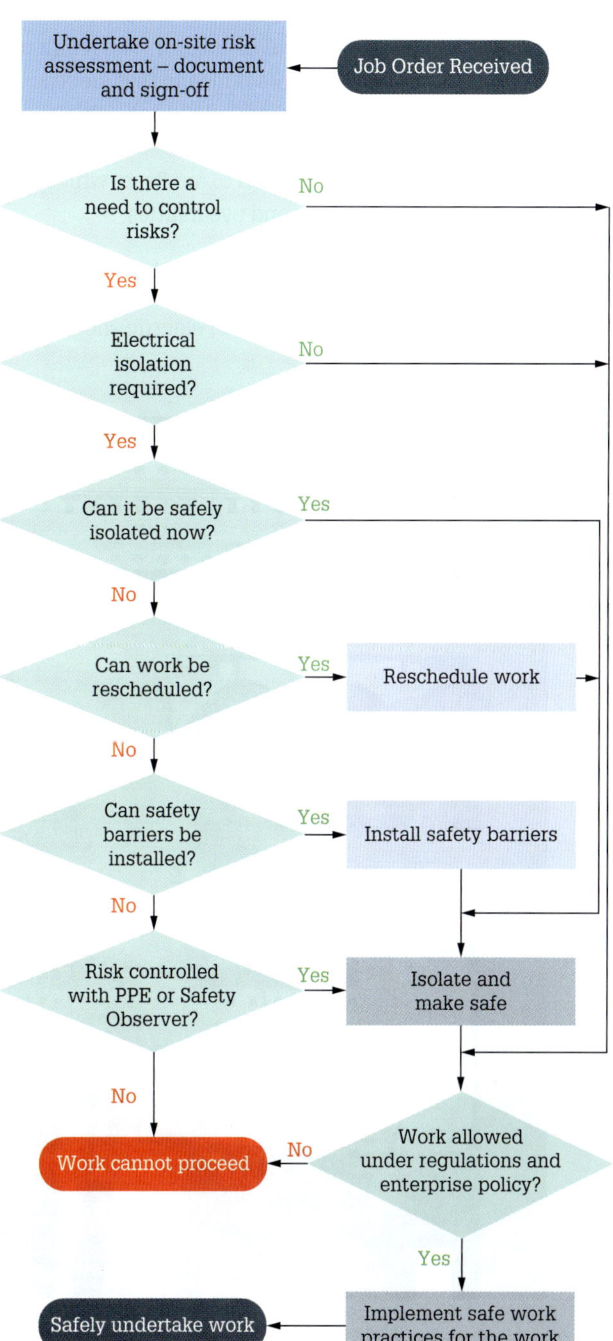

FIGURE 12.1 Principles of risk control for electrical isolation

Hazards

The major and perhaps most obvious hazard associated with working on or near electrical equipment is electric shock. Electric current has a tendency to follow a low impedance conductive path from and to the source of supply. If a human body forms part of this conductive path, the current will cause injury between where it enters and exits the body. The degree of injury depends on the magnitude of the voltage potential, the magnitude of the current, the path the current takes through the body, and the time for which the current flows. In an emergency situation, time is a critical factor that requires prompt action to minimise the degree of exposure.

The risk of serious injury or death is high if the electric current passes through major organs such as the heart. Contact with low-voltage alternating current can cause muscle contraction, which may lead to the casualty's hands fastening onto an electrical conductor. The result of them not being able to break contact means that electric current is flowing for a prolonged period of time, which increases the probability of death.

Another hazard is being exposed to arc flash, which can cause severe radiant burns and eye injuries.

Electric shock

There are numerous possible sources of electric shock, which include:

- potential difference between line conductors and neutral
- potential difference between line conductors and earth, which includes exposed metalwork, conductive surfaces and materials, and the like
- potential difference across open switch contacts, which can be equal to the supply voltage
- electrical energy supplied from other sources of supply such as photovoltaic arrays, wind generators, hydro generators, inverters, motor generator set, uninterruptible power supply
- potential difference due to undischarged capacitors
- potential difference impressed on conductors and conductive parts through static electricity, electromagnetic induction, and the like
- faulty electrical equipment or wiring, which may result in conductive parts attaining a potential above earth
- high resistance in the earth return path in an MEN installation causing a rise in potential on earth conductors.

A person exhibiting any of the following may be the victim of electric shock:

- not breathing
- cardiac arrest
- ventricular fibrillation
- involuntary muscle reaction
- entry and exit burns
- bone fractures caused by muscle spasm.

Source: Shutterstock.com/Nightman1965

FIGURE 12.2 Arc flash in control panel

Arc flash

An arc flash, as shown in **Figure 12.2**, is the heat and light produced as a result of the ionisation of air. Temperatures of an arc flash can be as high as 19 000 °C. Temperatures of this order can ignite clothing and burn the skin of anyone within close proximity. The temperatures generated by the arc flash can also melt metal. Other effects are damage to the lungs from breathing the vapours generated by the flash and eyesight damage from molten metal and the intense radiation produced by the arc flash.

The main cause of arc flash is from switching reactive loads or lightning strikes causing voltage transients (spikes). The duration of this transient might be in the order of microseconds but it can carry many thousands of amperes during this time. If this were to occur when an electrician is in the process of taking measurements, a plasma arc can form either inside or around the measurement instrument.

Arc flashes may be caused by:

- touching a test probe to the wrong surface
- worn or loose connections
- dust causing tracking
- gaps in conductor insulation
- improperly installed parts causing high resistance joints and connections
- corrosion causing high resistance joints and connections.

SWITCH ON

Switching operations can cause arc flashes.

Only operate a switch carrying load current if the switch is designed for such operation.

Arc blast

An arc blast is the pressure wave that an arc flash creates. The intensity of an arc blast may be sufficient to knock a person who is near the source of the blast to the ground, or cause extensive equipment damage. An arc blast can cause:

- damage to hearing or brain functions
- loose equipment, tools, machinery and debris to be projected through the air and cause further damage or injury.

Risks associated with alternative power sources

Alternative electricity power sources pose additional risks in that one or more of these generation systems, if installed and functioning, may still be live after isolating the normal supply, or the generator may start unexpectedly. Alternative electricity power sources and some associated risks are summarised in **Table 12.2**.

TABLE 12.2 Alternative electricity power sources and associated risks

Alternative energy source	Associated risk
Motor generator sets either diesel or petrol driven	▪ Switching on and off may be controlled manually or automatically. ▪ May connect via a plug and socket or may be hard-wired. ▪ May still supply electricity in the event of normal electrical supply being isolated. ▪ Chance of electric shock from the generator, connecting cables and the electrical installation to which the generator connects. ▪ Possibility of fire or explosion from fuel source igniting. ▪ Chance of oxygen depletion and breathing noxious fumes from unvented exhaust gases. An installation may have diesel generators or other continuous electrical power systems installed for critical processes, such as life-sustaining equipment at a hospital.
Grid connect solar photovoltaic (PV) systems	▪ Generation of electricity only occurs during daylight hours. ▪ Cables connecting solar panels, inverter and switchboard may be live during night hours. ▪ Chance of electric shock from the PV array, connecting cables, and the electrical installation to which the PV system connects.
Wind powered generator	▪ Wind is used to turn a turbine that drives a generator to produce electrical energy. ▪ Cables connecting the generator, inverter and switchboard are live at all times. ▪ Chance of electric shock from the wind generator, connecting cables, and the electrical installation to which the generator connects.
Hydro powered generator	▪ Water is used to turn a turbine that drives a generator to produce electrical energy. ▪ Cables connecting the generator, inverter and switchboard are live at all times. ▪ Chance of electric shock from the hydro generator, connecting cables, and the electrical installation to which the generator connects.
Uninterruptible power supply (UPS) or battery storage systems	▪ May connect via a plug and socket or may be hard-wired. ▪ Will supply electricity in the event of normal electrical supply being isolated. ▪ Chance of electric shock from the battery system, connecting cables and the installation to which the UPS or battery storage system connects. An installation may have a UPS or battery system installed for critical processes, such as life-sustaining equipment at a hospital.

For sites containing storage batteries, the following additional safety risks exist:

- Batteries present hazards similar to that of compressed gas cylinders or natural gas supply in that they can explode or combust.
- Charging or discharging of lithium-ion batteries, which are common storage batteries for alternate supplies, may emit noxious gases. These gases may not be flammable, but they pose a health risk in poorly ventilated environments.
- Storage batteries may leak corrosive chemicals.

REVIEW QUESTIONS

1 Describe the meaning of supplementary supply.
2 What are some of the risks to consider when undertaking a risk assessment for electrical work?
3 List two hazards associated with working on or near electrical equipment.
4 What are some of the injuries that can result from exposure to an arc flash?
5 How is an arc blast created?
6 Identify three alternative electricity power sources.
7 List two risks associated with motor generator sets as alternative electricity power sources.
8 What additional safety risks exist for sites containing storage batteries?

12.2 Types and configurations of alternate supplies

There are a number of technologies and methods used to produce usable electrical power. We are seeing a shift away from a reliance on burning fossil fuels as the source of this energy towards renewable energy sources using technologies such as wind generators and photovoltaic panels. Common technologies used for alternative electrical power supplies are:

- engine-driven alternators and generators
- photovoltaic panels
- wind-driven alternators and generators
- personal or micro hydroelectric systems.

Source: Shutterstock.com/MAXSHOT.PL

FIGURE 12.3 Portable petrol alternator

Engine-driven alternators and generators

Engine-driven alternators and generators have been the stalwart of the production of electrical energy utilising the basic concept of moving a conductor through a magnetic field. Engine-driven alternators and generators find application as:

- back-up supply to the electricity grid for critical installations such as hospitals and data centres
- standalone generation units to supply electrical power for remote communities and industries such as mining
- temporary power source during installation and maintenance work on the electricity network.

Source: Shutterstock.com/ballchadowdesign

FIGURE 12.4 Back-up diesel engine-driven alternator

Types of prime mover

The type of prime mover depends largely on the output rating, size, location and local requirements. Low VA rated portable alternators, like that shown in **Figure 12.3**, frequently utilise a small petrol-driven engine to provide the mechanical input. Larger VA alternators, like that shown in **Figure 12.4**, are capable of providing back-up power in the event of loss of normal supply. They have a diesel engine as the prime mover as these provide for greater fuel efficiency and reliability.

Natural gas is also used as an energy source and there is a range of gas turbine-driven generators on the market that are capable of producing high thermal efficiency (often in excess of 60%).

The choice and use of an engine-driven alternator depends on a range of factors. These factors include the required output of the alternator (or VA rating of the unit), physical location, whether portable or fixed, and available fuel source.

Photovoltaic (PV) arrays

Photovoltaic (PV) arrays are made of a number of individual cells, which use semiconductor materials to convert radiant energy in the form of sunlight directly into electricity. Silicon is the most common form of semiconductor material used in the manufacture of solar cells. The power produced by a single solar cell is in the order of 1.5 watts and the typical open circuit voltage of a single cell is about 0.58 volts when exposed to full sunlight. It is necessary to combine individual solar cells together in series and parallel combinations to obtain higher power capabilities by increasing both potential difference across the cell and the current capability. A commercially available photovoltaic panel, as shown in **Figure 12.5**, is constructed using between 32 and 48 individual solar cells in series to give a panel capable of charging a 12 V d.c. battery.

Source: Shutterstock.com/PainterMaster

FIGURE 12.5 Photovoltaic (PV) array or solar panel

PV panels utilise a modular technology that facilitates the interconnection of a number of panels to form solar arrays that can have ratings in hundreds of megawatts, as is the case of solar farms. Solar arrays generate direct current (d.c.) electricity. A single solar panel, depending on the technology and the module size, can generate between 12 and 100 volts and be capable of delivering a current in excess of 8 amperes. Solar panels connected together to form a solar array can have an open circuit potential difference in the order of hundreds of volts. **Figure 12.6** illustrates a residential solar array comprising 32 individual panels having a combined 'nominal power' rating of 8 kW (32 panels × 250 watts per panel). It is important, therefore, to exercise due care when working with or around solar panel installations.

The output of a solar panel is normally described using what is referred to as the 'nominal power' rating, which is the amount of power the panel will produce if operated at the standard test conditions. The standard test conditions relate to the panel being exposed to a solar radiation equivalent to 1000 W/m^2 at a temperature of 25 °C. An installed and operational panel will unlikely be subject to these conditions and therefore produce a power much less than the quoted 'nominal power'. This reduction is attributable to installation conditions having a temperature exceeding 25 °C and variations in the amount of solar radiation to which the panel is exposed.

Solar cells are a source of 'green energy' in that they do not produce any greenhouse gases (GHG) such as carbon dioxide (CO_2) or any other gases that are harmful to the environment while operating.

Source: Shutterstock.com/Slavun

FIGURE 12.6 Residential solar installation

Wind-driven alternators and generators

For centuries the power in wind was used by windmills to produce mechanical power in the form of torque, which was then used directly for pumping water or grinding corn. A recent adaptation of this concept comes from modifying the windmill to generate electrical power, which can be used for heating and lighting, by using the sails or blades of the windmill to drive an electrical generator via the rotating shaft connected to the windmill's sails.

As the sun heats the planet unevenly it makes the air hotter around the Equator and colder near the poles. As air is a gas, heating it causes it to expand and, conversely, cooling it causes air to contract. Wind is created from differences in atmospheric pressure. Warm air at the Equator rises higher into the atmosphere and travels toward the poles. This creates what is known as a low-pressure system. Conversely, cooler, denser air from the poles moves over Earth's surface toward the Equator to replace the heated air, creating a high-pressure system. It is the difference in air pressure that creates winds, which generally blow from high-pressure areas towards low-pressure areas. Wind is considered to be 'air in motion' and is extremely variable in velocity.

A modern-day windmill, known as a wind turbine, is capable of converting the kinetic energy contained in the wind into mechanical and hence electrical energy. Wind turbines are a form of wind generator that operate differently and more efficiently than a conventional sail windmill. Like solar panels, it is possible to cluster multiple wind turbines to capture large amounts of wind energy. A 'wind farm' is the name given to such a cluster. The wind energy is then converted into electrical energy and the combined electrical energy is fed into the electricity grid. Wind farms may be located on flat land, mountain tops or offshore.

While wind turbines look like an extremely large fan, they comprise many complex mechanical and electrical components. The wind turbine rotor blades rotate around a central hub, which turns a low-speed gearbox shaft that in turn drives a generator at a higher speed and generates electricity. The electrical generator is the component that converts the kinetic energy from the rotating blades into electrical energy. Wind turbines have a number of air foil shaped rotor blades that are similar to aeroplane propellers, which makes them more efficient than the original windmills that usually had several flat blades or sails. Blade rotation in a wind turbine is due to the wind creating both lift and drag.

Wind turbines are available in a number of different configurations but generally fall into one of two configurations:

1 vertical-axis wind turbines (VAWTs), which have blades that rotate about a vertical axis, as the left image in **Figure 12.7** shows

2 horizontal-axis wind turbines (HAWTs), which have blades that rotate about a horizontal axis parallel to the wind, as the right image in **Figure 12.7** shows.

Both configurations have their advantages and disadvantages, but each is available in designs capable of generating electrical power from a few hundred watts to many kilowatts. Both configurations comprise:

- a support mechanism such as a tower that supports the rotors, gearbox, generator and auxiliary equipment
- the wind turbine with two or more rotor blades which captures the wind energy
- a mechanical gearbox, which increases the rotational speed to the generator
- a generator or alternator that produces the electrical energy
- various sensors and control electronics to monitor and regulate the speed and output
- electrical cables that connect the generator or alternator to the load.

Source: Shutterstock.com/Praiwun Thungsarn

FIGURE 12.7 Vertical-axis wind turbine (left) and horizontal-axis wind turbine (right)

Electrical output

The output of a wind generator can be either a.c. (from an alternator) or d.c. (from a generator), depending on their design. The most common type is the wind-driven alternator, which produces a sinusoidal a.c. output. While this output is a.c. it is not generally of a quality suited for exporting to the electricity grid. It is therefore necessary to modify the output by feeding the alternator's a.c. output into a rectifier to convert the a.c. to d.c. for battery charging, or feeding the alternator's a.c. output through a converter to match the generated output to the grid specifications.

Small-scale generation

Many residential, roof-mounted wind turbines are able to operate with a wind velocity from around 15 km/h up to a maximum of around 100 km/h. On a calm windless day, the turbine sits idle and the blades do not rotate. As the wind velocity increases, it eventually reaches the cut-in velocity of the turbine of around 15 km/h. At this wind velocity, the turbine blades will spin up to their cut-in operating speed and allow the attached generator to commence generating electrical energy and as the wind velocity increases further, the rotor blade velocity increases and the generator output also increases.

Wind turbine generators deliver maximum power at a wind velocity of around 60 km/h. This means that a generator having a rated (nameplate) output of 50 kW, will be capable of supplying 50 kW of electrical power at the rated wind velocity. The same generator will deliver less than 25% of rated output power at the cut-in wind velocity of only 15 km/h. At a wind velocity exceeding 60 km/h the generator maintains its rated capacity until the wind velocity approaches 100 km/h where the turbine reaches its cut-out speed and the in-built protection will cease generating electrical power.

The wind velocity is a very important consideration for the safe and proper operation of a wind turbine generator. Historical wind velocity data is used in the calculation of wind power when determining proper placement of a wind turbine generator – whether this be on the ground or on a roof.

Advantages of wind-driven alternators and generators

The main advantages of wind-driven alternators and generators include:

- energy harnessed from the wind is clean and renewable energy that does not release pollutants, emissions or by-products into the atmosphere during operation
- wind turbine generators can be located in remote areas and on scales ranging from small personal and residential use to large full-sized wind farms, which means it is possible to have electrical energy available in locations that would normally be considered as 'off grid'
- the design of wind turbines allows them to operate within a wind velocity usually between 15 km/h and 100 km/h
- larger land-based installations can be used simultaneously for wind generation, crop cultivation, animal grazing, and the like.

Disadvantages of wind-driven alternators and generators

The main disadvantages of wind-driven alternators and generators include:

- wind farms are considered as unsightly man-made structures that have a negative visual impact – viewed as a form of visual pollution
- wind farms require large expanses of land or are placed in environmentally sensitive areas to take advantage of stronger and constant wind
- the wind turbine does not generate any useful electrical energy in calm conditions
- wind farms can injure and kill birds or disturb the flight patterns of migratory and predatory birds
- wind turbines can produce noise pollution as they produce a low frequency 'whooshing' sound as the blades rotate
- the best locations for siting wind farms are usually far from populated urban areas, which necessitates the transportation of electrical energy over long cable distances
- although annual winds and power output from a wind turbine are relatively predictable, hourly and daily wind energy output levels are not as wind speed does not remain constant giving little power output in low winds.

Personal or micro hydroelectric systems

Personal or micro hydroelectric generating systems utilise the kinetic energy of flowing water to produce localised electrical generation for particular situations. They are a low-maintenance and emission-free source of electrical energy. These small-scale hydro-electric generating systems can be used to supply electrical energy to nearby installations. **Figure 12.8** shows a 20 kW micro hydro-electric generator.

These systems are generally suited to localised applications that are close to the energy source. Micro hydro-electric systems have declined in popularity due to the significant and ongoing reduction in the cost of PV systems.

Source: Shutterstock.com/kaitong.yepoon

FIGURE 12.8 20 kW micro hydro-electric generator

Connection methods

Where an electrical installation integrates an alternative supply system with the normal (grid-connected) supply, there are a number of regulatory considerations. Section 7.3 of AS/NZS 3000:2018 *Wiring Rules* details the requirements for the connection of alternative systems of electrical supply. The following electricity generation systems are recognised as alternate supplies in the *Wiring Rules*:

- Alternative and supplementary supply where a motor-generator set:
  - provides an alternative or stand-by a.c. electrical supply if the normal electrical supply to the installation should fail
  - forms the primary electrical supply to an electrical installation
  - forms part of a standalone electrical system.
- Standalone power system that does not connect to the power distribution system of a network provider and may include:
  - PV array
  - wind turbine or mini-hydroelectric turbine
  - engine-driven generator set in the form of an a.c. supply or a d.c. supply.

- An inverter system that provides a.c. electrical power from an interactive inverter, which itself is supplied from a renewable energy source, such as PV array, wind turbine or mini-hydroelectric turbine.
- A battery system that is provided with an electrical supply from an alternative energy source, such as a PV array, wind turbine, generator set or mini-hydroelectric turbine, for the purpose of charging a battery bank and provide a d.c. supply to an electrical installation.

Figure 12.9 shows the arrangement for connecting an alternative supply to a switchboard having a local MEN connection.

Control

It is a requirement of the *Wiring Rules* that a main switch or switches can control the alternative supply system at the installation switchboard to which the alternative supply system connects. Any installed main switch must be appropriate for starting and stopping the alternative supply system. If control is via multiple devices these devices must be grouped and clearly identified as to what they control. Unless allowed by the electrical energy supplier, an alternative supply system will not supply energy upstream of the point of connection to the installation either directly or indirectly. This is known as co-generation, where an alternative supply system exports electrical energy into the upstream network under specific conditions. Accepted examples of co-generation include grid-connected inverter-based systems (such as PV systems) or engine-driven generating systems intended to operate in parallel with the network.

It is essential to ensure that if the alternative supply system is providing electrical energy to the installation that all necessary electrical connections required for basic and fault protection within the installation, such as the MEN connection, remain intact.

Every alternative supply system must have an isolating switch, which:

- is either installed adjacent to, or on, the source of alternative supply so that anyone operating the switch can see anyone who may be working on the alternative supply system; or has a securing method such that the device when in the isolated position requires a deliberate action to engage or disengage
- may incorporate overcurrent protection
- complies with Clause 4.13 if an electric motor is incorporated in the alternative supply system
- is under manual control
- cannot be overridden or bypassed by programmable control systems or the like.

Requirements for earthing

Clause 7.3.6 of the *Wiring Rules* requires all alternative supply systems rated at 25 kVA or greater to employ the MEN system of earthing and must comply with Section 5

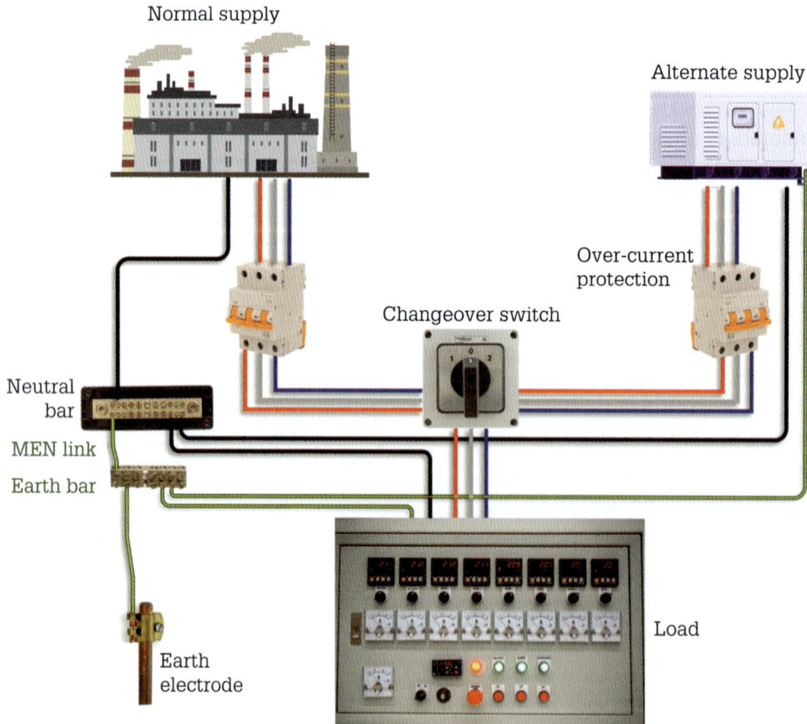

Sources: Thermal power plant, Shutterstock.com/Elegant Solution; Alternate supply, Shutterstock.com/Seahorse Vector; Control panel, Shutterstock.com/lowpower225

FIGURE 12.9 Connecting an alternative supply to a switchboard incorporating MEN connection

of the *Wiring Rules*. Alternative supply systems that are not rated more than 25 kVA can be arranged as a separated supply complying with Clause 7.4 of the *Wiring Rules*. Furthermore, all exposed conductive parts of an alternative supply system must connect to the main earthing conductor at the main switchboard.

Changeover switches

A changeover switch ensures effective electrical separation between the normal electrical supply to the installation and the alternative supply system. The purpose of the changeover switch is to ensure that electrical energy from each supply cannot connect to the installation at the same time as it is likely that the line voltages, frequencies and phase positions will be different. A changeover switch incorporates a mechanical interlock that prevents both supplies simultaneously connecting to the installation. The changeover switch can be of the manual or automatic switching type with some devices incorporating over-current protection in the form of a circuit breaker. **Figure 12.10** illustrates a three-pole/three-pole automatic transfer switch (ATS).

AS/NZS 3010:2017 is the applicable Standard, which requires that in the case of a three-pole/three-pole ATS:

- the MEN link is effected at the main switchboard, which houses the ATS
- the neutral conductors of the normal supply and alternative supply are not switched
- the neutral conductor of the normal (mains) supply connects directly to the MEN link
- the neutral conductor of the alternative supply connects directly to the MEN link
- over-current protection for the generator has three-phase protection without neutral protection.

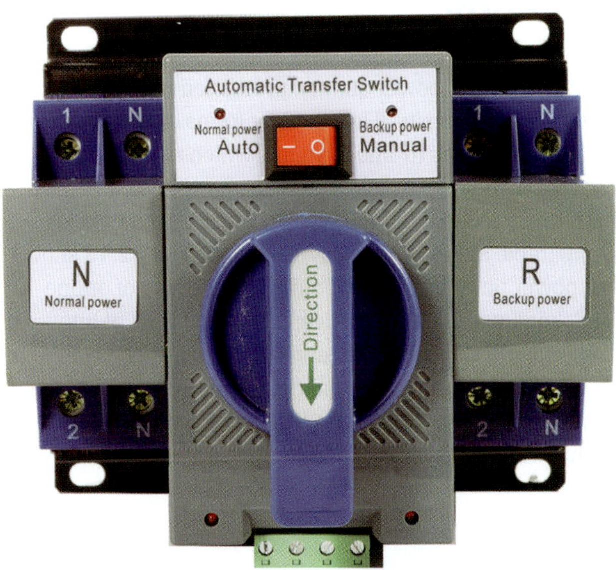

FIGURE 12.10 Two-pole automatic transfer switch

Requirements of local supply authority

Where an electrical installation integrates an alternative supply system with the normal (grid-connected) supply, it is necessary to consult the relevant Service and Installation Rules for the specific jurisdiction.

The Service and Installation Rules for most jurisdictions will detail:

- Stand-by generating plant
  - Conditions of use
  - Spacing for conductors
  - Changeover equipment for non-parallel operation
  - Switching the neutral
  - Multiple generators
- Requirements for stand-by generator synchronise close transfer trip
  - Operating procedure
  - Additional protection
- Requirements for generator parallel operation
  - Operating procedure
  - Additional protection
- Small-scale parallel customer generation (via inverters)
  - Responsibilities
  - Export limiting systems
  - Approvals documentation
  - Metering requirements
  - Construction permits
  - Islanding
  - Multi-phase generating systems
  - Power factor setting
  - d.c. isolation
  - Generator connection arrangement
  - Switching requirements
  - Conductor size requirements
  - Islanding protection equipment
  - Reconnection procedure
  - Generator supply main switch
  - Shutdown procedure
  - Storage systems
- Labelling.

The relevant Service and Installation Rules should be consulted in conjunction with Australian Standards.

Uninterruptible power supplies

An uninterruptible power supply, which is usually referred to as a UPS, is an item of electrical equipment that is capable of supplying electrical power to a load should the normal electrical power source fail. A UPS can range in capacity from a few VA, which is intended to supply electrical power to a personal computer for a sufficient period of time to allow it to shut down gracefully, to several

MVA, which is intended to keep operating theatres and other critical hospital areas functioning for extended periods of time.

Where auxiliary or emergency power systems or stand-by generators take a period of time to commence supplying an electrical installation, a UPS provides near-instantaneous protection from disruptions to the normal electrical power source. It does this by supplying energy that is stored in batteries, supercapacitors, or flywheels to the electrical installation as a.c. power.

Battery-based UPS

A battery-based UPS is typically used to protect hardware such as computers, data centres, telecommunication equipment or other critical electrical equipment where an unexpected power disruption may cause injuries, fatalities, serious business disruption or data loss. A UPS, like the one shown in **Figure 12.11** having a rating of about 200 VA, is suitable for operating a single personal computer without a visual display unit. A battery-based UPS located in Hornsdale, South Australia is touted as the world's first big battery capable of providing essential electricity grid-support services. The first 100 MW/129 MWh was completed in November 2017. In September 2020, a 50 MW/64.5 MWh expansion was completed. This 150 MW capacity is being upgraded to include Tesla's Virtual Machine Mode, which enables the battery to provide inertia support services to the national electricity grid.

Source: Shutterstock.com/Thitikorn Suksao

FIGURE 12.11 Small battery back-up UPS

Rotary UPS

A rotary UPS uses flywheel energy storage to provide short-term ride-through in the event of power loss. The rotating mass of the flywheel not only stores rotational energy but can act as a buffer to smooth spikes and sags in the normal power source. This is due to the fact that these power events of short duration do not significantly affect the rotational speed of the high-mass flywheel. This type of UPS has been around for several decades making it one of the oldest designs, predating electronic power devices.

The rotary UPS is considered 'on-line', as under normal operating conditions it is continuously rotating. The rotating UPS can generally maintain electrical supply for up to 20 seconds before inertia in the flywheel decreases to a point where electrical supply cannot be maintained. It is expected that a diesel engine or similar has started and is able to maintain rotational energy. **Figure 12.12** shows the functional blocks of a diesel rotary UPS (DRUPS).

The decision to install a rotary UPS is not taken lightly as these are expensive installations, which are generally reserved for applications needing more than 10 000 W of protected load capability. It is possible to increase the initial run-time by installing a larger flywheel or having multiple flywheels operating in parallel.

The flywheel provides a source of mechanical energy that can, when coupled through a gearbox, be used to start a diesel engine. Once running and up to speed, the diesel engine can spin the flywheel directly to provide emergency electrical power.

A rotary UPS has an expected operational life of about 30 years, which is far in excess of a purely electronic UPS. In order to achieve this longevity, they do, however, require periodic downtime for mechanical maintenance, such as replacing ball bearings. Battery-based UPS designs do not require downtime if it is possible to hot-swap the batteries. Advances in machine technology mean that technologies such as magnetic bearings and air-evacuated enclosures, when employed on rotary UPS, increase stand-by efficiency and reduce maintenance to very low levels.

Topology

There are many UPS topologies available and each provides varying levels of power quality, protection time, sensitivity, as well as complexity. The simplest forms of battery-based UPS are known as in-line systems, which use a battery charger to provide d.c. electrical energy to the battery or batteries; the d.c. from the battery is converted to grid-comparable alternating by an inverter. **Figure 12.13** shows the power flow through an online UPS. A type of online UPS, known as line interactive, provides a level of line conditioning where the battery source provides power and the unit disconnects from the normal supply if this supply voltage is too high or too low. When the normal supply is within range, it is fed through to the load.

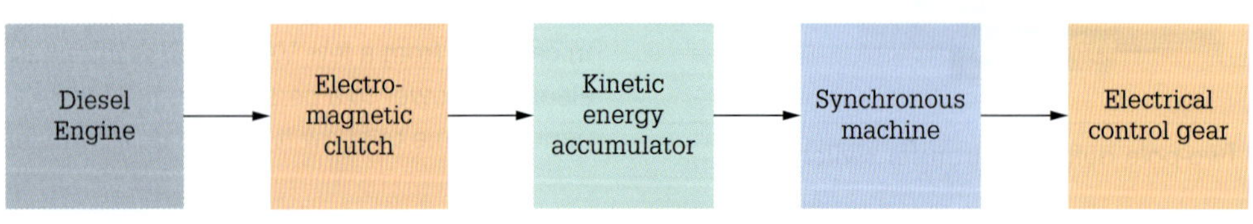

FIGURE 12.12 Functional blocks in a rotary UPS

Switched systems are similar but they switch to battery power after the normal supply has failed. This type of system provides a back-up power supply.

Rotary UPS is usually for installations requiring a minimum of 10 kVA back-up electrical power for large loads such as hospitals and commercial data centres.

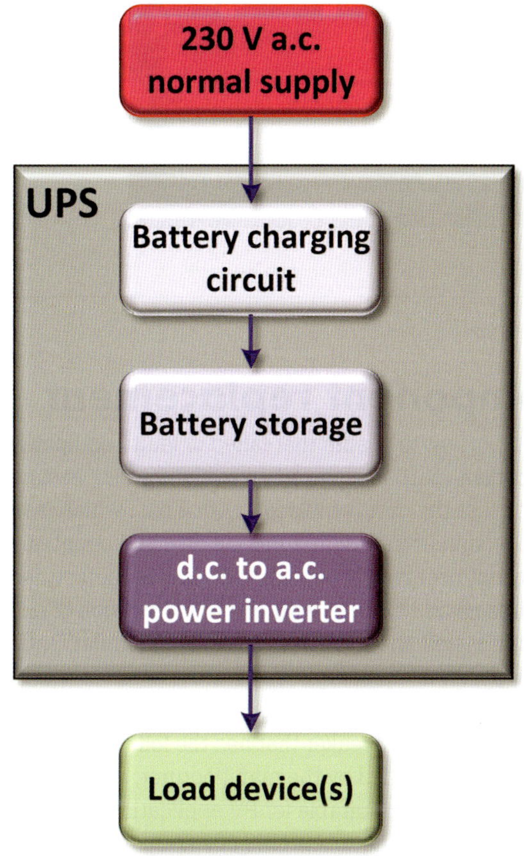

FIGURE 12.13 Online UPS power flow

Installation requirements

As with electrical installations, it is necessary to install UPS in accordance with the relevant Australian Standards and the requirements of Local Service and Installation Rules. For safety, it is essential that circuits within the electrical installation and the electricity grid do not become unintentionally 'live' in the event of failure of the normal electrical supply. It is imperative when designing a UPS installation to consider which loads will connect to the UPS and how the associated circuit wiring is configured to ensure that circuit wiring and attached equipment having the possibility of being live during failure or isolation of the normal supply are not inadvertently considered a conventional circuit.

All parts of a UPS must be suitably housed to protect them from the environment and protect end users from the possibility of receiving electric shock. This is particularly important for battery banks, which have the possibility of supplying high short-circuit energy. The applicable Australian Standard is AS/NZS 4777 series that requires that any grid-protection devices installed for a UPS do not isolate the neutral or earth conductors.

AS/NZS 4777.1:2016 and AS/NZS 4777.2:2020 identify additional requirements for UPS systems, which are additional to the requirements for standard grid-connected inverter systems. These requirements include:

- the UPS must only 'supply an identifiably separate set of circuits'
- these circuits should originate from a separate switchboard or load centre
- warning signs are provided at the main switchboard and any other relevant switchboards to indicate that some circuits within the installation may be live when the main switchboard is isolated, which may include any neutral or earth conductor within the installation.

Polarity requirements for d.c. components

Appendix Q of the *Wiring Rules* provides guidance on the polarity requirements for circuit protection and switching devices that operate on a d.c. supply. Installers should refer to this Appendix for detailed information.

Polarised and non-polarised isolators and circuit breakers used to isolate solar panels and d.c. generating equipment, are reasonable to use. There are, however, some issues that can lead to very dangerous conditions.

A polarised circuit breaker or isolator needs the electric current to flow with a specific polarity within the device. This is due to the method employed for arc suppression within the device where a permanent magnet is used to draw the arc from the opening contacts into the arc chute where it lengthens and subsequently weakens. If the device is connected with reverse polarity, then the electric current will flow in the wrong direction, causing the arc to move away from the arc chute and deeper into the body of the device. This may result in a fire within the device, which may occur regardless of the disconnection method – whether a deliberate manual switching operation or automatic should there be an overload.

This situation is not present in a non-polarised circuit breaker or isolator that employs a polarity-independent electro-magnetic blowout rather than polarity-dependent permanent magnet. Non-polarised circuit breakers and isolators are the preferred option. Installers must follow the manufacturer's instructions to ensure these devices work safely as intended. It is a requirement of AS/NZS 4509.1 that over-current protection of battery systems be of the non-polarised type.

Figure 12.14 illustrates common circuit breakers.

SWITCH ON

It is important to note that an a.c. rated circuit breaker or isolator **cannot** be used for d.c. applications.

Sources: Shutterstock.com/Somsit

FIGURE 12.14 Various circuit breakers (d.c. left and a.c. right)

A typical application of a polarised circuit breaker is a distribution switchboard that supplies d.c. operated equipment. With this arrangement current flow is unidirectional (one way).

A typical application of a non-polarised circuit breaker is for over-current protection of a battery system in which the current flow can be in opposite directions: from supply to the battery when being charged, and from the battery when being discharged by the load or when the d.c. charging source is not outputting.

In all jurisdictions it is required to locate the d.c. isolator or circuit breaker at the d.c. to a.c. inverter. In some jurisdictions there is an additional requirement to locate a d.c. isolator at the solar panels.

If you consider that solar panels are producing d.c. energy while exposed to solar radiation, isolating at the inverter means that the cables supplying the inverter from the solar panels are in a 'live' state. Coming into contact with conductors within these cables upstream of the inverter isolator presents an unacceptable risk of electric shock. This risk is greatly mitigated by having a d.c.-rated isolator at the solar panel end of the cables supplying the inverter. It is important to bear in mind that the cables interconnecting the individual panels are still 'live'. To date there is no satisfactory way of isolating the individual solar panels and making them safe.

Earthed d.c. supply

The d.c. supply may operate as a separated or isolated system, which requires all switching or overload devices to operate in all unearthed conductors. To this end all multiple pole devices should have their actuator linked so that all contacts will operate simultaneously.

It is also possible to operate a d.c. supply where one pole of the d.c. supply connects to earth. In this system, any switching device only needs to operate in the unearthed conductor.

Component replacement

It is important to ensure that failed or faulty equipment or component (part) is replaced on a 'like-for-like' basis. This means that the part is replaced with a part which is the same or similar in design, function, use and maintenance. It may not necessarily be from the same manufacturer and requires no additional alteration or modification of existing parts to install and occupies the same or similar footprint as the part it is replacing.

An example would be in replacing a circuit breaker. Consider a circuit breaker that has the following information supplied by the manufacturer:

- non-polarised, suitable for PV applications
- tested according to AS/NZS 60947-2
- rated short-circuit breaking capacity Icu 10 kA
- rated operating voltage Ue 180 V DC per pole
- tripping characteristics K.

A suitable replacement would have the same parameters. The critical specifications being Icu = 10 kA; Ue = 180 V DC per pole; and Tripping characteristic = K.

Inverter principles

The inverter is the item of equipment responsible for converting the direct current produced by an alternative energy source such as solar panels into alternating current for use in a home or business (normal single-phase supply is 230 V a.c). There are four types of inverter most commonly used for small-scale installations up to 5 kW per phase:

1. String inverters, as shown in **Figure 12.15**, are single units that connect to a string of solar panels, which are the most common type for households.
2. Micro inverters connect to each panel and represent a good option when some panels are frequently shaded. These are more expensive than string inverters.
3. Battery-only inverters are paired with battery chargers to store 'surplus' electrical energy in a

battery bank and release it later when it is more useful or economically viable.

4 Hybrid inverters combine a string inverter and battery inverter into a single unit.

Source: Shutterstock.com/Rene Notenbomer

FIGURE 12.15 String inverter

Standalone inverters

Standalone inverters range in size from as little as 100 watts to several kilowatts. Small-scale inverters are popular with campers who like the creature comforts of modern life. These inverters are able to convert the 12 V d.c. battery supply available from the vehicle's battery to 230 V 50 Hz a.c. to power low energy devices such as televisions. Specialised systems, specifically aimed at campers, connect to portable PV arrays and are capable of providing up to 2.4 kW.

Similar to battery-based UPS, standalone inverters are available with an output that is either a modified sine wave or a true sine wave. Modified sine wave inverters generally produce a very poor-shaped sine wave that is acceptable for equipment with switch-mode power supplies and lighting but is unsuitable for most applications including motors, transformers and voltage-sensitive equipment. Figure 12.16 shows oscilloscope graphs of a modified sine wave and a pure sine wave.

For standalone systems it is important to use a true sine wave inverter that is capable of supplying grid-quality sine wave output (230 V at 50 Hz).

SWITCH ON

Standalone inverters must never connect to the normal a.c. (grid) supply system.

Grid-connected inverters

Grid-connected inverters are intended for operation on the normal electricity grid supply and can supply electrical energy back into the network. The design of these systems ensures that they do not operate without first synchronising with the normal electrical supply. The

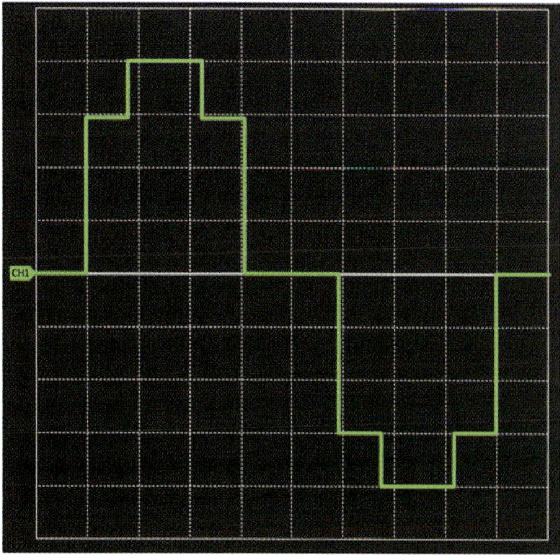

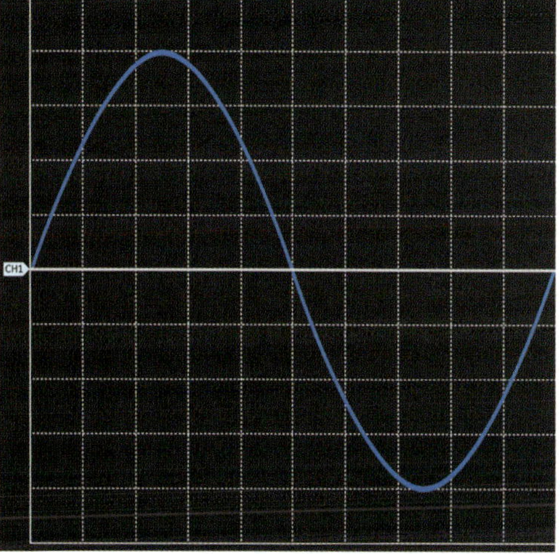

FIGURE 12.16 (a) Modified sine wave; (b) pure sine wave

output characteristics of the inverter are automatically matched to those of the normal electrical supply. In the event of the normal electrical supply failing, the inverter will automatically cease outputting energy. The inverter will generally wait a pre-determined time after restoration of the normal electrical supply before outputting energy to the 'grid'. This functionality is known as 'anti-islanding' and it is a requirement of AS/NZS 4777.1:2016.

A grid-connected inverter must comply with the requirements of AS/NZS 4777 series of standards. As such, a compliant inverter will have the following electrical characteristics:

- grid voltage of 230 V single-phase measured between line and neutral
- power factor within the range of 0.8 leading to 0.95 lagging (unity power factor is common)
- voltage tolerance within the range of +10% to −6%
- frequency of 50 Hz
- total harmonic distortion not exceeding 5%.

Multi-mode inverters

Multi-mode inverters are capable of providing both off-grid and on-grid solutions. When normal supply is available, they operate as a grid-connected inverter – importing and exporting electrical energy as the installation demands. When combined with an energy storage system such as a battery bank, multi-mode inverters function as a standalone or off-grid system.

They offer the flexibility to import and export electricity, depending on installation needs, as well as providing a complete standalone system when coupled with an energy storage system. This flexibility can also allow the unit to provide the installation with an uninterruptible power supply and load-shifting adjustable export control.

Regulatory requirements for inverters

The installation of grid-connected and multi-mode inverters with associated solar panels must comply with the relevant Australian Standard and, depending on the local supply authority, Local Service and Installation Rules. In most jurisdictions these installations require individual inspection, which may include the associated energy metering that allows both the import and export of energy associated with the electricity grid connection.

There are a number of applicable requirements detailed in the *Wiring Rules* pertaining to the safe and effective installation of inverters. In particular Section 4.12 relates to the selection, control, isolation, overcurrent protection and earthing. Section 7.3 has further requirements for inverters as energy-generation systems.

The AS/NZS 4777 series of standards details requirements for the installation of grid-connected inverters while AS/NZS 4509 details requirements for standalone power systems.

Appendix Q of the *Wiring Rules* also recommends that unless the inverter is marked as providing full electrical isolation, then all parts of the d.c. supply system should be insulated and screened.

Identification and labelling

The purpose of signs and labels is to provide clear indication that an electrical installation is served by multiple supplies and to identify the affected circuits. Signs pertaining to inverter energy systems (IES) need to be placed on the switchboard to which the IES connects. Additional signage may be needed to direct emergency services and these would be located on the fire indicator panel. If the IES connects to a switchboard other than the main switchboard, then additional signage is needed alerting to this fact.

Any signage must be legibly and indelibly printed in English and be of suitable size.

Figures 12.17 to **12.22** are sample labels and appropriate locations for fixing as identified in AS/NZS 4777.1:2016.

FIGURE 12.17 Example of label affixed to switchboard to which IES connects

MAIN SWITCH (INVERTER SUPPLY)	MAIN SWITCH (GRID SUPPLY)

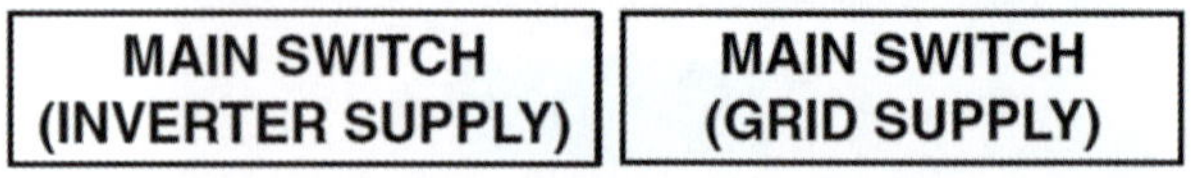

FIGURE 12.18 Example of main switch labels affixed to switchboard to which IES connects

FIGURE 12.19 Example of label affixed to main switchboard where IES connects to a downstream switchboard

FIGURE 12.20 Example of energy source labelling

Some isolators are not capable of operating on load, so it is necessary to remove the load before opening the isolator.

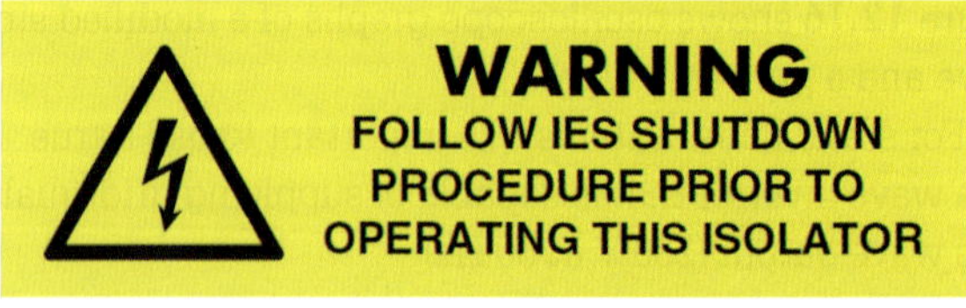

FIGURE 12.21 Example of shutdown procedure before isolating labelling

MULTIPLE SUPPLIES IES
SHUTDOWN PROCEDURE
LOCATED IN MAIN SWITCHBOARD

FIGURE 12.22 Example of shutdown procedure location labelling

REVIEW QUESTIONS

1 Describe some typical applications of engine-driven alternators and generators for the production of electrical energy.
2 What is meant by the 'nominal power' rating of a solar array?
3 What is the purpose of the mechanical gearbox in a wind turbine?
4 List two advantages of wind turbines and wind-driven alternators and generators.
5 In terms of an alternative supply system, what is meant by 'co-generation'?
6 What is the purpose of a changeover switch for an installation with an alternative supply system?
7 Identify two types of uninterruptible power supplies.
8 Which type of circuit breaker or isolator is required for over-current protection of battery systems?
9 Why would micro inverters be used for a solar array installation?
10 Describe the 'anti-islanding' functionality of grid-connected inverters.

12.3 Safe isolation

Being able to safely isolate a generator or inverter energy supply system is of paramount importance to facilitate safe working on an electrical installation.

Circuit isolation

An important work practice for an electrician carrying out electrical tasks is isolation and lockout. Correct isolation and lockout procedures will minimise the dangers of unintentional or unexpected energisation of circuits and the start-up of electrical equipment. All electrical equipment and conductors must be considered as live until isolated and proven de-energised. Work must not be carried out on, or near, electrical circuits or equipment until an electrical worker has:

1 Confidently identified the work task including the circuit and any connected electrical equipment. In addition, any additional energy sources (capacitors, batteries, UPS) for connected equipment must also be identified. Finally, suitable isolation points must be noted. Any test instruments used must be confirmed to be 'in-test'.
2 Turned OFF all sources of supply and securely locked out these supplies from providing service, and in addition, danger tagged all switches and controls for the equipment being worked on.
3 Secured the electrical inaccessibility of the circuit or equipment by testing for service at all identified isolation points. This step checks that the electrical circuit and equipment is de-energised and electrically safe. Then prove correct operation of any test instruments used.
4 Provided and identified a safe area of work for electrical workers. The safe area of work should be identified by erecting a barricade or warning signs or by other suitable means.

Testing tools and equipment

For all low-voltage electrical installation verification and testing work, electrical persons should, as a minimum, have the following range of test instruments: loop impedance tester, multimeter, clamp ammeter, insulation continuity tester and Category IV test leads.

It is suggested that all testers and leads should be Category IV and should be the preferred choice of electrical persons due to their higher fault capabilities.

Shutting down alternative energy source

The proper shutdown procedure should be located onsite. The location should be clearly indicated on a prominent sign similar to that shown in **Figure 12.22**. The generic shutdown process for an engine-driven generator set is outlined below.

Diesel generator set correct shutdown operation method

Following the correct procedure is an effective way to prevent the development of faults, and has direct impact on extending its service life.

1 Prior to stopping it is necessary to remove the electrical load and then reduce the speed to about 750 rpm.
2 After about 3 to 5 minutes of this load speed operation, turn the stop handle to stop and bring the system to rest.
3 For a 12-cylinder V-type diesel engine, turn the electric key to the intermediate position after shutdown to prevent battery current from flowing back.

Inverter system shutdown procedure

The following procedure is suitable for shutting down grid-connected inverter systems.

1 Locate the solar supply main switch and turn the switch to the off position.

2 If the solar power inverter is more than 3 metres away from the switchboard, locate the switch marked solar AC isolator. This will be located next to the inverter. If the inverter and switchboard are within 3 metres of each other, disregard this step.
3 Go to the inverter and locate the switch marked **PV Array and DC Isolator**. Turn this switch to the **off** position (in some cases there will be two switches).
4 The inverter may have a switch marked Inverter Isolator. If present, turn this switch to the off position. Skip this step if the system does not have this isolator.
5 The solar PV system should now be completely switched off. All lights and screen displays will be extinguished. Keep the system off for a minimum of five minutes.
6 To re-start the system, follow this guide in reverse order. That is, **DC isolator** on first, followed by **AC isolator**, followed by the **solar supply main switch**.

Figure 12.23 shows a solar inverter with shutdown procedure.

Source: Shutterstock.com/Douglas Cliff

FIGURE 12.23 Inverter showing shutdown procedure

Anti-islanding protection

Anti-islanding protection is an important safety feature integral to all grid-connected inverters. A grid tie inverter has sophisticated monitoring circuits capable of detecting the loss of normal power supply from the grid and within a few milliseconds automatically power down the inverter. Remember the inverter is still supplied with direct current electricity from the solar panels, wind turbine generator or similar.

The term 'islanding', refers to a condition where the electricity grid does not provide power but some installations are supplying electrical energy into the grid, transmitting electrical energy back into the otherwise inactive electricity supply lines. This creates a potentially hazardous situation:

- In the case of network failure, systems that are feeding energy back into the grid create a hazardous situation for service personnel trying to fix the cause of the network failure.
- In the event of a network fault, the grid infrastructure such as transformers shut down for their own protection. If inverters are feeding energy back into the network the equipment may suffer damage as a result of this back-feeding.

Automatic transfer switch

An automatic transfer switch (ATS) is an automatic power switching device that utilises dedicated control logic to perform intelligent switching operations. The primary function of an ATS is to ensure the continuous delivery of electrical energy from one of two sources of electrical energy to a connected electrical load, which is typically the critical portions of an electrical installation.

The control logic or automatic controller is microprocessor-based and constantly monitors the electrical parameters of voltage and frequency of primary and alternate power sources. If the normal (primary) supply fails, the ATS will automatically transfer (switch) the load circuit to the alternate energy source (if it is available). Most automatic transfer switches attempt to connect to the primary energy source (utility power) by default and will only attempt to connect to the alternate energy source (engine-generator, back-up utility) if the primary source were to fail or requested to do so through manual switching command.

Typically, the operation of an ATS is as follows:

1 The normal utility power source fails.
2 The ATS shifts the load to the emergency power source when power from the generator or back-up utility feed is stable and within prescribed voltage and frequency tolerances. The transfer process may be automatic or manually-initiated, depending on the requirements of the particular installation.
3 The ATS returns the load from the emergency power source to the normal power source when utility power is restored. Again, the process may be automatic or manually-initiated.

ATS switching configurations

Two common switching arrangements are:

1 Utility–Generator as shown in Figure 12.24.
2 Utility–Utility as shown in Figure 12.25.

Utility power to back-up generator transfer

The standard transfer switch configuration includes a supply from an electricity distributor's network and a back-up generator for normal and emergency power sources respectively. This system arrangement is typically referred to as an emergency standby generator system. The single generator depicted in the diagram may be several engine-driven generator sets operating in parallel.

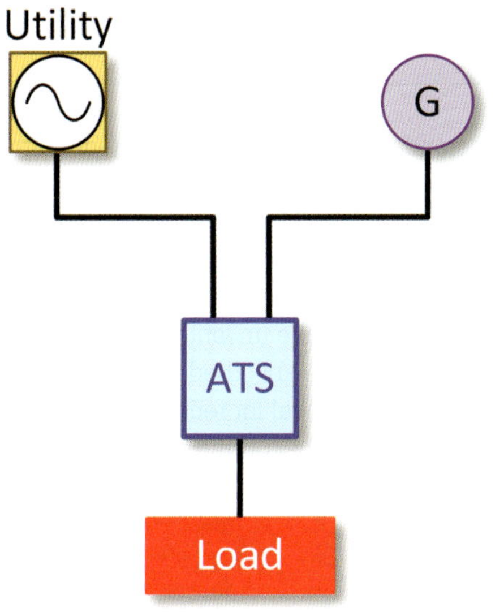

FIGURE 12.24 ATS switching between utility power and back-up generator

Utility power to utility power transfer

This configuration utilises electrical supply from electrical distribution networks that provide redundancy in the distribution system and allows for quick restoration of service to the installation should an upstream equipment failure occur. The two sources can be independent of each other, requiring the public utility company to provide dual electric services, or they can originate from a single electric service that is distributed through redundant paths within the facility.

Earth fault alarm

It is not usually the case to reference d.c. supplies to earth, so earth fault risk is commonly overlooked. Should the d.c. supply connect to an operating inverter that does not have a transformer, the potential difference between the d.c. supply and earth can exceed 400 V d.c. This means a d.c. supply cable that is secured to the rail, usually with sharp stainless steel cable ties, has a 400 V d.c. potential to the rail. A failure in this cable's insulation, wherever it occurs on the solar installation, creates an earth fault.

An earth fault may cause the solar generating system to stop operating. It also poses the risk of someone working on the roof receiving an electric shock. If multiple faults occur simultaneously, then a person working on a roof could receive an electric shock strong enough to throw them off the roof.

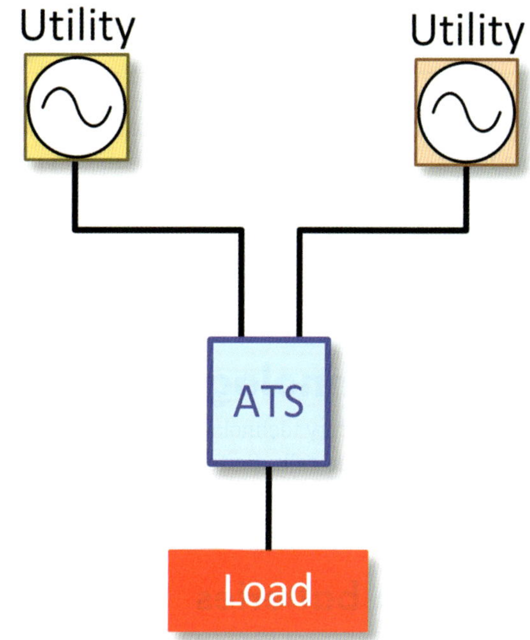

FIGURE 12.25 ATS switching between two utility power sources

An earth fault alarm is capable of detecting this type of fault and, depending on the system, instigating appropriate action, which can be combined with online inverter monitoring.

Solar voltage rise

The nominal grid potential in Australia is 230 volts. This potential will vary and fluctuate throughout the day, depending on the demand on the grid and how much energy from solar inverters is being fed into the grid. It is common to see this potential fluctuate 10 volts throughout the day. The electricity distributor will endeavour to maintain this potential between about 217 volts and 253 volts. Maintaining the potential below 253 volts is becoming more of a challenge for the electricity distributors.

When a solar system is producing more energy than is being used, it feeds the excess back into the grid. In order for power to flow from an electrical installation to the grid, the potential from the solar inverter has to be higher than the grid potential. This situation creates what is known as solar voltage rise.

REVIEW QUESTIONS

1 List the test instruments electrical persons should have as a minimum for all low-voltage electrical installation verification and testing work.
2 Describe the correct shutdown operation method for a diesel generator set being used as an alternative energy source.
3 What is the procedure for re-starting a grid-connected inverter?
4 Why can islanding, where the electricity grid does not provide power but some installations are supplying electrical energy into the grid, create a potentially hazardous situation?
5 What is the primary function of an ATS?
6 Describe two common ATS switching arrangements.
7 How can an earth fault arise in a solar generating system?
8 Explain the term 'solar voltage rise'.

12.4 Battery storage systems

A major drawback to solar power installations is that they only produce electrical energy in daylight hours and grid-connected systems only produce this energy when the grid is operating. It is therefore necessary to store surplus energy to use later. Battery systems are used for the storage of electrical energy for use at a later time.

Battery technology

The most prevalent battery technologies are lithium-ion and lead-acid storage cells. Battery technology is advancing at a great rate with emerging technologies, such as sodium nickel, making progress.

Lithium-ion batteries

This is the most popular grid-connected battery chemistry, and is the same type of technology as that used in mobile phone and laptop batteries. There are different types of lithium chemistry; common types are nickel-manganese-cobalt (NMC) or iron phosphate (LiFePO/LFP). LFP batteries are safer but less efficient than NMC batteries. Lithium batteries are popular because they:

- have a long lifespan (expected to be more than 10 years, and researchers are working to further extend this)
- can be used to almost their full capacity (discharged to almost zero from a full charge)
- work in a wide range of climates.

Lead-acid batteries

Lead-acid batteries, as shown in **Figure 12.26**, are commonly used in automobiles. They are cheaper than lithium-ion batteries but bulky and provide less flexibility. They have a slow charge cycle and are sensitive to high temperatures. Sometimes these batteries can be coupled with a supercapacitor for a faster charge cycle. This technology is often used in back-up power supplies, which cycle batteries only occasionally. Lead-acid batteries are still popular for standalone (off-grid) power systems, although lithium-ion batteries are taking over this role as their lifetime performance becomes better understood.

Source: Shutterstock.com/Vittee

FIGURE 12.26 Lead-acid battery bank

Lead-acid batteries may be either of the wet cell (vented) or sealed (valve-regulated) type. Vented batteries use liquid electrolyte while unvented batteries use either a gel or liquid electrolyte absorbed into fibreglass matt. Vented batteries are typical for renewable energy systems, but sealed batteries are becoming more common due to their safer operation and ease of maintenance.

Hazards

Batteries and battery banks present significant safety hazards. Battery storage systems have the ability to delivery very high current, which is the case of a short circuit occurring across the terminals of the battery or the battery is heavily loaded. While being able to supply high current to fluctuating loads may be desirable, it creates hazards such as fire, injury to personnel and magnetic stresses. Lead-acid batteries were for decades the battery of choice for providing back-up power to installations such as telephone exchanges, emergency lighting and other essential services. Some of the hazards associated with these installations include the production of hydrogen gas both during charging and discharging cycles, acid spillage and associated burns, and potential short-circuiting of the battery terminals. The rooms containing lead-acid batteries were normally negatively pressured to prevent the hydrogen gas escaping into the working environment. Visits to such facilities made it obvious how dangerous these were considered, with well-documented emergency procedures, chemical showers, and appropriate PPE.

While modern battery installations are less hazardous, they still harbour the potential for serious harm and should be treated as such. Lithium-ion batteries have received bad press in relation to batteries in mobile phones and other portable devices igniting. Most lithium-ion batteries have in-built protection for both over-current and over-temperature protection.

The applicable standards are AS 3011.1:2019 for vented cells and AS 3011.2:2019 for sealed cells, which provide guidance for the installation of battery systems.

Regulations

There are numerous Australian Standards and regulations that cover alternative energy and renewable energy systems. The purpose of these documents is to ensure the safe design and installation of these systems as well as stipulating required system performance. The Australian Clean Energy Council (CEC) publishes additional guidelines covering the design and installation of grid-connected PV and standalone power systems.

All electrical systems, whether connected to the normal supply grid or standalone electrical supply systems, must comply with the *Wiring Rules*. Furthermore, the installation of wiring and equipment must comply with the relevant WHS and safe work practice requirements.

Table 12.3 is not an exhaustive list and only shows some of the applicable Standards. Any installation involving alternate supply systems may be required to comply with any or all of these Standards. Those involved with the design, installation and maintenance of these systems have a responsibility to keep up to date with current requirements of their work practices.

TABLE 12.3 Australian Standards pertaining to alternate supplies

Standard	Application
AS/NZS 3000:2018 **Electrical installations (known as the Australian/New Zealand Wiring Rules)**	This Standard details the requirements for the design, construction and verification of electrical installations, which includes the selection and installation of electrical equipment that forms any part of the electrical installations. It comprises two parts: ▪ Part 1 details the minimum regulatory requirements to ensure that the electrical installation is safe. ▪ Part 2 details compliant work methods and installation practices aligned with the requirements of Part 1.
AS/NZS 4777.1:2016 **Grid connection of energy systems via inverters Installation requirements**	This Standard details the electrical and general safety installation requirements for inverter energy systems not exceeding 200 kVA intended to supply electrical energy to and from a low-voltage electrical installation that connects to the grid.
AS/NZS 4777.2:2020 **Grid connection of energy systems via inverters Inverter requirements**	This Standard details the specifications, functionality, testing and compliance requirements pertaining to electrical safety and performance of inverters that connect between low-voltage energy sources and/or energy storage systems. This includes electric vehicles that operate in vehicle to grid mode as well as standalone inverters that connect to a low-voltage electrical installation that may connect to the grid.
AS/NZS 5033:2021 **Installation and safety requirements for photovoltaic (PV) arrays**	This Standard details the general installation and safety requirements for PV arrays, which includes d.c. array wiring, electrical protection devices, switching, and earthing provisions.
AS/NZS 3010:2017 **Electrical installations – Generating sets**	This Standard details the minimum safety requirements for the use of generating sets that supply electrical energy at potentials exceeding 50 V a.c. or 120 V d.c.
AS/NZS 4509.2:2010 **Stand-alone power systems System design**	This Standard provides requirements and guidance pertaining to the design of standalone power systems having energy storage at extra-low voltage and used to supply extra-low and/or low-voltage electrical energy in a domestic installation.
AS 3011.1:2019 **Electrical installations – Secondary batteries installed in buildings Vented cells**	This Standard specifies minimum requirements to ensure safety from fire and electric shock from the installation of vented lead-acid batteries and vented alkaline batteries.
AS 3011.2:2019 **Electrical installations – Secondary batteries installed in buildings Sealed cells**	This Standard specifies minimum requirements to ensure safety from fire and electric shock from the installation of sealed secondary batteries.
AS/NZS 5139:2019 **Electrical installations – Safety of battery systems for use with power conversion equipment**	This Standard specifies general installation and safety requirements for battery energy storage systems located in a dedicated enclosure or room and supplies electrical energy to other parts of an electrical installation via power conversion equipment.

REVIEW QUESTIONS

1 What is the advantage of using a battery system with a solar power installation?

2 Identify the two most prevalent battery technologies for energy storage systems.

3 What are two advantages of lithium batteries over other battery technologies?

4 What are the two types of lead-acid batteries?

5 Describe some of the hazards associated with the use of lead-acid batteries.

6 Which Standard details the specifications, functionality, testing and compliance requirements pertaining to electrical safety and performance of inverters that connect between low-voltage energy sources and/or energy storage systems?

7 Which Standard specifies minimum requirements to ensure safety from fire and electric shock from the installation of vented lead-acid batteries?

8 Which body publishes additional guidelines covering the design and installation of grid-connected PV and standalone power systems?

CHAPTER REVIEW

12.1 Working safely with alternate supplies

- Before undertaking any electrical work, it is essential that a thorough risk assessment be undertaken. The purpose is to determine potential hazards associated with the work.
- After identifying possible hazards, it is necessary to put in place preventative control measures to mitigate the incidence of potential hazards.
- The major and perhaps most obvious hazard associated with working on or near electrical equipment is electric shock.
- Alternative electricity power sources pose additional risks in that one or more of these generation systems, if installed and functioning, may still be live after isolating the normal supply, or the generator may start unexpectedly.
- Every alternative supply system must have an isolating switch.

12.2 Types and configurations of alternate supplies

- There are a number of technologies and methods used to produce usable electrical power including engine-driven alternators and generators, PV panels, wind-driven alternators and generators, and personal or micro hydro-electric systems.
- Where an electrical installation integrates an alternative supply system with the normal (grid-connected) supply, there are a number of regulatory considerations.
- It is a requirement of the *Wiring Rules* that a main switch or switches can control the alternative supply system at the installation switchboard to which the alternative supply system connects.
- A changeover switch ensures effective electrical separation between the normal electrical supply to the installation and the alternative supply system.
- Where an electrical installation integrates an alternative supply system with the normal (grid-connected) supply, it is necessary to consult the relevant Service and Installation Rules for the specific jurisdiction.
- An uninterruptible power supply, which is usually referred to as a UPS, is an item of electrical equipment that is capable of supplying electrical power to a load should the normal electrical power source fail.
- The inverter is the item of equipment responsible for converting the direct current produced by an alternative energy source such as solar panels into alternating current for use in a home or business (normal single-phase supply is 230 V a.c).
- The purpose of signs and labels is to provide clear indication that an electrical installation is served by multiple supplies and to identify affected circuits.

12.3 Safe isolation

- Being able to safely isolate a generator or inverter energy supply system is of paramount importance to facilitate safe working on an electrical installation.
- Correct isolation and lockout procedures will minimise the dangers of unintentional or unexpected energisation of circuits and the start-up of electrical equipment.
- The proper shutdown procedure should be located onsite.
- The term 'islanding', refers to a condition where the electricity grid does not provide power but some installations are supplying electrical energy into the grid, transmitting electrical energy back into the otherwise inactive electricity supply lines.
- An automatic transfer switch (ATS) is an automatic power switching device that utilises dedicated control logic to perform intelligent switching operations.

12.4 Battery storage systems

- The most prevalent battery technologies are lithium-ion and lead-acid storage cells.
- Battery storage systems have the ability to deliver very high current, which is the case of a short circuit occurring across the terminals of the battery or the battery is heavily loaded.
- There are numerous Australian Standards and regulations that cover alternative energy and renewable energy systems.

TRIAL EXAM

For Chapter 12 knowledge assessment, please complete the following trial exam.

1 A system of electrical supply that is intended to maintain the functioning of all or part of an electrical installation in case of interruption of the normal electrical supply is:
   a the normal supply
   b a supplementary supply
   c grid-based supply
   d an alternative supply

2 Which of the following is an essential step prior to undertaking **any** electrical work?
   a a thorough risk assessment
   b preparation of safe work method statements
   c analysis of safety data sheets
   d a profit/loss assessment

3 An important operational principle is:
  a that all electrical conductors and equipment are isolated until proven otherwise
  b that all electrical conductors and equipment are live until proven otherwise
  c all electrical isolators and other isolation points are correctly labelled
  d electrical testing instruments always work as intended

4 The major hazard associated with working on or near electrical equipment is:
  a burns
  b slips, trips and falls
  c electric shock
  d lacerations

5 The heat and light produced as a result of the ionisation of air is:
  a discharge lighting
  b an arc flash
  c a normal switching operation
  d a product of electromagnetic induction

6 Solar arrays generate:
  a alternating current (a.c.) electricity
  b direct current (d.c.) electricity
  c either alternating current (a.c.) or direct current (d.c.) electricity
  d pulsating direct current (d.c.) electricity

7 Solar cells are a considered a source of 'green energy' because:
  a they do not produce any greenhouse gases
  b they are commonly used in greenhouses
  c their country of origin is Greenland
  d the underside of the panels has a green coating

8 The output of a wind generator is:
  a alternating current (a.c.) electricity
  b direct current (d.c.) electricity
  c either alternating current (a.c.) or direct current (d.c.) electricity
  d pulsating direct current (d.c.) electricity

9 Personal or micro hydroelectric generating systems utilise:
  a heat from the sun
  b the kinetic energy of wind
  c the potential energy stored in a head of water
  d the kinetic energy of flowing water

10 Every alternative supply system must have an isolating switch, which:
  a is under manual control
  b is under fully automatic control
  c is installed remote from the generation source
  d can be bypassed by programmable control

11 The *Wiring Rules* requires that all alternative supply systems employ the MEN system of earthing where the system is rated at:
  a 5 kVA or less
  b 15 kVA or greater
  c 25 kVA or greater
  d no more than 25 kVA

12 The device that ensures effective electrical separation between the normal electrical supply to the installation and the alternative supply system is:
  a an isolation switch
  b a changeover switch
  c a programmable controller
  d not required for installations with alternate supplies

13 An item of electrical equipment that is capable of supplying electrical power to a load should the normal electrical power source fail is:
  a a function of the normal supply system
  b only rated to 50 VA
  c not necessary for installations with underground electrical supply
  d an uninterruptible power supply

14 A device that uses flywheel energy storage to provide short-term ride-through in the event of power loss is:
  a a rotary UPS
  b a battery-based UPS
  c usually diesel engine-driven
  d is a cheap alternative to solar panels

15 An item of equipment responsible for converting the direct current produced by an alternative energy source such as solar panels into alternating current is:
  a a rotary amplifier
  b an inverter
  c a battery-based UPS
  d an engine-driven generator

16 An appropriate circuit breaker for over-current protection of battery systems is:
  a polarised
  b non-polarised
  c a residual current device
  d rated for a.c. operation

17 It is suggested that all testers and leads should be:
  a Category I
  b Category II
  c Category III
  d Category IV

18 The most popular grid-connected battery chemistry is:
  a lead-acid
  b nickel-cadmium
  c lithium-ion
  d carbon-zinc

19 Rooms containing lead-acid batteries are normally negatively pressurised to:
  a prevent the hydrogen gas escaping into the working environment
  b keep the batteries at a cooler operating temperature
  c ensure any fire can be extinguished quickly
  d maintain the electrolyte in solid (gel) consistency

20 The Standard that details the minimum safety requirements for the use of generating sets that supply electrical energy at potentials exceeding 50 V a.c. or 120 V d.c. is:
  a AS/NZS 3010
  b AS/NZS 5139
  c AS 3011.1
  d AS/NZS 5033

13 Rescue from a live LV panel

This chapter provides electrotechnology workers with knowledge and skills required to undertake the safe rescue of a person from a live low voltage panel. This chapter provides underpinning knowledge for the unit UETDRRF004 from the UEE training package.

LEARNING OBJECTIVES

Preparing for LV electrical work

- Outline the responsibilities of employers and employees when undertaking work on a live LV panel.
- Document the steps involved in planning to work on a live LV panel.
- Outline safe isolation procedures for working on a live LV panel.
- List the contents of an LV rescue kit.

General principles for an LV rescue

- Explain the process of effecting a rescue from a live LV panel.
- Outline the recommended treatment for burns.
- Describe the recommended action to control bleeding.

13.1 Preparing for LV electrical work

There are specific laws and regulations covering working safely with electricity. The most important of these is the management of the risks associated with any electrical work, which may involve electrical equipment or electrical installation work.

It is also important to ensure that equipment is de-energised prior to the commencement of work. In general, nobody is permitted to work on equipment that is energised. Test every circuit and every conductor and never assume that equipment is de-energised. Work health and safety (WHS) regulations in most jurisdictions do not permit work on live electrical equipment unless:

- In the interests of health and safety it is necessary to undertake the electrical work on energised equipment. For example, for the preservation of human life to undertake work on energised life-saving equipment.
- In order for the work to be effected it is necessary for the electrical equipment to remain energised.
- For the purposes of testing, the equipment must remain energised.
- There is no reasonable alternative to the performance of the work.

If work on live equipment is necessary, it is vital to ensure that:

- It is not possible for a person to accidentally contact any exposed part that is live.
- A competent person undertakes all the work using the appropriate tools, testing equipment and personal protective equipment (PPE).
- A safe work method statement (SWMS) is prepared before the work is undertaken and the work is performed in accordance with this SWMS.
- A competent safety observer is present. A safety observer is not, however, required for testing, or if a risk assessment identified no serious risks.
- Access to areas where work is being undertaken on energised equipment is not permitted.

Responsibilities of employers and employees

Both employers and their employees have legal obligations or responsibilities in the performance of electrical work involving live electrical equipment

Responsibilities of employers

Employers have a responsibility to provide appropriate training and all the necessary safety equipment to allow a person to safely effect a rescue. Depending on the particular jurisdiction, it may be a requirement of the applicable Electrical Regulations that an employer provides a safety observer if there is danger of a person accidentally coming in direct contact with exposed live conductors or exposed live parts of electrical equipment.

Responsibilities of employees

Employees have a responsibility to correctly wear the appropriate safety clothing and correctly use the provided safety equipment. The safety observer must not undertake any other duties or activities other than being a safety observer.

Persons undertaking the role of a safety observer must ensure that they possess the appropriate knowledge of rescue and resuscitation relevant to the type of work undertaken. They must have received training and been assessed or re-assessed as competent to perform the duties of a safety observer during the previous 12 months. Any member of the work team may request proof of this training or assessment.

Joint responsibilities

Both the employer and employee must complete a risk assessment and apply prior experience and knowledge of potential risk exposure in order to determine the most appropriate methods of effecting a safe rescue in the event of an emergency.

SWITCH ON

The safety observer:

- Has the explicit duty of observing and warning against unsafe approach to equipment and other potential hazards.
- May also implement any control measures prior to work being undertaken and respond to an incident when required.
- Must be competent in electrical rescue and administering CPR.
- Must be competent in performing the particular task involved.
- May assist the person undertaking electrical testing but must be aware of any hazards present.

Planning the work

After arriving at the worksite, it is necessary for all persons involved with the electrical work to complete an LV work safety checklist like that shown in **Figure 13.1**.

An important part of the pre-work activity is to complete a risk assessment and have this checked. Risk assessments are covered in Chapter 1 of this textbook.

Risk mitigation

The purpose of risk mitigation is to assess the likelihood of identified hazards occurring and to use appropriate control measures to reduce the level of risks to allow the work to be carried out safely. A risk assessment of the worksite should be undertaken by all those involved in the work. This assessment should include what is to be done in the event of an incident occurring when undertaking the work.

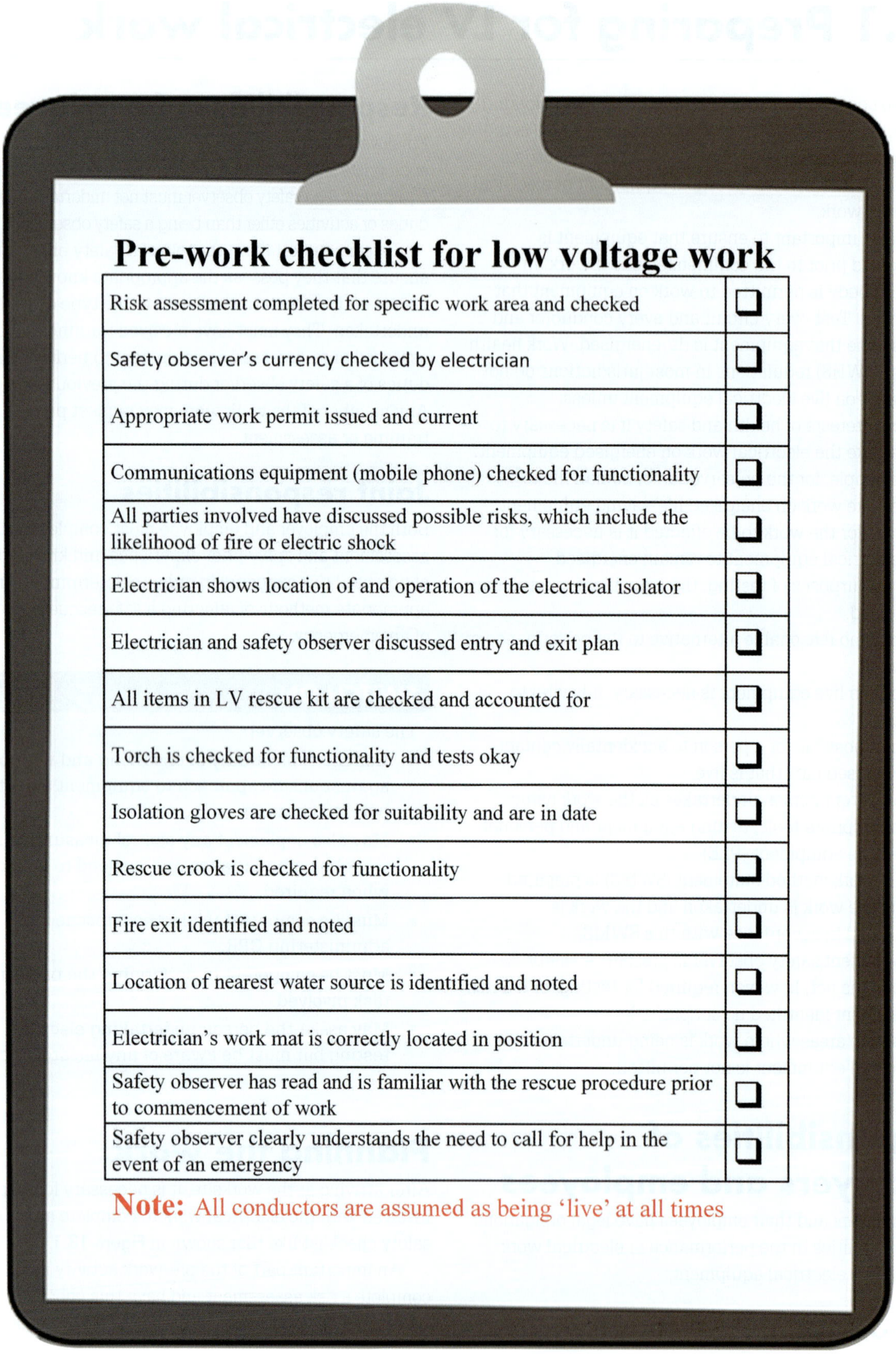

Pre-work checklist for low voltage work

Risk assessment completed for specific work area and checked	❑
Safety observer's currency checked by electrician	❑
Appropriate work permit issued and current	❑
Communications equipment (mobile phone) checked for functionality	❑
All parties involved have discussed possible risks, which include the likelihood of fire or electric shock	❑
Electrician shows location of and operation of the electrical isolator	❑
Electrician and safety observer discussed entry and exit plan	❑
All items in LV rescue kit are checked and accounted for	❑
Torch is checked for functionality and tests okay	❑
Isolation gloves are checked for suitability and are in date	❑
Rescue crook is checked for functionality	❑
Fire exit identified and noted	❑
Location of nearest water source is identified and noted	❑
Electrician's work mat is correctly located in position	❑
Safety observer has read and is familiar with the rescue procedure prior to commencement of work	❑
Safety observer clearly understands the need to call for help in the event of an emergency	❑

Note: All conductors are assumed as being 'live' at all times

FIGURE 13.1 LV work safety checklist

SWITCH ON

The basic principles of risk mitigation include:

- Identifying the hazards associated with performing the work or the work environment.
- Assessing the risk, using the method with which the work team is acquainted, to determine if controls are necessary to reduce the risk.
- Reducing the risk to an acceptable level using the identified controls.

Possible control measures for working on a live LV panel include:

- Eliminating the risk by not continuing with the activity until other controls are in place.
- Substituting the tests using equipment or plant that present a lower, more acceptable level of risk.
- Engineering controls through the use of isolating barriers or insulating mats or by improving the design of equipment or changing its location.
- Administration through rescheduling the activity, providing adequate training or using appropriate warning signs.
- Wearing PPE; for example, in the form of insulating gloves, protective clothing, face shields, safety footwear and the like.

After undertaking a risk assessment, it may be deemed necessary to implement one or more of the control measures detailed above to minimise the risks associated with the activities being performed.

Table 13.1 outlines some risk events and likely factors to consider when assessing the worksite.

TABLE 13.1 Risk event and associated considerations

Event	Considerations
Worker receives an electric shock	▪ Worker is thrown clear of the live electrical equipment ▪ Worker remains in contact with live electrical equipment
Fire develops as a result of an incident	▪ Worker receives flash burns to the eyes or to other parts of the body ▪ Worker receives contact burns to any part of the body or is engulfed in flames ▪ Presence of smoke, toxic fumes, extreme heat and/or poor visibility
Worker receives other injuries	▪ Head/spinal injuries ▪ Fractures ▪ Contusions ▪ Lacerations ▪ Burns

Safe work procedure

A safe work procedure (SWP) is an operational document that is created locally to describe the safest and most efficient way to perform a specific work activity. SWPs are usually the result of an administrative control identified in a risk assessment. Safe work procedures are needed for work identified as being medium to high risk. **Figure 13.2** shows part of an SWP for working on an LV panel.

Safe work method statement

A safe work method statement (SWMS) delineates the high-risk work activities to be undertaken along with the associated controls, codes or legislation. An SWMS is a statement detailing the risk control measures identified for the specific high-risk work activity. The SWMS shows that any high-risk hazards that may affect the work to be performed were identified and that there are suitable control measures in place to reduce or remove the risk.

In most jurisdictions, the relevant safety regulations require the preparation of an SWMS before commencing 'high-risk construction work' if that work may pose a risk to the health or safety of any person including other workers onsite or the public.

Isolation

The appropriate point of isolation must be identified (**Figure 13.3**) before commencing work and the isolation method explained to the safety observer. In an emergency situation it is important to be able to immediately disconnect the source of electrical supply to allow for a safe rescue. In some situations, however, it may be more advantageous to release the casualty without isolating the supply; if this is the case, it is imperative that the rescuer follow all safety precautions and only act if there is a low risk of the rescuer receiving an electric shock.

SWITCH ON

Prior to starting any work on a live LV panel, there are specific procedures to follow:

- Obtain a work order or work permit that states the electrical issue and the scope of the work.
- Locate the electrical schedule for the switchboard or panel to determine the isolator to be marked as the point of isolation.

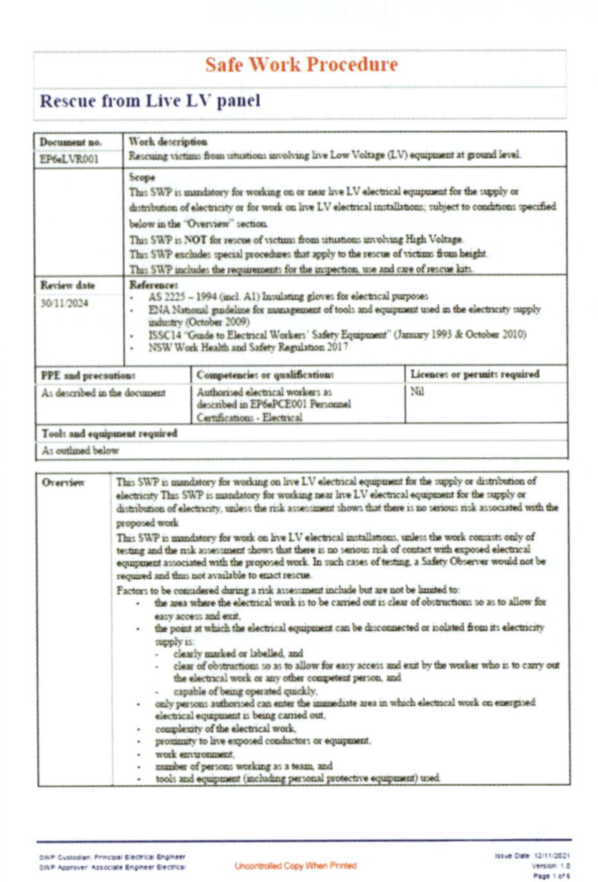

Safe Work Procedure

Rescue from Live LV panel

Document no.	**Work description**
EP6eLVR001	Rescuing victims from situations involving live Low Voltage (LV) equipment at ground level.
	Scope This SWP is mandatory for working on or near live LV electrical equipment for the supply or distribution of electricity or for work on live LV electrical installations; subject to conditions specified below in the "Overview" section. This SWP is **NOT** for rescue of victims from situations involving High Voltage. This SWP excludes special procedures that apply to the rescue of victims from height. This SWP includes the requirements for the inspection, use and care of rescue kits.
Review date 30/11/2024	**References** - AS 2225 – 1994 (incl. A1) Insulating gloves for electrical purposes - ENA National guideline for management of tools and equipment used in the electricity supply industry (October 2009) - ISSC14 "Guide to Electrical Workers' Safety Equipment" (January 1993 & October 2010) - NSW Work Health and Safety Regulation 2017

PPE and precautions	**Competencies or qualifications**	**Licences or permits required**
As described in the document	Authorised electrical workers as described in EP6ePCE001 Personnel Certifications - Electrical	Nil
Tools and equipment required		
As outlined below		

Overview	This SWP is mandatory for working on live LV electrical equipment for the supply or distribution of electricity This SWP is mandatory for working near live LV electrical equipment for the supply or distribution of electricity, unless the risk assessment shows that there is no serious risk associated with the proposed work
	This SWP is mandatory for work on live LV electrical installations, unless the work consists only of testing and the risk assessment shows that there is no serious risk of contact with exposed electrical equipment associated with the proposed work. In such cases of testing, a Safety Observer would not be required and thus not available to enact rescue.
	Factors to be considered during a risk assessment include but are not be limited to:
	- the area where the electrical work is to be carried out is clear of obstructions so as to allow for easy access and exit,
	- the point at which the electrical equipment can be disconnected or isolated from its electricity supply is:
	  - clearly marked or labelled, and
	  - clear of obstructions so as to allow for easy access and exit by the worker who is to carry out the electrical work or any other competent person, and
	  - capable of being operated quickly,
	- only persons authorised can enter the immediate area in which electrical work on energised electrical equipment is being carried out,
	- complexity of the electrical work,
	- proximity to live exposed conductors or equipment,
	- work environment,
	- number of persons working as a team, and
	- tools and equipment (including personal protective equipment) used.

SWP Custodian: Principal Electrical Engineer
SWP Approver: Associate Engineer Electrical
Uncontrolled Copy When Printed
Issue Date: 12/11/2021
Version: 1.0
Page 1 of 6

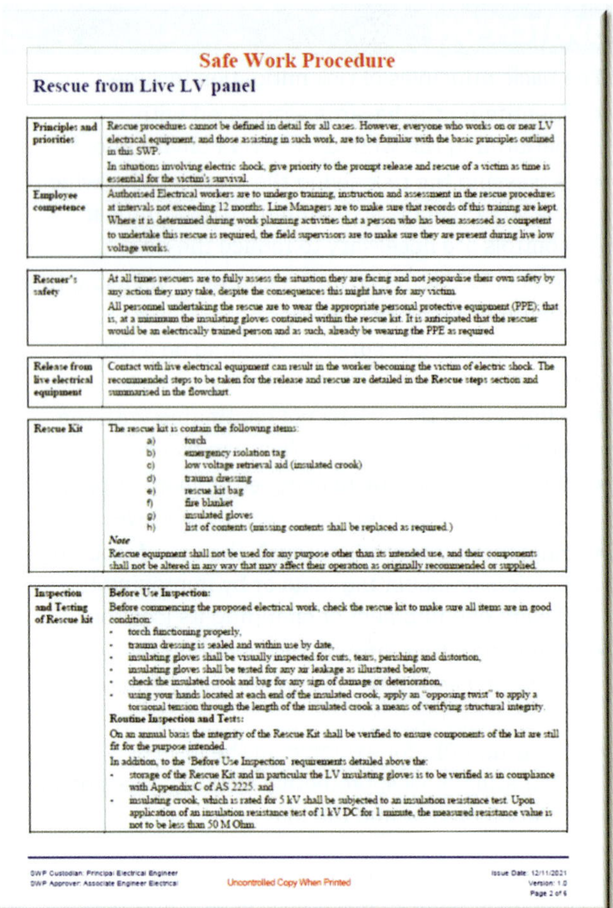

Safe Work Procedure

Rescue from Live LV panel

Principles and priorities	Rescue procedures cannot be defined in detail for all cases. However, everyone who works on or near LV electrical equipment, and those assisting in such work, are to be familiar with the basic principles outlined in this SWP. In situations involving electric shock, give priority to the prompt release and rescue of a victim as time is essential for the victim's survival.
Employee competence	Authorised Electrical workers are to undergo training, instruction and assessment in the rescue procedures at intervals not exceeding 12 months. Line Managers are to make sure that records of this training are kept. Where it is determined during work planning activities that a person who has been assessed as competent to undertake this rescue is required, the field supervisors are to make sure they are present during live low voltage works.

Rescuer's safety	At all times rescuers are to fully assess the situation they are facing and not jeopardise their own safety by any action they may take, despite the consequences this might have for any victim. All personnel undertaking the rescue are to wear the appropriate personal protective equipment (PPE); that is, at a minimum the insulating gloves contained within the rescue kit. It is anticipated that the rescuer would be an electrically trained person and as such, already be wearing the PPE as required

Release from live electrical equipment	Contact with live electrical equipment can result in the worker becoming the victim of electric shock. The recommended steps to be taken for the release and rescue are detailed in the **Rescue steps** section and summarised in the flowchart.

Rescue Kit	The rescue kit is contain the following items: a) torch b) emergency isolation tag c) low voltage retrieval aid (insulated crook) d) trauma dressing e) rescue kit bag f) fire blanket g) insulated gloves h) list of contents (missing contents shall be replaced as required.) *Note* Rescue equipment shall not be used for any purpose other than its intended use, and their components shall not be altered in any way that may affect their operation as originally recommended or supplied.

Inspection and Testing of Rescue kit	**Before Use Inspection:** Before commencing the proposed electrical work, check the rescue kit to make sure all items are in good condition: - torch functioning properly, - trauma dressing is sealed and within use by date, - insulating gloves shall be visually inspected for cuts, tears, perishing and distortion, - insulating gloves shall be tested for any air leakage as illustrated below, - check the insulated crook and bag for any sign of damage or deterioration, - using your hands located at each end of the insulated crook, apply an "opposing twist" to apply a torsional tension through the length of the insulated crook a means of verifying structural integrity. **Routine Inspection and Tests:** On an annual basis the integrity of the Rescue Kit shall be verified to ensure components of the kit are still fit for the purpose intended. In addition, to the 'Before Use Inspection' requirements detailed above the: - storage of the Rescue Kit and in particular the LV insulating gloves is to be verified as in compliance with Appendix C of AS 2225, and - insulating crook, which is rated for 5 kV shall be subjected to an insulation resistance test. Upon application of an insulation resistance test of 1 kV DC for 1 minute, the measured resistance value is not to be less than 50 M Ohm.

SWP Custodian: Principal Electrical Engineer
SWP Approver: Associate Engineer Electrical
Uncontrolled Copy When Printed
Issue Date: 12/11/2021
Version: 1.0
Page 2 of 6

FIGURE 13.2 Part of a safe work procedure for working on an LV panel

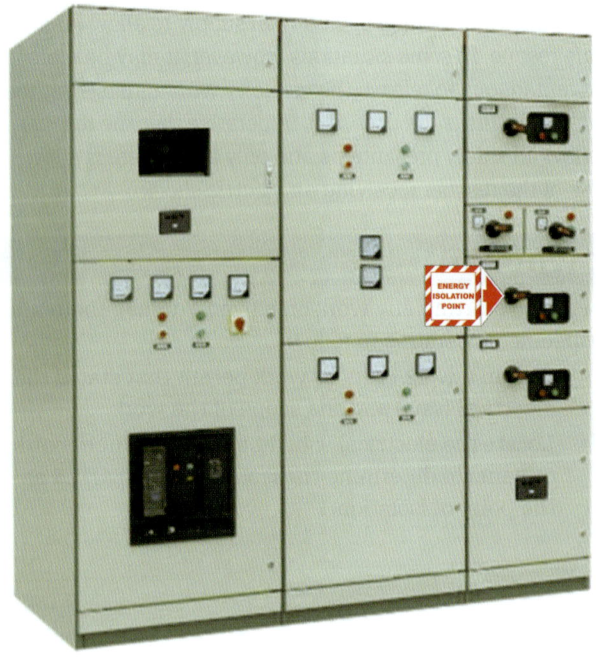

FIGURE 13.3 Identifying point of isolation

Rescue kit

A low voltage rescue kit, like that shown in **Figure 13.4**, is required when any work is carried out on a live LV panel. It is necessary to check a low voltage rescue kit prior to starting work, to ensure the contents are in good condition and suitable for the work environment. The rescue kit should be placed in a position which is accessible to the work area. Suitable harnesses or lifting equipment may also be needed to rescue a casualty from a confined space, such as when working in a cable pit.

The safety observer is the person who has responsibility for the LV rescue kit prior to commencement of any work, while the work is carried out, and after the completion of the work. It is essential that the safety observer:

- is familiar with the rescue kit equipment
- is rigorous with the checking of the rescue kit equipment
- places the rescue kit in an accessible location to enable fast response and rescue.

SWITCH ON

The contents of the LV rescue kit, as shown in Figure 13.4, should be checked as follows:

- bag or case
  - clearly marked as LV rescue kit
- insulated gloves
  - approved, test date stamped not more than 12 months old
  - air tested to ensure no holes or tears
- isolation tag
  - clearly marked
- torch
  - operational
  - non-conductive
- insulated crook
  - outer coating intact with no cracks or chips
  - within test date
- fire blanket
  - 1800 mm × 1200 mm
- burns dressing
  - in date
  - packaging intact.

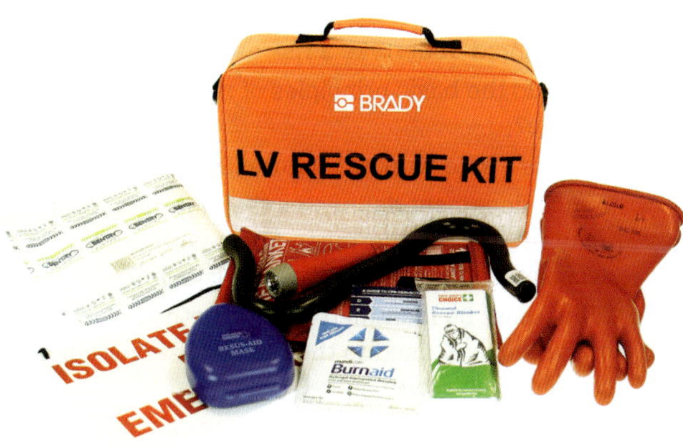

Source: Brady Australia Pty Ltd

FIGURE 13.4 Low voltage rescue kit

Personal protective equipment (PPE)

When undertaking work on a live LV electrical panel it is imperative that PPE, which is appropriate to the work task, is worn correctly. Generally, this would include wearing clothing of 100% cotton or clothing with flame retardant properties, which if covering all body surfaces (i.e. sleeves rolled down and buttoned at the wrist as well as having legs totally covered), will afford maximum protection. It is also important to remove all conductive jewellery and body piercings prior to commencement of work.

SWITCH ON

PPE must be suitable for the identified risk and suitable for the worker to use. The worker must have received appropriate training in how to:

- correctly use PPE provided for their protection
- store and maintain PPE
- keep PPE clean and ready for use.

The types of PPE available and suited to undertaking work activities on a live LV panel include:

- full face shield that is arc rated (may also be integrated with a hard hat)
- eye protection such as cover-all goggles; metal spectacle frames must not be worn
- insulated gloves must be rated to the highest potential voltage expected for the work; leather work gloves would be appropriate only for de-energised electrical work
- clothing should be high visibility non-synthetic, made of non-fusible material that is flame resistant; do not wear clothing made from conductive material or containing metal threads
- footwear must be non-conductive; for example, composite toe work boots or shoes manufactured to a suitable standard (steel-capped boots should not be worn)
- hard hat which complies with safety standards should be worn; this may also include a full arc-rated face shield
- safety belt and harness should be checked and inspected each time before use with particular attention to the condition of buckles, rings, hooks, clips and webbing.

Tools available and suited to undertaking work activities on a live LV panel include:

- ladder that is constructed from non-conductive fibreglass (aluminium can carry an electrical current)
- hand tools such as pliers and screwdrivers must be insulated to 1000 volts.

Other considerations

Other considerations include being able to promptly contact emergency services if needed and the possibility of a low oxygen environment.

Contacting emergency services

In the event of a major incident it is recommended to seek assistance from ambulance and/or medical personnel. The designated safety observer must be familiar with the means of calling for assistance such as the use of two-way radios, mobile phone checked for signal before commencement of work with appropriate emergency telephone numbers stored and ready to call relevant emergency services.

Low oxygen environment

A risk associated with working in a confined space such as a cable pit is oxygen depletion due to accumulation of toxic gases displacing breathable air, which may cause asphyxiation or poisoning of people working in the confined space. If the risk assessment identifies the possibility of toxic gas being present, it is prudent to have appropriate gas testing and rescue equipment onsite and ready for use.

Safe approach distance

Safe approach distance (SAD) is the distance required to be maintained between machinery and anything held by a person and the low voltage panel or electrical installation for the purpose of preventing electrical arcing. Recommended safe approach distances are shown in **Table 13.2**.

TABLE 13.2 Safe approach distance

Qualified electrical worker	Unqualified electrical worker	General public
500 mm	1000 mm	3000 mm

Emergency services

An important aspect of planning to work on a live LV panel is the preparation of a thorough emergency management plan, which documents the strategies to manage emergencies or disasters. Effective emergency management provides the ability to quickly and capably respond to an emergency situation. Creating a robust emergency response plan is critical and it must include unambiguous, succinct instructions on how those responding to an emergency situation are to act. Those who are likely to respond to an emergency situation include the fire, rescue and state emergency services, police, and ambulance service.

The ambulance service provides pre-hospital ambulance care for sick and injured patients, and transports patients to and from medical facilities.

The paramedics and ambulance officers constantly monitor and manage the patient's condition and apply advanced life support and medications when required. They also triage multiple casualty scenes.

To get external assistance in an emergency or disaster situation call:

- Police/Fire/Ambulance: 000
- SES assistance in floods and storms: 132 500
- Police attendance: 131 444 (all states except Victoria)
- International incident emergency helpline: 1300 555 135 (within Australia)

The Triple Zero (000) service is used to contact police, fire or ambulance services in life-threatening or emergency situations.

Emergency + is a useful smartphone app that uses the phone's GPS functionality, which enables callers to provide emergency service personnel with their location as determined by their smartphone. The app also includes SES, Police Assistance Line and other numbers as options, which allows non-emergency calls to be made to the most appropriate phone number.

REVIEW QUESTIONS

1. Outline the purpose of risk mitigation.
2. Who should undertake a risk assessment?
3. What do the basic principles of risk mitigation include?
4. What is a safe work procedure (SWP)?
5. Which document delineates the high-risk work activities to be undertaken along with the associated controls, codes or legislation?
6. What must be identified and its method of operation explained to the safety observer before commencing work?
7. List the required contents of a low voltage rescue kit.
8. What clothing should be worn when undertaking work on a live LV electrical panel?
9. From whom should assistance be sought in the event of a major incident involving physical injuries?
10. What is meant by safe approach distance?

13.2 General principles for an LV rescue

This section describes the principles of rescuing a casualty who has received an electric shock or other injuries from an LV electrical panel. LV is defined in the *Wiring Rules* as a voltage not exceeding 1000 volts a.c or 1500 volts d.c.

In the event of an incident, it is essential to rescue the victim from a live LV panel as quickly as possible and promptly administer treatment as necessary.

The steps involved in rescuing the victim are:

1. Send for help as soon as is practical and safe.
2. Assess the situation and rescue the casualty as quickly as possible.
3. Isolate the supply, and if this not possible or safe, then ensure all safety precautions are in place.
4. Avoid becoming another casualty through receiving an electric shock or being exposed to extreme heat, toxic gases or smoke.

5 Use the insulated crook to remove the casualty and at all times avoid direct skin-to-skin contact with the casualty.
6 In the event of the casualty being on fire, apply the fire blanket by:
   - holding the tabs of the blanket ensuring your hands are behind the blanket
   - holding the blanket between yourself and the casualty
   - placing the blanket over the casualty being sure to cover from the head downwards
   - patting-down to ensure the fire is extinguished.
7 Move the casualty to a clear, safe area to allow for assessment of their condition and treatment as necessary.
8 Assess the casualty's condition.
9 Perform cardiopulmonary resuscitation (CPR) if necessary.
10 Treat the casualty's injuries such as controlling bleeding, managing burns, and the like.
11 Confirm that emergency services were advised and are on their way.
12 Place the casualty in the recovery position awaiting further medical help.
13 Ask a bystander to meet and direct the medical professionals to the casualty.
14 Remain with the casualty until medical help arrives and continually monitor the casualty's pulse and breathing.

Placing the casualty in a safe area

If access is restricted or hazards exist, the casualty should be moved to a clear, safe area for treatment. The most effective way of moving the casualty is by the one person drag method. This is done by:

- crouching behind the casualty
- positioning your arms around the casualty's upper chest
- securely gripping one hand over the opposite wrist
- adopting the correct lifting procedure to avoid sustaining a back injury from lifting and dragging the casualty
- dragging the casualty to a clear and safe area.

Awaiting medical assistance

The types of injuries you may encounter from accidental contact with a live LV electrical panel include:

1 cardiac arrest
2 respiratory arrest
3 burns and/or tissue damage.

While waiting for medical assistance to arrive onsite, remember to follow the DRSABCD emergency action plan (this is covered in Chapter 2 of this textbook) and is summarised in **Figure 13.5**.

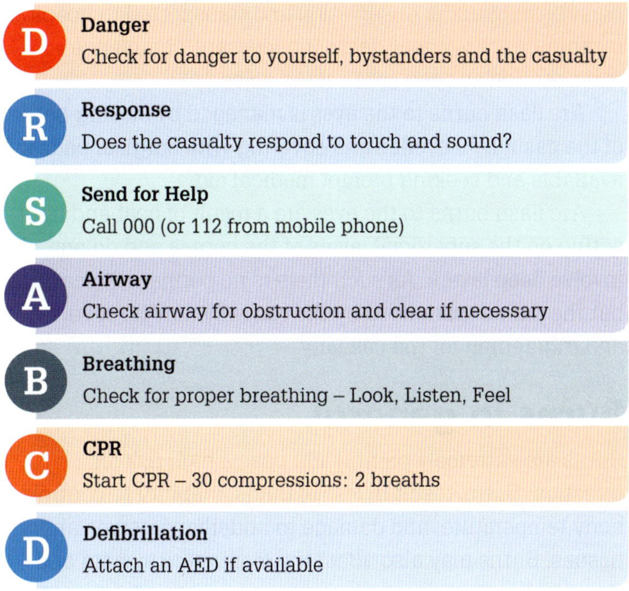

D — **Danger**
Check for danger to yourself, bystanders and the casualty

R — **Response**
Does the casualty respond to touch and sound?

S — **Send for Help**
Call 000 (or 112 from mobile phone)

A — **Airway**
Check airway for obstruction and clear if necessary

B — **Breathing**
Check for proper breathing – Look, Listen, Feel

C — **CPR**
Start CPR – 30 compressions: 2 breaths

D — **Defibrillation**
Attach an AED if available

FIGURE 13.5 DRSABCD emergency action plan

Remember CPR stands for **C**ardio**p**ulmonary **R**esuscitation

Burns from electrical contact

A burn from electrical contact, which causes current to pass through the body, may be more serious than it appears in that:

- the burn may be deep
- there may be damage to internal organs, depending on the path of current through the body
- current flow through the heart may cause cardiac arrest or cardiac arrhythmias
- an entry and an exit wound may be present; the entry wound is generally smaller than the exit wound.

Management of burns from electrical contact

The first step is to safely remove the casualty from the electrical energy and then:

1 Follow the DRSABCD emergency action plan (**Figure 13.5**).
2 Assess the casualty for both entrance and exit wounds.
3 Cool burned areas under running water for twenty (20) minutes.
4 Cover the burn with either a non-adherent burn dressing or loosely applied cling wrap.
   - Note apply cling wrap only after twenty (20) minutes of cooling.
5 Reassure the casualty.
6 Seek medical aid for burns from electrical contact.

Human skin offers a high resistance electrical path. Current flowing through the skin produces heat (I^2R), which results in burns. Burns from electrical contact will usually have an entry and exit point along with substantial internal tissue damage. Burns may also be as a result of arc flashes. Smoke and gas inhalation may cause the casualty to suffer burns to the airway. A casualty suffering airway burns must be kept under observation and transported to hospital by ambulance without delay. If necessary, commence resuscitation as soon as is safe to do so.

Arc flash burns to the eyes is managed by closing both of the casualty's eyelids and covering with surgical pads if available and seeking prompt medical aid.

Arc flash burns to the eyes are a result of heat and light acting on the superficial layers of the cornea and do not involve deep layers. As such there is no permanent scarring but the pain experienced by the casualty may be severe and frightening for the casualty.

Burns in general

Burns are extremely painful and pose a high risk for infection. Burns result in loss of fluids, inability to control body temperature, and damage to underlying nerves and tissues. Burns may also affect the respiratory system and the eyes.

Burns may be classified as:

- superficial:
  - skin is red and painful
  - skin may blister and swell, such as with sunburn
- deep:
  - skin is white, dark red, or charred
  - casualty will not experience pain where nerve endings were destroyed
  - usually surrounded by superficial burns.

Treatment for general burns

The recommended first aid treatment for burns from non-electrical contact is:

- Follow DRSABCD emergency action plan (**Figure 13.5**).
- Ensure that the scene is safe and poses no risk to those administering first aid.
- Extinguish burning clothing using a fire blanket.
- If clothing is stuck to the skin, do not attempt to remove the clothing – leave intact.
- Irrigate the area of the burn with a gentle stream of tap water for 20 minutes – do not use ice or iced water to cool a burn as this will result in further tissue damage.
- Gently remove any rings (if possible), watches, belts or tight clothing from burnt areas before the onset of swelling.
- Observe the casualty and ensure that they do not become too cold.
- Do not apply lotions, ointments or creams.
- Do not pierce any blisters that may have formed.
- For scalds, remove wet clothing from affected area.
- After cooling the burn for at least twenty (20) minutes cover it with a clean non-adherent burn dressing or loosely applied cling wrap.
  - This will minimise the risk of infection. Cling wrap must only be applied after twenty (20) minutes of cooling.
- Rest and reassure the casualty.
- Seek urgent medical attention for:
  - burns involving airways, hands, feet, face or genitals
  - deep burns
  - superficial burns larger than a twenty (20) cent piece on an adult or a ten (10) cent piece on a child.

Control severe bleeding

Bleeding is defined as loss of blood and can range from minor bleeding through to severe external and internal bleeding. When attending to any wound, it is important to take appropriate measures to avoid direct contact with blood and any bodily fluids.

The recommended treatment for the control of bleeding is:

- Wear gloves to minimise the risk of infection.
- Follow the DRSABCD emergency action plan (**Figure 13.5**).
- Help the casualty to lie down and if the wound is covered by clothing, remove or cut their clothing to expose the wound.
- Squeeze the edges of the wound together if possible.
- Apply direct pressure over the wound using a sterile pad or your hands (use gloves if available); if the casualty is able, ask them to do this.
- Elevate and support the injured limb above the level of the casualty's heart; if you suspect a broken bone then exercise care when elevating the limb.
- Apply a sterile wound dressing over the wound if not already in place.
- Secure the dressing by bandaging over the padded wound.
- If bleeding is not controlled by this action, then leave the initial dressing in place and apply a second dressing and secure it with a bandage.
- If bleeding continues through the second dressing, then replace only the second dressing and re-bandage.
- Check every fifteen (15) minutes to ensure that the bandages are not too tight and that there is circulation below the wound.
- Continue to monitor the casualty's breathing and pulse.
- Observe the casualty for shock and treat as necessary.
- Seek medical assistance.

Post-incident actions

All incidents require investigation to prevent recurrence. The degree of investigation is dependent on the seriousness of the incident and the relevant jurisdiction. In general, it is expected that an authority to release a worksite post-incident is required before work can continue. In the case of a minor injury then the employer has authority to release a worksite. In the case of an incident resulting in a fatality, then emergency services such as police and relevant fire services will close the incident area and suspend use of equipment until the conclusion of their investigation. WorkSafe may issue an improvement notice, which if not acted upon by the employer, can lead to prosecution.

All electric shocks, accidents and incidents must be reported to the relevant authorities.

REVIEW QUESTIONS

1 What does the *Wiring Rules* define as low voltage?
2 Outline the requirements for rescuing a victim from a live LV panel in the event of an incident.
3 Describe the one person drag method.
4 What are three types of injuries that may occur due to accidental contact with a live LV electrical panel?
5 In the DRSABCD emergency action plan, what does the letter 'S' represent?
6 In the DRSABCD emergency action plan, what does the letter 'C' represent?
7 Name the two classifications of burns.
8 What classification of burn has occurred when the skin is white, dark red, or charred and the casualty does not experience pain?
9 Outline the recommended treatment for a burn.
10 What is an important consideration when attempting to control bleeding?

CHAPTER REVIEW

13.1 Preparing for LV electrical work

- There are specific laws and regulations covering working safely with electricity.
- Both employers and their employees have legal obligations and responsibilities in the performance of electrical work involving live electrical equipment.
- After arriving at the worksite, it is necessary for all persons involved with the electrical work to complete an LV work safety checklist.
- The purpose of risk mitigation is to assess the likelihood of identified hazards occurring and to use appropriate control measures to reduce the level of risks to allow the work to be carried out safely.
- A safe work procedure (SWP) is an operational document that is created locally to describe the safest and most efficient way to perform a specific work activity.
- The appropriate point of isolation must be identified before commencing work and the isolation method must be explained to the safety observer.
- Low voltage rescue kits are required when any work is carried out on live LV panels, and should be placed in a suitable position, which is accessible to the work area.

13.2 General principles for an LV rescue

- Low voltage is defined in the *Wiring Rules* as a voltage not exceeding 1000 volts a.c or 1500 volts d.c.
- In the event of an incident, it is essential to rescue the victim from a live LV panel as quickly as possible and promptly administer treatment as necessary.
- If access is restricted or hazards exist, the casualty should be moved to a clear, safe area for treatment.
- The types of injuries you may encounter from accidental contact with a live LV electrical panel include cardiac arrest, respiratory arrest and burns and/or tissue damage.
- Current flowing through the skin produces heat (I^2R), which results in burns.
- All incidents require investigation to prevent recurrence.

TRIAL EXAM

For Chapter 13 knowledge assessment, please complete the following trial exam.

1 If work on live equipment is necessary, it is vital to ensure that:
  a an inexperienced person undertakes all the work
  b a competent safety observer is on call
  c it is not possible for a person to accidentally contact any exposed part that is live
  d emergency services are advised and are on standby
2 A safety observer must have received training and been assessed or re-assessed as competent to perform the duties of a safety observer during the previous:
  a 12 months
  b 2 years
  c 5 years
  d 10 years
3 After arriving at the worksite it is necessary for all persons involved with the electrical work to complete:
  a an attendance sheet
  b an LV work safety checklist
  c a CPR refresher
  d a skills test

4 A risk assessment of the worksite should be undertaken by:
a the safety observer
b the qualified electrician
c the site supervisor
d all those involved in the work

5 Control measures in order of priority are:
a elimination, substitution, controls, administration, and PPE
b PPE, substitution, controls, administration, and elimination
c elimination, administration, controls, substitution, and PPE
d substitution, elimination, administration, controls, and PPE

6 An operational document that is created locally to describe the safest and most efficient way to perform a specific work activity is:
a a risk assessment
b a safe work method statement
c a safe work procedure
d not required for LV electrical work on an isolated panel

7 A document detailing the risk control measures identified for the specific high-risk work activity is:
a a risk assessment
b a safe work method statement
c a safe work procedure
d not required for LV electrical work on an isolated panel

8 The appropriate point of isolation must be identified:
a at the entrance to the site
b on Google Maps
c through discussions with the site supervisor
d before commencing work

9 Low voltage rescue kits are required when:
a any work is carried out on live LV panels
b any work is carried out on isolated LV panels
c any work is carried out on live extra low voltage panels
d an inexperienced person is undertaking the work

10 Personal protective clothing that is appropriate for working on a live LV panel is:
a clothing of polyester and nylon blend for insulation properties
b clothing of synthetic cotton blend for coolness
c clothing of 100% cotton or clothing with flame retardant properties
d not required to be rated for arc flashes

11 A risk associated with working in a confined space such as a cable pit is:
a oxygen depletion
b diminished visibility
c exposure to ultraviolet radiation
d thermal stress

12 The recommended safe approach distance for a qualified electrical worker is:
a 500 mm
b 1000 mm
c 1500 mm
d 3000 mm

13 The service used to contact police, fire or ambulance services in life threatening or emergency situations is:
a Dial before you dig
b 911
c 000
d 13 123

14 In the DRSABCD emergency action plan, the letter 'B' represents:
a Below
b Breathing
c Bitten
d Brave

15 Arc flash burns to the eyes is managed by:
a irrigating the eyes with cool flowing water
b placing ice cubes on the eyelids and taping in place
c closing both of the casualty's eyelids and covering with surgical pads
d holding the eyelids open until medical help arrives

Electrical control circuits

This chapter provides electrotechnology workers with essential knowledge and skills in developing, connecting and testing electrical power and control circuits. This chapter provides underpinning knowledge for the unit UEEEL0005 from the UEE training package.

LEARNING OBJECTIVES

Relay circuits

- Explain the purpose of manufacturers' terminal numbers.
- Recognise common control devices.
- Draw various control circuits using correct symbols.

Remote stop–start control

- Describe the operation of remote stop–start control and interlocking.

Time delay relays

- Discuss the operating principles of time delay relays.

Circuits using contactors

- Identify circuits using contactors.

Jogging and interlocking

- Explain the purpose of jogging controls.

Control devices

- State the operating principles of various transducers.

Programmable relays

- State the purpose of microprocessor systems.
- Describe the operation of the programmable relay.
- Describe the operation of a programmable logic controller.

Three-phase induction motor starters

- State the operating characteristics of various types of motor starters.

Three-phase induction motor reversal and braking

- Explain various braking management principles.
- Explain reversing operating principles.

Three-phase induction motor speed control

- Describe several motor speed control methods.

14.1 Relay circuits

A relay is essentially an electrically operated switch that opens and closes in response to electrical signals from outside sources. Relays may be of electromechanical or solid state construction. The basic electromechanical relay comprises a coil, which receives an electric signal and converts it to a mechanical action causing contacts to either open or close an associated electric circuit. Symbols are used to represent the various parts of the relay in circuit diagrams.

Circuit symbols

A circuit component may be represented by a single graphical symbol or an arrangement of several symbols as illustrated in **Figure 14.1**.

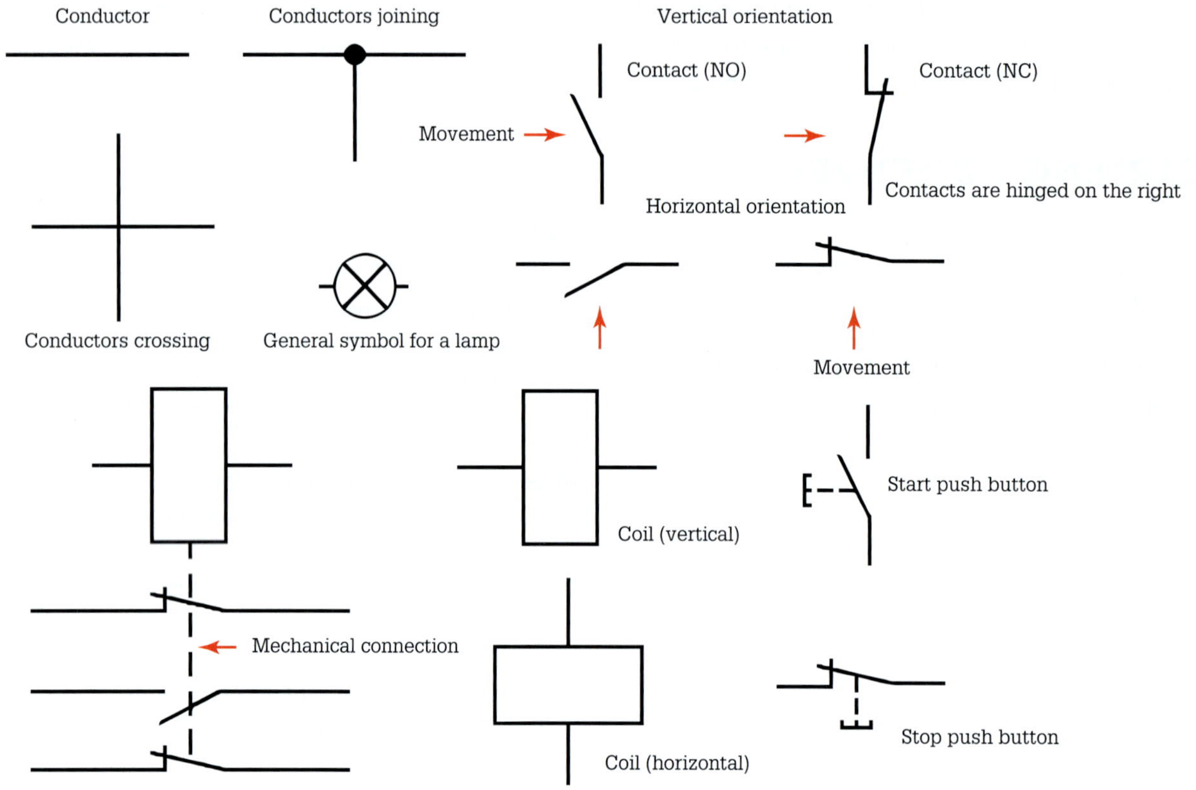

FIGURE 14.1 Circuit symbols (NO means normally open; NC means normally closed)

The meaning of the symbol is defined by its shape and its content. All circuit symbols should be drawn to a size suitable for understanding. However, the size of the symbol and its line width does not affect the meaning. Circuit symbols are used in the drawing of circuit diagrams such as control diagrams and power diagrams to show how an actual circuit operates. In the control and power circuits of electrical systems, it is frequently necessary to make intricate interconnections of relay contacts and switches. Circuit symbols of contacts and pushbuttons are always shown in their normal state as they would be in an actual un-energised circuit.

When the orientation of a symbol to be placed in a circuit diagram deviates from the orientation the symbol has in the AS/NZS 1102 *Graphical symbols for Electrotechnical documentation* series the symbol may be rotated or mirror-imaged if its meaning thereby will not be changed. (Refer to D2.4 SAA/SNZ HB3:1996 *Electrical and electronic drawing practice for students* and EN 61082-1:2006 *Preparation of documents used in electrotechnology* 5.12.3 Orientation of symbols.)

Three types of relays are used in switching and signal routing; each offers distinct advantages and disadvantages:

1. reed relays
2. electromechanical relays
3. solid-state relays.

Reed relays

Contacts of reed relays are made from a ferromagnetic material (reed) which is encapsulated in glass. The coil is wrapped around the glass and when energised the electromagnetic field produced brings the two reeds together, closing the contacts. Contacts are the point where a switch action (throw) makes or breaks a circuit. The contact that moves with the switch actuator is called the movable contact or lever; the contact that is fixed is called the stationary contact.

There are two types of reed relays: dry or wet. Both types are identical except that a wet reed relay has a small amount of mercury added inside the glass tube to provide a consistent contact resistance.

When a circuit needs to switch at high speeds, reed relays are used. Overall, reed relays switch quicker than electromechanical relays. In addition reeds have very low contact resistance and offer the added benefit of being hermetically sealed (contacts do not suffer from oxidation). However, they do not have the capacity to carry high voltages and currents as do electromechanical relays. They are capable of switching much more rapidly than electromechanical relays, up to several hundred times per second, but they can only switch low currents (up to 500 mA). A single-pole, single-throw (SPST), normally closed, 12 V 0.5 A d.c. reed relay is illustrated in **Figure 14.2**.

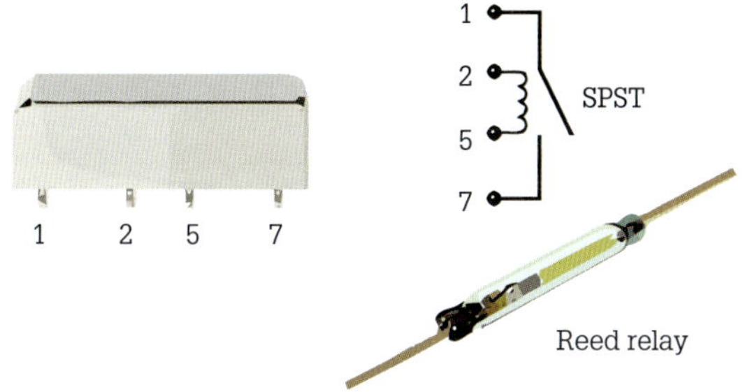

FIGURE 14.2 Reed relay

Reed relays are available in the following configurations:

1. normally open SPST or double-pole, single-throw (DPST)
2. normally closed SPST or DPST
3. SPDT.

Electromechanical relays

The purpose of contactors and contactor relays is to latch and unlatch associated contacts and to provide no-volt and under-volt protection. A letter or letter/number is used to designate the coil (e.g. 'K1'); the associated contacts have the same identifying letter (e.g. K1.1, K1.2).

Coils require a holding-in voltage to ensure that the contacts remain latched. If the supply voltage falls below the required holding-in voltage, the contacts unlatch. This is what is meant by the term 'no-volt protection'.

Electromechanical relays are electromagnetic devices consisting of an electromagnetic coil, an armature mechanism, and electrical contacts. A relay is a switch with normally open (NO) and normally closed (NC) contacts. When the coil is energised a moving contact 'breaks' away from the first fixed contact and moves to 'make' contact with a second fixed contact thereby creating a changeover switch. The function of the relay is to:

- convert the coil current to an electromagnetic field
- convert the electromagnetic field into a mechanical force
- use the force to operate the contacts
- conduct or stop electric current when the contacts switch.

An octal base control relay is illustrated in **Figures 14.3** and **14.4**.

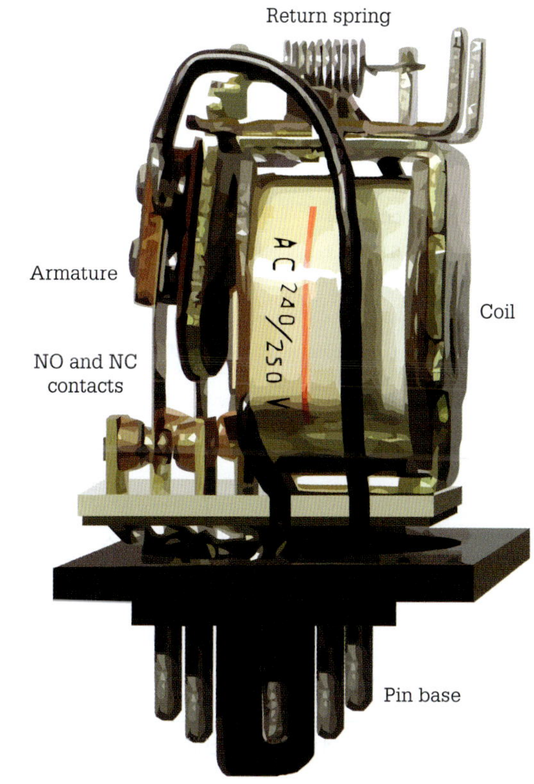

FIGURE 14.3 Octal base control relay

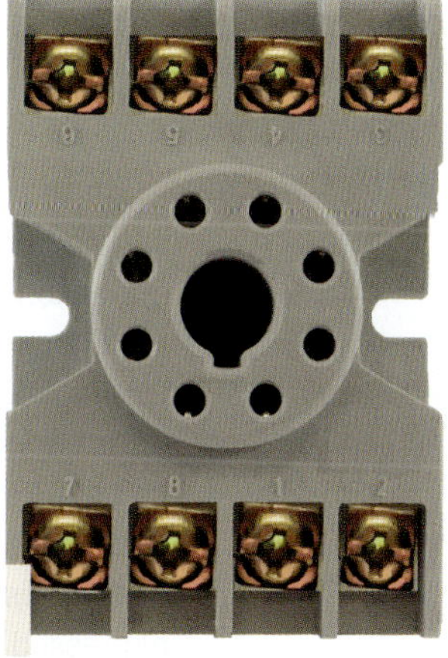

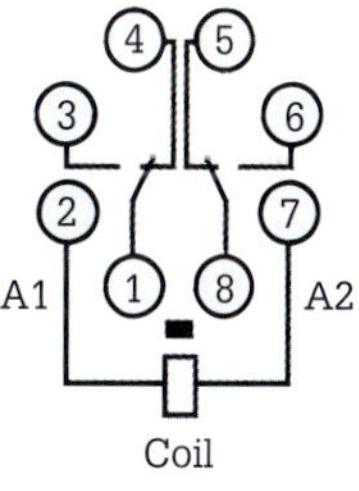

FIGURE 14.4 Octal relay, base, and circuit diagram

The energising of an electromagnetic coil with a small value of current and voltage effects the movement of these contacts. When the coil is energised, current passes through the coil, creating an electromagnetic field that is concentrated within the coil's soft-iron core. This soft iron attracts a flat metal strip called an armature. Because of the movement of the armature, the normally open (NO) contacts close and the normally closed (NC) contacts open. The NO contact closes instantaneously when the electromagnet is energised. On de-energising the coil, the NO contact returns to its normal state, that is, open.

The main operation of a relay occurs where only a low-voltage control signal can be used to control a larger power circuit. In **Figure 14.5**, we have a control circuit consisting of a 24 V d.c. supply feeding a relay coil (K1) through a single-pole switch (manually operated). For an explanation of the operation, this relay coil has one normally open (NO) contact (K1.1) which is connected to the 230 V power circuit. The energisation of the relay coil closes contact K1.1 (electromagnetically operated) which allows current to flow in the power circuit through the lamps.

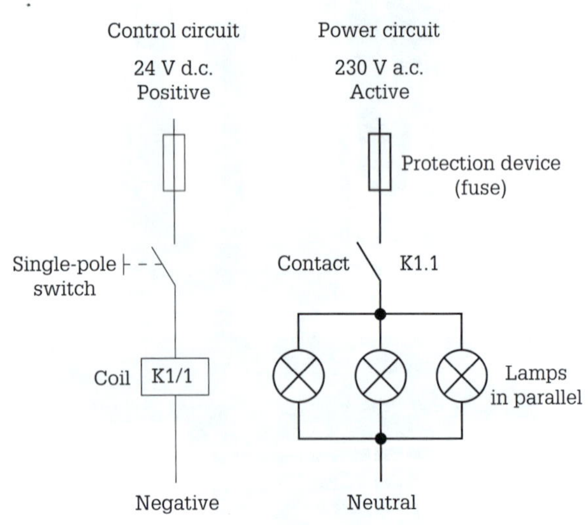

FIGURE 14.5 Control and power circuits

The power circuit provides the current for the lamps (load), while it is the control circuit that provides the logic behind the operation of the lamps. When the current to the coil is switched off, the armature is returned to its relaxed position via a spring.

An electromechanical relay can control two separate circuits. One circuit (control circuit) energises the coil, while the contacts switch the power (high current) or controlled circuit that is electrically isolated from the energy of the control circuit.

Relay switching routines are affected by high ambient temperatures, humidity, and dust and contaminant gases. The relay itself generates heat and oxidants as it operates, resulting in a reduced service life. However, the main influencing factor limiting a relay's service life is the arc formed when the contacts open and close.

General-purpose electromechanical relays are available in cube, DIN mounted or octal base and operate with a.c. or d.c. current, at voltages ranging from 6 V to 230 V and they can control power currents ranging from 2 A to 30 A. A 30 ampere relay is shown in **Figure 14.6**.

FIGURE 14.6 30 ampere relay (cover removed)

Relays can be connected to build very efficient control systems that can perform simple or complex logic functions. An example of a relay system exercising a control circuit function is illustrated in **Figure 14.7**.

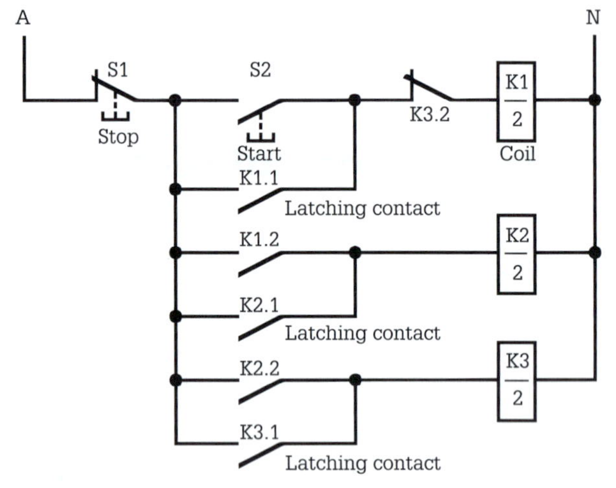

FIGURE 14.7 Circuit diagram showing use of relays

In this system, relay K1 becomes energised when a starting event is established. The initiating event is the activation of the pushbutton switch S2.

The normally open (NO) contacts (K1.1 and K1.2) of the relay K1 close and relay K2 is energised. Relay K2's normally open contacts (K2.1 and K2.2) close and relay coil K3 is energised. Relay K3's normally open contacts close and its normally closed (NC) contact (K3.2) opens, de-energising relay coil K1. The relay sequence can be expressed logically as follows: relay K3 energises if relay

K2 is energised. In order for K2 to energise, K1 must first energise. K1 will de-energise when K3 is energised.

Contacts K1.1, K2.1, and K3.1 are latching contacts. Latching or electrical interlocking is also included in the control circuit to ensure that the coils remain energised after the start button (S2) is released. This is not true for coil K1 once coil K3 becomes energised because contact K3.2 opens, de-energising coil K1 which unlatches contact K1.1.

Solid-state relay

A solid-state relay (SSR) provides electrical isolation between a control circuit and a switched or power circuit and may replace an electromechanical relay. Solid-state relays do not require any moving parts. Solid-state relays have advantages over electromechanical relays: for example, increased operational life, decreased electrical noise (no contact bounce and arcing) and high corrosion resistance. Solid-state relays are compatible with digital circuitry and have a wide variety of uses for such circuits. Switching is carried out using a power semiconductor device capable of handling high voltages and large currents. A solid-state relay may include a driver circuit and a field-effect transistor (FET) providing the output signal and having a control terminal or gate connected to the driver circuit. The driver circuit receives a control signal and operates or drives the output FET based upon the control signal.

Solid-state relays are applied to perform switching in telecommunication, battery-powered devices, programmable controllers, electronic instruments and industrial controls, such as microprocessor control of solenoids, lights, and motors. In general, in a relay-controlled circuit used for signalling or control functions, relatively small currents are switched. A solid-state relay is illustrated in **Figure 14.8**.

FIGURE 14.8 Solid-state relay

Relay testing

A multimeter with a continuity testing ability can be used to test the armature arm switching between the normally closed contact and the normally open contact. This test demonstrates a relay connecting and completing a circuit when the relay is activated. There are two ways to test the switching.

Method one

Set the multimeter to 'diode check'. Connect the common lead of the multimeter to the normally open contact pin of the relay and the other lead to the main contact pin of the relay. Energise the relay and observe the result. Alternatively, use the millivolt meter range on the multimeter to measure the voltage drop across the closed contacts while the relay is in circuit and energised.

Method two

Set the multimeter to the 'ohms' range. With the relay de-energised, connect the common lead of the multimeter to the normally open contact pin of the relay and the other lead to the main contact pin of the relay. Note the result. NO = infinity Ω, NC = zero Ω. An infinity Ω result indicates that the NO contact circuit is fit for service. With an NC contact, a 0 Ω result shows fitness for service.

Drawing conventions

For reference purposes, documents presenting information in diagram form on paper or equivalent should have a reference grid (A, 1) around the border as illustrated in **Figure 14.9**.

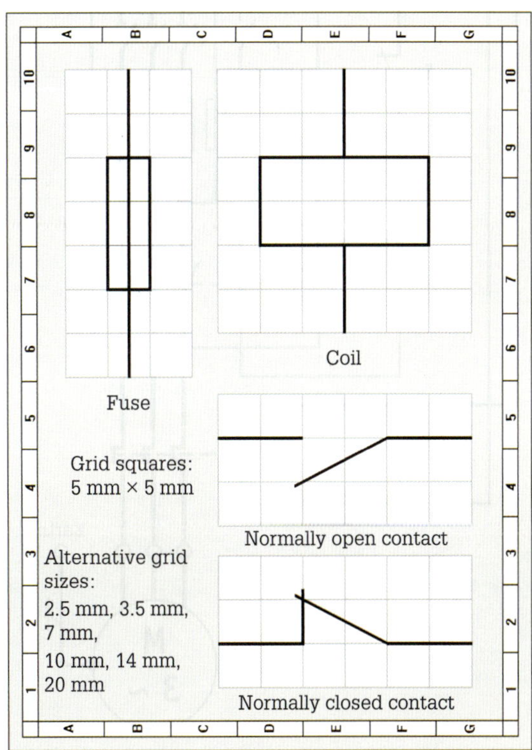

FIGURE 14.9 Document with reference grid and graphical symbols

Graphical symbols that are presented in an electrotechnical document should be taken from the AS/NZS 1102 *Graphical symbols for electrotechnical documentation* series. Symbols should be drawn to a size using an appropriate grid pattern suggested by the standard.

An explanation of the electrical system or circuit may begin with a circuit diagram. A circuit diagram is a basic predictable graphical representation, presented in the form of symbols to provide an outline of circuit and device functions. The circuit diagram illustrates the relationship of the features of the various devices and any physical links that connect them. By explicitly representing individual current paths, the circuit diagram also shows how the electrical circuit operates. The signal current flow in a circuit diagram should be from left to right and from top to bottom.

Circuit diagrams can involve two circuits, either separate or together. One circuit is called the control circuit and the other is the power circuit. Circuit diagrams that show both the control circuit and the power circuit can be represented in three forms – attached, semi-detached and detached as presented in **Figure 14.10 (a)**, **14.10 (b)** and **14.10 (c)**.

All connections in a circuit diagram should be made so that the functioning of the devices can be easily traced. The highlighting of control or power circuit diagrams may be achieved by the use of colours or scaling of symbols or increased line widths.

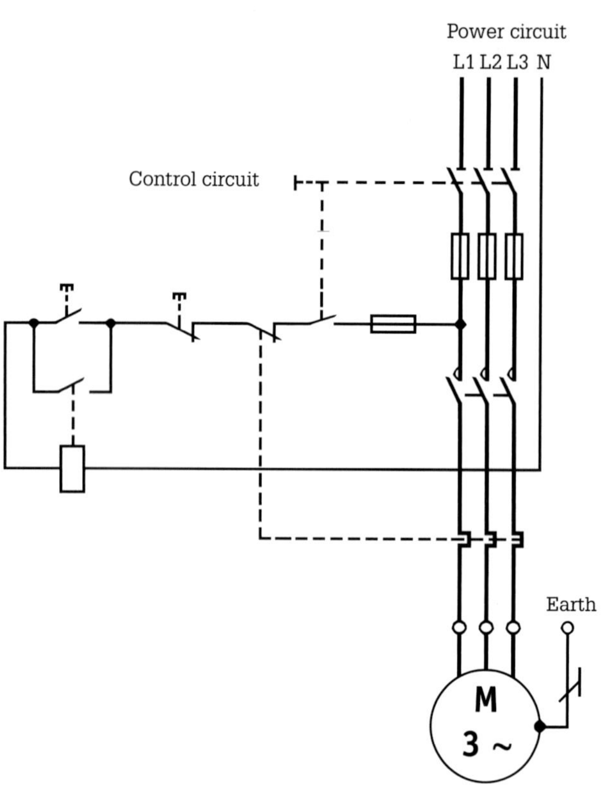

FIGURE 14.10 (B) Circuit diagram using semi-detached representation

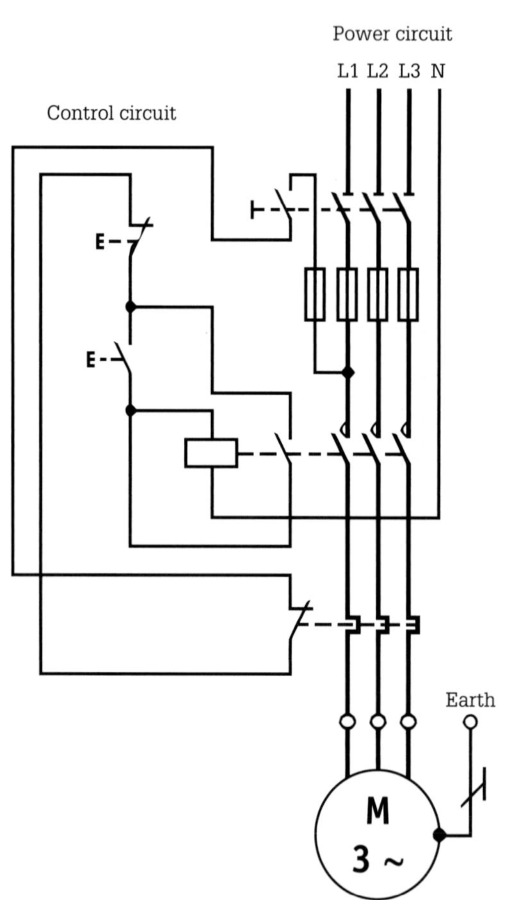

FIGURE 14.10 (A) Circuit diagram using attached representation

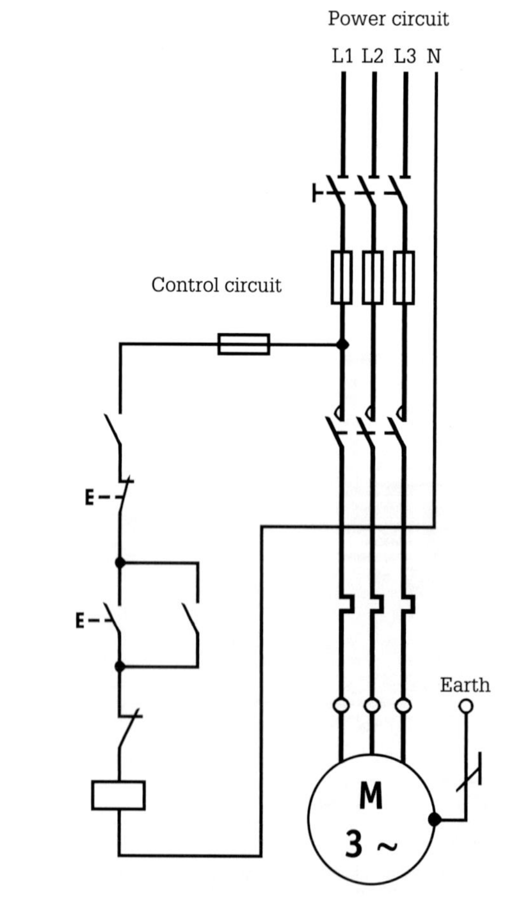

FIGURE 14.10 (C) Circuit diagram using detached representation

Referencing in diagrams

Reference designations (S1), as shown in **Figure 14.11**, associated with a symbol should be located to the left of the symbol when it is shown with vertical circuit drawing orientation or above the symbol when it is shown with horizontal circuit drawing orientation. Coil designations can be placed inside the coil symbol or outside it (shown in red in the diagram). Contact numbers (K1.1) are always opposite the moving contact.

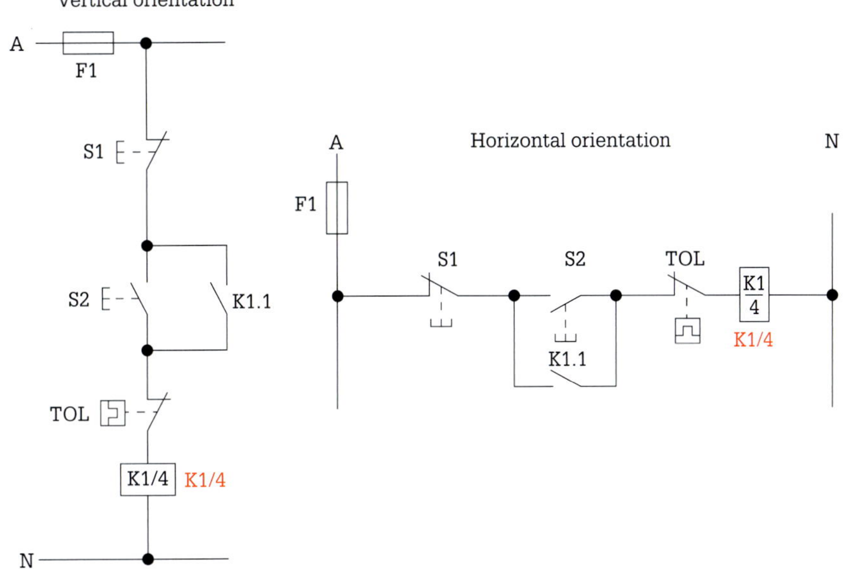

FIGURE 14.11 Reference designations

Terminal designations (1, 2, 13, 14) as illustrated in **Figure 14.12** should be located above horizontal connecting lines and to the left of vertical connecting lines.

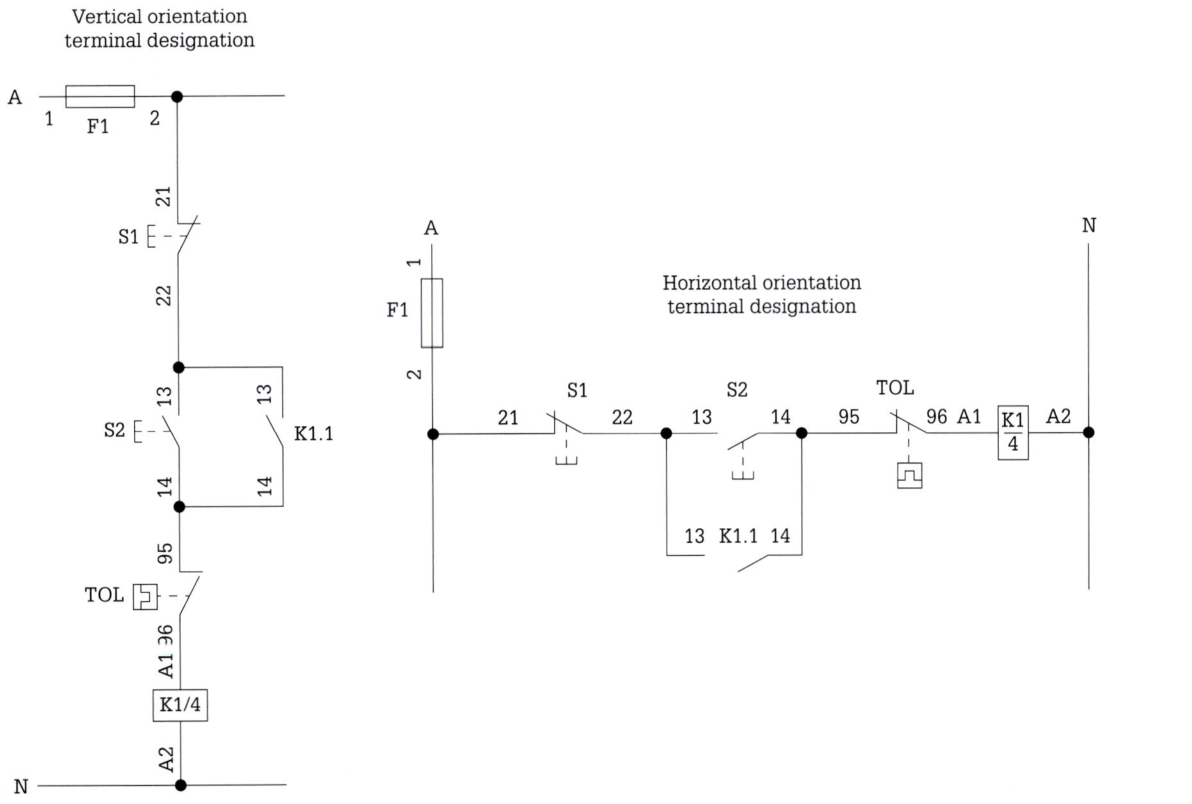

FIGURE 14.12 Terminal designations and reference designations

Line number reference

Each line in a circuit diagram should be numbered starting with the first line and reading down for horizontal orientation.

With vertical orientation, each vertical line should be numbered on the first vertical line and reading across to the right. Line number referencing is shown in **Figure 14.13**.

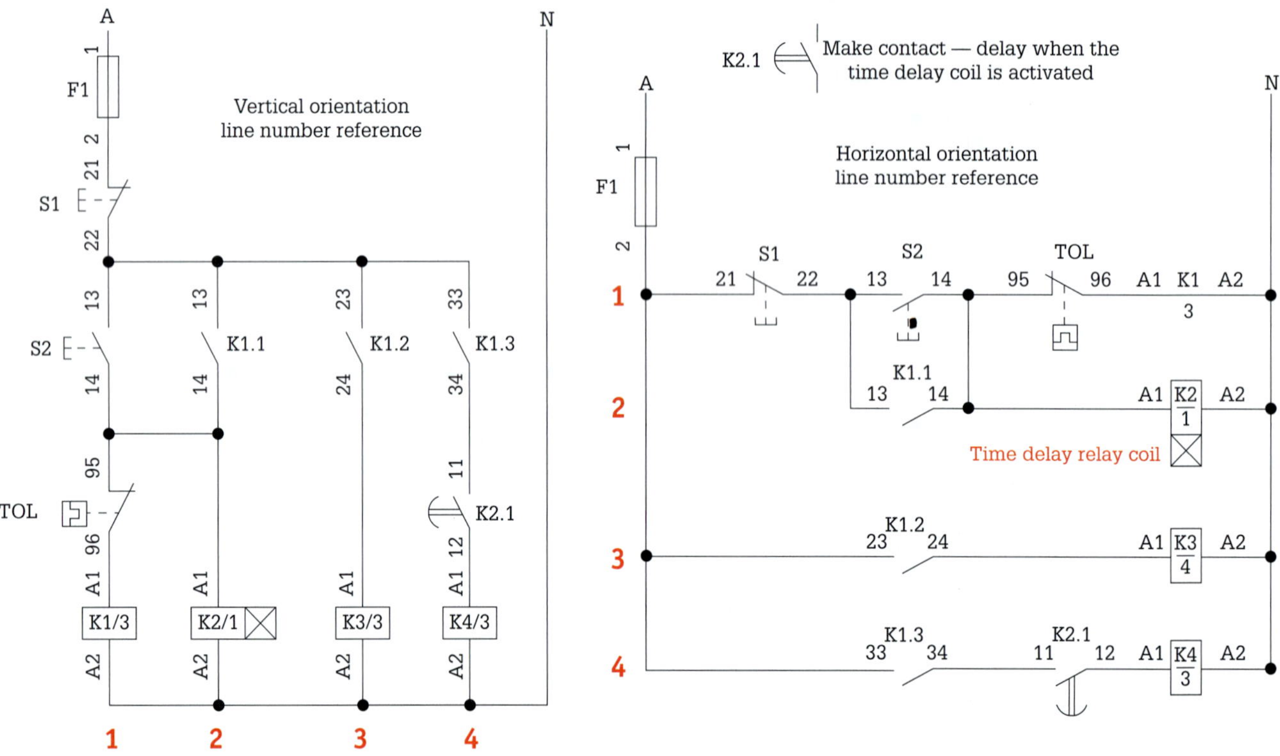

FIGURE 14.13 Line number reference

Starting at the top and reading down, all vertical orientation lines (except N) and all horizontal orientation lines are numbered 1, 2 and 3 in red. These figures serve as an aid in determining the overall logic of the control circuit.

Line number cross-reference

Line number cross-reference as shown in **Figure 14.14** identifies the location and type of contacts controlled by a given relay coil. A line number cross-reference

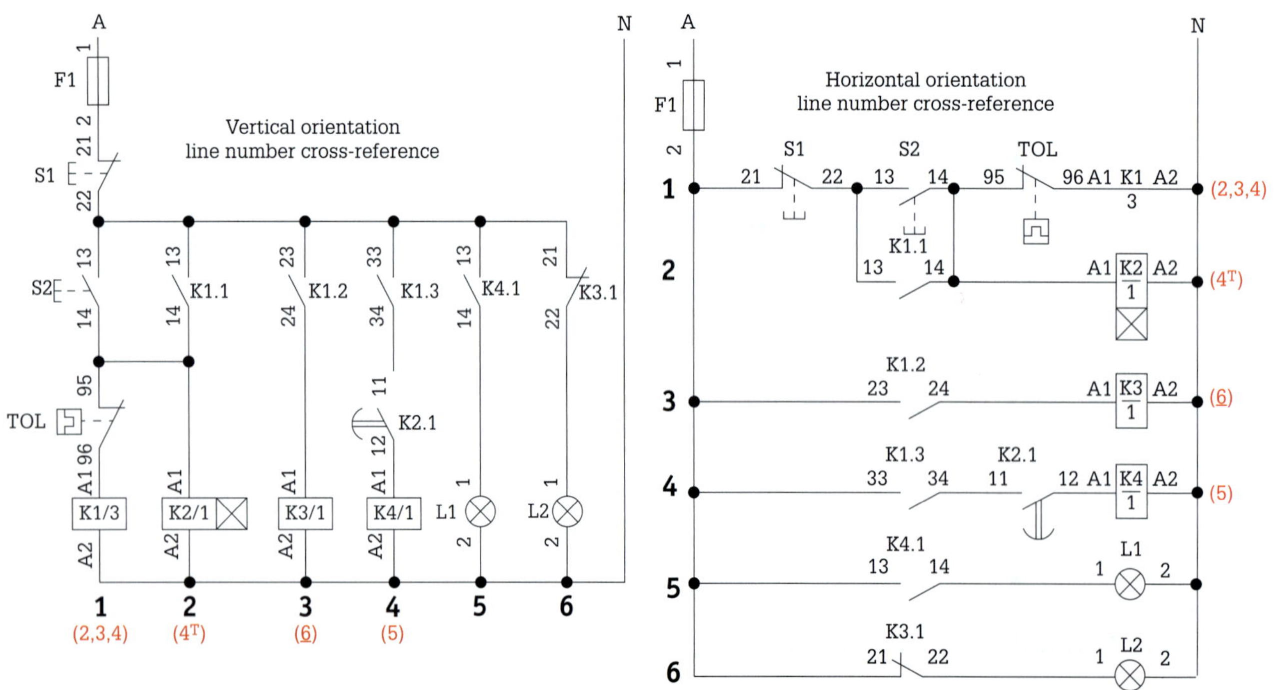

FIGURE 14.14 Line number cross-reference

system consists of numbers in parenthesis to the right of the horizontal orientation circuit diagram and at the bottom of the vertical orientation circuit diagram. To differentiate between normally open contacts and normally closed contacts, normally closed contacts are indicated by a number that is underlined. Timer contacts have a superscript T next to the line number reference (4^T).

Conductor reference numbers

Each conductor in the control circuit can be assigned a reference number (2, 3, and 4 in blue) as shown in **Figure 14.15** on a circuit diagram. Conductor reference numbers provide knowledge of the number of conductors that connect the devices to the circuit.

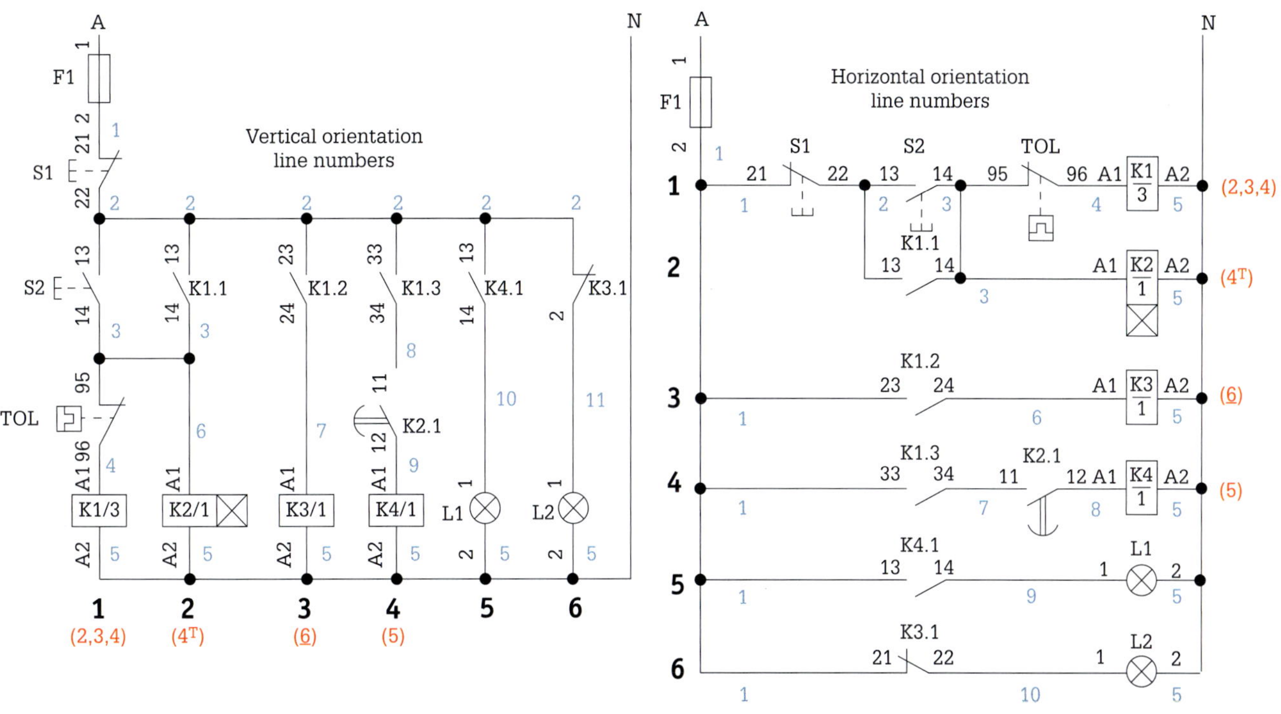

FIGURE 14.15 Conductor reference numbers

With this system of conductor identification, each conductor is given a number. Conductors that are permanently connected have the same number (e.g. number 5 in the figure). This method can be used instead of the numbering of terminal designations system.

Circuit diagrams showing the presentation of symbols for a direct-on-line contactor motor starter with attached, semi-attached and detached representation are shown in **Figure 14.16**

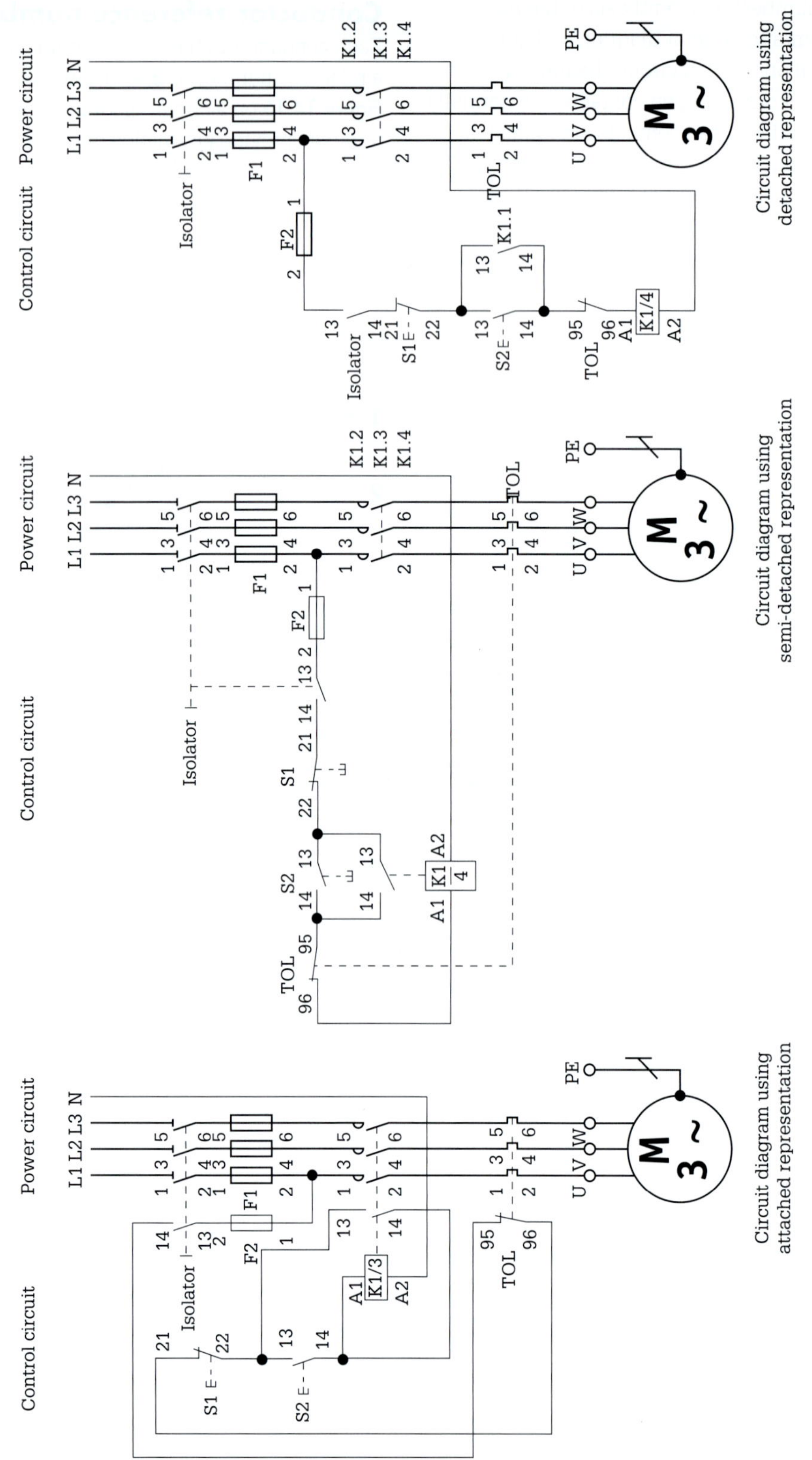

FIGURE 14.16 Circuit diagrams

Labelling wires and terminal (numbering systems)

Some wiring systems have many cables, and it can be hard to distinguish one from another. Consequently, wire labels or markers and terminal numbering systems exist.

A clear, meaningful label identifies cables, and a centralised location for the tag reduces the amount of time required for troubleshooting and maintenance. Labels are placed on wires at the start and end of each wire. Sleeve or shaped ring labels (24, A3) are illustrated in **Figure 14.17**.

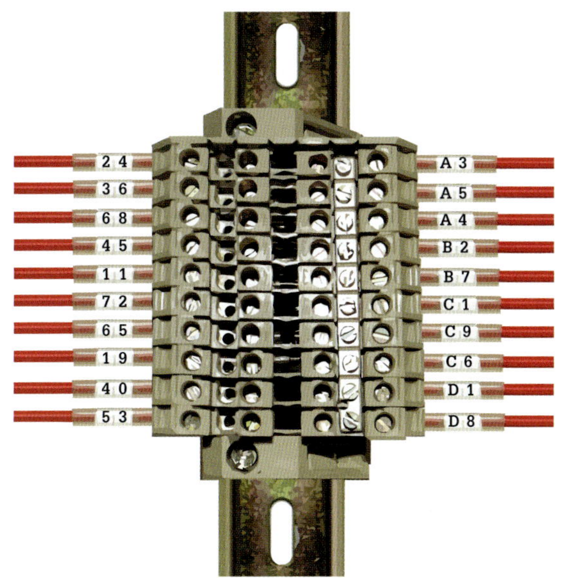

FIGURE 14.17 Sleeve or shaped ring markers

These types of markers provide a permanent cable identification system. The sleeve or shaped ring markers (**Figure 14.17**) and the terminal numbering system (**Figure 14.18**) are used in motor starter panel boards and cabinets where the control wiring is loomed. The markers and terminal numbers relate to the wire numbers that would be listed on an electrical circuit diagram.

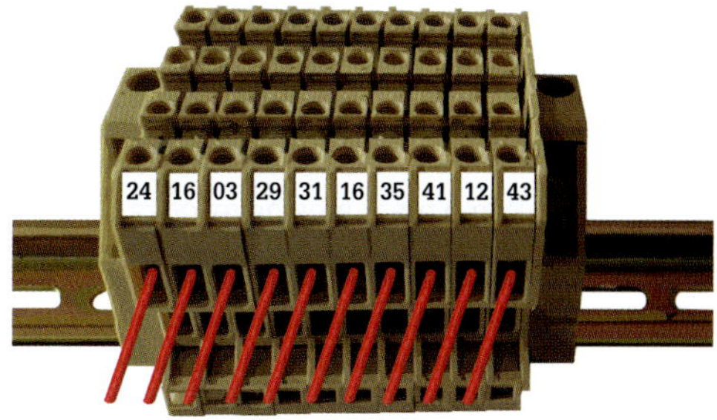

FIGURE 14.18 Terminal numbering

REVIEW QUESTIONS

1 What is a relay?
2 Name the properties that define the meaning of a symbol.
3 Name the two types of reed relays.
4 When would a reed relay be considered for use in a circuit?
5 State the main influencing factor limiting a relay's service life.
6 Where are labels placed on wires?
7 What type of instrument is used to test the armature arm switching?
8 How can distinguishing of control or power circuit diagrams be achieved?
9 How are contact numbers (K1.1) placed in the circuit diagram?
10 What is identified by line number cross-reference?
11 How are line numbers for timer contacts shown?
12 What system provides knowledge of the number of conductors that connect the devices in a circuit?
13 Name the two states for relay contacts.
14 Design a horizontal orientation control circuit diagram using the following circuit description:
- Include line numbers, wire number, and contact locations.
- When pushbutton S1 is pressed relay K1 will energise and latch.
- Once energised K1 causes an indicator lamp (L1) to illuminate.
- When pushbutton S2 is pressed relay K2 energises and latches.
- Once energised relay K2 will cause an indicator lamp (L2) to illuminate and turn off relay K1 and indicator lamp L1.
- Pushbutton S3 de-energises the control circuit.

14.2 Remote stop–start control

One major advantage of using relays in control circuits is that they allow remote switching of other circuits. This is often referred to as remote stop-start control.

Stop–start circuits

A starting device is a control device used primarily to start machinery and hold it in operation. A stopping device is a control device used to shut down machinery, while an isolator is a device that holds and locks out the machinery of operation. A start and stop pushbutton control device with legend plate is illustrated in **Figure 14.19**.

FIGURE 14.19 Start and stop buttons

When choosing a pushbutton control device, factors that should be considered include:

- mounting requirements
- number and contacts
- current and voltage rating

- colour of the button – red for stop, green or black for start, yellow for a return to the initial state, white for any function not covered by the other colours
- illumination requirement
- shape and physical size
- IP rating
- security requirements
- suitability for the environmental conditions.

Devices to start and stop machinery should be in a position where they can be readily and conveniently reached by the person who operates such machinery, and be fabricated and arranged so as to avoid the accidental starting of such machinery. Some machines have lockable isolating switches, shown in **Figure 14.20**, to avoid unauthorised use.

FIGURE 14.20 Lockable isolating switch

For safety, the start pushbuttons may have individual key locks to prevent unauthorised operation of the machinery. The use of a pilot light to indicate that the machinery is energised is another additional safety factor. A hold-in contact or latching contact is usually a normally open contact connected in parallel with a start button to allow a relay coil to remain energised once the start button is released. A three-wire start–stop control circuit diagram with an 'On' indicator lamp is illustrated in **Figure 14.21**.

The term 'three-wire control' is derived from the fact that in the control circuit at least three wires (1, 2, 3 in red) are required to connect the momentary pushbuttons to the coil. Coil K1.5 is energised by pressing the start button S2. An auxiliary contact K1.1 is connected in parallel with the start button contacts. This contact keeps the coil, energised when the start button is released and provides under-voltage protection. This means that if the supply voltage falls below the holding-in voltage of the coil, then the contacts release. When the voltage is restored the start button must be operated to energise the coil.

EXERCISE 14.1

An underground tunnel requires a 230 V lighting circuit to be controlled externally. A STOP/START station is to be used to control two lights. When the START button is pressed, the control relay energises and its associated contacts change state. A green indicator lamp is now ON, and the red indicator lamp is OFF. Two lamps in the tunnel are now ON. When the STOP button is pressed, the contacts return to their resting state; the red indicator lamp is ON, and the green indicator lamp OFF and the lamps in the tunnel are also OFF. Draw a circuit diagram using the horizontal orientation for this process.

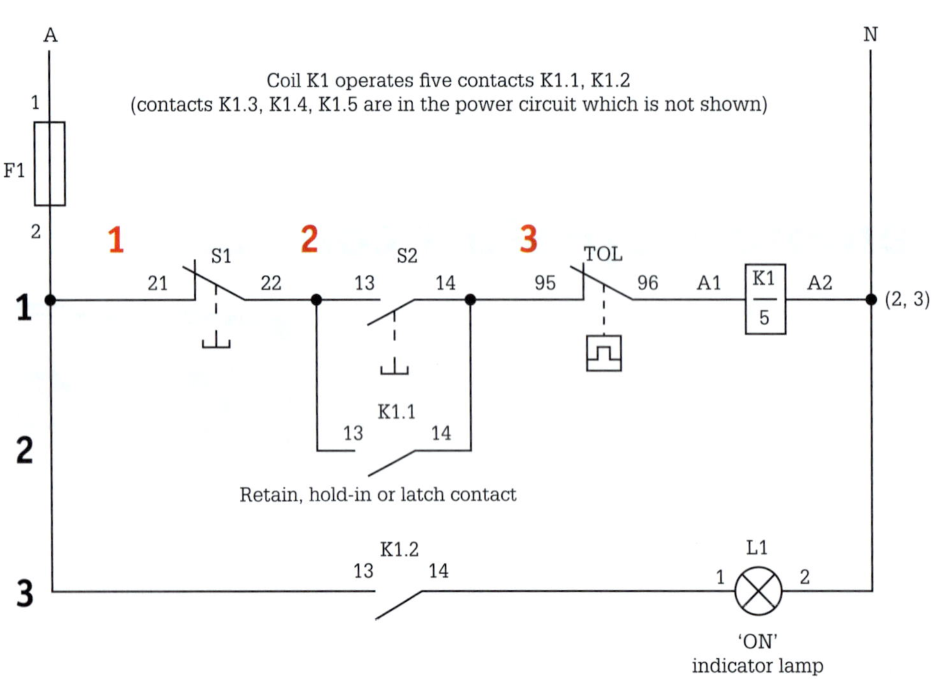

FIGURE 14.21 Three-wire start–stop control circuit diagram

Remote stop–start circuit

When equipment must be started and stopped at more than one location, any number of start and stop momentary contacts may be wired together. A set of start and stop pushbutton stations may be located in a position remote from the controlled machinery in order to energise or de-energise the equipment. All start contacts are connected in parallel with each other, while stop contacts are connected in series with each other. **Figure 14.22** shows how this might be implemented.

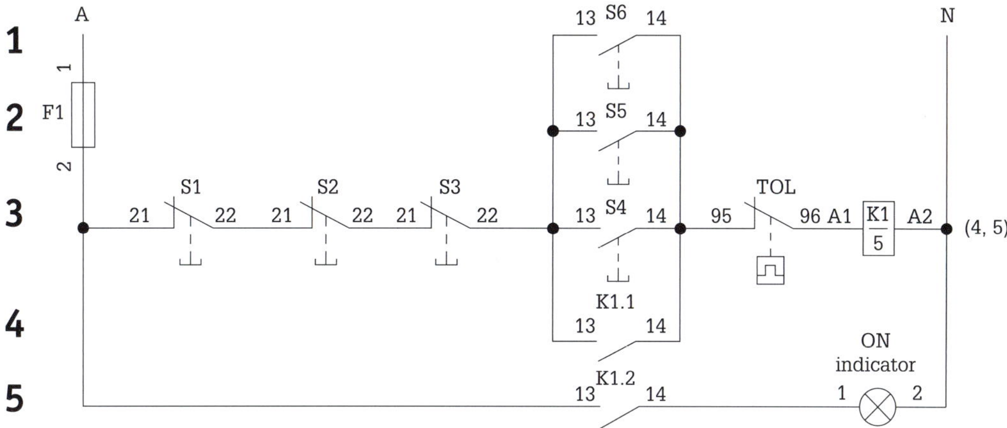

FIGURE 14.22 Local and remote start–stop stations

A remote stop–start station can be used with pumps, with the local stop–start station in the control room and the remote controls located next to the pump for maintenance applications. Remote devices to start and stop machinery should be in a location where they can freely and conveniently be reached by persons who need to operate the machinery; and be so constructed and arranged as to prevent the accidental starting or stopping of such machinery.

Emergency stop

Emergency stop circuits are one of the most standard safety systems found in commercial and industrial installations. Emergency stop buttons (see **Figure 14.23**) should be located within reach of any point where an operator of machinery is exposed to a hazard (e.g. cutting hazard). Emergency stop circuitry must override all machine functions and cause all moving parts to stop, and remove line power from the machine.

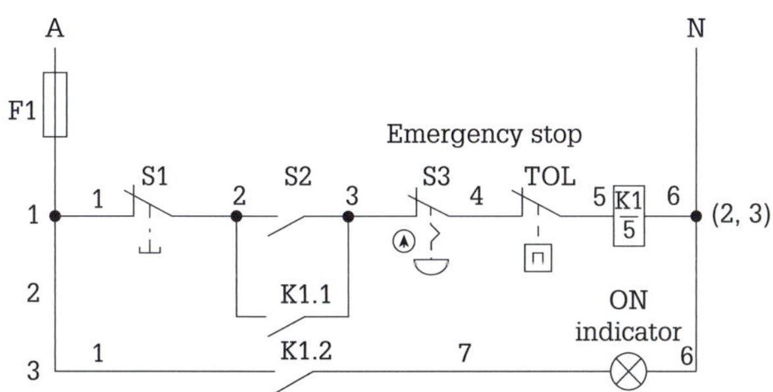

FIGURE 14.23 Emergency stop button

Emergency stop buttons provide alternative precautionary measures whereby the safety of the machinery operator is ensured as far as is practicable.

The emergency stop button is wired in series with the machinery stop button and requires a push for emergency stop and a twist and pull out to reset. This feature provides a positive means of rendering the controls of machinery inoperative (locks out) while the emergency is in place. The resetting of the emergency stop button must not initiate a restart of the machinery. The restart is achieved only through the local start button.

The safety requirements for emergency stop pushbuttons are that they must be red in colour with a yellow background as shown in **Figure 14.24** and be immediately accessible. Also, the pushbutton shape should be the palm or mushroom head type.

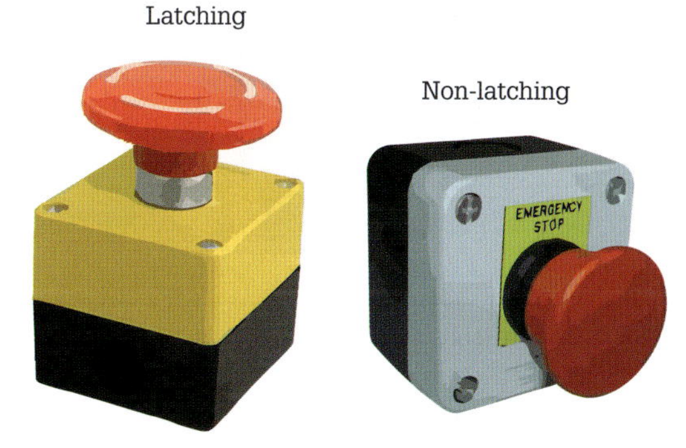

FIGURE 14.24 Emergency stop buttons

Two-wire control

The term 'two-wire control' earns this description because only two wires are required to connect to the control device. Two-wire control wiring is used for remote or inaccessible installations such as pumping stations, refrigeration plants, and compressors. In these applications, it is desirable to have an automatic start control. The control device used for the start control can be a float, thermostat, limit or pressure switch (P) (shown in **Figure 14.25**).

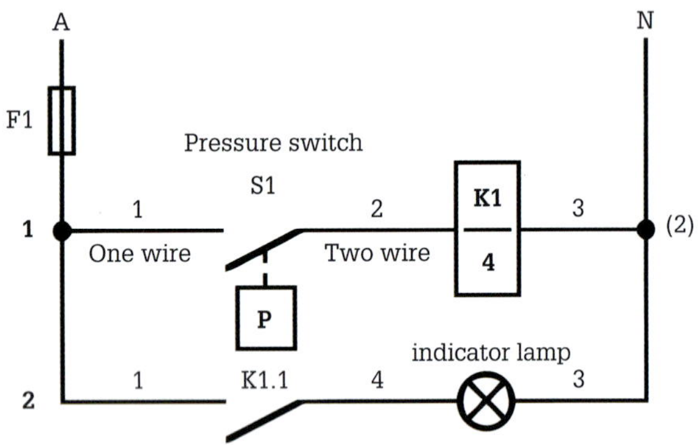

FIGURE 14.25 Pressure switch using a two-wire control system

When the contacts of the control device open, power is detached from the coil and the equipment it is controlling stops. When the control device closes, the coil restarts connected equipment without warning. A pressure switch is illustrated in **Figure 14.26**.

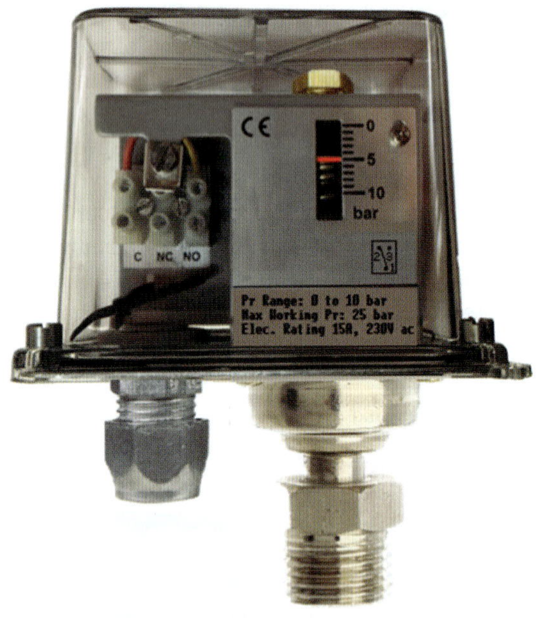

FIGURE 14.26 Pressure switch

Make-before-break

Make-before-break (MBB) switches are those that close a new connection path before another circuit breaks. With this type of function, the next contact is made or closed before the previous contact is broken or opened. This stops the switched path from ever seeing an open circuit. A make-before-break switch has two contacts; one set of normally open contacts makes before the normally closed set breaks when the switch is operated. The circuit symbol for a make-before-break switch is illustrated in **Figure 14.27**.

FIGURE 14.27 Make-before-break symbol

Make-before-break switches are sometimes applied to two generator sets for base load power use. In such applications, source power is intermittently alternated between the generator sets to share base load time equally. Because some loads are extremely critical and difficult to shut down, a manual transfer of power sources without causing an outage to the load is ideal.

Break-before-make

A break-before-make switch is valuable in applications where it is undesirable to have two independent sources short-circuited while switching. On actuation, the break-before-make switch has a movable contact which breaks contact with a fixed contact before making contact (non-shorting) with another fixed contact. The circuit symbol for a break-before-make switch showing one normally closed and one normally open contact is illustrated in **Figure 14.28**.

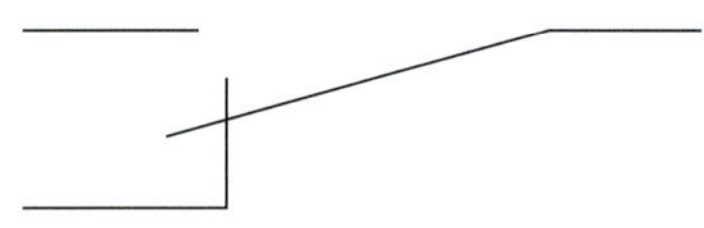

FIGURE 14.28 Break-before-make symbol

A break-before-make switch is used as a transfer switch which will 'break' the electrical connection with utility mains power before it 'makes' the link between an emergency alternative power supply such as a generator. The switch prevents back-feeding from the alternative power source into the utility mains. A break-before-make switch also prevents utility power from damaging the alternative power supply when mains service is restored.

Dead-man controls

The name derives from the concept that if the operator was to die, fall asleep, suffer a heart attack or become incapacitated the device the operator was controlling would stop. The dead-man system is an electrical control unit (handle, foot pedal and switch) designed to provide a means of switching off electrical equipment. The dead-man switch provides an ON or closed circuit when the enable button is fully depressed, and releases control when the button is released. The controlled equipment can be pumped motor starter coils, control valve solenoids and so on.

Dead-man controls are used with shot blast cleaning machines, cranes, trains, trams, and assembly lines. Dead-man switches are mandatory for power tools; for example, chainsaws, circular saws, planers and routers.

Other types of safety control circuits include through-beam light curtains (which act as an emergency stop) and safety door interlock switches. These devices ensure that the machine operator cannot run the system if the control circuit experiences a detectable fault.

REVIEW QUESTIONS

1 Explain the difference between a starting device, a stopping device and an isolator.
2 Outline the requirements for devices used to start and stop machinery.
3 Why do some machines have lockable control switches?
4 How are all stop contacts of remote start–stop stations connected?
5 What is the requirement for emergency stop circuitry?
6 What are the safety requirements for emergency stop pushbuttons?
7 Describe the operation of a break-before-make (BBM) switch.
8 When does a dead-man switch provide a closed circuit?
9 Name common applications for dead-man controls.

14.3 Time delay relays

Time delay relays, like other control relays, have a coil to control the operation of contacts. The difference between a control relay and a timer relay is that the contacts of the timer relay can change the state after a time delay when the coil is either energised or de-energised. Time delay relays can be separated into two general types: the on-delay relay and the off-delay relay.

The on-delay timer has contacts that delay changing their position when the coil is energised, but revert to their normal position immediately when the coil is de-energised. For example, assume that the timer that has normally open contacts has been set for a delay of 8 seconds. When the control voltage is applied to the coil, the contacts remain in the open position for 8 seconds and then close. When the control voltage is removed the coil de-energises and the contacts return immediately to their normally open position.

The off-delay timer has contacts that change position immediately when the coil is energised, but delay returning to their normal position when the coil is de-energised. For example, assume that the timer that has normally open contacts has been set for a delay of 8 seconds. When the control voltage is applied to the coil, the contacts change from open to closed. When the control voltage is removed the coil de-energises; however, the contacts remain in their closed position for 8 seconds before returning to their normally open position (for an electronic timer, a constant supply is required for this).

Timing diagrams are used to demonstrate the relative operation of the timer contacts as time progresses. Timelines are drawn to represent two possible conditions: ON or OFF. There are four options for time-delay relay operation, as shown in Figures 14.29 to 14.32.

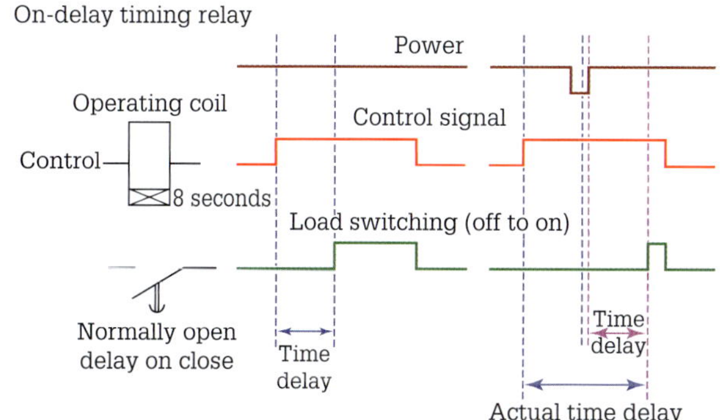

In an on-delay timing relay, the time delay occurs only after the control signal is energised. The load switching occurs at some time after the application of the signal power.

If constant power is normally required for the operation of the time delay relay then any interruption to this power source before load switching occurs will require a repeat of the timing cycle from zero.

FIGURE 14.29 Normally open, on-delay timer diagram

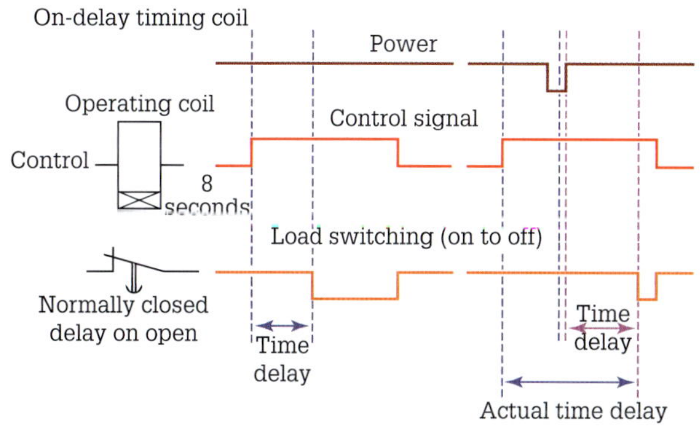

FIGURE 14.30 Normally closed, on-delay timer diagram

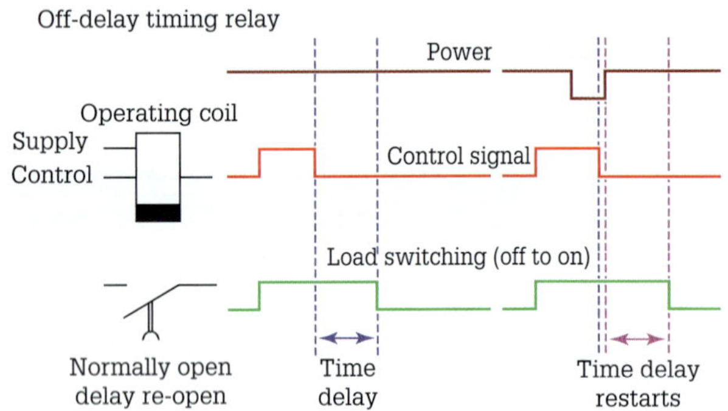

In an off-delay timing relay, the time delay occurs only after the control signal is de-energised. The load switching occurs at some time after the removal of the signal power. Off-delay timing relays usually require a constant power for correct operation. Any interruption to this power source before load switching occurs may result in inaccurate timing.

FIGURE 14.31 Normally open, on-delay timer diagram

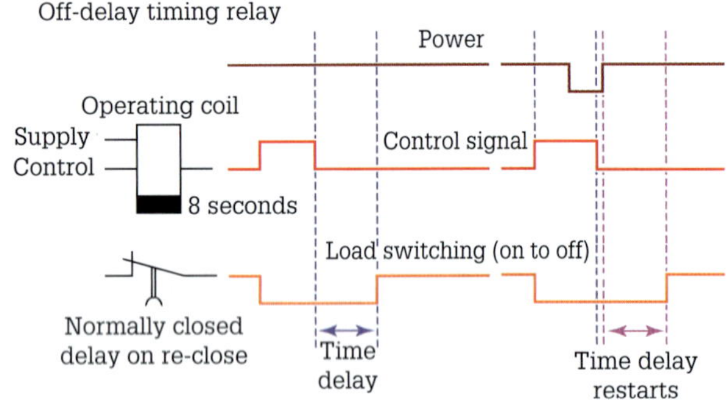

FIGURE 14.32 Normally closed, off-delay timer diagram

A timed function is useful in applications such as the starting system of a large motor in which a reduced voltage must be delivered to the windings of the motor for a set period of time before full-line voltage is connected. Note that large kW motors can be started direct-on-line (DOL) if written approval is obtained from the supply authority. Certain requirements must be satisfied such that the supply system can start the motor DOL, and that other customers of the supply authority will not experience voltage sag and flicker.

In pneumatic timers, the timing is accomplished by the relocation of air through a restricted orifice. The amount of restriction is controlled by an adjustable needle valve, permitting changes to be made during the timing period. A timing circuit is shown in **Figure 14.33**.

Pressing S2 (the start button) causes the coil of K1 to energise as described previously. Instead of illuminating a signal lamp, contact K1.2 energises a timing relay (K2). This is a delay-on-energise time delay relay.

This means that its contacts (K2.1 and K2.2) change state some time interval after energising the coil. The engaged indicator lamp is illuminated and extinguished after the delay while the run indicator lamp is extinguished and illuminates after the delay.

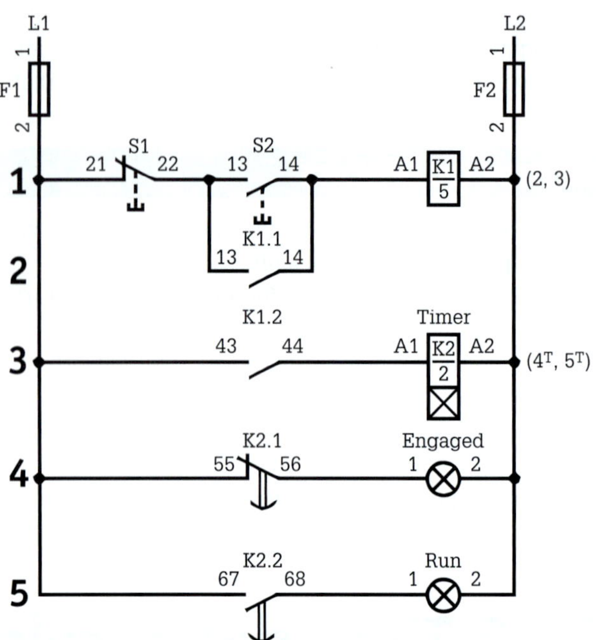

FIGURE 14.33 Timing circuit

Relay specifications

Pole and throw conditions are important to consider when selecting electromechanical relays. Pole choices include single pole (SP), double pole (DP), triple pole (TP), four poles (4P) and greater than four poles. Throw choices are single throw or double throw. Double-throw (DT) relays have a pair of contacts open in one position and closed in the other.

Essential contact specifications to consider when selecting electromechanical relays include the number of normally open contacts, normally closed contacts and the number of changeover contacts. Furthermore, contact ratings to consider include maximum switching current, maximum a.c. switching voltage, maximum d.c. switching voltage, maximum a.c. switching power and maximum d.c. switching power.

Key coil ratings to consider when selecting electromechanical relays include a.c. coil voltage, d.c. coil voltage, coil resistance, coil rated a.c. power and d.c. power. In addition, significant performance specifications to consider for electromechanical relays include making time (operate time) and break time (release time).

Consideration must also be given to what type of electrical load is being switched. An electrical load is defined in terms of voltage (a.c. or d.c.), current and type of loads such as resistive, inductive, electric motor and incandescent. The arc produced as the contacts open differs between a.c. and d.c. load switching. In addition, an inductive load creates an inrush current via the struck arc that is 10 times higher than the nominal load current. Consequently, this consideration of load determines the specification of a relay for a given application.

A relay may have supplementary features. Common examples include a manual test button which can operate the coil plunger so that the circuit can be tested without having to power the coil. An indicator light can be fitted to show if and when the coil is energised; this is often referred to as a coil LED indicator. Electrical protection devices can be integrated into the relay to protect the relay contacts from voltage peaks causing excessive arcing conditions.

REVIEW QUESTIONS

1 State the difference between a control relay and a timer relay.
2 Describe the operation of an on-delay timer relay.
3 Describe the operation of an off-delay timer relay.
4 What is the purpose of timing diagrams?
5 Name a function where a timed function is useful.
6 Identify two important considerations when selecting an electromechanical control relay.
7 What is the purpose of an indicator light fitted to a relay?

14.4 Circuits using contactors

An air break contactor (see **Figure 14.34**) is an electromagnetically operated device for repeatedly establishing and interrupting an electrical power circuit. A contactor contains a set of main (heavy-duty) contacts for switching high-load currents at load rated voltages and may have a set of auxiliary contacts (light duty) for the control circuit. The auxiliary contacts can be added to either side or both sides of the contactor. A top adder for auxiliary contacts is also available. It is always wise to have at least one NO and one NC spare auxiliary contact in addition to those specified for control purposes.

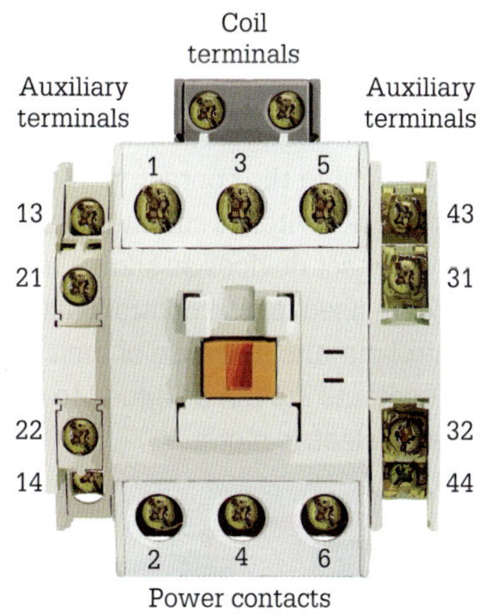

FIGURE 14.34 Contactor

Coils

A coil, illustrated in **Figure 14.35**, is used in a.c. electromagnetic contactors and contactor relays. The coil produces the electromagnetic flux needed to attract the armature of the electromagnet. The coil when energised either latches (closes) or unlatches (opens) the contacts.

FIGURE 14.35 Contactor coil (a.c.)

An a.c. coil draws more current when in the unlatched state than when in the latched state. This effect occurs because coils designed for alternating current have relatively little resistance (R). The little resistance causes a high pick-up (inrush, pull-in) current when the coil is energised. As the electromagnetic effect of the coil starts closing the armature, inductive reactance (X_L) starts to build up (the air gap is getting smaller). The coil now has impedance (Z) which is the phasor sum of its resistance and its inductive reactance. Maximum impedance occurs when the armature has closed (smaller air gap in its magnetic path). Maximum impedance limits the holding current of the coil.

Holding power

The Mitsubishi Electric SR-N4 (CX) magnetic contactor has an inrush value of 60 VA, and when sealed 10 VA. The Siemens 3RH1122-1AB00 contactor relay, 2NO + 2NC, has an apparent pull-in power of the solenoid for a.c. of 27 VA and an apparent holding power of the solenoid for a.c. of 4.6 VA. The ratio of apparent pull-in power to apparent holding power can be 6:1 or more. Coil voltage plays its part. The lower the coil voltage, the greater is the current consumption for both pick-up and holding.

AS 60947.4.1:2015 *Low-voltage switchgear and control gear – Contactors and motor-starters – Electromechanical contactors and motor-starters* sets parameters for dependable contactor operation. A contactor must pull in consistently at 85% of the nominal voltage which allows for a maximum volt drop of 15% during the pick-up stage. In practice, higher control voltages tend to be more dependable and have fewer operational problems than lower voltage control. On the other hand, lower voltages are considered safer.

A.C. contactors

Alternating current contactors have a laminated iron core called an armature with a shading ring embedded in the surface (see **Figure 14.36**) to reduce electromagnetic hum and chattering.

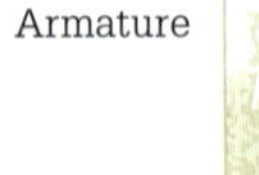

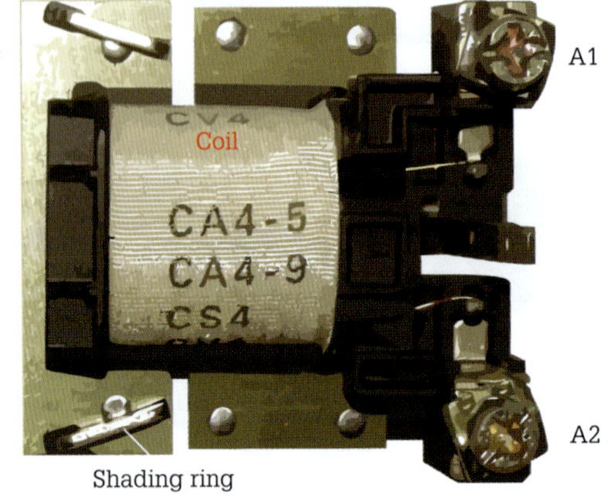

FIGURE 14.36 Shading ring, coil and the armature

When the coil is energised the magnetic field developed within the armature reverses in unison with the a.c. current that is applied. As the electromagnetic field reverses, it induces an electromagnetic field in the shading ring, causing a current to flow. This current develops a secondary reversing electromagnetic field. This field is out of phase with that of the applied coil power and holds the armature faces continuously closed between coil current reversals and minimises electromagnetic noise (chattering). Contactors are suited for starting three-phase a.c. motors ranging from 3 kW to 250 kW at 400 V. A utilisation category has been established for the switching capability of contactors. Types AC-3 (normal switching duty) and AC-4 (extreme switching duty) are the most often encountered contactors. These contactors can provide from 150 to 4000 on-load operations per hour. A contactor chosen for a particular application should be rated for the maximum fault current allowed by circuit protection devices at the point where the contactor is installed.

Contactors are essentially large-capacity relays specifically designed to make and break power circuits for loads such as motors, heaters, and lights. Contactors are multi-pole devices typically arranged to interrupt all energised conductors connected to an electrical load.

Contactors are controlled by control devices (pushbutton switches, float switches) which can turn the contactor coil on or off.

Contactors include normally closed and normally open contacts, but are most often of the single-throw, normally open, double-break configuration. The power contacts and coil are marked as shown using a semi-detached drawing orientation in **Figure 14.37** with a numerical system that is applied worldwide. With the semi-detached drawing style, the coil contacts are joined by a dotted line to indicate that they belong to that coil.

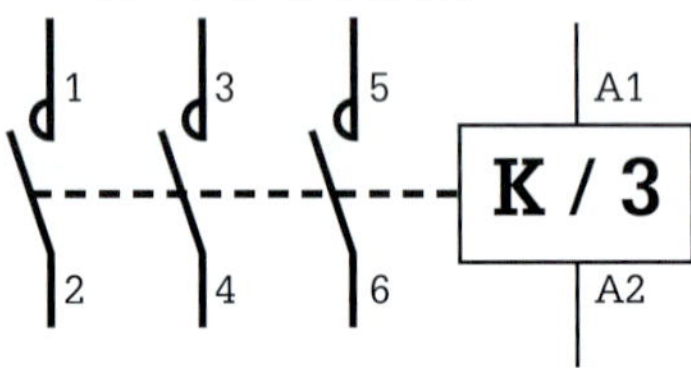

FIGURE 14.37 Contact marking for a contactor

The power contacts are numbered 1-2, 3-4 and 5-6 for each phase. Contactors do not include overload protection for the load they are serving. When contactors are used to control motors, the power circuit includes thermal overload protection for the motor.

Auxiliary contact devices

Additional contacts associated with a contactor may be made available by the use of auxiliary contact blocks. These contact blocks are available with one, two or four contacts and in various configurations of NO and/or NC contacts. When switching the contactor on and off, the auxiliary contact block switches simultaneously.

The auxiliary contact devices as shown in **Figure 14.38** and **Figure 14.39** are used for the control of auxiliary circuits and control circuits. Because of their low current rating auxiliary contacts are not designed to switch power to a load such as an electric motor.

FIGURE 14.38 Auxiliary contacts top adder

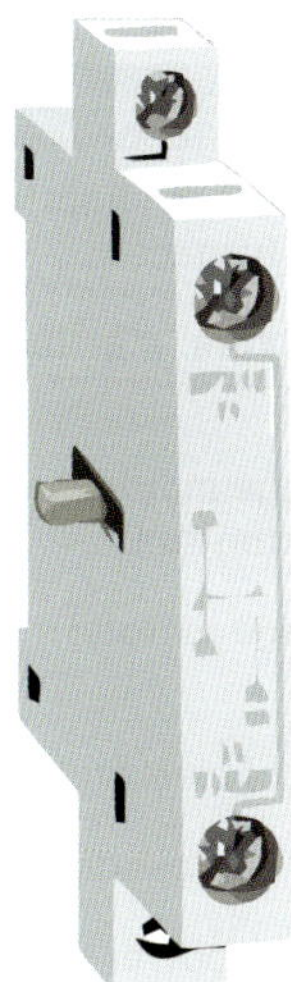

FIGURE 14.39 Auxiliary contact block side adder

A top-adder auxiliary contact block allows two or four auxiliary contacts to be added without increasing the mounting area of the magnetic contactors. The side-mounting auxiliary contact block allows auxiliary contacts to be added to the magnetic contactors without increasing the depth. The auxiliary contacts are also marked as shown in Figure 14.40 with a numerical system that is applied worldwide.

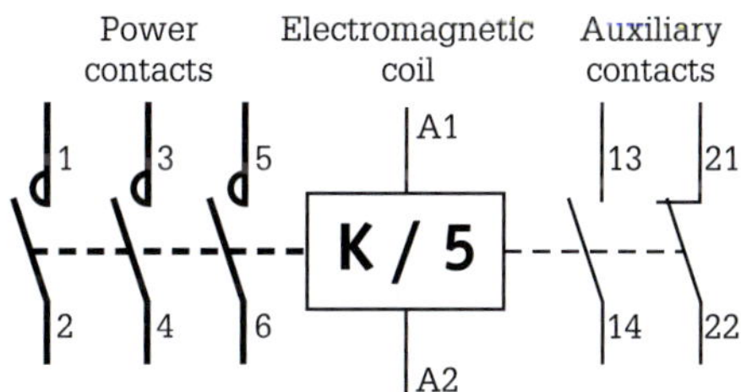

FIGURE 14.40 Contact marking for the coil, power and auxiliary contacts

Coil drive device

This device as illustrated in Figure 14.41 performs an on–off operation of the contactor with input signals from electronic equipment.

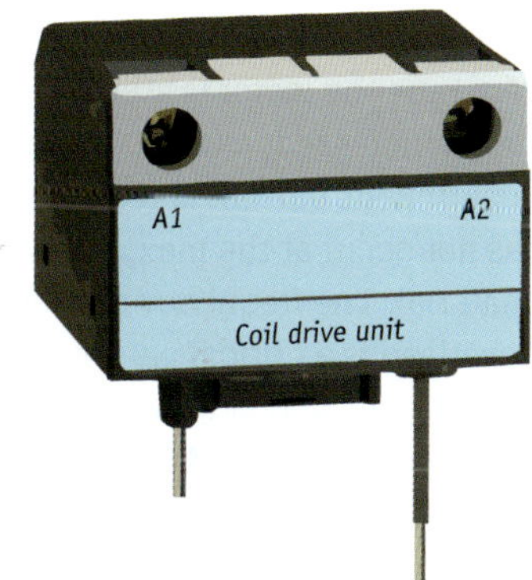

FIGURE 14.41 Coil drive device

The advantages of a coil drive device are:

- protection of main contacts welding and uncharacteristic power intake by voltage drop
- can be used for both d.c. and a.c. voltage
- filtering a.c. voltage interference
- integrated surge suppressor
- energy saving.

Coil surge suppression device

When current through a coil is disturbed, the sudden change of coil current induces a surge voltage due to the coil inductance. This voltage can produce noise that can damage or cause disruption to adjacent electronic devices.

Contactors like relays can be fitted with resistance/capacitance (RC) elements, varistors or Zener-diode assemblies for absorbing the coil surge voltage due to contactor on–off operations. The coil surge suppression unit (CSU) is connected across (in parallel) the contactor coil terminals. Some versions have an LED for 'ON' indication.

RC types suppress rapid increases of surge voltages by lowering the surge voltage oscillation frequency and varistor types amend the peak value of surge voltages. **Note:** The CSU as shown in Figure 14.42 is required for motor starter or contactor applications involving PLC output control.

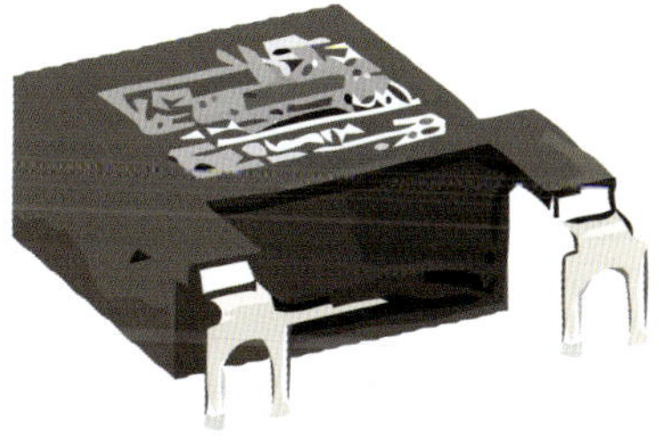

FIGURE 14.42 Coil surge suppression device

Contactor relay

A contactor relay (illustrated in Figure 14.43) is an electromagnetically controlled device that opens and closes electrical contacts to affect the operation of other devices in the same or another circuit. The contactor relay is suitable for switching small control or auxiliary circuit currents at low voltages.

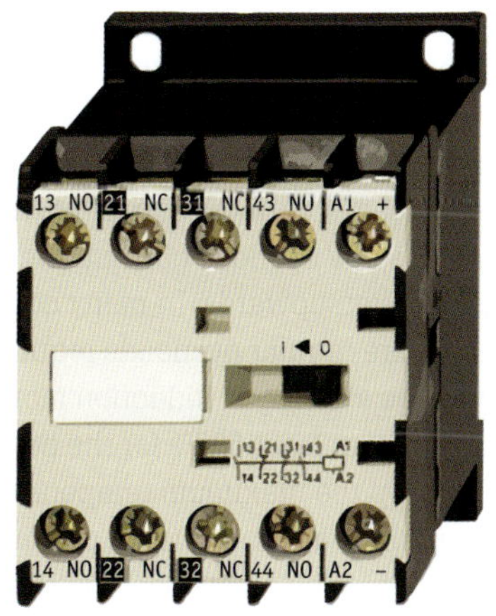

FIGURE 14.43 Contactor relay (a.c.)

Contactors and contactor relays are selected according to the type of coil voltage (a.c. or d.c.), voltage magnitude, their required contact current rating (heating current) when in service, and the number of poles. The load characteristics and their utilisation requirement also have a significant bearing on their selection, as does the service operating environment.

Direct current coils

Instead of using an a.c.-type contactor or contactor relay it is sometimes preferable to use a d.c.-type electromagnet. The power in a d.c. coil is independent of the position of the armature. Direct current coils compensate for this condition; they contain three to four times as much copper as an a.c. coil and more powerful solenoids (an a.c. magnet is laminated steel, a d.c. magnet is made of solid steel) are used. Conventional d.c. contactors are, therefore, able to withstand relatively high holding current without the coil becoming damaged by excess heat. Many d.c. contactors utilise economiser resistors to minimise power wastage in the coil when energised. Also, d.c. contactors or contactor relays have inherently greater mechanical life expectancy than their a.c. counterparts. The most common source of d.c. is rectified a.c.

Utilisation categories

Utilisation categories are important because they help the electrician identify the correct contactor for a particular application. The designation of the utilisation category is made up of three parts:

1 The prefix a.c. or d.c., which indicates the nature of the current.
2 A one-digit number or a two-digit number, which indicates the type of application the contactor is designed for.
3 The suffix ‘a’ or ‘b’, which means whether the contactor is suitable for (a) frequent operation or (b) infrequent operation.

Table 14.1 illustrates a.c. utilisation categories only.

The utilisation categories allow for initial selection of a contactor that can meet the demands of the circuit purpose. As can be seen from the utilisation categories, there are a number of ratings that can be applied to any given contactor. The thermal or heating current is the maximum continuous current that the contactor can carry. The rating of the contacts within the contactor is a thermal rating. As the current increases, the temperature of the contacts increases with the square of the increase in current. If there is any overload current, the temperature of the contacts is driven upwards. In many contactor applications, there are regular overloads required. For example, full-voltage starting of induction motors requires an overload current of at least 600% for the period of the start. In order to keep the maximum temperature of the contacts at acceptable values, the rated current is reduced for applications where such an overload occurs.

TABLE 14.1 Alternating current (a.c.) utilisation categories

Category	Typical applications
AC-1	Non-inductive or slightly inductive loads, resistance furnaces
AC-2	Slipring motors: starting, switching off motors during run
AC-3	Squirrel cage motors: starting, switching off motors during run
AC-4	Squirrel cage motors: starting, plugging, inching
AC-5a	Switching of electric discharge lamp control
AC-5b	Switching of incandescent lamps
AC-6a	Switching of power transformers
AC-6b	Switching of three-phase capacitors. Inductance of leads between capacitors in parallel: min. 6 mH
AC-7a	Slightly inductive loads in household appliances
AC-7b	Motor loads for household applications
AC-8a	Switching of hermetically sealed compressor motors (manual reset)
AC-8b	Switching of hermetically sealed compressor motors (auto reset)
AC-12	Control of resistive loads and solid-state loads with isolation by optocouplers
AC-13	Control of solid-state loads with transformer isolation
AC-14	Control of small electromechanical loads
AC-15	Electromagnets for contactors, valves, solenoid actuators
AC-20	Connecting and disconnecting under no-load condition
AC-21	Switching of resistive loads, including moderate overloads
AC-22	Switching of mixed resistive and inductive loads, including moderate overloads
AC-23	Switching of motor loads or other highly inductive loads

Contactors designed for controlling motors are rated AC-3 and are suitable for regular overloads of 600% of full-load current every time the motor is started. The contacts heat more due to an increase in current during the overload (I^2r) and so the continuous current is reduced to ensure that overheating does not occur at the maximum temperature of the contact material. The frequency of overload and the duration of the overload is a critical part of the AC-3 rating. If there is a very frequent overload, the AC-3 rating should be further reduced. Never use the AC-1 (thermal current) for starting motors. For arduous starting conditions, use AC-4 ratings. To size the contactor properly you need to know the voltage, current, kW load and use the appropriate AC rating.

Contact tips

Contact tips, illustrated in **Figure 14.44**, are stamped or are produced by powder metallurgy infiltration and manufactured by sintering from various combinations of the base material silver (Ag) and other materials. Examples are silver–nickel (Ag–Ni), silver–zinc oxide (Ag–ZnO), silver–cadmium oxide (Ag–CdO), silver–tin oxide (Ag–SnO_2) and silver–tin oxide–copper (Ag–SnO_2–Cu).

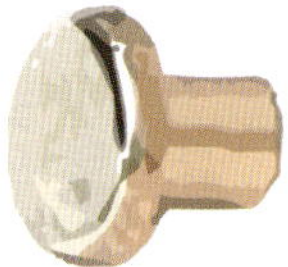

FIGURE 14.44 Silver alloy contact tips

The various combinations of materials for the contact tips result in different contact properties. These features include:

- flame resistance
- high resistance to arc erosion
- high resistance to contact welding
- cold contact surface with the arc extinguished automatically
- low contact resistance
- self-lubricating behaviour
- good electrical thermal conductivity.

Contact tips find application in relays, contactors, switches, controllers and circuit breakers.

Thermal overload

Thermal overload relays prevent an electric motor from drawing too much current and overheating. Thermal overload relays are three-pole bimetallic devices, as shown in **Figure 14.45**. The load current flows through the bimetal poles which are indirectly heated (heater is connected in series with the load current) or directly heated (the bimetal is the path for current). Under the effect of the heating, the bimetallic releases bend and cause the relay to trip. The relay setting range is graduated in amperes. Thermal overload relays provide protection against overload, phase failure imbalance and phase loss.

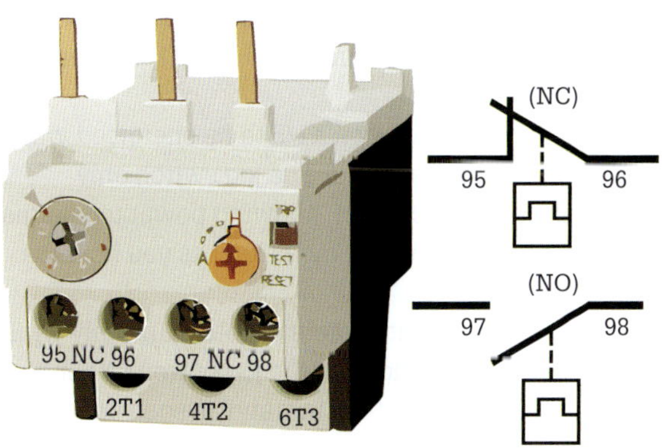

FIGURE 14.45 Thermal overload with NO and NC auxiliary contacts

To make an adjustment of the time, change the distance between the heater and the bimetal releases (indirect) or the distance the bimetal has to bend (direct) by adjusting the timer setting screw. An increase in this distance means that more load current must flow before sufficient heat develops to activate the bimetal releases. Many thermal overload relays have a differential trip mechanism which allows the relay to be phase-loss sensitive. When selecting a thermal overload you need to know:

- the full-load current rating of the motor
- the overload setting range in amperes (fixed or adjustable)
- whether there is or is not single-phase protection
- suitability for two- or three-phase thermal sensing
- whether resetting is automatic or manual
- whether there is direct on contactor (piggy back) or independent mounting
- direct sensing or current transformer (CT) type.

Main circuit suppression device

These devices consist of a capacitor and resistor for a coil and a delta-connected capacitor and resistor for an electric motor circuit. With coils, the operation of inductive circuits causes over-voltages when opening the contactor coil. Electromagnetic energy stored by the electromagnetic field is returned to the circuit in the form of voltage surges (may be several kilovolts) when the coil circuit is opened. When the contactor is used to control a motor, a surge voltage is generated from the motor circuit at 'on–off' conditions. Main circuit suppression devices as illustrated in **Figure 14.46** prevent the misoperation of electronic controllers by suppressing motor surge voltages when connected across the contactor terminals.

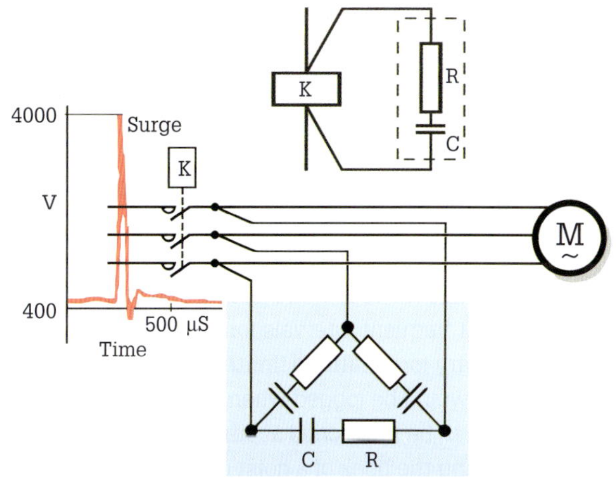

FIGURE 14.46 Main circuit suppression device and circuit diagram

The RC surge suppressor uses a resistor and a capacitor to minimise the surge voltage peak.

The capacitor is used to absorb the voltage spike. The resistor controls the rate of charge for the capacitor when the coil is turned off, and the capacitor limits the coil's inrush current when the coil is turned on. At the instant of switch opening, the RC combination engages and suppresses the energy of the arc by permitting it to bypass the switch.

Operation counter

An operation counter as shown in **Figure 14.47** counts the 'on–off' operation times of a contactor. These devices have application within a preventative maintenance program. By directly counting the operation times, the service life of the contacts can be estimated.

FIGURE 14.47 Operation counter

Mechanical latching relays

Once the contactor is closed, a mechanical latching device holds the contactor in the closed position should the supply voltage fail at the contactor terminals. The latching relay as shown in **Figure 14.48** can be controlled electrically by an a.c. or d.c. impulse or manually by pressing a pushbutton on the latching device.

FIGURE 14.48 Mechanical latching relay

REVIEW QUESTIONS

1. What type of contacts are available on contactors?
2. What is produced by the operating coil of a contactor?
3. When does maximum impedance occur with an a.c. contactor?
4. Why do a.c. contactors have a laminated iron core with a shading ring embedded in the surface?
5. What functions are provided by auxiliary contact blocks fitted to contactors?
6. What function does a coil drive device provide?
7. Identify the considerations for selecting contactors and contactor relay.
8. Why is it sometimes preferable to use a d.c.-type electromagnet?
9. What is a typical application of a contactor having a utilisation category of AC-6a?
10. What is the purpose of an operation counter?

14.5 Jogging and interlocking

Jogging describes the repeated starting and stopping of machinery at frequent intervals for any short periods of time necessary for obtaining limited movement. The electric motor would be jogged when a piece of driven machinery has to be positioned accurately, for example, when positioning the hook of a hoist during a lifting set-up.

A jogging function can make use of a pushbutton changeover switch with a normally closed and a normally open switch configuration. The circuit diagram of a jogging circuit is shown in **Figure 14.49**.

The circuit uses both the K1.1 contact and the changeover switch (break-before-make) to provide the electrical hold-in when used as a regular stop–start control. When the jogging pushbutton is pressed, contact K1.1 is isolated, and the start pushbutton is bypassed. When the jog pushbutton is released coil K1/5 de-energises and contact K1.1 opens.

The jog circuit allows coil K1/5 to be energised when the jog pushbutton is pressed. The jog switch only bypasses the holding contact, allowing power to energise the coil for a brief period (i.e. for as long as you hold the switch down).

Jogging is a capability where a motor runs only while the operator has their finger on the jog button. When the button is pressed, the motor runs, and as soon as the button is released the motor stops. Always avoid excessive use of jogging since it may cause reduced contactor life and motor overheating.

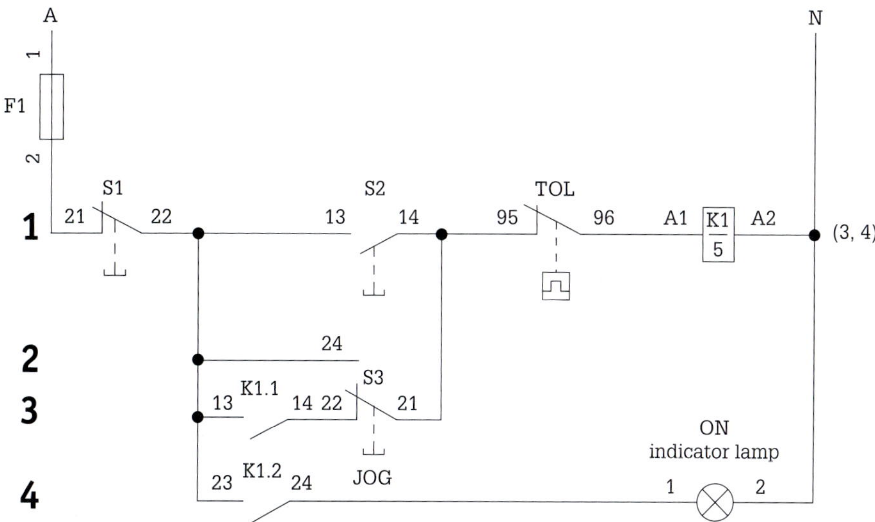

FIGURE 14.49 Jogging circuit using a changeover switch

Relay jogging control

A jogging relay eliminates the problem of the interlock contacts making connection before the NC contact of the make-before-break switch reconnects. In the relay jogging circuit of **Figure 14.50**, it is the control relay coil and not the contactor coil that provides the interlocking contacts.

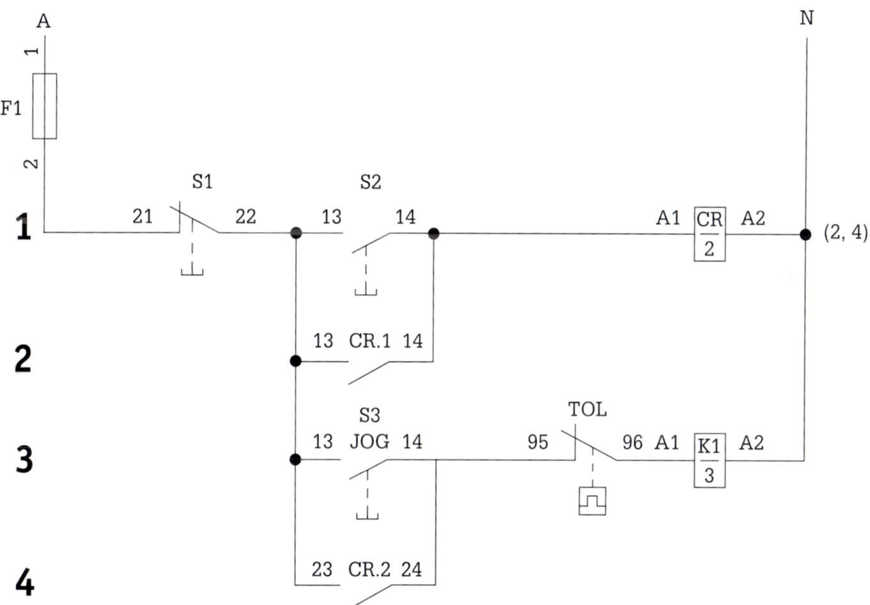

FIGURE 14.50 Jogging relay circuit

The jog pushbutton energises the contactor coil (K1) for the motor, but not the control relay (CR). The start pushbutton (S2) is used to energise the control relay. When energised the CR relay contacts provide connection to the contactor coil.

Two-hand control

For working environments where it is prudent for an operator to account for the position of both hands, some control devices provide for two start switches. This means that the controlled process is hazardous, and its operation requires concurrent pressure from both the operator's hands during a substantial part of the machine process.

Two-hand control is a method of machine safety where both the operator's hands must be located in a designated safe place to operate the machine. In general, it serves to ensure the location of both hands of a machine operator which gives a control signal for a movement that can be dangerous. The areas of application include the die-closing portion of the stroke of metal-shaping presses for metal working, printing and paper processing machines, croppers and guillotines. A two-hand control circuit diagram is illustrated in **Figure 14.51**.

On machines that are mechanically operated, an electric trip solenoid is added to the operating linkage to trip the machine.

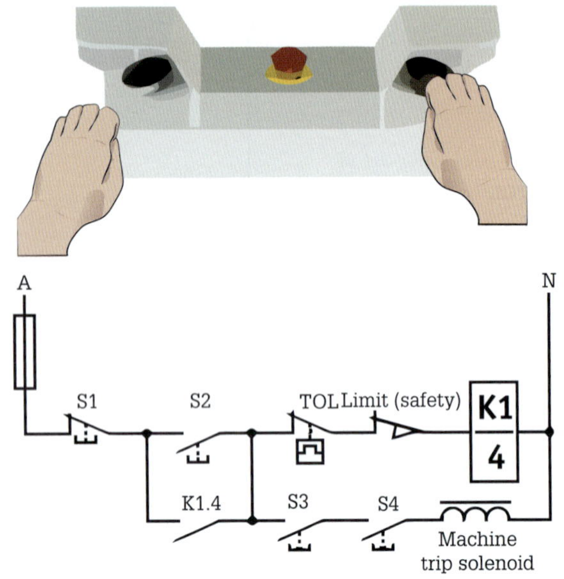

FIGURE 14.51 Two-hand control

Interlock

Electrical interlocking refers to the use of auxiliary contacts of one relay (NC or NO) in series with another relay coil to make certain that this second relay cannot be energised at the same time and vice versa. An example would be a motor-driven vibrating separator. To convert that rotary motion to a single direction of movement, two oppositely rotating vibrator motors are combined with a dual drive. In such a two-motor drive, electrical interlocking must ensure that whenever one motor is de-energised, the other automatically becomes de-energised at the same time. Otherwise, the vibratory force would cease to be unidirectional, which could severely damage the process equipment. Another example would be a reversing motor. Relay coils K1 and K2 are electrically interlocked by the normally closed contacts K1.2 and K2.1, as shown in the forward and reverse motor circuit of **Figure 14.52**.

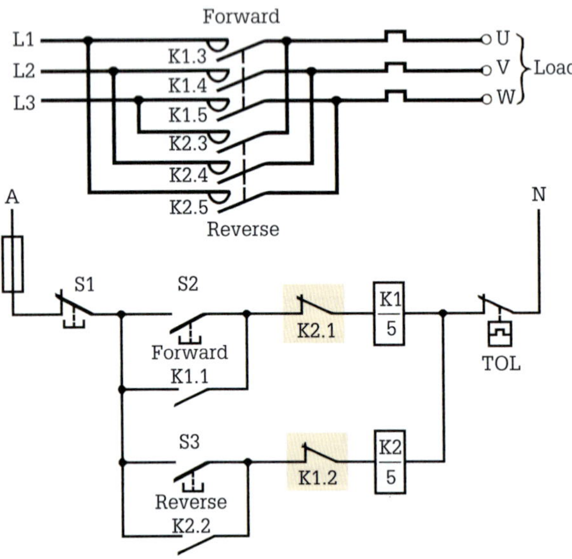

FIGURE 14.52 Electrical interlock circuit

Mechanical interlock

A mechanical interlock is used for side-by-side interlocking of two contactors. The mechanical interlock is either a front or top-mountable or side-mountable contactor accessory device as illustrated in **Figure 14.53** that interlocks two contactors of the same frame size.

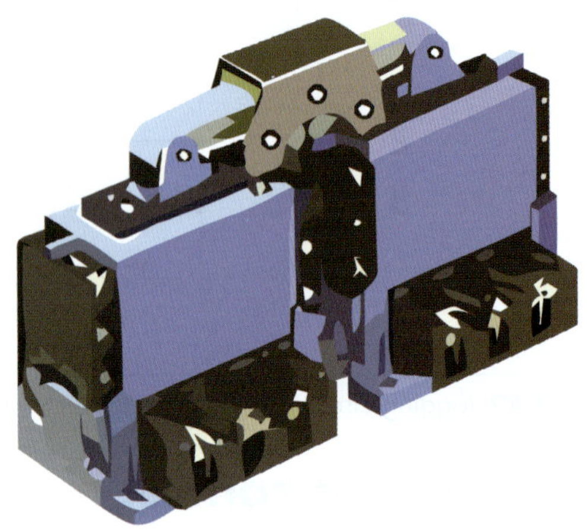

FIGURE 14.53 Mechanical interlock top-mountable

A mechanical interlock device is a lever joining the armatures of two contactors so that they are physically prevented from simultaneous closure that would cause a short-circuit. When fixed between two contactors, the mechanical interlock device stops one of the contactors closing as long as the other contactor is closed. The mechanical interlock is shown by an equilateral triangle symbol linking the two contactors in **Figure 14.54**.

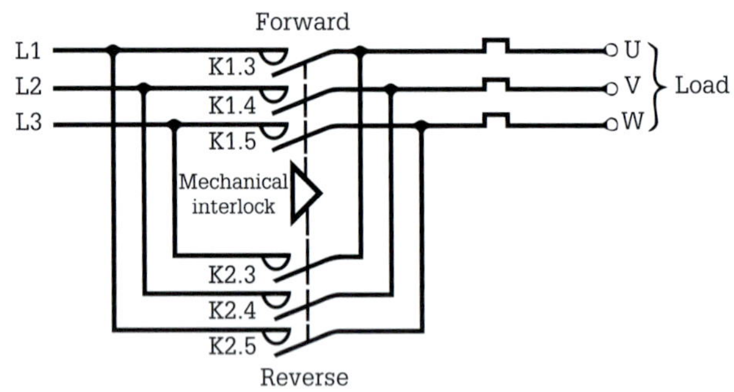

FIGURE 14.54 Mechanical interlock symbol

In **Figure 14.54**, contactor K1 can only energise if contactor K2 is de-energised and vice versa. In circuit diagrams, both contacts would be drawn in the open or non-energised position. The 'on' contactor must be switched 'off' before the other contactor can be switched 'on'. Simultaneous contactor positions are 'on-off', 'off-off' and 'off-on'. Mechanical interlocks are used with reversing contactors, reversing starters, two-speed starters, and star-delta starters.

REVIEW QUESTIONS

1 Describe the process of 'jogging' an electrical machine.
2 Describe with the aid of a diagram how a 'jogging' control function can be implemented.
3 What problem is overcome through the use of a control relay in a 'jogging' circuit?
4 State the purpose of two-hand control.
5 State the function of electrical interlocking.
6 What is a mechanical interlock?

14.6 Control devices

Control devices keep electric motors and other electrical equipment safe and operational.

Pushbuttons

A pushbutton as illustrated in **Figure 14.55** is a spring-loaded manual control device for an electric circuit that changes state when pressed. Pushbuttons are available in two states – normally open (NO) as is the case of the green button of **Figure 14.55** or normally closed (NC) as is the case of the other buttons in **Figure 14.55** or a multiple combination of both. Start and stop pushbuttons can have one state or they may be latching as in the case of a push on–push off or momentary action.

FIGURE 14.55 Pushbuttons

The current rating of a pushbutton must be suitable for the load current that the control device is switching. Pushbuttons usually have a legend plate attached stating the purpose (stop, start) of the control device.

A diagram illustrating the use of legend plates is shown in **Figure 14.56**.

FIGURE 14.56 Legend plate

Indicator lamps

Indicator lamps as illustrated in **Figure 14.57** display the principal state of relays: energised or de-energised. The different lamp lens colours signify different electrical conditions within a control wiring system. For example, red indicates an abnormal state, yellow means caution or attention, green means ready for operation, white indicates that the circuit is live or operating normally while blue represents any function not covered by the other colours.

FIGURE 14.57 Indicator lamps and their symbol

Transducers

Transducers can be used to sense a broad range of different energy forms such as movement, electrical signals, radiant energy, thermal or magnetic energy. There are various types of both analogue and digital input and output transducer devices available. The kind of input or output transducer being used really depends on the kind of signal or process being 'sensed' or 'controlled' but we can define a transducer as a device that converts one physical quantity into another.

Devices which perform an input function are commonly called sensors because they 'sense' a physical change in some characteristic that changes in response to some excitation, for example, heat or force, and convert that into an electrical signal. Devices that perform an output function (converting an electrical signal into a physical output) are called actuators and are used to control some external device, for example, movement.

Both sensors and actuators are collectively known as transducers because they are used to convert energy of one kind into energy of another kind. For example, a microphone (input device) as shown in **Figure 14.58** converts sound waves into electrical signals for the amplifier to amplify, and a loudspeaker (output device) converts the electrical signals back into sound waves.

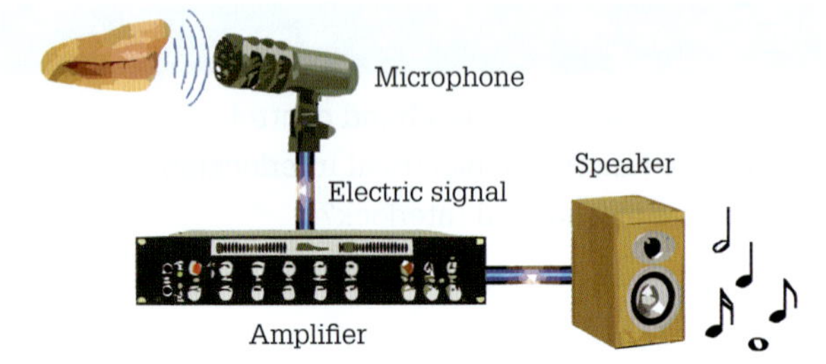

FIGURE 14.58 Transducers

Sound data exist as patterns of air pressure; the microphone sensor changes this information into patterns of electric signals. The amplifier receives the electric signals and boosts them before sending them to a loudspeaker actuator that transforms them into sound waves.

Sensors are used to acquire data concerning:

- pressure in solids, liquids and gases
- temperature
- levels of liquids in tanks
- flow rate in liquids and gases.

Other applications for sensors include humidity, light, mass, position, speed, velocity, acceleration and vibration measurement and control. Sensors, such as microphones, photovoltaic cells, and alternators, have an electric output converted from some other form of energy (sound, light, rotational).

Actuators, such as light bulbs, electric motors and cooktops, have an electric input transduced into some other form of useful energy (light, rotational, heat). All actuators alter electrical energy into some kind of useful work.

Sensors are divided into analogue or digital, depending on whether they provide a continuously varying signal (analogue) or only two possible states (digital). There are many different types of sensors so our discussion is limited to the following types of sensors.

Limit switches

A limit switch, as illustrated in **Figure 14.59**, is the simplest type of sensor. The function of a limit switch is to switch an electrical signal when physical contact occurs between the device to be detected and the operating head of the switch. The limit switch, which is a primary sensor, has either a normally open contact or a normally closed contact. The contacts are switched when the arm is rotated a few degrees.

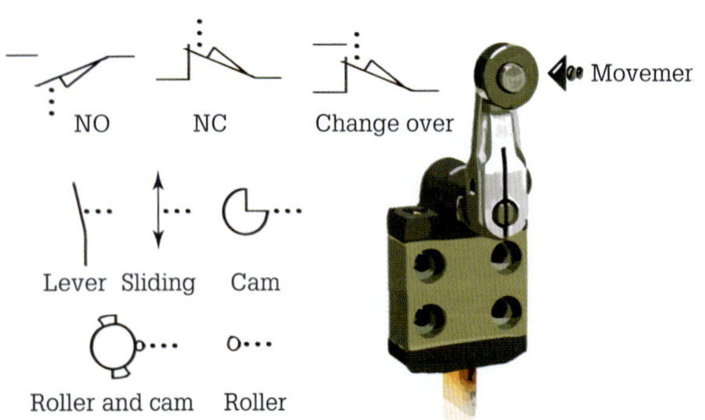

FIGURE 14.59 Limit switch and symbols

A limit switch is simply an electromechanical device that detects the position of an object by making direct physical contact with the object. The direct contact enables the limit switch to open or close a circuit.

Applications for limit switches as presence/absence detection devices include conveyor systems, transfer machines, automatic turret lathes, milling and boring machines and radial drills.

Temperature sensors

There are many sensors that can measure temperature. The choice depends on a number of factors, including the accuracy, the temperature range, access to the heat point, the speed of response, and the working environment.

Temperature switch

The bimetal temperature switch (shown in **Figure 14.60**) finds applications in temperature monitoring/control and signalling in cooling and heating systems, compressors and motors.

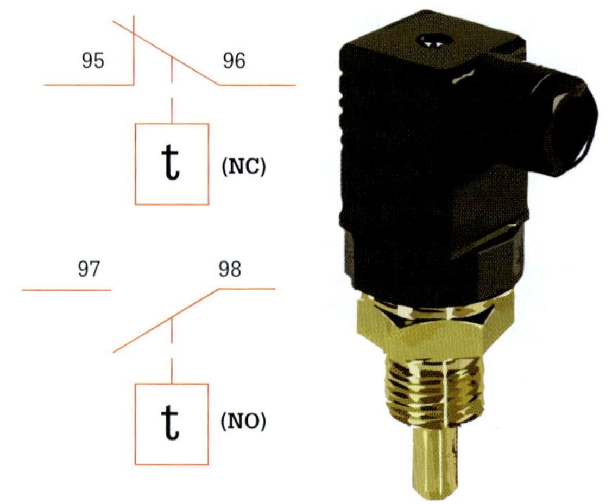

FIGURE 14.60 Temperature switch

If the temperature change occurs within the equipment that the temperature switch is monitoring, then the temperature difference is transferred to a bimetallic release mechanism via the fitting. When the pre-set temperature is reached, the temperature switch is activated.

Thermocouples

Thermocouples are made by connecting two conductors of different types of metal and work on the Seebeck principle. This principle explains how two dissimilar metals generate an emf that is a function of the temperature of the junction when connected in an external closed circuit. Thermocouples can be used to measure temperatures of up to 2320 °C.

A 2320 °C thermocouple would consist of tungsten–rhenium metals combined in different percentages for both the positive and negative elements. Such a combination would develop an emf between 0 and 37 mV over its temperature range. Usually, a reference junction is used in addition to the measurement junction. A thermocouple is illustrated in **Figure 14.61**.

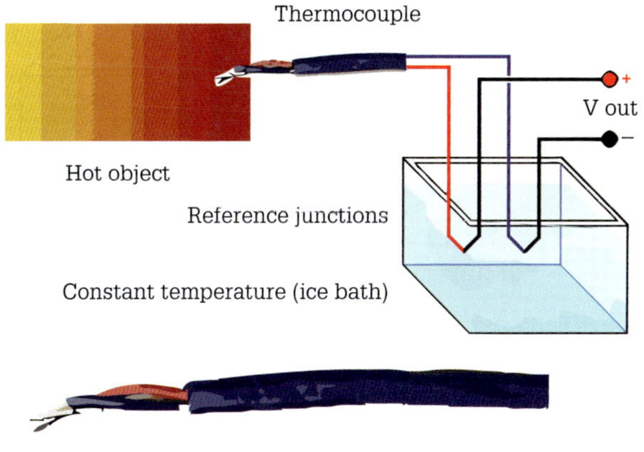

FIGURE 14.61 Thermocouple

When accurate thermocouple measurements are required, a practice that can be used is to reference both thermocouple leads to the output conductors at the ice point so that these conductors may be connected to the emf meter. This procedure avoids the generation of a thermal electromotive force (emf) at the terminals of the meter causing a measurement error. The emf generated is dependent on the difference in temperature, so in order to make an analysis the reference must be known. Knowledge of the reference can be accomplished by placing the reference junction in an ice water bath at a constant 0 °C. Applications for thermocouples include temperature measurement for engine control systems and industrial process control systems.

Thermistors

There are two types of thermistors: negative temperature coefficient (NTC) and positive temperature coefficient (PTC). Thermistors and their R–T characteristics are shown in **Figure 14.62**.

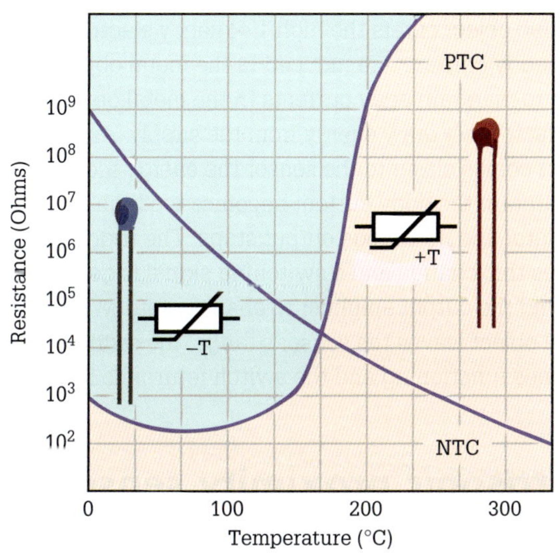

FIGURE 14.62 Thermistors and their characteristic curve including symbols

The shape of the thermistor probe can take the form of a bead, disc, rod or washer. Applications include temperature measurement, control and as a differential thermometer.

Pressure switch

Pressure switches provide electrical contact outputs to electromechanical or electronic devices signalling the presence or absence of pressure. Pressure switches are control devices that regulate the pressure of a fluid or gas. The pressure force is converted into a switchable state.

Apart from the usual monitoring and limitation of pressure, their task can also include keeping a particular pressure constant, to commence and end various control and regulating processes or transmit a signal (pressure transducer).

Pressure switches have two adjustable switching states: NC and NO (single pole or three pole). Pressure closes or opens the NC and NO electrical contacts when the monitored pressure reaches a predetermined level. These switching states can be used either as a switching signal for a control circuit (contactor control, programmable logic controller (PLC), alarm) or for direct switching of a load such as a pump, compressor or a motor starter.

Pressure switches as illustrated in **Figure 14.63** are used for automatic switching of motors on pumps and compressors. They are used to monitor the pressure conditions of liquids or gaseous mediums in pipelines, containers and tanks. They also have application for controlling the pressure in process control, lubrication, cooling, refrigeration, pneumatics and hydraulic systems of various machines.

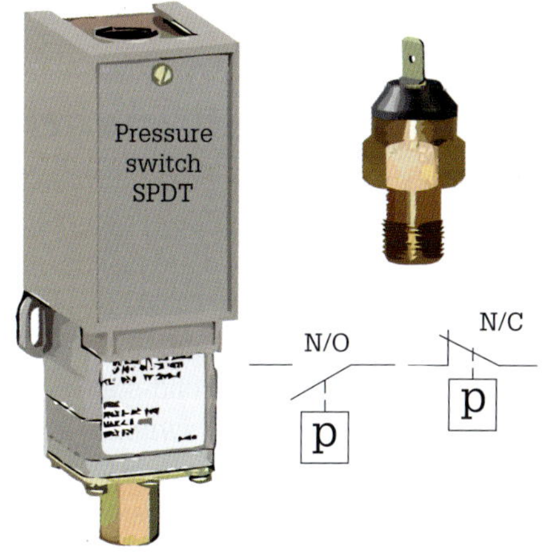

FIGURE 14.63 Pressure switches and symbols

Proximity switches

A proximity switch is the name given to a variety of sensors that produce an electrical signal without any mechanical interaction (physical contact) when a physical substance passes through their sensor field. Proximity switches can use a magnetic field, capacitive, inductive and ultrasonic techniques with which to detect a physical substance.

Magnetic field proximity switches

Magnetic field proximity switches consist of two elements: the sensor (can be a reed switch or movable contacts attached to a ferromagnetic material) and the actuator (magnet attached to some moving device). These switches are used to detect a magnetic field. The switch is contained in a fixed unit, and the magnet is mounted on a moving object to be detected.

Micro dry-reed switches are hermetically sealed in an inert gas-filled glass envelope and can consist of single-pole, single-throw (SPST) switch elements having normally open contacts. With **Figure 14.64**, when the magnet attached to the actuator moves close to the switch, the reeds become magnetised and the normally open contacts will close, completing the circuit.

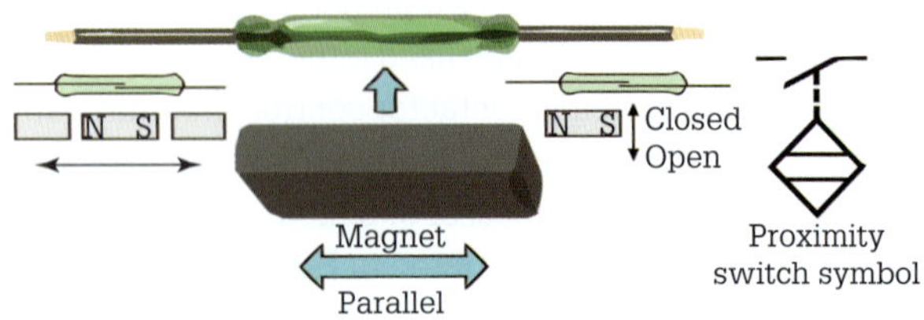

FIGURE 14.64 NO reed switch and the general symbol for a sensor

When the magnet moves away from the switch, the contacts open and disconnect the circuit. Some applications of magnetic proximity switches are machine tool position sensing, security door opening sensing and elevator position sensing.

Capacitive proximity switches

Capacitive proximity switches can be used when the object to be detected is either metallic or non-metallic. The most common materials sensed are plastic, glass, wood, paper, liquids and all metals. Capacitive proximity switches consist of a metal plate, a high-frequency oscillator, a detection circuit and a solid-state switch. The operating principle of this proximity switch is similar to that of a simple capacitor.

When the switch is energised via an external power supply the solid-state switch is ready to operate. Once energised, the oscillator detects the amount of capacitance (charge stored in an electrostatic field) between an exterior target and the metal plate in the sensor. As the target approaches, the capacitance increases. When the capacitance exceeds a pre-set threshold value a detection circuit is activated which sends a signal to the solid-state switch to change state. A capacitive proximity switch used for liquid level detection is shown in **Figure 14.65**.

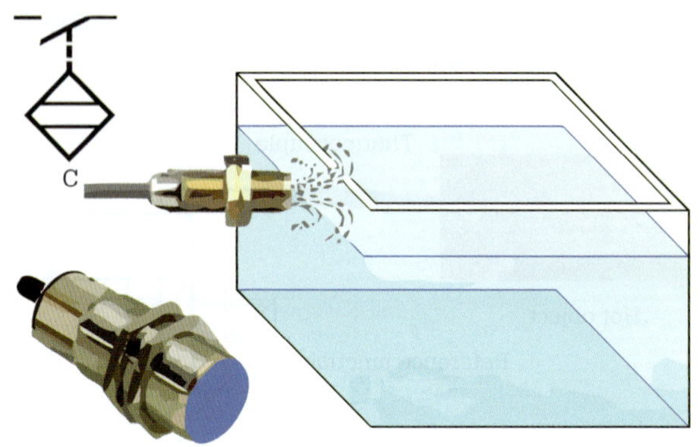

FIGURE 14.65 Capacitive proximity switch used for level detection

Inductive proximity sensors

An inductive proximity sensor as shown in **Figure 14.66** consists of four main sections: sensor coil, oscillator, trigger and amplifier. These devices are used to detect metal objects and are also used for mould-level detection.

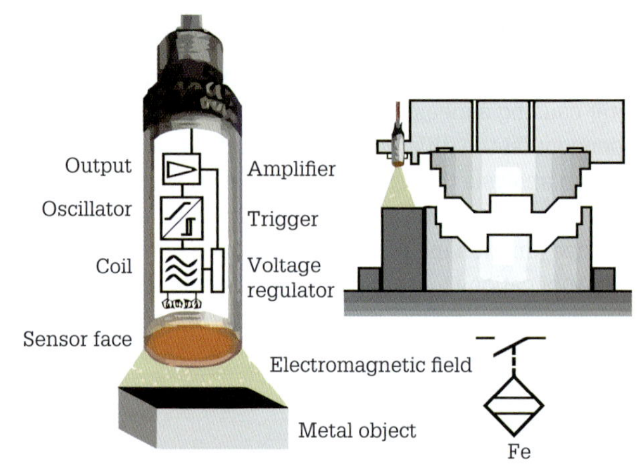

FIGURE 14.66 NO inductive proximity sensor

The oscillator creates an alternating high-frequency field that radiates from the coil in the sensor face. When a metallic object enters the high-frequency electromagnetic field, eddy currents are induced in the metal object. In order to maintain eddy currents in the metal object, the oscillator must draw energy from the supply. As the metal object comes closer to the sensor the energy required to maintain the eddy current becomes too high for the oscillator and its voltage output stops. The trigger circuit senses this change and a switching signal is generated causing the output amplifier to change state. When the metal object leaves the sensor's range, the oscillator resumes functioning and the switch returns to its normal state.

Ultrasonic proximity sensors

Ultrasonic proximity sensors as shown in **Figure 14.67** use a piezoelectric transducer to send and receive high-frequency sound waves.

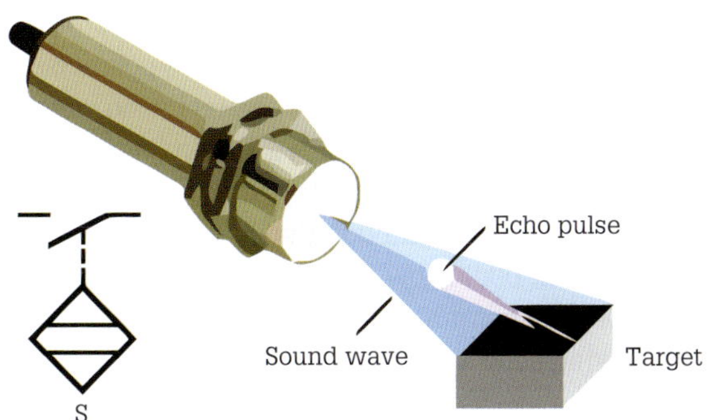

FIGURE 14.67 NO ultrasonic proximity sensor and symbol

When a target enters the wave, an echo pulse is reflected back to the transducer. The time it takes for the reflected pulse to return is evaluated by the transducer and when the target enters the sensor's pre-set operating range, the sensor switch changes states.

Photoelectric switches

Photoelectric switches contain the following components: a luminous radiation source called the emitter, a receiver unit to detect the emitted light, and an amplification circuit which converts, evaluates, amplifies and processes the detected light causing the switch to change states.

These switches use a light source generated from semiconductors such as LEDs that produce radiation within a narrow frequency band (called a modulated pulse) of the electromagnetic spectrum. The use of a narrow light frequency band limits the effect of ambient light on the receiver. The light sources utilised by photoelectric switches range from visible green to infrared. When in operation the radiated beam that is emitted is either reflected or broken with respect to the object to be sensed. A photoelectric switch with emitter and reflector is illustrated in **Figure 14.68**.

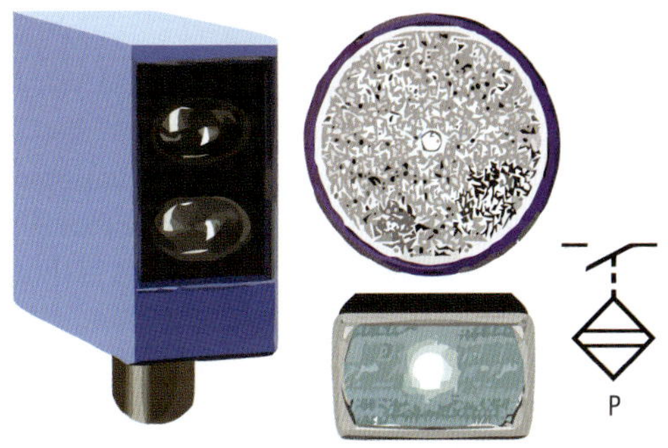

FIGURE 14.68 NO photoelectric switch and symbol

In **Figure 14.68**, the emitter and receiver are in the same device. Radiation from the emitter is transmitted in a straight line to the reflector and returns to the receiver as polarised light. The polarised light is the only light that the receiver responds to. When an object breaks the light path, the sensor switch changes its state. When an object moves away from the light beam the switch returns to its normal state. The operation of a photoelectric switch counting bolts on a conveyer belt is shown in **Figure 14.69**.

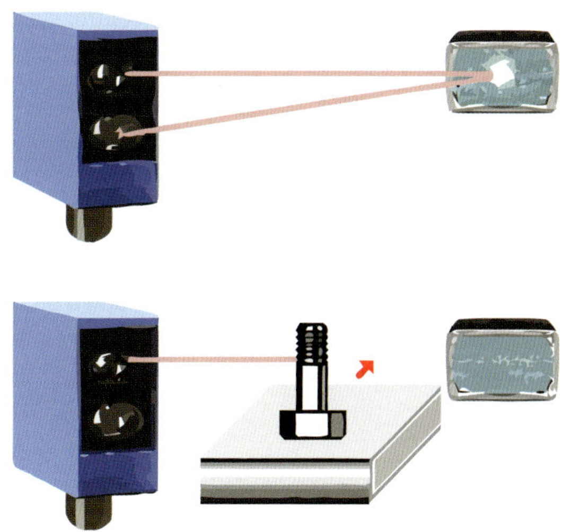

FIGURE 14.69 Operation of a photoelectric switch

Other photoelectric switches include fibre-optic sensors and lasers. Applications for these devices include positioning, speed detection and counting.

Tachogenerator

Devices that convert angular speed to some other form of reading or usable signal are known as tachometers. Some tachometer devices are direct contact units that are pressed against a rotating shaft to measure the rpm on a dial. Non-contact devices use strobe lights or infrared technology to sense the speed of rotating shafts. When the signal produced is an emf, the device is known as a tachogenerator. The tachogenerator is a precision d.c. generator, and the output voltage signal is a precise function of the rotational speed at which the tachogenerator is driven. Tachogenerator components are illustrated in **Figure 14.70**.

FIGURE 14.70 Tachogenerator armature and permanent magnetic field

Tachogenerators are frequently used to measure and control the rotational speeds of variable speed drives used in paper and steel mills, machine tools and process plants.

A tachogenerator gives a polarised output voltage: positive voltage in one direction and negative in the reverse direction, so the feedback control system has to be able to cope with either polarity if the controller is to reverse the motor. Information gained by a tachogenerator is sent to the operator or for use in a signal feedback control system such as a d.c. power drive or an a.c. inverter powering the drive motor.

Float switches

A float switch as illustrated in **Figure 14.71** is simply an electrical switching circuit controlled by the tilting movement as the float switch moves up and down on the surface of the liquid. The movement is detected by a built-in switch and triggers the switching action.

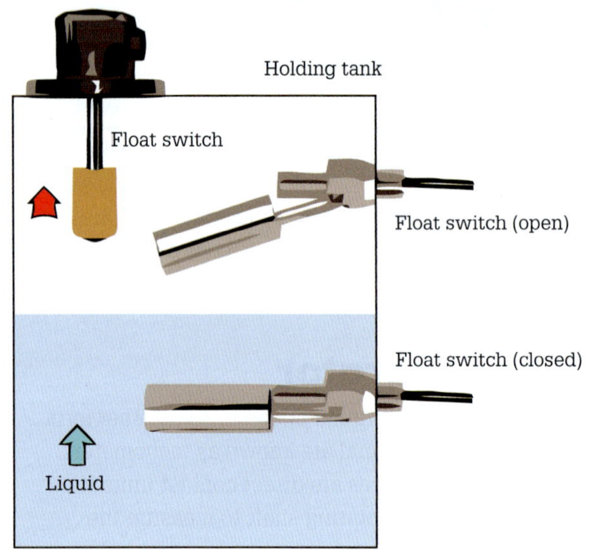

FIGURE 14.71 Float switch

The switching action can be used to activate alarms and controllers or complete a circuit that is used to control pumps in liquid systems, and levels in ponds, sumps and holding tanks.

Arc suppression devices

When an electromechanical relay is de-energised rapidly the collapsing magnetic field produces a voltage transient (high voltage spike) in the wires of the control circuit in its effort to disperse the stored energy and oppose the sudden change of current flow. It is common practice to suppress the voltage transient with other components that limit the transient voltage to a much smaller level. The suppression device may be in parallel with the relay coil or in parallel with the contacts used to control the relay. It is frequently preferred to have the suppression parallel to the coil since it can be located closer to the relay. When the suppression is in parallel with the relay coil, any of the following devices may be used:

1. Protection rectifier diode
2. Metal-oxide varistors
3. RC discrete devices
4. Suppressor diodes.

A protection rectifier diode as shown in **Figure 14.72** is suitable for d.c. coil voltage and includes a polarity safeguard. Conduction of the diode occurs in one direction only when the coil is switched off. At this moment, current tries to continue flowing through the coil, and it is harmlessly diverted through the diode. Without the diode, no current could flow, and the coil would produce an inductive voltage transient in its attempt to keep the current flowing across the contacts.

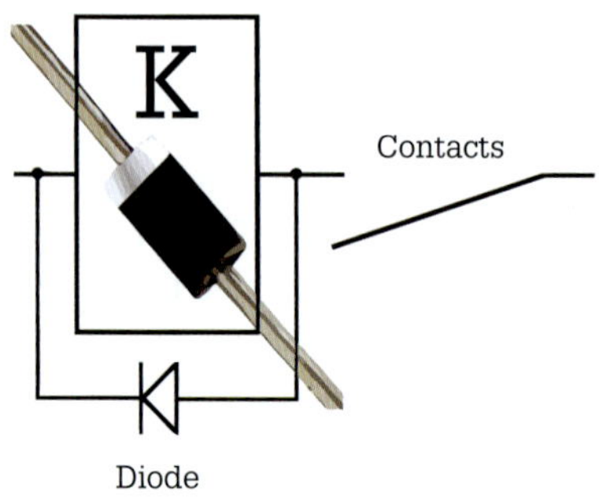

FIGURE 14.72 Protection diode

A metal-oxide varistor (voltage-sensitive resistor) as shown in **Figure 14.73** is suitable for both a.c. and d.c. coil voltage. Varistors have an advantage over diodes and transient suppressor diodes in that they can absorb much higher transient energies.

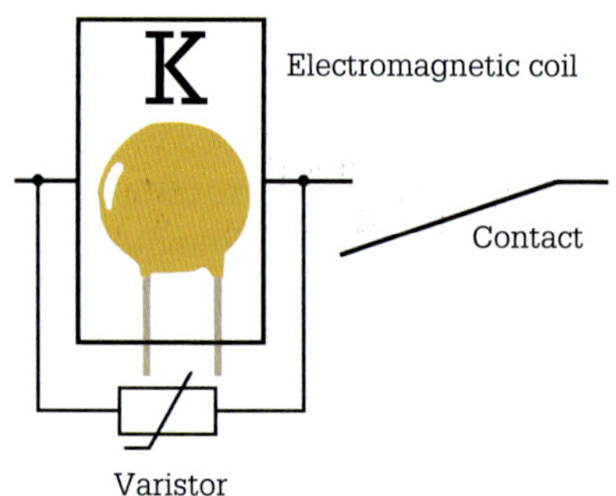

FIGURE 14.73 Varistor protection

When the voltage transient happens, the varistor changes its resistance from a very high value to a very small conducting value. The voltage transient is thus absorbed and clamped to a safe level, protecting circuit components.

RC discrete devices (connected to a snubber circuit) as shown in **Figure 14.74** apply to both a.c. and d.c. coil voltages, with minimal opening delay. The disadvantage is that the capacitor can lead to high currents across the contacts. An RC snubber limits fast load voltage changes (e.g. caused by inductive loads) at the load terminals. The capacitor is used to absorb the high transient voltage. The resistor controls the rate at which the capacitor charges when the coil is turned off and to limit the coil's inrush current when the coil is energised.

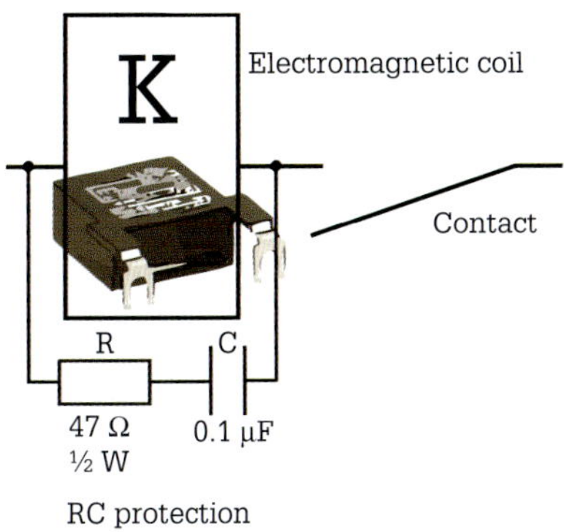

FIGURE 14.74 RC protection

Suppressor diodes as shown in **Figure 14.75** apply to both a.c. and d.c. coil voltages, with minimal dropout delay. A suppressor diode is a reverse-biased rectifier diode in series with a Zener diode such that their anodes (or cathodes) are common, and the rectifier diode prevents normal current flow. The diode surge suppressor clamps the transient voltage close to the coil's nominal or rated voltage. The energy caused by the voltage transient is dissipated by the coil itself.

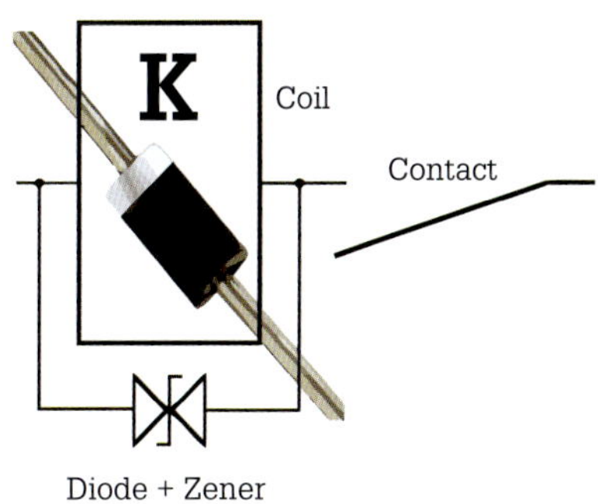

FIGURE 14.75 Suppressor diode protection

REVIEW QUESTIONS

1. What is a pushbutton?
2. Describe the purpose of a pushbutton legend plate.
3. What name is given to devices that perform an input function?
4. List three common actuators.
5. State the function of a limit switch.
6. Name the device that produces an electrical signal when a physical substance passes through their sensor field.
7. Name three types of temperature sensors.
8. List the components in a photoelectric switch.
9. What kind of device converts angular speed to some other form of usable signal?
10. What is a float switch?

14.7 Programmable relays

Programmable relays combine timers, relays, counters, special functions, inputs and outputs into one easily programmed compact device.

Microprocessor control systems

A microprocessor is an integrated system of electronic circuits and devices infused on a single chip. The integrated system is designed so that the microprocessor can perform logic and arithmetic operations. It needs memory such as RAMs and ROMs and input/output integrated circuits in order to work. An example of a microprocessor mounted on a circuit board is shown in **Figure 14.76**.

The microprocessor chip is a small piece of silicon that has been engineered, by a process akin to contact printing, to manipulate electrical signals. Microprocessors usually contain five essential principal components as part of their architecture (physical arrangement) as portrayed in **Figure 14.77**.

Source: Shutterstock.com/Poravute Siriphiroon

FIGURE 14.76 A microprocessor

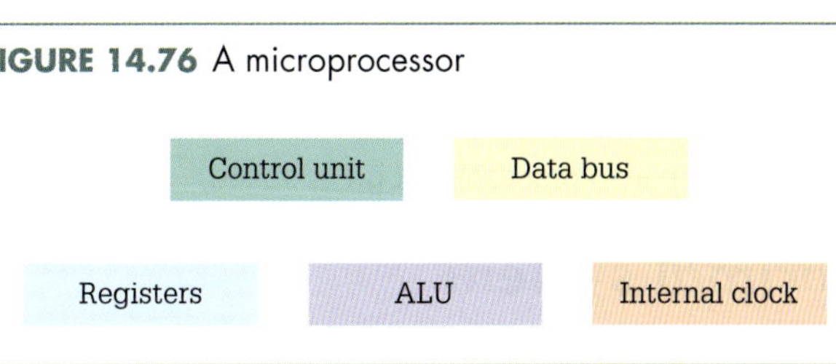

FIGURE 14.77 Microprocessor key components

One important element is the control unit consisting of many transistors connected in Boolean configurations whose purpose is to interpret data and execute programs stored in the registers (number holding areas). Another central element is the arithmetic and logic unit (ALU), made of transistors arranged in the form of Boolean logic gates, which executes mathematic (like addition, subtraction, multiplication and division) and logic functions (comparison of two numbers).

A third element, called the register, is a bank of internal storage data cells (small amounts of memory) which the ALU accesses when required. The clock is also a vital element as it directs and synchronises when the various processor operations (fetch and execute cycles) may be carried out.

The clock pulsates or cycles at a fixed rate and determines the speed (gigahertz, GHz) at which the microprocessor runs. Each gigahertz is equal to a million clock cycles per second. The last central element is the data bus or circuitry, which links all the various sections and components of the microprocessor chip.

Arithmetic operations

The mathematical operations that the microprocessor performs are in essence the repeated addition or the sum of two numbers. Microprocessors can, by adding two numbers, perform addition, multiplication, subtraction and division.

EXAMPLE 14.1

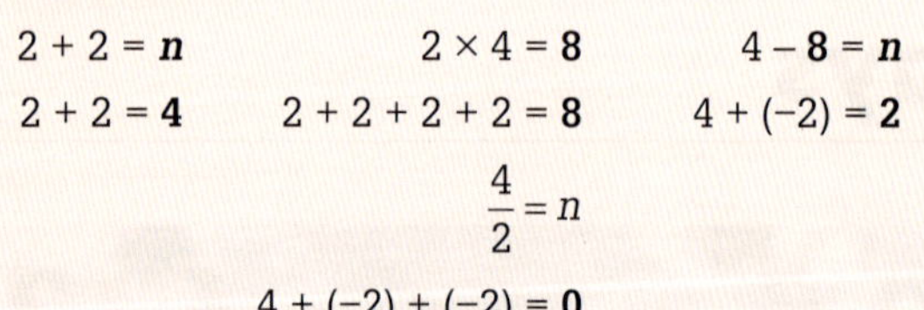

$2 + 2 = \mathbf{n}$ $\quad$ $2 \times 4 = \mathbf{8}$ $\quad$ $4 - \mathbf{8} = \mathbf{n}$

$2 + 2 = \mathbf{4}$ $\quad$ $2 + 2 + 2 + 2 = \mathbf{8}$ $\quad$ $4 + (-2) = \mathbf{2}$

$$\frac{4}{2} = n$$

$$4 + (-2) + (-2) = \mathbf{0}$$

Count the number of −2s that must be added before zero is reached: Answer = **2** (*n* is an unspecified number)

As the speed at which the microprocessor runs is very fast, complex mathematical operations can be processed in minuscule time frames.

Logic operations

The logic operation that a microprocessor performs is the contrast between two numbers to see if one number is the same as, less than or greater than the other number. For example, microprocessor logic can determine by sensors if the temperature of a heated room is the same as the required set temperature. If not, the heater turns on. Logic operations are also processed at great speed.

Decimal system

The decimal numbering system is used in most commonplace calculations. This method can be called base 10 counting which means that each digit can have 10 different states (0–9). With the decimal numbering system we count 0, 1, 2, 3, 4, 5, 6, 7, 8, 9 in the first digit location and instead of going to zero we place a 1 in the second digit position and begin counting again (10, 11, 12, etc.). After 9, we start a third digit position and start counting again (100, 101 and so on.).

Binary system

Using programmable relays requires us to become accustomed to other number systems besides decimal.

Binary is the numbering system computers, programmable relays, and PLCs use. The binary system uses the number 2 as the base. The only allowable digits are 0 and 1. With digital circuits, it is easy to distinguish between two voltage levels (e.g. 15 V and 0 V), which can be related to the binary digits 1 and 0. Therefore, the binary system can be applied quite easily to programmable relays and computer systems. Since the binary system uses only two digits, each position of a binary number can go through only two changes, and then a 1 is carried to the immediate left position.

Data from the real world have to be converted into a binary code. Binary code (base 2 numbering system) represents the microprocessor transistors' switching condition. A zero (0) represents an open switch, while a one (1) represents a closed switch. In binary, a single digit number is 1 or 0; there is no in-between state.

Electronically these states can be represented as 'on' or 'off' and 'voltage' or 'no voltage'. The binary number system works like the decimal number system except the number system uses base 2.

$$2^7\ 2^6\ 2^5\ 2^4\ 2^3\ 2^2\ 2^1\ 2^0$$

Each digit (0 or 1) of a binary number is called a bit. Eight bits is called one byte, 16 bits is called one binary word, and 32 bits is four bytes.

A group of 8 bits forms a pattern known as a byte. A byte can represent a single character. For example, in 7-bit ASCII (American Standard Code for Information Interchange) binary code the text 'on' is:

on = 0110 1111 0110 1110

The text 'off' in the same binary code has three bytes, an 'o' and two 'f's and is:

off = 0110 1111 0110 0110 0110 0110

ASCII is a binary code standard used to represent alphanumeric and control characters in 128 possible bit combinations (7 bit = 2^7 = 128 possibilities for encoding). Each bit combination can be represented by a 7-digit binary number: 0000000 to 1111111. The ASCII 7-bit code covering a portion of the 128-bit combinations available is shown in **Table 14.2 (a)**.

TABLE 14.2 (A) 128-bit combinations

Decimal	Binary	Value
048	0011 0000	0
049	0011 0001	1

»

050	0011 0010	2
051	0011 0011	3
052	0011 0100	4
053	0011 0101	5
054	0011 0110	6
055	0011 0111	7
056	0011 1000	8
057	0011 1001	9
097	0110 0001	a
098	0110 0010	b
099	0110 0011	c
100	0110 0100	d
101	0110 0101	e
102	0110 0110	f
103	0110 0111	g
104	0110 1000	h
105	0110 1001	i
106	0110 1010	j
107	0110 1011	k
108	0110 1100	l
109	0110 1101	m
110	0110 1110	n
111	0110 1111	o
112	0111 0000	p
113	0111 0001	q
114	0111 0010	r
115	0111 0011	s
116	0111 0100	t
117	0111 0101	u
118	0111 0110	v
119	0111 0111	w
120	0111 1000	x
121	0111 1001	y
122	0111 1010	z

Binary numbers and arithmetic allow the representation of any amount by using just two digits: 0 and 1. Each '1' and '0' in a binary number represents a particular power of two. See **Table 14.2 (b)**.

TABLE 14.2 (B) Binary as an exponent

0001 is 2 to the zero power or 1
0010 is 2 to the 1st power or 2
0100 is 2 to the 2nd power or 4
1000 is 2 to the 3rd power or 8

When you have a number like 11111111, you can work out what it means by adding the powers of 2 ($2^0 = 1$) as shown in the following:

2^7	2^6	2^5	2^4	2^3	2^2	2^1	2^0
1	1	1	1	1	1	1	1
128 +	64 +	32 +	16 +	8 +	4 +	2 +	1 = 255

Similarly, we can convert 10110010 to decimal:

1	0	1	1	0	0	1	0
128 +	0 +	32 +	16 +	0 +	0 +	2 +	0 = 178

To obtain binary numbers from a decimal number divide the number by 2. If there is no remainder the result is a zero, except for the last division. If there is a remainder, the result is a one. Read the outcome from the bottom to the top. For example, the decimal number to convert is 178_{10}:

178/2 = 89 with the remainder of 0

89/2 = 44 remainder of 1, etc.

2	178	0 Least significant bit
2	89	1
2	44	0
2	22	0
2	11	1
2	5	1
2	2	0
2	1	1 Most significant bit

$\mathbf{178_{10} = 10110010_2}$

Bus

Buses are groups of wires such as a twisted pair that carry data and control signals from one device to another. Most microprocessor systems contain several buses such as data, address and control. The bus that carries data to and from the microprocessor is called the data bus. The more information that can be sent at the same time, the more data can be transmitted in a specified interval and therefore the faster (and wider) the bus.

A data bus can be 4, 8, 16, 32 or 64 bits in width. A wider data bus allows for more data to be moved in or out of the microprocessor in one cycle. The size of the data bus is, therefore, an indication of the microprocessor's data-moving capability.

The address bus carries the address information used to describe the memory location to which the data is being sent or from which the data is being retrieved. As with the data bus, a wider bus allows a greater number of address locations to be accessed. The width of the address bus indicates the maximum amount of memory (RAM) that a microprocessor chip can address. A simple block diagram of the bus system is shown in **Figure 14.78**.

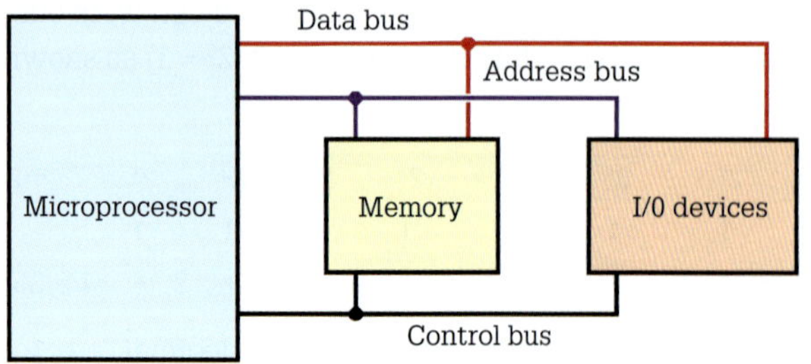

FIGURE 14.78 Bus block diagram (I/O = input/output)

SWITCH ON

Advantages of microprocessor systems

Microprocessors are used increasingly in many devices such as motorised vehicles, automated teller machines, and telephones to handle control tasks and data-processing tasks that would formerly have been carried out by analogue control circuits.

A microprocessor is programmable, so that alterations or new application provisions may be easily accomplished without a total redesign of the device or process it is controlling. Microprocessors are becoming very popular in the industry for various reasons. They help to minimise operator errors; they significantly increase the machine or process speed they are governing, and they enhance the overall energy efficiency of an electrical installation.

Systems software

Systems software is the set of instructions that govern the microprocessor operating system (e.g. C-Lotion or C-Gate server). Its purpose is to convert input data from peripheral devices (such as light switches) into data that the processor understands. It also decodes commands generated by a human interface into those that are interpreted by the microprocessor.

Energy management systems

There are several systems available, for example, Dynalite, Icontrol and C-Bus. These systems offer complete control of any load, either digital or analogue in design.

C-Bus (Clipsal Bus) is a fully programmable microprocessor-controlled system for industrial, commercial and domestic installations. Communication data is sent from touch screens, light level sensors, critical inputs and occupancy sensors via twisted pair cable to load controllers, which operate relays and dimmers. C-Bus technology is used to control all types of lighting loads, electronic dimmable fluorescent ballasts and other electrical loads such as electric motors, security and audio-visual devices, ventilation and heating. Each device within a C-Bus system has its microprocessor which provides for two-way communication between all system devices. Data is held within each microprocessor rather than at a central control point. C-Bus uses unshielded twisted-pair (UTP) category 5 data cable to carry the communication data (class D link = 100 Mb/s). Twisted-pair cables (8 conductors – 4 pairs), as shown in **Figure 14.79**, are so named because the uniquely colour-coded pairs of wires are twisted around one another. Each pair consists of two insulated copper wires twisted together. The wire pairs are twisted because it helps reduce the effect of electromagnetic induction (noise and crosstalk) and capacitive coupling between each wire.

FIGURE 14.79 UTP category 5 sheathed cable and a modular 8 × 8 plug

Therefore, each connection to twisted-pair cables requires both wires. A 36 V 300 mA d.c. supply provides the source current to the C-Bus twisted-pair network. With respect to the lighting system, C-Bus provides lighting switching flexibility when compared with conventional lighting systems. C-Bus replaces all mains switch wires with UTP cable. Microprocessors are an integral part of a C-Bus key input light switch. Their purpose is to send command signals to a load controller which in turn activates only the lights that are needed. Once set up C-Bus can monitor all lighting requirements for the installation. Lights can be adjusted to any desired level and can be programmed to turn on only when a person enters a room. A block diagram of this process is illustrated in **Figure 14.80**.

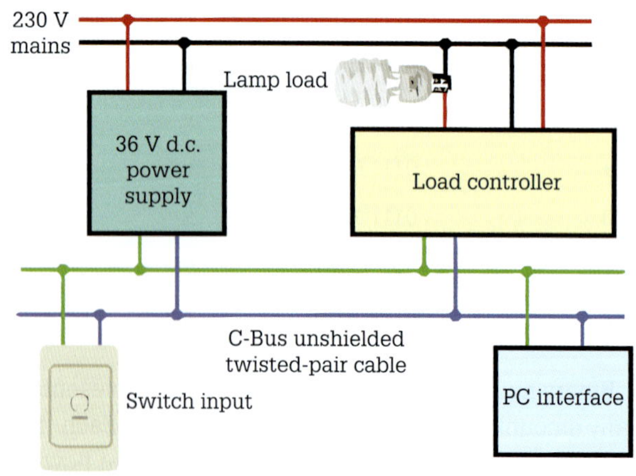

FIGURE 14.80 C-Bus wiring

Programmable logic controller (PLC)

A programmable logic controller (PLC) is a microprocessor-based device that accepts input data from various devices and processes that data by making decisions according to an installed program. The controller then generates output signals to devices that perform a specific function.

Typical input devices are pushbuttons, proximity sensors, liquid level sensors, photoelectric sensors, selector switches and pressure transducers. Typical output devices are contactors, magnetic motor starters, solenoids and indicator lamps.

PLCs, with their various memory locations acting as relays, and their programmable control device, allow a command sequence to be varied without major rewiring of an industrial system. Many PLCs using the relay programming method (control, input and output) called ladder logic language as their encoded system are activated by external sensory devices.

There are three main sections of a PLC:

1 Input interface
2 Central processing unit
3 Output interface.

The microprocessor unit receives communication input from the interface of entry, which then translates the data according to its programming into a signal that is fed into an output interface. A block diagram illustrating the communication patterns of a PLC is shown in **Figure 14.81**.

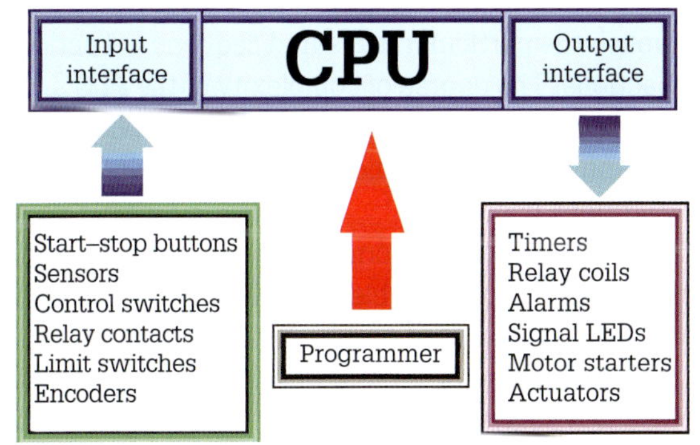

FIGURE 14.81 A block diagram for a PLC

A PLC with its input interface or rack, and central processing unit is shown in **Figure 14.82**.

FIGURE 14.82 A programmable controller

A PLC uses a programming language based upon recognisable symbols common to electric motor control. Hand-held PCs are commonly used for programming the PLC. A PLC performs three tasks: checking of the input status, execution of the program and updating of the output state.

First, the PLC checks the input status by scanning each input to determine if a connected device is on or off and then records that information in memory. Second, the PLC executes the operator's program one rung at a time to make decisions. For example, assume the operator's program instructs the PLC to turn on an output device if input number 1 is on, and then turn on another output device if input number 2 is off. The PLC analyses these conditions, executes the appropriate action and then stores that information in memory.

Last, the PLC updates the output status. Updating means it will send data to an output device, such as the coil of a magnetic motor starter, to enable some process to begin. The time taken for the PLC to go through this cycle is called the scan time.

Input interface

The input interface can be in modular form, which means that if more input devices are required it is a simple matter of just plugging in another module beside the existing one. When a process or a machine is controlled by a PLC, it uses inputs from various devices in order to make decisions and updates outputs to drive various pieces of equipment The input interface receives different digital or analogue communication signals from each of the input devices. The signal to an input module is in either digital or analogue form. If the input is an on/off signal (pushbutton, limit switch, photo cells) the signal is considered to be digital. If the input varies, as with temperature, pressure, humidity or level sensors, the signal received is analogue (range of values between on and off) in form. The interface input provides termination, isolation and indication for an input signal's state and it conditions these signals to a digital form readable by the microprocessor. An input interface module is shown in **Figure 14.83**.

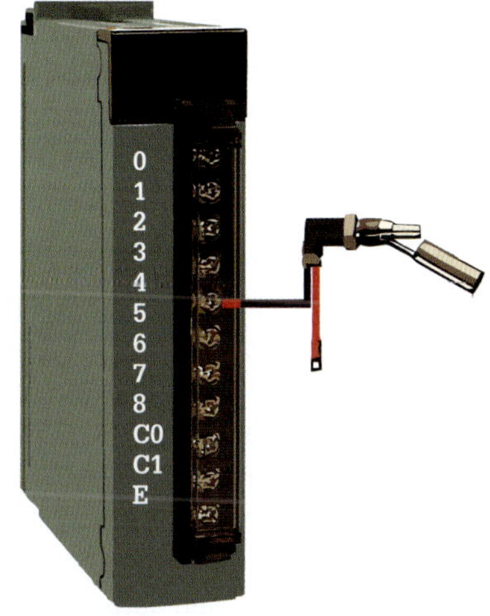

FIGURE 14.83 An input interface module with a float valve as the input device

Every input device has its set of terminals and circuitry thereby ensuring electrical isolation and signal integrity for each input signal. It also ensures that every 'field input' device has its unique address.

Central processing unit

The central processing unit (CPU) includes a microprocessor, RAM, and ROM memory. This section of the PLC scans, arrays, ensures, authorises and directs all other parts of the system. The CPU drives the program carrying out one instruction at a time in the order entered by the programmer. The CPU performs three essential functions:

1. Reads the state of all input signals from the interface of entry.
2. Executes the program that has been installed in its memory.
3. Sends updated information to the output interface.

The three functions have a repetitive nature ensuring that the CPU continuously reads all inputs in order to deliver the correct process or machine operation before it carries out the next programmed instruction. CPUs vary in design, and each have their array of components and features. A CPU module is shown in **Figure 14.84**.

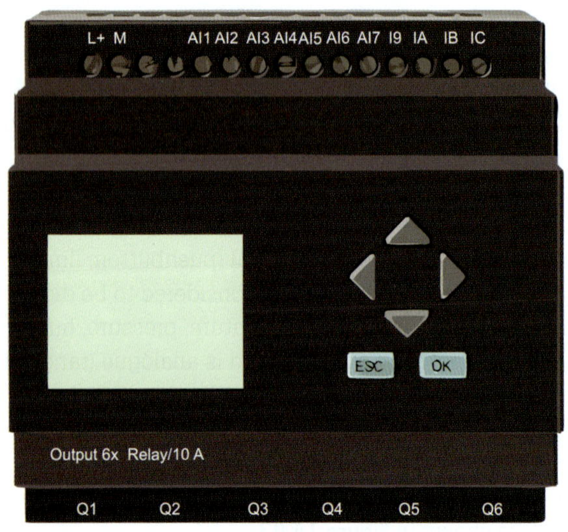

FIGURE 14.84 A CPU module

When a logic event has been programmed into the CPU then the event can be started, paused and stopped by the CPU itself.

Output interface

The output interface module is similar to the input module. This interface converts the digital communication of the CPU into usable electric signals to drive the output devices to new states or modes of operation. The output devices can be solid-state switches or relays. Some output interfaces also indicate the status of the connected output devices on an LED display. PLCs have a limited number of connections built in for input and output signals. Expansion units are available if the base model does not have sufficient I/O (input and output) interface.

'In-the-present-time' device

Programmable logic controllers are an 'in-the-present-time' device. They are capable of making decisions and initialising actions at the speed required by changing conditions, sometimes in nanoseconds. They are used for welding, mixing, weighing, sorting, painting and many other types of applications.

The PLC can be used to sort coins moving on a conveyor system. The coins are fed down an inclined chute where a size-measuring sensor views the coins as they pass. Using the light extinction principle, three sets of photodiodes see each coin from three directions. The coin size is determined by the average of the signals from the three photodiodes. By measuring the area of each coin, the sensors decide whether it is a five cent, 10 cent or one dollar coin and deliver an input signal to the CPU. Based on this input signal, the CPU instructs an output device to open and close several grids to direct the coins to their correct location.

Ladder logic diagrams

All industrial and machine processes follow a determined sequence of actions and reactions that must occur during a typical operation. A program consists of instructions that accomplish the particular measured sequence of actions and reactions. The degree of complexity of the PLC's program depends upon the difficulty of the application, the number and type of input and output devices and the types of instructions used.

This sequence can be formulated into a ladder logic diagram. Most programmable controllers are designed to receive input instructions using ladder logic diagrams. Ladder logic incorporates programming functions that are graphically displayed to resemble symbols used in hard-wired control diagrams.

A ladder diagram is a specific kind of schematic diagram that shows in simple terms how a circuit works. These diagrams use two vertical parallel lines (rails) to represent the primary power bus. Devices such as switches (inputs) and coils (outputs) are represented by their schematic symbol on horizontal logic lines called rungs drawn between the parallel bus lines. Each rung of the ladder diagram corresponds to a set of instructions (logic line) that tells the CPU what to do in response to the status of each device connected on the rung. Ladder diagrams are read from left to right and from top to bottom.

All 'field input' devices such as stop buttons, limit switches and thermal overload must be programmed as 'examine on' (NO when referring to circuit diagrams) while other devices follow logic. Programming is vital because the ladder logic diagram is a visual representation of a program – a set of logic instructions that the CPU must interpret and execute no matter how illogical they may be.

The symbol is the 'examine off' instruction and the logic diagram in **Figure 14.85** would be interpreted as: 'Activate the output at address *O*1 when input address *I*1 is on and input address *I*2 is off'. This means push the stop button (to the off position) while pushing the start button, which is of course a nonsense. However, this is a machine following instructions; it is not a physical wired circuit.

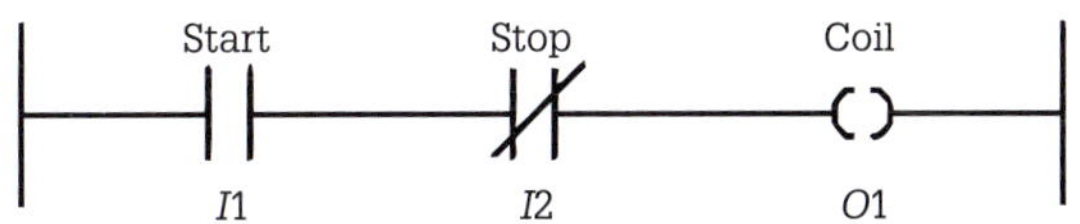

FIGURE 14.85 A PLC ladder logic diagram

Programmed correctly, as in **Figure 14.86**, the instructions will be interpreted as: 'Activate the output address *O*1 when input address *I*1 is on and input address *I*2 is on'. Now the stop button is always an NC device, so that when the CPU scans the input rack it finds *I*1 open (off), input *I*2 closed (on) and therefore does not activate *O*1. Pressing the start button turns *I*1 on, and the CPU now finds *I*1 on and therefore activates *O*1.

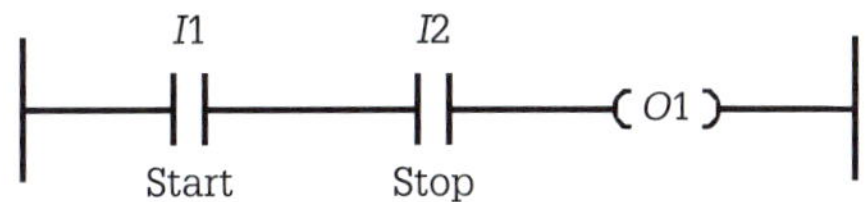

FIGURE 14.86 Correctly programmed instructions

'Internal' or 'virtual' contacts that have been assigned to an output address forming part of the program are programmed as 'examine off' or 'examine on' as required by the logic.

PLC application

A three-phase pump motor is only able to operate when certain conditions are met. A pushbutton must be energised to turn the system on and another pushbutton activated in order to operate the pump motor. A pump system is shown in **Figure 14.87**.

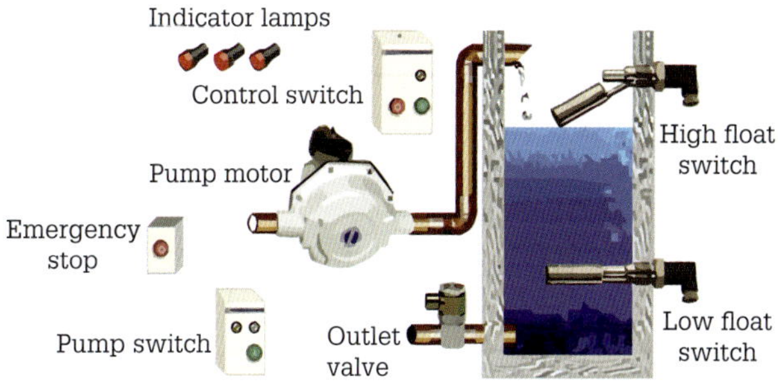

FIGURE 14.87 A pump system

A control circuit for the pump is designed to keep the water level in a tank between two defined levels. The tank fill cycle is operated by the low-level float switch and continues until the high-level float switch is activated. Once the high-level float switch is activated an outlet valve is energised allowing a measured amount of water to be released. The pump system in **Figure 14.88** indicates that there are six field input devices of various types and five physical output devices (motor coil, solenoid valve and three external lamp indicators). It shows them connected to the appropriate I/O racks with an assigned generic address for each of them: *I*1 to *I*5 as inputs and *O*1 to *O*5 as outputs (the actual address would depend on the PLC involved).

The ladder program is as follows. When the PLC is on, as in **Figure 14.89**, and set to 'run', the CPU scans the input and output racks and sets the status of each as shown, that is, inputs *I*2, *I*3 and *I*4 are all 'on', and because output *O*2 is 'off' the 'examine off' contacts assigned to it, on the third rung, are also in the 'on' state (the CPU examines *address*

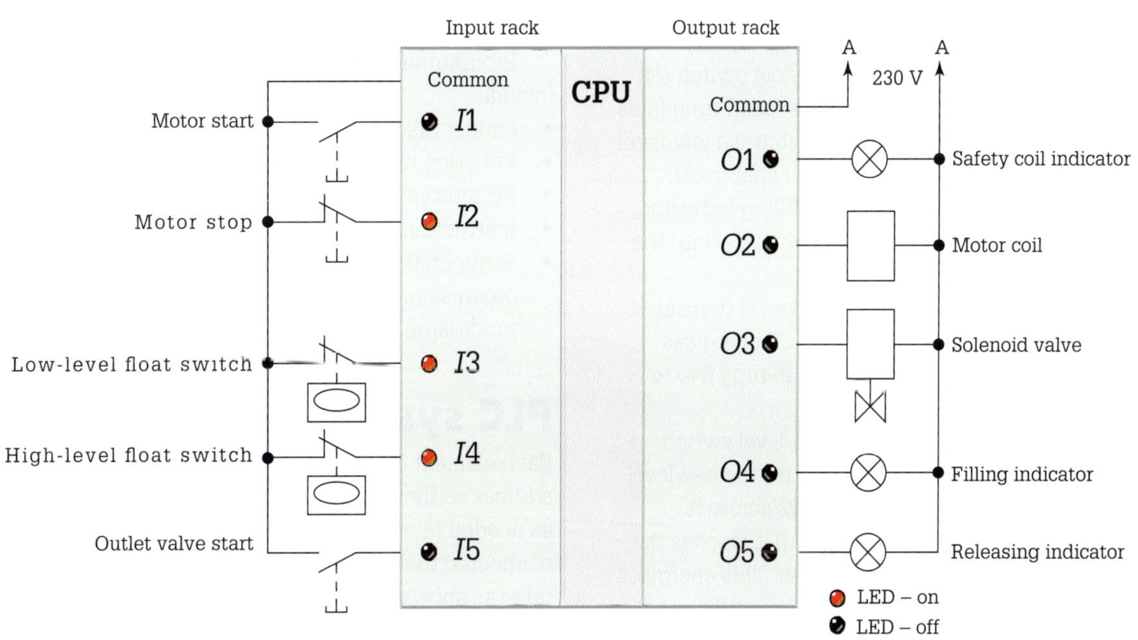

FIGURE 14.88 CPU and input and output racks

*O*2 and finding it off then assigns a *logic* 1 to the *examine off instruction* in the third rung).

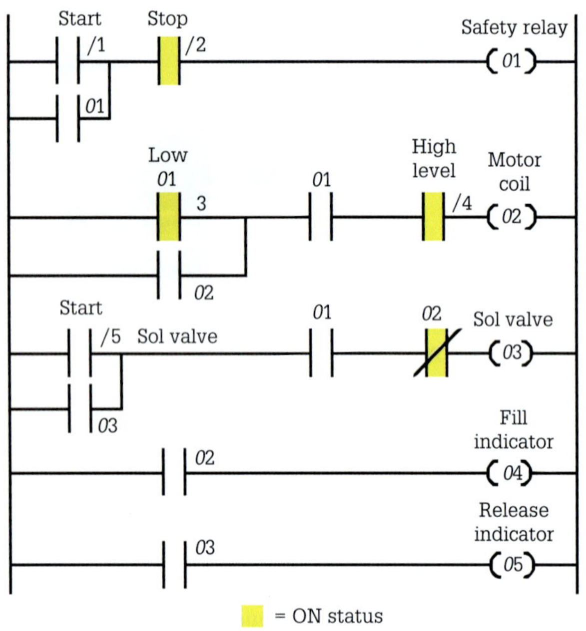

FIGURE 14.89 A PLC ladder diagram for the pump system

Pressing the motor start button makes the first rung true and the CPU energises output *O*1. *O*1 latches itself (*O*1 across *I*1) and closes *O*1 in the second rung, making this true also, and closes *O*1 in rung three (interlocking contacts to prevent the solenoid valve being activated before the whole system is started).

Output *O*2 energises the motor coil (and latches *I*4), and the motor begins to pump in water. Because *O*2 is on, its contacts in the third rung open to disable that rung, preventing *O*3 being energised so that the solenoid valve is held off and another set in rung four closed to bring on *O*4 and the filling indicator.

As the water rises to the low-level float switch (*I*3), it opens, but this has no effect as it is latched (by *O*2). The water continues to rise to the high-level float switch (*I*4) causing it to switch off. This makes the second rung false and *O*2 is de-energised, its contacts unlatch the low-level float switch (*I*3), close in the third rung to enable *O*3, and open in rung four to extinguish the filling indicator. The tank is now full and the process now waits until the solenoid valve start (*I*5) is pressed.

Pressing *I*5 allows the CPU to energise *O*3 (because contact *O*2 has re-closed as discussed). *O*3 energises the solenoid valve, latches *I*5 and closes in rung five to energise the releasing indicator.

The water level drops below the high-level switch (*I*4), which closes but has no effect because the low-level switch (*I*3) is still open as are contacts *O*2 across it.

The water level continues to drop until it reaches the low-level switch (*I*3), which now re-closes. This energises *O*2 (the pump coil) via *O*1 contacts (that are still closed) and *I*4 (high level).

The energising of *O*2 latches the low-level switch (*I*3) and disables rung three thus de-energising *O*3 (the solenoid valve). At the same time *O*2 contacts close in rung four to energise the filling indicator and *O*3 unlatches *I*5 and also de-energises *O*5 (release indicator). The pump refills the tank automatically, then becomes static and awaits further instruction.

Note: There is a set of 'examine on' contacts from output *O*1 in rung three. The purpose of this is to avoid a safety hazard – that is, pushing the solenoid valve start button (*I*5) while the tank is empty. This would allow *O*3 to activate and while nothing happens (the tank is empty) it does present a hazard that can only be cleared (*O*3 turned off) by pressing the pump start (*I*1). Contacts *O*1 prevent this.

In practice, the start–stop buttons are hard wired into a standard circuit external to the PLC and only the contacts of the safety relay that they control appear as part of the PLC program.

Entry of the ladder diagram via a notebook computer

The following is a very brief explanation of data entry. Before entering a program into the PLC ensure that the screen and PLC memory are clear. Once this occurs component symbols can be entered. Each component of a rung is introduced separately. The element, its name, and any relevant data are keyed into the data entry. When the write key is pressed, the symbol is added to the diagram at the current cursor position. When the output (coil) is entered the cursor automatically moves to the last status of the rung and inserts the output symbol. There can only be one output per rung. The rungs must also be filled from the top of the screen with no rungs missed. Any unused rungs must be left at the bottom of the screen where they do not take up storage in the PLC memory. The cursor is moved from rung to rung, and each required component added. Finally, the completed screen is sent to the PLC memory.

Programming languages that are used with PLCs include:

- Ladder diagram (LD), graphical
- Function block diagram (FBD), graphical
- Structured text (ST), textual
- Instruction list (IL), textual
- Sequential function chart (SFC), has elements to organise programs for sequential and parallel control processing.

PLC system addressing

Each element of data input or output has to have a unique address so that the processor can locate the information as needed in order to control the process or the machine connected to it. The addresses are contained in the data table as shown in **Figure 14.90** that is stored in the memory of the processor. I/O addresses describe the physical location that the input or output device is connected to.

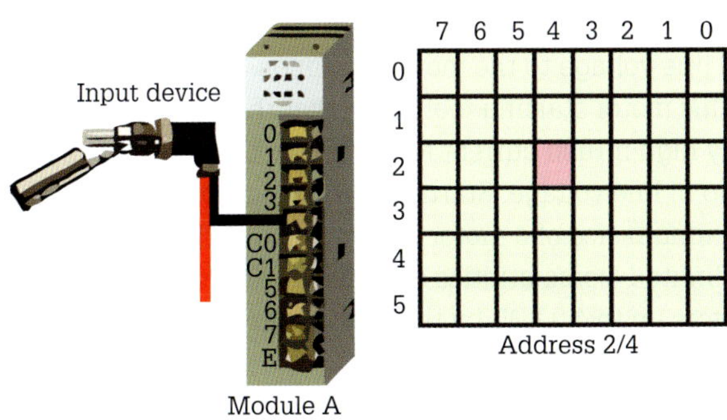

FIGURE 14.90 A data table

Referring to our pump system of **Figure 14.89**, the various components are assigned a particular address using the protocol relevant to the PLC to be used. All the pertinent information is in **Table 14.3**.

TABLE 14.3 PLC table for Figure 14.89

Field inputs	Address	Internal devices	Address	Field outputs	Address
Start button	*I*1	Safety coil	*O*1	Pump contactor	*O*2
Stop button	*I*2			Solenoid valve	*O*3
Low-level float	*I*3			Fill indicator	*O*4
High-level float	*I*4			Release indicator	*O*5
Valve start	*I*5				

REVIEW QUESTIONS

1 What is a microprocessor?
2 How many essential components are a part of a microprocessor's architecture?
3 What is a register?
4 Which numbering system is used by a computer's programmable relays and PLCs?
5 What is meant by 'systems software'?
6 What is a PLC?
7 Name the main sections of a PLC.
8 Name typical PLC input devices.
9 Explain the purpose of the CPU.
10 State the three essential functions that a CPU performs in association with a PLC.
11 What is the purpose of the 'rung' in a ladder diagram?
12 Why is it important that each element of data input or output has a unique address?

14.8 Three-phase induction motor starters

When an induction motor stator winding is energised, a rotating magnetic field is created which rotates at synchronous speed.

$$N_s = \frac{120f}{p}$$

where: N_s = synchronous speed in rpm
f = rotor frequency in Hz
p = number of stator poles

EXAMPLE 14.2

Calculate the synchronous speed of a two-pole induction motor operating at a supply frequency of 50 Hz.

Solution:

$$N_s = \frac{120f}{p} = \frac{120 \times 50}{2} = 3000 \text{ rpm}$$

EXERCISE 14.2

Calculate the synchronous speed of a four-pole induction motor operating at a supply frequency of 50 Hz.

The expanding rotating magnetic field passes through the air gap between the stator and the rotor and cuts the stationary rotor conductors. Owing to the synchronous speed between the rotating magnetic flux and the stationary rotor an electromotive force (emf) is induced in the rotor conductors. Because the rotor conductors are short-circuited, currents start flowing in these conductors. These currents create a rotor magnetic field which interacts with the stator magnetic field. Subsequently a mechanical force acts on each of the rotor conductors. The summation of the mechanical forces on the rotor conductors produces angular movement of the rotor called torque that tends to move a rotor in the same direction as the stator magnetic field. Maximum torque is necessary because it is needed to overcome the stationary inertia of the rotor and connected load it is trying to move.

The fact that the rotor moves in the direction of the rotating stator magnetic field can be explained by Lenz's law. This law states that the direction of the rotor currents are such that they tend to oppose the cause producing them. The cause producing the rotor currents is the relative speed between the rotating stator magnetic field and the stationary rotor conductors. Consequently, the rotor starts turning in the same direction as that of the stator field and attempts to catch it.

Starting a motor using reduced voltage causes the motor current to be reduced in direct proportion to the line voltage reduction. The torque, however, is reduced by the square of the voltage reduction. For example, when you start a motor at 50% of the line voltage the starting current would be 50% of the full line voltage value (0.5 × 600% = 300% full load current). The torque would then be 0.5^2 or 25% of the normal starting torque (0.25 × 150% = 37.5% full load torque). Therefore, reduced voltage starting would be a good method to reduce both inrush current and starting torque.

Motor starter circuits

Motor starter circuit diagrams show the connections of both the power and the control circuits as clearly as possible with all conductors drawn neatly as straight lines. The actual layout of the components is usually quite different from the circuit diagram.

There are two drawing conventions used for setting out motor starter circuit diagrams. They are:

1 horizontal layout
2 vertical layout.

The motor starter circuit diagrams are:

- simple direct-on-line (DOL) starter circuit
- contactor-controlled DOL starter
- star–delta starter
- autotransformer starter
- primary resistance starter
- secondary resistance starter.

Simple direct-on-line (DOL) starter circuit

The components of a DOL starter consist of only a manually operated disconnector as shown in **Figure 14.91**.

To operate the DOL, the disconnector is closed, applying full line voltage to the motor windings. The disadvantage of this motor starting method is that the motor draws a very high inrush current for a very short time causing a drop in line voltage. Starting torque is also very high. The maximum size of a motor allowed on a DOL starter may be limited (some utilities only allow DOL for three-phase motors up to 4.0 kW) by the supply utility because of the drop in line voltage.

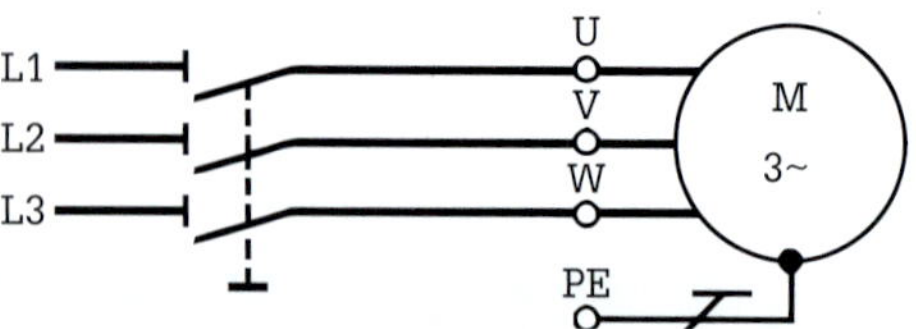

FIGURE 14.91 Manually operated DOL starter

Contactor-controlled DOL starter

A contactor-controlled DOL starter with overload protection circuit diagram drawn in a vertical orientation is illustrated in **Figure 14.92**.

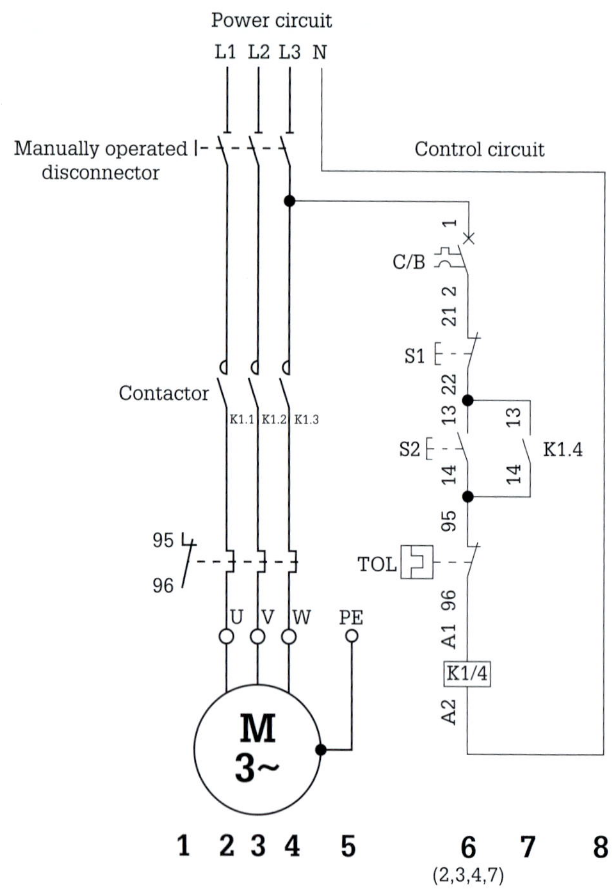

FIGURE 14.92 Contactor-controlled DOL starter vertical orientation circuit diagram

A contactor-controlled DOL starter with overload protection circuit diagram using conductor numbering and drawn in a horizontal orientation is shown in **Figure 14.93**.

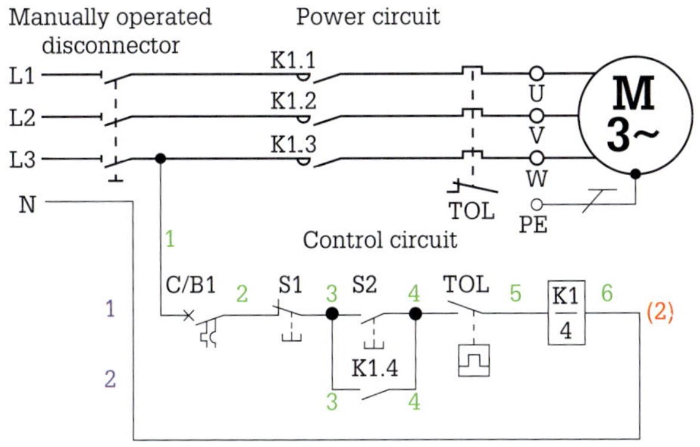

FIGURE 14.93 Contactor-controlled DOL starter horizontal orientation circuit diagram

Figure 14.93 shows a contactor-controlled DOL starter that utilises 230 V control. Pressing the start button (S2) completes the circuit from line L3 through the normally closed stop pushbutton and thermal overload contact to coil K1/4 and back to the neutral. The main contactor coil K1/4 closes contacts K1.1, K1.2 and K1.3, which places the motor across the supply and contact K1.4 is also closed. Contact K1.4's purpose is to provide an electrical interlock or latching (allows current flow through the coil) across the contacts of the start button once the start button is released. Pressing the stop button de-energises the main contactor coil, which opens all contacts it controls, and the motor is taken off line. This type of motor control provides immediate local or remote control of all kinds of machinery, fans and pumps.

All DOL starters should be capable of 15 starts per hour. However, where plug braking (immediate reversal of motor) is required; DOL starters should be capable of 40 starts per hour. Plug braking is an electric motor deceleration method that reverses the supply circuit connection so that the motor develops a counter-torque to stop the motor.

Star–delta starter

With the star–delta starting method, the motor is started in star configuration and then it is transferred to the delta configuration, allowing the full voltage to be applied to the motor at its regular speed so as to get the full torque output. This starting method reduces starting current and starting torque.

The components typically consist of three contactors, an overload relay and a timer for setting the time to the star position. The starting current is about 33% of the DOL value, and the starting torque is reduced to about 33% of the torque available for a DOL starter.

When starting, the load torque is low and increases with the square of the speed. When reaching approximately 80–85% of the motor rated speed, the load torque is equal to the motor torque and the acceleration ceases. To reach rated speed, a switch over to the delta is required. An illustration of a star–delta starter circuit diagram using a vertical orientation is shown in **Figure 14.94**.

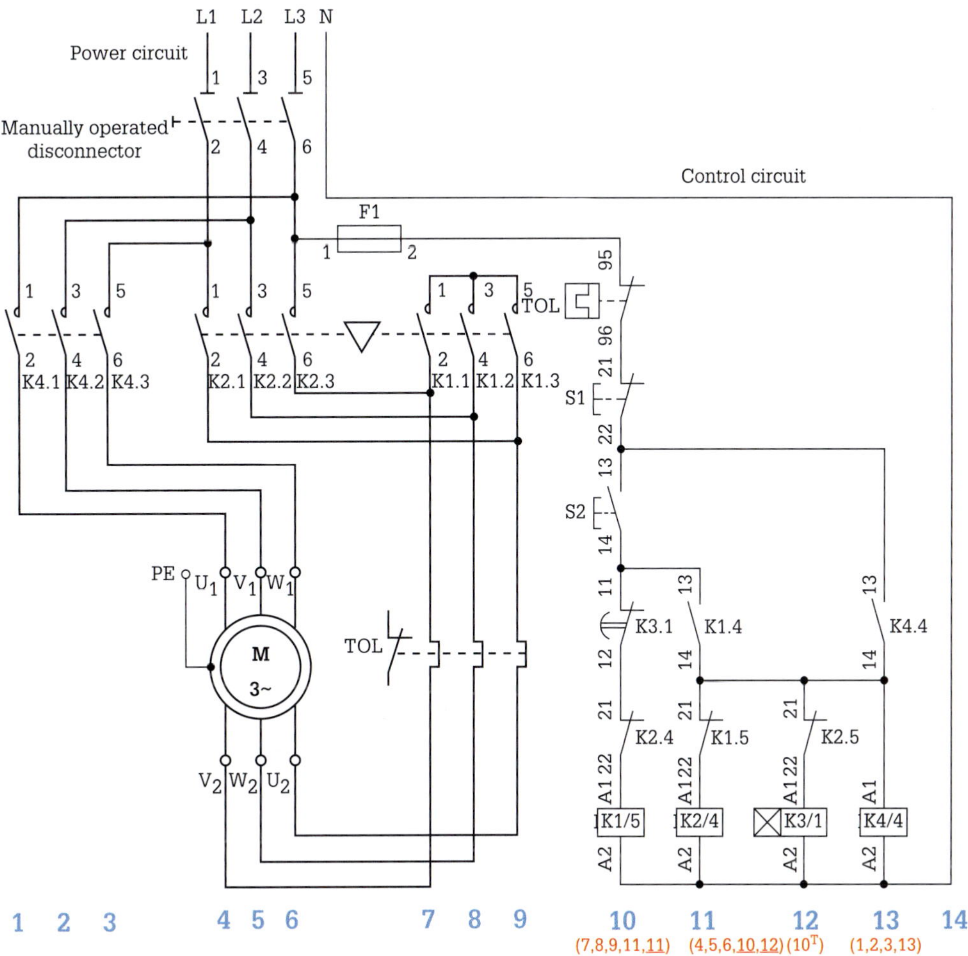

FIGURE 14.94 Star–delta starter circuit diagram using vertical orientation

Figure 14.94 shows a contactor-controlled star–delta starter that utilises 230 V control. Pressing the start button (S2) completes the circuit from line L3 through the thermal overload contact, the normally closed stop pushbutton and usually closed timer contact to coil K1/5 and back to the neutral.

The main contactor coil K1/5 closes contacts K1.1, K1.2 and K1.3, which joins the open ends of the stator winding (V_2, W_2 and U_2) to form a star point. The current in star is one-third of the current in delta, so this contactor is rated at one-third of the load current rating of the motor. Simultaneously, the delta connecting coil K2 is open-circuited by the normally closed contact K1.5. To provide safety, coils K1 and K2 are mechanically interlocked.

Contact K1.4 also closes to provide an electrical interlock or latching across the contacts of the start button once the start button is released. When contact K1.4 closes, voltage is applied to the time delay coil K3 and the main contactor K4/4 is placed on line. Placing on line causes contact K4.4 to close and bridge out the start button and contact K1.4. Contacts K4.1, K4.2 and K4.3 close providing a voltage to the stator winding leads (U_1, V_1 and W_1).

Because contactor K1 is closed, the stator is now connected in the star configuration. Star configuration means that the voltage across the stator windings is 58% of the full line voltage which in turn causes a reduction in starting current.

Once the time delay setting has elapsed, contact K3.1 opens causing coil K1 to open the star connection. Simultaneously, contact K1.5 closes and coil K2 is energised. This action causes contact K2.4 to open and electrically isolate coil K1. Contact K2.5 also opens and disconnects the timer coil K3. Contacts K2.1, K2.2 and K2.3 close and connect the stator winding in the delta configuration. The mechanical interlock is now controlled by the contactor K2/5 and prevents contactor K1/5 from closing. Full line voltage is now applied to the stator. Pressing the stop button de-energises the main contactor coil K4, which opens all contacts it controls, and the motor is taken off line.

The timers for star–delta starters must be break-before-make contacts with a minimum time of 50 ms between break and make in order to quench the arc developed on the star contactor. In addition, the switching action of the timer prevents the decay of the stator's electromagnetic flux with resulting high-voltage transients. A star–delta circuit diagram using a horizontal orientation is illustrated in **Figure 14.95**.

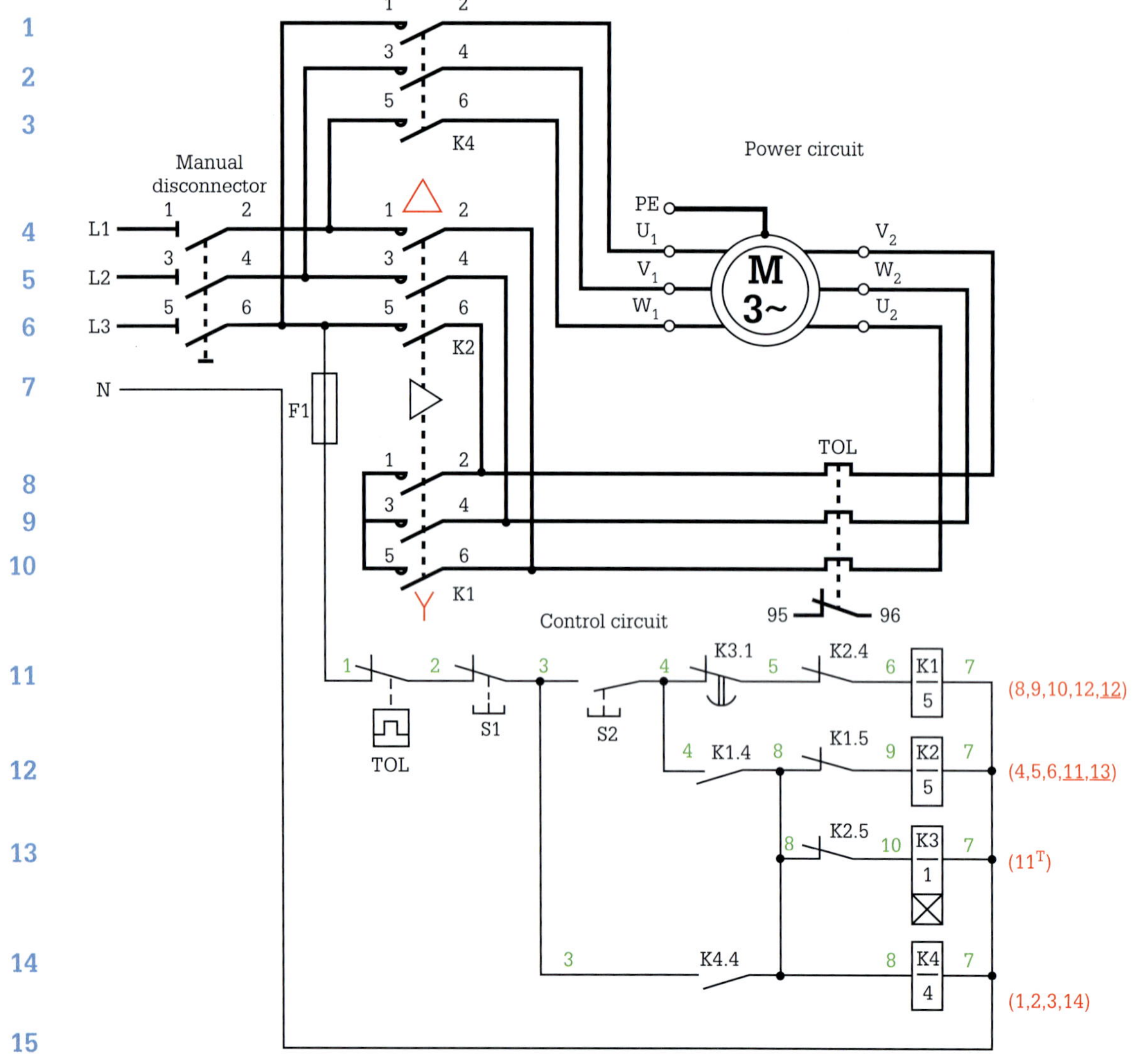

FIGURE 14.95 Star–delta starter circuit diagram using a horizontal orientation

Starting torque

Reducing the voltage applied to an induction motor will produce a reduction in the current drawn from the supply and a reduction in torque produced by the motor. The torque produced by the motor at the lower voltage is calculated from the following equation:

$$T_{reduced\ V} = T_{DOL} \times \left(\frac{V_{reduced}}{V_{DOL}}\right)^2$$

where $T_{\text{reduced V}}$ = torque produced at reduced voltage in newton metres (Nm)
T_{DOL} = torque produced when started DOL in newton metres (Nm)
V_{reduced} = reduced voltage in volts (V)
V_{DOL} = DOL voltage in volts (V)

EXAMPLE 14.3

A 400 V three-phase motor is started using a star–delta starter. Calculate the torque produced at the reduced voltage if the DOL starting torque is 40 Nm.

$$T_{reduced\ V} = T_{DOL} \times \left(\frac{V_{reduced}}{V_{DOL}}\right)^2$$

$$= 40 \times \left(\frac{230}{400}\right)^2$$

$$= \mathbf{13.2\ Nm}$$

This represents 1/3 of the torque produced at full voltage.

It follows then that for star–delta motor starters:

$$T_{star} = \frac{T_{DOL}}{3}$$

where T_{star} = torque produced in star connection in newton metres (Nm)
T_{DOL} = torque produced when started DOL in newton metres (Nm)

EXERCISE 14.3

A 400 V three-phase motor is started using a star–delta starter. Calculate the torque produced at the reduced voltage if the DOL starting torque is 30 Nm.

Starting current

The starting current is calculated from the equation:

$$I_{reduced\ V} = I_{DOL} \times \left(\frac{V_{reduced}}{V_{DOL}}\right)^2$$

where $I_{\text{reduced V}}$ = current drawn at reduced voltage in amperes (A)
I_{DOL} = current drawn when started DOL in amperes (A)
V_{reduced} = reduced voltage in volts (V)
V_{DOL} = DOL voltage in volts (V)

EXAMPLE 14.4

A 400 V three-phase motor is started using a star–delta starter. Calculate the current drawn at start if the DOL starting current is 42 A.

$$I_{reduced\ V} = I_{DOL} \left(\frac{V_{reduced}}{V_{DOL}}\right)^2$$

$$= 42 \times \left(\frac{230}{400}\right)^2$$

$$= \mathbf{14\ A}$$

This represents 1/3 of the current drawn at full voltage.

It follows then that for star–delta motor starters:

$$I_{star} = \frac{I_{DOL}}{3}$$

where I_{star} = current drawn at reduced voltage in amperes (A)
I_{DOL} = current drawn when started DOL in amperes (A)

EXERCISE 14.4

A 400 V three-phase motor is started using a star–delta starter. Calculate the current drawn at start if the DOL starting current is 60 A.

Autotransformer starter

The autotransformer starter is another starting method that reduces the starting current and starting torque. Autotransformers are wound with taps for each phase in order to allow the transformer to be adapted to the starting conditions that are required for the motor. While starting, the stator is connected to the autotransformer taps.

Taps on the autotransformer allow for selection of the motor with 50%, 65% or 80% of the current inrush seen during a DOL start. The resulting starting torque is 25%, 42% or 64% of full voltage values, and the current drawn by the stator reflects the same percentages.

An autotransformer motor starter allows reduced voltage starting of a motor while maintaining useful starting torque. **Figure 14.96** illustrates an autotransformer starter circuit diagram using a vertical orientation.

Power circuit operation

Manually close the disconnector. The disconnector is rated for motor full-load current. After pressing the start button S2, contactor K1 closes to connect the autotransformer into a star configuration. Contactor K3 closes to energise the autotransformer and start the three-phase motor. K1.5 electrically interlocks contactor K2.

Contactors K1 and K3 are mechanically and electrically interlocked to prevent them from closing simultaneously.

The sequence timer K4 is energised by contact K1.4. After a short period (between 7 to 20 seconds) K3 closes via timer contact K4.2. Contact K3.4 holds it in. Timer contact K4.1 opens and contact K2.4 closes.

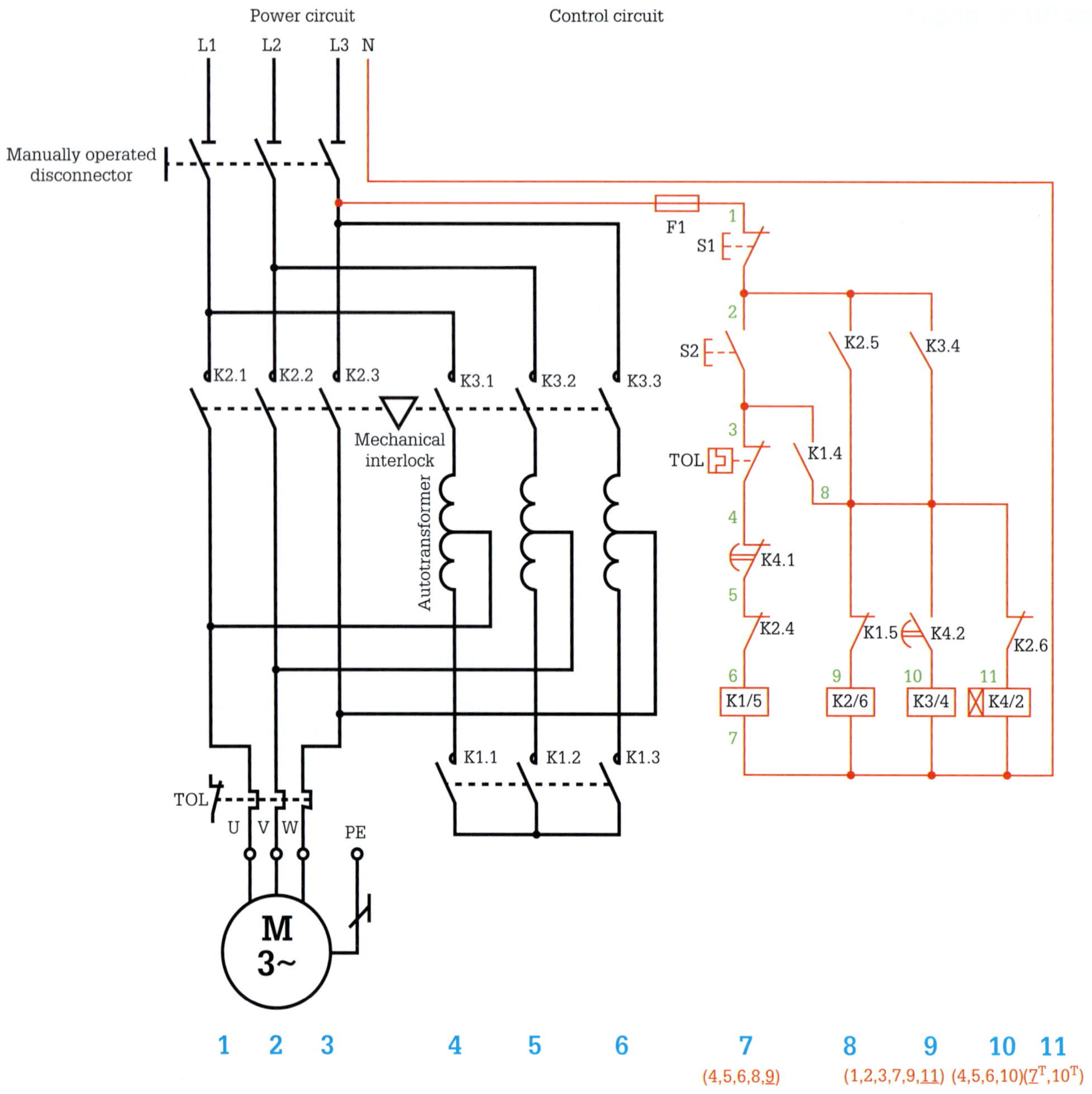

FIGURE 14.96 Autotransformer starter circuit diagram using vertical orientation

Contactor K1 is electrically interlocked by contact K2.4 and coil K2 holds it in. Timer relay K4 is de-energised by contact K2.6. Contactor K3 opens because contact K1.4 opens.

Contactor K1 opens after a short delay to disconnect the star contacts (K1.1, K1.2, K1.3). Contactor K2 closes to place the fully supplied voltage to the motor terminals. Contactor K3 opens to disconnect the autotransformer.

Contactors K1 and K3 have to be rated for the type of autotransformer used, the starting period and the number of starts per hour. Contactor K2 is rated for the full-load motor current.

The motor may be stopped by pressing S1.

Control circuit operation

Pressing the start button (S2) completes the circuit from line L3 through the normally closed stop pushbutton, the normally closed thermal overload contact, the closed timer contact to coil K1 and back to the neutral. Contactor coil K1 causes power contacts K1.1, K1.2 and K1.3 to close forming the star point for the autotransformer and contact K1.4 closes and contact K1.5 opens. Simultaneously timer coil K4 energises and closes timer contact K4.2.

The closing of timer contact K4.2 energises coil K3. The power contacts K3.1, K3.2 and K3.3 associated with contactor K3 connect the autotransformer to the motor terminals. This action allows a reduced voltage to be applied to the motor terminals.

The power contacts K2.1, K2.2 and K2.3 associated with contactor K2 connect the motor terminals to the line.

Pressing the stop button allows all contacts to be reset to their original configuration. Typical applications for this starting method are conveyors, compressors, and pumps. Some autotransformer starters include solid-state motor protection and vacuum contactors.

Autotransformer starters should have timers in the control circuit to limit the number of starts per hour to the rating of the transformer. Some starters may also incorporate a 'policeman timer' to disconnect the motor circuit from the supply if the allowable starting time is exceeded.

With the star K1 and autotransformer contactors K3 closed, the motor is under reduced voltage. Consequently, the torque is reduced as the square of the applied voltage. When the motor reaches 80–95% of the nominal speed, the star contactor opens. Then the line contactor closes K2 and the autotransformer contactor K3 opens. The motor is never disconnected from the power supply while starting. **Figure 14.97** shows an autotransformer starter circuit diagram using horizontal orientation and wire colours to distinguish between power and control circuits.

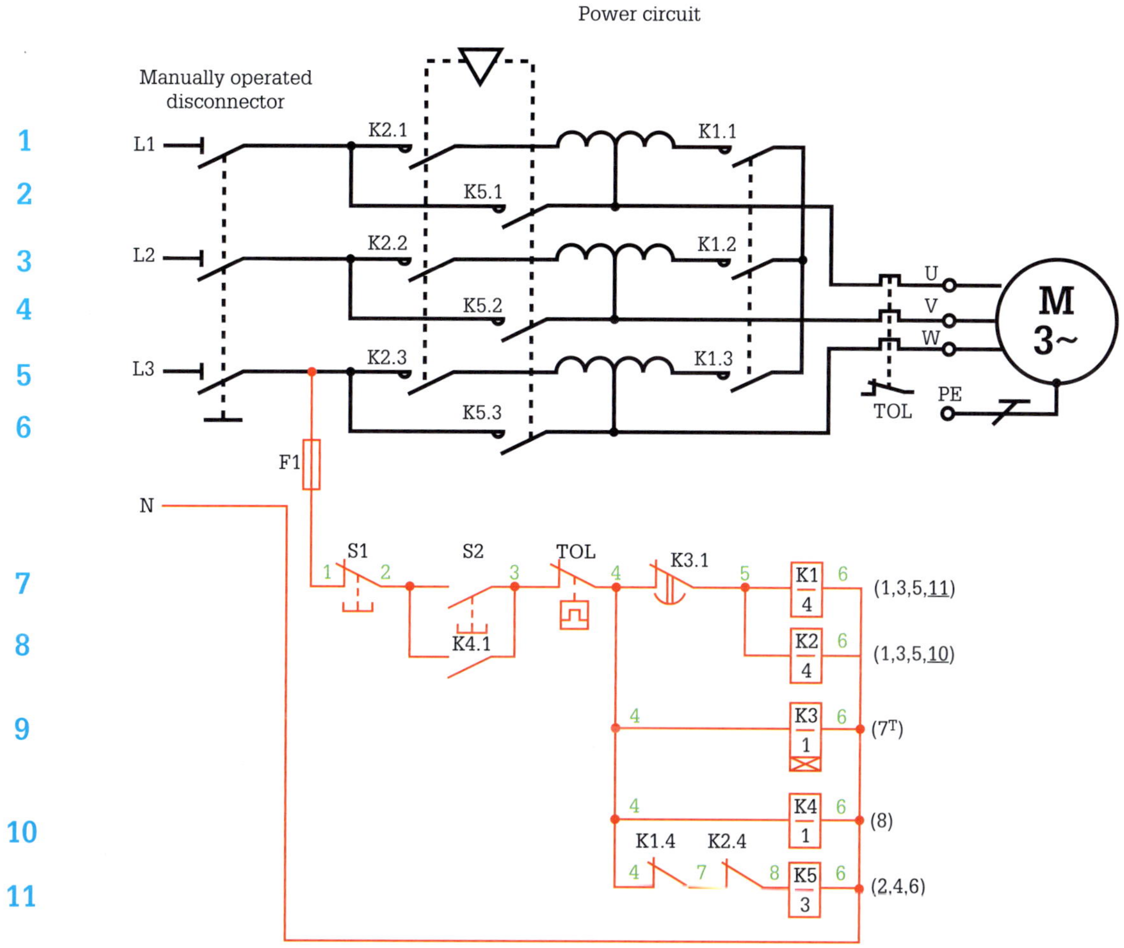

FIGURE 14.97 Autotransformer starter circuit diagram using horizontal orientation

Starting torque

If we consider the equation for the star–delta motor starter and apply it to the autotransformer starter, we can see that the reduced voltage equates to % tap × V_{DOL}. This results in the following equation:

$$T_{reduced\ V} = T_{DOL} \times (\%\ tap)^2$$

where $T_{reduced\ V}$ = torque produced at reduced voltage in newton metres (Nm)

T_{DOL} = torque produced when started DOL in newton metres (Nm)

% tap = autotransformer tapping % expressed as a decimal

EXAMPLE 14.5

The DOL starting torque for a 400 V three-phase induction motor is 86 Nm. Calculate the starting torque if the motor is started on an autotransformer using the 80% tap.

$$T_{reduced\ V} = T_{DOL} \times (\%\ tap)^2$$
$$= 86 \times (0.8)^2$$
$$= 55\ \text{Nm}$$

EXERCISE 14.5

The DOL starting torque for a 400 V three-phase induction motor is 86 Nm. Calculate the starting torque if the motor is started on an autotransformer using the 65% tap.

Starting current

The starting current drawn by the motor is calculated from the equation:

$$I_{motor} = I_{DOL} \times \% \; tap$$

where I_{motor} = current drawn by motor at reduced voltage in amperes (A)

I_{DOL} = current drawn when started DOL in amperes (A)

% tap = autotransformer tapping % expressed as a decimal

EXAMPLE 14.6

A 400 V three-phase motor is started on the 80% tap of an autotransformer starter. Calculate the current drawn by the motor at start if the DOL starting current is 60 A.

$$\begin{aligned} I_{motor} &= I_{DOL} \times \% \; tap \\ &= 60 \times 0.8 \\ &= \mathbf{48\ A} \end{aligned}$$

EXERCISE 14.6

A 400 V three-phase motor is started on the 65% tap of an autotransformer starter. Calculate the current drawn by the motor at start if the DOL starting current is 60 A.

The actual current drawn from the supply (line current) is calculated from the equation:

$$I_{motor} = I_{DOL} \times (\% \; tap)^2$$

where I_{motor} = current drawn by motor at reduced voltage in amperes (A)

I_{DOL} = current drawn when started DOL in amperes (A)

% tap = autotransformer tapping % expressed as a decimal

EXAMPLE 14.7

A 400 V three-phase motor is started on the 80% tap of an autotransformer starter. Calculate the current drawn by the motor at start if the DOL starting current is 200 A.

$$\begin{aligned} I_{motor} &= I_{DOL} \times (\% \; tap)^2 \\ &= 200 \times (0.8)^2 \\ &= \mathbf{128\ A} \end{aligned}$$

EXERCISE 14.7

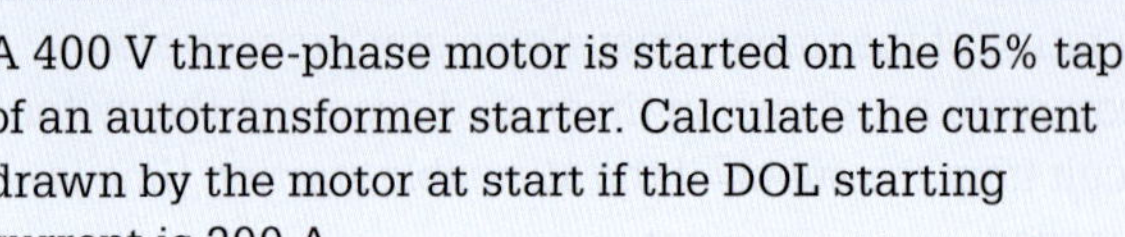

A 400 V three-phase motor is started on the 65% tap of an autotransformer starter. Calculate the current drawn by the motor at start if the DOL starting current is 200 A.

Primary resistance starter

Primary resistance starters are classified as closed transition starters because the motor is not disconnected from the supply during the starting sequence. These starters incorporate a resistor bank that is connected in series with the stator windings. The resistor bank limits the initial current surge drawn by the motor at starting. Once the motor draws current a voltage drop occurs across the resistor bank resulting in a reduced voltage across the motor terminals.

As motor speed increases, the back emf induced in the stator windings opposes the applied voltage and reduces further the starting current. While the voltage across the resistor bank decreases, the voltage across the motor terminals increases. This increase in voltage causes the motor torque to increase. The effect is a steady, even acceleration of the rotor without transient surges occurring on the supply line. Once the motor is nearly up to full-load speed a timing relay is connected to short out the resistors, allowing full line voltage to be applied across the motor terminals. The full line voltage brings the motor up to full-load speed. As the torque is proportional to the square of the voltage the starting torque is also reduced. The value of the resistor bank is chosen to reduce current at start-up while allowing sufficient voltage across the motor terminals for the required starting torque. The power resistors used for resistance starters can have a high wattage value and some resistance banks need forced air-cooling.

There are two types of resistors used for resistance starters – solid resistors such as wire-wound resistors, metal-plate resistors or cast-iron grid resistors and electrolyte resistors made of metal plates in a saline solution or caustic soda. Metal resistors have a positive temperature coefficient characteristic and as they heat up their resistance increases.

Electrolyte resistors have a negative temperature coefficient characteristic and as they heat up their resistance reduces. The chemistry of the electrolyte solution is based on a number of factors including tank volume/size required, starting current and starting torque, number of starts per hour and ambient temperature. The most frequent problem with liquid starters is the evaporation of the water that increases the starting resistance and affects the speed–torque curve, slowing the acceleration. A liquid starter operates with current flowing in an electrolyte solution between copper electrodes.

The electrical resistance between the electrodes varies with the amount of electrolyte in contact with the electrodes. The heat generated by solid and electrolyte resistors needs time to condense back into liquid form between starts during the start cycle. Their resistance characteristics make it necessary for them to cool down sufficiently between starts. The cooling down period restricts the number of stop–start operations and the minimum time between motor starts.

Primary resistor starters usually provide one step of resistance with about 65% to 70% of line voltage available at the motor terminals. Motor current is also reduced to about 65% to 70% of rated locked rotor current and about 42% to 49% of rated locked rotor torque is available for the load. Some of the disadvantages of primary resistance starting include a limited number of starts per hour, difficulty in starting high-inertia loads and the fact that a lot of heat needs to be dissipated. A contactor-controlled primary resistance starter circuit diagram using the horizontal orientation is shown in **Figure 14.98**.

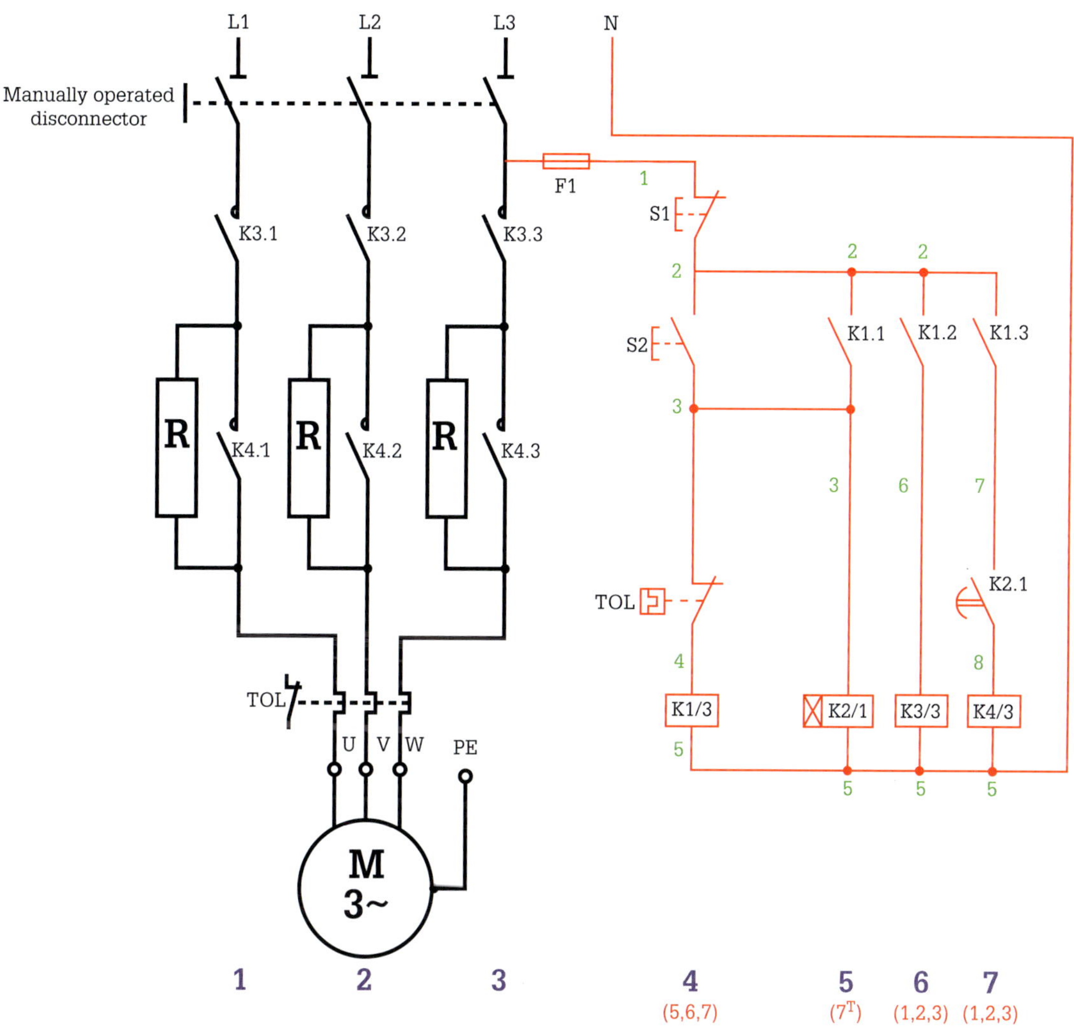

FIGURE 14.98 Primary resistance starter circuit diagram using horizontal orientation

Figure 14.98 illustrates one method of connecting a primary resistance contactor starter. When the start button is pressed the control relay K1 is energised. This relay closes contacts K1.1, K1.2 and K1.3.

Contact K1.1 is the electrical interlock for the start pushbutton that now can be released. The closing of K1.1 energises the time relay K2. Contact K1.2 energises the contactor coil K3, which closes contacts K3.1, K3.2 and K3.3. The closure of these contacts connects the motor to a reduced voltage supply through the starting resistors. After a pre-set time delay (usually 10–20 seconds) the time relay K2 closes contact K2.1, energising contactor coil K4, which closes contacts K4.1, K4.2 and K4.3.

The closing of these contacts shorts out the starting resistors and the motor is now of full line voltage. Pressing the stop button allows all contacts to be reset to their original configuration. Some starters may also incorporate a 'policeman timer' to disconnect the motor circuit from the supply if the allowable starting time is exceeded.

Secondary resistance starter

The name 'secondary resistance starter' refers to the fact that resistance is added to the rotor or secondary circuit of the motor. Only slip-ring (also called wound-rotor) induction motors have the rotor design necessary for this type of motor starting. Slip-ring or wound-rotor induction motors have performance characteristics that are very different to that of squirrel cage motors even though the operating principle is the same.

The maximum torque characteristic of squirrel cage motors cannot be altered as it depends upon rotor design and usually occurs in about 80% of rotor speed. The wound-rotor motor can have its rotor characteristic altered by inserting variable external resistance (metal or liquid) via slip rings in series with the rotor winding. Series connection allows the maximum developed torque to be realised at any point between zero and 80% of full-load speed.

The combination of this starter and motor means that a smooth acceleration and useful torque can be developed throughout the motor's speed range by reducing the value of the external resistance until the rotor is shorted out. It then operates as a standard squirrel cage motor. In addition, the maximum developed torque is directly proportional to the current drawn by the stator. Consequently, at 250% torque the line current is 250% of the motor's full-load value.

DOL starting is the only method of starting which can achieve this level of torque development, but requires a very high line current (up to 10 times full-load value). The torque and current characteristics offer distinct advantages over squirrel cage motors, and the motor and starter find application in providing for variable speed load demands.

The liquid secondary resistance starters have a stepless resistance characteristic, which does not cause transients. As the motor is run up to speed, there is an automatic decrease in resistance as the electrolyte is heated by the passage of rotor current inside its chamber.

The electrolyte becomes partially vaporised, creating a difference in resistivity between the vapour and the liquid. It is the thermal differences between these two resistance forms that enable the opposition to current flow to gradually fall. An adjustable timing relay then closes contacts that short out the liquid resistance. The number of starts per hour is limited because the vapour needs to cool and become liquid.

The slip-ring induction motor and secondary resistance starter combination with respect to some applications are under threat by electronic variable-voltage variable-frequency (VVVF) drives combined with standard squirrel cage motors. These starters are suitable for increasing load duty such as that produced by pumps and fans. A contactor-controlled secondary resistance starter circuit diagram is shown in **Figure 14.99**.

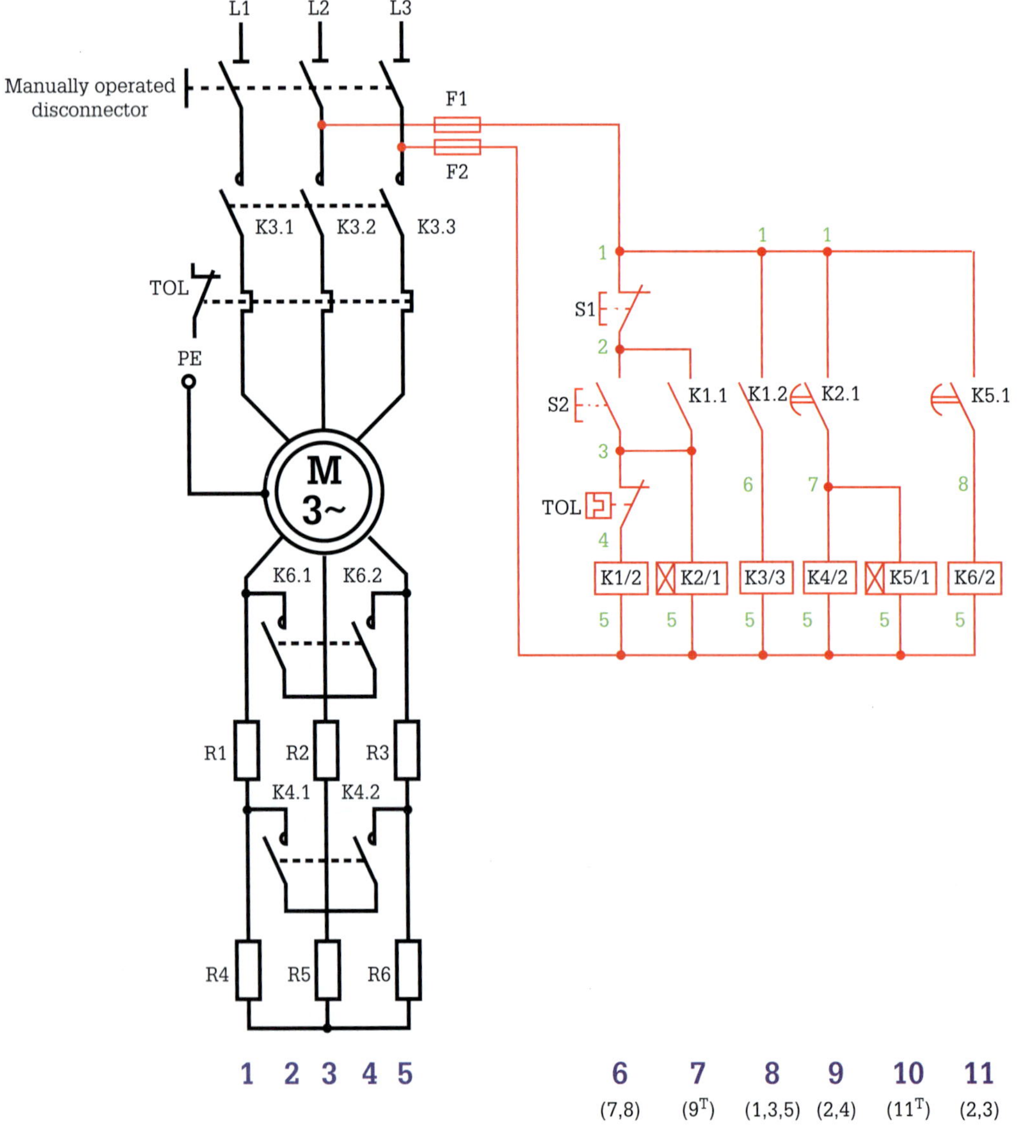

FIGURE 14.99 Secondary resistance starter

Figure 14.99 illustrates one method of connecting a secondary resistance contactor starter. When the start button is pressed the control relay K1 is energised. This relay closes contacts K1.1 and K1.2. Contact K1.1 is the electrical interlock for the start pushbutton that now can be released. The closing of K1.1 energises the time relay K2. Contact K1.2 energises the contactor coil K3 that closes the contacts K3.1, K3.2, and K3.3. The closure of these contacts connects the motor stator winding to line voltage while the rotor windings are connected in the star configuration through the starting resistors R1, R2, R3, R4, R5 and R6. The closure of these contacts allows the motor to develop torque and begin turning. After a pre-set time delay (usually 10–20 seconds) the time relay K2 closes the contact K2.1 energising contactor K4 that closes contacts K4.1 and K4.2. The closure of these contacts connects the motor across a reduced resistance R1, R2 and R3. The reduced resistance allows the motor to increase speed. After a pre-set time delay (usually 10–20 seconds) the time relay K5 closes the contact K5.1, energising contactor K6 that closes contacts K6.1 and K6.2. The closing of these contacts shorts out the remaining resistance in the rotor circuit by connecting the rotor windings in the star configuration and the motor now develops full-load speed. Pressing the stop button allows all contacts to be reset to their original settings.

SWITCH ON

***Wiring Rules* and service rule requirements**

Before installing three-phase motors and starters the Australian and New Zealand AS/NZS 3000:2018 *Wiring Rules* and local supply authority's requirements need to be consulted. Areas of interest include connection, control switches, isolating switches, limitation of transient current, automatic starting and protection against over-temperature.

REVIEW QUESTIONS

1. Calculate the synchronous speed of an eight-pole induction motor operating at a supply frequency of 50 Hz.
2. Name the two drawing conventions used for setting out motor starter circuit diagrams.
3. Draw a contactor-controlled DOL starter with electronic overload protection configuration using a vertical layout.
4. What is the major disadvantage of starting a large motor DOL?
5. Draw a star–delta starter control and power circuit using a vertical layout.
6. Draw an autotransformer starter control and power circuit using a vertical layout.
7. The DOL starting torque for a 400 V, three-phase induction motor is 86 Nm. Calculate the starting torque if the motor is started on an autotransformer using the 50% tap.
8. A 400 V three-phase motor is started on the 50% tap of an autotransformer starter. Calculate the current drawn by the motor at start if the DOL starting current is 60 A.
9. Name the two types of resistors used with primary resistance starters.
10. What type of starter adds resistance to the rotor circuit of a wound-rotor motor?

14.9 Three-phase induction motor reversal and braking

Reversing the direction of rotation of a three-phase motor is achieved by inter-changing the connections of any two of the three power conductors. The regular number of starts per day over a period of months or years influences the life of the motor and its power and control system.

Excessive cycling affects the life of control components such as starters, sensors, and relays. The number of starts and reverses that a motor sees (on and off) can also cause motor shaft damage (twisting stress), bearing damage, stressed insulation and motor overheating (for every 10°C rise the motor insulation life is reduced by half). These conditions can reduce motor life.

Intermittent operation of a reversible motor for a short period results in a significant flow of current when the motor is started or reversed, creating increased heat generation. Therefore, the motor should be allowed to run for a short period in order to dissipate heat build-up from starting current. Additionally, the current surge in the windings at start-up causes the stator windings to experience a mechanical stress. The current surge causes the winding to bend slightly at the point where the winding conductors leave the stator slot thereby placing mechanical pressure on the insulation at those points. The effects of heat and mechanical stresses mean that there are a limited number of starts per hour. Standard three-phase motors below 1 kW can deliver up to 12 starts while motors above 30 kW would be limited to four starts or less per hour. These values depend on the design and duty application of the motor. A reversing motor control circuit is illustrated in **Figure 14.100**.

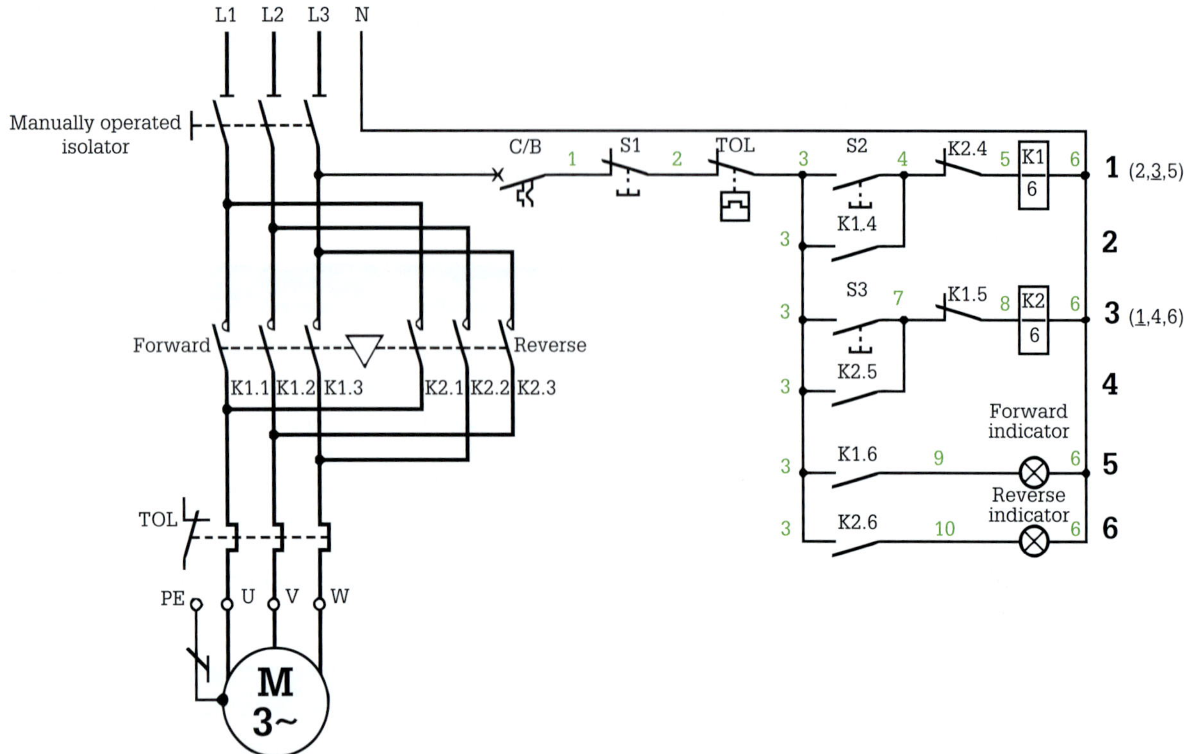

FIGURE 14.100 Reversing starter circuit diagram

Methods of braking

Braking is essentially the removal of stored kinetic energy in the form of motion from a motor-mechanical load. When the motor and attached load is brought up to speed, energy drawn from the supply is converted and stored as motion. To stop motion, the kinetic energy stored in the motor-load combination must be removed. The braking methods that are applied to motor-load combinations convert the kinetic energy into heat. Braking provides a means of stopping an a.c. motor and can be accomplished in several ways.

Electromechanical brake

The electromechanical brake applies a braking force directly to the brake disc mounted on the motor shaft as shown in **Figure 14.101**.

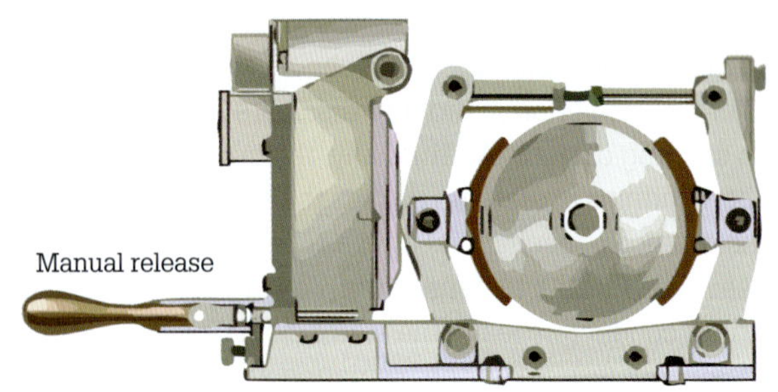

FIGURE 14.101 Electromechanical brake

Electromechanical-actuated friction brakes convert the kinetic energy into heat energy by pushing a friction disc via solenoid action against rotating or moving parts to stop the rotor. Therefore, without friction-generating components such as a drum, disc and shoe lining material the motor shaft cannot be stopped.

The armature of the electromagnet opposes the force of the spring when voltage is applied to the solenoid. The energised solenoid releases the brake and lets the rotor shaft freely rotate. When power is removed, the spring energises the brake and holds the load. Many hoisting systems rely solely upon the electromechanical brakes to stop the motor shaft movement when an emergency condition such as loss of power occurs. All emergency brakes are electromechanical.

Dynamic braking

Dynamic braking, also called d.c. injection braking, utilises the ability of the a.c. drive motor to act as a generator. Disconnecting the a.c. feed to the stator and injecting direct current across any pair of stator terminals obtains this effect. The injection of direct current into the stator windings creates a stationary d.c. magnetic field within the stator, and the kinetic energy of the revolving rotor is dissipated into the rotor cage by generator action. When the rotor, via its motion, cuts the magnetic field of the stator an emf is induced across the rotor bars.

This voltage develops circulating currents in the rotor cage. This current is dissipated as heat energy at the expense of the rotational energy stored in the rotor thereby slowing the rotor. Dynamic braking cannot provide the holding torque necessary to keep the motor at standstill, and an electromechanical brake is required to complete the stopping of the motor. The d.c. supply applied to the stator windings can be greater than three times the rated full-load value of current and must be removed immediately the rotor stops; otherwise the stator overheats. The braking torque developed by the d.c. injection method is proportional to the square of the applied injected d.c. current.

Dynamic braking can also be obtained through a braking circuit and phase capacitors. The capacitors are connected across the terminals of the stator when the line feed is removed. The discharging of the capacitor's energy enables the motor to operate as a self-excited generator being driven by the kinetic energy of the revolving load. After the capacitors are switched from the circuit, the generated energy can be connected across an electrical load such as dynamic load resistors to dissipate the generated power. Thus, the rotational energy of the load can be converted to electrical energy and then be dissipated as heat energy in the dynamic resistors. Because load torque is required for the motor to generate energy the retarding torque developed by the generator mode of the motor limits the speed of the load, bringing it to a complete stop.

Plug braking

Plug braking is the fast deceleration of a motor by reversing any two line feeds to the stator terminals while it is running at full speed. Although reversing draws a high line current, even exceeding the locked rotor current value of the motor, plug braking is the simplest and quickest braking method. Plug braking imposes a retarding force on the motor and places considerable stress on the stator and rotor windings and motor bearings. The generated heat losses in both stator and rotor windings can be up to three times that experienced during normal acceleration. Also, the contactor switching and out-of-phase re-closing can cause severe torque and current transients. The increased temperature can degrade stator insulation and melt rotor bars. The torque transients can loosen the rotor bars, twist or break the rotor shaft and create flat spots on the balls or rollers in the motor bearings. Therefore, a motor used for plug braking must be specially designed for this application.

Finally, plug braking cannot hold the motor at the standstill position by providing a positive stop. If the line voltage is not removed a little before zero speed is reached, the motor begins to re-accelerate in the reverse direction. Electromechanical brakes are needed to hold the rotor shaft in a stationary position while the line voltage is de-energised. A timer-type plug braking control circuit is illustrated in **Figure 14.102**.

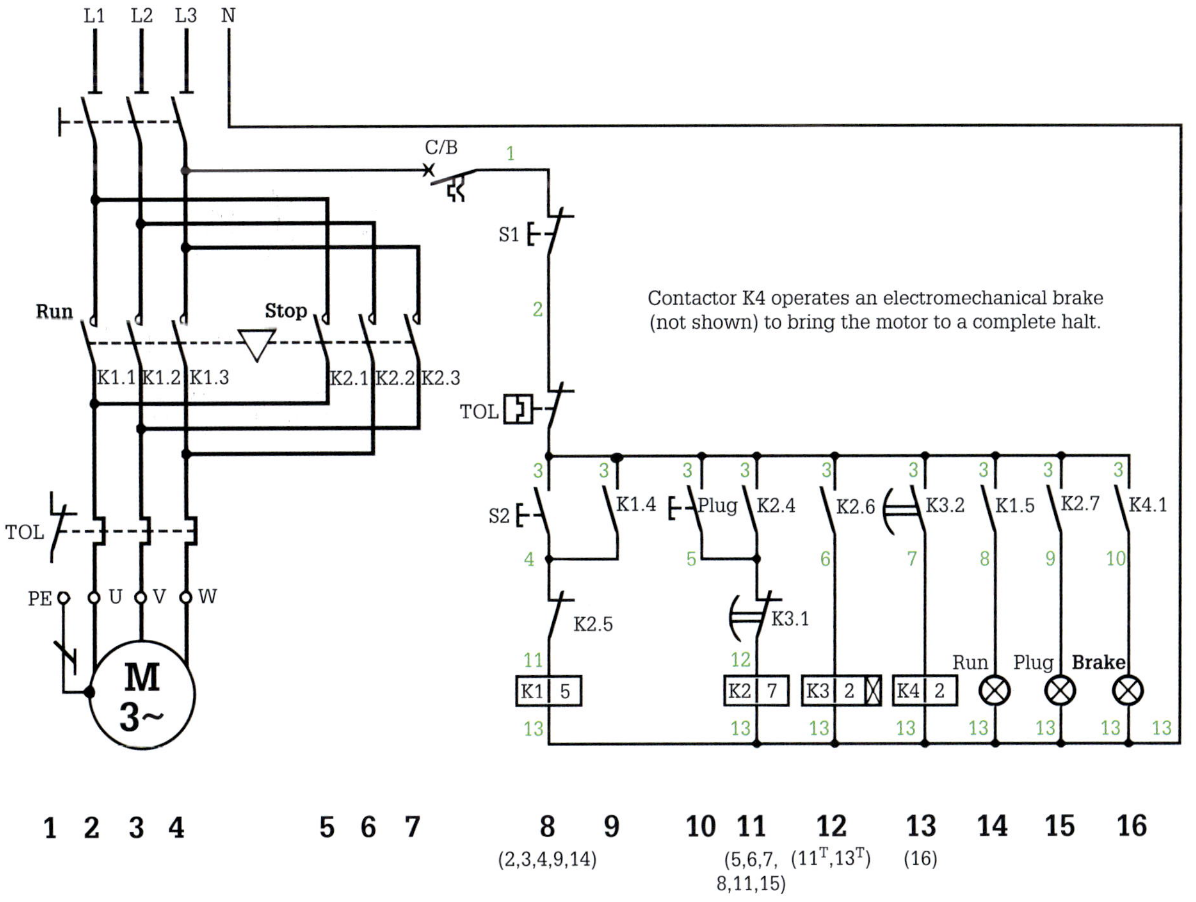

FIGURE 14.102 Timer-type plug braking circuit

Regenerative braking

Overhauling loads are the most demanding braking application. Examples of overhauling loads include the slowing of a flywheel, the slowing of trains and trams, the control of an elevator car or crane load under the force of gravity as it descends, and in drilling and sawing operations where a sudden drop in torque occurs when these machines complete their designated operation. These examples illustrate the need to make the driven load move more slowly than it would in an uncontrolled state.

When an overhauling load drives the motor above synchronous speed, the motor acts as an induction generator feeding energy back into the supply system and at the same time causing a reversal of torque. Regeneration takes place when the voltage generated (E_g) becomes greater than the supply voltage (V). Consequently, the direction of the stator current and hence rotor torque is reversed, and speed falls until E_g becomes lower than V. A reverse contactor that operates on a defined timing sequence controls this flow of energy. The energy that is fed back into the system declines as the overhauling load slows to synchronous speed and, at a predetermined value, the energy is prevented from entering the supply system.

With induction motors that have multi-pole connections available, the stator can be reconfigured externally to provide a larger number of poles. Reconfiguring changes the synchronous speed of the motor that is now lower than the overhauling speed of the load. The motor again acts as an induction generator and the overhauling load slows to the new synchronous speed. Regenerative braking, when applied to overhauling loads, can only be used to reduce the speed to synchronous value. When a little synchronous speed is reached, an electromechanical brake operates to hold the load in the stop position.

Eddy current braking

A typical eddy current brake as shown in **Figure 14.103** consists of a segmented (to dissipate heat) stationary field assembly, and a smooth-surface brake rotor coupled mechanically to the brake shaft that surrounds the stationary field assembly. A small air gap exists between the smooth-surface rotor and the stationary field assembly.

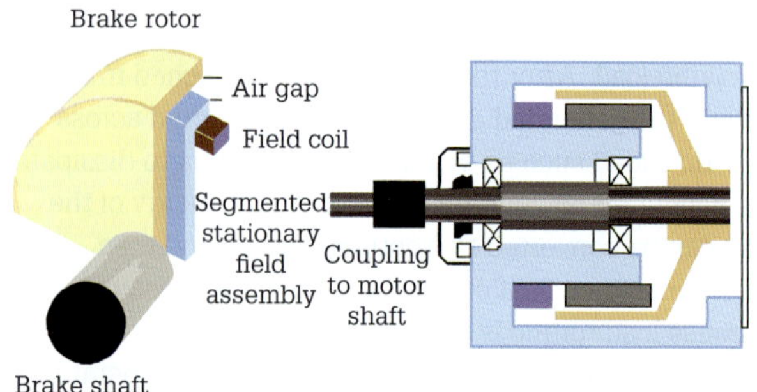

FIGURE 14.103 Eddy current brake

In operation, direct current is applied to the field coil, and an electromagnetic field is established in the stationary field assembly. If the brake rotor is turning through the magnetic field, eddy currents are induced in it. These eddy currents react with the magnetic field in the field assembly and produce a torque that opposes motion of the brake rotor. This torque is proportional to the square of the direct current applied to the field. The brake rotor rotates at the speed of the prime mover until the field coil is energised. Rotation of the rotor is slowed by controlling the current in the field coil.

The faster the brake rotor is turning; the stronger the eddy current effect (high slip), meaning that as the rotor slows the braking force is reduced, producing a smooth stopping motion. Note that at zero slip, the eddy current brake has no torque and therefore cannot be used where holding a load is required.

REVIEW QUESTIONS

1. Explain how to reverse the direction of rotation of a three-phase induction motor.
2. Why should motors be allowed to run for a short period after starting?
3. With induction motors what is meant by the term 'braking'?
4. What type of device has a brake disc mounted on a motor shaft?
5. Give another name for dynamic braking.
6. Which method of braking is the simplest and quickest?
7. What types of load are the most demanding for braking applications?
8. What type of braking is used as emergency brakes?

14.10 Three-phase induction motor speed control

For three-phase induction motors, there is a direct relationship between speed and torque.

Torque–speed relationships

The mechanical power (P) available from an induction motor can be expressed as:

$$P = \frac{2\pi NT}{60} \text{ in Watts}$$

where N = speed of the motor in rpm

T = torque developed in Newton metres (Nm)

As 60 / 2π equals 9.55 (a constant), the full-load torque of a motor in newton metres (Nm) can be calculated from the fundamental equation:

$$T = \frac{9.55 \times P}{N}$$

One of the elementary essentials of induction motor principles is that the torque is proportional to the square of the motor terminal voltage. If half of the available line voltage is applied to the motor terminals, the rotor produces only a quarter of the full-load torque. This is the principle on which all reduced-voltage starters operate. Many induction motors are designed to have a standard torque–speed curve that offers a compromise between starting torque and efficiency. The important fact is that an induction motor only increases the speed of the applied load when it produces more torque than the connected load can contain. This truth applies for all rotor speeds including a stationary rotor and a rotor at full-load speed.

EXAMPLE 14.8

Calculate the torque produced by a 2.4 kW induction motor having a shaft speed of 1440 rpm.

Solution:

$$T = \frac{9.55P}{N}$$

$$= \frac{9.55 \times (2.4 \times 10^3)}{1440}$$

$$= \mathbf{15.9\ Nm}$$

EXERCISE 14.8

Calculate the torque produced by a 1.8 kW induction motor having a shaft speed of 1445 rpm.

DOL torque–speed relationship at starting

DOL starters are 'full-voltage-across-the-line starters'. Full voltage means that the torque developed at the instant of starting is very high because torque is directly proportional to the square of the voltage. The starting torque of motors using this method of starting for a standard resistance rotor is 150% of full-load torque. An illustration of a typical DOL torque–speed characteristic is shown in **Figure 14.104**.

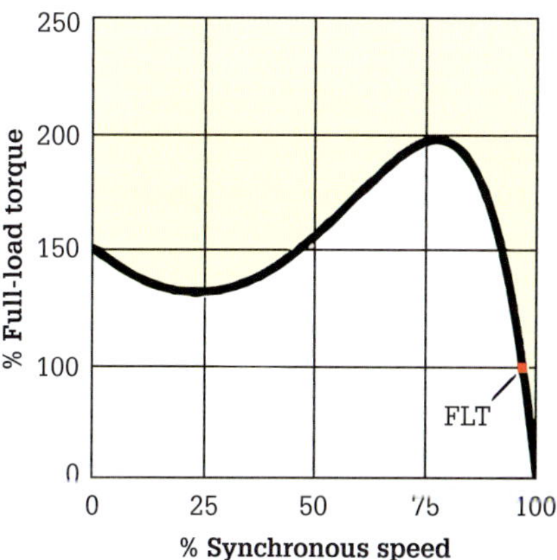

FIGURE 14.104 DOL torque–speed characteristics

Note: This figure represents a hypothetical squirrel cage motor started DOL, and the starting methods that follow are all starting this same motor.

The graph indicates that the starting torque is high. High starting torque means that the loads that are driven by a motor started by the DOL method must be able to withstand a severe starting shock.

Star–delta torque–speed relationship

A guide that is used for the maximum starting time of a star–delta starter is the rating of the motor in kilowatts divided by three, plus 9 seconds. For example, a 7.5 kW motor should be capable of attaining 80–90% of synchronous speed while driving the connected load in star before the changeover to delta occurs.

Consequently, an acceleration has to occur in 11.5 seconds or less. If this rate of acceleration of rotor speed is not obtained then the load is outside the capability of a star–delta starter.

All contactor-type star–delta starters use a timing relay for switching from star to delta, and the timer needs to be correctly set for each application. This type of starter provides very small torque (33% of DOL torque = 50% full-load torque (FLT) when compared to the graph of **Figure 14.104**) at the instant of starting. It is, therefore, suitable for no-load or very light load starting applications (lathes, drilling machines). An illustration of a typical star–delta speed torque characteristic is shown in **Figure 14.105**. Note that the motor torque peaks when the open transition changeover point is reached.

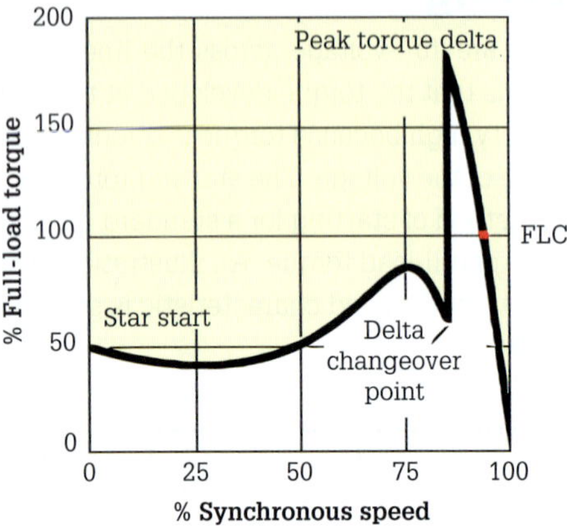

FIGURE 14.105 Star–delta torque–speed characteristics

Autotransformer torque–speed relationship at starting

Three starting torques are available with the standard tappings of 80%, 65% and 50% of line voltage, depending upon design requirements of the connected load. The tappings allow for easy adjustment of the starting torque and starting current to suit the load parameters. Starting torque is proportional to the square of the voltage so that the starting torque on these taps will be 64%, 42% and 25% respectively of the maximum DOL starting torque (64% of DOL = 96% FLT when related to the graph of **Figure 14.104**). An autotransformer torque–speed characteristic at the 80% tapping is shown in **Figure 14.106**.

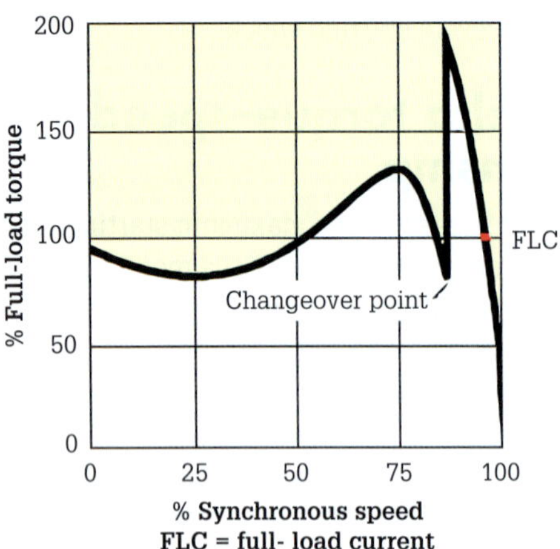

FIGURE 14.106 Autotransformer torque–speed characteristics;

Autotransformer starters are suitable for heavy loads requiring a high starting torque. Applications include centrifugal pumps, conveyors and air compressors.

Primary resistance torque–speed relationship at starting

The resistors with primary resistance starters are designed to allow approximately 60% of line voltage to be connected across the motor terminals at starting. Starting torque is proportional to the square of the voltage so that the starting torque developed at this voltage will be 36% DOL (= 54% FLT when related to the graph of **Figure 14.104**) that could be obtained with DOL starting. An illustration of a torque–speed characteristic at 60% of line voltage is shown in **Figure 14.107**.

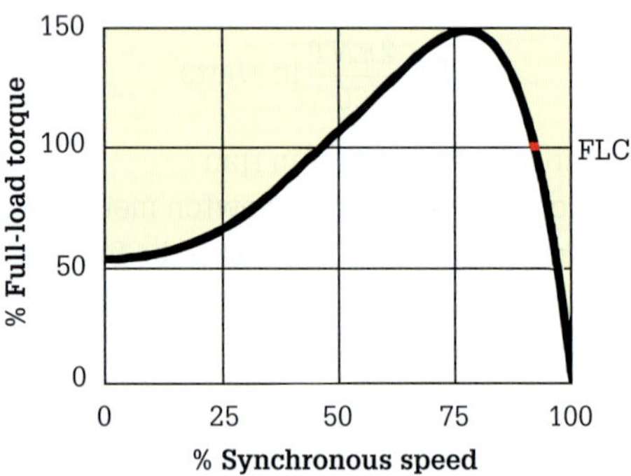

FIGURE 14.107 Primary resistance torque–speed characteristics

From the graph, it can be seen that the torque increases with acceleration. These starters are for small kW rated motors with light starting requirements. The starter resistors are easily adjustable to suit the load and give smooth acceleration with a steadily increasing torque.

Secondary resistor starter torque–speed relationship

The wound-rotor motor can develop a very high starting torque (up to 2.5 DOL value) from zero speed to full speed at a relatively low starting current. The starter has the capability of total control over the start characteristics of the wound-rotor motor by the selection of the rotor resistors. An illustration of a two-stage secondary resistor starter torque–speed characteristic is shown in **Figure 14.108**.

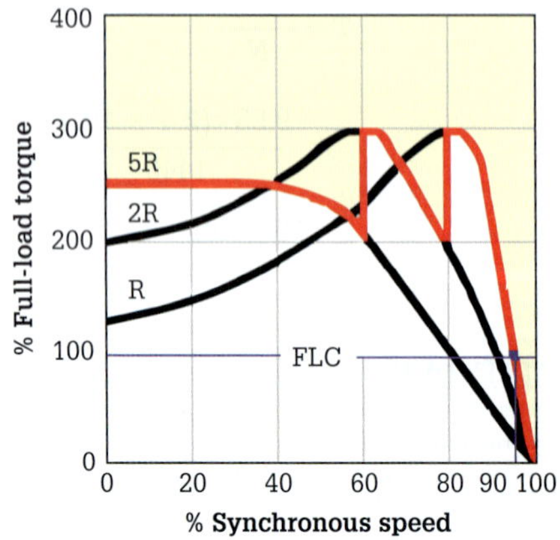

FIGURE 14.108 Secondary resistance torque–speed characteristics

The graph displays the change in torque–speed characteristic that is achieved by lowering the rotor resistance in steps from an initial high value to zero as the motor is accelerated. Note the equal peaks in each step. At 5R the full load speed is 80%, 2R = 90% and R = 96%.

This type of motor is only used where frequent starting, accelerating high-inertia loads, inching and plug braking is required.

Calculating the starting resistance

From the motor nameplate or the manufacturer, values for rotor current and rotor voltage, developed at locked rotor condition when the motor is connected across the supply, can be obtained. From these values a resistor value per phase can be calculated which, when connected in series with the rotor windings, allows the maximum locked rotor torque to be developed. Using Ohm's law,

$$R = \frac{V}{\sqrt{3} \times I} \text{ in ohms}$$

where R = resistance value in ohms (Ω)

V = rotor volts phase-to-phase in volts (V)

I = rotor current in amperes (A)

If the rotor values cannot be obtained then the rotor voltage is the open-circuit voltage at standstill and can be measured across any two slip rings when the starter is connected to the supply with the brushes on the slip rings lifted. Once this is obtained the approximate rotor current can be calculated.

$$\text{Rotor current} = \frac{\text{kW} \times 100 \times 1.1}{\sqrt{3}\,\text{rotor volts}}$$

With wound-rotor a.c. motors (slip-ring motors), if the stator voltage is held constant at line voltage, and the rotor resistance is varied, the torque capacity of the motor and hence the speed can be varied.

Since the rotor current is proportional to the relative motion between the rotating field and the rotor speed, the rotor current and hence the torque are both directly proportional to the slip.

The rotor current is proportional to the rotor resistance. Increasing the rotor resistance reduces the current and increases the slip; hence a form of speed and torque control is possible with wound-rotor motors. This speed-control technique is only useful over a range of 50% to 100% of full speed. An illustration of the wound-rotor circuit with external rotor speed-control resistors is shown in **Figure 14.109**.

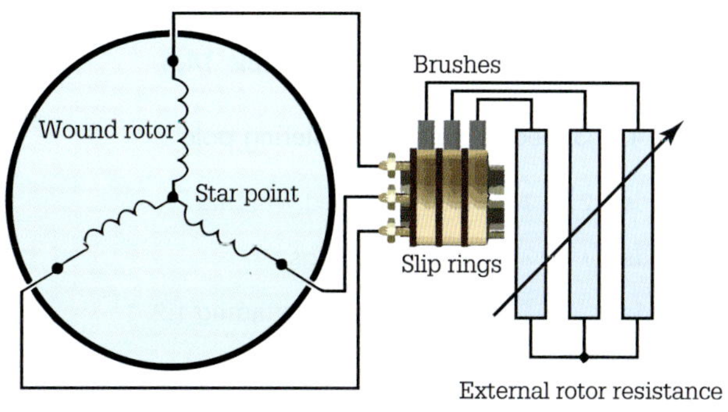

FIGURE 14.109 Wound-rotor and speed-control resistors

Electronic starters torque–speed relationship

The starter can be programmed to provide excellent step-less control of torque from zero to maximum speed depending upon the load requirements. An illustration of a possible electronic starter torque–speed characteristic is shown in **Figure 14.110**.

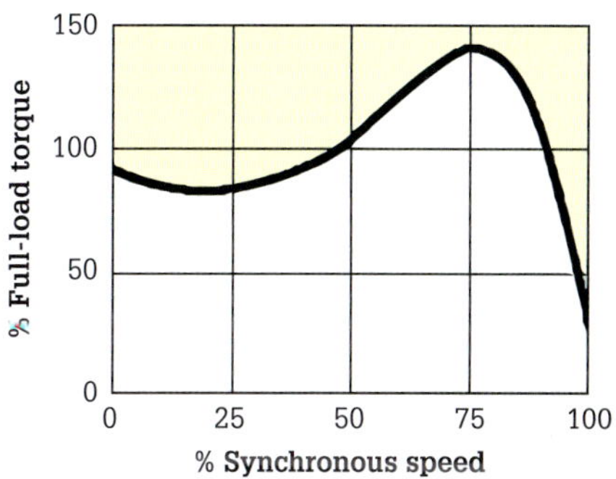

FIGURE 14.110 Electronic torque–speed characteristics

Variable-frequency drives

An electric motor turns at a given speed depending on the number of poles in the motor, and the frequency of the alternating current applied. Motor speed can be changed by changing the a.c. frequency. Nearly all variable-frequency drives (VFDs) manufactured today are referred to as pulse width modulation drives. These drives convert the incoming 50 Hz a.c. power to direct current and then re-transmit the new power signal after being filtered through a d.c. to a.c. inverter to the motor at varying frequencies and voltages. VFDs can enable a motor to develop shaft speeds ranging from near 0 rpm to as high as 150% of the rated speed of the motor.

Pole changing

In a squirrel cage induction motor, the speed is dependent upon the number of poles so this type of motor can have its speed varied by changing the stator winding pole connections to create a different pole configuration.

Several speeds can be obtained by altering the number of poles. The usual types are shown in **Table 14.4**.

TABLE 14.4 Speeds obtained by altering pole

2 speeds 1:2	1 reversible tapped winding
2 speeds	2 separate windings
3 speeds	1 reversible tapped winding 1:2, 1 separate winding
4 speeds	2 reversible tapped windings 1:2
2 speeds	Tapped winding

Preferred speed combinations are shown in **Table 14.5**.

By having two or more sets of winding leads brought out to the terminal connection box, the number of active poles can be changed. Consequently, the leads for each pole configuration are wired to a pole change switch external to the motor.

TABLE 14.5 Ideal speed combinations

Motors with tapped winding	1500/3000	–	720/1500	500/1000
Motors with separate windings		100/1500		
No. of poles	4/2	6/4	8/4	12/6

With the change in speed, the kilowatt rating of the motor also changes. For example, in a two-speed motor for which the slower speed is one-half the higher speed, the kilowatt rating is one-half the kilowatt rating of the higher speed. These pole-changing motors require two sets of overload relays in order to provide adequate protection. A typical power and control diagram for a two-speed separate-winding three-phase motor is illustrated in **Figure 14.111**.

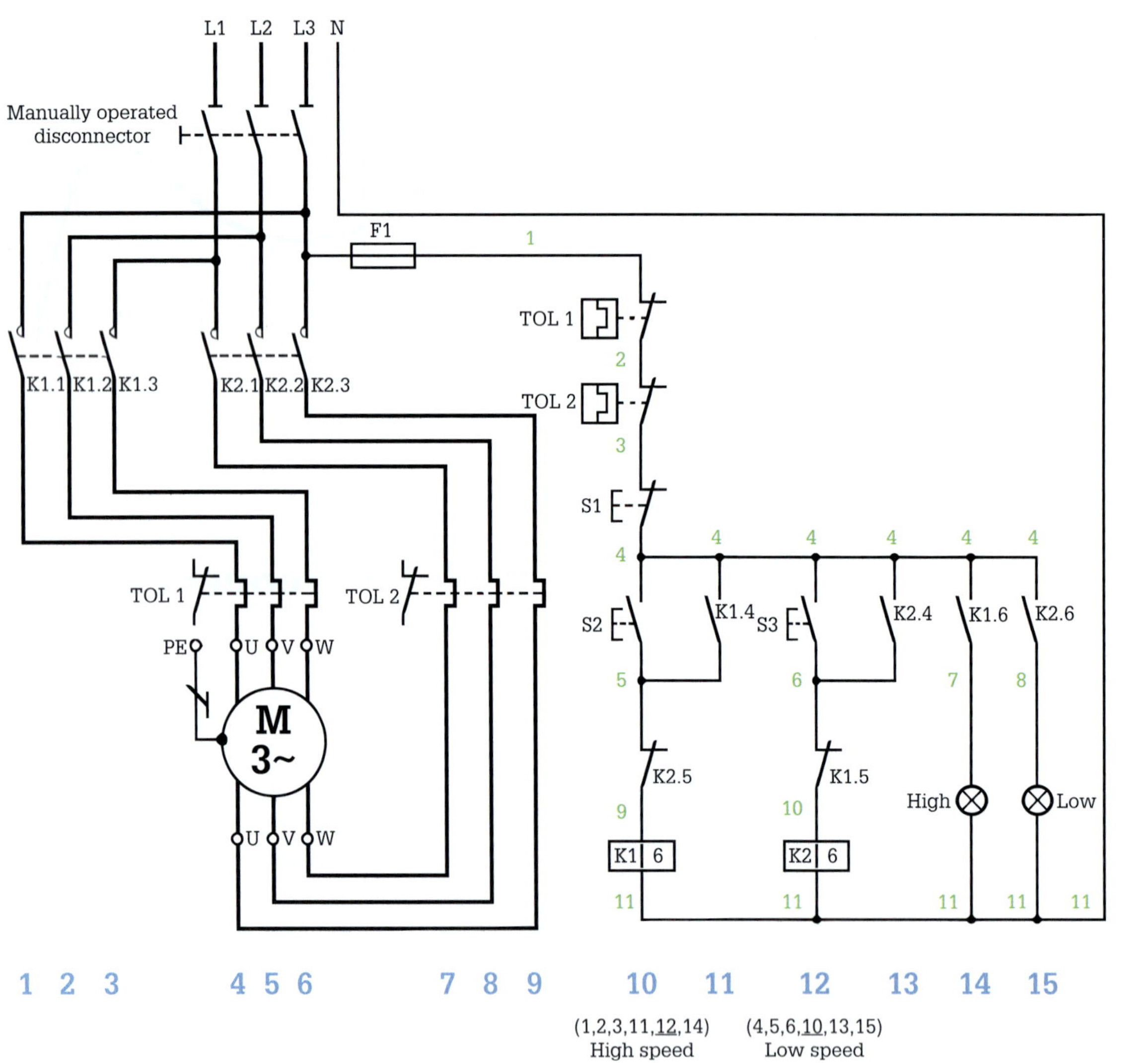

FIGURE 14.111 Two-speed separate-winding power and control circuit diagram

Speed changing for an induction motor with one stator winding can be performed with a consequent-pole connection. This type of connection always produces a speed ratio of 2:1. For example, the consequent-pole motor could provide synchronous speeds of 3000 and 1500 rpm, or 1500 and 750 rpm, or 1000 and 500 rpm. In addition, the stator connections can be configured to produce one of the following motor characteristics: constant kilowatt output, constant torque or variable torque. A connection diagram for a two-speed consequent-pole motor is illustrated in **Figure 14.112**; note that the stator coils are indicated in red.

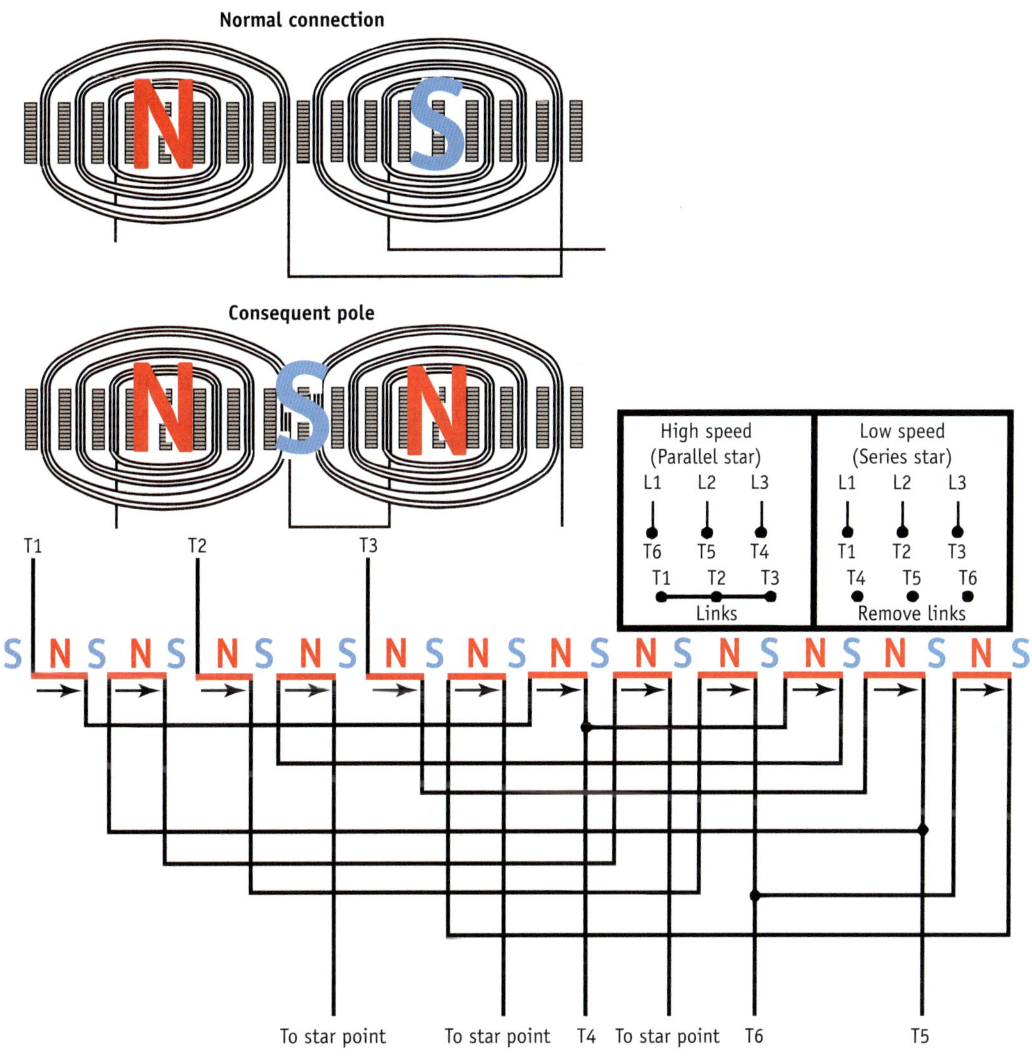

FIGURE 14.112 Connection diagram for a two-speed consequent-pole motor

REVIEW QUESTIONS

1. Calculate the torque produced by a 3.6 kW induction motor having a shaft speed of 1440 rpm.
2. State the principle on which all reduced-voltage starters operate.
3. Name the starter that is a full-voltage-across-the-line starter.
4. List the standard tappings for an autotransformer starter.
5. What starter allows approximately 58% of line voltage to be connected across the motor terminals at starting?
6. Which motor starter allows the motor to develop a very high starting torque (up to 2.5 DOL value) from zero speed to full speed at a relatively low starting current?
7. What is the purpose of variable-frequency drives?
8. Name the type of motor that can have its speed varied by changing the stator winding pole connections to create a different pole configuration.

CHAPTER REVIEW

14.1 Relay circuits

- A circuit component may be represented by a single graphical symbol or an arrangement of several symbols.
- Circuit symbols are used in the drawing of circuit diagrams such as control diagrams and power diagrams to show how an actual circuit operates.
- There are two types of reed relays: dry or wet.
- Electromechanical relays are electromagnetic devices consisting of an electromagnetic coil, an armature mechanism, and electrical contacts.
- Relays can be connected to build very efficient control systems that can perform simple or complex logic functions.
- A solid-state relay (SSR) provides electrical isolation between a control circuit and a switched or power circuit and may replace an electromechanical device.
- A circuit diagram is a basic predictable graphical representation, presented in the form of symbols to provide an outline of circuit and device functions.
- Circuit diagrams that show both the control circuit and the power circuit can be represented in three forms – attached, semi-detached and detached.
- Each line in a circuit diagram should be numbered starting with the top line and reading down for horizontal orientation. With vertical orientation, each vertical line should be numbered on the first vertical line and reading across to the right.
- Line number cross-reference identifies the location and type of contacts controlled by a given relay coil.

14.2 Remote stop–start control

- A starting device is a control device used primarily to start machinery and hold it in operation. A stopping device is a control device used to shut down machinery, while an isolator is a device that holds and locks out the machinery from operation.
- The term 'three-wire control' is derived from the fact that in the control circuit, at least three wires are required to connect the momentary pushbuttons to the coil.
- Emergency stop circuits are one of the most common safety systems found in commercial and industrial installations.
- The term 'two-wire control' obtains this description because only two wires are required to wire to the control device.
- Make-before-break (MBB) switches are those that close a new connection path before another circuit breaks.
- A break-before-make switch is valuable in applications where it is undesirable to have two independent sources short-circuited while switching.
- The dead-man system is an electrical control unit (handle, foot pedal and switch) designed to provide a means of switching off electrical equipment.

14.3 Time delay relays

- The on-delay timer has contacts that delay changing their position when the coil is energised, but revert to their normal position immediately the coil is de-energised.
- The off-delay timer has contacts that change position immediately when the coil is energised, but delay returning to their normal position when the coil is de-energised.
- In pneumatic timers, the timing is accomplished by the transfer of air through a restricted orifice.
- Pole and throw conditions are important to consider when selecting electromechanical relays.

14.4 Circuits using contactors

- An air break contactor is an electromagnetically operated device, for repeatedly establishing and interrupting an electrical power circuit.
- The purpose of contactors and contactor relays is to operate associated contacts and to provide no-volt and under-volt protection.
- Alternating current contactors have a laminated iron core called an armature with a shading ring embedded in the surface to reduce electromagnetic hum and chattering.
- Contactors include normally closed and normally open contacts, but are most often of the single-throw, normally open, double-break configuration.
- A coil drive device performs an on–off operation of the contactor with input signals from electronic equipment.
- Contactors like relays can be fitted with resistance/capacitance (RC) elements, varistors or Zener-diode assemblies for absorbing the coil surge voltage due to contactor on–off operations.
- A contactor relay is an electromagnetically controlled device that opens and closes electrical contacts to effect the operation of other devices in the same or another circuit.
- Contact tips are stamped or are produced by powder metallurgy infiltration and manufactured by sintering from various combinations of the base material silver (Ag) and other materials.
- Thermal overload relays prevent an electric motor from drawing too much current and overheating.
- A mechanical latching device holds the contactor in the closed position should the supply voltage fail at the contactor terminals.

14.5 Jogging and interlocking

- Jogging describes the repeated starting and stopping of machinery at frequent intervals for any short periods of time necessary for obtaining limited movement.
- For working environments where it is prudent for an operator to account for the position of both hands, some control devices provide for two start switches.

- Electrical interlocking refers to the use of auxiliary contacts of one relay (NC or NO) in series with another relay coil to make certain that this second relay cannot be energised at the same time and vice versa.

14.6 Control devices

- An arc suppression device may be in parallel with the relay coil or in parallel with the contacts used to control the relay.
- The function of a limit switch is to switch an electrical signal when physical contact occurs between the device to be detected and the operating head of the switch.
- Pressure switches are used to verify that the operational conditions of the process are being met.
- An RC snubber limits fast load voltage changes (e.g. caused by inductive loads) at the load terminals.
- Sensors may also be divided into analogue or digital, depending on whether they provide a continuously varying signal (analogue) or only two possible states (digital).
- Devices that convert angular speed to some other form of reading or usable signal are known as tachometers.
- A float switch is simply an electrical switching circuit controlled by the tilting movement as the float switch moves up and down on the surface of the liquid.

14.7 Programmable relays

- A microprocessor is an integrated system of electronic circuits and devices infused on a single chip.
- The mathematical operations that the microprocessor performs are in essence the repeated addition or the sum of two numbers.
- The logic operation that a microprocessor performs is the contrast between two numbers to see if one number is the same as, less or greater than the other number.
- All input to a microprocessor whether from sensors or from the keyboard is converted to zeros or ones.
- Buses are groups of wires such as a twisted pair that carry data and control signals from one device to another.
- C-Bus (Clipsal Bus) is a microprocessor fully programmable controlled wiring system for industrial, commercial and domestic installations.
- A programmable logic controller (PLC) is a microprocessor-based device that accepts input data from various devices and processes that data by making decisions according to an installed program.
- A PLC uses a programming language based upon recognisable symbols common to electric motor control.
- The input interface receives various digital or analogue communication signals from each of the input devices.
- The central processing unit (CPU) includes a microprocessor, RAM, and ROM memory.
- A ladder diagram is a specific kind of schematic diagram that shows in simple terms how a circuit works.

14.8 Three-phase induction motor starters

- Motor starter circuit diagrams show the connections of both the power and the control circuits as clearly as possible with all conductors drawn neatly as straight lines.
- The timers for open transition star–delta starters should be break-before-make contacts with a minimum time of 50 ms between break and make in order to quench the arc developed on the star contactor.
- Autotransformer starters should have timers in the control circuit to limit the number of starts per hour to the rating of the transformer.
- Primary resistance starters are classified as closed transition starters because the motor is not disconnected from the supply during the starting sequence.
- The wound-rotor motor can have its rotor characteristic altered by inserting variable external resistance (metal or liquid) via slip rings in series with the rotor winding.

14.9 Three-phase induction motor reversal and braking

- Braking is essentially the removal of stored kinetic energy in the form of motion from a motor-mechanical load.
- The electromechanical brake applies a braking force directly to the brake disc mounted on the motor shaft.
- Dynamic braking, also called d.c. injection braking, utilises the ability of the a.c. drive motor to act as a generator.
- Plug braking is the fast deceleration of a motor by reversing any two line feeds to the stator terminals while it is running at full speed.
- Regenerative braking takes advantage of the fact that when the motor stops being powered and just coasts, it acts as a generator and can feed energy back into the a.c. supply system.

14.10 Three-phase induction motor speed control

- When an induction motor stator winding is energised a rotating magnetic field is created which rotates at synchronous speed.
- The difference between the synchronous speed (N_s) of the stator magnetic field and the actual rotor speed (N) is called slip.
- DOL starters are a full-voltage-across-the-line starter.

- Three starting torques are available with the standard tappings of autotransformer starters: 80%, 65% and 50% of line voltage, depending upon design requirements with respect to the connected load.
- Primary resistance starters are designed to allow approximately 60% of line voltage to be connected across the motor terminals at starting.
- The wound-rotor motor can develop a very high starting torque (up to 2.5 DOL value) from zero speed to full speed at a relatively low starting current.
- With wound-rotor a.c. motors (slip-ring motor), if the stator voltage is held constant at line voltage, and the rotor resistance is varied, the torque capacity of the motor and hence the speed can be varied.
- Electronic starters can be programmed to provide excellent stepless control of torque from zero to maximum speed, depending upon the load requirements.
- In a squirrel cage induction motor, the speed is dependent upon the number of poles; therefore, this type of motor can have its speed varied by changing the stator winding pole connections to create a different pole configuration.

TRIAL EXAM

For Chapter 14 knowledge assessment, please complete the following trial exam.

1 The relay that does not contain moving parts is the:
  a solid-state relay
  b timer relay
  c contactor relay
  d thermal overload relay

2 The synchronous speed of a four-pole induction motor operating at a supply frequency of 50 Hz is:
  a 6000 rpm
  b 3600 rpm
  c 3000 rpm
  d 1500 rpm

3 What kind of diagram is a basic, predictable graphical representation of an electrical system?
  a wiring diagram
  b control diagram
  c circuit diagram
  d line diagram

4 A latching contact is a:
  a power contact
  b normally closed contact
  c stop contact
  d normally open contact

5 Stop buttons are connected:
  a in series with other stop buttons
  b in parallel with other stop buttons
  c in the power circuit
  d across the electromagnetic coil

6 Three-wire control is a control wiring method where:
  a the coil is connected at both the start and stop pushbuttons
  b the power circuit is connected across three-phase
  c there are 3 wires used to connect the momentary pushbuttons to the coil
  d a break-before-make switch is used

7 What type of device prevents a control circuit from operating out of sequence?
  a an interlock
  b a reed relay
  c a solid-state relay
  d a contactor

8 Why does an a.c. coil draw more current when in the unlatched state than when in the latched state?
  a because of its low resistance
  b because of the load current it is controlling
  c because of the loading effect of the devices connected in series with it
  d because of its low impedance

9 An AC-5a type contactor is used for:
  a switching of electric discharge lamps
  b squirrel cage motors: starting
  c squirrel cage motors: plug braking
  d slip-ring motors: starting

10 A pushbutton changeover switch with a normally closed and a normally open switch configuration is used as a:
  a mechanical latching device
  b surge suppression device
  c jogging device
  d high-speed switch

11 Name the type of switch that is used as a transfer switch between mains power and an emergency power supply.
  a break-before-make switch
  b temperature switch
  c limit switch
  d make-before-break switch

12 Name a transducer that produces an electrical signal without any mechanical interaction.
  a a proximity switch
  b the limit switch
  c the pressure switch
  d the float switch

13 What type of component can, by adding two numbers, perform addition, multiplication, subtraction and division?
- a a register
- b a microprocessor
- c the data bus
- d the twisted pair

14 Name the relay that checks the inputs, runs through the program and changes the outputs.
- a reed relay
- b an electromechanical relay
- c a programmable relay
- d a solid-state relay

15 Which of the following starters is an open transition starter?
- a liquid resistance starter
- b primary resistance starter
- c secondary resistance starter
- d star–delta starter

16 What is another name for an electronic starter?
- a step-less reduced-voltage starter
- b DOL starter
- c autotransformer starter
- d Wye-delta starter

17 Which of the following braking methods is also referred to as d.c. injection braking?
- a dynamic
- b electromechanical
- c plug
- d eddy current

18 Variable-frequency drives:
- a are dependent upon the number of poles
- b use tapped windings for speed changes
- c use pulse width modulation to vary the speed
- d use separate windings for speed changes

19 A 400 V three-phase motor is started using a star–delta starter. Calculate the current drawn at start if the DOL starting current is 45 A is:
- a 15 A
- b 22.7 A
- c 45 A
- d 135 A

20 The DOL starting torque for a 400 V three-phase induction motor is 45 Nm. The starting torque if the motor is started on an autotransformer using the 65% tap is:
- a 9.23 Nm
- b 19 Nm
- c 38 Nm
- d 135 Nm

15 Select wiring systems and cables

This chapter provides electrotechnology workers with essential knowledge and skills pertaining to the application of common wiring systems and cable types. It assists learners to develop skills and knowledge in selecting wiring systems that are compatible with the installation conditions, selecting cables that comply with required current-carrying capacity, voltage drop, and earth fault-loop impedance limitations, as well as coordinating between protective devices and conductors, and documenting selection decisions. This chapter provides underpinning knowledge for the unit UEEEL0018 from the UEE training package.

LEARNING OBJECTIVES

Design and safety performance requirements

- Explain factors that must be considered when designing or planning an electrical installation.

Final sub-circuit arrangements

- Understand the factors affecting the various circuits within an installation.
- Determine the number and types of circuits necessary for an installation.

Factors affecting the suitability of wiring systems

- Identify the factors influencing the installation of a wiring system.

Maximum demand

- Determine the maximum demand of various electrical installations.

Cable selection based on current-carrying capacity

- Select cable based on current-carrying capacity requirements.

Cable selection based on voltage drop

- Select cable based on voltage-drop requirements.

Cable selection based on fault-loop impedance

- Select cable based on fault-loop impedance requirements.

Selecting protection devices

- Select protection devices.
- Identify devices used for the automatic disconnection of supply.

Selecting devices for isolation and switching

- Select devices for isolation and switching.

Switchboards

- Apply AS/NZS 3000:2018 and local supply authority requirements for switchboards.
- Explain the various tariffs.

15.1 Design and safety performance requirements

An electrical installation consists of the fixed electrical wiring and all associated connection components and fittings, which includes permanently wired stationary appliances. An electrical installation does not include portable equipment and appliances. The design of an electrical installation must take into account the proposed electrical load, the circuit system design, environmental aspects in which the electrical system is to be installed, client requirements, type of service to be supplied, legislation, local service rules and AS/NZS 3000:2018 *Wiring Rules*, Section 1.6, 'Design of an electrical installation'.

The installation must be designed, constructed, installed and protected to minimise the risk of fire or property damage. In normal operation, the electrical installation must not create the risk of electric shock, excessive temperatures likely to cause burns, fires or other physical injury hazards to persons and livestock (see Section 1, 'Scope application and fundamental principles', AS/NZS 3000:2018).

The electrical installation must:

- safely deliver maximum demand
- include correctly rated protection devices for safeguarding against over-current, fault current and earth leakage conditions
- be controlled by a main switch or switches
- be certified that it is electrically safe.

These requirements reflect the fact that the layout, design, and testing of the electrical system is the responsibility of suitably qualified persons. It is essential, therefore, that the licensed electrical contractor document all design aspects of the installation to demonstrate its fitness for the intended purpose.

An installation meets performance standards if it has been inspected and tested by a licensed electrician and found to conform to specified rules such as the AS/NZS *Wiring Rules* and to procedures of applicable Electrical Acts and the regulations framed under them, local rules and codes of practice.

Design considerations

Important items to consider in the design of an electrical installation are:

- harmful effects
- supply characteristics
- supply availability
- earthing
- use of equipment
- harmonic currents.

Harmful effects

The design of every electrical installation must be such that it eliminates or minimises the harmful effects of electric current that may occur within the wiring system. The design of the installation must afford protection of the circuit conductors by disconnecting faulty circuits from the supply. If the installation design is not appropriate, then the consequences for the circuits are thermal overloading of the conductor insulation and fire.

Persons and livestock must be protected from electric shock hazards arising from direct and indirect contact with live conductors or exposed conductive parts of the installation.

Supply characteristics

Characteristics of the electrical supply that apply to an electrical installation include:

- frequency
- current
- voltage.

Frequency

Electricity supplies throughout Australia are usually an alternating-current (a.c.) sinusoidal waveform at a frequency of 50 Hz nominal (49.85 Hz to 50.15 Hz). Frequency control is a function of the generation process and is maintained to within ±0.2% of the nominal 50 Hz. However, on occasions a generation event or a load event can change the frequency by a significant amount (called an 'excursion').

SWITCH ON

A load event means an identifiable connection or disconnection of more than 50 MW of load. A generation event means a synchronisation of a generating unit of more than 50 MW.

Current

While the supply to most electrical installations in Australia is alternating current, some installations in remote areas use direct-current (d.c.) sources for electricity supplies. For example, energy from renewable energy sources, mainly solar and wind, relies heavily on energy storage technologies such as lead-acid batteries. Such is the case with Basslink, which is an undersea power cable linking Tasmania to mainland Australia's electricity network and carries d.c. electricity from a 480 MW monopolar HVDC converter station that has an a.c. input of 220 kV.

Voltage

Standard nominal voltages are: 230 V/400 V within a range of +10% and –6%, and 460 V, 6.6 kV, 11 kV, 22 kV, 33 kV, 44 kV and 66 kV. Standard low-voltage systems are three-phase four wire 230/400 V, single-phase 230 V and three wire 230/460 V. The highest voltage allowed for a single-phase 230 V system is 253 V (nominal +10%), and the lowest voltage is 216 V (nominal –6%).

Supply availability

To determine if an electrical installation (greater than 70 A) can be provided with the current required, an electrical design consultant needs to create the installation using the electrical distributor standards and guidelines. The design and application forms must then be sent to the electrical distributor to review its suitability and to provide a quote. Possible details required are:

- plan/layout of the installation (switchboards, cable route) including a single-line diagram of the installation
- network connection: domestic, commercial, industrial and so on
- total consumption and the method used to estimate load such as AS/NZ 3000:2018 *Wiring Rules* or VA/m^2
- load patterns
- tariff details (low- or high-voltage supply)
- CT metering (loads greater than 100 A)
- disturbing equipment (inverters, welders, etc.) and motor details
- transformer requirements (high voltage or low voltage)
- underground or overhead supply
- earthing details.

The electrical distributor needs these details to see how the installation affects the current-carrying capacity of its distribution system. In order to connect this new installation an increase in the electrical distributor's capacity may be required. Increased capacity could include larger distribution transformers, switchgear and larger conductors for which the owner of the intended electrical installation must pay a percentage of costs.

Earthing

All new electrical installations must use the multiple earthed neutral (MEN) system of earthing and comply with the requirements for this earthing system as specified in AS/NZS 3000:2018 *Wiring Rules*.

Use of equipment

The equipment in an electrical installation must be arranged and operated so as to minimise or prevent adverse effects on the distribution system and other electrical installations connected to the distribution system. For example, the starting current of motors, which is significantly higher than running current, can cause a substantial fluctuation of the supply voltage. Other types of equipment that can cause excessive voltage disturbances because of their load demands are welding machines and electric arc furnaces.

Harmonic currents

Harmonic currents are produced when a load draws a non-linear current from a sinusoidal supply. Switch-mode power supplies used extensively in computer and electronic equipment, or variable-frequency drives used to control electric motors can distort the voltage and current waveform. Harmonic currents combine in the neutral conductor, resulting in the current being well above the current-carrying capacity of the cable. Overload currents increase the temperature of conductors and their insulation, causing the insulation to break down (refer to AS/NZS 3000:2018 *Wiring Rules*, 'a neutral conductor', 'harmonic currents').

Prospective fault current

The electrical installation must be capable of enduring, without destruction, the prospective fault current that can be drawn under a fault condition. The prospective fault current refers to the magnitude of current that would flow at a particular point in the installation if a fault of negligible impedance occurred at that point. The prospective fault current depends on the voltage at the source and the impedance of the upstream circuit. Refer to Chapter 11 of this textbook for detailed information on prospective fault current.

Determining cable impedance

Cable impedance is calculated from the conductor resistance and reactance values provided in AS/NZS 3008.1.1:2017 *Electrical installations – Selection of cables Cables for alternating voltages up to and including 0.6/1 kV – Typical Australian installation conditions* and AS/NZS 3008.1.2:2017 *Electrical installations – Selection of cables Cables for alternating voltages up to and including 0.6/1 kV – Typical New Zealand conditions* using the equation:

$$Z_C = L \times \frac{\sqrt{R_C^{\,2} + X_C^{\,2}}}{1000}$$

where: L = cable length in metres

Z_C = Cable impedance in Ω

R_C = Cable resistance in Ω/km

X_C = Cable reactance in Ω/km

The values for cable reactance are obtained from Tables 30 to 33 inclusive and values for cable resistance are obtained from Tables 34 to 39 inclusive.

EXAMPLE 15.1

Calculate the impedance of a 20 m length of 16 mm^2, two-core and earth, V-90 insulated cable having copper conductors.

From Table 34 of AS/NZS 3008.1.1, the resistance of 16 mm^2 copper conductor at 75 °C (maximum operating temperature for V-90 cables) is 1.40 Ω/km and from Table 30 the reactance of 16 mm^2 copper conductor with PVC insulation is 0.0861 Ω/km (column 9).

$$Z_C = L \times \frac{\sqrt{Rc^2 + Xc^2}}{1000}$$

$$= 20 \times \frac{\sqrt{1.40^2 + 0.0861^2}}{1000}$$

$$= 0.0281\ \Omega$$

EXERCISE 15.1

Calculate the impedance of a 50 m length of 16 mm^2, two-core and earth, V-90 insulated cable having copper conductors.

Determining fault current level

It is important to be able to calculate fault current levels to enable the correct selection of protection devices. A circuit breaker, which is not able to safely isolate the fault current may suffer catastrophic failure. For example, if the fault level at a switchboard was determined to be 6 kA, it would be inappropriate to install 4.5 kA circuit breakers without installing additional upstream fault current limiters. Refer to Chapter 11 of this textbook for detailed information on determining fault current level.

Maximum demand of an installation

It is essential to know the magnitude of required load current in the installation before installation work commences. The maximum demand of the installation is defined as the electrical loading of the installation. Consequently, the conductor cross-sectional area (CSA) of consumer's mains and sub-mains must be able to carry the electrical demand.

Figure 15.1 illustrates how the consumer's mains supplies the sub-mains and sub-circuits within the electrical installation. The consumer's mains carry the sum of the installation current when the sub-circuits are energised. The sub-main, however, only carries the load current demanded by its connected sub-circuits.

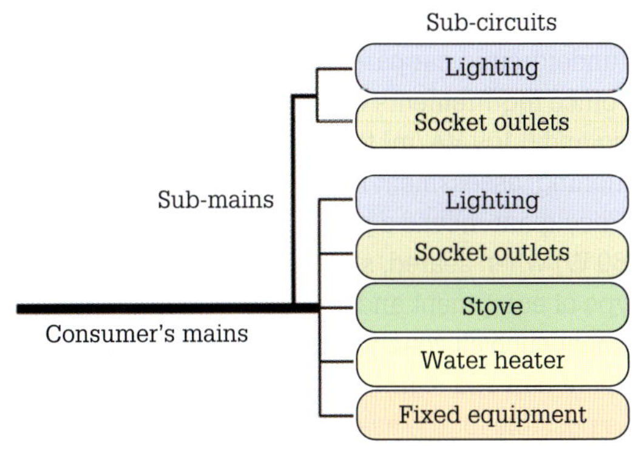

FIGURE 15.1 Basic installation diagram

Knowledge of the maximum demand of the installation is also necessary so that supply entities can organise a suitable supply for the electrical installation (single, two phase or three phase).

Methods for determining maximum demand

AS/NZS 3000:2018, Appendix C, recognises four methods for determining the maximum demand of consumer's mains or sub-mains:

1. Assessment
2. Limitation
3. Measurement
4. Calculation.

Assessment method

Assessment of maximum demand can be based on the duty cycle (the proportion of time during which the installed equipment operated) or the load characteristics of the installed equipment. For example, the maximum demand of a single phase 3.6 kW storage water heater can be calculated by dividing the power (3.6 kW) by the nominal equipment voltage (230 V) to give a current of 15.65 A.

In the larger commercial and industrial installations where the load characteristics of the type of occupancy or the floor area and system of climate are known, the maximum demand may be assessed. **Table 15.1** shows assessment values for different building categories. Refer to AS/NZS 3000:2018 for guidance on using the assessment method for determining maximum demand.

TABLE 15.1 Assessment values

Building categories	VA/m^2	kVA/m^2
Air-conditioned with non-electric heating	100	0.10
Air-conditioned with electric heating	120	0.12
Air-conditioned (reverse cycle)	110	0.11
Electric heating with no cooling	100	0.10
Non-electric heating with no cooling	60	0.06

Hospital developments might be assessed at 125 VA/m^2, while underground car parks may be evaluated at 15 VA/m^2. Small office requirements could use 0.065 VA/m^2 for lighting and 0.109 VA/m^2 for light power uses. For air-conditioning, assessment is based on typical occupancy levels (occupant means a person – a person has a heat load (60 W) when seated, standing or walking), lighting and type of equipment and optimum building criteria. Equipment such as vending machines and other similar equipment can add significantly to the heat load estimates.

It is possible for the assessment method to be conservative resulting in an over-design for consumer's mains or sub-mains. However, some installation designers using the evaluation method may use diversity. Diversity, when applied to installation design, takes into account the percentage of people occupying, lights, equipment and so on that are on simultaneously compared with the total demand used for design.

Assessment values can also be taken from a similar existing installation in which the maximum demand has been measured using a maximum demand indicator. The measured value provides the actual value for a particular period rather than a theoretical value obtained by calculation.

Limitation method

A fixed setting or an adjustable circuit breaker protecting all the associated final sub-circuits may be used to limit the maximum demand of consumer's mains and sub-mains. Alternatively, the maximum demand could be determined by the summation of the current settings of the circuit breakers protecting the final sub-circuits.

Measurement method

The maximum demand of an existing electrical installation may be measured by a maximum demand indicator (provides a guideline to the maximum current drawn) over 15 minutes or measured by the use of a data logger that measures in 15-minute intervals over a seven-day period. The maximum demand indicator and the data logger use current transformers (CTs) that are clipped around the cables to be measured. The measurement method is usually applied to existing electrical installations where the load profile needs to be changed. A maximum demand indicator (MDI), as illustrated in **Figure 15.2**, is a resettable, peak demand current-measuring device with a red maximum demand slave pointer. An MDI meter must be provided in each phase over the same 15-minute interval.

Care has to be taken when using the maximum demand indicator measurement method. For example, the maximum recorded demand for a major commercial complex was 512 kW that at a power factor of 0.85 equates to 602 kVA. As this recorded measurement took place during the day, outdoor car park lighting was omitted. A valid measurement must include all loads.

Calculation method

The calculation method may be used as a basis for determining the expected maximum demand of an electrical installation. The maximum demand may be calculated by following the directions in Appendix C of AS/NZS 3000:2018 for the suitable category of electrical installation. Maximum demand calculation is based on two tables: Table C1 is used for domestic installations and Table C2 is used for non-domestic installations. Tables C1 and C2 apply the principle of diversity, which takes into account the normal operating conditions of circuits during which all equipment is not operating simultaneously at full load or for periods exceeding 15 minutes.

FIGURE 15.2 Maximum demand indicator

For example, in a domestic electrical installation we might have 40 lighting points that are 60 W at each point. If we assume that the average full load for the points is 2400 W, then the current expected to flow in the final sub-circuit will be approximately 10 A. However, since the lighting points are separately controlled, we do not know exactly when each point turns on. The condition that only a few points are in use at the same time and that their total load is small is known as diversity.

Note: While the nominal supply voltage is 230 V, equipment is rated at 220–240 V. The table for maximum demand, domestic purposes, gives us a correction factor which results in a lower total current load for the 40 lighting points (3 A for 1 to 20 points + 2 A for each additional 20 points or part thereof). So instead of a load current of 10 A, the load current for maximum demand purposes, using the diversity factor, is only 5 A.

Diversity is applied in the calculation method on the basis that the percentage of people, lights, equipment and so on that are on (or present) simultaneously compared with the total design demand may not occur at the same time. As a consequence, the consumer's mains or sub-mains are not required to carry the real full-load current of the installation at any one time.

The AS/NZS prescribed calculation method accounts for all items of electrical equipment connected to the main or sub-main circuits together with the appropriate diversity factors for different types of loads. This method of assessment is used by installation designers for domestic installations.

REVIEW QUESTIONS

1 List the items that compromise an electrical installation.
2 What are the requirements for an electrical installation?
3 When is an electrical installation deemed to meet performance standards?
4 Name six important design considerations for an electrical installation.
5 What electrical supply characteristics apply to an electrical installation?
6 Which earthing system is used in all new electrical installations?
7 What causes harmonic currents in an electrical installation?
8 To what does 'prospective fault current' refer?
9 In which Standard are conductor resistance and reactance published?
10 What does the term 'maximum demand' mean?
11 Name four acceptable methods for determining the maximum demand of an electrical installation.
12 What is the function of a maximum demand indicator?
13 When is it appropriate to use the assessment methods for determining maximum demand?
14 To what does the term 'diversity' refer in relation to determining maximum demand?

15.2 Final sub-circuit arrangements

The creation of different circuits in an electrical installation makes it possible to:

- limit the effects of a fault to the circuit concerned
- simplify fault-finding
- carry out maintenance work, inspections, testing or circuit extensions without interrupting the whole electrical installation. In general, the following circuit groups are used:
  - lighting – indoor, outdoor, communal
  - socket outlets – 10 A, 15 A, 20 A
  - cooking – ranges, ovens, cooktops
  - fixed-space heating – air-conditioning, ventilation
  - water heating – instantaneous, storage, spa, swimming pool and laundry equipment
  - safety – emergency lighting, fire protection, security and for uninterruptible power supplies
  - power circuits for fixed plant – lifts, motors, kilns and welding machines.
- control circuits for fixed plant.

The number and type of circuits required are determined by the load on the circuit, the location of the loads, any seasonal variations affecting the loads and any special conditions.

Refer to AS/NZS 3000:2018 *Wiring Rules*, 'electrical installation circuit arrangement'.

Essential principles

When dividing an electrical installation into separate circuits, some important considerations are:

- external influences
- protection for safety
- protection against unwanted voltages
- protection against injury from mechanical movement.

External influences

The design and erection of an electrical installation must be appropriate for the external environmental influences. AS/NZS 3000:2018 requires that all parts of electrical installations be designed to afford protection against environmental influences that may cause damage.

External influences mean the surrounding environment to the extent that it has an impact on the electrical installation.

SWITCH ON

Such outdoor influences are: ambient temperature, atmospheric humidity, altitude, ultraviolet radiation, presence of water, presence of foreign solid bodies, presence of corrosive or polluting substances, mechanical stress impact, other mechanical stresses, presence of flora and/or mould growth, presence of fauna, and electromagnetic, electrostatic or ionising influences.

Other external influences that can affect parts of an electrical installation include:

- activities to be approved in and adjacent to the electrical installation
- a building's construction materials which may affect the temperature and fire rating of proposed circuit cables.

An example of where external influences affect the design of the installation is where an electrical installation has an aerial service. An aerial service must be designed and erected in such a way that its construction and location guards against dangers to persons, livestock and property. Therefore, an aerial service must be routed with an adequate clearance to ground, vegetation, other types of aerials, roadways and buildings. Consequently, external influences affect the choice of appropriate installation equipment and the selection of protective measures for the safety of persons, livestock and buildings (refer to AS/NZS 3000:2018 *Wiring Rules*, 'electrical installation circuit arrangement', external influences).

Protection for safety

Refer to AS/NZS 3000:2018 *Wiring Rules*, Clause 1.5 'fundamental principles'.

In electrical installations two major types of risk exist:

1. shock currents
2. excessive temperatures are likely to cause burns and fires and have other harmful effects.

In addition to the two main areas requiring protection for safety – direct and indirect contact – there are six other fundamental principles specified in AS/NZS 3000:2018 *Wiring Rules* where protection is needed to minimise the risk of electric shock and fire:

- Protection against thermal effects that may cause persons and livestock to be burnt or flammable materials to ignite. Methods used can be screening, enclosure or guarding.
- Protection against unwanted voltages via screening and removal of all unused conductors.
- Protection against over-current that could cause damage through excessive temperatures or electromechanical stresses. Refer to AS/NZS 3000:2018 *Wiring Rules*, '2.5 Protection against overcurrent'.
- Protection against the effects of fault currents (prospective short-circuit currents). Refer to AS/NZS 3000:2018 *Wiring Rules*, '2.5 Protection against overcurrent'.
- Protection against the effects of over-voltages (a fault between live parts that are supplied at different voltages). Refer to AS/NZS 3000:2018 *Wiring Rules*, '2.7 Protection against overvoltage'.
- Protection against the hazards from mechanical movement. Methods used can be disconnection or isolation. Refer to AS/NZS 3000:2018 *Wiring Rules*, 'Protection against injury from mechanical movement'.

The methods of achieving this protection in the six additional fundamental areas are established as specific rules in AS/NZS 3000:2018 *Wiring Rules*. When designing or installing an electrical installation, it is essential to document how the installation complies with the fundamental principles for the protection of safety.

Protection against unwanted voltages

It is necessary to provide protection from injury or property damage due to any harmful effects of voltage that may exist in unused conductors because of induction or any other cause. Refer to AS/NZS 3000:2018 *Wiring Rules*, 'voltage in unused conductors'.

Typical sources of unwanted voltages include:

- voltages between phases
- voltages between phases and earth – during single phase to earth faults, voltages of sufficient magnitude can exist on an unearthed metallic pipe
- live exposed conductors
- voltages between live exposed conductors and conductive materials
- voltages across undischarged capacitors
- voltages on disconnected conductors – mainly neutrals and unused conductors that have not been physically removed from service
- induced voltages – due to electromagnetic or capacitive coupling
- multiple-supply sources
- high impedance neutral (MEN system)
- electrical testing – routine and in damp situations and hazardous areas
- circuits becoming live – omission of safety tags and lockouts
- circuits with insulation damage – lack of mechanical protection.

To prevent unwanted voltages, all circuits must be placed into an electrically safe work condition. This means that all circuits must be:

- tested for voltage presence
- isolated from energised parts
- earthed if deemed necessary.

Protection against injury from mechanical movement

All parts of equipment or machines that move in the course of the performance of work may contribute to accidents causing injury and damage. All electrically actuated rotating, linear and reciprocating machine movements can be dangerous. Refer to Chapter 11 of this textbook for appropriate methods used to protect against injury from mechanical movement.

Number and types of circuits required

Electrical installations will comprise some or all of the following circuits:

- consumer's mains
- sub-mains
- final sub-circuits

Consumer's mains

The supply to a consumer's premises is a four-wire configuration, which comprises three phases and a neutral with 230 V between each phase and neutral and 400 V between each of the phases. There is no minimum conductor size for consumer's mains in AS/NZS 3000:2018

Wiring Rules. However, refer to the local utility service and installation rules for their requirements as some indicate 6 mm² for copper conductors while others require 4 × 16 mm². The final connection to the consumer from the mains in the road verge is made via a lead-in service cable or service line.

The term 'consumer's mains' describes a cable connected between the point of supply (POS) and the main switchboard in an installation (refer to AS/NZS 3000:2018 *Wiring Rules*, consumer's mains). These cables belong to the consumer. The POS is the junction at which the utility's service mains connect to the consumer's mains. There are three wiring systems for consumer's mains: underground, surface or overhead wiring. The consumer's mains connection can occur at underground POSs or overhead service line POSs.

Sub-circuits

Every electrical installation must be divided into circuits to avoid hazards and to contain the disruption in the event of a fault. In addition, the division of an installation into various circuits makes easier the safe operation of the installation, improves the quality of circuit inspection and aids reliability with testing and makes maintenance simpler. The electrical distribution within an installation ends with final sub-circuits (see AS/NZS 3000:2018). A final sub-circuit is that section of a wiring system that extends beyond the final circuit protection device. Final sub-circuits can originate from either a main switchboard or a sub-main distribution board.

Having ascertained the number of final sub-circuits, a decision can be made on the installation method, the type of cable and how to protect against electric shock and over-current.

Current requirements for final sub-circuits

The current requirements for final sub-circuits are determined by the full connected load of the equipment. However, Appendix C5.2 and Table C9 of the *Wiring Rules* provide guidance on the loading of the number of points allowable on a circuit and must be consulted. Cables selected to continuously carry the load current need to meet installing methods and environmental conditions such as ambient temperature.

EXAMPLE 15.2

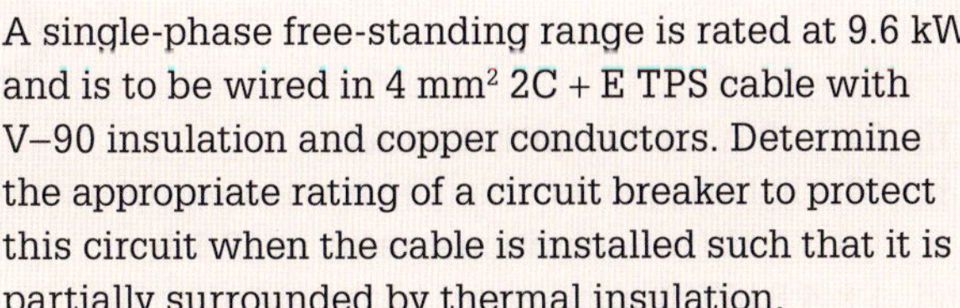

A single-phase free-standing range is rated at 9.6 kW and is to be wired in 4 mm² 2C + E TPS cable with V–90 insulation and copper conductors. Determine the appropriate rating of a circuit breaker to protect this circuit when the cable is installed such that it is partially surrounded by thermal insulation.

Solution:

From Table C9, 4 mm² can supply a 10 kW range when protected by a 25 A circuit breaker, which is supported by Table C6.

EXERCISE 15.2

A single-phase free-standing range is rated at 7.6 kW and is to be wired in 2.5 mm² 2C + E TPS cable with V–90 insulation and copper conductors. Determine the appropriate rating of a circuit breaker to protect this circuit when the cable is installed such that it is partially surrounded by thermal insulation.

Labelling and circuit schedules

Circuit reference labels of Traffolyte with black letters on a white background or other approved non-deteriorating material, fixed by means of screws, rivets or glue, must identify every relay, pushbutton, fuse, circuit breaker, switch, meter, indicator light, terminal block, motor and other apparatus including all socket outlets and light switches in commercial and industrial installations. With domestic installations, functional switches providing isolation for various types of loads such as cooktops must be identified.

Labels for power sockets and light switches should be located adjacent to the accessory and must refer to the switchboard origin and circuit breaker number (e.g. DB-1 CB12). An alternative accessory labelling system is the Clipsal C2000 Series. An approved label showing the outlet circuit ID number should also be mounted adjacent to each lighting socket in false ceilings.

All control circuit wiring should be labelled at each end and in junction boxes by an approved non-deteriorating means with circuit reference numbers. All wiring should be colour coded to maintain phase sequence identification. Typed circuit schedules, an example of which is shown in **Figure 15.3**, should be fitted on the doors of all switchboards behind Perspex panels, and should comprise the following:

- circuit reference number
- size and cable and rating of circuit breaker or fuse
- description of the circuit
- name of switchboard

Switchboard SB 12 **Creative Arts Bld**

CB	AMPS	CIRCUIT DETAILS	CB	AMPS	CIRCUIT DETAILS
1	10		2	16	
3	10	Lights	4	20	
5	—		6	20	
7	16		8	10	
9	20		10	16	
11	16	Socket outlets	12	16	
13	20		14	—	
15	20		16	32	
17	—		18	32	
19	20		20	16	

FIGURE 15.3 Typical switchboard circuit schedule

- name of switchboard feeding the board
- switchboard reference number if more than one switchboard.

The circuit schedule must be installed in each switchboard in a circuit schedule holder with a transparent plastic sheet. All main switchboards in commercial and industrial installations should also have a reduced copy of an up-to-date version of a single-line circuit diagram representing the circuits controlled by the switchboard.

Schedule of circuits

A schedule of circuits as shown in **Figure 15.3** must be placed on the main switchboard when installation work is completed so that any person wanting to operate or alter the installation can do so with reasonable safety. For the schedule illustrated in **Figure 15.4**, the connected load per point for lighting is calculated using a 60 W load (0.26 A per point). The connected load per point for socket outlets assumes a 250 W load (1.09 A per point) thereby allowing 15 points to be connected to a 2.5 mm² TPS cable. Table C9 in AS/NZS 3000:2018 provides guidance on suggested per-point loading to assist with determining of the number of socket outlets, lighting points and appliances connected to a circuit.

EXAMPLE 15.3

A single-phase domestic installation consisting of three bedrooms, lounge, dining room, kitchen, bath, WC, laundry, hall and double garage has the following requirements:

- 1 × 80 A isolator
- 1 × 8 kW range
- 11 × 10 A double socket outlets plus 1 × 10 A single socket outlet
- 16 × lighting points including two smoke detectors
- 1 × air-conditioner rated at 11 A (2.42 kW)
- 1 × 4.8 kW storage water heater rated at 20 A

The circuit schedule for this installation is shown in **Figure 15.4**.

Circuit Schedule

Supply	CCT Ref. No.	MCB RCBO rating (A)	Cable size mm²	Description	No. points	Load per point
80 Ampere Isolator	1	6 RCBO	1.5 mm²	Lighting – bed 1, 2, 3, kitchen, garage, hall, outside Smoke detectors 1 in hall – on CCT 1 1 in kitchen on CCT 2	9	0.06 kW
	2	6 RCBO	1.5 mm²	Lighting – lounge, dining, laundry, WC, bath	7	0.06 kW
	3	16 RCBO	2.5 mm²	Socket outlets, kitchen, laundry	3D + 1S	0.25 kW
	4	16 RCBO	2.5 mm²	Socket outlets, lounge, dining, bed 1, 2, 3	8D	0.25 kW
	5	20 MCB	4.0 mm²	Range	1	8.0 kW
	6	25 MCB	4.0 mm²	Water heater – controlled	1	4.8 kW
	7	12 RCBO	2.5 mm²	Air-conditioner	1	2.42 kW (heat)

FIGURE 15.4 Schedule of circuits

EXERCISE 15.3

a A single-phase domestic installation consisting of four bedrooms, lounge, dining room, kitchen, bath, WC, laundry, hall and double garage has the following requirements:

i 1 × 80 A isolator
ii 1 × 12.5 kW range
iii 24 × 10 A double socket outlets
iv 36 × lighting points including two smoke detectors
v 1 × 3.6 kW air-conditioner rated at 17.6 A
vi 1 × 4.8 kW storage water heater rated at 20 A

Create a circuit schedule using **Figure 15.5**.

FIGURE 15.5 Circuit Schedule 1

b A domestic installation consisting of three bedrooms, lounge, dining room, kitchen, bath, WC, laundry, hall and single garage has the following requirements:

i 1 × 80 A isolator

ii 1 × 9.6 kW range

iii 18 × 10 A double socket outlets and one single socket outlet for the fridge/freezer

iv 30 × lighting points including two smoke detectors

v 1 × 4.8 kW twin-element storage water heater rated at 20 A per element

Create a circuit schedule using **Figure 15.6**.

FIGURE 15.6 Circuit schedule 2

As there is no prescribed standard form, these tabulated schedules can be quite varied in their format, content, and layout. **Figure 15.7** is an illustration of an alternative circuit schedule.

Some schedules may use measured values or even ticks for test results to confirm that the installation conforms to specifications.

SCHEDULE OF CIRCUIT DETAILS								
Circuit No.	**Circuit description**	**Protective device**	**I rating**	**Cable size installation method**	**Cable length**	**Continuity**	**Ins Res**	**Z**
			A	mm^2	m	Ω	Ω	Ω
L1	Lighting (internal)	RCBO	10	TPS – cavity walls/ceiling 1.5 mm^2	30	0.3	∞	1.32
L2	Lighting (external)	RCBO	6	TPS – TPS in UG conduit 1.5 mm^2	15	0.26	∞	0.63
P1	Socket outlets (10A)	RCBO	16	TPS – cavity walls/ceiling 2.5 mm^2	30	0.52	∞	0.92
R	4.8 kW range	CB	16	TPS – cavity walls 2.5 mm^2	12	0.1	∞	0.56
HW	310 L 3.6 kW HWS	CB	16	TPS – cavity walls 4 mm^2 + 2.5 mm^2 E	14	0.2	∞	0.43
Mains				16 mm^2 SDI	9	0.2	∞	✓
Sub-mains				NA	NA	NA	NA	NA
Main earth				6 mm^2	2.5	0.2	∞	✓
Equipotential bonding (at HWS)				4 mm^2	2	0.2	∞	✓

FIGURE 15.7 Alternative circuit schedule

REVIEW QUESTIONS

1 Why are different circuits created within an electrical installation?
2 What are important considerations when dividing an electrical installation into circuits?
3 What are the two main risks that exist in an electrical installation?
4 List the six fundamental principles specified in AS/NZS 3000:2018 *Wiring Rules* where protection is needed to minimise the risk of electric shock and fire.
5 How can circuits be placed into an electrically safe work condition to prevent unwanted voltages?
6 Name the three broad types of circuits in an electrical installation.
7 Name five items that circuit reference labels should identify.
8 Where should the circuit schedule be installed?

15.3 Factors affecting the suitability of wiring systems

Wiring systems are an assembly comprising one or more conductors, cables or busbars and including fixings and when appropriate, their mechanical protection. It is necessary to install wiring systems and accessories in locations that prevent them from being subject to mechanical damage, dampness, corrosion, chemical attack, insect attack, heat and other damaging environmental conditions. Correct selection of the cable types and their accessories must consider these impacts as well as the utilisation and design of the building, the accessibility and ease by which increasing the capacity of the cable systems can be achieved and the possibility of electromagnetic interference. Where exposures to such conditions are unavoidable, protective measures must be used. Other factors influencing the choice of a wiring system include the planned use of the installation, and installation requirements for defined special locations (swimming pools, and the like) (refer to AS/NZS 3000:2018 *Wiring Rules*, '3.2 Types of wiring systems').

The types of wiring systems that can be used, depending upon the type of cable are:

- resting on a continuous surface fixed or without fixings, in free air or thermal insulation
- in a wiring enclosure such as a conduit, duct, trunking or other approved enclosure
- supported by a cable tray or ladder
- backed by a catenary system
- buried direct in the ground
- within a wiring enclosure buried in the ground.

When selecting a wiring system for specific applications, it is necessary to consider:

- external influences
- heat sources
- solar radiation
- effect of wiring system.

External influences

External influences include ambient temperature and soil conditions.

Ambient temperature

The current-carrying capacities of cables are based on a defined ambient air and soil temperature. AS/NZS 3008.1.1:2017 *Electrical installations – Selection of cables* is applicable to Australian installation conditions where the nominal ambient air and ground temperatures are 40 °C and 25 °C respectively.

V–90 rated conductor insulation has a conductor operating temperature rating of 75 °C. The temperature rating V–90 of the insulation is needed to handle a higher ambient temperature in the environment and not to increase the conductor's current rating.

Higher ambient temperatures prevent effective transference of heat from the cable to the surrounding environment. Therefore, cables used in ambient temperatures above the standard must have their current-carrying capacity reduced. Operating cables at their normal current-carrying capacity and above the allowable ambient temperature causes their electrical insulation to age at an accelerated rate. In this regard, attention should be given to cables that are run in confined areas such as roof spaces and building shafts.

Note: Cables used for wiring to and between light fittings must be suitable for the temperatures that apply to the terminations. For example, in some luminaires such as oyster fittings the heat from the lamp(s) increases the temperature within the light fittings and their enclosures and so cable with a higher temperature rating must be used to ensure safe operation.

Soil conditions

Extreme soil and environmental conditions exist where the ambient temperature exceeds 40 °C. Hot, dry conditions in the summer create an arid soil which produces a high thermal resistance. Under such conditions, the environmental and soil parameters significantly influence the current-carrying capability of the installed cable.

Current-capacity ratings of buried cables are, therefore, based on a number of variables including cable depth and local climatic conditions such as air temperature and soil type (different soils have different thermal properties). The recommended depth for underground cables in Australia is 0.5 m – measured from ground level to the top of the cable or wiring enclosure.

Note: Always refer to local requirements for the recommended depth of underground cables and enclosures.

Effect of ambient temperature and soil conditions on cable rating

High and low air and soil temperatures must be taken into account when selecting a suitable conductor installed in those two mediums. The ambient air and soil temperatures used by AS/NZS 2017 for Australia are 40 °C (at ground level) and 25 °C (at a depth of 0.5 m), while for New Zealand 30 °C and 15 °C applies (AS/NZS 3008.1.2).

Table 27(1) provides rating factors for cables installed in the air where the ambient temperature varies from the specified 40 °C.

EXAMPLE 15.4

A cable has a nominal current rating of 20 A for an ambient temperature of 40 °C. Determine the current-carrying capacity of this cable when installed in a location where the ambient temperature is:

a 30 °C

b 60 °C

i From Table 27(1), Column 1, locate 75 °C then under Column 5 (for 30 °C), you will find the rating of 1.14. This means for the nominal current rating of 20 A (at 40 °C) the conductor can safely carry 22.8 A (20 × 1.14) due to the lower ambient temperature.

ii From Table 27(1), Column 1, locate 75 °C then under Column 11 (for 60 °C), you will find the rating of 0.60. This means for the nominal current rating of 20 A (at 40 °C) the conductor can safely carry 12 A (20 × 0.60) due to the higher ambient temperature.

EXERCISE 15.4

A cable has a nominal current rating of 25 A for an ambient temperature of 40 °C. Determine the current-carrying capacity of this cable when installed in a location where the ambient temperature is:

a 30 °C

b 55 °C

Table 27(2) provides rating factors for cables installed in soil where the soil temperature varies from the specified 25 °C.

EXAMPLE 15.5

A cable has a nominal current rating of 20 A for a soil temperature of 25 °C. Determine the current-carrying capacity of this cable when installed in a location where the soil temperature is:

a 15 °C

b 35 °C

i From Table 27(2), Column 1, locate 75 °C then under Column 3 (for 15 °C), you will find the rating of 1.10. This means for the nominal current rating of 20 A (at 25 °C) the conductor can safely carry 22 A (20 × 1.1) due to the lower soil temperature.

ii From Table 27(2), Column 1, locate 75 °C then under Column 7 (for 35 °C), you will find the rating of 0.89. This means for the nominal current rating of 20 A (at 25 °C) the conductor can safely carry 17.8 A (20 × 0.89) due to the higher soil temperature.

EXERCISE 15.5

A cable has a nominal current rating of 25 A for a soil temperature of 25 °C. Determine the current-carrying capacity of this cable when installed in a location where the soil temperature is:

a 22 °C

b 35 °C

External heat sources

The cable must also be protected from the effects of external heat sources such as a heated concrete slab or an electric furnace.

Table 27(1) provides rating factors for cables installed in a heated concrete slab. For example, the rating factor of a V–90 TPS cable installed in the slab that is subjected to an ambient temperature of 40 °C can be determined.

Using Table 27(1), Column 1, locate 75 °C then find column 7, and you should be given the rating of 1.00.

EXAMPLE 15.6

A cable has a nominal current rating of 20 A for an ambient temperature of 40 °C. Determine the current-carrying capacity of this cable when installed in a concrete slab having an ambient temperature of 45 °C.

From Table 27(1), Column 1, locate 75 °C then under Column 8 (for 45 °C), you will find the rating of 0.91. This means for the nominal current rating of 20 A (at 40 °C) the conductor can safely carry 18.2 A (20 × 0.91) due to the higher ambient temperature.

EXERCISE 15.6

A cable has a nominal current rating of 20 A for an ambient temperature of 40 °C. Determine the current-carrying capacity of this cable when installed in a concrete slab having an ambient temperature of 55 °C.

Solar radiation

Cables may be directly exposed or contained within enclosures that are exposed to solar radiation. This exposure increases the temperature of the insulation surrounding the conductors thereby decreasing the effectiveness of the heat transference from the conductor to the surrounding environment. Refer to AS/NZS:2008 'Effect of direct sunlight'.

If 2C + E flat TPS 2.5 mm^2 copper stranded conductors as illustrated in **Figure 15.8** are unenclosed and fixed to a wall then from Table 10 of AS/NZS 2008.1.1, Column 5, the current-carrying capacity of the cable is 26 A. If the same cable is exposed to the sun then column 8 states that the cable is now rated at 20 A.

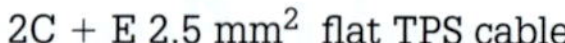

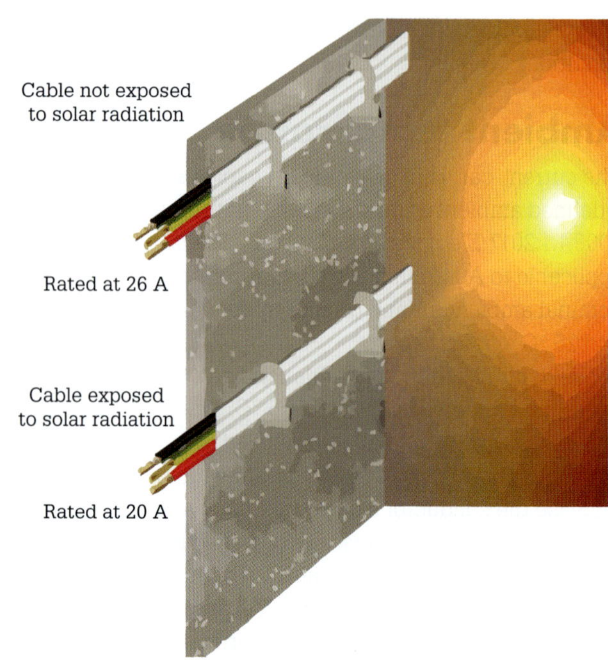

FIGURE 15.8 Flat TSP cable fixed to a wall

Effect of wiring system

The wiring system itself and the installation method have an impact on the current-carrying capacity of a given conductor. Some of these systems and conditions include:

- cables installed in the air spaced from supporting surface
- cables in air touching a continuous surface
- cables in a wiring enclosure in air
- cables installed in thermal insulation
- cables buried direct in the ground
- cables installed in underground wiring enclosure.

Cables installed in the air spaced from supporting surface

When cables are installed on cable tray or hung using insulated wire hangers with a catenary as illustrated in **Figure 15.9** they can carry current with no de-rating factors being applied. No de-rating occurs because the cables can dissipate their conductor heat quickly. From Table 10, a 2.5 mm^2 2C+ E TPS cable installed in air has a current-carrying capacity of 27 A (column 2).

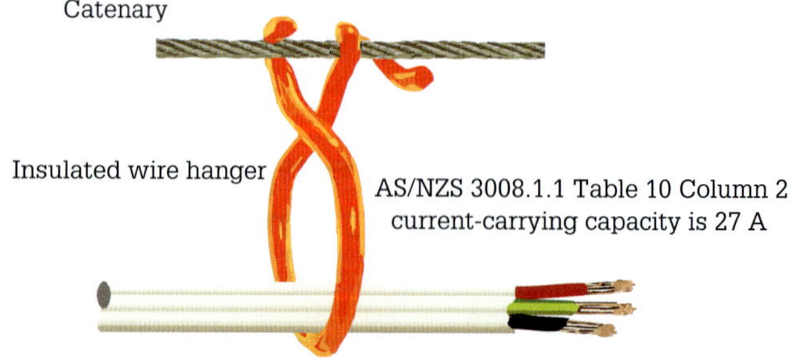

FIGURE 15.9 Cable installed in the air

Cables in air touching a continuous surface

When cables are fixed to a continuous surface, as illustrated in **Figure 15.10**, they can carry current with a slight de-rating factor being applied (compare with **Figure 15.8**). From Table 10, a 2.5 mm^2 2C + E TPS cable touching a continuous surface has a current-carrying capacity of 26 A (column 5). The slight reduction in current-carrying capacity is due to the fact that the cable is not entirely surrounded by air.

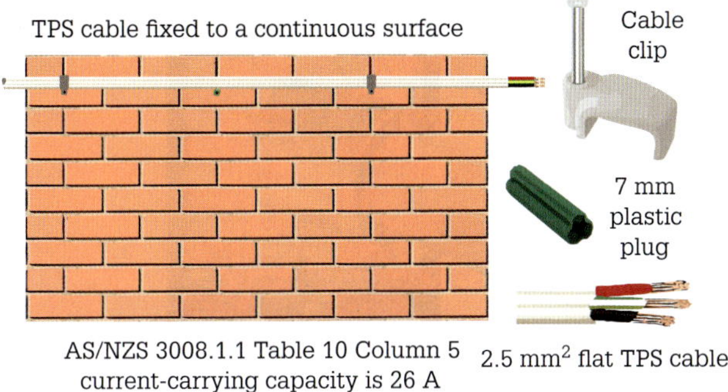

FIGURE 15.10 Cable fixed to a continuous surface

If fixing TPS directly to a masonry surface such as brick it is suggested not to place the cable clips in the mortar nibs (vertical cement between the bricks) as the clips will come away from the mortar. The professional approach is to drill holes in the brick and insert a 7 mm plastic plug into which the cable clip is fixed.

Cables in a wiring enclosure in air

When cables are installed in an enclosure in air as illustrated in **Figure 15.11** they can carry current with a significant de-rating factor being applied (compare with **Figure 15.10**). The reduction in current-carrying capacity is because the heat from the conductor is contained within the enclosure before it is dissipated to the surrounding air. From Table 10, a 2.5 mm^2 2C + E TPS cable installed in a wiring enclosure in air has a current-carrying capacity of 23 A (column 11).

FIGURE 15.11 Cable installed in an enclosure in air

Cables installed in thermal insulation

Cables as shown in **Figure 15.12** can be partially surrounded or completely surrounded by thermal insulation. Cables can also be installed in enclosures that may be partially or wholly surrounded by thermal insulation. Refer to AS/NZS 3008.1.1, 'Cables installed in thermal insulation' and 'Effect of thermal insulation'. From Table 10, a 2.5 mm^2 2C + E TPS cable installed so as to be partially surrounded by thermal insulation has a current-carrying capacity of 20 A (column 15). From Table 10, a 2.5 mm^2 2C + E TPS cable installed so as to be totally surrounded by thermal insulation has a current-carrying capacity of 13 A (column 19), which is less than half the current-carrying capacity of the cable installed in air. From Table 10, a 2.5 mm^2 2C + E TPS cable installed in a wiring enclosure which is totally surrounded by thermal insulation has a current-carrying capacity of 12 A (column 21).

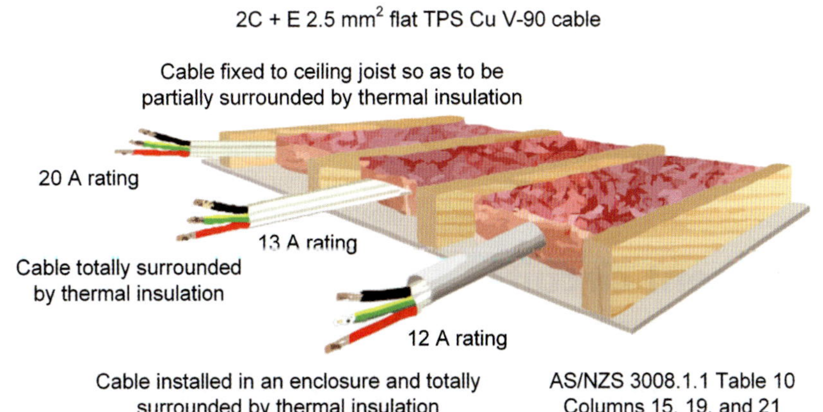

FIGURE 15.12 Cables with thermal insulation

Cables buried direct in the ground

Refer to AS/NZS 3008.1.1, 'Cables buried direct in the ground'. Because of the difference in ambient temperature between surface air and soil, cables installed in the ground are able to dissipate their conductor heat more efficiently. An illustration of a 2.5 mm^2 2C + E circular TPS cable installed in the ground is shown in **Figure 15.13**. From Table 10, a 2.5 mm^2 2C + E TPS cable installed buried direct in the ground has a current-carrying capacity of 30 A (column 23).

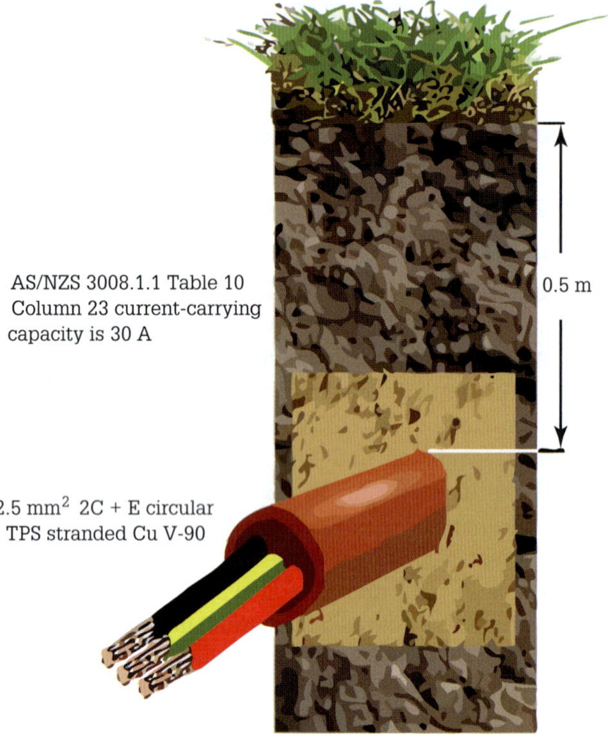

FIGURE 15.13 Cables installed in soil

Cables installed in underground wiring enclosure

Refer to 'Cables installed in underground wiring enclosures'. An illustration of cables within a metal pipe enclosure in soil is shown in **Figure 15.14**. From Table 10, a 2.5 mm^2 2C + E TPS cable installed in an underground enclosure has a current-carrying capacity of 30 A (column 25).

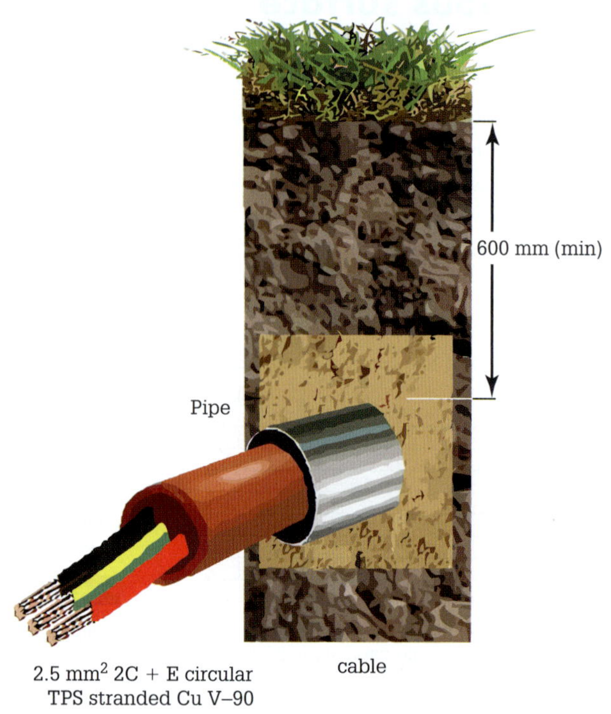

FIGURE 15.14 Cables installed in an underground wiring enclosure

Cables for different circuits

When cables for different circuits are bunched together, excessive heat becomes a problem. The problem exists because there is limited air surrounding each cable, and the heat from each cable heats an adjacent cable. As well as additional heating by conduction, convection, and radiation from other cables, induction heating can also occur.

Cables bunched in groups are de-rated according to the number of cables in the group. To determine the de-rated current-carrying capacity of cables grouped together, it is necessary to apply the de-rating factors listed in Tables 22 to 26 of AS/NZS 3008.1.1. Current-carrying capacities for cables in Tables 3 to 21, AS/NZS 3008.1.1, are based on cables installed in single circuit configuration. Circuit cables can be placed in different grouping formations. Typical formations include trefoil and flat formations as shown in **Figure 15.15.**

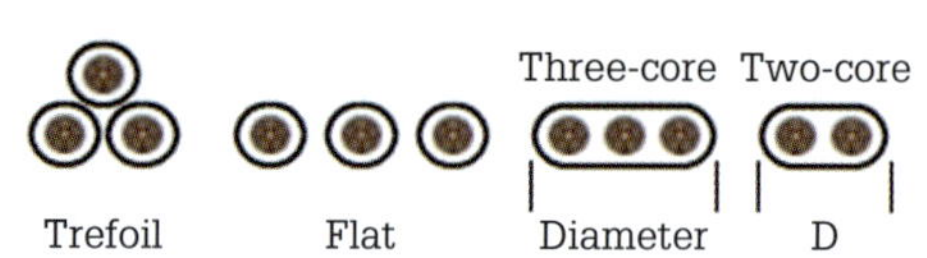

FIGURE 15.15 Cable formations

Refer to Figure 1 in AS/NZS for cable formations and spacing in air. If the spacing between cable surfaces is not greater than that specified, it is necessary to de-rate the nominal (single-circuit) current-carrying capacity of the cable.

Cables bunched in air

EXAMPLE 15.7

A wiring system comprises five circuits of 2.5 mm^2 2C+E flat V-90 TPS cable installed bunched and suspended from a catenary as shown in **Figure 15.16**. Determine the current-carrying capacity of each cable.

»

Using AS/NZS as reference:

- From Table 3(1), Column 1, Item 9 matches the cable type (column 2) and installation method (column 5). This directs us to Table 10, columns 2 to 4 for the current. This directs us to Table 14, columns 2 to 4 for the current-carrying capacity and Table 24 for the applicable de-rating factor.
- From Table 10, column 2, 2.5 mm² 2C has a current-carrying capacity of 27 A.
- From Table 22, Item 1, column 8 (for 5 circuits) the applicable de-rating factor is 0.70.
- This means for the nominal current rating of 27 A (at 40 °C) the conductor can safely carry 18.9 A (27 × 0.70) due to the bunching

Five circuits of 2.5 mm² TPS cable bunched in air supported by an insulated wire hanger

FIGURE 15.16 Five circuits bunched in air

EXERCISE 15.7

A wiring system comprises three circuits of 6 mm² 2C+E flat V–90 TPS cable installed bunched and suspended from a catenary. Determine the current-carrying capacity of each cable.

Cables installed on cable tray

Cables can also be installed in trays or other enclosures in air. **Figure 15.17** provides an example of this type of installation method. If the cables touch one another then their current-carrying capacity must be de-rated.

AS/NZS 3008.1.1 Table 24 Column 1 item 1, Column 4 item 1, Column 7 item 1 de-rating factor for 3 cables is 0.78

4C+ E circular cable PVC/PVC
4 × 6 mm² + 2.5 mm² earth stranded Cu V–90 cables

FIGURE 15.17 Multi-core cables touching on unperforated tray

EXAMPLE 15.8

A wiring system comprises four circuits of 4 mm² 4C+ E circular V–90 TPS cable installed flat, touching on an unperforated cable tray. Determine the current-carrying capacity of each cable.

Using AS/NZS as reference:

- From Table 3(1), Column 1, Item 10 matches the cable type (column 2) and installation method (column 5). This directs us to Table 14, columns 2 to 4 for the current-carrying capacity and Table 24 for the applicable de-rating factor.
- From Table 14, column 2, 4 mm² 4C has a current-carrying capacity of 38 A.
- From Table 24, Item 1, column 8 (for 4 cables) the applicable de-rating factor is 0.75.
- This means for the nominal current rating of 38 A (at 40 °C) the conductor can safely carry 28.5 A (38 × 0.75) due to the restricted air flow.

EXERCISE 15.8

A wiring system comprises four circuits of 6 mm² 4C+ E circular V–90 TPS cable installed flat, touching on an unperforated cable tray. Determine the current-carrying capacity of each cable.

REVIEW QUESTIONS

1. What is a wiring system?
2. What are the requirements for the installation of wiring systems?
3. Name four considerations for selecting a wiring system.
4. What are two external influences that impact a wiring system?
5. What is the nominal ambient air temperature for Australian installations?
6. What is the nominal soil temperature for Australian installations?
7. Why is it necessary to de-rate cables?
8. How is it possible to avoid cable de-rating?
9. A wiring system comprises six circuits of 1.5 mm² 2C+E flat V–90 TPS cable installed bunched and suspended from a catenary. Determine the current-carrying capacity of each cable.
10. A wiring system comprises four circuits of 2.5 mm² 2C+E circular V–90 TPS cable installed flat, touching on a perforated cable tray. Determine the current-carrying capacity of each cable.

15.4 Maximum demand

It is essential to know the magnitude of required load current in the installation before installation work commences. The maximum demand of the installation is defined as the electrical loading of the installation. Consequently, the conductor cross-sectional area (CSA) of consumer's mains and sub-mains must be able to carry the electrical demand.

The calculation method may be used as a basis for determining the expected maximum demand of an electrical installation. The maximum demand may be calculated by following the directions in Appendix C of AS/NZS 3000:2018 for the suitable category of electrical installation. Maximum demand calculation is based on two tables: Table C1 is used for domestic installations and Table C2 is used for non-domestic installations. Tables C1 and C2 apply the principle of diversity, which takes into account the normal operating conditions of circuits during which all equipment is not operating simultaneously at full load or for periods exceeding 15 minutes.

Single domestic installations

Consumer's mains (aerial or underground) and sub-mains in a domestic installation are illustrated in **Figure 15.18**. The consumer's mains commence at the consumer's terminals and terminate at the installation's main switchboard.

In general, the maximum capacity of single-phase two-wire domestic consumer's mains is 70 A. If the installation is larger than this, either a two-phase and neutral system (70 A per phase) or a three-phase and neutral four-wire system (when demand is over 140 A) will be used. However, it is the local distribution entity in each state or territory which determines the type of supply to be used (e.g. less than 100 A uses a single-phase two-wire system; greater than 100 A uses a three-phase and neutral four-wire system).

Prior to calculating maximum demand, the distribution of the loads across sub-circuits must be determined. Loading tells us the number of sub-circuits required. An estimation of the number of phases may also be required.

Using Table C1

Prior to using Table C1, Appendix C, AS/NZS 3000:2018, note the following data:

- **Column 1, load group A(i):** Within the calculation method, lighting loads represent only a small percentage of the mains/sub-mains loading. This is reflected in Column 2 where 20 points receive a demand value of only 3 A. The definition of a 'point' (Clause 1.4.92) in wiring is 'a termination of installation wiring' and Clause 1.4.80 states that a luminaire is 'a complete lighting assembly'. It is apparent that a loose diversity factor has been applied to load group A(i). This is because a lighting point may range from one luminaire with a lamp of 8 W to a luminaire containing 60 or more lamps of 15 W each (actual load is 15/230 × 60 =3.91 A). It is assumed that all luminaires will not be on at the same time, thus providing validity to the reasoning for Column 2, 1 to 20 points. Notes d, e and f in Table C1 are necessary and must be referred to when calculating lighting demand. Lighting track and socket outlets for low-wattage appliances (clocks, exhaust and ceiling fans) at a particular height are included as lighting points.
- **Column 1, load group A(ii):** Outdoor lighting is a separate category and must exceed 1000 W in total (75% connected load). Also refer to notes f and g to Table C1.

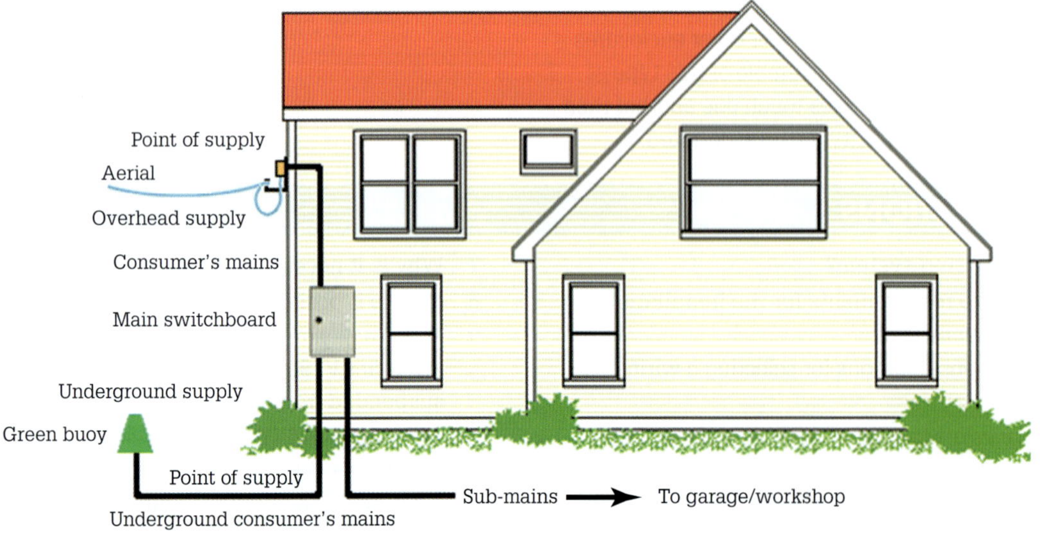

FIGURE 15.18 Domestic installation showing the two possible methods of supply

- **Column 1, load group B(i):** 10 A single-phase socket outlets. A loose diversity factor has been applied to socket outlets as well. A sub-circuit may have four socket outlets connected with a potential to draw 16 A. However, 1 to 20 socket outlets have a loading of 10 A. Also refer to notes e and h to Table C1.
- **Column 1, load group B(ii):** 15 A single-phase socket outlets. One or more have 10 A totals. Also refer to notes h and j to Table C1.

Moreover, diversity is also applied to other loads as Table C1 indicates.

Calculation method

A suggested approach to calculating maximum demand for a single domestic installation is as follows.

1 Determine the number of circuits the installation loads require and tabulate them into circuit number and function as illustrated in **Figure 15.19**.

Circuit No.	Function
1	Indoor lighting
2	Outdoor lighting

FIGURE 15.19 Circuit number and function

2 Using Table C1 tabulate each circuit on the load list as shown in **Figure 15.20**, according to its circuit number, load group, description of load, calculation and demand current.

Load group	Description of load	Calculation	Demand current (A)

FIGURE 15.20 Load list

3 Calculate the contribution each load group makes towards the maximum demand and then total the result.

Note: The current-carrying capacity of the consumer's mains should, when installed in accordance with AS/NZS 3000:2018 and AS 3008.1.1, exceed the rating of the circuit breaker main switch.

Domestic demand single-phase

The following examples outline how to use Table C1 of AS/NZS 3000:2018 to determine the maximum demand of a single-phase domestic residence. Note how the table detailed in **Figure 15.20** is used to document the calculations.

EXAMPLE 15.9

1 Determine the maximum demand for a single-phase single domestic residence comprising the following load:

i 25 × lighting points
ii 11 × double 10 A socket outlets
iii 1 × 15 A socket outlet
iv 1 × 7.36 kW range @ 230 V
v 1 × 20 A instantaneous water heater @ 230 V

TABLE 15.2 Solution to Example 15.9 Question 1

Load group	Description of load	Calculation	Demand current (A)
A(i)	25 lighting points	3 + 2	5.0
B(i)	11 double outlets = 22 points	10 + 5	15.0
B(ii)	15 A single-phase outlet	–	10.0
C	7.36 kW range	7360/230 × 0.5	16.0
E	instantaneous HWS	20 × 0.333	6.66
Total demand current (A)			**52.66**

From these load units, the maximum demand can now be calculated using Table C1 for domestic installations.

Double socket outlets count as two points.

Note: Local service utility rules may require a minimum size for consumer's mains that is larger than calculated sizes.

2 Determine the maximum demand for a single-phase domestic installation comprising:

i 3 × 1 metre lighting tracks
ii 2 × 150 W floodlights
iii 4 × twin 36 W fluorescent luminaires
iv 20 × lighting points
v 16 × double 10 A socket outlets
vi 2 × single 10 A socket outlets
vii 1 × 15 A socket outlet
viii 1× 9.71 kW range @ 230 V
ix 1 × 20 A storage water heater @ 230 V

»

TABLE 15.3 Solution to Example 15.9 Question 2

Load group	Description of load	Calculation	Demand current (A)
A(i)	Track lighting (3 × 1 m) = 6 points Floodlights × 2 = 2 points Fluorescent luminaire × 4 = 4 points Lighting points × 20 = 20 points = **32 points**	3 + 2	5.0
B(i)	10 A double outlets × 16 = 32 points 10 A single outlets × 2 = 2 points = **34 points**	10 + 5	15.0
B(ii)	15 A single-phase outlet × 1	–	10.0
C	9.71 kW range × 1	9710/230 × 0.5	21.1
F	Storage water heater	–	20.0
Total demand current (A)			**71.1**

Domestic demand three-phase

Example 15.10 outlines how to use Table C1 of AS/NZS 3000:2018 to determine the maximum demand of a three-phase domestic residence. Note how the tables detailed in **Figures 15.19** and **15.20** are modified to document the circuit arrangement and subsequent calculations.

EXAMPLE 15.10

Determine the maximum demand for the three-phase 230/400 V domestic installation detailed below.

- 38 × lighting points
- 4 × 200 W exterior lights
- 15 × single 10 A socket outlets
- 9 × double 10 A socket outlets
- 1 × 15 A single-phase socket outlet
- 1 × 4.2 kW cooktop @ 230 V
- 1 × 5.2 kW oven @ 230 V
- 1 × 15 A storage water heater
- 1 × 0.75 kW 230 V pool filter pump rated at 4.3 A
- 1 × 11.1 kW three-phase ducted air-conditioner rated at 23.6 A per phase

The installation is divided into the following circuits:

TABLE 15.4 Sub-circuit arrangement for Example 15.10

Circuit No.	Function	Line (phase)		
		1	2	3
1	13 indoor lighting points	✓		
2	12 indoor lights + 2 outdoor lights		✓	
3	13 indoor lights + 2 outdoor lights			✓
4	8 single 10 A outlets + 4 double 10 A outlets	✓		
5	7 single 10 A outlets + 5 double 10 A outlets		✓	
6	15 A single-phase outlet	✓		
7	cooktop rating 4.2 kW @ 230 V		✓	
8	oven rating 5.2 kW @ 230 V			✓
9	pool filter pump rating 4.3 A	✓		
10	air-conditioner rating 23.6 A per phase	✓	✓	✓
11	storage hot water unit rating 15 A			✓

Lighting should be divided over the three phases for safety reasons in the event that one phase is lost. The socket outlets can be rearranged to balance the phases.

The maximum demand can now be calculated using Table C1 for domestic installations.

TABLE 15.5 Maximum demand for Example 15.10

Circuit No.	Load group	Description of load	Calculation	Demand current (A)		
				L_1	L_2	L_3
1	A(i)	13 lighting points	–	3.0		
2	A(i)	14 lighting points	–		3.0	
3	A(i)	15 lighting points	–			3.0
4	B(i)	8 single + 4 double outlets	8+ (4 × 2) + 1 (pool filter pump) =17 points	10.0		
5	B(i)	7 single + 5 double outlets	7 + (5 × 2) = 17 points		10.0	
6	B(ii)	15 A single-phase outlet	–	10.0		
7	C	cooktop	(4200/230) × 0.5 = 9.13		9.13	
8	C	oven rating	(5200/230) × 0.5 = 11.3			11.3
9	B(i)	pool filter pump	Included in circuit 4			
10	D	air-conditioner	23.6 × 0.75	17.7	17.7	17.7
11	F	storage water heater	–			15.0
	Total demand current (A)			**40.7**	**39.83**	**47.0**

The maximum demand of this installation is the highest loaded phase, which is L3 with 47.0 A.

Note: Local service utility rules may require a minimum size for consumer's mains that is larger than calculated sizes.

EXERCISE 15.9

Determine the maximum demand for the three-phase 230/400 V domestic installation detailed below.

- **a** 54 × lighting points
- **b** 9 × single 10 A single-phase socket outlets
- **c** 42 × double 10 A single-phase socket outlets
- **d** 2 × 15 A single-phase socket outlets
- **e** 1 × 28 A single-phase cooktop
- **f** 1 × 18 A single-phase wall oven
- **g** 1 × 22 kW rated capacity three-phase ducted air-conditioner rated at 17.2 A per phase
- **h** 1 × 20 A storage hot water unit

As a preliminary, list the load units spread over three phases and create a load list for the installation as shown in the tables below.

Note that the loads may need to be rearranged to ensure the phases are balanced as much as possible.

TABLE 15.6 Sub-circuit arrangement for Exercise 15.9

Circuit No.	Function	Line (phase)		
		1	2	3
1	18 indoor lighting points	✓		
2	18 indoor lighting points		✓	
3	18 indoor lighting points			✓
4	2 single 10 A outlet + 7 double 10 A outlets	✓		
5	1 single 10 A outlet + 7 double 10 A outlets	✓		
6	2 single 10 A outlet + 7 double 10 A outlets		✓	
7	1 single 10 A outlet + 7 double 10 A outlets		✓	
8	2 single 10 A outlet + 7 double 10 A outlets			✓
9	1 single 10 A outlet + 7 double 10 A outlets			✓

»

TABLE 15.6 Sub-circuit arrangement for Exercise 15.9 (*Continued*)

Circuit No.	Function	Line (phase)		
		1	2	3
10	15 A single-phase outlet	✓		
11	15 A single-phase outlet		✓	
12	cooktop rating 28 A		✓	
13	oven rating 18 A	✓		
14	air-conditioner rating 17.2 A per phase	✓	✓	✓
15	storage hot water unit rating 20 A			✓

TABLE 15.7 Maximum demand for Exercise 15.9

Circuit No.	Load group	Description of load	Calculation	Demand current (A)		
				L_1	L_2	L_3
1		18 indoor lighting points				
2		18 indoor lighting points				
3		18 indoor lighting points				
4		2 single 10 A outlet + 7 double 10 A outlets				
5		1 single 10 A outlet + 7 double 10 A outlets				
6		2 single 10 A outlet + 7 double 10 A outlets				
7		1 single 10 A outlet + 7 double 10 A outlets				
8		2 single 10 A outlet + 7 double 10 A outlets				
9		1 single 10 A outlet + 7 double 10 A outlets				
10		15 A single-phase outlet				
11		15 A single-phase outlet				
12		cooktop rating 28 A				
13		oven rating 18 A				
14		air-conditioner rating 17.2 A per phase				
15		storage hot water unit rating 20 A				
			Total demand current (A)			

Multiple occupancy residential premises such as villas, units, townhouses

Where there are more than two individually metered units, the supply can be assumed to be three-phase. However, the local distribution entity must be consulted.

Using Table C1

Columns 3, 4 and 5 of Table C1 are used for blocks of residential home units, and it is necessary to determine the number of units per phase before proceeding. For example, if a block of units has 26 units on a site, then the number of units per supply phase would be nine on one phase, nine on the second phase and eight on the final phase (9 + 9 + 8 = 26). However, this may vary according to the type of equipment (three phase) and size of the individual units.

In the example just given, column 4 of Table C1 of AS/NZS 3000:2018 (6–20 units per phase) would be used. The calculations for maximum demand that would follow would be based on the highest-loaded phase (nine units).

Note that it is standard practice on unit developments to supply electrical loads that are shared by all occupants (e.g. security lighting, laundry and lifts) on separate meters, the cost being shared by all. Such loads are referred to as communal loads. About halfway down Table C1 a heading in bold type indicates the start of the section devoted to such communal loads.

The number of phases provided for communal loads can be determined by the same service rules applying to any single installation. Therefore, if the total communal loading is less than 70 A, the communal load is on single phase.

For example, if the only loading in the communal area of a block of units is 10 × 60 W lights and 15 × 10 A single-phase socket outlets, these must all be connected to the same phase as there will be only one kilowatt-hour meter for the community loading. If, however, there is three-phase loading connected to the community tariff, then the community load can be spread over the three phases.

The calculations listed under Table C1, Columns 3, 4 and 5, can be confusing. The following notes should help.

- **A(i) Column 3:** The value of 6 A does not mean 6 A per unit. It means a single allowance of 6 A, regardless of whether there are 2–5 units of that phase.
- **A(ii):** It appears at first that no allowance is made for outdoor lighting; however, there is provision further down the table under load group H, for communal lighting.
- **B(i):** Referring to Column 3, allow 10 A once then 5 A per unit. Use the other columns in a similar way.
- **B(ii) and B(iii):** Make the allowance once, not on a per unit basis.
- **C:** Using Column 3, make the allowance once, regardless of the number of units applicable, whereas the allowances under Columns 4 and 5 are on a per unit basis.
- **D and E:** Uncomplicated.
- **F:** Uncomplicated.
- **H, L and J:** Uncomplicated.
- **K–M:** These are simple in application.

Multiple domestic demands (three-phase)

The following examples outline how to use Table C1 of AS/NZS 3000:2018 to determine the maximum demand of a three-phase multiple domestic residence.

EXAMPLE 15.11

Determine the three-phase maximum demand for the consumer's mains for a block of 18 living units as illustrated in Figure 15.21, which comprises the following loads.

Each living unit is supplied via a single-phase sub-main

- Lighting – 10 points
- Socket outlets – 12 × 10 A double outlets
- Electric range – rated at 45 A
- Water heater –storage (quick recovery) rated at 20 A
- Air-conditioner – 6.8 kW cooling only rated at 12.8 A

Communal services

These are metered by a single, single-phase meter.

- 24 × 60 W lighting points
- 6 × 10 A single socket outlets

The number of living units per phase = 18 ÷ 3 = 6 living units. We are using Column 4 of Table C1.

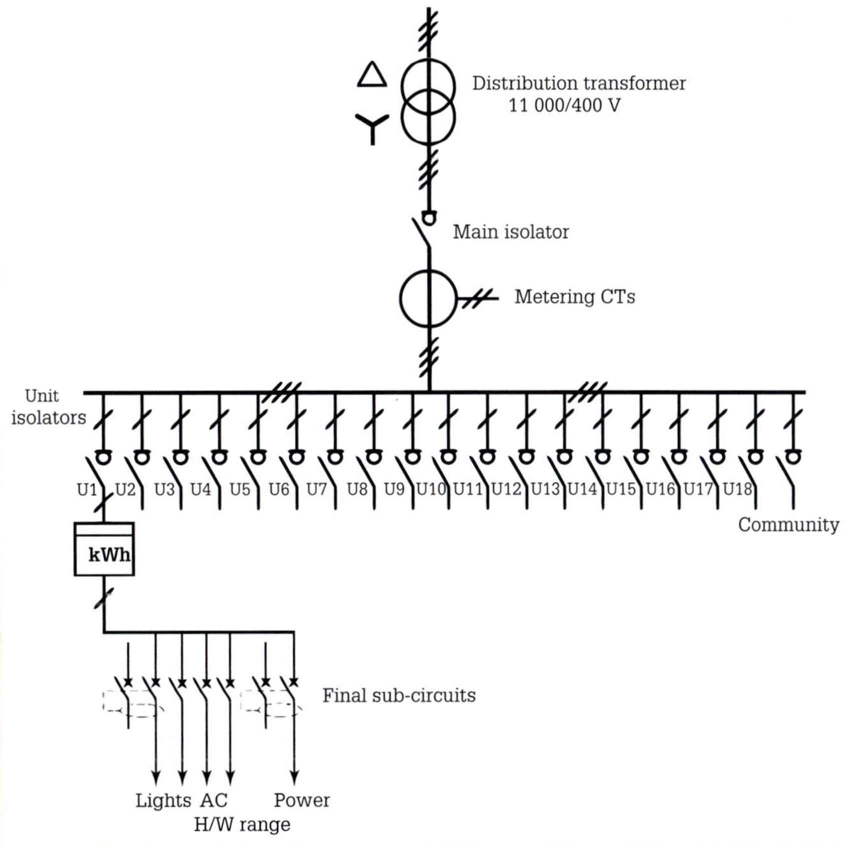

FIGURE 15.21 Domestic installation showing the two possible methods of supply

»

TABLE 15.8 Maximum demand for living units, Example 15.11

Load group	Description of load	Units / phase			Calculation	Demand (A)		
		L_1	L_2	L_3		L_1	L_2	L_3
A(i)	Lighting	6	6	6	5 + (6 × 0.25)	6.5	6.5	6.5
B(i)	Socket outlets 10 A	6	6	6	15 + (6 × 3.75)	37.5	37.5	37.5
C	Range	6	6	6	6 × 2.8	16.8	16.8	16.8
D	Air conditioning	6	6	6	6 × 12.8 × 0.75	57.6	57.6	57.6
F	Hot water QR	6	6	6	6 × 6	36.0	36.0	36.0
					Total demand current (A)	**154.4**	**154.4**	**154.4**

TABLE 15.9 Maximum demand for communal load, Example 15.11

Load group	Description of load	Qty / phase			Calculation	Demand (A)		
		L_1	L_2	L_3		L_1	L_2	L_3
H	Lighting	24	0	0	(24 × 60) / 230	6.3	0	0
I	Socket outlets 10 A	6	0	0	6 × 2	12.0	0	0
					Total demand current (A)	**18.3**	**0**	**0**

Total loading: 154.4+ 18.3 = **172.7 A**

Note: Local service utility rules may require a minimum size for consumer's mains that is larger than calculated sizes and may not allow unenclosed (buried direct) consumer's mains.

EXERCISE 15.10

Determine the three-phase maximum demand for the consumer's mains for a block of 24 living units, which comprises the following loads.

Each living unit is supplied via a single-phase sub-main

a Lighting – 16 points
b Socket outlets – 11 × 10 A double outlets
c Electric range – rated at 40 A
d Water heater – storage (quick recovery) rated at 20 A
e Air-conditioner – 6.8 kW cooling only rated at 12.8 A

Communal services

These are metered by a single, single-phase meter.

a 30 × 60 W lighting points
b 9 × 10 A single socket outlets

TABLE 15.10 Maximum demand for living units, Exercise 15.10

Load group	Description of load	Units / phase			Calculation	Demand (A)		
		L_1	L_2	L_3		L_1	L_2	L_3
	Lighting							
	Socket outlets 10 A							
	Range							
	Air conditioning							
	Hot water QR							
					Total demand current (A)			

»

TABLE 15.11 Maximum demand for communal load, Exercise 15.10

Load group	Description of load	Qty / phase			Calculation	Demand (A)		
		L_1	L_2	L_3		L_1	L_2	L_3
	Lighting							
	Socket outlets 10 A							
					Total demand current (A)			
					Total maximum demand for installation (A)			

Light industrial demand

Prior to calculating maximum demand, the distribution of the loads across sub-circuits must be determined. Loading tells us the number of sub-circuits required. An estimation of the number of phases may also be required.

Using Table C2

Table C2 of AS/NZS 3000:2018 considers two broad installation categories for the purpose of determining maximum demand. Essentially these categories are residential environments such as boarding houses and hotels, and non-residential such as factories, shops, schools and the like. Prior to using Table C2, Appendix C, AS/NZS 3000:2018, note the following data:

- **Column 1, load group A: Calculation is based on a minimum** 60 W (or the actual wattage of the lamp to be installed if greater) for incandescent lamps, full connected load for fluorescent and other discharge lamps, and 0.5 A per metre per phase of track lighting (or the actual connected load). Other items that may be considered as part of this load group are socket-outlets installed more than 2.3 m above a floor, which supply a luminaire or appliance rated at no more than 150 W and permanently connected appliances rated at not more than 150 W.
- **Column 1, load group B(i):** 10 A socket outlets in non-air-conditioned or heated portions of buildings are based on a wattage rating for each point. To obtain the current loading, simply divide the total wattage by the supply voltage of 230. The first point per phase is rated at 1000 W. Remember that this applies to each phase and not simply the first point. Note also that when determining maximum demand, a multiple combination socket outlet counts as the same number of points as the number of integral socket outlets (a double outlet counts as two points).
- **Column 1, load group B(ii):** 10 A socket outlets in air-conditioned or heated portions of buildings are based on a wattage rating for each point. To obtain the current loading, simply divide the total wattage by the supply voltage of 230. The first point per phase is rated at 1000 W.
- **Column 1, load group B(iii):** Socket outlets rated at more than 10 A are based on the full rating of the highest-rated outlet per phase plus a percentage of the sum of the remainder. If the installation had a number of 32 A outlets as the highest rated outlet, you would allow 32 A plus a percentage of the remainder. As an example, if the installation had six, 32 A outlets this would be $32 + (x\% \times 5 \times 32)$.
- **Column 1, load group C(i): Appliances for cooking, heating and cooling, including instantaneous water heaters** are based on the full rating of the highest-rated appliance per phase plus a percentage of the sum of the remainder.
- **Column 1, load group C(ii): Electric vehicle charging equipment** are based on the full rating of the highest-rated item of equipment per phase plus a percentage of the sum of the remainder.
- **Column 1, load group D: Motors for general use** are based on the full rating of the highest-rated motor per phase plus a percentage of the sum of the remainder.
- **Column 1, load group E: Lifts** are based on the full rating of the highest-rated motor per phase plus a percentage of the sum of the remainder.
- **Column 1, load group F: Fuel dispensing units** are based on the full rating of the highest-rated motor per phase plus a percentage of the sum of the remainder.
- **Column 1, load group G: Heating elements for thermal storage heaters, which includes water heaters, space heaters and the like (swimming pools, spas, saunas) is the full-connected load.**
- **Column 1, load group H: Welding machines require calculation in accordance with AS/NZS 3000:2018.**
- **Column 1, load group I: X-ray machines have diversity applied and only require 50% of the highest-rated machine with all others being ignored.**
- **Column 1, load group J: Is for any equipment not previously covered and requires assessment.**

Calculation method

A suggested approach to calculating maximum demand for a non-domestic installation is as follows.

1. Determine the number of circuits the installation loads require and tabulate them into circuit number and function.
2. Using Table C2 tabulate each circuit on the load list, according to its circuit number, load group, description of load, calculation and demand current.

3 Calculate the contribution each load group makes towards the maximum demand and then total the result.

Note: The current-carrying capacity of the consumer's mains should, when installed in accordance with AS/NZS 3000:2018 and AS 3008.1.1, exceed the rating of the circuit breaker main switch.

Non-domestic demand three-phase

Example 15.12 outlines how to use Table C2 of AS/NZS 3000:2018 to determine the maximum demand of a three-phase non-domestic installation.

EXAMPLE 15.12

Determine the maximum demand for a three-phase factory installation as detailed below.

- 24 × twin 36 W fluorescent lighting fittings @ 0.4 A/fitting
- 6 × 400 W highbay MV light fittings @ 2.2 A/fitting
- 20 × double 10 A socket outlets
- 4 × 15 A single-phase socket outlets
- 2 × 15 A three-phase socket outlets
- 1 × 6.4 kW single-phase range rated at 230 V
- 1 × 2.4 kW single-phase instantaneous water heater rated at 230 V
- 3 × 4 kW three-phase motors rated at 8.5 A per phase
- 3 × 5.5 kW three-phase motors rated at 11.7 A per phase

TABLE 15.12 Sub-circuit arrangement for Example 15.12

Circuit No.	Function	Line (phase)		
		1	2	3
1	12 twin 36 W fluorescent light fittings	✓		
2	12 twin 36 W fluorescent light fittings		✓	
3	3 highbay MV light fittings			✓
4	3 highbay MV light fittings	✓		
5	7 double 10 A outlets		✓	
6	7 double 10 A outlets			✓
7	6 double 10 A outlets	✓		
8	2 × 15 A single-phase outlets		✓	
9	2 × 15 A single-phase outlets			✓
10	15 A three-phase outlet	✓	✓	✓
11	15 A three-phase outlet	✓	✓	✓
12	single-phase range rated at 6.4 kW	✓		
13	instantaneous water heater rated at 2.4 kW		✓	
14	5.5 kW, three-phase motor @ 11.7 A per phase	✓	✓	✓
15	5.5 kW, three-phase motor @ 11.7 A per phase	✓	✓	✓
16	5.5 kW, three-phase motor @ 11.7 A per phase	✓	✓	✓
17	4 kW, three-phase motor @ 8.5 A per phase	✓	✓	✓
18	4 kW, three-phase motor @ 8.5 A per phase	✓	✓	✓
19	4 kW, three-phase motor @ 8.5 A per phase	✓	✓	✓

TABLE 15.13 Maximum demand for Example 15.12

Circuit No.	Load group	Description of load	Qty	Calculation	Demand (A)		
					L_1	L_2	L_3
1	A	twin 36 W fluorescent lights	12	12 × 0.4 = 4.8	4.8		
2	A	twin 36 W fluorescent lights	12	12 × 0.4 = 4.8		4.8	
3	A	highbay MV light fittings	3	3 × 2.2 = 6.6			6.6

»

TABLE 15.13 Maximum demand for Example 15.12 (*Continued*)

Circuit No.	Load group	Description of load	Qty	Calculation	Demand (A)		
					L_1	L_2	L_3
4	A	highbay MV light fittings	3	3 × 2.2 = 6.6	6.6		
5	B(ii)	10 A single-phase socket outlets	14	(1000+(13 × 750))/230		46.7	
6	B(ii)	10 A single-phase socket outlets	14	(1000 + (13 × 750))/230			46.7
7	B(ii)	10 A single-phase socket outlets	12	(1000 + (11 × 750))/230	40.2		
8	B(iii)	15 A single-phase outlets	2	2 × (0.75 × 15)		22.5	
9	B(iii)	15 A single-phase outlets	2	2 × (0.75 × 15)			22.5
10	B(iii)	15 A three-phase outlet	1	Highest rated	15.0	15.0	15.0
11	B(iii)	15 A three-phase outlet	1	0.75 × 15	11.3	11.3	11.3
12	C(i)	single-phase range	1	6400/230	27.8		
13	C(i)	instantaneous water heater	1	2400/230		10.4	
14	D	5.5 kW, three-phase motor	1	Highest rated motor	11.7	11.7	11.7
15	D	5.5 kW, three-phase motor	1	0.75 × 11.7 = 8.8	8.8	8.8	8.8
16	D	5.5 kW, three-phase motor	1	0.5 × 11.7	5.8	5.8	5.8
17	D	4 kW, three-phase motor	1	0.5 × 8.5 = 4.3	4.3	4.3	4.3
18	D	4 kW, three-phase motor	1	0.5 × 8.5 = 4.3	4.3	4.3	4.3
19	D	4 kW, three-phase motor	1	0.5 × 8.5 =4.3	4.3	4.3	4.3
				Total demand current (A)	**144.9**	**149.9**	**141.3**

The maximum demand of this installation is the highest loaded phase, which is L2 with 149.9 A. It is necessary to check local supply authority requirements for the permissible unbalance between phases.

EXERCISE 15.11

Determine the maximum demand for a three-phase factory installation as detailed below.

- **a** 54 × twin 36 W fluorescent lighting fittings @ 0.35 A/fitting
- **b** 21 × 500 W highbay MV light fittings @ 2.5 A/fitting
- **c** 18 × double 10 A socket outlets
- **d** 3 × 20 A single-phase socket outlets
- **e** 2 × 32 A three-phase socket outlets
- **f** 1 × 9.6 kW two-phase range rated at 230 V
- **g** 2 × 4.8 kW single-phase quick recovery water heater rated at 230 V
- **h** 2 × 6 kW three-phase motors rated at 12.5 A per phase
- **j** 3 × 4 kW three-phase motors rated at 7.7 A per phase

TABLE 15.14 Sub-circuit arrangement for Exercise 15.11

✓	Function	Line (phase)		
		1	2	3
1	18 twin 36 W fluorescent light fittings	✓		
2	18 twin 36 W fluorescent light fittings		✓	
3	18 twin 36 W fluorescent light fittings			✓
4	7 highbay MV light fittings	✓		
5	7 highbay MV light fittings		✓	
6	7 highbay MV light fittings			✓
7	6 double 10 A outlets	✓		

»

TABLE 15.14 Sub-circuit arrangement for Exercise 15.11 *(Continued)*

Circuit No.	Function	Line (phase)		
		1	2	3
8	6 double 10 A outlets		✓	
9	6 double 10 A outlets			✓
10	1 × 20 A single-phase outlet	✓		
11	1 × 20 A single-phase outlet		✓	
12	1 × 20 A single-phase outlet			✓
13	32 A three-phase outlet	✓	✓	✓
14	32 A three-phase outlet	✓	✓	✓
15	two-phase range rated at 9.6 kW (4.8 kW /phase)	✓	✓	
16	QR water heater rated at 4.8 kW			✓
17	6 kW, three-phase motor @ 12.5 A per phase	✓	✓	✓
18	6 kW, three-phase motor @ 12.5 A per phase	✓	✓	✓
19	4 kW, three-phase motor @ 7.7 A per phase	✓	✓	✓
20	4 kW, three-phase motor @ 7.7 A per phase	✓	✓	✓
21	4 kW, three-phase motor @ 7.7 A per phase	✓	✓	✓

TABLE 15.15 Maximum demand for Exercise 15.11

Circuit No.	Load group	Description of load	Qty	Calculation	Demand (A)		
					L_1	L_2	L_3
1		fluorescent light fittings					
2		fluorescent light fittings					
3		fluorescent light fittings					
4		highbay MV light fittings					
5		highbay MV light fittings					
6		highbay MV light fittings					
7		6 double 10 A outlets					
8		6 double 10 A outlets					
9		6 double 10 A outlets					
10		1 × 20 A single-phase outlet					
11		1 × 20 A single-phase outlet					
12		1 × 20 A single-phase outlet					
13		32 A three-phase outlet					
14		32 A three-phase outlet					
15		range					
16		QR water heater					
17		6 kW, three-phase motor					
18		6 kW, three-phase motor					
19		4 kW, three-phase motor					
20		4 kW, three-phase motor					
21		4 kW, three-phase motor					
				Total demand current (A)			

REVIEW QUESTIONS

1 What is meant by the term 'maximum demand of the installation'?

2 What factor with consumer's mains must exceed the rating of the circuit breaker main switch?

3 Why should lighting be spread over three phases?

4 Determine the maximum demand of a single-phase 230 V single domestic installation that has the following connected load:
- 24 lighting points
- 500 W outdoor lighting
- 8 single and 14 double socket outlets
- 1 × 15 A socket outlet
- range rated at 44 A
- 20 A controlled load water heater

5 Determine the maximum demand of a three-phase single domestic installation that has the following load groups connected:
- 45 lighting points
- 12 × 30 W LED floodlights in warm white
- 21 double socket outlets
- 4.5 kW hard-wired 60 cm oven and 6.5 kW hard-wired hotplates
- Three-phase ducted air-conditioner rated at 11.2 A per phase

6 Determine the maximum demand of a three-phase block of living units and the maximum demand of the sub-mains to each unit. In the block there are:
- 3 × three-bedroom units
- 6 × two-bedroom units
- 6 × one-bedroom units

Each unit is connected to a single-phase supply at their distribution switchboard via their sub-mains originating at the main switchboard. A three-phase supply feeds the main switchboard. The individual units have the following load groups connected. The load on each phase comprises 1 × three-bedroom unit, 2 × two-bedroom units, and 2 × one-bedroom units. The communal load is single-phase and metered separately.

Three-bedroom units
- 18 lighting points
- 14 double socket outlets
- 2.8 kW hard-wired oven and 6.8 kW hard-wired cooktop
- 4.8 kW 315 litre quick recovery storage-type water heater

Two-bedroom units
- 16 lighting points
- 10 double socket outlets
- 2.8 kW hard-wired oven and 6.8 kW hard-wired cooktop
- 3.6 kW 250 litre quick recovery storage type water heater

One-bedroom units
- 12 lighting points
- 8 double socket outlets
- 2.8 kW hard-wired oven and 6.8 kW hard-wired cooktop
- 2.4 kW 125 litre quick recovery storage type water heater

Communal loads
- 9 socket outlets
- 8 × 8 W CFL lights (0.05 A each) and 4 × 30 W LED floodlights (0.2 A each)

7 An industrial workshop is supplied by a single-phase 230 V mains and has the following load groups connected:
- lighting points consisting of 24 × 36 W fluorescent lamps each of which draws 0.65 A
- 16 double socket outlets
- 2.3 kW hard-wired oven
- 3.6 kW 250 litre storage-type water heater

Determine the maximum demand for the installation.

8 An industrial installation has a three-phase 230/400 V supply. Balance the proposed load group over the three phases and determine the maximum demand.
- 48 lighting points each drawing 0.56 A
- 20 double 10 A socket outlets and 5 × 15 A socket outlets (single-phase)
- 4.6 kW 230 V storage water heater
- 1 × 5 A/phase three-phase motor
- 1 × 15 A/phase three-phase motor
- 2 × 22.5 A/phase three-phase motors
- 2 × 2.6 A/phase three-phase motors

9 Determine the maximum demand for the single-phase 230 V single domestic installation detailed below.
- 26 × lighting points
- 15 × double 10 A socket outlets
- 2 × 15 A socket outlets
- 1 × 3.6 kW range at 230 V
- 1 × 3.6 kW at 230 V instantaneous water heater

10 Determine the maximum demand for the consumer's mains for the three-phase 400 V single domestic installation detailed below.
- 42 × lighting points
- 8 × 200 W exterior lights
- 28 × double 10 A socket outlets
- 2 × 15 A socket outlets
- 1 × 7.6 kW range
- 1 × 3.6 kW instantaneous water heater @ 240 V
- 1 × 1.1 kW 230 V pool filter pump rated at 4.78 A

15.5 Cable selection based on current-carrying capacity

There are three types of cable wiring systems. The first are bare conductors that are used as out-of-reach aerials/lines and are supported using insulators. The second are unsheathed thermoplastic insulated (TPI) conductors that can be installed in various types of enclosures to provide supplementary insulation and mechanical protection. Third are the sheathed cables which can be installed using any suitable method.

The most important element of any cable is its current-carrying capacity. This capability is a factor of the maximum permissible insulation temperature allowed. The way the cable is installed can have a significant effect on a cable's ability to dissipate heat from the conductor. Heat produced by a current-carrying conductor is an effect caused by the conductor's resistance. The heat generated is directly proportional to the square of the current. Therefore, if the current has doubled, the heat will quadruple. While the conductor of the cables has a very high melting point (copper melts at 1083 °C), it is the insulation surrounding the conductor that is the main limiting factor for heat dissipation. A 2C + E flat, 450/750 V PVC-insulated, PVC white-sheathed cable rated at 90 °C with copper conductors is shown in **Figure 15.22**.

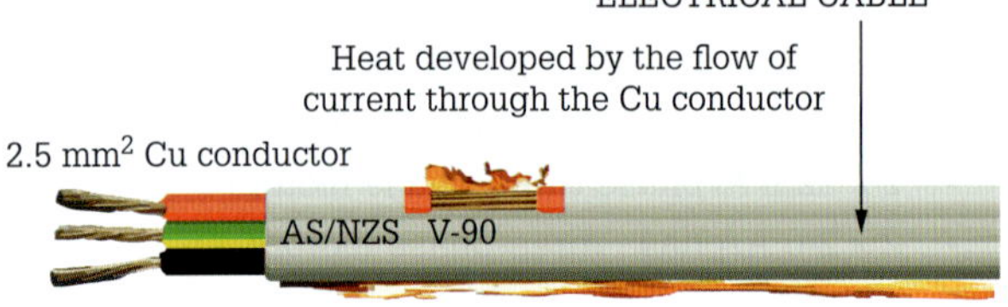

FIGURE 15.22 Heat being emitted by a copper conductor into its solid polymer insulation

Even though the TPS cable in **Figure 15.22** is rated at 90 °C, for purposes of calculating current-carrying capacity, a temperature of 75 °C is used. Refer to Tables 4, 7, 10 and 13 in Section 3, AS/NZS 2009 *Electrical installations – Selection of cables*. These tables enable the selection of the minimum size of conductor necessary to carry the load current under the installed conditions. Note that a larger size conductor can always be used.

When determining the current requirements of the circuit:

$$I_B \leq I_Z$$

where I_B = the current for which the circuit is designed

I_Z = the continuous current-carrying capacity of the cable

The establishment of current-carrying capacities for cables provides for a safe installation. They ensure that the temperature of the cable is not exceeded under normal load conditions.

Refer to AS/NZS 3008.1.1:2017 (Australia) *Electrical installations – Selection of cables*, Section 3, 'Current-carrying capacity'.

Factors affecting current-carrying capacity

The current-carrying capability of a cable is dependent upon the type of conductor material, its cross-sectional area (CSA), method of installation (may require a de-rating factor) and external influences that affect the allowable operating temperature of the cable insulation.

The types of conductor materials used are high-conductivity copper or aluminium. All conductors have electrical resistance and they experience an energy loss when they carry current. This loss appears as heat and causes the resistance of the cable to rise. As the temperature of the conductor rises, the heat it loses to its surroundings by conduction, convection and radiation also increases.

Therefore, different methods of installation affect the rate at which the heat generated by the current flowing through a conductor is dissipated to the surrounding environment. A cable installed in the air and clipped to a flat surface is able to dissipate heat more easily than a similar cable which is within a wiring enclosure buried in the ground.

The resistance of a conductor is inversely proportional to its CSA. Therefore, any increase in the CSA of any conductor material significantly decreases the conductor resistance and increases the current-carrying capacity of the conductor. The lower the resistance, the more current a conductor can carry at a given temperature.

However, even though copper has a melting point of 1083 °C and pure aluminium has a melting point of 660 °C, the current a conductor of a given CSA is capable of carrying safely depends upon its own insulation and sheath temperature limitations.

PVC is probably the most usual form of cable insulation and is very susceptible to damage by temperatures above 90 °C. Therefore, the current ratings of the various CSA conductors using PVC as an insulant are designed to ensure that this will not happen.

AS/NZS 3008.1.1:2017 *Electrical installations – Selection of cables* has separate tables for different types of cables, with differing conditions of installation, grouping factors and so on. As an example, Table 14 states that a V–90 PVC three-core 0.6/1 kV insulated and sheathed 2.5 mm² stranded copper cable has a current-carrying capacity of 13 A if enclosed and completely surrounded by thermal insulation.

If the same cable is unenclosed and fixed directly to the wall, then its current-carrying capacity is 26 A.

Coordination between protection devices and circuit conductors

Refer to AS/NZS 3000:2018 *Wiring Rules*, Section 2, 'General arrangement, control, and protection'.

A commonly overloaded component of an electrical system is the wiring. The correct current-carrying capacity of a cable is a function of its insulation, temperature rating, the type of enclosure in which it is installed, number of conductors being installed in a common enclosure and whether it is above ground or underground. Protective devices must be provided to break any overload current flowing in the circuit conductors before the current causes a temperature rise that harms the insulation, terminations, joins or surrounding environment. Therefore, there has to be a 'condition of coordination' between conductors and protective devices.

Coordination and overload protection requirements ensure that the life of the cable's insulation is not significantly shortened. **Figure 15.23** illustrates the relationship between the four current conditions that exist when a load is connected to a protective device (see Clause 2.5.3, 'Protection against overload current').

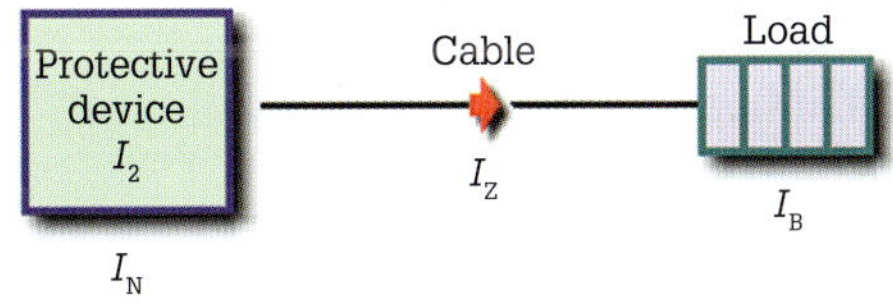

FIGURE 15.23 Four current conditions

I_N = the nominal current rating or current setting of the device protecting the circuit against over-current

I_Z = the current-carrying capacity of the cable for continuous service with respect to a particular installation condition concerned

I_B = the designed current of the circuit (the current carried by the circuit in normal service)

I_2 = the minimum operating current of the protective device.

The operating characteristic of each protective device protecting a cable from overload and short-circuit should satisfy the following two conditions:

1. The normal current or current setting (I_N) of the protective device should not be less than the design current (I_B) of the circuit. To avoid overheating the cable insulation it is essential that the maximum sustained current (I_B) carried by a cable was less than or equal to its current-carrying capacity (I_Z).
2. The normal current or current setting (I_N) of the protective device should not exceed the lowest of the current-carrying capacities (I_Z) of any of the cables of the circuit.

   Therefore:

$$I_B \le I_N \le I_Z \text{ Condition 1 (for coordination).}$$

To ensure that the life of the cable insulation is not significantly shortened, protective devices should have an adequate current tripping rating (I_2) of:

$$I_2 \le 1.45 \times I_Z \text{ Condition 2 (for overload)}$$

The factor 1.45 is the fusing factor of the protective device (fusing factor $= \frac{I_2}{I_N}$).

Miniature circuit breakers and moulded case circuit breakers normally have tripping factors equal to or below the 1.45 factor. Consequently:

$$1.45 I_N = I_2$$
$$\text{and } 1.45 I_N \le 1.45 I_Z$$
$$\text{so } I_N < I_Z$$

Therefore, if either of these devices is used in compliance with condition 1, condition 2 is also met, thus providing overload protection for the cables concerned.

For HRC fuses, the nominal fusing current rating is $I_N > 1.6$, the current-carrying capacity (I_Z) of the cable they protect within one hour. Therefore, to satisfy condition 2 the nominal tripping current rating (I_N) is:

$$I_N = \frac{1.45}{1.6}$$
$$= 0.9$$
$$I_N = 0.9 I_Z$$

Coordination example

Let us take as our example a single circuit for a 20 kW split-ducted reverse-cycle air-conditioner with the ratings in **Table 15.16**.

TABLE 15.16 Capacity ratings examples

Cooling capacity 18000 W	Power supply 400 V three-phase plus neutral 50 Hz
Heating capacity 20000 W	Operating voltage range 342–418 V
Rated input power	Rated input current
Cooling 6500 W	Cooling 11.5 A
Heating 6200 W	Heating 11 A

The cable is copper 4.0 mm^2 four-core plus earth V–90 TPS installed enclosed in PVC conduit on a surface in air with nominal ambient temperature not exceeding 40 °C.

Determine the size of the miniature circuit breaker (MCB) and HRC fuse protection devices that satisfy the overload requirements.

The first important factor that must be considered is the current-carrying capacity of the cable to be protected. The current-carrying capacity is clearly dependent on the conductor's insulation materials and its CSA. In addition, the current-carrying capacity is affected by the ambient temperature of the environment in which the cable operates and on the installation arrangements, including the spacing and adequacy of air circulation.

- The current-carrying capacity of the cable installed enclosed in free air, I_Z, is 25 A (from Table 13, column 11, AS/NZS 3008.1.1:2017).
- The design current for the circuit (I_B) is 11.5 A.
- The nominal commercial current rating of the protection device (I_N) is 20 A.

These values satisfy the coordination condition 1.

$$I_B \leq I_N \leq I_Z$$
$$11.5 < 20 < 25$$

Overload protection for condition 2:

$$I_2 \leq 1.45 I_Z$$
$$= 1.45 \times 25$$
$$= 36.2\,\text{A}$$

For an MCB:

$$I_N \leq I_Z \leq 25\,\text{A}$$

For an HRC fuse:

$$I_N \leq 0.91_Z$$
$$\leq 0.9 \times 25$$
$$\leq 22.5\,\text{A}$$

The nominal manufacturer's rating of overload protection is: the nominal current of the MCB and HRC fuse must be less than the current rating of the cable it is protecting (I_Z), but higher than the current that it will carry continuously (I_B).

Therefore, for an MCB with $I_N = 20$ A, a B-type MCB would be suitable, and for an HRC fuse with $I_N = 20$ A, a general purpose 'gG' category high-rupturing fuse would be appropriate as this category provides complete protection.

These two protection devices satisfy the overload requirements of $I_2 \leq 1.45 I_Z$ using a 20 A MCB or $I_N = 0.9 I_Z$ using a 20 A HRC fuse.

SWITCH ON

HRC fuse links are defined by a two-letter code.

The first letter indicates the breaking range of the fuse link, as follows:

'g' Full-range breaking capacity fuse link

'a' Partial-range breaking capacity fuse link

The second letter indicates utilisation category, as follows:

'G' Fuse link for general application, including the protection of motor circuits

'M' Fuse link for protection of motor circuits

These two letters are combined to recognise three HRC fuse link classes: gG, gM and aM. A standard fuse link classified as type 'gG' will protect an associated PVC-insulated cable against both overload and short-circuit if its current rating (I_N) is equal to or less than 90% of the current rating of the cable (I_Z).

Using AS/NZS 3000:2018 to determine current-carrying capacity

Reference to current-carrying capacity is found in Clause 3.4.1, AS/NZS 3000:2018.

All current-carrying capacity calculations are made using AS/NZS 3008.1.1. However, there are several clauses in AS/NZS 3000:2018 that deal with conductor size. These are Clause 3.4.3, which provides the minimum size of conductors in parallel, Clause 3.5.1, which gives the nominal CSA of conductors, and Clause 3.5.2, which covers the minimum size of the neutral conductor in single-phase two-wire and multiphase circuits.

There are two ways in which cable selection using AS/NZS 3000 can be implemented. One way is selecting a cable size and then determining the maximum current-carrying capacity. The second way is by choosing a cable size capable of carrying a known current demand.

Current-carrying capacity of consumer's mains

Clause 1.4.37, AS/NZS 3000:2018, provides a definition for consumer's mains while AS/NZS 3008.1.1 provides elements that are to be considered when determining the minimum size of consumer's mains. These features include:

- type of conductor
- type of cable
- installation conditions
- external influences on cables
- voltage drop
- short-circuit performance.

In addition, the requirements of the local electricity entity must be followed.

To determine the current-carrying capacity of a cable refer to AS/NZS 3008.1.1, Columns 3 and 5 of Table 3. These columns list the various installation methods that could be employed in the installation. Once the method has been chosen, then refer to Table 3, Column 4, to obtain the relevant table and column for the current-carrying capacity. See column 6 for any applicable de-rating factors.

In simple terms, the selection of a cable involves the selection of the correct table and then applying any necessary de-rating factors to the current-carrying capacity of the cable. In order to understand 'de-rating', read Clauses 3.1 to 3.5 of AS/NZS 3008.1.1.

When de-rating is required

De-rating factors must be applied when:

- cables are installed in the air, and the ambient temperature exceeds 40 °C
- cables are grouped.

De-rating factors must be implemented to underground cables if the:

- ambient soil temperature exceeds 25 °C
- depth of laying exceeds 0.5 m

- thermal resistivity of the soil exceeds 1.2 °C m/w
- cables are contained within wiring enclosures (see Clause 3.4.5d of AS/NZS 3008.1.1).

The de-rating factors inform us that the current-carrying capacity of the cable depends on the type of insulation, the insulation's temperature rating and all other factors which influence the cable's operating temperature.

The effect of the installation methods, such as unenclosed, enclosed, buried direct and installed within non-metallic enclosures, can be clearly observed by considering the current-carrying capacity of consumer's mains consisting of two copper 16 mm^2 single-core 0.6/1 kV sheathed cables with V–90 insulation in Table 4 of AS/NZS 3008.1.1.

From Table 4, if the consumer's mains were connected to an aerial and were installed unenclosed within the eaves space then Column 8 conditions would apply. Consequently, each 16 mm^2 copper conductor has a current rating of 72 amperes. If these same conductors were installed on a ceiling with one side of the conductors in continuous contact with the ceiling material then the conditions of Column 18, Table 4, apply. Consequently, the current capacity of the consumer's mains is now 56 A. If the consumer's mains were protected by a 65 A type C circuit breaker then the 72 ampere-rated cable condition is suitable. Where the conductors were installed on a ceiling with one side of the conductors in continuous contact with the ceiling material, the cable is hazardous because we have a 65 A circuit breaker protecting a 56 ampere-rated cable.

Current-carrying capacity of sub-circuits

The cable selection procedure for sub-circuits requires application of one of the following.

Using AS/NZS 3008.1.1 to determine maximum current-carrying capacity for a predetermined cable size or select a cable size to deliver a known demand current capacity

To determine the current-carrying capacity of a given cable apply the following:

- Select the chosen installation method by reference to Table 3 (1 to 4), Columns 3 and 5.
- Refer to Column 4 in Table 3 to obtain the reference table and column for the given current-carrying capacity of the cable.
- Note the de-rating requirements of Table 3, Column 6, and the corresponding footnotes to the referenced table.
- After going to the referenced table, record the current-carrying capacity of the known cable.
- Apply any de-rating that the referenced table requires (refer to table notes). To avoid de-rating of cables, refer to Clause 3 of AS/NZS 3008.1.1.
- Record the current-carrying capacity of the known cable.

EXAMPLE 15.13

Flat TPS with four possible de-rating requirements

Determine the current-carrying capacity of flat TPS 0.6/1 kV 2.5 mm^2, two-core plus earth V–90 PVC/PVC cable with stranded copper conductors for four possible 'de-rating' requirements as shown in Figure 15.24.

Spaced on horizontal surface

Table 3(1), Item 12, Columns 3 and 5, applies to configuration.

Tables 10 or 11, Columns 5, 6 or 7 (or Table 12, Columns 4 and 5).

No de-rating for single-circuit configuration as per footnote 4a of Table 10 (5a of Table 11 or 3a of Table 12).

From Table 22, Item 3 and Column 5, a de-rating factor of 0.85 applies.

From Table 10, Column 5: 2.5 mm^2 is rated at **26 A.**

Touching on horizontal surface

Table 3(1), Item 12, Columns 3 and 5, applies to configuration.

Tables 10 or 11, Column 5, 6 or 7 (Table 12, Columns 4 and 5).

De-rating Table 22 applies.

From Table 22, Item 3 and Column 5, a de-rating factor of 0.85 applies.

From Table 10, Column 5: 2.5 mm^2 is rated at 26 A, which is de-rated to 22.1 A.

Partially surrounded by thermal insulation

Table 3(2), Item 9, applies to configuration.

Tables 10 or 11, Column 15 or 16 (Table 12, Column 11).

No de-rating for single-circuit configuration as per footnote 4a of Table 10 (5a of Table 11 or 3a of Table 12).

From Table 10, Column 15: 2.5 mm^2 is rated at **20 A**.

Surrounded by thermal insulation

Table 3(2), Item 11, Columns 3 and 5, applies to configuration

Tables 10 or 11, Column 19 or 20 (Table 12, Column 12).

No de-rating for single circuit configuration as per footnote 4a of Table 10 (5a of Table 11 or 3a of Table 12).

From Table 10, Column 19: 2.5 mm^2 is rated at 13 A.

»

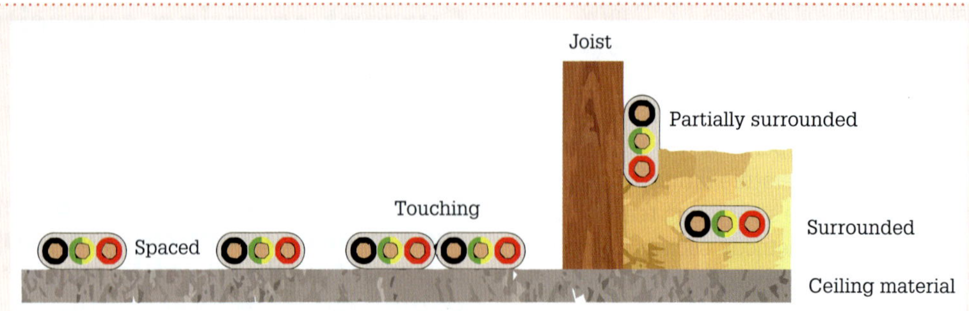

FIGURE 15.24 Flat TPS 2.5 mm² copper conductors two-core plus earth V–90 cable

EXERCISE 15.12

Determine the current-carrying capacity of flat TPS 0.6/1 kV 6 mm², two-core plus earth V–90 PVC/PVC cable with stranded copper conductors for four possible 'de-rating' requirements as follows:

- **a** Spaced on horizontal surface
- **b** Touching on horizontal surface
- **c** Partially surrounded by thermal insulation
- **d** Surrounded by thermal insulation

EXAMPLE 15.14

Single-circuit cable clipped to a concrete wall

Determine the current-carrying capacity of a 25 mm², four-core plus earth V–90 PVC/PVC cable with aluminium conductors as shown in Figure 15.25. The method of installation has the single-circuit cable clipped to a concrete wall.

Table 3(1), Item 13, Columns 3 and 5, applies to configuration.

Table 13 or 14, Column 5, 6 or 7 (Table 15, Columns 4 and 5).

No de-rating for single-circuit configuration.

From Table 13, Column 7: 25 mm² is rated at **71 A**.

FIGURE 15.25 25 mm², four-core plus earth V–90 PVC/PVC cable with aluminium conductors

EXAMPLE 15.15

Two separate circuits unenclosed clipped to the wall and touching

Determine the current-carrying capacity of two separate circuits consisting of two-core plus earth 0.6/1 kV insulated and sheathed stranded copper 4 mm² V–90 PVC/PVC cables as shown in Figure 15.26. The cables are unenclosed, clipped directly to the wall and touching each other.

Table 3(1), Item 12, Columns 3 and 5, applies to configuration.

Table 10 or 11, Column 5, 6 or 7 (Table 10, Column 5, for V–90 copper).

Table 10, Column 5: 4 mm² is rated at 34 A.

De-rating factors apply for two circuits. Table 22, Item 3, Column 5: applicable de-rating is 0.85: $34 \times 0.85 =$ **28.9 A**.

FIGURE 15.26 Two-core plus earth 0.6/1 kV copper 4 mm² V–90 PVC/PVC cables

EXAMPLE 15.16

Multi-core cable circuits supported on an unperforated cable tray

Determine the current-carrying capacity of the three touching four-core and earth 35 mm^2 0.6/1 kV insulated and sheathed V–90 PVC/PVC multi-core stranded copper cable circuits supported on an unperforated cable tray as shown in Figure 15.27.

Table 3(1), Item 10, Column 2, 3 or 4, multi-core cables in unperforated cable tray.

Table 13 or 14, Column 2, 4 or 5 (Table 15, Columns 2 and 3).

Table 13, Column 2: 35 mm^2 is rated at 120 A.

From Table 3(1), Item 10, Column 6: de-rating for more than one circuit is found in Table 24.

Table 24, Item 1, Column 7: de-rating factor of 0.78 applies to three circuits (refer to footnote 3 to determine for which columns the de-rating is applicable).

The 35 mm^2 is de-rated to: $120 \times 0.78 =$ **93.6 A**.

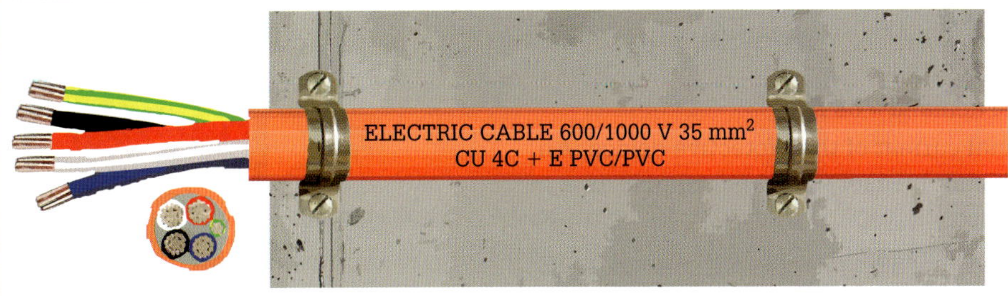

FIGURE 15.27 Four-core and earth 35 mm^2 0.6/1 kV insulated and sheathed V–90 PVC/PVC

EXAMPLE 15.17

Copper cables on perforated cable tray in a horizontal formation

Determine the current-carrying capacity of the two three-phase circuits, consisting of 25 mm^2 copper stranded V–90 three single-core (SDI) PVC/PVC 0.6/1 kV copper cables supported on perforated cable tray in a horizontal formation as shown in Figure 15.28.

Table 3(1), Item 5, Columns 3 and 5, SDI cables on perforated cable tray.

Table 7 or 8, Column 5, 6 or 7 (Table 9, Columns 6 and 7).

Table 7, Column 5: 25 mm^2 is rated at 103 A.

From Table 3(1), for Item 5, Column 6: de-rating for more than one circuit is found in Table 23.

Table 23, Item 4: de-rating factor of 0.89 applies to two circuits.

The 25 mm^2 is de-rated to: $103 \times 0.89 =$ **91.6 A**.

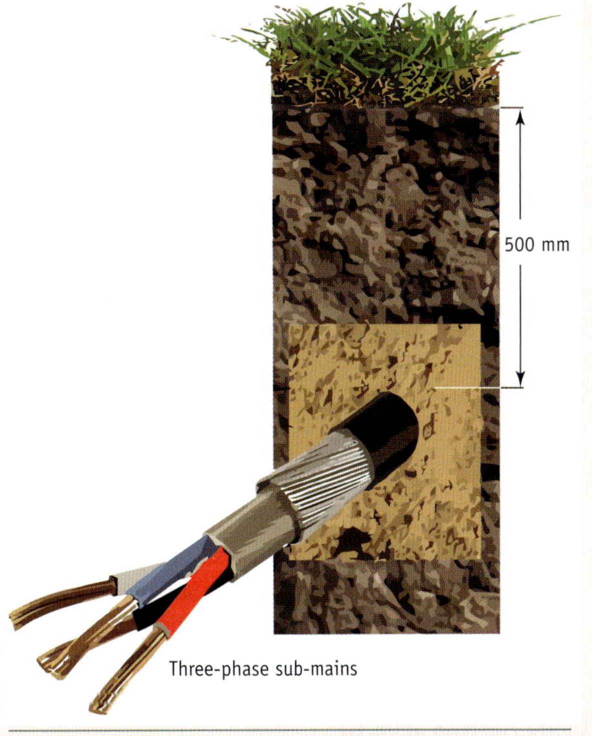

FIGURE 15.28 25 mm^2 copper V–90 three single-core (SDI) PVC/PVC 0.6/1 kV copper cables

Using AS/NZS 3008.1.1 to determine the CSA of a cable that relates to the known maximum demand current

To determine the CSA of a cable, apply the following:

- Circuit protection must be addressed before the cable is chosen.
- Locate the correct table and columns using the installation methods shown in Table 3.

- Note any de-rating applicable from Table 3 and relevant footnotes.
- Apply any de-ratings required to the current-carrying capacity.
- Note total current-carrying capacity from the correct table.
- Record the CSA of the cable that relates to the current-carrying capacity.

EXAMPLE 15.18

Consumer's mains in a buried non-metallic wiring enclosure

Two single-core stranded PVC/PVC V–90 0.6/1 kV sheathed consumer's mains with a connected load of 68 A are to be installed in a non-metallic wiring enclosure as shown in Figure 15.29. Determine the minimum CSA conductor size that can carry the maximum demand current.

- I_B = maximum demand = 68 A
- I_N = the nominal current of the protective device. An 80 A type C circuit breaker would be suitable as a protective device.
- I_Z can now be considered.

Using AS/NZS 3008.1.1

Table 3(4), Item 1, Columns 3 and 5, show installation methods.

Table 4 or 5, Columns 24 to 26 (Table 6, Columns 16 and 17), show the current-carrying capacities.

From Table 4, Column 24: a 16 mm² conductor is rated at 89 A.

No de-rating is required as the wiring enclosure contains only one circuit.

Therefore, $I_B \leq I_N \leq I_Z$ is satisfied as $68 \leq 80 \leq 89$.

FIGURE 15.29 Single-core PVC/PVC V–90 0.6/1 kV sheathed consumer's mains

EXAMPLE 15.19

Three-phase plus neutral sub-main unenclosed and buried direct

Select the minimum CSA for a three-phase plus neutral sub-main consisting of a four-core stranded copper 0.6/1 kV armoured and sheathed PVC/PVC cable supplying a load demand of 46 A, as shown in Figure 15.30. The cable must be V–90 insulated and contain copper conductors. The installation method chosen is unenclosed and buried direct.

- I_B = maximum demand = 46 A
- I_N = the nominal current of the protective device. A 50 A type C circuit breaker would be suitable as a protective device.
- I_Z can now be considered.

Using AS/NZS 3008.1.1

Table 3(3), Item 4, Columns 3 and 5, show installation methods.

Table 13 or 14, Column 23 or 24 (Table 15, Column 13), show the current-carrying capacities.

Table 25(2) is the de-rating table in Column 6 of Table 3(3).

Table 13, Column 23: 10 mm² is rated at 55 A (the CSA chosen must be rated as 50 A).

Therefore, $I_B \leq I_N \leq I_Z$ is satisfied as $46 \leq 50 \leq 55$.

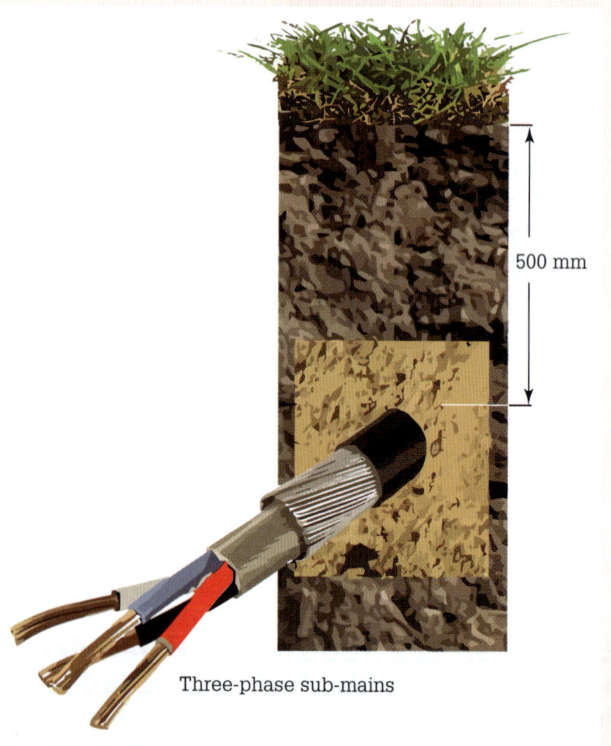

FIGURE 15.30 Four-core 0.6/1 kV armoured and sheathed PVC/PVC cable

EXAMPLE 15.20

Three-phase sub-main supported by a catenary

Select the minimum CSA for a three-phase sub-main consisting of a four-core stranded copper 0.6/1 kV armoured and sheathed circular PVC/PVC cable supplying a load demand of 95 A, as shown in Figure 15.31. The cable, which is supported by a catenary, is V–90 insulated and contains copper conductors.

- I_B = maximum demand = 95 A
- I_N = the nominal current of the protective device. A 100 A type C circuit breaker would be suitable as a protective device.
- I_Z can now be considered.

Using AS/NZS 3008.1.1

Table 3(1), Item 11, Columns 3 and 5, show the installation methods.

Table 13 or 14, Columns 2 to 4 (Table 15, Columns 2 and 3), show the current-carrying capacities.

Table 22 is the de-rating table in Column 6 of Table 3(1).

No de-rating is required because there is only one circuit.

Table 13, Column 2: 120 A = 35 mm^2

Therefore, $I_B \le I_N \le I_Z$ is satisfied as 95 ≤ 100 ≤ 120.

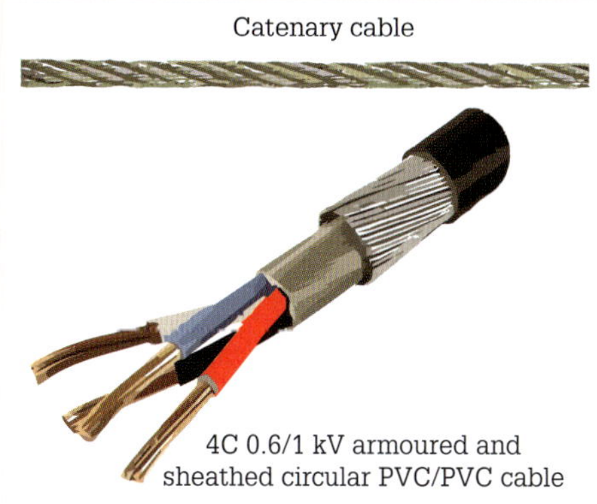

FIGURE 15.31 Four-core 0.6/1 kV armoured and sheathed circular PVC/PVC

EXAMPLE 15.21

Four-core aluminium aerial bundled cable

Select the minimum CSA for a three-phase circuit consisting of a four-core aluminium aerial bundled cable (ABC) 0.6/1 kV XLPE X–90UV cable supplying a load demand of 97 A, as shown in Figure 15.32. The cable is installed in the air and experiences a wind speed of 2 m/s.

- I_B = maximum demand = 97 A
- I_N = the nominal current of the protective device. A 100 A type C circuit breaker would be suitable as a protective device.
- I_Z can now be considered.

Using AS/NZS 3008.1.1

In AS/NZS 3008.1.1 the current-carrying capacity of aerial cables is determined from Tables 20 and 21.

Table 21 is for aluminium ABC cables. Column 20: 113 A = 25 mm^2 (the CSA curent-carrying capacity chosen must be rated as ≥ 100 A).

Therefore, $I_B \le I_N \le I_Z$ is satisfied as 97 ≤ 100 ≤ 113.

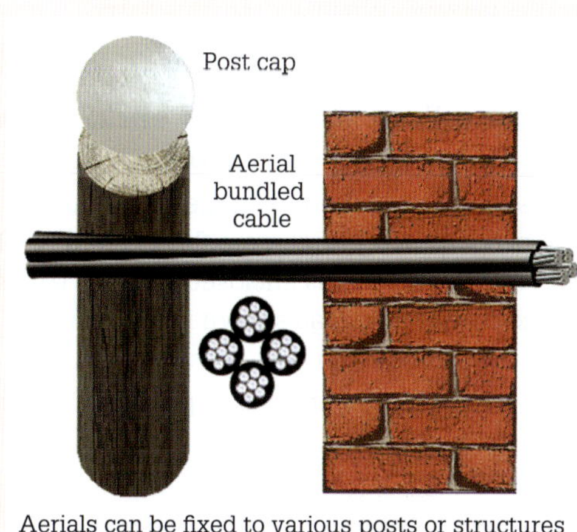

FIGURE 15.32 Four-core aluminium aerial bundled cable (ABC) 0.6/1 kV XLPE X–90UV cable

EXAMPLE 15.22

MIMS in free air and clipped to a concrete wall

Select the minimum CSA for a three-phase final sub-circuit consisting of a bare multi-core (4) mineral-insulated 1/1 kV copper-sheathed cable with copper conductors and a sheath temperature of 100 °C supplying a load demand of 24 A, as shown in Figure 15.33. The cable is installed in free air and clipped to a concrete wall.

- I_B = maximum demand = 24 A
- I_N = the nominal current of the protective device. A 25 A type C circuit breaker would be suitable as a protective device.
- I_Z can now be considered.

Using AS/NZS 3008.1.1

For mineral-insulated metal-sheathed (MIMS) cables, the current-carrying capacity of bare or served copper MIMS cables is determined from Tables 18 and 19.

Table 19 is for bare multi-core MIMS cables.

Column 5: 28 A = 2.5 mm^2 (the CSA chosen must be rated as ≥ 24 A).

Therefore, $I_B \leq I_N \leq I_Z$ is satisfied as $24 \leq 25 \leq 28$.

FIGURE 15.33 Bare multi-core (4) mineral-insulated 1/1kV copper-sheathed cable

EXAMPLE 15.23

Single-phase final sub-circuits enclosed in cable trunking fixed on a wall

There are six (6) only separate single-phase final sub-circuits comprising three-core flat 2.5 mm^2 V–90 stranded copper TPS cables enclosed in cable trunking, which is fixed on the wall as shown in Figure 15.34. Determine the current-carrying capacity of each circuit.

Note: As the cables are not fixed in position they are considered as bunched.

Using AS/NZS 3008.1.1

Table 3(2), Item 7, Columns 3 and 5, show the installation methods.

Table 10 or 11, Columns 11 to 13 (Table 12, Column 9 or 10) show the current-carrying capacities.

Table 22 is the de-rating table in Column 6 of Table 3(2).

Table 10, Column 11: 2.5 mm^2 = 23 A

De-rating: Table 22, Item 2, Column 9: 0.57: 23×0.57 =13.1 A.

Appropriate circuit protection would be a 10 A type C circuit breaker.

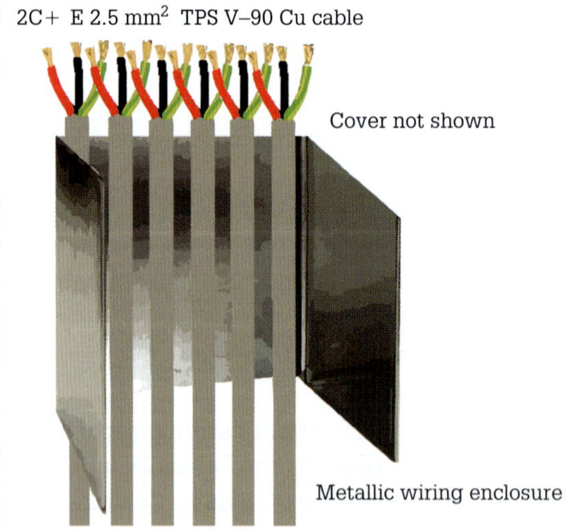

FIGURE 15.34 Three-core flat 2.5 mm^2 V–90 copper TPS cables

EXAMPLE 15.24

Single-phase sub-main enclosed in a non-metallic duct in free air

Select the minimum CSA for a single-phase sub-main consisting of a two-core and earth 0.6/1 kV circular sheathed PVC/PVC cable supplying a load demand of 45 A, as shown in Figure 15.35. The cable is enclosed in a non-metallic duct in free air and must be V–90 insulated and contain copper conductors.

- I_B = maximum demand = 45 A
- I_N = the nominal current of the protective device. A 50 A type C circuit breaker would be suitable as a protective device.
- I_Z can now be considered.

Using AS/NZS 3008.1.1

Table 3(2), Item 7, Columns 3 and 5, show the installation methods.

Table 10 or 11, Columns 11 to 13 (Table 12, Column 9 or 10) show the current-carrying capacities.

Table 22 is the de-rating table in Column 6 of Table 3(2).

Table 10, Column 11: 52 A = 10 mm² (the CSA chosen must be rated as ≥ 50 A).

Therefore, $I_B \le I_N \le I_Z$ is satisfied as $45 \le 50 \le 52$.

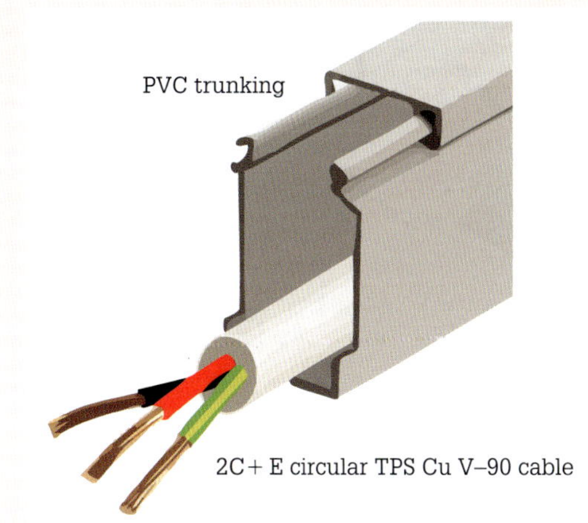

FIGURE 15.35 Two-core and earth 0.6/1 kV circular sheathed PVC/PVC

EXERCISE 15.13

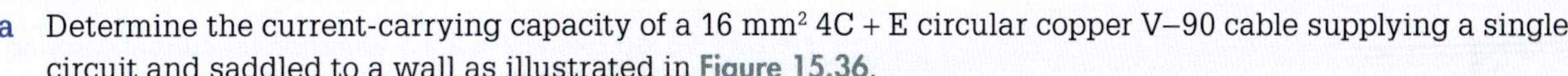

a Determine the current-carrying capacity of a 16 mm² 4C + E circular copper V–90 cable supplying a single circuit and saddled to a wall as illustrated in Figure 15.36.

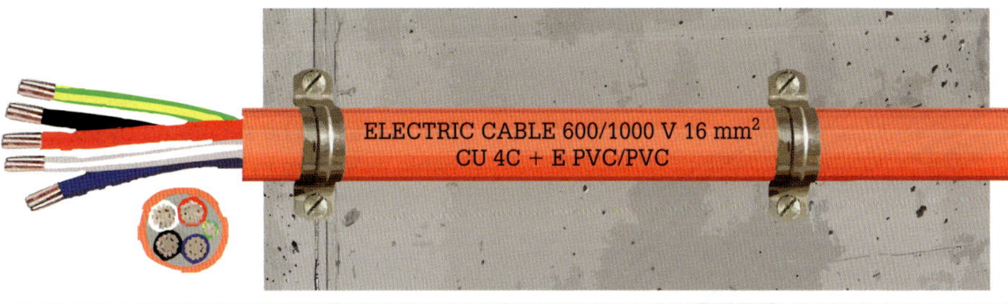

FIGURE 15.36 Exercise 15.13 Question (a) current-carrying capacity

b Determine the current-carrying capacity of three 6 mm² copper 2C + E flat TPS V–90 cables, which are to supply three individual circuits and are installed unenclosed in air and clipped to a timber beam as illustrated in Figure 15.37.

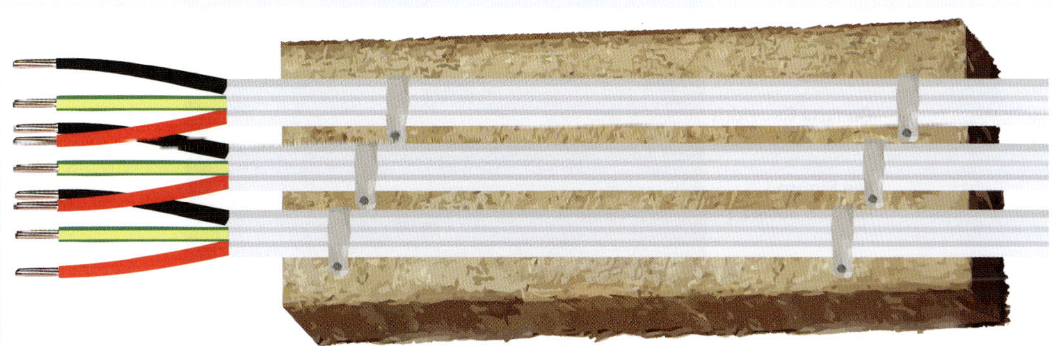

FIGURE 15.37 Exercise 15.13 Question (b) current-carrying capacity

c Determine the current-carrying capacity of three 10 mm² copper 4C + E V–90 circular multi-core cables, which are fixed to an unperforated cable tray as illustrated in **Figure 15.38**.

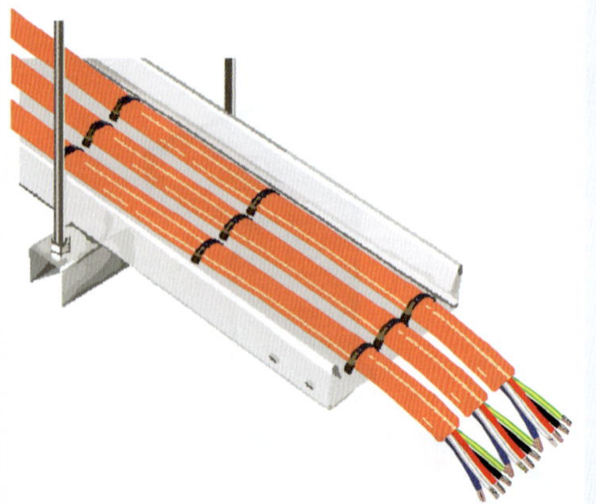

FIGURE 15.38 Exercise 15.13 Question (c) current-carrying capacity

d Determine the current-carrying capacity of two three-phase circuits consisting of 16 mm² copper V–90 SDI cables, which are fixed to a perforated cable tray as shown in **Figure 15.39**.

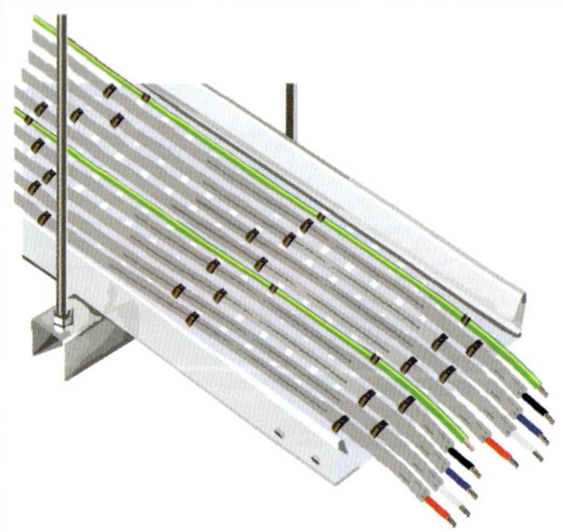

FIGURE 15.39 Exercise 15.13 Question (d) current-carrying capacity

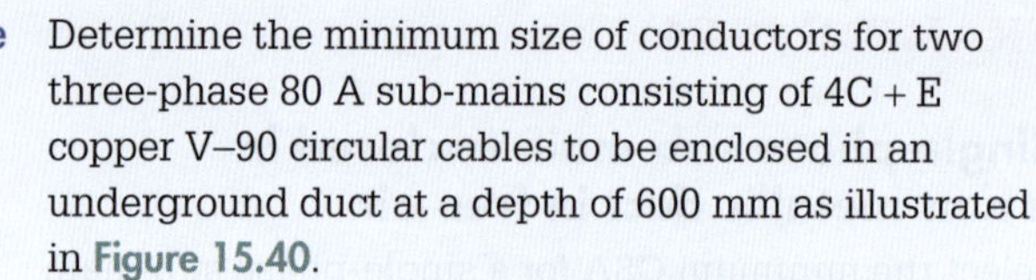

e Determine the minimum size of conductors for two three-phase 80 A sub-mains consisting of 4C + E copper V–90 circular cables to be enclosed in an underground duct at a depth of 600 mm as illustrated in **Figure 15.40**.

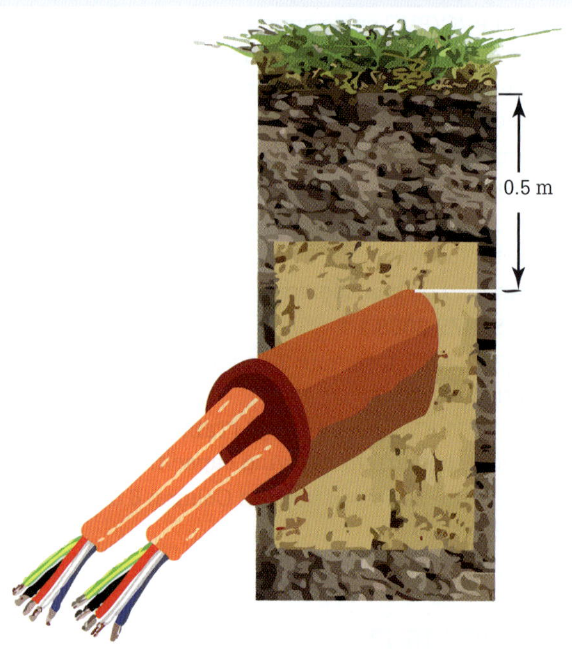

FIGURE 15.40 Exercise 15.13 Question (e) current-carrying capacity

f Determine the minimum size of two sub-main conductors having a maximum demand of 45 A each and comprising flat 2C + E V–90 PVC/PVC copper conductor cables fixed flat that are to be separated by six times the width of the cable as shown in **Figure 15.41**.

FIGURE 15.41 Exercise 15.13 Question (f) current-carrying capacity

Parallel current-carrying conductors

It is possible to determine the nominal current-carrying capacity of a single conductor within a parallel group using the equation:

$$I_Z = \frac{I_N}{\text{conductors in parallel} \times \text{product of derating factors}}$$

where I_Z is the continuous current-carrying capacity of the conductor

I_N is the nominal current rating of the protective device

EXAMPLE 15.25

Sub-mains conductors fixed to a ladder

Select the minimum CSA for a three-phase sub-main consisting of six single-core PVC/PVC 0.6/1 kV sheathed cables supplying a load demand of 470 A as shown in Figure 15.42. The cables must be V–90 insulated and contain copper conductors. The sub-main consists of two conductors in parallel per phase fixed to the ladder. All conductors are laid flat and are touching.

- I_B = maximum demand = 470 A
- I_N = the nominal current of the protective device. A 500 ampere HRC fuse would be suitable as a protective device.
- I_Z can now be considered.

Note: With some commercial and industrial installations two or more cables are often run in parallel as mains or sub-mains because current loads are large.

Using AS/NZS 3008.1.1

Table 3(1), Item 5, Columns 3 and 5, show installation methods. Note that the installation method shown is for one three-phase circuit consisting of three only SDI cables.

Table 7 or 8, Columns 5 to 7 (Table 9, Columns 4 and 5), show the current-carrying capacities.

Table 23 is the de-rating table in Column 6 of Table 3(1). Note two conductors per phase are considered two circuits for de-rating purposes (see footnote 3 to Table 23).

Table 23, Item 7, Column 7: de-rating is 0.95 for two circuits.

To obtain the required current-carrying capacity per conductor it is necessary to re-rate the load current by 0.95 for multiple circuits and 0.9 for protection by HRC fuses (see Note 1 to Clause 2 of AS/NZS 3000:2018).

So the cables must be capable of supplying a continuous current of 585 A (500/[0.95 × 0.9]).

As there are two conductors in parallel per phase, each conductor carries half the load current:

$$585/2 = 292.5 \text{ A}$$

For cables touching:

- Table 7, Column 5 (refer to footnote 3 of Table 23):
- 150 mm^2 is rated at 330 A (the CSA chosen must be rated as ≥ 292.5 A).

$I_B \le I_N \le 0.9\, I_Z$ is satisfied as $470 \le 500 \le 564$.

Note: Figure 1 on page 22 of AS/NZS shows that, with a space equal to the outside diameter of the cable between the two sets of conductors, de-rating can be avoided.

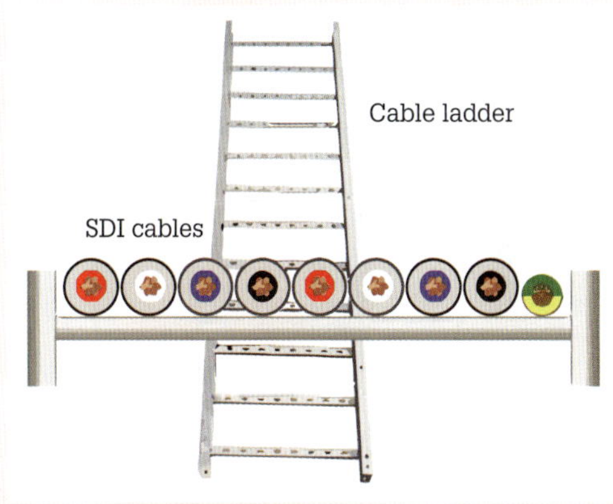

FIGURE 15.42 Six single-core PVC/PVC 0.6/1 kV sheathed cables

EXAMPLE 15.26

Three-phase SDI main on a perforated tray

Select the minimum CSA for a three-phase main consisting of three SDI cables in parallel per phase 0.6/1 kV PVC/PVC cable installed on perforated tray supplying a load demand of 492 A, as shown in Figure 15.43. The cable must be XLPE (X–90) insulated and contain copper conductors.

- I_B = maximum demand = 492 A
- I_N = the nominal current of the protective device. A 500 A HRC fuse would be suitable as a protective device.
- I_Z can now be considered.

Using AS/NZS 3008.1.1

Table 3(1), Item 5, Columns 3 and 5, show installation methods.

Table 7 or 8, Columns 5 to 7 (Table 9, Columns 4 and 5) show the current-carrying capacities.

Table 23 is the de-rating table in Column 6 of Table 3(1). Note three parallel conductors per phase are considered three circuits (footnote 3 to Table 23).

Table 23, Item 15, Column 8: Table 23, number of tiers 1: de-rating is 0.96 for three circuits.

To obtain the required current-carrying capacity per conductor, de-rate the load current by 0.96.

However, with HRC fuses I_Z needs to be de-rated to 90%.

»

Now find the current-carrying capacity of the conductor.

$$I_Z = \frac{I_N}{\text{conductors in parallel} \times \text{derating factors}}$$
$$= \frac{500}{3 \times 0.9 \times 0.96} = 192.9\text{ A}$$

From Table 8, Column 5, a 193 A current capacity means that a 70 mm² (240 A) cable must be used. Also, $I_B \le I_N \le I_Z$ is satisfied as 492 ≤ 500 ≤ 622 (3 × 240 A × 0.96 × 0.9).

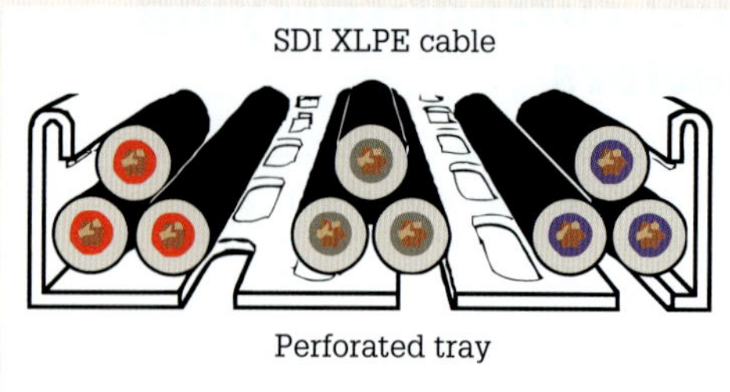

FIGURE 15.43 XLPE (X–90) SDI 0.6/1 kV PVC/PVC cable

EXERCISE 15.14

Determine the minimum size copper conductor for a 320 A load to be supplied by two parallel TPI V–90 PVC/PVC cables per phase fixed flat and touching to a cable ladder as shown in **Figure 15.44**. Note that cables in parallel are often used for mains and sub-mains because of the ease of handling and lower cost. The disadvantage is that two cables in parallel means two circuits and therefore the cables have to be de-rated. Refer to Clause 3 of AS/NZS 2008.1.1.

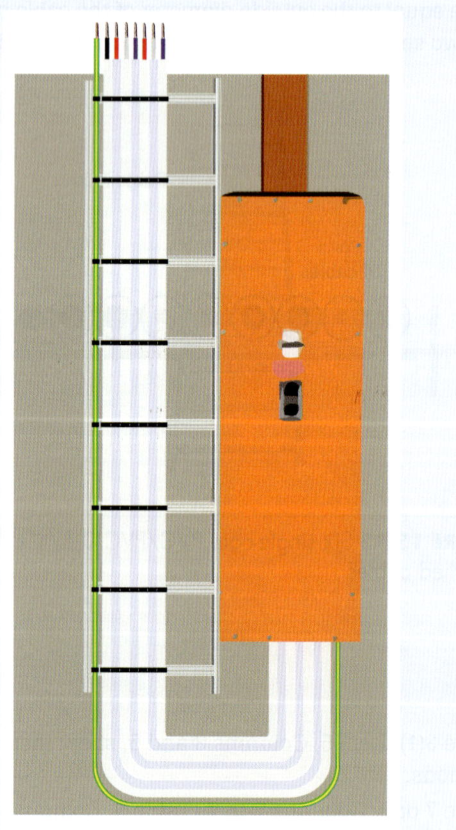

FIGURE 15.44 Exercise 15.14 current-carrying capacity

Protection of cables insulation from heat

Heat can have an adverse effect on cable insulation. Temperatures in excess of that for which the insulation is designed to safely operate can cause the insulation to soften and suffer a reduction in dielectric strength. Major causes of heat in a cable are short-circuit currents and solar radiation.

Heat from short-circuit currents

Section 2 of AS/NZS 3000:2018 *Wiring Rules* and Section 5 of AS/NZS 3008.1.1:2017 *Electrical installations – Selection of cables* takes account of the clearance time of the protective device under short-circuit conditions by applying what is known as the adiabatic equation. This equation states that the time in which a given short-circuit current will raise the temperature of conductors in normal operation to their limiting temperature can be calculated from the adiabatic equation:

$$t = \frac{K^2 \times S^2}{I_2}$$

where t = duration in seconds

K = a factor taking into account various criteria of a conductor such as conductor material, insulation and final temperature

S = cable cross-section in mm²

I = effective short-circuit current in amperes (A_{rms})

Values of K for typical materials for calculation of the effects of short-circuit current are given in the AS/NZS series. **Table 15.17** illustrates some of these values for 450/750 V cables up to 300 mm².

Assessment of protection under short-circuit condition, when based on the adiabatic equation, is accurate for faults of short duration only, for example, less than 0.1 second, as the adiabatic equation assumes no heat loss from the cable.

EXAMPLE 15.27

Calculate the time in which a 2.5 kA short-circuit current will raise the temperature of 2.5 mm² copper V–90 thermoplastic cables installed as a final sub-circuit from an assumed initial temperature of 75 °C to their limiting final temperature. From **Table 15.17**, K = 111.

$$t = \frac{K^2 \times S^2}{I_2}$$
$$= \frac{111^2 \times 2.5^2}{(2.5 \times 10^3)^2}$$
$$= 12.3\text{ ms}$$

TABLE 15.17 K values for 450/750 V cables up to 300 mm^2

Conductor material	Insulation material	Assumed initial temperature	Limiting final temperature	*K*
Copper	Cross-linked elastomer			
	R–EP–90	90 °C	250 °C	143
	R–CSP–90	90 °C	250 °C	143
	R–S–150	130 °C	350 °C	155
	Thermoplastic			
	V–75	75 °C	160 °C	111
	V–90	75 °C	160 °C	111
	V–90HT	90 °C	160 °C	99.9
	XPLE			
	X–90	90 °C	250 °C	143
Aluminium	Cross-linked elastomer			
	R–EP–90	90 °C	250 °C	143
	R–CSP–90	90 °C	250 °C	143
	R–S–150	130 °C	350 °C	155
	Thermoplastic			
	V–75	75 °C	160 °C	111
	V–90	90 °C	160 °C	99.9
	V–90	90 °C	160 °C	99.9
	XLPE			
	X–90	90 °C	250 °C	143

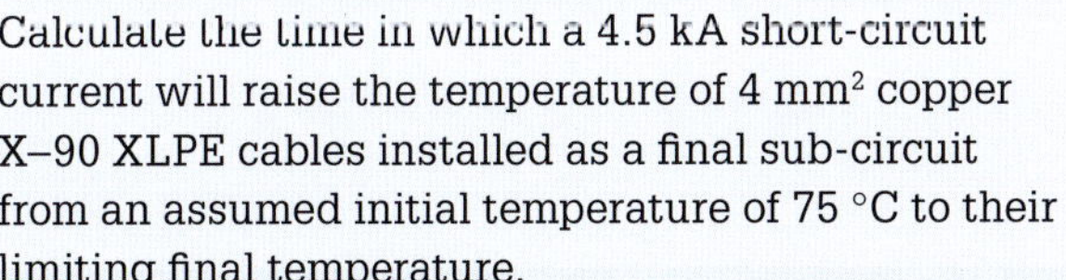
EXERCISE 15.15

Calculate the time in which a 4.5 kA short-circuit current will raise the temperature of 4 mm^2 copper X–90 XLPE cables installed as a final sub-circuit from an assumed initial temperature of 75 °C to their limiting final temperature.

For a short-circuit having a duration of less than 0.1 second, where the irregularity of the current is of less importance, the value of K^2S^2 for the cable should be greater than the manufacturer's value of energy let-through (I^2t) of the short-circuit protective device. Let-through energy, I^2t, is an expression related to energy that passes through an over-current protective device during an interruption.

$$I^2t^2 \le K^2S^2$$
$$K^2S^2 = 111^2 \times 2.5^2$$
$$= 7.7006 \times 10^4 \text{ A}^2\text{S}$$

Effective short-circuit protection is only possible if the let-through energy (I^2t) is equal to or less than the maximum allowable short-time energy (K^2S^2) for the conductor insulation to remain unaffected by heat.

The let-through energy for the protective device can be obtained from the let-through energy curves or tables provided by the individual cable manufacturer. The allowable short-time energy, K^2S^2, for the cable to be protected, can be determined from the rated short-time current density *K* and its cross-sectional area *S*. The TPS insulated copper cable in the final sub-circuit to be protected in our example above possesses a permissible K^2S^2 of 7.7006 × 104 A^2 s. From the I^2t characteristics graph for C-type MCBs in **Figure 15.45** a 10 A MCB experiencing a fault of 3 kA lets through 7.5 × 10^4 A^2 seconds of energy.

A 10 A circuit breaker protecting the cable operates in less time than that required for the cable to reach its temperature limit, so the cable is protected.

Solar radiation

It is necessary to consider the effects of solar radiation when calculating the maximum current-carrying capacity of aerials, flexible cables and other types of cables for given environmental conditions, or when finding the time over which mitigation actions may take place. In order to find the maximum current-carrying capacity of these cables it is necessary to consider thermal effects such as ambient temperature (°C), wind speed (m/s) and intensity of sunlight (W/m^2). Tables 4 to 15, 20 and 21 in AS/NZS 3008.1.1:2017 *Electrical installations – Selection of cables* provide guidance on the current-carrying capacity of flexible cables and aerials. Refer to Tables 4 to 17.

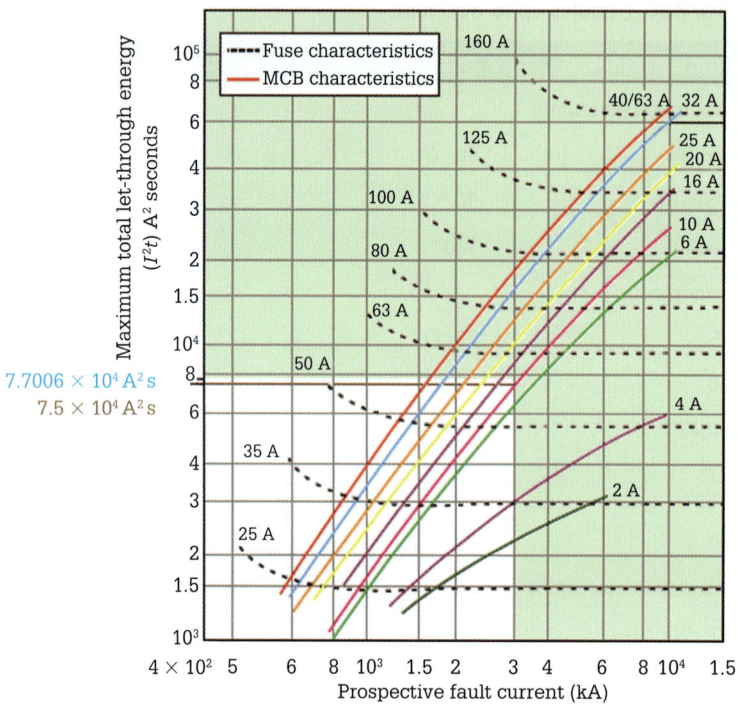

FIGURE 15.45 I^2t characteristics graph for C-type MCBs

EXAMPLE 15.28

a Determine the current-carrying capacity of PVC V–90 copper, 25 mm² 450/750 V two-core flexible cable exposed to the sun.
From AS/NZS 3008.1.1, Table 10, Column 9, a 25 mm² conductor can safely carry 77 A.

b Determine the current-carrying capacity of a 16 mm² PVC-insulated two-core parallel webbed aerial with copper conductors in a 2 m/s air movement.
From AS/NZS 3008.1.1, Table 20, Column 10, a 16 mm² conductor can safely carry 107 A.

EXERCISE 15.16

Determine the current-carrying capacity of a 25 mm² PVC-insulated two-core parallel webbed aerial with copper conductors in still air.

REVIEW QUESTIONS

1 What are the three types of cable wiring systems?
2 The maximum permissible cable insulation temperature allowed has direct bearing on what cable parameter?
3 Which cable parameter provides for a safe electrical installation?
4 Why are installation methods important considerations when selecting conductor size?
5 Why is it necessary to coordinate protection devices and conductor current-carrying capacity?
6 Name four features to consider when determining the minimum size of consumer's mains.
7 Determine the current-carrying capacity of two separate circuits consisting of two-core plus earth 0.6/1 kV insulated and sheathed stranded copper 6 mm² V–90 PVC/PVC cables installed unenclosed, clipped directly to the wall and touching each other.
8 Determine the current-carrying capacity of three 6 mm² 4C + E V–90 circular multi-core cables, which are fixed to a perforated cable tray.
9 Determine the minimum size copper conductor for a 220 A load to be supplied by two parallel TPI V–90 PVC/PVC cables per phase fixed flat and touching to a cable ladder.
10 Calculate the time in which a 2.5 kA short-circuit current will raise the temperature of 4 mm² copper X–90 XLPE cables installed as a final sub-circuit from an assumed initial temperature of 75 °C to their limiting final temperature.

15.6 Cable selection based on voltage drop

The voltage at a load device will always be lower than that at the point of supply (POS). This is known as voltage drop and is due to the potential across a length of conductor. The magnitude of the potential is a factor of load current and the impedance of the circuit conductors. The impedance of the circuit conductors is dependent on the type of conductor material (resistivity), the length of the conductor, the conductor's CSA and the type and thickness of conductor insulation. For most cases, resistance is the major component of cable impedance, which allows us to approximate voltage drop as:

$$\text{from Ohm's Law } V_d = I \times R$$

$$\text{and as } R = \frac{\rho^l}{A}$$

$$\text{gives } R = I\frac{\rho^l}{A}$$

Voltage drop is an important consideration when selecting cables for an electrical installation. While the resistance of circuit conductors within an installation is of a small value when carrying circuit current, a voltage drop does occur between the origin of the circuit and the load terminals.

Another important consideration is the heat developed by a current-carrying conductor. Heat is proportional to the square of the value of the current and this is the most significant factor when selecting cables for an installation.

$$H = I^2Rt$$

AS/NZS 3000:2018 *Wiring Rules* places a limit to the magnitude of the voltage drop within an electrical installation. For low-voltage installations, the voltage drop between the POS and any point in the installation must not exceed 5% of the nominal supply voltage when maximum current is being drawn from the service mains (see **Figure 15.46**).

The CSA of each series-connected circuit cable as illustrated in **Figure 15.47** should be such that the voltage drop from the POS to the load does not exceed the statutory requirement.

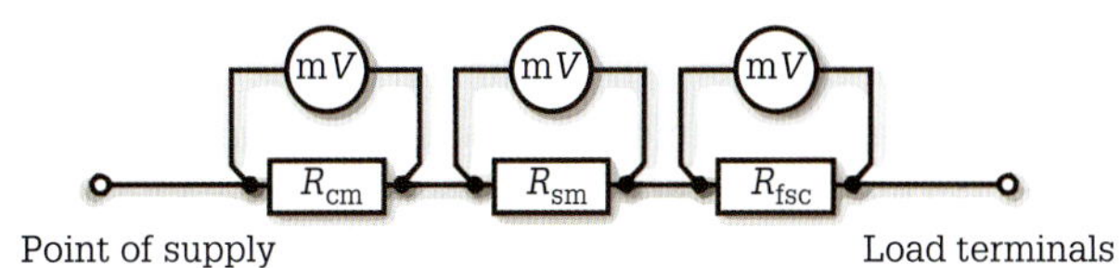

FIGURE 15.47 Voltage drop from the POS

Effect of reduced voltage

Reduced voltage can cause some types of lamps to flicker. Power circuits may experience unreliable performance from electromechanical devices such as relays and contactors. Induction motors will operate at reduced torque. Electronic equipment such as programmable logic controllers and electronic relays may have sensitivity to short-duration voltage sags and cannot ride through the reduced voltage event. This is more of a problem where the voltage at the point of origin could already be at the lower limit of 218.5 V and this, combined with a 5% circuit voltage drop, means that the voltage across the load terminals could be as low as 207.6 V. With lower voltage, equipment power output falls and induction motors may stall at start due to the reduction in starting torque (torque is proportional to the square of the supply voltage).

Note: Voltage sags are a partial reduction in rms voltage that usually lasts for 0.5 to 30 cycles, while a momentary interruption is a complete loss of a.c. power which can be 0.5 cycle to minutes in duration.

Installation circuits working on the fringes of operating conditions can result in increased heat generation.

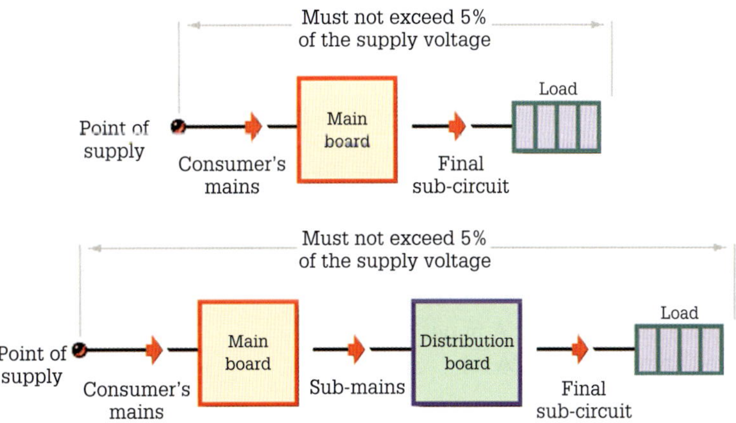

FIGURE 15.46 Series arrangement of installation circuits

Current capacity of the circuit cables changes drastically as temperatures rise and a reduction in capacity begins to generate heat on its own while also causing a drop in voltage. Under short-circuit conditions, the fault current may not be of sufficient magnitude, due to reduced voltage, to operate a circuit protective device before the cable insulation is damaged from overheating.

The voltage drop for any specified electrical installation is the sum of the voltage drops across each series-connected circuit cable from the POS to the load. In a 230 V wiring system, the maximum voltage drop is 11.5 V (5% of 230 V). For a 400 V wiring system the maximum voltage drop is 20 V.

Cable size is sometimes governed by voltage drop rather than by heating effect. The voltage drop method occurs when the current demand is close to the allowable level of the conductor and the circuit route length is long. Therefore, these long circuit lengths may require larger CSA conductors for a given current load. Hence, the CSA of each series-connected circuit cable should be chosen so that the voltage drop from the POS to the load does not exceed the statutory requirement.

Determination of voltage drop

Determination of voltage drop is usually implemented using a calculation known as the millivolt per ampere metre (mV/A.m) method. The voltage drop (V_d) for each circuit cable in an installation can be calculated using the following equation found in AS/NZS 3008.1.1:2017 *Electrical installations – Selection of cables*.

$$V_d = \frac{L \times I \times V_c}{1000}$$

where V_d = actual voltage drop in volts

L = route length in metres

I = current to be carried by the cable

V_c = unit voltage drop in millivolts per ampere metre (mV/A.m) (values of V_c are tabulated in AS/NZS 3008.1.1:2017)

Alternatively, a simplified voltage drop using percentage terms (see AS/NZS 3000:2018, Appendix C) can be used for PVC cables operating at 75 °C when route length and circuit current are known.

$$\% V_d = \frac{L \times I \times V_c}{10 \times V_o}$$

where $\%V_d$ = actual voltage drop on the circuit that represents 1% of supply voltage

L = route length in metres

I = current to be carried by the cable

V_c = unit voltage drop in millivolts per ampere metre (mV/A.m) (values of V_c are tabulated in AS/NZS 3008.1.1:2017)

V_o = circuit supply voltage in volts

EXAMPLE 15.29

This example uses Table C8 of AS/NZS 3000:2018.

Determine the percentage voltage drop for multi-core PVC V–90 insulated and sheathed copper conductors laid flat for a 230 V final sub-circuit carrying 20 A over a route length of 30 m as shown in Figure 15.48.

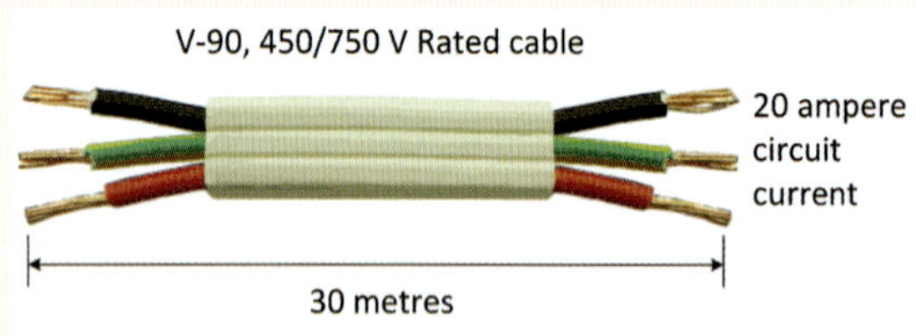

FIGURE 15.48 Percentage voltage drop

1. Required ampere metres (A.m) = $I \times L = 20 \times 30 =$ 600 A.m
2. From Table C8 A.m per % of V_d (230 V circuit), the possible cable sizes are:
   - i 6 mm² at 306 A.m per % of V_d
   - ii 4 mm² at 205 A.m per % of V_d
   - iii 2.5 mm² at 128 A.m per % of V_d

   The cable sizes selected must have their A.m per % of V_d less than the required ampere metres (A.m).
3. Divide required A.m by A.m per % of V_d for each possible cable size to obtain percentages.

$$6\,\text{mm}^2 = \frac{600}{306} = 1.96\%$$

$$4\,\text{mm}^2 = \frac{600}{205} = 2.93\%$$

$$2.5\,\text{mm}^2 = \frac{600}{128} = 4.69\%$$

The selected cable depends upon the percentage voltage drop occurring across the consumer's mains. If the percentage V_d across the consumer's mains or consumer's mains plus sub-mains was 2%, only the 6 mm² cable and a 4 mm² cable would be suitable.

- i (6 mm²) 2% + 1.96% = 3.96% which is less than the 5% statutory requirement.
- ii (4 mm²) 2% + 2.93% = 4.93% which is less than the 5% statutory requirement.
- iii (2.5 mm²) 2% + 4.69% = 6.69% which is more than the 5% statutory requirement.

The actual voltage drops (V_d) across these cables with a 230 V supply are:

1.96% × 230 V = **4.51 V**

2.93% × 230 V = **6.74 V**

4.69% × 230 V = **10.79 V**

EXERCISE 15.17

Use Table C8 of AS/NZS 3000:2018 to determine the actual voltage drop for multi-core PVC V–90 insulated and sheathed copper conductors laid flat for a 230 V final sub-circuit carrying 16 A over a route length of 40 m, as shown in Figure 15.49.

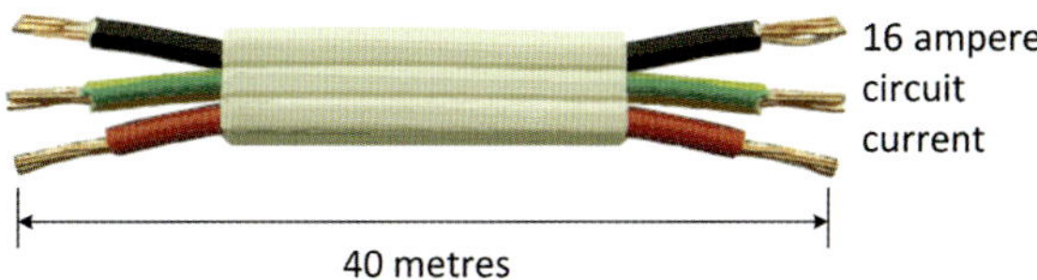

FIGURE 15.49 Percentage voltage drop in cable segment

EXAMPLE 15.30

This example uses Table 42, AS/NZS 3008.1.1.

When using Tables 40–50 in AS/NZS for voltage drop calculations it will be necessary to convert the three-phase voltage drop values (V_c) given in the tables to a single-phase value for single-phase installations. To do this the following conversion factor is applied:

$$\text{Single-phase } V_c \text{ (mV/A.m)} = \text{three-phase } V_c \times 1.155$$

Determine the voltage drop for multi-core PVC V–90 insulated and sheathed copper conductors laid flat for a single-phase 230 V final sub-circuit carrying 20 A over a route length of 30 m for 6 mm², 4 mm² and 2.5 mm² cable.

Using Table 42 the three-phase per unit voltage drops (V_c) for V–90 cables in normal use for conductors at 75 °C are:

6.49, 9.71 and 15.6 mV/A.m respectively

Converting these to single phase we obtain:

For 6 mm², $V_c = 6.49 \times 1.155 = \mathbf{7.50\ mV/A.m}$

For 4 mm², $V_c = 9.71 \times 1.155 = \mathbf{11.21\ mV/A.m}$

For 2.5 mm², $V_c = 15.6 \times 1.155 = \mathbf{18.0\ mV/A.m}$

$$\text{Using } V_d = \frac{L \times I \times V_C}{1000}$$

For 6 mm²

$$V_d = \frac{L \times I \times V_C}{1000} = \frac{30 \times 20 \times 7.50}{1000} = 4.50\,V$$

For 4 mm²

$$V_d = \frac{L \times I \times V_C}{1000} = \frac{30 \times 20 \times 11.21}{1000} = 6.73\,V$$

For 2.5 mm²

$$V_d = \frac{L \times I \times V_C}{1000} = \frac{30 \times 20 \times 18.0}{1000} = 10.80\,V$$

EXERCISE 15.18

Use Table 42 of AS/NZS 3008.1.1:2017 to determine the actual voltage drop for 2.5 mm² copper conductor, multi-core PVC V–90 insulated and sheathed copper conductors laid flat for a single-phase 230 V final sub-circuit carrying 18 A over a route length of 45 m. Is this arrangement suitable?

EXAMPLE 15.31

This example uses Table C8, AS/NZS 3000:2018.

Determine the minimum size of a multi-core cable having PVC-insulated and sheathed copper conductors and required to carry 60 A over a route length of 40 m if the maximum voltage drop for this single-phase circuit is 1.5%.

1 Required ampere metres (A.m)

$$\begin{aligned}\text{Required A.m} &= I \times L\\ &= 60 \times 40\\ &= 2400\ \text{A.m}\end{aligned}$$

2 Required capacity for allowable voltage drop (A.m per % of V_d)

$$\begin{aligned}V_d &= \frac{\text{A.m}}{\%\,V_d}\\ &= \frac{2400}{1.5}\\ &= 1600\ \text{A.m}/\%\,V_d\end{aligned}$$

3 From Table C8, possible cable sizes that satisfy the 1600 A.m / % V_d are:

i 35 mm² at 1773 A.m per % of V_d

ii 50 mm² at 2377 A.m per % of V_d

iii 70 mm² at 3342 A.m per % of V_d

The cable sizes selected must have their A.m per % of V_d not less than the required ampere metres (A.m).

Note: Even though cables meet voltage drop requirements the selected cable size must also be able to carry the current.

EXERCISE 15.19

Use Table C8 of AS/NZS 3000:2018 to determine the minimum size of a multi-core cable having PVC-insulated and sheathed copper conductors and required to carry 45 A over a route length of 35 m if the maximum allowable voltage drop for this single-phase circuit is 1.8%.

Sub-mains

A sub-main circuit can be defined as a circuit connected directly from the main low-voltage switchboard to any other sub-main distribution board or from one sub-main switchboard to another. The CSA of any sub-main cable must be able to carry the maximum demand of the portion of the installation it is supplying.

When an electrical supply is required for outbuildings, such as a detached workshop, garage or garden shed, sub-mains are used to supply the power from the main switchboard. A sub-main distribution board is required to be installed in an outbuilding. Sub-circuits are then run from the sub-main distribution board to lights, socket outlets and appliances.

Each sub-main circuit is required to have its protection device at the main switchboard. In addition, each sub-main circuit may have an earthing conductor connected to the main switchboard earthing bar and taken to the earthing bar on the sub-mains distribution board. Alternatively, an MEN system may be established at the outbuilding with a separate earth stake. RCD protection at outbuildings or on the sub-main circuit is recommended.

Determining the cable size for a sub-main based on voltage drop

Similar to determining cable size based on voltage drop for final sub-circuits, the size of sub-mains must also take into account voltage drop requirements. Example 15.32 demonstrates cable selection based on voltage drop limitations.

EXAMPLE 15.32

An 80 A type C circuit breaker protects an X–HF–90 insulated multi-core aluminium sub-main cable in an industrial installation. The maximum demand for this portion of the installation is 70 A per phase. The cable, which has a route length of 60 m, is clipped to a perforated tray as shown in Figure 15.50. Determine the minimum conductor size for this cable segment if the maximum allowable voltage drop is 0.8%.

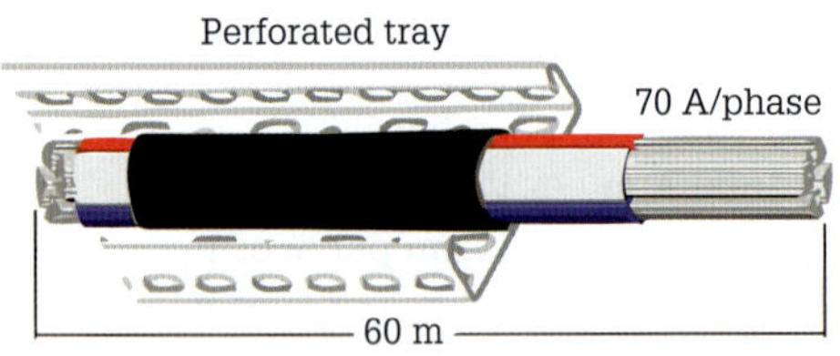

FIGURE 15.50 X–HF–90 cable clipped to perforated tray

It is necessary to calculate the allowable value for V_c, which is based on circuit length, maximum demand current, and allowable voltage drop. In this case the allowable voltage drop is 3.2 V (0.8% × 400).

$$\text{Using } V_d = \frac{L \times I \times V_C}{1000}$$

Then

$$V_C = \frac{1000 \times V_d}{L \times I} = \frac{1000 \times 3.2}{60 \times 70} = 0.762\,\text{m}V/\text{A.m}$$

Optimum conductor size: from Table 45 of AS/NZS 3008.1.1, a 95 mm² X–HF–90 aluminium conductor has a three-phase voltage drop of 0.723 mV/A.m which is less than the calculated 0.762 mV/A.m.

Now it is necessary to ensure that the selected cable has sufficient current-carrying capacity. From AS/NZS 3008.1.1, Table 14, Column 4, a 95 mm² X–HF–90 aluminium conductor can carry 263 A. Therefore, $I_B \leq I_N \leq I_Z$ is satisfied as $70 \leq 80 \leq 263$.

If HRC fuses protected the sub-main the current-carrying capacity would have to be multiplied by 0.9 to de-rate it. In this case:

$$263 \times 0.9 = 236 \text{ A}$$

Consequently, the cable size would still be suitable as $I_B \leq I_N \leq I_Z$ is satisfied as $70 \leq 80 \leq 236$.

EXERCISE 15.20

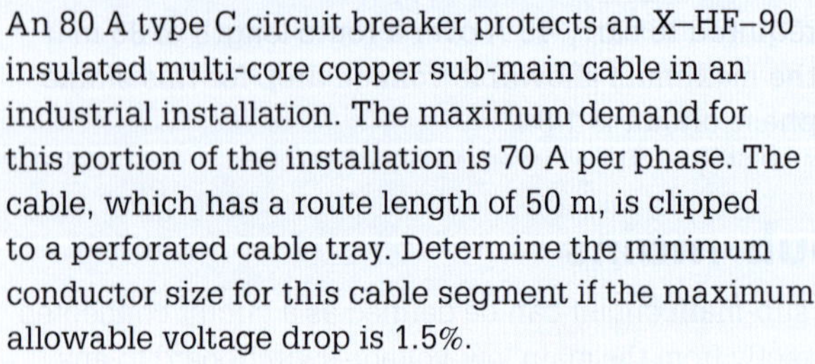

An 80 A type C circuit breaker protects an X–HF–90 insulated multi-core copper sub-main cable in an industrial installation. The maximum demand for this portion of the installation is 70 A per phase. The cable, which has a route length of 50 m, is clipped to a perforated cable tray. Determine the minimum conductor size for this cable segment if the maximum allowable voltage drop is 1.5%.

Determining the cable size for an installation based on voltage drop

As the 5% voltage drop limit applies from the POS to the load terminals of any consuming device, it is necessary to consider the voltage drop for the entire installation. Example 15.33 shows this.

EXAMPLE 15.33

A commercial installation with a maximum demand of 80 A is supplied via an underground consumer's mains having a route length of 25 m and comprising a three-phase, four-core cable having 16 mm² copper conductors and enclosed in conduit. A single-phase 15 A socket outlet circuit protected by a 16 A circuit breaker using V–90 insulated copper cable with a route length of 30 m is to be installed. The cable is to be clipped to steel battens in the insulated ceiling space. Determine a suitable cable size for this final sub-circuit.

1 Determine voltage drop in the consumer's mains.

From Table 42 of AS/NZS 3008.1.1, a 16 mm², V–90 (Table 1, AS/NZS 3008.1.1, recommends 75 °C for regular use) has a three-phase per unit voltage drop of 2.43 mV/A.m.

$$V_d = \frac{L \times I \times V_c}{1000} = \frac{25 \times 80 \times 2.43}{1000} = 4.86\,V$$

»

2 As the final-sub-circuit is single-phase it is convenient to express the voltage drop in the consumer's mains as a percentage.

$$V_{\text{d con mains}}\% = \frac{V_{\text{d con mains}}}{V} \times 100 = \frac{4.86 \times 100}{400} = 1.21\%$$

3 Determine allowable voltage drop in final sub-circuit.

$$V_{\text{d sub-cct allowable}}\% = 5 - 1.21 = 3.79\%$$

4 Express allowable voltage drop percentage for final sub-circuit as a voltage.

$$V_{\text{d sub-cct allowable}} = \frac{3.79}{100} \times 230 = 8.72\,V$$

5 Determine the optimum conductor for the final sub-circuit.

$$V_c = \frac{1000 \times V_d}{L \times I} = \frac{1000 \times 8.72}{30 \times 16} = 18.17\text{m}V/\text{A.m}$$

6 Determine optimum conductor size.

As the final-sub-circuit is single-phase, it is necessary to convert the calculated value for V_c to a three-phase value to use Table 42 in AS/NZS 3008.1.1.

$$V_{\text{c3-phase}} = 0.866 \times V_{\text{c1-phase}} = 0.866 \times 18.17$$
$$= \mathbf{15.73\,mV / A.m}$$

Optimum conductor size based on voltage drop: from Table 42 of AS/NZS 3008.1.1, a 2.5 mm² V–90 copper conductor has a three-phase voltage drop of 15.6 mV/A.m which is less than the calculated 15.73 mV/A.m.

7 Check current-carrying capacity of selected conductor.

From Table 10, Column 15 of AS/NZS 3008.1.1, a 2.5 mm² copper conductor has a current-carrying capacity of 20 A and is the minimum-sized conductor that can be used. $I_B < I_N < I_Z$ is satisfied as 15 < 16 < 20.

Note: Clause 3.6.2c of AS/NZS 3008.1.1:2017, allows a voltage drop dispensation for final sub-circuits, with distributed load (such as socket outlets or lighting), where half the current rating of the protective device may be used in the voltage drop determination. Only those circuits with the load distributed across the entire circuit would be considered to have a distributed load. For a hot water service, stove, air-conditioner, 15 A socket outlet and so on, no such dispensation is provided.

EXERCISE 15.21

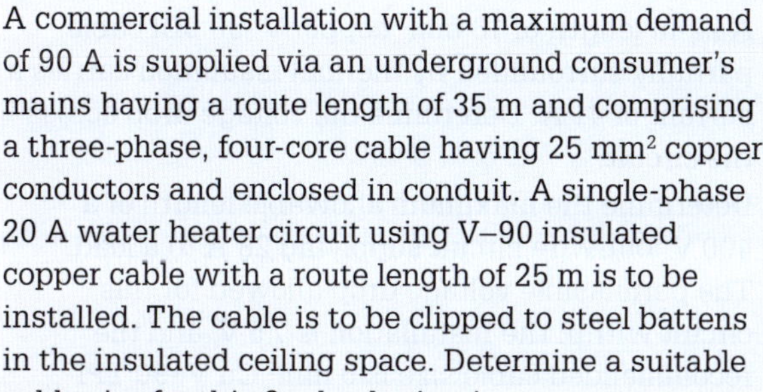

A commercial installation with a maximum demand of 90 A is supplied via an underground consumer's mains having a route length of 35 m and comprising a three-phase, four-core cable having 25 mm² copper conductors and enclosed in conduit. A single-phase 20 A water heater circuit using V–90 insulated copper cable with a route length of 25 m is to be installed. The cable is to be clipped to steel battens in the insulated ceiling space. Determine a suitable cable size for this final sub-circuit.

Determining voltage drop within an installation

An electrical installation will usually comprise consumer's mains, sub-mains and final sub-circuits. As the 5% voltage drop limit applies from the POS to the load terminals of any consuming device, it is necessary to consider the voltage drop for the entire installation. Example 15.34 shows this.

EXAMPLE 15.34

Figure 15.51 illustrates the design parameters of a single-phase domestic installation. The proposed installation has a calculated maximum demand of 60 A and has the following requirements:

- The consumer's mains are 16 mm² V–90 copper SDI cable installed over 18 m in PVC conduit underground and feed the main switchboard (MSB).
- A 40 A sub-main installed over a 20 m length consists of 2C + E 6.0 mm² copper V–90 flat TPS cable which feeds a distribution switchboard (DSB).
- A final sub-circuit using 2C + E 4.0 mm² copper V–90 flat TPS cable from the DSB feeds a single-phase ducted air-conditioner with an operating current of 11.1 A which is located 16 m from the DSB.

Determine the voltage drop in the consumer's mains, sub-mains, and the final sub-circuit.

»

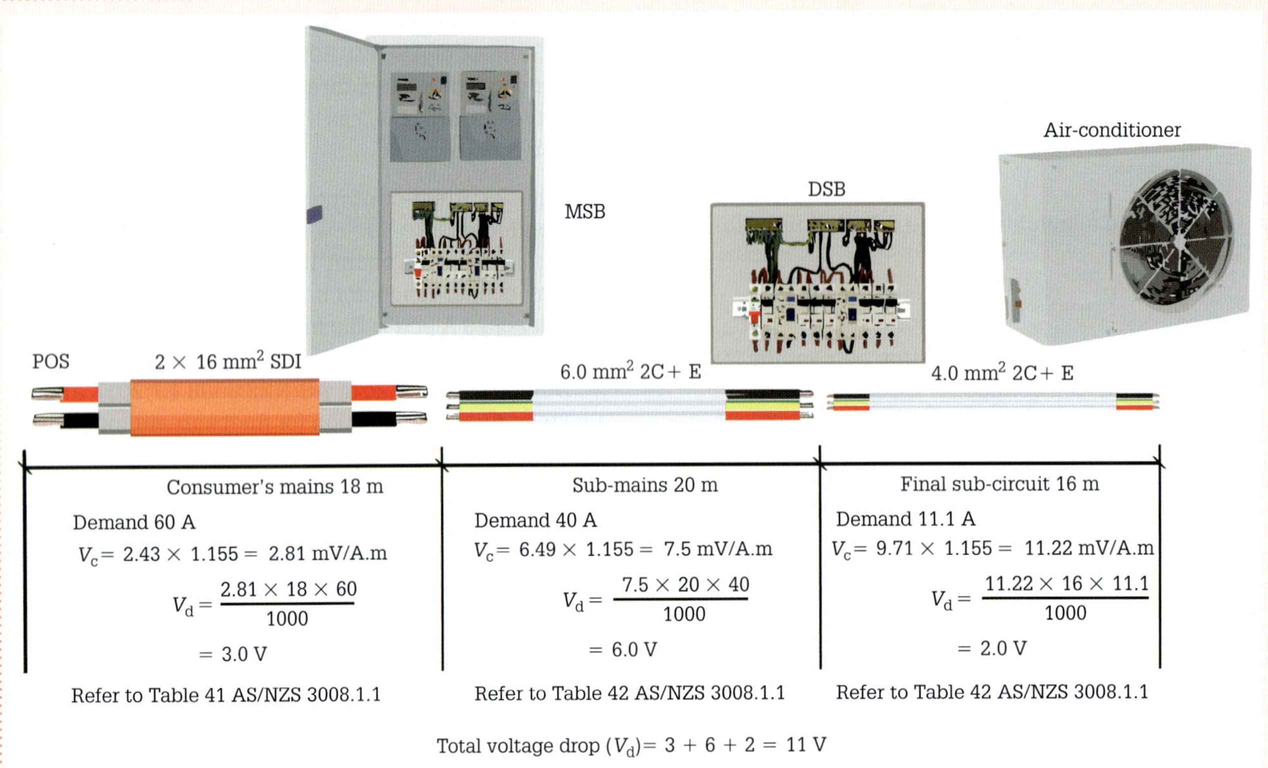

FIGURE 15.51 Design parameters of a domestic installation

EXERCISE 15.22

a An inverter-type split-system single-phase air-conditioner having an operating current of 14.5 A is protected by a 20 A type C circuit breaker and has been hard wired using 2C + E 4 mm² copper V–90 flat TPS cable. The measured route length for the circuit is 23 m. Determine the voltage drop across the circuit.

b A single-phase 250 litre electric storage heater rated at 20 A is protected by a 20 A type C circuit breaker and has been hard wired using 2C + E 2.5 mm² Cu V–90 flat TPS cable. The measured route length for the circuit is 25 m. Determine the voltage drop across the circuit.

c A 30-m length of 2C + E 1.5 mm² copper V–90 flat TPS cable protected by a 16 A type C circuit breaker in a single-phase installation with distributed loads provides a final sub-circuit for a 14 A lighting load. Determine the voltage drop across the circuit.

d A 40-m length of 4 mm² copper V–90 thermoplastic insulated (TPI) cable supplying a 10 A three-phase load is installed in PVC conduit fixed to the wall. Determine the voltage drop across the circuit.

e A 22-m length of 16 mm² copper V–90 SDI cable partially surrounded by thermal insulation carries a current of 41 A. Determine the voltage drop across the circuit.

f Determine the maximum allowable length of a 400 V four-wire circuit supplying 28 A to a load. The permissible voltage drop allowed for this circuit within the installation is 7.5 V, and the recommended cable size is 6 mm² Cu V–90 TPI cable installed in PVC fixed to the ceiling.

g An industrial installation has a low-voltage three-phase standard range TEFC cast-iron metric frame induction motor. This 15 kW, 2935 rpm motor draws 27.4 A at full load supplied by three-core 10 mm² multi-core circular V–90 TPS cables clipped to a perforated tray. The motor is protected by a type D 30 A circuit breaker and the motor is installed 80 m from its related switchboard. It is proposed to replace the installed motor with a larger high-efficiency 18.5 kW 2925 rpm TEFC cast-iron metric frame induction motor. This motor draws 30.5 A at full load. Can the existing cables and the existing 30 A circuit breaker be used if the voltage drop allowance for this circuit is 9 V?

REVIEW QUESTIONS

1 State the AS/NZS 3000:2018 *Wiring Rules* limit on the voltage drop within an electrical installation.
2 How can reduced voltage affect an installation? (Provide two examples.)
3 What is the maximum allowable voltage drop for a single-phase 230 V electrical installation?
4 Between which two points in an electrical installation is voltage drop considered?
5 Determine the percentage voltage drop for multi-core PVC V–90 insulated and sheathed copper conductors laid flat touching for a 230 V final sub-circuit carrying 20 A over a route length of 30 m for 6 mm², 4 mm² and 2.5 mm² cables.
6 What size PVC-insulated and sheathed multi-core cable would be required to carry 55 A over a route length of 40 m with a maximum voltage drop of 1.2% in a single-phase circuit?

15.7 Cable selection based on fault-loop impedance

Earth-fault-loop impedance is the impedance of the earth fault-current loop, which comprises the active-to-earth loop that commences and concludes at the point-of-earth fault. It is essential that this impedance is sufficiently low to ensure that the supply is automatically disconnected within the specified time if a fault of negligible impedance occurs anywhere in the electrical installation between an active conductor and a protective earthing conductor or exposed conductive part. See Chapter 11 of this textbook for detailed information on fault-loop impedance and devices for automatic disconnection of supply.

Cable selection based on fault-loop impedance

The objective of the earth-fault-loop impedance calculation is to determine accurate circuit earth cable size. The earth cable size needs to be sufficient to ensure:

1 Suitable earth-fault-loop impedance (Z_{int}).
2 Satisfactory current-carrying capacity for the prospective earth fault currents for a time that relates to the tripping time of the associated circuit protection (adiabatic equation).
3 Suitable mechanical strength.

The selection of the earth cable size is determined from either:

- tables in AS/NZS 3000:2018 which provide conventional earth cable sizes in relation to the largest CSA active cable size (or summation where there are parallel circuits)
- by calculation – protective device details need to be known.

An electrical installation can be viewed as a network of different impedances all connected in series as shown in **Figure 15.52**.

Maximum circuit impedance – tabulated values

Table 8.2 of AS/NZS 3000:2018 provides guidance on the maximum permissible resistance of final sub-circuits for a range of cable sizes and circuit protective device ratings.

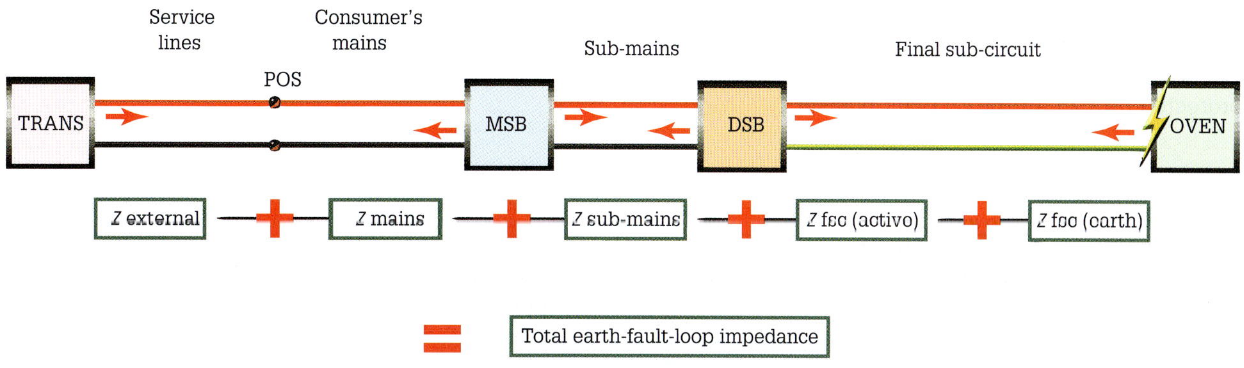

FIGURE 15.52 Total earth-fault-loop impedance

EXAMPLE 15.35

Determine the maximum permissible resistance (R_{phe}) for a circuit comprising 4 mm² active conductors with a 2.5 mm² protective earthing conductor that supplies a 20 A socket outlet when protected by a 25 A type C circuit breaker.

From Table 8.2, the maximum permissible value of R_{phe} is 0.8 Ω.

EXERCISE 15.23

Determine the maximum permissible resistance (R_{phe}) for a circuit comprising 2.5 mm² active conductors with a 2.5 mm² protective earthing conductor that supplies a 15 A socket outlet when protected by a 20 A type C circuit breaker.

Maximum circuit impedance – calculated values

The measured value of R_{phe} assumes that 80% of the earth-fault-loop impedance is internal to the electrical installation due to the reduced conductor size when compared to the size of the distribution conductors. A further assumption is that as the conductors are not carrying fault current when testing the earth-fault-loop impedance using an ohmmeter, the conductors are at ambient temperature and will have a lower resistance. As such, the value of R_{phe} is corrected to 80%. This results in the following equation:

$$R_{phe} = 0.64 \times \frac{U_0}{I_a}$$

where R_{phe} = fault-loop impedance

U_0 = nominal phase voltage (230 V)

I_a = prospective short-circuit current causing automatic tripping of the protective device

EXAMPLE 15.36

Calculate the maximum permissible resistance (R_{phe}) for a single-phase 230 V circuit comprising 4 mm² active conductors with a 2.5 mm² protective earthing conductor that supplies a 20 A socket outlet when protected by a 25 A type C circuit breaker.

$$R_{phe} = 0.64 \times \frac{U_0}{I_a}$$
$$= 0.64 \times \frac{230}{(7.5 \times 25)}$$
$$= 0.785\ \Omega$$

EXERCISE 15.24

Determine the maximum permissible resistance (R_{phe}) for a single-phase 230 V circuit comprising 2.5 mm² active conductors with a 2.5 mm² protective earthing conductor that supplies a 15 A socket outlet when protected by a 20 A type C circuit breaker.

Calculating the impedance of cables using the reactance and resistance tables in AS/NZS 3008.1.1

Cable impedance is calculated from the conductor resistance and reactance values provided in AS/NZS 3008.1.1:2017 *Electrical installations – Selection of cables Cables for alternating voltages up to and including 0.6/1 kV – Typical Australian installation conditions* and AS/NZS 3008.1.2:2017 *Electrical installations – Selection of cables Cables for alternating voltages up to and including 0.6/1 kV – Typical New Zealand conditions* using the equation:

$$Z_C = L \times \frac{\sqrt{R_C{}^2 + X_C{}^2}}{1000}$$

where: L = cable length in metres

Z_C = cable impedance in Ω

R_C = cable resistance in Ω/km

X_C = cable reactance in Ω/km

The values for cable reactance are obtained from Tables 30 to 33 inclusive and values for cable resistance are obtained from Tables 34 to 39 inclusive.

Using these tables and calculation, it is possible to estimate the impedance of a cable run prior to installation and subsequent testing to ensure compliance with earth-fault-loop impedance requirements.

EXAMPLE 15.37

Calculate the impedance of a 35 m length of 16 mm², two-core and earth, V–90 insulated cable having copper conductors.

From Table 34 of AS/NZS 3008.1.1, the resistance of 16 mm² copper conductor at 75 °C (maximum operating temperature for V–90 cables) is 1.40 Ω/km and from Table 30 the reactance of 16 mm² copper conductor with PVC insulation is 0.0861 Ω/km (column 9).

$$Z_C = L \times \frac{\sqrt{R_C{}^2 + X_C{}^2}}{1000}$$
$$= 35 \times \frac{\sqrt{1.40^2 + 0.0861^2}}{1000}$$
$$= 0.0491\ \Omega$$

EXERCISE 15.25

Calculate the impedance of a 90 m length of 25 mm², two-core and earth, V–90 insulated cable having copper conductors.

Prospective touch voltage

Prospective touch voltage (set by convention at 50 V) is concerned with electric shock as a result of contact with persons or livestock with exposed conductive parts made live by a fault. The prospective touch voltage is the magnitude of the voltage to which the person or livestock at risk would be subjected in the event of an earth fault occurring in an installation.

If the touch voltage is 50 V or less, automatic disconnection of the supply is not necessary for protection against electric shock. However, disconnection is required to protect cables from thermal effects.

Touch voltages well in excess of 50 V can occur under earth fault conditions in the installation. However, compliance with the requirements of AS/NZS 3000 should mean that the magnitude and duration of the touch voltages is such as not to cause danger.

The extent of the earth fault current determines the time of disconnection of the over-current protective device being used to provide protection against indirect contact.

$$I_{fault} = \frac{V_{supply}}{Z_{FL}}$$

The touch voltage can be calculated by applying the following equation.

$$\text{Touch voltage} = I_{fault} \times Z_{PE}$$

EXAMPLE 15.38

If the total measured fault-loop impedance of an installation is 0.5213 Ω and the impedance of the final sub-circuit earth is 0.125 Ω for a 230 V supply calculate the touch voltage.

First, we must calculate the fault current.

$$I_{fault} = \frac{V_{supply}}{Z_{FL}} = \frac{230}{0.5213} = 441.2\text{ A}$$

By knowing the fault current, touch voltage can be calculated.

$$\begin{aligned}\text{Touch voltage} &= I_{fault} \times Z_{PE} \\ &= 441.2 \times 0.125 \\ &= 55.15\text{ V}\end{aligned}$$

As this circuit does not comply with the 50 V touch voltage requirement, a check of the disconnection time for the protection device must be carried out. If the circuit was protected by a 32 A type D circuit breaker then the current required for instantaneous operation (I_a) is:

$$32 \times 12.5 = 400\text{ A}$$

The 441.2 A fault current means that the circuit complies as this value allows automatic tripping of the protective device. Any value less than 400 A means that the protective device takes longer to trip and therefore does not provide protection against touch voltage. For a protective device to disconnect within the required time then the measured Z_{FL} must be less than the total fault loop impedance Z_s:

$$\begin{aligned}Z_s &= \frac{V_s}{I_a} = \frac{230}{400} = 0.575\,\Omega \\ &= 0.5213 < 0.575\end{aligned}$$

EXERCISE 15.26

If the total measured fault-loop impedance of an installation is 0.562 Ω and the impedance of the final sub-circuit earth is 0.145 Ω for a 230 V supply, calculate the touch voltage.

REVIEW QUESTIONS

1. What is earth-fault-loop impedance?
2. What is the purpose of calculating earth-fault-loop impedance?
3. Name three important considerations for ensuring a safe electrical installation.
4. Prospective touch voltage should not exceed what value?
5. If the total measured fault-loop impedance of an installation is 0.45 Ω and the impedance of the final sub-circuit earth is 0.25 Ω for a 230 V supply, calculate the touch voltage.

15.8 Selecting protection devices

It is a requirement of the *Wiring Rules* that electrical circuits be protected from over-current conditions. See also Chapter 11 of this textbook for detailed information about circuit protection devices.

Over-current protection

An over-current protective device or devices that ensure protection against overload and short-circuit current must be installed at the origin of each circuit and at every point where a reduction occurs in the current-carrying capacity of the conductors. Protective devices on installations using a selective coordination protection arrangement, which is referred to as discrimination, may be all circuit breakers (CBs) incorporating short-circuit and overload release, circuit breaker–fuse combinations or all fuses, as **Figure 15.53** shows.

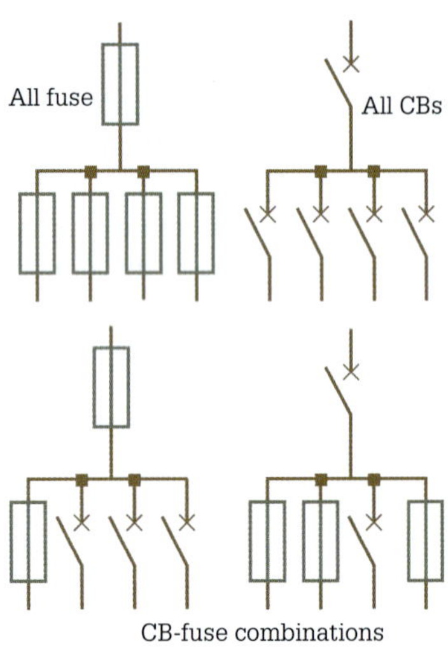

FIGURE 15.53 Possible arrangement of over-current protection devices

An installation with some protective devices in series is considered a selective coordination arrangement when, in the event of short-circuit or overload, the installation is interrupted only by the device that is immediately upstream of the fault point. The term 'upstream' refers to a device being closer to the origin of the installation and 'downstream' refers to a device being closer to the load. Selectivity is ensured when the time–current characteristic of the upstream protective device is above the time–current characteristic of the downstream protective device. Selectivity may be total or partial.

All of these types of protective devices must be able to interrupt any over-current including the prospective short-circuit current at the location where the devices are installed. 'Interrupting rating' is defined as the prospective short-circuit current at rated operating voltage that a device is intended to interrupt under load conditions. Where fuses or circuit breakers without effective interrupting ratings are used, they can rupture aggressively with consequent hazards to equipment and personnel. Refer to Clause 2.5.2 in AS/NZS 3000:2018 *Wiring Rules*.

Series rated

Circuit protection, which is series rated, is a combination of two short-circuit protective devices that have been tested under the manufacturer's test conditions to work together to isolate a fault. Only tested combinations can be used.

Series rated devices can be a combination of circuit breakers or fuses and circuit breakers that can be used at short-circuit levels above the interrupting rating of the load-side circuit breaker, but not above the interrupting rating of the main or supply-side device. A series rated combination can consist of fuses protecting circuit breakers, or circuit breakers protecting other circuit breakers.

For short-circuit protection the series combination interrupting rating must be higher than the available short-circuit current at the load-side circuit breaker. In addition, the interrupting rating of the supply-side fuse or circuit breaker must be greater than the available short-circuit current at its line terminals.

A disadvantage of a series-rated combination (see **Figure 15.54**) is that for fault conditions the supply-side circuit breaker must interrupt at the same time as, and in harmony with, the load-side circuit breaker. This means that the entire circuit branch loses power because the protecting circuit breaker feeding the branch must open under medium- to high-level short-circuit conditions.

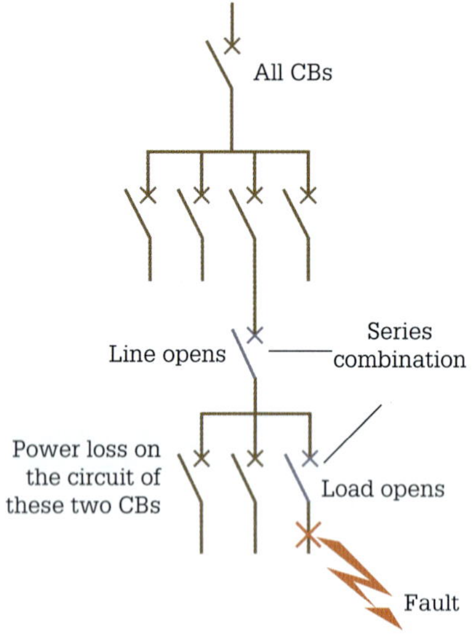

FIGURE 15.54 Series-rated circuit combination

Figure 15.55 shows an installation system with a 100 A HRC fuse and a 32 A sub-main circuit breaker feeding a distribution board containing 16 A circuit breakers. If a fault occurs at point A, the 100 A HRC fuse will operate and the total installation will be disconnected. If the fault is at point C, the 16 A circuit breaker should operate and not the 32 A circuit breaker or 100 A HRC fuse. A fault at point B should operate the 32 A circuit breaker and not the 100 A HRC fuse. If this takes place, the system has discriminated properly.

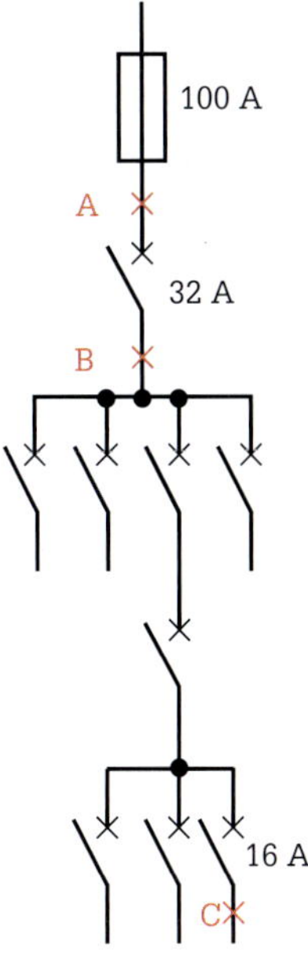

FIGURE 15.55 Discrimination

Circuit protection devices

The continuous load ratings of all electrical equipment are based on thermal considerations. The limiting factor is the maximum temperature at which the conductor insulation can be stressed without suffering deterioration. Circuit protection devices are designed to protect insulating materials by stopping current flow by opening the circuit. To do this, a circuit protection device must always connect in series with the circuit it is protecting. Essentially these devices are over-current protection devices designed to protect only the conductors from over-temperature damage by short-circuits and current surges. There are two types of over-current protection devices: fuses and circuit breakers. Fuses and circuit breakers also protect against short-circuit currents.

A third protection device is the core balance earth leakage device. Its role is to protect small appliances used in earthed situations and indirectly this device also protects people. It does not provide protection against over-current.

Selecting over-current protection

Circuits in low-voltage electrical installations follow what is known as a 'radial branched distribution' method as shown in **Figures 15.53** to **15.55**. As stated previously, it is imperative that over-current protection operate in a coordinated manner that provides effective discrimination to minimise inconvenience in the event of a fault. The *Wiring Rules* include both overload current and short-circuit current when referring to 'over-current'.

The steps taken to ensure correct discrimination are:

1. Ascertain the required current-carrying capacity for the circuit conductors, in accordance with the AS/NZS 3008.1 series, dependent on the installation method and taking into account any external influences.
2. Ascertain the over-current requirements, in accordance with Clause 2.5 of the *Wiring Rules*.
3. Determine voltage drop requirements, in accordance with Clause 3.6 of the *Wiring Rules*.
4. Determine the requirements for automatic disconnection of supply, in accordance with Clause 5.7 of the *Wiring Rules*.

It is essential to disconnect over-current conditions from the electrical supply to prevent the temperature of conductors and their associated insulation from rising to levels that will reduce the effectiveness of the insulation thereby reducing its expected service life.

Currents due to short-circuit conditions

Short-circuit currents may be in the magnitude of several thousand times nominal circuit current. Currents of this magnitude can cause overheating as well as mechanical stresses on conductors and their associated connections.

Clause 3.4 of the *Wiring Rules* requires current-carrying capacity of cables to be determined from the AS/NZS 3008.1 series of Standards, which provide a comprehensive set of tables and methods of calculation that take into consideration various cable/conductor types, installation methods and external influences.

Referencing the AS/NZS 3008.1 series of Standards may not be necessary for many typical electrical installations such as residential (domestic) installations. Tables C6 and C7 in AS/NZS 3000:2018 provide guidance on the selection of protective devices suitable for use with cables of CSA from 1 mm^2 to 25 mm^2. These tables address single-phase and three-phase cable applications respectively, for a restricted range of installation conditions. These tables reflect a more conservative approach to cable selection by limiting the circuit current through the selection of appropriately rated protective devices.

Tables C6 and C7 represent a simplified selection method as they only consider single-circuit installation methods and as such they do not address de-rating factors for groups of cables. To this end, these tables presume that the circuits will:

- be separated from each other
- operate below maximum current in lower ambient temperature or for cables with a nominal rating of 75 °C, will not experience a rise in cable temperature above 90 °C due to grouping.

EXAMPLE 15.39 

A three-phase air conditioning unit is rated at 18 A per phase. Choose an appropriate cable and circuit breaker combination if the 4C + E TPS cable with V–90 insulation and copper conductors is installed such that it is unenclosed and partially surrounded by thermal insulation.

Solution:

From Table C7, 4 mm^2 can supply 20 A when protected by a 20 A circuit breaker.

EXERCISE 15.27

a A three-phase sub-main to a detached garage needs to supply 30 A. Choose an appropriate cable and circuit breaker combination if the 4C + E TPS cable with V–90 insulation and copper conductors is installed such that it is enclosed in an underground enclosure.

b A three-phase air conditioning unit is rated at 16 A per phase. Choose an appropriate cable and circuit breaker combination if the 4C + E TPS cable with V–90 insulation and copper conductors is installed such that it is unenclosed and completely surrounded by thermal insulation.

Protection by use of RCDs and use of extra-low voltage

Protection from harm can be achieved through the use of residual current devices (RCDs) and supplying circuits at extra-low voltage (ELV).

Core balance earth leakage

A core balance earth leakage (CBEL) device is also called a residual current device (RCD) or a safety switch and is able to detect very small earth leakage currents (10 mA). Its purpose is to protect portable appliances in earthed situations. Because of this ability it is the only protection device that affords protection to people and livestock. Note that RCDs do not protect against active to neutral faults. This type of fault is cleared by a fuse or circuit breaker. The RCD essentially operates on the principle that the magnetic flux in a differential transformer is balanced under normal conditions.

The transformer uses the effects of magnetism to detect any out-of-balance current in the system. If the current in the return conductor, the neutral, does not equal the current entering the active conductor, due to an earth fault, then the system is out of balance. A magnetic field is created that induces a voltage in the secondary winding of the toroidal transformer. A semiconductor circuit amplifies this voltage to provide sufficient energy to trip the circuit breaker.

These devices are mandatory in Australia for all circuits in new domestic wiring installations. **Figure 15.56** shows the use of combination MCBs/RCDs to protect all circuits.

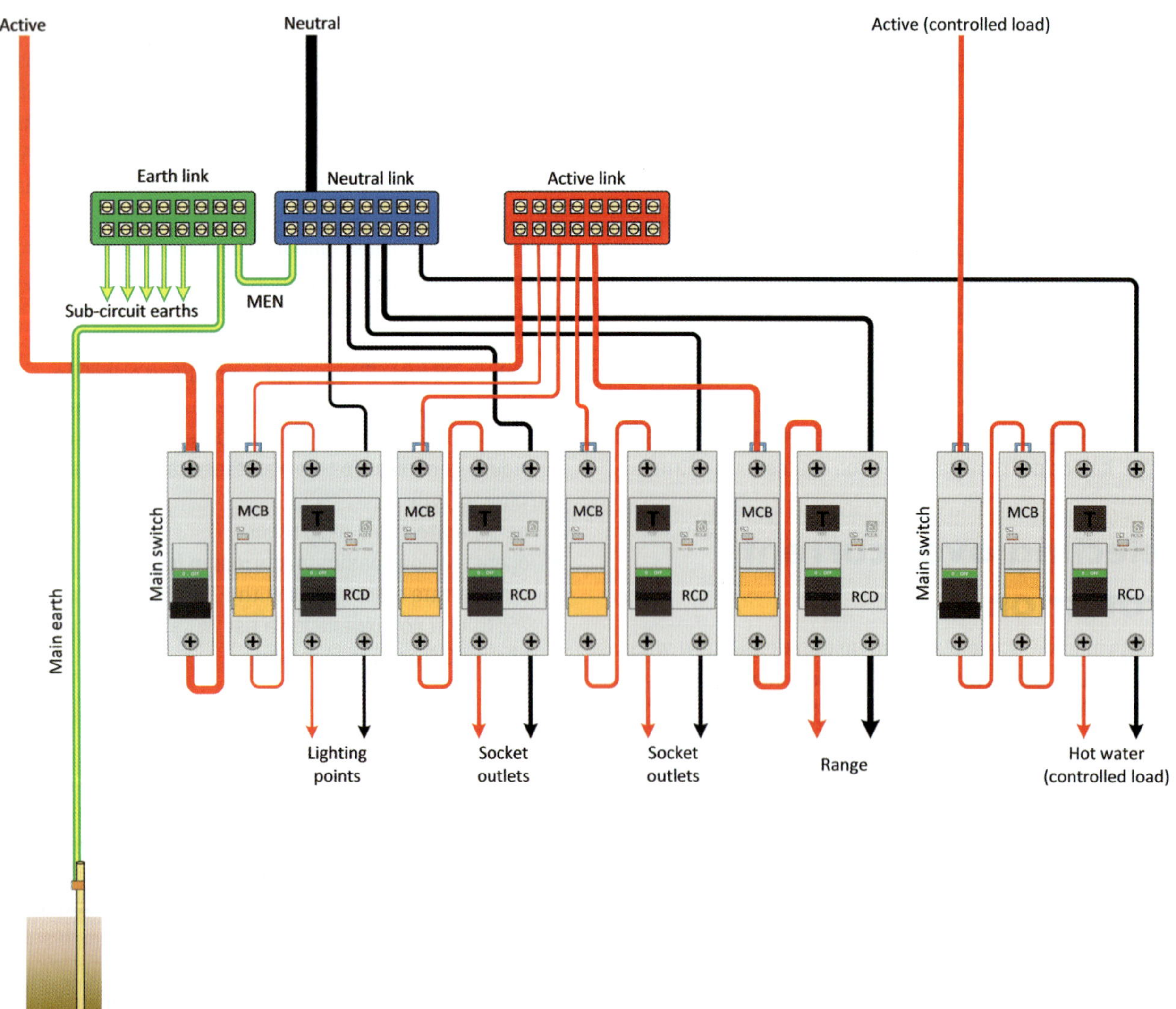

FIGURE 15.56 Domestic installation protected by combination CMCB/RCD

The 10 mA sensitive RCD with 10 A rated outlets is used in extremely hazardous areas, including hospitals.

For fixed domestic installations with 10 A rated outlets, a 30 mA RCD is recommended. The RCD is either double-pole or four-pole allowing it to isolate actives and the neutral.

Extra-low voltage

Voltage is regarded as a difference of potential that exists between all active conductors and between all active conductors and earth. Extra-low voltage is a voltage not exceeding 50 V a.c. or 120 V ripple-free d.c. Extra-low voltage is a measure designed to protect people and livestock from direct and indirect contact with live parts without automatic disconnection of the supply.

Two systems are used for this purpose – separated extra-low voltage and protected extra-low voltage. The source of these types of supply must comply with the requirements of AS/NZS 3000:2018 *Wiring Rules*.

Separated extra-low voltage (SELV)

SELV systems are used where an inherently safe system of supply is required. This safe system ensures that a failure of primary insulation will not result in an increased risk of electric shock. The safe system is achieved by electrical separation of the SELV output from the source of energy, typically using a safety isolating transformer.

Protection by SELV is used in high-risk situations where the operation of electrical equipment presents a serious hazard, for example, around swimming pools. This method of protection requires supplying power from a safety isolating transformer, which uses double or reinforced insulation. AS/NZS 3000:2018 *Wiring Rules* requires that the nominal voltage must not exceed 12 V a.c. or 30 V ripple-free d.c. in areas containing water vessels such as would be the case of underwater pool lighting.

Protected extra-low voltage (PELV)

A PELV wiring system is used where low voltage is required but the situation is not as high risk as with SELV wiring situations. A PELV power supply includes double insulation to protect against direct and indirect contact with exposed conductive parts by means of a safe separation of the primary and secondary transformer windings. A conductive shield placed between the primary and secondary windings with a protective earth connection can also be used. The shield collects any leakage current from the primary winding and transfers it to the earth. This prevents any current leaked from the primary entering the secondary winding. For this reason no separate protective earth conductor is required in a PELV system. Connected equipment and exposed metallic parts in a PELV wiring system can be earthed but it is not essential.

If the circuits of a PELV system require earthing, then the secondary circuit of the transformer cannot exceed 25 V a.c. or 60 V ripple-free d.c. and must have an integrated earth connection to enable connection of the protective earth conductors. PELV wiring systems are used in current supply in automation, machine and process control applications.

REVIEW QUESTIONS

1 Where are protective devices placed in an electrical installation?
2 Name the types of protective devices that provide a selective coordination protection arrangement.
3 When is an installation, with some protective devices in series, considered a selective coordination arrangement?
4 Outline the steps taken to ensure appropriate discrimination of circuit protection devices.
5 What rating is recommended for an RCD protecting 10 A rated outlets in a fixed domestic installation?
6 What are the voltage limits for extra-low voltage?
7 When would an SELV system be used?
8 What protective measures does a PELV supply offer?

15.9 Selecting devices for isolation and switching

Devices used for isolation and switching are broadly classified as:

- isolators
- switches.

Control of installations

Control of an installation is achieved by means of a mechanical switching device (see Clause 2.3 of AS/NZS 3000:2018). This device is capable of making, carrying and breaking an electrical current under normal installation conditions. There are three important features to consider when selecting a switch:

1 contacts; for example, single pole, double throw
2 ratings, such as maximum voltage and current
3 method of operation; for example, toggle, slide, key and the like.

Devices providing control of an electrical installation must comply with the appropriate requirements of Clause 2.3 in the *Wiring Rules*. These devices are classified according to one of the following functions:

- Isolation (see Clause 2.3.2.2)
- Emergency (see Clause 2.3.5.2)
- Mechanical maintenance (see Clause 2.3.6.2)
- Functional [control] (see Clause 2.3.7.2).

Switching devices for isolation need to effectively isolate all active conductors from the circuit. A semiconductor (solid-state) device is not suitable for isolation purposes.

Emergency switching devices can be a single switching device that directly interrupts the incoming supply, or a combination of several items of electrical equipment that are operated in one single action that results in interrupting the appropriate supply.

Switching devices used for shutting down for mechanical maintenance need to be manually operated, clearly and reliably indicate the 'OFF' position, and designed or installed so as to prevent unintentional closure.

Devices for functional switching must be suitable for the most demanding of the duties required of them.

Switch terminology

Several terms are used to describe switch contacts:

- Pole – number of switch contact sets
- Throw – number of conducting positions, such as single or double
- Way – number of conducting positions, such as three or more
- Momentary – switch returns to its normal position when released
- Open – off position, contacts not conducting
- Closed – on position, contacts conducting; there may be several on positions.

For example, the simplest on–off switch has one set of contacts (single pole) and one switching position which conducts (single throw). The switch mechanism has two positions: open (off) and closed (on), but it is called 'single throw' because only one position conducts.

Switch contacts are rated with a maximum voltage and current, and there may be different ratings for a.c. and d.c. The a.c. values are higher because the current falls to zero many times each second and an arc is less likely to form across the switch contacts.

For extra-low voltage electronic tasks, the voltage rating will not matter, but the current rating is important. The maximum current is less for inductive loads such as coils and motors because they cause more sparking at the contacts when switched off.

REVIEW QUESTIONS

1 What are the two broad classifications of devices used to isolate an electrical supply?
2 How is control of an electrical installation achieved?
3 What are the three important features to consider when selecting a switch?
4 Name the four types of control devices used in an electrical installation.
5 Is a solid-state switch permitted for use as an isolating switch?

15.10 Switchboards

AS/NZS 3000:2018, Clause 2.10, 'Switchboards' specifies requirements for location, construction and mounting of switchboards. Regardless of the number or type of switchboards within an installation, they must all:

- Comply with the unique requirements of the relevant supply authority. Complying ensures that the installation coordinates with their protective and control equipment.
- Be readily accessible in well-ventilated areas, free from excess humidity, dust, and dirt and away from hazardous materials. A main switchboard should be located within easy access to the main entrance of the building.
- Be weatherproof; if installed outdoors, the switchboard must be protected to prevent moisture from entering the enclosure.
- Be constructed of materials (all steel construction or all insulated polycarbonate) able to withstand the environmental constraints and severe operating conditions that may be expected at the installation location.
- Be safely and securely mounted to a permanent structure such as a wall or a floor.
- Have a means to prevent strain/damage to cables (see Clause 2.10.6).
- Have no exposed live parts if the switchboard is domestic. Note Clause 2 for non-domestic boards.
- Include protective and metering devices positioned so as to provide safe and easy access for operation and maintenance.

All circuit breakers are mounted so that the ON/OFF and current rating indications are clearly visible with covers or escutcheons in position. In addition, align operating toggles of each circuit breaker in the same plane.

Types of switchboards

Main switchboards for industrial and commercial installations are of the modular type while the smaller sub-boards are usually of steel construction with insulated busbars systems. See Chapter 11 of this textbook for details on switchboard construction and arrangement.

Switchboards at shows, carnivals and displays

Switchboards for a show, carnival or display are required to meet certain requirements.

Where a temporary switchboard having connection facilities is installed within 2.5 m of the ground in an area accessible to unauthorised persons the switchboard shall:

1 Have no exposed live parts.
2 Be enclosed within a cabinet fitted with a door or lid that can be locked. The cabinet shall be:
   a constructed to permit the opening or closing of the door without the removal of, or damage to, any cables or flexible cords attached to the connection facilities
   b provided with a tie-bar or similar arrangement for the anchoring of the cables or flexible cords in order to prevent strain at the termination of the cables or cords.

   Each socket outlet within the cabinet shall be:
   a individually controlled by a switch that interrupts all live conductors
   b protected by a 30 mA RCD.

Every outgoing connection shall be connected to a separate RCD-protected final sub-circuit.

Metering

The wiring of each installation must have the metered and un-metered conductors installed, prepared, connected and arranged to conform to the metering arrangements to suit the required tariffs. A simple single electronic kilowatt hour meter for domestic general light and power supply is shown in **Figure 15.57**.

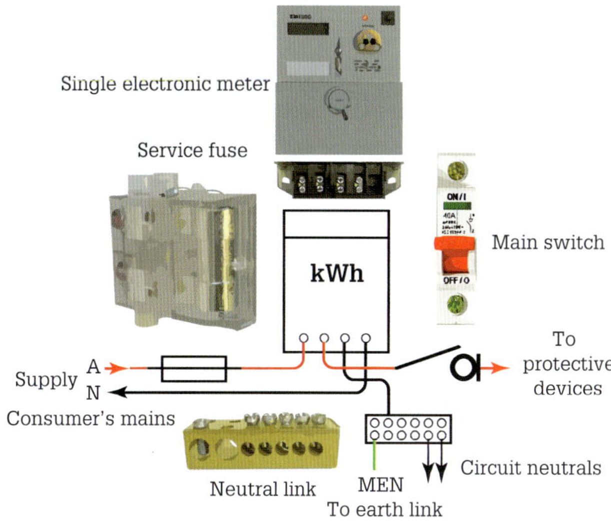

FIGURE 15.57 Single electronic kilowatt hour

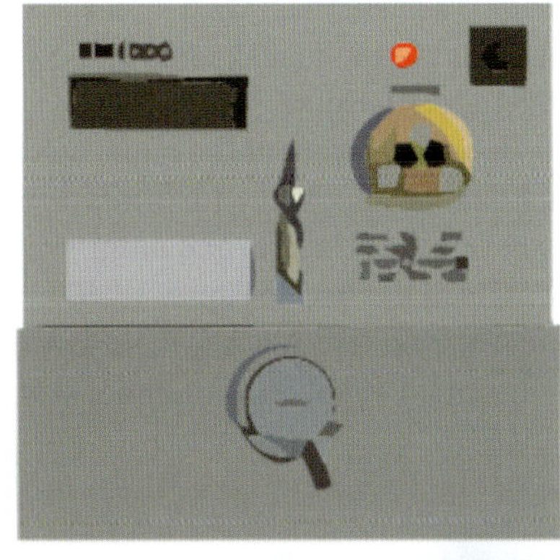

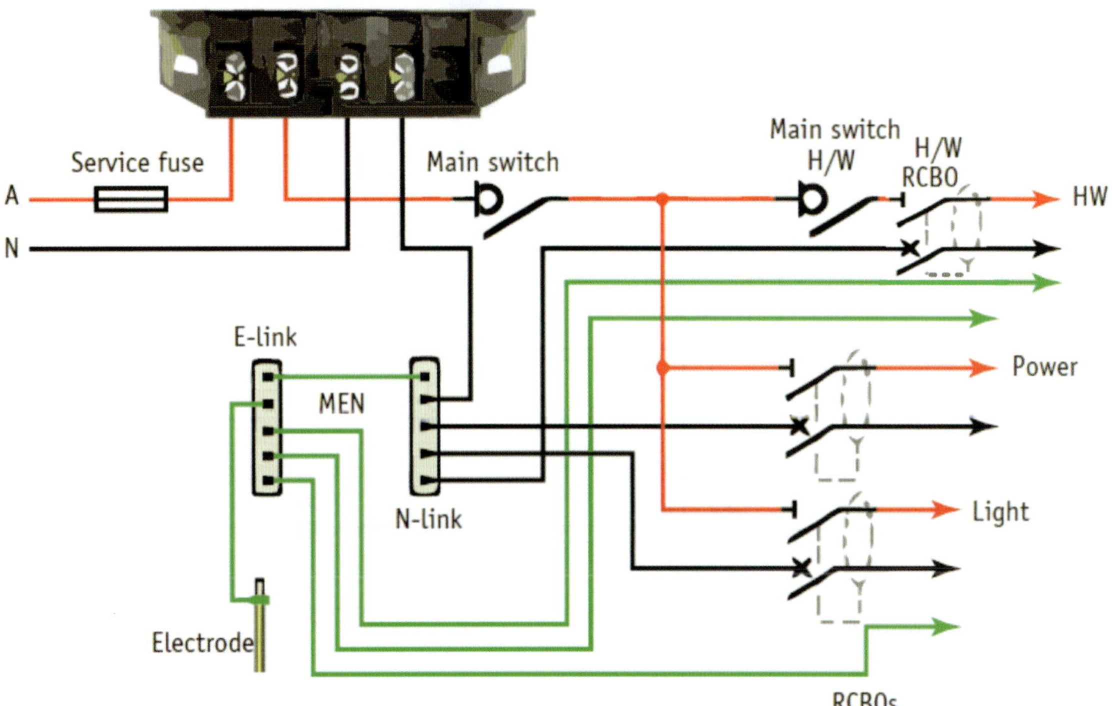

FIGURE 15.58 Single-phase single-tariff entire current metering

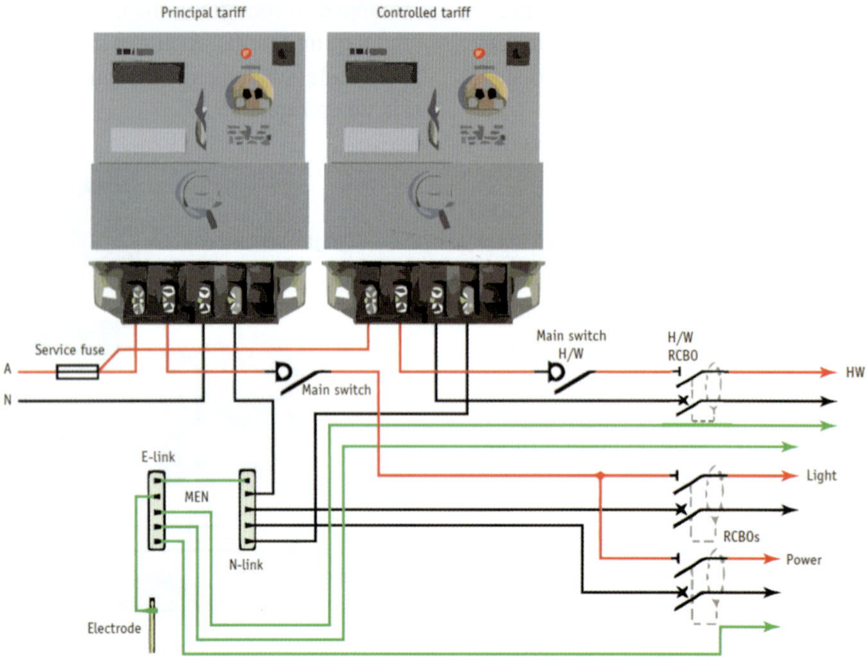

FIGURE 15.59 Direct-connected single-phase metering with controlled load

Figures 15.58 to **15.60** illustrate possible wiring for various metering arrangements. Local supply authorities require meters to be installed to their regulations, and local rules must be followed.

A wiring layout of the main switchboard for an installation supplied with single-phase single-tariff whole current metering is illustrated in **Figure 15.58**.

In domestic installations where parallel consumer's mains are used, the meter neutral conductor should be soldered to one conductor only.

A wiring layout of the main switchboard for an installation supplied with direct-connected single-phase metering for principal and controlled load tariff is illustrated in **Figure 15.59**.

A wiring layout of the main switchboard for an installation supplied with direct-connected single-phase metering for principal and multi-switch controlled load tariff is illustrated in **Figure 15.60**.

A wiring layout of the main switchboard for an installation supplied with direct-connected three-phase

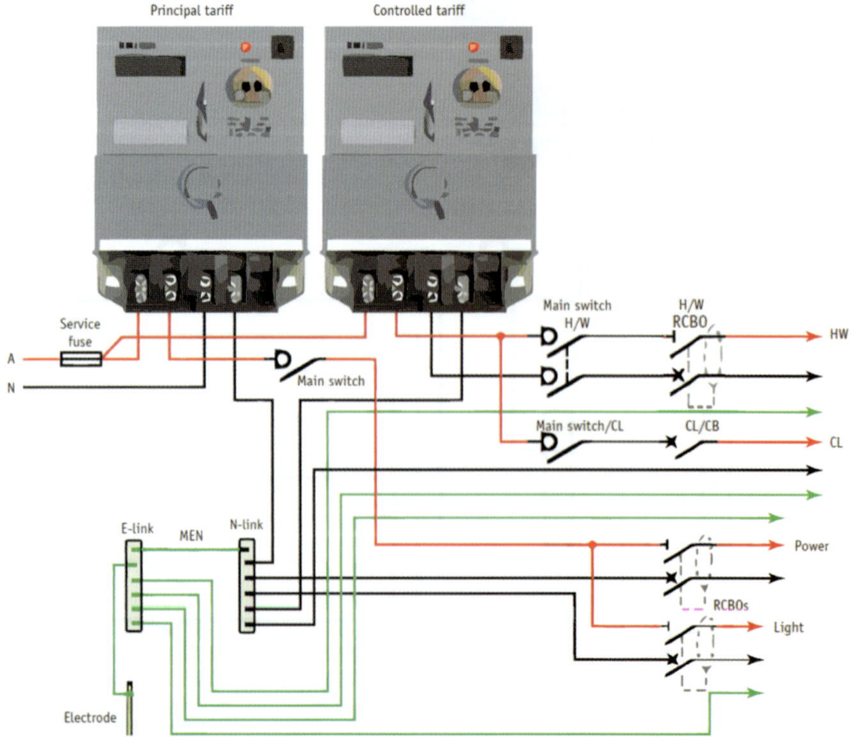

FIGURE 15.60 Direct-connected single-phase metering with multi-switch controlled load

metering for principal and multi-switch controlled load tariff is illustrated in **Figure 15.61**.

A wiring layout of the main switchboard for an installation supplied with low-voltage three-phase current transformer (CT) metering is illustrated in **Figure 15.62**.

Energy suppliers require a low-voltage CT metering equipment order form, a single line diagram and a switchboard layout diagram to be submitted for approval.

Tariffs

A tariff is the agenda of an energy distributor, containing all rates and charges stated separately by type of service, the rules and regulations of the utility and any contracts that affect rates, charges, terms or conditions of service. The kilowatt hour (kWh) is the unit by which electricity is sold to the buyer, with different price tariffs for peak and off-peak usage. The peak period may be between 6 a.m. to 6 p.m. inclusive on any day.

Generated energy, unlike other economic commodities, cannot be stored. So the interrelationship between the local supply authorities and the end user is a distinctive feature of the electrical energy sector. Therefore, tariff pricing plays a significant role in maintaining discipline in the system. If the prices are very high there is every possibility that some of the consumers may withdraw from the system and purchase electricity from another distributor or generate their own electricity.

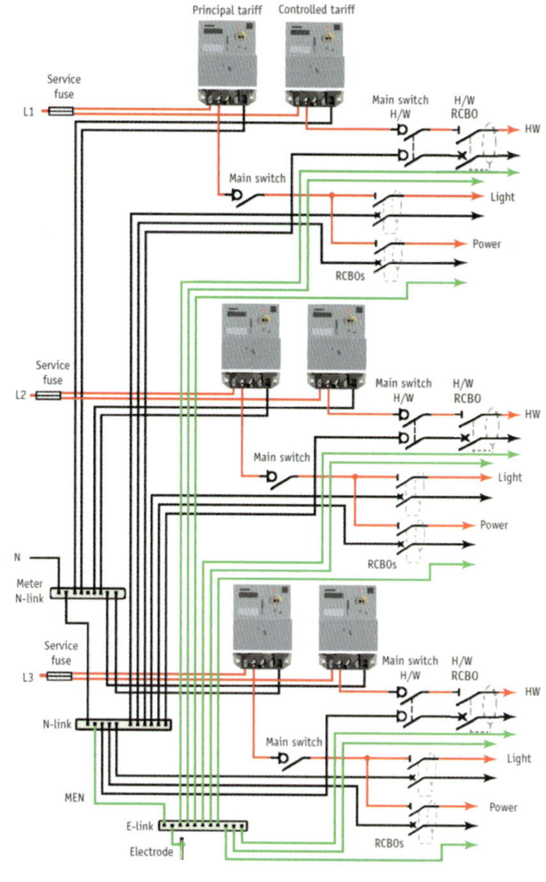

FIGURE 15.61 Direct-connected three-phase metering with multi-switch controlled load

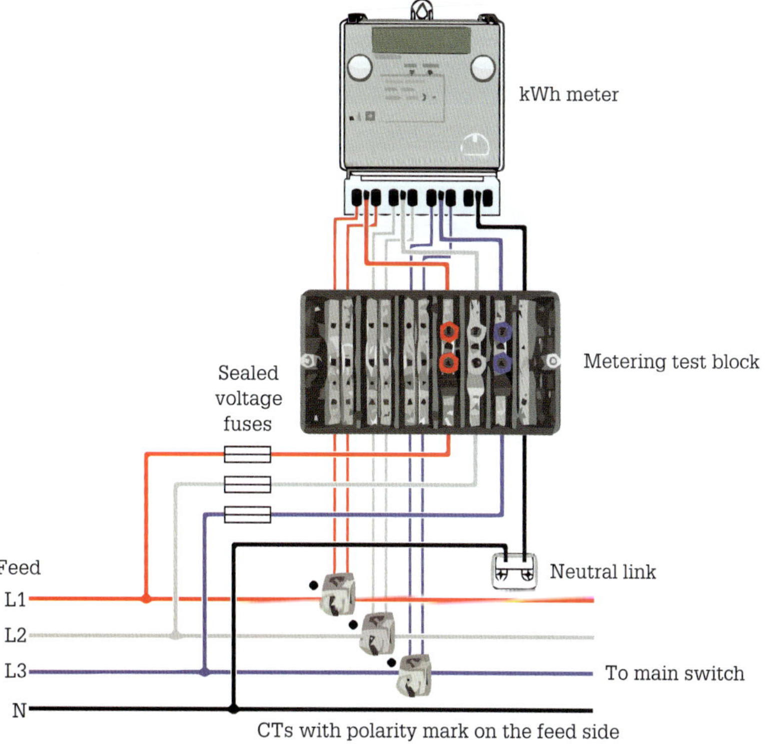

FIGURE 15.62 Low-voltage CT metering

TABLE 15.18 (A) General domestic tariffs

Domestic time-of-use (tariff 12)	Unit	Excl. GST	Inc. GST
Peak consumption (Weekdays 4 p.m.–8 p.m.)	c/kWh	35.0	38.5
Shoulder consumption			
(Weekdays 7 a.m.–4 p.m., 8 p.m.–10 p.m., Weekends 7 a.m.–10 p.m.)	c/kWh	21.5	23.6
Off-peak consumption (Weekdays and Weekends 10 p.m.–7 a.m.)	c/kWh	17.2	19.0
Daily supply charge	c/day	79.0	87.0

Tariff structures

Tariffs are charged to recover costs associated with the generation and for the general provision of supply and maintenance of the network of poles, cables and electrical equipment which distribute power to consumers' premises. The objectives of the electricity tariff structure are:

1. To have a tariff pricing that actually reflects the economic costs and to promote efficient use of electricity, in particular to encourage less consumption during the peak consumption periods of the distribution system.
2. To secure the financial status of the electrical energy distributors.
3. To provide equity for all energy consumer classes (residential, industrial and rural) by reducing cross-subsidisation from one consumer class to another. One approach used to create equity in tariff pricing is the establishment of different tariff rates at various intervals of the day.
4. To attain a method for tariff adjustment that is flexible and regular, corresponding with changing production and distribution costs in a competitive market.

Domestic light and power tariff

This is the electricity tariff for all light and power in general domestic (homes, units and flats) usage and is in the form of progressive rates. (The rates in **Table 15.18** are illustrative only.)

The daily supply charge is a charge that applies for supplying electricity to the premises for each day of the billing period, regardless of how much electricity is used.

TABLE 15.18 (B) Daily supply charges

Controlled Load 1 (tariff 31)	Unit	Excl. GST	Inc. GST
Controlled Load 1 consumption – Night rate 10 p.m.–7 a.m. (super economy)	c/kWh	11.0	12.2
Controlled Load 2 (tariff 33)	**Unit**	**Excl. GST**	**Inc. GST**
Controlled Load 2 consumption – Controlled supply (economy)	c/kWh	15.6	17.2

REVIEW QUESTIONS

1. Which AS/NZS 3000:2018 clause specifies requirements for location, construction and mounting of switchboards?
2. What are the requirements for the orientation of circuit breakers on a switchboard?
3. What type of switchboard is commonly used for a main switchboard in industrial and commercial installations?
4. What are the requirements for every outgoing connection from a switchboard at a carnival?
5. What is a tariff?
6. Why do electricity distributors charge tariffs?
7. What is a daily supply charge?

RULE BOOK QUESTIONS ON SWITCHBOARDS

Refer to AS/NZS 3000:2018, Part 2, *Wiring Rules*, 'General arrangement, control, and protection: Switchboards and emergency systems', Clause 2.10, Switchboards.

1 When can a main switch not be located on a switchboard, or be readily accessible?
Answer____________________
Clause number____________________

2 What are three requirements for the location of switchboards?
Answer____________________
Clause____________________

3 How can sufficient access and exit facilities be achieved?
Answer____________________
Clause number____________________

4 What are the requirements for the identification of the 'main switchboard'?
Answer____________________
Clause number____________________

5 When can a switchboard be installed in the vicinity of an automatic fire sprinkler system?
Answer____________________
Clause number____________________

6 When can live parts be exposed in a non-domestic electrical installation?
Answer____________________
Clause number____________________

7 What are the specific requirements for a bar installed on a switchboard to which a neutral connects?
Answer____________________
Clause number____________________

8 What are the requirements for identifying the relationship of electrical equipment on a switchboard?
Answer____________________
Clause number____________________

9 What are the requirements for identifying terminals of bars, circuit-breakers, fuses and other electrical equipment mounted on a switchboard?
Answer____________________
Clause number____________________

10 What is required where the marking of the fuse base does not correctly indicate the rating of the associated fuse-element?
Answer____________________
Clause number____________________

CHAPTER REVIEW

15.1 Design and safety performance requirements

- An electrical installation consists of the electrical wiring and associated components and fittings, including permanently wired and large stationary appliances, but excluding portable equipment and appliances.
- An electrical installation must be designed, constructed, installed and protected to minimise the risk of fire or property damage.
- All new installations must comply with the requirements for the multiple earthed neutral (MEN) system of earthing as laid down in AS/NZS 3000:2018 *Wiring Rules*.
- The equipment in an electrical installation must be arranged and operated so as to minimise or prevent adverse effects on the distribution system and other electrical installations connected to the distribution system.
- AS/NZS 3000:2018, Appendix C, recognises four methods for determining the maximum demand of consumer's mains or sub-mains.
- Assessment of maximum demand can be based on the duty cycle (the proportion of time during which the installed equipment operates) or the load characteristics of the installed equipment.
- The maximum demand of an existing electrical installation may be measured by a maximum demand indicator.

15.2 Final sub-circuit arrangements

- The equipment in an electrical installation must be arranged and operated so as to minimise or prevent adverse effects on the distribution system and other electrical installations connected to the distribution system.

- The term 'consumer's mains' describes a cable connected between the point of supply and the main switchboard in an installation.
- External influences means the surrounding environment to the extent that it has an impact on the electrical installation.
- The circuit schedule must be installed in each switchboard in a circuit schedule holder with a clear plastic sheet.

15.3 Factors affecting the suitability of wiring systems

- Wiring systems and accessories should be installed in locations that prevent them from being subject to mechanical damage, dampness, corrosion, chemical attack, insect attack, heat and other damaging environmental conditions.
- Higher ambient temperatures prevent effective transference of heat from the cable to the surrounding environment.
- Extreme soil and environmental conditions exist where the ambient temperature exceeds 40 °C. Hot, dry conditions in the summer create an arid soil which produces a high thermal resistance.
- The cable must also be protected from the effects of external heat sources such as a heated concrete slab or an electric furnace.
- Cables may be directly exposed or contained within enclosures that are exposed to solar radiation.
- When cables are fixed to a continuous surface, they can carry current with a slight de-rating factor being applied.
- When cables are installed in an enclosure in air they can carry current with a significant de-rating factor being used.
- Cables can be partially surrounded or completely surrounded by thermal insulation.
- Because of the difference in ambient temperature between surface air and soil, cables installed in ground are able to dissipate their conductor heat more efficiently.
- When cables for different circuits are bunched together, excessive heat becomes a problem.

15.4 Maximum demand

- The maximum demand may be calculated by following the directions in Appendix C of AS/NZS 3000:2018 for the suitable category of electrical installation.
- Diversity is applied in the calculation method on the basis that the percentage of people, lights, equipment, etc. that are on (or present) simultaneously compared with the total design demand may not occur at the same time.
- Where there are more than two individually metered units, the supply can be assumed to be three-phase. However, the local distribution entity must be consulted.
- Prior to calculating maximum demand, the distribution of the loads across sub-circuits must be determined. Loading tells us the number of sub-circuits required.

15.5 Cable selection based on current-carrying capacity

- The most important element of any cable is its current-carrying capacity.
- The correct current-carrying capacity of a cable is a function of its insulation, temperature rating, the type of enclosure in which it is installed, number of conductors being installed in a standard enclosure and whether it is above ground or underground.
- Coordination and overload protection requirements ensure that the life of the cable insulation is not significantly shortened.
- Cable selection can be done by selecting a cable size and then determining the maximum current-carrying capacity, or by choosing a cable size capable of carrying a known current demand.
- Section 2 of AS/NZS 3000:2018 *Wiring Rules* and Section 5 of AS/NZS 3008.1.1:2017 *Electrical installations – Selection of cables* takes account of the clearance time of the protective device under short-circuit conditions by applying what is known as the adiabatic equation.
- Where a number of circuits are installed together, de-rating factors must apply because the cooling of the cables within the surrounding environment has been affected.

15.6 Cable selection based on voltage drop

- Voltage drop is dependent upon the impedance of the cable, the magnitude of the circuit current and the circuit power factor.
- The voltage drop between the point of supply and any point in the installation must not exceed 5% of the nominal supply voltage.
- Voltage sags are a partial reduction in rms voltage that usually lasts for 0.5 to 30 cycles, while a momentary interruption is a complete loss of a.c. power which can be 0.5 cycle to minutes in duration.
- Cable size is sometimes governed by voltage drop rather than by heating. The voltage drop method occurs when the circuit route length is long, and these long circuit lengths may require larger CSA conductors for a given current.
- The voltage drop (V_d) for each circuit cable in an installation can be calculated using the equation found in AS/NZS 3008.1.1:2017 *Electrical installations – Selection of cables.*
- When using Tables 40–50 in AS/NZS 3008.1.1:2017 for voltage drop calculations it will be necessary to convert the three-phase voltage drop values (V_c) given in the tables to a single-phase value for single-phase installations.
- Even though cables meet voltage drop requirements, the selected cable size must also be able to carry the circuit current.
- AS/NZS 3000:2018, Clause 3.6.2c, allows a voltage drop dispensation for final sub-circuits, with distributed load (such as socket outlets or lighting), where half the current rating of the protective device may be used in the voltage drop determination.

15.7 Cable selection based on fault-loop impedance

- An essential safety requirement demanded by AS/NZS 3000:2018, Clause 1.5, is the protection of persons and livestock from 'indirect contact' by touching exposed metallic components that have become 'live' due to an earth fault.
- The earth-fault-loop impedance is the impedance of the phase-to-earth fault current loop starting and ending at the point of the earth fault.
- The main concern with earth-fault impedance is the need to disconnect the circuit within a required time when the touch voltage exceeds 50 V a.c.

15.8 Selecting protection devices

- A protective device must be placed at the point where a reduction of the CSA of the circuit conductors or another change causes alteration in the characteristics of the installation.
- Series rated is a combination of circuit breakers or fuses and circuit breakers that can be used at short-circuit levels above the interrupting rating of the load-side (protected) circuit breaker, but not above the interrupting rating of the main or supply-side device.

15.9 Selecting devices for isolation and switching

- Isolators or switch disconnectors are used for the electrical isolation of electrical systems from all sources of power and to conduct continuous current up to the level of their defined rated current level.

15.10 Switchboards

- Switchboard design and arrangement of equipment on the panel board depend on several factors.
- A tariff is the agenda of an energy distributor, containing all rates and charges stated separately by type of service, the rules and regulations of the utility and any contracts that affect rates, charges, terms or conditions of service.
- Tariffs are charged to recover costs associated with the generation and for the general provision of supply and maintenance of the network of poles, cables and electrical equipment which distribute power to consumers' premises.

TRIAL EXAM

For Chapter 15 knowledge assessment, please complete the following trial exam.

1 The prospective touch voltage should not exceed:
 a 25 V
 b 50 V
 c 100 V
 d 230 V

2 The objective of the earth-fault-loop impedance calculation is to determine accurately:
 a the circuit earth cable size
 b the consumer's main size
 c the main earth size
 d the current rating of the protective device

3 In what location must protective devices be placed with the circuits of an installation?
 a where voltage drops occur
 b where a reduction of the CSA of the circuit conductors occurs
 c where the supply authority regulations suggest
 d where a potential short-circuit exists

4 A circuit protection device must be:
 a connected at the end of all circuits
 b connected to the earthing circuit
 c connected across the active and neutral circuit conductors
 d connected in series with the circuit it is protecting

5 Over-current protection is provided by a:
 a circuit breaker
 b rocker switch
 c isolating switch
 d functional switch

6 What is a circuit connected directly from the main low-voltage switchboard to any other sub-main distribution board called?
 a final sub-circuit
 b sub-mains
 c consumer's mains
 d switchboard wiring

7 When does an installation meet performance standards?
 a after it has been inspected, tested and found to be safe by a licensed electrician
 b at the design stage
 c after the layout has been finalised
 d after the final fit-out

8 The range of standard 230 V nominal voltages is:
 a +20% and −5%
 b +7% and −4%
 c +10% and −6%
 d +5% and −12%

9 The maximum short-circuit current that can flow under short-circuit conditions at the POS within an electrical installation is called:
 a earth fault current
 b the main earth current
 c the prospective fault current
 d impedance current

10 To prevent unwanted voltages:
 a all active conductors must be solidly bonded to earth
 b all neutral conductors must be bonded to earth
 c all circuits must be placed into an electrically safe work condition
 d unused protective earth conductors must be isolated

11 Which method of determining the maximum demand is based on looking up values in a Table for specific items of equipment?
a estimation
b limitation
c assessment
d calculation

12 In low-voltage installations, the voltage drop between the POS and any point in the installation must not exceed:
a 1% of the nominal supply voltage
b 2% of the nominal supply voltage
c 5% of the nominal supply voltage
d 12 V

13 In electrical installations, two major types of risk exist. Shock current is one; name the other.
a excessive temperatures
b unwanted voltages
c fault currents
d mechanical movement

14 A maximum demand indicator can be used to:
a measure maximum demand
b determine load types
c determine number of final sub-circuits
d graph voltage drop over time

15 For the purpose of maximum demand in a single domestic residence, 15 double 10 A socket outlets contributes:
a 5 A
b 10 A
c 15 A
d 20 A

16 A commonly overloaded component of an electrical system is the:
a earth electrode
b circuit wiring
c bonding conductors
d main isolator

17 The current-carrying capacity of the consumer's mains should, when installed in accordance with AS/NZS 3000:2018 and AS 3008.1.1:
a not exceed the rating of the circuit breaker main switch
b exceed the rating of the circuit breaker main switch
c exceed the rating of the main earthing conductor
d not exceed the rating of the main earthing conductor

18 Why is it necessary to de-rate cables installed in thermal insulation?
a because a smaller CSA cable is required
b because of a reduced ability to dissipate heat
c because the value of the protective device for circuits installed in thermal insulation must be reduced
d because the cable insulation may be chemically changed by being in contact with the thermal insulation

19 Using Table 42 of AS/NZS 3008.1.1:2017, determine the actual voltage drop for multi-core PVC V–90 insulated and sheathed cable having 2.5 mm² copper conductors laid flat for a single-phase 230 V final sub-circuit carrying 20 A over a route length of 30 m.
a 4.5 V
b 6.73 V
c 8.16 V
d 10.8 V

20 A cable has a nominal current rating of 20 A for an ambient temperature of 40 °C. The current-carrying capacity of this cable when installed in a concrete slab having an ambient temperature of 45 °C is:
a 22 A
b 20 A
c 18.2 A
d indeterminate

21 A wiring system comprises five circuits of 2.5 mm² 2C + E flat V–90 TPS cable installed bunched and suspended from a catenary. The current-carrying capacity for each cable is:
a 25 A
b 22 A
c 18.9 A
d 16 A

22 What electrical rating for buried cables is affected by cable depth and local climatic conditions such as air temperature and soil type?
a voltage rating
b ohmic resistance of the conductors
c current capacity
d insulation resistance

23 Using AS/NZS 3008.1.1, Table 10, what is the current rating for a 2C + E 4 mm² flat TPS stranded copper V–90 cable totally surrounded by thermal insulation?
a 27 A
b 24 A
c 17 A
d 12 A

24 Name the principle that takes into account the normal operating conditions of circuits during which all equipment is not operating simultaneously.
a variety
b diversity
c selection
d range

25 Select the minimum CSA for a three-phase final sub-circuit consisting of a bare multi-core (4) mineral-insulated 1/1 kV copper-sheathed cable with copper conductors and a sheath temperature of 100 °C supplying a load demand of 24 A. The cable is installed in free air and clipped to a concrete wall.
a 1.5 mm²
b 2.5 mm²
c 4 mm²
d 6 mm²

Answers to the exercises

Chapter 1

Exercise 1.1

a single 9 m industrial

b single 5 m domestic

Exercise 1.2

a 2.5 m

b 3.75 m

Exercise 1.3

Arm paralysis

Chapter 3

Exercise 3.1

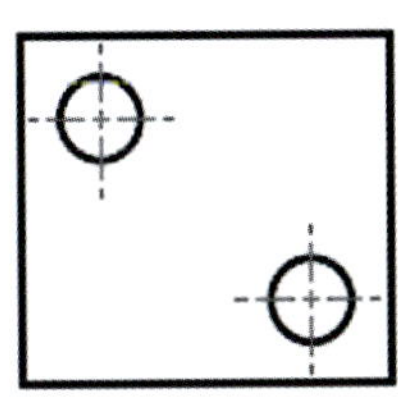

Centre lines

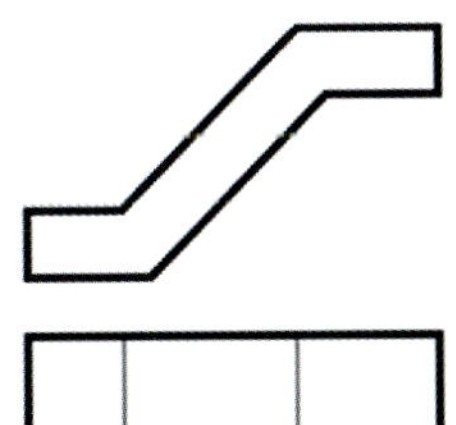

Bend lines

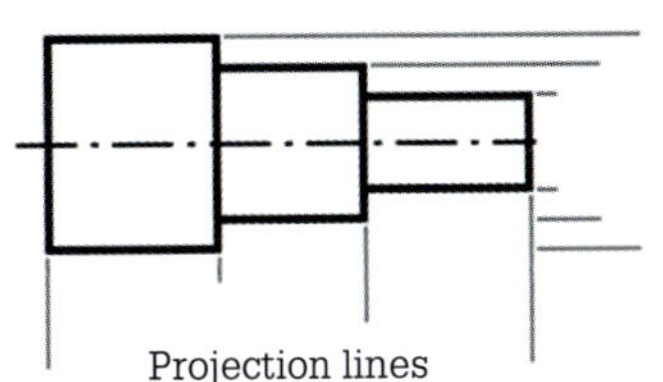

Projection lines

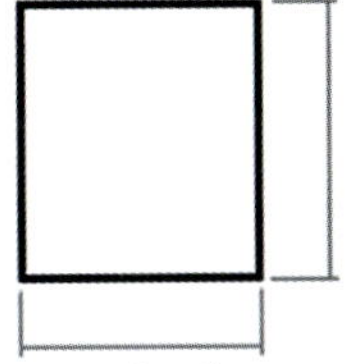

Dimension lines

Exercise 3.2

a 550 mm

b 350 mm

c 240 mm

Exercise 3.3

a **i** kilometres

ii metres

iii millimetres

iv millimetres

b Approximately:

- width: 17 mm
- depth: 70.0 mm
- height: 80.0 mm

c 49 m–51 m

d ± 2.78 %

Exercise 3.4

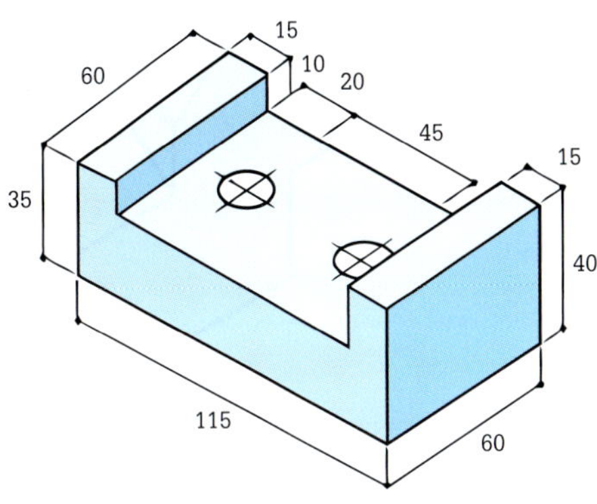

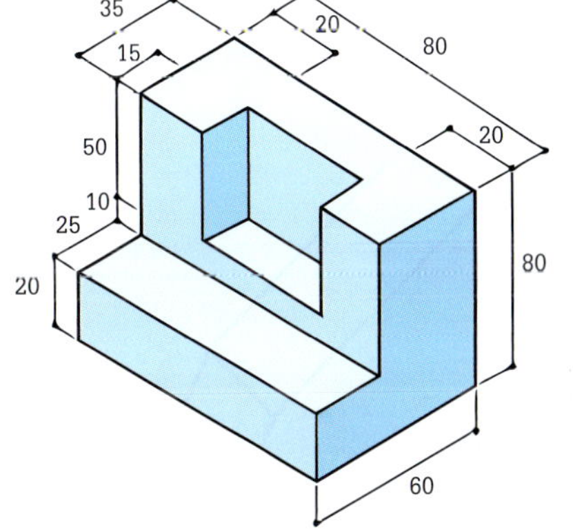

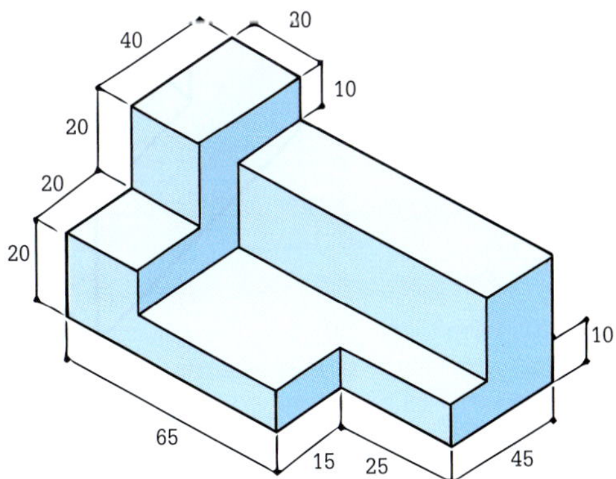

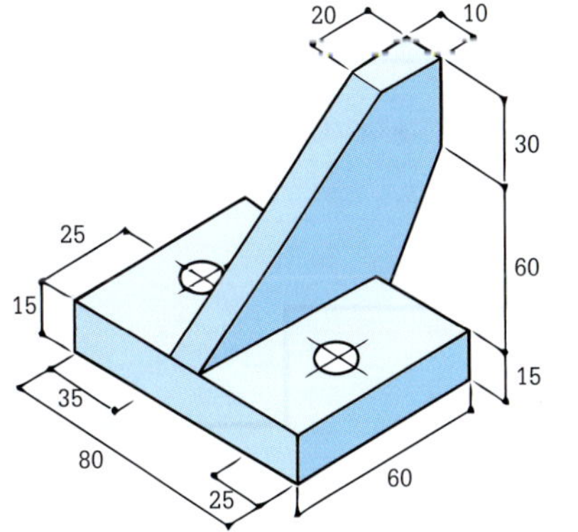

Exercise 3.5

a

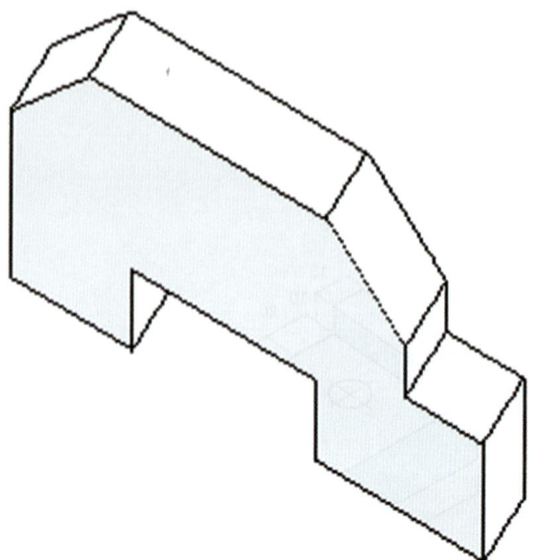

b

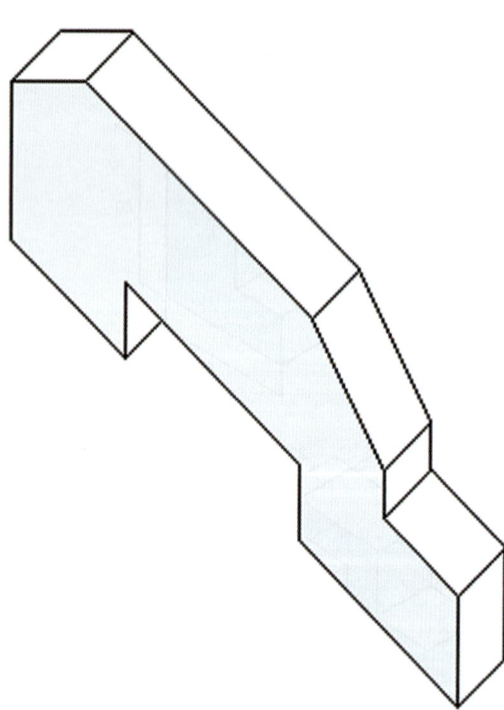

c

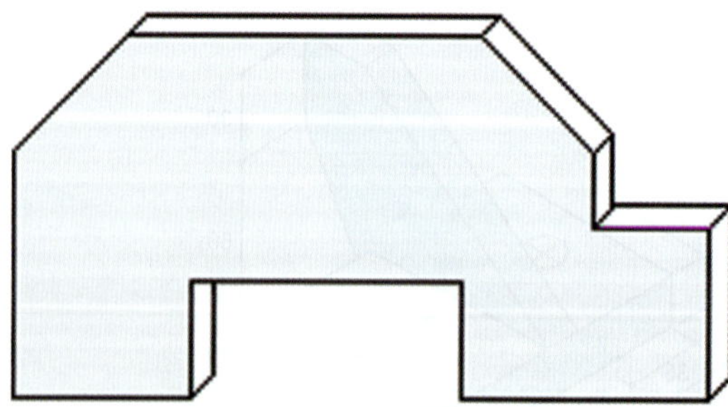

Exercise 3.6

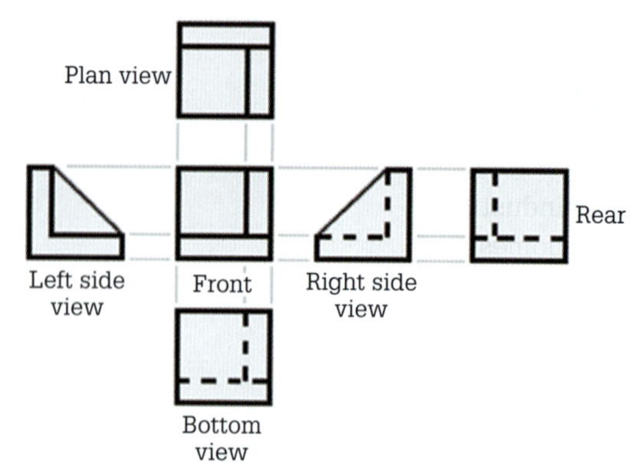

Exercise 3.7

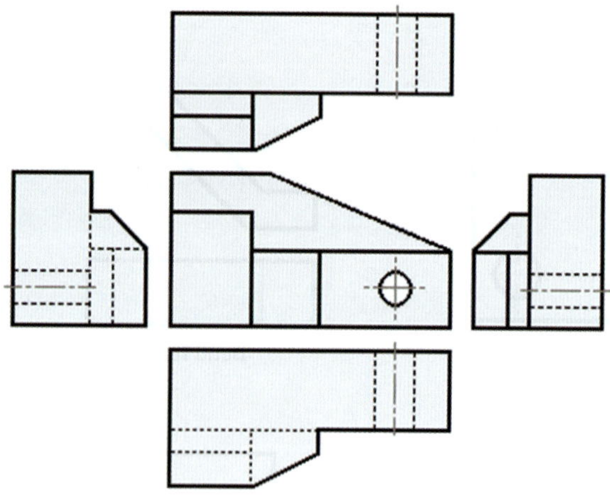

Exercise 3.8

a 2.5 mm
b 8.8 mm
c 17.5 mm

Exercise 3.9

5.00 mm

Exercise 3.10

93.80 mm

Chapter 4

Exercise 4.1

60 mm (55 + 5)

Exercise 4.2

Pilot hole size of 6 mm for softwood and 8 mm for hardwood.

Exercise 4.3

Tensile stress area (TSA) of the anchor is 50.3 mm^2
A Class 5.8 bolt has a tensile strength of 500 MPa, so UTS = 25.15 kN
kgf = 2564 kg

Applying a safety margin of 4:1 the anchor can carry a load of 641 kg shear weight vertical force flat to the wall. The horizontal force taken at 0.6 times the vertical force would be 384 kg.

The actual tension force on the bolt is the height divided by the width, multiplied by the mass of the object. Therefore, the tension created by the mass:

$$0.5 \div 1 = 0.5$$

$$0.5 \times 100 \text{ kg} = 50 \text{ kg}$$

Now 384 divided by 50 = 7.68. To support the load securely eight M8 masonry anchors would be suitable. Additional anchors may be warranted, not because of weight, but because of the size of the fixture.

Chapter 5

Exercise 5.1

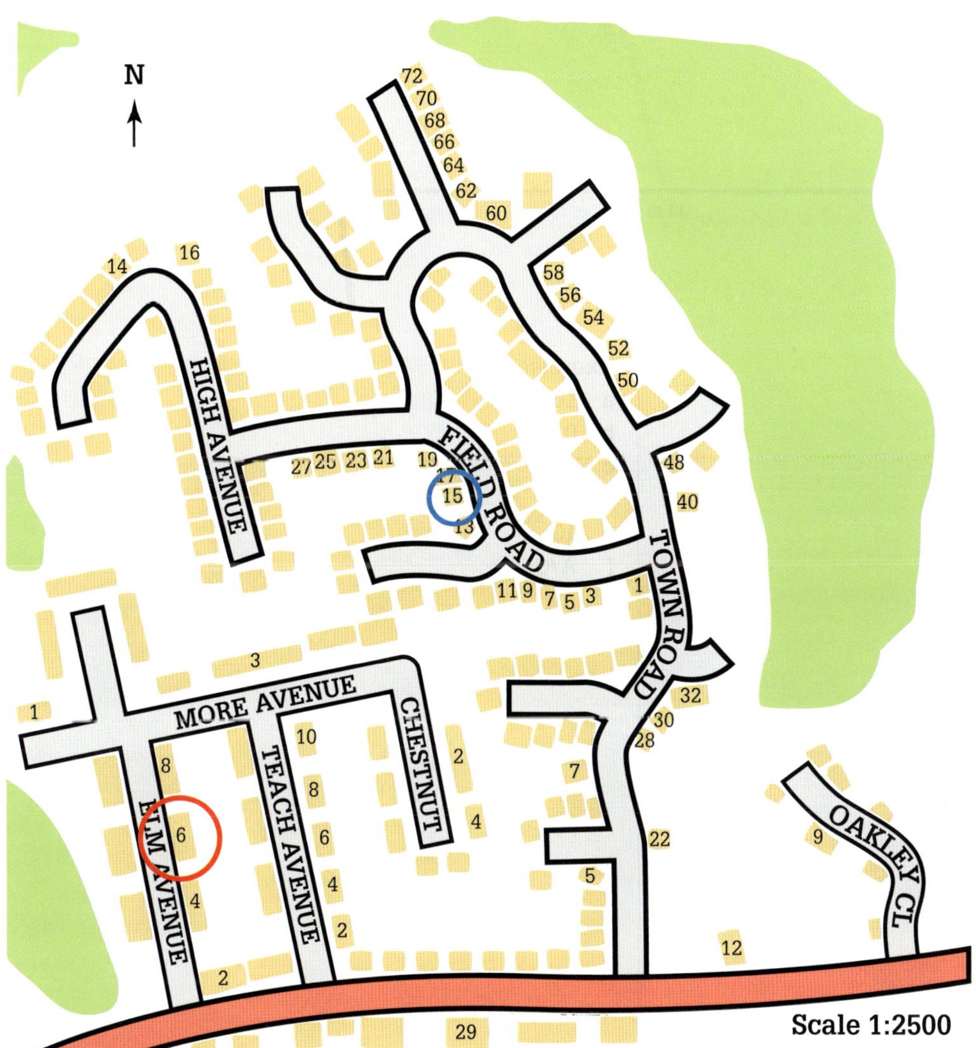

Exercise 5.2

E 3° N for 21.8 m

Exercise 5.3

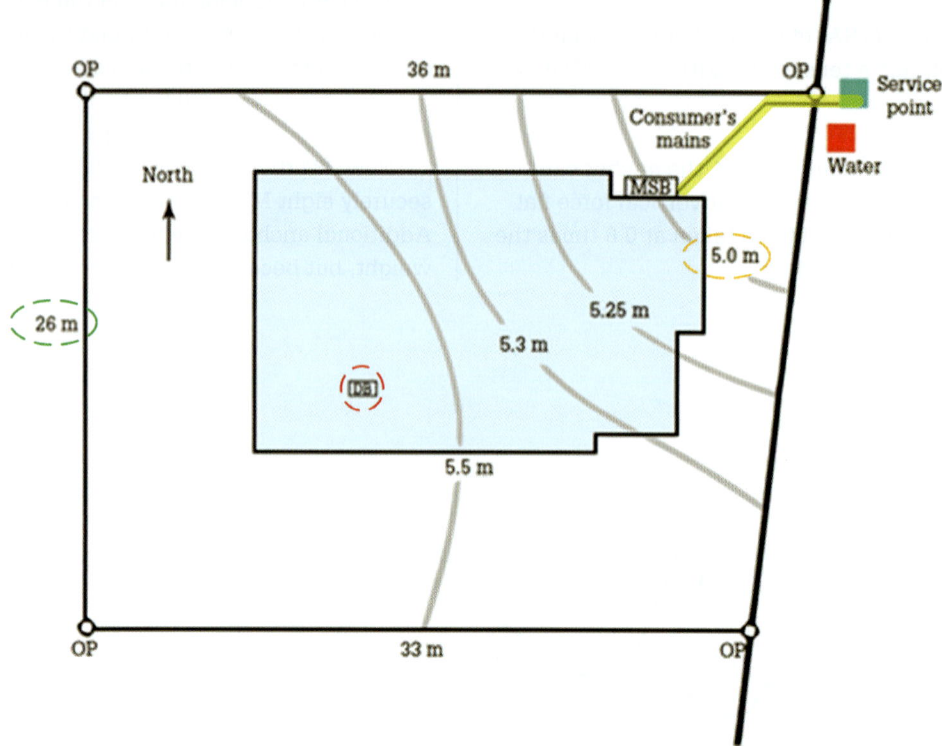

Exercise 5.4

- a site plan
- b floor plan standard drawing
- c floor plan detailed drawing
- d location plan

Exercise 5.5

- a 21.2 m
- b 14.8 m
- c 296 m^2
- d 26 m^2
- e 3000 mm
- f 9.2 m × 5.4 m

Exercise 5.6

- a 10
- b 0
- c 60 A
- d 3
- e 10
- f A/C

Exercise 5.7

8 Pole Chassis
Main Switchboard
Domestic Installation

Spare

25 A RCBO
P5 – 2 x 2.5 mm^2
TPS
HWS
N5

32 A RCBO
P4 – 2 x 6 mm^2
TPS
Range
N4

16 A RCBO
P3 – 2 x 2.5 mm^2
TPS
Power
N3

16 A RCBO
P2 – 2 x 2.5 mm^2
TPS
Power
N2

10 A RCBO
P1 – 2 x 1.5 mm^2
TPS
Light
N1

Main CB
60A SP

Exercise 5.8

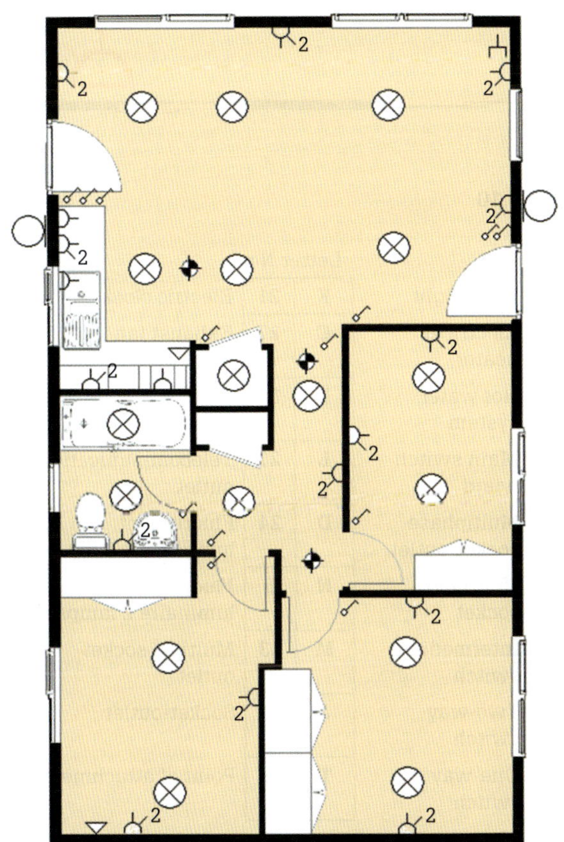

Exercise 5.9

a

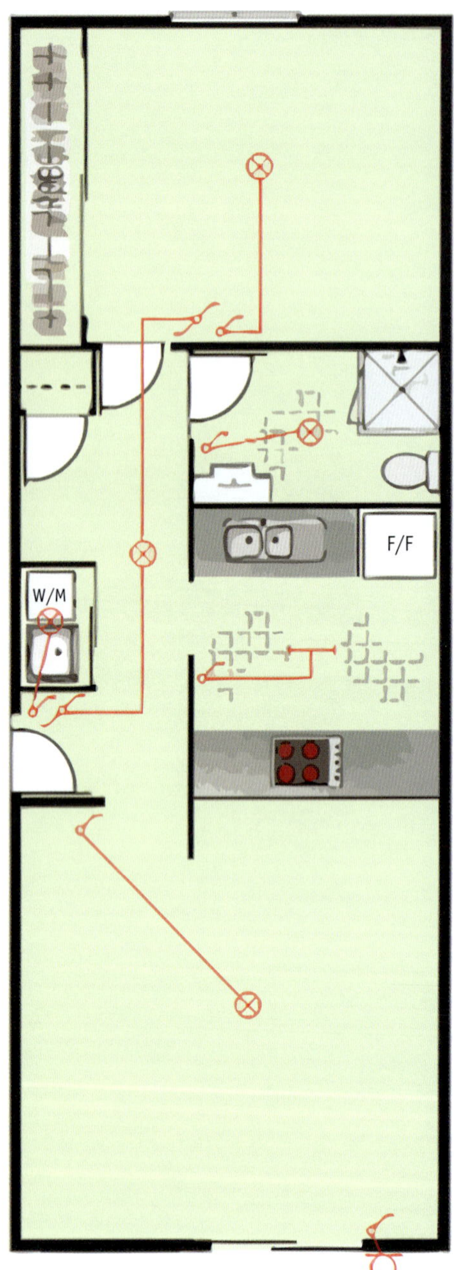

b

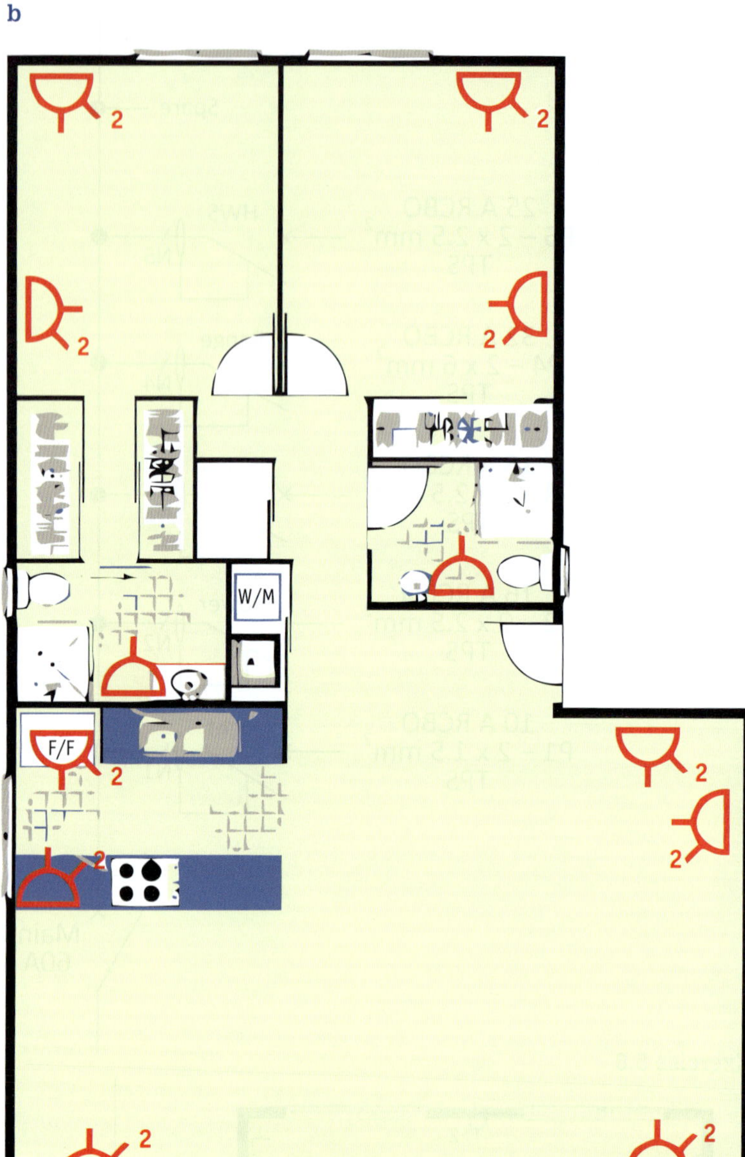

Exercise 5.10

Letter	No		Letter	No	
T	31	Flood light	K	21	Electric clock
A	7	Electrical heater	C	27	Exhaust fan
S	23	Hot water system	L	31	Wall telephone outlet
S	28	Main switch board	L	21	Telecommunications outlet
S	19	Multiphase socket-outlet	D	24	Fluorescent luminaire 1 lamp
A	10	15 A plug socket	N	5	Fluorescent luminaire 2 lamps
S	3	Intermediate switch	M	23	Multiple socket-outlet
A	3	Two-way switch	J	31	Socket-outlet
O	25	One-way switch	T	28	Point of attachment

Exercise 5.11

Circuit no.	Core size mm^2	Current (A)	Route length	Application
MSB C/B 1	1.0 T&E flat	10	130 m	Lighting
MSB C/B 2	1.0 T&E flat	10	170 m	Lighting

Exercise 5.12

- **a** Protection against overcurrent
- **b** Maximum demand
- **c** Functional (control) switching
- **d** Protection against overvoltage
- **e** Limitation of circulating and eddy currents

Chapter 6

Exercise 6.1

- **a** WS24
- **b** WS35

Exercise 6.2

- **a** 3
- **b** 25 mm

Exercise 6.3

240 mm

Exercise 6.4

- **a** $\frac{23 \times 30}{100} = \mathbf{6.9\ kg}$
- **b** 15 A

Chapter 8

Exercise 8.1

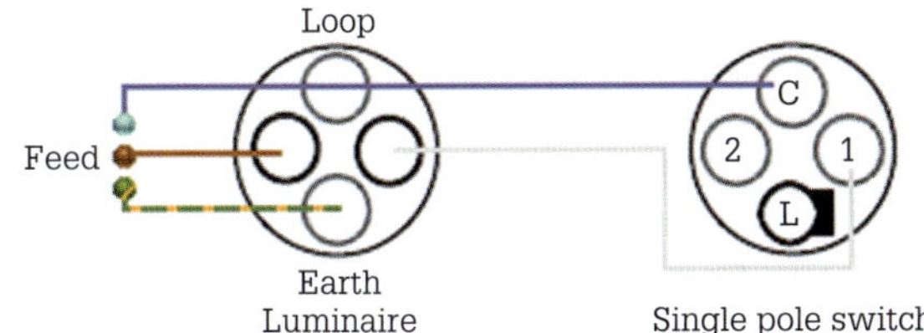

Exercise 8.2

200 lx

Exercise 8.3

13.33 lm/W

Exercise 8.4

7.2 lx

Chapter 9

Exercise 9.1

a
$$R = \frac{V^2}{P} = \frac{240^2}{(2.4 \times 10^3)} = \mathbf{24\ \Omega}$$

b
$$I = \frac{V}{R} = \frac{230}{24} = \mathbf{9.58\ A}$$

Exercise 9.2

10.58 Ω

Exercise 9.3

- **a** 393.15 K
- **b** 553.15 K
- **c** −23.0 °C
- **d** −153.0 °C

Exercise 9.4

- **a** 103.3 kJ
- **b** 110.9 °C

Exercise 9.5

No. Appliance 2 has a higher power rating. It is not a direct replacement.

Chapter 10

Exercise 10.1

From Table C9 a 20 A socket outlet is rated at 20 A for all installations. As the circuit is rated at 20 A, it is possible to install only one 20 A socket outlet on this circuit.

Exercise 10.2

From Table C7, 2.5 mm^2 copper conductors with V-90 PVC insulated and sheathed cables are rated at 16 A when installed partially surrounded by thermal insulation.

Exercise 10.3

From Table C9 a lighting point is rated at 0.5 A and a socket outlet is rated at 1 A for all domestic installations.

The contribution of the lighting points is:

$$\text{lighting demand} = \text{number of points} \times 0.5\text{ A} = 12 \times 0.5\text{ A} = 6\text{ A}$$

As the circuit is rated at 20 A, the available demand for socket outlets is:

available demand = circuit demand – lighting demand
= 20 – 6
= 14 A

it is possible to install 14 socket outlets on this circuit. This is calculated from:

$$\text{Number of points} = \frac{\text{available demand}}{\text{rating per point}}$$

Chapter 11

Exercise 11.1

1 Answer
   a Enable automatic disconnection of supply in the event of a short-circuit to earth fault or excessive earth leakage current in the protected part of the installation through protective earthing arrangements.
   b Enable equipment requiring an earth reference to function correctly through functional earth (FE) arrangements.
   c Mitigate voltage differences appearing between exposed conductive parts of equipment and extraneous conductive parts through equipotential bonding arrangements.
   d Provide an effective and reliable low impedance fault path capable of carrying earth fault and earth leakage currents without danger or failure from thermal, electromechanical, mechanical, environmental and other external influences.
   e Provide measures for the connection of exposed conductive parts and extraneous conductive parts.

   Clause number 5.1.2

2 Under the MEN system, the neutral conductor (PEN) of the distribution system is earthed at the source of supply at regular intervals throughout the system and at each electrical installation connected to the system. Within the electrical installation, the earthing system is separated from the neutral conductor and is arranged for the connection of the exposed conductive parts of equipment.
   Clause number 5.1.3

3 When a fault occurs between a live part and an exposed conductive part or parts of the protective earthing system, a prospective touch voltage may arise between simultaneously accessible conductive parts. Fault protection by means of automatic disconnection of supply is intended to limit this voltage.
   Clause number 5.2.1

4 4 mm^2; Clause number 5.3.3.2

5 Conductive sheaths, armours and screens of cables may be regarded as a protective earthing conductor, provided that the electrical equipment to be earthed is supplied only by live conductors incorporated in the cable.
   Clause number 5.3.2.3 (b)

6 A catenary wire may be regarded as a protective earthing conductor, provided that it:
   (i) has not less than seven strands;
   (ii) is supported by means of suitable anchorages;
   (iii) has a nominal cross-sectional area of not less than 8.5 mm^2 if constructed of hard-drawn copper or galvanised low carbon (mild) steel;
   (iv) has a resistance in accordance with the requirements of Clause 5.3.3; and
   (v) is identified as an earthing conductor, in accordance with Clause 3.8.3.4, and for aerial earthing conductors, at each anchorage point or catenary support.

   Clause number 5.3.2.3 (d)

7 The cross-sectional area of a protective earthing conductor shall ensure:
   a adequate current-carrying capacity for prospective fault currents for a time at least equal to the operating time of the associated overcurrent protective device;
   b appropriate earth fault-loop impedance (see Clause 5.7);
   c adequate mechanical strength and resistance to external influences; and
   d for parts of the protective earthing conductor that do not consist of cables, or parts of cables, that there is an allowance for the subsequent deterioration in conductivity that may reasonably be expected.

   Clause number 5.3.3.1.1

8 Where the active conductor comprises a number of conductors, connected in parallel, the earthing conductor shall be determined in relation to the summation of the cross-sectional areas of the individual conductors forming the largest active conductor to be protected.
   Clause number 5.3.3.1.2

9 10 mm^2; Clause number 5.3.3.1.2 and Table 5.1

10 120 mm^2; Clause number 5.3.3.2

11 The function of the MEN connection is to connect the earthing system within the electrical installation to the supply neutral conductor by means of a connection from the main earthing terminal/connection or bar to the earthing terminal on the main neutral bar
   Clause number 5.3.5.1 (Note)

12 The MEN connection shall be a conductor complying with Clause 5.3.2 and have a cross-sectional area capable of carrying the maximum current that it may be required to carry under short-circuit conditions
   Clause number 5.3.5.2

13 The location of the earth electrode shall be identified at the main switchboard; Clause number 5.3.6.4

14 Yes; Clause number 5.4.2

15 Yes; Clause number 5.4.3

16 An earthing conductor, deemed to be the main earthing conductor, shall be taken from the main earthing terminal/connection or bar at the main

switchboard to an earth electrode complying with Clause 5.3.6
Clause number 5.5.1.1

17 The main earthing conductor shall have a permanent label attached at the connection to the earth electrode with a legible warning against disconnection in the following form:
WARNING: MAIN ELECTRICAL EARTHING CONDUCTOR—DO NOT DISCONNECT.
Clause number 5.5.1.3

18 0.5 Ω; Clause number 5.5.1.4

19 Exposed conductive parts of wiring enclosures shall be earthed at the end adjacent to the switchboard or accessory at which the wiring enclosure originates
Clause number 5.5.3.2 (a)

Exercise 11.2

1 Yes; Clause number 5.6.1

2 Maximum resistance shall not exceed 0.5 Ω.
Clause number 5.6.2.6.1

3 Equipotential bonding is intended to minimise the risks associated with the occurrence of voltage differences between exposed conductive parts of electrical equipment and extraneous conductive parts
Clause number 5.6.1

4 Conductive water piping that is both:
   a installed and accessible within the building containing the electrical installation; and
   b continuously conductive from inside the building to a point of contact with the ground, shall be bonded to the earthing system of the electrical installation.

   Clause number 5.6.2.2

5 Any conductive reinforcing within a concrete floor or wall of a room containing a shower or bath shall be bonded to the earthing system of the electrical installation
Clause number 5.6.2.5

6 a The conductive pool structure and the pool equipotential bonding conductor connection point specified in Clauses 5.6.2.6.2 and 5.6.2.6.3
   b The items of electrical equipment specified in Clause 5.6.2.6.4
   c The conductive fixtures and fittings specified in Clause 5.6.2.6.5
   d The earthing conductors associated with each circuit supplying the pool or spa, or the earthing bar at the switchboard at which the circuits originate.

   Clause number 5.6.2.6.1

7 An equipotential bonding conductor, in accordance with Clause 5.6.3, shall be connected between:
   a the conductive pool structure and the pool equipotential bonding conductor connection point specified in Clauses 5.6.2.6.2 and 5.6.2.6.3;
   b the items of electrical equipment specified in Clause 5.6.2.6.4;
   c the conductive fixtures and fittings specified in Clause 5.6.2.6.5; and
   d the earthing conductors associated with each circuit supplying the pool or spa, or the earthing bar at the switchboard at which the circuits originate.

   Clause number 5.6.2.6.1

Exercise 11.3

a 27.9 kA
b 39.2 kA

Exercise 11.4

0.0092 Ω

Exercise 11.5

9.75 kA

Exercise 11.6

5.54 kA

Exercise 11.7

$I_{PFC} = 20.1$ kA
$Z_S = 0.011\ 44\ \Omega$
$Z_{cma} = 0.004\ \Omega$
$Z_{sma} = 0.011\ 76\ \Omega$
$Z_{smpe} = 0.031\ 08\ \Omega$
$I_{Fault\ SB} = 3.95$ kA

Exercise 11.8

a $Z_S = 1.53\ \Omega$, $I = 150$ A, $I/I_N = 7.5$, time = 0.005 to 4.8 sec (from Figure 9.56)
b $Z_S = 0.958\ \Omega$, $I = 240$ A, $I/I_N = 7.5$, time = 0.005 to 4.8 sec (from Figure 9.56)

Exercise 11.9

4 mm^2 protected by 25 A CB has maximum length of 67 m

Exercise 11.10

104.8 m

Exercise 11.11

About 0.5 sec

Exercise 11.12

20 A type D circuit breaker would be suitable

Chapter 14

Exercise 14.1

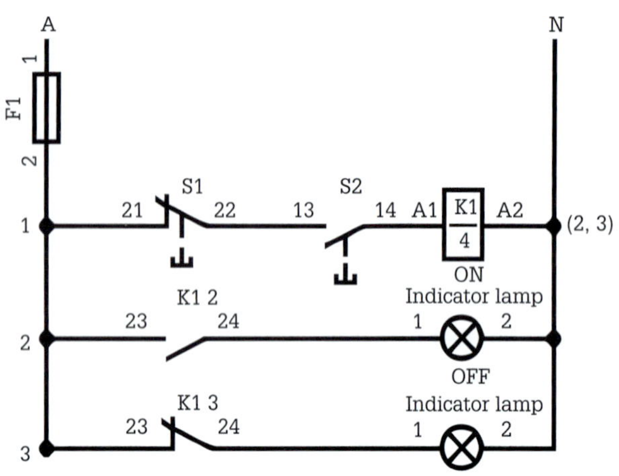

Exercise 14.2

1800 rpm

Exercise 14.3

15.1 Nm

Exercise 14.4

$$I_{star} = \frac{I_{DOL}}{3} = \frac{60}{3} = \mathbf{20\ A}$$

Exercise 14.5

36.3 Nm

Exercise 14.6

39 A

Exercise 14.7

84.5 A

Exercise 14.8

11.9 Nm

Chapter 15

Exercise 15.1

$$Z_C = L \times \frac{\sqrt{R_C^2 \pm X_C^2}}{1000} = 50 \times \frac{\sqrt{1.40^2 + 0.0861^2}}{1000} = 0.701\ \Omega$$

Exercise 15.2

From Table C9, 2.5 mm^2 can supply an 8 kW range when protected by a 20 A circuit breaker, which is supported by Table C6.

Exercise 15.3

a

Circuit schedule						
Supply	Cct ref. no.	MCB RCBO rating (A)	Cable size (mm^2)	Description	No. points	Load per point
80 Ampere Isolator	1	6 RCBO	1.5 mm^2	Lighting – bed 1, 2, 3, kitchen, garage, hall, outside Smoke detector x1	18	0.06 kW
	2	6 RCBO	1.5 mm^2	Lighting – lounge, dining, laundry, WC, bath Smoke detector x1	18	0.06 kW
	3	16 RCBO	2.5 mm^2	Socket-outlets, ½ kitchen, lounge, bed 3	12D	0.25 kW
	4	16 RCBO	2.5 mm^2	Socket-outlets ½ kitchen, laundry, dining, bed 1, 2	12D	0.25 kW
	5	55 RCBO	10 mm^2	Range	1	12.5 kW
	6	25 RCBO	4.0 mm^2	Water heater – controlled	1	4.8 kW
	7	16 RCBO	2.5 mm^2	Air-conditioner	1	3.6 kW

b

Circuit schedule						
Supply	Cct ref. no.	MCB RCBO rating (A)	Cable size (mm^2)	Description	No. points	Load per point
80 Ampere Isolator	1	6 RCBO	1.5 mm^2	Lighting – ½ kitchen, bed 1, 2, 3, garage, hall, outside Smoke detector x1	15	0.06 kW
	2	6 RCBO	1.5 mm^2	Lighting – ½ kitchen, lounge, dining, laundry, WC, bath Smoke detector x1	15	0.06 kW
	3	16 RCBO	2.5 mm^2	Socket-outlets, kitchen, laundry	9D + 1s	0.25 kW
	4	16 RCBO	2.5 mm^2	Socket-outlets lounge, dining, bed 1, 2, 3	9D	0.25 kW
	5	45 RCBO	6 mm^2	Range	1	9.6 kW
	6	25 RCBO	4.0 mm^2	Water heater – controlled	1	4.8 kW
	7	16 RCBO	2.5 mm^2	Air-conditioner	1	3.6 kW

Exercise 15.4

a From Table 27(1), Column 1, locate 75 °C then under Column 5 (for 30 °C), you will find the rating of 1.14. This means for the nominal current rating of 25 A (at 40 °C) the conductor can safely carry 28.5 A (25 × 1.14) due to the lower ambient temperature

b From Table 27(1), Column 1, locate 75 °C then under Column 10 (for 55 °C), you will find the rating of 0.72. This means for the nominal current rating of 25 A (at 40 °C) the conductor can safely carry 18 A (25 × 0.72) due to the higher ambient temperature

Exercise 15.5

a From Table 27(2), Column 1, locate 75 °C then under Column 5 (for 25 °C which is the closest to 22°C), you will find the rating of 1.0. This means for the nominal current rating of 25 A (at 25 °C) the conductor can safely carry 25 A (25 × 1.0) due to the lower soil temperature.

b From Table 27(2), Column 1, locate 75 °C then under Column 7 (for 35 °C), you will find the rating of 0.89. This means for the nominal current rating of 25 A (at 25 °C) the conductor can safely carry 22.2 A (25 × 0.89) due to the higher soil temperature.

Exercise 15.6

From Table 27(1), Column 1, locate 75 °C then under Column 10 (for 55 °C), you will find the rating of 0.72. This means for the nominal current rating of 20 A (at 40 °C) the conductor can safely carry 14.4 A (20 × 0.72) due to the higher ambient temperature.

Exercise 15.7

From Table 3(1), Column 1, Item 9 matches the cable type (column 2) and installation method (column 5). This directs us to Table 10, columns 2 to 4 for the current. This directs us to Table 14, columns 2 to 4 for the current-carrying capacity and Table 24 for the applicable de-rating factor.

From Table 10, column 2, 6 mm^2 2C has a current-carrying capacity of 46 A.

From Table 22, Item 1, column 6 (for 3 circuits) the applicable de-rating factor is 0.75.

This means for the nominal current rating of 46 A (at 40 °C) the conductor can safely carry 34.5 A (46 × 0.75) due to the bunching.

Exercise 15.8

From Table 3(1), Column 1, Item 10 matches the cable type (column 2) and installation method (column 5). This directs us to Table 14, columns 2 to 4 for the current-carrying capacity and Table 24 for the applicable de-rating factor.

From Table 14, column 2, 6 mm^2 4C has a current-carrying capacity of 48 A.

From Table 24, Item 1, column 8 (for 4 cables) the applicable de-rating factor is 0.75.

This means for the nominal current rating of 48 A (at 40 °C) the conductor can safely carry 36 A (48 × 0.75) due to the restricted air flow.

Exercise 15.9

Circuit No.	Load group	Description of load	Calculation	Demand current (A)		
				L_1	L_2	L_3
1	A(i)	18 indoor lighting points		3.0		
2	A(i)	18 indoor lighting points			3.0	
3	A(i)	18 indoor lighting points				3.0
4	B(i)	2 single 10 A outlet + 7 double 10 A outlets	$2 + (7 \times 2) = 16$	---		
5	B(i)	1 single 10 A outlet + 7 double 10 A outlets	$1 + (7 \times 2) = 15$ Cct 4 & 5 = 31 points 10 + 5	15.0		
6	B(i)	2 single 10 A outlet + 7 double 10 A outlets	$2 + (7 \times 2) = 16$		---	
7	B(i)	1 single 10 A outlet + 7 double 10 A outlets	$1 + (7 \times 2) = 15$ Cct 6 & 7 = 31 points 10 + 5		15.0	
8	B(i)	2 single 10 A outlet + 7 double 10 A outlets	$2 + (7 \times 2) = 16$			---
9	B(i)	1 single 10 A outlet + 7 double 10 A outlets	$1 + (7 \times 2) = 15$ Cct 8 & 9 = 31 points 10 + 5			15.0
10	B(ii)	15 A single-phase outlet		10.0		
11	B(ii)	15 A single-phase outlet			10.0	
12	C	cooktop rating 28 A	$28 \times 0.5 = 14$		14.0	
13	C	oven rating 18 A	$18 \times 0.5 = 9$	9.0		
14	D	air-conditioner rating 17.2 A per phase	17.2×0.75	12.9	12.9	12.9
15	F	storage hot water unit rating 20 A	---			20.0
			Total demand current (A)	49.9	54.9	50.9

Exercise 15.10

Load group	Description of load	Units / phase			Calculation	Demand (A)		
		L_1	L_2	L_3		L_1	L_2	L_3
A(i)	Lighting	8	8	8	$5 + (8 \times 0.25)$	7.0	7.0	7.0
B(i)	Socket outlets 10 A	8	8	8	$15 + (8 \times 3.75)$	45.0	45.0	45.0
C	Range	8	8	8	8×2.8	22.4	22.4	22.4
D	Air conditioning	8	8	8	$8 \times 12.8 \times 0.75$	76.8	76.8	76.8
F	Hot water QR	8	8	8	8×6	48.0	48.0	48.0
					Total demand current (A)	**199.2**	**199.2**	**199.2**

Load group	Description of load	Qty / phase			Calculation	Demand (A)		
		L_1	L_2	L_3	L_1	L_1	L_2	L_3
H	Lighting	30			$30 \times 60 / 230$	7.8		
I	Socket outlets 10 A	9			$9 \times 2 = 18$ but 15 max	15.0		
					Total demand current (A)	**22.8**		
					Total maximum demand for installation (A)	**222**		

Exercise 15.11

Circuit No.	Load group	Description of load	Qty	Calculation	Demand (A)		
					L_1	L_2	L_3
1	A	fluorescent light fittings	18	18 × 0.35 = 6.3	6.3		
2	A	fluorescent light fittings	18	18 × 0.35 = 6.3		6.3	
3	A	fluorescent light fittings	18	18 × 0.35 = 6.3			6.3
4	A	highbay MV light fittings	7	7 × 2.5 = 17.5	17.5		
5	A	highbay MV light fittings	7	7 × 2.5 = 17.5		17.5	
6	A	highbay MV light fittings	7	7 × 2.5 = 17.5			17.5
7	B(i)	6 double 10 A outlets	12	(1000 + (11 × 750))/230	40.2		
8	B(i)	6 double 10 A outlets	12	(1000 + (11 × 750))/230		40.2	
9	B(i)	6 double 10 A outlets	12	(1000 + (11 × 750))/230			40.2
10	B(iii)	1 × 20 A single-phase outlet	1	20 × 0.75 = 15.0	15.0		
11	B(iii)	1 × 20 A single-phase outlet	1	20 × 0.75 = 15.0		15.0	
12	B(iii)	1 × 20 A single-phase outlet	1	20 × 0.75 = 15.0			15.0
13	B(iii)	32 A three-phase outlet	1	---	32.0	32.0	32.0
14	B(iii)	32 A three-phase outlet	1	32 × 0.75 = 24.0	24.0	24.0	24.0
15	C(i)	range	1	9600/(2 × 230)	20.9	20.9	
16	C(i)	QR water heater	1	4800/230 + (0.75 × 4800/230) = 36.5			36.5
17	D	6 kW, three-phase motor	1	---	12.5	12.5	12.5
18	D	6 kW, three-phase motor	1	0.75 × 12.5	9.4	9.4	9.4
19	D	4 kW, three-phase motor	1	0.5 × 7.7	3.8	3.8	3.8
20	D	4 kW, three-phase motor	1	0.5 × 7.7	3.8	3.8	3.8
21	D	4 kW, three-phase motor	1	0.5 × 7.7	3.8	3.8	3.8
				Total demand current (A)	**189.2**	**189.2**	**204.8**

Exercise 15.12

a Table 3(1), Item 12, Columns 3 and 5, applies to configuration.
Tables 10 or 11, Columns 5, 6 or 7 (or Table 12, Columns 4 and 5).
No de-rating for single-circuit configuration as per footnote 4a of Table 10 (5a of Table 11 or 3a of Table 12).
From Table 22, Item 3 and Column 5, a de-rating factor of 0.85 applies.
From Table 10, Column 5: 6 mm^2 is rated at **44 A**.

b Table 3(1), Item 12, Columns 3 and 5, applies to configuration.
Tables 10 or 11, Column 5, 6 or 7 (Table 12, Columns 4 and 5).
De-rating Table 22 applies.
From Table 22, Item 3 and Column 5, a de-rating factor of 0.85 applies.
From Table 10, Column 5: 6 mm^2 is rated at 44 A, which is de-rated to **37.4 A**.

c Table 3(1), Item 12, Columns 3 and 5, applies to configuration.
Tables 10 or 11, Column 15 or 16 (Table 12, Column 11).
No de-rating for single-circuit configuration as per footnote 4a of Table 10 (5a of Table 11 or 3a of Table 12).
From Table 10, Column 15: 6 mm^2 is rated at **35 A**.

d Table 3(1), Item 12, Columns 3 and 5, applies to configuration
Table 10 or 11, Column 19 or 20 (Table 12, Column 12).
No de-rating for single-circuit configuration as per footnote 4a of Table 10 (5a of Table 11 or 3a of Table 12).
From Table 10, Column 19: 6 mm^2 is rated at **22 A**.

Exercise 15.13

a From Table 13, Column 5: 16 mm^2 Cu is rated at **37 A**.

b Table 10, Column 5: 6 mm^2 is rated at 44 A.
De-rating factors apply for three circuits. Table 22, Item 3, Column 5: applicable de-rating is 0.79: 44 × 0.79 = **34.7 A**.

c Table 13, Column 2: 10 mm^2 is rated at 54 A.
Table 24, Item 1, Column 7: de-rating factor of 0.78 applies to three circuits (refer to footnote 3 to determine for which columns the de-rating is applicable).
The 10 mm^2 is de-rated to: 54 × 0.78 = **40.5 A**.

d Table 7, Column 5: 16 mm^2 is rated at 77 A.
Table 23, Item 4: de-rating factor of 0.89 applies to two circuits.
The 16 mm^2 is de-rated to: $77 \times 0.89 = \mathbf{68.5\ A}$.

e In AS/NZS 3008.1.1:
Table 13, Column 25 and De-rating Table 22
To obtain the required current-carrying capacity per cable up-rate the current by 0.80
$80 \div 0.8 = 100$ A
Table 13, Column 25; 35 mm^2 can carry 114 A.

f In AS/NZS 3008.1.1:
Table 10 for V–90 Cu Column 5 Conductor size for 45 A is 10 mm^2.

Exercise 15.14

In AS/NZS 3008.1.1:

Table 3(1), Item 5, Columns 3 and 5 show installation methods. Note that the installation method shown is for one three-phase circuit consisting of three only SDI cables.

Table 7, Column 5 for V–90 copper

Table 23 is suggested as the de-rating table in Column 6 of Table 3(1). Note two conductors per phase are two circuits.

Table 23, Item 7, Column 7: de-rating is 0.95 for two circuits

To obtain the required current-carrying capacity per conductor up-rate the current by 0.95

Because there are two conductors per phase, each conductor carries half the load current.

$320 / 2 = 160$ A

Up-rated current requirements $= 160 \div 0.95 = 168.4$ A

For cables touching: Table 7, Column 5: 70 mm^2 can carry 197 A.

Exercise 15.15

16.1 ms

Exercise 15.16

From AS/NZS 3008.1.1, Table 20, Column 8, a 25 mm^2 conductor can safely carry 52 A.

Exercise 15.17

a Required ampere metres (A.m) $= I \times L = 16 \times 40 = 640$ A.m

b From Table C8 A.m per % of V_d (230 V circuit), the possible cable sizes are:
- 6 mm^2 at 306 A.m per % of V_d
- 4 mm^2 at 205 A.m per % of V_d
- 2.5 mm^2 at 128 A.m per % of V_d

The cable sizes selected must have their A.m per % of V_d less than the required ampere metres (A.m).

c Divide required A.m by A.m per % of V_d for each possible cable size to obtain percentages.
- 6 mm^2 = 2.09%
- 4 mm^2 = 3.12%
- 2.5 mm^2 = 5%

The actual voltage drops (V_d) across these cables with a 230 V supply are:
- $2.09\% \times 230\ V = 4.81\ V$
- $3.12\% \times 230\ V = 7.18\ V$
- $5\% \times 230\ V = 11.5\ V$

Exercise 15.18

$$V_d = \frac{L \times I \times V_C}{1000} = \frac{45 \times 18 \times 18}{1000} = 14.6\ V$$

This arrangement is not suitable as it exceeds the allowable volt-drop.

Exercise 15.19

a Required ampere metres (A.m) is 1575 A.m

b Required capacity for allowable voltage drop (A.m per % of V_d) = 875

c From Table C8, possible cable size that satisfies the 800 A.m /%V_d is:
25 mm^2 at 1289 A.m per % of V_d

Exercise 15.20

Allowable voltage drop is 6 V ($1.5\% \times 400$)
Optimum conductor size is 25 mm^2 from Table 42, column 8

Exercise 15.21

Volt drop in consumer mains = 4.85 V which represents 1.21%
Allowable volt drop in sub-circuit is 8.72 V
[(5% − 1.21% × 230)]
V_c = 17.43 mV/A.m converted to 3-phase =15.1 mV/A.m
Optimum conductor size is 4 mm^2 from Table 42 column 6

Exercise 15.22

a 3.7 V

b 9 V

c 13.9 V

d 3.88 V

e 2.2 V

f 41 m

g Using Table 42 (AS/NZS 3008.1.1) the three-phase per unit voltage drops (V_c) for V-90 10 mm^2 cables in normal use is conductors at 75° C = 3.86 mV/A.m.
Voltage drop old motor

$$V_d = \frac{L \times I \times V_C}{1000} = \frac{80 \times 27.4 \times 2.81}{1000} = \mathbf{6.159\ V}$$

Voltage drop new motor

$$V_d = \frac{L \times I \times V_C}{1000} = \frac{80 \times 30.5 \times 2.81}{1000} = \mathbf{6.86\ V}$$

10 mm^2 multicore circular V90 TPS cables clipped to a perforated tray AS/NZS 3008.1.1 2009 Table 13 column 5 = 51 A. The existing cables and CB are suitable.

Exercise 15.23

1.0 Ω

Exercise 15.24

0.981 Ω

Exercise 15.25

0.000 888 Ω

Exercise 15.26

59 V

Exercise 15.27

a From Table C7, 4 mm^2 can supply 32 A when protected by a 32 A circuit breaker.

b From Table C7, 4 mm^2 can supply 16 A when protected by a 16 A circuit breaker.

Index